Handbook of Experimental Pharmacology

Volume 224

Editor-in-Chief
W. Rosenthal, Jena

Editorial Board
J.E. Barrett, Philadelphia
V. Flockerzi, Homburg
M.A. Frohman, Stony Brook, NY
P. Geppetti, Florence
F.B. Hofmann, München
M.C. Michel, Ingelheim
P. Moore, Singapore
C.P. Page, London
A.M. Thorburn, Aurora, CO
K. Wang, Beijing

More information about this series at
http://www.springer.com/series/164

Arnold von Eckardstein · Dimitris Kardassis
Editors

High Density Lipoproteins

From Biological Understanding
to Clinical Exploitation

Editors
Arnold von Eckardstein
University Hospital Zurich
Institute of Clinical Chemistry
Zurich
Switzerland

Dimitris Kardassis
University of Crete Medical School
Iraklion, Crete
Greece

ISSN 0171-2004 ISSN 1865-0325 (electronic)
ISBN 978-3-319-34861-2 ISBN 978-3-319-09665-0 (eBook)
DOI 10.1007/978-3-319-09665-0
Springer Cham Heidelberg New York Dordrecht London

© The Editor(s) and the Author(s) 2015
Softcover reprint of the hardcover 1st edition 2015
Open Access This book is distributed under the terms of the Creative Commons Attribution Noncommercial License, which permits any noncommercial use, distribution, and reproduction in any medium, provided the original author(s) and source are credited.
All commercial rights are reserved by the Publisher, whether the whole or part of the material is concerned, specifically the rights of translation, reprinting, re-use of illustrations, recitation, broadcasting, reproduction on microfilms or in any other way, and storage in data banks. Duplication of this publication or parts thereof is permitted only under the provisions of the Copyright Law of the Publisher's location, in its current version, and permission for commercial use must always be obtained from Springer. Permissions for commercial use may be obtained through RightsLink at the Copyright Clearance Center. Violations are liable to prosecution under the respective Copyright Law.
The use of general descriptive names, registered names, trademarks, service marks, etc. in this publication does not imply, even in the absence of a specific statement, that such names are exempt from the relevant protective laws and regulations and therefore free for general use.
While the advice and information in this book are believed to be true and accurate at the date of publication, neither the authors nor the editors nor the publisher can accept any legal responsibility for any errors or omissions that may be made. The publisher makes no warranty, express or implied, with respect to the material contained herein.

Printed on acid-free paper

Springer is part of Springer Science+Business Media (www.springer.com)

Preface

In both epidemiological and clinical studies as well as the meta-analyses thereof, low plasma levels of high-density lipoprotein (HDL) cholesterol (HDL-C) identified individuals at increased risk of major coronary events. Observational studies also found inverse associations between HDL-C and risks of ischemic stroke, diabetes mellitus type 2, and various cancers. In addition, HDLs exert many effects in vitro and in vivo which protect the organism from chemical or biological harm and thereby may interfere with the pathogenesis of atherosclerosis, diabetes, and cancer but also other inflammatory diseases. Moreover, in several animal models transgenic overexpression or exogenous application of apolipoprotein A-I (apoA-I), the most abundant protein of HDL, decreased or prevented the development of atherosclerosis, glucose intolerance, or tissue damage induced by ischemia or mechanical injury.

However, as yet drugs increasing HDL-C such as fibrates, niacin, or inhibitors of cholesteryl ester transfer protein have failed to consistently and significantly reduce the risk of major cardiovascular events, especially when combined with statins. Moreover, mutations in several human genes as well as targeting of several murine genes were found to modulate HDL-C levels without changing cardiovascular risk and atherosclerotic plaque load, respectively, into the opposite direction as expected from the inverse correlation of HDL-C levels and cardiovascular risk in epidemiological studies. Because of these controversial data, the pathogenic role, and, hence, the suitability of HDL as a therapeutic target, has been increasingly questioned. Because of the frequent confounding of low HDL-C with hypertriglyceridemia, it has been argued that low HDL-C is an innocent bystander of (postprandial) hypertriglyceridemia or another culprit related to insulin resistance or inflammation.

These complex relationships are depicted in Fig. 1. It is important to note that previous intervention and genetic studies targeted HDL-C, i.e., the cholesterol measured by clinical laboratories in HDL. By contrast to the pro-atherogenic and, hence, disease causing cholesterol in LDL (measured or estimated by clinical laboratories as LDL cholesterol, LDL-C) which after internalization turns macrophages of the arterial intima into pro-inflammatory foam cells, the cholesterol in HDL (i.e., HDL-C) neither exerts nor reflects any of the potentially anti-atherogenic activities of HDL. By contrast to LDL-C, HDL-C is only a nonfunctional surrogate marker for estimating HDL particle number and size without

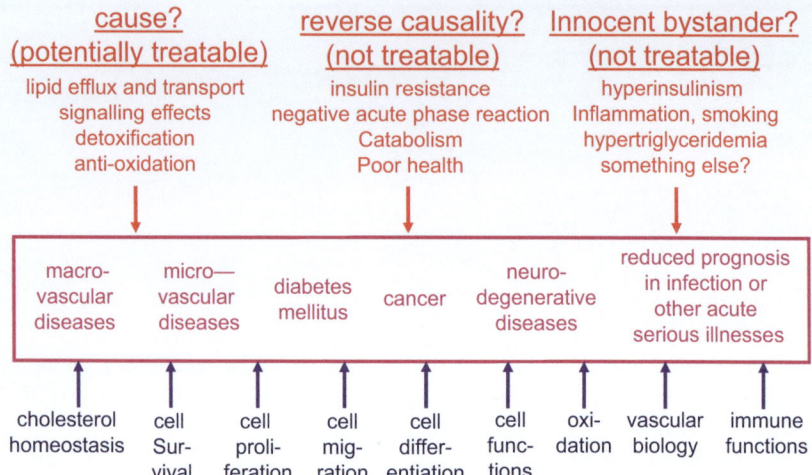

Fig. 1 Possible pathophysiological relationships of low HDL cholesterol with its associated diseases

deciphering the heterogeneous composition and, hence, functionality of HDL. HDL particles are heterogeneous and complex macromolecules carrying hundreds of lipid species and dozens of proteins as well as microRNAs. This physiological heterogeneity is further increased in pathological conditions due to additional quantitative and qualitative molecular changes of HDL components which have been associated with both loss of physiological function and gain of pathological dysfunction. This structural and functional complexity of HDL has prevented clear assignments of molecules to the many functions of HDL. Detailed knowledge of structure–function relationships of HDL-associated molecules is a prerequisite to test them for their relative importance in the pathogenesis of HDL-associated diseases. The identification of the most relevant biological activities of HDL and their mediating molecules within HDL, as well as their cellular interaction partners, is pivotal for the successful development of anti-atherogenic and anti-diabetogenic drugs as well as of diagnostic biomarkers for the identification, treatment stratification, and monitoring of patients at increased risk for cardiovascular diseases or diabetes mellitus but also other diseases which show associations with HDL.

This *Handbook of Experimental Pharmacology* on HDL emerged from the European Cooperation in Science and Technology (*COST*) *Action BM0904* entitled "*HDL—from biological understanding to clinical exploitation*" (HDL*net*: http://cost-bm0904.gr/). This COST Action was run from 2010 to 2014 and involved more than 200 senior and junior scientists from 16 European countries. HDL*net* has been a scientific network dedicated to the study of HDL in health and disease, to the identification of targets for novel HDL-based therapies, and to the discovery of biomarkers which can be used for diagnostics, prevention, and therapy of cardiovascular disease. HDL*net* fostered the cooperation and interaction of European HDL-researchers, the exchange of information and materials, the training and promotion of early career scientists, the gain of technological know-how, and the dissemination of old and new knowledge on HDL to the scientific and medical community as well as the lay public. In this setting, the chapters of this handbook have been written by cooperative and interactive efforts of many senior scientists of the HDL*net* consortium and colleagues from the United States. It is published as open access through COST funding so that the knowledge on HDL can be spread without limitation.

As the chairman and vice-chairman of HDL*net*, the editors of this Handbook of Experimental Pharmacology issue like to thank not only the authors of the 22 chapters of this handbook but all members of the COST Action for their engaged participation and cooperation. We thank Ms. Zinovia Papatheodorou (senior Administrative Officer of the grant holder FORTH, Heraklion) for excellent grant administrative work in HDL*net*, the Science Officers Dr. Magdalena Radwanska and Dr Inga Dadeshidze, the Administrative Officers Ms Anja van der Snickt and Ms Jeannette Nchung (all from COST Office, Brussels, Belgium), as well as the DC Rapporteur, Prof. Marieta Costache (Bucharest, Romania), for their excellent support and sustained interest in our Action. We gratefully acknowledge Andrea Bardelli and Giulia Miotto from COST Publications Office for their help in publishing this book as an open access Final Action Publication (FAP). Finally we wish to thank Prof. Martin Michel for his interest and guidance as well as Susanne Dathe and Wilma McHugh from Springer who supported us with patience and enthusiasm in the production of this book.

Zurich	Arnold von Eckardstein
Iraklion	Dimitris Kardassis

Acknowledgement

This publication is supported by COST

COST is supported by the EU Framework Programme Horizon 2020

COST—European Cooperation in Science and Technology is an intergovernmental framework aimed at facilitating the collaboration and networking of scientists and researchers at European level. It was established in 1971 by 19 member countries and currently includes 35 member countries across Europe, and Israel as a cooperating state.

COST funds pan-European, bottom-up networks of scientists and researchers across all science and technology fields. These networks, called "COST Actions", promote international coordination of nationally funded research.

By fostering the networking of researchers at an international level, COST enables break-through scientific developments leading to new concepts and products, thereby contributing to strengthening Europe's research and innovation capacities.

COST's mission focuses in particular on:

- Building capacity by connecting high-quality scientific communities throughout Europe and worldwide
- Providing networking opportunities for early career investigators
- Increasing the impact of research on policy makers, regulatory bodies, and national decision makers as well as the private sector

Through its inclusiveness policy, COST supports the integration of research communities in less research-intensive countries across Europe, leverages national research investments, and addresses societal issues.

Over 45,000 European scientists benefit from their involvement in COST Actions on a yearly basis. This allows the pooling of national research funding and helps countries' research communities achieve common goals.

As a precursor of advanced multidisciplinary research, COST anticipates and complements the activities of EU Framework Programmes, constituting a "bridge" towards the scientific communities of emerging countries.

Traditionally, COST draws its budget for networking activities from successive EU RTD Framework Programmes.

COST Mission: COST aims to enable breakthrough scientific developments leading to new concepts and products. It thereby contributes to strengthening Europe's research and innovation capacities.

Contents

Part I Physiology of HDL

Structure of HDL: Particle Subclasses and Molecular Components 3
Anatol Kontush, Mats Lindahl, Marie Lhomme, Laura Calabresi, M. John Chapman, and W. Sean Davidson

HDL Biogenesis, Remodeling, and Catabolism 53
Vassilis I. Zannis, Panagiotis Fotakis, Georgios Koukos, Dimitris Kardassis, Christian Ehnholm, Matti Jauhiainen, and Angeliki Chroni

Regulation of HDL Genes: Transcriptional, Posttranscriptional, and Posttranslational 113
Dimitris Kardassis, Anca Gafencu, Vassilis I. Zannis, and Alberto Davalos

Cholesterol Efflux and Reverse Cholesterol Transport 181
Elda Favari, Angelika Chroni, Uwe J.F. Tietge, Ilaria Zanotti, Joan Carles Escolà-Gil, and Franco Bernini

Functionality of HDL: Antioxidation and Detoxifying Effects 207
Helen Karlsson, Anatol Kontush, and Richard W. James

Signal Transduction by HDL: Agonists, Receptors, and Signaling Cascades 229
Jerzy-Roch Nofer

Part II Pathology of HDL

Epidemiology: Disease Associations and Modulators of HDL-Related Biomarkers 259
Markku J. Savolainen

Beyond the Genetics of HDL: Why Is HDL Cholesterol Inversely Related to Cardiovascular Disease? 285
J.A. Kuivenhoven and A.K. Groen

Mouse Models of Disturbed HDL Metabolism................... 301
Menno Hoekstra and Miranda Van Eck

**Dysfunctional HDL: From Structure-Function-Relationships
to Biomarkers**.. 337
Meliana Riwanto, Lucia Rohrer, Arnold von Eckardstein, and Ulf Landmesser

Part III Possible Indications and Target Mechanisms of HDL Therapy

HDL and Atherothrombotic Vascular Disease.................... 369
Wijtske Annema, Arnold von Eckardstein, and Petri T. Kovanen

HDLs, Diabetes, and Metabolic Syndrome....................... 405
Peter Vollenweider, Arnold von Eckardstein, and Christian Widmann

**High-Density Lipoprotein: Structural and Functional Changes Under
Uremic Conditions and the Therapeutic Consequences**.............. 423
Mirjam Schuchardt, Markus Tölle, and Markus van der Giet

**Impact of Systemic Inflammation and Autoimmune Diseases
on apoA-I and HDL Plasma Levels and Functions**................. 455
Fabrizio Montecucco, Elda Favari, Giuseppe Danilo Norata,
Nicoletta Ronda, Jerzy-Roch Nofer, and Nicolas Vuilleumier

HDL in Infectious Diseases and Sepsis.......................... 483
Angela Pirillo, Alberico Luigi Catapano, and Giuseppe Danilo Norata

High-Density Lipoproteins in Stroke........................... 509
Olivier Meilhac

Therapeutic Potential of HDL in Cardioprotection and Tissue Repair.... 527
Sophie Van Linthout, Miguel Frias, Neha Singh, and Bart De Geest

Part IV Treatments for Dyslipidemias and Dysfunction of HDL

HDL and Lifestyle Interventions............................... 569
Joan Carles Escolà-Gil, Josep Julve, Bruce A. Griffin, Dilys Freeman,
and Francisco Blanco-Vaca

**Effects of Established Hypolipidemic Drugs on HDL Concentration,
Subclass Distribution, and Function**............................ 593
Monica Gomaraschi, Maria Pia Adorni, Maciej Banach, Franco Bernini,
Guido Franceschini, and Laura Calabresi

Emerging Small Molecule Drugs................................ 617
Sophie Colin, Giulia Chinetti-Gbaguidi, Jan A. Kuivenhoven,
and Bart Staels

ApoA-I Mimetics 631
R.M. Stoekenbroek, E.S. Stroes, and G.K. Hovingh

Antisense Oligonucleotides, microRNAs, and Antibodies 649
Alberto Dávalos and Angeliki Chroni

Index 691

Part I
Physiology of HDL

Structure of HDL: Particle Subclasses and Molecular Components

Anatol Kontush, Mats Lindahl, Marie Lhomme, Laura Calabresi, M. John Chapman, and W. Sean Davidson

Contents

1 HDL Subclasses 5
2 Molecular Components of HDL 7
   2.1 Proteome 7
      2.1.1 Major Protein Components 7
      2.1.2 Protein Isoforms, Translational and Posttranslational Modifications 19
   2.2 Lipidome 23
      2.2.1 Phospholipids 23
      2.2.2 Sphingolipids 27
      2.2.3 Neutral Lipids 27

A. Kontush (✉) • M. Lhomme • M.J. Chapman
National Institute for Health and Medical Research (INSERM), UMR-ICAN 1166, Paris, France

University of Pierre and Marie Curie - Paris 6, Paris, France

Pitié – Salpêtrière University Hospital, Paris, France

ICAN, Paris, France
e-mail: anatol.kontush@upmc.fr; m.lhomme@ican-institute.org; john.chapman@upmc.fr

M. Lindahl
Department of Clinical and Experimental Medicine, Linköping University, Linköping, Sweden
e-mail: mats.lindahl@liu.se

L. Calabresi
Department of Pharmacological and Biomolecular Sciences, Center E. Grossi Paoletti, University of Milan, Milan, Italy
e-mail: laura.calabresi@unimi.it

W.S. Davidson
Department of Pathology and Laboratory Medicine, University of Cincinnati, Cincinnati, OH 45237, USA
e-mail: davidswm@ucmail.uc.edu

3 The Structure of HDL 28
   3.1 Introduction/Brief History 28
   3.2 HDL in the Test Tube 31
        3.2.1 Discoid HDL 31
        3.2.2 Spherical rHDL 33
   3.3 "Real" HDL Particles 35
References 41

Abstract

A molecular understanding of high-density lipoprotein (HDL) will allow a more complete grasp of its interactions with key plasma remodelling factors and with cell-surface proteins that mediate HDL assembly and clearance. However, these particles are notoriously heterogeneous in terms of almost every physical, chemical and biological property. Furthermore, HDL particles have not lent themselves to high-resolution structural study through mainstream techniques like nuclear magnetic resonance and X-ray crystallography; investigators have therefore had to use a series of lower resolution methods to derive a general structural understanding of these enigmatic particles. This chapter reviews current knowledge of the composition, structure and heterogeneity of human plasma HDL. The multifaceted composition of the HDL proteome, the multiple major protein isoforms involving translational and posttranslational modifications, the rapidly expanding knowledge of the HDL lipidome, the highly complex world of HDL subclasses and putative models of HDL particle structure are extensively discussed. A brief history of structural studies of both plasma-derived and recombinant forms of HDL is presented with a focus on detailed structural models that have been derived from a range of techniques spanning mass spectrometry to molecular dynamics.

Keywords

HDL • Composition • Structure • Heterogeneity • Proteomics • Lipidomics • Proteome • Lipidome • Post-translational • Modifications

High-density lipoprotein (HDL) is a small, dense, protein-rich lipoprotein as compared to other lipoprotein classes, with a mean size of 8–10 nm and density of 1.063–1.21 g/ml (Kontush and Chapman 2012). HDL particles are plurimolecular, quasi-spherical or discoid, pseudomicellar complexes composed predominantly of polar lipids solubilised by apolipoproteins. HDL also contains numerous other proteins, including enzymes and acute-phase proteins, and may contain small amounts of nonpolar lipids. Furthermore, HDL proteins often exist in multiple isoforms and readily undergo posttranslational modification. As a consequence of such diverse compositional features, HDL particles are highly heterogeneous in their structural, chemical and biological properties. This chapter reviews current knowledge of the composition, structure and heterogeneity of human plasma HDL.

1 HDL Subclasses

Human plasma HDLs are a highly heterogeneous lipoprotein family consisting of several subclasses differing in density, size, shape and lipid and protein composition (Table 1).

Differences in HDL subclass distribution were first described by Gofman and colleagues in the early 1950s by using analytic ultracentrifugation (De Lalla and Gofman 1954), the gold standard technique for HDL separation. Two HDL subclasses were identified: the less dense (1.063–1.125 g/mL), relatively lipid-rich form was classified as **HDL2** and the more dense (1.125–1.21 g/mL), relatively protein-rich form as **HDL3**. The two major HDL subclasses can be separated by other ultracentrifugation methods, such as rate-zonal ultracentrifugation (Franceschini et al. 1985) or single vertical spin ultracentrifugation (Kulkarni et al. 1997). Ultracentrifugation methods are accurate and precise but require expensive instruments, time and technical skills. A precipitation method has been proposed for HDL2 and HDL3 separation and quantitation (Gidez et al. 1982), which is inexpensive and easier, but with a high degree of interlaboratory variability. HDL2 and HDL3 can be further fractionated in distinct subclasses with different electrophoretic mobilities by non-denaturing polyacrylamide gradient gel electrophoresis (GGE) (Nichols et al. 1986), which separates HDL subclasses on the basis of particle size. Two HDL2 and three HDL3 subclasses have been identified and their particle size characterised by this method: **HDL3c**, 7.2–7.8 nm diameter; **HDL3b**, 7.8–8.2 nm; **HDL3a**, 8.2–8.8 nm; **HDL2a**, 8.8.–9.7 nm; and **HDL2b**, 9.7–12.0 nm. The equivalent subclasses of HDL with similar size distribution may be preparatively isolated by isopycnic density gradient ultracentrifugation (Chapman et al. 1981; Kontush et al. 2003).

Agarose gel electrophoresis allows analytical separation of HDL according to surface charge and shape into **α-migrating particles**, which represent the majority of circulating HDL, and **preβ-migrating particles**, consisting of nascent discoidal and poorly lipidated HDL. Agarose gels can be stained with Coomassie blue or with anti-apolipoprotein A-I (apoA-I) antibodies, and the relative protein content of the two HDL subclasses can be determined (Favari et al. 2004). The plasma preβ-HDL concentration can be also quantified using a sandwich enzyme immunoassay (Miida et al. 2003). The assay utilises a monoclonal antibody which specifically recognises apoA-I bound to preβ-HDL. The agarose gel and the GGE can be combined into a 2-dimensional (2D) electrophoretic method, which separates HDL according to charge in the first run and according to size in the second run. Gels can be stained with apolipoprotein-specific antibodies, typically with anti-apoA-I antibodies, allowing the detection of distinct HDL subclasses (Asztalos et al. 2007). This is by far the method with the highest resolving power: up to 12 distinct apoA-I-containing HDL subclasses can be identified, referred to as preβ (preβ$_1$ and preβ$_2$), α ($α_1$, $α_2$, $α_3$ and $α_4$) and preα (preα$_1$, preα$_2$, preα$_3$) according to their mobility and size (Asztalos and Schaefer 2003a, b).

According to the protein component, HDL can be separated into **particles containing apoA-I with (LpA-I:A-II) or without apoA-II (LpA-I)** by an

Table 1 Major HDL subclasses according to different isolation/separation techniques

Density (ultracentrifugation)
HDL2 (1.063–1.125 g/mL)
HDL3 (1.125–1.21 g/mL)
Size (GGE)
HDL2b (9.7–12.0 nm)
HDL2a (8.8–9.7 nm)
HDL3a (8.2–8.8 nm)
HDL3b (7.8–8.2 nm)
HDL3c (7.2–7.8 nm)
Size (NMR)
Large HDL (8.8–13.0 nm)
Medium HDL (8.2–8.8 nm)
Small HDL (7.3–8.2 nm)
Shape and charge (agarose gel)
α-HDL (spherical)
Preβ-HDL (discoidal)
Charge and size (2D electrophoresis)
Preβ-HDL (preβ$_1$ and preβ$_2$)
α-HDL (α$_1$, α$_2$, α$_3$ and α$_4$)
Preα-HDL (preα$_1$, preα$_2$, preα$_3$)
Protein composition (electroimmunodiffusion)
LpA-I
LpA-I:A-II

electroimmunodiffusion technique in agarose gels; plasma concentration of LpA-I and LpA-I:A-II can be determined from a calibration curve (Franceschini et al. 2007).

More recently, a nuclear magnetic resonance (NMR) method for HDL subclass analysis has been proposed, based on the concept that each lipoprotein particle of a given size has its own characteristic lipid methyl group NMR signal (Otvos et al. 1992). According to the NMR signals, three HDL subclasses can be identified: **large HDL** (8.8–13.0 nm diameter), **medium HDL** (8.2–8.8 nm) and **small HDL** (7.3–8.2 nm); results are expressed as plasma particle concentration. The relative plasma content of small, medium and large HDL (according to the same cut-off) can also be determined by GGE, dividing the absorbance profile into the three size intervals (Franceschini et al. 2007).

HDL subfractionation on an analytical scale has been generally accomplished by the different techniques in academic laboratories; however, clinical interest in HDL heterogeneity has been growing in the last 10 years and a number of laboratory tests for determining HDL subclass distribution are now available, including GGE, NMR and 2D gel electrophoresis (Mora 2009). Whether evaluation of HDL subfractions is performed by academic or commercial laboratories, there are a number of factors that confound the interpretation of the results of such analyses. The number and nomenclature of HDL subclasses are not uniform among the different techniques;

moreover, each subclass contains distinct subpopulations, as identified by, e.g. 2D electrophoresis. In addition, whereas some methodologies measure HDL subclass concentrations, others describe the percent distribution of the HDL subclasses relative to the total or characterise the HDL distribution by average particle diameter. As a consequence, there is little relation among HDL subfractionation data produced by different analytical techniques. A panel of experts has recently proposed a classification of HDL by physical properties, which integrates terminology from several methods and defines five HDL subclasses, termed very large, large, medium, small and very small HDL (Rosenson et al. 2011). The proposed nomenclature, possibly together with widely accepted standards and quality controls, should help in defining the relationship between HDL subclasses and cardiovascular risk as well as in assessing the clinical effects of HDL modifying drugs.

2 Molecular Components of HDL

2.1 Proteome

2.1.1 Major Protein Components

Proteins form the major structural and functional component of HDL particles. HDL carries a large number of different proteins as compared to other lipoprotein classes (Table 2). HDL proteins can be divided into several major subgroups which include apolipoproteins, enzymes, lipid transfer proteins, acute-phase response proteins, complement components, proteinase inhibitors and other protein components. Whereas apolipoproteins and enzymes are widely recognised as key functional HDL components, the role of minor proteins, primarily those involved in complement regulation, protection from infections and the acute-phase response, has received increasing attention only in recent years, mainly as a result of advances in proteomic technologies (Heinecke 2009; Hoofnagle and Heinecke 2009; Davidsson et al. 2010; Shah et al. 2013). These studies have allowed reproducible identification of more than 80 proteins in human HDL (Heinecke 2009; Hoofnagle and Heinecke 2009; Davidsson et al. 2010; Shah et al. 2013) (for more details see the HDL Proteome Watch at http://homepages.uc.edu/~davidswm/HDLproteome.html). Numerous proteins involved in the acute-phase response, complement regulation, proteinase inhibition, immune response and haemostasis were unexpectedly found as components of normal human plasma HDL, raising the possibility that HDL may play previously unsuspected roles in host defence mechanisms and inflammation (Hoofnagle and Heinecke 2009).

Importantly, the composition of the HDL proteome may depend on the method of HDL isolation. Indeed, ultracentrifugation in highly concentrated salt solutions of high ionic strength can remove some proteins from HDL, whereas other methods of HDL isolation (gel filtration, immunoaffinity chromatography, precipitation) provide HDL extensively contaminated with plasma proteins or subject HDL to unphysiological conditions capable of modifying its structure and/or composition

Table 2 Major components of the HDL proteome

Protein	M_r, kDa	Major function	Number of proteomic studies in which the protein was detected[a]
Apolipoproteins			
ApoA-I	28	Major structural and functional apolipoprotein, LCAT activator	14
ApoA-II	17	Structural and functional apolipoprotein	13
ApoA-IV	46	Structural and functional apolipoprotein	14
ApoC-I	6.6	Modulator of CETP activity, LCAT activator	12
ApoC-II	8.8	Activator of LPL	12
ApoC-III	8.8	Inhibitor of LPL	14
ApoC-IV	11	Regulates TG metabolism	6
ApoD	19	Binding of small hydrophobic molecules	11
ApoE	34	Structural and functional apolipoprotein, ligand for LDL-R and LRP	13
ApoF	29	Inhibitor of CETP	8
ApoH	38	Binding of negatively charged molecules	8
ApoJ	70	Binding of hydrophobic molecules, interaction with cell receptors	11
ApoL-I	44/46	Trypanolytic factor of human serum	14
ApoM	25	Binding of small hydrophobic molecules	12
Enzymes			
LCAT	63	Esterification of cholesterol to cholesteryl esters	4
PON1	43	Calcium-dependent lactonase	12
PAF-AH (LpPLA$_2$)	53	Hydrolysis of short-chain oxidised phospholipids	
GSPx-3	22	Reduction of hydroperoxides by glutathione	
Lipid transfer proteins			
PLTP	78	Conversion of HDL into larger and smaller particles, transport of LPS	5
CETP	74	Heteroexchange of CE and TG and homoexchange of PL between HDL and apoB-containing lipoproteins	3
Acute-phase proteins			
SAA1	12	Major acute-phase reactant	10
SAA4	15	Minor acute-phase reactant	10
Alpha-2-HS-glycoprotein	39	Negative acute-phase reactant	9

(continued)

Table 2 (continued)

Protein	M_r, kDa	Major function	Number of proteomic studies in which the protein was detected[a]
Fibrinogen alpha chain	95	Precursor of fibrin, cofactor in platelet aggregation	10
Complement components			
C3	187	Complement activation	9
Proteinase inhibitors			
Alpha-1-antitrypsin	52	Inhibitor of serine proteinases	11
Hrp	39	Decoy substrate to prevent proteolysis	10
Other proteins			
Transthyretin	55	Thyroid hormone binding and transport	12
Serotransferrin	75	Iron binding and transport	10
Vitamin D-binding protein	58	Vitamin D binding and transport	10
Alpha-1B-glycoprotein	54	Unknown	9
Hemopexin	52	Heme binding and transport	8

[a]Out of total of 14 proteomic studies according to Shah et al. (2013). Only proteins detected in more than 50 % of the studies are listed, together with seven others previously known to be associated with HDL (apoC-IV, apoH, LCAT, PAF-AH, GSPx-3, PLTP, CETP)

CE cholesteryl ester, *CETP* cholesteryl ester transfer protein, GSPx-3 glutathione selenoperoxidase 3, *Hrp* haptoglobin-related protein, *LDL-R* LDL receptor, *LCAT* lecithin/cholesterol acyltransferase, *LPL* lipoprotein lipase, *LpPLA2* lipoprotein-associated phospholipase A2, *LRP* LDL receptor-related protein, *PAF-AH* platelet-activating factor acetyl hydrolase, *PL* phospholipid, *PLTP* phospholipid transfer protein, *PON1* paraoxonase 1, *SAA* serum amuloid A, *TG* triglyceride

(e.g. extreme pH and ionic strength involved in immunoaffinity separation). Thus, proteomics of apoA-I-containing fractions isolated from human plasma by a non-denaturing approach of fast protein liquid chromatography (FPLC) reveal the presence of up to 115 individual proteins per fraction, only up to 32 of which were identified as HDL-associated proteins in ultracentrifugally isolated HDL (Collins et al. 2010). Indeed, co-elution with HDL of plasma proteins of matching size is inevitable in FPLC-based separation; the presence of a particular protein across a range of HDL-containing fractions of different size isolated by FPLC on the basis of their association with phospholipid would however suggest that such a protein is indeed associated with HDL (Gordon et al. 2010). Remarkably, several of the most abundant plasma proteins, including albumin, haptoglobin and alpha-2-macroglobulin, are indeed present in all apoA-I-containing fractions isolated by FPLC (Collins et al. 2010), suggesting their partial association with HDL by a non-specific, low-affinity binding.

Apolipoproteins

Apolipoprotein A-I is the major structural and functional HDL protein which accounts for approximately 70 % of total HDL protein. Almost all HDL particles are believed to contain apoA-I (Asztalos and Schaefer 2003a, b; Schaefer et al. 2010). Major functions of apoA-I involve interaction with cellular receptors, activation of lecithin/cholesterol acyltransferase (LCAT) and endowing HDL with multiple anti-atherogenic activities. Circulating apoA-I represents a typical amphipathic protein that lacks glycosylation or disulfide linkages and contains eight alpha-helical amphipathic domains of 22 amino acids and two repeats of 11 amino acids. As a consequence, apoA-I binds avidly to lipids and possesses potent detergent-like properties. ApoA-I readily moves between lipoprotein particles and is also found in chylomicrons and very low-density lipoprotein (VLDL). As for many plasma apolipoproteins, the main sites for apoA-I synthesis and secretion are the liver and small intestine.

ApoA-II is the second major HDL apolipoprotein which represents approximately 15–20 % of total HDL protein. About a half of HDL particles may contain apoA-II (Duriez and Fruchart 1999). ApoA-II is more hydrophobic than apoA-I and circulates as a homodimer composed of two identical polypeptide chains (Shimano 2009; Puppione et al. 2010) connected by a disulfide bridge at position 6 (Brewer et al. 1972). ApoA-II equally forms heterodimers with other cysteine-containing apolipoproteins (Hennessy et al. 1997) and is predominantly synthesised in the liver but also in the intestine (Gordon et al. 1983).

ApoA-IV, an O-linked glycoprotein, is the most hydrophilic apolipoprotein which readily exchanges between lipoproteins and also circulates in a free form. ApoA-IV contains thirteen 22-amino acid tandem repeats, nine of which are highly alpha-helical; many of these helices are amphipathic and may serve as lipid-binding domains. In man, apoA-IV is synthesised in the intestine and is secreted into the circulation with chylomicrons.

ApoCs form a family of small exchangeable apolipoproteins primarily synthesised in the liver. **ApoC-I** is the smallest apolipoprotein which associates with both HDL and VLDL and can readily exchange between them. ApoC-I carries a strong positive charge and can thereby bind free fatty acids and modulate activities of several proteins involved in HDL metabolism. Thus, apoC-I is involved in the activation of LCAT and inhibition of hepatic lipase and cholesteryl ester transfer protein (CETP). **ApoC-II** functions as an activator of several triacylglycerol lipases and is associated with HDL and VLDL. **ApoC-III** is predominantly present in VLDL with small amounts found in HDL. The protein inhibits lipoprotein lipase (LPL) and hepatic lipase and decreases the uptake of chylomicrons by hepatic cells. **ApoC-IV** induces hypertriglyceridemia when overexpressed in mice (Allan and Taylor 1996; Kim et al. 2008). In normolipidemic plasma, greater than 80 % of the protein resides in VLDL, with most of the remainder in HDL. The HDL content of apoC-IV is much lower as compared to the other apoC proteins.

ApoD is a glycoprotein mainly associated with HDL (McConathy and Alaupovic 1973). The protein is expressed in many tissues, including liver and intestine. ApoD does not possess a typical apolipoprotein structure and belongs to the lipocalin family which also includes retinol-binding protein, lactoglobulin and uteroglobulin. Lipocalins are small lipid transfer proteins with a limited amino acid sequence identity but with a common tertiary structure. Lipocalins share a structurally conserved beta-barrel fold, which in many lipocalins bind hydrophobic ligands. As a result, apoD transports small hydrophobic ligands, with a high affinity for arachidonic acid (Rassart et al. 2000). In plasma, apoD forms disulfide-linked homodimers and heterodimers with apoA-II.

ApoE is a key structural and functional glycoprotein component of HDL despite its much lower content in HDL particles as compared to apoA-I (Utermann 1975). The major fraction of circulating apoE is carried by triglyceride-containing lipoproteins where it serves as a ligand for apoB/apoE receptors and ensures lipoprotein binding to cell-surface glycosaminoglycans. Similar to apoA-I and apoA-II, apoE contains eight amphipathic alpha-helical repeats and displays detergent-like properties towards phospholipids (Lund-Katz and Phillips 2010). ApoE is synthesised in multiple tissues and cell types, including liver, endocrine tissues, central nervous system and macrophages.

ApoF is a sialoglycoprotein present in human HDL and low-density lipoprotein (LDL) (Olofsson et al. 1978), also known as lipid transfer inhibitor protein (LTIP) as a consequence of its ability to inhibit CETP. ApoF is synthesised in the liver and heavily glycosylated with both O- and N-linked sugar groups. Such glycosylation renders the protein highly acidic and results in a molecular mass some 40 % greater than predicted (Lagor et al. 2009).

ApoH, also known as beta-2-glycoprotein 1, is a multifunctional N- and O-glycosylated protein. ApoH binds to various kinds of negatively charged molecules, primarily to cardiolipin, and may prevent activation of the intrinsic blood coagulation cascade by binding to phospholipids on the surface of damaged cells. Such binding properties are ensured by a positively charged domain. ApoH regulates platelet aggregation and is expressed by the liver.

ApoJ (also called **clusterin** and complement-associated protein SP-40,40) is an antiparallel disulfide-linked heterodimeric glycoprotein. Human apoJ consists of two subunits designated alpha (34–36 kDa) and beta (36–39 kDa) which share limited homology (de Silva et al. 1990a, b) and are linked by five disulfide bonds. The distinct structure of apoJ allows binding of both a wide spectrum of hydrophobic molecules on the one hand and of specific cell-surface receptors on the other.

ApoL-I is a key component of the trypanolytic factor of human serum associated with HDL (Duchateau et al. 1997). ApoL-I possesses a glycosylation site and shares structural and functional similarities with intracellular apoptosis-regulating proteins of the Bcl-2 family. ApoL-I displays high affinity for phosphatidic acid and cardiolipin (Zhaorigetu et al. 2008) and is expressed in pancreas, lung, prostate, liver, placenta and spleen.

ApoM is an apolipoprotein found mainly in HDL (Axler et al. 2007) which possesses an eight stranded antiparallel beta-barrel lipocalin fold and a hydrophobic pocket that ensures binding of small hydrophobic molecules, primarily sphingosine-1-phosphate (S1P) (Ahnstrom et al. 2007; Christoffersen et al. 2011). ApoM reveals 19 % homology with apoD, another apolipoprotein member of the lipocalin family (Sevvana et al. 2009), and is synthesised in the liver and kidney. The binding of apoM to lipoproteins is assured by its hydrophobic N-terminal signal peptide which is retained on secreted apoM, a phenomenon atypical for plasma apolipoproteins (Axler et al. 2008; Christoffersen et al. 2008; Dahlback and Nielsen 2009).

ApoO, a minor HDL component expressed in several human tissues (Lamant et al. 2006), is present in HDL, LDL and VLDL, belongs to the proteoglycan family and contains chondroitin sulphate chains, a unique feature distinguishing it from other apolipoproteins. The physiological function of apoO remains unknown (Nijstad et al. 2011).

Minor apolipoprotein components isolated within the density range of HDL are also exemplified by **apoB** and **apo(a),** which reflect the presence of lipoprotein (a) and result from overlap in the hydrated densities of large, light HDL2 and lipoprotein (a) (Davidson et al. 2009).

Enzymes

LCAT catalyses the esterification of cholesterol to cholesteryl esters in plasma lipoproteins, primarily in HDL but also in apoB-containing particles. Approximately 75 % of plasma LCAT activity is associated with HDL. In plasma, LCAT is closely associated with apoD, which frequently co-purify (Holmquist 2002). The LCAT gene is primarily expressed in the liver and, to a lesser extent, in the brain and testes. The LCAT protein is heavily N-glycosylated. The tertiary structure of LCAT is maintained by two disulfide bridges and involves an active site covered by a lid (Rousset et al. 2009). LCAT contains two free cysteine residues at positions 31 and 184.

Human **paraoxonases** (PON) are calcium-dependent lactonases PON1, PON2 and PON3 (Goswami et al. 2009). In the circulation, **PON1** is almost exclusively associated with HDL; such association is mediated by HDL surface phospholipids and requires the hydrophobic leader sequence retained in the secreted PON1. Human PON1 is largely synthesised in the liver but also in the kidney and colon (Mackness et al. 2010). Hydrolysis of homocysteine thiolactone has been proposed to represent a major physiologic function of PON1 (Jakubowski 2000). The name "PON" however reflects the ability of PON1 to hydrolyse the organophosphate substrate paraoxone together with other organophosphate substrates and aromatic carboxylic acid esters. Catalytic activities of the enzyme involve reversible binding to the substrate as the first step of hydrolytic cleavage. PON1 is structurally organised as a six-bladed beta-propeller, with each blade consisting of four beta-sheets (Harel et al. 2004). Two calcium atoms needed for the stabilisation of the structure and the catalytic activity of the enzyme are located in the central tunnel of the enzyme. Three helices, located at the top of the propeller, are involved in the anchoring of PON1 to HDL. The enzyme is N-glycosylated and may contain a disulfide bond. **PON2**, another member of the PON family, is an intracellular enzyme not detectable in serum despite its expression in many tissues, including the brain, liver, kidney and testis. The enzyme hydrolyses organophosphate substrates and aromatic carboxylic acid esters. **PON3** possesses properties which are similar to those of PON1, such as requirement for calcium, N-glycosylation, secretion in the circulation with retained signal peptide and association with HDL. PON3 displays potent lactonase, limited arylesterase and no PON activities and is predominantly expressed in the liver.

Platelet-activating factor acetyl hydrolase (PAF-AH) equally termed lipoprotein-associated phospholipase A_2 (**LpPLA$_2$**) is a calcium-independent, N-glycosylated enzyme, which degrades PAF by hydrolysing the sn-2 ester bond to yield biologically inactive lyso-PAF (Mallat et al. 2010). The enzyme cleaves phospholipid substrates with a short residue at the sn-2 position and can therefore hydrolyse proinflammatory oxidised short-chain phospholipids; however, it is inactive against long-chain non-oxidised phospholipids. PAF-AH is synthesised throughout the brain, white adipose tissue and placenta. Macrophages represent the most important source of the circulating enzyme (McIntyre et al. 2009). Plasma PAF-AH circulates in association with LDL and HDL particles, with the majority of the enzyme bound to small, dense LDL and to lipoprotein (a) (Tselepis et al. 1995). The crystal structure of PAF-AH reveals a typical lipase alpha/beta-hydrolase fold and a catalytic triad (Samanta and Bahnson 2008). The active site is close to the lipoprotein surface and at the same time accessible to the aqueous phase. Two clusters of hydrophobic residues build a lipid-binding domain ensuring association with lipoproteins.

Plasma glutathione selenoperoxidase 3 (GSPx-3), also called **glutathione peroxidase 3**, is distinct from two other members of the GSPx family termed GSPx-1 and GSPx-2 which represent erythrocyte and liver cytosolic enzymes. All GSPx enzymes protect biomolecules from oxidative damage by catalysing the reduction of hydrogen peroxide, lipid peroxides and organic hydroperoxide, in a reaction involving glutathione. Human GSPx-3 is a homotetrameric protein containing selenium as a selenocysteine residue at position 73. Human GSPx-3 is synthesised in the liver, kidney, heart, lung, breast and placenta. In plasma, GSPx-3 is exclusively associated with HDL (Chen et al. 2000).

Lipid Transfer Proteins

Phospholipid transfer protein (PLTP) belongs to the bactericidal permeability-increasing protein (BPI)/lipopolysaccharide (LPS)-binding protein (LBP)/Plunc superfamily of proteins. PLTP is synthesised in the placenta, pancreas, lung, kidney, heart, liver, skeletal muscle and brain. In the circulation, PLTP is primarily associated with HDL and converts it into larger and smaller particles. PLTP also plays a role in extracellular phospholipid transport and can bind LPS. The protein contains multiple glycosylation sites and is stabilised by a disulfide bond. PLTP is a positive acute-phase reactant with a potential role in the innate immune system.

CETP equally belongs to the BPI/LBP/Plunc superfamily and contains multiple N-glycosylation sites. It is primarily expressed by the liver and adipose tissue. In the circulation, CETP shuttles between HDL and apoB-containing lipoproteins and facilitates the bidirectional transfer of cholesteryl esters and triglycerides between them. The structure of CETP includes a hydrophobic tunnel filled with two cholesteryl ester molecules and plugged by an amphiphilic phosphatidylcholine

(PC) molecule at each end (Qiu et al. 2007). Such interactions additionally endow CETP with PC transfer activity. CETP may also undergo conformational changes to accommodate lipoprotein particles of different sizes (Qiu et al. 2007).

Acute-Phase Response Proteins

Positive acute-phase response proteins, whose plasma concentrations are markedly elevated by acute inflammation, form a large family of HDL-associated proteins (Vaisar et al. 2007; Heinecke 2009). Under normal conditions, the content of such proteins in HDL is however much lower as compared to apolipoproteins. On the other hand, plasma levels of several HDL apolipoproteins, such as apoA-I and apoA-IV, are reduced during the acute-phase response (Navab et al. 2004); such proteins can therefore be considered as negative acute-phase response proteins.

Serum amyloid A (SAA) proteins, major acute-phase reactants, are secreted during the acute phase of the inflammatory response. In humans, three SAA isoforms, SAA1, SAA2 and SAA4, are produced predominantly by the liver. Hepatic expression of SAA1 and SAA2 in the liver is induced during the acute-phase reaction, resulting in increase in their circulating levels by as much as 1,000-fold from basal concentrations of about 1–5 mg/l (Khovidhunkit et al. 2004). By contrast, SAA4 is expressed constitutively in the liver and is therefore termed constitutive SAA. **SAA1**, the major member of this family, is predominantly carried by HDL in human, rabbit and murine plasma (Hoffman and Benditt 1982; Marhaug et al. 1982; Cabana et al. 1996). In the circulation, SAA1 does not exist in a free form and associates with non-HDL lipoproteins in the absence of HDL (Cabana et al. 2004).

LBP is an acute-phase glycoprotein capable of binding the lipid A moiety of LPS of Gram-negative bacteria and facilitating LPS diffusion (Wurfel et al. 1994). LBP/LPS complexes appear to interact with the CD14 receptor to enhance cellular responses to LPS. LBP also binds phospholipids, thereby acting as a lipid exchange protein (Yu et al. 1997), and belongs to the same BPI/LBP/Plunc protein superfamily as PLTP and CETP.

Fibrinogen is a common acute-phase protein and a cofactor in platelet aggregation synthesised by the liver, which is converted by thrombin into fibrin during blood coagulation. Fibrinogen is a disulfide-linked heterohexamer which contains two sets of three non-identical chains (alpha, beta and gamma).

Alpha-1-acid glycoprotein 2, equally termed orosomucoid-2, belongs to the calycin protein superfamily which also includes lipocalins and fatty acid-binding proteins. The protein appears to modulate activity of the immune system during the acute-phase reaction. In plasma, the protein is N-glycosylated and stabilised by disulfide bonds.

Alpha-2-HS-glycoprotein (fetuin-A) promotes endocytosis, possesses opsonic properties and influences the mineral phase of bone. The protein shows affinity for calcium and barium ions and contains two chains, A and B, which are held together by a single disulfide bond. Alpha-2-HS-glycoprotein is synthesised in the liver and secreted into plasma.

Complement Components

Several complement components associate with HDL. **Complement component 3 (C3)** plays a central role in the activation of the complement system through both classical and alternative activation pathways. C3 exists in a form of two chains, beta and alpha, linked by a disulfide bond. **C4** is a key component involved in the activation of the classical pathway of the complement system, which circulates as a disulfide-linked trimer of alpha, beta and gamma chains. **C4b-binding protein** controls the classical pathway of complement activation and binds as a cofactor to C3b/C4b inactivator, which then hydrolyses complement fragment C4b. **C9** is a pore-forming subunit of the membrane attack complex that provides an essential contribution to the innate and adaptive immune response by forming pores in the plasma membrane of target cells.

Vitronectin is another HDL-associated protein involved in complement regulation. The protein belongs to cell-to-substrate adhesion molecules present in serum and tissues, which interact with glycosaminoglycans and proteoglycans and can be recognised by members of the integrin family. In complement regulation, vitronectin serves as an inhibitor of the membrane-damaging effect of the terminal cytolytic pathway. Vitronectin is largely expressed in the liver but also in visceral tissue and adrenals. The presence of vitronectin in HDL raises the possibility that certain HDL components can be derived from non-cellular sources or cells distinct from those that synthesise apoA-I in the liver and intestine (Heinecke 2009).

Proteinase Inhibitors

A family of proteins in HDL contains serine proteinase inhibitor domains (Vaisar et al. 2007). Serine protease inhibitors (serpins) are important regulators of biological pathways involved in inflammation, coagulation, angiogenesis and matrix degradation. HDL-associated serpins are exemplified by **alpha-1-antitrypsin** which in the circulation is exclusively present in HDL (Karlsson et al. 2005; Ortiz-Munoz et al. 2009). **Alpha-2-antiplasmin**, a serpin that inhibits plasmin and trypsin and inactivates chymotrypsin, is another key proteinase inhibitor which circulates in part associated with HDL. HDL also carries inter-alpha-trypsin inhibitor heavy chain H4 and bikunin, two components of inter-alpha-trypsin inhibitors consisting of three of four heavy chains selected from the groups 1 to 4 and one light chain selected from the alpha-1-microglobulin/bikunin precursor or Kunitz-type protease inhibitor 2 groups. The full complex inhibits trypsin, plasmin and lysosomal granulocytic elastase. **Inter-alpha-trypsin inhibitor heavy chain H4** is produced in the liver and is cleaved by kallikrein to yield 100 and 35 kDa fragments in plasma, and the resulting 100 kDa fragment is further converted to a 70 kDa fragment. **Bikunin** is a light chain of the inter-alpha-trypsin inhibitor, which is produced via proteolytic cleavage of the alpha-1-microglobulin/bikunin precursor protein together with alpha-1-microglobulin and trypstatin.

Other proteolysis-related proteins which are detected on HDL include haptoglobin-related protein (Hrp), kininogen-1, prothrombin, angiotensinogen and procollagen C-proteinase enhancer-2 (PCPE2). **Hrp** contains a crippled catalytic triad residue that may allow it to act as a decoy substrate to prevent proteolysis.

Kininogen-1 (alpha-2-thiol proteinase inhibitor) plays an important role in blood coagulation and inhibits the thrombin- and plasmin-induced aggregation of thrombocytes. **Prothrombin** is a precursor of thrombin, a key serine protease of the coagulation pathway. **Angiotensinogen** is a an alpha-2-globulin that is produced constitutively mainly by the liver and represents a substrate for renin whose action forms angiotensin I. **PCPE2** binds to the C-terminal propeptide of type I or II procollagens and enhances the cleavage of the propeptide by bone morphogenetic protein 1 (BMP-1, also termed procollagen C-proteinase).

Other Protein Components

HDL equally transports distinct proteins displaying highly specialised functions. The metabolic purpose of such association is unclear; it might prolong the residence time of a protein or represent a mechanism for protein conservation in the circulation. For example, plasma **retinol-binding protein**, which delivers retinol from the liver stores to peripheral tissues, co-isolates with HDL3 (Vaisar et al. 2007). In plasma, the complex of retinol-binding protein and retinol interacts with **transthyretin**, thereby preventing loss of retinol-binding protein by filtration through the kidney glomeruli. As a corollary, transthyretin, a homotetrameric thyroid hormone-binding protein, is equally present on HDL (Hortin et al. 2006; Vaisar et al. 2007; Davidson et al. 2009). **Serotransferrin,** an iron-transport glycoprotein largely produced in the liver, is also in part associated with HDL. **Hemopexin**, an iron-binding protein that binds heme and transports it to the liver for breakdown and iron recovery, equally co-isolates with HDL3 (Vaisar et al. 2007).

HDL also carries proteins involved in the regulation of various biological functions, such as **Wnt signalling molecules,** which participate in cell-to-cell signalling (Neumann et al. 2009), and **progranulin**, a precursor of granulins which play a role in inflammation, wound repair and tissue remodelling.

In addition, HDL transports lysosomal proteins, such as **prenylcysteine oxidase,** which is involved in the degradation of prenylated proteins (Vaisar et al. 2007). Other minor abundance proteins reported to be associated with HDL are **albumin**, **alpha-1B-glycoprotein**, **alpha-amylase**, **vitamin D-binding protein** and **platelet basic protein** (Vaisar et al. 2007; Davidson et al. 2009).

In addition to proteins, HDL carries a large number of small **peptides** in the mass range from 1 to 5 kDa (Hortin et al. 2006). These peptides are present in HDL at low concentrations of about 1 % of total HDL protein, with some representing fragments of larger proteins, such as apoB, fibrinogen and transthyretin (Hortin et al. 2006). The association of small peptides with HDL as a vehicle may represent a pathway for peptide delivery or scavenging, in order to slow renal clearance and proteolysis (Hortin et al. 2006).

Heterogeneity in HDL Proteins

HDL proteins are non-uniformly distributed across HDL subpopulations. Indeed, proteomic analysis of five HDL subpopulations isolated from normolipidemic subjects by isopycnic density gradient ultracentrifugation identifies five distinct patterns of distribution of individual protein components across the HDL density subfractions (Davidson et al. 2009) (Fig. 1). The most interesting of these distributions identifies small, dense HDL3b and 3c as particle subpopulations in which seven proteins occur predominantly, notably apoJ, apoL-1, apoF, PON1/3, PLTP and PAF-AH. Activities of HDL-associated enzymes (LCAT, PON1, PAF-AH) are equally elevated in small, dense HDL3c (Kontush et al. 2003; Kontush and Chapman 2010). The HDL3c proteome also contains apoA-I; apoA-II; apoD; apoM; SAA 1, 2, and 4; apoC-I; apoC-II; and apoE (Davidson et al. 2009). Consistent with these data, apoL-I (Hajduk et al. 1989), apoF (He et al. 2008; Lagor et al. 2009), apoJ (de Silva et al. 1990a, b; Bergmeier et al. 2004), PON1 (Kontush et al. 2003; Bergmeier et al. 2004), apoA-IV (Bisgaier et al. 1985; Ohta et al. 1985), apoM (Wolfrum et al. 2005), apoD (Campos and McConathy 1986) and SAA1/2 (Benditt and Eriksen 1977; Coetzee et al. 1986) are known to preferentially co-isolate with dense HDL3. Furthermore, small, dense HDL may also represent a preferential carrier for human CETP (Marcel et al. 1990). On the other hand, apoE, apoC-I, apoC-II and apoC-III preferentially localise to large, light HDL2 (Schaefer et al. 1979; Cheung and Albers 1982; Schaefer and Asztalos 2007; Davidson et al. 2009) (Fig. 1). Importantly, these associations can in part be confirmed using an alternative approach of gel filtration subfractionation of HDL particles (Gordon et al. 2010).

The low HDL content of the majority of these proteins of less than one copy per HDL particle suggests internal heterogeneity of the HDL3c subfraction. This conclusion is further consistent with the isolation of a unique particle containing the trypanosome lytic factor apoL-I, plus apoA-I and Hrp, in the HDL3 density range (Shiflett et al. 2005).

In a similar fashion, apoF co-isolates with dense HDL (mean density, 1.134 g/ml) which also contains apoA-I, apoC-II, apoE, apoJ, apoD and PON1 (He et al. 2008). Complexes formed by PON1 with human phosphate-binding protein and apoJ represent another example of protein–protein interactions occurring within the HDL particle spectrum (Vaisar 2009).

Specific protein–protein interactions should thus drive the formation of such complexes in the circulation. In support of such a mechanism, PLTP in human plasma resides on lipid-poor complexes dominated by apoJ and proteins implicated in host defence and inflammation (Cheung et al. 2010).

In addition and as mentioned above (Sect. 1), immunoaffinity technique allows separating HDL particles containing only apoA-I (LpA-I) from those containing both apoA-I and apoA-II (LpA-I:A-II) (Duriez and Fruchart 1999). ApoA-I is typically distributed approximately equally between LpA-I and LpA-I:A-II, whereas virtually all apoA-II is in LpA-I:A-II (Duriez and Fruchart 1999). LpA-I and LpA-I:A-II contain approximately 35 and 65 % of plasma apoA-I, respectively (James et al. 1988). On the other hand, approximately half of HDL particles contain apoA-II (Wroblewska et al. 2009). In addition to apoA-II, the two subclasses may differ in their content of other proteins, as exemplified by PON1 which preferably associates with LpA-I (Moren et al. 2008).

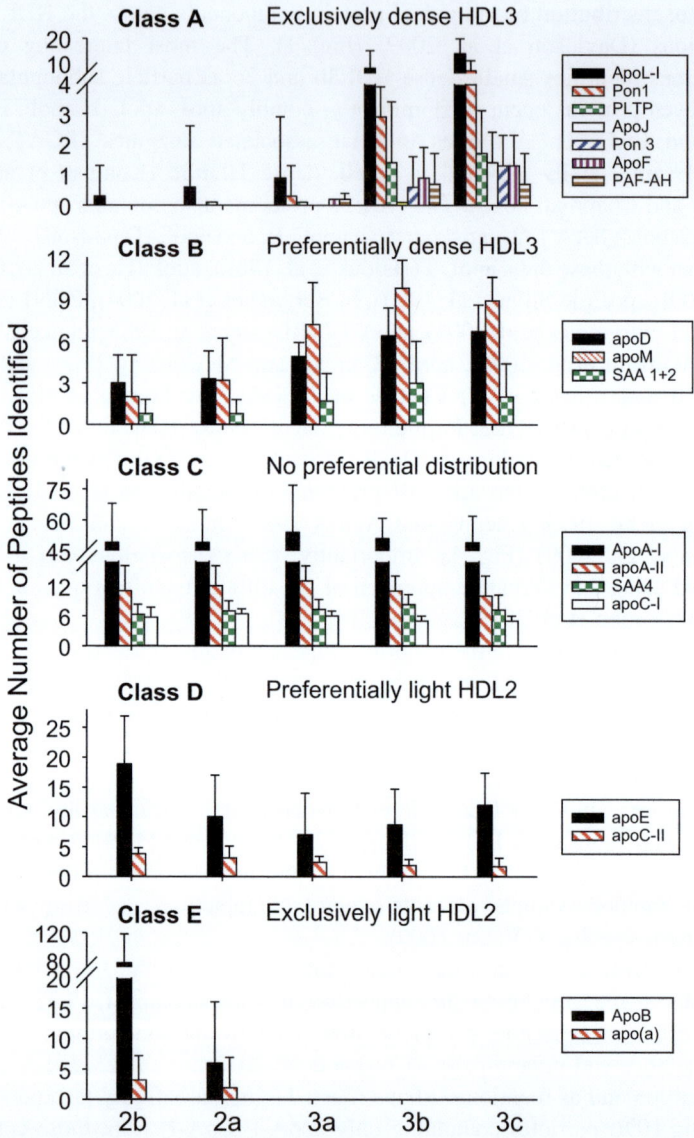

Fig. 1 Abundance pattern of proteins across healthy normolipidemic HDL subpopulations. Class A: exclusively present in small, dense HDL3b and 3c. Class B: enriched in small, dense HDL3b and 3c. Class C: equally abundant across HDL subpopulations. Class D: enriched in large, light HDL2b and 2a. Class E: exclusively present in large, light HDL2b and 2a [modified from (Davidson et al. 2009)]

LpA-I particles can be further subfractionated according to size. The number of apoA-I molecules in such subpopulations is increased from two to three to four with an increase in the particle size (Gauthamadasa et al. 2010). On the other hand, the entire population of LpA-I:A-II demonstrates the presence of only two apoA-I molecules per particle, while the number of apoA-II molecules varies from one dimeric apoA-II to two and then to three. Upon compositional analyses of individual subpopulations, LpA-I:A-II exhibits comparable proportions for major lipid classes across subfractions, while LpA-I components show significant variability (Gauthamadasa et al. 2010).

Another important subpopulation of HDL particles is formed by apoE-containing HDL. The presence of apoE facilitates expansion of the lipid core as a result of the accumulation of cholesteryl ester, with formation of large, lipid-rich HDL; these particles represent an excellent ligand for the LDL receptor (Hatters et al. 2006).

The diversity of molecules which bind to HDL suggests that the lipoprotein can serve as a versatile adsorptive surface for proteins and peptides to form complexes playing roles not only in lipid metabolism but equally in acute-phase response, innate immune response, complement activation, plaque stability and proteolysis inhibition (Heinecke 2009). As the abundance of the most of HDL-associated proteins is below 1 mol/mol HDL (i.e. less than 1 copy per HDL particle), it remains however unclear as to how they are distributed among minor HDL subpopulations of potentially distinct origin and function. Specific proteins may therefore be confined to distinct HDL subpopulations of distinct origin and function, which are differentially distributed across the HDL particle spectrum (Davidson et al. 2009). The HDL fraction as a whole therefore appears to represent "a collection of individualised species with distinct functionalities that happen to have similar physicochemical properties" (Shah et al. 2013) and are primarily defined by specific protein–protein interactions facilitated by the presence of phospholipid, rather than "a transient ensemble of randomly exchanging proteins" (Gordon et al. 2013; Shah et al. 2013).

2.1.2 Protein Isoforms, Translational and Posttranslational Modifications

In line with the evolvement of proteomics, a large number of proteins have been identified in HDL as described in the previous section (Vaisar et al. 2007). In addition, most proteins are expressed as different isoforms due to co- and posttranslational modifications (Karlsson et al. 2005; Candiano et al. 2008). This makes the proteome of HDL both complex and dynamic, which most likely result in various HDL particles with different protein composition in respect to the environment. Common posttranslational modifications (PTMs) such as glycosylations, truncations and phosphorylations change the charge and/or the size of the protein, which can be utilised for separation of the isoforms, and modern mass spectrometric techniques can be used to detect, characterise and nowadays also measure even small mass differences in proteins. However, although isoforms of major HDL proteins have been known for decades (Zannis et al. 1980; Hussain and Zannis 1990), surprisingly little is still known on how these variations of the HDL proteome affect the functionality. The following is a comprehensive review of isoforms patterns described in common human HDL apolipoproteins, apoA-I, apoA-II, apoC-III and SAA, and their possible functional relevance.

ApoA-I (p*l* 5.3/28 kDa, accession no. P02647) is normally found as different charge isoforms; besides the major isoform (70–75 % of total apoA-I) and the slightly more basic pre-apoA-I (5–10 % of total apoA-I), also two more acidic isoforms are generally detected by isoelectric focusing (IEF) and 2D gel electrophoresis (2-DE) (Contiero et al. 1997; Karlsson et al. 2005). The nature of these acidic isoforms is still unclear. An early report suggested deamidation of Gln or Asn residues, resulting in a +1 charge shift, which could be formed during the analytical procedure (Ghiselli et al. 1985). At the same time, a few reports indicated the importance of acidic apoA-I in vivo; increased levels of acidic apoA-I, while decreased levels of the major form, were found in LDL from obese subjects, especially in women (Karlsson et al. 2009). Also, higher degree of deamidated apoA-I has been shown in relation to diabetes (Jaleel et al. 2010) and acidic apoA-I may be more vulnerable to methionine oxidation (Fernandez-Irigoyen et al. 2005). In 2-DE HDL protein patterns also 30–35 kDa variants of apoA-I are usually detected (Karlsson et al. 2005). Although mass spectrometry (MS) analysis was in agreement with O-glycosylation at two potential sites (Thr78 or Thr92), this has not been confirmed by others, and it is generally regarded that apoA-I is not N-linked or O-linked glycosylated. In contrast, non-enzymatically glycation of apoA-I has been found in association to diabetes and believed to affect apoA-I functions, such as LCAT activation (Fievet et al. 1995; Nobecourt et al. 2007; Park et al. 2010).

Another potentially important PTM of apoA-I is truncation. During atherosclerotic inflammation, apoA-I might be N-terminally and C-terminally truncated by released proteases. Specific cleavage sites at Tyr42, Phe57, Tyr216 and Phe253 for chymase have been identified that in reconstituted HDL reduces its ability to promote cholesterol efflux (Lee et al. 2003; Usami et al. 2013). Low amounts of C-terminally truncated apoA-I can be measured in normal serum (Usami et al. 2011) and fragmented apoA-I is a feature in plasma from children with nephrotic syndrome, a condition linked to higher risk of atherosclerosis (Santucci et al. 2011). Truncation has also been implicated in apoA-I dimerisation as studied in apoA-I Milano (R197C) and apoA-I Paris (R175C) (Calabresi et al. 2001; Favari et al. 2007; Gursky et al. 2013). Notably, apoA-I with an apparent molecular mass of 50 kDa, consistent with dimeric apoA-I, has been found in patients with myocardial infarction (Majek et al. 2011) but also appears to be present in HDL from healthy individuals (Karlsson et al. 2005). Finally, oxidatively modified apoA-I have been extensively studied during the recent years by the help of MS techniques, as described in detail in several reviews elsewhere (e.g. Nicholls and Hazen 2009; Shao 2012). It has been proposed that myeloperoxidase-mediated inflammation results in oxidation of apoA-I. Specific sites have been identified for methionine oxidation, for nitrated/chlorinated tyrosines and for lysines modified by reactive carbonyls. Importantly, the modifications have been coupled to functional impairment in apoA-I activity such as ABCA1-mediated cholesterol efflux and are linked to cardiovascular disease.

ApoA-II (pI 5.0/8.7 kDa, accession no. P02652) is mostly found as two isoforms in HDL that differ slightly according to pI and molecular mass, probably due to O-linked glycosylation/sialylation (Karlsson et al. 2005; Halim et al. 2013). Similar to apoA-I, the protein is produced as a more basic pro-form (Hussain and Zannis 1990). In contrast, apoA-II appears to be quickly processed to the mature form, as the pro-form is not found in the circulation. In addition to glycosylation, phosphorylation at Ser68, C-terminal truncated variants (des-Gln and des-Thr-Gln) and a cysteinylated variant has been detected in the circulation (Jin and Manabe 2005; Nelsestuen et al. 2008; Zhou et al. 2009). ApoA-II also forms a homodimer at Cys29 that is abundant in plasma (Gillard et al. 2005; Jin and Manabe 2005). Overall, more than ten different variants of apoA-II are present in humans, but the physiological relevance of this heterogeneity is unclear. However, sialylated apoA-II appear to be selectively associated to HDL3 (Remaley et al. 1993; Karlsson et al. 2005), and elevated levels of modified apoA-II isoforms have been linked to premature delivery in pregnant women (Flood-Nichols et al. 2013).

ApoC-III (pI 4.7/8.8 kDa, accession no. P02656) is generally found as three charge isoforms depending on O-linked glycosylation (GalGalNAc) at Thr94 with or without sialylation; disialylated apoC-III$_2$, monosialylated apoCIII$_1$ and non-sialylated apoC-III$_0$ (Karlsson et al. 2005; Bruneel et al. 2008). An early report showed that glycosylation is not necessary for apoC-III secretion and does not affect its relative affinity to different lipoprotein particles (Roghani and Zannis 1988), and, as judged by gel electrophoresis and MS analysis of HDL and plasma, the non-sialylated variant is least abundant, usually less than 5 % of total apoC-III in normal individuals (Wopereis et al. 2003; Bruneel et al. 2008; Mazur et al. 2010; Holleboom et al. 2011). In addition to glycosylation, apoC-III can also be C-terminal truncated (des-Ala and des-Ala-Ala), which further increases the number of isoforms (Bondarenko et al. 1999; Jin and Manabe 2005; Nicolardi et al. 2013a). Interestingly, novel results strongly suggest that glycosylation of apoC-III is an important event in the regulation of lipid metabolism (Holleboom et al. 2011; Baenziger 2012). Thus, apoC-III is exclusively glycosylated by GalNAc transferase 2 (GALNT2) (Holleboom et al. 2011; Schjoldager et al. 2012), and heterozygotes with a loss-of-function mutation in GALNT2 present with an altered apoC-III isoform pattern with more of the non-sialylated variant and less of the monosialylated variant, while the total apoC-III plasma concentration was about the same as compared to wild-type controls (Holleboom et al. 2011). This is then linked to reduced inhibition of lipoprotein lipase and improved triglyceride clearance. In line, the production rate of apoC-III$_1$ and -III$_2$ is more strongly correlated with plasma triglyceride levels than apoC-III$_0$ (Mauger et al. 2006), increased apoC-III$_1$/apoC-III$_0$ ratio has been found in diabetic subjects (Jian et al. 2013), and HDL3 from subjects with low HDL-C is characterised by higher levels of monosialylated apoC-III than subjects with high HDL-C (Mazur et al. 2010). The evaluation of apoC-III isoforms is complicated by the fact that apoC-III$_0$ can be separated into a non-glycosylated form and glycosylated but non-sialylated forms (Bruneel et al. 2008; Holleboom et al. 2011; Nicolardi et al. 2013a). Moreover, a recent MS study of 96 serum samples showed that 30 % of the individuals displayed an apoC-III pattern with additional glycosylated variants, characterised by fucosylation (Nicolardi et al. 2013b). The relevance of these glycosylated non-sialylated variants of apoC-III, as of the C-terminal truncated forms, is yet unclear, but may explain somewhat contradictory results showing higher relative levels of non- and less-sialylated apoC-III in obese subjects than in lean subjects (Harvey et al. 2009; Karlsson et al. 2009), although obesity is generally associated with high triglyceride levels.

SAA exists in a form of SAA1 (pI 5.9/11.7 kDa, accession no. P0DJI8) and SAA2 (pI 8.3/11.6 kDa, accession no. P0DJI9). The two proteins display about 93 % sequence homology, and depending on natural variation in alleles, SAA1 is separated into five isoforms, SAA1.1 to SAA1.5, and SAA2 is separated into two isoforms SAA2.1 and SAA2.2 (often also denoted as alpha, beta, etc.), with one to three amino acid difference between the isoforms. In addition, both N-terminal and C-terminal truncated variants of SAA1/SAA2 are detected in serum (Ducret et al. 1996; Kiernan et al. 2003; de Seny et al. 2008). By using a combined 2-DE/MS and SELDI-TOF approach, eight isoforms were identified in HDL after LPS infusion in healthy individuals; besides native SAA1.1 and SAA2.1, N-terminal truncations (des-R, -RS and -RSFF) of each variant were also found (Levels et al. 2011). Interestingly, a subgroup, based on HDL protein profile, characterised by elevated antioxidative PON1 activity showed a delayed response of SAA to LPS in particular for the most truncated (des-RSFF) variants. Otherwise, very little is known about differential physiological relevance of the SAA isoforms. However, SAA2.1 but not SAA1.1 has been shown to promote cholesterol efflux from macrophages (Kisilevsky and Tam 2003). Today, contradictory results make it unclear whether the increased level of SAA in inflammation, believed to replace apoA-I in HDL, actually is a mechanism in atherosclerosis or merely is a marker for inflammation (de Beer et al. 2010; Chiba et al. 2011; Kisilevsky and Manley 2012). Future differential quantitative MS analysis of the highly homologous SAA isoforms such as described by Sung et al. (2012) may resolve this controversy.

Sensitive MS techniques have revealed a large number of proteins associated to HDL. Most of them are also expressed as different isoforms depending on translational and posttranslational modifications. This leads to a need to develop MS-based methods for specific and reliable quantification of protein isoforms. With appropriate standards, measurements can be performed with low coefficient of variation and with a specificity superior to, e.g. immunoassays. Consequently, such applications in the field of HDL are being presented by using, e.g. multiple reaction monitoring and top-down proteomics with high-resolution MS (Mazur et al. 2010; Sung et al. 2012). Another interesting and fairly simple approach is to use ratio determinations, e.g. modified/native protein expression (Nelsestuen et al. 2008). As these measures are concentration-independent, they bear a potential to reduce individual variations. Furthermore, such MS approaches are not only useful for PTMs but also of value to understand the impact of protein variations caused by genetic polymorphism. For example, a recent study of heterozygotes with an apoA-I mutation (K131Del) showed that, in contrast to what could be expected, the mutant protein was more abundantly expressed in HDL than the native protein (Ljunggren et al. 2013). Herein four HDL proteins that are all expressed as different isoforms have been discussed: two (apoA-I and apoC-III) in which PTMs have been shown to be important for lipid metabolism and two (apoA-II and SAA) in which the role of PTMs is still unclear. In light of the vital importance of carboxylations and truncations in the processes of haemostasis, which also involves other PTMs such as phosphorylation, hydroxylation, glycosylation and sulphation, it appears highly unlikely that the diversity in the "HDLome" would not be relevant for lipid metabolism and cardiovascular disease. Therefore, characterisation of PTMs is probably one of the most challenging but also one of the most important tasks in order to understand the complex function of HDL.

2.2 Lipidome

The real power of lipidomic technologies involving mass spectrometry results from their ability to provide quantitative data on individual molecular species of lipids and on low-abundance lipid molecules. The pioneering study of Wiesner and colleagues published in 2009 (Wiesner et al. 2009) provided reference values for the lipidome of HDL isolated from healthy normolipidemic controls by FPLC. In an attempt to further characterise HDL composition and address its inherent heterogeneity, we recently reported the phospho- and sphingolipidome of five major HDL subpopulations isolated from healthy normolipidemic subjects (Camont et al. 2013).

2.2.1 Phospholipids

Phosphatidylcholine is the principal plasma phospholipid that accounts for 32–35 mol % of total lipids in HDL (Wiesner et al. 2009) (Table 3). PC is a structural lipid, consistent with its even distribution across HDL subpopulations (Fig. 2). Major molecular species of PC are represented by the 16:0/18:2, 18:0/18;2 and 16:0/20:4 species (Lhomme et al. 2012). As compared to other lipoproteins, HDL is enriched in PC containing polyunsaturated fatty acid moieties (Wiesner et al. 2009).

LysoPC is an important phospholipid subclass in HDL (1.4–8.1 mol % of total lipids; Table 3). It is derived from the regulated degradation of PC by phospholipases, including LCAT, consistent with the preferential association of the latter with HDL particles (Kontush et al. 2007). More specifically, LCAT was reported earlier to associate mainly with small, dense HDL particles, which are also enriched in lysoPC by approximately twofold as compared to large, light HDL (Camont et al. 2013) (Fig. 2). LysoPC is also produced by the hydrolytic action of Lp-PLA$_2$ on oxidised PC or by secreted PLA$_2$ under pro-atherogenic conditions, such as oxidative stress and inflammation, and constitutes therefore a potential biomarker of inflammation. Major molecular species of HDL lysoPC contain saturated fatty acid moieties of predominantly 16 and 18 carbon atoms, reflecting LCAT preference for 16 and 18 carbon atom long PCs (Lhomme et al. 2012). As

Table 3 Major components of the HDL lipidome

Lipid class	HDL content in mol % of total lipids
Phospholipids	*37.4–49.3*
Phosphatidylcholine	32–35
PC-plasmalogen	2.2–3.5
LysoPC	1.4–8.1
Phosphatidylethanolamine	0.70–0.87
PE-plasmalogen	0.54–0.87
Phosphatidylinositol	0.47–0.76
Cardiolipin	0.077–0.201
Phosphatidylserine	0.016–0.030
Phosphatidylglycerol	0.004–0.006
Phosphatidic acid	0.006–0.009
Sphingolipids	*5.7–6.9*
Sphingomyelin	5.6–6.6
Ceramide	0.022–0.097
Hexosyl Cer	0.075–0.123
Lactosyl Cer	0.037–0.060
S1P d18:1	0.015–0.046
S1P d18:0	0.007
SPC d18:1	0.001
Neutral lipids	*46.7–54.0*
Cholesteryl esters	35–37
Free cholesterol	8.7–13.5
Triacylglycerides	2.8–3.2
Diacylglycerides	0.17–0.28
Minor lipids	
Free fatty acids	16:0, 18:0, 18:1[a]
Isoprostane-containing PC	ND (IPGE2/D2-PC (36:4))[a]

Data are shown for HDL obtained from normolipidemic healthy subjects according to Deguchi et al. (2000), Kontush et al. (2007), Wiesner et al. (2009), Camont et al. (2013), Stahlman et al. (2013), Pruzanski et al. (2000), Sattler et al. (2010), Argraves et al. (2011)
SPC sphingosylphosphorylcholine, *S1P* sphingosine-1-phosphate, *IPGE2* isoprostaglandin E2
[a] no quantitative data available, major molecular species identified

considerable amounts of serum lysoPC are also associated with albumin (Wiesner et al. 2009), HDL contamination by the both compounds is typical for FPLC isolation. However, in HDL isolated by isopycnic density gradient ultracentrifugation, lysoPC content in HDL is two- to tenfold lower (Camont et al. 2013; Stahlman et al. 2013).

Phosphatidylethanolamine (PE) is moderately abundant in HDL (0.7–0.9 mol % of total lipids; Table 3), and its content tends to increase with increasing HDL hydrated density (Wiesner et al. 2009; Camont et al. 2013) (Fig. 2). PE principal molecular species are represented by the 36:2 and 38:4 fatty acid residues in HDL (Kontush et al. 2007).

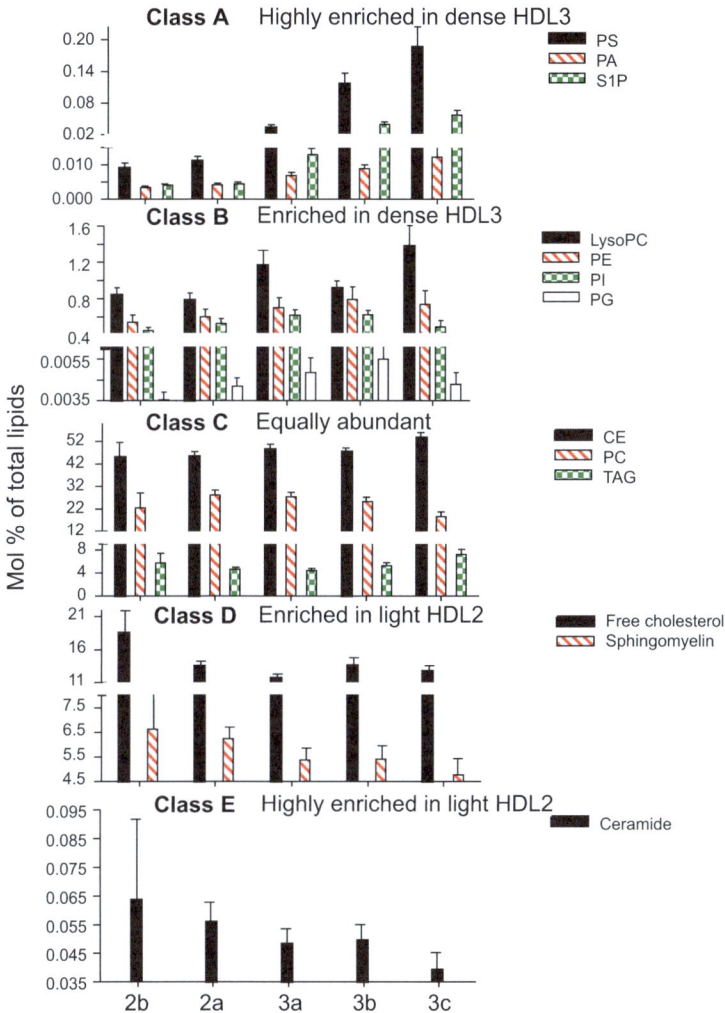

Fig. 2 Abundance pattern of lipids across healthy normolipidemic HDL subpopulations. Class A: highly enriched in small, dense HDL3b and 3c (>1.5-fold relative to HDL2b). Class B: enriched in small, dense HDL3b and 3c (1.2–1.5-fold relative to HDL2b). Class C: equally abundant across HDL subpopulations (<1.2-fold variations between HDL2b and HDL3b + 3c). Class D: enriched in large, light HDL2b (1.2–1.5-fold relative to HDL3b + 3c). Class E: highly enriched in large, light HDL2b (>1.5-fold relative to HDL3b + 3c). S1P, sphingosine-1-phosphate

Plasmalogens contain a vinyl ether-linked fatty acid essential for their specific antioxidative properties (Maeba and Ueta 2003). PC-plasmalogens are the most abundant species in HDL (2.2–3.5 mol %) but represent less than 10 % of total PC (Stahlman et al. 2013). On the contrary, PE-plasmalogens and PE are equally abundant in HDL (0.6–0.9 mol %; Table 3). PC- and PE-plasmalogens contain mainly polyunsaturated species: 38:5 and 36:2 in PC-plasmalogens (Stahlman et al. 2013) and 18:0/20:4 and 16:0/20:4 in PE-plasmalogens (Wiesner et al. 2009).

Phosphatidylinositol (PI), **phosphatidylserine** (PS), **phosphatidylglycerol** (PG), **phosphatidic acid** (PA) and **cardiolipin** are negatively charged phospholipids present in HDL (Table 3) which may significantly impact on its net surface charge (Rosenson et al. 2011; Lhomme et al. 2012). The content of these lipids can thereby modulate lipoprotein interactions with lipases, membrane proteins, extracellular matrix and other protein components; indeed, such interactions are largely charge-dependent.

PI, similarly to PE, is moderately abundant in HDL (0.5–0.8 mol %; Table 3) and tends to be enriched in small, dense HDL (Fig. 2). Major molecular species of PI in HDL include the 18:0/20:3 and 18:0/20:4 species (Lee et al. 2010).

PS is a minor negatively charged phospholipid component of HDL (0.016–0.030 mol %; Table 3). This phospholipid was very recently reported to be highly enriched (34-fold) in the small, dense HDL3c subpopulation relative to large, light HDL2 (Fig. 2) (Camont et al. 2013) as well as in small discoid preβ HDL and small nascent HDL formed by ABCA1 (up to 2.5 mol % of total lipids) (see Kontush and Chapman 2012 for review). Interestingly, small, dense HDL also displayed potent biological activities which correlated positively with PS content in HDL (Camont et al. 2013). This lipid could therefore, in part, account for enhanced functionality of HDL3c.

PA, a second messenger, is both a common metabolic precursor and an enzymatic product of phospholipid metabolism. This negatively charged lipid is present in very low abundance in HDL (0.006–0.009 mol %; Table 3) but, similarly to PS, is enriched in small, dense HDL (by more than threefold) (Camont et al. 2013) (Table 3). This observation might reflect preferential association of PA with apoL-I which is equally enriched in small, dense HDL (Kontush and Chapman 2012).

PG is a metabolic precursor of cardiolipin present in HDL in very low amounts (0.004–0.006 mol %; Table 3). PG tends to be enriched in small, dense particles (Camont et al. 2013) (Fig. 2).

Cardiolipin is a minor anionic phospholipid present in trace amounts in HDL (0.08–0.2 mol %; Table 3). This lipid with potent anticoagulant properties may contribute to the effects of lipoproteins on coagulation and platelet aggregation (Deguchi et al. 2000).

Together, these data indicate that although negatively charged lipids represent minor HDL constituents (0.8 mol % of total lipids), they are highly enriched in small, dense HDL, consistent with the elevated surface electronegativity of this subpopulation (Rosenson et al. 2011).

Isoprostanes are well established as biomarkers of oxidative stress and are predominantly associated with HDL (see Kontush and Chapman 2012 for review). Major molecular species of isoprostane-containing PCs include 5,6-epoxy-isoprostaglandine A2-PC (EIPGA2-PC) 36:3, 5,6 EIPGE2-PC 36:4, IPGE2/D2-PC 36:4, IPGF-PC 36:4, IPGE2/D2-PC 38:4 and IPGF-PC 38:4 (Pruzanski et al. 2000) (Table 3).

2.2.2 Sphingolipids

Sphingomyelin, a structural lipid which enhances surface lipid rigidity (Rye et al. 1996; Saito et al. 2000), is the major sphingolipid in circulating HDL (5.6–6.6 mol % of total lipids) (Wiesner et al. 2009; Camont et al. 2013; Stahlman et al. 2013) (Table 3), which largely originates from triacylglyceride-rich lipoproteins and only to a minor extent from nascent HDL (Nilsson and Duan 2006). Major molecular species of sphingomyelin are the 16:0 and 24:1 species (Lhomme et al. 2012). Unlike negatively charged PL, sphingomyelin is depleted by up to 30 % in small, dense relative to large, light HDL (Kontush et al. 2007; Camont et al. 2013) (Fig. 2). This result may, in part, reflect the low abundance of sphingomyelin in nascent HDL, a metabolic precursor of HDL3c, and suggest distinct metabolic pathways for HDL subpopulations.

Among lysosphingolipids, **S1P** is particularly interesting as this bioactive lipid plays key roles in vascular biology (Lucke and Levkau 2010). More than 90 % of circulating sphingoid base phosphates are found in HDL and albumin-containing fractions (Table 3) (Kontush and Chapman 2012). Interestingly, S1P associates preferentially with small, dense HDL particles (up to tenfold enrichment compared to large, light HDL) (Kontush et al. 2007) (Fig. 2) consistent with the high content in apoM, a specific carrier of S1P, in small, dense particles (Davidson et al. 2009). Other biologically active lysosphingolipids carried by HDL are represented by lysosphingomyelin and lysosulfatide (Lhomme et al. 2012).

Ceramide is a sphingolipid intermediate implicated in cell signalling, apoptosis, inflammatory responses, mitochondrial function and insulin sensitivity (Lipina and Hundal 2011). This lipid is poorly transported by HDL, which carries only 25 mol % of total plasma ceramide (Wiesner et al. 2009), and constitutes only between 0.022 and 0.097 mol % of total HDL lipids (Wiesner et al. 2009; Argraves et al. 2011; Camont et al. 2013; Stahlman et al. 2013) (Table 3). Similarly to sphingomyelin, this product of sphingomyelin hydrolysis is enriched in large, light HDL (Fig. 2), suggesting common metabolic pathways for these lipids. This hypothesis is however not supported by the pattern of major molecular species of ceramide observed in HDL, which are the 24:0 and 24:1 species (Wiesner et al. 2009; Stahlman et al. 2013).

Lipidomic data on glycosphingolipids, gangliosides and sulfatides are scarce (Lhomme et al. 2012). Hexosyl and lactosyl species constitute the major glycosphingolipids in plasma lipoproteins (Scherer et al. 2010) (Table 3).

2.2.3 Neutral Lipids

Unesterified (free) **sterols** are located in the surface lipid monolayer of HDL particles and regulate its fluidity. HDL sterols are dominated by cholesterol, reflecting the key role of lipoproteins in cholesterol transport through the body. Other sterols are present in lipoproteins at much lower levels as exemplified by minor amounts of lathosterol, ergosterol, phytosterols (β-sitosterol, campesterol), oxysterols and estrogens (largely circulating as esters) (Kontush and Chapman 2012). Free cholesterol, whose affinity for sphingomyelin is now well established, tends to preferentially associate with large, light HDL (Fig. 2).

Cholesteryl esters (CE) are largely (up to 80 %) formed in plasma HDL (Fig. 2), as a result of transesterification of PL and cholesterol catalysed by LCAT. These highly hydrophobic lipids form the lipid core of HDL and contribute up to 36 mol % of total HDL lipid (Wiesner et al. 2009; Camont et al. 2013; Stahlman et al. 2013) (Table 3). Most of HDL CE is accounted for by cholesteryl linoleate (Lhomme et al. 2012). A pioneering work on CE molecular species distribution across HDL subpopulations using gas chromatography showed very similar profiles between HDL2 and HDL3 particles (Vieu et al. 1996).

HDL-associated **triacylglycerides** (TAG) are dominated by species containing oleic, palmitic and linoleic acid moieties (Lhomme et al. 2012) and represent around 3 mol % of total HDL lipids (Vieu et al. 1996; Kontush et al. 2007; Wiesner et al. 2009; Camont et al. 2013; Stahlman et al. 2013) (Table 3). Similarly to CE, TAG species profile is conserved between HDL2 and HDL3 (Vieu et al. 1996).

Minor bioactive lipids present in HDL include **diacylglycerides** (DAG), **monoacylglycerides** (MAG) and **free fatty acids** (Lhomme et al. 2012). In the study of Vieu et al., sn-1,2 and sn-1,3 isomers of DAG species were identified in HDL (Vieu et al. 1996). More recently, detailed characterisation of DAG molecular species in HDL revealed 16:0/18:1 and 18:1/18:1 as major species, estimating their content at 0.2 mol % of total lipids (Stahlman et al. 2013) (Table 3).

HDL composition in free fatty acids shows a predominance of palmitic, stearic and oleic acid-containing species (Lhomme et al. 2012).

Together, available lipidomic studies have already provided detailed characterisation of HDL isolated using two complementary techniques, FPLC (Wiesner et al. 2009) and ultracentrifugation (Kontush et al. 2007; Camont et al. 2013). The latter studies demonstrated preferential association of negatively charged lipids (PS, PA) and S1P with small, dense HDL together with preferential association of sphingomyelin and ceramide with large, light particles, consistent with distinct metabolic origins and potent biological activities of small, dense HDL (Camont et al. 2013). These data illustrate the power of lipidomics to provide crucial information on the metabolism and function of lipoproteins relevant for the development of cardiovascular disease, which can in turn deliver novel biomarkers of cardiovascular risk.

3 The Structure of HDL

3.1 Introduction/Brief History

When it was becoming clear that plasma levels of HDL were inversely correlated with cardiovascular disease, numerous laboratories were already doing pioneering work on the structure of human plasma HDL (Edelstein et al. 1972; Jonas 1972; Laggner et al. 1973; Atkinson et al. 1974; Schonfeld et al. 1976; Tardieu et al. 1976). Knowing little about the nature of the stabilising proteins, Scanu and colleagues used detailed compositional studies to build models of human HDL3 particles (Shen et al. 1977; Scanu 1978). Concepts derived from these studies with respect to fundamental packing of the protein and various lipid components still apply today. After the sequences of the major HDL proteins, apolipoprotein A-I and apoA-II, were reported in the 1970s (Brewer et al. 1972, 1978), Segrest (Segrest et al. 1974) and others (McLachlan 1977) quickly noted periodically repeating units that, when mapped on a helical wheel plot, indicated the presence of amphipathic alpha helices. With hydrophobic faces mediating lipid interactions and polar faces interacting with water, these structures turned out to be responsible for the detergent-like ability of these proteins to solubilise lipids into stable lipoprotein particles. At around the same time, Jonas and colleagues were learning how to combine purified apolipoproteins with lipids under control of detergents to produce recombinant forms of HDL (Matz and Jonas 1982). Furthermore, the work of Nichols et al was exploring new ways to characterise these particles by native electrophoresis (Nichols et al. 1983). Because the electron microscopy work of Forte indicated that they likely have a discoid shape (Forte and Nordhausen 1986), these particles were referred to as reconstituted (r)HDL discs. Their compositional simplicity and homogeneity quickly made them a mainstay for HDL structural studies.

Segrest et al. and others (Wlodawer et al. 1979) proposed in the late 1970s that the α-helices of apoA-I were arranged around the circumference of discoidal HDL with the long axis of the helices perpendicular to the acyl chains (Fig. 3). This became known as the "belt" or "bicycle wheel" model. Alternatively, other investigators theorised that the 22-amino acid helical repeats were an ideal length to traverse the bilayer edge with the helices parallel to the acyl chains (Jonas et al. 1989; Nolte and Atkinson 1992). The intervening proline residues were thought to induce hairpin turns that punctuated the repeats. This "picket fence" model was favoured during the early 1990s because of supporting infrared (IR) spectroscopy studies (Brasseur et al. 1990) and the fact that the model accounted nicely for the experimentally observed size classes of apoA-I containing HDL (Wald et al. 1990). Furthermore, the plausibility of the picket fence model was supported by Phillips et al. using molecular modelling techniques (Phillips et al. 1997).

However, the publication of the first successful X-ray crystal structure of a lipid-free fragment of apoA-I by Borhani et al. (1997) brought fresh energy to the debate. The crystal structure showed a tetramer of highly α-helical apoA-I molecules arranged in a ring-shaped complex with no evidence of hairpin turns that would be prevalent in the picket fence. The belt model picked up more support with methodologically updated IR experiments performed by Koppaka et al. (1999) that contradicted the earlier IR studies supporting the picket fence in rHDL discs. Over the next few years, several laboratories reported results that supported the belt orientation using methodologies like fluorescence energy transfer (Li et al. 2000) and lipid-based fluorescence quenching (Panagotopulos et al. 2001). With the question of helical orientation largely addressed, much of the focus for the first decade of the 2000s centred on determining the spatial relationships between

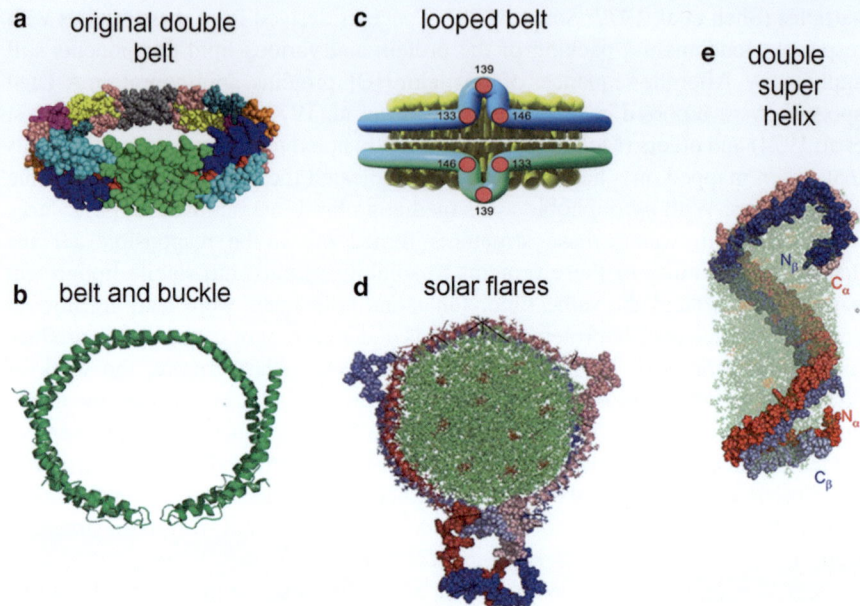

Fig. 3 The double belt model for a rHDL discoidal particle and its various refinements. The classic LL 5/5 double belt model for apoA-I as proposed by Segrest (Segrest et al. 2000) is shown in panel (**a**). Each amphipathic helical domain is shown in a different colour with helix 5 in *green* at the front. This model did not include the N-terminal 43 a.a. The "belt and buckle" model (**b**) does not form a *continuous circle* around the particle but has its N-terminal 43 a.a. folding back across the belts (40). The "looped belt" (**c**) features a localised separation between residues 133 and 146 (helix 5) (41). The "solar flares" model (**d**) encapsulates the lipid with an antiparallel double belt but has disorganised domains that resemble solar flares erupting between residues 165 and 180. The double superhelix model (**e**) maintains similar intermolecular contact among the molecules of apoA-I as the double belt, but the lipid is stabilised as an elongated micelle and no intramolecular contact is made between the N- and C-termini of apoA-I

apoA-I molecules on these discs. This effort was aided by the detailed "double belt" computer model put forth in the classic paper by Segrest (Segrest et al. 1999).

The preceding paragraphs quickly summarise a tremendous amount of work between 1970 and 2000 aimed at understanding the structure of HDL and its recombinant forms, and by no means does it do justice to all contributions. For more complete discussions of this period of HDL structural research, the reader is referred to the classic review series by Brouillette (Brouillette and Anantharamaiah 1995; Brouillette et al. 2001). In addition, due to space limitations, this section will not focus on the structure of lipid-free apolipoproteins nor their proposed mechanisms for lipid binding and particle generation. The reader is directed to the recent review by Phillips (Phillips 2013) for an excellent discussion of these issues. For the purposes of this section, we will focus on work done over the last 14 years that has built upon the concept of the double belt model in both reconstituted and "real" circulating plasma HDL.

3.2 HDL in the Test Tube

3.2.1 Discoid HDL

Discoid HDLs probably exist in human plasma but are extremely short-lived due to the fact that they are excellent substrates for LCAT (Jonas 2002). HDL discs have been detected in sequestered compartments like peripheral lymph and interstitial fluid which likely have lower LCAT activity (Sloop et al. 1983). As stated above, discs can be reconstituted in vitro, and these particles exhibit many of the traits of native HDL such as LCAT activation (Jonas and McHugh 1984), lipid transfer (Davidson et al. 1995) and receptor binding (Liadaki et al. 2000).

The most well-studied rHDL discs consist of two apoA-I molecules, are ~96 Å in diameter and ~47 Å in thickness and contain 150–200 phospholipid molecules (Jonas 1986; Jonas et al. 1989). This "benchmark" particle was used as a basis for the original **double belt** molecular model (Segrest et al. 1999). In this model, two ring-shaped apoA-I molecules encapsulate a lipid membrane leaflet in an antiparallel orientation (Segrest 1977; Segrest et al. 1999; Brouillette et al. 2001; Klon et al. 2002a) in which helix 5 of both apoA-I molecules directly oppose each other (Fig. 3a). In this configuration, there are two possible interfaces between the molecules depending on how they are stacked: left to left (LL) and right to right (RR). Computer analysis indicated that an LL interface in which helix 5 (more specifically, glycine 129) of both molecules are directly opposed produced the highest weighted score of potential salt bridges between the apoA-I molecules (Segrest et al. 1999). This is called the LL5/5 (G129j) registry and, interestingly, is the same orientation found in the Borhani crystal structure (Borhani et al. 1997). This orientation was experimentally validated in three studies that utilised chemical cross-linking to derive the molecular orientation of the two apoA-I molecules in these particles. Two from our laboratory were highly consistent with the 5/5 orientation across most of the molecule, although certain cross-links were also consistent with a second 5/2 registry (Davidson and Hilliard 2003; Silva et al. 2005). Cross-linking studies by Bhat et al. (2005) were also consistent with the general 5/5 belt model for most of the molecule, but these investigators proposed that the N- and C-terminal 40–50 residues doubled back on the molecule (Fig. 3b). This refinement was called the "belt and buckle". Electron paramagnetic resonance studies performed by Martin and colleagues (Martin et al. 2006) on 96 Å rHDL particles were also consistent with a 5/5 antiparallel double belt orientation over most of the particle circumference. However, they observed that the spin coupling signatures of residues 134–145 (within helix 5) were consistent with a looping region that causes a localised opening between the parallel belts. This "looped belt" was proposed to be a potential site by which LCAT could gain access to the cholesterol and phospholipid acyl chains that are otherwise buried in the lipid bilayer (Fig. 3c). Using hydrogen/deuterium exchange (HDX) experiments, Wu and colleagues proposed an alternative refinement of the double belt model which included the full-length apoA-I (Wu et al. 2007). This "solar flares" model again featured a 5/5 double belt for most of the particle, but the N-termini were modelled as globular nodules (Fig. 3d). The HDX profile of the region between 165 and

180 suggested that this region loops off the particle edge and could be a site for LCAT interaction. This argument was bolstered by results showing that added LCAT could alter the exchange patterns of these residues and that mutations in this region affected LCAT activity. The conclusions of this study were called into some question by a more recent HDX study performed by Chetty et al. on similar particles that failed to support the existence of the looped solar flares (Chetty et al. 2013). They showed that almost all of apoA-I was helical in the discs with the exception of the first 6 and last 7 residues. Interestingly, the region between residues 125 and 158 exhibited bimodal exchange kinetics, suggesting that this region flits between a helical and less ordered structure. This is consistent with the looped belt idea (Martin et al. 2006), but the putative solar flares region (Wu et al. 2007) was not found to be particularly dynamic in this study. The authors suggested that this region is responsible for absorbing changes in particle size by unfolding under conditions when lipid surface area is limiting. This is consistent with previous studies from our laboratory which used lipid-based fluorescence quenching agents to show that this same region changes its exposure to lipid when the particle diameter is reduced (Maiorano et al. 2004).

The recent publication of a high-resolution (2.2 Å) crystal structure of a lipid-free truncation mutant of apoA-I (Δ185–243) also strongly supports the double belt model (Mei and Atkinson 2011). Although lipid-free, the dimeric apoA-I formed a curved structure derived from long, kinked alpha helices. Interestingly, much of the structure resembled the 5/5 double belt, even down to a similar degree of curvature. The model also suggested increased temperature factors in helix 5 hinting that there could be a separation of the helices at this site, consistent with the looped belt idea. This advance should prove to be a valuable guide for future modelling of both the lipid-free and lipid-bound forms of apoA-I.

The models discussed above conform to the generally accepted idea that these rHDLs are discoid in shape. However, Hazen and colleagues proposed a model that departed from this idea (Wu et al. 2009). Based on small-angle neutron scattering (SANS), they proposed that antiparallel apoA-I molecules form a **double superhelix** (DSH) that encapsulates a cigar-shaped micelle of phospholipids (Fig. 3e). Thus, the particles resemble an ellipsoid rather than a disc. Despite the different shape, the overall molecular interactions along the interface of the two apoA-I molecules are the same as the 5/5 double belt model. However, unlike in the planar disc, the N- and C-termini are not predicted interact in the DSH. The viability of this model remains under debate as molecular dynamics studies suggest that this model rapidly collapses to a disc-shaped structure (Jones et al. 2010). Even simulations done by the group that proposed the model seem to indicate that the elongated structure collapses to an oblate spheroid, though they argue it remains micellar (Gogonea et al. 2010). Going forward, this model must be squared with observations of disc-like shapes by electron microscopy (Zhang et al. 2011), EPR data (Jones et al. 2010), and the quantised changes in particle size that can be readily demonstrated by reconstituting different lipid to protein ratios (Wald et al. 1990).

In addition to experimental work, significant progress has been made using molecular dynamics to simulate the structure of rHDL discs. The earliest simulations by Klon et al. utilised apoA-I lacking the N-terminal 43 amino acids (Klon et al. 2002b). The short 1 ns simulations were performed using the 5/5 double belt model, encapsulating simulated phospholipids in a standard ~96 Å particle as a starting point. The simulations provided four major conclusions. First, the overall structure remained intact over the course of the simulation, arguing the physical plausibility of the structure. Second, specific salt bridges were found to be important with regard to the molecular registry of the two apoA-I molecules. Third, the punctuating proline residues were found to induce peri-helical "kinks" that also played a role in molecular registry. Finally, the motions of the molecule suggested that apoA-I may have the potential to change its molecular registry. This concept was first proposed in the fluorescence resonance energy transfer experiments of Li et al. (2002), and the concept was later supported by certain chemical cross-links (Silva et al. 2005). This idea of molecular registry shifts on the HDL particle surface raises fascinating implications with respect to alternative interactions with the wide array of HDL docking proteins.

Further simulations in which lipids were added to or subtracted from the system showed that apoA-I can twist into a saddle-shaped structure in response to lipid removal (Gu et al. 2010). These results were supported by independent studies (Miyazaki et al. 2010) and in subsequent temperature jump studies using full-length apoA-I (Jones et al. 2011). Segrest and colleagues concluded that apoA-I maintains minimal surface bilayers and responds to changes in surface lipid content by winding or unwinding of this twisted saddle shape. It is important to emphasise, however, that these twisted models of dimeric apoA-I maintain similar intermolecular contacts as the planar double belt model and thus are consistent with the fluorescence, EPR and cross-linking data described above.

3.2.2 Spherical rHDL

Although the majority of HDL particles in human plasma are spherical, these particles are extremely understudied from a structural point of view. Spherical HDL particles contain a neutral lipid core composed of cholesteryl ester and triglyceride and thus lack a particle "edge" to constrain the apolipoproteins as in the discs. The helical domains have the potential to spread across the phospholipid surface penetrating past the phosphate group to interact with the acyl chains. Segrest et al. (1994) argued early on that if apoA-I exists in a belt-like model in discoidal particles, the fundamental interactions of apoA-I helices with the phospholipid acyl chains should not change significantly in spheres. Brasseur (Brasseur et al. 1990) and Borhani (Borhani et al. 1997) have also argued the similarity in apoA-I structure between the two shapes. These arguments were supported by circular dichroism and fluorescence studies showing that apoA-I secondary structure content and the exposure of Trp residues do not undergo gross changes when a reconstituted discoidal particle converts to a sphere in the presence of LCAT (Jonas et al. 1990). In addition, careful NMR studies suggested that the changes in conformation of apoA-I between reconstituted discs and spheres are primarily limited to the apoA-I N-terminus (Sparks et al. 1992).

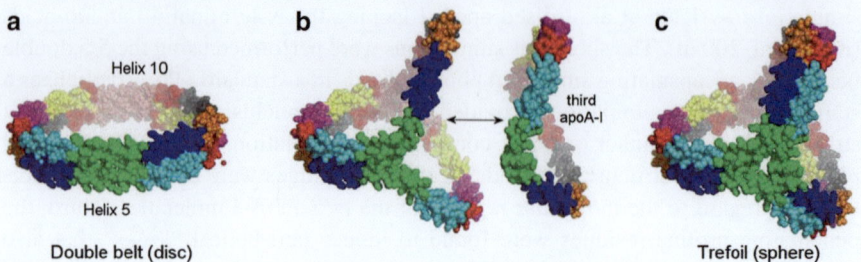

Fig. 4 The development of the trefoil model for a *spherical particle* from the double belt. The classic LL 5/5 double belt for apoA-I is shown in panel (**a**) coloured as in Fig. 3. (**b**) Kinks are introduced within helix 5 and helix 10 and the molecules are bent 60° out of the plane of the disc. A third molecule of apoA-I, bent the same way, can now be introduced. (**c**) The trefoil model demonstrates similar intermolecular contacts between all three apoA-I molecules

Using the cross-linking approach, we studied the structure of reconstituted apoA-I-containing spherical particles generated by incubating rHDL discs with LCAT under controlled conditions (Silva et al. 2008). These spheres contained a largely cholesteryl ester core and were of roughly similar diameter to the "benchmark" discs described above. Cross-linking experiments showed that these spheres contained 3 molecules of apoA-I, vs. only two in the discs. Quite surprisingly, the cross-linking pattern derived in the spheres was almost identical to that of the discs. Our solution to the problem of how the exact same molecular interactions occur between two apoA-Is in a disc vs. three in a sphere was the "trefoil model" (Fig. 4). In this scheme, kinks were added in helix 5 and 10 of apoA-I. Two antiparallel belts as per the double belt model are then bent 120° on these kinks and a third molecule, bent the same way, was inserted. This formed a three-dimensional cage-like structure in which the helical domains in each of the three apoA-I molecules made exactly the same salt bridge interactions postulated for the original double belt. This cage can be envisioned to encapsulate a neutral lipid ester core and support the surface polar lipids in the intervening open spaces, or lines, between the molecules.

Studying the exact same particles using SANS, Wu and colleagues came to a different conclusion (Wu et al. 2011). The low-resolution molecular envelope of the protein component was consistent with three different architectures: (1) two apoA-I molecules in a double belt with the third as a separate hairpin, called the **helical dimer with hairpin** (HdHp), (2) three separate hairpins (3Hp), or (3) an integrated trimer (iT) which is similar to the trefoil arrangement. Based on chemical cross-linking, the authors argued that the data was most consistent with the HdHp model. Despite the differences in the HdHp and the trefoil arrangements from a conceptual point of view, the models are quite similar from a molecular perspective in that the interactions in both models are predicted to be similar. All contacts in the trefoil are intermolecular where the interactions within the hairpined apoA-I are intramolecular, but the same sequences interact in all molecules. Thus, the epitopes on the HDL particles should be similar in both models and therefore might be expected to interact similarly with HDL docking proteins. It should also be noted that an alternative spherical model that is related to the trefoil organisation was proposed by Gursky (Gursky 2013) that invokes different **hinge domains** as extrapolated from the crystal structure of the lipid-free apoA-I fragment (Mei and Atkinson 2011). Distinction between these models awaits the development of strategies that allow for the higher resolution determination of cross-link span on particles containing more than two molecules of apoA-I.

Catte et al. made initial explorations into the structure of reconstituted spherical particles using a combination of atomistic and course-grained molecular dynamics simulations (Catte et al. 2008). The simulated particle consisted of a saddle-shaped rHDL particle with two molecules of apoA-I in which surface phospholipids were replaced in silico with cholesteryl ester (CE) molecules. During the simulations, the CE is sequestered into a core that was well covered by the surface protein and phospholipid, forming a prolate ellipsoid shape. The overall helicity and salt bridge orientation between the resident apoA-I molecules remained similar to the disc models, though there was an increase in overall salt bridge number. Despite an uncertain relationship with experimentally derived spherical particles, these simulations are consistent with the idea that apoA-I structures in discs and spheres are generally similar. The same group later simulated particles that more closely resembled those appearing in plasma (Segrest et al. 2013). Their data suggested that spheres created under artificial conditions in the test tube may represent kinetically trapped species with artificially high ratios of surface lipids with respect to the scaffold proteins. However, again, the resident apoA-I particles exhibited overall similar structures.

3.3 "Real" HDL Particles

Reconstituted HDL discs and spheres have been a critical tool for understanding the fundamental arrangements of apoA-I on the HDL particle surface. The homogeneity of these preparations allowed the use of traditional spectrophotometric approaches such as circular dichroism, NMR, EPR, etc. However, the ultimate goal has always been to understand HDL in its natural plasma form. As outlined in other sections of this chapter, human plasma HDL as isolated by density gradient ultracentrifugation is heterogeneous in almost every property: protein composition, lipid composition, particle size, particle shape and particle charge. This heterogeneity precludes the use of most traditional structural techniques. For example, circular dichroism studies of a mixed system in which apoA-I is present either in several different conformations or with other apolipoproteins will produce averaged data that will likely not be applicable to any of the particles present in the sample. Even attempts to simplify these particles, such as by immunoisolating only those particles that contain apoA-I but not apoA-II (LpA-I), still result in preparations with unacceptable compositional and size heterogeneity for informative structural studies. For these reasons, our understanding of HDL structure in authentic plasma HDL particles has lagged far behind the reconstituted forms.

The use of mass spectrometry as a structural tool has finally allowed this barrier to be breached. With the mass accuracy and resolution of modern mass spectrometers, it is possible to identify cross-links that are specific for a given protein, even in the presence of many others. Taking advantage of this, we studied 5 density subfractions of normal human plasma HDL as delineated by ultracentrifugation. In order to simplify the particles to the extent possible, we isolated those particles containing predominantly apoA-I (Huang et al. 2011). Surprisingly, we noted cross-linking patterns that were remarkably similar to those found in the reconstituted spheres and discs, irrespective of the particle size or density. This strongly suggested that the general features of apoA-I structure are related among lipid-associated forms of all sizes, shapes and origin. Careful measurements of the protein/lipid components indicated a range 3–5 apoA-I molecules per particle, depending on size. In fact, an important outcome of this study was the realisation that the surface of HDL particle is dominated by protein. For example, we determined that apoA-I accounts for 87 % of the surface of an LpA-I$_{3c}$ particle. This only reduces to 71 % in the larger LpA-I$_{2b}$ particles. This is in agreement with the assertions of Segrest et al. (2013) whose compositional calculations indicate that apoA-I is much more tightly packed on native spheres than it is on typical reconstituted particles. Figure 5 shows one proposal for how 4 and 5 apoA-Is can be accommodated by changing the hinge bend angles of the trefoil while maintaining the same intermolecular interactions, and hence cross-linking pattern. Furthermore, we proposed that HDL particle size is modulated via a twisting motion of the resident apoA-Is, consistent with the MD studies of the Segrest lab (Catte et al. 2008). Combining these data, we developed the first detailed molecular models for native LpA-I HDL particles, exemplified by that for LpA-I$_{2b}$ in Fig. 6. In this model, apoA-I forms an 11 nm cage stabilising ~170 molecules of phospholipid at a physiological packing density. The cutaway shows a cavity of correct volume for the ~143 CEs and ~23 TGs measured in the core. These are the first detailed, molecular-scale models of native plasma HDL to be proposed.

Taking another MS approach, Chetty et al. used HDX to study spherical LpA-I particles isolated from human plasma by ultracentrifugation and further isolated by high-resolution gel filtration chromatography (Chetty et al. 2013). These particles were approximately 10 nm in diameter and contained on average 5 molecules of apoA-I per particle as measured by cross-linking. Comparing these directly to a benchmark rHDL disc, they found that the total helical content was similar between the native spheres and the reconstituted discs. Strikingly, the HDX profiles were also quite similar between the two with both showing near-complete helicity with the exception of the first and last 6 or so residues. Like the discs, the plasma spheres showed bimodal HDX kinetics indicating a possible loop region in helix 5 of apoA-I. Interestingly, the length of this partially disordered segment was similar

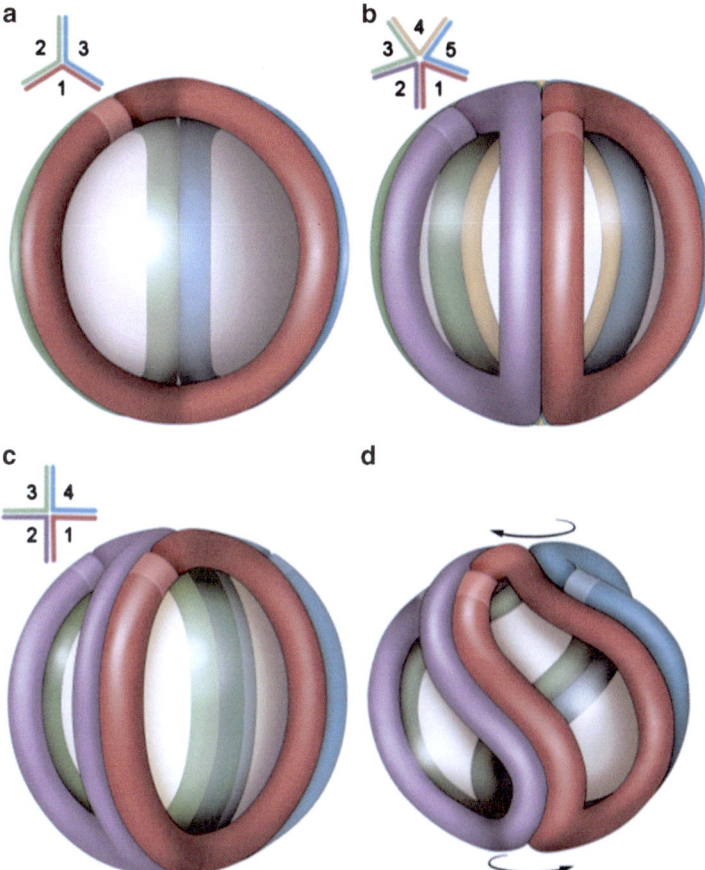

Fig. 5 Models of apoA-I in human plasma HDL particles of various size and number of apoA-I molecules. (**a**) The trefoil model is shown as a schematic with each molecule of apoA-I in a *different colour*. (**b**) LpA-I$_{2b}$ contains 5 apoA-I molecules, but shows a similar cross-linking patterns to reconstituted particles that have only 2 and 3 molecules of apoA-I. This figure shows that more apoA-I molecules can be added to the trefoil framework by increasing the hinge bend angle and adding more apoA-I molecules. (**c**) A possible model for LpA-I$_{2a}$ that has four apoA-I molecules on average. (**d**) Further reductions in HDL particle size may be accomplished by a twisting action of the resident apoA-I molecules

to that of rHDL discs of 8 nm in diameter, suggesting that apoA-I on these spheres is more tightly packed than in the 96 Å discs. This is consistent with higher number of apoA-I molecules on the surface. Overall, these studies reinforce the idea that the global apoA-I organisation does not change significantly between particles of different shape or origin; however, more subtle conformational adaptations may occur in localised regions of the protein in response to changes in either particle diameter or surface packing density.

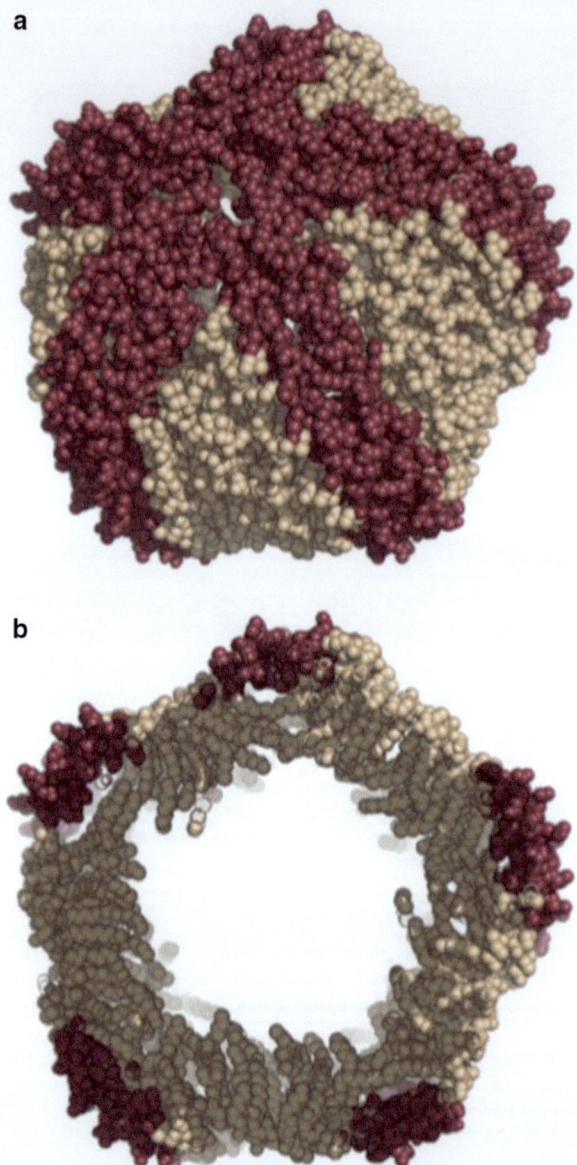

Fig. 6 Molecular model of a human plasma LpA-I$_{2b}$ particle. (**a**) Five molecules of apoA-I are shown in an organisation shown schematically in Fig. 5b. Phospholipid headgroups (tan) are present in the intervening spaces (*lunes*) between the apoA-I belts (*purple*). (**b**) A cross section of the same particle showing the helical belts floating atop the acyl chains of the phospholipids (*brown*). Note the large internal cavity that is of sufficient volume to contain the number of triglycerides and cholesteryl ester molecules that have been measured in these particles (Huang et al. 2011). (Colour figure online)

Conclusions and Perspectives

HDL, a lipoprotein class historically and classically defined and isolated on the basis of its hydrated density alone, is a highly heterogeneous particle family consisting of multiple subclasses differing in hydrated density, size, shape and lipid and protein composition. HDL-associated proteins have been considered until now to predominate in determining the particle structure and biological function of all HDL particles. Indeed, HDL is distinguished from other lipoprotein classes by its cargo of a large number (>80) of different proteins whose abundance may vary by more than 1,000-fold and which fall into several major subgroups, including apolipoproteins, diverse enzymes, lipid transfer proteins, acute-phase response proteins, complement components and proteinase inhibitors; additional functions, such as host defence and inflammation, cannot be excluded. Many of the individual proteins are present as distinct isoforms due to co- and posttranslational modification, thereby adding a further dimension to the complex and dynamic HDL structure.

It would be erroneous however to consider that the proteome(s) of HDL particles solely determines their respective functions, particularly as cutting edge lipidomic technologies have recently revealed significant—and indeed determinant—roles of specific HDL lipid components in particle function. In this regard, negatively charged phospholipids, and notably molecular species of phosphatidylserines, are intimately associated with the cellular cholesterol efflux, antioxidative, antiapoptotic and anti-inflammatory actions of small dense HDL3 particles and equally modulate platelet activity (Camont et al. 2013). In addition, earlier studies have suggested that both core neutral lipids in HDL (cholesteryl esters and triacylglycerols) and surface lipids can impact surface fluidity and, in turn, antioxidative and other functions (Zerrad-Saadi et al. 2009; Camont et al. 2013).

Increasingly then, a new concept is emerging, one in which specific protein and lipid components interact closely to form individual HDL particle species, such as overall compositional and structural features translating into expression of highly defined biological functions (Camont et al. 2011; Shah et al. 2013). Clearly we are at the beginning of a new era in our understanding of HDL structure and function, with the promise of exciting mechanistic insights into function, which, in turn, will translate into new therapeutic strategies to harness the anti-atherosclerotic potential of HDL.

The central thesis that specific protein–protein interactions may drive the formation of lipoprotein complexes in the HDL particle spectrum is equally emergent. Indeed, the diversity of molecules which bind to HDL suggests that the lipoprotein can serve as a versatile adsorptive surface for proteins to form complexes playing roles not only in lipid metabolism but equally in the acute-phase response, innate immune response, complement activation, proteolysis inhibition and plaque stability.

More comprehensive understanding of HDL structure is critical to future progress in HDL biology. A molecular understanding of HDL-bound apoA-I will allow further insight into its interactions with key plasma remodelling factors, such as LCAT, PLTP or CETP, and in addition with cell-surface proteins that mediate HDL assembly and clearance, such as ABCA1 and scavenger receptor B-I. A firm handle on these interactions would contribute to therapeutic manipulation that could enhance the protective features of HDL with respect to cardiovascular disease, or perhaps other pathologies (Gordon et al. 2011). Because HDL and its recombinant analogues have been resistant to traditional high-resolution structural approaches such as NMR or X-ray crystallography, investigators have been forced to apply a series of lower resolution techniques to the problem. As described above, these approaches have led to a general understanding of apoA-I structure in HDL that can be summarised by the following statements: (1) apoA-I pairs with itself in a double belt-like arrangement in almost all of its lipidated states; (2) apoA-I interacts via intermolecular salt bridges that commonly stabilise the registry in the 5/5 orientation, though there may be a possibility of dynamic shifts to other orientations; and (3) despite a similar general organisation among different types of HDL particles, specific regions within the scaffold can undergo conformational changes depending on particle size or perhaps composition—possible sites include the N- and C-termini and central region in helix 5 or between residues 165 and 180. These conformationally dynamic sites are attractive possibilities for the interaction of lipid-modifying docking proteins such as LCAT and CETP.

Now that these conformationally dynamic sites within apoA-I have been proposed, it will be important to specifically define their importance for key HDL remodelling reactions. This will require well-designed site-directed mutagenesis experiments in reconstituted particles and antibody blocking experiments in native plasma particles. Furthermore, cross-linking studies hold much promise for directly mapping the sites of interaction for plasma remodelling factors on HDL-bound apoA-I.

Finally, yet another key unanswered question relates to the unknown effects of other HDL structural proteins on apoA-I structure. ApoA-II is commonly found with apoA-I on many HDL particles and would be expected to have a significant impact not only on properties such as protein packing on the particle surface but also on the conformation of apoA-I in a functionally significant way. For example, does apoA-II perform distinct structural and protein docking roles vs. apoA-I? How does each of these proteins determine the components of the bound lipidome in individual particle species? Resolution of these and other questions will continue to require the use of both reconstituted and native, authentic HDL particles combined with the appropriate application of multiple structural techniques (including continued attempts at X-ray crystallography).

It is poignant to consider that in 2014 we still lack critically important insights into HDL structure as it relates to the intravascular metabolism and atheroprotective function of these pseudomicellar, quasi-spherical particles. Had sufficient effort been invested in solving these unknowns heretofore, then perhaps our costly forays—and failures—into HDL-targeted therapeutics over the past 15 years could at least partially have been avoided.

Acknowledgements This article results from individual contributions of all the co-authors. The highly complex world of HDL subclasses is presented by Laura Calabresi. The multifaceted composition of the HDL proteome is discussed by Anatol Kontush. Protein isoforms, including translational and posttranslational modifications, are reviewed by Mats Lindahl. Rapidly growing knowledge of the HDL lipidome is discussed by Marie Lhomme. W. Sean Davidson presents a brief history of structural studies of HDL, critically reviews existing models of HDL structure, and together with M. John Chapman draws conclusions and discusses future perspectives.

Open Access This chapter is distributed under the terms of the Creative Commons Attribution Noncommercial License, which permits any noncommercial use, distribution, and reproduction in any medium, provided the original author(s) and source are credited.

References

Ahnstrom J et al (2007) Hydrophobic ligand binding properties of the human lipocalin apolipoprotein M. J Lipid Res 48(8):1754–1762

Allan CM, Taylor JM (1996) Expression of a novel human apolipoprotein (apoC-IV) causes hypertriglyceridemia in transgenic mice. J Lipid Res 37(7):1510–1518

Argraves KM et al (2011) S1P, dihydro-S1P and C24:1-ceramide levels in the HDL-containing fraction of serum inversely correlate with occurrence of ischemic heart disease. Lipids Health Dis 10:70

Asztalos BF, Schaefer EJ (2003a) High-density lipoprotein subpopulations in pathologic conditions. Am J Cardiol 91(7A):12E–17E

Asztalos BF, Schaefer EJ (2003b) HDL in atherosclerosis: actor or bystander? Atheroscler Suppl 4(1):21–29

Asztalos BF et al (2007) Role of LCAT in HDL remodeling: investigation of LCAT deficiency states. J Lipid Res 48(3):592–599

Atkinson D et al (1974) The structure of a high density lipoprotein (HDL3) from porcine plasma. Proc R Soc Lond B Biol Sci 186(083):165–180

Axler O et al (2007) An ELISA for apolipoprotein M reveals a strong correlation to total cholesterol in human plasma. J Lipid Res 48(8):1772–1780

Axler O et al (2008) Apolipoprotein M associates to lipoproteins through its retained signal peptide. FEBS Lett 582(5):826–828

Baenziger JU (2012) Moving the O-glycoproteome from form to function. Proc Natl Acad Sci USA 109(25):9672–9673

Benditt EP, Eriksen N (1977) Amyloid protein SAA is associated with high density lipoprotein from human serum. Proc Natl Acad Sci USA 74(9):4025–4028

Bergmeier C et al (2004) Distribution spectrum of paraoxonase activity in HDL fractions. Clin Chem 50(12):2309–2315

Bhat S et al (2005) Intermolecular contact between globular N-terminal fold and C-terminal domain of ApoA-I stabilizes its lipid-bound conformation: studies employing chemical cross-linking and mass spectrometry. J Biol Chem 280(38):33015–33025

Bisgaier CL et al (1985) Distribution of apolipoprotein A-IV in human plasma. J Lipid Res 26(1):11–25

Bondarenko PV et al (1999) Mass spectral study of polymorphism of the apolipoproteins of very low density lipoprotein. J Lipid Res 40(3):543–555

Borhani DW et al (1997) Crystal structure of truncated human apolipoprotein A-I suggests a lipid-bound conformation. Proc Natl Acad Sci USA 94(23):12291–12296

Brasseur R et al (1990) Mode of assembly of amphipathic helical segments in model high-density lipoproteins. Biochim Biophys Acta 1043(3):245–252

Brewer HB Jr et al (1972) Amino acid sequence of human apoLp-Gln-II (apoA-II), an apolipoprotein isolated from the high-density lipoprotein complex. Proc Natl Acad Sci USA 69(5):1304–1308

Brewer HB Jr et al (1978) The amino acid sequence of human APOA-I, an apolipoprotein isolated from high density lipoproteins. Biochem Biophys Res Commun 80(3):623–630

Brouillette CG, Anantharamaiah GM (1995) Structural models of human apolipoprotein A-I. Biochim Biophys Acta 1256(2):103–129

Brouillette CG et al (2001) Structural models of human apolipoprotein A-I: a critical analysis and review. Biochim Biophys Acta 1531(1–2):4–46

Bruneel A et al (2008) Two dimensional gel electrophoresis of apolipoprotein C-III and MALDI-TOF MS are complementary techniques for the study of combined defects in N- and mucin type O-glycan biosynthesis. Proteomics Clin Appl 2:1670–1674

Cabana VG et al (1996) HDL content and composition in acute phase response in three species: triglyceride enrichment of HDL a factor in its decrease. J Lipid Res 37(12):2662–2674

Cabana VG et al (2004) Influence of apoA-I and apoE on the formation of serum amyloid A-containing lipoproteins in vivo and in vitro. J Lipid Res 45(2):317–325

Calabresi L et al (2001) Limited proteolysis of a disulfide-linked apoA-I dimer in reconstituted HDL. J Lipid Res 42(6):935–942

Camont L et al (2011) Biological activities of HDL subpopulations and their relevance to cardiovascular disease. Trends Mol Med 17(10):594–603

Camont L et al (2013) Small, dense high-density lipoprotein-3 particles are enriched in negatively charged phospholipids: relevance to cellular cholesterol efflux, antioxidative, antithrombotic, anti-inflammatory, and antiapoptotic functionalities. Arterioscler Thromb Vasc Biol 33(12):2715–2723

Campos E, McConathy WJ (1986) Distribution of lipids and apolipoproteins in human plasma by vertical spin ultracentrifugation. Arch Biochem Biophys 249(2):455–463

Candiano G et al (2008) High-resolution 2-DE for resolving proteins, protein adducts and complexes in plasma. Electrophoresis 29(3):682–694

Catte A et al (2008) Structure of spheroidal HDL particles revealed by combined atomistic and coarse-grained simulations. Biophys J 94(6):2306–2319

Chapman MJ et al (1981) A density gradient ultracentrifugal procedure for the isolation of the major lipoprotein classes from human serum. J Lipid Res 22(2):339–358

Chen N et al (2000) Physiologic concentrations of homocysteine inhibit the human plasma GSH peroxidase that reduces organic hydroperoxides. J Lab Clin Med 136(1):58–65

Chetty PS et al (2013) Comparison of apoA-I helical structure and stability in discoidal and spherical HDL particles by HX and mass spectrometry. J Lipid Res 54(6):1589–1597

Cheung MC, Albers JJ (1982) Distribution of high density lipoprotein particles with different apoprotein composition: particles with A-I and A-II and particles with A-I but no A-II. J Lipid Res 23(5):747–753

Cheung MC et al (2010) Phospholipid transfer protein in human plasma associates with proteins linked to immunity and inflammation. Biochemistry 49(34):7314–7322

Chiba T et al (2011) Serum amyloid A facilitates the binding of high-density lipoprotein from mice injected with lipopolysaccharide to vascular proteoglycans. Arterioscler Thromb Vasc Biol 31(6):1326–1332

Christoffersen C et al (2008) The signal peptide anchors apolipoprotein M in plasma lipoproteins and prevents rapid clearance of apolipoprotein M from plasma. J Biol Chem 283(27):18765–18772

Christoffersen C et al (2011) Endothelium-protective sphingosine-1-phosphate provided by HDL-associated apolipoprotein M. Proc Natl Acad Sci USA 108(23):9613–9618

Coetzee GA et al (1986) Serum amyloid A-containing human high density lipoprotein 3. Density, size, and apolipoprotein composition. J Biol Chem 261(21):9644–9651

Collins LA et al (2010) Integrated approach for the comprehensive characterization of lipoproteins from human plasma using FPLC and nano-HPLC-tandem mass spectrometry. Physiol Genomics 40(3):208–215

Contiero E et al (1997) Apolipoprotein AI isoforms in serum determined by isoelectric focusing and immunoblotting. Electrophoresis 18(1):122–126

Dahlback B, Nielsen LB (2009) Apolipoprotein M affecting lipid metabolism or just catching a ride with lipoproteins in the circulation? Cell Mol Life Sci 66(4):559–564

Davidson WS, Hilliard GM (2003) The spatial organization of apolipoprotein A-I on the edge of discoidal high density lipoprotein particles: a mass specrometry study. J Biol Chem 278 (29):27199–27207

Davidson WS et al (1995) Effects of acceptor particle size on the efflux of cellular free cholesterol. J Biol Chem 270(29):17106–17113

Davidson WS et al (2009) Proteomic analysis of defined HDL subpopulations reveals particle-specific protein clusters: relevance to antioxidative function. Arterioscler Thromb Vasc Biol 29 (6):870–876

Davidsson P et al (2010) Proteomics of apolipoproteins and associated proteins from plasma high-density lipoproteins. Arterioscler Thromb Vasc Biol 30(2):156–163

de Beer MC et al (2010) Impact of serum amyloid A on high density lipoprotein composition and levels. J Lipid Res 51(11):3117–3125

De Lalla OF, Gofman JW (1954) Ultracentrifugal analysis of serum lipoproteins. Methods Biochem Anal 1:459–478

de Seny D et al (2008) Monomeric calgranulins measured by SELDI-TOF mass spectrometry and calprotectin measured by ELISA as biomarkers in arthritis. Clin Chem 54(6):1066–1075

de Silva HV et al (1990a) Purification and characterization of apolipoprotein J. J Biol Chem 265 (24):14292–14297

de Silva HV et al (1990b) A 70-kDa apolipoprotein designated ApoJ is a marker for subclasses of human plasma high density lipoproteins. J Biol Chem 265(22):13240–13247

Deguchi H et al (2000) Cardiolipin is a normal component of human plasma lipoproteins. Proc Natl Acad Sci USA 97(4):1743–1748

Duchateau PN et al (1997) Apolipoprotein L, a new human high density lipoprotein apolipoprotein expressed by the pancreas. Identification, cloning, characterization, and plasma distribution of apolipoprotein L. J Biol Chem 272(41):25576–25582

Ducret A et al (1996) Characterization of human serum amyloid A protein isoforms separated by two-dimensional electrophoresis by liquid chromatography/electrospray ionization tandem mass spectrometry. Electrophoresis 17(5):866–876

Duriez P, Fruchart JC (1999) High-density lipoprotein subclasses and apolipoprotein A-I. Clin Chim Acta 286(1–2):97–114

Edelstein C et al (1972) On the subunit structure of the protein of human serum high density lipoprotein. I. A study of its major polypeptide component (Sephadex, fraction 3). J Biol Chem 247(18):5842–5849

Favari E et al (2004) Depletion of pre-beta-high density lipoprotein by human chymase impairs ATP-binding cassette transporter A1- but not scavenger receptor class B type I-mediated lipid efflux to high density lipoprotein. J Biol Chem 279(11):9930–9936

Favari E et al (2007) A unique protease-sensitive high density lipoprotein particle containing the apolipoprotein A-I Milano dimer effectively promotes ATP-binding Cassette A1-mediated cell cholesterol efflux. J Biol Chem 282(8):5125–5132

Fernandez-Irigoyen J et al (2005) Oxidation of specific methionine and tryptophan residues of apolipoprotein A-I in hepatocarcinogenesis. Proteomics 5(18):4964–4972

Fievet C et al (1995) Non-enzymatic glycosylation of apolipoprotein A-I and its functional consequences. Diabete Metab 21(2):95–98

Flood-Nichols SK et al (2013) Longitudinal analysis of maternal plasma apolipoproteins in pregnancy: a targeted proteomics approach. Mol Cell Proteomics 12(1):55–64

Forte TM, Nordhausen RW (1986) Electron microscopy of negatively stained lipoproteins. Methods Enzymol 128:442–457

Franceschini G et al (1985) Effects of storage on the distribution of high density lipoprotein subfractions in human sera. J Lipid Res 26(11):1368–1373

Franceschini G et al (2007) Effects of fenofibrate and simvastatin on HDL-related biomarkers in low-HDL patients. Atherosclerosis 195(2):385–391

Gauthamadasa K et al (2010) Speciated human high-density lipoprotein protein proximity profiles. Biochemistry 49(50):10656–10665

Ghiselli G et al (1985) Origin of apolipoprotein A-I polymorphism in plasma. J Biol Chem 260 (29):15662–15668

Gidez LI et al (1982) Separation and quantitation of subclasses of human plasma high density lipoproteins by a simple precipitation procedure. J Lipid Res 23(8):1206–1223

Gillard BK et al (2005) Plasma factors required for human apolipoprotein A-II dimerization. Biochemistry 44(2):471–479

Gogonea V et al (2010) Congruency between biophysical data from multiple platforms and molecular dynamics simulation of the double-super helix model of nascent high-density lipoprotein. Biochemistry 49(34):7323–7343

Gordon JI et al (1983) Biosynthesis of human preproapolipoprotein A-II. J Biol Chem 258 (22):14054–14059

Gordon SM et al (2010) Proteomic characterization of human plasma high density lipoprotein fractionated by gel filtration chromatography. J Proteome Res 9(10):5239–5249

Gordon SM et al (2011) High density lipoprotein: it's not just about lipid transport anymore. Trends Endocrinol Metab 22(1):9–15

Gordon SM et al (2013) Multi-dimensional co-separation analysis reveals protein-protein interactions defining plasma lipoprotein subspecies. Mol Cell Proteomics 12(11):3123–3134

Goswami B et al (2009) Paraoxonase: a multifaceted biomolecule. Clin Chim Acta 410(1–2):1–12

Gu F et al (2010) Structures of discoidal high density lipoproteins: a combined computational-experimental approach. J Biol Chem 285(7):4652–4665

Gursky O (2013) Crystal structure of Delta(185–243)ApoA-I suggests a mechanistic framework for the protein adaptation to the changing lipid load in good cholesterol: from flatland to sphereland via double belt, belt buckle, double hairpin and trefoil/tetrafoil. J Mol Biol 425 (1):1–16

Gursky O et al (2013) Structural basis for distinct functions of the naturally occurring Cys mutants of human apolipoprotein A-I. J Lipid Res 54(12):3244–3257

Hajduk SL et al (1989) Lysis of Trypanosoma brucei by a toxic subspecies of human high density lipoprotein. J Biol Chem 264(9):5210–5217

Halim A et al (2013) LC-MS/MS characterization of O-glycosylation sites and glycan structures of human cerebrospinal fluid glycoproteins. J Proteome Res 12(2):573–584

Harel M et al (2004) Structure and evolution of the serum paraoxonase family of detoxifying and anti-atherosclerotic enzymes. Nat Struct Mol Biol 11(5):412–419

Harvey SB et al (2009) O-glycoside biomarker of apolipoprotein C3: responsiveness to obesity, bariatric surgery, and therapy with metformin, to chronic or severe liver disease and to mortality in severe sepsis and graft vs host disease. J Proteome Res 8(2):603–612

Hatters DM et al (2006) Apolipoprotein E structure: insights into function. Trends Biochem Sci 31 (8):445–454

He Y et al (2008) Control of cholesteryl ester transfer protein activity by sequestration of lipid transfer inhibitor protein in an inactive complex. J Lipid Res 49(7):1529–1537

Heinecke JW (2009) The HDL proteome: a marker - and perhaps mediator - of coronary artery disease. J Lipid Res Supp 50:S167–S171

Hennessy LK et al (1997) Isolation of subpopulations of high density lipoproteins: three particle species containing apoE and two species devoid of apoE that have affinity for heparin. J Lipid Res 38(9):1859–1868

Hoffman JS, Benditt EP (1982) Changes in high density lipoprotein content following endotoxin administration in the mouse. Formation of serum amyloid protein-rich subfractions. J Biol Chem 257(17):10510–10517

Holleboom AG et al (2011) Heterozygosity for a loss-of-function mutation in GALNT2 improves plasma triglyceride clearance in man. Cell Metab 14(6):811–818

Holmquist L (2002) Selective extraction of lecithin:cholesterol acyltransferase (EC 2.3.1.43) from human plasma. J Biochem Biophys Methods 52(1):63–68

Hoofnagle AN, Heinecke JW (2009) Lipoproteomics: using mass spectrometry-based proteomics to explore the assembly, structure, and function of lipoproteins. J Lipid Res 50(10):1967–1975

Hortin GL et al (2006) Diverse range of small peptides associated with high-density lipoprotein. Biochem Biophys Res Commun 340(3):909–915

Huang R et al (2011) Apolipoprotein A-I structural organization in high-density lipoproteins isolated from human plasma. Nat Struct Mol Biol 18(4):416–422

Hussain MM, Zannis VI (1990) Intracellular modification of human apolipoprotein AII (apoAII) and sites of apoAII mRNA synthesis: comparison of apoAII with apoCII and apoCIII isoproteins. Biochemistry 29(1):209–217

Jakubowski H (2000) Calcium-dependent human serum homocysteine thiolactone hydrolase. A protective mechanism against protein N-homocysteinylation. J Biol Chem 275(6):3957–3962

Jaleel A et al (2010) Identification of de novo synthesized and relatively older proteins: accelerated oxidative damage to de novo synthesized apolipoprotein A-1 in type 1 diabetes. Diabetes 59 (10):2366–2374

James RW et al (1988) Protein heterogeneity of lipoprotein particles containing apolipoprotein A-I without apolipoprotein A-II and apolipoprotein A-I with apolipoprotein A-II isolated from human plasma. J Lipid Res 29(12):1557–1571

Jian W et al (2013) Relative quantitation of glycoisoforms of intact apolipoprotein C3 in human plasma by liquid chromatography-high-resolution mass spectrometry. Anal Chem 85(5):2867–2874

Jin Y, Manabe T (2005) Direct targeting of human plasma for matrix-assisted laser desorption/ionization and analysis of plasma proteins by time of flight-mass spectrometry. Electrophoresis 26(14):2823–2834

Jonas A (1972) Studies on the structure of bovine serum high density lipoprotein using covalently bound fluorescent probes. J Biol Chem 247(23):7773–7778

Jonas A (1986) Reconstitution of high-density lipoproteins. Methods Enzymol 128:553–582

Jonas A (2002) Lipoprotein structure. In: Vance DE, Vance JE (eds) Biochemistry of lipids, lipoproteins and membranes. Elsevier, Amsterdam, pp 483–504

Jonas A, McHugh HT (1984) Reaction of lecithin: cholesterol acyltransferase with micellar substrates. Effect of particle sizes. Biochim Biophys Acta 794(3):361–372

Jonas A et al (1989) Defined apolipoprotein A-I conformations in reconstituted high density lipoprotein discs. J Biol Chem 264(9):4818–4824

Jonas A et al (1990) Apolipoprotein A-I structure and lipid properties in homogeneous, reconstituted spherical and discoidal high density lipoproteins. J Biol Chem 265(36):22123–22129

Jones MK et al (2010) Assessment of the validity of the double superhelix model for reconstituted high density lipoproteins: a combined computational-experimental approach. J Biol Chem 285 (52):41161–41171

Jones MK et al (2011) "Sticky" and "promiscuous", the Yin and Yang of apolipoprotein A-I termini in discoidal high-density lipoproteins: a combined computational-experimental approach. Biochemistry 50(12):2249–2263

Karlsson H et al (2005) Lipoproteomics II: mapping of proteins in high-density lipoprotein using two-dimensional gel electrophoresis and mass spectrometry. Proteomics 5(5):1431–1445

Karlsson H et al (2009) Protein profiling of low-density lipoprotein from obese subjects. Proteomics Clin Appl 3(6):663–671

Khovidhunkit W et al (2004) Effects of infection and inflammation on lipid and lipoprotein metabolism: mechanisms and consequences to the host. J Lipid Res 45(7):1169–1196

Kiernan UA et al (2003) Detection of novel truncated forms of human serum amyloid A protein in human plasma. FEBS Lett 537(1–3):166–170

Kim E et al (2008) Expression of apolipoprotein C-IV is regulated by Ku antigen/peroxisome proliferator-activated receptor gamma complex and correlates with liver steatosis. J Hepatol 49(5):787–798

Kisilevsky R, Manley PN (2012) Acute-phase serum amyloid A: perspectives on its physiological and pathological roles. Amyloid 19(1):5–14

Kisilevsky R, Tam SP (2003) Macrophage cholesterol efflux and the active domains of serum amyloid A 2.1. J Lipid Res 44(12):2257–2269

Klon AE et al (2002a) Comparative models for human apolipoprotein A-I bound to lipid in discoidal high-density lipoprotein particles. Biochemistry 41(36):10895–10905

Klon AE et al (2002b) Molecular dynamics simulations on discoidal HDL particles suggest a mechanism for rotation in the apo A-I belt model. J Mol Biol 324(4):703–721

Kontush A, Chapman MJ (2010) Antiatherogenic function of HDL particle subpopulations: focus on antioxidative activities. Curr Opin Lipidol 21(4):312–318

Kontush A, Chapman MJ (2012) High-density lipoproteins: structure, metabolism, function and therapeutics. Wiley, New York

Kontush A et al (2003) Small, dense HDL particles exert potent protection of atherogenic LDL against oxidative stress. Arterioscler Thromb Vasc Biol 23(10):1881–1888

Kontush A et al (2007) Preferential sphingosine-1-phosphate enrichment and sphingomyelin depletion are key features of small dense HDL3 particles: relevance to antiapoptotic and antioxidative activities. Arterioscler Thromb Vasc Biol 27(8):1843–1849

Koppaka V et al (1999) The structure of human lipoprotein A-I. Evidence for the "belt" model. J Biol Chem 274(21):14541–14544

Kulkarni KR et al (1997) Quantification of HDL2 and HDL3 cholesterol by the vertical auto profile-II (VAP-II) methodology. J Lipid Res 38(11):2353–2364

Laggner P et al (1973) Studies on the structure of lipoprotein A of human high density lipoprotein HDL3: the spherically averaged electron density distribution. FEBS Lett 33(1):77–80

Lagor WR et al (2009) Overexpression of apolipoprotein F reduces HDL cholesterol levels in vivo. Arterioscler Thromb Vasc Biol 29(1):40–46

Lamant M et al (2006) ApoO, a novel apolipoprotein, is an original glycoprotein up-regulated by diabetes in human heart. J Biol Chem 281(47):36289–36302

Lee M et al (2003) Apolipoprotein composition and particle size affect HDL degradation by chymase: effect on cellular cholesterol efflux. J Lipid Res 44(3):539–546

Lee JY et al (2010) Profiling of phospholipids in lipoproteins by multiplexed hollow fiber flow field-flow fractionation and nanoflow liquid chromatography-tandem mass spectrometry. J Chromatogr A 1217(10):1660–1666

Levels JH et al (2011) High-density lipoprotein proteome dynamics in human endotoxemia. Proteome Sci 9(1):34

Lhomme M et al (2012) Lipidomics in lipoprotein biology. In: Ekroos K (ed) Lipidomics: technologies and applications. Wiley, Weinheim, pp 197–231

Li H et al (2000) Structural determination of lipid-bound ApoA-I using fluorescence resonance energy transfer. J Biol Chem 275(47):37048–37054

Li HH et al (2002) ApoA-I structure on discs and spheres. Variable helix registry and conformational states. J Biol Chem 277(42):39093–39101

Liadaki KN et al (2000) Binding of high density lipoprotein (HDL) and discoidal reconstituted HDL to the HDL receptor scavenger receptor class B type I. Effect of lipid association and APOA-I mutations on receptor binding. J Biol Chem 275(28):21262–21271

Lipina C, Hundal HS (2011) Sphingolipids: agents provocateurs in the pathogenesis of insulin resistance. Diabetologia 54(7):1596–1607

Ljunggren S et al (2013) APOA-I mutations, L202P and K131DEL, IN HDL from heterozygotes with low HDL-cholesterol. Proteomics Clin Appl 8(3–4):241–250

Lucke S, Levkau B (2010) Endothelial functions of sphingosine-1-phosphate. Cell Physiol Biochem 26(1):87–96

Lund-Katz S, Phillips MC (2010) High density lipoprotein structure-function and role in reverse cholesterol transport. Subcell Biochem 51:183–227

Mackness B et al (2010) Human tissue distribution of paraoxonases 1 and 2 mRNA. IUBMB Life 62(6):480–482

Maeba R, Ueta N (2003) Ethanolamine plasmalogen and cholesterol reduce the total membrane oxidizability measured by the oxygen uptake method. Biochem Biophys Res Commun 302(2):265–270

Maiorano JN et al (2004) Identification and structural ramifications of a hinge domain in apolipoprotein A-I discoidal high-density lipoproteins of different size. Biochemistry 43(37):11717–11726

Majek P et al (2011) Plasma proteome changes in cardiovascular disease patients: novel isoforms of apolipoprotein A1. J Transl Med 9:84

Mallat Z et al (2010) Lipoprotein-associated and secreted phospholipases A in cardiovascular disease: roles as biological effectors and biomarkers. Circulation 122(21):2183–2200

Marcel YL et al (1990) Distribution and concentration of cholesteryl ester transfer protein in plasma of normolipemic subjects. J Clin Invest 85(1):10–17

Marhaug G et al (1982) Characterization of amyloid related protein SAA complexed with serum lipoproteins (apoSAA). Clin Exp Immunol 50(2):382–389

Martin DD et al (2006) Apolipoprotein A-I assumes a "looped belt" conformation on reconstituted high density lipoprotein. J Biol Chem 281(29):20418–20426

Matz CE, Jonas A (1982) Micellar complexes of human apolipoprotein A-I with phosphatidylcholines and cholesterol prepared from cholate-lipid dispersions. J Biol Chem 257(8):4535–4540

Mauger JF et al (2006) Apolipoprotein C-III isoforms: kinetics and relative implication in lipid metabolism. J Lipid Res 47(6):1212–1218

Mazur MT et al (2010) Quantitative analysis of intact apolipoproteins in human HDL by top-down differential mass spectrometry. Proc Natl Acad Sci USA 107(17):7728–7733

McConathy WJ, Alaupovic P (1973) Isolation and partial characterization of apolipoprotein D: a new protein moiety of the human plasma lipoprotein system. FEBS Lett 37(2):178–182

McIntyre TM et al (2009) The emerging roles of PAF acetylhydrolase. J Lipid Res Supp 50:S255–S259

McLachlan AD (1977) Repeated helical pattern in apolipoprotein-A-I. Nature 267(5610):465–466

Mei X, Atkinson D (2011) Crystal structure of C-terminal truncated apolipoprotein A-I reveals the assembly of high density lipoprotein (HDL) by dimerization. J Biol Chem 286(44):38570–38582

Miida T et al (2003) Analytical performance of a sandwich enzyme immunoassay for pre beta 1-HDL in stabilized plasma. J Lipid Res 44(3):645–650

Miyazaki M et al (2010) Static and dynamic characterization of nanodiscs with apolipoprotein A-I and its model peptide. J Phys Chem B 114(38):12376–12382

Mora S (2009) Advanced lipoprotein testing and subfractionation are not (yet) ready for routine clinical use. Circulation 119(17):2396–2404

Moren X et al (2008) HDL subfraction distribution of paraoxonase-1 and its relevance to enzyme activity and resistance to oxidative stress. J Lipid Res 49(6):1246–1253

Navab M et al (2004) The oxidation hypothesis of atherogenesis: the role of oxidized phospholipids and HDL. J Lipid Res 45(6):993–1007

Nelsestuen GL et al (2008) Top-down proteomic analysis by MALDI-TOF profiling: concentration-independent biomarkers. Proteomics Clin Appl 2(2):158–166

Neumann S et al (2009) Mammalian Wnt3a is released on lipoprotein particles. Traffic 10(3):334–343

Nicholls SJ, Hazen SL (2009) Myeloperoxidase, modified lipoproteins, and atherogenesis. J Lipid Res Supp 50:S346–S351

Nichols AV et al (1983) Characterization of discoidal complexes of phosphatidylcholine, apolipoprotein A-I and cholesterol by gradient gel electrophoresis. Biochim Biophys Acta 750(2):353–364

Nichols AV et al (1986) Nondenaturing polyacrylamide gradient gel electrophoresis. Methods Enzymol 128:417–431

Nicolardi S et al (2013a) Mapping O-glycosylation of apolipoprotein C-III in MALDI-FT-ICR protein profiles. Proteomics 13(6):992–1001

Nicolardi S et al (2013b) Identification of new apolipoprotein-CIII glycoforms with ultrahigh resolution MALDI-FTICR mass spectrometry of human sera. J Proteome Res 12(5):2260–2268

Nijstad N et al (2011) Overexpression of apolipoprotein O does not impact on plasma HDL levels or functionality in human apolipoprotein A-I transgenic mice. Biochim Biophys Acta 1811(4):294–299

Nilsson A, Duan RD (2006) Absorption and lipoprotein transport of sphingomyelin. J Lipid Res 47(1):154–171

Nobecourt E et al (2007) The impact of glycation on apolipoprotein A-I structure and its ability to activate lecithin:cholesterol acyltransferase. Diabetologia 50(3):643–653

Nolte RT, Atkinson D (1992) Conformational analysis of apolipoprotein A-I and E-3 based on primary sequence and circular dichroism. Biophys J 63(5):1221–1239

Ohta T et al (1985) Studies on the in vivo and in vitro distribution of apolipoprotein A-IV in human plasma and lymph. J Clin Invest 76(3):1252–1260

Olofsson SO et al (1978) Isolation and partial characterization of a new acidic apolipoprotein (apolipoprotein F) from high density lipoproteins of human plasma. Biochemistry 17(6):1032–1036

Ortiz-Munoz G et al (2009) HDL antielastase activity prevents smooth muscle cell anoikis, a potential new antiatherogenic property. FASEB J 23(9):3129–3139

Otvos JD et al (1992) Development of a proton nuclear magnetic resonance spectroscopic method for determining plasma lipoprotein concentrations and subspecies distributions from a single, rapid measurement. Clin Chem 38(9):1632–1638

Panagotopulos SE et al (2001) Apolipoprotein A-I adopts a belt-like orientation in reconstituted high density lipoproteins. J Biol Chem 276(46):42965–42970

Park KH et al (2010) Fructated apolipoprotein A-I showed severe structural modification and loss of beneficial functions in lipid-free and lipid-bound state with acceleration of atherosclerosis and senescence. Biochem Biophys Res Commun 392(3):295–300

Phillips MC (2013) New insights into the determination of HDL structure by apolipoproteins: thematic review series: high density lipoprotein structure, function, and metabolism. J Lipid Res 54(8):2034–2048

Phillips JC et al (1997) Predicting the structure of apolipoprotein A-I in reconstituted high-density lipoprotein disks. Biophys J 73(5):2337–2346

Pruzanski W et al (2000) Comparative analysis of lipid composition of normal and acute-phase high density lipoproteins. J Lipid Res 41(7):1035–1047

Puppione DL et al (2010) Mass spectral analyses of the two major apolipoproteins of great ape high density lipoproteins. Comp Biochem Physiol Part D Genomics Proteomics 4(4):305–309

Qiu X et al (2007) Crystal structure of cholesteryl ester transfer protein reveals a long tunnel and four bound lipid molecules. Nat Struct Mol Biol 14(2):106–113

Rassart E et al (2000) Apolipoprotein D. Biochim Biophys Acta 1482(1–2):185–198

Remaley AT et al (1993) O-linked glycosylation modifies the association of apolipoprotein A-II to high density lipoproteins. J Biol Chem 268(9):6785–6790

Roghani A, Zannis VI (1988) Mutagenesis of the glycosylation site of human ApoCIII. O-linked glycosylation is not required for ApoCIII secretion and lipid binding. J Biol Chem 263(34):17925–17932

Rosenson RS et al (2011) HDL measures, particle heterogeneity, proposed nomenclature, and relation to atherosclerotic cardiovascular events. Clin Chem 57(3):392–410

Rousset X et al (2009) Lecithin: cholesterol acyltransferase–from biochemistry to role in cardiovascular disease. Curr Opin Endocrinol Diabetes Obes 16(2):163–171

Rye KA et al (1996) The influence of sphingomyelin on the structure and function of reconstituted high density lipoproteins. J Biol Chem 271(8):4243–4250

Saito H et al (2000) Inhibition of lipoprotein lipase activity by sphingomyelin: role of membrane surface structure. Biochim Biophys Acta 1486(2–3):312–320

Samanta U, Bahnson BJ (2008) Crystal structure of human plasma platelet-activating factor acetylhydrolase: structural implication to lipoprotein binding and catalysis. J Biol Chem 283 (46):31617–31624

Santucci L et al (2011) Protein-protein interaction heterogeneity of plasma apolipoprotein A1 in nephrotic syndrome. Mol Biosyst 7(3):659–666

Sattler KJ et al (2010) Sphingosine 1-phosphate levels in plasma and HDL are altered in coronary artery disease. Basic Res Cardiol 105(6):821–832

Scanu AM (1978) Ultrastructure of serum high density lipoproteins: facts and models. Lipids 13 (12):920–925

Schaefer EJ, Asztalos BF (2007) Increasing high-density lipoprotein cholesterol, inhibition of cholesteryl ester transfer protein, and heart disease risk reduction. Am J Cardiol 100(11A): S25–S31

Schaefer EJ et al (1979) The composition and metabolism of high density lipoprotein subfractions. Lipids 14(5):511–522

Schaefer EJ et al (2010) Marked HDL deficiency and premature coronary heart disease. Curr Opin Lipidol 21(4):289–297

Scherer M et al (2010) Sphingolipid profiling of human plasma and FPLC-separated lipoprotein fractions by hydrophilic interaction chromatography tandem mass spectrometry. Biochim Biophys Acta 1811(2):68–75

Schjoldager KT et al (2012) Probing isoform-specific functions of polypeptide GalNAc-transferases using zinc finger nuclease glycoengineered SimpleCells. Proc Natl Acad Sci USA 109(25):9893–9898

Schonfeld G et al (1976) Structure of high density lipoprotein. The immunologic reactivities of the COOH- and NH2-terminal regions of apolipoprotein A-I. J Biol Chem 251(13):3921–3926

Segrest JP (1977) Amphipathic helixes and plasma lipoproteins: thermodynamic and geometric considerations. Chem Phys Lipids 18(1):7–22

Segrest JP et al (1974) A molecular theory of lipid-protein interactions in the plasma lipoproteins. FEBS Lett 38(3):247–258

Segrest JP et al (1994) The amphipathic alpha helix: a multifunctional structural motif in plasma apolipoproteins. Adv Protein Chem 45:303–369

Segrest JP et al (1999) A detailed molecular belt model for apolipoprotein A-I in discoidal high density lipoprotein. J Biol Chem 274(45):31755–31758

Segrest JP et al (2000) Detailed molecular model of apolipoprotein A-I on the surface of high-density lipoproteins and its functional implications. Trends Cardiovasc Med 10(6):246–252

Segrest JP et al (2013) MD simulations suggest important surface differences between reconstituted and circulating spherical HDL. J Lipid Res 54(10):2718–2732

Sevvana M et al (2009) Serendipitous fatty acid binding reveals the structural determinants for ligand recognition in apolipoprotein M. J Mol Biol 393(4):920–936

Shah AS et al (2013) Proteomic diversity of high density lipoproteins: our emerging understanding of its importance in lipid transport and beyond. J Lipid Res 54(10):2575–2585

Shao B (2012) Site-specific oxidation of apolipoprotein A-I impairs cholesterol export by ABCA1, a key cardioprotective function of HDL. Biochim Biophys Acta 1821(3):490–501

Shen BW et al (1977) Structure of human serum lipoproteins inferred from compositional analysis. Proc Natl Acad Sci USA 74(3):837–841

Shiflett AM et al (2005) Human high density lipoproteins are platforms for the assembly of multi-component innate immune complexes. J Biol Chem 280(38):32578–32585

Shimano H (2009) ApoAII controversy still in rabbit? Arterioscler Thromb Vasc Biol 29 (12):1984–1985

Silva RA et al (2005) A mass spectrometric determination of the conformation of dimeric apolipoprotein A-I in discoidal high density lipoproteins. Biochemistry 44(24):8600–8607

Silva RA et al (2008) Structure of apolipoprotein A-I in spherical high density lipoproteins of different sizes. Proc Natl Acad Sci USA 105(34):12176–12181

Sloop CH et al (1983) Characterization of dog peripheral lymph lipoproteins: the presence of a disc-shaped "nascent" high density lipoprotein. J Lipid Res 24(11):1429–1440

Sparks DL et al (1992) The conformation of apolipoprotein A-I in discoidal and spherical recombinant high density lipoprotein particles. 13C NMR studies of lysine ionization behavior. J Biol Chem 267(36):25830–25838

Stahlman M et al (2013) Dyslipidemia, but not hyperglycemia and insulin resistance, is associated with marked alterations in the HDL lipidome in type 2 diabetic subjects in the DIWA cohort: impact on small HDL particles. Biochim Biophys Acta 1831(11):1609–1617

Sung HJ et al (2012) Large-scale isotype-specific quantification of Serum amyloid A 1/2 by multiple reaction monitoring in crude sera. J Proteomics 75(7):2170–2180

Tardieu A et al (1976) Structure of human serum lipoproteins in solution. II. Small-angle x-ray scattering study of HDL and LDL. J Mol Biol 101(2):129–153

Tselepis AD et al (1995) PAF-degrading acetylhydrolase is preferentially associated with dense LDL and VHDL-1 in human plasma. Catalytic characteristics and relation to the monocyte-derived enzyme. Arterioscler Thromb Vasc Biol 15(10):1764–1773

Usami Y et al (2011) Detection of chymase-digested C-terminally truncated apolipoprotein A-I in normal human serum. J Immunol Methods 369(1–2):51–58

Usami Y et al (2013) Identification of sites in apolipoprotein A-I susceptible to chymase and carboxypeptidase A digestion. Biosci Rep 33(1):49–56

Utermann G (1975) Isolation and partial characterization of an arginine-rich apolipoprotein from human plasma very-low-density lipoproteins: apolipoprotein E. Hoppe Seylers Z Physiol Chem 356(7):1113–1121

Vaisar T (2009) Thematic review series: proteomics. Proteomic analysis of lipid-protein complexes. J Lipid Res 50(5):781–786

Vaisar T et al (2007) Shotgun proteomics implicates protease inhibition and complement activation in the antiinflammatory properties of HDL. J Clin Invest 117(3):746–756

Vieu C et al (1996) Identification and quantification of diacylglycerols in HDL and accessibility to lipase. J Lipid Res 37(5):1153–1161

Wald JH et al (1990) Structure of apolipoprotein A-I in three homogeneous, reconstituted high density lipoprotein particles. J Biol Chem 265(32):20037–20043

Wiesner P et al (2009) Lipid profiling of FPLC-separated lipoprotein fractions by electrospray ionization tandem mass spectrometry. J Lipid Res 50(3):574–585

Wlodawer A et al (1979) High-density lipoprotein recombinants: evidence for a bicycle tire micelle structure obtained by neutron scattering and electron microscopy. FEBS Lett 104 (2):231–235

Wolfrum C et al (2005) Apolipoprotein M is required for prebeta-HDL formation and cholesterol efflux to HDL and protects against atherosclerosis. Nat Med 11(4):418–422

Wopereis S et al (2003) Apolipoprotein C-III isofocusing in the diagnosis of genetic defects in O-glycan biosynthesis. Clin Chem 49(11):1839–1845

Wroblewska M et al (2009) Phospholipids mediated conversion of HDLs generates specific apoA-II pre-beta mobility particles. J Lipid Res 50(4):667–675

Wu Z et al (2007) The refined structure of nascent HDL reveals a key functional domain for particle maturation and dysfunction. Nat Struct Mol Biol 14(9):861–868

Wu Z et al (2009) Double superhelix model of high density lipoprotein. J Biol Chem 284(52): 36605–36619

Wu Z et al (2011) The low resolution structure of ApoA1 in spherical high density lipoprotein revealed by small angle neutron scattering. J Biol Chem 286(14):12495–12508

Wurfel MM et al (1994) Lipopolysaccharide (LPS)-binding protein is carried on lipoproteins and acts as a cofactor in the neutralization of LPS. J Exp Med 180(3):1025–1035

Yu B et al (1997) Lipopolysaccharide binding protein and soluble CD14 catalyze exchange of phospholipids. J Clin Invest 99(2):315–324

Zannis VI et al (1980) Isoproteins of human apolipoprotein A-I demonstrated in plasma and intestinal organ culture. J Biol Chem 255(18):8612–8617

Zerrad-Saadi A et al (2009) HDL3-mediated inactivation of LDL-associated phospholipid hydroperoxides is determined by the redox status of apolipoprotein A-I and HDL particle surface lipid rigidity: relevance to inflammation and atherogenesis. Arterioscler Thromb Vasc Biol 29(12):2169–2175

Zhang L et al (2011) Morphology and structure of lipoproteins revealed by an optimized negative-staining protocol of electron microscopy. J Lipid Res 52(1):175–184

Zhaorigetu S et al (2008) ApoL1, a BH3-only lipid-binding protein, induces autophagic cell death. Autophagy 4(8):1079–1082

Zhou W et al (2009) An initial characterization of the serum phosphoproteome. J Proteome Res 8(12):5523–5531

HDL Biogenesis, Remodeling, and Catabolism

Vassilis I. Zannis, Panagiotis Fotakis, Georgios Koukos, Dimitris Kardassis, Christian Ehnholm, Matti Jauhiainen, and Angeliki Chroni

Contents

1	Biogenesis of HDL	56
	1.1 ATP-Binding Cassette Transporter A1 (ABCA1)	57
	1.1.1 Structure of apoA-I and Its Interactions with ABCA1 In Vitro	57
	1.1.2 Interaction of apoA-I with ABCA1 In Vivo Initiates the Biogenesis of HDL	58
	1.1.3 Unique Mutations in apoA-I May Affect apoA-I/ABCA1 Interactions and Inhibit the First Step in the Pathway of HDL Biogenesis	60
	1.2 Lecithin/Cholesterol Acyltransferase (LCAT)	63
	1.2.1 Interactions of Lipid-Bound ApoA-I with LCAT	63
	1.2.2 ApoA-I Mutations that Affect apoA-I/LCAT Interactions	64

V.I. Zannis (✉) • P. Fotakis
Molecular Genetics, Whitaker Cardiovascular Institute, Boston University School of Medicine, Boston, MA 02118, USA

Department of Biochemistry, University of Crete Medical School and Institute of Molecular Biology and Biotechnology, Foundation of Research and Technology of Hellas, Heraklion, Crete 71110, Greece
e-mail: vzannis@bu.edu

G. Koukos
Molecular Genetics, Whitaker Cardiovascular Institute, Boston University School of Medicine, Boston, MA 02118, USA

D. Kardassis
Department of Biochemistry, University of Crete Medical School and Institute of Molecular Biology and Biotechnology, Foundation of Research and Technology of Hellas, Heraklion, Crete 71110, Greece

C. Ehnholm • M. Jauhiainen
National Institute for Health and Welfare, Public Health Genomics Unit, Biomedicum, Helsinki FI-00251, Finland

A. Chroni
Institute of Biosciences and Applications, National Center for Scientific Research "Demokritos", Athens 15310, Greece

© The Author(s) 2015
A. von Eckardstein, D. Kardassis (eds.), *High Density Lipoproteins*, Handbook of Experimental Pharmacology 224, DOI 10.1007/978-3-319-09665-0_2

1.3 ApoA-I Mutations May Induce Hypertriglyceridemia
and/or Hypercholesterolemia 68
  1.3.1 Potential Mechanism of Dyslipidemia Resulting from apoA-I
  Mutations 71
1.4 ApoE and apoA-IV Participate in the Biogenesis of HDL Particles Containing
the Corresponding Proteins 72
1.5 Clinical Relevance of the Aberrant HDL Phenotypes 74
2 Remodeling and Catabolism of HDL 74
  2.1 ATP-Binding Cassette Transporter G1 76
  2.2 Phospholipid Transfer Protein 78
  2.3 apoM 79
  2.4 Hepatic Lipase and Endothelial Lipase 80
  2.5 Cholesteryl Ester Transfer Protein 82
  2.6 Scavenger Receptor BI 83
    2.6.1 Role of SR-BI in HDL Remodeling Based on Its In Vitro Interactions
    with Its Ligands 83
    2.6.2 In Vivo Functions of SR-BI 84
  2.7 Role of Ecto-F_1-ATPase/$P2Y_{13}$ Pathway in Hepatic HDL Clearance 86
  2.8 Transcytosis of apoA-I and HDL by Endothelial Cells 86
  2.9 The Role of Cubilin in apoA-I and HDL Catabolism by the Kidney 87
3 HDL Subclasses 88
  3.1 The Origin and Metabolism of Preβ-HDL Subpopulations 88
  3.2 Complexity of HDL 89
4 Sources of Funding 90
References 91

Abstract

In this chapter, we review how HDL is generated, remodeled, and catabolized in plasma. We describe key features of the proteins that participate in these processes, emphasizing how mutations in apolipoprotein A-I (apoA-I) and the other proteins affect HDL metabolism.

The biogenesis of HDL initially requires functional interaction of apoA-I with the ATP-binding cassette transporter A1 (ABCA1) and subsequently interactions of the lipidated apoA-I forms with lecithin/cholesterol acyltransferase (LCAT). Mutations in these proteins either prevent or impair the formation and possibly the functionality of HDL.

Remodeling and catabolism of HDL is the result of interactions of HDL with cell receptors and other membrane and plasma proteins including hepatic lipase (HL), endothelial lipase (EL), phospholipid transfer protein (PLTP), cholesteryl ester transfer protein (CETP), apolipoprotein M (apoM), scavenger receptor class B type I (SR-BI), ATP-binding cassette transporter G1 (ABCG1), the F1 subunit of ATPase (Ecto F1-ATPase), and the cubulin/megalin receptor.

Similarly to apoA-I, apolipoprotein E and apolipoprotein A-IV were shown to form discrete HDL particles containing these apolipoproteins which may have important but still unexplored functions. Furthermore, several plasma proteins were found associated with HDL and may modulate its biological functions. The effect of these proteins on the functionality of HDL is the topic of ongoing research.

HDL Biogenesis, Remodeling, and Catabolism 55

Keywords

HDL biogenesis • HDL remodeling • HDL catabolism • HDL phenotypes • Apolipoprotein A-I mutations • Apolipoprotein E • Apolipoprotein A-IV • ATP-binding cassette transporter A1 (ABCA1) • Lecithin/cholesterol acyltransferase (LCAT) • HDL subclasses • Preβ- and α-HDL particles • Dyslipidemia • Hypertriglyceridemia • Lipoprotein lipase (LPL) • Hepatic lipase (HL) • Endothelial lipase (EL) • Phospholipid transfer protein (PLTP) • Apolipoprotein M • Cholesteryl ester transfer protein (CETP) • Scavenger receptor class B type I (SR-BI) • ATP-binding cassette transporter G1 (ABCG1) • Ecto-F_1-ATPase • Cubilin • Transcytosis • Clinical phenotypes

Abbreviations

ABCA1	ATP-binding cassette transporter A1
ABCG1	ATP-binding cassette transporter G1
apoA-I	Apolipoprotein A-I
apoA-IV	Apolipoprotein A-IV
apoB	Apolipoprotein B
apoE	Apolipoprotein E
apoJ	Apolipoprotein J
apoM	Apolipoprotein M
BHK cells	Baby hamster kidney cells
Caco-2 cells	Human epithelial colorectal adenocarcinoma cells
CAD	Coronary artery disease
CD36	Cluster of differentiation 36
CE	Cholesterol ester
CETP	Cholesteryl ester transfer protein
CHO cells	Chinese hamster ovary cells
eNOS	Endothelial nitric oxide synthase
EL	Endothelial lipase
EM	Electron microscopy
FED	Fish eye disease
FLD	Familial LCAT deficiency
FPLC	Fast protein liquid chromatography
HA-PLTP	High-activity PLTP
HDL	High-density lipoprotein
HDL-A-IV	apoA-IV-containing HDL
HDL-C	HDL cholesterol
HDL-E	apoE-containing HDL
HEK293 cells	Human embryonic kidney 293 cells

HepG2 cells	Liver hepatocellular carcinoma cells
HL	Hepatic lipase
IDL	Intermediate density lipoprotein
LA-PLTP	Low-activity PLTP
LCAT	Lecithin/cholesterol acyltransferase
LDL	Low-density lipoprotein
LDLr	Low-density lipoprotein receptor
MPO	Myeloperoxidase
PLA1	Phospholipase A1
PLTP	Phospholipid transfer protein
RCT	Reverse cholesterol transport
rHDL	Recombinant HDL
SNP	Single-nucleotide polymorphism
SR-BI	Scavenger receptor class B type I
SREBPs	Sterol regulatory element-binding proteins
TC	Total cholesterol
VEGF-A	Vascular endothelial growth factor A
VLDL	Very low-density Lipoprotein
WT	Wild type

1   Biogenesis of HDL

The biogenesis of HDL is a complex process and involves several membrane bound and plasma proteins (Zannis et al. 2004a). The first step in HDL biogenesis involves secretion of apoA-I mainly by the liver and the intestine (Zannis et al. 1985). Secreted apoA-I interacts functionally with ABCA1, and this interaction leads to the transfer of cellular phospholipids and cholesterol to lipid-poor apoA-I. The lipidated apoA-I is gradually converted to discoidal particles enriched in unesterified cholesterol. The esterification of free cholesterol by the enzyme lecithin/cholesterol acyltransferase (LCAT) (Zannis et al. 2006a) converts the discoidal to spherical HDL particles (Fig. 1).

The absence or inactivating mutations in apoA-I, ABCA1, and LCAT prevent the formation of apoA-I-containing HDL (Daniil et al. 2011). For this reason, we classify the apoA-I, ABCA1, and LCAT interactions that will be discussed in this chapter as early steps in the biogenesis of HDL. Following a similar pathway, apoE and apoA-IV can also synthesize HDL particles that contain these proteins (Duka et al. 2013; Kypreos and Zannis 2007). The first part of this review provides important information on the unique properties of apoA-I that permits it to acquire lipids via interactions with ABCA1 and LCAT. It also provides examples of how specific mutations in apoA-I disrupt specific steps in the pathway of HDL biogenesis and generate distinct aberrant HDL phenotypes. The HDL phenotypes described here can serve as molecular markers that could be used for the diagnosis, prognosis,

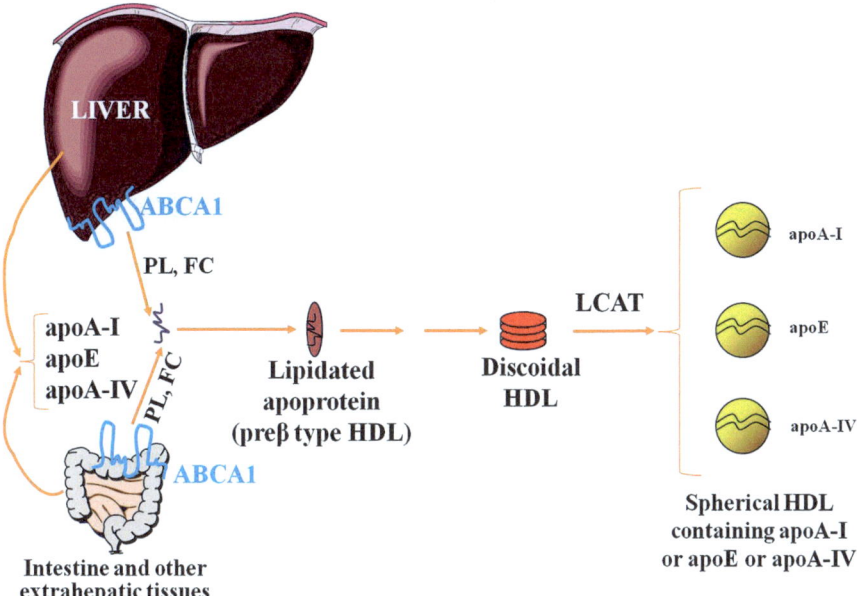

Fig. 1 Schematic representation of the pathway of the biogenesis of HDL containing apoA-I or apoE or apoA-IV

and potential treatment of HDL abnormalities or dyslipidemias associated with HDL.

1.1 ATP-Binding Cassette Transporter A1 (ABCA1)

1.1.1 Structure of apoA-I and Its Interactions with ABCA1 In Vitro

apoA-I contains 22 or 11 amino acid repeats which, according to the models of Nolte and Atkinson (1992), are organized in amphipathic a-helices (Segrest et al. 1974). Based on the crystal structure of apoA-I in solution (Borhani et al. 1997, 1999), a belt model was proposed to explain the structure of apoA-I on discoidal HDL particles (Segrest et al. 1999). Various models have been proposed to explain the arrangement of apoA-I on spherical HDL particles based on structural work and cross-linking(Wu et al. 2011; Silva et al. 2008; Huang et al. 2011). Details on the structure of apoA-I and HDL are provided in chapter "Structure of HDL: Particle Subclasses and Molecular Components."

ABCA1 is a ubiquitous protein that belongs to the ABC family of transporters and is expressed abundantly in the liver, macrophages, brain, and various other tissues (Langmann et al. 1999; Kielar et al. 2001). ABCA1 is localized only on the basolateral surface of the hepatocytes (Neufeld et al. 2002); it is also found on endocytic vesicles and was shown to travel between late endocytic vesicles and the cell surface (Neufeld et al. 2001). ABCA1 promotes efflux of cellular phospholipids and cholesterol to lipid-free or minimally lipidated apoA-I and other apolipoproteins, but not to spherical HDL particles (Wang et al. 2000; Remaley et al. 2001).

Studies in HeLa cells that expressed an ABCA1 green fluorescence fusion protein showed the intracellular trafficking of ABCA1 complexed to apoA-I (Neufeld et al. 2001, 2002). Other studies showed that in macrophages ABCA1 associates with apoA-I in the coated pits, is internalized, interacts with intracellular lipid pools, and is re-secreted as a lipidated particle (Takahashi and Smith 1999; Smith et al. 2002; Lorenzi et al. 2008). A similar pathway that leads to transcytosis has been described in endothelial cells (Cavelier et al. 2006; Ohnsorg et al. 2011).

A series of cell culture and in vitro experiments investigated the ability of apoA-I mutants to promote ABCA1-mediated efflux of cholesterol and phospholipids and to cross-link to ABCA1. These mutants had amino-terminal deletions, carboxy-terminal deletions that removed the 220-231 region, carboxy-terminal deletions that maintained the 220-231 region, and double deletions of the amino- and carboxy-terminal regions (Chroni et al. 2003).

These studies presented in Chroni et al. (2003) showed that wild-type (WT) - ABCA1-mediated cholesterol and phospholipid efflux was not affected by amino-terminal apoA-I deletions, but it was diminished by carboxy-terminal deletions in which residues 220-231 were removed. Efflux was not affected by deletion of the carboxy-terminal 232-243 region, and it was restored to 80 % of WT control by double deletions of both the amino- and carboxy-termini (Zannis et al. 2004a, 2006a; Chroni et al. 2003; Reardon et al. 2001). Lipid efflux was either unaffected or moderately reduced by a variety of point mutations or deletions of internal helices 2–7. The findings indicated that different combinations of central helices can promote lipid efflux (Chroni et al. 2004a, b). Chemical cross-linking/immunoprecipitation studies showed that the ability of apoA-I mutants to promote ABCA1-depended lipid efflux is correlated with the ability of these mutants to be cross-linked efficiently to ABCA1 (Chroni et al. 2004b). Cross-linking between apoA-I and ABCA1 and cholesterol efflux was also affected by mutations in ABCA1 that are found in patients with Tangier disease. The majority of the ABCA1 mutants cross-link poorly to WT apoA-I and have diminished capacity to promote cholesterol efflux (Bodzioch et al. 1999; Fitzgerald et al. 2002). A notable exemption is the ABCA1[W590S] mutant which cross-linked stronger to apoA-I than to WT ABCA1 but had diminished capacity to promote cholesterol efflux and to promote formation of HDL (Bodzioch et al. 1999; Fitzgerald et al. 2001, 2002). We suggested that this ABCA1 mutation may have altered the environment of the binding site of ABCA1 in such a way that the binding of apoA-I is strong but not productive and thus prevented efficient lipid efflux (Chroni et al. 2004b).

1.1.2 Interaction of apoA-I with ABCA1 In Vivo Initiates the Biogenesis of HDL

Inactivating mutations in ABCA1 found in patients with Tangier disease are associated with very low levels of total plasma and HDL cholesterol, diminished capacity to promote cholesterol efflux, formation of preβ-migrating particles, and

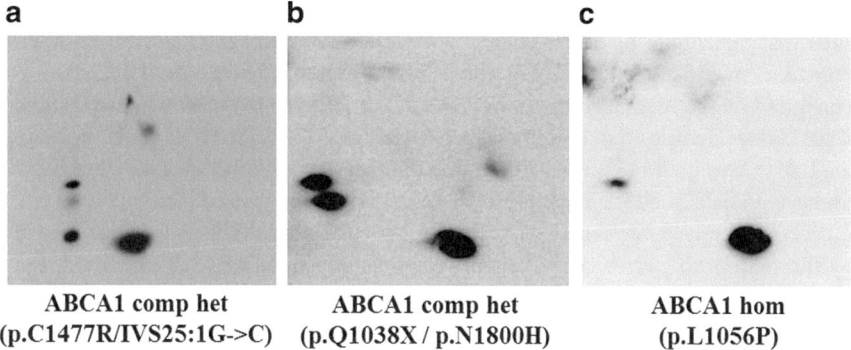

Fig. 2 Two-dimensional gel electrophoresis of plasma obtained from homozygotes or compound heterozygotes ABCA1-deficient human subjects with Tangier disease as indicated [Adapted from Daniil et al. (2011)]

abnormal lipid deposition in various tissues (Daniil et al. 2011; Brunham et al. 2006b; Orso et al. 2000; Assmann et al. 2001; Timmins et al. 2005).

Analysis of the serum of carriers with ABCA1 mutations by two-dimensional gel electrophoresis and western blotting using a rabbit polyclonal anti-human apoA-I antibody showed that subjects homozygous or compound mutant heterozygous for ABCA1 fail to form α-HDL particles, but instead they form preβ- and other small-size particles (Fig. 2).

As explained later, such particles are found in the plasma of mice expressing C-terminal mutants and may be created by mechanisms that involve nonproductive interactions between ABCA1 and apoA-I (Chroni et al. 2007; Fotakis et al. 2013a).

Several ABCA1 mutations in humans that alter the functions of ABCA1 are associated with increased susceptibility to atherosclerosis (Singaraja et al. 2003). Specific amino-acid substitutions found in the Danish general population were associated with increased risk for ischemic heart disease and reduced life expectancy through unknown mechanisms (Frikke-Schmidt et al. 2008). Inactivation of the ABCA1 gene in mice leads to low total serum cholesterol levels, lipid deposition in various tissues, impaired growth, and neuronal development and mimics the phenotype described for patients with Tangier disease (Orso et al. 2000). In addition ABCA1-deficient mice exhibit moderate increase in cholesterol absorption in response to high cholesterol diet (McNeish et al. 2000; Aiello et al. 2002). The role of ABCA1 on the lipid content of bile salts and cholesterol secretion is not clear (Vaisman et al. 2001; Groen et al. 2001).

Bone marrow transplantation experiments indicated that ABCA1 plays an important role in the control of macrophage recruitment to the tissues (Van Eck et al. 2002, 2006).

The contribution of ABCA1 in the pathogenesis of atherosclerosis in mice is presented in chapter "Mouse Models of Disturbed HDL Metabolism" (Hoesha and Van Eck). In men and mice, the majority of HDL is produced by the liver (Brunham et al. 2006a). When the liver and intestinal ABCA1 genes were inactivated in mice,

HDL was not found in plasma, indicating that the liver and the intestine are the only sites that contribute to the production of HDL (Timmins et al. 2005). Following intestinal-specific inactivation of the ABCA1 gene in mice, the HDL that was generated by the liver accounted for 70 % of the HDL found in WT mice (Brunham et al. 2006a). In mice that do not express hepatic ABCA1, the HDL concentration in the lymph was greatly diminished despite the fact that the intestine contributes 30 % to the synthesis of HDL. This implies that the HDL that is produced in the intestine is secreted directly into the plasma (Brunham et al. 2006a). This is further supported by the finding that in mice that do not express intestinal ABCA1, the apoA-I and cholesterol concentration of lymph was not affected (Brunham et al. 2006a).

In liver-specific or whole-body ABCA1 knockout mice, the plasma HDL catabolism and the fractional catabolic rate of HDL by the liver and to a lesser extent by the kidney and the adrenal is increased (Timmins et al. 2005; Singaraja et al. 2006). In ABCA1-deficient mice, lipidated apoA-I particles or preβ-HDL failed to mature and are rapidly catabolized by the kidney (Timmins et al. 2005).

1.1.3 Unique Mutations in apoA-I May Affect apoA-I/ABCA1 Interactions and Inhibit the First Step in the Pathway of HDL Biogenesis

The in vivo interactions of apoA-I with ABCA1 were studied systematically by adenovirus-mediated gene transfer of WT and mutant apoA-I forms. Similar studies were performed to probe the interactions of lipid-bound apoA-I with LCAT. Four to five days postinfection plasma was collected and analyzed for lipids and lipoproteins and by two-dimensional gel electrophoresis to identify the HDL subpopulations. The plasma was fractionated by density gradient ultracentrifugation and fast protein liquid chromatography (FPLC), and the HDL fraction was analyzed by electron microscopy (EM) to assess the size and shape of HDL (Chroni et al. 2003; Reardon et al. 2001). Also the hepatic mRNA levels of apoA-I were determined to ensure that there was comparable expression of the WT and the mutant apoA-I forms (Zannis et al. 2004a; Chroni et al. 2003, 2007; Reardon et al. 2001).

We have studied most recently the effect of two sets of point mutations in the 218-222 and 225-230 region of apoA-I that affects apoA-I/ABCA1 interactions on the biogenesis of HDL. Adenovirus-mediated gene transfer of these mutants in apoA-I$^{-/-}$ x apoE$^{-/-}$ mice (Fotakis et al. 2013a, b) showed that compared to the WT apoA-I, the expression of an apoA-I[L218A/L219A/V221A/L222A] mutant decreased plasma cholesterol, apoA-I, and HDL cholesterol levels and generated preβ- and α4 HDL subpopulations (Fotakis et al. 2013a) (Table 1 and Fig. 3a–g).

To eliminate the involvement of apoE in the generation of apoE-containing HDL particles (Kypreos and Zannis 2007), the apoA-I[L218A/L219A/V221A/L222A] mutant was expressed in apoA-I$^{-/-}$ x apoE$^{-/-}$ mice via adenovirus-mediated gene transfer (Fotakis et al. 2013a). In this mouse background, the FPLC fractionation of the plasma showed the near absence of an HDL cholesterol peak. Density gradient ultracentrifugation of the plasma showed small amount of the apoA-I in HDL3 and in $d < 1.21$ g/ml fractions (Fig. 3a). EM analysis showed the presence of discoidal particles along with larger particles corresponding in size to IDL/LDL (Fig. 3b).

HDL Biogenesis, Remodeling, and Catabolism

Table 1 Plasma lipids and hepatic mRNA levels of apoA-I$^{-/-}$ or apoA-I$^{-/-}$ x apoE$^{-/-}$ mice expressing WT and mutant forms of apoA-I in the presence and absence of LCAT as indicated

Protein expressed	Total cholesterol (mg/dL)	Triglycerides (mg/dL)	Relative apoA-I mRNA (%)	apoA-I plasma levels (mg/dL)
WT apoA-I in apoA-I$^{-/-}$ mice	278 ± 74	78 ± 24	100 ± 26	260 ± 40
apoA-I[L218A/L219A/V221A/L222A] in apoA-I$^{-/-}$ mice	45 ± 14	50 ± 20	95 ± 24	41 ± 5
WT apoA-I in apoA-I$^{-/-}$ x apoE$^{-/-}$ mice	1,343 ± 104	294 ± 129	100 ± 13	–
apoA-I[L218A/L219A/V221A/L222A] in apoA-I$^{-/-}$ x apoE$^{-/-}$ mice	778 ± 52	18 ± 2	92 ± 23	–
apoA-I[L218A/L219A/V221A/L222A] plus LCAT in apoA-I$^{-/-}$ x apoE$^{-/-}$ mice	754 ± 122	37 ± 10	90 ± 3	–

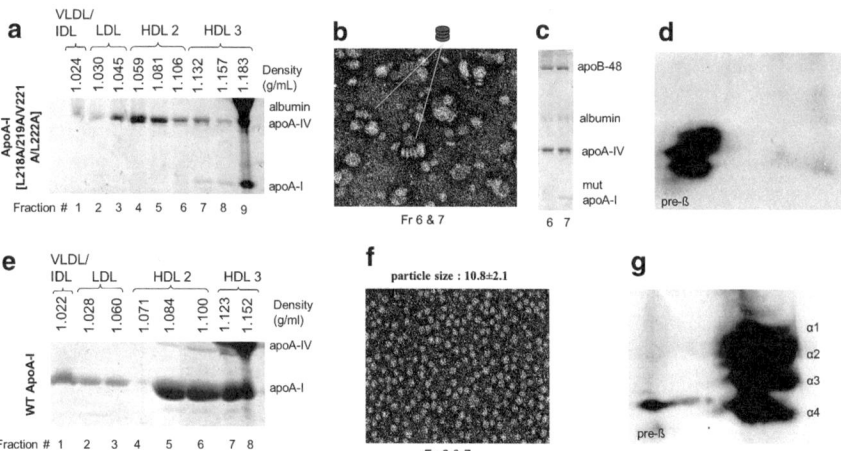

Fig. 3 (**a–g**) ApoA-I mutations and their effect on the HDL phenotypes. Analysis of plasma of apoA-I$^{-/-}$ x apoE$^{-/-}$ mice infected with adenoviruses expressing the WT apoA-I or the apoA-I [L218A/L219A/V221A/L222A] mutant, as indicated, by density gradient ultracentrifugation and SDS-PAGE (**a, e**). EM analysis of HDL fractions 6–7 obtained from apoA-I$^{-/-}$ x apoE$^{-/-}$ mice expressing the WT apoA-I or the apoA-I[L218A/L219A/V221A/L222A] mutant, as indicated, following density gradient ultracentrifugation of plasma (**b, f**). SDS gel electrophoresis showing apoprotein composition of fractions 6 and 7 (**c**) used for the EM analysis in panel **b**. Two-dimensional gel electrophoresis of plasma of apoA-I$^{-/-}$ x apoE$^{-/-}$ mice infected with adenoviruses expressing WT apoA-I or the apoA-I[L218A/L219A/V221A/L222A] mutant, as indicated (**d, g**)

These particles may originate from apoB-48 containing lipoproteins that are found in the HDL fractions (Fig. 3c). Two-dimensional gel electrophoresis showed the presence of only preβ-HDL particles (Fig. 3d). Control experiments showed that WT apoA-I when expressed in apoA-I$^{-/-}$ x apoE$^{-/-}$ mice floated predominantly in the HDL2/HDL3 region and generated spherical particles and normal preβ- and α-HDL subpopulation (Fig. 3e–g).

Co-expression of the apoA-I[L218A/L219A/V221A/L222A] mutant and human LCAT in apoA-I$^{-/-}$ x apoE$^{-/-}$ mice did not correct the plasma apoA-I levels and did not correct the aberrant HDL phenotype.

The findings shed light to previous studies which showed that carboxy-terminal deletion mutants that lacked the 220-231 region of apoA-I prevented the biogenesis of normal α-HDL particles but allowed the formation of preβ-HDL particles (Chroni et al. 2003, 2007).

Studies discussed in a later section showed that naturally occurring point mutations in apoA-I when expressed in mouse models activate LCAT insufficiently and in some instances lead to the accumulation of discoidal HDL particles in plasma (Koukos et al. 2007a, b; Chroni et al. 2005a). A characteristic feature of these mutations is that they could be corrected in vivo by the co-expression of the mutant apoA-I and LCAT.

The phenotype produced by the apoA-I[L218A/L219A/V221A/L222A] mutations is distinct from all previously described phenotypes and cannot be corrected by overexpression of LCAT. In addition, the mutant protein had reduced capability to promote the ABCA1-mediated cholesterol efflux. Although other interpretations are possible, the in vivo and in vitro data suggest that the interaction of the apoA-I [L218A/L219A/V221A/L222A] mutant with ABCA1 results in defective lipidation of apoA-I that leads to the generation of preβ-HDL particles that are not a good substrate for LCAT. If this interpretation is correct, the normal lipidation of apoA-I may require a precise initial orientation of the apoA-I ligand within the binding site of ABCA1 that is similar to that described for enzyme-substrate interactions. Such a configuration would allow correct lipidation of apoA-I that could subsequently undergo cholesterol esterification and formation of mature HDL, whereas incorrectly lipidated apoA-I becomes a poor substrate for LCAT.

In addition to the unique role of the L218/L219/V221/L222 residues in the biogenesis of HDL, the same residues are also required to confer trans-endothelial transport capacity (Ohnsorg et al. 2011) and bactericidal activity (Biedzka-Sarek et al. 2011) to apoA-I. These overall properties suggest that the L218/L219/V221/L222 residues represent an effector domain for several activities of apoA-I.

The phenotype generated by the expression of the apoA-I[F225A/V227A/F229A/L230A] mutant was similar to that obtained with the apoA-I[L218A/L219A/V221A/L222A] mutant. However co-expression of the apoA-I[F225A/V227A/F229A/L230A] mutant and human LCAT corrected the abnormal HDL levels, created normal preβ- and α-HDL subpopulations, and generated spherical HDL particles (Fotakis et al. 2013b).

1.2 Lecithin/Cholesterol Acyltransferase (LCAT)

1.2.1 Interactions of Lipid-Bound ApoA-I with LCAT

Plasma LCAT is a 416 amino acid long plasma protein that is synthesized and secreted primarily by the liver and to a much lesser extent by the brain and the testis (Warden et al. 1989; Simon and Boyer 1970). LCAT interacts with discoidal and spherical HDL and catalyzes the transfer of the 2-acyl group of lecithin or phosphatidylethanolamine to the free hydroxyl residue of cholesterol to form cholesteryl ester, using apoA-I as an activator (Fielding et al. 1972; Zannis et al. 2004b). It also catalyzes the reverse reaction of esterification of lysolecithin to lecithin (Subbaiah et al. 1980). The esterification of free cholesterol of HDL in vivo converts the discoidal to mature spherical HDL particles (Jonas 2000; Chroni et al. 2005a). Regarding the mechanism of activation of LCAT by apoA-I, it has been proposed that residues R130 and K133 play an important role in the formation of an amphipathic presentation tunnel located between helices 5-5 in the double belt model. Such a tunnel could allow migration of the hydrophobic acyl chains of phospholipids and the amphipathic unesterified cholesterol from the bilayer to the active site of LCAT that contains sites for phospholipase activity and esterification activity (Jones et al. 2009). The esterification of the cholesterol converts the 3.67 residues per turn helices to an idealized 3.6 residues per turn helix (Borhani et al. 1997).

Mutations in LCAT are associated with two phenotypes. The familial LCAT deficiency (FLD) is characterized by the inability of the mutant LCAT to esterify cholesterol on HDL and LDL and the accumulation of discoidal HDL in the plasma. The fish eye disease (FED) is characterized by the inability of mutant LCAT to esterify cholesterol on HDL only. Both diseases are characterized by low HDL levels (Santamarina-Fojo et al. 2001) and formation of preβ- and α4 subpopulations.

Analysis of plasma of patients with complete LCAT deficiency (homozygous or compound heterozygous carriers of functional LCAT mutations) by two-dimensional electrophoresis shows mostly the presence of small α-HDL subpopulation (Daniil et al. 2011) (Fig. 4a, b). Analysis of plasma of LCAT-deficient mice by two-dimensional electrophoresis shows the presence of preβ- and small-size α4 HDL particles (Fig. 4c). Expression of human LCAT by adenovirus-mediated gene transfer generated large-size α-HDL subpopulations (Fig. 4d). In contrast, expression of a LCAT mutant generated preβ-HDL along with small-size α4, α3, and α2 HDL subpopulations (Fig. 4e).

Sera obtained from LCAT heterozygotes had increased capacity to promote ABCA1-mediated cholesterol efflux and decreased capacity to promote ABCG1- and SR-BI-mediated cholesterol efflux from macrophages as compared with sera obtained from normal subjects. These properties were attributed to the increased preβ- and decreased α-HDL subpopulations in the sera of the LCAT heterozygotes (Calabresi et al. 2009b). Heterozygosity for LCAT mutations in the Italian population was not associated with increased preclinical atherosclerosis (Calabresi et al. 2009a).

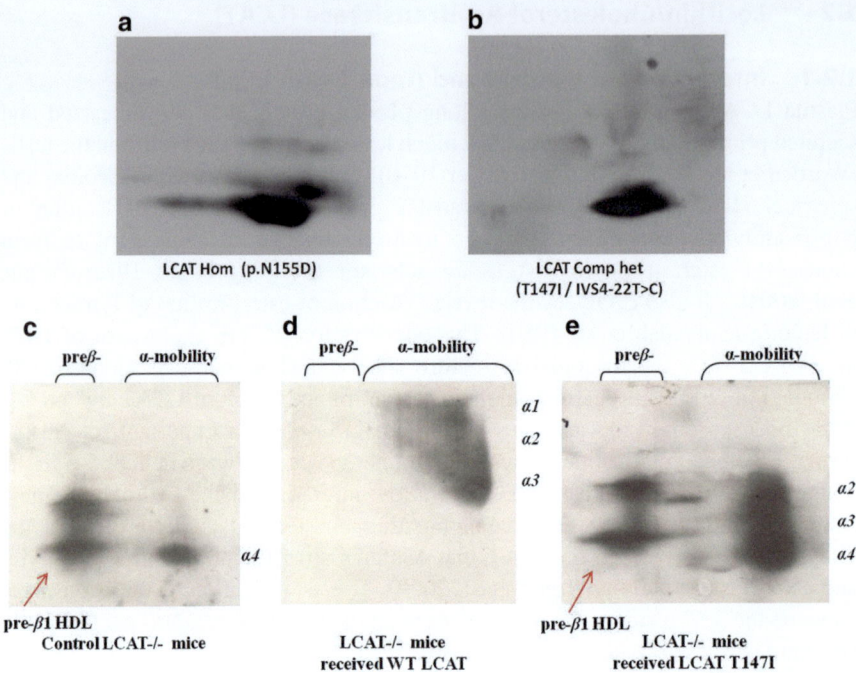

Fig. 4 (**a–e**) Two-dimensional gel electrophoresis from plasma obtained from one homozygote and one compound heterozygote for LCAT deficiency as indicated (**a, b**) (Daniil et al. 2011). Two-dimensional gel electrophoresis of plasma obtained from plasma of control LCAT$^{-/-}$ mice (**a**) and LCAT$^{-/-}$ mice infected with an adenovirus expressing either the WT human LCAT (**d**) or human LCAT[T147I] found in a patient with fish eye disease (**e**). LCAT cDNA probes of WT and LCAT mutants were provided by Dr. J.A. Kuivenhoven (Medical Center of Groningen)

The LCAT activation ability of apoA-I is inhibited following treatment with MPO (Shao et al. 2008). In addition myeloperoxidase modification alters several functions of HDL and generates pro-inflammatory HDL particles (Undurti et al. 2009).

1.2.2 ApoA-I Mutations that Affect apoA-I/LCAT Interactions

Several naturally occurring apoA-I mutations that produce pathological phenotypes have been described (Zannis et al. 2006b; Sorci-Thomas and Thomas 2002; Miettinen et al. 1997b). It has been estimated that structural mutations of apoA-I occur in 0.3 % of the Japanese population and may affect the plasma HDL levels (Yamakawa-Kobayashi et al. 1999). From a total of 46 natural mutations of apoA-I, 25 are associated with low HDL levels, and 17 of these mutants reduce the capacity of apoA-I to activate LCAT (Sorci-Thomas and Thomas 2002; Zannis et al. 2006b). These mutations are clustered predominantly in or at the vicinity of helix 6 of apoA-I and some of them predispose to atherosclerosis (Huang et al. 1998; Miller et al. 1998; Miettinen et al. 1997b; Miccoli et al. 1996). Here we describe how some representative mutations of this category affect the biogenesis and maturation of HDL.

Previous studies showed that hemizygotes (compound heterozygotes for an apoA-I null allele and an apoA-I(L141R)$_{Pisa}$ allele) had greatly decreased plasma apoA-I levels and near absence of HDL cholesterol. Plasma from hemizygotes contained preβ1-HDL and low concentration of small particles with alpha electrophoretic mobility (Miccoli et al. 1997). Heterozygotes for apoA-I(L141R)$_{Pisa}$ had approximately half-normal values for HDL cholesterol and plasma apoA-I (Miccoli et al. 1996; Pisciotta et al. 2003). Three male hemizygote patients and one heterozygote patient developed coronary stenosis (Miccoli et al. 1996).

Other studies also showed that heterozygotes for apoA-I(L159R)$_{FIN}$ mutation had greatly reduced plasma levels of HDL cholesterol and apoA-I (Miettinen et al. 1997a) that was mainly distributed in the HDL3 region and had abnormal electrophoretic mobility (Miettinen et al. 1997b; McManus et al. 2001). They also had small-size (8–9 nm) HDL particles and decreased plasma and HDL cholesteryl ester levels (Miettinen et al. 1997b; McManus et al. 2001). Human HDL containing apoA-I(L159R)$_{FIN}$ had increased fractional catabolic rate compared to normal HDL, indicating increased catabolism of the mutant apoA-I protein (Miettinen et al. 1997a, b). Only one affected patient with this mutation had clinically manifested atherosclerosis (Miettinen et al. 1997b).

To explain the etiology and potential therapy of genetically determined low levels of HDL resulting from natural apoA-I mutations, we have studied the in vitro and in vivo properties of the naturally occurring mutants, apoA-I[L141A]$_{Pisa}$ and apoA-I[L159R]FIN (Koukos et al. 2007a). In vitro studies showed that both mutants were secreted efficiently from cells, had normal ability to promote ABCA1-mediated cholesterol efflux, but greatly diminished capacity to activate LCAT (0.4–2 % of WT apoA-I). Adenovirus-mediated gene transfer showed that compared to WT apoA-I, expression of either of the two mutants in apoA-I-deficient (apoA-I$^{-/-}$) mice greatly decreased total plasma cholesterol and apoA-I levels as well as the CE/TC ratio compared to WT apoA-I (Table 2). Another change that was associated with differences between the WT apoA-I and either of

Table 2 Plasma lipids and hepatic mRNA levels of apoA-I$^{-/-}$ mice expressing WT and the mutant forms of apoA-I in the presence and absence of LCAT as indicated

Protein expressed	Cholesterol (mg/dl)	CE/TC	Triglycerides (mg/dl)	Relative ApoA-I mRNA (%)	Apo A-I Protein (mg/dl)
apo A-I WT	148 ± 11	0.78 ± 0.01	63 ± 1	100 ± 32	186 ± 34
apoA-I (L141R)$_{Pisa}$	23 ± 0.4	0.44 ± 0.03	11 ± 2.8	88 ± 9	17 ± 4
apoA-I (L141R)$_{Pisa}$ + LCAT	184 ± 53	0.68 ± 0.01	41 ± 0.3	91 ± 2	224 ± 7
apoA-I (L159R)$_{FIN}$	16 ± 5	0.13 ± 0.04	25 ± 4	216 ± 32	25 ± 9
apoA-I (L159R)$_{FIN}$ + LCAT	224 ± 22	0.73 ± 0.01	53 ± 15	63 ± 9	190 ± 20
apo A-I (R159L)$_{Oslo}$	43 ± 13	0.23 ± 0.01	36 ± 4	117 ± 30	66 ± 31
apo A-I (R160L)$_{Oslo}$	250 ± 47	0.082 ± 0.01	62 ± 11	60 ± 1	127 ± 26

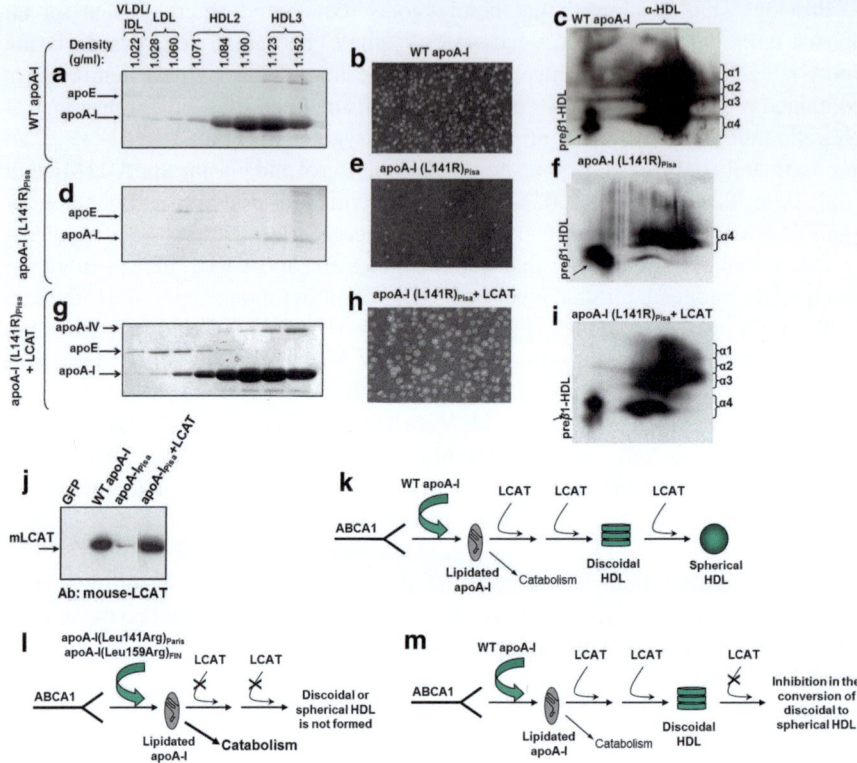

Fig. 5 (**a–m**) ApoA-I mutation that influence activity of LCAT. Analysis of plasma of apoA-I$^{-/-}$ mice infected with adenoviruses expressing the WT apoA-I or the apoA-I[L141]$_{Pisa}$ mutant alone or in combination with human LCAT by density gradient ultracentrifugation and SDS-PAGE (**a, d, g**) as indicated. EM analysis of HDL fractions 6–7 obtained from apoA-I$^{-/-}$ mice expressing the WT apoA-I or the apoA-I[L141R]$_{Pisa}$ or mutant alone or in combination with human LCAT, following density gradient ultracentrifugation of plasma, as indicated (**b, e, h**). Two-dimensional gel electrophoresis of plasma of apoA-I$^{-/-}$ mice infected with adenoviruses expressing WT apoA-I or the apoA-I[L141q]$_{Pisa}$ or mutant alone or in combination with human LCAT, as indicated (**c, f, i**). Western blot analysis of plasma from apoA-I$^{-/-}$ mice infected with adenoviruses expressing either the control protein, GFP, or the WT apoA-I or the apoA-I(L141R)$_{Pisa}$ alone or in combination with human LCAT, as indicated at the *top of the figure* (**j**). Schematic representation showing the pathway of biogenesis of HDL (**k**) and how the apoA-I(L141R)$_{Pisa}$ and apoA-I(L159R)$_{FIN}$ mutants affect the esterification of cholesterol of the pre-HDL particles and prevent their conversion to discoidal and spherical HDL, thus promoting their catabolism (**l**). Schematic representation showing the inability of the apoA-I(R160L)$_{Oslo}$ mutant to convert the discoidal to spherical HDL particles (**m**)

the two mutants was the greatly decreased HDL cholesterol peak as determined by FPLC fractionation of the plasma. Density gradient ultracentrifugation of plasma showed great reduction of the amount of apoA-I that floated in the HDL region of the apoA-I[L141R]$_{Pisa}$ mutant as compared to WT apoA-I (Fig. 5a, d). EM analysis of the HDL fractions obtained by density gradient ultracentrifugation showed the presence of a large number of spherical HDL for WT apoA-I but only a few spherical HDL particles for the apoA-I[L141R]$_{Pisa}$ mutant (Fig. 5b, e). Two-dimensional gel electrophoresis of the plasma showed the formation of small amount of preβ-HDL and large amount of α1, α2, α3, α4 HDL subpopulations for the WT apoA-I and only preβ- and small-size α4 HDL subpopulations for the apoA-I[L141R]$_{Pisa}$ mutant (Fig. 5c, f). Similar results were observed for apoA-I [L159R]$_{FIN}$ (Koukos et al. 2007a).

Coinfection of apoA-I$^{-/-}$ mice with adenoviruses expressing either of the two mutants and human LCAT normalized the plasma apoA-I, the total plasma cholesterol levels, and the CE/TC ratio (Table 2) and increased the HDL cholesterol peak and the amount of apoA-I that floated in the HDL region (Fig. 5g). It also generated large amount of spherical HDL (Fig. 5h) and restored the normal preβ- and α-HDL subpopulations (Fig. 5i). Similar results were observed for apoA-I[L159R]$_{FIN}$ (Koukos et al. 2007a).

Another interesting naturally occurring apoA-I mutation is the apoA-I [R160L]$_{Oslo}$. Previous studies showed that heterozygotes of apoA-I[R160L]$_{Oslo}$ have approximately 60 and 70 % of normal HDL and apoA-I levels, respectively, from preβ1-and small-size α-HDL particles and have a 30 % reduction in their plasma LCAT activity (Leren et al. 1997). Gene transfer of the apoA-I[R160L]$_{Oslo}$ mutant in apoA-I$^{-/-}$ mice resulted in low plasma cholesterol and apoA-I levels (Table 2) and generated discoidal particles with α4 electrophoretic mobility. The aberrant phenotype could be corrected by co-expression of this mutant with human LCAT (Koukos et al. 2007b).

Similar but not identical phenotypes were produced by expressing the bioengineered apoA-I[R160V/H162A] and apoA-I[R149A] mutants and the naturally occurring mutants apoA-I[R151C]$_{Paris}$ and apoA-I[L144R]$_{Zaragosa}$ (Haase et al. 2011; Chroni et al. 2005a; Koukos et al. 2007b). This phenotype could be corrected by co-expression of the mutant with human LCAT.

The last two mutations have not been associated with incidence of atherosclerosis in humans.

The apoA-I mutations discussed here offer a valuable tool to dissect the molecular events which lead to the biogenesis of HDL and possibly to understand the types of molecular interactions between apoA-I and LCAT which lead to the activation of the enzyme.

In our case, residues R149, R153, and R160 were reported to create a positive electrostatic potential around apoA-I. Mutations in these residues reduced drastically the ability of rHDL particles containing these apoA-I mutants to activate LCAT in vitro (Roosbeek et al. 2001). Based on the "belt" model for discoidal rHDL, these residues are located on the hydrophilic face of the apoA-I helices and do not form intramolecular salt bridges in the antiparallel apoA-I dimer that covers the fatty acyl chain of the discoidal particle. This arrangement allows in principle these apoA-I residues to form salt bridges or hydrogen bonds with appropriate residues of LCAT and thus contribute to LCAT activation.

To explain the low HDL levels and the abnormal HDL phenotype of the apoA-I$^{-/-}$ mice expressing the apoA-I(L141R)$_{Pisa}$, we analyzed the relative abundance of the endogenous mouse LCAT following gene transfer of the apoA-I(L141R)$_{Pisa}$ mutant alone or in the presence of LCAT. This analysis showed a dramatic increase of the mouse LCAT in mice expressing the apoA-I(L141R)$_{Pisa}$ mutant as compared to mice expressing the WT apoA-I. Coinfection of apoA-I$^{-/-}$ mice with the apoA-I (L141R)$_{Pisa}$ mutant and human LCAT restored the mouse LCAT to normal levels (those observed in the presence of WT apoA-I) (Fig. 5j). The depletion of the endogenous LCAT in mice expressing the mutant forms of apoA-I could be the result of rapid degradation of endogenous mouse LCAT bound to minimally lipidated apoA-I mutants possibly by the kidney.

The ability of the apoA-I[L141R]$_{Pisa}$ and apoA-I(L159R)$_{FIN}$ mutants to be secreted efficiently from cells and to promote ABCA1-mediated cholesterol efflux suggests that the functional interactions between apoA-I and ABCA1 that lead to the lipidation of apoA-I are normal and the low apoA-I and HDL levels caused by these two mutants are the result of fast removal of the lipidated nascent HDL particles from the plasma compartment. This interpretation is supported by the increased catabolic rate of HDL containing apoA-I(L159R)$_{FIN}$ (Miettinen et al. 1997b) and the accumulation of proapoA-I in the plasma of hemizygotes for apoA-I(L141R)$_{Pisa}$ (Miccoli et al. 1996). Accumulation of proapoA-I has been previously observed in patients with Tangier disease (Zannis et al. 1982) that are characterized by increased catabolic rate of HDL (Assmann et al. 2001). It has been also shown previously that cubilin, a 600 KDa membrane protein, binds both apoA-I and HDL and promotes their catabolism by the kidney (Kozyraki et al. 1999; Hammad et al. 1999).

Previous studies showed that preβ-HDL is an efficient substrate of LCAT (Nakamura et al. 2004b). In the presence of excess LCAT, the esterification of the cholesterol of the newly formed preβ-particles appears to prevent their fast catabolism and allows them to proceed in the formation of discoidal and spherical HDL. In the case of the apoA-I[R160L]$_{Oslo}$, the HDL pathway appears to be inhibited in the step of the conversion of the discoidal to spherical HDL particles. Figure 5k–m depicts the normal pathway of the biogenesis of HDL (Fig. 5k) and the disruption of this pathway by the apoA-I(L141R)$_{Pisa}$ and apoA-I(L159R)$_{FIN}$ (Fig. 5l) and apoA-I[R160L]$_{Oslo}$ mutants (Fig. 5m).

1.3 ApoA-I Mutations May Induce Hypertriglyceridemia and/or Hypercholesterolemia

A series of apoA-I mutations had resulted in severe hypertriglyceridemia (Chroni et al. 2004a; 2005b) (Table 3). The most recently studied case was apoA-I[D89A/E91A/E92A] mutant where the charged residues were substituted by alanines. The capacity of the apoA-I[D89A/E91A/E92A] mutant to promote ABCA1-mediated cholesterol efflux and activate LCAT in vitro was approximately 2/3 of that of WT apoA-I (Kateifides et al. 2011).

Table 3 Plasma lipids and hepatic mRNA levels of apoA-I$^{-/-}$ mice expressing WT and the mutant forms of apoA-I as indicated

Protein expressed	Cholesterol (mg/dL)	CE/TC	Triglycerides (mg/dL)	Relative apo A-I mRNA (%)	Plasma apo A-I (mg/dL)
apo A-I$^{-/-}$	33 ± 6	–	42 ± 7	–	–
WT apoA-I	268 ± 55	0.72 ± 0.06	70 ± 11	100 ± 32	283 ± 84
apo A-I [D89A/E91A/E92A]	497 ± 139	0.36 ± 0.31	2,106 ± 1,629	101 ± 24	235 ± 106
apo A-I [D89A/E91A/E92A] +hLPL	122 ± 56	0.44 ± 0.14	49 ± 16	41 ± 6	99 ± 18
apo A-I [Δ(62–78)]	220 ± 16	–	986 ± 289	130 ± 5	265 ± 36
apo A-I [E110A/E111A]	520 ± 45	–	1,510 ± 590	69 ± 23	204 ± 27

In vivo studies using adenovirus-mediated gene transfer in apoA-I-deficient mice showed that compared to WT apoA-I, the apoA-I[D89A/E91A/E92A] mutant increased plasma and HDL cholesterol, reduced the CE/TC ratio, and caused severe hypertriglyceridemia (Table 3) (Kateifides et al. 2011). Following density gradient ultracentrifugation of plasma, approximately 40 % of the apoA-I mutant was distributed in the VLDL/IDL region. In contrast, the WT apoA-I was distributed in the HDL2/HDL3 region (Fig. 6a). Whereas the WT apoA-I formed spherical HDL (Fig. 5b), the apoA-I[D89A/E91A/E92A] mutant formed mostly spherical and few discoidal HDL particles as determined by EM (Fig. 6b). Two-dimensional gel electrophoresis showed that WT apoA-I formed normal preβ- and α-HDL subpopulations, whereas the apoA-I[D89A/E91A/E92A] mutants formed preβ- and α4 HDL subpopulations (Fig. 6c) (Kateifides et al. 2011).

Co-expression of apoA-I[D89A/E91A/E92A] mutants and human lipoprotein lipase in apoA-I-deficient mice abolished hypertriglyceridemia (Table 3), redistributed apoA-I in the HDL2/HDL3 regions (Fig. 6d), restored in part the α1,2,3,4 HDL subpopulations (Fig. 6f), but did not change significantly the cholesterol ester to total cholesterol ratio (Table 3) or the formation of discoidal HDL particles (Fig. 6e) (Kateifides et al. 2011).

The findings indicate that residues D89, E91, and E92 of apoA-I are important for plasma cholesterol and triglyceride homeostasis as well as for the maturation of HDL.

The lipid, lipoprotein, and HDL profiles generated by another mutant, apoA-I [K94A/K96A], where the charged residues were changed to Alanines, were similar to those of WT apoA-I, indicating that the observed changes on the HDL phenotype were unique for the charged residues D89, E91, and E92 (Kateifides et al. 2011). Expression of a deletion mutant, apoA-I[Δ89-99], in apoA-I$^{-/-}$ mice, increased plasma cholesterol levels, increased the plasma preβ-HDL subpopulation, generated discoidal HDL particles, but did not induce hypertriglyceridemia (Chroni et al. 2005b).

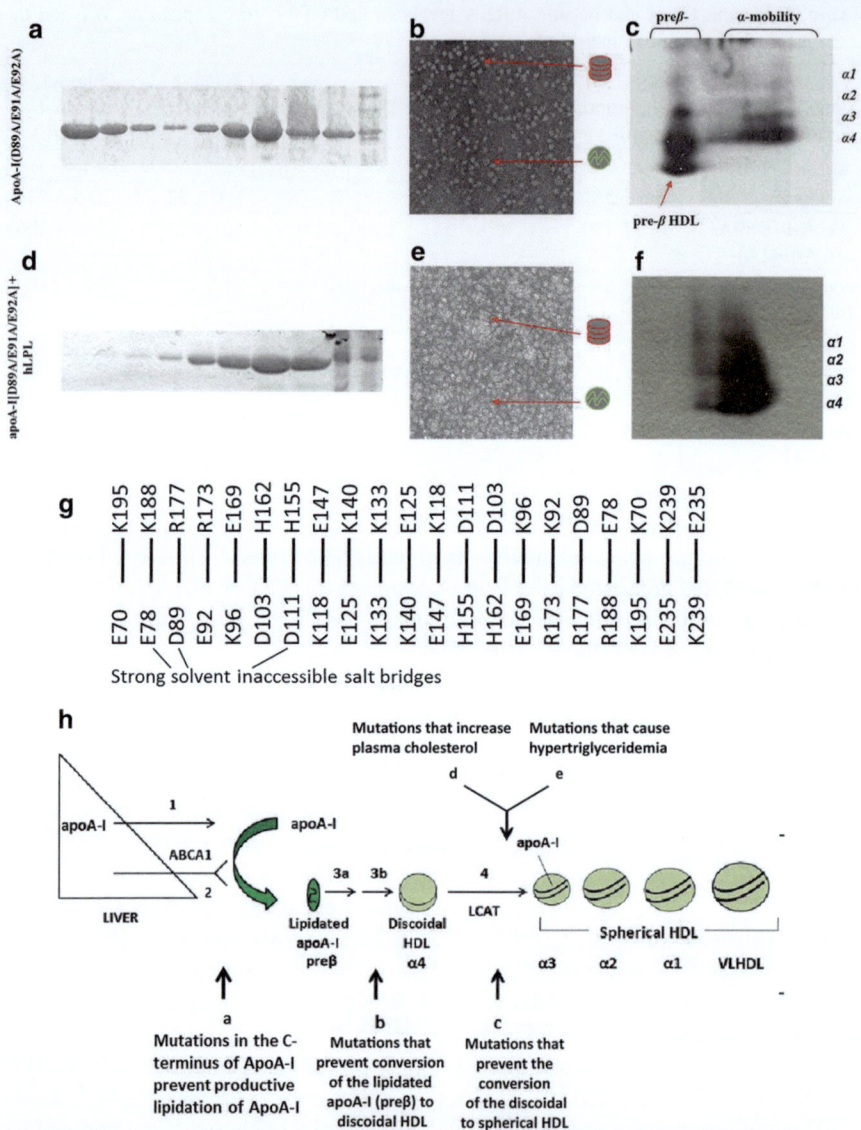

Fig. 6 (**a–h**) Effects of apoA-I mutations on the induction of dyslipidemia. Location of the apoA-I mutations that cause hypertriglyceridemia as indicated. Analysis of plasma of apoA-I$^{-/-}$ mice infected with adenoviruses expressing the apoA-I[D89A/E91A/E92A] mutant alone or in combination with human LPL, as indicated, by density gradient ultracentrifugation and SDS-PAGE (**a, d**). EM analysis of HDL fractions 6–7 obtained from apoA-I$^{-/-}$ mice expressing the apoA-I [D89A/E91A/E92A] mutant alone or in combination with LPL, as indicated, following density gradient ultracentrifugation of plasma (**b, e**). Two-dimensional gel electrophoresis of plasma of apoA-I$^{-/-}$ mice infected with adenoviruses expressing the apoA-I[D89A/E91A/E92A] mutant alone or in combination with LPL, as indicated (**c, f**). Schematic representation of the solvent inaccessible interhelical charged interactions of apoA-I dimers arranged in an antiparallel

1.3.1 Potential Mechanism of Dyslipidemia Resulting from apoA-I Mutations

The apoA-I[D89A/E91A/E92A] mutant has two similar characteristics with two other mutants in different regions of apoA-I, the apoA-I[Δ(61-78)] and the apoA-I [E110A/E111A] (Chroni et al. 2004a; 2005b), that cause hypertriglyceridemia (Table 3). The first characteristic is that all three mutants caused accumulation of apoA-I in the VLDL/IDL region. As shown previously, the accumulation of apoA-I in the lower densities affects the in vitro lipolysis of the VLDL/IDL fraction by exogenous lipoprotein lipase (Chroni et al. 2004a; 2005b). The second characteristic is that the three apoA-I mutants have lost negative-charged residues that are present in the WT sequence. The E78, D89, and E111 residues have the ability to form solvent inaccessible salt bridges with positively charged residues present in the antiparallel apoA-I molecule of a discoidal HDL particle (Segrest et al. 1999) (Fig. 6g).

In these arrangements of the apoA-I molecules on the HDL particle, residues E78 in helix 2, D89 in helix 3, and E111 in helix 4 can form solvent inaccessible salt bridges with residues R188 in helix 8, R177 in helix 7, and H155 in helix 6, respectively, of the antiparallel strand. The affinity of all three mutants for triglyceride-rich lipoprotein particles is further supported by binding studies to triglyceride-rich emulsion particles (Gorshkova and Atkinson 2011).

It is interesting that in the 11/3 α-helical wheel residues E78, D89, and E111 are all located in wheel position 2. With the exception of R188, all other five residues involved in salt bridges are conserved in mammals.

The lipid and lipoprotein abnormalities observed in this mutant suggest that the increased abundance of apoA-I in the VLDL/IDL region may create lipoprotein lipase insufficiency that is responsible for the induction of hypertriglyceridemia.

The persistence of discoidal particle following the lipoprotein lipase treatment indicates a direct effect of the [D89A/E91A/E92A] mutation in the activation of LCAT in vivo. Previous studies showed that discoidal and small-size HDL particles and LCAT associated with them may be catabolized fast by the kidney and thus lead to LCAT insufficiency and reduced plasma HDL levels (Koukos et al. 2007a; Timmins et al. 2005; Miettinen et al. 1997a).

It is conceivable that loosening of the structure of apoA-I around the D89 or E92 area due to the substitution of the original residues by alanines may provide new surfaces for interaction of HDL with other proteins or lipoprotein particles such as VLDL in ways that inhibit triglyceride hydrolysis. Furthermore, the accumulation of discoidal HDL as well as the formation of preβ- and small α4 HDL particles as shown by the in vivo experiments indicate that replacement of D89, E91, and E92 by A has a direct impact on the activation of LCAT.

The preceding analyses described in Figs. 3, 4, 5, and 6 demonstrate that expression of mutant apoA-I forms in different mouse models disrupted specific steps along the pathway of the biogenesis of HDL and generated discrete lipid and HDL

Fig. 6 (continued) orientation in the belt model of rHDL (**g**). The pathway of HDL biogenesis. Superimposed on the pathway are defects that inhibit different steps of this pathway (**h**)

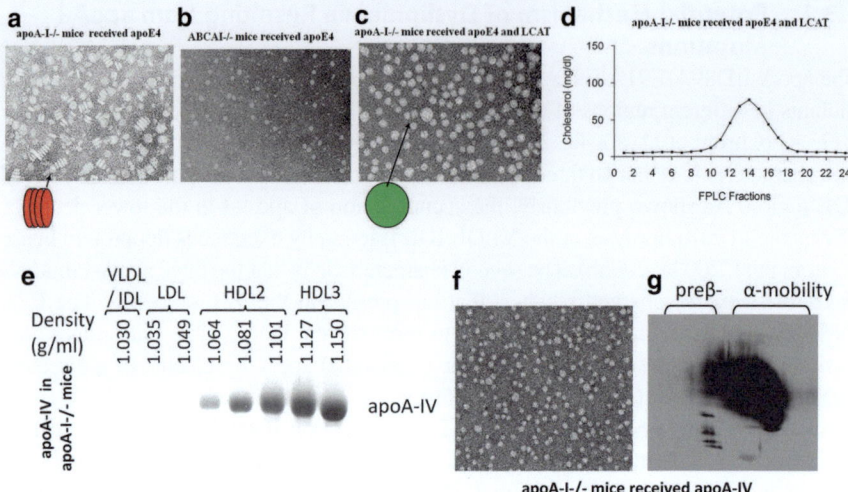

Fig. 7 EM analysis of apoA-I$^{-/-}$ and ABCA1$^{-/-}$ mice following infection with adenovirus expressing human apoE4 as indicated (**a, b**). EM analysis of apoA-I$^{-/-}$ mice following infection with adenovirus expressing human apoE4 and LCAT (**c**). FPLC profile of apoA-I$^{-/-}$ mice expressing apoE4 and LCAT (**d**). Analyses of the plasma of apoA-I$^{-/-}$ mice following infection with adenovirus expressing human apoA-IV by density gradient ultracentrifugation and SDS-PAGE (**e**), EM (**f**), and two-dimensional gel electrophoresis (**g**)

phenotypes (Zannis et al. 2006b). The phenotypes generated included inhibition of the formation of HDL (Chroni et al. 2003, 2007; Fotakis et al. 2013a, b), generation of unstable intermediates (Koukos et al. 2007a), inhibition of the activation of LCAT (Chroni et al. 2005a; Koukos et al. 2007b), and increase in plasma cholesterol or increase in both plasma cholesterol and triglycerides (Chroni et al. 2004a, 2005b; Kateifides et al. 2011) (Fig. 6h).

1.4 ApoE and apoA-IV Participate in the Biogenesis of HDL Particles Containing the Corresponding Proteins

Using adenovirus-mediated gene transfer in apoA-I- or ABCA1-deficient mice, we obtained unequivocal evidence that apoE of any phenotype participates in the biogenesis of apoE-containing HDL particles (HDL-E) using a similar pathway that is used for the biogenesis of apoA-I-containing HDL particles (Kypreos and Zannis 2007). In the initial experiments, gene transfer of an apoE4-expressing adenovirus increased both HDL and the triglyceride-rich VLDL/IDL/LDL fraction and generated discoidal HDL particles (Fig. 7a). Control experiments showed the absence of discoidal or spherical HDL-size particles in the plasma of apoA-I-deficient mice. The involvement of ABCA1 which was established by gene transfer of apoE in ABCA1$^{-/-}$ mice prior to and after treatment with apoE4, and indicated that apoE4 could not promote formation of HDL particles in ABCA1$^{-/-}$ mice

(Fig. 7b). Other experiments in apoA-I$^{-/-}$ x apoE$^{-/-}$ mice established that residues 1-202 of apoE are sufficient to promote biogenesis of apoE-containing HDL (Vezeridis et al. 2011). Coinfection of apoA-I$^{-/-}$ mice with a mixture of adenoviruses expressing both apoE4 and human LCAT converted the discoidal to spherical HDL (Fig. 7c), suggesting that LCAT is essential for the maturation of the discoidal apoE-containing HDL to spherical particles (Kypreos and Zannis 2007). The LCAT treatment also cleared the triglyceride-rich lipoproteins and increased the HDL cholesterol peak as determined by FPLC (Fig. 7d).

These findings suggest that in contrast to apoA-I where the C-terminal domain is required for the biogenesis of HDL (Chroni et al. 2007), the carboxy-terminal domain of apoE is not required for HDL formation. Overall, the findings indicate that apoE has a dual functionality. In addition to its documented role in the clearance of triglyceride-rich lipoproteins, it participates in the biogenesis of HDL-E in a process that is similar to that of apoA-I.

HDL-E thus formed may have antioxidant and anti-inflammatory functions similar to those described for apoA-I-containing HDL, which may contribute to the atheroprotective properties of apoE (Mineo et al. 2003; Plump et al. 1992; Schaefer et al. 1986; Navab et al. 2000). ApoE-containing HDL may also have important biological functions in the brain (Li et al. 2003).

A similar set of gene transfer experiments in apoA-I$^{-/-}$ and apoA-I$^{-/-}$ x apoE$^{-/-}$ mice also established that similar to apoE, apoA-IV also participates in the biogenesis of apoA-IV containing HDL (HDL-A-IV) and requires for this purpose the activity of ABCA1 and LCAT (Duka et al. 2013).

Gene transfer of apoA-IV in apoA-I$^{-/-}$ mice did not change plasma lipid levels. Density gradient ultracentrifugation showed that apoA-IV floated in the HDL2/HDL3 region (Fig. 7e), promoted the formation of spherical HDL particles as determined by electron microscopy (Fig. 7f), and generated mostly α- and a few preβ-HDL subpopulations as determined by two-dimensional gel electrophoresis (Fig. 7g). When expressed in apoA-I$^{-/-}$ × apoE$^{-/-}$ mice, apoA-IV increased plasma cholesterol and triglyceride levels and shifted the distribution of the apoA-IV protein in the lower density fractions. This treatment likewise generated spherical particles and α- and preβ-like HDL subpopulations. Co-expression of apoA-IV and LCAT in apoA-I$^{-/-}$ mice restored the formation of HDL-A-IV. Spherical and α-migrating HDL particles were not detectable following gene transfer of apoA-IV in ABCA1$^{-/-}$ or LCAT$^{-/-}$ mice (Duka et al. 2013). The ability of apoA-IV to promote biogenesis of HDL may explain previously reported anti-inflammatory and atheroprotective properties of apoA-IV.

In vitro studies showed that lipid-free apoA-IV and reconstituted HDL-A-IV promoted ABCA1- and scavenger receptor BI (SR-BI)-mediated cholesterol efflux, with the same efficiency as apoA-I and apoE (Chroni et al. 2005c; Duka et al. 2013).

1.5 Clinical Relevance of the Aberrant HDL Phenotypes

Genome-wide association studies indicated that specific gene loci were associated with low or high HDL cholesterol and triglyceride levels and could in principle affect the risk for coronary artery disease (CAD) (Teslovich et al. 2010). Prospective population studies have also shown that HDL has a protective role against CAD (Gordon et al. 1989). The beneficial functions of HDL are also supported by the atheroprotective effect of apoA-I overexpression in transgenic mice (Rubin et al. 1991; Paszty et al. 1994) or rabbits (Emmanuel et al. 1996) or following adenovirus-mediated gene transfer in mice (Belalcazar et al. 2003; Benoit et al. 1999; Tangirala et al. 1999). The studies described above provide molecular markers that could be used for the diagnosis, prognosis, and potential treatment of HDL abnormalities or dyslipidemias associated with the biogenesis and remodeling of HDL. Diagnostic phenotypes such as those depicted in Figs. 2a–c, 3a–d, 4a, b, 5d–f, and 6a–c can be used to assess defects in apoA-I, ABCA1, and LCAT respectively.

The HDL phenotypes observed in human patients carrying the apoA-I [L141R]$_{Pisa}$ and apoA-I[L159R]$_{Fin}$ mutations resemble closely the phenotypes observed in apoA-I-deficient mice expressing these mutants and indicate the validity of the gene transfer studies in mice to establish defects in HDL biogenesis. It is possible that phenotypes generated by mutagenesis of apoA-I may exist in the human population and can be detected by one or more of the analyses described previously. The correction of the aberrant HDL phenotypes by treatment with LCAT suggests a potential therapeutic intervention for HDL abnormalities that result from specific mutations in apoA-I or conditions that result in low HDL levels. Additional supporting evidence has been obtained by adenovirus-mediated gene transfer of human LCAT in squirrel monkeys. This treatment increased two-fold the HDL levels without affecting apoA-I levels, increased the size of HDL, and decreased apoB levels (Amar et al. 2009).

The potential contribution of apoA-I mutations to hypertriglyceridemia in humans is interesting. Hypertriglyceridemia resulting from apoA-I mutations may be further aggravated by other genetic and environment factors such as diabetes and thyroid status. The contribution of apoA-I mutations to hypertriglyceridemia could be addressed in future studies in selected populations of patients with hypertriglyceridemia of unknown etiology.

2 Remodeling and Catabolism of HDL

Following synthesis by the liver and the intestine, HDL is remodeled by various plasma proteins and is subsequently catabolized in the plasma by cell receptors and other plasma proteins (Fig. 8a–d).

Turnover studies showed that the mean plasma residence time of radiolabeled ^{125}I-HDL2 and HDL3 was 6 days for normal subjects (Schaefer et al. 1979) and 0.22 days for patients with Tangier disease (Schaefer et al. 1981). Using stable

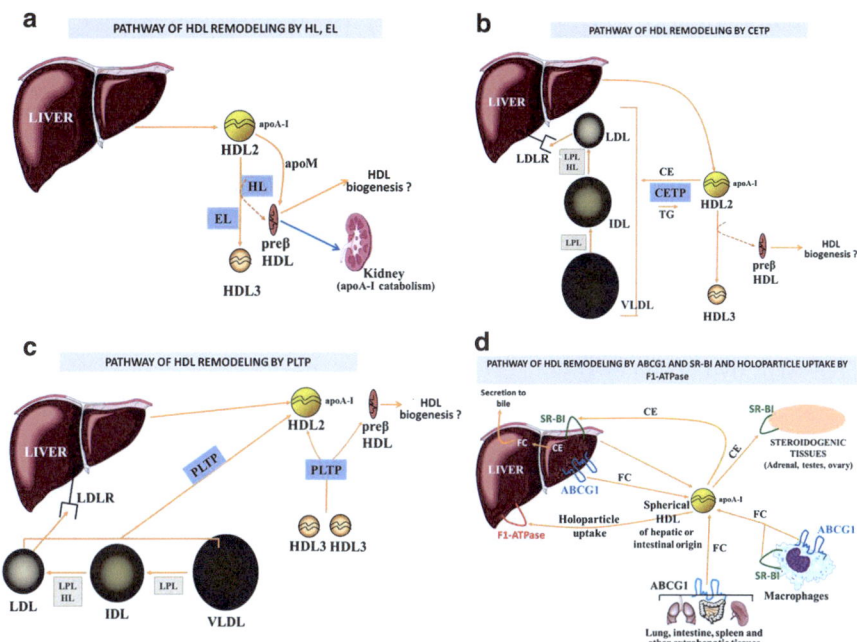

Fig. 8 (a–d) Schematic representation of the pathway of HDL remodeling by the action of hepatic and endothelial lipase (a), CETP (b), PLTP (c), and SR-BI, ABCG1, and HDL holoparticle uptake by F1-ATPase (d)

isotope turnover kinetic measurements, the resident time was estimated to be 4–5 days. The fractional catabolic rates (FCR) were expressed as pools per day, and the secretion rates that were determined by this method for men and women were similar. The FCR were not affected significantly by diet, diabetes, or LDL receptor defects, but it was increased in subjects with increased glucose tolerance (Marsh et al. 2000). HDL remodeling affects the structure and metabolic turnover of HDL and generates a dynamic mixture of discrete HDL subfractions that vary in size, shape, apolipoprotein, and lipid composition and functions (Siggins et al. 2007; Fielding and Fielding 2007; Xu and Nilsson-Ehle 2007; Harder and McPherson 2007).

Remodeling of HDL by the action of hepatic lipase (HL) and endothelial lipase (EL) involves hydrolysis of residual triglycerides and some phospholipids of HDL (Maugeais et al. 2003a; Santamarina-Fojo et al. 2004), leads to the conversion of HDL2 to HDL3 and preβ-HDL (Ishida et al. 2003; Krauss et al. 1974; Breckenridge et al. 1982; Brunzell and Deeb 2001), and accelerates the catabolism of HDL. Preβ-HDL formation also requires the functions of apolipoprotein M (apoM) (Wolfrum et al. 2005).

Portion of the cholesteryl esters formed by the actions of LCAT can be transferred to VLDL/IDL/LDL by the cholesteryl ester transfer protein (CETP)

(Hopkins and Barter 1980; Barter et al. 2003) The phospholipid transfer protein (PLTP) can transfer the phospholipids from VLDL/IDL to the HDL particle during lipolysis to generate HDL2 and can also convert HDL3 particles to HDL2 and preβ-HDL (Tall et al. 1983; Lusa et al. 1996). HDL-binding proteins/receptors or transporters have been documented at all steps of HDL metabolism and involve the SR-BI, which is mostly expressed in hepatocytes, macrophages, and steroidogenic tissues and mediates selective CE uptake by the cells and tissues and cholesterol efflux (Gu et al. 1998; Krieger 1999; Rohrl and Stangl 2013; Pagler et al. 2006a); the ABCG1, which mediates cholesterol efflux (Nakamura et al. 2004a); the ecto-F_1-ATPase subunit, which mediates HDL holoparticle uptake by the liver (Martinez et al. 2003, 2007); and the cubilin/megalin receptors for removal of apoA-I and preβ-HDL by the kidney (Martinez et al. 2007).

2.1 ATP-Binding Cassette Transporter G1

HDL can be remodeled following interactions with ABCG1, a 67 kDa protein, which is a member of the ABC family of half transporters. ABCG1 is expressed in the spleen, thymus, lung, and brain (Savary et al. 1996; Croop et al. 1997; Nakamura et al. 2004a) and was reported to be localized on plasma membrane, the Golgi, and recycling endosomes (Wang et al. 2006; Vaughan and Oram 2005; Sturek et al. 2010a; Xie et al. 2006a). The expression of ABCG1 is induced by LXR agonists in macrophages and the liver or by cholesterol loading in macrophages (Klucken et al. 2000; Venkateswaran et al. 2000; Wang et al. 2006). Overexpression of ABCG1 promotes cholesterol efflux from different cells to HDL but not to lipid-free apoA-I (Wang et al. 2004, 2006; Nakamura et al. 2004a; Vaughan and Oram 2005) (Fig. 8d). HDL obtained from CETP-deficient subjects or patients treated with the CETP inhibitors torcetrapib or anacetrapib was shown to have enhanced ability to promote ABCG1-dependent cholesterol efflux from macrophages (Matsuura et al. 2006; Yvan-Charvet et al. 2007, 2010). ABCG1-mediated cholesterol efflux to HDL is abolished by mutations in the ATP-binding Walker A motif indicating that the ATP-binding domain in ABCG1 is essential for both lipid transport activity and protein trafficking (Vaughan and Oram 2005). In addition, ABCG1 was shown to promote efflux of 7-ketocholesterol and related oxysterols from macrophages and endothelial cells to HDL, thus protecting cells from apoptosis (Terasaka et al. 2007; Li et al. 2010). In macrophages, ABCG1, and not SR-BI or ABCA1, has been shown recently to be primarily responsible for free cholesterol mobilization to rHDL (Cuchel et al. 2010a).

Studies in ABCG1-deficient mice also suggested that ABCG1 plays a critical role in the efflux of cellular cholesterol to HDL (Kennedy et al. 2005). Studies using intraperitoneal injection of mice with [^{3}H]cholesterol-labeled J774 macrophages with either increased or reduced ABCG1 expression, as well as primary macrophages lacking ABCG1 expression, and measurement of the macrophage-derived [^{3}H]cholesterol levels in plasma and feces, showed that ABCG1 plays a critical role in promoting macrophage reverse cholesterol transport in vivo (Wang et al. 2007). Nevertheless, the studies in ABCG1-deficient and ABCG1 transgenic mice showed that plasma lipids, HDL, and other lipoprotein levels were not affected (Kennedy et al. 2005; Out et al. 2007; Burgess et al. 2008). Other studies showed that loss of ABCG1 gene in mice results in massive lipid accumulation in hepatocytes and in macrophages within multiple tissues, with the more marked accumulation in pulmonary macrophages (Kennedy et al. 2005; Out et al. 2006, 2007; Burgess et al. 2008; Baldan et al. 2006; Ranalletta et al. 2006; Wojcik et al. 2008). In addition, ABCG1 has been suggested to mediate cholesterol efflux to HDL particles from other cell types than macrophages, including adipocytes (Zhang et al. 2010) and human placental endothelial cells, where it may facilitate the transfer of maternal cholesterol to the fetus (Stefulj et al. 2009).

Previous studies indicated that there is little or no specificity of ABCG1 for the acceptor of cholesterol since LDL, HDL_2, HDL_3, phospholipid/apoA-I particles of various sizes and small unilamellar particles can function as acceptors for cholesterol from cells in an ABCG1-mediated manner (Sankaranarayanan et al. 2009; Favari et al. 2009). However, a recent study by us showed that the ABCG1-mediated efflux of cholesterol to rHDL containing different apoA-I mutants is diminished by deletion of the carboxyl-terminal domain 185-243 from full-length apoA-I (Daniil et al. 2013). Analysis of rHDL used in these studies suggested that the impairment of ABCG1-mediated cholesterol efflux is not due to major differences in particle composition or size between rHDL particles containing WT apoA-I or apoA-I[$\Delta(185$-$243)$].

The mechanism by which ABCG1 promotes sterol efflux to extracellular acceptors has not been resolved. The earlier studies failed to detect specific HDL association in BHK or HEK293 cells overexpressing the human ABCG1 (Wang et al. 2004; Sankaranarayanan et al. 2009). Also the initial studies had suggested that ABCG1 is localized to both the plasma membrane and internal membrane structures (Vaughan and Oram 2005; Wang et al. 2006; Xie et al. 2006b), while more recent studies suggested that ABCG1 is localized to endosomes and recycling endosomes (Sturek et al. 2010b; Tarling and Edwards 2011). It has been proposed that ABCG1 could transport sterols across the bilayer of endosomes before their fusion with the plasma membrane and thus redistribute these sterols to the outer leaflet of the plasma membrane and facilitate their subsequent efflux of sterols to HDL or other acceptors (Tarling and Edwards 2011; Vaughan and Oram 2005). However, the similar pattern of lipid-free and rHDL-bound apoA-I mutants to promote ABCA1- and ABCG1-mediated cholesterol efflux is compatible with a transient localization of ABCG1 in the plasma membrane that will allow its interaction with lipoprotein acceptors (Chroni et al. 2003, 2004b; Daniil et al. 2013). The similar cholesterol efflux capacity of lipid-free and lipidated apoA-I mutants could favor a model where lipid-free apoA-I is lipidated in an ABCA1-mediated process, changes its conformation, and subsequently accepts more cholesterol from membrane pools generated by ABCA1 or ABCG1.

The different capacity of rHDL-associated apoA-I[$\Delta(185$-$243)$] mutant to promote ABCG1-mediated cholesterol and 7-ketocholesterol efflux (Daniil et al. 2013) may be related to the looser association of 7-ketocholesterol with the plasma membrane compared to cholesterol (Kan et al. 1992).

The finding that lipid-free and lipid-bound apoA-I[Δ(185-243)] has reduced capacity to promote ABCA1- and ABCG1-mediated cholesterol efflux, respectively, may have physiological significance since proteolysis of HDL-associated apoA-I in vivo may affect its ability to promote cholesterol efflux from macrophages. In this regard, proteolysis of apoA-I by metalloproteinases present in the arterial wall or alveolar macrophages (Russell et al. 2002; Galis et al. 1995) produces various fragments that correspond to apoA-I cleaved after residues 191 or 188 and are compatible in size with apoA-I[Δ(185-243)] (Lindstedt et al. 1999; Eberini et al. 2002). This may explain the accumulation of foam cells containing high cholesterol levels in alveolar macrophages of ABCG1-deficient mice (Kennedy et al. 2005; Out et al. 2006).

A recent study using high-density genotyping arrays containing single-nucleotide polymorphisms suggested an association between HDL cholesterol levels in humans and ABCG1 (Edmondson et al. 2011). Recent genetic association studies in humans identified functional variants in ABCG1 associated with increased risk of coronary artery disease (Xu et al. 2011; Schou et al. 2012), suggesting an important role of ABCG1 in the protection from atherosclerosis and cardiovascular disease.

2.2 Phospholipid Transfer Protein

Phospholipid transfer protein (PLTP) transfers phospholipids, diacylglycerol (Rao et al. 1997), free cholesterol (Nishida and Nishida 1997), R-tocopherol (vitamin E) (Kostner et al. 1995), and lipopolysaccharide among lipoproteins and between lipoproteins and cells (Hailman et al. 1996; Levels et al. 2005) (Fig. 8c). In vitro studies have identified a number of functions for PLTP in HDL metabolism (Albers and Cheung 2004; Siggins et al. 2007). PLTP displays two major functions in circulation: (1) it transforms HDL particles in a conversion or fusion process whereby small HDL3 particles are fused leading to the generation of large fused HDL particles and preβ-HDL that can participate in cholesterol removal from cells (Vikstedt et al. 2007a), and (2) it transfers post-lipolytic VLDL surface phospholipids to HDL (Albers and Cheung 2004; Siggins et al. 2007). Functions of PLTP which may influence the formation of atherosclerotic lesions include the generation of acceptors for lipid efflux from cells, regulation of plasma HDL levels, protection of lipoproteins from oxidation, and regulation of production of atherogenic lipoproteins (Jiang et al. 2001).

In human plasma, two distinct forms of PLTP are present, one with high activity (HA-PLTP) and the other with low activity (LA-PLTP) (Oka et al. 2000; Karkkainen et al. 2002). It was reported that phospholipid transfer activity is a prerequisite for efficient PLTP-mediated HDL enlargement (Huuskonen et al. 2000) and that enrichment of triglyceride in the HDL core could promote such fusion (Rye et al. 1998). Of these two forms, only the high specific activity

PLTP promotes macrophage cholesterol efflux via fusion of HDL particles that leads to the release of lipid-poor preβ-mobile apoA-I particles which act as efficient cholesterol acceptors (Vikstedt et al. 2007a). The mechanisms by which LA-PLTP is generated and its physiological functions are currently unknown. However, since apoE is able to interact with PLTP, and apoE-containing proteoliposomes can activate inactive or low active PLTP, the presence of apoE in PLTP complexes is expected to enhance PLTP activity. This is consistent with the suggestion that apoE may play a role in regulating the PLTP activity level in plasma (Janis et al. 2005). HA- and LA-PLTP forms are surface-active proteins, and the low active form was demonstrated to dock more strongly onto a phospholipid monolayer surface as compared to HA-PLTP form (Setala et al. 2007). It is therefore possible that LA-PLTP form could play other important lipid transfer-independent functions such as signaling on cell surface as suggested (Albers et al. 2012).

Although the role of PLTP in lipoprotein metabolism and atherogenesis has been intensively studied in gene-targeted mouse models and using in vitro experiments, the physiological role of PLTP in human metabolism is far from being resolved. Genetic approach has provided some evidence that genetic variation at the PLTP locus affects its phospholipid transfer activity and HDL particle size and might highlight its relevance in cholesterol efflux process (Vergeer et al. 2010a). PLTP-deficient mice have a marked decrease in HDL and apoA-I (Jiang et al. 1999) but reduced atherosclerosis in the background of apoE$^{-/-}$ or apoB-transgenic mice (Jiang et al. 2001). It has also been reported that macrophage-derived PLTP contributes significantly to total plasma PLTP activity and deficiency of PLTP in macrophages leads to reduced atherosclerosis in LDLr$^{-/-}$ mice (Vikstedt et al. 2007b). There is also an interesting interaction between PLTP and CETP since it was demonstrated that purified PLTP enhances cholesteryl ester transfer from HDL$_3$ to VLDL (Tollefson et al. 1988), although PLTP has no cholesterol ester transfer activity of its own. Moreover, CETP transgenic/PLTP KO mice have significantly lower plasma CETP activity as compared to that of CETP transgenic mice (Kawano et al. 2000). Currently, the physiological relevance of this PLTP-CETP interaction in HDL metabolism is poorly understood.

2.3 apoM

Apolipoprotein M (apoM), which is also involved in HDL remodeling (Fig. 8a), is a 26 kDa glycoprotein that belongs to the lipocalin protein superfamily and has been shown to bind lipophilic ligands in its hydrophobic binding pocket (Xu and Dahlback 1999; Nielsen et al. 2009; Dahlback and Nielsen 2009; Hu et al. 2010). It is secreted by the liver and to a lesser extent by the kidney and associates with HDL through its retained N-terminal signal peptide (Christoffersen et al. 2008a; Axler et al. 2008) and to a lesser extent with other lipoproteins. ApoM is involved in the recycling of small lipophilic ligands via the multi-ligand receptor megalin (Nielsen et al. 2009) and has been shown to participate in the remodeling and maturation of HDL in plasma.

Studies in humans and in mice overexpressing or lacking apoM have shown a positive association between plasma apoM levels and total as well as HDL and LDL cholesterol concentrations (Axler et al. 2007; Plomgaard et al. 2009; Christoffersen et al. 2008b). Lack of hepatocyte nuclear factor-1α (HNF-1α) (Shih et al. 2001) or inhibition of apoM expression in C57BL/6 mice injected with small interfering RNA for apoM (Wolfrum et al. 2005) is characterized by diminished concentration of preβ-HDL particles and presence of large-size HDL particles. In vitro experiments using plasma obtained from WT mice, apoM knockout, and apoM transgenic mice showed that apoM increases the formation of preβ-HDL particles following incubation of the plasma at 37 °C (Christoffersen et al. 2008b).

These studies also indicate that apoM-containing HDL particles isolated from human plasma and the plasma of apoM transgenic mice have increased capacity to stimulate cholesterol efflux from macrophage foam cells and are more efficient in protecting against LDL oxidation (Wolfrum et al. 2005; Christoffersen et al. 2008b). Cell culture studies indicated that expression of apoM in cells transfected with ABCA1 increases the size of preβ-HDL (Mulya et al. 2010).

Reduced plasma apoM levels have been reported in animal models of diabetes and some patients with diabetes and metabolic syndrome (Plomgaard et al. 2009; Dullaart et al. 2009; Xu et al. 2006), indicating potential involvement of apoM in the development of diabetes. Genetic linkage studies in Chinese populations have also associated two single-nucleotide polymorphisms (SNPs) located in the apoM proximal promoter region (SNP T-778C and SNP T-855C) with the development of coronary artery disease (Xu et al. 2008; Jiao et al. 2007) and one of them (the T-778C) associated with susceptibility to coronary artery disease.

Adenovirus-mediated gene transfer of apoM in LDL-receptor-deficient mice or hepatic overexpression of apoM in apoM transgenic mice partially protected the mice from atherosclerosis development (Christoffersen et al. 2008b; Wolfrum et al. 2005). Data accumulating until now strongly suggest a protective role for apoM, and the protection might be mediated via HDL.

2.4 Hepatic Lipase and Endothelial Lipase

HDL is first remodeled in the circulation and subsequently catabolized by cells and tissues. Hepatic lipase (HL) and endothelial lipase (EL) are two plasma lipases playing an important role in HDL remodeling (Fig. 8a). HL and EL have specificity primarily for phospholipids and to a lesser extend for triglycerides of apoB-containing lipoprotein remnants and large HDL (Maugeais et al. 2003a; Santamarina-Fojo et al. 2004). HL-deficient mice exhibit elevated levels of large HDL particles enriched in phospholipids and apoE (Homanics et al. 1995) and reduced atherosclerosis in the background of apoE$^{-/-}$ mice (Karackattu et al. 2006; Mezdour et al. 1997). In contrast, overexpression of HL in mice reduced plasma HDL levels (Braschi et al. 1998). A rat liver perfusion of human native HDL2 or triglyceride-enriched HDL promoted the formation of the preβ1-HDL subspecies and a reduction of the α-HDL2 (Barrans et al. 1994). These changes were attributed to the triglyceride lipase activity of HL (Barrans et al. 1994). Characterization of preβ1-HDL showed that these particles contain one to two molecules of apoA-I, associated with phospholipids, and some free and esterified cholesterol (Guendouzi et al. 1999). When compared to triglyceride-rich HDL2, remnant-HDL2 had lost on the average one molecule of apoA-I, 60 % of triglycerides, and 15 % of phospholipids. The estimated composition supported the hypothesis that HL had splitted the initial particle into one preβ1-HDL and one remnant-HDL2. Remnant-HDL2 had different composition and properties from HDL3, suggesting that HL did not promote the direct conversion of HDL2 to HDL3 (Guendouzi et al. 1999). Analysis of HL transgenic rabbits suggested that HL reduces the size of α-migrating HDL and increases the rate of catabolism of apoA-I (Kee et al. 2002). Cell studies showed that HL promotes selective HDL3 cholesterol ester uptake independent from SR-BI and that proteoglycans are needed for the HL action on selective CE uptake (Brundert et al. 2003). Earlier studies in mice deficient in both HL and EL suggested an additive effect of HL and EL on plasma HDL levels but not on macrophage-mediated reverse cholesterol transport in mice (Brown et al. 2010). However, a recent study demonstrated that targeted inactivation of both HL and EL in mice promoted macrophage-to-feces RCT and enhanced HDL antioxidant properties (Escola-Gil et al. 2013).

HL-deficient patients have elevated plasma concentrations of cholesterol in the HDL and β-VLDL and increased concentration of triglycerides and phospholipids in the LDL and HDL (Breckenridge et al. 1982). Analyses carried out in complete and partial HL-deficient subjects as well as in normotriglyceridemic and hypertriglyceridemic controls suggested that HL activity is important for physiologically balanced HDL metabolism (Ruel et al. 2004). However, the presence of HL may not be necessary for normal HDL-mediated reverse cholesterol transport process and is not associated with pro-atherogenic changes in HDL composition and metabolism (Ruel et al. 2004). In addition, another Mendelian randomization study showed that subjects with loss-of-function genetic variants of HL have elevated levels of HDL cholesterol, but are not associated with risk of ischemic cardiovascular disease and therefore may not be protected against ischemic cardiovascular disease (Johannsen et al. 2009).

Endothelial lipase (EL) has phospholipase activity (mostly PLA1 activity) and low levels of triglyceride lipase activity (Jaye et al. 1999). Overexpression of EL in mice markedly decreased plasma HDL cholesterol and apoA-I levels, had a modest effect on apoB-containing lipoproteins, and increased 2.5–3-fold the uptake of the HDL by the kidney and the liver (Ishida et al. 2003; Maugeais et al. 2003a). In contrast, the EL deficiency in mice increased HDL cholesterol levels (Ishida et al. 2003; Ma et al. 2003) and reduced atherosclerosis in the background of apoE$^{-/-}$ mice (Ishida et al. 2004). Analysis of atherosclerosis prone LDLR$^{-/-}$ × ApoB(100/100) mice suggested that EL and the HDL cholesterol levels were regulated by SREBPs and VEGF-A (Kivela et al. 2012). Overexpression of EL in mice markedly decreased plasma HDL cholesterol and apoA-I levels and had a modest effect on apoB-containing lipoproteins (Maugeais et al. 2003b; Ishida et al. 2003). Furthermore, the HDL phospholipid and cholesteryl ester content decreased, while HDL triglyceride content increased (Nijstad et al. 2009) and the free cholesterol content remained unaltered. Fast protein liquid chromatography analysis and agarose gel electrophoresis showed that the expression of EL resulted in the generation of small preβ-HDL particles (Nijstad et al. 2009). In addition, overexpression of EL increased the selective uptake of hepatic HDL cholesteryl ester by SR-BI as well as hepatic holoparticle uptake. This resulted in a dramatic increase in the uptake of the HDL protein, but not the cholesteryl ester moieties, into the kidneys (Nijstad et al. 2009). These data support a model in which EL-mediated phospholipid hydrolysis of HDL destabilizes the particle, resulting in the shedding of poorly lipidated apoA-I from the particle surface, which are preferentially cleared by the kidneys and via increased selective uptake by SR-BI.

Several genetic EL variants have been reported to be associated with plasma HDL-C levels (deLemos et al. 2002; Edmondson et al. 2009), and genome-wide association studies have shown that single-nucleotide polymorphisms (SNPs) near *LIPG* (EL) are associated with plasma HDL-C levels (Kathiresan et al. 2008a, 2009; Teslovich et al. 2010). However, the relationship of genetic variation in the EL locus with the risk for coronary artery disease remains uncertain (Vergeer et al. 2010b). A newer study showed that carriers of an EL mutant characterized by complete loss of function had significantly higher plasma HDL cholesterol levels compared to carriers having partial loss-of-function mutations (Singaraja et al. 2013). Apolipoprotein B-depleted serum from carriers of HL with complete loss of function had significantly enhanced capacity to promote cholesterol efflux as compared to apoB-depleted serum obtained from HL carriers with partial loss of function (Singaraja et al. 2013). In the same study, it was reported that carriers of certain EL mutations exhibited trends toward reduced coronary artery disease in four independent cohorts (Singaraja et al. 2013).

2.5 Cholesteryl Ester Transfer Protein

Cholesteryl ester transfer protein (CETP) promotes the transfer of cholesteryl esters from HDL to VLDL, IDL, and LDL in exchange for triglycerides (Fig. 8b). It was estimated that 66 % of the cholesteryl esters of HDL return to the liver by the action of CETP, indicating an important role of CETP in reverse cholesterol transport (Barter et al. 2003), and 33 % by the action of SR-BI (Fielding and Fielding 2007). Deficiency in CETP in humans is associated with increased plasma levels of HDL (hyperalphalipoproteinemia) (Inazu et al. 1990; Maruyama et al. 2003) and decreased levels of small preβ1-HDL particles (Arai et al. 2000; Asztalos 2004). An early study showed that the hyperalphalipoproteinemia and high plasma HDL cholesterol levels in a Japanese family without incidence of atherosclerosis was the result of deficiency in CETP (Koizumi et al. 1985). Inhibition of CETP activity by CETP inhibitors increased HDL cholesterol levels and the size of HDL particle and decreased LDL cholesterol levels in human subjects, but did not increase atheroprotection (Landmesser et al. 2012; Brousseau et al. 2004; de Grooth et al. 2002). However, subsequent studies indicated that heterozygous mutations in CETP increase the risk for CAD (Hirano et al. 1995, 1997; Zhong et al. 1996). The effect of CETP on the HDL pathway was also studied in mice expressing the human CETP gene. CETP transgenic mice have a significant decrease in apoA-I and HDL levels (Melchior et al. 1994) and increased preβ-HDL levels (Francone et al. 1996) and are susceptible to atherosclerosis in the background of $apoE^{-/-}$ or $LDLr^{-/-}$ mice (Plump et al. 1999).

2.6 Scavenger Receptor BI

2.6.1 Role of SR-BI in HDL Remodeling Based on Its In Vitro Interactions with Its Ligands

SR-BI is an 82 kDa membrane glycoprotein consisting of a large extracellular domain, two transmembrane domains, and two cytoplasmic amino and carboxy-terminal domains (Krieger 1999). SR-BI is primarily expressed in the liver, steroidogenic tissues, and endothelial cells but is also found in other tissues (Acton et al. 1996), and it binds a variety of ligands including HDL, LDL, VLDL, and modified lipoproteins. SR-BI has also been shown to affect chylomicron metabolism in vivo and bind non-HDL lipoproteins in vitro (Out et al. 2004b, 2005; Krieger 1999, 2001; Acton et al. 1994, 1996; Murao et al. 1997). The most important property of SR-BI is considered its ability to act as the HDL receptor (Fig. 8d).

It has been shown that SR-BI binds to native HDL and discoidal reconstituted HDL containing apoA-I or apoE, through their apolipoprotein moieties (Krieger 2001; Chroni et al. 2005c; Liadaki et al. 2000; Xu et al. 1997). When it is bound to HDL, SR-BI mediates selective uptake of cholesteryl ester, triglycerides, phospholipids, and vitamin E from HDL to cells (Acton et al. 1996; Greene et al. 2001; Thuahnai et al. 2001; Stangl et al. 1999; Gu et al. 1998, 2000b; Urban et al. 2000). It also promotes bidirectional movement of unesterified cholesterol (Ji et al. 1997; Gu et al. 2000a). Interactions of HDL with SR-BI are responsible for mobilization of free cholesterol from the whole body (Ji et al. 1997; Gu et al. 2000a; Cuchel et al. 2010b). SR-BI-mediated HDL holoparticle endocytosis may also be involved in SR-BI-mediated selective CE uptake under certain conditions in some types of cells (Pagler et al. 2006b; Ahras et al. 2008). To understand the molecular interaction of SR-BI with HDL, SR-BI mutants which display altered biological functions were generated by in vitro mutagenesis. A SR-BI[M158R] mutant does not bind HDL (Gu et al. 2000a). A SR-BI[Q402R/Q418R] mutant also does not bind HDL, but in contrast with the first mutant, it binds LDL (Gu et al. 2000a; b). A SR-BI[G420H] mutant has normal selective cholesteryl ester uptake but reduced cholesterol efflux to HDL and reduced hydrolysis of internalized cholesteryl esters (Parathath et al. 2004). Cell culture cholesterol efflux studies using rHDL containing mutated apoA-I and these SR-BI mutants showed that the greater reduction of cholesterol efflux in cells expressing WT SR-BI was with the mutants apoA-I[D102A/D103A] and apoA-I [R160V/H162A] (21 % and 49 %, respectively) (Liu et al. 2002).

Follow-up in vivo studies showed apoA-I-deficient mice infected with an adenovirus expressing the apoA-I[D102A/D103A] had an HDL phenotype that resembled that of WT apoA-I (Chroni et al. 2005a). In contrast, mice expressing the apoA-I[R160V/H162A] had a phenotype similar to that described for apoA-I [R160L]$_{Oslo}$ that could be corrected by co-expression of the apoA-I[R160L]$_{Oslo}$ mutant and human LCAT (Chroni et al. 2005a). Following density gradient ultracentrifugation, the apoA-I[R160V/H162A] mutant that floated in the HDL region was decreased relatively to WT apoA-I, and it was shifted toward the HDL3 region (Chroni et al. 2005a).

When the mutant SR-BI[M158R] was examined, several apoA-I mutants tested had reduced efflux and bound less tightly compared to WT apoA-I with the exception of rHDL that contained the mutant apoA-I[A160V/H162A]. The binding of this mutant was almost as tight to the cells that expressed SR-BI[M158R] mutant as it was for the cells that expressed WT SR-BI (Liu et al. 2002). Based on these findings, it was suggested that efficient SR-BI-mediated cholesterol efflux requires not only direct binding (Gu et al. 2000a) but also the formation of a productive complex between SR-BI and the rHDL particle (Liu et al. 2002).

2.6.2 In Vivo Functions of SR-BI

Expression of SR-BI in the liver was shown to be critical for the control of plasma levels of HDL cholesterol (HDL-C) (Leiva et al. 2011; Zhang et al. 2007), and its expression in the steroidogenic tissues is important for synthesis of steroid hormones (Landschulz et al. 1996; Krieger 1999). Transgenic mice expressing SR-BI in the liver had decreased apoA-I and HDL cholesterol levels and increased clearance of VLDL and LDL (Wang et al. 1998; Ueda et al. 1999). SR-BI-deficient mice had decreased HDL cholesterol clearance (Out et al. 2004a), twofold increased plasma cholesterol, and presence of large-size abnormal apoE-enriched particles that were distributed in the HDL/IDL/LDL region (Rigotti et al. 1997). The in vivo phenotypes generated by overexpression or deficiency of SR-BI are consistent with its in vitro functions to promote selectively lipid transport from HDL to cells and efflux of free cholesterol from cells. SR-BI has also been shown to affect chylomicron metabolism in vivo (Out et al. 2004b; 2005).

Deficiency of SR-BI in mice reduced greatly the cholesteryl ester stores of steroidogenic tissues and decreased the secretion of biliary cholesterol by approximately 50 %. However, the SR-BI deficiency did not affect the secretion of the pool size of bile acids or the fecal secretion of bile acids and the intestinal cholesterol absorption (Rigotti et al. 1997; Mardones et al. 2001). These findings established that two important functions of SR-BI are the transfer of the CE of HDL to the liver and subsequent incorporation into the bile for excretion (Rigotti et al. 1997; Mardones et al. 2001) and the delivery of cholesteryl esters to the steroidogenic tissues where it is utilized for synthesis of steroid hormones (Ji et al. 1999). Furthermore SR-BI controls the concentrations and composition of plasma HDL (Krieger 2001; Wang et al. 1998; Ueda et al. 1999; Rigotti et al. 1997; Webb et al. 2002) and protects different mouse models from atherosclerosis (Hildebrand et al. 2010; Arai et al. 1999; Ueda et al. 2000; Kozarsky et al. 2000; Huszar et al. 2000;

Trigatti et al. 1999; Braun et al. 2002, 2003; Karackattu et al. 2005; Zhang et al. 2005).

SR-BI deficiency also caused defective maturation of oocytes and red blood cells due to accumulation of cholesterol in the plasma membrane of progenitor cells (Trigatti et al. 1999; Holm et al. 2002) and caused infertility in the female but not the male mice (Trigatti et al. 1999; Yesilaltay et al. 2006a). The infertility could be corrected by restoration of SR-BI gene by adenovirus-mediated gene transfer (Yesilaltay et al. 2006b). Subsequent experiments showed a negative correlation of follicular HDL cholesterol levels in women and embryo fragmentation during in vitro fertilization (Browne et al. 2009). Taken together these data suggest a role of HDL in oocyte development and embryogenesis. The SR-BI-mediated selective uptake of the CE of HDL by the liver is a complex process and requires the functions of a liver-specific protein, PDZK1, that contains four PDZ domains that can recognize the C-terminal region of SR-BI. Interaction of PDZK1 with the C-terminal region of SR-BI, posttranscriptionally, regulates localization and stability of SR-BI (Fenske et al. 2009). Inactivation of hepatic PDZK1 significantly affected plasma HDL metabolism and structure and caused occlusive atherosclerosis in double-deficient mice for apoE and PDKZ (Kocher et al. 2008; Yesilaltay et al. 2009). The detailed mechanism of SR-BI-facilitated selective uptake of the CE of HDL is not yet clear. It has been suggested that HDL binding to hepatic SR-BI allows the entry of cholesteryl esters into a channel that is generated by SR-BI and along which cholesteryl esters move down their concentration gradient into the cell membrane. During this movement, HDL particles donate their CE to hepatocytes without the simultaneous uptake and degradation of the whole HDL particle.

Recent findings reviewed in Meyer et al. (2013) indicated that SR-BI-independent cholesterol ester uptake processes may also operate in macrophages. Liver-specific or whole-body ABCA1 deficiency in mice accelerated HDL catabolism in plasma without changing the hepatic expression of SR-BI, suggesting that other membrane proteins, such as those involved in the hepatic F_1-ATPase/$P2Y_{13}$ pathway (Martinez et al. 2003) (see below for more detail) and CD36 (Brundert et al. 2011), may be involved in the selective cholesteryl ester uptake. In a large-scale human study, several CD36 SNPs were strongly associated with HDL cholesterol levels, thus pointing to a potential role of CD36 in the regulation of human HDL metabolism (Love-Gregory et al. 2008).

Interactions of HDL with SR-BI in endothelial cells trigger signaling mechanisms that involve activation of eNOS and release of nitric oxide that causes vasodilation (Mineo et al. 2003; Yuhanna et al. 2001; Li et al. 2002; Gong et al. 2003).

Human subjects have been identified with a P297S substitution in SR-BI. Heterozygote carriers for this mutation had increased HDL levels and decreased adrenal steroidogenesis and dysfunctional platelets, but did not develop atherosclerosis. HDL derived from these subjects had decreased ability to promote cholesterol efflux from macrophages (Vergeer et al. 2011). A recent study has shown the impact of SR-BI SNPs on female fertility (Yates et al. 2011).

2.7 Role of Ecto-F_1-ATPase/$P2Y_{13}$ Pathway in Hepatic HDL Clearance

Based on Biacore's surface plasmon resonance studies of hepatic membranes, Martinez and his colleagues (Martinez et al. 2003) demonstrated the presence of a 50 kDa apoA-I binding protein that was identical to the subunit of the β-chain of ATP synthase (Boyer 1997). The HDL-binding protein was identified as ecto-F_1-ATPase that recognizes apoA-I (Fig. 8d). The multi-subunit ATPase complex consists of two major domains called F_0 and F_1 (Boyer 1997). The ecto-F_1-ATPase protein, which resides on cell membranes, hydrolyzes ATP to ADP and phosphate and can be inhibited by the mitochondrial inhibitor protein IF_1 (Cabezon et al. 2003). It was recently shown that IF1 is present in the serum, and its concentration correlates negatively with HDL-C levels and the risk for coronary heart disease (Genoux et al. 2013). Binding of lipid-free apoA-I to the high affinity side of ecto-F_1-ATPase enhances binding of HDL to the low-affinity binding sites. The apoA-I binding to the ecto-F_1-ATPase also increases the production of ADP that associates with its receptor, purinergic $P2Y_{13}$ (Jacquet et al. 2005). The ecto-F_1-ATPase/$P2Y_{13}$-mediated HDL uptake pathway is under careful control. Adenylate kinase and niacin are important factors that regulate HDL metabolism and plasma levels via ecto-F1-ATPase (Fabre et al. 2006; Zhang et al. 2008). Inhibitors of ecto-F1-ATPase or adenylate kinase activity that consume ADP generated by ecto-F1-ATPase downregulate holo-HDL particle uptake (Genoux et al. 2013). In vivo studies using a $P2Y_{13}$-deficient mouse model also indicated that the $P2Y_{13}$ ADP receptor may have an important role in HDL-mediated reverse cholesterol transport (Fabre et al. 2010). It is possible that induction of hepatic ecto-F_1-ATPase/$P2Y_{13}$ pathway might enhance hepatic HDL endocytosis and turnover and accelerate cholesterol removal from cholesterol-laden macrophages and other tissues and cells of the body.

2.8 Transcytosis of apoA-I and HDL by Endothelial Cells

It has been shown that endothelial cells have the ability to bind and transcytose lipid-free apoA-I in a specific manner. This process depends on ABCA1 and leads to the generation of a lipidated apoA-I particle that is secreted (Cavelier et al. 2006; Rohrer et al. 2006). Endothelial cells can also transcytose HDL and this process required the functions of SR-BI and ABCG1 (Rohrer et al. 2009). ApoA-I mutants with defective C-terminal apoA-I[Δ(185-243)] and apoA-I[L218A/L219A/V221A/L222A] had 80 % decreased specific binding and 90 % decreased specific transport by aortic endothelial cells. Following lipidation of these mutants, the rHDL particles formed were transported through endothelial cells by an ABCG1- and SR-BI-dependent process. Amino and combined amino- and carboxy-terminal apoA-I deletion mutants displayed increased nonspecific binding, but the specific binding or transport remained absent (Ohnsorg et al. 2011). These data support the model in which apoA-I is initially lipidated by ABCA1 and subsequently processed by ABCA1-independent mechanisms. Transcytosis of apoA-I and HDL may provide a mechanism for transfer of HDL into the subendothelial space.

2.9 The Role of Cubilin in apoA-I and HDL Catabolism by the Kidney

It has been shown that in humans, impaired cubilin and amnionless function results in the Imerslunds-Gräsbeck syndrome, which is characterized by intestinal vitamin B12 malabsorption and proteinuria (Fyfe et al. 2004). Cubilin is a 460 kDa endocytic receptor which is co-expressed with megalin, a 600 kDa multi-ligand receptor belonging to the LDL receptor gene family. It is localized in the apical membranes of epithelial cells in the proximal tubules in the kidney cortex (Kozyraki 2001). In addition to co-localization of cubilin with megalin, the transmembrane protein amnionless is a renal protein that interacts with cubilin and forms a large cubilin/amnionless complex. In this complex, cubilin plays a role as a ligand-binding domain, whereas amnionless is essential for subcellular localization and endocytosis of cubilin bound to its ligand (Strope et al. 2004; Fyfe et al. 2004). An important function of cubilin is related to its ability to bind apoA-I or HDL (Kozyraki et al. 1999). In the kidney, however, HDL particles are too large to cross the glomerular filtration barrier, and therefore megalin and cubilin/amnionless protein receptor system is only exposed to filtered lipid-free or poorly lipidated apoA-I, thereby affecting the overall HDL metabolism (Kozyraki et al. 1999; Moestrup and Nielsen 2005). In physiological terms, it is considered that the kidney cortex is a major site of catabolism for lipid-free and poorly lipidated apoA-I and that this uptake is a concerted action of glomerular filtration, tubular reabsorption, and intracellular degradation of free apoA-I (Woollett and Spady 1997). Graversen et al. (2008) analyzed urine samples from patients with Fanconi syndrome. This is a rare renal proximal tubular reabsorption failure and also includes dysfunction of cubilin. A high urinary excretion of both apoA-I and apoA-IV but not apoA-II was evident. This study demonstrated that the human kidney is a major site for filtered apoA-I and A-IV but not for HDL particles since urinary excretion of all major lipid classes (phospholipids, triglycerides, cholesterol, and cholesterol esters) in Fanconi patients was as low as in control subjects (Graversen et al. 2008). Although the kidney is not considered a central organ in lipoprotein catabolism, it plays an important role in the degradation of lipid-poor apoA-I via the cubilin function.

3 HDL Subclasses

3.1 The Origin and Metabolism of Preβ-HDL Subpopulations

Several preβ- and α-HDL subpopulations exist in plasma and are generated as a consequence of the pathway of biogenesis and remodeling of HDL. These subpopulations can be separated based on different fractionation procedures (Fielding and Fielding 1996; Chung et al. 1986; Nichols et al. 1986; Davidson et al. 1994). The precursor-product relationship between preβ- and α-HDL particles as well as the precise origin and functions of the preβ-HDL particles is still a matter of investigation.

It has been reported earlier that preβ-HDL comprises approximately 5 % of total plasma apoA-I level. It is heterogeneous in size and contains several species of 5–6 nm in diameter (Fielding and Fielding 1995; Nanjee et al. 2000). The best characterized species are preβ1 and preβ2 (Fielding and Fielding 1995). The concentration of preβ1-HDL is increased in large lymph vessels (Asztalos et al. 1993) and in aortic intima (Heideman and Hoff 1982).

Preβ-HDL particles can be formed by two different routes. The first is de novo synthesis by the HDL biogenesis pathway (Fig. 8a). The second is generation of preβ-HDL particles from α-HDL particles by reactions catalyzed by CETP, PLTP, HL, EL, and apoM discussed earlier (Barrans et al. 1994; Maugeais et al. 2003c; Arai et al. 2000; Huuskonen et al. 2001; Christoffersen et al. 2008b) (Fig. 8a–c).

Cell culture studies showed that lipid-free apoA-I added to a culture medium of CHO cells can recruit phospholipids and cholesterol, initially to form small 73 Å particles, and subsequently larger apoA-I-containing particles by the action of LCAT that have a precursor-product relationship (Forte et al. 1993, 1995).

Subsequent studies showed that a large proportion of apoA-I is secreted from HepG-2, CaCo-2, or apoA-I expressing CHO cells in lipid-free monomeric form, with a Stokes radius of 2.6 nm and preα electrophoretic mobility that is unable to promote efflux of phospholipids and cholesterol. It was suggested that in a reaction dependent on ABCA1, the 2.6 nm form was converted into a 3.6 nm monomeric apoA-I form with preβ electrophoretic mobility that was able to promote efflux of phospholipids and cholesterol from cells and thus increase its size (Chau et al. 2006). Expression of apoM in cells transfected with ABCA1 can also increase the size of preβ-HDL (Mulya et al. 2010). Other studies have shown that some types of preβ-HDL particles can be formed independently of apoA-I/ABCA1 interactions in the plasma of humans with Tangier disease and the plasma of apoA-I-deficient mice expressing mutant apoA-I forms (Chroni et al. 2007; Fotakis et al. 2013a; Asztalos et al. 2001). Furthermore, inhibition of ABCA1 in HepG2 cells and macrophage cultures by glyburide inhibited the formation of α-HDL particles but did not affect the formation of preβ-HDL particles (Krimbou et al. 2005).

The presence of increased concentrations of preβ1-HDL in the vascular bed suggests that these particles may be generated locally by gradual lipidation of lipid-poor apoA-I (Heideman and Hoff 1982; Smith et al. 1984). Preβ-HDL particles are typically lipid-poor and therefore they are efficient in promoting ABCA1-mediated cholesterol efflux. The ABCA1/preβ1-HDL interaction provides phospholipids and cholesterol and thereby converts the preβ-HDL to α-HDL-migrating particles. These particles may be enlarged further by recruitment of phospholipids and cholesterol from cell membranes (Fielding and Fielding 2007). In addition, esterification of the cholesterol of preβ1-HDL by LCAT contributes to their gradual conversion into spherical HDL without prior formation of discoidal HDL particles (Nakamura et al. 2004b; Fielding and Fielding 2007).

3.2 Complexity of HDL

Genome-wide association studies demonstrated that new genes and the corresponding proteins affect plasma HDL levels by unknown mechanisms (Holleboom et al. 2008; Kathiresan et al. 2008b, 2009; Sabatti et al. 2009; Aulchenko et al. 2009; Teslovich et al. 2010; Richards et al. 2009; Willer et al. 2008; Chasman et al. 2009; Waterworth et al. 2010; Laurila et al. 2013). In parallel, proteomic analysis showed that a large number of plasma proteins can associate with HDL and this may affect the HDL structure and functions (Fig. 9) (Gordon et al. 2010). The proteins associated with HDL can be classified in six major categories and include proteins involved in lipid, lipoprotein, and HDL biogenesis and metabolism, acute phase proteins, protease inhibitors, complement regulatory proteins, and a few others (albumin, fibrinogen a chain platelet basic protein) (Vaisar et al. 2007; Davidson et al. 2009b). Differences were observed in

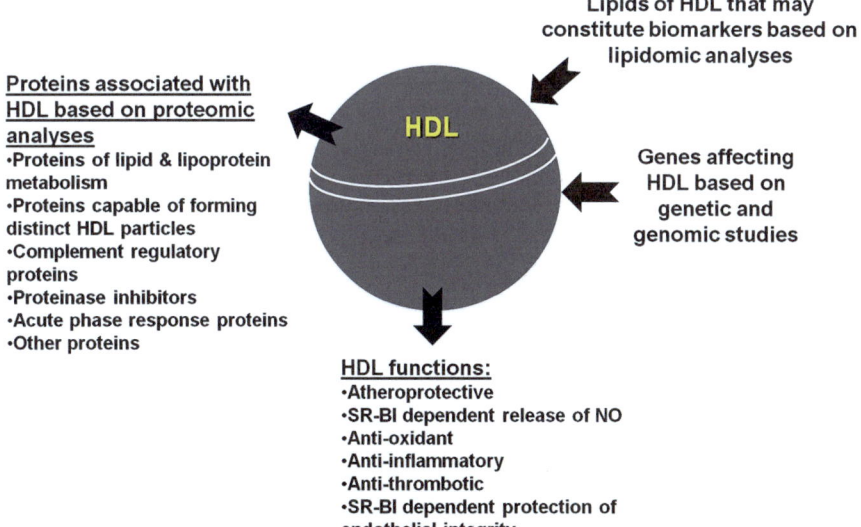

Fig. 9 Schematic representation of proteins associated with HDL and factors that may affect HDL levels and functions

the proteomic composition of HDL subpopulations derived from HDL particles of different sizes (Davidson et al. 2009a). Furthermore the HDL proteome could be altered by pharmacological treatments (Green et al. 2008).

The modulation of the concentration of various proteins associated with HDL has been studied in more detail under conditions of infection, inflammation, or tissue injury. Under these conditions, acute phase response is triggered that causes huge alterations in hepatic protein synthesis in response to cytokines that alter HDL protein composition (Rohrer et al. 2004; Shah et al. 2013). These changes include the abrupt increase in serum amyloid A, apoJ, and secretory non-pancreatic phospholipase A2 and a decrease in apoA-I, paraoxonase-1, LCAT, and PLTP. These changes affect the capacity of HDL to promote cellular cholesterol efflux as well as other HDL functions (Rohrer et al. 2004; Shah et al. 2013).

In addition to the variety of proteins, a variety of lipids are also carried by HDL, and some of them are or can be transformed to potent bioactive molecules (Vickers and Remaley 2014). Furthermore, HDL carries and transports fat soluble vitamins, steroid hormones, carotenoids, as well as numerous more polar metabolites such as heneicosanoic acid, pentitol, and oxalic acid which were found to be significantly correlated with insulin resistance (Vickers and Remaley 2014). Finally, it has been reported that HDL also transports small RNAs, including microRNAs, tRNA-derived RNA fragments, and RNase P-derived RNA fragments (Vickers et al. 2011). How all this protein decoration affects HDL metabolism and HDL particle function remain to be studied.

4   Sources of Funding

Collaboration among the groups of Drs Vassilis I. Zannis, Dimitris Kardassis, Matti Jauhiainen, and Angeliki Chroni is supported by COST Action BM0904. Dr Vassilis I. Zannis' research is supported by a grant from the National Institutes of Health HL48739. Drs Dimitris Kardassis and Angeliki Chroni's research is supported by grants of the General Secretariat of Research and Technology of Greece (Synergasia 09SYN-12-897) and by and the Ministry of Education of Greece (Thalis MIS 377286). P. Fotakis has been supported by pre-doctoral training Fellowship HERACLEITUS II by the European Union and Greek national funds through the Operational Program "Education and Lifelong Learning" of the National Strategic Reference Framework (NSRF). The work was also financially supported by the Academy of Finland (grant # 257545 to MJ).

Open Access This chapter is distributed under the terms of the Creative Commons Attribution Noncommercial License, which permits any noncommercial use, distribution, and reproduction in any medium, provided the original author(s) and source are credited.

References

Acton SL, Scherer PE, Lodish HF, Krieger M (1994) Expression cloning of SR-BI, a CD36-related class B scavenger receptor. J Biol Chem 269:21003–21009

Acton S, Rigotti A, Landschulz KT, Xu S, Hobbs HH, Krieger M (1996) Identification of scavenger receptor SR-BI as a high density lipoprotein receptor. Science 271:518–520

Ahras M, Naing T, McPherson R (2008) Scavenger receptor class B type I localizes to a late endosomal compartment. J Lipid Res 49:1569–1576

Aiello RJ, Brees D, Bourassa PA, Royer L, Lindsey S, Coskran T, Haghpassand M, Francone OL (2002) Increased atherosclerosis in hyperlipidemic mice with inactivation of ABCA1 in macrophages. Arterioscler Thromb Vasc Biol 22:630–637

Albers JJ, Cheung MC (2004) Emerging roles for phospholipid transfer protein in lipid and lipoprotein metabolism. Curr Opin Lipidol 15:255–260

Albers JJ, Vuletic S, Cheung MC (2012) Role of plasma phospholipid transfer protein in lipid and lipoprotein metabolism. Biochim Biophys Acta 1821:345–357

Amar MJ, Shamburek RD, Vaisman B, Knapper CL, Foger B, Hoyt RF Jr, Santamarina-Fojo S, Brewer HB Jr, Remaley AT (2009) Adenoviral expression of human lecithin-cholesterol acyltransferase in nonhuman primates leads to an antiatherogenic lipoprotein phenotype by increasing high-density lipoprotein and lowering low-density lipoprotein. Metabolism 58:568–575

Arai T, Wang N, Bezouevski M, Welch C, Tall AR (1999) Decreased atherosclerosis in heterozygous low density lipoprotein receptor-deficient mice expressing the scavenger receptor BI transgene. J Biol Chem 274:2366–2371

Arai T, Tsukada T, Murase T, Matsumoto K (2000) Particle size analysis of high density lipoproteins in patients with genetic cholesteryl ester transfer protein deficiency. Clin Chim Acta 301:103–117

Assmann G, von Eckardstein A, Brewer HB (2001) Familial analphalipoproteinemia: Tangier disease. In: Scriver CR, Beaudet AL, Sly WS, Valle D (eds) The metabolic and molecular basis of inherited disease. McGraw-Hill, New York, pp 2937–2960

Asztalos BF (2004) High-density lipoprotein metabolism and progression of atherosclerosis: new insights from the HDL Atherosclerosis Treatment Study. Curr Opin Cardiol 19:385–391

Asztalos BF, Sloop CH, Wong L, Roheim PS (1993) Comparison of apo A-I-containing subpopulations of dog plasma and prenodal peripheral lymph: evidence for alteration in subpopulations in the interstitial space. Biochim Biophys Acta 1169:301–304

Asztalos BF, Brousseau ME, McNamara JR, Horvath KV, Roheim PS, Schaefer EJ (2001) Subpopulations of high density lipoproteins in homozygous and heterozygous Tangier disease. Atherosclerosis 156:217–225

Aulchenko YS, Ripatti S, Lindqvist I, Boomsma D, Heid IM, Pramstaller PP, Penninx BW, Janssens AC, Wilson JF, Spector T, Martin NG, Pedersen NL, Kyvik KO, Kaprio J, Hofman A, Freimer NB, Jarvelin MR, Gyllensten U, Campbell H, Rudan I, Johansson A, Marroni F, Hayward C, Vitart V, Jonasson I, Pattaro C, Wright A, Hastie N, Pichler I, Hicks AA, Falchi M, Willemsen G, Hottenga JJ, de Geus EJ, Montgomery GW, Whitfield J, Magnusson P, Saharinen J, Perola M, Silander K, Isaacs A, Sijbrands EJ, Uitterlinden AG, Witteman JC, Oostra BA, Elliott P, Ruokonen A, Sabatti C, Gieger C, Meitinger T, Kronenberg F, Doring A, Wichmann HE, Smit JH, McCarthy MI, van Duijn CM, Peltonen L (2009) Loci influencing lipid levels and coronary heart disease risk in 16 European population cohorts. Nat Genet 41:47–55

Axler O, Ahnstrom J, Dahlback B (2007) An ELISA for apolipoprotein M reveals a strong correlation to total cholesterol in human plasma. J Lipid Res 48:1772–1780

Axler O, Ahnstrom J, Dahlback B (2008) Apolipoprotein M associates to lipoproteins through its retained signal peptide. FEBS Lett 582:826–828

Baldan A, Pei L, Lee R, Tarr P, Tangirala RK, Weinstein MM, Frank J, Li AC, Tontonoz P, Edwards PA (2006) Impaired development of atherosclerosis in hyperlipidemic Ldlr$^{-/-}$ and

ApoE$^{-/-}$ mice transplanted with Abcg1$^{-/-}$ bone marrow. Arterioscler Thromb Vasc Biol 26: 2301–2307

Barrans A, Collet X, Barbaras R, Jaspard B, Manent J, Vieu C, Chap H, Perret B (1994) Hepatic lipase induces the formation of pre-beta 1 high density lipoprotein (HDL) from triacylglycerol-rich HDL2. A study comparing liver perfusion to in vitro incubation with lipases. J Biol Chem 269:11572–11577

Barter PJ, Brewer HB Jr, Chapman MJ, Hennekens CH, Rader DJ, Tall AR (2003) Cholesteryl ester transfer protein: a novel target for raising HDL and inhibiting atherosclerosis. Arterioscler Thromb Vasc Biol 23:160–167

Belalcazar LM, Merched A, Carr B, Oka K, Chen KH, Pastore L, Beaudet A, Chan L (2003) Long-term stable expression of human apolipoprotein A-I mediated by helper-dependent adenovirus gene transfer inhibits atherosclerosis progression and remodels atherosclerotic plaques in a mouse model of familial hypercholesterolemia. Circulation 107:2726–2732

Benoit P, Emmanuel F, Caillaud JM, Bassinet L, Castro G, Gallix P, Fruchart JC, Branellec D, Denefle P, Duverger N (1999) Somatic gene transfer of human ApoA-I inhibits atherosclerosis progression in mouse models. Circulation 99:105–110

Biedzka-Sarek M, Metso J, Kateifides A, Meri T, Jokiranta TS, Muszynski A, Radziejewska-Lebrecht J, Zannis V, Skurnik M, Jauhiainen M (2011) Apolipoprotein A-I exerts bactericidal activity against *Yersinia enterocolitica* serotype O:3. J Biol Chem 286:38211–38219

Bodzioch M, Orso E, Klucken J, Langmann T, Bottcher A, Diederich W, Drobnik W, Barlage S, Buchler C, Porsch-Ozcurumez M, Kaminski WE, Hahmann HW, Oette K, Rothe G, Aslanidis C, Lackner KJ, Schmitz G (1999) The gene encoding ATP-binding cassette transporter 1 is mutated in Tangier disease. Nat Genet 22:347–351

Borhani DW, Rogers DP, Engler JA, Brouillette CG (1997) Crystal structure of truncated human apolipoprotein A-I suggests a lipid-bound conformation. Proc Natl Acad Sci USA 94: 12291–12296

Borhani DW, Engler JA, Brouillette CG (1999) Crystallization of truncated human apolipoprotein A-I in a novel conformation. Acta Crystallogr D Biol Crystallogr 55(Pt 9):1578–1583

Boyer PD (1997) The ATP synthase–a splendid molecular machine. Annu Rev Biochem 66: 717–749

Braschi S, Couture N, Gambarotta A, Gauthier BR, Coffill CR, Sparks DL, Maeda N, Schultz JR (1998) Hepatic lipase affects both HDL and ApoB-containing lipoprotein levels in the mouse. Biochim Biophys Acta 1392:276–290

Braun A, Trigatti BL, Post MJ, Sato K, Simons M, Edelberg JM, Rosenberg RD, Schrenzel M, Krieger M (2002) Loss of SR-BI expression leads to the early onset of occlusive atherosclerotic coronary artery disease, spontaneous myocardial infarctions, severe cardiac dysfunction, and premature death in apolipoprotein E-deficient mice. Circ Res 90:270–276

Braun A, Zhang S, Miettinen HE, Ebrahim S, Holm TM, Vasile E, Post MJ, Yoerger DM, Picard MH, Krieger JL, Andrews NC, Simons M, Krieger M (2003) Probucol prevents early coronary heart disease and death in the high-density lipoprotein receptor SR-BI/apolipoprotein E double knockout mouse. Proc Natl Acad Sci USA 100:7283–7288

Breckenridge WC, Little JA, Alaupovic P, Wang CS, Kuksis A, Kakis G, Lindgren F, Gardiner G (1982) Lipoprotein abnormalities associated with a familial deficiency of hepatic lipase. Atherosclerosis 45:161–179

Brousseau ME, Schaefer EJ, Wolfe ML, Bloedon LT, Digenio AG, Clark RW, Mancuso JP, Rader DJ (2004) Effects of an inhibitor of cholesteryl ester transfer protein on HDL cholesterol. N Engl J Med 350:1505–1515

Brown RJ, Lagor WR, Sankaranaravanan S, Yasuda T, Quertermous T, Rothblat GH, Rader DJ (2010) Impact of combined deficiency of hepatic lipase and endothelial lipase on the metabolism of both high-density lipoproteins and apolipoprotein B-containing lipoproteins. Circ Res 107:357–364

Browne RW, Bloom MS, Shelly WB, Ocque AJ, Huddleston HG, Fujimoto VY (2009) Follicular fluid high density lipoprotein-associated micronutrient levels are associated with embryo fragmentation during IVF. J Assist Reprod Genet 26:557–560

Brundert M, Heeren J, Greten H, Rinninger F (2003) Hepatic lipase mediates an increase in selective uptake of HDL-associated cholesteryl esters by cells in culture independent from SR-BI. J Lipid Res 44:1020–1032

Brundert M, Heeren J, Merkel M, Carambia A, Herkel J, Groitl P, Dobner T, Ramakrishnan R, Moore KJ, Rinninger F (2011) Scavenger receptor CD36 mediates uptake of high density lipoproteins in mice and by cultured cells. J Lipid Res 52:745–758

Brunham LR, Kruit JK, Iqbal J, Fievet C, Timmins JM, Pape TD, Coburn BA, Bissada N, Staels B, Groen AK, Hussain MM, Parks JS, Kuipers F, Hayden MR (2006a) Intestinal ABCA1 directly contributes to HDL biogenesis in vivo. J Clin Invest 116:1052–1062

Brunham LR, Singaraja RR, Hayden MR (2006b) Variations on a gene: rare and common variants in ABCA1 and their impact on HDL cholesterol levels and atherosclerosis. Annu Rev Nutr 26: 105–129

Brunzell JD, Deeb SS (2001) Familial lipoprotein lipase deficiency, apoC-II deficiency, and hepatic lipase deficiency. In: Scriver CR, Beaudet AL, Valle D, Sly WS (eds) The metabolic and molecular bases of inherited disease. McGraw-Hill, New York, pp 2789–2816

Burgess B, Naus K, Chan J, Hirsch-Reinshagen V, Tansley G, Matzke L, Chan B, Wilkinson A, Fan J, Donkin J, Balik D, Tanaka T, Ou G, Dyer R, Innis S, McManus B, Lutjohann D, Wellington C (2008) Overexpression of human ABCG1 does not affect atherosclerosis in fat-fed ApoE-deficient mice. Arterioscler Thromb Vasc Biol 28:1731–1737

Cabezon E, Montgomery MG, Leslie AG, Walker JE (2003) The structure of bovine F1-ATPase in complex with its regulatory protein IF1. Nat Struct Biol 10:744–750

Calabresi L, Baldassarre D, Castelnuovo S, Conca P, Bocchi L, Candini C, Frigerio B, Amato M, Sirtori CR, Alessandrini P, Arca M, Boscutti G, Cattin L, Gesualdo L, Sampietro T, Vaudo G, Veglia F, Calandra S, Franceschini G (2009a) Functional lecithin: cholesterol acyltransferase is not required for efficient atheroprotection in humans. Circulation 120:628–635

Calabresi L, Favari E, Moleri E, Adorni MP, Pedrelli M, Costa S, Jessup W, Gelissen IC, Kovanen PT, Bernini F, Franceschini G (2009b) Functional LCAT is not required for macrophage cholesterol efflux to human serum. Atherosclerosis 204:141–146

Cavelier C, Rohrer L, von Eckardstein A (2006) ATP-binding cassette transporter a1 modulates apolipoprotein A1 transcytosis through aortic endothelial cells. Circ Res 99:1060–1066

Chasman DI, Pare G, Mora S, Hopewell JC, Peloso G, Clarke R, Cupples LA, Hamsten A, Kathiresan S, Malarstig A, Ordovas JM, Ripatti S, Parker AN, Miletich JP, Ridker PM (2009) Forty-three loci associated with plasma lipoprotein size, concentration, and cholesterol content in genome-wide analysis. PLoS Genet 5

Chau P, Nakamura Y, Fielding CJ, Fielding PE (2006) Mechanism of prebeta-HDL formation and activation. Biochemistry 45:3981–3987

Christoffersen C, Ahnstrom J, Axler O, Christensen EI, Dahlback B, Nielsen LB (2008a) The signal peptide anchors apolipoprotein M in plasma lipoproteins and prevents rapid clearance of apolipoprotein M from plasma. J Biol Chem 283:18765–18772

Christoffersen C, Jauhiainen M, Moser M, Porse B, Ehnholm C, Boesl M, Dahlback B, Nielsen LB (2008b) Effect of apolipoprotein M on high density lipoprotein metabolism and atherosclerosis in low density lipoprotein receptor knock-out mice. J Biol Chem 283:1839–1847

Chroni A, Liu T, Gorshkova I, Kan HY, Uehara Y, von Eckardstein A, Zannis VI (2003) The central helices of apoA-I can promote ATP-binding cassette transporter A1 (ABCA1)-mediated lipid efflux. Amino acid residues 220-231 of the wild-type apoA-I are required for lipid efflux in vitro and high density lipoprotein formation in vivo. J Biol Chem 278: 6719–6730

Chroni A, Kan HY, Kypreos KE, Gorshkova IN, Shkodrani A, Zannis VI (2004a) Substitutions of glutamate 110 and 111 in the middle helix 4 of human apolipoprotein A-I (apoA-I) by alanine affect the structure and in vitro functions of apoA-I and induce severe hypertriglyceridemia in apoA-I-deficient mice. Biochemistry 43:10442–10457

Chroni A, Liu T, Fitzgerald ML, Freeman MW, Zannis VI (2004b) Cross-linking and lipid efflux properties of apoA-I mutants suggest direct association between apoA-I helices and ABCA1. Biochemistry 43:2126–2139

Chroni A, Duka A, Kan HY, Liu T, Zannis VI (2005a) Point mutations in apolipoprotein a-I mimic the phenotype observed in patients with classical lecithin:cholesterol acyltransferase deficiency. Biochemistry 44:14353–14366

Chroni A, Kan HY, Shkodrani A, Liu T, Zannis VI (2005b) Deletions of helices 2 and 3 of human apoA-I are associated with severe dyslipidemia following adenovirus-mediated gene transfer in apoA-I-deficient mice. Biochemistry 44:4108–4117

Chroni A, Nieland TJ, Kypreos KE, Krieger M, Zannis VI (2005c) SR-BI mediates cholesterol efflux via its interactions with lipid-bound ApoE. Structural mutations in SR-BI diminish cholesterol efflux. Biochemistry 44:13132–13143

Chroni A, Koukos G, Duka A, Zannis VI (2007) The carboxy-terminal region of apoA-I is required for the ABCA1-dependent formation of alpha-HDL but not prebeta-HDL particles in vivo. Biochemistry 46:5697–5708

Chung BH, Segrest JP, Ray MJ, Brunzell JD, Hokanson JE, Krauss RM, Beaudrie K, Cone JT (1986) Single vertical spin density gradient ultracentrifugation. Methods Enzymol 128:181–209

Croop JM, Tiller GE, Fletcher JA, Lux ML, Raab E, Goldenson D, Son D, Arciniegas S, Wu RL (1997) Isolation and characterization of a mammalian homolog of the Drosophila white gene. Gene 185:77–85

Cuchel M, Lund-Katz S, Llera-Moya M, Millar JS, Chang D, Fuki I, Rothblat GH, Phillips MC, Rader DJ (2010) Pathways by which reconstituted high-density lipoprotein mobilizes free cholesterol from whole body and from macrophages. Arterioscler Thromb Vasc Biol 30:526–532

Dahlback B, Nielsen LB (2009) Apolipoprotein M affecting lipid metabolism or just catching a ride with lipoproteins in the circulation? Cell Mol Life Sci 66:559–564

Daniil G, Phedonos AA, Holleboom AG, Motazacker MM, Argyri L, Kuivenhoven JA, Chroni A (2011) Characterization of antioxidant/anti-inflammatory properties and apoA-I-containing subpopulations of HDL from family subjects with monogenic low HDL disorders. Clin Chim Acta 412:1213–1220

Daniil G, Zannis VI, Chroni A (2013) Effect of apoA-I mutations in the capacity of reconstituted HDL to promote ABCG1-mediated cholesterol efflux. PLoS ONE 8:e67993

Davidson WS, Sparks DL, Lund-Katz S, Phillips MC (1994) The molecular basis for the difference in charge between pre-beta- and alpha-migrating high density lipoproteins. J Biol Chem 269:8959–8965

Davidson WS, Silva RA, Chantepie S, Lagor WR, Chapman MJ, Kontush A (2009) Proteomic analysis of defined HDL subpopulations reveals particle-specific protein clusters: relevance to antioxidative function. Arterioscler Thromb Vasc Biol 29:870–876

de Grooth GJ, Kuivenhoven JA, Stalenhoef AF, de Graaf J, Zwinderman AH, Posma JL, van Tol A, Kastelein JJ (2002) Efficacy and safety of a novel cholesteryl ester transfer protein inhibitor, JTT-705, in humans: a randomized phase II dose-response study. Circulation 105:2159–2165

deLemos AS, Wolfe ML, Long CJ, Sivapackianathan R, Rader DJ (2002) Identification of genetic variants in endothelial lipase in persons with elevated high-density lipoprotein cholesterol. Circulation 106:1321–1326

Duka A, Fotakis P, Georgiadou D, Kateifides A, Tzavlaki K, von Eckardstein L, Stratikos E, Kardassis D, Zannis VI (2013) ApoA-IV promotes the biogenesis of apoA-IV-containing HDL particles with the participation of ABCA1 and LCAT. J Lipid Res 54:107–115

Dullaart RP, Plomgaard P, de Vries R, Dahlback B, Nielsen LB (2009) Plasma apolipoprotein M is reduced in metabolic syndrome but does not predict intima media thickness. Clin Chim Acta 406:129–133

Eberini I, Calabresi L, Wait R, Tedeschi G, Pirillo A, Puglisi L, Sirtori CR, Gianazza E (2002) Macrophage metalloproteinases degrade high-density-lipoprotein-associated apolipoprotein A-I at both the N- and C-termini. Biochem J 362:627–634

Edmondson AC, Brown RJ, Kathiresan S, Cupples LA, Demissie S, Manning AK, Jensen MK, Rimm EB, Wang J, Rodrigues A, Bamba V, Khetarpal SA, Wolfe ML, Derohannessian S, Li M, Reilly MP, Aberle J, Evans D, Hegele RA, Rader DJ (2009) Loss-of-function variants in endothelial lipase are a cause of elevated HDL cholesterol in humans. J Clin Invest 119: 1042–1050

Edmondson AC, Braund PS, Stylianou IM, Khera AV, Nelson CP, Wolfe ML, Derohannessian SL, Keating BJ, Qu L, He J, Tobin MD, Tomaszewski M, Baumert J, Klopp N, Doring A, Thorand B, Li M, Reilly MP, Koenig W, Samani NJ, Rader DJ (2011) Dense genotyping of candidate gene loci identifies variants associated with high-density lipoprotein cholesterol. Circ Cardiovasc Genet 4:145–155

Emmanuel F, Caillaud JM, Hennuyer N, Fievet C, Viry I, Houdebine JC, Fruchart JC, Denefle P, Duverger N (1996) Overexpression of human apolipoprotein A-I inhibits atherosclerosis development in Watanabe rabbits. Circulation 94:632

Escola-Gil JC, Chen X, Julve J, Quesada H, Santos D, Metso J, Tous M, Jauhiainen M, Blanco-Vaca F (2013) Hepatic lipase- and endothelial lipase-deficiency in mice promotes macrophage-to-feces RCT and HDL antioxidant properties. Biochim Biophys Acta 1831:691–697

Fabre AC, Vantourout P, Champagne E, Terce F, Rolland C, Perret B, Collet X, Barbaras R, Martinez LO (2006) Cell surface adenylate kinase activity regulates the F(1)-ATPase/P2Y (13)-mediated HDL endocytosis pathway on human hepatocytes. Cell Mol Life Sci 63: 2829–2837

Fabre AC, Malaval C, Ben AA, Verdier C, Pons V, Serhan N, Lichtenstein L, Combes G, Huby T, Briand F, Collet X, Nijstad N, Tietge UJ, Robaye B, Perret B, Boeynaems JM, Martinez LO (2010) P2Y13 receptor is critical for reverse cholesterol transport. Hepatology 52:1477–1483

Favari E, Calabresi L, Adorni MP, Jessup W, Simonelli S, Franceschini G, Bernini F (2009) Small discoidal pre-beta1 HDL particles are efficient acceptors of cell cholesterol via ABCA1 and ABCG1. Biochemistry 48:11067–11074

Fenske SA, Yesilaltay A, Pal R, Daniels K, Barker C, Quinones V, Rigotti A, Krieger M, Kocher O (2009) Normal hepatic cell surface localization of the high density lipoprotein receptor, scavenger receptor class B, type I, depends on all four PDZ domains of PDZK1. J Biol Chem 284:5797–5806

Fielding CJ, Fielding PE (1995) Molecular physiology of reverse cholesterol transport. J Lipid Res 36:211–228

Fielding CJ, Fielding PE (1996) Two-dimensional nondenaturing electrophoresis of lipoproteins: applications to high-density lipoprotein speciation. Methods Enzymol 263:251–259

Fielding CJ, Fielding PE (2007) Reverse cholesterol transport – new roles for preb1-HDL and lecithin: cholesterol acyltransferase. In: Fielding CJ (ed) High-density lipoproteins: from basic biology to clinical aspects. Wiley, Weinheim, pp 143–162

Fielding CJ, Shore VG, Fielding PE (1972) A protein cofactor of lecithin:cholesterol acyltransferase. Biochem Biophys Res Commun 46:1493–1498

Fitzgerald ML, Mendez AJ, Moore KJ, Andersson LP, Panjeton HA, Freeman MW (2001) ATP-binding cassette transporter A1 contains an NH2-terminal signal anchor sequence that translocates the protein's first hydrophilic domain to the exoplasmic space. J Biol Chem 276: 15137–15145

Fitzgerald ML, Morris AL, Rhee JS, Andersson LP, Mendez AJ, Freeman MW (2002) Naturally occurring mutations in the largest extracellular loops of ABCA1 can disrupt its direct interaction with apolipoprotein A-I. J Biol Chem 277:33178–33187

Forte TM, Goth-Goldstein R, Nordhausen RW, McCall MR (1993) Apolipoprotein A-I-cell membrane interaction: extracellular assembly of heterogeneous nascent HDL particles. J Lipid Res 34:317–324

Forte TM, Bielicki JK, Goth-Goldstein R, Selmek J, McCall MR (1995) Recruitment of cell phospholipids and cholesterol by apolipoproteins A-II and A-I: formation of nascent apolipoprotein-specific HDL that differ in size, phospholipid composition, and reactivity with LCAT. J Lipid Res 36:148–157

Fotakis P, Kateifides A, Gkolfinopoulou C, Georgiadou D, Beck M, Grundler K, Chroni A, Stratikos E, Kardassis D, Zannis VI (2013a) Role of the hydrophobic and charged residues in the 218 to 226 region of apoA-I in the biogenesis of HDL. J Lipid Res 54:3281–3292

Fotakis P, Tiniakou I, Kateifides AK, Gkolfinopoulou C, Chroni A, Stratikos E, Zannis VI, Kardassis D (2013b) Significance of the hydrophobic residues 225-230 of apoA-I for the biogenesis of HDL. J Lipid Res 54:3293–3302

Francone OL, Royer L, Haghpassand M (1996) Increased prebeta-HDL levels, cholesterol efflux, and LCAT-mediated esterification in mice expressing the human cholesteryl ester transfer protein (CETP) and human apolipoprotein A-I (apoA-I) transgenes. J Lipid Res 37:1268–1277

Frikke-Schmidt R, Nordestgaard BG, Jensen GB, Steffensen R, Tybjaerg-Hansen A (2008) Genetic variation in ABCA1 predicts ischemic heart disease in the general population. Arterioscler Thromb Vasc Biol 28:180–186

Fyfe JC, Madsen M, Hojrup P, Christensen EI, Tanner SM, de la Chapelle A, He Q, Moestrup SK (2004) The functional cobalamin (vitamin B12)-intrinsic factor receptor is a novel complex of cubilin and amnionless. Blood 103:1573–1579

Galis ZS, Muszynski M, Sukhova GK, Simon-Morrissey E, Libby P (1995) Enhanced expression of vascular matrix metalloproteinases induced in vitro by cytokines and in regions of human atherosclerotic lesions. Ann N Y Acad Sci 748:501–507

Genoux A, Ruidavets JB, Ferrieres J, Combes G, Lichtenstein L, Pons V, Laffargue M, Taraszkiewicz D, Carrie D, Elbaz M, Perret B, Martinez LO (2013) Serum IF1 concentration is independently associated to HDL levels and to coronary heart disease: the GENES study. J Lipid Res 54:2550–2558

Gong M, Wilson M, Kelly T, Su W, Dressman J, Kincer J, Matveev SV, Guo L, Guerin T, Li XA, Zhu W, Uittenbogaard A, Smart EJ (2003) HDL-associated estradiol stimulates endothelial NO synthase and vasodilation in an SR-BI-dependent manner. J Clin Invest 111:1579–1587

Gordon DJ, Probstfield JL, Garrison RJ, Neaton JD, Castelli WP, Knoke JD, Jacobs DR Jr, Bangdiwala S, Tyroler HA (1989) High-density lipoprotein cholesterol and cardiovascular disease. Four prospective American studies. Circulation 79:8–15

Gordon SM, Deng JY, Lu LJ, Davidson WS (2010) Proteomic characterization of human plasma high density lipoprotein fractionated by gel filtration chromatography. J Proteome Res 9: 5239–5249

Gorshkova IN, Atkinson D (2011) Enhanced binding of apolipoprotein A-I variants associated with hypertriglyceridemia to triglyceride-rich particles. Biochemistry 50:2040–2047

Graversen JH, Castro G, Kandoussi A, Nielsen H, Christensen EI, Norden A, Moestrup SK (2008) A pivotal role of the human kidney in catabolism of HDL protein components apolipoprotein A-I and A-IV but not of A-II. Lipids 43:467–470

Green PS, Vaisar T, Pennathur S, Kulstad JJ, Moore AB, Marcovina S, Brunzell J, Knopp RH, Zhao XQ, Heinecke JW (2008) Combined statin and niacin therapy remodels the high-density lipoprotein proteome. Circulation 118:1259–1267

Greene DJ, Skeggs JW, Morton RE (2001) Elevated triglyceride content diminishes the capacity of high density lipoprotein to deliver cholesteryl esters via the scavenger receptor class B type I (SR-BI). J Biol Chem 276:4804–4811

Groen AK, Bloks VW, Bandsma RH, Ottenhoff R, Chimini G, Kuipers F (2001) Hepatobiliary cholesterol transport is not impaired in Abca1-null mice lacking HDL. J Clin Invest 108: 843–850

Gu X, Trigatti B, Xu S, Acton S, Babitt J, Krieger M (1998) The efficient cellular uptake of high density lipoprotein lipids via scavenger receptor class B type I requires not only receptor-mediated surface binding but also receptor-specific lipid transfer mediated by its extracellular domain. J Biol Chem 273:26338–26348

Gu X, Kozarsky K, Krieger M (2000a) Scavenger receptor class B, type I-mediated [3H]cholesterol efflux to high and low density lipoproteins is dependent on lipoprotein binding to the receptor. J Biol Chem 275:29993–30001

Gu X, Lawrence R, Krieger M (2000b) Dissociation of the high density lipoprotein and low density lipoprotein binding activities of murine scavenger receptor class B type I (mSR-BI) using retrovirus library-based activity dissection. J Biol Chem 275:9120–9130

Guendouzi K, Jaspard B, Barbaras R, Motta C, Vieu C, Marcel Y, Chap H, Perret B, Collet X (1999) Biochemical and physical properties of remnant-HDL2 and of pre beta 1-HDL produced by hepatic lipase. Biochemistry 38:2762–2768

Haase CL, Frikke-Schmidt R, Nordestgaard BG, Kateifides AK, Kardassis D, Nielsen LB, Andersen CB, Kober L, Johnsen AH, Grande P, Zannis VI, Tybjaerg-Hansen A (2011) Mutation in APOA1 predicts increased risk of ischaemic heart disease and total mortality without low HDL cholesterol levels. J Intern Med 270:136–146

Hailman E, Albers JJ, Wolfbauer G, Tu AY, Wright SD (1996) Neutralization and transfer of lipopolysaccharide by phospholipid transfer protein. J Biol Chem 271:12172–12178

Hammad SM, Stefansson S, Twal WO, Drake CJ, Fleming P, Remaley A, Brewer HB Jr, Argraves WS (1999) Cubilin, the endocytic receptor for intrinsic factor-vitamin B(12) complex, mediates high-density lipoprotein holoparticle endocytosis. Proc Natl Acad Sci USA 96: 10158–10163

Harder CJ, McPherson R (2007) HDL remodeling by CETP and SR-BI. In: Fielding CJ (ed) High-density lipoproteins. From basic biology to clinical aspects. Wiley, Weinheim, pp 163–182

Heideman CL, Hoff HF (1982) Lipoproteins containing apolipoprotein A-I extracted from human aortas. Biochim Biophys Acta 711:431–444

Hildebrand RB, Lammers B, Meurs I, Korporaal SJ, De Haan W, Zhao Y, Kruijt JK, Pratico D, Schimmel AW, Holleboom AG, Hoekstra M, Kuivenhoven JA, van Berkel TJ, Rensen PC, Van Eck M (2010) Restoration of high-density lipoprotein levels by cholesteryl ester transfer protein expression in scavenger receptor class B type I (SR-BI) knockout mice does not normalize pathologies associated with SR-BI deficiency. Arterioscler Thromb Vasc Biol 30: 1439–1445

Hirano K, Yamashita S, Kuga Y, Sakai N, Nozaki S, Kihara S, Arai T, Yanagi K, Takami S, Menju M et al (1995) Atherosclerotic disease in marked hyperalphalipoproteinemia. Combined reduction of cholesteryl ester transfer protein and hepatic triglyceride lipase. Arterioscler Thromb Vasc Biol 15:1849–1856

Hirano K, Yamashita S, Nakajima N, Arai T, Maruyama T, Yoshida Y, Ishigami M, Sakai N, Kameda-Takemura K, Matsuzawa Y (1997) Genetic cholesteryl ester transfer protein deficiency is extremely frequent in the Omagari area of Japan. Marked hyperalphalipoproteinemia caused by CETP gene mutation is not associated with longevity. Arterioscler Thromb Vasc Biol 17:1053–1059

Holleboom AG, Vergeer M, Hovingh GK, Kastelein JJ, Kuivenhoven JA (2008) The value of HDL genetics. Curr Opin Lipidol 19:385–394

Holm TM, Braun A, Trigatti BL, Brugnara C, Sakamoto M, Krieger M, Andrews NC (2002) Failure of red blood cell maturation in mice with defects in the high-density lipoprotein receptor SR-BI. Blood 99:1817–1824

Homanics GE, de Silva HV, Osada J, Zhang SH, Wong H, Borensztajn J, Maeda N (1995) Mild dyslipidemia in mice following targeted inactivation of the hepatic lipase gene. J Biol Chem 270:2974–2980

Hopkins GJ, Barter PJ (1980) Transfers of esterified cholesterol and triglyceride between high density and very low density lipoproteins: in vitro studies of rabbits and humans. Metabolism 29:546–550

Hu YW, Zheng L, Wang Q (2010) Characteristics of apolipoprotein M and its relation to atherosclerosis and diabetes. Biochim Biophys Acta 1801:100–105

Huang W, Sasaki J, Matsunaga A, Nanimatsu H, Moriyama K, Han H, Kugi M, Koga T, Yamaguchi K, Arakawa K (1998) A novel homozygous missense mutation in the apo A-I gene with apo A-I deficiency. Arterioscler Thromb Vasc Biol 18:389–396

Huang R, Silva RA, Jerome WG, Kontush A, Chapman MJ, Curtiss LK, Hodges TJ, Davidson WS (2011) Apolipoprotein A-I structural organization in high-density lipoproteins isolated from human plasma. Nat Struct Mol Biol 18:416–422

Huszar D, Varban ML, Rinninger F, Feeley R, Arai T, Fairchild-Huntress V, Donovan MJ, Tall AR (2000) Increased LDL cholesterol and atherosclerosis in LDL receptor-deficient mice with attenuated expression of scavenger receptor B1. Arterioscler Thromb Vasc Biol 20:1068–1073

Huuskonen J, Olkkonen VM, Ehnholm C, Metso J, Julkunen I, Jauhiainen M (2000) Phospholipid transfer is a prerequisite for PLTP-mediated HDL conversion. Biochemistry 39:16092–16098

Huuskonen J, Olkkonen VM, Jauhiainen M, Ehnholm C (2001) The impact of phospholipid transfer protein (PLTP) on HDL metabolism. Atherosclerosis 155:269–281

Inazu A, Brown ML, Hesler CB, Agellon LB, Koizumi J, Takata K, Maruhama Y, Mabuchi H, Tall AR (1990) Increased high-density lipoprotein levels caused by a common cholesteryl-ester transfer protein gene mutation. N Engl J Med 323:1234–1238

Ishida T, Choi S, Kundu RK, Hirata K, Rubin EM, Cooper AD, Quertermous T (2003) Endothelial lipase is a major determinant of HDL level. J Clin Invest 111:347–355

Ishida T, Choi SY, Kundu RK, Spin J, Yamashita T, Hirata K, Kojima Y, Yokoyama M, Cooper AD, Quertermous T (2004) Endothelial lipase modulates susceptibility to atherosclerosis in apolipoprotein-E-deficient mice. J Biol Chem 279:45085–45092

Jacquet S, Malaval C, Martinez LO, Sak K, Rolland C, Perez C, Nauze M, Champagne E, Terce F, Gachet C, Perret B, Collet X, Boeynaems JM, Barbaras R (2005) The nucleotide receptor P2Y13 is a key regulator of hepatic high-density lipoprotein (HDL) endocytosis. Cell Mol Life Sci 62:2508–2515

Janis MT, Metso J, Lankinen H, Strandin T, Olkkonen VM, Rye KA, Jauhiainen M, Ehnholm C (2005) Apolipoprotein E activates the low-activity form of human phospholipid transfer protein. Biochem Biophys Res Commun 331:333–340

Jaye M, Lynch KJ, Krawiec J, Marchadier D, Maugeais C, Doan K, South V, Amin D, Perrone M, Rader DJ (1999) A novel endothelial-derived lipase that modulates HDL metabolism. Nat Genet 21:424–428

Ji Y, Jian B, Wang N, Sun Y, Moya ML, Phillips MC, Rothblat GH, Swaney JB, Tall AR (1997) Scavenger receptor BI promotes high density lipoprotein-mediated cellular cholesterol efflux. J Biol Chem 272:20982–20985

Ji Y, Wang N, Ramakrishnan R, Sehayek E, Huszar D, Breslow JL, Tall AR (1999) Hepatic scavenger receptor BI promotes rapid clearance of high density lipoprotein free cholesterol and its transport into bile. J Biol Chem 274:33398–33402

Jiang XC, Bruce C, Mar J, Lin M, Ji Y, Francone OL, Tall AR (1999) Targeted mutation of plasma phospholipid transfer protein gene markedly reduces high-density lipoprotein levels. J Clin Invest 103:907–914

Jiang XC, Qin S, Qiao C, Kawano K, Lin M, Skold A, Xiao X, Tall AR (2001) Apolipoprotein B secretion and atherosclerosis are decreased in mice with phospholipid-transfer protein deficiency. Nat Med 7:847–852

Jiao GQ, Yuan ZX, Xue YS, Yang CJ, Lu CB, Lu ZQ, Xiao MD (2007) A prospective evaluation of apolipoprotein M gene T-778C polymorphism in relation to coronary artery disease in Han Chinese. Clin Biochem 40:1108–1112

Johannsen TH, Kamstrup PR, Andersen RV, Jensen GB, Sillesen H, Tybjaerg-Hansen A, Nordestgaard BG (2009) Hepatic lipase, genetically elevated high-density lipoprotein, and risk of ischemic cardiovascular disease. J Clin Endocrinol Metab 94:1264–1273

Jonas A (2000) Lecithin cholesterol acyltransferase. Biochim Biophys Acta 1529:245–256

Jones MK, Catte A, Li L, Segrest JP (2009) Dynamics of activation of lecithin:cholesterol acyltransferase by apolipoprotein A-I. Biochemistry 48:11196–11210

Kan CC, Yan J, Bittman R (1992) Rates of spontaneous exchange of synthetic radiolabeled sterols between lipid vesicles. Biochemistry 31:1866–1874

Karackattu SL, Picard MH, Krieger M (2005) Lymphocytes are not required for the rapid onset of coronary heart disease in scavenger receptor class B type I/apolipoprotein E double knockout mice. Arterioscler Thromb Vasc Biol 25:803–808

Karackattu SL, Trigatti B, Krieger M (2006) Hepatic lipase deficiency delays atherosclerosis, myocardial infarction, and cardiac dysfunction and extends lifespan in SR-BI/apolipoprotein E double knockout mice. Arterioscler Thromb Vasc Biol 26:548–554

Karkkainen M, Oka T, Olkkonen VM, Metso J, Hattori H, Jauhiainen M, Ehnholm C (2002) Isolation and partial characterization of the inactive and active forms of human plasma phospholipid transfer protein (PLTP). J Biol Chem 277:15413–15418

Kateifides AK, Gorshkova IN, Duka A, Chroni A, Kardassis D, Zannis VI (2011) Alteration of negatively charged residues in the 89 to 99 domain of apoA-I affects lipid homeostasis and maturation of HDL. J Lipid Res 52:1363–1372

Kathiresan S, Melander O, Guiducci C, Surti A, Burtt NP, Rieder MJ, Cooper GM, Roos C, Voight BF, Havulinna AS, Wahlstrand B, Hedner T, Corella D, Tai ES, Ordovas JM, Berglund G, Vartiainen E, Jousilahti P, Hedblad B, Taskinen MR, Newton-Cheh C, Salomaa V, Peltonen L, Groop L, Altshuler DM, Orho-Melander M (2008a) Six new loci associated with blood low-density lipoprotein cholesterol, high-density lipoprotein cholesterol or triglycerides in humans. Nat Genet 40:189–197

Kathiresan S, Musunuru K, Orho-Melander M (2008b) Defining the spectrum of alleles that contribute to blood lipid concentrations in humans. Curr Opin Lipidol 19:122–127

Kathiresan S, Willer CJ, Peloso GM, Demissie S, Musunuru K, Schadt EE, Kaplan L, Bennett D, Li Y, Tanaka T, Voight BF, Bonnycastle LL, Jackson AU, Crawford G, Surti A, Guiducci C, Burtt NP, Parish S, Clarke R, Zelenika D, Kubalanza KA, Morken MA, Scott LJ, Stringham HM, Galan P, Swift AJ, Kuusisto J, Bergman RN, Sundvall J, Laakso M, Ferrucci L, Scheet P, Sanna S, Uda M, Yang Q, Lunetta KL, Dupuis J, de Bakker PI, O'Donnell CJ, Chambers JC, Kooner JS, Hercberg S, Meneton P, Lakatta EG, Scuteri A, Schlessinger D, Tuomilehto J, Collins FS, Groop L, Altshuler D, Collins R, Lathrop GM, Melander O, Salomaa V, Peltonen L, Orho-Melander M, Ordovas JM, Boehnke M, Abecasis GR, Mohlke KL, Cupples LA (2009) Common variants at 30 loci contribute to polygenic dyslipidemia. Nat Genet 41:56–65

Kawano K, Qin SC, Lin M, Tall AR, Jiang XC (2000) Cholesteryl ester transfer protein and phospholipid transfer protein have nonoverlapping functions in vivo. J Biol Chem 275:29477–29481

Kee P, Rye KA, Taylor JL, Barrett PH, Barter PJ (2002) Metabolism of apoA-I as lipid-free protein or as component of discoidal and spherical reconstituted HDLs: studies in wild-type and hepatic lipase transgenic rabbits. Arterioscler Thromb Vasc Biol 22:1912–1917

Kennedy MA, Barrera GC, Nakamura K, Baldan A, Tarr P, Fishbein MC, Frank J, Francone OL, Edwards PA (2005) ABCG1 has a critical role in mediating cholesterol efflux to HDL and preventing cellular lipid accumulation. Cell Metab 1:121–131

Kielar D, Dietmaier W, Langmann T, Aslanidis C, Probst M, Naruszewicz M, Schmitz G (2001) Rapid quantification of human ABCA1 mRNA in various cell types and tissues by real-time reverse transcription-PCR. Clin Chem 47:2089–2097

Kivela AM, Dijkstra MH, Heinonen SE, Gurzeler E, Jauhiainen S, Levonen AL, Yla-Herttuala S (2012) Regulation of endothelial lipase and systemic HDL cholesterol levels by SREBPs and VEGF-A. Atherosclerosis 225:335–340

Klucken J, Buchler C, Orso E, Kaminski WE, Porsch-Ozcurumez M, Liebisch G, Kapinsky M, Diederich W, Drobnik W, Dean M, Allikmets R, Schmitz G (2000) ABCG1 (ABC8), the human homolog of the Drosophila white gene, is a regulator of macrophage cholesterol and phospholipid transport. Proc Natl Acad Sci USA 97:817–822

Kocher O, Yesilaltay A, Shen CH, Zhang S, Daniels K, Pal R, Chen J, Krieger M (2008) Influence of PDZK1 on lipoprotein metabolism and atherosclerosis. Biochim Biophys Acta 1782: 310–316

Koizumi J, Mabuchi H, Yoshimura A, Michishita I, Takeda M, Itoh H, Sakai Y, Sakai T, Ueda K, Takeda R (1985) Deficiency of serum cholesteryl-ester transfer activity in patients with familial hyperalphalipoproteinaemia. Atherosclerosis 58:175–186

Kostner GM, Oettl K, Jauhiainen M, Ehnholm C, Esterbauer H, Dieplinger H (1995) Human plasma phospholipid transfer protein accelerates exchange/transfer of alpha-tocopherol between lipoproteins and cells. Biochem J 305(Pt 2):659–667

Koukos G, Chroni A, Duka A, Kardassis D, Zannis VI (2007a) LCAT can rescue the abnormal phenotype produced by the natural ApoA-I mutations (Leu141Arg)Pisa and (Leu159Arg)FIN. Biochemistry 46:10713–10721

Koukos G, Chroni A, Duka A, Kardassis D, Zannis VI (2007b) Naturally occurring and bioengineered apoA-I mutations that inhibit the conversion of discoidal to spherical HDL: the abnormal HDL phenotypes can be corrected by treatment with LCAT. Biochem J 406: 167–174

Kozarsky KF, Donahee MH, Glick JM, Krieger M, Rader DJ (2000) Gene transfer and hepatic overexpression of the HDL receptor SR-BI reduces atherosclerosis in the cholesterol-fed LDL receptor-deficient mouse. Arterioscler Thromb Vasc Biol 20:721–727

Kozyraki R (2001) Cubilin, a multifunctional epithelial receptor: an overview. J Mol Med (Berl) 79:161–167

Kozyraki R, Fyfe J, Kristiansen M, Gerdes C, Jacobsen C, Cui S, Christensen EI, Aminoff M, de la Chapelle A, Krahe R, Verroust PJ, Moestrup SK (1999) The intrinsic factor-vitamin B12 receptor, cubilin, is a high-affinity apolipoprotein A-I receptor facilitating endocytosis of high-density lipoprotein. Nat Med 5:656–661

Krauss RM, Levy RI, Fredrickson DS (1974) Selective measurement of two lipase activities in postheparin plasma from normal subjects and patients with hyperlipoproteinemia. J Clin Invest 54:1107–1124

Krieger M (1999) Charting the fate of the "good cholesterol": identification and characterization of the high-density lipoprotein receptor SR-BI. Annu Rev Biochem 68:523–558

Krieger M (2001) Scavenger receptor class B type I is a multiligand HDL receptor that influences diverse physiologic systems. J Clin Invest 108:793–797

Krimbou L, Hajj HH, Blain S, Rashid S, Denis M, Marcil M, Genest J (2005) Biogenesis and speciation of nascent apoA-I-containing particles in various cell lines. J Lipid Res 46: 1668–1677

Kypreos KE, Zannis VI (2007) Pathway of biogenesis of apolipoprotein E-containing HDL in vivo with the participation of ABCA1 and LCAT. Biochem J 403:359–367

Landmesser U, von Eckardstein A, Kastelein J, Deanfield J, Luscher TF (2012) Increasing high-density lipoprotein cholesterol by cholesteryl ester transfer protein-inhibition: a rocky road and lessons learned? The early demise of the dal-HEART programme. Eur Heart J 33:1712–1715

Landschulz KT, Pathak RK, Rigotti A, Krieger M, Hobbs HH (1996) Regulation of scavenger receptor, class B, type I, a high density lipoprotein receptor, in liver and steroidogenic tissues of the rat. J Clin Invest 98:984–995

Langmann T, Klucken J, Reil M, Liebisch G, Luciani MF, Chimini G, Kaminski WE, Schmitz G (1999) Molecular cloning of the human ATP-binding cassette transporter 1 (hABC1): evidence for sterol-dependent regulation in macrophages. Biochem Biophys Res Commun 257:29–33

Laurila PP, Surakka I, Sarin AP, Yetukuri L, Hyotylainen T, Soderlund S, Naukkarinen J, Tang J, Kettunen J, Mirel DB, Soronen J, Lehtimaki T, Ruokonen A, Ehnholm C, Eriksson JG, Salomaa V, Jula A, Raitakari OT, Jarvelin MR, Palotie A, Peltonen L, Oresic M, Jauhiainen M, Taskinen MR, Ripatti S (2013) Genomic, transcriptomic, and lipidomic profiling highlights the role of inflammation in individuals with low high-density lipoprotein cholesterol. Arterioscler Thromb Vasc Biol 33:847–857

Leiva A, Verdejo H, Benitez ML, Martinez A, Busso D, Rigotti A (2011) Mechanisms regulating hepatic SR-BI expression and their impact on HDL metabolism. Atherosclerosis 217:299–307

Leren TP, Bakken KS, Daum U, Ose L, Berg K, Assmann G, von Eckardstein A (1997) Heterozygosity for apolipoprotein A-I(R160L)Oslo is associated with low levels of high density lipoprotein cholesterol and HDL-subclass LpA-I/A- II but normal levels of HDL-subclass LpA-I. J Lipid Res 38:121–131

Levels JH, Marquart JA, Abraham PR, van den Ende AE, Molhuizen HO, van Deventer SJ, Meijers JC (2005) Lipopolysaccharide is transferred from high-density to low-density lipoproteins by lipopolysaccharide-binding protein and phospholipid transfer protein. Infect Immun 73:2321–2326

Li XA, Titlow WB, Jackson BA, Giltiay N, Nikolova-Karakashian M, Uittenbogaard A, Smart EJ (2002) High density lipoprotein binding to scavenger receptor, Class B, type I activates endothelial nitric-oxide synthase in a ceramide-dependent manner. J Biol Chem 277: 11058–11063

Li X, Kypreos K, Zanni EE, Zannis V (2003) Domains of apoE required for binding to apoE receptor 2 and to phospholipids: implications for the functions of apoE in the brain. Biochemistry 42:10406–10417

Li D, Zhang Y, Ma J, Ling W, Xia M (2010) Adenosine monophosphate activated protein kinase regulates ABCG1-mediated oxysterol efflux from endothelial cells and protects against hypercholesterolemia-induced endothelial dysfunction. Arterioscler Thromb Vasc Biol 30: 1354–1362

Liadaki KN, Liu T, Xu S, Ishida BY, Duchateaux PN, Krieger JP, Kane J, Krieger M, Zannis VI (2000) Binding of high density lipoprotein (HDL) and discoidal reconstituted HDL to the HDL receptor scavenger receptor class B type I. Effect of lipid association and APOA-I mutations on receptor binding. J Biol Chem 275:21262–21271

Lindstedt L, Saarinen J, Kalkkinen N, Welgus H, Kovanen PT (1999) Matrix metalloproteinases-3, -7, and -12, but not -9, reduce high density lipoprotein-induced cholesterol efflux from human macrophage foam cells by truncation of the carboxyl terminus of apolipoprotein A-I. Parallel losses of pre-beta particles and the high affinity component of efflux. J Biol Chem 274:22627–22634

Liu T, Krieger M, Kan HY, Zannis VI (2002) The effects of mutations in helices 4 and 6 of apoA-I on scavenger receptor class B type I (SR-BI)-mediated cholesterol efflux suggest that formation of a productive complex between reconstituted high density lipoprotein and SR-BI is required for efficient lipid transport. J Biol Chem 277:21576–21584

Lorenzi I, von Eckardstein A, Cavelier C, Radosavljevic S, Rohrer L (2008) Apolipoprotein A-I but not high-density lipoproteins are internalised by RAW macrophages: roles of ATP-binding cassette transporter A1 and scavenger receptor BI. J Mol Med 86:171–183

Love-Gregory L, Sherva R, Sun L, Wasson J, Schappe T, Doria A, Rao DC, Hunt SC, Klein S, Neuman RJ, Permutt MA, Abumrad NA (2008) Variants in the CD36 gene associate with the metabolic syndrome and high-density lipoprotein cholesterol. Hum Mol Genet 17:1695–1704

Lusa S, Jauhiainen M, Metso J, Somerharju P, Ehnholm C (1996) The mechanism of human plasma phospholipid transfer protein-induced enlargement of high-density lipoprotein particles: evidence for particle fusion. Biochem J 313(Pt 1):275–282

Ma K, Cilingiroglu M, Otvos JD, Ballantyne CM, Marian AJ, Chan L (2003) Endothelial lipase is a major genetic determinant for high-density lipoprotein concentration, structure, and metabolism. Proc Natl Acad Sci USA 100:2748–2753

Mardones P, Quinones V, Amigo L, Moreno M, Miquel JF, Schwarz M, Miettinen HE, Trigatti B, Krieger M, VanPatten S, Cohen DE, Rigotti A (2001) Hepatic cholesterol and bile acid metabolism and intestinal cholesterol absorption in scavenger receptor class B type I-deficient mice. J Lipid Res 42:170–180

Marsh JB, Welty FK, Schaefer EJ (2000) Stable isotope turnover of apolipoproteins of high-density lipoproteins in humans. Curr Opin Lipidol 11:261–266

Martinez LO, Jacquet S, Esteve JP, Rolland C, Cabezon E, Champagne E, Pineau T, Georgeaud V, Walker JE, Terce F, Collet X, Perret B, Barbaras R (2003) Ectopic beta-chain of ATP synthase is an apolipoprotein A-I receptor in hepatic HDL endocytosis. Nature 421:75–79

Martinez LO, Perretv B, Barbaras R, Terce F, Collet X (2007) Hepatic and renal HDL receptors. In: Fielding CJ (ed) High-density lipoproteins. From basic biology to clinical aspects. Wiley, Weinheim, pp 307–338

Maruyama T, Sakai N, Ishigami M, Hirano K, Arai T, Okada S, Okuda E, Ohya A, Nakajima N, Kadowaki K, Fushimi E, Yamashita S, Matsuzawa Y (2003) Prevalence and phenotypic spectrum of cholesteryl ester transfer protein gene mutations in Japanese hyperalphalipoproteinemia. Atherosclerosis 166:177–185

Matsuura F, Wang N, Chen W, Jiang XC, Tall AR (2006) HDL from CETP-deficient subjects shows enhanced ability to promote cholesterol efflux from macrophages in an apoE- and ABCG1-dependent pathway. J Clin Invest 116:1435–1442

Maugeais C, Tietge UJ, Broedl UC, Marchadier D, Cain W, McCoy MG, Lund-Katz S, Glick JM, Rader DJ (2003) Dose-dependent acceleration of high-density lipoprotein catabolism by endothelial lipase. Circulation 108:2121–2126

McManus DC, Scott BR, Franklin V, Sparks DL, Marcel YL (2001) Proteolytic degradation and impaired secretion of an apolipoprotein A-I mutant associated with dominantly inherited hypoalphalipoproteinemia. J Biol Chem 276:21292–21302

McNeish J, Aiello RJ, Guyot D, Turi T, Gabel C, Aldinger C, Hoppe KL, Roach ML, Royer LJ, de Wet J, Broccardo C, Chimini G, Francone OL (2000) High density lipoprotein deficiency and foam cell accumulation in mice with targeted disruption of ATP-binding cassette transporter-1. Proc Natl Acad Sci USA 97:4245–4250

Melchior GW, Castle CK, Murray RW, Blake WL, Dinh DM, Marotti KR (1994) Apolipoprotein A-I metabolism in cholesteryl ester transfer protein transgenic mice. Insights into the mechanisms responsible for low plasma high density lipoprotein levels. J Biol Chem 269: 8044–8051

Meyer JM, Graf GA, van der Westhuyzen DR (2013) New developments in selective cholesteryl ester uptake. Curr Opin Lipidol 24:386–392

Mezdour H, Jones R, Dengremont C, Castro G, Maeda N (1997) Hepatic lipase deficiency increases plasma cholesterol but reduces susceptibility to atherosclerosis in apolipoprotein E-deficient mice. J Biol Chem 272:13570–13575

Miccoli R, Bertolotto A, Navalesi R, Odoguardi L, Boni A, Wessling J, Funke H, Wiebusch H, Eckardstein A, Assmann G (1996) Compound heterozygosity for a structural apolipoprotein A-I variant, apo A-I(L141R)Pisa, and an apolipoprotein A-I null allele in patients with absence of HDL cholesterol, corneal opacifications, and coronary heart disease. Circulation 94: 1622–1628

Miccoli R, Zhu Y, Daum U, Wessling J, Huang Y, Navalesi R, Assmann G, von Eckardstein A (1997) A natural apolipoprotein A-I variant, apoA-I (L141R)Pisa, interferes with the formation of alpha-high density lipoproteins (HDL) but not with the formation of pre beta 1-HDL and influences efflux of cholesterol into plasma. J Lipid Res 38:1242

Miettinen HE, Gylling H, Miettinen TA, Viikari J, Paulin L, Kontula K (1997a) Apolipoprotein A-IFin. Dominantly inherited hypoalphalipoproteinemia due to a single base substitution in the apolipoprotein A-I gene. Arterioscler Thromb Vasc Biol 17:83–90

Miettinen HE, Jauhiainen M, Gylling H, Ehnholm S, Palomaki A, Miettinen TA, Kontula K (1997b) Apolipoprotein A-IFIN (Leu159−>Arg) mutation affects lecithin cholesterol acyltransferase activation and subclass distribution of HDL but not cholesterol efflux from fibroblasts. Arterioscler Thromb Vasc Biol 17:3021–3032

Miller M, Aiello D, Pritchard H, Friel G, Zeller K (1998) Apolipoprotein A-I(Zavalla) (Leu159−>Pro): HDL cholesterol deficiency in a kindred associated with premature coronary artery disease. Arterioscler Thromb Vasc Biol 18:1242–1247

Mineo C, Yuhanna IS, Quon MJ, Shaul PW (2003) High density lipoprotein-induced endothelial nitric-oxide synthase activation is mediated by Akt and MAP kinases. J Biol Chem 278: 9142–9149

Moestrup SK, Nielsen LB (2005) The role of the kidney in lipid metabolism. Curr Opin Lipidol 16: 301–306

Mulya A, Seo J, Brown AL, Gebre AK, Boudyguina E, Shelness GS, Parks JS (2010) Apolipoprotein M expression increases the size of nascent pre beta HDL formed by ATP binding cassette transporter A1. J Lipid Res 51:514–524

Murao K, Terpstra V, Green SR, Kondratenko N, Steinberg D, Quehenberger O (1997) Characterization of CLA-1, a human homologue of rodent scavenger receptor BI, as a receptor for high density lipoprotein and apoptotic thymocytes. J Biol Chem 272:17551–17557

Nakamura K, Kennedy MA, Baldan A, Bojanic DD, Lyons K, Edwards PA (2004a) Expression and regulation of multiple murine ATP-binding cassette transporter G1 mRNAs/isoforms that stimulate cellular cholesterol efflux to high density lipoprotein. J Biol Chem 279:45980–45989

Nakamura Y, Kotite L, Gan Y, Spencer TA, Fielding CJ, Fielding PE (2004b) Molecular mechanism of reverse cholesterol transport: reaction of pre-beta-migrating high-density lipoprotein with plasma lecithin/cholesterol acyltransferase. Biochemistry 43:14811–14820

Nanjee MN, Cooke CJ, Olszewski WL, Miller NE (2000) Concentrations of electrophoretic and size subclasses of apolipoprotein A-I-containing particles in human peripheral lymph. Arterioscler Thromb Vasc Biol 20:2148–2155

Navab M, Hama SY, Anantharamaiah GM, Hassan K, Hough GP, Watson AD, Reddy ST, Sevanian A, Fonarow GC, Fogelman AM (2000) Normal high density lipoprotein inhibits three steps in the formation of mildly oxidized low density lipoprotein: steps 2 and 3. J Lipid Res 41:1495–1508

Neufeld EB, Remaley AT, Demosky SJ, Stonik JA, Cooney AM, Comly M, Dwyer NK, Zhang M, Blanchette-Mackie J, Santamarina-Fojo S, Brewer HB Jr (2001) Cellular localization and trafficking of the human ABCA1 transporter. J Biol Chem 276:27584–27590

Neufeld EB, Demosky SJ Jr, Stonik JA, Combs C, Remaley AT, Duverger N, Santamarina-Fojo S, Brewer HB Jr (2002) The ABCA1 transporter functions on the basolateral surface of hepatocytes. Biochem Biophys Res Commun 297:974–979

Nichols AV, Krauss RM, Musliner TA (1986) Nondenaturing polyacrylamide gradient gel electrophoresis. Methods Enzymol 128:417–431

Nielsen LB, Christoffersen C, Ahnstrom J, Dahlback B (2009) ApoM: gene regulation and effects on HDL metabolism. Trends Endocrinol Metab 20:66–71

Nijstad N, Wiersma H, Gautier T, van der Giet M, Maugeais C, Tietge UJ (2009) Scavenger receptor BI-mediated selective uptake is required for the remodeling of high density lipoprotein by endothelial lipase. J Biol Chem 284:6093–6100

Nishida HI, Nishida T (1997) Phospholipid transfer protein mediates transfer of not only phosphatidylcholine but also cholesterol from phosphatidylcholine-cholesterol vesicles to high density lipoproteins. J Biol Chem 272:6959–6964

Nolte RT, Atkinson D (1992) Conformational analysis of apolipoprotein A-I and E-3 based on primary sequence and circular dichroism. Biophys J 63:1221–1239

Ohnsorg PM, Rohrer L, Perisa D, Kateifides A, Chroni A, Kardassis D, Zannis VI, von Eckardstein A (2011) Carboxyl terminus of apolipoprotein A-I (ApoA-I) is necessary for the transport of lipid-free ApoA-I but not prelipidated ApoA-I particles through aortic endothelial cells. J Biol Chem 286:7744–7754

Oka T, Kujiraoka T, Ito M, Egashira T, Takahashi S, Nanjee MN, Miller NE, Metso J, Olkkonen VM, Ehnholm C, Jauhiainen M, Hattori H (2000) Distribution of phospholipid transfer protein in human plasma: presence of two forms of phospholipid transfer protein, one catalytically active and the other inactive. J Lipid Res 41:1651–1657

Orso E, Broccardo C, Kaminski WE, Bottcher A, Liebisch G, Drobnik W, Gotz A, Chambenoit O, Diederich W, Langmann T, Spruss T, Luciani MF, Rothe G, Lackner KJ, Chimini G,

Schmitz G (2000) Transport of lipids from golgi to plasma membrane is defective in tangier disease patients and Abc1-deficient mice. Nat Genet 24:192–196

Out R, Hoekstra M, Spijkers JA, Kruijt JK, Van Eck M, Bos IS, Twisk J, van Berkel TJ (2004a) Scavenger receptor class B type I is solely responsible for the selective uptake of cholesteryl esters from HDL by the liver and the adrenals in mice. J Lipid Res 45:2088–2095

Out R, Kruijt JK, Rensen PC, Hildebrand RB, De Vos P, Van Eck M, van Berkel TJ (2004b) Scavenger receptor BI plays a role in facilitating chylomicron metabolism. J Biol Chem 279: 18401–18406

Out R, Hoekstra M, de Jager SC, Vos PD, van der Westhuyzen DR, Webb NR, Van Eck M, Biessen EA, van Berkel TJ (2005) Adenovirus-mediated hepatic overexpression of scavenger receptor BI accelerates chylomicron metabolism in C57BL/6J mice. J Lipid Res 46:1172–1181

Out R, Hoekstra M, Hildebrand RB, Kruit JK, Meurs I, Li Z, Kuipers F, van Berkel TJ, Van Eck M (2006) Macrophage ABCG1 deletion disrupts lipid homeostasis in alveolar macrophages and moderately influences atherosclerotic lesion development in LDL receptor-deficient mice. Arterioscler Thromb Vasc Biol 26:2295–2300

Out R, Hoekstra M, Meurs I, de Vos P, Kuiper J, Van Eck M, van Berkel TJ (2007) Total body ABCG1 expression protects against early atherosclerotic lesion development in mice. Arterioscler Thromb Vasc Biol 27:594–599

Pagler TA, Rhode S, Neuhofer A, Laggner H, Strobl W, Hinterndorfer C, Volf I, Pavelka M, Eckhardt ER, van der Westhuyzen DR, Schutz GJ, Stangl H (2006) SR-BI-mediated high density lipoprotein (HDL) endocytosis leads to HDL resecretion facilitating cholesterol efflux. J Biol Chem 281:11193–11204

Parathath S, Sahoo D, Darlington YF, Peng Y, Collins HL, Rothblat GH, Williams DL, Connelly MA (2004) Glycine 420 near the C-terminal transmembrane domain of SR-BI is critical for proper delivery and metabolism of high density lipoprotein cholesteryl ester. J Biol Chem 279:24976–24985

Paszty C, Maeda N, Verstuyft J, Rubin EM (1994) Apolipoprotein AI transgene corrects apolipoprotein E deficiency-induced atherosclerosis in mice. J Clin Invest 94:899–903

Pisciotta L, Miccoli R, Cantafora A, Calabresi L, Tarugi P, Alessandrini P, Bittolo BG, Franceschini G, Cortese C, Calandra S, Bertolini S (2003) Recurrent mutations of the apolipoprotein A-I gene in three kindreds with severe HDL deficiency. Atherosclerosis 167:335–345

Plomgaard P, Dullaart RP, de Vries R, Groen AK, Dahlback B, Nielsen LB (2009) Apolipoprotein M predicts pre-beta-HDL formation: studies in type 2 diabetic and nondiabetic subjects. J Intern Med 266:258–267

Plump AS, Smith JD, Hayek T, Aalto-Setala K, Walsh A, Verstuyft JG, Rubin EM, Breslow JL (1992) Severe hypercholesterolemia and atherosclerosis in apolipoprotein E-deficient mice created by homologous recombination in ES cells. Cell 71:343–353

Plump AS, Masucci-Magoulas L, Bruce C, Bisgaier CL, Breslow JL, Tall AR (1999) Increased atherosclerosis in ApoE and LDL receptor gene knock-out mice as a result of human cholesteryl ester transfer protein transgene expression. Arterioscler Thromb Vasc Biol 19: 1105–1110

Ranalletta M, Wang N, Han S, Yvan-Charvet L, Welch C, Tall AR (2006) Decreased atherosclerosis in low-density lipoprotein receptor knockout mice transplanted with Abcg1$^{-/-}$ bone marrow. Arterioscler Thromb Vasc Biol 26:2308–2315

Rao R, Albers JJ, Wolfbauer G, Pownall HJ (1997) Molecular and macromolecular specificity of human plasma phospholipid transfer protein. Biochemistry 36:3645–3653

Reardon CA, Kan HY, Cabana V, Blachowicz L, Lukens JR, Wu Q, Liadaki K, Getz GS, Zannis VI (2001) In vivo studies of HDL assembly and metabolism using adenovirus-mediated transfer of ApoA-I mutants in ApoA-I-deficient mice. Biochemistry 40:13670–13680

Remaley AT, Stonik JA, Demosky SJ, Neufeld EB, Bocharov AV, Vishnyakova TG, Eggerman TL, Patterson AP, Duverger NJ, Santamarina-Fojo S, Brewer HB Jr (2001) Apolipoprotein specificity for lipid efflux by the human ABCAI transporter. Biochem Biophys Res Commun 280:818–823

Richards JB, Waterworth D, O'Rahilly S, Hivert MF, Loos RJ, Perry JR, Tanaka T, Timpson NJ, Semple RK, Soranzo N, Song K, Rocha N, Grundberg E, Dupuis J, Florez JC, Langenberg C, Prokopenko I, Saxena R, Sladek R, Aulchenko Y, Evans D, Waeber G, Erdmann J, Burnett MS, Sattar N, Devaney J, Willenborg C, Hingorani A, Witteman JC, Vollenweider P, Glaser B, Hengstenberg C, Ferrucci L, Melzer D, Stark K, Deanfield J, Winogradow J, Grassl M, Hall AS, Egan JM, Thompson JR, Ricketts SL, Konig IR, Reinhard W, Grundy S, Wichmann HE, Barter P, Mahley R, Kesaniemi YA, Rader DJ, Reilly MP, Epstein SE, Stewart AF, van Duijn CM, Schunkert H, Burling K, Deloukas P, Pastinen T, Samani NJ, McPherson R, Davey SG, Frayling TM, Wareham NJ, Meigs JB, Mooser V, Spector TD (2009) A genome-wide association study reveals variants in ARL15 that influence adiponectin levels. PLoS Genet 5:e1000768

Rigotti A, Trigatti BL, Penman M, Rayburn H, Herz J, Krieger M (1997) A targeted mutation in the murine gene encoding the high density lipoprotein (HDL) receptor scavenger receptor class B type I reveals its key role in HDL metabolism. Proc Natl Acad Sci USA 94:12610–12615

Rohrer L, Hersberger M, von Eckardstein A (2004) High density lipoproteins in the intersection of diabetes mellitus, inflammation and cardiovascular disease. Curr Opin Lipidol 15:269–278

Rohrer L, Cavelier C, Fuchs S, Schluter MA, Volker W, von Eckardstein A (2006) Binding, internalization and transport of apolipoprotein A-I by vascular endothelial cells. Biochim Biophys Acta 1761:186–194

Rohrer L, Ohnsorg PM, Lehner M, Landolt F, Rinninger F, von Eckardstein A (2009) High-density lipoprotein transport through aortic endothelial cells involves scavenger receptor BI and ATP-binding cassette transporter G1. Circ Res 104:1142–1150

Rohrl C, Stangl H (2013) HDL endocytosis and resecretion. Biochim Biophys Acta 1831: 1626–1633

Roosbeek S, Vanloo B, Duverger N, Caster H, Breyne J, De Beun I, Patel H, Vandekerckhove J, Shoulders C, Rosseneu M, Peelman F (2001) Three arginine residues in apolipoprotein A-I are critical for activation of lecithin:cholesterol acyltransferase. J Lipid Res 42:31–40

Rubin EM, Krauss RM, Spangler EA, Verstuyft JG, Clift SM (1991) Inhibition of early atherogenesis in transgenic mice by human apolipoprotein AI. Nature 353:265–267

Ruel IL, Couture P, Cohn JS, Bensadoun A, Marcil M, Lamarche B (2004) Evidence that hepatic lipase deficiency in humans is not associated with proatherogenic changes in HDL composition and metabolism. J Lipid Res 45:1528–1537

Russell RE, Culpitt SV, DeMatos C, Donnelly L, Smith M, Wiggins J, Barnes PJ (2002) Release and activity of matrix metalloproteinase-9 and tissue inhibitor of metalloproteinase-1 by alveolar macrophages from patients with chronic obstructive pulmonary disease. Am J Respir Cell Mol Biol 26:602–609

Rye KA, Jauhiainen M, Barter PJ, Ehnholm C (1998) Triglyceride-enrichment of high density lipoproteins enhances their remodelling by phospholipid transfer protein. J Lipid Res 39: 613–622

Sabatti C, Service SK, Hartikainen AL, Pouta A, Ripatti S, Brodsky J, Jones CG, Zaitlen NA, Varilo T, Kaakinen M, Sovio U, Ruokonen A, Laitinen J, Jakkula E, Coin L, Hoggart C, Collins A, Turunen H, Gabriel S, Elliot P, McCarthy MI, Daly MJ, Jarvelin MR, Freimer NB, Peltonen L (2009) Genome-wide association analysis of metabolic traits in a birth cohort from a founder population. Nat Genet 41:35–46

Sankaranarayanan S, Oram JF, Asztalos BF, Vaughan AM, Lund-Katz S, Adorni MP, Phillips MC, Rothblat GH (2009) Effects of acceptor composition and mechanism of ABCG1-mediated cellular free cholesterol efflux. J Lipid Res 50:275–284

Santamarina-Fojo S, Hoeg JM, Assmann G, Brewer HB Jr (2001) Lecithin cholesterol acyltransferase deficiency and fish eye disease. In: Scriver CR, Beaudet AL, Sly WS, Valle D (eds) The metabolic and molecular bases of inherited disease. McGraw-Hill, New York, pp 2817–2834

Santamarina-Fojo S, Gonzalez-Navarro H, Freeman L, Wagner E, Nong Z (2004) Hepatic lipase, lipoprotein metabolism, and atherogenesis. Arterioscler Thromb Vasc Biol 24:1750–1754

Savary S, Denizot F, Luciani M, Mattei M, Chimini G (1996) Molecular cloning of a mammalian ABC transporter homologous to Drosophila white gene. Mamm Genome 7:673–676

Schaefer EJ, Foster DM, Jenkins LL, Lindgren FT, Berman M, Levy RI, Brewer HB Jr (1979) The composition and metabolism of high density lipoprotein subfractions. Lipids 14:511–522

Schaefer EJ, Anderson DW, Zech LA, Lindgren FT, Bronzert TB, Rubalcaba EA, Brewer HB Jr (1981) Metabolism of high density lipoprotein subfractions and constituents in Tangier disease following the infusion of high density lipoproteins. J Lipid Res 22:217–228

Schaefer EJ, Gregg RE, Ghiselli G, Forte TM, Ordovas JM, Zech LA, Brewer HB Jr (1986) Familial apolipoprotein E deficiency. J Clin Invest 78:1206–1219

Schou J, Frikke-Schmidt R, Kardassis D, Thymiakou E, Nordestgaard BG, Jensen G, Grande P, Tybjaerg-Hansen A (2012) Genetic variation in ABCG1 and risk of myocardial infarction and ischemic heart disease. Arterioscler Thromb Vasc Biol 32:506–515

Segrest JP, Jackson RL, Morrisett JD, Gotto AM Jr (1974) A molecular theory of lipid-protein interactions in the plasma lipoproteins. FEBS Lett 38:247–258

Segrest JP, Jones MK, Klon AE, Sheldahl CJ, Hellinger M, De Loof H, Harvey SC (1999) A detailed molecular belt model for apolipoprotein A-I in discoidal high density lipoprotein. J Biol Chem 274:31755–31758

Setala NL, Holopainen JM, Metso J, Wiedmer SK, Yohannes G, Kinnunen PK, Ehnholm C, Jauhiainen M (2007) Interfacial and lipid transfer properties of human phospholipid transfer protein: implications for the transfer mechanism of phospholipids. Biochemistry 46: 1312–1319

Shah AS, Tan L, Long JL, Davidson WS (2013) Proteomic diversity of high density lipoproteins: our emerging understanding of its importance in lipid transport and beyond. J Lipid Res 54: 2575–2585

Shao B, Cavigiolio G, Brot N, Oda MN, Heinecke JW (2008) Methionine oxidation impairs reverse cholesterol transport by apolipoprotein A-I. Proc Natl Acad Sci USA 105: 12224–12229

Shih DQ, Bussen M, Sehayek E, Ananthanarayanan M, Shneider BL, Suchy FJ, Shefer S, Bollileni JS, Gonzalez FJ, Breslow JL, Stoffel M (2001) Hepatocyte nuclear factor-1alpha is an essential regulator of bile acid and plasma cholesterol metabolism. Nat Genet 27:375–382

Siggins S, Rye K-A, Olkkonen VM, Jauhiainen M, Ehnholm C (2007) Human plasma phospholipid transfer protein (PLTP) – structural and functional features. In: Fielding CJ (ed) Highdensity lipoproteins. From basic biology to clinical aspects. Wiley, Weinheim, pp 183–206

Silva RA, Huang R, Morris J, Fang J, Gracheva EO, Ren G, Kontush A, Jerome WG, Rye KA, Davidson WS (2008) Structure of apolipoprotein A-I in spherical high density lipoproteins of different sizes. Proc Natl Acad Sci USA 105:12176–12181

Simon JB, Boyer JL (1970) Production of lecithin: cholesterol acyltransferase by the isolated perfused rat liver. Biochim Biophys Acta 218:549–551

Singaraja RR, Brunham LR, Visscher H, Kastelein JJ, Hayden MR (2003) Efflux and atherosclerosis: the clinical and biochemical impact of variations in the ABCA1 gene. Arterioscler Thromb Vasc Biol 23:1322–1332

Singaraja RR, Stahmer B, Brundert M, Merkel M, Heeren J, Bissada N, Kang M, Timmins JM, Ramakrishnan R, Parks JS, Hayden MR, Rinninger F (2006) Hepatic ATP-binding cassette transporter A1 is a key molecule in high-density lipoprotein cholesteryl ester metabolism in mice. Arterioscler Thromb Vasc Biol 26:1821–1827

Singaraja RR, Sivapalaratnam S, Hovingh K, Dube MP, Castro-Perez J, Collins HL, Adelman SJ, Riwanto M, Manz J, Hubbard B, Tietjen I, Wong K, Mitnaul LJ, van Heek M, Lin L, Roddy TA, McEwen J, Dallinge-Thie G, van Vark-van der Zee L, Verwoert G, Winther M, van Duijn C, Hofman A, Trip MD, Marais AD, Asztalos B, Landmesser U, Sijbrands E, Kastelein JJ, Hayden MR (2013) The impact of partial and complete loss-of-function mutations in endothelial lipase on high-density lipoprotein levels and functionality in humans. Circ Cardiovasc Genet 6:54–62

Smith EB, Ashall C, Walker JE (1984) High density lipoprotein (HDL) subfractions in interstitial fluid from human aortic intima and atherosclerotic lesions. Biochem Soc Trans 12:843–844

Smith JD, Waelde C, Horwitz A, Zheng P (2002) Evaluation of the role of phosphatidylserine translocase activity in ABCA1-mediated lipid efflux. J Biol Chem 277:17797–17803

Sorci-Thomas MG, Thomas MJ (2002) The effects of altered apolipoprotein A-I structure on plasma HDL concentration. Trends Cardiovasc Med 12:121–128

Stangl H, Hyatt M, Hobbs HH (1999) Transport of lipids from high and low density lipoproteins via scavenger receptor-BI. J Biol Chem 274:32692–32698

Stefulj J, Panzenboeck U, Becker T, Hirschmugl B, Schweinzer C, Lang I, Marsche G, Sadjak A, Lang U, Desoye G, Wadsack C (2009) Human endothelial cells of the placental barrier efficiently deliver cholesterol to the fetal circulation via ABCA1 and ABCG1. Circ Res 104: 600–608

Strope S, Rivi R, Metzger T, Manova K, Lacy E (2004) Mouse amnionless, which is required for primitive streak assembly, mediates cell-surface localization and endocytic function of cubilin on visceral endoderm and kidney proximal tubules. Development 131:4787–4795

Sturek JM, Castle JD, Trace AP, Page LC, Castle AM, Evans-Molina C, Parks JS, Mirmira RG, Hedrick CC (2010) An intracellular role for ABCG1-mediated cholesterol transport in the regulated secretory pathway of mouse pancreatic beta cells. J Clin Invest 120:2575–2589

Subbaiah PV, Albers JJ, Chen CH, Bagdade JD (1980) Low density lipoprotein-activated lysolecithin acylation by human plasma lecithin-cholesterol acyltransferase. Identity of lysolecithin acyltransferase and lecithin-cholesterol acyltransferase. J Biol Chem 255:9275–9280

Takahashi Y, Smith JD (1999) Cholesterol efflux to apolipoprotein AI involves endocytosis and resecretion in a calcium-dependent pathway. Proc Natl Acad Sci USA 96:11358–11363

Tall AR, Forester LR, Bongiovanni GL (1983) Facilitation of phosphatidylcholine transfer into high density lipoproteins by an apolipoprotein in the density 1.20-1.26 g/ml fraction of plasma. J Lipid Res 24:277–289

Tangirala RK, Tsukamoto K, Chun SH, Usher D, Pure E, Rader DJ (1999) Regression of atherosclerosis induced by liver-directed gene transfer of apolipoprotein A-I in mice. Circulation 100:1816–1822

Tarling EJ, Edwards PA (2011) ATP binding cassette transporter G1 (ABCG1) is an intracellular sterol transporter. Proc Natl Acad Sci USA 108:19719–19724

Terasaka N, Wang N, Yvan-Charvet L, Tall AR (2007) High-density lipoprotein protects macrophages from oxidized low-density lipoprotein-induced apoptosis by promoting efflux of 7-ketocholesterol via ABCG1. Proc Natl Acad Sci USA 104:15093–15098

Teslovich TM et al (2010) Biological, clinical and population relevance of 95 loci for blood lipids. Nature 466:707–713

Thuahnai ST, Lund-Katz S, Williams DL, Phillips MC (2001) Scavenger receptor class B, type I-mediated uptake of various lipids into cells. Influence of the nature of the donor particle interaction with the receptor. J Biol Chem 276:43801–43808

Timmins JM, Lee JY, Boudyguina E, Kluckman KD, Brunham LR, Mulya A, Gebre AK, Coutinho JM, Colvin PL, Smith TL, Hayden MR, Maeda N, Parks JS (2005) Targeted inactivation of hepatic Abca1 causes profound hypoalphalipoproteinemia and kidney hypercatabolism of apoA-I. J Clin Invest 115:1333–1342

Tollefson JH, Ravnik S, Albers JJ (1988) Isolation and characterization of a phospholipid transfer protein (LTP-II) from human plasma. J Lipid Res 29:1593–1602

Trigatti B, Rayburn H, Vinals M, Braun A, Miettinen H, Penman M, Hertz M, Schrenzel M, Amigo L, Rigotti A, Krieger M (1999) Influence of the high density lipoprotein receptor SR-BI on reproductive and cardiovascular pathophysiology. Proc Natl Acad Sci USA 96:9322–9327

Ueda Y, Royer L, Gong E, Zhang J, Cooper PN, Francone O, Rubin EM (1999) Lower plasma levels and accelerated clearance of high density lipoprotein (HDL) and non-HDL cholesterol in scavenger receptor class B type I transgenic mice. J Biol Chem 274:7165–7171

Ueda Y, Gong E, Royer L, Cooper PN, Francone OL, Rubin EM (2000) Relationship between expression levels and atherogenesis in scavenger receptor class B, type I transgenics. J Biol Chem 275:20368–20373

Undurti A, Huang Y, Lupica JA, Smith JD, Didonato JA, Hazen SL (2009) Modification of high density lipoprotein by myeloperoxidase generates a pro-inflammatory particle. J Biol Chem 284:30825–30835

Urban S, Zieseniss S, Werder M, Hauser H, Budzinski R, Engelmann B (2000) Scavenger receptor BI transfers major lipoprotein-associated phospholipids into the cells. J Biol Chem 275: 33409–33415

Vaisar T, Pennathur S, Green PS, Gharib SA, Hoofnagle AN, Cheung MC, Byun J, Vuletic S, Kassim S, Singh P, Chea H, Knopp RH, Brunzell J, Geary R, Chait A, Zhao XQ, Elkon K, Marcovina S, Ridker P, Oram JF, Heinecke JW (2007) Shotgun proteomics implicates protease inhibition and complement activation in the antiinflammatory properties of HDL. J Clin Invest 117:746–756

Vaisman BL, Lambert G, Amar M, Joyce C, Ito T, Shamburek RD, Cain WJ, Fruchart-Najib J, Neufeld ED, Remaley AT, Brewer HB Jr, Santamarina-Fojo S (2001) ABCA1 overexpression leads to hyperalphalipoproteinemia and increased biliary cholesterol excretion in transgenic mice. J Clin Invest 108:303–309

Van Eck M, Bos IS, Kaminski WE, Orso E, Rothe G, Twisk J, Bottcher A, Van Amersfoort ES, Christiansen-Weber TA, Fung-Leung WP, van Berkel TJ, Schmitz G (2002) Leukocyte ABCA1 controls susceptibility to atherosclerosis and macrophage recruitment into tissues. Proc Natl Acad Sci USA 99:6298–6303

Van Eck M, Singaraja RR, Ye D, Hildebrand RB, James ER, Hayden MR, van Berkel TJ (2006) Macrophage ATP-binding cassette transporter A1 overexpression inhibits atherosclerotic lesion progression in low-density lipoprotein receptor knockout mice. Arterioscler Thromb Vasc Biol 26:929–934

Vaughan AM, Oram JF (2005) ABCG1 redistributes cell cholesterol to domains removable by HDL but not by lipid-depleted apolipoproteins. J Biol Chem 280:30150–30157

Venkateswaran A, Repa JJ, Lobaccaro JM, Bronson A, Mangelsdorf DJ, Edwards PA (2000) Human white/murine ABC8 mRNA levels are highly induced in lipid-loaded macrophages. A transcriptional role for specific oxysterols. J Biol Chem 275:14700–14707

Vergeer M, Boekholdt SM, Sandhu MS, Ricketts SL, Wareham NJ, Brown MJ, de Faire U, Leander K, Gigante B, Kavousi M, Hofman A, Uitterlinden AG, van Duijn CM, Witteman JC, Jukema JW, Schadt EE, van der Schoot E, Kastelein JJ, Khaw KT, Dullaart RP, van Tol A, Trip MD, Dallinga-Thie GM (2010a) Genetic variation at the phospholipid transfer protein locus affects its activity and high-density lipoprotein size and is a novel marker of cardiovascular disease susceptibility. Circulation 122:470–477

Vergeer M, Cohn DM, Boekholdt SM, Sandhu MS, Prins HM, Ricketts SL, Wareham NJ, Kastelein JJ, Khaw KT, Kamphuisen PW, Dallinga-Thie GM (2010b) Lack of association between common genetic variation in endothelial lipase (LIPG) and the risk for CAD and DVT. Atherosclerosis 211:558–564

Vergeer M, Korporaal SJA, Franssen R, Meurs I, Out R, Hovingh GK, Hoekstra M, Sierts JA, Dallinga-Thie GM, Motazacker MM, Holleboom AG, Van Berkel TJC, Kastelein JJP, Van Eck M, Kuivenhoven JA (2011) Genetic variant of the scavenger receptor BI in humans. N Engl J Med 364:136–145

Vezeridis AM, Chroni A, Zannis VI (2011) Domains of apoE4 required for the biogenesis of apoE-containing HDL. Ann Med 43:302–311

Vickers KC, Remaley AT (2014) HDL and cholesterol: life after the divorce? J Lipid Res 55:4–12

Vickers KC, Palmisano BT, Shoucri BM, Shamburek RD, Remaley AT (2011) MicroRNAs are transported in plasma and delivered to recipient cells by high-density lipoproteins. Nat Cell Biol 13:423–433

Vikstedt R, Metso J, Hakala J, Olkkonen VM, Ehnholm C, Jauhiainen M (2007a) Cholesterol efflux from macrophage foam cells is enhanced by active phospholipid transfer protein through generation of two types of acceptor particles. Biochemistry 46:11979–11986

Vikstedt R, Ye D, Metso J, Hildebrand RB, van Berkel TJ, Ehnholm C, Jauhiainen M, Van EM (2007b) Macrophage phospholipid transfer protein contributes significantly to total plasma phospholipid transfer activity and its deficiency leads to diminished atherosclerotic lesion development. Arterioscler Thromb Vasc Biol 27:578–586

Wang N, Arai T, Ji Y, Rinninger F, Tall AR (1998) Liver-specific overexpression of scavenger receptor BI decreases levels of very low density lipoprotein ApoB, low density lipoprotein ApoB, and high density lipoprotein in transgenic mice. J Biol Chem 273:32920–32926

Wang N, Silver DL, Costet P, Tall AR (2000) Specific binding of ApoA-I, enhanced cholesterol efflux, and altered plasma membrane morphology in cells expressing ABC1. J Biol Chem 275: 33053–33058

Wang N, Lan D, Chen W, Matsuura F, Tall AR (2004) ATP-binding cassette transporters G1 and G4 mediate cellular cholesterol efflux to high-density lipoproteins. Proc Natl Acad Sci USA 101:9774–9779

Wang N, Ranalletta M, Matsuura F, Peng F, Tall AR (2006) LXR-induced redistribution of ABCG1 to plasma membrane in macrophages enhances cholesterol mass efflux to HDL. Arterioscler Thromb Vasc Biol 26:1310–1316

Wang X, Collins HL, Ranalletta M, Fuki IV, Billheimer JT, Rothblat GH, Tall AR, Rader DJ (2007) Macrophage ABCA1 and ABCG1, but not SR-BI, promote macrophage reverse cholesterol transport in vivo. J Clin Invest 117:2216–2224

Warden CH, Langner CA, Gordon JI, Taylor BA, McLean JW, Lusis AJ (1989) Tissue-specific expression, developmental regulation, and chromosomal mapping of the lecithin: cholesterol acyltransferase gene. Evidence for expression in brain and testes as well as liver. J Biol Chem 264:21573–21581

Waterworth DM, Ricketts SL, Song K, Chen L, Zhao JH, Ripatti S, Aulchenko YS, Zhang W, Yuan X, Lim N, Luan J, Ashford S, Wheeler E, Young EH, Hadley D, Thompson JR, Braund PS, Johnson T, Struchalin M, Surakka I, Luben R, Khaw KT, Rodwell SA, Loos RJ, Boekholdt SM, Inouye M, Deloukas P, Elliott P, Schlessinger D, Sanna S, Scuteri A, Jackson A, Mohlke KL, Tuomilehto J, Roberts R, Stewart A, Kesaniemi YA, Mahley RW, Grundy SM, McArdle W, Cardon L, Waeber G, Vollenweider P, Chambers JC, Boehnke M, Abecasis GR, Salomaa V, Jarvelin MR, Ruokonen A, Barroso I, Epstein SE, Hakonarson HH, Rader DJ, Reilly MP, Witteman JC, Hall AS, Samani NJ, Strachan DP, Barter P, van Duijn CM, Kooner JS, Peltonen L, Wareham NJ, McPherson R, Mooser V, Sandhu MS (2010) Genetic variants influencing circulating lipid levels and risk of coronary artery disease. Arterioscler Thromb Vasc Biol 30:2264–2276

Webb NR, de Beer MC, Yu J, Kindy MS, Daugherty A, van der Westhuyzen DR, de Beer FC (2002) Overexpression of SR-BI by adenoviral vector promotes clearance of apoA-I, but not apoB, in human apoB transgenic mice. J Lipid Res 43:1421–1428

Willer CJ, Sanna S, Jackson AU, Scuteri A, Bonnycastle LL, Clarke R, Heath SC, Timpson NJ, Najjar SS, Stringham HM, Strait J, Duren WL, Maschio A, Busonero F, Mulas A, Albai G, Swift AJ, Morken MA, Narisu N, Bennett D, Parish S, Shen H, Galan P, Meneton P, Hercberg S, Zelenika D, Chen WM, Li Y, Scott LJ, Scheet PA, Sundvall J, Watanabe RM, Nagaraja R, Ebrahim S, Lawlor DA, Ben Shlomo Y, Davey-Smith G, Shuldiner AR, Collins R, Bergman RN, Uda M, Tuomilehto J, Cao A, Collins FS, Lakatta E, Lathrop GM, Boehnke M, Schlessinger D, Mohlke KL, Abecasis GR (2008) Newly identified loci that influence lipid concentrations and risk of coronary artery disease. Nat Genet 40:161–169

Wojcik AJ, Skaflen MD, Srinivasan S, Hedrick CC (2008) A critical role for ABCG1 in macrophage inflammation and lung homeostasis. J Immunol 180:4273–4282

Wolfrum C, Poy MN, Stoffel M (2005) Apolipoprotein M is required for prebeta-HDL formation and cholesterol efflux to HDL and protects against atherosclerosis. Nat Med 11:418–422

Woollett LA, Spady DK (1997) Kinetic parameters for high density lipoprotein apoprotein AI and cholesteryl ester transport in the hamster. J Clin Invest 99:1704–1713

Wu Z, Gogonea V, Lee X, May RP, Pipich V, Wagner MA, Undurti A, Tallant TC, Baleanu-Gogonea C, Charlton F, Ioffe A, Didonato JA, Rye KA, Hazen SL (2011) The low resolution structure of ApoA1 in spherical high density lipoprotein revealed by small angle neutron scattering. J Biol Chem 286:12495–12508

Xie Q, Engel T, Schnoor M, Niehaus J, Hofnagel O, Buers I, Cullen P, Seedorf U, Assmann G, Lorkowski S (2006) Cell surface localization of ABCG1 does not require LXR activation. Arterioscler Thromb Vasc Biol 26:e143–e144

Xu N, Dahlback B (1999) A novel human apolipoprotein (apoM). J Biol Chem 274:31286–31290

Xu N, Nilsson-Ehle P (2007) ApoM – a novel apolipoprotein with antiatherogenic properties. In: Fielding CJ (ed) High-density lipoproteins. From basic biology to clinical aspects. Wiley, Weinheim, pp 89–110

Xu S, Laccotripe M, Huang X, Rigotti A, Zannis VI, Krieger M (1997) Apolipoproteins of HDL can directly mediate binding to the scavenger receptor SR-BI, an HDL receptor that mediates selective lipid uptake. J Lipid Res 38:1289–1298

Xu N, Nilsson-Ehle P, Ahren B (2006) Suppression of apolipoprotein M expression and secretion in alloxan-diabetic mouse: partial reversal by insulin. Biochem Biophys Res Commun 342: 1174–1177

Xu WW, Zhang Y, Tang YB, Xu YL, Zhu HZ, Ferro A, Ji Y, Chen Q, Fan LM (2008) A genetic variant of apolipoprotein M increases susceptibility to coronary artery disease in a Chinese population. Clin Exp Pharmacol Physiol 35:546–551

Xu Y, Wang W, Zhang L, Qi LP, Li LY, Chen LF, Fang Q, Dang AM, Yan XW (2011) A polymorphism in the ABCG1 promoter is functionally associated with coronary artery disease in a Chinese Han population. Atherosclerosis 219:648–654

Yamakawa-Kobayashi K, Yanagi H, Fukayama H, Hirano C, Shimakura Y, Yamamoto N, Arinami T, Tsuchiya S, Hamaguchi H (1999) Frequent occurrence of hypoalphalipoproteinemia due to mutant apolipoprotein A-I gene in the population: a population-based survey. Hum Mol Genet 8:331–336

Yates M, Kolmakova A, Zhao Y, Rodriguez A (2011) Clinical impact of scavenger receptor class B type I gene polymorphisms on human female fertility. Hum Reprod 26:1910–1916

Yesilaltay A, Morales MG, Amigo L, Zanlungo S, Rigotti A, Karackattu SL, Donahee MH, Kozarsky KF, Krieger M (2006) Effects of hepatic expression of the high-density lipoprotein receptor SR-BI on lipoprotein metabolism and female fertility. Endocrinology 147:1577–1588

Yesilaltay A, Daniels K, Pal R, Krieger M, Kocher O (2009) Loss of PDZK1 causes coronary artery occlusion and myocardial infarction in Paigen diet-fed apolipoprotein E deficient mice. PLoS ONE 4:e8103

Yuhanna IS, Zhu Y, Cox BE, Hahner LD, Osborne-Lawrence S, Lu P, Marcel YL, Anderson RG, Mendelsohn ME, Hobbs HH, Shaul PW (2001) High-density lipoprotein binding to scavenger receptor-BI activates endothelial nitric oxide synthase. Nat Med 7:853–857

Yvan-Charvet L, Matsuura F, Wang N, Bamberger MJ, Nguyen T, Rinninger F, Jiang XC, Shear CL, Tall AR (2007) Inhibition of cholesteryl ester transfer protein by torcetrapib modestly increases macrophage cholesterol efflux to HDL. Arterioscler Thromb Vasc Biol 27:1132–1138

Yvan-Charvet L, Kling J, Pagler T, Li H, Hubbard B, Fisher T, Sparrow CP, Taggart AK, Tall AR (2010) Cholesterol efflux potential and antiinflammatory properties of high-density lipoprotein after treatment with niacin or anacetrapib. Arterioscler Thromb Vasc Biol 30:1430–1438

Zannis VI, Lees AM, Lees RS, Breslow JL (1982) Abnormal apoprotein A-I isoprotein composition in patients with Tangier disease. J Biol Chem 257:4978–4986

Zannis VI, Cole FS, Jackson CL, Kurnit DM, Karathanasis SK (1985) Distribution of apolipoprotein A-I, C-II, C-III, and E mRNA in fetal human tissues. Time-dependent induction of apolipoprotein E mRNA by cultures of human monocyte-macrophages. Biochemistry 24: 4450–4455

Zannis VI, Chroni A, Kypreos KE, Kan HY, Cesar TB, Zanni EE, Kardassis D (2004a) Probing the pathways of chylomicron and HDL metabolism using adenovirus-mediated gene transfer. Curr Opin Lipidol 15:151–166

Zannis VI, Chroni A, Liu T, Liadaki KN, Laccotripe M (2004) New insights on the roles of apolipoprotein A-I, the ABCA1 lipid transporter, and the HDL receptor SR-BI in the biogenesis and the functions of HDL. In: Simionescu M, Sima A, Popov D (eds) Cellular dysfunction in atherosclerosis and diabetes. Romanian Academy Publishing House, Bucharest, pp 33–72

Zannis VI, Chroni A, Krieger M (2006a) Role of apoA-I, ABCA1, LCAT, and SR-BI in the biogenesis of HDL. J Mol Med 84:276–294

Zannis VI, Zanni EE, Papapanagiotou A, Kardassis D, Chroni A (2006b) ApoA-I functions and synthesis of HDL: insights from mouse models of human HDL metabolism. In: Fielding CJ (ed) High-density lipoproteins. From basic biology to clinical aspects. Wiley, Weinheim, pp 237–265

Zhang S, Picard MH, Vasile E, Zhu Y, Raffai RL, Weisgraber KH, Krieger M (2005) Diet-induced occlusive coronary atherosclerosis, myocardial infarction, cardiac dysfunction and premature death in SR-BI-deficient, hypomorphic apolipoprotein ER61 mice. Circulation 111:3457–3464

Zhang Y, Ahmed AM, Tran TL, Lin J, McFarlane N, Boreham DR, Igdoura SA, Truant R, Trigatti BL (2007) The inhibition of endocytosis affects HDL-lipid uptake mediated by the human scavenger receptor class B type I. Mol Membr Biol 24:442–454

Zhang LH, Kamanna VS, Zhang MC, Kashyap ML (2008) Niacin inhibits surface expression of ATP synthase beta chain in HepG2 cells: implications for raising HDL. J Lipid Res 49:1195–1201

Zhang Y, McGillicuddy FC, Hinkle CC, O'Neill S, Glick JM, Rothblat GH, Reilly MP (2010) Adipocyte modulation of high-density lipoprotein cholesterol. Circulation 121:1347–1355

Zhong S, Sharp DS, Grove JS, Bruce C, Yano K, Curb JD, Tall AR (1996) Increased coronary heart disease in Japanese-American men with mutation in the cholesteryl ester transfer protein gene despite increased HDL levels. J Clin Invest 97:2917–2923

Regulation of HDL Genes: Transcriptional, Posttranscriptional, and Posttranslational

Dimitris Kardassis, Anca Gafencu, Vassilis I. Zannis, and Alberto Davalos

Contents

1 Regulation of Genes Involved in HDL Metabolism at the Transcriptional Level 117
   1.1 General Introduction to Hormone Nuclear Receptors 118
   1.2 Transcriptional Regulation of the apoA-I Gene in the Liver 119
      1.2.1 The Role of the Distal Enhancer in apoA-I Gene Transcription 123
      1.2.2 Other Factors Regulating apoA-I Gene Transcription 124
   1.3 Transcriptional Regulation of the ABCA1 Gene 125
      1.3.1 Upregulatory Mechanisms of ABCA1 Gene Expression 125
      1.3.2 Negative Regulation of ABCA1 Gene Transcription 129
   1.4 Transcriptional Regulation of the ABCG1 Gene 130
   1.5 Transcriptional Regulation of the Apolipoprotein E Gene 132
      1.5.1 Proximal Regulatory Binding Sites Involved in the apoE Gene Expression 134
      1.5.2 Distal Regulatory Binding Sites That Modulate apoE Gene Expression in Macrophages 135
   1.6 Transcriptional Regulation of the Human apoM Gene in the Liver 136
   1.7 Transcriptional Regulation of the CETP Gene 138
   1.8 Transcriptional Regulation of the PLTP Gene 139
   1.9 Transcriptional Regulation of the Bile Acid Transporters ABCG5/ABCG8 140
   1.10 Transcriptional Regulation of the HDL Receptor SR-BI 141

D. Kardassis (✉)
Department of Biochemistry, University of Crete Medical School and Institute of Molecular Biology and Biotechnology, Foundation of Research and Technology of Hellas, Heraklion, Crete 71110, Greece
e-mail: kardasis@imbb.forth.gr

A. Gafencu
Institute of Cellular Biology and Pathology "Nicolae Simionescu", Bucharest, Romania

V.I. Zannis
Section of Molecular Genetics, Boston University Medical Center, Boston, MA 02118, USA

A. Davalos
Laboratory of Disorders of Lipid Metabolism and Molecular Nutrition, Madrid Institute for Advanced Studies (IMDEA)-Food, Ctra. de Cantoblanco 8, 28049 Madrid, Spain

© The Author(s) 2015
A. von Eckardstein, D. Kardassis (eds.), *High Density Lipoproteins*, Handbook of Experimental Pharmacology 224, DOI 10.1007/978-3-319-09665-0_3

2   Posttranscriptional Regulation of HDL Genes by Noncoding RNAs and microRNAs . .   143
    2.1   miRNAs: Biogenesis and Function   144
    2.2   Posttranscriptional Modulation of HDL Metabolism by miRNAs   146
          2.2.1   Targeting ABCA1 and ABCG1   146
          2.2.2   Targeting SR-BI   149
          2.2.3   Targeting Other miRNAs Related to HDL Biogenesis and Function   150
3   Posttranslational Mechanisms of HDL Regulation   150
    3.1   ABCA1   150
    3.2   ABCG1   154
    3.3   SR-BI   155
References   157

Abstract

HDL regulation is exerted at multiple levels including regulation at the level of transcription initiation by transcription factors and signal transduction cascades; regulation at the posttranscriptional level by microRNAs and other noncoding RNAs which bind to the coding or noncoding regions of HDL genes regulating mRNA stability and translation; as well as regulation at the posttranslational level by protein modifications, intracellular trafficking, and degradation. The above mechanisms have drastic effects on several HDL-mediated processes including HDL biogenesis, remodeling, cholesterol efflux and uptake, as well as atheroprotective functions on the cells of the arterial wall. The emphasis is on mechanisms that operate in physiologically relevant tissues such as the liver (which accounts for 80 % of the total HDL-C levels in the plasma), the macrophages, the adrenals, and the endothelium. Transcription factors that have a significant impact on HDL regulation such as hormone nuclear receptors and hepatocyte nuclear factors are extensively discussed both in terms of gene promoter recognition and regulation but also in terms of their impact on plasma HDL levels as was revealed by knockout studies. Understanding the different modes of regulation of this complex lipoprotein may provide useful insights for the development of novel HDL-raising therapies that could be used to fight against atherosclerosis which is the underlying cause of coronary heart disease.

Keywords

High-density lipoprotein • Regulation • Transcriptional • Posttranscriptional • Posttranslational • miRNAs • Protein stability • Hormone nuclear receptors • Hepatocyte nuclear factors • apoA-I • ABCA1 • ABCG1 • ABCG5 • ABCG8 • apoE • SR-BI • CETP

List of Abbreviations

HMG-CoA	3-Hydroxy-3-methylglutaryl-coenzyme A
12/15LO	12/15-Lipoxygenase
AP1	Activator protein 1
AF1	Activation function 1
Ang	Angiotensin
apoA-I	Apolipoprotein A-I
apoB	Apolipoprotein B
apoE	Apolipoprotein E
apoM	Apolipoprotein M
ARP-1	Apolipoprotein A-I-regulated protein 1
DHHC8	Asp-His-His-Cys 8
ABCA1	ATP-binding cassette transporter A1
ABCG1	ATP-binding cassette transporter G1
BIG1	Brefeldin A-inhibited guanine nucleotide-exchange protein 1
JNK	c-Jun N-terminal kinase
C/EBP	CCAAT/enhancer-binding protein
CAV-1	Caveolin 1
CXCL	Chemokine (C-X-C motif) ligand
COUP-TFI	Chicken ovalbumin upstream promoter transcription factor I
CEH	Cholesterol ester hydrolase
CETP	Cholesterol ester transfer protein
3C	Chromosome conformational capture
CLA	Conjugated linoleic acid
CAR	Constitutive androstane receptor
CREB	Cyclic AMP response element-binding protein
DNase	Deoxyribonuclease
DR4	Direct repeat with 4 nucleotides in the spacer region
DBD	DNA-binding domain
EGR-1	Early growth response protein 1
LIPG	Endothelial lipase
EE	Early endosome
ER	Endoplasmic reticulum
ESCRT	Endosomal sorting complex required for transport
ERα and β	Estrogen receptors α and β
ERE	Estrogen response element
ESRRG	Estrogen receptor-related gamma
FXR	Farnesoid X receptor
FXRE	Farnesoid X receptor-responsive element
FF	Fenofibrate
FOXA2	Forkhead box A2
FOXO1	Forkhead box O1
GF	Gemfibrozil

GR	Glucocorticoid receptor
GSK3β	Glycogen synthase kinase 3β
HNF-4	Hepatocyte nuclear factor-4
HDL	High-density lipoprotein
HDL-C	High-density lipoprotein cholesterol
Hcy	Homocysteine
HRE	Hormone response element
IRAK-1	IL-1 receptor-associated kinase 1
IDOL	Inducible degrader of the LDLR
IFN	Interferon
IL	Interleukin
KO	Knockout
KLF4	Kruppel-like factor 4
LCAT	Lecithin-cholesterol acyltransferase
LXR	Liver X receptor
LXRE	Liver X receptor response element
LRH-1	Liver receptor homologue-1
LBD	Ligand-binding domain
LPS	Lipopolysaccharide
LDLR	Low-density lipoprotein receptor
LAL	Lysosomal acid lipase
MCSF	Macrophage colony-stimulating factor
miRNAs	microRNAs
MEKK	Mitogen-activated protein kinase/ERK kinase kinase
MCP-1	Monocyte chemoattractant protein-1
ME	Multienhancer
MVB	Multivesicular bodies
DMHCA	N,N-Dimethyl-3β-hydroxycholenamide
NHERF	Na^+/H^+ exchanger regulator factor
NF-κB	Nuclear factor kappa-light-chain-enhancer of activated B cells
NFY	Nuclear factor Y
ORP1S	Oxysterol-binding protein-related protein 1S
PDZK1	PDZ domain-containing adaptor protein
PPAR	Peroxisome proliferator-activated receptors
PPRE	Peroxisome proliferator-activated receptor-responsive element
PMA	Phorbol 12-myristate 13-acetate
PI3K	phosphatidylinositol 3- kinase
PLD	Phospholipase D
PGC-1	PPARgamma coactivator 1
PLTP	Phospholipid transfer protein
pri-miRNA	Primary long miRNA
PREB	Prolactin regulatory element-binding
PEST	Proline, glutamic acid, serine, threonine
PKC	Protein kinase C
PCA	Protocatechuic acid

PCSK9	Proprotein convertase subtilisin/kexin type 9
RAR	Retinoic acid receptor
RORα	Retinoic acid receptor-related orphan receptor α
RXRα	Retinoid X receptor α
RISC	RNA-induced silencing complex
SR-BI	Scavenger receptor class B type I
SPTLC1	Serine palmitoyltransferase 1
STAT	Signal transducer and activator of transcription
SHP	Small heterodimer partner
siRNAs	Small interfering RNAs
SP1	Specificity protein 1
S1P	Sphingosine 1 phosphate
SRC-1	Steroid receptor coactivator-1
SF-1	Steroidogenic factor-1
SREBP	Sterol regulatory element-binding protein
SREs	Sterol-responsive elements
TRβ	Thyroid hormone receptor β
TAD	Transactivation domain
TGFβ1	Transforming growth factor β1
TNFα	Tumor necrosis factor α
Tpl2	Tumor progression locus 2
UTR	Untranslated region
URE	Upstream regulatory element
USF	Upstream stimulatory factors
VDR	Vitamin D receptor
WT	Wild type
YY1	Yin Yang 1
ZIC	Zinc finger of the cerebellum
ZNF202	Zinc finger protein 202

1 Regulation of Genes Involved in HDL Metabolism at the Transcriptional Level

A large body of work generated over the past four decades has revealed that eukaryotic gene transcription is a remarkably intricate biochemical process that is tightly regulated at many levels by the ordered assembly of multiprotein transcription initiation complexes to specific regulatory regions in the promoters of genes (Roeder 1998, 2005; Lemon and Tjian 2000). Despite the progress made, still limited knowledge regarding the details exists. It is believed that specificity in gene regulation is determined by the unique order of *cis*-acting regulatory regions which are recognized by sequence-specific DNA-binding transcription factors. Recent advances in gene regulation technologies including the powerful chromatin immunoprecipitation assay have enabled the monitoring in real time of the ordered assembly and the disassembly of transcription factor complexes on the promoters and the enhancers of genes in response to extracellular or intracellular cues (Christova 2013; Rodriguez-Ubreva and Ballestar 2014). High-throughput sequencing technologies have revolutionized the fields of genomics, epigenomics, and transcriptomics and have provided novel insights into the transcription signatures of human diseases (Churko et al. 2013). Furthermore, using new powerful methodologies such as chromosome conformation capture (3C) and its derivatives, we are at a position to monitor dynamic intra- and interchromosomal interactions that allow the optimal expression of genes at a given time and space (Gavrilov et al. 2009; Wei et al. 2013).

Transcription factors may be constitutively active in a cell or work in an inducible mode in response to various ligands and signal transduction pathways. The cross talk between different signaling pathways which orchestrate the cellular responses can be facilitated by the physical and functional interactions between transcription factors, and these interactions can be monitored by various methods both in vivo and in vitro. All known transcription factors are modular in nature and contain a DNA-binding domain and a transcriptional activation domain (Mitchell and Tjian 1989; Lemon and Tjian 2000). In addition, several factors contain a dimerization domain that permits them to form homodimers and/or heterodimers. A variety of nuclear receptors for steroids, thyroids, retinoids, etc. contain a ligand binding site. Via their transcription activation domains, transcription factors appear to facilitate the recruitment of the proteins of the coactivator complex and the basal transcription complex to the transcription initiation site of each gene and thus initiate transcription (Roeder 2005). Importantly, the activity of transcription factors can be modulated by drugs against diseases such as cancer and cardiovascular disease as exemplified by the drugs that activate or repress the hormone nuclear receptors (Gronemeyer et al. 2004).

It is beyond the scope of this chapter to provide a thorough review of the different mechanisms of transcriptional regulation of eukaryotic genes or to describe extensively the different classes of transcription factors, their structures, and their mode of regulation. We will only focus on those classes of transcription factors that have been shown to play key roles in the regulation of the genes involved in lipid and lipoprotein metabolism and more specifically on those involved in the metabolism of high-density lipoproteins (HDL) such as the hormone nuclear receptors.

1.1 General Introduction to Hormone Nuclear Receptors

Hormone nuclear receptors belong to a superfamily of transcription factors that are activated by steroid hormones (estrogens, androgens, glucocorticoids, etc.), retinoids, thyroids, and products of intermediate metabolism such as bile acids, fatty acids, and cholesterol derivatives, among others (Gronemeyer et al. 2004). Some members of this family do not need ligand binding to regulate transcription and are classified as "orphans" (Blumberg and Evans 1998). Nuclear receptors are structurally highly conserved. In terms of primary structure, the highest degree of homology among family members is in the DNA-binding domain that contains two zinc fingers (Helsen et al. 2012). Nuclear receptors also contain two transactivation domains (TADs), one N-terminal ligand-independent TAD called activation function 1 (AF1) and a ligand-dependent TAD called AF2 located close to the ligand-binding domain (LBD) (Rochel et al. 2011). Nuclear receptors bind to hormone response elements (HREs) on the promoters of target genes either as homodimers or as heterodimers with the retinoid X receptor (RXR). The HREs consist of direct repeats (DRs), inverted repeats (IRs), or palindromic repeats (PRs) of the consensus sequence 5′ AG(G/T)TCA 3′. The repeats are separated by 1, 2, 3, 4, or 5 nucleotides and are designated DR1, DR2, etc. (for the direct repeats); IR1, IR2, etc. (for the inverted repeats); and PR1, PR2, etc. (for the palindromic repeats) as described previously (Kardassis et al. 2007; Helsen et al. 2012). The HRE type and inter-repeat spacing determine to a large degree the specificity in nuclear receptor binding, but this rule is not strict at all. For instance, both LXR/RXR and T3R/RXR heterodimers prefer to bind to DR4 HREs, whereas RAR/RXR heterodimers bind to DR5. The direct repeats with one base spacing (DR1 type) appear to be very promiscuous as they bind RXR, COUP-TFI, ARP-1, and HNF-4 homodimers and PPAR/RXR, RAR/RXR, COUP-TFI/RXR, and ARP-1/RXR heterodimers (Nakshatri and Bhat-Nakshatri 1998). The elucidation of the three-dimensional structure of the ligand-binding domain of several nuclear receptors by X-ray crystallography in the absence and in the presence of ligands has allowed a good understanding of the modulation of nuclear receptor action by ligands and the development of very potent agonists and antagonists, some of which have been used therapeutically (Bourguet et al. 2000). Chromatin immunoprecipitation studies usually reveal that nuclear receptors are constitutively nuclear and bound to chromatin but they are transcriptionally silent in the absence of ligand. The binding of the ligand to the LBD causes a major conformational change to this domain which culminates in the recruitment of nuclear receptor coactivators such as PGC-1 and CBP/p300 and the displacement of corepressors (Chen and Li 1998; Liu and Lin 2011). Nuclear receptors can cross talk with other transcription factors in a positive or a negative manner as exemplified by the negative regulation of Jun or NF-κB transcription factors by the glucocorticoid receptors during inflammation, a mechanism termed *trans*-repression (Adcock and Caramori 2001).

1.2 Transcriptional Regulation of the apoA-I Gene in the Liver

The hypothesis that apolipoprotein (apo) A-I overexpression positively influences plasma concentrations of HDL cholesterol (HDL-C) has been validated experimentally in transgenic mice expressing human apoA-I under homologous or heterologous regulatory sequences. These mice have significantly elevated plasma levels of HDL-C and human apoA-I (Rubin et al. 1991; Kan et al. 2000). These "humanized" apoA-I transgenic mice are valuable tools for the study of apoA-I gene regulation in vivo. Furthermore, it was demonstrated that the overexpression of apoA-I in apoE KO or LDLR KO mice via transgene- or adenovirus-mediated gene transfer reduced atherosclerosis development confirming the anti-atherogenic role of apoA-I upregulation (Paszty et al. 1994; Belalcazar et al. 2003; Valenta et al. 2006).

In humans, the apoA-I gene is expressed abundantly in the liver and intestine and to a lesser extent in other tissues (Zannis et al. 1985). Early studies had established that the human apoA-I promoter containing 250 bp upstream from the transcription start site of the gene is sufficient to drive liver-specific gene expression both in cell cultures and in mice (Walsh et al. 1989; Tzameli and Zannis 1996; Hu et al. 2010a). This promoter region is rich in nuclear factor binding sites and responds to various intracellular as well as extracellular ligands (Zannis et al. 2001a; Haas and Mooradian 2010). As shown in Fig. 1, prominent role in the regulation of the apoA-I promoter play two hormone response elements (HREs) located at positions $-210/-190$ and $-132/-120$ that bind members of the hormone nuclear receptor superfamily in a competitive manner (Tzameli and Zannis 1996).

One of the nuclear receptors that plays a prominent role in apoA-I gene regulation in the liver and the intestine is the hepatocyte nuclear factor-4 (HNF-4).

HNF-4 was discovered as a rat liver nuclear protein that binds to the promoters of liver-specific genes such as transthyretin and apolipoprotein C-III (Sladek 1994). In the adult organism, HNF-4 is expressed in the liver, kidney, intestine, and pancreas (Sladek 1994). The total disruption of the HNF-4 gene in mice leads to an embryonic lethal phenotype due to the impairment of endodermal differentiation and gastrulation (Chen et al. 1994). This early developmental arrest was rescued by the complementation of the HNF-4$\alpha^{-/-}$ embryos with a tetraploid embryo-derived

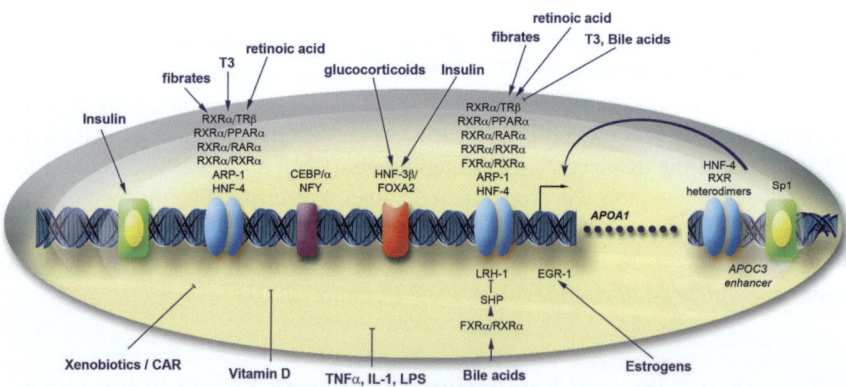

Fig. 1 Regulatory elements and transcription factors that control the expression of the apoA-I gene. *Arrows* and *block lines* denote activation and repression, respectively. The mechanisms are described in detail in the text. Abbreviations: retinoic acid receptor (RAR); retinoid X receptor (RXR); thyroid receptor β (TRβ); farnesoid X receptor (FXR); hepatocyte nuclear factor (HNF); apoA-I regulatory protein 1 (ARP-1); forkhead box 2 (FOXA2); nuclear factor Y (NFY); CCAAT/ enhancer-binding protein (CEBP); liver receptor homologue-1 (LRH-1); small heterodimer partner (SHP); specificity protein 1 (Sp1); tumor necrosis factor α (TNFα), interleukin-1 (IL-1); lipopolysaccharide (LPS)

wild-type visceral endoderm (Li et al. 2000). The analysis of the rescued mice showed that the expression of the apoA-I gene as well as of other apolipoprotein genes, shown previously to be regulated by HNF-4 including apoA-II, apoB, apoC-III, and apoC-II, was abolished confirming the cell culture data (Li et al. 2000). Experiments in mice in which the HNF-4 gene was disrupted in the adult liver using Alb-Cre revealed that HNF-4 is essential not only for the establishment but also for the maintenance of hepatic differentiation status (Hayhurst et al. 2001). Lipid and lipoprotein analysis of plasma of these mice revealed a dramatic reduction in total cholesterol, HDL cholesterol, and triglycerides as well as a dramatic increase in the concentration of bile acids (Hayhurst et al. 2001). Furthermore, FPLC analysis showed that HDL cholesterol from the HNF-4 Liv KO mice eluted later than that from controls indicative of the presence of smaller HDL populations. Interestingly, the expression of the two essential genes for HDL biogenesis, namely apoA-I and ABCA1, was not affected in the livers of the HNF-4 Liv KO, suggesting that the reduction in the plasma HDL levels was the result of altered HDL remodeling rather than reduced biosynthesis. In agreement with this, the expression level of the HDL receptor SR-BI gene was dramatically increased (Hayhurst et al. 2001).

Studies in transgenic mice expressing the human apoA-I gene under its own regulatory sequences and clinical studies in humans have shown that fibrates have a positive effect on apoA-I gene transcription as well as on plasma HDL levels. The increase in human apoA-I gene transcription by fibrates is mediated by peroxisome proliferator-activated receptor α (PPARα) which binds to a PPRE on the proximal apoA-I promoter as a heterodimer with RXRα (Tzameli and Zannis 1996; Staels and Auwerx 1998). This was confirmed by in vivo experiments performed in mice that express human apoA-I under the control of its own promoter but lack the expression of PPARα. When these mice were given fenofibrate (FF) or gemfibrozil (GF) for 17 days, an increase in plasma HDL-C levels was observed by FF and to a lesser extent by GF only in the mice that express endogenous PPARα (Duez et al. 2005). The fibrate-treated mice had larger HDL particles possibly due to the upregulation of phospholipid transfer protein and downregulation of SR-BI (Duez et al. 2005). Interestingly, the apoA-I gene cannot be upregulated by fibrates in rodents due to a three base pair difference in the PPRE rendering the rodent apoA-I PPRE nonfunctional (Vu-Dac et al. 1994). In line with the above findings, liver-specific inactivation of the PPARα heterodimer partner retinoid X receptor α (RXRα) gene in mice was associated with increased expression of the apoA-I gene (Wan et al. 2000).

In a clinical study involving 234 patients with combined hyperlipidemia, both FF and GF reduced triglycerides and increased HDL-C to a similar extent, but only FF treatment increased apoA-I plasma levels, and this was in agreement with the previous clinical trials (Schaefer et al. 1996; Durrington et al. 1998; Sakai et al. 2001; Duez et al. 2005).

A humanized apoA-I transgenic mouse model expressing human apoA-I under its own regulatory sequences in a mouse apoA-I null background was recently used to identify by global gene expression profiling candidate genes that affect lipid and lipoprotein metabolism in response to fenofibrate treatment (Sanoudou et al. 2009).

Bioinformatical analysis and stringent selection criteria (twofold change, 0 % false discovery rate) identified 267 significantly changed genes. In contrast to the study by Duez et al. discussed above (Duez et al. 2005), fenofibrates (FF) did not significantly alter the levels of hepatic human apoA-I mRNA and plasma apoA-I protein. This could be due to differences either in the mouse models used (for instance, the humanized apoA-I mouse of Sanoudou et al. has 2.1 kb apoA-I promoter fragment, whereas the model of Duez et al. has a 5.5 kb apoA-I promoter) or the doses of fibrates (0.2 % w/w in the paper of Duez et al. and 0.03 % in the paper by Sanoudou et al.). Despite the lack of apoA-I responsiveness, the FF treatment increased cholesterol levels 1.95-fold mainly due to the increase in HDL-C. The observed changes in HDL were associated with the upregulation of genes involved in phospholipid biosynthesis and lipid hydrolysis, as well as in the phospholipid transfer protein. The gene encoding the estrogen receptor-related gamma (ESRRG) transcription factor was upregulated 2.36-fold by FF and had a significant positive correlation with genes of lipid and lipoprotein metabolism and mitochondrial functions, indicating an important role of this orphan receptor in mediating the FF-induced activation of a specific subset of its target genes (Sanoudou et al. 2009).

In addition to HNF-4 and PPARα, the two HREs of the proximal human apoA-I promoter bind apoA-I regulatory protein 1 (ARP-1) and liver receptor homologue-1 (LRH-1) which repress and activate the apoA-I promoter, respectively, as illustrated in Fig. 1 (Ladias and Karathanasis 1991; Delerive et al. 2004). LRH-1 is a member of the *fushi tarazu* subfamily of nuclear receptors that is highly expressed in the liver, intestine, pancreas, and ovary (Fayard et al. 2004). In the liver, LRH-1 plays a key role in cholesterol homeostasis, through the control of the expression of genes that are implicated in bile acid biosynthesis and enterohepatic circulation such as CYP7A1, CYP8B1, and ABCG5/8 (del Castillo-Olivares and Gil 2000; Freeman et al. 2004; Kir et al. 2012; Back et al. 2013), reverse cholesterol transport (SR-BI, apoA-I) (Schoonjans et al. 2002; Delerive et al. 2004), and HDL remodeling (CETP) (Luo et al. 2001). However, mice with targeted inactivation of the LRH-1 gene in the liver are characterized by physiological levels of HDL cholesterol, LDL cholesterol, and triglycerides but have a profound effect on bile acid composition in the liver which leads to reduced intestinal reuptake of bile acids and to the enhanced removal of lipids from the body (Mataki et al. 2007). Recent data suggest that LRH-1 functions in a compensatory safeguard mechanism for adequate induction of bile salt synthesis under conditions of high bile salt loss (Out et al. 2011).

The two HREs of the apoA-I promoter also mediate the response of apoA-I to thyroids, retinoids, and bile acids via heterodimers of RXRα with thyroid hormone receptor β (TRβ), retinoic acid receptor α (RARα), and farnesoid X receptor α (FXRα), respectively (Rottman et al. 1991; Hargrove et al. 1999). Although retinoids activate apoA-I gene expression, thyroids have dual effects on apoA-I promoter activity, whereas bile acids inhibit apoA-I gene expression (Taylor et al. 1996; Tzameli and Zannis 1996; Srivastava et al. 2000; Claudel et al. 2002). As shown in Fig. 1, in response to bile acids, FXR downregulates apoA-I gene transcription by two complementary mechanisms: (a) a direct binding to the apoA-I HRE and (b) an indirect mechanism via the induction of small heterodimer partner (Bavner et al. 2005) which, in turn, represses the activity of LRH-1 (Delerive et al. 2004).

The nuclear receptor constitutive androstane receptor (CAR) regulates the detoxification of xenobiotics and endogenous molecules. In mice, the specific CAR agonist 1,4-bis[2-(3,5-dichloropyridyloxy)]benzene (TCPOBOP) decreased HDL cholesterol and plasma apoA-I levels in a CAR-specific manner (Masson et al. 2008). In transient transfections, CAR decreased the activity of the human apoA-I promoter in the presence of TCPOBOP, but the mechanism by which this repression is facilitated remains unknown (Masson et al. 2008).

Ligands of the vitamin D receptor (VDR) were also shown to affect negatively apoA-I gene expression in hepatic cells (Wehmeier et al. 2005). In VDR KO mice, serum HDL-C levels were 22 % higher and the mRNA levels of apoA-I were 49.2 % higher compared with WT mice. The mechanisms by which VDR ligands affect HDL levels remain unclear (Wang et al. 2009).

1.2.1 The Role of the Distal Enhancer in apoA-I Gene Transcription

In addition to its own promoter, optimal expression of the human apoA-I gene in hepatic and intestinal cells requires the presence of a 200 bp transcriptional enhancer element located downstream of the apoA-I gene, 500 bp upstream of the first exon of the adjacent apoC-III gene. This regulatory region, which coordinates the expression of all three genes of the apoA-I/C-III/A-IV cluster, contains two hormone response elements that bind HNF-4 and different combinations of ligand-dependent nuclear receptors as well as two binding sites for the ubiquitous transcription factor specificity protein 1 (Sp1) (Kardassis et al. 1997; Lavrentiadou et al. 1999). The mutagenesis of the HREs and of the Sp1 sites reduced the activity of the apoA-I promoter/C-III enhancer cassette in cell cultures and abolished the binding of the corresponding factors (Kardassis et al. 1997; Lavrentiadou et al. 1999).

The contribution of the HREs and the Sp1 binding sites to the tissue-specific expression of the apoA-I gene in vivo was addressed using transgenic mice bearing the WT apoA-I/apoC-III gene cluster under the control of their regulatory regions or the same cluster bearing mutations in different regulatory elements (Georgopoulos et al. 2000; Kan et al. 2000, 2004). It was shown that mutations in one of the two HREs of the enhancer (element I4) abolished the intestinal expression and reduced the hepatic expression of the adjacent apoA-I gene to 20 % of the control. Mutations in the two HREs of the proximal apoA-I promoter reduced the hepatic and intestinal expression of the apoA-I gene to approximately 15 % of the control, whereas combined mutations in all three HREs totally eliminated the intestinal and hepatic expression of the apoA-I gene (Kan et al. 2000). Studies in cell cultures established that HNF-4 and Sp1 factors are both required for the synergy between the apoA-I promoter and the enhancer by physically interacting with each other and forming transcriptional complexes in order to facilitate the recruitment of the basal transcriptional machinery (Kardassis et al. 2002). The aforementioned mouse model that expresses human apoA-I under its own promoter and enhancer (Kan et al. 2000) is very useful for the in vivo characterization of the mechanisms that regulate the expression of the apoA-I gene under physiological or pathological conditions as well as for the identification and validation of novel compounds that are designed to upregulate human apoA-I gene transcription and serve as HDL-raising drugs. This is especially important in light of the differences between the mouse and the human apoA-I promoters. For instance, the mouse gene cannot be upregulated by fibrates due to a three base pair difference in the PPRE compared to the human promoter which responds to 0.2 % fibrates but not to 0.03 % fibrates as mentioned above.

1.2.2 Other Factors Regulating apoA-I Gene Transcription

Further upstream from the two apoA-I HREs, an insulin response core element (IRCE) was identified and shown to bind Sp1 (Murao et al. 1998). Insulin-activated signaling pathways including the Ras/raf and the phosphatidylinositol 3-kinase (PI3K) have been shown to posttranslationally modify Sp1, and this leads to increased apoA-I promoter activity (Mooradian et al. 2004).

Early growth response protein 1 (EGR-1) is another transcription factor that regulates apoA-I expression via the proximal HREs (Kilbourne et al. 1995; Cui et al. 2002). Mice with experimental nephrotic syndrome are characterized by a fivefold increase in the levels of EGR-1, and these changes were associated with high plasma apoA-I and HDL-C levels as well as apoA-I gene transcription in the liver (Zaiou et al. 1998). In line with these findings, mice deficient in EGR-1 have reduced plasma HDL-C and apoA-I as well as hepatic apoA-I mRNA levels (Zaiou et al. 1998). EGR-1 was shown to mediate the response of the apoA-I promoter to estrogens (Hargrove et al. 1999).

The proximal apoA-I promoter also contains one element that binds the basic leucine zipper (bZip) factor CCAAT/enhancer-binding protein (C/EBP) and nuclear factor Y (NFY) and another element that binds hepatocyte nuclear factor-3β/FOXA2 (Papazafiri et al. 1991; Novak and Bydlowski 1997). Nuclear factor HNF-3β was shown to mediate the response of the apoA-I promoter to glucocorticoids (Hargrove et al. 1999).

Several natural compounds with antioxidant, pro-estrogenic, or other activities were shown to affect apoA-I and HDL-C levels, and these studies are summarized in Haas and Mooradian (2010).

Pro-inflammatory cytokines including tumor necrosis factor α (TNFα) and interleukin-1β (IL-1β) were previously shown to inhibit apoA-I gene expression both in cell cultures and in animals (Ettinger et al. 1994; Song et al. 1998). Furthermore, plasma levels of HDL-C and apoA-I were shown to be highly increased in mice deficient in the p50 subunit of the pro-inflammatory transcription factor NF-κB (Morishima et al. 2003). In agreement with this observation, the activation of NF-κB by lipopolysaccharide (LPS) caused a reduction in apoA-I mRNA and protein levels in HepG2 cells, whereas the inhibition of NF-κB via adenovirus-mediated overexpression of IκBα abolished the reduction (Morishima et al. 2003). This IκBα-induced apoA-I increase was blocked by preincubation with MK886, a selective inhibitor of peroxisome proliferator-activated receptor α, and mutations in the PPARα binding site in the apoA-I promoter abrogated these changes (Morishima et al. 2003). In a recent study, it was shown that apoA-I promoter activity in HepG2 cells is inhibited by TNFα in a c-Jun-dependent manner but no AP1-responsive element within the apoA-I promoter was reported to mediate this effect (Parseghian et al. 2013). The inhibition of the expression of apoA-I and other HDL genes in hepatocytes during inflammation could also be mediated by HNF-4 which was previously shown to be negatively regulated by the TNFα/NF-κB signaling pathway by physically interacting with NF-κB (Nikolaidou-Neokosmidou et al. 2006).

1.3 Transcriptional Regulation of the ABCA1 Gene

The gene encoding the ATP-binding cassette transporter A1 (ABCA1) is expressed in the liver, small intestine, macrophages, kidney, and various other tissues (Langmann et al. 1999, 2003; Kielar et al. 2001; Wellington et al. 2002). ABCA1 is an important regulator of HDL biogenesis in the liver and facilitates the removal of excess cholesterol from macrophages.

ABCA1 is particularly abundant in macrophages (Langmann et al. 1999). ABCA1 expression in macrophages has little influence on HDL-C plasma levels (Haghpassand et al. 2001) but is an important factor in the prevention of cholesterol accumulation in the macrophages found in the atherosclerotic plaque and their transformation into foam cells (Aiello et al. 2002). ABCA1 mRNA and protein are very unstable, having a half-life of 1–2 h in murine macrophages (Wang and Oram 2002). Fine-tuning regulatory mechanisms (transcriptional regulation as well as posttranscriptional and posttranslational modifications) are involved to ensure the constant and inducible ABCA1 expression in macrophages. In this section, we will focus on the transcriptional regulation of the ABCA1 gene with emphasis on macrophages. Posttranscriptional and posttranslational regulation of this gene will be discussed in later sections.

The human ABCA1 gene mapped to chromosome 9q31.1 is composed of 50 exons, which encode 2261-amino-acid residues (Santamarina-Fojo et al. 2000). The ABCA1 gene promoter contains a TATA box localized 24 bp upstream of the transcription initiation site, essential for promoter activity in macrophages as well as in hepatocytes (Langmann et al. 2002). The engagement of alternative promoters and transcription initiation sites localized upstream of the first exon or inside the first intron of the gene enables the inducible and tissue-specific expression regulation of ABCA1 gene (Huuskonen et al. 2003; Singaraja et al. 2005). In addition, other transcriptional response elements of the promoter influence the constitutive and tissue-specific expression of ABCA1 (Fig. 2).

1.3.1 Upregulatory Mechanisms of ABCA1 Gene Expression

The major transcription factors that upregulate ABCA1 gene expression in macrophages are the nuclear receptors liver X receptors α and β (LXRα and LXRβ), both expressed by this cell type. LXRs heterodimerize with the retinoic

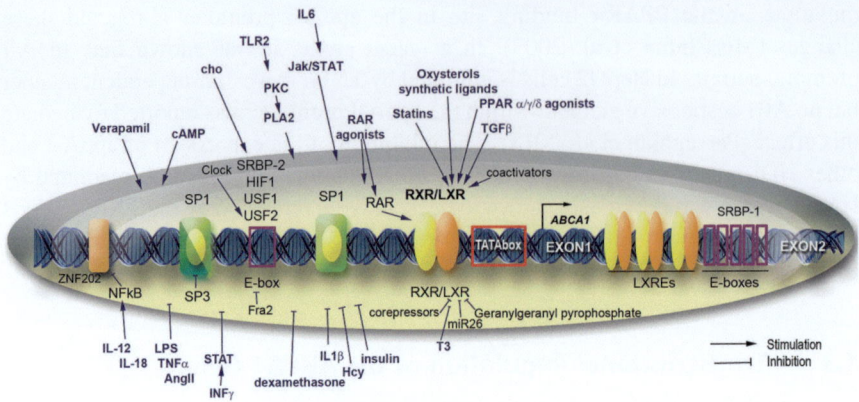

Fig. 2 Regulatory regions, transcription factors, and signaling molecules (cytokines, growth factors, metabolites, drugs) that modulate the expression of the ABCA1 gene in macrophages and other tissues. *Arrows* and *block lines* denote activation and repression, respectively. The mechanisms are described in detail in the text. Abbreviations: retinoic acid receptor (RAR); retinoid X receptor (RXR); liver X receptor (LXR); sterol regulatory element-binding protein (SREBP); specificity protein 1 (Sp1); tumor necrosis factor α (TNFα), interleukin (IL); lipopolysaccharide (LPS); interferon γ (IFNγ); angiotensin II (AngII); nuclear factor kappa beta (NF-κB); upstream stimulatory factor (USF); hypoxia-inducible factor (HIF); transforming growth factor β (TGFβ); peroxisome proliferator-activated receptor (PPAR); Janus kinase (Jak); signal transducer and activator of transcription (STAT)

X receptor (RXRα) to bind to the direct repeats separated by four nucleotides (direct repeat 4; DR4 elements) found at position −62/−47 of the ABCA1 promoter (Costet et al. 2000). Additional but not well-characterized LXREs are present at the intron 1 promoter of the ABCA1 gene (Singaraja et al. 2001). Since the LXRα promoter is subject to autoregulation, the LXR-mediated gene expression is auto-amplified (Laffitte et al. 2001a). The natural ligands of LXR are several hydroxylated derivatives of cholesterol (oxysterols) and include 27-hydroxycholesterol (Schwartz et al. 2000; Fu et al. 2001), 22(R)-hydroxycholesterol (Costet et al. 2000), 20(S)-hydroxycholesterol (Schwartz et al. 2000), and desmosterol (Yang et al. 2006). Among these ligands, 27-hydroxycholesterol which is endogenously produced by the action of CYP27A1 enzyme represents the sensor of cholesterol loading in macrophages. In addition to these natural ligands, synthetic LXR ligands such as TO901317 have also been developed. These synthetic LXR agonists upregulate ABCA1 in cultured macrophages more efficiently than cholesterol loading via modified LDL particles (Sparrow et al. 2002).

LXR tissue-selective gene transcription is dependent on co-regulatory proteins. For example, data showed that the activation of the ABCA1 promoter by LXRα/RXRα heterodimers and their ligands require Sp1 (Thymiakou et al. 2007). The overexpression of Sp1 increased ABCA1 mRNA level in HeLa cells and enhanced cellular cholesterol and phospholipid efflux in RAW 246.7 macrophages. Gel shift experiments revealed in vitro binding of Sp1 to −85/−91 and −151/−156 GC boxes. Moreover, it was shown that Sp3 competed with Sp1 for binding to the latter GC box, acting as a repressor (Langmann et al. 2002). Physical interactions between Sp1 and LXRα require the N-terminal region of LXRα, which includes the DNA-binding domain and two different domains of Sp1: the transactivation domain B and the DNA-binding domain (Thymiakou et al. 2007). LXR agonists, such as the natural steroidal ligand 22(R)-hydroxycholesterol and the weak nonsteroidal ligand GSK418224, differentially recruit coactivators and corepressors compared with full LXR agonists, such as the nonsteroidal ligand T0901317 (Albers et al. 2006; Peng et al. 2008, 2011; Phelan et al. 2008). The synthetic oxysterol N, N-dimethyl-3β-hydroxycholenamide (DMHCA) caused a differential induction of the ABCA1 and the sterol regulatory element-binding protein (SREBP)-1c genes in hepatic and macrophage cell lines, as well as in mice (Quinet et al. 2004). In cholesterol-loaded or unloaded peritoneal macrophages, DMHCA increased ABCA1 mRNA, whereas SREBP-1c mRNA levels were downregulated (Quinet et al. 2004). Cineole, a small aromatic compound found in teas and herbs, considerably stimulated the transactivation potential of LXRα and LXRβ and induced ABCA1 expression in macrophages but significantly reduced the expression of LXRα- and LXRα-responsive genes in hepatocytes (Jun et al. 2013). Another LXR agonist, ATI-111, had a strong effect on ABCA1 expression in macrophages as well as in the intestine and small effect on ABCA1 expression in the liver. ATI-111 significantly stimulated SREBP-1c mRNA in some tissues but inhibited the conversion of SREBP-1c precursor form into its active form (Peng et al. 2011).

These findings revealed that LXR agonists have a promising potential for the upregulation of the ABCA1 transporter and the promotion of the cellular lipid efflux capacity of macrophages. Due to the concomitant LXR-mediated upregulation of two genes involved in the fatty acid biosynthesis, fatty acid synthase (Joseph et al. 2002a), and SREBP-1c (Yoshikawa et al. 2001), the development of LXR agonists for therapeutic uses has been limited by their adverse effects that include hepatic steatosis and hypertriglyceridemia. In order to dissociate the positive effects of LXR agonists on cholesterol homeostasis from the adverse effects on fatty acid metabolism, the next step will be the discovery of new LXRβ-selective agonists and the synthesis of novel tissue-specific LXR ligands with weaker transcriptional effects on SREBP-1c. SREBP-1a, a different member of the SREBP family of proteins, binds to several sites present inside intron 1; however, the role of these elements in ABCA1 gene regulation is still unknown (Thymiakou et al., unpublished observations) (Fig. 2).

Besides LXR, other nuclear receptors and transcription factors are involved in ABCA1 gene regulation in macrophages. Retinoic acid receptor (RAR) activators such as all-*trans*-retinoic acid and 4-[(E)-2-(5,6,7,8-tetrahydro-5,5,8,8-tetramethyl-2-naphthalenyl)-1-propenyl]benzoic acid (arotinoid acid) were found to increase ABCA1 mRNA and protein levels in macrophages (Costet et al. 2003). Co-transfection experiments showed that the same DR4 promoter element of the ABCA1 promoter binds RXR heterodimers in the following order: RARγ/RXRα bound stronger and activated the human ABCA1 promoter, and RARα/RXRα bound weaker, while no RARβ/RXRα binding was detected. However, in macrophages from RARγ$^{-/-}$ mice, arotinoid acid still induced ABCA1 gene expression and caused marked upregulation of RARα, suggesting that high levels of RARα can compensate for the absence of RARγ (Costet et al. 2003).

Peroxisome proliferator-activated receptors α (PPARα) and γ (PPARγ) are both expressed in human macrophages where they exert anti-inflammatory effects. The hydroxylated derivative of linoleic acid, 13-hydroxy linoleic acid, a natural PPAR agonist, and pioglitazone (PPARγ agonist) increased PPAR transcriptional activity and induced ABCA1 gene expression in macrophages (Kammerer et al. 2011; Ozasa et al. 2011). However, data suggest that these effects are indirect and most probably are mediated by LXRα (Chinetti et al. 2001). PPARδ activators appeared to induce ABCA1 gene expression and cholesterol efflux moderately and to increase HDL levels in an obese-monkey model (Oliver et al. 2001).

Recent data showed that Clock, a key transcription factor that controls circadian rhythm, is involved in ABCA1 regulation in macrophages (Pan et al. 2013). The dominant-negative Clock mutant protein (ClockΔ19/Δ19) enhanced plasma cholesterol and atherosclerosis. Mutant ClockΔ19/19/apoE$^{-/-}$ mice had macrophage dysfunction, expressed low levels of ABCA1, and had higher levels of scavenger receptors. Molecular studies revealed that Clock regulated ABCA1 expression in macrophages by modulating the activity of upstream stimulatory factor 2 (Pan et al. 2013).

Experimental data showed that signaling molecules upregulate ABCA1 expression in macrophages, but the fine-tuning mechanism of modulation and the regulatory regions involved remain to be elucidated. Transforming growth factor β1 (TGFβ1) increased ABCA1 mRNA levels in cholesterol-loaded macrophages (Argmann et al. 2001; Panousis et al. 2001). It was demonstrated that in THP-1 macrophage-derived foam cells, the LXRα pathway is involved in TGFβ-mediated upregulation of ABCA1 expression (Hu et al. 2010b). Interleukin-6 (IL-6) significantly increased ABCA1 at both the mRNA and protein levels. This effect was abolished by selective inhibition of the JAK2/STAT3 signaling pathway (Frisdal et al. 2011). ABCA1 mRNA levels were significantly increased by estradiol treatment of macrophages for a short period of time, suggesting a direct activation of the ABCA1 promoter via the estrogen receptor β (Schmitz and Langmann 2005). Membrane-permeable analogues of cyclic adenosine monophosphate (cAMP) induced the ABCA1 mRNA in macrophages and other cells by an unknown mechanism (Lin and Bornfeldt 2002). Inhibitors that are able to block the action of phosphodiesterase 4 on cAMP have been found to increase ABCA1 mRNA and cellular cholesterol efflux (Lin and Bornfeldt 2002). Verapamil, a calcium channel blocker, enhanced ABCA1 transcription by an LXR-independent process (Suzuki et al. 2004). Toll-like receptor 2 agonist Pam(3)CSK(4) upregulated ABCA1 gene expression in RAW 264.7 macrophages via the activation of the PKCη/phospholipase D2 signaling pathway (Park et al. 2013). S-Allylcysteine, the most abundant organosulfur compound in aged garlic extract, also elevated ABCA1 content in human THP-1 macrophages (Malekpour-Dehkordi et al. 2013). The data showed that stimulation with CXCL5 that has a protective role in atherosclerosis (Rousselle et al. 2013) induced ABCA1 expression in alternatively activated (M2) macrophages but not in classically activated (M1) macrophages.

1.3.2 Negative Regulation of ABCA1 Gene Transcription

The thyroid hormone T3 strongly suppressed ABCA1 gene transcription (Huuskonen et al. 2004). It was demonstrated that T3 significantly inhibited the ability of oxysterols to activate LXR. Moreover, the TR/RXR heterodimers competed with LXR/RXR for the DR4 element in the ABCA1 promoter (Huuskonen et al. 2004). A reciprocal negative cross talk exists between LXRs and STAT1 based on the competition for CREB-binding protein (Pascual-Garcia et al. 2013). This may explain the IFNγ-mediated downregulation of ABCA1 in cholesterol-loaded macrophages (Panousis and Zuckerman 2000; Argmann et al. 2001; Ma et al. 2013). Moreover, TNFα downregulates ABCA1 as well as LXRα expression in macrophages via a PKCθ-dependent pathway (Ma et al. 2013).

Unsaturated fatty acids decrease the expression of ABCA1 in RAW 264.7 macrophages by a mechanism that involves LXR/RXR binding to the promoter (Uehara et al. 2007) and by modulation of the histone acetylation state (Ku et al. 2012). In MCSF-activated human monocytes, linoleic acid decreases ABCA1 gene expression (Mauerer et al. 2009). Geranylgeranyl pyrophosphate, a product of the mevalonate pathway that is used for protein isoprenylation, suppresses LXR-induced ABCA1 synthesis in two ways: as an antagonist of the LXR interaction with the steroid receptor coactivator-1 (SRC-1) and as an activator of Rho GTP-binding proteins (Gan et al. 2001).

Among the transcription factors that downregulate ABCA1 expression in macrophages in an LXR-independent manner is the SCAN domain-containing zinc finger transcription factor ZNF202 which binds to a GnT motif in the region −234/−215 of the ABCA1 promoter (Porsch-Ozcurumez et al. 2001). High intracellular levels of ZNF202 prevented LXR/RXR-mediated induction of the ABCA1 promoter in response to oxysterols (Porsch-Ozcurumez et al. 2001).

The conserved E-box at position −140 binds transcription factors that modulate the ABCA1 gene expression. Experimental data indicated that upstream stimulatory factors (USF) 1 and 2, hepatocyte nuclear factor-1α (HNF-1α), and fos-related antigen (Fra)2 bind to the intact E-box of the human ABCA1 promoter and differentially modulate the gene expression: USF1 and USF2 enhanced and Fra2 repressed ABCA1 promoter activity (Langmann et al. 2002; Yang et al. 2002). The same E-box element also binds the sterol regulatory element-binding protein 2 (SREBP-2), a key regulator of cholesterol metabolism, which suppresses ABCA1 gene transcription in response to cholesterol depletion (73).

ABCA1 gene expression was severely decreased in the liver and peritoneal macrophages of diabetic mice (Uehara et al. 2002). This observation was explained using in vitro models in which it was revealed that acetoacetate downregulates ABCA1 mRNA and protein in HepG2 hepatocytes and RAW 264.7 macrophages (Uehara et al. 2002) and thus high glucose concentration decreases ABCA1 gene expression in MCSF-activated monocytes (Mauerer et al. 2009). Data showed that the treatment of THP-1 macrophages with 100 nM dexamethasone, a potent synthetic ligand of the glucocorticoid receptor, decreases the expression of the ABCA1 gene (Sporstol et al. 2007).

LPS downregulates ABCA1 in macrophages; this inhibition was reverted by the treatment with betulinic acid acting via the downregulation of miR-33 and suppression of NF-κB pathway (Zhao et al. 2013). IL-18 and IL-12 synergistically decrease ABCA1 levels in THP-1 macrophage-derived foam cells through the IL-18 receptor/NF-κB/ZNF202 signaling pathway (Yu et al. 2012). ABCA1 expression was strongly suppressed by angiotensin (Ang) II at both mRNA and protein levels in a dose-dependent manner in THP-1-derived macrophages, whereas ABCG1 expression was not affected. The effect of Ang II on ABCA1 expression could be mediated by the angiotensin II type 1 (AT1) receptor (Chen et al. 2012a). It was demonstrated that clinically relevant concentrations of homocysteine (Hcy) decreased the mRNA and protein expression levels of ABCA1 in macrophages. It was revealed that mRNA expression and the activity of DNA methyltransferase were increased by Hcy, which may explain the higher DNA methylation level of ABCA1 gene in macrophages incubated with Hcy (Liang et al. 2013).

1.4 Transcriptional Regulation of the ABCG1 Gene

ABCG1 mediates cholesterol removal from macrophages to HDL particles, but not to lipid-free apoA-I (Kennedy et al. 2005; Fitzgerald et al. 2010). Although recent data showed that the combined macrophage deficiency of ABCA1/G1 is pro-atherogenic, probably by promoting plaque inflammation (Westerterp et al. 2013), the data concerning the role of ABCG1 expression in macrophages is controversial. Two independent groups reported that $LDLR^{-/-}$ mice lacking macrophage ABCG1 show decreased atherosclerotic lesions (Baldan et al. 2006; Ranalletta et al. 2006), while others reported that the absence of macrophage ABCG1 causes a modest increase in atherosclerotic lesions (Out et al. 2006). These contradictory results may be explained by recent data showing that the absence of ABCG1 leads to increased lesions in early stages of atherosclerosis but causes retarded lesion progression in more advanced stages of atherosclerosis in $LDLR^{-/-}$ mice, suggesting that the influence of ABCG1 deficiency on lesion development depends on the stage of atherogenesis (Meurs et al. 2012).

The human ABCG1 gene has been mapped to chromosome 21q22.3 and encodes for a 678-amino-acid protein of 75.5 kDa molecular mass (Chen et al. 1996). The ABCG1 gene spans more than 70 kb and includes 15 exons, each containing between 30 and 1,081 bp, while the intron size is between 137 bp and more than 45 kb. All exon-intron boundaries display the canonical GT/AG sequences. In contrast to the ABCA1 gene, the ABCG1 gene does not contain a canonical TATA box in the promoter (Langmann et al. 2000).

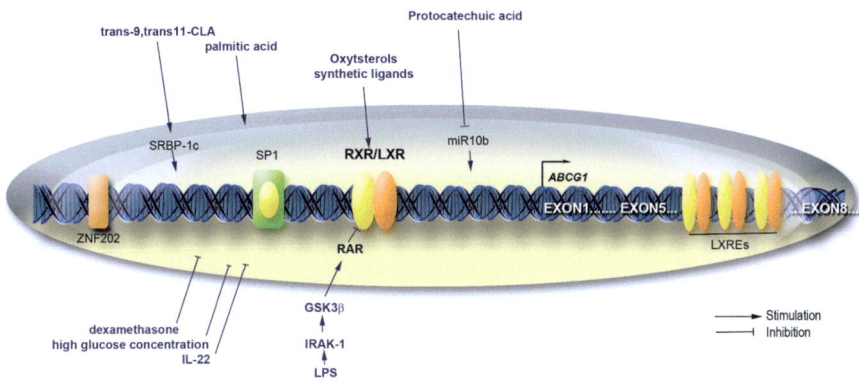

Fig. 3 Regulatory regions, transcription factors, and signaling molecules (cytokines, growth factors, metabolites, drugs) that modulate the expression of the ABCG1 gene. *Arrows* and *block lines* denote activation and repression, respectively. The mechanisms are described in detail in the text. Abbreviations: conjugated linoleic acid (CLA); glycogen synthase kinase 3β (GSK3β); sterol regulatory element-binding protein (SREBP); specificity protein 1 (Sp1)

The regulation of ABCG1 expression has similarities with that of ABCA1 (Fig. 3). Thus, LXR plays an important role in ABCG1 promoter activation. Experimental data indicated that various LXR ligands upregulate ABCG1 expression. ATI-111, a novel steroidal LXR agonist, induces ABCG1 mRNA expression in peritoneal macrophages more potently than T0901317 and inhibits its expression in the liver, suggesting tissue selectivity (Peng et al. 2011). Different natural compounds, such as cineole and fucosterol that are LXR activators, are able to increase ABCG1 levels in macrophages (Hoang et al. 2012; Jun et al. 2013). Recent data showed that the knockdown of LXRα impaired cholesterol efflux in human primary macrophages, while LXRβ silencing had no detectable impact on the expression of LXR target genes such as ABCA1 and ABCG1 and did not affect cholesterol efflux (Ishibashi et al. 2013). The indirect effects of LXR in ABCG1 regulation were recently shown. Adiponectin treatment significantly increased ABCG1 mRNA and protein levels in macrophages from diabetic patients, whereas the pharmacological or genetic inhibition of LXR abrogated this enhancement; these data demonstrated that the mechanism of adiponectin-mediated upregulation of ABCG1 includes LXRα (Wang et al. 2013b). Similar with ABCA1, TGFβ upregulates the expression of ABCG1, while unsaturated fatty acids suppress ABCG1 expression via the LXR pathway (Uehara et al. 2007; Hu et al. 2010b).

The analysis of potential regulatory elements in the promoter region carried out using the MatInspector program identified multiple Sp1 sites at positions −184, −382, and −566, an AP2 binding site at position −222, a NF-κB site at position −338, an E-box motif at position −233, a sterol regulatory element at position −660, and the NFY binding site at position −198 in ABCG1 proximal promoter (Langmann et al. 2000). A functional genetic variant of the ABCG1 promoter associated with an increased risk of myocardial infarction and ischemic heart disease in the general population was revealed (Schou et al. 2012). This study showed that the ABCG1 expression was decreased by approximately 40 % in g.-376C>T heterozygotes versus noncarriers. This gene polymorphism is included in a Sp1 binding site located at position −382/−373 in the ABCG1 promoter. Thus, the presence of the −376 T allele reduced the binding and transactivation of the promoter by Sp1, leading to a decreased ABCG1 expression (Schou et al. 2012).

ZNF202 was identified as a transcriptional repressor of ABCG1 gene which binds at position −560 in the ABCG1 promoter (Porsch-Ozcurumez et al. 2001).

Recently, microbiotic and dietary factors were shown to regulate the ABCG1 expression. Protocatechuic acid (PCA), a gut microbiota metabolite of cyanidin-3 formed by 0-β-glucoside, exerts an anti-atherogenic effect partially through the inhibition of miR-10b-mediated downregulation of ABCG1 expression (Wang et al. 2012a). Extra-virgin olive oil intake has been shown to improve the capacity of HDL to mediate cholesterol efflux and increased ABCG1 and ABCA1 expression in human macrophages (Helal et al. 2013). Conjugated linoleic acids (CLAs) are minor components of the diet with many reported biological activities. It was revealed that in MCSF-differentiated monocytes, *trans*-9,*trans*-11-CLA, but not *cis*-9,*trans*-11-CLA and *trans*-10,*cis*-12-CLA, activated ABCG1 via SREBP-1c (Ecker et al. 2007). In addition, it was demonstrated that palmitic acid upregulates the ABCG1 gene, while high glucose concentration decreased ABCG1 gene expression in MCSF-activated human monocytes (Mauerer et al. 2009).

Among the downregulators of ABCG1, low doses of LPS strongly reduce the expression of ABCG1 in bone marrow-derived macrophages through IL-1 receptor-associated kinase 1 (IRAK-1)/glycogen synthase kinase 3β (GSK3β)/retinoic acid receptor α (RARα) signaling pathway (Maitra and Li 2013). HMG-CoA reductase inhibitors, simvastatin and atorvastatin, decreased ABCG1-mediated cholesterol efflux in human macrophages, despite the fact that the protein expression remained unaltered (Wang et al. 2013c). In THP-1 monocytes, 100 nM dexamethasone, a synthetic glucocorticoid, inhibited the mRNA expression of ABCG1, although the glucocorticoid receptor expression was very low in this cell line (Sporstol et al. 2007). IL-22, a member of the IL-10 cytokine family secreted primarily by Th17 and Th22 subsets of T lymphocytes, was induced by S100/calgranulin and impaired cholesterol efflux in macrophages by downregulation of ABCG1 (Chellan et al. 2013).

1.5 Transcriptional Regulation of the Apolipoprotein E Gene

Apolipoprotein E (apoE), a glycoprotein of 35 kDa, plays an important role in plasma cholesterol level regulation and in cholesterol efflux, as documented by studies in patients and animal models with apoE deficiency or mutated apoE genes (Nakashima et al. 1994; Linton et al. 1995; von Eckardstein 1996; Van Eck et al. 1997; Grainger et al. 2004; Ali et al. 2005; Davignon 2005; Raffai et al. 2005). ApoE is mainly synthesized by the liver but also by various cells and peripheral tissues (Zannis et al. 2001b). ApoE is a marker for the developmental state of macrophages; the culture of mouse bone marrow cells in vitro showed that mature macrophages, but not their monocytic precursors, synthesized apoE (Werb and Chin 1983c). At the site of atherosclerotic lesion, apoE is provided by infiltrated macrophages. Transgenic mice expressing apoE only in macrophages are protected against atherosclerosis, even though the plasma levels of apoE are exceedingly low and the animals are hypercholesterolemic (Bellosta et al. 1995). In contrast, transgenic mice with normal levels of apoE in plasma, but not in macrophages, are more susceptible to atherosclerosis (Fazio et al. 1997). ApoE secreted by macrophages within the atherosclerotic plaque facilitates the cholesterol efflux to exogenous acceptors (such as HDL), thus assisting the reverse cholesterol transport to the liver. The uptake of acetylated LDL or cholesterol ester-rich β-VLDL into peritoneal macrophages stimulates apoE synthesis and secretion (Basu et al. 1981).

The human apoE gene is located on chromosome 19 at the 5′ end of a cluster containing also apoC-I, apoC-IV, and apoC-II genes (Myklebost and Rogne 1988; Smit et al. 1988; Allan et al. 1995a, b). The regulation of apolipoprotein E gene transcription is a highly complex process and requires the interaction of transcription factors with the proximal promoters but also with the distal regulatory regions (Fig. 4).

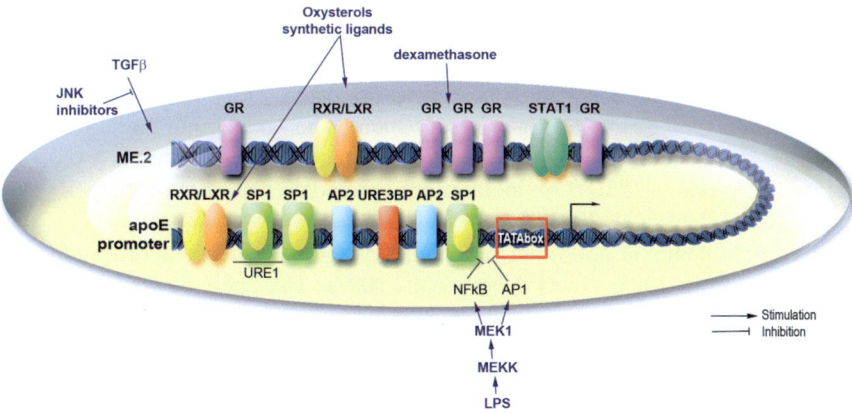

Fig. 4 Regulatory regions, transcription factors, and signaling molecules (cytokines, growth factors, metabolites, drugs) that modulate the expression of the apoE gene. *Arrows* and *block lines* denote activation and repression, respectively. The mechanisms are described in detail in the text. Abbreviations: glucocorticoid receptor (GR); signal transducer and activator of transcription (STAT); Jun N-terminal kinase (JNK); retinoid X receptor (RXR); liver X receptor (LXR); specificity protein 1 (Sp1); activator protein 2 (AP2); nuclear factor kappa beta (NF-κB); upstream regulatory region 3 binding protein (URE3BP); lipopolysaccharide (LPS); multienhancer 2 (ME.2); transforming growth factor β (TGFβ)

1.5.1 Proximal Regulatory Binding Sites Involved in the apoE Gene Expression

The proximal apoE promoter is well conserved in humans and mice, having the same localization of TATA box and GC box (Rajavashisth et al. 1985; Horiuchi et al. 1989). Multiple positive and negative elements that modulate apoE gene expression have been detected on the apoE promoter, using different in vitro systems (Larkin et al. 2000; Zannis et al. 2001b). Smith et al. analyzed the apoE promoter in both expressing (HepG2) and non-expressing (HeLa) cells (Smith et al. 1988). Within the proximal 5′-flanking sequence and the first intron, eight regions were identified which had a positive effect and three regions with a negative effect on apoE expression, in both HepG2 and HeLa cells (Smith et al. 1988). The proximal apoE promoter contains a GC box transcriptional control element at $-59/-45$, a nonspecific enhancer element at $-366/-246$, an upstream regulatory element (URE1) at $-193/-124$, and a downstream regulatory element at $+44/+262$ (Paik et al. 1988). Within URE1, a sequence spanning $-161/-141$, defined as a positive element for transcription, has the ability to act alone as an enhancer element (Chang et al. 1990). This element interacts with Sp1 transcription factor that constitutively binds the GC box motif, suggesting that Sp1 may play an important role in the basal level of apoE expression, as well as in the activity of this enhancer element. Another regulatory element, termed URE3, was identified at position $-101/-89$ and found to bind a 300 kDa protein from placental nuclear extracts termed URE3 BP (Jo et al. 1995). DNase I footprinting revealed the existence of two binding sites for recombinant AP2 in the regions from $-48/-74$ and from $-107/-135$ of the apoE promoter (Olaisen et al. 1982; Smith et al. 1988; Garcia et al. 1996; Salero et al. 2001, 2003). Gel mobility-shift assays showed the direct binding of LXRα/RXRα and LXRβ/RXRα to a low-affinity LXRE present in the region $-494/-465$ of the proximal promoter (Laffitte et al. 2001b). Other studies revealed that USF binds to an atypical E-box located in the $-101/-91$ region of the apoE promoter (Salero et al. 2003). The same group found that Zic1 and Zic2 transcription factors can bind to three binding sites located at $-65/-54$, $-136/-125$, and $-185/-174$ in the apoE promoter and stimulate apoE gene expression (Salero et al. 2001).

Bacterial endotoxin and other inflammatory agents decrease apoE production (Werb and Chin 1983a, b; Gafencu et al. 2007). The apoE downregulation in macrophages impaired the local beneficial effect of apoE during the plaque development. As a result, despite the fact that macrophages are present in the lesion, their ability to regress atherosclerosis is seriously compromised. We have previously reported the mechanisms of apoE downregulation in macrophages exposed to inflammatory conditions, similar to those found at the atherosclerotic site (Gafencu et al. 2007). Tumor progression locus 2 (Tpl2) and mitogen-activated protein kinase/ERK kinase kinase 1 (MEKK1) were identified as the kinases that are primarily responsible for the downregulation of apoE promoter activity by LPS. Tpl2 and MEKK1 signaling pathways converge to NF-κB and AP1, acting on the apoE core promoter $-55/+73$ (Gafencu et al. 2007).

1.5.2 Distal Regulatory Binding Sites That Modulate apoE Gene Expression in Macrophages

Despite this complex transcription factor machinery that may be targeted to the apoE promoter, the promoter itself lacks the ability to direct gene transcription in vivo in any cells, in the absence of the distal enhancers (Shih et al. 2000). In many tissues, cell-specific distal enhancers regulate the expression of genes in the apoE/apoC-I/apoC-I'/apoC-IV/apoC-II gene cluster (Shih et al. 2000). The expression of apoE in macrophages is controlled by two homologous enhancers (95 % identical in sequence), designated as multienhancer 1 (ME.1) and multienhancer 2 (ME.2), containing 620 and 619 nucleotides, respectively (Shih et al. 2000). These enhancers are located at 3.3 and 15.9 kb downstream of the apoE gene, respectively. We demonstrated by chromosome conformational capture (3C) and transient transfections that both ME.1 and ME.2 can interact with the apoE promoter only in phorbol 12-myristate 13-acetate (PMA)-differentiated macrophages, but not in undifferentiated monocytes (Trusca et al. 2011). The results showed that the interactions take place in antisense orientation of the promoter and ME.1/2. Our data obtained using a series of deletion mutants of the promoter or of the ME.2 identified the fragment $-100/+73$ as the minimal region of the apoE promoter that is activated by the ME.2. We showed that the entire sequence of ME.2 is necessary for an optimal interaction with the apoE promoter, but the 5' region of ME.2 is more important than 3' region for enhancing apoE promoter activity (Trusca et al. 2011). The interaction of the apoE promoter with ME.1/2 facilitates the transcriptional enhancement of the apoE gene by various transcription factors.

LXRα and LXRβ and their oxysterol ligands are key regulators of apoE expression in macrophages (Laffitte et al. 2001b; Joseph et al. 2002b; Mak et al. 2002b). The ability of oxysterols and synthetic ligands to regulate apoE expression in peritoneal macrophages as well as in adipose tissue is reduced in LXRα$^{-/-}$ or LXRβ$^{-/-}$ mice and abolished in double knockouts. However, basal expression of apoE is not compromised in LXR null mice, suggesting that LXRs mediate lipid-inducible expression rather than tissue-specific expression of this gene (Laffitte et al. 2001b). Data revealed that LXR/RXR binds to a low-affinity LXRE present in the apoE promoter as well as a high-affinity site conserved in both human ME.1 and ME.2 (Laffitte et al. 2001b). Experimental data revealed that the ligand activation of the LXR/RXR heterodimer enhanced the activity of the reporter constructs under the control of human ME.1 or ME.2 fused to the apoE proximal promoter (Laffitte et al. 2001b). Oxysterol-binding protein-related protein 1S (ORP1S) translocates from the cytoplasm to the nucleus in response to sterol binding and then binds to LXRs, promoting the binding of LXRs to LXREs. Thus, ORP1S mediates the LXR-dependent transcription via the ME.1 and ME.2 of the apoE gene (Lee et al. 2012). An interesting finding was that the induction of apoE gene expression by LXR agonists is attenuated by inhibitors of JNK and PI3K pathways (Huwait et al. 2011). A similar inhibition was noticed in the case of TGFβ-induced expression of apoE, which was prevented by pharmacological inhibitors of JNK, p38 kinase, and casein kinase 2 (Singh and Ramji 2006).

The synthetic glucocorticoid receptor (GR) ligand, dexamethasone, increased apoE mRNA levels in mature macrophages up to sixfold over basal levels (Zuckerman et al. 1993). In silico analysis of the ME.1 and ME.2 revealed some transcription factor binding motifs for the GR (Shih et al. 2000). The presence of these GR binding sites in the multienhancers may explain the apoE upregulation by GR ligands, but the biological activity of these GR biding sites remains to be revealed.

TRANSFAC analysis of the apoE promoter, ME.1, and ME.2 showed that STAT1 transcription factor has a binding site only on the ME.2. Our data showed that this binding site is biologically active and STAT1 specifically upregulates apoE gene expression via ME.2, in macrophages, but not in hepatocytes. The STAT1 binding site was located in the 174/182 region of ME.2 (Trusca et al. 2011). Interestingly, a simultaneous increase in the expression of apoE and STAT1 was recorded after monocyte differentiation with PMA treatment (for 4 h). Our model proposed that after DNA bending, which probably takes place during monocyte differentiation, STAT1 bound on ME.2 interacts with the transcription initiation complex, leading to the activation of apoE expression. In addition, STAT1 can interact and cooperate with other transcription factors bound on the ME.2 or on the apoE promoter, for the modulation of apoE gene expression. Recently, we have revealed that STAT1 can interact with RXR and modulate gene expression of the apoC-II gene (Trusca et al. 2012). Since RXRα binds to the ME.2, we can speculate that apoE expression in macrophages may be modulated by the STAT1-RXRα interactions, similarly with apoC-II.

1.6 Transcriptional Regulation of the Human apoM Gene in the Liver

Apolipoprotein M (apoM) belongs to the lipocalin protein superfamily and differs from typical water-soluble apolipoproteins by its tertiary structure (Dahlback and Nielsen 2009; Nielsen et al. 2009). ApoM is secreted primarily by the liver and associates with HDL particles through its retained N-terminal signal peptide (Axler et al. 2008; Christoffersen et al. 2008). The silencing of the endogenous apoM gene in mice showed a loss of pre-β-HDL particles and formation of large HDL particles (Wolfrum et al. 2005). In addition to its role in HDL remodeling, it was shown that apoM is the sole carrier of the bioactive lipid sphingosine 1 phosphate (S1P) in HDL, thus mediating many of the atheroprotective properties of HDL in the endothelium (Christoffersen et al. 2011; Arkensteijn et al. 2013; Christoffersen and Nielsen 2013).

The expression of apoM in the liver is primarily controlled by hepatocyte nuclear factor-1α (HNF-1α) (Richter et al. 2003). HNF-1α$^{-/-}$ mice are characterized by the complete absence of apoM from plasma. The plasma concentrations of other apolipoproteins in HNF-1α$^{-/-}$ mice were either similar (apoA-II, apoB, apoC) or increased (apoA-I, apoE) compared to wild-type mice. This was not due to the absence of apoM since restoration of apoM gene expression in the liver via adenovirus-mediated gene transfer could not rescue the abnormal apolipoprotein profile (Wolfrum et al. 2005). The analysis of the plasma lipoprotein profile of HNF-1$\alpha^{-/-}$ mice showed that similar to the apoM gene-silenced mice, plasma cholesterol was primarily associated with the HDL fraction. In addition, an abnormal large apoE-enriched HDL fraction that was identified as HDLc or HDL1 was observed, suggesting that this abnormal lipid profile in HNF-1$\alpha^{-/-}$ mice may be caused by the lack of apoM (Shih et al. 2001). In humans, HNF-1α regulates apoM gene expression through direct binding to a conserved DNA element located in the proximal apoM promoter region between nucleotides -55 and -41 (Richter et al. 2003).

ApoM gene expression in the liver is negatively regulated during inflammation or infection via pro-inflammatory cytokines such as TNFα or IL-1β (Feingold et al. 2008). The HNF-1α binding element in the proximal human apoM promoter is a dual-specificity regulatory element that mediates the activation or repression of apoM promoter activity by HNF-1 and by activator protein 1 (AP1) proteins (c-Jun and JunB), respectively, in hepatic cells (Mosialou et al. 2011). Competition experiments showed that the binding of Jun proteins and HNF-1α to the apoM promoter is mutually exclusive and chromatin immunoprecipitation assays established that AP1 activation leads to the recruitment of c-Jun and JunB proteins to the proximal apoM promoter with the simultaneous displacement of HNF-1 (Mosialou et al. 2011). A similar mechanism of transcriptional repression via dual-specificity AP1-/HNF-1-responsive elements has been demonstrated in the case of the promoter of the human apolipoprotein A-II gene (Mosialou et al. 2011). AP1 factors were shown to inhibit the promoters of the apolipoprotein C-III (Hadzopoulou-Cladaras et al. 1998) and ABCA1 (Mosialou and Kardassis unpublished) genes in hepatic cells suggesting a broader role of AP1 factors in lipoprotein metabolism in the liver during inflammation.

Besides HNF-1α, apoM gene transcription in the liver has been shown to be controlled positively by liver receptor homologue-1 (LRH-1) and forkhead box A2 (FOXA2) transcription factors which bind to distinct sites on the proximal apoM promoter (Venteclef et al. 2008; Wolfrum et al. 2008). Bile acids suppress apoM expression in vivo by inhibiting LRH-1 transcriptional activity via the recruitment of small heterodimer partner (SHP) to the apoM promoter (Venteclef et al. 2008).

Insulin, insulin-like growth factor I (IGF-I), and IGF-I potential peptide (IGF-IPP) were all shown to inhibit apoM gene expression in a dose- and time-dependent manner in primary human and murine hepatocytes via a signal transduction pathway that involves the serial activation of phosphatidylinositol 3-kinase (PI3K) and protein kinase B (PKB) and the inactivation of Foxa2 (Xu et al. 2006). In HepG2 cells, glucose and insulin inhibited apoM gene expression in an additive manner, while in hyperglycemic rats, serum apoM concentrations and hepatic apoM mRNA levels were significantly reduced (Zhang et al. 2007).

The human apoM gene is under the control of various orphan- and ligand-dependent nuclear receptors (Mosialou et al. 2010). The overexpression via adenovirus and silencing via siRNA established that HNF-4 is an important regulator of apoM gene transcription in hepatic cells (Mosialou et al. 2010). In addition to

HNF-4, homodimers of retinoid X receptor and heterodimers of retinoid X receptor with receptors for retinoic acid, thyroid hormone, fibrates (peroxisome proliferator-activated receptor), and oxysterols (liver X receptor) were shown to bind with different affinities to the proximal HRE in vitro and in vivo (Mosialou et al. 2010). These findings provide novel insights into the role of apoM in the regulation of HDL by steroid hormones and into the development of novel HDL-based therapies for diseases such as diabetes, obesity, metabolic syndrome, and coronary artery disease that affect a large proportion of the population in Western countries.

1.7 Transcriptional Regulation of the CETP Gene

The gene encoding the cholesterol ester transfer protein (CETP) plays an important role in human HDL metabolism because it facilitates the transfer of cholesteryl esters from mature spherical HDL particles to VLDL/IDL lipoproteins in exchange of triglycerides and its activity determines the plasma levels of HDL cholesterol (von Eckardstein et al. 2005; Tall et al. 2008).

The CETP gene is expressed mainly in the liver, adipose tissue, and spleen and at lower levels in the small intestine, adrenal, kidneys, and heart (Jiang et al. 1991; Radeau et al. 1995). Atherogenic diets were shown to increase CETP mRNA levels in rabbits and in human CETP transgenic mice (Quinet et al. 1990; Jiang et al. 1992). Both LXRs and SREBPs were shown to bind to regulatory elements on the promoter of the CETP gene and regulate its transcription in response to intracellular cholesterol levels (Gauthier et al. 1999; Luo and Tall 2000). It was shown recently that synthetic LXR agonists enhanced plasma CETP activity and decreased HDL-C levels in cynomolgus monkeys and human CETP transgenic mice (Honzumi et al. 2010). The induction of CETP gene expression by the LXR agonist was significantly reduced by knocking down the expression of LXRα but not LXRβ both ex vivo and in mice (Honzumi et al. 2010). In another study, it was shown that the LXR agonist T0901317 markedly increased CETP mRNA levels and CETP production in human differentiated macrophages but not in human peripheral blood monocytes (Lakomy et al. 2009). In inflammatory mouse and human macrophages, LXR-mediated CETP gene upregulation was inhibited and this inhibition was independent of lipid loading. It was concluded that LXR-mediated induction of human CETP expression is switched on during monocyte-to-macrophage differentiation and is abrogated in inflammatory macrophages (Lakomy et al. 2009).

Other factors that regulate CETP promoter activity include Yin Yang 1 (YY1) that binds to the same element as SREBPs (Gauthier et al. 1999), the LRH-1 that potentiates the sterol-mediated induction of the CETP gene by LXRs (Luo et al. 2001), the orphan nuclear receptor ARP-1 (Gaudet and Ginsburg 1995) that inhibits CETP promoter activity, the retinoic acid receptor (RAR) which regulates CETP gene expression in response to all-*trans*-retinoic acid (Jeoung et al. 1999), and the CCAAT/enhancer-binding protein (C/EBP) which is an activator of CETP gene expression (Agellon et al. 1992). Binding sites for the ubiquitous transcription factors SP1 and SP3 have been identified on the CETP promoter at positions −690, −623, and −37 and seem to be essential for the basal CETP promoter activity (Le Goff et al. 2003).

CETP gene expression was shown recently to be under regulation by bile acids and their nuclear receptor farnesoid X receptor (FXR) (Gautier et al. 2013). It was shown that plasma CETP activity and mass was higher in patients with cholestasis than controls and this was associated with lower HDL-C levels (Gautier et al. 2013). In agreement with this observation, bile acid feeding of APOE3*Leiden mice expressing the human CETP transgene controlled by its endogenous promoter decreased HDL-C and increased plasma CETP activity and mass. An FXR response element (FXRE) was identified in the first intron of the human CETP gene which could be responsible for the upregulation of CETP gene expression in response to bile acids (Gautier et al. 2013). In another study, it was shown that FXRα binds to DR4 LXRE that is present in the proximal CETP promoter and represses LXR-mediated transactivation of the CETP promoter by a competition mechanism (Park et al. 2008).

1.8 Transcriptional Regulation of the PLTP Gene

Phospholipid transfer protein (PLTP) belongs to the lipopolysaccharide (LPS) binding/lipid transfer gene family that includes the LPS-binding protein (LBP), the neutrophil bactericidal/permeability-increasing protein (BPI), and the cholesteryl ester transfer protein (CETP). PLTP is essential in the transfer of very low-density lipoprotein phospholipids into HDL (Jiang et al. 2012).

PLTP is expressed ubiquitously, but the highest expression levels in human tissues were observed in the ovary, thymus, placenta, and lung (Day et al. 1994). Taking into account the organ size involved, the liver and small intestine appear to be important sites for the overall PLTP expression. A high-fat, high-cholesterol diet causes a significant increase in PLTP activity and in mRNA levels. Plasma PLTP activity and PLTP mRNA levels in the liver and adipose tissues were significantly decreased following LPS administration (Jiang and Bruce 1995).

An FXR-responsive element (FXRE) has been found in the proximal PLTP promoter that binds FXRα/RXRα heterodimers and mediates the response of the PLTP promoter to bile acids (Urizar et al. 2000). Fibrates were shown to increase PLTP gene expression by activating PPARs which bind to three PPAR-responsive elements on the PLTP promoter (Tu and Albers 1999; Bouly et al. 2001). Two of these PPAR-responsive elements also seem to be responsible for the induction of PLTP expression by high glucose (Tu and Albers 2001).

The human PLTP promoter contains at least two LXR-responsive elements, one in the proximal and one in the distal region, that were shown to mediate PLTP gene regulation by oxysterols ex vivo and in vivo (Cao et al. 2002; Mak et al. 2002a; Laffitte et al. 2003). It was recently demonstrated that LXR agonists activate triglyceride synthesis and PLTP gene transcription by activating SREBP-1c

(Okazaki et al. 2010). In concert with the increase in triglyceride synthesis, the increased PLTP caused triglyceride incorporation into abnormally large VLDL particles which were removed from plasma by LDL receptors, whereas in the absence of LDL receptors, the large VLDLs accumulated and caused massive hypertriglyceridemia (Okazaki et al. 2010).

Recently, microarray analysis following alteration of p53 status in several human- and mouse-derived cells identified a group of 341 genes whose expression was induced by p53 in the liver-derived cell line HepG2 (Goldstein et al. 2012). Twenty of these genes encode proteins involved in many aspects of lipid homeostasis including PLTP (Goldstein et al. 2012).

1.9 Transcriptional Regulation of the Bile Acid Transporters ABCG5/ABCG8

ATP-binding cassette half-transporters G5 and G8 (ABCG5 and ABCG8) play important roles in the control of sterol excretion from the liver (Fitzgerald et al. 2010; Tarling and Edwards 2012; Li et al. 2013; Yu et al. 2014). Mutations in either of these transporters leads to β-sitosterolemia, an autosomal recessive disease characterized by premature coronary atherosclerosis and elevated levels of phytosterols in plasma (Fitzgerald et al. 2010; Tarling and Edwards 2012; Li et al. 2013; Yu et al. 2014). Mice lacking ABCG5 and ABCG8 proteins have decreased ability to secrete sterols into the bile (Yu et al. 2002a). The overexpression of ABCG5 and ABCG8 in the liver increases biliary cholesterol secretion and decreases dietary cholesterol absorption (Yu et al. 2002b). The human ABCG5 and ABCG8 genes are oriented in a head-to-head configuration, they are transcribed in opposite directions, and their transcription is coordinated by a short 374 bp bidirectional promoter in the intergenic region (Remaley et al. 2002).

The bidirectional promoter of ABCG5/ABCG8 genes contains a binding site for LRH-1 at positions 134–142 which is required for the activity of both the ABCG5 and ABCG8 promoters (Freeman et al. 2004). Mutating this LRH-1 binding site reduced promoter activity of the human ABCG5/ABCG8 intergenic region in HepG2 and Caco2 cells. Bile acids such as deoxycholic acid repressed ABCG5 and ABCG8 promoters via the FXR-SHP-LXR pathway that was described above (Sumi et al. 2007).

Dietary cholesterol feeding was shown to increase duodenal, jejunal, and hepatic expression levels of ABCG5 and ABCG8 mRNA in wild-type mice (Berge et al. 2000). The increase in ABCG5 or ABCG8 gene expression by diet was compromised in mice lacking either LXRα or both LXRα and LXRβ (Repa et al. 2002). Both the RXR-specific agonist LG268 and the LXR-specific agonist T0901317 caused upregulation of ABCG5 and ABCG8 mRNA expression in the liver and intestine of wild-type mice but not in LXRα/β$^{-/-}$ mice (Repa et al. 2002). To identify functional LXREs that control the expression of the ABCG5/ABCG8 genes in response to oxysterols, a recent study searched for evolutionarily conserved regions (ECRs) between the human and the mouse genes and identified

23 ECRs which were studied by luciferase assays for LXR responsiveness (Back et al. 2013). Two ECRs were found to be responsive to the LXR and binding of LXRα to these regions was verified (Back et al. 2013).

The bidirectional promoter of the ABCG5/G8 genes was shown to bind HNF-4 and GATA transcription factors and to be regulated by these factors in a cooperative manner and independent of the orientation of the bidirectional promoter (Sumi et al. 2007).

It was shown that the expression of both ABCG5 and ABCG8 genes is upregulated in the livers of mice with genetic ablation of the insulin receptor gene (LIRKO mice) both at the mRNA and the protein levels (Biddinger et al. 2008). In agreement with these findings, insulin suppressed the expression of ABCG5 and ABCG8 genes at subnanomolar concentrations and in a dose-responsive manner in rat hepatoma cells (Biddinger et al. 2008). The observation that the short intergenic region responded to insulin in both the ABCG5 and ABCG8 orientations suggested the presence of an element in the intragenic region of the ABCG5 and ABCG8 genes that responds to insulin. Using ex vivo and in vivo approaches, it was shown that insulin resistance leads to the activation of the forkhead box 1 (FOXO1) transcription factor which binds to the bidirectional promoter and activates the transcription of both genes severalfold (Biddinger et al. 2008).

1.10 Transcriptional Regulation of the HDL Receptor SR-BI

The gene encoding the HDL receptor scavenger receptor class B type I (SR-BI) is expressed at high levels in the liver and steroidogenic tissues. Several transcription factors have been shown to bind to the human or rodent SR-BI promoter and to regulate SR-BI gene transcription in a positive or negative manner.

The steroidogenic factor-1 (SF-1) has been shown to regulate both the human and rat SR-BI promoters and to serve as mediator of the cAMP-dependent regulation of the SR-BI gene in response to steroidogenic hormones (Lopez et al. 1999).

Liver X receptors α and β and PPARα and γ were shown to bind to distal LXRE and PPARE, respectively, on the human and rat SR-BI promoters and regulate the expression of the human SR-BI gene in response to oxysterols and fibrates (Lopez and McLean 1999; Malerod et al. 2002, 2003), whereas HNF-4 enhances the PPARγ-mediated SR-BI gene transcription (Malerod et al. 2003; Zhang et al. 2011). As discussed above, conditional inactivation of the HNF-4 gene in the liver of adult mice was associated with a significant increase in hepatic SR-BI mRNA levels and a decrease in plasma HDL-C levels, suggesting that HNF-4 influences negatively the expression of the HDL receptor (Hayhurst et al. 2001).

LRH-1 binds to a proximal response element on the human SR-BI promoter in an overlapping manner with SF-1 and activates the SR-BI promoter (Schoonjans et al. 2002). Retrovirus-mediated overexpression of LRH-1 in hepatic cells induced SR-BI gene expression, and this was associated with histone H3 acetylation on the SR-BI promoter. In agreement with these findings, the SR-BI mRNA levels were decreased in the livers of LRH-1(+/−) animals providing evidence that LRH-1 regulates SR-BI gene expression in vivo (Schoonjans et al. 2002).

Estrogens regulate the activity of the rat SR-BI promoter via estrogen receptors α and β (ERα and β) which bind to three different estrogen response elements (ERE) on the SR-BI promoter (Lopez et al. 2002; Lopez and McLean 2006). In endothelial cells, 17beta-estradiol (E2) increased the mRNA levels of the human SR-BI gene and the activity of the hSR-BI promoter, and this upregulation was protein kinase C (PKC) dependent since it was blocked by the PKC inhibitor bisindolylmaleimide I and a dominant-negative mutant of PKC (Fukata et al. 2013).

The mRNA levels of the mouse SR-BI gene were decreased in mice lacking the FXR nuclear receptor ($FXR^{-/-}$ mice) (Lambert et al. 2003). When WT mice were placed on a diet containing 0.4 % of the FXR agonist cholic acid, the hepatic SR-BI mRNA and protein levels increased in the wild-type but not in the $FXR^{-/-}$ mice, indicating that bile acids positively regulate SR-BI gene expression via FXRs (Lambert et al. 2003). In agreement with these findings, treatment of human hepatoma HepG2 cells with FXR ligands resulted in the upregulation of SR-BI both at the mRNA and protein levels via FXR binding to a novel FXRE, a direct repeat 8 at position −703/−684 of the promoter (Chao et al. 2010). A natural ligand of FXR administered to mice increased hepatic SR-BI expression (Chao et al. 2010). However, in another study, it was reported that bile acids inhibit SR-BI gene expression in the liver of mice and reduce the SR-BI promoter activity (Malerod et al. 2005). It was proposed that this inhibition was due to the FXR-mediated activation of SHP, which repressed the activity of LRH-1 that binds to the proximal SR-BI promoter (Malerod et al. 2005).

The zinc finger transcription factor Kruppel-like factor 4 (KLF4) was shown to bind to a putative KLF4 element on the SR-BI promoter at position −342/−329 and upregulate its activity in peripheral blood mononuclear cells and PMA-differentiated THP-1 cells treated with HDL (Yang et al. 2010).

Using high-throughput screening of 6,000 microbial secondary metabolite crude extracts for the identification of compounds that upregulate the SR-BI promoter in HepG2 cells, several putative SR-BI upregulators were identified: hoxyl-3′,5,7-hydroxyl isoflavone; (9R,13S)-7-deoxy-13-dihydrodaunomycinone; pratensein; the isoflavones formononetin, genistein, and daidzein; and the histone deacetylase inhibitor trichostatin A (Yang et al. 2007, 2009; Bao et al. 2009). Some of these compounds were shown to increase the uptake of DiI-labeled HDL and the efflux of cholesterol to HDL by cells.

In steroidogenic tissues, SR-BI supplies the cells with exogenous cholesterol for storage or for the synthesis of steroid hormones. In these tissues, SR-BI expression was shown to be upregulated by adrenocorticotropic hormone (ACTH) (Sun et al. 1999). The suppression of ACTH by the synthetic corticosteroid dexamethasone (which inhibits the hypothalamic-pituitary axis and decreases ACTH secretion) decreased SR-BI levels. However, the mechanism by which ACTH and glucocorticoid regulate the expression of the SR-BI gene in steroidogenic tissues is unclear. It was shown that the transcription of the human SR-BI gene is subject to feedback inhibition by glucocorticoids in adrenal and ovarian cells. SR-BI mRNA levels were increased in adrenals from corticosterone-insufficient $Crh^{-/-}$ mice, whereas corticosterone replacement by oral administration inhibited SR-BI gene expression in these mice (Mavridou et al. 2010). The glucocorticoid-mediated inhibition of SR-BI gene transcription required de novo protein synthesis and the glucocorticoid receptor (GR). No direct binding of GR to the SR-BI promoter could be demonstrated in vitro and in vivo, suggesting an indirect mechanism of repression of SR-BI gene transcription by GR in adrenal cells (Mavridou et al. 2010).

In the rat ovary, the uptake of cholesterol by the theca-interstitial cells is mediated by SR-BI. It was shown that insulin and the trophic hormone LH/cGH stimulate the expression of SR-BI in theca-interstitial cells and increase intracellular cholesterol, which is subsequently mobilized for androgen biosynthesis (Li et al. 2001; Towns et al. 2005).

In the adrenocortical cell line Y-1, adenovirus-mediated overexpression of prolactin regulatory element-binding (PREB) protein, a transcription factor that regulates prolactin expression in the anterior pituitary and is induced by cAMP, increased the levels of SR-BI protein and the activity of the SR-BI promoter by binding to a PREB-responsive *cis*-element of the SR-BI promoter. Using small interfering RNA against PREB in Y-1 cells, the effects of cAMP on SR-BI expression were attenuated. It was concluded that in the adrenal gland under conditions of cAMP increase, PREB regulates the transcription of the SR-BI gene (Murao et al. 2008).

Intracellular sterol levels regulate SR-BI gene expression via SREBP-1a which binds to two sterol-responsive elements (SREs) present on the rat SR-BI promoter (Cao et al. 1997; Lopez and McLean 1999). The ubiquitous transcription factors SP1 and SP3 bind to several GC-rich boxes present on the proximal SR-BI promoter and have been shown to be important for the basal activity as well as the SREBP-1a-mediated transactivation of the SR-BI promoter (Shea-Eaton et al. 2001).

The expression of the SR-BI gene is also subject to negative regulation by transcription factors including the orphan nuclear receptor dorsal-sensitive sex adrenal hypoplasia congenital critical region on the X chromosome gene 1 (DAX-1), which was shown to repress the rat SR-BI promoter by interfering with the SF-1- and SREBP-1a-mediated transactivation of the this promoter (Lopez et al. 2001). In addition, YY1 transcription factor binds directly to two sites on the SR-BI promoter and downregulates its activity by interfering with the binding of SREBP-1a to the SR-BI promoter (Shea-Eaton et al. 2001).

2 Posttranscriptional Regulation of HDL Genes by Noncoding RNAs and microRNAs

According to the Encyclopedia of DNA Elements (ENCODE project), 76 % of the human genome is transcribed (Bernstein et al. 2012; Djebali et al. 2012). While novel promoter and enhancer regions have been described (Bernstein et al. 2012; Sanyal et al. 2012), some of the newly described genes encode for noncoding RNAs including microRNAs (miRNAs), small interfering RNAs (siRNAs), piwi-interacting RNAs (piRNAs), circular RNAs (circRNAs), long noncoding RNAs (lncRNAs), *trans*-acting siRNAs (tasiRNAs), and several other noncoding RNAs. Their abundance and their recognized roles in transcriptional and epigenetic gene regulation and their involvement in several diseases suggest the existence of an extensive regulatory network on the basis of RNA signaling (Mattick and Makunin 2005). An important class of these endogenous noncoding RNAs is miRNAs. miRNAs are small noncoding RNAs (~22 nt) that have emerged as important posttranscriptional regulators of different protein-coding genes (Bartel 2009) including those related to HDL metabolism (Davalos and Fernandez-Hernando 2013).

2.1 miRNAs: Biogenesis and Function

Although originally described as regulators of developmental timing in *Caenorhabditis elegans* (Lee et al. 1993; Wightman et al. 1993), miRNAs did not receive special attention until their widespread identification in different species (Lagos-Quintana et al. 2002; Reinhart et al. 2002) and their role in human diseases was uncovered (Calin et al. 2002). miRNAs are important posttranscriptional regulators of gene expression through sequence-specific complementary binding to the 3′ untranslated region (UTR) of the target mRNA (Bartel 2009), even though certain miRNAs can interact with other target mRNA regions including the 5′ UTR, coding region, or intron–exon junction and even increase rather than decrease target mRNA expression (Vasudevan et al. 2007; Orom et al. 2008; Tay et al. 2008; Schnall-Levin et al. 2010). The interaction of a miRNA with its target mRNA results in the inhibition of translation and/or degradation of mRNAs (Guo et al. 2010; Krol et al. 2010).

miRNA biogenesis and function have been intensively studied in recent years (Bartel 2009; Krol et al. 2010). Briefly, in most mammals, these small RNAs are transcribed by RNA polymerase II into a primary long miRNA (pri-miRNA). The pri-miRNA is then processed in the nucleus into an ~70 nucleotide precursor hairpin (pre-miRNA) by a multiprotein complex containing different cofactors and two core components, a ribonuclease III (Drosha) and a double-stranded RNA-binding domain protein (DGCR8/Pasha). Some intronic miRNAs that bypass Drosha-mediated processing are produced by splicing and debranching (miRtrons). Pre-miRNAs are transported to the cytoplasm by Exportin 5 in a Ran-GTP-dependent manner (Lund et al. 2004) where they are further cleaved by an RNAse III enzyme called Dicer, generating the mature ~22 nt long miRNA/miRNA* duplex. Dicer-independent miRNA biogenesis has also been described (Cifuentes et al. 2010). The duplex miRNA is separated and one of the strands associates with an argonaute protein (Ago) within the RNA-induced silencing complex (RISC), guiding the complex to the complementary sites within the 3′ UTR of the target mRNA to induce mRNA repression and/or degradation. Although the miRNA/miRNA* duplex is initially produced in equal amounts, the miRNA* strand (passenger strand) is usually degraded. However, it also contains target recognition sites and is thus functional (Yang et al. 2011). Argonaute proteins are major players in miRNA-mediated gene regulation (Meister 2013). Although several aspects of miRNA biogenesis and function are well characterized, the different factors that regulate their decay or turnover are not well known (Krol et al. 2010).

Sequencing studies have identified ~2,000 miRNAs encoded in our human genome, and prediction algorithms suggest that 60 % of human protein-coding genes have conserved targets for pairing with miRNAs within their 3′ UTR (Friedman et al. 2009). Factors that influence miRNA target selection include the "seed" sequence which consists of nucleotides 2–8 at the 5′ end of the mature miRNA (Bartel 2009) and tissue distribution or developmental stage, as certain miRNAs are highly expressed or even restricted to certain cell types and can only target their mRNA target if they are co-expressed in the same tissue at the same time (Lagos-Quintana et al. 2002; Small and Olson 2011). In contrast to most plant miRNAs, which bind with near perfect complementarity to their targets, most mammalian miRNAs bind with mismatches and bulges and the mode of binding determines the type of posttranscriptional repression (Carthew and Sontheimer 2009). Recent experimental data suggest that around 60 % of seed interactions are noncanonical, containing bulged or mismatched nucleotides which are accompanied by specific, non-seed base pairing (Helwak et al. 2013). Some miRNA-mRNA interactions also involve the 3′ end of the miRNA (Helwak et al. 2013). Moreover, pseudogenes (Poliseno et al. 2010), long noncoding RNAs (lncRNAs) (Cesana et al. 2011), and circular RNAs (Hansen et al. 2013) that contain miRNA binding sites also influence miRNA activity, acting as competing endogenous RNAs (ceRNAs) and thus sequestering miRNAs and preventing them from binding to their mRNA targets (Salmena et al. 2011).

While the primary role of miRNAs seems to be the "fine-tuning" of gene expression, their capacity to simultaneously bind and repress multiple target genes and similarly the possibility of a single mRNA to be targeted by multiple miRNAs provide a mechanism to synchronize the coordinated regulation of a large number of transcripts that govern an entire biological process, thus resulting in strong phenotypic output (Mendell and Olson 2012). The high redundancy among related and non-related miRNAs in regulating gene expression reduces the importance of a particular miRNA under conditions of normal cellular homeostasis. However, compelling evidence suggests that under conditions of stress the function of miRNAs become especially pronounced (Mendell and Olson 2012) and their pharmacological modulation represents a novel approach to treat disease by modulating entire biological pathways as described in Davalos and Chroni (2014).

Compelling evidences have shown that miRNAs are present in the systemic circulation and other human biological fluids associate with extracellular microvesicles, exosomes (Valadi et al. 2007; Hunter et al. 2008), Ago2 complex (Arroyo et al. 2011), or HDL (Vickers et al. 2011). Even exogenous miRNAs have been described in our circulating plasma (Wang et al. 2012b; Zhang et al. 2012). Although several circulating miRNAs have been implicated as disease biomarkers in several diseases (Allegra et al. 2012; Kinet et al. 2013) or secreted for intercellular communication (Chen et al. 2012b), the biological significance, the factors that modulate their extracellular secretion, and the mechanisms by which they reach the target tissue still remain elusive. HDL has been shown to transport and deliver endogenous miRNAs to recipient cells with functional targeting capabilities (Vickers et al. 2011). Although the HDL miRNA signature from normal subjects was found to be different from hypercholesterolemic subjects (Vickers et al. 2011), the extent by which HDL-bound miRNAs contribute to total circulating miRNAs and thus regulate target cell gene function is not clear (Wagner et al. 2013).

In summary, miRNAs play a major role in almost every aspect of cellular function by posttranscriptional regulation of gene expression. Whether other noncoding RNAs (that modulate miRNA activity) or circulating or extracellular miRNAs might modulate HDL metabolism is not known. However, it seems obvious that by modulating miRNA activity, they might also have a role in HDL metabolism.

2.2 Posttranscriptional Modulation of HDL Metabolism by miRNAs

For normal function, mammalian cells must maintain cellular cholesterol levels within tight limits. Thus, complex mechanisms have been evolutionarily developed to control both cellular input from endogenous cholesterol biosynthesis or the uptake from circulating lipoproteins and cellular output by controlling cholesterol efflux. miRNAs have emerged as important regulators of HDL and cholesterol metabolism, and their functions are summarized in Fig. 5 and described in detail below. Although miRNAs have increased the complexity of HDL metabolism regulation, our understanding of every single step in their regulation will provide better opportunities to develop novel targets for their therapeutic modulation.

2.2.1 Targeting ABCA1 and ABCG1

ABCA1 is one of the most important proteins directly involved in the elimination of excess cholesterol from cells (cholesterol efflux). As described in this chapter, ABCA1 is regulated at the transcriptional, posttranscriptional, and posttranslational level. Its regulation is in accordance with the cellular need to handle cholesterol levels within tight limits, as excess of free cholesterol is deleterious to cells. ABCA1 mRNA contains a particularly long $3'$ UTR (>3.3 kb) as compared to other common genes involved in cholesterol and HDL metabolism including LCAT (20 bp), apoA-I (55 bp), ApoA-II (112 bp), CETP (178 bp), apoB (301 bp), PDZK1 (583 bp), ABCG1 (852 bp), SR-BI (959 bp), PCSK9 (1,269 bp), IDOL (1,496 bp), CAV-1 (1,898 bp), LIPG (2,386 bp), and LDLR (2,513 bp) (Davalos and Fernandez-Hernando 2013). This unusually long $3'$ UTR of ABCA1 clearly raises the probability of posttranscriptional regulation by miRNAs (and probably other noncoding RNAs). Different prediction algorithms indicate that ABCA1 can potentially be targeted by ~100 miRs. Some of them have been experimentally validated

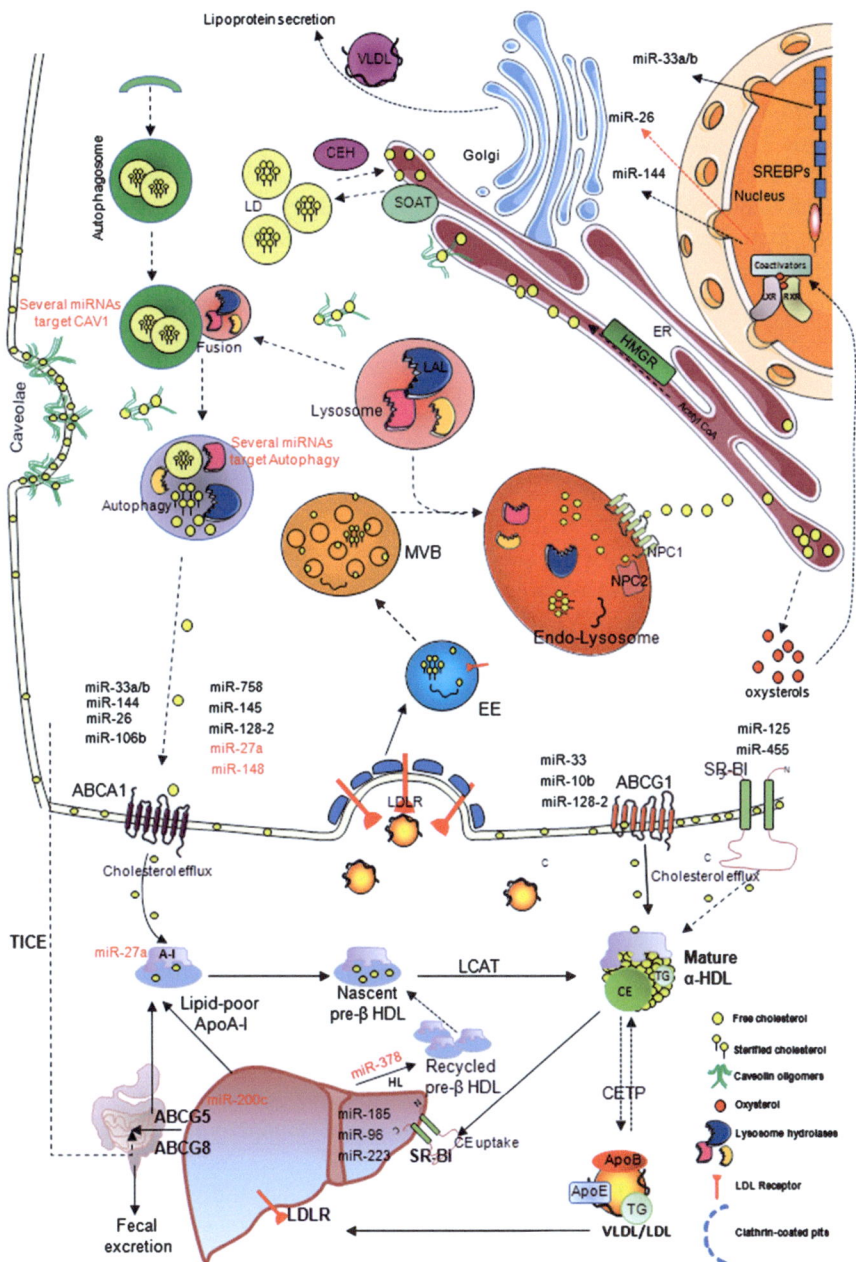

Fig. 5 Overview of posttranscriptional regulation by miRNAs of proteins involved in cholesterol efflux, RCT, and HDL metabolism. HDL metabolism and the role of individual proteins such as plasma enzymes, membrane transporters, and receptors are described in the text and more extensively in other chapters. Autophagy, the cell catabolism process through the lysosomal machinery, has been proposed to participate in cholesterol efflux. The induction of SREBPs induces the expression of miR-33 family. The induction of LXR induces the expression of miR-144 and represses miR-26. The induction of FXR also induces miR-144. miRNAs in black

for their importance in cholesterol efflux, reverse cholesterol transport, and cardiovascular disease, whereas other miRNAs still remain to be elucidated. Validated miRNAs that directly target ABCA1 are miR-33 family, miR-758, miR-106b, miR-26, miR-144, miR-10b, miR-128-2, and miR-145 (Fig. 5).

miR-33a and miR-33b play a crucial role in controlling cholesterol efflux and HDL function in concert with their host genes, the SREBP transcription factors (Najafi-Shoushtari et al. 2010; Horton et al. 2002; Horie et al. 2010; Marquart et al. 2010; Rayner et al. 2010). While both mature miRNAs only differ in two nucleotides, they are predicted to have largely overlapping sets of target genes in rodents (Rayner et al. 2011b; Horie et al. 2012) and nonhuman primates (Rayner et al. 2011a; Rottiers et al. 2013). The inhibition of the two miR-33 isoforms, either genetically or therapeutically, resulted in increased cholesterol efflux, increased HDL levels, increased reverse cholesterol transport, reduced atherosclerosis, and reduced VLDL triglyceride levels. Although miR-33a and miR-33b target different genes involved in lipid homeostasis, fatty acid β-oxidation, insulin signaling, and biliary transporters (Gerin et al. 2010; Davalos et al. 2011; Rayner et al. 2011a; Allen et al. 2012; Horie et al. 2013), their effects are mediated mainly by ABCA1. The human 3′ UTR of ABCA1 has three functional binding sites for miR-33. While most of the studies have been performed in tissues/cells directly related to lipoprotein metabolism and CVD (i.e., hepatocytes, macrophages), it is possible that from these ~100 miRs that potentially target ABCA1 3′ UTR, some of them might be relevant to other cells and tissues as well.

Regarding other tissues, miR-106b (Kim et al. 2012) and miR-758 (Ramirez et al. 2011) were found to regulate neuronal cholesterol excess. While miR-758 expression is particularly high in the brain and its expression is mediated by high cholesterol levels, the expression of miR-106 in brain tissues is low. However, their role in neuronal cholesterol efflux suggests that these miRNAs may contribute to the regulation of cholesterol levels in the brain. miR-758 and miR-106 directly target the 3′ UTR of ABCA1 by binding to two sites and one site, respectively. Moreover, miR-758 has other targets involved in neurological functions (Ramirez et al. 2011). miR-106b also targets the amyloid precursor protein (APP) and increases the amyloid β (Aβ) peptide secretion and clearance (Kim et al. 2012).

HDL metabolism is regulated by the LXRs which control the transcription of several genes related to lipid metabolism in response to hydroxylated products of cholesterol as described above (Calkin and Tontonoz 2012). Recent data suggest that LXRs also regulate HDL metabolism posttranscriptionally (Sun et al. 2012; de Aguiar Vallim et al. 2013; Ramirez et al. 2013; Vickers and Rader 2013). The induction of LXR resulted in the repression of miR-26 and induction of miR-26-a-1,

Fig. 5 (continued) labels are validated miRNAs for HDL metabolism. miRNAs in red labels are suggested miRNAs, but not fully validated, in HDL metabolism. All biological processes may not necessarily happen in the same cell type. Abbreviations: endoplasmic reticulum (ER); early endosome (EE); cholesterol ester hydrolase (CEH); multivesicular bodies (MVB); lysosomal acid lipase (LAL) (Figure courtesy of Dr. Alberto Canfrán-Duque)

and miR-144 that directly target the 3' UTR of ABCA1 and thus regulate cholesterol efflux, RCT, and HDL levels. While having one binding site for miR-26-a-1 (Sun et al. 2012), the human 3' UTR of ABCA1 has several (up to seven) predicted binding sites for miR-144 (Ramirez et al. 2013). miR-26-a-1 also targets the ADP-ribosylation factor-like 7 (Arl7), which participates in cellular cholesterol efflux (Engel et al. 2004). miR-144 expression is also induced by the nuclear receptor FXR (Vickers and Rader 2013). By controlling bile acid levels, FXR activation will not only repress the expression of ABCA1 posttranscriptionally but will also induce the expression of SR-BI, thereby resulting in an increased uptake of plasma HDL cholesterol and increased cholesterol biliary excretion via ABCG5/ABCG8 (de Aguiar Vallim et al. 2013).

Evolutionarily, the dual effect of LXR activation on miR-144 induction and mR-26-a-1 repression seems obvious, as cells must maintain cellular cholesterol levels within tight limits and these miRNAs might work as buffers against deleterious variation in gene expression programs.

Other miRNAs including miR-10b, miR-128-2, and miR-145 also regulate cellular cholesterol efflux by directly targeting the 3' UTR of ABCA1. Protocatechuic acid, an intestinal microbiota metabolite of cyanidin-3-O-glucoside, was found to repress miR-10b and thus regulate cholesterol efflux (Wang et al. 2012a). While in hepatocytes the inhibition of miR145 increases cholesterol efflux, in pancreatic islet beta cells it improves glucose-stimulated insulin secretion (Kang et al. 2013). The proapoptotic miRNA miR-128-2 was also found to regulate cholesterol efflux by targeting (one binding site) the 3' UTR of ABCA1 (Adlakha et al. 2013). Recent data suggest that this miRNA family governs neuronal excitability and motor behavior in mice (Tan et al. 2013) and that miR-128-2 might control neuronal cholesterol levels. RXRα is also a direct target of miR-128-2 (Adlakha et al. 2013). miR-27a and miR-148 were found to repress an ABCA1 3' UTR luciferase reporter construct; however, their physiological role in regulating cholesterol efflux has not been described (Kang et al. 2013).

ABCG1, the other member of the ATP-binding cassette transporter family, which also participates in cellular cholesterol efflux to HDL, was found to be directly regulated by some of these miRNAs. miR-33, miR-10b, and miR-128-2 regulate cholesterol efflux by posttranscriptionally regulating ABCG1 through binding to its 3' UTR (Rayner et al. 2010; Wang et al. 2012a; Adlakha et al. 2013).

2.2.2 Targeting SR-BI

Liver regulation of SR-BI is primarily achieved posttranslationally through a PDZ domain-containing adaptor protein (PDZK1) (Kocher et al. 2003; Kocher and Krieger 2009) as described in detail in a later section of this chapter. Increasing evidence suggests that other posttranscriptional mechanisms, such as targeting by miRNAs, also regulate SR-BI protein levels. Prediction miRNA analysis (www.targetscan.org) suggests that ~25 miRNA families might target SR-BI, most of which are poorly conserved among mammals and vertebrates. miR-125a and miR-455 were found to repress the lipoprotein-supported steroidogenesis by directly targeting the 3' UTR of SR-BI (Hu et al. 2012). In hepatic cells, where

SR-BI is expressed at higher levels, the expression of miR-125a was also high and its overexpression resulted in a reduced SR-BI-mediated selective cholesterol ester uptake. In contrast, miR-455 did not show any effect on hepatic SR-BI expression (Hu et al. 2012). miR-185, miR-96, and miR-223 were also found to directly target the 3′ UTR of SR-BI and to repress HDL cholesterol uptake in hepatic cells (Wang et al. 2013a).

2.2.3 Targeting Other miRNAs Related to HDL Biogenesis and Function

miR-27a was suggested to regulate apoA-I plasma levels, but the direct interaction to its 3′ UTR was not evaluated (Shirasaki et al. 2013). Several other miRNAs were proposed to regulate genes involved in cholesterol efflux, RCT, or HDL metabolism, including ABCG5 (Liu et al. 2012) or endothelial lipase (Kulyte et al. 2013), but either their direct interaction with the 3′ UTR of these genes or their physiological effects on HDL metabolism were not directly assessed.

Autophagy is the catabolic process by which unnecessary or dysfunctional cellular components are degraded in the lysosome. During starvation, this breakdown of cellular components contributes to the maintenance of cellular energy levels. The hydrolysis of cytoplasmic cholesteryl ester of lipid droplets is normally mediated by the action of neutral cholesteryl ester hydrolases. It has been recently recognized that lipid droplets can also be delivered to lysosomes via autophagy, where lysosomal acid lipase can hydrolyze lipid droplet cholesteryl esters to generate free cholesterol for cholesterol efflux (Ouimet et al. 2011).

Several miRNAs have been described that regulate different targets in autophagy (Frankel and Lund 2012). miRNAs that target key pathways in lipid-loaded macrophage autophagy genes and/or cholesterol ester hydrolases might be promising targets to promote cholesterol efflux (Davalos and Fernandez-Hernando 2013). Caveolin, the major protein coat of caveolae, has been proposed to contribute to cellular cholesterol efflux (Truong et al. 2010; Kuo et al. 2011). There is accumulating evidence that several miRNAs including miR-103, miR-107, miR-133a, miR-192, miR-802, and others target caveolin (Nohata et al. 2011; Trajkovski et al. 2011) but their contribution to cholesterol efflux, RCT, and HDL metabolism and their real physiological contribution to HDL function remain to be elucidated.

3 Posttranslational Mechanisms of HDL Regulation

3.1 ABCA1

In addition to the transcriptional and posttranscriptional mechanisms described above, the expression of ABCA1 is also regulated at the posttranslational level. Following its synthesis, ABCA1 is inserted into the ER where it undergoes proper folding, N-glycosylation, dimerization, and disulfide bond formation and transport to the plasma membrane via the ER-Golgi system (Fig. 6) (Kang et al. 2010).

Regulation of HDL Genes: Transcriptional, Posttranscriptional, and... 151

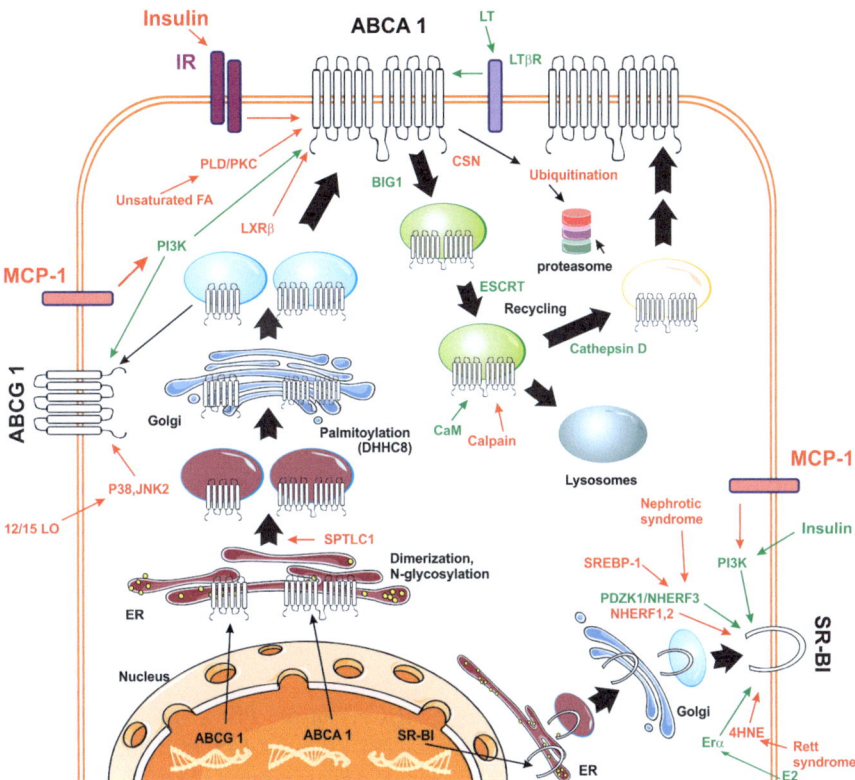

Fig. 6 Overview of the mechanisms that regulate the expression of ABCA1, ABCG1, and SR-BI genes at the posttranslational level. The mechanisms are described in the text. Abbreviations: endoplasmic reticulum (ER); Asp-His-His-Cys 8 (DHHC8); serine palmitoyltransferase 1 (SPTLC1); guanine nucleotide-exchange protein 1 (BIG1); endosomal sorting complex required for transport (ESCRT); Pro-GLu-Ser-Thr (PEST); calmodulin (Cam); lymphotoxin (LT); LT-β receptor (LTβR); phospholipase D (PLD); protein kinase C δ (PKCδ); insulin receptor (IR); phosphatidylinositol 3-kinase (PI3K); 12/15-lipoxygenase (12/15LO); PDZ-containing kidney protein 1 (PDZK1); Na^+/H^+ exchanger regulator factor-3 (NHERF3); 4-hydroxy-2-nonenal (4HNE); monocyte chemoattractant protein-1 (MCP-1)

Transport from the ER to the Golgi is prevented when ABCA1 interacts with the enzyme serine palmitoyltransferase 1 (SPTLC1) which participates in the biosynthetic pathway of sphingomyelin, a major phospholipid component of membranes. The ABCA1/SCPTL1 interaction may regulate the levels of intracellular pools of sphingolipids when the cellular demands for these lipids are high. Pharmacological inhibition of SPTLC1 with myriocin, an atypical amino acid and antibiotic derived from thermophilic fungi (Miyake et al. 1995), increases both the ABCA1 levels at the plasma membrane and the ABCA1-mediated cholesterol efflux (Tamehiro et al. 2008). Mutations in ABCA1 that prevent normal glycosylation cause failure of ABCA1 to exit the ER (Singaraja et al. 2006). In the Golgi compartment, ABCA1 undergoes palmitoylation at cysteine residues 3, 23, 1110, and 1111 by the palmitoyltransferase Asp-His-His-Cys 8 (DHHC8) (Singaraja et al. 2009). Inhibiting this enzyme by drugs or preventing ABCA1 palmitoylation by site-specific mutagenesis reduced ABCA1 levels at the plasma membrane and decreased efflux to apoA-I (Singaraja et al. 2009). The gene encoding cathepsin D, a lysosomal protease, was identified by comparative transcriptomic analysis to be associated with low HDL-C levels in humans (Haidar et al. 2006). Blocking the activity or expression of cathepsin D reduced ABCA1 expression and protein abundance as well as apoA-I-mediated lipid efflux by more than 70 % and caused retention of ABCA1 in lysosomal compartments (Haidar et al. 2006). Very recently, BIG1 (brefeldin A-inhibited guanine nucleotide-exchange protein 1) was found to modulate ABCA1 trafficking and functions in the liver cells (Lin et al. 2013). BIG1 depletion reduced surface ABCA1 on HepG2 cells and inhibited by 60 % cholesterol efflux, whereas BIG1 overexpression had the opposite effects. RNA interference with BIG1 dramatically decreased the internalization and recycling of ABCA1. This novel function of BIG1 was dependent on the guanine nucleotide-exchange activity and achieved through the activation of ADP-ribosylation factor 1 (Lin et al. 2013).

The transport of ABCA1 to the plasma membrane is also facilitated by certain members of the family of small GTPase including Rab8, Rab4A, and Rab4B (Linder et al. 2009) (Jean and Kiger 2012).

Calpain, a cysteine protease, cleaves ABCA1 at the PEST (proline, glutamic acid, serine, threonine) sequence (amino acids 1283–1306) that is located in the cytosolic loop of the molecule, and this cleavage takes place at the early endosomes (Fig. 6) (Yokoyama et al. 2012; Miyazaki et al. 2013). The deletion of the PEST sequence increases ABCA1 levels on the plasma membrane and apoA-I binding (Chen et al. 2005). The interaction with apoA-I stabilizes ABCA1 against this degradation (Arakawa and Yokoyama 2002; Wang et al. 2003; Arakawa et al. 2004). Importantly, the inhibition of calpain was shown to increase HDL biogenesis in cultured cells, indicating that the inhibition of proteolytic degradation of ABCA1 could be a promising strategy for increasing HDL (Lu et al. 2008). The phosphorylation of the PEST sequence of ABCA1 at Thr-1286 and Thr-1305 regulates calpain-mediated proteolysis of ABCA1 since the ABCA1-T1286A/T1305A mutant could not be degraded by calpain (Martinez et al. 2003). The interaction of ABCA1 with apoA-I results in the dephosphorylation of the ABCA1 PEST sequence and inhibition of calpain degradation, leading to an increase of ABCA1 cell-surface expression (Martinez et al. 2003). Calmodulin was also shown to interact with ABCA1 in the presence of Ca^{2+} and to protect from calpain-mediated degradation (Iwamoto et al. 2010).

In addition to the calpain-mediated proteolytic degradation of ABCA1 described above, ABCA1 is subject to ubiquitin-mediated proteolysis. The presence of ubiquitinated ABCA1 in the plasma membrane of several cell lines was recently demonstrated (Mizuno et al. 2011). In HuH-7 cells, the degradation of cell-surface-resident ABCA1 was inhibited by the overexpression of a dominant-negative form of ubiquitin. Moreover, the disruption of the endosomal sorting complex required for transport (ESCRT) pathway by the knockdown of hepatocyte growth factor-regulated tyrosine kinase substrate (HRS) delayed the degradation of ABCA1 (Mizuno et al. 2011). These findings suggested that ubiquitination mediates the lysosomal degradation of plasma membrane ABCA1 and thereby regulates the expression and cholesterol efflux functions of ABCA1 independently of the calpain-mediated pathway.

Unsaturated fatty acids such as oleate and linoleate were shown to destabilize ABCA1 protein and to inhibit ABCA1-mediated cholesterol efflux in macrophages by a mechanism that depends on the activity of the enzyme acyl-CoA synthetase 1 which is responsible for their activation to their CoA derivatives (Kanter et al. 2012). At the same time, unsaturated fatty acids increase the Ser phosphorylation of ABCA1 via a phospholipase D (PLD)/protein kinase C δ pathway which contributes to ABCA1 destabilization (Fig. 6) (Wang and Oram 2007).

Two additional proteins that were identified by two-hybrid screen to physically interact with ABCA1 and to regulate ABCA1 intracellular localization and turnover by interacting with its PDZ (PSD-95, Dlg, ZO-1) domain are the a1 and b1 syntrophins (Munehira et al. 2004). Given the short half-life of ABCA1 ($t_{1/2} = 1$–2 h) (Wang et al. 2003), interfering with the accessory proteins or the enzymes that posttranslationally modify ABCA1 may prove a valuable strategy to regulate ABCA1-mediated HDL biogenesis in the liver and the intestine or to enhance cholesterol efflux in peripheral cells.

Insulin enhances ABCA1 protein degradation in HepG2 cells via PI3K. In addition, it inhibits ABCA1 activity by phosphorylation at Tyr1206 (Nonomura et al. 2011). The kinase that is responsible for this Tyr phosphorylation of ABCA1 is not known, but it was hypothesized that it is the insulin receptor itself.

TNFα and lymphotoxin α (LT) are key inflammatory mediators which also contribute to the atherogenic process (Schreyer et al. 1996, 2002; Pamir et al. 2012). TNF induces ABCA1 mRNA and protein levels as well as cholesterol efflux from cultured macrophages to apoA-I (Gerbod-Giannone et al. 2006). The induction of ABCA1 by TNFα depended primarily on NF-κB (Gerbod-Giannone et al. 2006). It was also shown that the expression of the two TNF receptors is required to mediate full ABCA1 induction by TNFα. In addition, LT increased ABCA1 protein levels by inhibiting protein degradation through the LT-β receptor (LTβR) (Edgel et al. 2010).

ABCA1-mediated cholesterol efflux and ABCA1 protein levels were shown to be decreased by interferon γ (IFNγ) in murine macrophages and macrophage-derived foam cells (Panousis and Zuckerman 2000). This ABCA1 downregulation was an early event in IFNγ-mediated activation of macrophages (Alfaro Leon et al. 2005) and was dependent on signal transducer and activator of transcription 1 (STAT1) since similar effects were not observed in macrophages from STAT1 KO mice (Wang et al. 2002).

Additional protein kinases were shown to affect the activity and stability of ABCA1 at the posttranslational level including protein kinase A (PKA), protein kinase C (PKC), Janus kinase 2 (JAK2), and casein kinase (CK2) (Tang et al. 2004).

ABCA1 activity was shown to be regulated by the nuclear receptor LXRβ by a mechanism that is distinct from the transcriptional induction of the ABCA1 gene caused by cholesterol accumulation. It was shown that at low cholesterol levels, LXRβ binds to ABCA1 and the ABCA1-LXRβ complex is stably localized at the plasma membrane but is unable to facilitate apoA-I-mediated cholesterol efflux. Exogenously added LXR ligands, which mimic cholesterol accumulation, cause LXRβ to dissociate from ABCA1, thus freeing ABCA1 for apoA-I binding and subsequent cholesterol efflux (Hozoji et al. 2008; Hozoji-Inada et al. 2011).

3.2 ABCG1

Similar to ABCA1, the levels of ABCG1 transporter were shown to be subject to regulation at the posttranslational level (Fig. 6) (Tarling and Edwards 2012). Recent studies showed that the presence of additional 12 amino acids between the Walker B motif and the first transmembrane domain of ABCG1 can affect ABCG1 protein stability (Engel et al. 2006). It was shown that stable overexpression of 12/15-lipoxygenase (12/15LO) in macrophages was associated with a 30 % reduction in HDL-mediated cholesterol efflux and reduced ABCG1 protein expression (Nagelin et al. 2008). Treatment of macrophages with the 12/15LO eicosanoid product 12SHETE increased serine phosphorylation of ABCG1 and affected the stability of the protein (Nagelin et al. 2008). Proteasomal inhibitors blocked the downregulation of ABCG1 and resulted in the accumulation of phosphorylated ABCG1 (Nagelin et al. 2009). Macrophages that lack 12/15LO had enhanced ABCG1 expression, reduced ABCG1 phosphorylation, and increased cholesterol efflux. Conversely, macrophages that overexpress 12/15LO have reduced ABCG1 expression, increased transporter phosphorylation, and reduced cholesterol efflux (Nagelin et al. 2009). It was also shown that 12/15LO regulates ABCG1 expression and function through p38- and JNK2-dependent mechanisms (Nagelin et al. 2009).

The activation of adenosine monophosphate-activated protein kinase (AMPK) in human aortic endothelial cells resulted in increased levels of ABCG1 protein via a posttranscriptional mechanism that involved increased stability of ABCG1 mRNA (Li et al. 2010a, b). The aminoimidazole carboxamide ribonucleotide (AICAR)-dependent induction of ABCG1 was associated with increased efflux of cellular cholesterol and 7-ketocholesterol to HDL and protected nitric oxide synthase activity and vascular reactivity (Li et al. 2010a, b).

The protein levels of ABCG1 were shown to be controlled by palmitoylation in both human embryonic kidney 293 cells and in mouse macrophage, J774 (Gu et al. 2013). Five cysteine residues located at positions 26, 150, 311, 390, and 402 in the NH2-terminal cytoplasmic region of ABCG1 were shown to be palmitoylated. The removal of palmitoylation at Cys311 by mutating the residue to Ala (C311A) or Ser significantly decreased ABCG1-mediated cholesterol efflux. On the other hand, the removal of palmitoylation at sites 26, 150, 390, and 402 had no significant effect (Gu et al. 2013).

3.3 SR-BI

The stability of the HDL receptor SR-BI in the hepatocyte plasma membrane is very important for its functions and is subject to posttranscriptional regulation (Fig. 6). SR-BI stability is primarily controlled by its adapter protein, PDZ-containing kidney protein 1 (PDZK1 or NHERF3 for Na^+/H^+ exchanger regulator factor-3), since PZK1-knockout mice exhibit a >95 % reduction in hepatic SR-BI protein (but not mRNA) and are characterized by hypercholesterolemia and the presence of large cholesterol-rich HDL particles in the plasma (Kocher et al. 2003). In adrenal cells where the PDZK1 levels are very low, additional members of the NHERF family such as NHERF1 and NHERF2 (but not NHERF4) were shown to specifically interact with SR-BI and reduce its protein levels (Hu et al. 2013). The data showed that the downregulation of SR-BI by these NHERF1/2 factors significantly inhibits both HDL-CE uptake and steroid hormone production by adrenal cells. It was concluded that the PDZK1 homologues act as physiological translational/posttranslational regulators of the functional expression of SR-BI (Hu et al. 2013).

Feeding mice with an atherogenic diet was accompanied by a threefold post-translational downregulation of hepatic SR-BI at the protein level (Niemeier et al. 2009). A similar SR-BI downregulation was observed in transgenic mice overexpressing SREBP-1a and SREBP-1c on chow diet, and it was associated with a decrease in the expression levels of PDZK1 (Niemeier et al. 2009).

The levels of SR-BI protein were shown to be affected in two syndromes: the nephrotic syndrome and the Rett syndrome. It was shown that animals with nephrotic syndrome which is characterized by dyslipidemia, impaired high-density lipoprotein (HDL)-mediated reverse cholesterol transport, and atherosclerosis (Vaziri et al. 2003) exhibited severe hypercholesterolemia, hypertriglyceridemia, reduced HDL/total cholesterol ratio, significant upregulation of the endocytic HDL receptor ATP synthase mRNA and protein, and significant reduction of SR-BI protein despite its normal mRNA abundance. The reduction in SR-BI protein levels in animals with NS was accompanied by parallel reductions in PDZK1 mRNA and protein levels (Vaziri et al. 2011). Rett syndrome, a genetic form of infantile autism, is the second most common cause of mental retardation in women (Amir et al. 1999). When compared to healthy subjects, patients with Rett syndrome present with significant increases in total cholesterol (12 %), LDL cholesterol (15 %), and HDL-C (18 %) (Sticozzi et al. 2013). Skin fibroblasts isolated from patients with Rett syndrome exhibited low levels of SR-BI as a consequence of its association with 4-hydroxy-2-nonenal (4HNE), a product of lipid peroxidation that is increased in patients with Rett syndrome, and increased ubiquitination (Sticozzi et al. 2013). The role of the proteasome in SR-BI stability was confirmed using the proteasomal inhibitor (MG132). When Rett syndrome fibroblasts were pretreated with MG132, the loss of SR-BI was reversed demonstrating that SR-BI is degraded via the proteasome machinery (Sticozzi et al. 2013).

In hepatic cells and fibroblasts, SR-BI protein levels and SR-BI-mediated cholesterol transport (export and uptake) were shown to be affected by the Ras/MEK/ERK signaling cascade. This effect was mediated via PPARα-inducible degradation pathways (Wood et al. 2011). Metabolic labeling experiments in primary hepatocytes from mice demonstrated that fenofibrate enhances the degradation of SR-BI in a post-endoplasmic reticulum compartment (Lan and Silver 2005). Moreover, fenofibrate-induced degradation of SR-BI was independent of the proteasome, the calpain protease, or the lysosome (Lan and Silver 2005).

In primary macrophages and cell lines derived from female but not from male mice, tamoxifen and 4-hydroxytamoxifen increased SR-BI protein expression via the estrogen receptor α (Dong et al. 2011). Because SR-BI mRNA expression and promoter activity were not influenced by tamoxifen and 4-hydroxytamoxifen, it was demonstrated that the regulation of SR-BI by these substances takes place at the level of protein stability (Dong et al. 2011).

The phosphatidylinositol 3-kinase (PI3K) pathway was shown to affect positively the SR-BI-mediated HDL selective cholesterol ester uptake into human HepG2 cells as this process was compromised in the presence of the PI3K inhibitors wortmannin and LY294002 (Shetty et al. 2006). These inhibitors also blocked the positive effect of insulin on the SR-BI-dependent selective uptake. HDL cell-surface binding, receptor biotinylation studies, and confocal fluorescence microscopy of HepG2 cells expressing green fluorescent protein-tagged SR-BI demonstrated changes in SR-BI subcellular localization and cell-surface expression as a result of PI3K activation (Shetty et al. 2006). These data indicate that PI3K activation stimulates hepatic SR-BI function posttranslationally by regulating the subcellular localization of SR-BI. It was recently shown that monocyte chemo-attractant protein-1 (MCP-1), a protein expressed by endothelial cells under inflammatory conditions, decreases the cell-surface protein expression of ABCA1, ABCG1, and SR-BI in a dose-dependent and time-dependent manner (Huang et al. 2013). It was shown that PI3K activation corrected the MCP-1-induced decreases in the cell-surface protein expression of the three transporters. MCP-1 decreased the lipid uptake by HepG2 cells and the ABCA1-mediated cholesterol efflux to apoA-I, and this was reversed by the activation of PI3K. These data suggested that MCP-1 impairs RCT activity in HepG2 cells by a PI3K-/AKT-mediated posttranslational regulation of ABCA1, ABCG1, and SR-BI cell-surface expression (Huang et al. 2013).

Finally, it was shown that hormones such as triiodothyronine (T3) and thyromimetics and some pharmacological agents such as probucol increase SR-BI levels posttranscriptionally (Leiva et al. 2011).

Conclusions

As discussed earlier in this chapter as well as in other chapters of this handbook, plasma HDL cholesterol levels are determined by the relative rates of HDL biogenesis and HDL catabolism which in turn are mediated by several lipid-carrying proteins, plasma enzymes, and membrane transporters. The deregulation of the activity or the expression of these proteins by mutations or during diseases is associated with non-physiological levels of HDL cholesterol, HDL dysfunction, and predisposition to coronary heart disease. Epidemiological studies as well as experiments in genetically modified animals have provided unequivocal proof that raising HDL cholesterol levels may be therapeutic. Drugs that could specifically increase the rates of HDL biogenesis by inducing the levels of apoA-I, ABCA1, or LCAT are anticipated to be of great clinical benefit. Other HDL proteins such as ABCG1, ABCG5/G8, CETP, and LPL could also provide useful targets for HDL-based therapies that will interfere with HDL maturation or catabolism without compromising HDL functionality. The development of novel HDL-raising drugs that will capitalize on HDL biogenesis will certainly require a thorough understanding of the mechanisms and the molecules that are involved in HDL regulation at the transcriptional, posttranscriptional, or posttranslational level in hepatic cells as well as in other physiologically relevant cells such as the macrophages. For instance, understanding ABCA1 gene regulation in the macrophage has led to the elucidation of the important role of LXRs and their ligands in the reverse cholesterol transport and the development of novel LXR ligands which however have severe lipogenic effects in the liver and are thus of no clinical value today so the challenge here is to develop more specific drugs that will avoid lipogenesis. Novel single- or dual-specificity agonists of peroxisome proliferator-activated receptor isoforms are also anticipated. Understanding how the levels of HDL proteins and their functionality are affected by inflammatory factors will enable us to develop novel drugs to restore HDL levels and functionality and thus reduce the atherosclerotic burden in patients with chronic inflammatory diseases.

In summary, the current challenges in the HDL field are to develop novel therapies that would raise HDL cholesterol levels without compromising HDL functionality and to identify novel HDL-based biomarkers. The advances in the field of HDL regulation may provide crucial insights and tools to achieve this goal.

Acknowledgments This work was supported by the European Cooperation in Science and Technology (COST) Action BM0904; by grants from the Greek Secretariat for Research and Technology (SYNERGASIA-09 and THALIS MIS377286) to Dimitris Kardassis; by grants from the Spanish Instituto de Salud Carlos III (FIS, PI11/00315) to Alberto Dávalos; by a grant of the Romanian National Authority for Scientific Research, CNCS – UEFISCDI, project number PN-II-ID-PCE-2011-3-0591, and the Romanian Academy to Anca Gafencu; and by an NIH grant HL-48739 to Vassilis I. Zannis. We thank Dr. Ovidiu Croitoru for the graphic design.

Open Access This chapter is distributed under the terms of the Creative Commons Attribution Noncommercial License, which permits any noncommercial use, distribution, and reproduction in any medium, provided the original author(s) and source are credited.

References

Adcock IM, Caramori G (2001) Cross-talk between pro-inflammatory transcription factors and glucocorticoids. Immunol Cell Biol 79:376–384
Adlakha YK, Khanna S, Singh R, Singh VP, Agrawal A, Saini N (2013) Pro-apoptotic miRNA-128-2 modulates ABCA1, ABCG1 and RXRalpha expression and cholesterol homeostasis. Cell Death Dis 4:e780

Agellon LB, Zhang P, Jiang XC, Mendelsohn L, Tall AR (1992) The CCAAT/enhancer-binding protein trans-activates the human cholesteryl ester transfer protein gene promoter. J Biol Chem 267:22336–22339

Aiello RJ, Brees D, Bourassa PA, Royer L, Lindsey S, Coskran T et al (2002) Increased atherosclerosis in hyperlipidemic mice with inactivation of ABCA1 in macrophages. Arterioscler Thromb Vasc Biol 22:630–637

Albers M, Blume B, Schlueter T, Wright MB, Kober I, Kremoser C et al (2006) A novel principle for partial agonism of liver X receptor ligands. Competitive recruitment of activators and repressors. J Biol Chem 281:4920–4930

Alfaro Leon ML, Evans GF, Farmen MW, Zuckerman SH (2005) Post-transcriptional regulation of macrophage ABCA1, an early response gene to IFN-gamma. Biochem Biophys Res Commun 333:596–602

Ali K, Middleton M, Pure E, Rader DJ (2005) Apolipoprotein E suppresses the type I inflammatory response in vivo. Circ Res 97:922–927

Allan CM, Walker D, Segrest JP, Taylor JM (1995a) Identification and characterization of a new human gene (APOC4) in the apolipoprotein E, C-I, and C-II gene locus. Genomics 28:291–300

Allan CM, Walker D, Taylor JM (1995b) Evolutionary duplication of a hepatic control region in the human apolipoprotein E gene locus. Identification of a second region that confers high level and liver-specific expression of the human apolipoprotein E gene in transgenic mice. J Biol Chem 270:26278–26281

Allegra A, Alonci A, Campo S, Penna G, Petrungaro A, Gerace D et al (2012) Circulating microRNAs: new biomarkers in diagnosis, prognosis and treatment of cancer (review). Int J Oncol 41:1897–1912

Allen RM, Marquart TJ, Albert CJ, Suchy FJ, Wang DQ, Ananthanarayanan M et al (2012) miR-33 controls the expression of biliary transporters, and mediates statin- and diet-induced hepatotoxicity. EMBO Mol Med 4:882–895

Amir RE, Van den Veyver IB, Wan M, Tran CQ, Francke U, Zoghbi HY (1999) Rett syndrome is caused by mutations in X-linked MECP2, encoding methyl-CpG-binding protein 2. Nat Genet 23:185–188

Arakawa R, Yokoyama S (2002) Helical apolipoproteins stabilize ATP-binding cassette transporter A1 by protecting it from thiol protease-mediated degradation. J Biol Chem 277:22426–22429

Arakawa R, Hayashi M, Remaley AT, Brewer BH, Yamauchi Y, Yokoyama S (2004) Phosphorylation and stabilization of ATP binding cassette transporter A1 by synthetic amphiphilic helical peptides. J Biol Chem 279:6217–6220

Argmann CA, Van Den Diepstraten CH, Sawyez CG, Edwards JY, Hegele RA, Wolfe BM et al (2001) Transforming growth factor-beta1 inhibits macrophage cholesteryl ester accumulation induced by native and oxidized VLDL remnants. Arterioscler Thromb Vasc Biol 21:2011–2018

Arkensteijn BW, Berbee JF, Rensen PC, Nielsen LB, Christoffersen C (2013) The apolipoprotein m-sphingosine-1-phosphate axis: biological relevance in lipoprotein metabolism, lipid disorders and atherosclerosis. Int J Mol Sci 14:4419–4431

Arroyo JD, Chevillet JR, Kroh EM, Ruf IK, Pritchard CC, Gibson DF et al (2011) Argonaute2 complexes carry a population of circulating microRNAs independent of vesicles in human plasma. Proc Natl Acad Sci USA 108:5003–5008

Axler O, Ahnstrom J, Dahlback B (2008) Apolipoprotein M associates to lipoproteins through its retained signal peptide. FEBS Lett 582:826–828

Back SS, Kim J, Choi D, Lee ES, Choi SY, Han K (2013) Cooperative transcriptional activation of ATP-binding cassette sterol transporters ABCG5 and ABCG8 genes by nuclear receptors including Liver-X-Receptor. BMB Rep 46:322–327

Baldan A, Pei L, Lee R, Tarr P, Tangirala RK, Weinstein MM et al (2006) Impaired development of atherosclerosis in hyperlipidemic Ldlr−/− and ApoE−/− mice transplanted with Abcg1−/− bone marrow. Arterioscler Thromb Vasc Biol 26:2301–2307

Bao Y, Yang Y, Wang L, Gao L, Jiang W, Wang L et al (2009) Identification of trichostatin A as a novel transcriptional up-regulator of scavenger receptor BI both in HepG2 and RAW 264.7 cells. Atherosclerosis 204:127–135

Bartel DP (2009) MicroRNAs: target recognition and regulatory functions. Cell 136:215–233

Basu SK, Brown MS, Ho YK, Havel RJ, Goldstein JL (1981) Mouse macrophages synthesize and secrete a protein resembling apolipoprotein E. Proc Natl Acad Sci USA 78:7545–7549

Bavner A, Sanyal S, Gustafsson JA, Treuter E (2005) Transcriptional corepression by SHP: molecular mechanisms and physiological consequences. Trends Endocrinol Metab 16: 478–488

Belalcazar LM, Merched A, Carr B, Oka K, Chen KH, Pastore L et al (2003) Long-term stable expression of human apolipoprotein A-I mediated by helper-dependent adenovirus gene transfer inhibits atherosclerosis progression and remodels atherosclerotic plaques in a mouse model of familial hypercholesterolemia. Circulation 107:2726–2732

Bellosta S, Mahley RW, Sanan DA, Murata J, Newland DL, Taylor JM et al (1995) Macrophage-specific expression of human apolipoprotein E reduces atherosclerosis in hypercholesterolemic apolipoprotein E-null mice. J Clin Invest 96:2170–2179

Berge KE, Tian H, Graf GA, Yu L, Grishin NV, Schultz J et al (2000) Accumulation of dietary cholesterol in sitosterolemia caused by mutations in adjacent ABC transporters. Science 290: 1771–1775

Bernstein BE, Birney E, Dunham I, Green ED, Gunter C, Snyder M (2012) An integrated encyclopedia of DNA elements in the human genome. Nature 489:57–74

Biddinger SB, Haas JT, Yu BB, Bezy O, Jing E, Zhang W et al (2008) Hepatic insulin resistance directly promotes formation of cholesterol gallstones. Nat Med 14:778–782

Blumberg B, Evans RM (1998) Orphan nuclear receptors – new ligands and new possibilities. Genes Dev 12:3149–3155

Bouly M, Masson D, Gross B, Jiang XC, Fievet C, Castro G et al (2001) Induction of the phospholipid transfer protein gene accounts for the high density lipoprotein enlargement in mice treated with fenofibrate. J Biol Chem 276:25841–25847

Bourguet W, Germain P, Gronemeyer H (2000) Nuclear receptor ligand-binding domains: three-dimensional structures, molecular interactions and pharmacological implications. Trends Pharmacol Sci 21:381–388

Calin GA, Dumitru CD, Shimizu M, Bichi R, Zupo S, Noch E et al (2002) Frequent deletions and down-regulation of micro-RNA genes miR15 and miR16 at 13q14 in chronic lymphocytic leukemia. Proc Natl Acad Sci USA 99:15524–15529

Calkin AC, Tontonoz P (2012) Transcriptional integration of metabolism by the nuclear sterol-activated receptors LXR and FXR. Nat Rev Mol Cell Biol 13:213–224

Cao G, Garcia CK, Wyne KL, Schultz RA, Parker KL, Hobbs HH (1997) Structure and localization of the human gene encoding SR-BI/CLA-1. Evidence for transcriptional control by steroidogenic factor 1. J Biol Chem 272:33068–33076

Cao G, Beyer TP, Yang XP, Schmidt RJ, Zhang Y, Bensch WR et al (2002) Phospholipid transfer protein is regulated by liver X receptors in vivo. J Biol Chem 277:39561–39565

Carthew RW, Sontheimer EJ (2009) Origins and mechanisms of miRNAs and siRNAs. Cell 136:642–655

Cesana M, Cacchiarelli D, Legnini I, Santini T, Sthandier O, Chinappi M et al (2011) A long noncoding RNA controls muscle differentiation by functioning as a competing endogenous RNA. Cell 147:358–369

Chang DJ, Paik YK, Leren TP, Walker DW, Howlett GJ, Taylor JM (1990) Characterization of a human apolipoprotein E gene enhancer element and its associated protein factors. J Biol Chem 265:9496–9504

Chao F, Gong W, Zheng Y, Li Y, Huang G, Gao M et al (2010) Upregulation of scavenger receptor class B type I expression by activation of FXR in hepatocyte. Atherosclerosis 213:443–448

Chellan B, Yan L, Sontag TJ, Reardon CA, Hofmann Bowman MA (2013) IL-22 is induced by S100/calgranulin and impairs cholesterol efflux in macrophages by down regulating ABCG1. J Lipid Res 55:443–454

Chen JD, Li H (1998) Coactivation and corepression in transcriptional regulation by steroid/nuclear hormone receptors. Crit Rev Eukaryot Gene Expr 8:169–190

Chen WS, Manova K, Weinstein DC, Duncan SA, Plump AS, Prezioso VR et al (1994) Disruption of the HNF-4 gene, expressed in visceral endoderm, leads to cell death in embryonic ectoderm and impaired gastrulation of mouse embryos. Genes Dev 8:2466–2477

Chen H, Rossier C, Lalioti MD, Lynn A, Chakravarti A, Perrin G et al (1996) Cloning of the cDNA for a human homologue of the Drosophila white gene and mapping to chromosome 21q22.3. Am J Hum Genet 59:66–75

Chen W, Wang N, Tall AR (2005) A PEST deletion mutant of ABCA1 shows impaired internalization and defective cholesterol efflux from late endosomes. J Biol Chem 280:29277–29281

Chen HY, Xu Z, Chen LF, Wang W, Fang Q, Yan XW (2012a) Valsartan and telmisartan abrogate angiotensin II-induced downregulation of ABCA1 expression via AT1 receptor, rather than AT2 receptor or PPARgamma activation. J Cardiovasc Pharmacol 59:570–575

Chen X, Liang H, Zhang J, Zen K, Zhang CY (2012b) Secreted microRNAs: a new form of intercellular communication. Trends Cell Biol 22:125–132

Chinetti G, Lestavel S, Bocher V, Remaley AT, Neve B, Torra IP et al (2001) PPAR-alpha and PPAR-gamma activators induce cholesterol removal from human macrophage foam cells through stimulation of the ABCA1 pathway. Nat Med 7:53–58

Christoffersen C, Nielsen LB (2013) Apolipoprotein M: bridging HDL and endothelial function. Curr Opin Lipidol 24:295–300

Christoffersen C, Ahnstrom J, Axler O, Christensen EI, Dahlback B, Nielsen LB (2008) The signal peptide anchors apolipoprotein M in plasma lipoproteins and prevents rapid clearance of apolipoprotein M from plasma. J Biol Chem 283:18765–18772

Christoffersen C, Obinata H, Kumaraswamy SB, Galvani S, Ahnstrom J, Sevvana M et al (2011) Endothelium-protective sphingosine-1-phosphate provided by HDL-associated apolipoprotein M. Proc Natl Acad Sci USA 108:9613–9618

Christova R (2013) Detecting DNA-protein interactions in living cells-ChIP approach. Adv Protein Chem Struct Biol 91:101–133

Churko JM, Mantalas GL, Snyder MP, Wu JC (2013) Overview of high throughput sequencing technologies to elucidate molecular pathways in cardiovascular diseases. Circ Res 112:1613–1623

Cifuentes D, Xue H, Taylor DW, Patnode H, Mishima Y, Cheloufi S et al (2010) A novel miRNA processing pathway independent of Dicer requires Argonaute2 catalytic activity. Science 328:1694–1698

Claudel T, Sturm E, Duez H, Torra IP, Sirvent A, Kosykh V et al (2002) Bile acid-activated nuclear receptor FXR suppresses apolipoprotein A-I transcription via a negative FXR response element. J Clin Invest 109:961–971

Costet P, Luo Y, Wang N, Tall AR (2000) Sterol-dependent transactivation of the ABC1 promoter by the liver X receptor/retinoid X receptor. J Biol Chem 275:28240–28245

Costet P, Lalanne F, Gerbod-Giannone MC, Molina JR, Fu X, Lund EG et al (2003) Retinoic acid receptor-mediated induction of ABCA1 in macrophages. Mol Cell Biol 23:7756–7766

Cui L, Schoene NW, Zhu L, Fanzo JC, Alshatwi A, Lei KY (2002) Zinc depletion reduced Egr-1 and HNF-3beta expression and apolipoprotein A-I promoter activity in Hep G2 cells. Am J Physiol Cell Physiol 283:C623–C630

Dahlback B, Nielsen LB (2009) Apolipoprotein M affecting lipid metabolism or just catching a ride with lipoproteins in the circulation? Cell Mol Life Sci 66:559–564

Davalos A, Chroni A (2014) Antisense oligonucleotides, microRNAs and antibodies. In: von Eckardstein A, Kardassis D (eds) High density lipoproteins – from biological understanding to clinical exploitation. Springer, Heidelberg

Davalos A, Fernandez-Hernando C (2013) From evolution to revolution: miRNAs as pharmacological targets for modulating cholesterol efflux and reverse cholesterol transport. Pharmacol Res 75:60–72

Davalos A, Goedeke L, Smibert P, Ramirez CM, Warrier NP, Andreo U et al (2011) miR-33a/b contribute to the regulation of fatty acid metabolism and insulin signaling. Proc Natl Acad Sci USA 108:9232–9237

Davignon J (2005) Apolipoprotein E and atherosclerosis: beyond lipid effect. Arterioscler Thromb Vasc Biol 25:267–269

Day JR, Albers JJ, Lofton-Day CE, Gilbert TL, Ching AF, Grant FJ et al (1994) Complete cDNA encoding human phospholipid transfer protein from human endothelial cells. J Biol Chem 269: 9388–9391

de Aguiar Vallim TQ, Tarling EJ, Kim T, Civelek M, Baldan A, Esau C et al (2013) MicroRNA-144 regulates hepatic ATP binding cassette transporter A1 and plasma high-density lipoprotein after activation of the nuclear receptor farnesoid X receptor. Circ Res 112:1602–1612

del Castillo-Olivares A, Gil G (2000) Alpha 1-fetoprotein transcription factor is required for the expression of sterol 12alpha-hydroxylase, the specific enzyme for cholic acid synthesis. Potential role in the bile acid-mediated regulation of gene transcription. J Biol Chem 275: 17793–17799

Delerive P, Galardi CM, Bisi JE, Nicodeme E, Goodwin B (2004) Identification of liver receptor homolog-1 as a novel regulator of apolipoprotein AI gene transcription. Mol Endocrinol 18: 2378–2387

Djebali S, Davis CA, Merkel A, Dobin A, Lassmann T, Mortazavi A et al (2012) Landscape of transcription in human cells. Nature 489:101–108

Dong P, Xie T, Zhou X, Hu W, Chen Y, Duan Y et al (2011) Induction of macrophage scavenger receptor type BI expression by tamoxifen and 4-hydroxytamoxifen. Atherosclerosis 218:435–442

Duez H, Lefebvre B, Poulain P, Torra IP, Percevault F, Luc G et al (2005) Regulation of human apoA-I by gemfibrozil and fenofibrate through selective peroxisome proliferator-activated receptor alpha modulation. Arterioscler Thromb Vasc Biol 25:585–591

Durrington PN, Mackness MI, Bhatnagar D, Julier K, Prais H, Arrol S et al (1998) Effects of two different fibric acid derivatives on lipoproteins, cholesteryl ester transfer, fibrinogen, plasminogen activator inhibitor and paraoxonase activity in type IIb hyperlipoproteinaemia. Atherosclerosis 138:217–225

Ecker J, Langmann T, Moehle C, Schmitz G (2007) Isomer specific effects of Conjugated Linoleic Acid on macrophage ABCG1 transcription by a SREBP-1c dependent mechanism. Biochem Biophys Res Commun 352:805–811

Edgel KA, Leboeuf RC, Oram JF (2010) Tumor necrosis factor-alpha and lymphotoxin-alpha increase macrophage ABCA1 by gene expression and protein stabilization via different receptors. Atherosclerosis 209:387–392

Engel T, Lueken A, Bode G, Hobohm U, Lorkowski S, Schlueter B et al (2004) ADP-ribosylation factor (ARF)-like 7 (ARL7) is induced by cholesterol loading and participates in apolipoprotein AI-dependent cholesterol export. FEBS Lett 566:241–246

Engel T, Bode G, Lueken A, Knop M, Kannenberg F, Nofer JR et al (2006) Expression and functional characterization of ABCG1 splice variant ABCG1(666). FEBS Lett 580:4551–4559

Ettinger WH, Varma VK, Sorci-Thomas M, Parks JS, Sigmon RC, Smith TK et al (1994) Cytokines decrease apolipoprotein accumulation in medium from Hep G2 cells. Arterioscler Thromb 14:8–13

Fayard E, Auwerx J, Schoonjans K (2004) LRH-1: an orphan nuclear receptor involved in development, metabolism and steroidogenesis. Trends Cell Biol 14:250–260

Fazio S, Babaev VR, Murray AB, Hasty AH, Carter KJ, Gleaves LA et al (1997) Increased atherosclerosis in mice reconstituted with apolipoprotein E null macrophages. Proc Natl Acad Sci USA 94:4647–4652

Feingold KR, Shigenaga JK, Chui LG, Moser A, Khovidhunkit W, Grunfeld C (2008) Infection and inflammation decrease apolipoprotein M expression. Atherosclerosis 199:19–26

Fitzgerald ML, Mujawar Z, Tamehiro N (2010) ABC transporters, atherosclerosis and inflammation. Atherosclerosis 211:361–370

Frankel LB, Lund AH (2012) MicroRNA regulation of autophagy. Carcinogenesis 33:2018–2025

Freeman LA, Kennedy A, Wu J, Bark S, Remaley AT, Santamarina-Fojo S et al (2004) The orphan nuclear receptor LRH-1 activates the ABCG5/ABCG8 intergenic promoter. J Lipid Res 45:1197–1206

Friedman RC, Farh KK, Burge CB, Bartel DP (2009) Most mammalian mRNAs are conserved targets of microRNAs. Genome Res 19:92–105

Frisdal E, Lesnik P, Olivier M, Robillard P, Chapman MJ, Huby T et al (2011) Interleukin-6 protects human macrophages from cellular cholesterol accumulation and attenuates the proinflammatory response. J Biol Chem 286:30926–30936

Fu X, Menke JG, Chen Y, Zhou G, MacNaul KL, Wright SD et al (2001) 27-hydroxycholesterol is an endogenous ligand for liver X receptor in cholesterol-loaded cells. J Biol Chem 276:38378–38387

Fukata Y, Yu X, Imachi H, Nishiuchi T, Lyu J, Seo K et al (2013) 17β-Estradiol regulates scavenger receptor class BI gene expression via protein kinase C in vascular endothelial cells. Endocrine 46:644–650

Gafencu AV, Robciuc MR, Fuior E, Zannis VI, Kardassis D, Simionescu M (2007) Inflammatory signaling pathways regulating ApoE gene expression in macrophages. J Biol Chem 282:21776–21785

Gan X, Kaplan R, Menke JG, MacNaul K, Chen Y, Sparrow CP et al (2001) Dual mechanisms of ABCA1 regulation by geranylgeranyl pyrophosphate. J Biol Chem 276:48702–48708

Garcia MA, Vazquez J, Gimenez C, Valdivieso F, Zafra F (1996) Transcription factor AP-2 regulates human apolipoprotein E gene expression in astrocytoma cells. J Neurosci 16:7550–7556

Gaudet F, Ginsburg GS (1995) Transcriptional regulation of the cholesteryl ester transfer protein gene by the orphan nuclear hormone receptor apolipoprotein AI regulatory protein-1. J Biol Chem 270:29916–29922

Gauthier B, Robb M, Gaudet F, Ginsburg GS, McPherson R (1999) Characterization of a cholesterol response element (CRE) in the promoter of the cholesteryl ester transfer protein gene: functional role of the transcription factors SREBP-1a, -2, and YY1. J Lipid Res 40:1284–1293

Gautier T, de Haan W, Grober J, Ye D, Bahr MJ, Claudel T et al (2013) Farnesoid X receptor activation increases cholesteryl ester transfer protein expression in humans and transgenic mice. J Lipid Res 54:2195–2205

Gavrilov A, Eivazova E, Priozhkova I, Lipinski M, Razin S, Vassetzky Y (2009) Chromosome conformation capture (from 3C to 5C) and its ChIP-based modification. Methods Mol Biol 567:171–188

Georgopoulos S, Kan HY, Reardon-Alulis C, Zannis V (2000) The SP1 sites of the human apoCIII enhancer are essential for the expression of the apoCIII gene and contribute to the hepatic and intestinal expression of the apoA-I gene in transgenic mice. Nucleic Acids Res 28:4919–4929

Gerbod-Giannone MC, Li Y, Holleboom A, Han S, Hsu LC, Tabas I et al (2006) TNFalpha induces ABCA1 through NF-kappaB in macrophages and in phagocytes ingesting apoptotic cells. Proc Natl Acad Sci USA 103:3112–3117

Gerin I, Clerbaux LA, Haumont O, Lanthier N, Das AK, Burant CF et al (2010) Expression of miR-33 from an SREBP2 intron inhibits cholesterol export and fatty acid oxidation. J Biol Chem 285:33652–33661

Goldstein I, Ezra O, Rivlin N, Molchadsky A, Madar S, Goldfinger N et al (2012) p53, a novel regulator of lipid metabolism pathways. J Hepatol 56:656–662

Grainger DJ, Reckless J, McKilligin E (2004) Apolipoprotein E modulates clearance of apoptotic bodies in vitro and in vivo, resulting in a systemic proinflammatory state in apolipoprotein E-deficient mice. J Immunol 173:6366–6375

Gronemeyer H, Gustafsson JA, Laudet V (2004) Principles for modulation of the nuclear receptor superfamily. Nat Rev Drug Discov 3:950–964

Gu HM, Li G, Gao X, Berthiaume LG, Zhang DW (2013) Characterization of palmitoylation of ATP binding cassette transporter G1: effect on protein trafficking and function. Biochim Biophys Acta 1831:1067–1078

Guo H, Ingolia NT, Weissman JS, Bartel DP (2010) Mammalian microRNAs predominantly act to decrease target mRNA levels. Nature 466:835–840

Haas MJ, Mooradian AD (2010) Therapeutic interventions to enhance apolipoprotein A-I-mediated cardioprotection. Drugs 70:805–821

Hadzopoulou-Cladaras M, Lavrentiadou SN, Zannis VI, Kardassis D (1998) Transactivation of the ApoCIII promoter by ATF-2 and repression by members of the Jun family. Biochemistry 37:14078–14087

Haghpassand M, Bourassa PA, Francone OL, Aiello RJ (2001) Monocyte/macrophage expression of ABCA1 has minimal contribution to plasma HDL levels. J Clin Invest 108:1315–1320

Haidar B, Kiss RS, Sarov-Blat L, Brunet R, Harder C, McPherson R et al (2006) Cathepsin D, a lysosomal protease, regulates ABCA1-mediated lipid efflux. J Biol Chem 281:39971–39981

Hansen TB, Jensen TI, Clausen BH, Bramsen JB, Finsen B, Damgaard CK et al (2013) Natural RNA circles function as efficient microRNA sponges. Nature 495:384–388

Hargrove GM, Junco A, Wong NC (1999) Hormonal regulation of apolipoprotein AI. J Mol Endocrinol 22:103–111

Hayhurst GP, Lee YH, Lambert G, Ward JM, Gonzalez FJ (2001) Hepatocyte nuclear factor 4alpha (nuclear receptor 2A1) is essential for maintenance of hepatic gene expression and lipid homeostasis. Mol Cell Biol 21:1393–1403

Helal O, Berrougui H, Loued S, Khalil A (2013) Extra-virgin olive oil consumption improves the capacity of HDL to mediate cholesterol efflux and increases ABCA1 and ABCG1 expression in human macrophages. Br J Nutr 109:1844–1855

Helsen C, Kerkhofs S, Clinckemalie L, Spans L, Laurent M, Boonen S et al (2012) Structural basis for nuclear hormone receptor DNA binding. Mol Cell Endocrinol 348:411–417

Helwak A, Kudla G, Dudnakova T, Tollervey D (2013) Mapping the human miRNA interactome by CLASH reveals frequent noncanonical binding. Cell 153:654–665

Hoang MH, Jia Y, Jun HJ, Lee JH, Lee BY, Lee SJ (2012) Fucosterol is a selective liver X receptor modulator that regulates the expression of key genes in cholesterol homeostasis in macrophages, hepatocytes, and intestinal cells. J Agric Food Chem 60:11567–11575

Honzumi S, Shima A, Hiroshima A, Koieyama T, Ubukata N, Terasaka N (2010) LXRalpha regulates human CETP expression in vitro and in transgenic mice. Atherosclerosis 212:139–145

Horie T, Ono K, Horiguchi M, Nishi H, Nakamura T, Nagao K et al (2010) MicroRNA-33 encoded by an intron of sterol regulatory element-binding protein 2 (Srebp2) regulates HDL in vivo. Proc Natl Acad Sci USA 107:17321–17326

Horie T, Baba O, Kuwabara Y, Chujo Y, Watanabe S, Kinoshita M et al (2012) MicroRNA-33 deficiency reduces the progression of atherosclerotic plaque in ApoE−/− mice. J Am Heart Assoc 1:e003376

Horie T, Nishino T, Baba O, Kuwabara Y, Nakao T, Nishiga M et al (2013) MicroRNA-33 regulates sterol regulatory element-binding protein 1 expression in mice. Nat Commun 4:2883

Horiuchi K, Tajima S, Menju M, Yamamoto A (1989) Structure and expression of mouse apolipoprotein E gene. J Biochem 106:98–103

Horton JD, Goldstein JL, Brown MS (2002) SREBPs: activators of the complete program of cholesterol and fatty acid synthesis in the liver. J Clin Invest 109:1125–1131

Hozoji M, Munehira Y, Ikeda Y, Makishima M, Matsuo M, Kioka N et al (2008) Direct interaction of nuclear liver X receptor-beta with ABCA1 modulates cholesterol efflux. J Biol Chem 283: 30057–30063

Hozoji-Inada M, Munehira Y, Nagao K, Kioka N, Ueda K (2011) Liver X receptor beta (LXRbeta) interacts directly with ATP-binding cassette A1 (ABCA1) to promote high density lipoprotein formation during acute cholesterol accumulation. J Biol Chem 286:20117–20124

Hu Y, Ren X, Wang H, Ma Y, Wang L, Shen Y et al (2010a) Liver-specific expression of an exogenous gene controlled by human apolipoprotein A-I promoter. Int J Pharm 398:161–164

Hu YW, Wang Q, Ma X, Li XX, Liu XH, Xiao J et al (2010b) TGF-beta1 up-regulates expression of ABCA1, ABCG1 and SR-BI through liver X receptor alpha signaling pathway in THP-1 macrophage-derived foam cells. J Atheroscler Thromb 17:493–502

Hu Z, Shen WJ, Kraemer FB, Azhar S (2012) MicroRNAs 125a and 455 repress lipoprotein-supported steroidogenesis by targeting scavenger receptor class B type I in steroidogenic cells. Mol Cell Biol 32:5035–5045

Hu Z, Hu J, Zhang Z, Shen WJ, Yun CC, Berlot CH et al (2013) Regulation of expression and function of scavenger receptor class B, type I (SR-BI) by Na^+/H^+ exchanger regulatory factors (NHERFs). J Biol Chem 288:11416–11435

Huang CX, Zhang YL, Wang JF, Jiang JY, Bao JL (2013) MCP-1 impacts RCT by repressing ABCA1, ABCG1, and SR-BI through PI3K/Akt posttranslational regulation in HepG2 cells. J Lipid Res 54:1231–1240

Hunter MP, Ismail N, Zhang X, Aguda BD, Lee EJ, Yu L et al (2008) Detection of microRNA expression in human peripheral blood microvesicles. PLoS ONE 3:e3694

Huuskonen J, Abedin M, Vishnu M, Pullinger CR, Baranzini SE, Kane JP et al (2003) Dynamic regulation of alternative ATP-binding cassette transporter A1 transcripts. Biochem Biophys Res Commun 306:463–468

Huuskonen J, Vishnu M, Pullinger CR, Fielding PE, Fielding CJ (2004) Regulation of ATP-binding cassette transporter A1 transcription by thyroid hormone receptor. Biochemistry 43:1626–1632

Huwait EA, Greenow KR, Singh NN, Ramji DP (2011) A novel role for c-Jun N-terminal kinase and phosphoinositide 3-kinase in the liver X receptor-mediated induction of macrophage gene expression. Cell Signal 23:542–549

Ishibashi M, Filomenko R, Rebe C, Chevriaux A, Varin A, Derangere V et al (2013) Knock-down of the oxysterol receptor LXRalpha impairs cholesterol efflux in human primary macrophages: lack of compensation by LXRbeta activation. Biochem Pharmacol 86:122–129

Iwamoto N, Lu R, Tanaka N, Abe-Dohmae S, Yokoyama S (2010) Calmodulin interacts with ATP binding cassette transporter A1 to protect from calpain-mediated degradation and upregulates high-density lipoprotein generation. Arterioscler Thromb Vasc Biol 30:1446–1452

Jean S, Kiger AA (2012) Coordination between RAB GTPase and phosphoinositide regulation and functions. Nat Rev Mol Cell Biol 13:463–470

Jeoung NH, Jang WG, Nam JI, Pak YK, Park YB (1999) Identification of retinoic acid receptor element in human cholesteryl ester transfer protein gene. Biochem Biophys Res Commun 258: 411–415

Jiang XC, Bruce C (1995) Regulation of murine plasma phospholipid transfer protein activity and mRNA levels by lipopolysaccharide and high cholesterol diet. J Biol Chem 270:17133–17138

Jiang XC, Moulin P, Quinet E, Goldberg IJ, Yacoub LK, Agellon LB et al (1991) Mammalian adipose tissue and muscle are major sources of lipid transfer protein mRNA. J Biol Chem 266:4631–4639

Jiang XC, Agellon LB, Walsh A, Breslow JL, Tall A (1992) Dietary cholesterol increases transcription of the human cholesteryl ester transfer protein gene in transgenic mice. Dependence on natural flanking sequences. J Clin Invest 90:1290–1295

Jiang XC, Jin W, Hussain MM (2012) The impact of phospholipid transfer protein (PLTP) on lipoprotein metabolism. Nutr Metab (Lond) 9:75

Jo DW, Leren TP, Yang ZY, Chung YH, Taylor JM, Paik YK (1995) Characterization of an upstream regulatory element of the human apolipoprotein E gene, and purification of its binding protein from the human placenta. J Biochem 117:915–922

Joseph SB, Laffitte BA, Patel PH, Watson MA, Matsukuma KE, Walczak R et al (2002a) Direct and indirect mechanisms for regulation of fatty acid synthase gene expression by liver X receptors. J Biol Chem 277:11019–11025

Joseph SB, McKilligin E, Pei L, Watson MA, Collins AR, Laffitte BA et al (2002b) Synthetic LXR ligand inhibits the development of atherosclerosis in mice. Proc Natl Acad Sci USA 99: 7604–7609

Jun HJ, Hoang MH, Yeo SK, Jia Y, Lee SJ (2013) Induction of ABCA1 and ABCG1 expression by the liver X receptor modulator cineole in macrophages. Bioorg Med Chem Lett 23:579–583

Kammerer I, Ringseis R, Biemann R, Wen G, Eder K (2011) 13-hydroxy linoleic acid increases expression of the cholesterol transporters ABCA1, ABCG1 and SR-BI and stimulates apoA-I-dependent cholesterol efflux in RAW264.7 macrophages. Lipids Health Dis 10:222

Kan HY, Georgopoulos S, Zannis V (2000) A hormone response element in the human apolipoprotein CIII (ApoCIII) enhancer is essential for intestinal expression of the ApoA-I and ApoCIII genes and contributes to the hepatic expression of the two linked genes in transgenic mice. J Biol Chem 275:30423–30431

Kan HY, Georgopoulos S, Zanni M, Shkodrani A, Tzatsos A, Xie HX et al (2004) Contribution of the hormone-response elements of the proximal ApoA-I promoter, ApoCIII enhancer, and C/EBP binding site of the proximal ApoA-I promoter to the hepatic and intestinal expression of the ApoA-I and ApoCIII genes in transgenic mice. Biochemistry 43:5084–5093

Kang MH, Singaraja R, Hayden MR (2010) Adenosine-triphosphate-binding cassette transporter-1 trafficking and function. Trends Cardiovasc Med 20:41–49

Kang MH, Zhang LH, Wijesekara N, de Haan W, Butland S, Bhattacharjee A et al (2013) Regulation of ABCA1 protein expression and function in hepatic and pancreatic islet cells by miR-145. Arterioscler Thromb Vasc Biol 33:2724–2732

Kanter JE, Tang C, Oram JF, Bornfeldt KE (2012) Acyl-CoA synthetase 1 is required for oleate and linoleate mediated inhibition of cholesterol efflux through ATP-binding cassette transporter A1 in macrophages. Biochim Biophys Acta 1821:358–364

Kardassis D, Tzameli I, Hadzopoulou-Cladaras M, Talianidis I, Zannis V (1997) Distal apolipoprotein C-III regulatory elements F to J act as a general modular enhancer for proximal promoters that contain hormone response elements. Synergism between hepatic nuclear factor-4 molecules bound to the proximal promoter and distal enhancer sites. Arterioscler Thromb Vasc Biol 17:222–232

Kardassis D, Falvey E, Tsantili P, Hadzopoulou-Cladaras M, Zannis V (2002) Direct physical interactions between HNF-4 and Sp1 mediate synergistic transactivation of the apolipoprotein CIII promoter. Biochemistry 41:1217–1228

Kardassis D, Drosatos C, Zannis V (2007) Regulation of genes involved in the biogenesis and the remodeling of HDL. In: Fielding CJ (ed) High-density lipoproteins from basic biology to clinical aspects. Wiley-VCH, Weinheim, pp 237–253

Kennedy MA, Barrera GC, Nakamura K, Baldan A, Tarr P, Fishbein MC et al (2005) ABCG1 has a critical role in mediating cholesterol efflux to HDL and preventing cellular lipid accumulation. Cell Metab 1:121–131

Kielar D, Dietmaier W, Langmann T, Aslanidis C, Probst M, Naruszewicz M et al (2001) Rapid quantification of human ABCA1 mRNA in various cell types and tissues by real-time reverse transcription-PCR. Clin Chem 47:2089–2097

Kilbourne EJ, Widom R, Harnish DC, Malik S, Karathanasis SK (1995) Involvement of early growth response factor Egr-1 in apolipoprotein AI gene transcription. J Biol Chem 270: 7004–7010

Kim J, Yoon H, Ramirez CM, Lee SM, Hoe HS, Fernandez-Hernando C (2012) MiR-106b impairs cholesterol efflux and increases Aβ levels by repressing ABCA1 expression. Exp Neurol 235: 476–483

Kinet V, Halkein J, Dirkx E, Windt LJ (2013) Cardiovascular extracellular microRNAs: emerging diagnostic markers and mechanisms of cell-to-cell RNA communication. Front Genet 4:214

Kir S, Zhang Y, Gerard RD, Kliewer SA, Mangelsdorf DJ (2012) Nuclear receptors HNF4alpha and LRH-1 cooperate in regulating Cyp7a1 in vivo. J Biol Chem 287:41334–41341

Kocher O, Krieger M (2009) Role of the adaptor protein PDZK1 in controlling the HDL receptor SR-BI. Curr Opin Lipidol 20:236–241

Kocher O, Yesilaltay A, Cirovic C, Pal R, Rigotti A, Krieger M (2003) Targeted disruption of the PDZK1 gene in mice causes tissue-specific depletion of the high density lipoprotein receptor scavenger receptor class B type I and altered lipoprotein metabolism. J Biol Chem 278: 52820–52825

Krol J, Loedige I, Filipowicz W (2010) The widespread regulation of microRNA biogenesis, function and decay. Nat Rev Genet 11:597–610

Ku CS, Park Y, Coleman SL, Lee J (2012) Unsaturated fatty acids repress expression of ATP binding cassette transporter A1 and G1 in RAW 264.7 macrophages. J Nutr Biochem 23: 1271–1276

Kulyte A, Lorente-Cebrian S, Gao H, Mejhert N, Agustsson T, Arner P et al (2013) MicroRNA profiling links miR-378 to enhanced adipocyte lipolysis in human cancer cachexia. Am J Physiol Endocrinol Metab 306:E267–E274

Kuo CY, Lin YC, Yang JJ, Yang VC (2011) Interaction abolishment between mutant caveolin-1 (Delta62-100) and ABCA1 reduces HDL-mediated cellular cholesterol efflux. Biochem Biophys Res Commun 414:337–343

Ladias JA, Karathanasis SK (1991) Regulation of the apolipoprotein AI gene by ARP-1, a novel member of the steroid receptor superfamily. Science 251:561–565

Laffitte BA, Joseph SB, Walczak R, Pei L, Wilpitz DC, Collins JL et al (2001a) Autoregulation of the human liver X receptor alpha promoter. Mol Cell Biol 21:7558–7568

Laffitte BA, Repa JJ, Joseph SB, Wilpitz DC, Kast HR, Mangelsdorf DJ et al (2001b) LXRs control lipid-inducible expression of the apolipoprotein E gene in macrophages and adipocytes. Proc Natl Acad Sci USA 98:507–512

Laffitte BA, Joseph SB, Chen M, Castrillo A, Repa J, Wilpitz D et al (2003) The phospholipid transfer protein gene is a liver X receptor target expressed by macrophages in atherosclerotic lesions. Mol Cell Biol 23:2182–2191

Lagos-Quintana M, Rauhut R, Yalcin A, Meyer J, Lendeckel W, Tuschl T (2002) Identification of tissue-specific microRNAs from mouse. Curr Biol 12:735–739

Lakomy D, Rebe C, Sberna AL, Masson D, Gautier T, Chevriaux A et al (2009) Liver X receptor-mediated induction of cholesteryl ester transfer protein expression is selectively impaired in inflammatory macrophages. Arterioscler Thromb Vasc Biol 29:1923–1929

Lambert G, Amar MJ, Guo G, Brewer HB Jr, Gonzalez FJ, Sinal CJ (2003) The farnesoid X-receptor is an essential regulator of cholesterol homeostasis. J Biol Chem 278:2563–2570

Lan D, Silver DL (2005) Fenofibrate induces a novel degradation pathway for scavenger receptor B-I independent of PDZK1. J Biol Chem 280:23390–23396

Langmann T, Klucken J, Reil M, Liebisch G, Luciani MF, Chimini G et al (1999) Molecular cloning of the human ATP-binding cassette transporter 1 (hABC1): evidence for sterol-dependent regulation in macrophages. Biochem Biophys Res Commun 257:29–33

Langmann T, Porsch-Ozcurumez M, Unkelbach U, Klucken J, Schmitz G (2000) Genomic organization and characterization of the promoter of the human ATP-binding cassette transporter-G1 (ABCG1) gene. Biochim Biophys Acta 1494:175–180

Langmann T, Porsch-Ozcurumez M, Heimerl S, Probst M, Moehle C, Taher M et al (2002) Identification of sterol-independent regulatory elements in the human ATP-binding cassette transporter A1 promoter: role of Sp1/3, E-box binding factors, and an oncostatin M-responsive element. J Biol Chem 277:14443–14450

Langmann T, Mauerer R, Zahn A, Moehle C, Probst M, Stremmel W et al (2003) Real-time reverse transcription-PCR expression profiling of the complete human ATP-binding cassette transporter superfamily in various tissues. Clin Chem 49:230–238

Larkin L, Khachigian LM, Jessup W (2000) Regulation of apolipoprotein E production in macrophages (review). Int J Mol Med 6:253–258

Lavrentiadou SN, Hadzopoulou-Cladaras M, Kardassis D, Zannis VI (1999) Binding specificity and modulation of the human ApoCIII promoter activity by heterodimers of ligand-dependent nuclear receptors. Biochemistry 38:964–975

Le Goff W, Guerin M, Petit L, Chapman MJ, Thillet J (2003) Regulation of human CETP gene expression: role of SP1 and SP3 transcription factors at promoter sites −690, −629, and −37. J Lipid Res 44:1322–1331

Lee RC, Feinbaum RL, Ambros V (1993) The C. elegans heterochronic gene lin-4 encodes small RNAs with antisense complementarity to lin-14. Cell 75:843–854

Lee S, Wang PY, Jeong Y, Mangelsdorf DJ, Anderson RG, Michaely P (2012) Sterol-dependent nuclear import of ORP1S promotes LXR regulated trans-activation of apoE. Exp Cell Res 318:2128–2142

Leiva A, Verdejo H, Benitez ML, Martinez A, Busso D, Rigotti A (2011) Mechanisms regulating hepatic SR-BI expression and their impact on HDL metabolism. Atherosclerosis 217:299–307

Lemon B, Tjian R (2000) Orchestrated response: a symphony of transcription factors for gene control. Genes Dev 14:2551–2569

Li J, Ning G, Duncan SA (2000) Mammalian hepatocyte differentiation requires the transcription factor HNF-4alpha. Genes Dev 14:464–474

Li X, Peegel H, Menon KM (2001) Regulation of high density lipoprotein receptor messenger ribonucleic acid expression and cholesterol transport in theca-interstitial cells by insulin and human chorionic gonadotropin. Endocrinology 142:174–181

Li D, Wang D, Wang Y, Ling W, Feng X, Xia M (2010a) Adenosine monophosphate-activated protein kinase induces cholesterol efflux from macrophage-derived foam cells and alleviates atherosclerosis in apolipoprotein E-deficient mice. J Biol Chem 285:33499–33509

Li D, Zhang Y, Ma J, Ling W, Xia M (2010b) Adenosine monophosphate activated protein kinase regulates ABCG1-mediated oxysterol efflux from endothelial cells and protects against hypercholesterolemia-induced endothelial dysfunction. Arterioscler Thromb Vasc Biol 30:1354–1362

Li G, Gu HM, Zhang DW (2013) ATP-binding cassette transporters and cholesterol translocation. IUBMB Life 65:505–512

Liang Y, Yang X, Ma L, Cai X, Wang L, Yang C et al (2013) Homocysteine-mediated cholesterol efflux via ABCA1 and ACAT1 DNA methylation in THP-1 monocyte-derived foam cells. Acta Biochim Biophys Sin (Shanghai) 45:220–228

Lin G, Bornfeldt KE (2002) Cyclic AMP-specific phosphodiesterase 4 inhibitors promote ABCA1 expression and cholesterol efflux. Biochem Biophys Res Commun 290:663–669

Lin S, Zhou C, Neufeld E, Wang YH, Xu SW, Lu L et al (2013) BIG1, a brefeldin A-inhibited guanine nucleotide-exchange protein modulates ATP-binding cassette transporter A-1 trafficking and function. Arterioscler Thromb Vasc Biol 33:e31–e38

Linder MD, Mayranpaa MI, Peranen J, Pietila TE, Pietiainen VM, Uronen RL et al (2009) Rab8 regulates ABCA1 cell surface expression and facilitates cholesterol efflux in primary human macrophages. Arterioscler Thromb Vasc Biol 29:883–888

Linton MF, Atkinson JB, Fazio S (1995) Prevention of atherosclerosis in apolipoprotein E-deficient mice by bone marrow transplantation. Science 267:1034–1037

Liu C, Lin JD (2011) PGC-1 coactivators in the control of energy metabolism. Acta Biochim Biophys Sin (Shanghai) 43:248–257

Liu S, Tetzlaff MT, Cui R, Xu X (2012) miR-200c inhibits melanoma progression and drug resistance through down-regulation of BMI-1. Am J Pathol 181:1823–1835

Lopez D, McLean MP (1999) Sterol regulatory element-binding protein-1a binds to cis elements in the promoter of the rat high density lipoprotein receptor SR-BI gene. Endocrinology 140:5669–5681

Lopez D, McLean MP (2006) Estrogen regulation of the scavenger receptor class B gene: anti-atherogenic or steroidogenic, is there a priority? Mol Cell Endocrinol 247:22–33

Lopez D, Sandhoff TW, McLean MP (1999) Steroidogenic factor-1 mediates cyclic 3′,5-′-adenosine monophosphate regulation of the high density lipoprotein receptor. Endocrinology 140:3034–3044

Lopez D, Shea-Eaton W, Sanchez MD, McLean MP (2001) DAX-1 represses the high-density lipoprotein receptor through interaction with positive regulators sterol regulatory element-binding protein-1a and steroidogenic factor-1. Endocrinology 142:5097–5106

Lopez D, Sanchez MD, Shea-Eaton W, McLean MP (2002) Estrogen activates the high-density lipoprotein receptor gene via binding to estrogen response elements and interaction with sterol regulatory element binding protein-1A. Endocrinology 143:2155–2168

Lu R, Arakawa R, Ito-Osumi C, Iwamoto N, Yokoyama S (2008) ApoA-I facilitates ABCA1 recycle/accumulation to cell surface by inhibiting its intracellular degradation and increases HDL generation. Arterioscler Thromb Vasc Biol 28:1820–1824

Lund E, Guttinger S, Calado A, Dahlberg JE, Kutay U (2004) Nuclear export of microRNA precursors. Science 303:95–98

Luo Y, Tall AR (2000) Sterol upregulation of human CETP expression in vitro and in transgenic mice by an LXR element. J Clin Invest 105:513–520

Luo Y, Liang CP, Tall AR (2001) The orphan nuclear receptor LRH-1 potentiates the sterol-mediated induction of the human CETP gene by liver X receptor. J Biol Chem 276:24767–24773

Ma AZ, Zhang Q, Song ZY (2013) TNFa alter cholesterol metabolism in human macrophages via PKC-theta-dependent pathway. BMC Biochem 14:20

Maitra U, Li L (2013) Molecular mechanisms responsible for the reduced expression of cholesterol transporters from macrophages by low-dose endotoxin. Arterioscler Thromb Vasc Biol 33:24–33

Mak PA, Kast-Woelbern HR, Anisfeld AM, Edwards PA (2002a) Identification of PLTP as an LXR target gene and apoE as an FXR target gene reveals overlapping targets for the two nuclear receptors. J Lipid Res 43:2037–2041

Mak PA, Laffitte BA, Desrumaux C, Joseph SB, Curtiss LK, Mangelsdorf DJ et al (2002b) Regulated expression of the apolipoprotein E/C-I/C-IV/C-II gene cluster in murine and human macrophages. A critical role for nuclear liver X receptors alpha and beta. J Biol Chem 277:31900–31908

Malekpour-Dehkordi Z, Javadi E, Doosti M, Paknejad M, Nourbakhsh M, Yassa N et al (2013) S-Allylcysteine, a garlic compound, increases ABCA1 expression in human THP-1 macrophages. Phytother Res 27:357–361

Malerod L, Juvet LK, Hanssen-Bauer A, Eskild W, Berg T (2002) Oxysterol-activated LXRalpha/RXR induces hSR-BI-promoter activity in hepatoma cells and preadipocytes. Biochem Biophys Res Commun 299:916–923

Malerod L, Sporstol M, Juvet LK, Mousavi A, Gjoen T, Berg T (2003) Hepatic scavenger receptor class B, type I is stimulated by peroxisome proliferator-activated receptor gamma and hepatocyte nuclear factor 4alpha. Biochem Biophys Res Commun 305:557–565

Malerod L, Sporstol M, Juvet LK, Mousavi SA, Gjoen T, Berg T et al (2005) Bile acids reduce SR-BI expression in hepatocytes by a pathway involving FXR/RXR, SHP, and LRH-1. Biochem Biophys Res Commun 336:1096–1105

Marquart TJ, Allen RM, Ory DS, Baldan A (2010) miR-33 links SREBP-2 induction to repression of sterol transporters. Proc Natl Acad Sci USA 107:12228–12232

Martinez LO, Agerholm-Larsen B, Wang N, Chen W, Tall AR (2003) Phosphorylation of a pest sequence in ABCA1 promotes calpain degradation and is reversed by ApoA-I. J Biol Chem 278:37368–37374

Masson D, Qatanani M, Sberna AL, Xiao R, Pais de Barros JP, Grober J et al (2008) Activation of the constitutive androstane receptor decreases HDL in wild-type and human apoA-I transgenic mice. J Lipid Res 49:1682–1691

Mataki C, Magnier BC, Houten SM, Annicotte JS, Argmann C, Thomas C et al (2007) Compromised intestinal lipid absorption in mice with a liver-specific deficiency of liver receptor homolog 1. Mol Cell Biol 27:8330–8339

Mattick JS, Makunin IV (2005) Small regulatory RNAs in mammals. Hum Mol Genet 14(Spec No 1):R121–R132

Mauerer R, Ebert S, Langmann T (2009) High glucose, unsaturated and saturated fatty acids differentially regulate expression of ATP-binding cassette transporters ABCA1 and ABCG1 in human macrophages. Exp Mol Med 41:126–132

Mavridou S, Venihaki M, Rassouli O, Tsatsanis C, Kardassis D (2010) Feedback inhibition of human scavenger receptor class B type I gene expression by glucocorticoid in adrenal and ovarian cells. Endocrinology 151:3214–3224

Meister G (2013) Argonaute proteins: functional insights and emerging roles. Nat Rev Genet 14:447–459

Mendell JT, Olson EN (2012) MicroRNAs in stress signaling and human disease. Cell 148:1172–1187

Meurs I, Lammers B, Zhao Y, Out R, Hildebrand RB, Hoekstra M et al (2012) The effect of ABCG1 deficiency on atherosclerotic lesion development in LDL receptor knockout mice depends on the stage of atherogenesis. Atherosclerosis 221:41–47

Mitchell PJ, Tjian R (1989) Transcriptional regulation in mammalian cells by sequence-specific DNA binding proteins. Science 245:371–378

Miyake Y, Kozutsumi Y, Nakamura S, Fujita T, Kawasaki T (1995) Serine palmitoyltransferase is the primary target of a sphingosine-like immunosuppressant, ISP-1/myriocin. Biochem Biophys Res Commun 211:396–403

Miyazaki T, Koya T, Kigawa Y, Oguchi T, Lei XF, Kim-Kaneyama JR et al (2013) Calpain and atherosclerosis. J Atheroscler Thromb 20:228–237

Mizuno T, Hayashi H, Naoi S, Sugiyama Y (2011) Ubiquitination is associated with lysosomal degradation of cell surface-resident ATP-binding cassette transporter A1 (ABCA1) through the endosomal sorting complex required for transport (ESCRT) pathway. Hepatology 54:631–643

Mooradian AD, Haas MJ, Wong NC (2004) Transcriptional control of apolipoprotein A-I gene expression in diabetes. Diabetes 53:513–520

Morishima A, Ohkubo N, Maeda N, Miki T, Mitsuda N (2003) NFkappaB regulates plasma apolipoprotein A-I and high density lipoprotein cholesterol through inhibition of peroxisome proliferator-activated receptor alpha. J Biol Chem 278:38188–38193

Mosialou I, Zannis VI, Kardassis D (2010) Regulation of human apolipoprotein m gene expression by orphan and ligand-dependent nuclear receptors. J Biol Chem 285:30719–30730

Mosialou I, Krasagakis K, Kardassis D (2011) Opposite regulation of the human apolipoprotein M gene by hepatocyte nuclear factor 1 and Jun transcription factors. J Biol Chem 286:17259–17269

Munehira Y, Ohnishi T, Kawamoto S, Furuya A, Shitara K, Imamura M et al (2004) Alpha1-syntrophin modulates turnover of ABCA1. J Biol Chem 279:15091–15095

Murao K, Wada Y, Nakamura T, Taylor AH, Mooradian AD, Wong NC (1998) Effects of glucose and insulin on rat apolipoprotein A-I gene expression. J Biol Chem 273:18959–18965

Murao K, Imachi H, Yu X, Cao WM, Muraoka T, Dobashi H et al (2008) The transcriptional factor prolactin regulatory element-binding protein mediates the gene transcription of adrenal scavenger receptor class B type I via $3'$,$5'$-cyclic adenosine $5'$-monophosphate. Endocrinology 149:6103–6112

Myklebost O, Rogne S (1988) A physical map of the apolipoprotein gene cluster on human chromosome 19. Hum Genet 78:244–247

Nagelin MH, Srinivasan S, Lee J, Nadler JL, Hedrick CC (2008) 12/15-Lipoxygenase activity increases the degradation of macrophage ATP-binding cassette transporter G1. Arterioscler Thromb Vasc Biol 28:1811–1819

Nagelin MH, Srinivasan S, Nadler JL, Hedrick CC (2009) Murine 12/15-lipoxygenase regulates ATP-binding cassette transporter G1 protein degradation through p38- and JNK2-dependent pathways. J Biol Chem 284:31303–31314

Najafi-Shoushtari SH, Kristo F, Li Y, Shioda T, Cohen DE, Gerszten RE et al (2010) MicroRNA-33 and the SREBP host genes cooperate to control cholesterol homeostasis. Science 328: 1566–1569

Nakashima Y, Plump AS, Raines EW, Breslow JL, Ross R (1994) ApoE-deficient mice develop lesions of all phases of atherosclerosis throughout the arterial tree. Arterioscler Thromb 14: 133–140

Nakshatri H, Bhat-Nakshatri P (1998) Multiple parameters determine the specificity of transcriptional response by nuclear receptors HNF-4, ARP-1, PPAR, RAR and RXR through common response elements. Nucleic Acids Res 26:2491–2499

Nielsen LB, Christoffersen C, Ahnstrom J, Dahlback B (2009) ApoM: gene regulation and effects on HDL metabolism. Trends Endocrinol Metab 20:66–71

Niemeier A, Kovacs WJ, Strobl W, Stangl H (2009) Atherogenic diet leads to posttranslational down-regulation of murine hepatocyte SR-BI expression. Atherosclerosis 202:169–175

Nikolaidou-Neokosmidou V, Zannis VI, Kardassis D (2006) Inhibition of hepatocyte nuclear factor 4 transcriptional activity by the nuclear factor kappaB pathway. Biochem J 398:439–450

Nohata N, Hanazawa T, Kikkawa N, Mutallip M, Fujimura L, Yoshino H et al (2011) Caveolin-1 mediates tumor cell migration and invasion and its regulation by miR-133a in head and neck squamous cell carcinoma. Int J Oncol 38:209–217

Nonomura K, Arai Y, Mitani H, Abe-Dohmae S, Yokoyama S (2011) Insulin down-regulates specific activity of ATP-binding cassette transporter A1 for high density lipoprotein biogenesis through its specific phosphorylation. Atherosclerosis 216:334–341

Novak EM, Bydlowski SP (1997) NFY transcription factor binds to regulatory element AIC and transactivates the human apolipoprotein A-I promoter in HEPG2 cells. Biochem Biophys Res Commun 231:140–143

Okazaki H, Goldstein JL, Brown MS, Liang G (2010) LXR-SREBP-1c-phospholipid transfer protein axis controls very low density lipoprotein (VLDL) particle size. J Biol Chem 285: 6801–6810

Olaisen B, Teisberg P, Gedde-Dahl T Jr (1982) The locus for apolipoprotein E (apoE) is linked to the complement component C3 (C3) locus on chromosome 19 in man. Hum Genet 62:233–236

Oliver WR Jr, Shenk JL, Snaith MR, Russell CS, Plunket KD, Bodkin NL et al (2001) A selective peroxisome proliferator-activated receptor delta agonist promotes reverse cholesterol transport. Proc Natl Acad Sci USA 98:5306–5311

Orom UA, Nielsen FC, Lund AH (2008) MicroRNA-10a binds the 5'UTR of ribosomal protein mRNAs and enhances their translation. Mol Cell 30:460–471

Ouimet M, Franklin V, Mak E, Liao X, Tabas I, Marcel YL (2011) Autophagy regulates cholesterol efflux from macrophage foam cells via lysosomal acid lipase. Cell Metab 13: 655–667

Out R, Hoekstra M, Hildebrand RB, Kruit JK, Meurs I, Li Z et al (2006) Macrophage ABCG1 deletion disrupts lipid homeostasis in alveolar macrophages and moderately influences atherosclerotic lesion development in LDL receptor-deficient mice. Arterioscler Thromb Vasc Biol 26:2295–2300

Out C, Hageman J, Bloks VW, Gerrits H, Sollewijn Gelpke MD, Bos T et al (2011) Liver receptor homolog-1 is critical for adequate up-regulation of Cyp7a1 gene transcription and bile salt synthesis during bile salt sequestration. Hepatology 53:2075–2085

Ozasa H, Ayaori M, Iizuka M, Terao Y, Uto-Kondo H, Yakushiji E et al (2011) Pioglitazone enhances cholesterol efflux from macrophages by increasing ABCA1/ABCG1 expressions via PPARgamma/LXRalpha pathway: findings from in vitro and ex vivo studies. Atherosclerosis 219:141–150

Paik YK, Chang DJ, Reardon CA, Walker MD, Taxman E, Taylor JM (1988) Identification and characterization of transcriptional regulatory regions associated with expression of the human apolipoprotein E gene. J Biol Chem 263:13340–13349

Pamir N, McMillen TS, Edgel KA, Kim F, LeBoeuf RC (2012) Deficiency of lymphotoxin-alpha does not exacerbate high-fat diet-induced obesity but does enhance inflammation in mice. Am J Physiol Endocrinol Metab 302:E961–E971

Pan X, Jiang XC, Hussain MM (2013) Impaired cholesterol metabolism and enhanced atherosclerosis in clock mutant mice. Circulation 128:1758–1769

Panousis CG, Zuckerman SH (2000) Interferon-gamma induces downregulation of Tangier disease gene (ATP-binding-cassette transporter 1) in macrophage-derived foam cells. Arterioscler Thromb Vasc Biol 20:1565–1571

Panousis CG, Evans G, Zuckerman SH (2001) TGF-beta increases cholesterol efflux and ABC-1 expression in macrophage-derived foam cells: opposing the effects of IFN-gamma. J Lipid Res 42:856–863

Papazafiri P, Ogami K, Ramji DP, Nicosia A, Monaci P, Cladaras C et al (1991) Promoter elements and factors involved in hepatic transcription of the human ApoA-I gene positive and negative regulators bind to overlapping sites. J Biol Chem 266:5790–5797

Park SS, Choi H, Kim SJ, Kim OJ, Chae KS, Kim E (2008) FXRalpha down-regulates LXRalpha signaling at the CETP promoter via a common element. Mol Cells 26:409–414

Park DW, Lee HK, Lyu JH, Chin H, Kang SW, Kim YJ et al (2013) TLR2 stimulates ABCA1 expression via PKC-eta and PLD2 pathway. Biochem Biophys Res Commun 430:933–937

Parseghian S, Onstead-Haas LM, Wong NC, Mooradian AD, Haas MJ (2013) Inhibition of apolipoprotein A-I gene expression by tumor necrosis factor alpha in HepG2 cells: requirement for c-jun. J Cell Biochem 115:253–260

Pascual-Garcia M, Rue L, Leon T, Julve J, Carbo JM, Matalonga J et al (2013) Reciprocal negative cross-talk between liver X receptors (LXRs) and STAT1: effects on IFN-gamma-induced inflammatory responses and LXR-dependent gene expression. J Immunol 190:6520–6532

Paszty C, Maeda N, Verstuyft J, Rubin EM (1994) Apolipoprotein AI transgene corrects apolipoprotein E deficiency-induced atherosclerosis in mice. J Clin Invest 94:899–903

Peng D, Hiipakka RA, Dai Q, Guo J, Reardon CA, Getz GS et al (2008) Antiatherosclerotic effects of a novel synthetic tissue-selective steroidal liver X receptor agonist in low-density lipoprotein receptor-deficient mice. J Pharmacol Exp Ther 327:332–342

Peng D, Hiipakka RA, Xie JT, Dai Q, Kokontis JM, Reardon CA et al (2011) A novel potent synthetic steroidal liver X receptor agonist lowers plasma cholesterol and triglycerides and reduces atherosclerosis in LDLR(−/−) mice. Br J Pharmacol 162:1792–1804

Phelan CA, Weaver JM, Steger DJ, Joshi S, Maslany JT, Collins JL et al (2008) Selective partial agonism of liver X receptor alpha is related to differential corepressor recruitment. Mol Endocrinol 22:2241–2249

Poliseno L, Salmena L, Zhang J, Carver B, Haveman WJ, Pandolfi PP (2010) A coding-independent function of gene and pseudogene mRNAs regulates tumour biology. Nature 465:1033–1038

Porsch-Ozcurumez M, Langmann T, Heimerl S, Borsukova H, Kaminski WE, Drobnik W et al (2001) The zinc finger protein 202 (ZNF202) is a transcriptional repressor of ATP binding cassette transporter A1 (ABCA1) and ABCG1 gene expression and a modulator of cellular lipid efflux. J Biol Chem 276:12427–12433

Quinet EM, Agellon LB, Kroon PA, Marcel YL, Lee YC, Whitlock ME et al (1990) Atherogenic diet increases cholesteryl ester transfer protein messenger RNA levels in rabbit liver. J Clin Invest 85:357–363

Quinet EM, Savio DA, Halpern AR, Chen L, Miller CP, Nambi P (2004) Gene-selective modulation by a synthetic oxysterol ligand of the liver X receptor. J Lipid Res 45:1929–1942

Radeau T, Lau P, Robb M, McDonnell M, Ailhaud G, McPherson R (1995) Cholesteryl ester transfer protein (CETP) mRNA abundance in human adipose tissue: relationship to cell size and membrane cholesterol content. J Lipid Res 36:2552–2561

Raffai RL, Loeb SM, Weisgraber KH (2005) Apolipoprotein E promotes the regression of atherosclerosis independently of lowering plasma cholesterol levels. Arterioscler Thromb Vasc Biol 25:436–441

Rajavashisth TB, Kaptein JS, Reue KL, Lusis AJ (1985) Evolution of apolipoprotein E: mouse sequence and evidence for an 11-nucleotide ancestral unit. Proc Natl Acad Sci USA 82: 8085–8089

Ramirez CM, Davalos A, Goedeke L, Salerno AG, Warrier N, Cirera-Salinas D et al (2011) MicroRNA-758 regulates cholesterol efflux through posttranscriptional repression of ATP-binding cassette transporter A1. Arterioscler Thromb Vasc Biol 31:2707–2714

Ramirez CM, Rotllan N, Vlassov AV, Davalos A, Li M, Goedeke L et al (2013) Control of cholesterol metabolism and plasma high-density lipoprotein levels by microRNA-144. Circ Res 112:1592–1601

Ranalletta M, Wang N, Han S, Yvan-Charvet L, Welch C, Tall AR (2006) Decreased atherosclerosis in low-density lipoprotein receptor knockout mice transplanted with Abcg1−/− bone marrow. Arterioscler Thromb Vasc Biol 26:2308–2315

Rayner KJ, Suarez Y, Davalos A, Parathath S, Fitzgerald ML, Tamehiro N et al (2010) MiR-33 contributes to the regulation of cholesterol homeostasis. Science 328:1570–1573

Rayner KJ, Esau CC, Hussain FN, McDaniel AL, Marshall SM, van Gils JM et al (2011a) Inhibition of miR-33a/b in non-human primates raises plasma HDL and lowers VLDL triglycerides. Nature 478:404–407

Rayner KJ, Sheedy FJ, Esau CC, Hussain FN, Temel RE, Parathath S et al (2011b) Antagonism of miR-33 in mice promotes reverse cholesterol transport and regression of atherosclerosis. J Clin Invest 121:2921–2931

Reinhart BJ, Weinstein EG, Rhoades MW, Bartel B, Bartel DP (2002) MicroRNAs in plants. Genes Dev 16:1616–1626

Remaley AT, Bark S, Walts AD, Freeman L, Shulenin S, Annilo T et al (2002) Comparative genome analysis of potential regulatory elements in the ABCG5-ABCG8 gene cluster. Biochem Biophys Res Commun 295:276–282

Repa JJ, Berge KE, Pomajzl C, Richardson JA, Hobbs H, Mangelsdorf DJ (2002) Regulation of ATP-binding cassette sterol transporters ABCG5 and ABCG8 by the liver X receptors alpha and beta. J Biol Chem 277:18793–18800

Richter S, Shih DQ, Pearson ER, Wolfrum C, Fajans SS, Hattersley AT et al (2003) Regulation of apolipoprotein M gene expression by MODY3 gene hepatocyte nuclear factor-1alpha: haploinsufficiency is associated with reduced serum apolipoprotein M levels. Diabetes 52: 2989–2995

Rochel N, Ciesielski F, Godet J, Moman E, Roessle M, Peluso-Iltis C et al (2011) Common architecture of nuclear receptor heterodimers on DNA direct repeat elements with different spacings. Nat Struct Mol Biol 18:564–570

Rodriguez-Ubreva J, Ballestar E (2014) Chromatin immunoprecipitation. Methods Mol Biol 1094: 309–318

Roeder RG (1998) Role of general and gene-specific cofactors in the regulation of eukaryotic transcription. Cold Spring Harb Symp Quant Biol 63:201–218

Roeder RG (2005) Transcriptional regulation and the role of diverse coactivators in animal cells. FEBS Lett 579:909–915

Rottiers V, Obad S, Petri A, McGarrah R, Lindholm MW, Black JC et al (2013) Pharmacological inhibition of a microRNA family in nonhuman primates by a seed-targeting 8-mer antimiR. Sci Transl Med 5:212ra162

Rottman JN, Widom RL, Nadal-Ginard B, Mahdavi V, Karathanasis SK (1991) A retinoic acid-responsive element in the apolipoprotein AI gene distinguishes between two different retinoic acid response pathways. Mol Cell Biol 11:3814–3820

Rousselle A, Qadri F, Leukel L, Yilmaz R, Fontaine JF, Sihn G et al (2013) CXCL5 limits macrophage foam cell formation in atherosclerosis. J Clin Invest 123:1343–1347

Rubin EM, Ishida BY, Clift SM, Krauss RM (1991) Expression of human apolipoprotein A-I in transgenic mice results in reduced plasma levels of murine apolipoprotein A-I and the appearance of two new high density lipoprotein size subclasses. Proc Natl Acad Sci USA 88: 434–438

Sakai T, Kamanna VS, Kashyap ML (2001) Niacin, but not gemfibrozil, selectively increases LP-AI, a cardioprotective subfraction of HDL, in patients with low HDL cholesterol. Arterioscler Thromb Vasc Biol 21:1783–1789

Salero E, Perez-Sen R, Aruga J, Gimenez C, Zafra F (2001) Transcription factors Zic1 and Zic2 bind and transactivate the apolipoprotein E gene promoter. J Biol Chem 276:1881–1888

Salero E, Gimenez C, Zafra F (2003) Identification of a non-canonical E-box motif as a regulatory element in the proximal promoter region of the apolipoprotein E gene. Biochem J 370:979–986

Salmena L, Poliseno L, Tay Y, Kats L, Pandolfi PP (2011) A ceRNA hypothesis: the Rosetta Stone of a hidden RNA language? Cell 146:353–358

Sanoudou D, Duka A, Drosatos K, Hayes KC, Zannis VI (2009) Role of Esrrg in the fibrate-mediated regulation of lipid metabolism genes in human ApoA-I transgenic mice. Pharmacogenomics J 10:165–179

Santamarina-Fojo S, Peterson K, Knapper C, Qiu Y, Freeman L, Cheng JF et al (2000) Complete genomic sequence of the human ABCA1 gene: analysis of the human and mouse ATP-binding cassette A promoter. Proc Natl Acad Sci USA 97:7987–7992

Sanyal A, Lajoie BR, Jain G, Dekker J (2012) The long-range interaction landscape of gene promoters. Nature 489:109–113

Schaefer EJ, Lamon-Fava S, Cole T, Sprecher DL, Cilla DD Jr, Balagtas CC et al (1996) Effects of regular and extended-release gemfibrozil on plasma lipoproteins and apolipoproteins in hypercholesterolemic patients with decreased HDL cholesterol levels. Atherosclerosis 127: 113–122

Schmitz G, Langmann T (2005) Transcriptional regulatory networks in lipid metabolism control ABCA1 expression. Biochim Biophys Acta 1735:1–19

Schnall-Levin M, Zhao Y, Perrimon N, Berger B (2010) Conserved microRNA targeting in Drosophila is as widespread in coding regions as in 3'UTRs. Proc Natl Acad Sci USA 107: 15751–15756

Schoonjans K, Annicotte JS, Huby T, Botrugno OA, Fayard E, Ueda Y et al (2002) Liver receptor homolog 1 controls the expression of the scavenger receptor class B type I. EMBO Rep 3: 1181–1187

Schou J, Frikke-Schmidt R, Kardassis D, Thymiakou E, Nordestgaard BG, Jensen G et al (2012) Genetic variation in ABCG1 and risk of myocardial infarction and ischemic heart disease. Arterioscler Thromb Vasc Biol 32:506–515

Schreyer SA, Peschon JJ, LeBoeuf RC (1996) Accelerated atherosclerosis in mice lacking tumor necrosis factor receptor p55. J Biol Chem 271:26174–26178

Schreyer SA, Vick CM, LeBoeuf RC (2002) Loss of lymphotoxin-alpha but not tumor necrosis factor-alpha reduces atherosclerosis in mice. J Biol Chem 277:12364–12368

Schwartz K, Lawn RM, Wade DP (2000) ABC1 gene expression and ApoA-I-mediated cholesterol efflux are regulated by LXR. Biochem Biophys Res Commun 274:794–802

Shea-Eaton W, Lopez D, McLean MP (2001) Yin yang 1 protein negatively regulates high-density lipoprotein receptor gene transcription by disrupting binding of sterol regulatory element binding protein to the sterol regulatory element. Endocrinology 142:49–58

Shetty S, Eckhardt ER, Post SR, van der Westhuyzen DR (2006) Phosphatidylinositol-3-kinase regulates scavenger receptor class B type I subcellular localization and selective lipid uptake in hepatocytes. Arterioscler Thromb Vasc Biol 26:2125–2131

Shih SJ, Allan C, Grehan S, Tse E, Moran C, Taylor JM (2000) Duplicated downstream enhancers control expression of the human apolipoprotein E gene in macrophages and adipose tissue. J Biol Chem 275:31567–31572

Shih DQ, Bussen M, Sehayek E, Ananthanarayanan M, Shneider BL, Suchy FJ et al (2001) Hepatocyte nuclear factor-1alpha is an essential regulator of bile acid and plasma cholesterol metabolism. Nat Genet 27:375–382

Shirasaki T, Honda M, Shimakami T, Horii R, Yamashita T, Sakai Y et al (2013) MicroRNA-27a regulates lipid metabolism and inhibits hepatitis C virus replication in human hepatoma cells. J Virol 87:5270–5286

Singaraja RR, Bocher V, James ER, Clee SM, Zhang LH, Leavitt BR et al (2001) Human ABCA1 BAC transgenic mice show increased high density lipoprotein cholesterol and ApoAI-dependent efflux stimulated by an internal promoter containing liver X receptor response elements in intron 1. J Biol Chem 276:33969–33979

Singaraja RR, James ER, Crim J, Visscher H, Chatterjee A, Hayden MR (2005) Alternate transcripts expressed in response to diet reflect tissue-specific regulation of ABCA1. J Lipid Res 46:2061–2071

Singaraja RR, Visscher H, James ER, Chroni A, Coutinho JM, Brunham LR et al (2006) Specific mutations in ABCA1 have discrete effects on ABCA1 function and lipid phenotypes both in vivo and in vitro. Circ Res 99:389–397

Singaraja RR, Kang MH, Vaid K, Sanders SS, Vilas GL, Arstikaitis P et al (2009) Palmitoylation of ATP-binding cassette transporter A1 is essential for its trafficking and function. Circ Res 105:138–147

Singh NN, Ramji DP (2006) Transforming growth factor-beta-induced expression of the apolipoprotein E gene requires c-Jun N-terminal kinase, p38 kinase, and casein kinase 2. Arterioscler Thromb Vasc Biol 26:1323–1329

Sladek FM (1994) Orphan receptor HNF-4 and liver-specific gene expression. Receptor 4:64

Small EM, Olson EN (2011) Pervasive roles of microRNAs in cardiovascular biology. Nature 469: 336–342

Smit M, van der Kooij-Meijs E, Frants RR, Havekes L, Klasen EC (1988) Apolipoprotein gene cluster on chromosome 19. Definite localization of the APOC2 gene and the polymorphic Hpa I site associated with type III hyperlipoproteinemia. Hum Genet 78:90–93

Smith JD, Melian A, Leff T, Breslow JL (1988) Expression of the human apolipoprotein E gene is regulated by multiple positive and negative elements. J Biol Chem 263:8300–8308

Song H, Saito K, Fujigaki S, Noma A, Ishiguro H, Nagatsu T et al (1998) IL-1 beta and TNF-alpha suppress apolipoprotein (apo) E secretion and apo A-I expression in HepG2 cells. Cytokine 10: 275–280

Sparrow CP, Baffic J, Lam MH, Lund EG, Adams AD, Fu X et al (2002) A potent synthetic LXR agonist is more effective than cholesterol loading at inducing ABCA1 mRNA and stimulating cholesterol efflux. J Biol Chem 277:10021–10027

Sporstol M, Mousavi SA, Eskild W, Roos N, Berg T (2007) ABCA1, ABCG1 and SR-BI: hormonal regulation in primary rat hepatocytes and human cell lines. BMC Mol Biol 8:5

Srivastava RA, Srivastava N, Averna M (2000) Dietary cholic acid lowers plasma levels of mouse and human apolipoprotein A-I primarily via a transcriptional mechanism. Eur J Biochem 267: 4272–4280

Staels B, Auwerx J (1998) Regulation of apo A-I gene expression by fibrates. Atherosclerosis 137(Suppl):S19–S23

Sticozzi C, Belmonte G, Pecorelli A, Cervellati F, Leoncini S, Signorini C et al (2013) Scavenger receptor B1 post-translational modifications in Rett syndrome. FEBS Lett 587:2199–2204

Sumi K, Tanaka T, Uchida A, Magoori K, Urashima Y, Ohashi R et al (2007) Cooperative interaction between hepatocyte nuclear factor 4 alpha and GATA transcription factors regulates ATP-binding cassette sterol transporters ABCG5 and ABCG8. Mol Cell Biol 27: 4248–4260

Sun Y, Wang N, Tall AR (1999) Regulation of adrenal scavenger receptor-BI expression by ACTH and cellular cholesterol pools. J Lipid Res 40:1799–1805

Sun D, Zhang J, Xie J, Wei W, Chen M, Zhao X (2012) MiR-26 controls LXR-dependent cholesterol efflux by targeting ABCA1 and ARL7. FEBS Lett 586:1472–1479

Suzuki S, Nishimaki-Mogami T, Tamehiro N, Inoue K, Arakawa R, Abe-Dohmae S et al (2004) Verapamil increases the apolipoprotein-mediated release of cellular cholesterol by induction of ABCA1 expression via Liver X receptor-independent mechanism. Arterioscler Thromb Vasc Biol 24:519–525

Tall AR, Yvan-Charvet L, Terasaka N, Pagler T, Wang N (2008) HDL, ABC transporters, and cholesterol efflux: implications for the treatment of atherosclerosis. Cell Metab 7:365–375

Tamehiro N, Zhou S, Okuhira K, Benita Y, Brown CE, Zhuang DZ et al (2008) SPTLC1 binds ABCA1 to negatively regulate trafficking and cholesterol efflux activity of the transporter. Biochemistry 47:6138–6147

Tan CL, Plotkin JL, Veno MT, von Schimmelmann M, Feinberg P, Mann S et al (2013) MicroRNA-128 governs neuronal excitability and motor behavior in mice. Science 342: 1254–1258

Tang C, Vaughan AM, Oram JF (2004) Janus kinase 2 modulates the apolipoprotein interactions with ABCA1 required for removing cellular cholesterol. J Biol Chem 279:7622–7628

Tarling EJ, Edwards PA (2012) Dancing with the sterols: critical roles for ABCG1, ABCA1, miRNAs, and nuclear and cell surface receptors in controlling cellular sterol homeostasis. Biochim Biophys Acta 1821:386–395

Tay Y, Zhang J, Thomson AM, Lim B, Rigoutsos I (2008) MicroRNAs to Nanog, Oct4 and Sox2 coding regions modulate embryonic stem cell differentiation. Nature 455:1124–1128

Taylor AH, Wishart P, Lawless DE, Raymond J, Wong NC (1996) Identification of functional positive and negative thyroid hormone-responsive elements in the rat apolipoprotein AI promoter. Biochemistry 35:8281–8288

Thymiakou E, Zannis VI, Kardassis D (2007) Physical and functional interactions between liver X receptor/retinoid X receptor and Sp1 modulate the transcriptional induction of the human ATP binding cassette transporter A1 gene by oxysterols and retinoids. Biochemistry 46: 11473–11483

Towns R, Azhar S, Peegel H, Menon KM (2005) LH/hCG-stimulated androgen production and selective HDL-cholesterol transport are inhibited by a dominant-negative CREB construct in primary cultures of rat theca-interstitial cells. Endocrine 27:269–277

Trajkovski M, Hausser J, Soutschek J, Bhat B, Akin A, Zavolan M et al (2011) MicroRNAs 103 and 107 regulate insulin sensitivity. Nature 474:649–653

Truong TQ, Aubin D, Falstrault L, Brodeur MR, Brissette L (2010) SR-BI, CD36, and caveolin-1 contribute positively to cholesterol efflux in hepatic cells. Cell Biochem Funct 28:480–489

Trusca VG, Fuior EV, Florea IC, Kardassis D, Simionescu M, Gafencu AV (2011) Macrophage-specific up-regulation of apolipoprotein E gene expression by STAT1 is achieved via long range genomic interactions. J Biol Chem 286:13891–13904

Trusca VG, Florea IC, Kardassis D, Gafencu AV (2012) STAT1 interacts with RXRalpha to upregulate ApoCII gene expression in macrophages. PLoS ONE 7:e40463

Tu AY, Albers JJ (1999) DNA sequences responsible for reduced promoter activity of human phospholipid transfer protein by fibrate. Biochem Biophys Res Commun 264:802–807

Tu AY, Albers JJ (2001) Glucose regulates the transcription of human genes relevant to HDL metabolism: responsive elements for peroxisome proliferator-activated receptor are involved in the regulation of phospholipid transfer protein. Diabetes 50:1851–1856

Tzameli I, Zannis VI (1996) Binding specificity and modulation of the ApoA-I promoter activity by homo- and heterodimers of nuclear receptors. J Biol Chem 271:8402–8415

Uehara Y, Engel T, Li Z, Goepfert C, Rust S, Zhou X et al (2002) Polyunsaturated fatty acids and acetoacetate downregulate the expression of the ATP-binding cassette transporter A1. Diabetes 51:2922–2928

Uehara Y, Miura S, von Eckardstein A, Abe S, Fujii A, Matsuo Y et al (2007) Unsaturated fatty acids suppress the expression of the ATP-binding cassette transporter G1 (ABCG1) and ABCA1 genes via an LXR/RXR responsive element. Atherosclerosis 191:11–21

Urizar NL, Dowhan DH, Moore DD (2000) The farnesoid X-activated receptor mediates bile acid activation of phospholipid transfer protein gene expression. J Biol Chem 275:39313–39317

Valadi H, Ekstrom K, Bossios A, Sjostrand M, Lee JJ, Lotvall JO (2007) Exosome-mediated transfer of mRNAs and microRNAs is a novel mechanism of genetic exchange between cells. Nat Cell Biol 9:654–659

Valenta DT, Bulgrien JJ, Banka CL, Curtiss LK (2006) Overexpression of human ApoAI transgene provides long-term atheroprotection in LDL receptor-deficient mice. Atherosclerosis 189: 255–263

Van Eck M, Herijgers N, Yates J, Pearce NJ, Hoogerbrugge PM, Groot PH et al (1997) Bone marrow transplantation in apolipoprotein E-deficient mice. Effect of ApoE gene dosage on serum lipid concentrations, (beta)VLDL catabolism, and atherosclerosis. Arterioscler Thromb Vasc Biol 17:3117–3126

Vasudevan S, Tong Y, Steitz JA (2007) Switching from repression to activation: microRNAs can up-regulate translation. Science 318:1931–1934

Vaziri ND, Sato T, Liang K (2003) Molecular mechanisms of altered cholesterol metabolism in rats with spontaneous focal glomerulosclerosis. Kidney Int 63:1756–1763

Vaziri ND, Gollapudi P, Han S, Farahmand G, Yuan J, Rahimi A et al (2011) Nephrotic syndrome causes upregulation of HDL endocytic receptor and PDZK-1-dependent downregulation of HDL docking receptor. Nephrol Dial Transplant 26:3118–3123

Venteclef N, Haroniti A, Tousaint JJ, Talianidis I, Delerive P (2008) Regulation of antiatherogenic apolipoprotein M gene expression by the orphan nuclear receptor LRH-1. J Biol Chem 283:3694–3701

Vickers KC, Rader DJ (2013) Nuclear receptors and microRNA-144 coordinately regulate cholesterol efflux. Circ Res 112:1529–1531

Vickers KC, Palmisano BT, Shoucri BM, Shamburek RD, Remaley AT (2011) MicroRNAs are transported in plasma and delivered to recipient cells by high-density lipoproteins. Nat Cell Biol 13:423–433

von Eckardstein A (1996) Cholesterol efflux from macrophages and other cells. Curr Opin Lipidol 7:308–319

von Eckardstein A, Hersberger M, Rohrer L (2005) Current understanding of the metabolism and biological actions of HDL. Curr Opin Clin Nutr Metab Care 8:147–152

Vu-Dac N, Schoonjans K, Laine B, Fruchart JC, Auwerx J, Staels B (1994) Negative regulation of the human apolipoprotein A-I promoter by fibrates can be attenuated by the interaction of the peroxisome proliferator-activated receptor with its response element. J Biol Chem 269: 31012–31018

Wagner J, Riwanto M, Besler C, Knau A, Fichtlscherer S, Roxe T et al (2013) Characterization of levels and cellular transfer of circulating lipoprotein-bound microRNAs. Arterioscler Thromb Vasc Biol 33:1392–1400

Walsh A, Ito Y, Breslow JL (1989) High levels of human apolipoprotein A-I in transgenic mice result in increased plasma levels of small high density lipoprotein (HDL) particles comparable to human HDL3. J Biol Chem 264:6488–6494

Wan YJ, An D, Cai Y, Repa JJ, Hung-Po Chen T, Flores M et al (2000) Hepatocyte-specific mutation establishes retinoid X receptor alpha as a heterodimeric integrator of multiple physiological processes in the liver. Mol Cell Biol 20:4436–4444

Wang Y, Oram JF (2002) Unsaturated fatty acids inhibit cholesterol efflux from macrophages by increasing degradation of ATP-binding cassette transporter A1. J Biol Chem 277:5692–5697

Wang Y, Oram JF (2007) Unsaturated fatty acids phosphorylate and destabilize ABCA1 through a protein kinase C delta pathway. J Lipid Res 48:1062–1068

Wang XQ, Panousis CG, Alfaro ML, Evans GF, Zuckerman SH (2002) Interferon-gamma-mediated downregulation of cholesterol efflux and ABC1 expression is by the Stat1 pathway. Arterioscler Thromb Vasc Biol 22:e5–e9

Wang N, Chen W, Linsel-Nitschke P, Martinez LO, Agerholm-Larsen B, Silver DL et al (2003) A PEST sequence in ABCA1 regulates degradation by calpain protease and stabilization of ABCA1 by apoA-I. J Clin Invest 111:99–107

Wang JH, Keisala T, Solakivi T, Minasyan A, Kalueff AV, Tuohimaa P (2009) Serum cholesterol and expression of ApoAI, LXRbeta and SREBP2 in vitamin D receptor knock-out mice. J Steroid Biochem Mol Biol 113:222–226

Wang D, Xia M, Yan X, Li D, Wang L, Xu Y et al (2012a) Gut microbiota metabolism of anthocyanin promotes reverse cholesterol transport in mice via repressing miRNA-10b. Circ Res 111:967–981

Wang K, Li H, Yuan Y, Etheridge A, Zhou Y, Huang D et al (2012b) The complex exogenous RNA spectra in human plasma: an interface with human gut biota? PLoS ONE 7:e51009

Wang L, Jia XJ, Jiang HJ, Du Y, Yang F, Si SY et al (2013a) MicroRNAs 185, 96, and 223 repress selective high-density lipoprotein cholesterol uptake through posttranscriptional inhibition. Mol Cell Biol 33:1956–1964

Wang M, Wang D, Zhang Y, Wang X, Liu Y, Xia M (2013b) Adiponectin increases macrophages cholesterol efflux and suppresses foam cell formation in patients with type 2 diabetes mellitus. Atherosclerosis 229:62–70

Wang W, Song W, Wang Y, Chen L, Yan X (2013c) HMG-CoA reductase inhibitors, simvastatin and atorvastatin, downregulate ABCG1-mediated cholesterol efflux in human macrophages. J Cardiovasc Pharmacol 62:90–98

Wehmeier K, Beers A, Haas MJ, Wong NC, Steinmeyer A, Zugel U et al (2005) Inhibition of apolipoprotein AI gene expression by 1, 25-dihydroxyvitamin D3. Biochim Biophys Acta 1737:16–26

Wei Z, Huang D, Gao F, Chang WH, An W, Coetzee GA et al (2013) Biological implications and regulatory mechanisms of long-range chromosomal interactions. J Biol Chem 288:22369–22377

Wellington CL, Walker EK, Suarez A, Kwok A, Bissada N, Singaraja R et al (2002) ABCA1 mRNA and protein distribution patterns predict multiple different roles and levels of regulation. Lab Invest 82:273–283

Werb Z, Chin JR (1983a) Apoprotein E is synthesized and secreted by resident and thioglycollate-elicited macrophages but not by pyran copolymer- or bacillus Calmette-Guerin-activated macrophages. J Exp Med 158:1272–1293

Werb Z, Chin JR (1983b) Endotoxin suppresses expression of apoprotein E by mouse macrophages in vivo and in culture. A biochemical and genetic study. J Biol Chem 258:10642–10648

Werb Z, Chin JR (1983c) Onset of apoprotein E secretion during differentiation of mouse bone marrow-derived mononuclear phagocytes. J Cell Biol 97:1113–1118

Westerterp M, Murphy AJ, Wang M, Pagler TA, Vengrenyuk Y, Kappus MS et al (2013) Deficiency of ATP-binding cassette transporters A1 and G1 in macrophages increases inflammation and accelerates atherosclerosis in mice. Circ Res 112:1456–1465

Wightman B, Ha I, Ruvkun G (1993) Posttranscriptional regulation of the heterochronic gene lin-14 by lin-4 mediates temporal pattern formation in C. elegans. Cell 75:855–862

Wolfrum C, Poy MN, Stoffel M (2005) Apolipoprotein M is required for prebeta-HDL formation and cholesterol efflux to HDL and protects against atherosclerosis. Nat Med 11:418–422

Wolfrum C, Howell JJ, Ndungo E, Stoffel M (2008) Foxa2 activity increases plasma high density lipoprotein levels by regulating apolipoprotein M. J Biol Chem 283:16940–16949

Wood P, Mulay V, Darabi M, Chan KC, Heeren J, Pol A et al (2011) Ras/mitogen-activated protein kinase (MAPK) signaling modulates protein stability and cell surface expression of scavenger receptor SR-BI. J Biol Chem 286:23077–23092

Xu N, Ahren B, Jiang J, Nilsson-Ehle P (2006) Down-regulation of apolipoprotein M expression is mediated by phosphatidylinositol 3-kinase in HepG2 cells. Biochim Biophys Acta 1761:256–260

Yang XP, Freeman LA, Knapper CL, Amar MJ, Remaley A, Brewer HB Jr et al (2002) The E-box motif in the proximal ABCA1 promoter mediates transcriptional repression of the ABCA1 gene. J Lipid Res 43:297–306

Yang C, McDonald JG, Patel A, Zhang Y, Umetani M, Xu F et al (2006) Sterol intermediates from cholesterol biosynthetic pathway as liver X receptor ligands. J Biol Chem 281:27816–27826

Yang Y, Zhang Z, Jiang W, Gao L, Zhao G, Zheng Z et al (2007) Identification of novel human high-density lipoprotein receptor Up-regulators using a cell-based high-throughput screening assay. J Biomol Screen 12:211–219

Yang Y, Jiang W, Wang L, Zhang ZB, Si SY, Hong B (2009) Characterization of the isoflavone pratensein as a novel transcriptional up-regulator of scavenger receptor class B type I in HepG2 cells. Biol Pharm Bull 32:1289–1294

Yang T, Chen C, Zhang B, Huang H, Wu G, Wen J et al (2010) Induction of Kruppel-like factor 4 by high-density lipoproteins promotes the expression of scavenger receptor class B type I. FEBS J 277:3780–3788

Yang JS, Phillips MD, Betel D, Mu P, Ventura A, Siepel AC et al (2011) Widespread regulatory activity of vertebrate microRNA* species. RNA 17:312–326

Yokoyama S, Arakawa R, Wu CA, Iwamoto N, Lu R, Tsujita M et al (2012) Calpain-mediated ABCA1 degradation: post-translational regulation of ABCA1 for HDL biogenesis. Biochim Biophys Acta 1821:547–551

Yoshikawa T, Shimano H, Amemiya-Kudo M, Yahagi N, Hasty AH, Matsuzaka T et al (2001) Identification of liver X receptor-retinoid X receptor as an activator of the sterol regulatory element-binding protein 1c gene promoter. Mol Cell Biol 21:2991–3000

Yu L, Hammer RE, Li-Hawkins J, Von Bergmann K, Lutjohann D, Cohen JC et al (2002a) Disruption of Abcg5 and Abcg8 in mice reveals their crucial role in biliary cholesterol secretion. Proc Natl Acad Sci USA 99:16237–16242

Yu L, Li-Hawkins J, Hammer RE, Berge KE, Horton JD, Cohen JC et al (2002b) Overexpression of ABCG5 and ABCG8 promotes biliary cholesterol secretion and reduces fractional absorption of dietary cholesterol. J Clin Invest 110:671–680

Yu XH, Jiang HL, Chen WJ, Yin K, Zhao GJ, Mo ZC et al (2012) Interleukin-18 and interleukin-12 together downregulate ATP-binding cassette transporter A1 expression through the interleukin-18R/nuclear factor-kappaB signaling pathway in THP-1 macrophage-derived foam cells. Circ J 76:1780–1791

Yu XH, Qian K, Jiang N, Zheng XL, Cayabyab FS, Tang CK (2014) ABCG5/ABCG8 in cholesterol excretion and atherosclerosis. Clin Chim Acta 428:82–88

Zaiou M, Azrolan N, Hayek T, Wang H, Wu L, Haghpassand M et al (1998) The full induction of human apoprotein A-I gene expression by the experimental nephrotic syndrome in transgenic mice depends on cis-acting elements in the proximal 256 base-pair promoter region and the trans-acting factor early growth response factor 1. J Clin Invest 101:1699–1707

Zannis VI, Cole FS, Jackson CL, Kurnit DM, Karathanasis SK (1985) Distribution of apolipoprotein A-I, C-II, C-III, and E mRNA in fetal human tissues. Time-dependent induction of apolipoprotein E mRNA by cultures of human monocyte-macrophages. Biochemistry 24: 4450–4455

Zannis VI, Kan HY, Kritis A, Zanni E, Kardassis D (2001a) Transcriptional regulation of the human apolipoprotein genes. Front Biosci 6:D456–D504

Zannis VI, Kan HY, Kritis A, Zanni EE, Kardassis D (2001b) Transcriptional regulatory mechanisms of the human apolipoprotein genes in vitro and in vivo. Curr Opin Lipidol 12: 181–207

Zhang X, Jiang B, Luo G, Nilsson-Ehle P, Xu N (2007) Hyperglycemia down-regulates apolipoprotein M expression in vivo and in vitro. Biochim Biophys Acta 1771:879–882

Zhang Y, Shen C, Ai D, Xie X, Zhu Y (2011) Upregulation of scavenger receptor B1 by hepatic nuclear factor 4alpha through a peroxisome proliferator-activated receptor gamma-dependent mechanism in liver. PPAR Res 2011:164925

Zhang L, Hou D, Chen X, Li D, Zhu L, Zhang Y et al (2012) Exogenous plant MIR168a specifically targets mammalian LDLRAP1: evidence of cross-kingdom regulation by microRNA. Cell Res 22:107–126

Zhao GJ, Tang SL, Lv YC, Ouyang XP, He PP, Yao F et al (2013) Antagonism of betulinic acid on LPS-mediated inhibition of ABCA1 and cholesterol efflux through inhibiting nuclear factor-kappaB signaling pathway and miR-33 expression. PLoS ONE 8:e74782

Zuckerman SH, Evans GF, O'Neal L (1993) Exogenous glucocorticoids increase macrophage secretion of apo E by cholesterol-independent pathways. Atherosclerosis 103:43–54

Cholesterol Efflux and Reverse Cholesterol Transport

Elda Favari, Angelika Chroni, Uwe J.F. Tietge, Ilaria Zanotti, Joan Carles Escolà-Gil, and Franco Bernini

Contents

1 Cholesterol Efflux as the First Step of Reverse Cholesterol Transport (RCT) 183
   1.1 ABCA1-Mediated Lipid Efflux to Lipid-Poor apoA-I 184
   1.2 Cholesterol Efflux to Lipidated HDL 185
2 HDL Quality and Cholesterol Efflux 187
3 RCT in Animal Models 190
   3.1 Physiology 190
      3.1.1 Methodological Approaches to Quantify RCT In Vivo 191
      3.1.2 Factors Impacting In Vivo RCT 192
   3.2 Pharmacology 196
      3.2.1 CETP Inhibitors 196
      3.2.2 Nuclear Receptor Activation 197
      3.2.3 Cholesterol Absorption Inhibitors 197
      3.2.4 Augmenting or Mimicking apoA-I 197
4 Serum Cholesterol Efflux Capacity (CEC) 198
References 199

E. Favari • I. Zanotti • F. Bernini (✉)
Department of Pharmacy, University of Parma, Parco Area delle Scienze 27/A, 43124 Parma, Italy
e-mail: fbernini@unipr.it

A. Chroni
Institute of Biosciences and Applications, National Center for Scientific Research "Demokritos", Agia Paraskevi, Athens, Greece

U.J.F. Tietge
Department of Pediatrics, University of Groningen, University Medical Center Groningen, Groningen, The Netherlands

J.C. Escolà-Gil
IIB Sant Pau-CIBER de Diabetes y Enfermedades Metabolicas Asociadas, Barcelona, Spain

© The Author(s) 2015
A. von Eckardstein, D. Kardassis (eds.), *High Density Lipoproteins*, Handbook of Experimental Pharmacology 224, DOI 10.1007/978-3-319-09665-0_4

Abstract

Both alterations of lipid/lipoprotein metabolism and inflammatory events contribute to the formation of the atherosclerotic plaque, characterized by the accumulation of abnormal amounts of cholesterol and macrophages in the artery wall. Reverse cholesterol transport (RCT) may counteract the pathogenic events leading to the formation and development of atheroma, by promoting the high-density lipoprotein (HDL)-mediated removal of cholesterol from the artery wall. Recent in vivo studies established the inverse relationship between RCT efficiency and atherosclerotic cardiovascular diseases (CVD), thus suggesting that the promotion of this process may represent a novel strategy to reduce atherosclerotic plaque burden and subsequent cardiovascular events. HDL plays a primary role in all stages of RCT: (1) cholesterol efflux, where these lipoproteins remove excess cholesterol from cells; (2) lipoprotein remodeling, where HDL undergo structural modifications with possible impact on their function; and (3) hepatic lipid uptake, where HDL releases cholesterol to the liver, for the final excretion into bile and feces. Although the inverse association between HDL plasma levels and CVD risk has been postulated for years, recently this concept has been challenged by studies reporting that HDL antiatherogenic functions may be independent of their plasma levels. Therefore, assessment of HDL function, evaluated as the capacity to promote cell cholesterol efflux may offer a better prediction of CVD than HDL levels alone. Consistent with this idea, it has been recently demonstrated that the evaluation of serum cholesterol efflux capacity (CEC) is a predictor of atherosclerosis extent in humans.

Keywords

HDL • RCT • Cholesterol efflux • CEC

Abbreviations

15-LO-1	15-Lipoxygenase-1
ABCA1	ATP-binding cassette transporter A1
ABCG1	ATP-binding cassette transporter G1
apoB	Apolipoprotein B
apoE	Apolipoprotein E
CEC	Cholesterol efflux capacity
CETP	Cholesteryl ester transfer protein
CVD	Cardiovascular diseases
EL	Endothelial lipase
HDL	High-density lipoproteins
LCAT	Lecithin-cholesterol acyltransferase

LDL	Low-density lipoproteins
LXR	Liver X receptor
LXR/RXR	Liver X receptor/retinoid X receptor
MPO	Myeloperoxidase
NPC1L1	Niemann-Pick type C like 1
PLA_2	Phospholipase A_2
PLTP	Phospholipid transfer protein
PPAR	Peroxisome proliferator-activated receptor
RCT	Reverse cholesterol transport
SR-BI	Scavenger receptor class B type I

1 Cholesterol Efflux as the First Step of Reverse Cholesterol Transport (RCT)

Lipid accumulation within macrophages, leading to the formation of foam cells, represents a main feature of atherogenesis (Wynn et al. 2013). Among the strategies by which cells may counteract this process, the release of excess cholesterol to extracellular lipid acceptors has been matter of several studies in the recent years. Although cholesterol efflux from macrophages has a minor contribution to the whole body cholesterol transported by plasma high-density lipoproteins (HDL), it is the most important for the antiatherosclerotic extent. For this reason, the macrophage RCT concept has been proposed, and it was recently identified as a novel therapeutic target (Rosenson et al. 2012).

Both active and passive mechanisms are responsible for cholesterol efflux, and several components, such as cell cholesterol status, lipid transporter activity, and the nature of extracellular acceptors, have been shown to impact its efficiency (Zanotti et al. 2012).

The efflux process involves cholesterol localized on the plasma membrane that in turn may derive from intracellular sites, such as the late endosomal/lysosomal compartment and from the Golgi apparatus. Cholesterol localized in the endoplasmic reticulum originates from endogenous synthesis and can be delivered to intracellular organelles mainly through a nonvesicular pathway involving protein carriers. Cholesterol in endosomes and lysosomes derives from lipoprotein uptake and undergoes hydrolyzation by acid hydrolase (Maxfield and Tabas 2005). This lipoprotein-derived cholesterol is then rapidly released into the cytoplasm and delivered throughout the cells by Niemann-Pick type C 1 and 2. These proteins, expressed on the membrane of late endosomes and in lysosomes, respectively, facilitate lipid mobilization from these compartments to the plasma membranes via the trans-Golgi network (Boadu and Francis 2006). In physiological conditions cholesterol content in intracellular locations is low, because most of it (about 80 %) is transferred to the plasma membrane, where it establishes a dynamic equilibrium with endoplasmic reticulum and the Golgi system pools (Huang et al. 2003). The plasma membrane bilayer contains distinct lipid environments in a steady state equilibrium. Lipid rafts are characterized by a tightly packed, liquid-ordered state, where cholesterol associates with sphingolipids and caveolin, playing a crucial role in cell signaling (Gargalovic and Dory 2003). Several studies carried out in macrophages, demonstrated that the major mechanisms of cholesterol efflux are raft-independent (Gargalovic and Dory 2003). The non-raft membrane microdomains serve as the principal source of cholesterol available for interaction with extracellular acceptors and subsequent efflux via ATP-binding cassette transporter A1 (ABCA1) that is localized in these portions of the membrane (Landry et al. 2006; Vaughan and Oram 2005). Interestingly, RCT can be triggered also from extracellular matrix-associated cholesterol microdomains, whose formation is mediated by ATP-binding cassette transporter G1 (ABCG1) (Freeman et al. 2013).

Mechanisms accounting for cholesterol efflux include passive diffusion processes as well as active pathways mediated by ABCA1, ABCG1, and scavenger receptor class B type I (SR-BI). Aqueous diffusion mainly involves free cholesterol in the plasma membrane and HDL as lipid acceptors. The nature of this process is a matter of debate: whereas this mechanism appears to be a relevant contributor of lipid removal in foam cell macrophages (Adorni et al. 2007), the involvement of a still unknown transporter is not completely ruled out.

1.1 ABCA1-Mediated Lipid Efflux to Lipid-Poor apoA-I

ABCA1 is a 2,261-amino acid integral membrane protein, member of the large superfamily of ABC transporters that use ATP as an energy source to transport lipids across membranes. ABC transporters are characterized by the presence of nucleotide-binding domains containing two conserved peptide motifs known as Walker A and Walker B, a unique amino acid signature between the two Walker motifs, which defines the family. ABC transporters are integrated into the membrane by domains containing six transmembrane helices. The minimum requirement for an active ABC transporter is two nucleotide-binding and two 6-helix transmembrane domains (Oram and Lawn 2001). At the cellular level ABCA1 is localized both on the plasma membrane and intracellular compartments, the Golgi complex, and the late endosome/lysosomes, cycling between these loci and promoting a flow of intracellular cholesterol from the late endosomes/lysosomes, through the trans-Golgi complex, to the plasma membrane (Boadu and Francis 2006). This movement produces an ABCA1-dependent depletion of intracellular pools of cholesterol that affects both the "regulatory" pool of cholesterol and the LDL-derived cholesterol. These activities result in the modulation of a number of cellular events including: endogenous cholesterol synthesis, LDL receptor expression, cholesteryl ester turnover, and nascent HDL formation (Boadu and Francis 2006). The last process is driven by ABCA1-mediated removal of lipids from the cell membrane to extracellular acceptors represented by lipid-free or lipid-poor apolipoproteins. ABCA1 expression on the plasma membrane leads to the generation of non-raft microdomains and enlargement of cholesterol and phospholipid domains in the outer leaflet, thus facilitating the interaction with apoA-I and subsequent cholesterol efflux. It is worth noting that these pools of cholesterol created by ABCA1 are specific substrates for ABCA1-mediated lipid release because they selectively interact with lipid-free or lipid-poor apolipoproteins (Landry et al. 2006). ABCA1 exerts its function through a floppase activity, which drives cholesterol and phospholipids from the inner leaflet of the membrane to cell surface domains and eventually back to intracellular compartments. In addition, ABCA1 promotes vesicular trafficking of cholesterol, phospholipids and ABCA1 itself between the plasma membrane and intracellular compartments (Landry et al. 2006). Cholesterol efflux via ABCA1 is related to the protein level, in turn regulated by transcriptional or posttranslational mechanisms. Abca1 gene expression is primarily induced by the stimulation of liver X receptor/retinoid X receptor (LXR/RXR) axis, stimulated by cholesterol accumulation in the cells (Larrede et al. 2009). Recently, the role of microRNAs in the inhibitory control of Abca1 expression has emerged (Sun et al. 2012; Rayner et al. 2010; Ramirez et al. 2011). The posttranscriptional regulation includes mechanisms that involve (1) the stabilization of ABCA1 protein by apoA-I and (2) the acceleration of its turnover by calpain-mediated proteolysis or polyunsaturated fatty acids (Oram and Vaughan 2006).

1.2 Cholesterol Efflux to Lipidated HDL

ABCA1-mediated generation of nascent HDL particles may in turn promote cholesterol efflux via ABCG1 and SR-BI. The former is a half-size ABC protein, where the nucleotide-binding domain at the N-terminus is followed by six transmembrane-spanning domains. Interestingly, various transcripts of ABCG1 have been detected in different cells, possibly arising from alternative splicing events (Schmitz et al. 2001). Early studies reported many similarities between ABCG1 and ABCA1, including the cellular localization, the translocation from intracellular compartments to the plasma membrane, the floppase activity, and the expression promoted by cholesterol enrichment via LXR. A main feature of ABCG1-mediated efflux is the specific interaction with HDL and low-density lipoproteins (LDL) thus accounting for the elimination of cholesterol and toxic oxysterols from the macrophages (Vaughan and Oram 2005). It is important to note that under physiologic conditions ABCA1 and ABCG1 can act in a sequential fashion, with ABCA1 generating, which then promote lipid release via ABCG1 (Gelissen et al. 2006). However, recent data challenge these concepts, suggesting that, unlike ABCA1, ABCG1 effluxes cellular cholesterol by a process that is not dependent upon interaction with an extracellular protein (Tarling and Edwards 2012).

Although several cholesterol-responsive ABC transporters, other than ABCA1 and ABCG1, have been described in the macrophages, their potential relevance for the process of foam cell formation and RCT needs further investigations (Fu et al. 2013).

SR-BI is an 82-kDa integral membrane protein, belonging to the CD36 family, whose physiological role is related to the selective uptake of HDL cholesteryl ester, the process by which the core cholesteryl ester is taken into the cell without the endocytic uptake and degradation of the whole HDL. Since this pathway is the major route for the delivery of HDL cholesteryl ester to the liver, the role of SR-BI in determining the plasma levels of HDL is of major importance. Importantly, SR-BI has shown to stimulate free cholesterol efflux facilitating the aqueous diffusion pathway to phospholipid-enriched acceptors. Thus, the expression of this receptor establishes a bidirectional flux between cells and HDL, whose net effect will be related to cell cholesterol status, as well as the composition and concentration of the acceptor in the extracellular environment. Differently from the ABC transporters, SR-BI is localized in caveolae, a subset of lipid rafts, cell surface invaginations enriched in free cholesterol (Rosenson et al. 2012). Despite the significant role of SR-BI in cell cholesterol metabolism, its role in cholesterol efflux from macrophages is still unclear. It is important to note that the relative contribution of a single pathway to cholesterol export is species specific: whereas ABCA1-mediated efflux is the predominant mechanism both in human and murine cultured foam cells, the role of ABCG1 is elusive in the former, but not in the latter. Conversely, SR-BI (Cla-1 in humans) plays a pivotal role in human, but not murine cells (Adorni et al. 2007; Sun et al. 2012). Moreover, the antiatherosclerotic properties of ABCA1 and ABCG1 have recently been challenged by the demonstration that some efflux-independent activities could be related to deleterious effects on macrophage function (Adorni et al. 2011; Olivier et al. 2012).

A relevant role in cholesterol efflux from macrophages and RCT has been also attributed to apolipoprotein E (apoE) (Zanotti et al. 2011a, b). This 34-kDa protein is synthesized by many cell types, including macrophages, upon different stimuli, such as differentiation, cytokines, and lipid enrichment. The last process activates the LXR pathway, as described for ABCA1 and ABCG1. ApoE synthesis in the endoplasmic reticulum is followed by its movement to the Golgi and trans-Golgi network and incorporation into vesicular structures, before being transported to the plasma membrane for the final secretion. Secreted apoE can be released into the extracellular medium or alternatively can bind to the cell surface, particularly in association with heparan sulfate proteoglycans. Cell surface pools may be reinternalized and subsequently degraded, transported to the Golgi network for posttranslational modifications, or released into the extracellular medium (Kockx et al. 2008). When ApoE is secreted from cholesterol-enriched macrophages, this process can promote cholesterol efflux in the absence of added cholesterol acceptors or in presence of exogenous HDL, causing the generation of nascent HDL particles. There is evidence that apoE can promote cholesterol release by both ABCA1-dependent and independent mechanisms, whereas ABCG1 may contribute by driving cholesterol efflux to apoE-enriched particles (Huang et al. 2001).

2 HDL Quality and Cholesterol Efflux

HDL comprises several subclasses that differ in composition and size and exhibit a series of atheroprotective and other properties, including the ability to efflux cholesterol from various cell types, as well as antioxidative, anti-inflammatory, anticoagulatory, and anti-aggregatory properties (Annema and von Eckardstein 2013; Vickers and Remaley 2014). These properties are exerted by the various protein and lipid components of HDL (Annema and von Eckardstein 2013; Vickers and Remaley 2014). It is increasingly accepted that the quality rather than quantity of HDL is more relevant for its atheroprotective activity (see chapter "Dysfunctional HDL: From Structure-Function-Relationships to Biomarkers" for more details).

As anticipated, the ability of HDL to promote cholesterol efflux from lipid-laden macrophages is thought to be important for the atheroprotective function of HDL. The capacity of HDL to promote cholesterol efflux from macrophages was shown to have a strong inverse association with both carotid intima–media thickness and the likelihood of angiographic coronary artery disease, independently of the HDL-C level (Khera et al. 2011). As mentioned before, cellular cholesterol efflux is mediated by a number of pathways, with the various HDL subpopulations to display a varied capacity to promote cholesterol efflux via each of these pathways (Annema and von Eckardstein 2013; Vickers and Remaley 2014). Therefore, the efficiency of serum from an individual to accept cellular cholesterol can be affected by the distribution and composition of HDL particles acting as cholesterol acceptors. Indeed, it has been shown that apolipoprotein B (apoB)-depleted sera from subjects with similar HDL-C or apoA-I can have higher total macrophage efflux capacity due to significantly higher ABCA1-mediated efflux, and this efflux is significantly correlated with the levels of preβ1-HDL (de la Llera-Moya et al. 2010; Calabresi et al. 2009). Small discoidal preβ1-HDL particles are also efficient acceptors of cell cholesterol via ABCG1 (Favari et al. 2009). In another study, apoB-depleted sera from patients treated with the cholesteryl ester transfer protein (CETP) inhibitor anacetrapib were shown to have enhanced ability to promote ABCG1-mediated cholesterol efflux from macrophages, which was associated with increased lecithin-cholesterol acyltransferase (LCAT) and apoE mass in HDL (Yvan-Charvet et al. 2010).

The various components of HDL particles can undergo structural or chemical modifications during atherogenesis or other pathologic processes, having as a result adverse effects on HDL functionality, including the cholesterol efflux capacity. In vitro and in vivo studies have shown that enzymatic oxidation, lipolysis, and proteolysis can modify HDL and affect the HDL capacity to promote cellular cholesterol efflux. HDL isolated from humans with established CVD was found to contain higher levels of protein-bound 3-chlorotyrosine and protein-bound 3-nitrotyrosine compared to HDL from controls (Bergt et al. 2004; Pennathur et al. 2004; Zheng et al. 2004). In addition, the levels of both 3-chlorotyrosine and 3-nitrotyrosine were higher in HDL isolated from atherosclerotic lesions compared with plasma HDL. One pathway that generates such species involves myeloperoxidase (MPO), a major constituent of artery wall macrophages (Bergt et al. 2004; Pennathur et al. 2004; Zheng et al. 2004). It was proposed that apoA-I is a selective target for MPO-catalyzed nitration and chlorination in vivo and that MPO-catalyzed oxidation of HDL and apoA-I results in selective inhibition of ABCA1-dependent cholesterol efflux from macrophages (Zheng et al. 2004). More specifically, it was shown that chlorination of apoA-I impaired the ability of the protein to promote cholesterol efflux by MPO, while nitration had a much lesser effect (Bergt et al. 2004; Zheng et al. 2005; Shao et al. 2005) and that a combination of Tyr-192 chlorination and methionine oxidation is necessary for depriving apoA-I of its ABCA1-dependent cholesterol transport activity (Shao et al. 2006). More recently it was suggested that oxidative damage to apoA-I by MPO limits the ability of apoA-I to be liberated in a lipid-free form from HDL and that this impairment of apoA-I exchange reaction may contribute to reduced ABCA1-mediated cholesterol efflux (Cavigiolio et al. 2010).

Treatment of HDL_3 with the reticulocyte-type 15-lipoxygenase-1 (15-LO-1), which has been suggested to play a pathophysiological role in atherosclerosis, induced HDL apolipoprotein cross-linking and reduced cholesterol efflux from lipid-laden J774 cells (Pirillo et al. 2006). A reduced binding of 15-LO-modified HDL_3 to SR-BI explained, in part, the observed reduction of cholesterol efflux. In addition, the ABCA1-mediated cholesterol efflux was also reduced, as a consequence of loss of preβ-particles after HDL_3 modification (Pirillo et al. 2006).

The capacity of HDL to promote cellular cholesterol efflux from lipid-loaded mouse peritoneal macrophages was significantly decreased after treatment of HDL with secretory phospholipase A_2 (PLA_2) group X or V (Zanotti et al. 2011a, b). $sPLA_2$-X and $sPLA_2$-V that have been associated with the pathogenesis of atherosclerosis affect the capacity of HDL to promote cholesterol efflux by catalyzing the hydrolysis of phosphatidylcholine in HDL without any modification of apoA-I (Ishimoto et al. 2003).

Several proteases, including metalloproteinases, cathepsins, chymase, tryptase, kallikrein, neutrophil elastase, and plasmin, that are secreted from various types of cells of human atherosclerotic arterial intima can proteolytically modify HDL in vitro (Lee-Rueckert et al. 2011). It has been shown that the various proteases proteolyze apoA-I that is present in preβ-HDL and therefore reduce cholesterol efflux from macrophage foam cells (Favari et al. 2004). A series of studies aiming to elucidate the effect of proteolytic modification of HDL on cholesterol efflux and RCT used mast cells and the neutral serine proteases chymase and tryptase, which are secreted from mast cells. Treatment of HDL_3 with human chymase resulted in rapid depletion of preβ-HDL and a concomitant decrease in the efflux of cholesterol and phospholipids by an ABCA1-dependent pathway, while aqueous or SR-BI-facilitated diffusion of cholesterol was not affected (Favari et al. 2004). Furthermore, local activation of mast cells by the specific mast cell degranulating compound 48/80 in the mouse peritoneal cavity, which causes acute release of active chymase, resulted in a 90 % reduction of human apoA-I injected into the peritoneal cavity. This reduction reflected the reduction in preβ-HDL particles and resulted in attenuation of cholesterol efflux from intraperitoneally co-injected J774 macrophages and a reduced rate of macrophage RCT. In addition, pretreatment of apoA-I with chymase also fully abolished the stimulatory effect of untreated apoA-I to promote the transfer of macrophage-derived cholesterol to the intestine (Lee-Rueckert, Silvennoinen et al. 2011).

Proteolysis of lipid-free apoA-I or preβ-HDL- or HDL_3-associated apoA-I by intima proteases may result in the formation of carboxy-terminal truncated apoA-I fragments. Specifically, it has been shown recently that lipid-free apoA-I is preferentially digested by chymase at the C-terminus rather than the N-terminus and that the Phe_{229} and Tyr_{192} residues are the main cleavage sites, while the Phe_{225} residue is a minor cleavage site (Usami et al. 2013). C-terminally truncated apoA-I was detected in normal human serum using a specific monoclonal antibody (16-4mAb) recognizing C-terminally truncated apoA-I that has been cleaved after Phe_{225} by chymase (Usami et al. 2011). In addition, it has been shown that proteolysis of apoA-I in preβ-HDL by plasmin generated apoA-I fragments lacking the C-terminal region (Kunitake et al. 1990). Such C-terminal truncation of apoA-I can affect its capacity to promote ABCA1-mediated cholesterol efflux and the biogenesis of HDL. Previous studies showed that when C-terminal segments that contain residues 220–231 are deleted, the apoA-I cannot associate with ABCA1 and has a diminished capacity to promote ABCA1-mediated phospholipid cholesterol efflux (Favari et al. 2002; Chroni et al. 2003, 2004). In another study, the treatment of lipid-free human apoA-I by chymase-containing lysate derived from mouse peritoneal mast cells or recombinant human chymase generated apoA-I fragments, lacking the C-terminal site (Usami et al. 2013), which had diminished capacity to promote cholesterol efflux from mouse peritoneal macrophage foam cells (Lee-Rueckert, Silvennoinen et al. 2011). Furthermore, adenovirus-mediated gene transfer in apoA-I-deficient mice showed that the apoA-I mutants that lack C-terminal residues 220–231 fail to form spherical αHDL in vivo (Chroni et al. 2003, 2007).

Other proteolysis studies showed that apoA-I in HDL_3 by various recombinant metalloproteinases in vitro or by metalloproteinases secreted from macrophages generated apoA-I fragments also lacking the carboxyl-terminal region (cleavage after residues 213, 202, 215, 225,199, 191, or 188) (Lindstedt et al. 1999; Eberini et al. 2002). In addition, chymase, in vitro, cleaves apoA-I in reconstituted HDL at the C-terminus (after Phe_{225}) (Lee et al. 2003). These C-terminal truncated lipoprotein-associated apoA-I displayed diminished capacity to promote cholesterol efflux from mouse macrophage foam cells (Lindstedt et al. 1999; Lee et al. 2003). A recent study showed that reconstituted HDL containing the C-terminal deletion mutants apoA-I(Δ(185–243)) or apoA-I(Δ(220–243)) had strongly impaired capacity to promote ABCG1-mediated cholesterol efflux (Daniil et al. 2013). In addition, limited proteolysis of rHDL containing wild-type apoA-I by plasmin resulted in 66 % decrease of ABCG1-dependent cholesterol efflux (Daniil et al. 2013). It is therefore possible that proteolysis of HDL-associated apoA-I in vivo by proteases, which are present in the human arterial intima, could yield apoA-I fragments similar to the apoA-I(Δ(185–243)) or apoA-I(Δ(220–243)) mutants and thus may impair their capacity to promote ABCG1-mediated cholesterol efflux from macrophages. The structure-function relationship seen in this study between rHDL-associated apoA-I mutants and ABCG1-mediated cholesterol efflux closely resembles that seen before in lipid-free apoA-I mutants and ABCA1-dependent cholesterol efflux, suggesting that both processes depend on the same structural determinants of apoA-I.

The levels, composition, and the antiatherogenic properties, including the cholesterol efflux capacity, of HDL has been shown to be affected in patients suffering from chronic inflammatory rheumatic diseases (Onat and Direskeneli 2012; Watanabe et al. 2012). Specifically, cholesterol efflux capacity of HDL was impaired in rheumatoid arthritis and systemic lupus erythematosus patients with high disease activity and was correlated with systemic inflammation, higher plasma MPO activity, and HDL's antioxidant capacity (Ronda et al. 2013; Charles-Schoeman et al. 2012). In addition, HDL cholesterol efflux capacity was lower in patients with psoriasis (Watanabe et al. 2012; Holzer et al. 2013), while antipsoriatic therapy significantly improved the HDL cholesterol efflux capacity, along with improved serum LCAT activity and without any effect on serum HDL cholesterol levels (Holzer et al. 2013).

Overall, modifications of HDL particles have been demonstrated to affect the atheroprotective properties of HDL, including its capacity to promote cellular cholesterol efflux. Elucidation of the conditions and processes that affect the compositional, structural, and functional intactness of HDL may become an important tool for the assessment of cardiovascular risk and may provide us with novel therapeutic approaches. In addition, monitoring of the capacity of HDL to promote cholesterol efflux could be useful to evaluate novel therapeutic approaches for the reduction of cardiovascular risk, as it will be discussed later.

3 RCT in Animal Models

3.1 Physiology

The term RCT summarizes the transport of cholesterol from macrophage foam cells within atherosclerotic lesions through the aqueous compartment of the blood for final excretion into the feces, which could either occur directly as cholesterol or after metabolic conversion into bile acids (Annema and Tietge 2012). As detailed above, cholesterol efflux from lipid-laden macrophages is mediated by the transporters ABCA1, ABCG1, and SR-BI with HDL as acceptor, and this process is facilitated by apoE. Within the blood HDL can be remodeled in several ways. LCAT esterifies free cholesterol in the particles resulting in larger HDL and increased plasma HDL-C levels (Annema and Tietge 2012). CETP transfers cholesterol out of HDL toward apoB-containing lipoproteins in exchange for triglycerides, which are then rapidly hydrolyzed by hepatic lipase (HL) (Annema and Tietge 2012). Thereby CETP and also HL, either alone or in combination, lower plasma HDL-C levels. Other HDL remodeling proteins that increase the catabolic rate of HDL and lower circulating HDL-C are the phospholipid transfer protein (PLTP) (Annema and Tietge 2012) and the phospholipases endothelial lipase (EL) (Annema and Tietge 2011) and group IIA secretory phospholipase A_2 (sPLA_2-IIA) (Rosenson and Hurt-Camejo 2012) . Uptake of HDL-C into hepatocytes can be accomplished in two distinct ways, via selective uptake mediated by SR-BI (Annema and Tietge 2012) or via holoparticle uptake in a process that has not been fully elucidated yet (Vantourout et al. 2010). However, it has been shown that via ectopic localization of the mitochondrial F1-beta-ATPase on the cell membrane, ADP is generated which then stimulates via the ADP receptor P2Y13 holoparticle uptake of HDL by still elusive receptor mechanisms; thereby, HDL holoparticle uptake is fully dependent on the expression of the ADP receptor P2Y13 (Vantourout et al. 2010). In hepatocytes, cholesterol is either secreted directly into the bile, which is mediated in terms of mass mainly by ABCG5/G8 and to a lesser extent by SR-BI, or following conversion into bile acids, for which ABCB11 is the critical transporter into bile (Dikkers and Tietge 2010). Cholesterol can be reabsorbed in the intestine by Niemann-Pick type C like 1 (NPC1L1) but also (re)secreted by the ABCG5/G8 heterodimer, whereby a low activity of NPC1L1 would increase and a low activity of ABCG5/G8 would decrease cholesterol excretion into the feces (Annema and Tietge 2012). Bile acids can be reabsorbed in the terminal ileum by ASBT (Dawson 2011). In addition to its role in sterol absorption, the intestine has also been indicated to mediate direct secretion of cholesterol, a process termed transintestinal cholesterol excretion [TICE, current knowledge summarized in Tietge and Groen (2013)]. All different steps discussed in this paragraph have the potential to impact on and modulate RCT.

3.1.1 Methodological Approaches to Quantify RCT In Vivo

The key methodological problem for quantifying RCT specifically from macrophages, which is most relevant for atherosclerotic disease, is that the macrophage cholesterol pool is rather small. Initially employed techniques such as mass determinations of centripetal cholesterol fluxes to the liver or isotope dilution methods were therefore not able to accurately allow a conclusion of specific cholesterol fluxes from macrophages (Annema and Tietge 2012). Recently, a now widely used and accepted technique had been developed that overcomes such methodological drawbacks (Zhang et al. 2003). Thereby, macrophages, either primary or cell lines, are loaded with radiolabeled cholesterol and then injected intraperitoneally into recipient animals. After injection, the appearance of the tracer is determined in plasma at different time points and, most importantly, in the feces that are collected continuously. The time course of such experiments is usually 24–48 h; in the feces distinguishing between labels in neutral sterols and bile acids is in our view preferable.

Although this method allows tracing of cholesterol from macrophages to feces, there are certain limitations that need to be taken into account when interpreting results from such studies. One is that in the assay a potential influx of cholesterol from the plasma compartment into the macrophage is not taken into account. As first steps to overcome this, two approaches were reported, in which the application procedure of the labeled macrophages is modified in a way that allows reisolation of the cells to measure label as well as mass cholesterol content. One of these approaches makes use of Matrigel plugs that are implanted subcutaneously (Malik and Smith 2009), while the other another, entrapment of the macrophages in semipermeable holofibers (Weibel et al. 2011). Another limitation of in vivo RCT assays is that all approaches place the labeled macrophages outside the vascular compartment, which supposedly is a good surrogate but might also not accurately reflect the situation in an atherosclerotic lesion in terms of oxygen tension, pH, or accessibility by the HDL particles. In the following, results obtained from the above-described method will be summarized. It is relevant to point out that although initial steps toward a macrophage RCT assay in humans have been reported at conferences (Dunbar et al. 2013), current knowledge on the regulation of in vivo RCT is based on studies in animal models that differ in several aspects of their lipoprotein metabolism.

3.1.2 Factors Impacting In Vivo RCT

Looking at the level of the macrophage, available studies consistently indicate that expression of ABCA1 and ABCG1 is associated with increased RCT (Wang et al. 2007a, b; Out et al. 2008). However, SR-BI deficiency in macrophages does not impact RCT (Wang et al. 2007b; Zhao et al. 2011). On the other hand, macrophage apoE was shown to stimulate RCT (Zanotti et al. 2011a, b).

Regarding plasma proteins that have a role in HDL metabolism, for LCAT surprisingly no consistent effects on RCT were observed using several different models and overexpression as well as knockout strategies resulting in the conclusion that LCAT only minimally contributes to RCT although it has a major role in determining plasma HDL-C levels (Tanigawa et al. 2009). In the case of the lipid transfer proteins, PLTP overexpression resulted in increased RCT (Samyn et al. 2009), while for CETP (over)expressing models, either an increase (Tchoua et al. 2008; Tanigawa et al. 2007), dependent on functional expression of the LDL receptor (Tanigawa et al. 2007), or no effect (Rotllan et al. 2008) was observed. With respect to (phospho)lipases, $sPLA_2$-IIA had no impact on RCT (Annema et al. 2010). In case of HL and EL, conflicting data have been reported. While one group found no effect of knocking out either lipase separate or both in combination (Brown et al. 2010), others reported increased RCT for EL as well as HL (Escola-Gil et al. 2013). In contrast to apoE expression in macrophages, systemic overexpression of apoE had no effect on RCT either in wild-type or CETP transgenic mice (Annema et al. 2012).

In the case of the hepatic component of RCT, SR-BI expression in hepatocytes has a clear increasing effect on RCT as indicated in studies using both knockout and overexpression (Zhao et al. 2011; El Bouhassani et al. 2011; Zhang et al. 2005). Interestingly, it has been shown that in the liver the impact of SR-BI is independent of ABCG5/G8 expression, since double-knockout mice for both transporters had a significant decrease in RCT compared with ABCG5 knockout mice alone (Dikkers et al. 2013). ABCG5/G8 knockouts on the other hand exhibited no change in RCT (Calpe-Berdiel et al. 2008). In addition, also the consequences of abolishing functional HDL holoparticle uptake into hepatocytes on RCT were assessed. In

P_2Y_{13} knockout mice, this pathway is completely absent translating into a significant reduction in macrophage-to-feces RCT (Fabre et al. 2010; Lichtenstein et al. 2013). These combined studies indicate that both selective uptake and holoparticle uptake of HDL are critical for RCT. Regarding expression of ABCA1 in hepatocytes, liver supposedly contributes around 70 % to total HDL formation (Timmins et al. 2005); the picture is not so clear. While in wild-type and SR-BI knockout mice blocking hepatic ABCA1 with probucol enhanced RCT (Annema et al. 2012; Yamamoto et al. 2011), no such effect was seen in liver-specific ABCA1 knockout mice on a LDLR-deficient background (Bi et al. 2013).

With respect to the final step of RCT, sterol excretion from the body, the differential contribution of the biliary pathway and the intestine is a subject of active study; it might be important to note that no unequivocally established concept has emerged thus far. The classical view on the RCT pathway puts biliary secretion of RCT-relevant cholesterol central (Annema and Tietge 2012). This view is supported by studies using either surgical disruption of biliary secretion or Abcb4 knockout mice that have a genetic defect in cholesterol secretion into the bile secondary to their inability to secrete phospholipids and form mixed micelles (Nijstad et al. 2011). In the surgical model RCT was virtually absent, while in the genetic model, RCT was strongly reduced (Nijstad et al. 2011). Studies in ezetimibe-treated mice that express NPC1L1 only in hepatocytes reached a similar conclusion, namely, that functional RCT in this model depended on efficient biliary sterol secretion (Xie et al. 2013). On the other hand, RCT studies in NPC1L1 liver-transgenic mice on a wild-type background as well as a short-term experiment in bile-diverted mice resulted in opposite findings, namely, that RCT can proceed when the biliary secretion pathway is impaired (Temel et al. 2010). Further studies are clearly required to assess the role of the intestine and a possible contribution of TICE to macrophage RCT. However, currently such experiments are hampered by the lack of information how TICE is precisely mediated (Tietge and Groen 2013). Commonly accepted on the other hand is the role of intestinal sterol absorption in RCT, and several pharmacological intervention studies were carried out to show that blocking sterol absorption increases RCT (for details please see Table 1 and text above).

In addition, there are also more complex systemic pathophysiological states that are clinically associated with an increased atherosclerosis incidence and in which also alterations in RCT have been observed in mouse models. One example is an inflammatory response. Consistently, RCT was severely impaired in LPS- (Annema et al. 2010; McGillicuddy et al. 2009) and zymosan-induced (Malik et al. 2011) models of acute inflammation. Another example is diabetes. In both, an insulin-deficient model of type 1 diabetes (de Boer et al. 2012) as well as in db/db mice (Low et al. 2012), which lack the leptin receptor and serve as model of type 2 diabetes, RCT was decreased. Interestingly, the first study indicated that the selective uptake step of HDL into the liver is defective in type 1 diabetic mice (de Boer et al. 2012), while the latter study provided evidence that advanced glycation end products are not likely to be pathophysiologically involved (Low et al. 2012). Furthermore, acute psychological stress has been shown to increase

Table 1 Effects of therapeutic agents on in vivo macrophage-to-feces RCT

Treatment	Agent	Animal model	Macrophage-to-feces RCT	References
CETP inhibitors	Torcetrapib	Wild-type mice, human CETP-expressing mice	↓	Tchoua et al. (2008)
	Torcetrapib	Human CETP/apoB100-expressing mice	↑	Briand et al. (2011)
	Torcetrapib	Hamster	↑	Tchoua et al. (2008), Niesor et al. (2010)
	Dalcetrapib	Hamster	↑	Niesor et al. (2010)
	Anacetrapib	Hamster	= or ↑	Niesor et al. (2010), Castro-Perez et al. (2011)
	Anacetrapib	Human CETP-expressing/LDLR-deficient mice	=	Bell et al. (2013)
	CETP antisense oligonucleotide	Human CETP-expressing/LDLR-deficient mice	↑	Bell et al. (2013)
Nuclear receptors	Systemic LXR agonists T0901317 and GW3965	Wild-type mice, LDLR/apobec-1 double-knockout mice, human CETP/apoB100-expressing mice	↑	Naik et al. (2006), Zanotti et al. (2008), Calpe-Berdiel et al. (2008), Yasuda et al. (2010), Nijstad et al. (2011)
	Systemic LXR agonist GW3965	Hamster	↑	Briand et al. (2010)
	Intestinal-specific LXR agonist GW6340	Wild-type mice	↑	Yasuda et al. (2010)
	PPARβ/δ agonists GW0742	Wild-type mice	↑	Briand et al. (2009), Silvennoinen et al. (2012)
	Fenofibrate	Human apoA-I transgenic mice	↑	Rotllan et al. (2011)
	Gemfibrozil	Human apoA-I transgenic mice	=	Rotllan et al. (2011)
	PPARα agonists GW7647	Wild-type mice, LDLR/apobec-1 double-knockout mice overexpressing human apoA-I	↑	Nakaya et al. (2011)
	Anti-miR33 oligonucleotide	LDLR-deficient mice	↑	Rayner et al. (2011)
	PPARγ agonist GW7845	Wild-type mice	↓	Toh et al. (2011)
	FXR agonist GW4064	Wild-type mice, SR-BI-deficient mice	↑	Zhang et al. (2010)

Cholesterol and bile acid absorption inhibitors	Ezetimibe	Wild-type mice	↑	Briand et al. (2009), Sehayek and Hazen (2008), Silvennoinen et al. (2012), Maugeais et al. (2013)
	Cholestyramine	Wild-type mice	↑	Maugeais et al. (2013)
Injection of reconstituted HDL and apoA-I forms	Human apoA-I	Wild-type mice	↑	Lee-Rueckert et al. (2011)
	Reconstituted HDL	Wild-type mice	↑	Maugeais et al. (2013)
	Mimetic D-4 F	ApoE-deficient mice	↑	Navab et al. (2004)
	Mimetic ATI-5261	ApoE-deficient mice	↑	Bielicki et al. (2010)
	Mimetic 5A	Wild-type mice	↑	Amar et al. (2010)
Upregulation of apoA-I production	Thienotriazolodiazepine Ro 11-1464	Human apoA-I transgenic mice	↑	Zanotti et al. (2011a, b)

↑ or ↓ denotes that the measured parameter increased or decreased, respectively
=, the parameter remained unchanged

RCT in mice, mainly by a corticosterone-mediated downregulation of NPC1L1 and subsequent reduction in intestinal cholesterol absorption (Silvennoinen et al. 2012). These data underline the important role of intestinal NPC1L1 in RCT but are counterintuitive to the increased atherosclerosis risk attributed to psychological stress in clinical settings.

Finally, two recent additions of pathways that are linked to RCT have emerged which are not so obvious given the classical view on the RCT pathway but are worth mentioning. On the one hand, red blood cells were shown to contribute to RCT in a way that functional RCT was decreased in anemic mice (Hung et al. 2012). On the other hand, lymphatic drainage has been indicated to be involved in RCT (Martel et al. 2013; Lim et al. 2013). Hereby, blocking lymphatic transport or lymph vessel regrowth inhibited RCT, while enabling it increased RCT (Martel et al. 2013; Lim et al. 2013). Interestingly, for this route of RCT, transcytosis of HDL through lymphatic endothelium in a SR-BI-dependent fashion was critical (Lim et al. 2013), findings very similar to the role of SR-BI in HDL transcytosis through vascular endothelial cells (Rohrer et al. 2009).

3.2 Pharmacology

RCT-enhancing therapies are currently considered a promising strategy for the prevention and treatment of atherosclerotic CVD. An important number of RCT-targeted drugs have been used in preclinical animal models (mainly mice and hamsters) to test their effects on in vivo RCT from labeled cholesterol macrophages to feces. These RCT-targeted drugs can be classified among four different therapeutic approaches: CETP inhibition, nuclear receptors activation, cholesterol absorption inhibition, and directly augmenting or mimicking apoA-I. Most of these drugs are being used in clinical practice or tested in clinical trials in phases I, II, or III. This section discusses recent findings indicating that some of these therapies may be atheroprotective by promoting RCT in vivo (Table 1 shows a summary of available data).

3.2.1 CETP Inhibitors

CETP inhibition presents a preferential target for raising HDL-C and enhancing RCT. However, available data on the effect of the CETP inhibitors torcetrapib, dalcetrapib, and anacetrapib in macrophage-to-feces RCT have produced divergent results (Tchoua et al. 2008; Briand et al. 2011; Niesor et al. 2010; Castro-Perez et al. 2011). These controversial effects, together with the disappointing results of two large clinical trials using torcetrapib and dalcetrapib, have raised reasonable doubts regarding the clinical use of CETP inhibitors (Barter and Rye 2012). The positive effects of CETP antisense oligonucleotides on macrophage RCT (Bell et al. 2013) open up an alternative pathway to further evaluate whether the inhibition of CETP may improve cardiovascular risk.

3.2.2 Nuclear Receptor Activation

Liver X receptor (LXR) is considered an attractive target for therapeutic strategies aimed at stimulating RCT since it promotes HDL biogenesis, macrophage cholesterol efflux, and biliary cholesterol excretion. A considerable number of studies tested the effect of systemic LXR agonists T0901317 and GW3965 on macrophage-to-feces RCT and consistently found a higher flux through this pathway (Naik et al. 2006; Zanotti et al. 2008; Calpe-Berdiel et al. 2008; Yasuda et al. 2010; Nijstad et al. 2011; Briand et al. 2010). However, systemic LXR activators have detrimental consequences for the liver such as the induction of lipogenesis. As discussed in Sect. 3.1.2, multiple evidences indicate that excretion of macrophage-derived cholesterol can be modulated in the last step of RCT pathway which occurs in the small intestine. Indeed, intestine-specific LXR activator GW6340 also enhances macrophage RCT but avoids the lipogenic toxicity associated with liver LXR activation in mice (Yasuda et al. 2010). Peroxisome proliferator-activated receptor (PPAR) α agonists such as GW7647 and fenofibrate promote macrophage RCT (Rotllan et al. 2011; Nakaya et al. 2011). This effect is strongly correlated with the positive effects on apoA-I levels; more importantly, macrophage PPARα and LXR expression are required for the PPARα-mediated enhancement of macrophage RCT (Nakaya et al. 2011). ABCA1 and G1 are targeted for degradation by microRNA (miR)-33, an intronic microRNA located within the SREBF2 gene; anti–miR-33 therapy enhances macrophage RCT (Rayner et al. 2011). In contrast, the PPARγ agonist GW7845 reduces macrophage RCT. The authors hypothesize that GW7845 redirects macrophage-derived cholesterol to adipose tissue via SR-BI, thereby reducing its biliary excretion (Toh et al. 2011). Beyond PPAR-LXR activation, the farnesoid X receptor (FXR) agonist GW4064 also promotes macrophage-to-feces RCT; this effect is related to liver SR-BI upregulation and reduced intestinal cholesterol absorption (Zhang et al. 2010).

3.2.3 Cholesterol Absorption Inhibitors

Interventions that inhibit cholesterol absorption including ezetimibe administration and PPAR β/δ activation with GW0742 increase the excretion of macrophage-derived cholesterol in feces by reducing intestinal NPC1L1 activity (Briand et al. 2009; Sehayek and Hazen 2008; Silvennoinen et al. 2012; Maugeais et al. 2013). The bile acid sequestrant cholestyramine also promotes macrophage-to-feces RCT (Maugeais et al. 2013).

3.2.4 Augmenting or Mimicking apoA-I

The use of apoA-I and its different forms for the prevention and treatment of atherosclerosis is a long-term goal of many laboratories. The full-length apoA-I, reconstituted apoA-I-containing HDL, or apoA-I mimetics D4-F, ATI-5261, and 5A have been demonstrated to be effective for enhancing macrophage-to-feces RCT in different mouse models (Maugeais et al. 2013; Lee-Rueckert and Kovanen 2011; Navab et al. 2004; Bielicki et al. 2010; Amar et al. 2010). The upregulation of liver apoA-I production, as occurred with the thienotriazolodiazepine Ro 11-1464, also enhances macrophage RCT (Zanotti et al. 2011a, b).

4 Serum Cholesterol Efflux Capacity (CEC)

Several epidemiological studies define HDL as the most powerful plasmatic factor with atheroprotective activity in humans (Di Angelantonio et al. 2009); a 1 mg/dl increase of plasma HDL-C is associated with a 3–4 % reduction in cardiovascular mortality (Assmann et al. 2002). Some post hoc analyses from randomized controlled trials also suggest that raising HDL-C beneficially affects the risk of CVD (Toth et al. 2013). However, the clinical efficacy of raising plasma HDL-C levels to achieve cardiovascular risk reduction has been difficult to prove. Recently published outcome trials involving the addition of niacin or dalcetrapib to standard low-density lipoprotein cholesterol reduction therapy failed to demonstrate clinical benefit despite increases in HDL-C ((AIM-HIGH) trial 2011; Schwartz et al. 2012). Furthermore, genetic variants associated with increased HDL-C, thus conferring lifelong exposure to higher circulating levels, are not consistently associated with improved vascular outcomes (Voight et al. 2012). These findings have reinforced the idea that changes in HDL-C levels are an inadequate surrogate for therapeutic use. Therefore, an emerging concept is that of the quality of HDL, which are heterogeneous in terms of size, charge, and lipid content (Calabresi et al. 2010) and display functional differences, such as cell cholesterol efflux promotion. Animal studies have suggested that HDL-mediated cholesterol flux through the different RCT steps is a better predictor of the atherosclerotic impact of various genetic and pharmacological perturbations than static, mass-based quantification of circulating HDL-C (Rader et al. 2009). The efficacy of such HDL function in a single individual may be estimated by measuring the cholesterol efflux capacity (CEC) using widely standardized techniques that allow distinguishing between the various mechanisms involved (Adorni et al. 2007). Several lines of evidence suggest that CEC is sensitive to HDL composition rather than HDL-C plasma levels; for instance, it has been shown that the serum capacity to promote cholesterol efflux via ABCA1 strictly depends on the nascent (preβ) HDL plasma levels (de la Llera-Moya et al. 2010; Favari et al. 2004). Subjects with the apoA-I$_{Milano}$ mutation or LCAT deficiency have high levels of circulating particles and efficient serum ability to induce macrophage cholesterol depletion despite very low HDL levels (Favari et al. 2007; Calabresi et al. 2009). Thus, there is increasing evidence that the measure of HDL functionality in different populations may be a better predictor of coronary artery disease than the measure of absolute HDL-C levels (Khera et al. 2011). As mentioned before, this concept is consistent with studies revealing that subjects affected by CVD not only have HDL deficiency but also major rearrangements of their composition (Campos et al. 1995; Sweetnam et al. 1994) and consequently function. The role of serum CEC as an index of cardiovascular protection has been suggested by a recent study demonstrating that in two distinct cohorts of subjects, the CEC variable has a stronger predictive power of the carotid intima-media thickness, an index of subclinical atherosclerosis, than plasma HDL-C levels (Khera et al. 2011). In addition, in a population of healthy subjects, ABCA1-mediated serum CEC was inversely correlated with pulse wave velocity, an index of arterial stiffness, independent of HDL-C serum levels (Favari et al. 2013). A work measuring the flow-mediated dilation, as a parameter of endothelial function, showed a positive correlation with the ABCA1-mediated efflux pathway confirming the concept that the functional measures of HDL might be a better marker for cardiovascular risk rather than HDL cholesterol levels (Vazquez et al. 2012). Recently, Li and colleagues provided data indicating that individuals in the top tertile of CEC had a moderately increased risk of a composite cardiovascular end point of incident myocardial infarction, stroke, or death during 3 years of follow-up (Li et al. 2013). However, in the same study, CEC was inversely associated with coronary artery disease. The reasons for this apparently contradictory findings require further evaluation, but may be related to the characteristics of the population involved in the study (Khera and Rader 2013). Overall, the available data suggest that evaluation of CEC may better correlate with coronary artery diseases than HDL-C; however, further studies are required to demonstrate its role as predictor of cardiovascular event risk.

Conclusions

RCT provides a physiological strategy of protection from atherosclerosis. HDL plays a leading role by promoting the removal of excess cholesterol in the arterial wall through the induction of cellular cholesterol efflux from cells. Cholesterol efflux represents the first and perhaps most important step of RCT. The ability of HDL to promote cholesterol efflux depends on their quality and the pathologic processes that induce their modifications. Many of the available data on the physiology and pharmacology of RCT depend on studies in animal models that have demonstrated that the enhancement of RCT is inversely correlated with the development of atherosclerosis. In humans, the efficiency of the RCT can be evaluated with the surrogate parameter of serum cholesterol efflux capacity (CEC) that indicates the ability of HDL to promote efflux of cholesterol in the individual patient. Recent clinical data suggest that the evaluation of CEC is a strong predictor of atherosclerosis extent in humans and may represent in the future a useful biomarker of cardiovascular risk.

Open Access This chapter is distributed under the terms of the Creative Commons Attribution Noncommercial License, which permits any noncommercial use, distribution, and reproduction in any medium, provided the original author(s) and source are credited.

References

Adorni MP, Zimetti F et al (2007) The roles of different pathways in the release of cholesterol from macrophages. J Lipid Res 48(11):2453–2462

Adorni MP, Favari E et al (2011) Free cholesterol alters macrophage morphology and mobility by an ABCA1 dependent mechanism. Atherosclerosis 215(1):70–76

AIM-HIGH (2011) The role of niacin in raising high-density lipoprotein cholesterol to reduce cardiovascular events in patients with atherosclerotic cardiovascular disease and optimally treated low-density lipoprotein cholesterol: baseline characteristics of study participants. The

Atherothrombosis Intervention in Metabolic syndrome with low HDL/high triglycerides: impact on Global Health outcomes (AIM-HIGH) trial. Am Heart J 161(3):538–543

Amar MJ, D'Souza W et al (2010) 5A apolipoprotein mimetic peptide promotes cholesterol efflux and reduces atherosclerosis in mice. J Pharmacol Exp Ther 334(2):634–641

Annema W, Tietge UJ (2011) Role of hepatic lipase and endothelial lipase in high-density lipoprotein-mediated reverse cholesterol transport. Curr Atheroscler Rep 13(3):257–265

Annema W, Tietge UJ (2012) Regulation of reverse cholesterol transport - a comprehensive appraisal of available animal studies. Nutr Metab (Lond) 9(1):25

Annema W, von Eckardstein A (2013) High-density lipoproteins. Multifunctional but vulnerable protections from atherosclerosis. Circ J 77(10):2432–2448

Annema W, Nijstad N et al (2010) Myeloperoxidase and serum amyloid A contribute to impaired in vivo reverse cholesterol transport during the acute phase response but not group IIA secretory phospholipase A(2). J Lipid Res 51(4):743–754

Annema W, Dikkers A et al (2012) ApoE promotes hepatic selective uptake but not RCT due to increased ABCA1-mediated cholesterol efflux to plasma. J Lipid Res 53(5):929–940

Assmann G, Cullen P et al (2002) Simple scoring scheme for calculating the risk of acute coronary events based on the 10-year follow-up of the prospective cardiovascular Munster (PROCAM) study. Circulation 105(3):310–315

Barter PJ, Rye KA (2012) Cholesteryl ester transfer protein inhibition as a strategy to reduce cardiovascular risk. J Lipid Res 53(9):1755–1766

Bell TA 3rd, Graham MJ et al (2013) Antisense oligonucleotide inhibition of cholesteryl ester transfer protein enhances RCT in hyperlipidemic, CETP transgenic, LDLr-/- mice. J Lipid Res 54(10):2647–2657

Bergt C, Pennathur S et al (2004) The myeloperoxidase product hypochlorous acid oxidizes HDL in the human artery wall and impairs ABCA1-dependent cholesterol transport. Proc Natl Acad Sci U S A 101(35):13032–13037

Bi X, Zhu X et al (2013) Liver ABCA1 deletion in LDLrKO mice does not impair macrophage reverse cholesterol transport or exacerbate atherogenesis. Arterioscler Thromb Vasc Biol 33(10):2288–2296

Bielicki JK, Zhang H et al (2010) A new HDL mimetic peptide that stimulates cellular cholesterol efflux with high efficiency greatly reduces atherosclerosis in mice. J Lipid Res 51(6):1496–1503

Boadu E, Francis GA (2006) The role of vesicular transport in ABCA1-dependent lipid efflux and its connection with NPC pathways. J Mol Med (Berl) 84(4):266–275

Briand F, Naik SU et al (2009) Both the peroxisome proliferator-activated receptor delta agonist, GW0742, and ezetimibe promote reverse cholesterol transport in mice by reducing intestinal reabsorption of HDL-derived cholesterol. Clin Transl Sci 2(2):127–133

Briand F, Treguier M et al (2010) Liver X receptor activation promotes macrophage-to-feces reverse cholesterol transport in a dyslipidemic hamster model. J Lipid Res 51(4):763–770

Briand F, Thieblemont Q et al (2011) CETP inhibitor torcetrapib promotes reverse cholesterol transport in obese insulin-resistant CETP-ApoB100 transgenic mice. Clin Transl Sci 4(6):414–420

Brown RJ, Lagor WR et al (2010) Impact of combined deficiency of hepatic lipase and endothelial lipase on the metabolism of both high-density lipoproteins and apolipoprotein B-containing lipoproteins. Circ Res 107(3):357–364

Calabresi L, Favari E et al (2009) Functional LCAT is not required for macrophage cholesterol efflux to human serum. Atherosclerosis 204(1):141–146

Calabresi L, Gomaraschi M et al (2010) High-density lipoprotein quantity or quality for cardiovascular prevention? Curr Pharm Des 16(13):1494–1503

Calpe-Berdiel L, Rotllan N et al (2008) Liver X receptor-mediated activation of reverse cholesterol transport from macrophages to feces in vivo requires ABCG5/G8. J Lipid Res 49(9):1904–1911

Campos H, Roederer GO et al (1995) Predominance of large LDL and reduced HDL2 cholesterol in normolipidemic men with coronary artery disease. Arterioscler Thromb Vasc Biol 15(8): 1043–1048

Castro-Perez J, Briand F et al (2011) Anacetrapib promotes reverse cholesterol transport and bulk cholesterol excretion in Syrian golden hamsters. J Lipid Res 52(11):1965–1973

Cavigiolio G, Geier EG et al (2010) Exchange of apolipoprotein A-I between lipid-associated and lipid-free states: a potential target for oxidative generation of dysfunctional high density lipoproteins. J Biol Chem 285(24):18847–18857

Charles-Schoeman C, Lee YY et al (2012) Cholesterol efflux by high density lipoproteins is impaired in patients with active rheumatoid arthritis. Ann Rheum Dis 71(7):1157–1162

Chroni A, Liu T et al (2003) The central helices of ApoA-I can promote ATP-binding cassette transporter A1 (ABCA1)-mediated lipid efflux. Amino acid residues 220-231 of the wild-type ApoA-I are required for lipid efflux in vitro and high density lipoprotein formation in vivo. J Biol Chem 278(9):6719–6730

Chroni A, Liu T et al (2004) Cross-linking and lipid efflux properties of apoA-I mutants suggest direct association between apoA-I helices and ABCA1. Biochemistry 43(7):2126–2139

Chroni A, Koukos G et al (2007) The carboxy-terminal region of apoA-I is required for the ABCA1-dependent formation of alpha-HDL but not prebeta-HDL particles in vivo. Biochemistry 46(19):5697–5708

Daniil G, Zannis VI et al (2013) Effect of apoA-I mutations in the capacity of reconstituted HDL to promote ABCG1-mediated cholesterol efflux. PLoS ONE 8(6):e67993

Dawson PA (2011) Role of the intestinal bile acid transporters in bile acid and drug disposition. Handb Exp Pharmacol 201:169–203

de Boer JF, Annema W et al (2012) Type I diabetes mellitus decreases in vivo macrophage-to-feces reverse cholesterol transport despite increased biliary sterol secretion in mice. J Lipid Res 53(3):348–357

de la Llera-Moya M, Drazul-Schrader D et al (2010) The ability to promote efflux via ABCA1 determines the capacity of serum specimens with similar high-density lipoprotein cholesterol to remove cholesterol from macrophages. Arterioscler Thromb Vasc Biol 30(4):796–801

Di Angelantonio E, Sarwar N et al (2009) Major lipids, apolipoproteins, and risk of vascular disease. JAMA 302(18):1993–2000

Dikkers A, Tietge UJ (2010) Biliary cholesterol secretion: more than a simple ABC. World J Gastroenterol 16(47):5936–5945

Dikkers A, Freak de Boer J et al (2013) Scavenger receptor BI and ABCG5/G8 differentially impact biliary sterol secretion and reverse cholesterol transport in mice. Hepatology 58(1): 293–303

Dunbar RL, Cuchel M, Millar JS, Baer A, Poria R, Freifelder RH, Pryma DA, Billheimer JJ, Rader DJ (2013) Abstract 439: niacin does not accelerate reverse cholesterol transport in man. Arterioscler Thromb Vasc Biol 33:A439

Eberini I, Calabresi L et al (2002) Macrophage metalloproteinases degrade high-density-lipoprotein-associated apolipoprotein A-I at both the N- and C-termini. Biochem J 362(Pt 3):627–634

El Bouhassani M, Gilibert S et al (2011) Cholesteryl ester transfer protein expression partially attenuates the adverse effects of SR-BI receptor deficiency on cholesterol metabolism and atherosclerosis. J Biol Chem 286(19):17227–17238

Escola-Gil JC, Chen X et al (2013) Hepatic lipase- and endothelial lipase-deficiency in mice promotes macrophage-to-feces RCT and HDL antioxidant properties. Biochim Biophys Acta 1831(4):691–697

Fabre AC, Malaval C et al (2010) P2Y13 receptor is critical for reverse cholesterol transport. Hepatology 52(4):1477–1483

Favari E, Bernini F et al (2002) The C-terminal domain of apolipoprotein A-I is involved in ABCA1-driven phospholipid and cholesterol efflux. Biochem Biophys Res Commun 299(5): 801–805

Favari E, Lee M et al (2004) Depletion of pre-beta-high density lipoprotein by human chymase impairs ATP-binding cassette transporter A1- but not scavenger receptor class B type I-mediated lipid efflux to high density lipoprotein. J Biol Chem 279(11):9930–9936

Favari E, Gomaraschi M et al (2007) A unique protease-sensitive high density lipoprotein particle containing the apolipoprotein A-I(Milano) dimer effectively promotes ATP-binding cassette A1-mediated cell cholesterol efflux. J Biol Chem 282(8):5125–5132

Favari E, Calabresi L et al (2009) Small discoidal pre-beta1 HDL particles are efficient acceptors of cell cholesterol via ABCA1 and ABCG1. Biochemistry 48(46):11067–11074

Favari E, Ronda N et al (2013) ABCA1-dependent serum cholesterol efflux capacity inversely correlates with pulse wave velocity in healthy subjects. J Lipid Res 54(1):238–243

Freeman SR, Jin X et al (2013) ABCG1-mediated generation of extracellular cholesterol microdomains. J Lipid Res 55(1):115–127

Fu Y, Mukhamedova N et al (2013) ABCA12 regulates ABCA1-dependent cholesterol efflux from macrophages and the development of atherosclerosis. Cell Metab 18(2):225–238

Gargalovic P, Dory L (2003) Caveolins and macrophage lipid metabolism. J Lipid Res 44(1):11–21

Gelissen IC, Harris M et al (2006) ABCA1 and ABCG1 synergize to mediate cholesterol export to apoA-I. Arterioscler Thromb Vasc Biol 26(3):534–540

Holzer M, Wolf P et al (2013) Anti-psoriatic therapy recovers high-density lipoprotein composition and function. J Invest Dermatol 134(3):635–642

Huang ZH, Lin CY et al (2001) Sterol efflux mediated by endogenous macrophage ApoE expression is independent of ABCA1. Arterioscler Thromb Vasc Biol 21(12):2019–2025

Huang ZH, Gu D et al (2003) Expression of scavenger receptor BI facilitates sterol movement between the plasma membrane and the endoplasmic reticulum in macrophages. Biochemistry 42(13):3949–3955

Hung KT, Berisha SZ et al (2012) Red blood cells play a role in reverse cholesterol transport. Arterioscler Thromb Vasc Biol 32(6):1460–1465

Ishimoto Y, Yamada K et al (2003) Group V and X secretory phospholipase A(2)s-induced modification of high-density lipoprotein linked to the reduction of its antiatherogenic functions. Biochim Biophys Acta 1642(3):129–138

Khera AV, Rader DJ (2013) Cholesterol efflux capacity: full steam ahead or a bump in the road? Arterioscler Thromb Vasc Biol 33(7):1449–1451

Khera AV, Cuchel M et al (2011) Cholesterol efflux capacity, high-density lipoprotein function, and atherosclerosis. N Engl J Med 364(2):127–135

Kockx M, Jessup W et al (2008) Regulation of endogenous apolipoprotein E secretion by macrophages. Arterioscler Thromb Vasc Biol 28(6):1060–1067

Kunitake ST, Chen GC et al (1990) Pre-beta high density lipoprotein. Unique disposition of apolipoprotein A-I increases susceptibility to proteolysis. Arteriosclerosis 10(1):25–30

Landry YD, Denis M et al (2006) ATP-binding cassette transporter A1 expression disrupts raft membrane microdomains through its ATPase-related functions. J Biol Chem 281(47):36091–36101

Larrede S, Quinn CM et al (2009) Stimulation of cholesterol efflux by LXR agonists in cholesterol-loaded human macrophages is ABCA1-dependent but ABCG1-independent. Arterioscler Thromb Vasc Biol 29(11):1930–1936

Lee M, Kovanen PT et al (2003) Apolipoprotein composition and particle size affect HDL degradation by chymase: effect on cellular cholesterol efflux. J Lipid Res 44(3):539–546

Lee-Rueckert M, Kovanen PT (2011) Extracellular modifications of HDL in vivo and the emerging concept of proteolytic inactivation of prebeta-HDL. Curr Opin Lipidol 22(5):394–402

Lee-Rueckert M, Silvennoinen R et al (2011) Mast cell activation in vivo impairs the macrophage reverse cholesterol transport pathway in the mouse. Arterioscler Thromb Vasc Biol 31(3):520–527

Li XM, Tang WH et al (2013) Paradoxical association of enhanced cholesterol efflux with increased incident cardiovascular risks. Arterioscler Thromb Vasc Biol 33(7):1696–1705

Lichtenstein L, Serhan N et al (2013) Lack of P2Y13 in mice fed a high cholesterol diet results in decreased hepatic cholesterol content, biliary lipid secretion and reverse cholesterol transport. Nutr Metab (Lond) 10(1):67

Lim HY, Thiam CH et al (2013) Lymphatic vessels are essential for the removal of cholesterol from peripheral tissues by SR-BI-mediated transport of HDL. Cell Metab 17(5):671–684

Lindstedt L, Saarinen J et al (1999) Matrix metalloproteinases-3, -7, and -12, but not -9, reduce high density lipoprotein-induced cholesterol efflux from human macrophage foam cells by truncation of the carboxyl terminus of apolipoprotein A-I. Parallel losses of pre-beta particles and the high affinity component of efflux. J Biol Chem 274(32):22627–22634

Low H, Hoang A et al (2012) Advanced glycation end-products (AGEs) and functionality of reverse cholesterol transport in patients with type 2 diabetes and in mouse models. Diabetologia 55(9):2513–2521

Malik P, Smith JP (2009) A novel in vivo assay for reverse cholesterol transport. Arterioscler Thromb Vasc Biol 29:E46

Malik P, Berisha SZ et al (2011) Zymosan-mediated inflammation impairs in vivo reverse cholesterol transport. J Lipid Res 52(5):951–957

Martel C, Li W et al (2013) Lymphatic vasculature mediates macrophage reverse cholesterol transport in mice. J Clin Invest 123(4):1571–1579

Maugeais C, Annema W et al (2013) rHDL administration increases reverse cholesterol transport in mice, but is not additive on top of ezetimibe or cholestyramine treatment. Atherosclerosis 229(1):94–101

Maxfield FR, Tabas I (2005) Role of cholesterol and lipid organization in disease. Nature 438 (7068):612–621

McGillicuddy FC, de la Llera Moya M et al (2009) Inflammation impairs reverse cholesterol transport in vivo. Circulation 119(8):1135–1145

Naik SU, Wang X et al (2006) Pharmacological activation of liver X receptors promotes reverse cholesterol transport in vivo. Circulation 113(1):90–97

Nakaya K, Tohyama J et al (2011) Peroxisome proliferator-activated receptor-alpha activation promotes macrophage reverse cholesterol transport through a liver X receptor-dependent pathway. Arterioscler Thromb Vasc Biol 31(6):1276–1282

Navab M, Anantharamaiah GM et al (2004) Oral D-4 F causes formation of pre-beta high-density lipoprotein and improves high-density lipoprotein-mediated cholesterol efflux and reverse cholesterol transport from macrophages in apolipoprotein E-null mice. Circulation 109(25): 3215–3220

Niesor EJ, Magg C et al (2010) Modulating cholesteryl ester transfer protein activity maintains efficient pre-beta-HDL formation and increases reverse cholesterol transport. J Lipid Res 51(12):3443–3454

Nijstad N, Gautier T et al (2011) Biliary sterol secretion is required for functional in vivo reverse cholesterol transport in mice. Gastroenterology 140(3):1043–1051

Olivier M, Tanck MW et al (2012) Human ATP-binding cassette G1 controls macrophage lipoprotein lipase bioavailability and promotes foam cell formation. Arterioscler Thromb Vasc Biol 32(9):2223–2231

Onat A, Direskeneli H (2012) Excess cardiovascular risk in inflammatory rheumatic diseases: pathophysiology and targeted therapy. Curr Pharm Des 18(11):1465–1477

Oram JF, Lawn RM (2001) ABCA1. The gatekeeper for eliminating excess tissue cholesterol. J Lipid Res 42(8):1173–1179

Oram JF, Vaughan AM (2006) ATP-binding cassette cholesterol transporters and cardiovascular disease. Circ Res 99(10):1031–1043

Out R, Jessup W et al (2008) Coexistence of foam cells and hypocholesterolemia in mice lacking the ABC transporters A1 and G1. Circ Res 102(1):113–120

Pennathur S, Bergt C et al (2004) Human atherosclerotic intima and blood of patients with established coronary artery disease contain high density lipoprotein damaged by reactive nitrogen species. J Biol Chem 279(41):42977–42983

Pirillo A, Uboldi P et al (2006) 15-Lipoxygenase-mediated modification of high-density lipoproteins impairs SR-BI- and ABCA1-dependent cholesterol efflux from macrophages. Biochim Biophys Acta 1761(3):292–300

Rader DJ, Alexander ET et al (2009) The role of reverse cholesterol transport in animals and humans and relationship to atherosclerosis. J Lipid Res 50(Suppl):S189–S194

Ramirez CM, Davalos A et al (2011) MicroRNA-758 regulates cholesterol efflux through post-transcriptional repression of ATP-binding cassette transporter A1. Arterioscler Thromb Vasc Biol 31(11):2707–2714

Rayner KJ, Suarez Y et al (2010) MiR-33 contributes to the regulation of cholesterol homeostasis. Science 328(5985):1570–1573

Rayner KJ, Sheedy FJ et al (2011) Antagonism of miR-33 in mice promotes reverse cholesterol transport and regression of atherosclerosis. J Clin Invest 121(7):2921–2931

Rohrer L, Ohnsorg PM et al (2009) High-density lipoprotein transport through aortic endothelial cells involves scavenger receptor BI and ATP-binding cassette transporter G1. Circ Res 104 (10):1142–1150

Ronda N, Favari E et al (2013) Impaired serum cholesterol efflux capacity in rheumatoid arthritis and systemic lupus erythematosus. Ann Rheum Dis 73(3):609–615

Rosenson RS, Hurt-Camejo E (2012) Phospholipase A2 enzymes and the risk of atherosclerosis. Eur Heart J 33(23):2899–2909

Rosenson RS, Brewer HB Jr et al (2012) Cholesterol efflux and atheroprotection: advancing the concept of reverse cholesterol transport. Circulation 125(15):1905–1919

Rotllan N, Calpe-Berdiel L et al (2008) CETP activity variation in mice does not affect two major HDL antiatherogenic properties: macrophage-specific reverse cholesterol transport and LDL antioxidant protection. Atherosclerosis 196(2):505–513

Rotllan N, Llaverias G et al (2011) Differential effects of gemfibrozil and fenofibrate on reverse cholesterol transport from macrophages to feces in vivo. Biochim Biophys Acta 1811 (2):104–110

Samyn H, Moerland M et al (2009) Elevation of systemic PLTP, but not macrophage-PLTP, impairs macrophage reverse cholesterol transport in transgenic mice. Atherosclerosis 204(2): 429–434

Schmitz G, Langmann T et al (2001) Role of ABCG1 and other ABCG family members in lipid metabolism. J Lipid Res 42(10):1513–1520

Schwartz GG, Olsson AG et al (2012) Effects of dalcetrapib in patients with a recent acute coronary syndrome. N Engl J Med 367(22):2089–2099

Sehayek E, Hazen SL (2008) Cholesterol absorption from the intestine is a major determinant of reverse cholesterol transport from peripheral tissue macrophages. Arterioscler Thromb Vasc Biol 28(7):1296–1297

Shao B, Bergt C et al (2005) Tyrosine 192 in apolipoprotein A-I is the major site of nitration and chlorination by myeloperoxidase, but only chlorination markedly impairs ABCA1-dependent cholesterol transport. J Biol Chem 280(7):5983–5993

Shao B, Oda MN et al (2006) Myeloperoxidase: an inflammatory enzyme for generating dysfunctional high density lipoprotein. Curr Opin Cardiol 21(4):322–328

Silvennoinen R, Escola-Gil JC et al (2012) Acute psychological stress accelerates reverse cholesterol transport via corticosterone-dependent inhibition of intestinal cholesterol absorption. Circ Res 111(11):1459–1469

Sun D, Zhang J et al (2012) MiR-26 controls LXR-dependent cholesterol efflux by targeting ABCA1 and ARL7. FEBS Lett 586(10):1472–1479

Sweetnam PM, Bolton CH et al (1994) Associations of the HDL2 and HDL3 cholesterol subfractions with the development of ischemic heart disease in British men. The Caerphilly and Speedwell collaborative heart disease studies. Circulation 90(2):769–774

Tanigawa H, Billheimer JT et al (2007) Expression of cholesteryl ester transfer protein in mice promotes macrophage reverse cholesterol transport. Circulation 116(11):1267–1273

Tanigawa H, Billheimer JT et al (2009) Lecithin: cholesterol acyltransferase expression has minimal effects on macrophage reverse cholesterol transport in vivo. Circulation 120(2):160–169

Tarling EJ, Edwards PA (2012) Dancing with the sterols: critical roles for ABCG1, ABCA1, miRNAs, and nuclear and cell surface receptors in controlling cellular sterol homeostasis. Biochim Biophys Acta 1821(3):386–395

Tchoua U, D'Souza W et al (2008) The effect of cholesteryl ester transfer protein overexpression and inhibition on reverse cholesterol transport. Cardiovasc Res 77(4):732–739

Temel RE, Sawyer JK et al (2010) Biliary sterol secretion is not required for macrophage reverse cholesterol transport. Cell Metab 12(1):96–102

Tietge UJ, Groen AK (2013) Role the TICE?: advancing the concept of transintestinal cholesterol excretion. Arterioscler Thromb Vasc Biol 33(7):1452–1453

Timmins JM, Lee JY et al (2005) Targeted inactivation of hepatic Abca1 causes profound hypoalphalipoproteinemia and kidney hypercatabolism of apoA-I. J Clin Invest 115(5):1333–1342

Toh SA, Millar JS et al (2011) PPARgamma activation redirects macrophage cholesterol from fecal excretion to adipose tissue uptake in mice via SR-BI. Biochem Pharmacol 81(7):934–941

Toth PP, Barter PJ et al (2013) High-density lipoproteins: a consensus statement from the national lipid association. J Clin Lipidol 7(5):484–525

Usami Y, Matsuda K et al (2011) Detection of chymase-digested C-terminally truncated apolipoprotein A-I in normal human serum. J Immunol Methods 369(1–2):51–58

Usami Y, Kobayashi Y, Kameda T, Miyazaki A, Matsuda K, Sugano M, Kawasaki K, Kurihara Y, Kasama T, Tozuka M (2013) Identification of sites in apolipoprotein A-I susceptible to chymase and carboxypeptidase A digestion. Biosci Rep 33:49–56

Vantourout P, Radojkovic C et al (2010) Ecto-F(1)-ATPase: a moonlighting protein complex and an unexpected apoA-I receptor. World J Gastroenterol 16(47):5925–5935

Vaughan AM, Oram JF (2005) ABCG1 redistributes cell cholesterol to domains removable by high density lipoprotein but not by lipid-depleted apolipoproteins. J Biol Chem 280(34):30150–30157

Vazquez E, Sethi AA et al (2012) High-density lipoprotein cholesterol efflux, nitration of apolipoprotein A-I, and endothelial function in obese women. Am J Cardiol 109(4):527–532

Vickers KC, Remaley AT (2014) HDL and cholesterol: life after the divorce? J Lipid Res 55(1):4–12

Voight BF, Peloso GM et al (2012) Plasma HDL cholesterol and risk of myocardial infarction: a mendelian randomisation study. Lancet 380(9841):572–580

Wang MD, Franklin V et al (2007a) In vivo reverse cholesterol transport from macrophages lacking ABCA1 expression is impaired. Arterioscler Thromb Vasc Biol 27(8):1837–1842

Wang X, Collins HL et al (2007b) Macrophage ABCA1 and ABCG1, but not SR-BI, promote macrophage reverse cholesterol transport in vivo. J Clin Invest 117(8):2216–2224

Watanabe J, Charles-Schoeman C et al (2012) Proteomic profiling following immunoaffinity capture of high-density lipoprotein: association of acute-phase proteins and complement factors with proinflammatory high-density lipoprotein in rheumatoid arthritis. Arthritis Rheum 64(6):1828–1837

Weibel GL, Hayes S et al (2011) Novel in vivo method for measuring cholesterol mass flux in peripheral macrophages. Arterioscler Thromb Vasc Biol 31(12):2865–2871

Wynn TA, Chawla A et al (2013) Macrophage biology in development, homeostasis and disease. Nature 496(7446):445–455

Xie P, Jia L et al (2013) Ezetimibe inhibits hepatic Niemann-Pick C1-Like 1 to facilitate macrophage reverse cholesterol transport in mice. Arterioscler Thromb Vasc Biol 33(5):920–925

Yamamoto S, Tanigawa H et al (2011) Pharmacologic suppression of hepatic ATP-binding cassette transporter 1 activity in mice reduces high-density lipoprotein cholesterol levels but promotes reverse cholesterol transport. Circulation 124(12):1382–1390

Yasuda T, Grillot D et al (2010) Tissue-specific liver X receptor activation promotes macrophage reverse cholesterol transport in vivo. Arterioscler Thromb Vasc Biol 30(4):781–786

Yvan-Charvet L, Kling J et al (2010) Cholesterol efflux potential and antiinflammatory properties of high-density lipoprotein after treatment with niacin or anacetrapib. Arterioscler Thromb Vasc Biol 30(7):1430–1438

Zanotti I, Poti F et al (2008) The LXR agonist T0901317 promotes the reverse cholesterol transport from macrophages by increasing plasma efflux potential. J Lipid Res 49(5):954–960

Zanotti I, Maugeais C et al (2011a) The thienotriazolodiazepine Ro 11-1464 increases plasma apoA-I and promotes reverse cholesterol transport in human apoA-I transgenic mice. Br J Pharmacol 164(6):1642–1651

Zanotti I, Pedrelli M et al (2011b) Macrophage, but not systemic, apolipoprotein E is necessary for macrophage reverse cholesterol transport in vivo. Arterioscler Thromb Vasc Biol 31(1):74–80

Zanotti I, Favari E et al (2012) Cellular cholesterol efflux pathways: impact on intracellular lipid trafficking and methodological considerations. Curr Pharm Biotechnol 13(2):292–302

Zhang Y, Zanotti I et al (2003) Overexpression of apolipoprotein A-I promotes reverse transport of cholesterol from macrophages to feces in vivo. Circulation 108(6):661–663

Zhang Y, Da Silva JR et al (2005) Hepatic expression of scavenger receptor class B type I (SR-BI) is a positive regulator of macrophage reverse cholesterol transport in vivo. J Clin Invest 115(10):2870–2874

Zhang Y, Yin L et al (2010) Identification of novel pathways that control farnesoid X receptor-mediated hypocholesterolemia. J Biol Chem 285(5):3035–3043

Zhao Y, Pennings M et al (2011) Hypocholesterolemia, foam cell accumulation, but no atherosclerosis in mice lacking ABC-transporter A1 and scavenger receptor BI. Atherosclerosis 218(2):314–322

Zheng L, Nukuna B et al (2004) Apolipoprotein A-I is a selective target for myeloperoxidase-catalyzed oxidation and functional impairment in subjects with cardiovascular disease. J Clin Invest 114(4):529–541

Zheng L, Settle M et al (2005) Localization of nitration and chlorination sites on apolipoprotein A-I catalyzed by myeloperoxidase in human atheroma and associated oxidative impairment in ABCA1-dependent cholesterol efflux from macrophages. J Biol Chem 280(1):38–47

Functionality of HDL: Antioxidation and Detoxifying Effects

Helen Karlsson, Anatol Kontush, and Richard W. James

Contents

1. High-Density Lipoproteins and Oxidative Stress 209
   1.1 High-Density Lipoproteins: Antioxidative Function 210
   1.2 Mechanisms of Protection 210
   1.3 Heterogeneity of Antioxidant Activity of HDL Particles 213
2. High-Density Lipoproteins, Paraoxonase-1 214
   2.1 PON1 as an Antioxidant 214
   2.2 PON1 and Bacterial Pathogens 216
3. High-Density Lipoproteins, Environmental Pathogens and Toxins 217
   3.1 Bacterial Pathogens 217
   3.2 Parasites 218
   3.3 Hepatitis, Dengue and Other Viruses 219
   3.4 Metal Oxides, Carbon Nanotubes and PLGA Nanoparticles 219
   3.5 PON1 and Organophosphates 220
   3.6 Detoxification of Plasma and External Fluids 220
References 221

H. Karlsson
Occupational and Environmental Medicine, Heart Medical Centre, County Council of Ostergotland, Linkoping University, SE-58185 Linkoping, Sweden
e-mail: helen.m.karlsson@liu.se

A. Kontush
INSERM UMRS 939, University of Pierre and Marie Curie (UPMC) Paris 6, Hôpital de la Pitié, Paris, France
e-mail: anatol.kontush@upmc.fr

R.W. James (✉)
Clinical Diabetes Unit, Department of Internal Medicine, University of Geneva, Geneva, Switzerland
e-mail: Richard.James@hcuge.ch

© The Author(s) 2015
A. von Eckardstein, D. Kardassis (eds.), *High Density Lipoproteins*, Handbook of Experimental Pharmacology 224, DOI 10.1007/978-3-319-09665-0_5

Abstract

High-density lipoproteins (HDL) are complexes of multiple talents, some of which have only recently been recognised but all of which are under active investigation. Clinical interest initially arose from their amply demonstrated role in atherosclerotic disease with their consequent designation as a major cardiovascular disease (CVD) risk factor. However, interest is no longer confined to vascular tissues, with the reports of impacts of the lipoprotein on pancreatic, renal and nervous tissues, amongst other possible targets. The ever-widening scope of HDL talents also encompasses environmental hazards, including infectious agents and environmental toxins. In almost all cases, HDL would appear to have a beneficial impact on health. It raises the intriguing question of whether these various talents emanate from a basic ancestral function to protect the cell.

The following chapter will illustrate and review our current understanding of some of the functions attributed to HDL. The first section will look at the antioxidative functions of HDL and possible mechanisms that are involved. The second section will focus specifically on paraoxonase-1 (PON1), which appears to bridge the divide between the two HDL functions discussed herein. This will lead into the final section dealing with HDL as a detoxifying agent protecting against exposure to environmental pathogens and other toxins.

Keywords

HDL • Lipoproteins • Oxidative stress • Paraoxonase • Organophosphates • Bacterial pathogen • Antiviral activity • Nanoparticles • Bisphenol

Abbreviations

AAPH	Azo-initiator (2,2′-azobis-(2-amidinopropane) hydrochloride)
Apo	Apolipoprotein
BPA	Bisphenol A
CEOOH	Cholesteryl ester hydroperoxide
CETP	Cholesterol ester transfer protein
Cu	Copper
CVD	Cardiovascular disease
Cys	Cysteine
GSP	Glutathione peroxidase
Hb	Haemoglobin
Hbr	Haptoglobin-related protein
HBV	Hepatitis B virus
HDL	High-density lipoproteins
HETE	Hydroxy-eicosatetraenoic acid

HODE	Hydroxyoctadecadienoic acid
HIV	Human immunodeficiency virus
HSL	Homoserine lactones
LCAT	Lecithin-cholesterol acyltransferase
LDL	Low-density lipoproteins
LOOH	Lipid hydroperoxide
LPS	Lipopolysaccharides
LSI	Liver somatic index
LTA	Lipoteichoic acid
Met	Methionine
MPO	Myeloperoxidase
NADPH	Nicotinamide adenine dinucleotide phosphate
NLF	Nasal lavage fluid
OP	Organophosphate
PAF-AH	Platelet-activating factor acetylhydrolase
PCB	Polychlorinated biphenyl
PLGA	Poly(lactic-co-glycolic acid)
PLOOH	Phospholipid hydroperoxide
PLTP	Phospholipid transfer protein
PON1	Paraoxonase-1
POP	Persistent organic pollutant
POPC	1-palmitoyl-2-oleoyl-sn-glycero-3-phosphocholine
rHDL	Reconstituted HDL
SAA	Serum amyloid A
SRA	Serum resistance-associated
SR-BI	Scavenger receptor class B-I
TLF	Trypanolytic factor
TLR	Toll-like receptor
TNF	Tumour necrosis factor
Tyr	Tyrosine
VLDL	Low-density lipoproteins

1 High-Density Lipoproteins and Oxidative Stress

The role of high-density lipoproteins (HDL) in the vascular system has been the primary focus of clinical interest. It arose from early studies of the influence of the lipoprotein on cholesterol metabolism and atherosclerosis. With increasing understanding of the complexity of the atherosclerotic process and involvement of other pathological mechanisms, attention has logically progressed to the impact of HDL on such mechanisms. One process is oxidative stress and there is now persuasive evidence that the lipoprotein can attenuate its consequences by a number of mechanisms. These are discussed in this section.

1.1 High-Density Lipoproteins: Antioxidative Function

The response-to-retention hypothesis of atherosclerosis (Williams and Tabas 1995) postulates that cholesterol-rich lipoproteins, primarily low-density lipoproteins (LDL), are retained in the arterial wall and oxidatively modified under the action of resident cells (Stocker and Keaney 2004). Oxidation in the arterial intima results from local oxidative stress, which represents an imbalance between prooxidants and antioxidants in favour of the former. Cellular oxidative systems involved in vivo include myeloperoxidase (MPO), NADPH oxidase, nitric oxide synthase and lipoxygenase. They produce a variety of reactive chlorine, nitrogen and oxygen species in the form of one-electron (free radical) and two-electron oxidants (Gaut and Heinecke 2001).

HDL can protect LDL and other lipoproteins from oxidative stress induced by both one- and two-electron species. It can be observed in vitro on their co-incubation (Parthasarathy et al. 1990) and in vivo upon HDL supplementation (Klimov et al. 1993). One-electron oxidants modify both lipid and protein moieties of HDL with formation of lipid and protein radicals. It is followed by accumulation of primary oxidation products, initially lipid hydroperoxides (LOOH), which in turn propagate further oxidation of HDL components to stable termination products (Stocker and Keaney 2004). HDL particles potently protect both lipid and protein moieties of LDL from free radical-induced oxidation, inhibiting accumulation of both primary and secondary oxidation products (Kontush and Chapman 2010). The following overview will focus on the impact of HDL on lipid hydroperoxides and other primary products of lipid peroxidation.

1.2 Mechanisms of Protection

HDL potently inhibits accumulation of LOOH in LDL. Removal of LOOH from LDL (or cells) represents a key step of HDL-mediated protection from oxidative damage induced by free radicals. Indeed, phospholipid hydroperoxides (PLOOHs) are rapidly transferred from LDL to HDL upon their co-incubation (Zerrad-Saadi et al. 2009). Addition of a hydroperoxyl group to a phospholipid or cholesteryl ester molecule strongly increases hydrophilicity. As a result, LOOH molecules are more surface active than their non-peroxidised counterparts (Nuchi et al. 2002). Their exposure to the aqueous phase at the HDL surface facilitates their removal from the lipoprotein. Transfer from LDL to HDL can occur directly between lipoprotein phospholipid monolayers, either spontaneously or mediated by lipid transfer proteins, including cholesterol ester transfer protein (CETP). The latter can accelerate the transfer of both cholesteryl ester hydroperoxides (CEOOHs) and PLOOHs (Christison et al. 1995). Thus, CETP can enhance the antioxidative activity of HDL towards LDL-derived LOOH. Removal of LOOH molecules from LDL can also be mediated by lipid-free apoA-I. As a consequence, HDL particles constitute a major transport vehicle of LOOH in human plasma, effectively functioning as a "sink" for oxidised lipids (Bowry et al. 1992). They can accumulate in the particle when its

LOOH-inactivating capacity is overwhelmed. Furthermore, HDL represents a major plasma carrier of F2-isoprostanes, stable final products of lipid peroxidation (Proudfoot et al. 2009). The ability of HDL to remove LOOH from cell membranes of erythrocytes and astrocytes has also been reported (Ferretti et al. 2003; Klimov et al. 2001). Thus, the accumulation of oxidised lipids in HDL most probably results not only from their transfer from LDL but also from remnant triglyceride-rich lipoproteins and/or arterial wall cells, mediated in part by lipid transfer proteins (Christison et al. 1995). Subsequently, LOOHs and their corresponding hydroxides can be rapidly removed from HDL via scavenger receptor class B-I (SR-BI)-mediated selective uptake by the liver (Christison et al. 1996). This pathway may significantly contribute to the removal of toxic, oxidised lipids from the body. As CEOOH are removed from the circulation more rapidly than the corresponding nonoxidised cholesteryl ester, addition of a hydroperoxyl group may accelerate cholesterol excretion. Following their transfer, HDL-associated LOOH are inactivated by reduction to the corresponding hydroxides in a two-electron redox reaction with HDL proteins (Garner et al. 1998; Zerrad-Saadi et al. 2009). ApoA-I plays a central role in redox inactivation of LOOH and may account for a large part of the antioxidation effects of the lipoprotein. ApoA-I Met residues 112 and 148 reduce LOOH to redox-inactive lipid hydroxides, thereby terminating chain reactions of lipid peroxidation (Garner et al. 1998; Zerrad-Saadi et al. 2009). Simulation by molecular dynamics shows interaction of Met112 and Met148 with Tyr115, creating a microenvironment unique to human apoA-I (Bashtovyy et al. 2011) that may be optimal for such redox reactions. LOOH is also inactivated by apoA-II, although to a lesser extent. The key role of apoA-I in the HDL-mediated protection of LDL from one-electron oxidants is supported by the observation that reconstituted HDL (rHDL) containing only purified apoA-I and phospholipid (POPC), but devoid of other protein and lipid components, are comparable to natural, small, dense HDL3b + 3c subfractions in their capacities to delay lipid peroxidation (Zerrad-Saadi et al. 2009). Concentrations of redox-active Met residues in apoA-I and of PLOOH suggest a 1:1 reaction stoichiometry (Garner et al. 1998; Zerrad-Saadi et al. 2009). HDL content of apoA-I and oxidative status of apoA-I Met residues are therefore important determinants of the capacity of HDL to inactivate LOOH and protect LDL from free radical-induced oxidation (Zerrad-Saadi et al. 2009). The redox reaction between HDL Met residues and LOOH is paralleled by the formation of oxidised forms of apoA-I and apoA-II which contain methionine sulfoxides. These methionine sulfoxide moieties can be reduced back to methionine by methionine sulfoxide reductases. In addition to Met, Cys residues may be important for the antioxidative properties of apoA-I, as demonstrated by the elevated inhibitory potency of the N74C mutant of apoA-I towards LDL oxidation (Zhang et al. 2010). His residues may also contribute to the apoA-I-mediated inhibition of LOOH accumulation in LDL, as a result of their metal-chelating properties (Nguyen et al. 2006). Indeed, transition metal ions are well-established catalysers of lipid peroxidation.

HDL particles contain other apolipoproteins that may contribute to inhibition of LOOH accumulation (Davidson et al. 2009; Ostos et al. 2001). Such antioxidative properties have been reported for apoA-IV (Ostos et al. 2001) and apoE (Miyata and Smith 1996). In addition, apoM displays antioxidative properties in transgenic mice where possible mechanisms may involve binding of oxidised phospholipids, including LOOH (Elsoe et al. 2012). Enzymatic components potentially contributing to antioxidative properties of HDL include paraoxonase-1 (PON1), platelet-activating factor acetylhydrolase (PAF-AH) and lecithin-cholesterol acyltransferase (LCAT), all of which have been proposed to hydrolyse oxidised phospholipids (Kontush and Chapman 2006). In addition, HDL carries glutathione peroxidase (GSPx) which can detoxify LOOH by reducing them to the corresponding hydroxides (Maddipati and Marnett 1987). Finally, the trypanosome lytic factor present in very high-density subpopulations of human HDL exhibits a peroxidase-like activity (Molina Portela et al. 2000) that may provide a minor contribution to LOOH-inactivating properties of HDL. With respect to PON1, PAF-AH and LCAT, these are weakly reactive towards LOOH (Goyal et al. 1997; Kriska et al. 2007; Teiber et al. 2004). HDL-associated PON1, which has been attributed a major antioxidant role in HDL (see Sect. 2.1), does not appear to contribute significantly to the inactivation of LDL-derived PLOOH (Garner et al. 1998; Zerrad-Saadi et al. 2009). Such activity that has been previously reported has been ascribed to the presence of detergents or unidentified proteins (Teiber et al. 2004). The major activity of PON1 is that of a calcium-dependent lactonase rather than a peroxidase (Khersonsky and Tawfik 2005); its affinity for LOOH is several orders of magnitude lower than its affinity for lactones (see Sect. 2.1 for further discussion of PON1). It is more probable that PAF-AH, rather than PON1, represents the hydrolase for PLOOH in HDL (Kriska et al. 2007). LCAT may equally hydrolyse PLOOH generated during lipoprotein oxidation (Goyal et al. 1997). In addition, it can delay LDL oxidation, acting as a chain-breaking antioxidant, most likely via its Cys residues (McPherson et al. 2007). Inhibition of LCAT does not, however, significantly influence HDL-mediated inactivation of LDL-derived PLOOH (Zerrad-Saadi et al. 2009), thereby indicating that the enzyme is at most a minor factor in PLOOH inactivation. Irrespective of the enzyme involved, PLOOH hydrolysis releases lysophosphatidylcholine and a free fatty acid hydroperoxide. Interestingly, PAF-AH is capable of hydrolysing PLOOH within oxidised LDL; in this case, free fatty acid hydroperoxides are transferred to HDL for subsequent two-electron reduction to corresponding hydroxides by apoA-I (Kotosai et al. 2013). As a result of elevated hydrophilicity, free fatty acid hydroperoxides, as compared to PLOOH and CEOOH (Kotosai et al. 2013), are preferentially transported to HDL and reduced to hydroxides.

In contrast to its effects on free radical-induced oxidation, HDL weakly protects LDL from oxidation by two-electron oxidants, such as hypochlorite, which mainly modify the protein moiety of LDL. Such antioxidative action appears largely unspecific, reflecting direct oxidant scavenging by HDL. A rare example of HDL-mediated protection from two-electron oxidants is inactivation by enzymes such as PAF-AH and/or PON1 with participation of apoA-I (Ahmed et al. 2001), of phospholipid core aldehydes generated upon HDL oxidation by peroxynitrite.

Finally, HDL lipids can significantly modulate antioxidative activities displayed by the protein components. First, HDL carries small amounts of lipophilic antioxidants, primarily tocopherols, which may provide a minor contribution to its LOOH-inactivating properties (Bowry et al. 1992; Goulinet and Chapman 1997). Second, the rigidity of the phospholipid monolayer of HDL particles is a key negative modulator of PLOOH transfer efficiency from LDL and cell membranes (Vila et al. 2002; Zerrad-Saadi et al. 2009). The surface monolayer rigidity of HDL is primarily determined by the relative content of such lipids (sphingomyelin, free cholesterol, and saturated and monounsaturated fatty acids). Increasing the content of each rigidifies the surface monolayer of HDL particles at a given protein/lipid ratio (Zerrad-Saadi et al. 2009).

1.3 Heterogeneity of Antioxidant Activity of HDL Particles

Apolipoproteins, enzymes and lipids which determine the antioxidative activities of HDL are non-uniformly distributed across the spectrum of HDL subpopulations. ApoA-I, the major protein component of HDL participating in the reduction of LOOHs, is enriched relative to apoA-II in small, dense HDL3c and in large, less dense HDL2b as compared to HDL2a, 3a and 3b subpopulations (Kontush et al. 2007). LCAT, PAF-AH, PON1 (Kontush et al. 2003) and apoA-IV (Bisgaier et al. 1985) are enriched in small, dense HDL. As a consequence, HDL particles are heterogeneous in their capacity to protect LDL from oxidative damage induced by one-electron oxidants. Small, dense, protein-rich HDL particles in particular potently protect LDL from mild oxidative stress induced by azo-initiator (2,2-′-azobis-(2-amidinopropane) hydrochloride; AAPH) or Cu^{2+} (Kontush et al. 2003; Yoshikawa et al. 1997). It involves inactivation of LOOHs via a two-step mechanism with transfer of LOOHs to HDL (facilitated by enhanced fluidity of the HDL surface lipid monolayer) and subsequent reduction to redox-inactive hydroxides by Met residues of apoA-I (Christison et al. 1995; Garner et al. 1998; Zerrad-Saadi et al. 2009). Small, dense HDL3 may be superior to large, less dense HDL2 in terms of their capacity to remove oxidised lipids from other lipoproteins and cellular membranes. Structural defects in the packing of surface lipids, which allow insertion of exogenous molecules, become more pronounced with decreasing HDL particle size and may account for this property (Kumpula et al. 2008). Consistent with this possibility, small discoid rHDL complexes appear to remove negatively charged lipids from oxidised LDL to a greater degree than native, spherical HDL (Miyazaki et al. 1994). In addition, a diminished content of sphingomyelin and free cholesterol in small, dense HDL may result in an increased fluidity of the surface lipid monolayer, thereby facilitating incorporation of oxidised lipids of exogenous origin (Kontush and Chapman 2010).

Inactivation of oxidised lipid molecules following their transfer to HDL may also occur more rapidly in small, dense particles. First, reduction of LOOHs to hydroxides is more efficient in HDL3 as compared to HDL2 (Zerrad-Saadi et al. 2009). Relative enrichment of small, dense HDL in apoA-I may underlie this feature (Kontush et al. 2007). Moreover, the distinctly low lipid content of small, dense HDL can induce conformational changes in apoA-I relative to large, less dense HDL. It results in enhanced exposure to the aqueous phase, as reflected in modified reactivity towards monoclonal antibodies (Sparks et al. 1995), which might in turn facilitate the redox reaction between Met residues of apoA-I and LOOHs. Second, LOOH hydrolysis by HDL-associated hydrolytic enzymes appears to be predominantly associated with small, dense HDL3 (Kontush et al. 2003). Preferential localisation of enzymatic activities of PAF-AH within HDL3 is probably responsible for such association. Furthermore, enzymatic activities might be beneficially influenced by the lipidome of small, dense HDL3, which is distinct in displaying a low sphingomyelin to phosphatidylcholine ratio (Kontush et al. 2007). Indeed, sphingomyelin belongs to a class of structural lipids that exert a positive impact on surface rigidity and a negative impact on LCAT activity (Subbaiah and Liu 1993).

Finally, the unique proteome of HDL3c may have implications for its antioxidative activity. Several proteins, including apoJ, apoM, serum amyloid A4 (SAA4), apoD, apoL-1, PON1 and PON3 and phospholipid transfer protein (PLTP) occur predominantly in this subfraction (Davidson et al. 2009). Indeed, PLTP is present on HDL together with apoJ, apoL-I, apoD, apoA-I, apoA-II and several other proteins, suggesting a direct, functional interaction between them.

2   High-Density Lipoproteins, Paraoxonase-1

PON1 is an intriguing example of the functional flexibility of HDL components as it bridges the apparent divide between systemic oxidative stress and environmental toxicology. Originally identified as an enzyme neutralising environmental toxins, it progressed to an antioxidant role before revealing its capacity to combat pathogens: the functional complexity and versatility of HDL reflected in a single, peptide component. This section will focus on two aspects of PON1 function. It will expand on the antioxidant capacity of PON1 that was referred to in the preceeding section. Subsequently the ability of PON1 to limit bacterial virulence, notably with respect to *Pseudomonas aeruginosa*, will be addressed.

2.1   PON1 as an Antioxidant

Early, in vitro studies suggested that purified PON1 and HDL-associated PON1 could protect LDL from oxidation (Mackness et al. 2003), although these conclusions were questioned in subsequent reports (James 2006). However, studies from several animal models (James 2006) have provided much stronger support for the conclusion that PON1 protects against atherosclerotic disease by a process that involves reduction of oxidative stress. Thus PON1 knockout in mice was associated with a significant increase in the development of atherosclerotic lesions, whilst HDL from such mice were less able to protect LDL from oxidation (Shih et al. 1998, 2000). The latter was corrected by addition of purified PON1 to the HDL. Conversely, over-expression of PON1 protected against lesion development, with HDL showing a greater capacity to protect against oxidative stress (Tward et al. 2002). A consistent observation from these animal models is that PON1 activity is a negative correlate of markers of oxidative stress.

Studies in man are also consistent with a protective role of PON1, involving an impact on oxidative stress. In a number of disease states that increase risk of atherosclerotic disease and where increased oxidative stress is noted, PON1 serum activity is significantly reduced. These include diabetes, the metabolic syndrome, familial hypercholesterolaemia and renal dysfunction (Abbott et al. 1995; James 2006). Enzyme activity was also reduced in coronary disease patients, with the extent of coronary lesions correlating inversely with the PON1 activity (Graner et al. 2006). The relationship was independent of other risk factors. Other case-control investigations confirm that PON1 is a risk factor for coronary disease (Soran et al. 2009), although it is not a consistent observation (Troughton et al. 2008). Whilst the above studies focus on enzyme activity, there have also been a number of studies examining the relationship of PON1 coding region polymorphisms to coronary risk. The rationale is that a polymorphism at amino acid 192 gives rise to two isoforms with strikingly different activities towards paraoxon, one of the substrates used to analyse PON1 activity. Such studies have given somewhat equivocal results, as illustrated in a meta-analysis by Wheeler et al. (2004). However, this data should be interpreted with reserve as the range of activities is wide within each isoform. Moreover, the isoforms manifest much smaller differences in activities towards a second substrate, phenylacetate, frequently used to analyse PON1. It has led to suggestions that PON1 phenotype should be used to analyse the relationship of the enzyme to vascular risk (Jarvik et al. 2000).

More persuasive evidence of a role for PON1 in CVD comes from prospective studies. In the Caerphilly prospective study, serum PON1 activity was shown to be an independent determinant of future coronary events (Mackness et al. 2003). No measures of oxidative stress were undertaken. In two prospective studies, Hazen and colleagues (Bhattacharyya et al. 2008; Tang et al. 2012) firmly established PON1 as an independent risk factor for vascular disease, in both primary and secondary disease settings. They also analysed an impressive range of systemic indices of lipid oxidation (HETE, HODE, isoprostane) and showed a strong negative correlation between PON1 activity and serum concentrations of these markers.

These studies clearly suggest that it is the ability of PON1 to limit systemic oxidative stress and the generation of oxidised lipids that underlies its links to atherosclerotic disease. Alternative or complementary mechanisms by which PON1 could reduce risk have also been proposed (Aviram 2012; Rosenblat et al. 2013). The authors suggest that the enzyme can increase the free radical scavenging capacity of monocyte/macrophages and lower oxidative stress. The same group also proposed that PON1 increased the ability of HDL to remove cholesterol from macrophages (Rosenblat et al. 2011). The latter is thought to be an important step in reverse cholesterol transport. Data from the same study indicated that PON1 may increase the ability of HDL to prevent macrophage-mediated oxidation of LDL. In another study, the group proposed that PON1 could reduce the risk of diabetes, a potent CVD risk factor (Koren-Gluzer et al. 2011). The antioxidant function, acting via sulphydryl groups PON1, was suggested to play a role, but not enzyme activity.

Whilst there is persuasive evidence that PON1 prevents systemic lipid oxidation, how and what substrates it may act on are still unclear. As discussed above (Sect. 1.2), LOOH would appear to be poor substrates for the enzyme. Nevertheless, a feature of the PON1 knockout mice was an accumulation of LOOH in HDL (Shih et al. 1998; Tward et al. 2002). It is possible that short-chain oxidised phospholipids may be a better substrate for the enzyme. Increased lipoperoxides as well as lipid oxidation end products were also a feature of reduced PON1 activity in human studies (Bhattacharyya et al. 2008; Tang et al. 2012). One possible explanation may be furnished by a recent, intriguing proposal that PON1 can modulate the activity of myeloperoxidase (MPO). The latter is an important source of reactive oxygen species and is associated with HDL. Huang et al. (2013) reported that PON1 and MPO form a ternary complex with HDL. In this constellation, PON1 was able to modulate MPO activity (and conversely MPO inhibited PON1). Confirmation was obtained in vitro by showing that adding PON1 to active MPO significantly reduced lipid peroxidation by the latter.

At present, the question remains open as to how exactly PON1 functions, with possible explanations being prevention of ROS formation, curtailing propagation of ROS or hydrolysing oxidised lipids.

2.2 PON1 and Bacterial Pathogens

A major advance in our appreciation of the physiological relevance of PON1 was the demonstration that its principal enzyme activity is that of a lactonase (Khersonsky and Tawfik 2005). Indeed, lactonase activity would appear to be the initial, ancestral enzyme activity of the PON gene family. These observations considerably broadened the scope of HDL functions as lactonases are instrumental in neutralising bacterial virulence factors.

Pseudomonas aeruginosa is an opportunist bacterium and a common cause of disease in immunocompromised patients. It is also a major contributor to nosocomial infections. Bacterial virulence is governed by homoserine lactones (HSL) synthesised and secreted by the bacterium. They are central to the phenomenon of quorum sensing that allows communication between bacteria. The HSLs give a measure of bacterial density and when a threshold bacteria count is achieved, they activate gene programmes that greatly increase the rate of Pseudomonas infection (Juhas et al. 2005). The ability of the PON enzymes to hydrolyse the HSL interrupts bacterial communication and reduces virulence. Whilst the intracellular PON2 variant shows the strongest lactonase activity, PON1is also active towards HSL (Ozer et al. 2005) and is able to protect against Pseudomonas infection (Stoltz et al. 2008). Indeed, PON1 is the major source of HSL neutralising activity in serum (Teiber et al. 2008).

Discovery of its role in combating bacterial infection and knowledge that the paraoxonases are a highly conserved gene family have led to suggestions that they may play a role in innate immunity (Shih and Lusis 2009).

3 High-Density Lipoproteins, Environmental Pathogens and Toxins

In recent studies, the molecular composition of the HDL particle has been assigned significant, functional importance (Besler et al. 2012), which is of particular relevance with respect to reported changes in HDL composition in disease. Interestingly, alterations in HDL composition have also been observed in man during endotoxemia (Levels et al. 2011) and in endotoxin-treated mice (Chiba et al. 2011). As suggested above, data indicate that HDL has an important role in neutralising and detoxifying invading pathogens to prevent systemic inflammatory responses or sepsis, which is the leading cause of death in intensive care units (ICUs) of high-income countries (Levels et al. 2011; Russell 2006). Concurrently, the interaction of HDL with viruses, parasites and environmental toxins has also been a focus of recent interest. This is reflected in reviews of the impact of gram-positive and gram-negative bacterial toxins in sepsis (Ramachandran 2013), the role of HDL in innate immunity (Feingold and Grunfeld 2011) and the immune system (Kaji 2013) and xenobiotic metabolism, disposition and regulation of receptors (Omiecinski et al. 2011) to which the reader is referred. In this chapter, the role(s) of HDL as a detoxifying agent during exposure to different environmental pathogens or toxins will be discussed.

3.1 Bacterial Pathogens

Multiple alterations in lipid and lipoprotein metabolism occur during the acute phase response. Within HDL, apoA-I, cholesterol and phospholipid levels decrease, as do certain enzyme activities, e.g. PON1 (Khovidhunkit et al. 2004), whilst free apoA-I (Cabana et al. 1997) and triglyceride levels increase (Cabana et al. 1996). As a consequence, reverse cholesterol transport decreases, accompanied by increased cholesterol delivery to immune cells. Such alterations initially protect the host from the harmful effects of invading pathogens but will contribute to atherogenesis if prolonged (Khovidhunkit et al. 2004).

A crucial factor in the detoxifying properties attributed to HDL is the ability to neutralise the invading pathogen. A number of human studies have shown that circulating lipoproteins, principally HDL, are able to bind lipopolysaccharides (LPS) from gram-negative bacteria as well as lipoteichoic acid (LTA) from gram-positive bacteria (Khovidhunkit et al. 2004; Levels et al. 2003, 2011). Binding to HDL inhibits LPS interaction with cell surface toll-like receptor 4 (TLR-4) (Underhill and Ozinsky 2002) and LTA interaction with TLR-2 (Flo et al. 2000), receptors known to mediate inflammatory responses (macrophage activation, cytokine release). These neutralising effects of HDL have been confirmed in humans with low HDL levels who show a more robust inflammatory response to LPS administration (Birjmohun et al. 2007).

During the acute phase response, HDL protein and lipid compositions change (Khovidhunkit et al. 2004). Observations indicate that apoA-I can be replaced by the positive acute phase protein SAA (Artl et al. 2000). SAA is able to influence HDL-mediated cholesterol metabolism through its inhibitory effects on SR-BI-mediated selective cholesterol uptake (Cai et al. 2005). It has recently been suggested to facilitate the binding of HDL, isolated from mice injected with LPS, to vascular proteoglycans (Chiba et al. 2011). The relevance of these observations is important when taken together with the up to 1,000-fold increase in concentrations of plasma SAA during acute phase reactions such as inflammation or infection. In addition, several isoforms of SAA, still with unknown functions, have recently been detected in acute phase HDL (Levels et al. 2011).

The neutralising effects of HDL during LPS exposure are dependent on phospholipids as well as proteins. ApoA-I is suggested to be the major neutralising factor (Massamiri et al. 1997). Thus, direct interaction between LPS and the C-terminus of apoA-I can decrease TNF-α release from macrophages in vitro (Henning et al. 2011). The role of apoA-I as a neutralising agent is further strengthened by the beneficial effects of apoA-I on LPS-induced acute lung injury and endotoxemia, as shown in mice (Yan et al. 2006).

The HDL receptor, SR-BI, is highly expressed in hepatocytes and steroid-producing cells and has an important role regarding the fate of the neutralised pathogens. In mice, it is suggested to protect against endotoxemia through its roles in facilitating glucocorticoid production as well as hepatic clearance of LPS. In this respect, SR-BI has been suggested to mediate the binding and uptake of LPS as well as LTA (Vishnyakova et al. 2006). Additionally, an alternatively spliced variant, SR-BII, has been shown to bind directly a variety of bacteria suggesting a conserved role of these receptors in pattern recognition and innate immunity (Webb et al. 1998).

3.2 Parasites

Several species of African trypanosomes cause fatal disease in livestock, but most are unable to infect humans due to innate trypanolytic factors (TLFs). It has been shown that in human serum, TLFs consist of two minor HDL subfractions characterised by the presence of haptoglobin-related protein (Hpr) and apoL-I (Vanhamme et al. 2003). The underlying mechanisms of TLFs action during parasite infection involve their endocytosis by trypanosomes where, subsequently, apoL-I forms membrane pores within the acidic lysosome resulting in ion dysregulation that leads to osmotic imbalance, parasite swelling and lysis (Molina-Portela Mdel et al. 2005). Endocytosis of apoL-I is facilitated by a complex formed by Hpr and haemoglobin that interacts with the haptoglobin-haemoglobin receptor on the trypanosome cell surface (Vanhollebeke et al. 2008). However, *Trypanosoma brucei rhodesiense*, carried by tsetse flies and causing sleeping sickness in man, is able to escape lysis by TLFs. It is achieved mainly by the expression of the serum resistance-associated (SRA) protein, which binds to and neutralises apoL-I (Vanhamme et al. 2003). It should be noted however that recent population-based observations, focused on individual variations in Hpr, indicate that a more dynamic view of the relative roles of Hpr and Hpr-Hb complexes needs to be considered for understanding innate immunity to African trypanosomes. Other pathogens including the newly discovered Plasmodium (Imrie et al. 2012) should possibly be considered.

3.3 Hepatitis, Dengue and Other Viruses

HDL account for part of the broad, non-specific antiviral activity in human serum (Singh et al. 1999). Thus, apoA-I was found to prevent cell penetration by inhibiting fusion of the herpes virus as well as the human immunodeficiency virus (HIV) (Srinivas et al. 1990). Interestingly, HIV infection has also been linked to dysfunctional HDL with reduced antioxidant properties, which may be associated to progression of subclinical atherosclerosis (Kelesidis et al. 2013). In addition, enhanced levels of human apoM, which is mainly associated with HDL, were observed during hepatitis B virus (HBV) infection and may reflect feedback suppression of HBV replication (Gu et al. 2011). Conversely, HDL may also promote virulence by facilitating virus entry via the SR-BI receptor. This was first demonstrated for the hepatitis C viruses (Voisset et al. 2006) and recently extended to dengue viruses that directly associate with apoA-I of HDL (Li et al. 2013). It is perhaps unsurprising that SR-BI has recently been proposed as a therapeutic target that warrants further research, based on reports indicating its involvement in the capture and cross-presentation of antigens from viruses, bacteria and parasites.

3.4 Metal Oxides, Carbon Nanotubes and PLGA Nanoparticles

Nanotechnology is an emerging industry that involves the creation of new materials with a variety of useful functions. However, concerns are growing that some of them may have toxic effects. This has promoted interest in the role of HDL both as a neutralising agent and carrier of nanosize particles. Thus the biodistribution of nanoparticles is significantly influenced by their interaction with plasma proteins, notably HDL proteins apoA-I and apoE. These, amongst others, have been found to recognise and interact with carbon nanotubes, a number of metal oxides (Karlsson et al. 2012) and PLGA (poly(lactic-co-glycolic acid)) nanoparticles (Sempf et al. 2013) in in vitro plasma models. There are, as yet, no detailed studies confirming HDL neutralisation and clearance of engineered nanoparticles. However, the concept of incorporating lipophilic drugs assembled with apoA-I and phospholipids into soluble, HDL-like nanoparticles for drug delivery is presently widely explored (Marrache and Dhar 2013; Shin et al. 2013). They could serve a number of therapeutic purposes, for example, targeting tissues expressing high amounts of SR-BI, such as cancer cells (Marrache and Dhar 2013; Shin et al. 2013).

3.5 PON1 and Organophosphates

A crucial function of HDL-associated PON1, in addition to its potential antioxidant role, is neutralisation of toxic organophosphate (OP) derivatives. Such an activity was described in human serum many years ago (Aldridge 1953). The activity was attributed to the enzyme PON1, a name derived from the substrate paraoxon, initially employed in such studies. Its involvement in the detoxification of OPs has since been well documented, encompassing chlorpyrifos, oxon and diazoxon as well as nerve agents (Costa et al. 2013). Indeed, HDL-associated PON1 has been shown to be a key determinant in detoxification of OPs (Shih et al. 1998) in mouse models. Intriguingly, PON1 was shown to be associated with HDL at a time when its antioxidant activity was unknown (James 2006).

A growing area of research is the importance of PON1 in environment toxicology and its role in susceptibility to OP exposure. Serum concentrations of the enzyme, as well as certain polymorphisms affecting activity, appear important for risk prediction (Androutsopoulos et al. 2011; Costa et al. 2013). Consequently, children may be more sensitive to environmental OPs since they have reportedly lower PON1 levels than adults (Huen et al. 2009). Finally, its impact on environmental toxins, as well as its antioxidant function, has led to suggestions that PON1 is linked to neurological disorders (Androutsopoulos et al. 2011), an area that requires considerably more research.

3.6 Detoxification of Plasma and External Fluids

As discussed in previous sections, HDL and apo A-I have a role in recognition, neutralisation and elimination of xenobiotics. Thus, elevated levels of plasma apoA-I in female rats have been found after exposure to the plastic chemical bisphenol A (BPA) (Ronn et al. 2013). The molecular mechanisms behind the increase in apoA-I may reflect the effects of estrogens, although conflicting data have been reported (Teeguarden et al. 2013). It could also be related to apoA-I/HDL-controlled clearance mechanisms that involve transport of xenobiotics to the liver. Interestingly, an increase in liver somatic index (LSI) accompanied the apoA-I increase in the exposed rats, indicating less favourable metabolic alterations. Since the structure of BPA is very similar to a number of recently proposed apoA-I enhancing drugs (Du et al. 2012), it may be relevant to investigate these mechanisms further.

Another group of environmental contaminants, present at low levels in most living organisms, are persistent organic pollutants (POPs). Due to their hydrophobic character, they can be transported by lipoprotein particles. Interestingly, long-term effects of POPs in human plasma have recently been highlighted in a number of epidemiological studies (Lind et al. 2012). In a recent study of an exposed Swedish population, significantly higher concentrations of POPs were found amongst individuals with CVD or cancer compared to controls (Ljunggren et al. 2014). Principal component analyses showed that POP concentrations in HDL were more associated with CVD, whilst POP concentrations in LDL/VLDL were more associated with cancer. Interestingly, PON-1 activity was negatively correlated to the sum of polychlorinated biphenyls (PCBs), and covariation between decreased arylesterase activity, increased PCB concentrations and CVD was found.

To expand the knowledge regarding the role of HDL as detoxifier, it may be beneficial exploring external fluids. Lipids are present in human tears (Rantamaki et al. 2011), and apoA-I has been detected in nasal lavage fluid (NLF) (Ghafouri et al. 2002) and saliva (Ghafouri et al. 2003). Interestingly, in NLF, apoA-I was shown to increase in hairdressers after 20 min of exposure to persulphates (Karedal et al. 2010), whilst apoA-I in gingival crevicular fluid has been suggested to be a novel periodontal disease marker (Tsuchida et al. 2012). In addition, decreased levels of apoA-I have been reported in bronchio-alveolar lavage fluids of patients with idiopathic pulmonary fibrosis compared to healthy controls (Kim et al. 2010). Together with lipocalin, apoA-I has been suggested as a potential biomarker for chronic obstructive pulmonary disease (Nicholas et al. 2010). The origin of apoA-I in fluids of the airways has not been extensively studied, but trans-endothelial transport has been investigated (Ohnsorg et al. 2011). The presence, composition and possible function of apoA-I containing lipoprotein particles in external fluids remain to be investigated.

Acknowledgements RWJ gratefully acknowledges support received from the Swiss National Research Foundation, the Novartis Consumer Health Foundation and Unitec, University of Geneva. AK acknowledges financial support from the National Institute for Health and Medical Research (INSERM).

HK, AK and RWJ are members of the COST Action BM0904.

Open Access This chapter is distributed under the terms of the Creative Commons Attribution Noncommercial License, which permits any noncommercial use, distribution, and reproduction in any medium, provided the original author(s) and source are credited.

References

Abbott CA, Mackness MI, Kumar S, Boulton AJ, Durrington PN (1995) Serum paraoxonase activity, concentration, and phenotype distribution in diabetes mellitus and its relationship to serum lipids and lipoproteins. Arterioscler Thromb Vasc Biol 15(11):1812–1818

Ahmed Z, Ravandi A, Maguire GF, Emili A, Draganov D, La Du BN, Kuksis A, Connelly PW (2001) Apolipoprotein A-I promotes the formation of phosphatidylcholine core aldehydes that are hydrolyzed by paraoxonase (PON-1) during high density lipoprotein oxidation with a peroxynitrite donor. J Biol Chem 276(27):24473–24481

Aldridge WN (1953) Serum esterases. II. An enzyme hydrolysing diethyl p-nitrophenyl phosphate (E600) and its identity with the A-esterase of mammalian sera. Biochem J 53(1):117–124

Androutsopoulos VP, Kanavouras K, Tsatsakis AM (2011) Role of paraoxonase 1 (PON1) in organophosphate metabolism: implications in neurodegenerative diseases. Toxicol Appl Pharmacol 256(3):418–424

Artl A, Marsche G, Lestavel S, Sattler W, Malle E (2000) Role of serum amyloid A during metabolism of acute-phase HDL by macrophages. Arterioscler Thromb Vasc Biol 20 (3):763–772

Aviram M (2012) Atherosclerosis: cell biology and lipoproteins – paraoxonases protect against atherosclerosis and diabetes development. Curr Opin Lipidol 23(2):169–171

Bashtovyy D, Jones MK, Anantharamaiah GM, Segrest JP (2011) Sequence conservation of apolipoprotein A-I affords novel insights into HDL structure-function. J Lipid Res 52 (3):435–450

Besler C, Luscher TF, Landmesser U (2012) Molecular mechanisms of vascular effects of high-density lipoprotein: alterations in cardiovascular disease. EMBO Mol Med 4(4):251–268

Bhattacharyya T, Nicholls SJ, Topol EJ, Zhang R, Yang X, Schmitt D, Fu X, Shao M, Brennan DM, Ellis SG, Brennan ML, Allayee H, Lusis AJ, Hazen SL (2008) Relationship of paraoxonase 1 (PON1) gene polymorphisms and functional activity with systemic oxidative stress and cardiovascular risk. JAMA 299(11):1265–1276

Birjmohun RS, van Leuven SI, Levels JH, van't Veer C, Kuivenhoven JA, Meijers JC, Levi M, Kastelein JJ, van der Poll T, Stroes ES (2007) High-density lipoprotein attenuates inflammation and coagulation response on endotoxin challenge in humans. Arterioscler Thromb Vasc Biol 27(5):1153–1158

Bisgaier CL, Sachdev OP, Megna L, Glickman RM (1985) Distribution of apolipoprotein A-IV in human plasma. J Lipid Res 26(1):11–25

Bowry VW, Stanley KK, Stocker R (1992) High density lipoprotein is the major carrier of lipid hydroperoxides in human blood plasma from fasting donors. Proc Natl Acad Sci U S A 89 (21):10316–10320

Cabana VG, Lukens JR, Rice KS, Hawkins TJ, Getz GS (1996) HDL content and composition in acute phase response in three species: triglyceride enrichment of HDL a factor in its decrease. J Lipid Res 37(12):2662–2674

Cabana VG, Gidding SS, Getz GS, Chapman J, Shulman ST (1997) Serum amyloid A and high density lipoprotein participate in the acute phase response of Kawasaki disease. Pediatr Res 42 (5):651–655

Cai L, de Beer MC, de Beer FC, van der Westhuyzen DR (2005) Serum amyloid A is a ligand for scavenger receptor class B type I and inhibits high density lipoprotein binding and selective lipid uptake. J Biol Chem 280(4):2954–2961

Chiba T, Chang MY, Wang S, Wight TN, McMillen TS, Oram JF, Vaisar T, Heinecke JW, De Beer FC, De Beer MC, Chait A (2011) Serum amyloid A facilitates the binding of high-density lipoprotein from mice injected with lipopolysaccharide to vascular proteoglycans. Arterioscler Thromb Vasc Biol 31(6):1326–1332

Christison JK, Rye KA, Stocker R (1995) Exchange of oxidized cholesteryl linoleate between LDL and HDL mediated by cholesteryl ester transfer protein. J Lipid Res 36(9):2017–2026

Christison J, Karjalainen A, Brauman J, Bygrave F, Stocker R (1996) Rapid reduction and removal of HDL- but not LDL-associated cholesteryl ester hydroperoxides by rat liver perfused in situ. Biochem J 314(Pt 3):739–742

Costa LG, Giordano G, Cole TB, Marsillach J, Furlong CE (2013) Paraoxonase 1 (PON1) as a genetic determinant of susceptibility to organophosphate toxicity. Toxicology 307:115–122

Davidson WS, Silva RA, Chantepie S, Lagor WR, Chapman MJ, Kontush A (2009) Proteomic analysis of defined HDL subpopulations reveals particle-specific protein clusters: relevance to antioxidative function. Arterioscler Thromb Vasc Biol 29(6):870–876

Du Y, Yang Y, Jiang W, Wang L, Jia XJ, Si SY, Chen XF, Hong B (2012) Substituted benzamides containing azaspiro rings as upregulators of apolipoprotein A-I transcription. Molecules 17

Elsoe S, Ahnstrom J, Christoffersen C, Hoofnagle AN, Plomgaard P, Heinecke JW, Binder CJ, Bjorkbacka H, Dahlback B, Nielsen LB (2012) Apolipoprotein M binds oxidized phospholipids and increases the antioxidant effect of HDL. Atherosclerosis 221(1):91–97

Feingold KR, Grunfeld C (2011) The role of HDL in innate immunity. J Lipid Res 52(1):1–3

Ferretti G, Bacchetti T, Moroni C, Vignini A, Curatola G (2003) Copper-induced oxidative damage on astrocytes: protective effect exerted by human high density lipoproteins. Biochim Biophys Acta 1635(1):48–54

Flo TH, Halaas O, Lien E, Ryan L, Teti G, Golenbock DT, Sundan A, Espevik T (2000) Human toll-like receptor 2 mediates monocyte activation by *Listeria monocytogenes*, but not by group B streptococci or lipopolysaccharide. J Immunol 164(4):2064–2069

Garner B, Waldeck AR, Witting PK, Rye KA, Stocker R (1998) Oxidation of high density lipoproteins. II. Evidence for direct reduction of lipid hydroperoxides by methionine residues of apolipoproteins AI and AII. J Biol Chem 273(11):6088–6095

Gaut JP, Heinecke JW (2001) Mechanisms for oxidizing low-density lipoprotein. Insights from patterns of oxidation products in the artery wall and from mouse models of atherosclerosis. Trends Cardiovasc Med 11(3–4):103–112. doi:10.1016/S1050-1738(01)00101-3

Ghafouri B, Stahlbom B, Tagesson C, Lindahl M (2002) Newly identified proteins in human nasal lavage fluid from non-smokers and smokers using two-dimensional gel electrophoresis and peptide mass fingerprinting. Proteomics 2(1):112–120

Ghafouri B, Tagesson C, Lindahl M (2003) Mapping of proteins in human saliva using two-dimensional gel electrophoresis and peptide mass fingerprinting. Proteomics 3 (6):1003–1015

Goulinet S, Chapman MJ (1997) Plasma LDL and HDL subspecies are heterogenous in particle content of tocopherols and oxygenated and hydrocarbon carotenoids. Relevance to oxidative resistance and atherogenesis. Arterioscler Thromb Vasc Biol 17(4):786–796

Goyal J, Wang K, Liu M, Subbaiah PV (1997) Novel function of lecithin-cholesterol acyltransferase. Hydrolysis of oxidized polar phospholipids generated during lipoprotein oxidation. J Biol Chem 272(26):16231–16239

Graner M, James RW, Kahri J, Nieminen MS, Syvanne M, Taskinen MR (2006) Association of paraoxonase-1 activity and concentration with angiographic severity and extent of coronary artery disease. J Am Coll Cardiol 47(12):2429–2435

Gu JG, Zhu CL, Cheng DZ, Xie Y, Liu F, Zhou X (2011) Enchanced levels of apolipoprotein M during HBV infection feedback suppresses HBV replication. Lipids Health Dis 10:154

Henning MF, Herlax V, Bakas L (2011) Contribution of the C-terminal end of apolipoprotein AI to neutralization of lipopolysaccharide endotoxic effect. Innate Immun 17(3):327–337

Huang Y, Wu Z, Riwanto M, Gao S, Levison BS, Gu X, Fu X, Wagner MA, Besler C, Gerstenecker G, Zhang R, Li XM, DiDonato AJ, Gogonea V, Tang WH, Smith JD, Plow EF, Fox PL, Shih DM, Lusis AJ, Fisher EA, DiDonato JA, Landmesser U, Hazen SL (2013) Myeloperoxidase, paraoxonase-1, and HDL form a functional ternary complex. J Clin Invest 123(9):3815–3828

Huen K, Harley K, Brooks J, Hubbard A, Bradman A, Eskenazi B, Holland N (2009) Developmental changes in PON1 enzyme activity in young children and effects of PON1 polymorphisms. Environ Health Perspect 117(10):1632–1638

Imrie HJ, Fowkes FJ, Migot-Nabias F, Luty AJ, Deloron P, Hajduk SL, Day KP (2012) Individual variation in levels of haptoglobin-related protein in children from Gabon. PLoS ONE 7(11): e49816

James RW (2006) A long and winding road: defining the biological role and clinical importance of paraoxonases. Clin Chem Lab Med 44(9):1052–1059. doi:10.1515/CCLM.2006.207

Jarvik GP, Rozek LS, Brophy VH, Hatsukami TS, Richter RJ, Schellenberg GD, Furlong CE (2000) Paraoxonase (PON1) phenotype is a better predictor of vascular disease than is PON1 (192) or PON1(55) genotype. Arterioscler Thromb Vasc Biol 20:2441–2447

Juhas M, Eberl L, Tummler B (2005) Quorum sensing: the power of cooperation in the world of Pseudomonas. Environ Microbiol 7(4):459–471

Kaji H (2013) High-density lipoproteins and the immune system. J Lipids 2013:684903

Karedal MH, Mortstedt H, Jeppsson MC, Kronholm Diab K, Nielsen J, Jonsson BA, Lindh CH (2010) Time-dependent proteomic iTRAQ analysis of nasal lavage of hairdressers challenged by persulfate. J Proteome Res 9(11):5620–5628

Karlsson H, Ljunggren S, Ahrén M, Ghafouri B, Uvdahl K, Lindahl M, Ljungman A (2012) Two-dimensional gel electrophoresis and mass spectrometry in studies of nanoparticle-protein interactions. In: Magdeldin S (ed) Gel glectrophoresis – advanced techniques. In Tech, Rijeka

Kelesidis T, Yang OO, Kendall MA, Hodis HN, Currier JS (2013) Dysfunctional HDL and progression of atherosclerosis in HIV-1-infected and -uninfected adults. Lipids Health Dis 12:23

Khersonsky O, Tawfik DS (2005) Structure-reactivity studies of serum paraoxonase PON1 suggest that its native activity is lactonase. Biochemistry 44(16):6371–6382

Khovidhunkit W, Kim MS, Memon RA, Shigenaga JK, Moser AH, Feingold KR, Grunfeld C (2004) Effects of infection and inflammation on lipid and lipoprotein metabolism: mechanisms and consequences to the host. J Lipid Res 45(7):1169–1196

Kim TH, Lee YH, Kim KH, Lee SH, Cha JY, Shin EK, Jung S, Jang AS, Park SW, Uh ST, Kim YH, Park JS, Sin HG, Youm W, Koh ES, Cho SY, Paik YK, Rhim TY, Park CS (2010) Role of lung apolipoprotein A-I in idiopathic pulmonary fibrosis: antiinflammatory and antifibrotic effect on experimental lung injury and fibrosis. Am J Respir Crit Care Med 182(5):633–642

Klimov AN, Gurevich VS, Nikiforova AA, Shatilina LV, Kuzmin AA, Plavinsky SL, Teryukova NP (1993) Antioxidative activity of high density lipoproteins in vivo. Atherosclerosis 100(1):13–18

Klimov AN, Kozhevnikova KA, Kuzmin AA, Kuznetsov AS, Belova EV (2001) On the ability of high density lipoproteins to remove phospholipid peroxidation products from erythrocyte membranes. Biochemistry (Mosc) 66(3):300–304

Kontush A, Chapman MJ (2006) Functionally defective high-density lipoprotein: a new therapeutic target at the crossroads of dyslipidemia, inflammation, and atherosclerosis. Pharmacol Rev 58(3):342–374

Kontush A, Chapman MJ (2010) Antiatherogenic function of HDL particle subpopulations: focus on antioxidative activities. Curr Opin Lipidol 21(4):312–318

Kontush A, Chantepie S, Chapman MJ (2003) Small, dense HDL particles exert potent protection of atherogenic LDL against oxidative stress. Arterioscler Thromb Vasc Biol 23(10):1881–1888

Kontush A, Therond P, Zerrad A, Couturier M, Negre-Salvayre A, de Souza JA, Chantepie S, Chapman MJ (2007) Preferential sphingosine-1-phosphate enrichment and sphingomyelin depletion are key features of small dense HDL3 particles: relevance to antiapoptotic and antioxidative activities. Arterioscler Thromb Vasc Biol 27(8):1843–1849

Koren-Gluzer M, Aviram M, Meilin E, Hayek T (2011) The antioxidant HDL-associated paraoxonase-1 (PON1) attenuates diabetes development and stimulates beta-cell insulin release. Atherosclerosis 219(2):510–518

Kotosai M, Shimada S, Kanda M, Matsuda N, Sekido K, Shimizu Y, Tokumura A, Nakamura T, Murota K, Kawai Y, Terao J (2013) Plasma HDL reduces nonesterified fatty acid hydroperoxides originating from oxidized LDL: a mechanism for its antioxidant ability. Lipids 48(6):569–578

Kriska T, Marathe GK, Schmidt JC, McIntyre TM, Girotti AW (2007) Phospholipase action of platelet-activating factor acetylhydrolase, but not paraoxonase-1, on long fatty acyl chain phospholipid hydroperoxides. J Biol Chem 282(1):100–108

Kumpula LS, Kumpula JM, Taskinen MR, Jauhiainen M, Kaski K, Ala-Korpela M (2008) Reconsideration of hydrophobic lipid distributions in lipoprotein particles. Chem Phys Lipids 155(1):57–62

Levels JH, Abraham PR, van Barreveld EP, Meijers JC, van Deventer SJ (2003) Distribution and kinetics of lipoprotein-bound lipoteichoic acid. Infect Immun 71(6):3280–3284

Levels JH, Geurts P, Karlsson H, Maree R, Ljunggren S, Fornander L, Wehenkel L, Lindahl M, Stroes ES, Kuivenhoven JA, Meijers JC (2011) High-density lipoprotein proteome dynamics in human endotoxemia. Proteome Sci 9(1):34

Li Y, Kakinami C, Li Q, Yang B, Li H (2013) Human apolipoprotein A-I is associated with dengue virus and enhances virus infection through SR-BI. PLoS ONE 8(7):e70390

Lind PM, van Bavel B, Salihovic S, Lind L (2012) Circulating levels of persistent organic pollutants (POPs) and carotid atherosclerosis in the elderly. Environ Health Perspect 120 (1):38–43

Ljunggren S, Helmfrid I, Salihovic S, van Bavel B, Wingren G, Lindahl M, Karlsson H (2014) Persistent organic pollutants distribution in lipoprotein fractions in relation to cardiovascular disease and cancer. Environ Int 65:93–99

Mackness B, Durrington P, McElduff P, Yarnell J, Azam N, Watt M, Mackness M (2003) Low paraoxonase activity predicts coronary events in the Caerphilly Prospective Study. Circulation 107(22):2775–2779

Maddipati KR, Marnett LJ (1987) Characterization of the major hydroperoxide-reducing activity of human plasma. Purification and properties of a selenium-dependent glutathione peroxidase. J Biol Chem 262(36):17398–17403

Marrache S, Dhar S (2013) Biodegradable synthetic high-density lipoprotein nanoparticles for atherosclerosis. Proc Natl Acad Sci U S A 110(23):9445–9450

Massamiri T, Tobias PS, Curtiss LK (1997) Structural determinants for the interaction of lipopolysaccharide binding protein with purified high density lipoproteins: role of apolipoprotein A-I. J Lipid Res 38(3):516–525

McPherson PA, Young IS, McEneny J (2007) A dual role for lecithin:cholesterol acyltransferase (EC 2.3.1.43) in lipoprotein oxidation. Free Radic Biol Med 43(11):1484–1493

Miyata M, Smith JD (1996) Apolipoprotein E allele-specific antioxidant activity and effects on cytotoxicity by oxidative insults and beta-amyloid peptides. Nat Genet 14(1):55–61

Miyazaki A, Sakai M, Suginohara Y, Hakamata H, Sakamoto Y, Morikawa W, Horiuchi S (1994) Acetylated low density lipoprotein reduces its ligand activity for the scavenger receptor after interaction with reconstituted high density lipoprotein. J Biol Chem 269(7):5264–5269

Molina Portela MP, Raper J, Tomlinson S (2000) An investigation into the mechanism of trypanosome lysis by human serum factors. Mol Biochem Parasitol 110(2):273–282

Molina-Portela Mdel P, Lugli EB, Recio-Pinto E, Raper J (2005) Trypanosome lytic factor, a subclass of high-density lipoprotein, forms cation-selective pores in membranes. Mol Biochem Parasitol 144(2):218–226

Nguyen SD, Jeong TS, Sok DE (2006) Apolipoprotein A-I-mimetic peptides with antioxidant actions. Arch Biochem Biophys 451(1):34–42

Nicholas BL, Skipp P, Barton S, Singh D, Bagmane D, Mould R, Angco G, Ward J, Guha-Niyogi B, Wilson S, Howarth P, Davies DE, Rennard S, O'Connor CD, Djukanovic R (2010) Identification of lipocalin and apolipoprotein A1 as biomarkers of chronic obstructive pulmonary disease. Am J Respir Crit Care Med 181(10):1049–1060

Nuchi CD, Hernandez P, McClements DJ, Decker EA (2002) Ability of lipid hydroperoxides to partition into surfactant micelles and alter lipid oxidation rates in emulsions. J Agric Food Chem 50(19):5445–5449

Ohnsorg PM, Rohrer L, Perisa D, Kateifides A, Chroni A, Kardassis D, Zannis VI, von Eckardstein A (2011) Carboxyl terminus of apolipoprotein A-I (ApoA-I) is necessary for the transport of lipid-free ApoA-I but not prelipidated ApoA-I particles through aortic endothelial cells. J Biol Chem 286(10):7744–7754

Omiecinski CJ, Vanden Heuvel JP, Perdew GH, Peters JM (2011) Xenobiotic metabolism, disposition, and regulation by receptors: from biochemical phenomenon to predictors of major toxicities. Toxicol Sci 120(Suppl 1):S49–S75

Ostos MA, Conconi M, Vergnes L, Baroukh N, Ribalta J, Girona J, Caillaud JM, Ochoa A, Zakin MM (2001) Antioxidative and antiatherosclerotic effects of human apolipoprotein A-IV in apolipoprotein E-deficient mice. Arterioscler Thromb Vasc Biol 21(6):1023–1028

Ozer EA, Pezzulo A, Shih DM, Chun C, Furlong C, Lusis AJ, Greenberg EP, Zabner J (2005) Human and murine paraoxonase 1 are host modulators of *Pseudomonas aeruginosa* quorum-sensing. FEMS Microbiol Lett 253(1):29–37

Parthasarathy S, Barnett J, Fong LG (1990) High-density lipoprotein inhibits the oxidative modification of low-density lipoprotein. Biochim Biophys Acta 1044(2):275–283

Proudfoot JM, Barden AE, Loke WM, Croft KD, Puddey IB, Mori TA (2009) HDL is the major lipoprotein carrier of plasma F2-isoprostanes. J Lipid Res 50(4):716–722

Ramachandran G (2013) Gram-positive and gram-negative bacterial toxins in sepsis: a brief review. Virulence 5(1)

Rantamaki AH, Seppanen-Laakso T, Oresic M, Jauhiainen M, Holopainen JM (2011) Human tear fluid lipidome: from composition to function. PLoS ONE 6(5):e19553

Ronn M, Lind PM, Karlsson H, Cvek K, Berglund J, Malmberg F, Orberg J, Lind L, Ortiz-Nieto F, Kullberg J (2013) Quantification of total and visceral adipose tissue in fructose-fed rats using water-fat separated single echo MRI. Obesity (Silver Spring) 21(9):E388–E395

Rosenblat M, Volkova N, Aviram M (2011) Injection of paraoxonase 1 (PON1) to mice stimulates their HDL and macrophage antiatherogenicity. Biofactors 37(6):462–467

Rosenblat M, Elias A, Volkova N, Aviram M (2013) Monocyte-macrophage membrane possesses free radicals scavenging activity: stimulation by polyphenols or by paraoxonase 1 (PON1). Free Radic Res 47(4):257–267

Russell JA (2006) Management of sepsis. N Engl J Med 355(16):1699–1713

Sempf K, Arrey T, Gelperina S, Schorge T, Meyer B, Karas M, Kreuter J (2013) Adsorption of plasma proteins on uncoated PLGA nanoparticles. Eur J Pharm Biopharm 85(1):53–60

Shih DM, Lusis AJ (2009) The roles of PON1 and PON2 in cardiovascular disease and innate immunity. Curr Opin Lipidol 20(4):288–292

Shih DM, Gu L, Xia YR, Navab M, Li WF, Hama S, Castellani LW, Furlong CE, Costa LG, Fogelman AM, Lusis AJ (1998) Mice lacking serum paraoxonase are susceptible to organophosphate toxicity and atherosclerosis. Nature 394(6690):284–287

Shih DM, Xia YR, Wang XP, Miller E, Castellani LW, Subbanagounder G, Cheroutre H, Faull KF, Berliner JA, Witztum JL, Lusis AJ (2000) Combined serum paraoxonase knockout/apolipoprotein E knockout mice exhibit increased lipoprotein oxidation and atherosclerosis. J Biol Chem 275(23):17527–17535

Shin JY, Yang Y, Heo P, Lee JC, Kong B, Cho JY, Yoon K, Shin CS, Seo JH, Kim SG, Kweon DH (2013) pH-responsive high-density lipoprotein-like nanoparticles to release paclitaxel at acidic pH in cancer chemotherapy. Int J Nanomed 7:2805–2816

Singh IP, Chopra AK, Coppenhaver DH, Anantharamaiah GM, Baron S (1999) Lipoproteins account for part of the broad non-specific antiviral activity of human serum. Antiviral Res 42(3):211–218

Soran H, Younis NN, Charlton-Menys V, Durrington P (2009) Variation in paraoxonase-1 activity and atherosclerosis. Curr Opin Lipidol 20(4):265–274

Sparks DL, Davidson WS, Lund-Katz S, Phillips MC (1995) Effects of the neutral lipid content of high density lipoprotein on apolipoprotein A-I structure and particle stability. J Biol Chem 270(45):26910–26917

Srinivas RV, Birkedal B, Owens RJ, Anantharamaiah GM, Segrest JP, Compans RW (1990) Antiviral effects of apolipoprotein A-I and its synthetic amphipathic peptide analogs. Virology 176(1):48–57

Stocker R, Keaney JF Jr (2004) Role of oxidative modifications in atherosclerosis. Physiol Rev 84(4):1381–1478

Stoltz DA, Ozer EA, Taft PJ, Barry M, Liu L, Kiss PJ, Moninger TO, Parsek MR, Zabner J (2008) Drosophila are protected from *Pseudomonas aeruginosa* lethality by transgenic expression of paraoxonase-1. J Clin Invest 118(9):3123–3131

Subbaiah PV, Liu M (1993) Role of sphingomyelin in the regulation of cholesterol esterification in the plasma lipoproteins. Inhibition of lecithin-cholesterol acyltransferase reaction. J Biol Chem 268(27):20156–20163

Tang WH, Hartiala J, Fan Y, Wu Y, Stewart AF, Erdmann J, Kathiresan S, Roberts R, McPherson R, Allayee H, Hazen SL (2012) Clinical and genetic association of serum paraoxonase and arylesterase activities with cardiovascular risk. Arterioscler Thromb Vasc Biol 32(11):2803–2812

Teeguarden J, Hanson-Drury S, Fisher JW, Doerge DR (2013) Are typical human serum BPA concentrations measurable and sufficient to be estrogenic in the general population? Food Chem Toxicol 62:949–963

Teiber JF, Draganov DI, La Du BN (2004) Purified human serum PON1 does not protect LDL against oxidation in the in vitro assays initiated with copper or AAPH. J Lipid Res 45(12):2260–2268

Teiber JF, Horke S, Haines DC, Chowdhary PK, Xiao J, Kramer GL, Haley RW, Draganov DI (2008) Dominant role of paraoxonases in inactivation of the *Pseudomonas aeruginosa* quorum-sensing signal N-(3-oxododecanoyl)-L-homoserine lactone. Infect Immun 76(6):2512–2519

Troughton JA, Woodside JV, Yarnell JW, Arveiler D, Amouyel P, Ferrieres J, Ducimetiere P, Patterson CC, Luc G (2008) Paraoxonase activity and coronary heart disease risk in healthy middle-aged males: the PRIME study. Atherosclerosis 197(2):556–563

Tsuchida S, Satoh M, Umemura H, Sogawa K, Kawashima Y, Kado S, Sawai S, Nishimura M, Kodera Y, Matsushita K, Nomura F (2012) Proteomic analysis of gingival crevicular fluid for discovery of novel periodontal disease markers. Proteomics 12(13):2190–2202

Tward A, Xia YR, Wang XP, Shi YS, Park C, Castellani LW, Lusis AJ, Shih DM (2002) Decreased atherosclerotic lesion formation in human serum paraoxonase transgenic mice. Circulation 106(4):484–490

Underhill DM, Ozinsky A (2002) Toll-like receptors: key mediators of microbe detection. Curr Opin Immunol 14(1):103–110

Vanhamme L, Paturiaux-Hanocq F, Poelvoorde P, Nolan DP, Lins L, Van Den Abbeele J, Pays A, Tebabi P, Van Xong H, Jacquet A, Moguilevsky N, Dieu M, Kane JP, De Baetselier P, Brasseur R, Pays E (2003) Apolipoprotein L-I is the trypanosome lytic factor of human serum. Nature 422(6927):83–87

Vanhollebeke B, De Muylder G, Nielsen MJ, Pays A, Tebabi P, Dieu M, Raes M, Moestrup SK, Pays E (2008) A haptoglobin-hemoglobin receptor conveys innate immunity to *Trypanosoma brucei* in humans. Science 320(5876):677–681

Vila A, Korytowski W, Girotti AW (2002) Spontaneous transfer of phospholipid and cholesterol hydroperoxides between cell membranes and low-density lipoprotein: assessment of reaction kinetics and prooxidant effects. Biochemistry 41(46):13705–13716

Vishnyakova TG, Kurlander R, Bocharov AV, Baranova IN, Chen Z, Abu-Asab MS, Tsokos M, Malide D, Basso F, Remaley A, Csako G, Eggerman TL, Patterson AP (2006) CLA-1 and its splicing variant CLA-2 mediate bacterial adhesion and cytosolic bacterial invasion in mammalian cells. Proc Natl Acad Sci U S A 103(45):16888–16893

Voisset C, Op de Beeck A, Horellou P, Dreux M, Gustot T, Duverlie G, Cosset FL, Vu-Dac N, Dubuisson J (2006) High-density lipoproteins reduce the neutralizing effect of hepatitis C virus (HCV)-infected patient antibodies by promoting HCV entry. J Gen Virol 87(Pt 9):2577–2581

Webb NR, Connell PM, Graf GA, Smart EJ, de Villiers WJ, de Beer FC, van der Westhuyzen DR (1998) SR-BII, an isoform of the scavenger receptor BI containing an alternate cytoplasmic tail, mediates lipid transfer between high density lipoprotein and cells. J Biol Chem 273(24):15241–15248

Wheeler JG, Keavney BD, Watkins H, Collins R, Danesh J (2004) Four paraoxonase gene polymorphisms in 11212 cases of coronary heart disease and 12786 controls: meta-analysis of 43 studies. Lancet 363(9410):689–695

Williams KJ, Tabas I (1995) The response-to-retention hypothesis of early atherogenesis. Arterioscler Thromb Vasc Biol 15(5):551–561

Yan YJ, Li Y, Lou B, Wu MP (2006) Beneficial effects of ApoA-I on LPS-induced acute lung injury and endotoxemia in mice. Life Sci 79(2):210–215

Yoshikawa M, Sakuma N, Hibino T, Sato T, Fujinami T (1997) HDL3 exerts more powerful antioxidative, protective effects against copper-catalyzed LDL oxidation than HDL2. Clin Biochem 30(3):221–225

Zerrad-Saadi A, Therond P, Chantepie S, Couturier M, Rye KA, Chapman MJ, Kontush A (2009) HDL3-mediated inactivation of LDL-associated phospholipid hydroperoxides is determined by the redox status of apolipoprotein A-I and HDL particle surface lipid rigidity: relevance to inflammation and atherogenesis. Arterioscler Thromb Vasc Biol 29(12):2169–2175

Zhang X, Zhu X, Chen B (2010) Inhibition of collar-induced carotid atherosclerosis by recombinant apoA-I cysteine mutants in apoE-deficient mice. J Lipid Res 51(12):3434–3442

Signal Transduction by HDL: Agonists, Receptors, and Signaling Cascades

Jerzy-Roch Nofer

Contents

1	Introduction	231
2	ApoA-I-Induced Cell Signaling Directly Mediated by ABCA1	232
3	ApoA-I-Induced Cell Signaling Indirectly Mediated by β-ATPase and P2Y$_{12/13}$ ADP Receptor	237
4	ApoA-I- and HDL-Induced Cell Signaling Indirectly Mediated by ABCA1 and/or ABCG1	238
5	HDL-Induced Cell Signaling Mediated by SR-BI	239
6	HDL-Induced Cell Signaling Mediated by S1P	243
7	HDL-Induced Cell Signaling: Future Challenges and Opportunities	247
References		249

Abstract

Numerous epidemiologic studies revealed that high-density lipoprotein (HDL) is an important risk factor for coronary heart disease. There are several well-documented HDL functions such as reversed cholesterol transport, inhibition of inflammation, or inhibition of platelet activation that may account for the atheroprotective effects of this lipoprotein. Mechanistically, these functions are carried out by a direct interaction of HDL particle or its components with receptors localized on the cell surface followed by generation of intracellular signals. Several HDL-associated receptor ligands such as apolipoprotein A-I (apoA-I) or sphingosine-1-phosphate (S1P) have been identified in addition to HDL holoparticles, which interact with surface receptors such as ATP-binding cassette transporter A1 (ABCA1); S1P receptor types 1, 2, and 3 (S1P$_1$, S1P$_2$, and S1P$_3$); or scavenger receptor type I (SR-BI) and activate intracellular

---

J.-R. Nofer (✉)
Center for Laboratory Medicine, University Hospital Münster, Albert-Schweizer-Campus 1, Geb. A1, D-48149 Münster, Germany
e-mail: nofer@uni-muenster.de

© The Author(s) 2015
A. von Eckardstein, D. Kardassis (eds.), *High Density Lipoproteins*, Handbook of Experimental Pharmacology 224, DOI 10.1007/978-3-319-09665-0_6

signaling cascades encompassing kinases, phospholipases, trimeric and small G-proteins, and cytoskeletal proteins such as actin or junctional protein such as connexin43. In addition, depletion of plasma cell cholesterol mediated by ABCA1, ATP-binding cassette transporter G1 (ABCG1), or SR-BI was demonstrated to indirectly inhibit signaling over proinflammatory or proliferation-stimulating receptors such as Toll-like or growth factor receptors. The present review summarizes the current knowledge regarding the HDL-induced signal transduction and its relevance to athero- and cardioprotective effects as well as other physiological effects exerted by HDL.

Keywords

High-density lipoprotein (HDL) • Apolipoprotein A-I (apoA-I) • ATP-binding cassette transporter A1 (ABCA1) • Scavenger receptor type I (SR-BI) • Sphingosine-1-phosphate (S1P) • Signal transduction

Abbreviations

ABCA1	ATP-binding cassette transporter A1
ABCG1	ATP-binding cassette transporter G1
AMPK	AMP-activated protein kinase
Apo	Apolipoprotein
CAMK	Calcium/calmodulin-dependent protein kinase
cAMP	Cyclic adenosine monophosphate
Cdc42	Cell division control protein 42
CHD	Coronary heart disease
CKD	Chronic kidney disease
COX-2	Cyclooxygenase-2
DAG	diacylglycerol
DHEA	Dehydroepiandrosterone
EL	Endothelial lipase
eNOS	Endothelial nitric oxide synthase
GLUT4	Glucose transporter type 4
GM-CSF	Granulocyte macrophage colony-stimulating factor
HDL	High-density lipoprotein
HSPC	Hematopoietic stem progenitor cells
IL	Interleukin
JAK2	Janus kinase 2
JNK	c-Jun N-terminal kinase
LCAT	Lecithin:cholesterol acyltransferase
LDL	Low-density lipoprotein

LPS	Lipopolysaccharide
MAPK	Mitogen-activated protein kinase
MCP1	Monocyte chemotactic protein 1
MDCK	Madin-Darby canine kidney
PC	Phosphatidylcholine
PC-PLC	PC-specific phospholipase C
PC-PLD	PC-specific phospholipase D
PDZK1	PDZ domain-containing adaptor protein
PGI_2	Prostacyclin
PHD	Prolyl hydroxylase
PI3K	Phosphatidylinositol 3-kinase
PI-PLC	Phosphatidylinositol-specific phospholipase C
PKA	Protein kinase A
PKC	Protein kinase C
S1P	Sphingosine-1-phosphate
SAA	Serum amyloid A
SDMA	Symmetric dimethylarginine
SK	Sphingosine kinase
SPC	Sphingosylphosphorylcholine
SPM	Sphingomyelin
SR-BI	Scavenger receptor type I
STAT3	Signal transducer and activator of transcription 3
TGFβ	Transforming growth factor-β
TLR	Toll-like receptor
TNFα	Tumor necrosis factor α
β-ATPase	F1F0 ATP synthase β-subunit

1 Introduction

Innumerable epidemiological studies document the inverse relationship between plasma levels of high-density lipoprotein (HDL) cholesterol and cardiovascular risk. The first hypothesis aiming to mechanistically explain this relationship has been formulated by John Glomset (1968), who proposed that HDL particles take up excess cholesterol in peripheral tissues including atherosclerotically changed arterial walls and transport it to the liver for excretion with bile. The first step in this process termed reverse cholesterol transport encompassed direct interaction between HDL and cell membrane with the subsequent efflux of cholesterol to lipoprotein particles. Early concepts envisioned cholesterol efflux as a purely physicochemical phenomenon with the unidirectional flux of free cholesterol perpetuated by a steady concentration gradient maintained by the activity of lecithin:cholesterol acyltransferase (LCAT) in HDL particles. John F. Oram was the first to note that HDL binding to specific binding sites precedes cholesterol efflux and is required for the effective mobilization of free cholesterol from internal pools and its transport to the plasma membrane, where it is available for further transfer to lipoprotein particles (Graham and Oram 1987; Slotte et al. 1987). The seminal discovery of Oram put HDL function in a fully new perspective: ever since then HDL has been perceived not only as a cholesterol transporter but also as hormone-like particle exerting its activities through binding to specific receptors.

Signal transduction occurs when an extracellular signaling molecule activates a surface receptor, which in turn alters intracellular molecules creating a response, which ultimately modifies cellular functions. The recognition of specific HDL binding sites raised immediately a question, whether HDL-cell interaction is linked to signaling events relevant to intracellular cholesterol transport and/or cholesterol efflux. It was Oram again who demonstrated that the exposure of fibroblasts to HDL causes activation of protein kinase C (PKC), which in turn contributes to translocation of free cholesterol between intracellular membranes and plasmalemma (Mendez et al. 1991). This finding opened a new field of research focusing on dissection of intracellular signaling pathways that are induced by HDL or its constituents and mediate cellular responses to lipoprotein particles. In the following years, several ligands present in HDL and their receptors expressed on the cell surface could be identified that mediate one or more physiological effects attributed to this lipoprotein. In addition, the interaction of HDL with the cell membrane was found to potently modify signal transduction induced by other ligands such as pathogen-associated molecular patterns (PAMPs) or growth factors. This review attempts to summarize the current state of knowledge on HDL-induced signal transduction and its physiological relevance.

2   ApoA-I-Induced Cell Signaling Directly Mediated by ABCA1

ATP-binding cassette transporter A1 (ABCA1) has been originally identified as a protein defective in familial HDL deficiency (Tangier disease), which is characterized on the molecular level by the failure of apoA-I to mobilize cholesterol from intracellular stores, to induce cholesterol and phospholipid efflux from cells, and ultimately to initiate the maturation of HDL particles and the HDL-mediated reversed cholesterol transport (Bodzioch et al. 1999; Brooks-Wilson et al. 1999; Rust et al. 1999). Early models explaining ABCA1-mediated lipid efflux from cells assumed that this transporter facilitates translocation of cholesterol and phospholipids to the exofacial leaflet of the cell membrane, where they became accessible for the passive transfer to apoA-I. Further studies, however, provided several pieces of evidence suggesting that apoA-I may directly interact with ABCA1 for the effective unloading of cell cholesterol. For instance, cells obtained from patients with Tangier disease displayed reduced number of apoA-I binding sites, whereas increased apoA-I binding to the cell surface could be observed in cells overexpressing full-length cDNA of ABCA1 (Francis et al. 1995; Wang et al. 2000). Moreover, studies utilizing chemical cross-linking indicated that apoA-I and ABCA1 are in a very close proximity (<7 Å) characteristic of ligand-receptor interaction, while naturally occurring mutations in extracellular ABCA1 loops profoundly impaired the cross-linking efficiency (Wang et al. 2000; Fitzgerald et al. 2002, 2004; Chroni et al. 2004). The binding of apoA-I to ABCA1 was subsequently found to increase the cell content of ABCA1 protein but not the ABCA1 mRNA abundance suggesting that apoA-I-ABCA1 interaction might be critical to ABCA1 stabilization (Wang et al. 2003). Actually, apoA-I was found to promote both the dephosphorylation of the proline-, glutamic acid-, serine-, and threonine-rich (PEST) sequence in the cytoplasmatic domain of ABCA1 and the recruitment of calmodulin to the specific 1-5-8-14 binding motif localized in a close vicinity to the PEST sequence (Wang et al. 2003; Martinez et al. 2003a; Iwamoto et al. 2010). In this manner, apoA-I binding appears to protect ABCA1 against calpain-mediated proteolysis. In addition, lipid-free apoA-I was found to retard ABCA1 degradation mediated by thiol proteases and thereby to facilitate ABCA1 accumulation on the cell surface (Arakawa and Yokoyama 2002).

The recognition of the apoA-I-ABCA1 complex formation and its stabilization as a mechanism protecting against protease-mediated degradation gave rise to the hypothesis that the interaction of apoA-I with ABCA1 may trigger intracellular signaling cascades that modulate ABCA1 levels and/or ABCA1-mediated lipid transport activity in a posttranslational manner (see Fig. 1 for schematic presentation of ABCA1-induced signaling). Early studies documented that apoA-I enhances phosphatidylcholine (PC) turnover in a specific process mediated by a G-protein-dependent activation of PC-specific phospholipases C and D (PC-PLC and PC-PLD) (Walter et al. 1995, 1996). Later investigations attributed the apoA-I-mediated activation of PC-PLC rather to ABCA1-dependent efflux of sphingomyelin (SPM) and the unspecific depletion of plasma membrane SPM content (Yamauchi et al. 2003). In both cases, stimulation of PC-PLC liberates diacylglycerol (DAG), which is the primary physiological activator of PKC. The activation of both PC-specific phospholipases and PKC seems to be critical for ABCA1-mediated lipid transport. Actually, pharmacological interference with PC-PLC and PC-PLD activation reduced apoA-I-induced cholesterol efflux, whereas exogenous application of their products DAG or phosphatidic acid (PA) partially corrected the cholesterol efflux defect in cells obtained from patients with Tangier disease (Walter et al. 1996; Haidar et al. 2001). Similarly, PKC inhibitors and activators were found, respectively, to block or to enhance apoA-I-induced cholesterol efflux from various cell types including macrophages, smooth muscle cells, and fibroblasts (Mendez et al. 1991; Li and Yokoyama 1995; Li et al. 1997). More recently, it has been suggested that PKC specifically controls the translocation of cholesterol from acylCoA:cholesterol acyltransferase (ACAT)-accessible pools to the as yet poorly defined intracellular cholesterol compartment, where it is available for the ABCA1-mediated release (Yamauchi et al. 2004). In addition, PKC isoform alpha (PKCα) was found to increase ABCA1 phosphorylation at as yet unknown serine sites and thereby to enhance it protection against calpain-mediated proteolysis (Yamauchi et al. 2003). Apelin, a bioactive peptide modulating vascular tone and blood pressure and displaying strong antiatherogenic

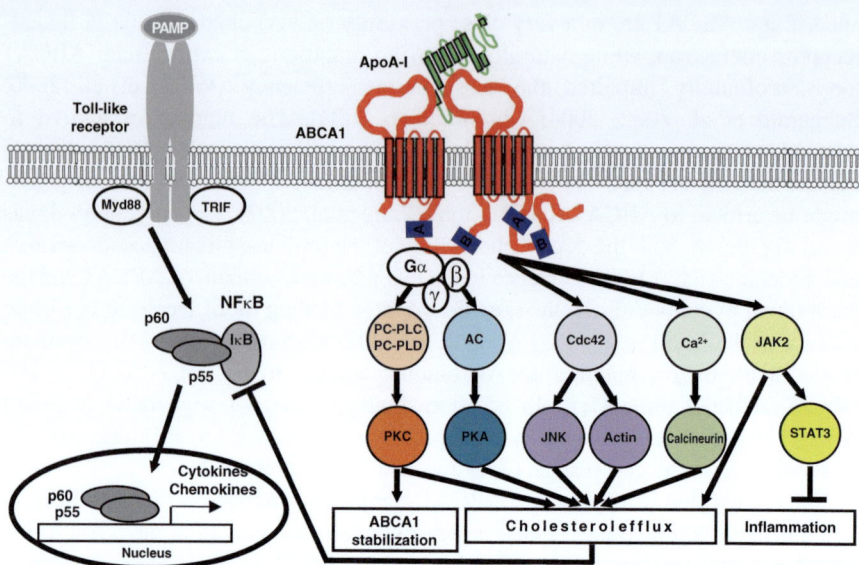

Fig. 1 ApoA-I-induced cell signaling mediated by ABCA1. The binding of apolipoprotein A-I (apoA-I) to lipid transporter ABCA1 induces the trimeric G-protein-mediated activation of phosphatidylcholine-specific phospholipases C and/or D (PC-PLC, PC-PLD) and adenylate cyclase (AC) and the ensuing activation of protein kinases C and A (PKC, PKA). Direct interaction of the small G-protein Cdc42 with ABCA1 induces activation of protein kinase JNK and actin polymerization. Direct interaction of Janus kinase-2 (JAK2) with ABCA1 leads to its autophosphorylation and activation as well as activation of transcription factor signal transducer and activator of transcription 3 (STAT3). These effects account for cholesterol efflux-inducing, anti-inflammatory, and ABCA1-stabilizing properties of apoA-I. In addition, ABCA1-mediated cholesterol efflux to apoA-I exerts indirect negative effect on signal transduction mediated by proinflammatory Toll-like receptors and the resulting activation of the transcription factor NF-κB and expression of cyto- and chemokines. *PAMP* Pathogen-associated molecular patterns

properties, has been recently found to phosphorylate serine residues in ABCA1 through the PKCα pathway and to inhibit calpain-mediated proteolysis, thereby promoting cholesterol efflux and reducing foam cell formation (Liu et al. 2013a).

ABCA1 is well known to be regulated by cyclic adenosine monophosphate (cAMP) at the transcriptional level, and cAMP responsive elements have been localized in the ABCA1 promoter. The interaction of apoA-I with ABCA1, in turn, was demonstrated to activate adenylate cyclase and to increase cAMP content, while severely reduced apoA-I-induced cAMP production was found in cells obtained from carriers of ABCA1 mutations (Haidar et al. 2004). These observations strongly suggested that cAMP and its downstream target protein kinase A (PKA) may represent an important component of apoA-I/ABCA1-induced signaling cascade regulating intracellular lipid translocation and/or their removal from cells. Actually, PKA phosphorylation sites were identified in the nucleotide binding domains of ABCA1, and the targeted mutation of one of these sites, S2054, significantly impaired apoA-I-mediated lipid efflux (See et al. 2002). Moreover, pharmacological modulation of cAMP/PKA signaling pathway exerted profound effects on apoA-I-induced cholesterol efflux. For instance, forskolin, an adenyl cyclase activator, and 8-Br-cAMP, a stable cAMP analog, substantially increased both ABCA1 phosphorylation and cholesterol efflux to apoA-I in wild-type cells and partly corrected the impaired efflux in cells obtained from patients with Tangier disease (Haidar et al. 2002). Similar effects were more recently observed in macrophages treated with Ht31, a PKA-anchoring inhibitor, which robustly elevated PKA activity in the cytoplasm, elevated ABCA1-dependent cholesterol export from cells, and completely reversed foam cell formation (Ma et al. 2011). Conversely, interference with cAMP/PKA signaling accomplished by treating cells with adenyl cyclase inhibitors, PKA inhibitors, or eicosapentaenoic acid, which decreased both cAMP levels and PKA activity, led to concomitant reduction of ABCA1 phosphorylation and ABCA1-mediated cholesterol efflux (Haidar et al. 2002, 2004; Hu et al. 2009). The molecular mechanisms underlying the stimulatory effect of PKA-mediated ABCA1 phosphorylation on the function of this transporter remain unclear. In contrast to PKC, PKA neither improves ABCA1 stability nor increases apoA-I binding. It seems more likely that phosphorylation of ABCA1 by the cAMP/PKA pathway allows the transporter to assume the conformation favoring more effective lipid translocation across the cell membrane (See et al. 2002).

Janus kinase 2 (JAK2) represents the third kinase, the activity of which is induced by ApoA-I/ABCA1 interaction. Even short exposure of cells expressing ABCA1 to apoA-I produces substantial autophosphorylation of JAK2, which is a prerequisite for further signal transmission to downstream located targets (Tang et al. 2004). Active JAK2 does not seem to directly phosphorylate ABCA1, and it is more likely that it targets an as yet unknown ABCA1 partner protein. Nevertheless, several experimental approaches clearly documented that JAK2 signaling is required for ABCA1 to effectively mediate lipid export from cells. For instance, JAK2-specific inhibitors substantially reduced apoA-I-induced cholesterol efflux, and mutant cells lacking JAK2 exhibited a severely impaired lipid transport (Tang et al. 2004, 2006). In addition, studies exploiting various ABCA1 variants revealed close interdependency between lipid export and JAK2 activation (Vaughan et al. 2009). In contrast to PKC and PKA, JAK2 activation by apoA-I neither promotes ABCA1 stability nor enhances intrinsic lipid translocase activity of the transporter. However, JAK2 activation seems to be critical for the optimum formation of apoA-I/ABCA1 complex, as JAK2-deficient cells exhibit severely impaired apoA-I binding despite normal ABCA1 protein levels (Tang et al. 2006).

JAK2 autophosphorylation is commonly followed by phosphorylation of several downstream targets, out of which signal transducer and activator of transcription 3 (STAT3) represents its canonical target. As ABCA1 possesses STAT3 docking sites, this molecule was an obvious candidate for linking JAK2 with ABCA1 activation. As a matter of fact, apoA-I induces STAT3 phosphorylation (which is tantamount to its activation) in an ABCA1-dependent fashion, and this effect is attenuated in cells expressing ABCA1, in which STAT3 docking sites have been mutated (Tang et al. 2009). However, neither complete STAT3 deficiency nor interruption of ABCA1-STAT3 interaction altered ABCA1 capacity to bind apoA-I and to efflux cholesterol from cells (Tang et al. 2009). Hence, apoA-I-induced STAT3 activation seems dispensable for the ABCA1-mediated lipid export. STAT3 represents an intracellular target of IL-10—a cytokine with well-established anti-inflammatory properties—and activation of STAT3 was demonstrated to suppress several aspects of inflammatory macrophage activation. These observations raised the interesting possibility that apoA-I might produce anti-inflammatory effects by activating STAT3 in ABCA1-dependent fashion. Actually, exposure of ABCA1-expressing macrophages to apoA-I suppressed lipopolysaccharide (LPS)-induced production of proinflammatory cytokines such as interleukin-6 (IL-6), and this effects were abolished after pharmacological STAT3 blockade or knocking down STAT3 expression with siRNA (Tang et al. 2009). Further investigations demonstrated that the suppressing effect of apoA-I on LPS-induced cytokine production was mediated by mRNA destabilization brought about by mRNA-destabilizing protein tristetraprolin, the expression of which was induced in cells exposed to apoA-I in an ABCA1- and JAK2/STAT3-dependent manner (Yin et al. 2011). Recently, apoA-I was demonstrated to promote cyclooxygenase-2 (COX-2) activation and prostacyclin (PGI_2) production in endothelial cells, and these anti-inflammatory effects were also found to depend on ABCA1 and JAK2 activation (Liu et al. 2011).

Cell division control protein 42 (Cdc42) is a member of a small G-protein family involved in the regulation of cytoskeleton organization and intracellular vesicular trafficking. Early studies revealed decreased Cdc42 expression and abnormal cytoskeleton architecture in cells obtained from Tangier patients (Hirano et al. 2000). Moreover, impaired cholesterol efflux observed in aged human fibroblasts as well as fibroblasts from Werner syndrome characterized by the early onset of senescent phenotypes including premature atherosclerotic cardiovascular disease could be corrected by forced expression of wild-type Cdc42 (Tsukamoto et al. 2002; Zhang et al. 2005). Expression of constitutively active Cdc42 variant in normal Madin-Darby canine kidney (MDCK) cells potentiated apoA-I-induced cholesterol efflux, whereas expression of dominant negative Cdc42 variants in normal MDCK cells or fibroblasts exerted opposite effects (Hirano et al. 2000; Nofer et al. 2003). ApoA-I-induced cholesterol efflux was also reduced in cells pretreated with *Clostridium difficile* toxin, which irreversibly modifies and disables Cdc42 (Nofer et al. 2003). Both Cdc42 and ABCA1 were found to co-segregate into lubrol-insoluble, triton-soluble lipid rafts (Drobnik et al. 2002). In addition, Cdc42 could be immunoprecipitated with wild-type ABCA1, but not with C-terminally truncated variant of this transporter indicating that both proteins interact with each other and suggesting that C-terminus of ABCA1 harbors a potential docking site for Cdc42 (Tsukamoto et al. 2001; Nofer et al. 2006). Exposure of cells expressing wild-type ABCA, but not cells obtained from Tangier patients or expressing C-terminally truncated variant of ABCA1 to apoA-I, induced Cdc42 activation and subsequent actin polymerization (Nofer et al. 2006). Taken together, these findings suggest that apoA-I signals through ABCA1 to activate Cdc42 and thereby to promote cholesterol efflux. How exactly Cdc42 activation promotes lipid export from cells remains to be elucidated. One possibility is that Cdc42 modulates intracellular vesicular traffic and thereby helps to supply cholesterol and/or phospholipids to an intracellular compartment, where they are available for the ABCA1-mediated release. The retardation of intracellular lipid transport in cells with reduced Cdc42 levels and its intensification after enforced expression of Cdc42 appears to support this concept (Tsukamoto et al. 2002). Alternatively, components of the signaling machinery localized downstream to Cdc42 might accelerate intracellular lipid transfer or the ABCA1-dependent lipid translocation across the cell membrane. Actually, apoA-I was demonstrated to produce consecutive activation of two serine/threonine kinases, namely, p21-activated kinase 1 (PAK1) and c-Jun N-terminal kinase (JNK) in ABCA1- and Cdc42-dependent manner (Nofer et al. 2003). The potential involvement of these two kinases in the processes of lipid export from cells has been not clarified to date.

Mobilization of intracellular Ca(2+) from intracellular stores or Ca(2+) influx from extracellular space represents an integral component of diverse intracellular signaling pathways. In one single report, evidence has been provided that apoA-I provokes extracellular Ca(2+) influx in an ABCA1-dependent manner and that inhibition of calcineurin, the downstream target of Ca(2+) influx, with cyclosporine A or FK506 completely abolished apoA-I lipidation and also interfered with JAK2 phosphorylation (Karwatsky et al. 2010). However, other authors failed to observe Ca(2+) mobilization in fibroblasts or smooth muscle cells after exposure to apoA-I or other apolipoproteins (apoA-II, apoC-III) (Nofer et al. 2000 and unpublished observations). Further studies are necessary, therefore, to fully understand the contribution of Ca(2+) mobilization to ABCA1-mediated signaling and lipid export.

3 ApoA-I-Induced Cell Signaling Indirectly Mediated by β-ATPase and P2Y$_{12/13}$ ADP Receptor

F1F0 ATP synthase is the terminal enzyme of the oxidative phosphorylation pathway responsible for the lion part of ATP synthesis in humans and other living being except in Archaea and present in mitochondria, chloroplasts, and prokaryotic membranes. Unexpectedly, the F1F0 ATP synthase (with F1 domain facing outside) has been found on the surface of several cells including hepatocytes, adipocytes, and endothelial cells, and its β-chain (β-ATPase) has been characterized as a receptor for apoA-I (Martinez et al. 2003b; Fabre et al. 2006; Radojkovic et al. 2009; Howard et al. 2011). The binding of apoA-I to β-ATPase induces ATP hydrolysis and promotes extracellular ADP generation, which in turn stimulates intracellular signaling by activating purinergic receptors, which belong to the family of G-protein-coupled receptors stimulated by extracellular nucleotides (Martinez et al. 2003b; Fabre et al. 2006). In hepatocytes, the β-ATPase/P2Y$_{13}$-induced signaling was reported to activate the small G-protein RhoA and the downstream-located Rho-associated protein kinase (ROCK-I) and thereby to promote HDL endocytosis by as yet poorly defined mechanism (Fabre et al. 2006;

Malaval et al. 2009). In endothelial cells, binding of apoA-I to β-ATPase and the subsequent generation of extracellular ADP stimulated endothelial proliferation and survival, but components of intracellular signaling cascade responsible for these effects remained uncharacterized (Radojkovic et al. 2009). In a recent study, the activation of β-ATPase/$P2Y_{12}$ was found to precede HDL binding and cell association as well as ABCG1- and SR-BI-dependent internalization, and transcytosis in endothelial cells (Cavelier et al. 2012). The identity of intracellular signaling pathway linking $P2Y_{12}$ activation to HDL internalization has not been addressed in this study.

4 ApoA-I- and HDL-Induced Cell Signaling Indirectly Mediated by ABCA1 and/or ABCG1

ABCA1 is unique among ABC transporters in that it requires the direct binding of acceptors for the transported substances. ATP-binding cassette transporter G1 (ABCG1) was also found to be a crucial mediator of lipid export. Studies by Wang et al. demonstrated that ABCG1 promotes efflux to a variety of acceptors, including HDL, low-density lipoprotein (LDL), phospholipid vesicles, and cyclodextrin (Wang et al. 2004, 2006). In major contrast to ABCA1, ABCG1 does not engage into direct interaction with lipid acceptors and for this reason cannot serve as a triggering element of intracellular second messenger cascade. However, recent studies document that ABCG1 (and to lesser extent ABCA1) profoundly impacts on signaling processes taking place in contiguity. Several studies showed that ABCA1 regulates cholesterol distribution between raft and non-raft membrane fractions and pointed to increased lipid raft formation in macrophages and other cells deficient in ABCG1 and ABCA1 (Landry et al. 2006; Sano et al. 2007; Koseki et al. 2007; Zhu et al. 2008a, Yvan-Charvet et al. 2008). Lipid rafts provide a molecular platform securing efficient signal transduction through several receptor families including proinflammatory Toll-like receptors (TLRs). Not surprisingly, therefore, both ABCA1-deficient cells and ABCG1-deficient cells, and in particular cells lacking both ABC transporters, exhibited increased cell surface expression of TLR4 and sensitivity to LPS stimulation (Koseki et al. 2007; Zhu et al. 2008a, Yvan-Charvet et al. 2008; Mogilenko et al. 2012). Also the NF-κB activation-dependent cytokine and chemokine release in response to stimulation with agonists of other TLR receptors (TLR2, TLR3) or other proinflammatory receptors (CD40) revealed to be increased in macrophages deficient in ABCA1 and/or ABCG1 (Sun et al. 2009; Yin et al. 2012). Such cells were also found to be prone to apoptosis during efferocytosis (phagocytosis of apoptotic cells) as compared to wild-type cells (Yvan-Charvet et al. 2010a). The enhanced apoptotic response observed in ABCA1- and ABCG1-deficient macrophages was dependent on an excessive assembly of NADPH oxidase/NOX2 complexes within lipid rafts followed by oxidative burst and ensuing sustained JNK activation, which turned on the apoptotic cell death program. It is very likely that similar mechanisms consisting in amplification of transmembrane signaling under conditions of decreased ABCG1 and/or ABCA1 activity and the resulting accumulation of lipid rafts may substantially contribute to the enhancement of several proinflammatory processes. For instance, cholesterol accumulation within lipid rafts was suggested to enhance monocyte adhesion and neutrophil activation, and both processes could be reversed by cholesterol removal from cells carried out either with help of synthetic cholesterol acceptors (e.g., cyclodextrin) or through induction of ABCA1-dependent cholesterol efflux (Murphy et al. 2008, 2011). In addition, ABCG1- and ABCA1-dependent modulation of cell membrane cholesterol content was demonstrated to critically regulate cellular signaling related to proliferatory responses. For instance, mice deficient in ABCG1- and ABCA1 developed severe monocytosis and neutrophilia, which was dependent on uncontrolled proliferation and expansion of hematopoietic stem progenitor cells (HSPC) (Yvan-Charvet et al. 2010b). The mechanistic analysis revealed increased plasma membrane lipid raft formation in these cells accompanied by elevated cell surface presence of interleukin IL-3/granulocyte macrophage colony-stimulating factor (GM-CSF) receptor. Addition of IL-3 or GM-CSF to ABCA1/G1-deficient bone marrow led to increased proliferation of HSPCs, reflecting activation of signaling pathways downstream of the IL-3/GM-CSF receptor. Similarly, defects in cholesterol efflux pathways in macrophages and dendritic cells in spleens of ABCA1/G1-deficient mice led to increased production of IL-23 and activation of a proinflammatory and proatherogenic signaling axis involving IL-23/IL-17 (Westerterp et al. 2012). Other studies showed increased lipid raft content in plasma membrane and enhanced proliferation in response to anti-CD3 antibody stimulation in T cells obtained from ABCG1-deficient animals (Bensinger et al. 2008). Similar increase in lymph node T-cell proliferation has been noted after disruption of cholesterol efflux pathway by deletion of apoA-I on a hypercholesterolemic background (Wilhelm et al. 2009).

5   HDL-Induced Cell Signaling Mediated by SR-BI

Scavenger receptor type I (SR-BI) is a cell surface glycoprotein originally identified by its homology to the scavenger receptor CD36 and capacity to bind acetylated LDL. Later studies in vitro revealed that SR-BI serves as the binding partner for a variety of ligands including native and oxidized LDL, very-low-density lipoproteins (VLDL), Lp(a), modified serum proteins such as maleylated albumin, lipid vesicles containing negatively charged phospholipids, apoptotic cells, and HDL. Despite this promiscuous binding behavior SR-BI is believed to primarily function as a physiologically relevant HDL receptor because of its tissue expression pattern reflecting sites of HDL cholesterol uptake and because of the profound impact of complete or tissue-specific SR-BI deficiency on HDL metabolism. The classic cellular function of SR-BI is to mediate the selective uptake of lipids from HDL and to transfer them into cells. Cholesteryl esters represent the primary substrate for SR-BI, but this receptor also facilitates transfer of various other lipid species including unesterified cholesterol, phospholipids, triglycerides, and lipid-soluble vitamins. SR-BI also mediates the efflux of cholesterol from cells to HDL, which in contrast to ABCA1- and ABCG1-mediated efflux is a passive, concentration gradient-dependent, and energy-independent process. Importantly, SR-BI-mediated lipid transfer from and to HDL occurs in the absence of consecutive internalization and degradation of HDL particle, which is a precedent condition for effective signal transduction (for recent reviews, see Hoekstra et al. 2010; Mineo and Shaul 2012).

Binding of HDL to SR-BI has been demonstrated to trigger several intracellular signaling events (see Fig. 2). The recruitment of non-receptor tyrosine kinase src to the C-terminal cytoplasmic tail of SR-BI followed by src autophosphorylation and phosphorylation and activation of AMP-activated protein kinase (AMPK) mediated through calcium/calmodulin-dependent protein kinase (CAMK) and/or serine/threonine liver kinase B1 (LKB1) appear to constitute initial steps in SR-BI signaling

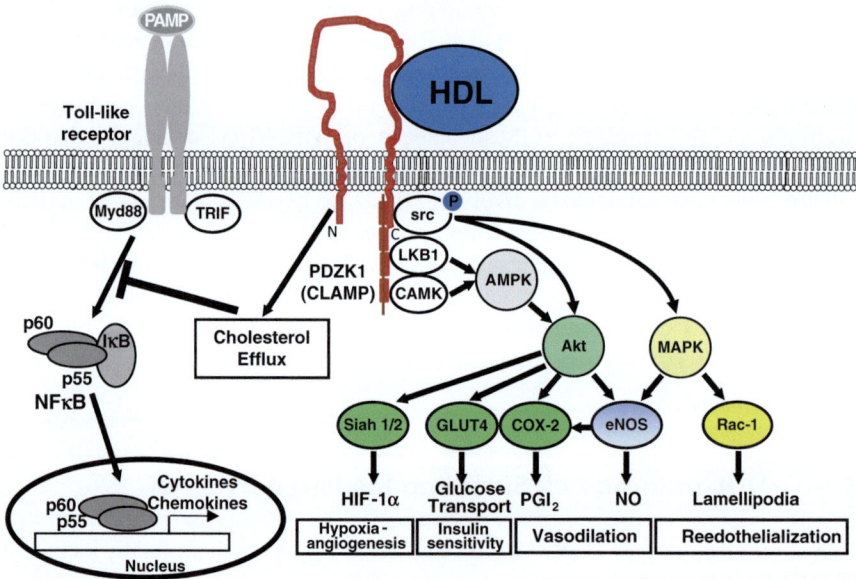

Fig. 2 HDL-induced cell signaling mediated by SR-BI. The binding of HDL to scavenger receptor type 1 (SR-BI) induces activation of protein kinases src, liver kinase B1 (LKB1), and calcium calmodulin-dependent protein kinase (CAMK) in a process dependent on an adapter protein PDZK1 (CLAMP). The resulting activation of AMP-activated protein kinase (AMPK) and protein kinase Akt as well as mitogen-activated protein kinase (MAPK) transduces signal to downstream effectors including endothelial nitric oxide synthase (eNOS) generating NO, cyclooxygenase-2 (COX-2) producing prostacyclin (PGI$_2$), glucose transporter GLUT4 mediating cellular glucose uptake, ubiquitin ligases Siah 1 and 2 stabilizing hypoxia-inducible factor-1α (HIF-1α), and small G-protein Rac-1 regulating formation of lamellipodia. These effects account for vasodilatory, insulin-sensitizing, proangiogenic, and endothelial regenerating properties of HDL. In addition, SR-BI-mediated cholesterol efflux to HDL exerts indirect negative effects on signal transduction mediated by proinflammatory Toll-like receptors and the resulting activation of the transcription factor NF-κB and expression of cyto- and chemokines. Pathogen-associated molecular patterns (PAMP)

(Mineo et al. 2003; Kimura et al. 2010). Detailed studies revealed that phosphorylations of src (though not its interaction with SR-BI) and AMPK are critically dependent on the presence of the PDZ domain-containing adaptor protein, PDZK1 (also termed CLAMP), which interacts with extreme C-terminal residues of SR-BI and is essential for its stability in the plasma membrane (Kimura et al. 2010; Assanasen et al. 2005; Zhu et al. 2008a, b). Src and AMPK activations lead to the activation of phosphatidylinositol 3-kinase (PI3K), which in turn induces parallel activation of protein kinase B (Akt) and mitogen-activated protein kinase (MAPK) (Kimura et al. 2010; Assanasen et al. 2005; Zhu et al. 2008b). The latter event seems to be required for the subsequent stimulation of the activity of small G-protein Rac1, which promotes formation of lamellipodia and induces cell shape change (Seetharam et al. 2006). It remains currently unclear, whether SR-BI-mediated lipid transfer is required for the effective initiation of signaling by HDL. The C-terminal transmembrane domain of SR-BI contains a cholesterol-binding site, which interacts with plasma cholesterol, as could be documented in photocholesterol binding experiments (Assanasen et al. 2005). Studies utilizing SR-BI/CD36 chimeras showed that only transfer of both C-terminal domains (cytoplasmatic and transmembrane) endows CD36 with signaling properties of SR-BI, whereas the transfer of the C-terminal cytoplasmatic domain only is insufficient (Assanasen et al. 2005). In addition, introduction of mutations into the transmembrane C-terminal domain severely impeded SR-BI-mediated signal transduction (Saddar et al. 2013). Other studies pointed to the enhancement of SR-BI signaling in the presence of extracellular cholesterol acceptors such as cyclodextrins or reconstituted HDL particles containing 2 molecules of apoA-I but not cholesterol (Assanasen et al. 2005). Furthermore, HDL enriched in phosphatidylcholine, which increases its cholesterol acceptor capacity, induces signal transduction at concentrations that are otherwise insufficient to invoke signaling. Collectively, these observations suggest that SR-BI-mediated cholesterol efflux exerts at least potentiating effects on HDL-triggered signaling transduced by this receptor. Recently, Al-Jarallah and Trigatti hypothesized that SR-BI-mediated removal of cholesterol from caveolae might suppress the activity of high molecular weight phosphatase complex, which would in turn help to maintain protein kinase targets such as MAPK in the phosphorylated state and to prolong their activation (Al-Jarallah and Trigatti 2010). Whether this mechanism indeed contributes to HDL-induced signaling mediated by SR-BI remains to be clarified.

The concerted activation of both Akt and MAPK appears to constitute the centerpiece of SR-BI-triggered signaling cascades. Both kinases are required for phosphorylation and full activation of endothelial nitric oxide synthase (eNOS), nitric oxide generation, and vasorelaxation, which are observed in endothelial cells or isolated aortas in the presence of HDL (Yuhanna et al. 2001; Mineo et al. 2003; Assanasen et al. 2005). In addition, PI3K/Akt/eNOS signaling was found to be involved in the HDL-induced COX-2 expression and PGI_2 release in endothelial cells (Zhang et al. 2012). The proliferation of bone-derived mesenchymal stem cells in response to HDL was likewise mediated via its binding to scavenger receptor-B type I and activation of PI3K/Akt/MAPK pathways (Xu et al. 2012). Activation of

MAPK but not Akt with subsequent induction of Rac1-GTPase was demonstrated to promote HDL-induced endothelial cell migration, which is critical for carotid artery reendothelialization after perivascular electric injury (Seetharam et al. 2006). Conversely, activation of Akt but not MAPK was necessary to sustain the migration and the proliferation of breast cancer cells (Danilo et al. 2013). In adipocytes, SR-BI-mediated activations of Akt and AMPK were found to promote glucose uptake and translocation of the glucose transporter GLUT4 to the plasma membrane suggesting that SR-BI signaling might contribute to insulin-sensitizing effects of HDL (Zhang et al. 2011). Finally, augmentation of hypoxia-induced angiogenesis, which is observed in the presence of HDL, was demonstrated to depend on SR-BI-mediated activation of Akt linked to increased expression of E3 ubiquitin ligases Siah1 and Siah2 and accelerated degradation of prolyl hydroxylases (PHD1-PHD3) combined with stabilization of hypoxia-inducible factor 1α (HIF-1α) (Tan et al. 2013). Recent studies provide evidence that similar to ABCG1 also SR-BI may indirectly affect proinflammatory signaling by reducing membrane cholesterol content. Actually, macrophages obtained from SR-BI-deficient animals were characterized by increased inflammatory responses to Toll-like receptor agonists (Guo et al. 2009; Cai et al. 2012). Feng et al. found that SR-BI deficiency enhanced lymphocyte proliferation, caused imbalanced interferon-γ (IFNγ) and IL-4 production in lymphocytes, and led to elevated inflammatory cytokine production in macrophages (Feng et al. 2011). Umemoto et al. demonstrated that HDL-induced disruption of lipid raft in adipocytes, which is accompanied by reduced activation of NADPH oxidase and production of monocyte chemotactic protein 1 (MCP1), was reversed by silencing both ABCG1 and SR-BI expression (Umemoto et al. 2013).

It is well established that HDL inhibits the activation of platelets by strong agonists such as thrombin and collagen (for review, see Nofer and van Eck 2011). The presence of SR-BI has been identified on platelet surface (Imachi et al. 2003). Recent investigations demonstrated that HDL binding to platelets was reduced in a concentration-dependent fashion by SR-BI ligands such as negatively charged liposomes, which in addition potently inhibited thrombin-induced platelet aggregation, granule secretion, fibrinogen binding, and mobilization of intracellular Ca^{2+} (Nofer et al. 2011). Furthermore, both native HDL and other SR-BI ligands failed to inhibit thrombin-induced activation of platelets obtained from SR-BI-deficient mice. These findings together with the results of Imachi et al. (2003), who observed an inverse relationship between the levels of SR-BI expression and the platelet aggregation in response to ADP, corroborate the hypothesis that SR-BI is a true functional HDL receptor on platelets. The likely consequence of the HDL binding to SR-BI on platelets is the induction of intracellular second messenger cascade. It has been previously demonstrated that incubation of platelets with HDL leads to the release of DAG from the plasma membrane phosphatidylcholine and to subsequent activation of PKC, which in turn stimulates the Na^+/H^+ antiport, promotes the alkalization of the cytoplasmic compartment, and inhibits the release of calcium ions from intracellular storage sites (Nofer et al. 1996). In addition, activation of PKC inhibits the activity of phosphatidylinositol-specific phospholipase C (PI-PLC), which is one of the most important signal transduction mediators of agonists such as thrombin and collagen (Nofer et al. 1998). It is conceivable that this signaling cascade is launched by interaction of HDL with SR-BI on platelet surface.

6   HDL-Induced Cell Signaling Mediated by S1P

Early studies on HDL-induced signal transduction focused on apoA-I or other apolipoproteins as potential binding partners for HDL receptor responsible for initiating the cascade of intracellular signaling events. However, it has soon become clear that neither apoA-I nor other proteins encountered in HDL particles can exclusively account for HDL signaling activities. Studies utilizing fibroblasts demonstrated that PI-PLC activation, which is observed in cells exposed to HDL and accompanied by increased phosphatidylinositol bisphosphate (PIP_2) turnover and the mobilization of intracellular calcium, is not emulated by purified apoA-I or apoA-I-containing proteoliposomes, but occurs in the presence of lipids extracted from HDL holoparticles (Nofer et al. 2000). Likewise, certain cellular responses triggered by HDL in fibroblasts of smooth muscle cells such as cell proliferation or proliferation-linked activation of MAPK could be observed in the presence of HDL lipids or HDL particles, in which proteins were covalently modified or enzymatically degraded, but not in the presence of proteins isolated from HDL (Deeg et al. 1997; Sachinidis et al. 1999; Nofer et al. 2001a). The intense search for lipid species accounting for the signal-inducing capacity of HDL revealed that these lipoproteins serve as a carrier for bioactive lysosphingolipids such as sphingosine-1-phosphate (S1P), sphingosylphosphorylcholine (SPC), and lysosulfatide (LSF) and that signaling events observed in cells exposed to native HDL or isolated lysosphingolipids largely overlap (Sachinidis et al. 1999; Nofer et al. 2000, 2004; Kimura et al. 2001).

S1P is the most prominent and best characterized member of the lysosphingolipid family. It is produced by phosphorylation of sphingosine by sphingosine kinases 1 and 2 (SK1 and SK2) in response to a variety of stimuli and degraded to phosphoethanolamine and hexadecanal by S1P-lyase or to sphingosine and organic phosphate by S1P phosphatase. S1P interacts with five related G-protein-coupled receptors termed $S1P_{1-5}$, which regulate a wide spectrum of cellular functions, including proliferation and survival, cytoskeletal rearrangements, and cell motility, and exert potent cytoprotective effects. In the vasculature, S1P receptors were identified on endothelial and smooth muscle cells, and S1P provides essential contribution to the new vessel formation and the maintenance of vascular barrier integrity. In addition, S1P was demonstrated to interfere with proliferation, migration, and activation of monocytes/macrophages and to regulate their recruitment to sites of inflammation (for recent reviews, see Maceyka et al. 2012; Levkau 2013). Erythrocytes and platelets are major sources of S1P in plasma, where it is preferentially (up to 60 %) associated with HDL subfraction rich in apolipoprotein M (apoM) (Dahm et al. 2006; Hänel et al. 2007; Christoffersen et al. 2011). Recent studies demonstrate that apoM specifically binds S1P and stimulates its biosynthesis for secretion (Liu et al. 2013b). There is substantial evidence suggesting that S1P may directly contribute to atheroprotective effects attributed to HDL. For instance, S1P closely correlates with HDL in a concentration range, in which HDL most effectively protects against atherosclerosis, and decreased HDL-bound S1P levels were noted in patients with coronary artery disease and myocardial infarction (Karuna et al. 2011; Sattler et al. 2010; Argraves et al. 2011). In addition, S1P was found to emulate in vitro several atheroprotective effects of HDL, which themselves could be attenuated by interrupting signaling through S1P receptors (Levkau et al. 2013).

Protein kinases MAPK, AMPK, and Akt are critical for the S1P-dependent signaling cascade, which unfolds in endothelial cells in the presence of HDL (see Fig. 3 for schematic presentation of S1P-mediated signaling). Kimura and

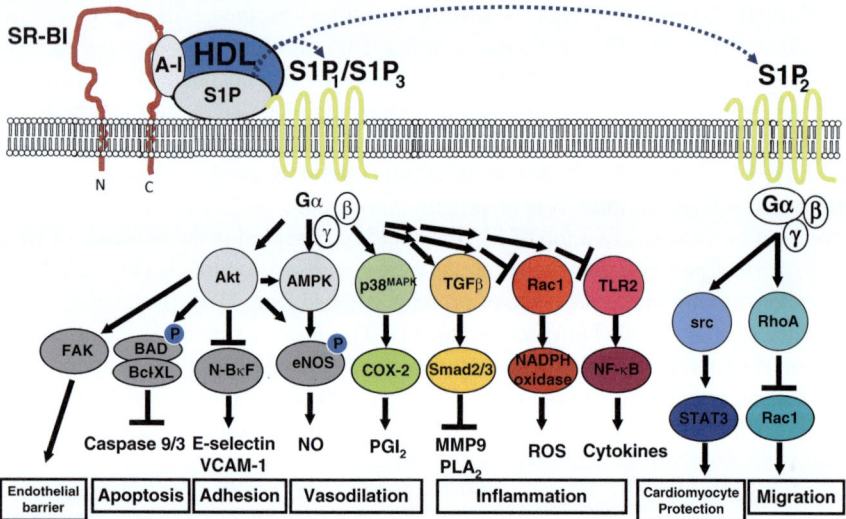

Fig. 3 HDL-induced cell signaling mediated by S1P. The binding of HDL to scavenger receptor type I (SR-BI) may provide a molecular platform for efficient receptor stimulation by sphingosine-1-phosphate (S1P) present in HDL molecule. Activation of AMP-activated protein kinase (AMPK) and protein kinase Akt via $S1P_1$ and/or $S1P_3$ receptors leads to the stimulation of focal adhesion kinase (FAK), the inactivation of proapoptotic protein BAD and the inhibition of caspases 9 and 3, the inhibition of the transcription factor NF-κB and the expression of adhesive molecules E-selectin and vascular cell adhesion molecule 1 (VCAM-1), as well as the activation of endothelial nitric oxide synthase (eNOS) and the generation of NO. Activation of the mitogen-activated protein kinase p38MAPK results in activation of cyclooxygenase-2 (COX-2) and production of prostacyclin (PGI_2). Increased expression of transforming growth factor β (TGFβ) contributes to the increased activation of transcription factors Smad 2 and 3, which suppress production of proinflammatory matrix metalloproteinases such as MMP9 as well as phospholipase A_2 (PLA_2). Inhibition of small G-protein Rac1 suppresses activity of NADPH oxidase and generation of reactive oxygen species. Inhibition of Toll-like receptor 2 (TLR2) reduces NF-κB activation and production of proinflammatory cytokines. These effects account for endothelial barrier-enhancing, anti-inflammatory, antiapoptotic, anti-adhesive, and vasodilatory properties of HDL. Activation of protein kinase src and transcription factor signal transducer and activator of transcription 3 (STAT3) via $S1P_2$ protects against cardiomyocyte apoptosis. Activation of the small G-protein RhoA via $S1P_2$ leads to inhibition of the small G-protein Rac1, which regulates smooth muscle cell migration

colleagues demonstrated that activation of MAPK is essential for endothelial proliferation and survival mediated by HDL-associated lysosphingolipids (Kimura et al. 2001). Using antisense oligonucleotides and siRNA strategy combined with application of pharmacological G-protein inhibitors, they further show that S1P receptor types 1 and 3 and inhibitory G-protein (Gi) contribute to HDL-mediated induction of endothelial migration, whereas the augmentation of endothelial survival in endothelial cells exposed to HDL is exclusively dependent on $S1P_1$. Miura et al. demonstrated that HDL promotes endothelial cell tube formation via activation of MAPK and suggested that this effect is mediated by S1P present in HDL particles (Miura et al. 2003). Later these authors showed that reconstituted HDL particles enriched in S1P are particularly effective in stimulating angiogenic response (Matsuo et al. 2007). HDL-like lipoproteins are also present in follicular fluid and induce angiogenesis in the S1P- and MAPK-dependent manner (von Otte et al. 2006). The activation of AMPK, which itself was entirely dependent on the parallel activation of CAMK, was localized upstream to Akt in the signaling sequel (Kimura et al. 2010). The stimulation of Akt by HDL-associated lysosphingolipids was—depending on the experimental system used—mediated by either $S1P_1$ or $S1P_3$ and controlled an array of important antiatherogenic activities attributed to HDL. Nofer et al. (2001b) found that activation of Akt by HDL and its associated lysosphingolipids disabled a proapoptotic protein BAD and thereby inhibited the collapse of mitochondrial potential, the cytochrome C release from mitochondria, and the ensuing activation of apoptosis-executing caspases 3 and 9. Nofer et al. (2004) and Kimura et al. (2006) demonstrated that the S1P content in HDL at least partially accounts for the HDL-stimulated activation of eNOS, generation of NO, and vasorelaxation. Stimulation of endothelial cells with statins such as pitavastatin or simvaststin increased the expression of $S1P_1$ and potentiated the stimulatory effect of HDL on eNOS activation (Igarashi et al. 2007; Kimura et al. 2008). Likewise, upregulated $S1P_1$ and $S1P_3$ expression and increased Akt and eNOS activation were found in fenofibrate-treated mice, which were characterized by elevated plasma levels of HDL and S1P (Krishna et al. 2012). Kimura et al. (2006) reported that signaling through $S1P_1$ receptor partly accounts for the inhibitory effects exerted by HDL on tumor necrosis factor α (TNFα)-induced NF-κB activation as well as endothelial expression of adhesins such as vascular cell adhesion molecule-1 (VCAM-1) and intercellular adhesion molecule-1 (ICAM-1). Argraves et al. (2008) found that HDL increased endothelial barrier integrity as measured by electric cell substrate impedance sensing in a process that was entirely dependent on S1P, $S1P_1$ expression, and Akt activation. The favorable effect of S1P on endothelial barrier was sustained for considerably longer time with HDL-associated rather than albumin-associated S1P, and this was related to the inhibitory effect of HDL on $S1P_1$ degradation (Wilkerson et al. 2012). Norata et al. reported that HDL potently stimulates endothelial expression of long pentraxin 3—a key component of innate immunity—as well as transforming growth factor-β (TGFβ), a compound with potent anti-inflammatory and immunomodulatory activities, in Akt-dependent manner. They attributed these effects to the presence of S1P and SPC in HDL particles and $S1P_1$ and $S1P_3$ receptors on endothelial cells (Norata et al. 2005, 2008). Increased activation of Smad 2/3, a transcription factor controlled by TGFβ, found in the latter study suggests that HDL-associated S1P may transactivate TGFβ signaling pathways in an autocrine fashion. Most recently, Tatematsu et al. reported that endothelial lipase (EL) is a critical determinant of HDL-stimulated S1P-dependent signaling (Tatematsu et al. 2013). Actually, HDL-induced endothelial cell migration and angiogenic responses as well as Akt activation were markedly decreased in EL-deficient endothelial cells, but restored in the presence of exogenous S1P. Basing on application of pharmacological inhibitors, Tatematsu and colleagues concluded that the effects of HDL on endothelial cell migration and Akt activation are mediated by $S1P_1$.

As pointed out above, HDL is strongly mitogenic for smooth muscle cells, and HDL-associated lysosphingolipids appear to account for this effect. On the other hand, HDL was identified as a potent inhibitor of smooth muscle cell migration, which is a hallmark of the development of advanced atherosclerotic plaque and restenosis (Tamama et al. 2005; Damirin et al. 2007). The inhibitory effect of HDL was likely mediated by its S1P content, as it was abolished in the presence of $S1P_2$-specific siRNA and following desensitization of S1P receptors and much enhanced in cells overexpressing $S1P_2$ receptor. Further effects exerted by HDL on smooth muscle cell through its S1P content include modulation of the activity of several vasoactive factors. Chrisman et al. found that HDL-associated S1P desensitizes guanylyl cyclase B, a receptor for C-type natriuretic peptide (CNP), and thereby inhibits CNP-induced cyclic guanosine monophosphate (cGMP) accumulation in vascular smooth muscle cells (Chrisman et al. 2003). Gonzalez-Diez and colleagues demonstrated induction of COX-2 expression and PGI_2 production in vascular smooth muscle cells exposed to HDL, which was mediated through S1P receptors 2 and 3 and protein kinase p38MAPK (González-Díez et al. 2008). They have also shown that simvastatin potentiates the HDL- and S1P-induced COX-2 expression by upregulating cellular levels of $S1P_3$. Finally, HDL-associated lysosphingolipids S1P and SPC were found to suppress the proinflammatory activation of smooth muscle cells brought about by stimulation with thrombin and characterized by the increased release of inflammatory cyto- and chemokines such as MCP1 (Tölle et al. 2008). The detailed examination of signaling pathways accounting for inhibitory effects exerted by HDL-associated lysosphingolipids revealed that they were related to the inhibition of Rac1-mediated activation of NADPH oxidase and generation of reactive oxygen species. $S1P_3$ was identified to mediate anti-inflammatory effects of HDL, S1P, and SPC in smooth muscle cells.

Cardiomyocytes express several S1P receptors, and HDL was found to exert protective effects on these cells through its S1P content. HDL and S1P showed potent capacity to protect cardiomyocytes against hypoxia-reoxygenation- or doxorubicin-induced apoptosis, and these effects were attributed to the induction of protein kinases Src, MAPK, and Akt and the transcription factor STAT3 (Frias et al. 2009, 2010; Tao et al. 2010). However, investigations concerning the identity of receptor-mediating protective effects of HDL-associated S1P were inconclusive, and equally supportive evidence for the involvement of either $S1P_2$

or S1P$_1$ and S1P$_3$ has been provided. Recently, short-term treatment of cardiomyocytes with HDL or S1P was found to induce a PKC-dependent phosphorylation of connexin43, which is a gap junction protein present in ventricular cardiomyocytes and involved in cardioprotection by modulating ischemic preconditioning (Morel et al. 2012). In addition, HDL and S1P improved gap junctional communication, but only incrementally affected conduction velocities in cardiomyocytes. The physiological relevance of these effects has been clearly documented using an ex vivo approach, in which both HDL and S1P protected perfused hearts against ischemia/reperfusion-induced cell death.

Several other cell types were found to be responsive to stimulation with HDL-associated lysosphingolipids, but the physiological relevance of these effects remains obscure. For instance, HDL and S1P were demonstrated to stimulate prostate cancer cell migration and invasion through S1P receptors 2 and 3 and activation of MAPK and STAT3, to stimulate plasminogen activator inhibitor-1 (PAI-1) release from adipocytes through S1P$_2$ and activation of PKC and Rho kinase, and to promote release of fibroblast growth factor-2 (FGF2), a potent neurotrophic factor, from astroglial cells through activation of PI-PLC and MAPK (Malchinkhuu et al. 2003; Lee et al. 2010; Sekine et al. 2011). HDL particles isolated from body fluids other than plasma were also found to contain S1P and to stimulate intracellular signaling through their S1P content. For instance, HDL-like particles obtained from cerebrospinal fluid stimulated astrocyte migration and oligodendrocyte retraction in a manner sensitive to inhibitors of S1P receptors 1 and 3 (Sato et al. 2007). Follicular fluid-derived HDL was found to promote human granulosa lutein cell migration through S1P$_3$ and activation of small G-protein Rac1 (Becker et al. 2011). Despite the prominent role played by monocytes and macrophages in the pathogenesis of atherosclerosis, very little is known about the modulatory effects exerted by S1P cargo of HDL particles on these cells. In one study, both HDL and S1P were found to blunt monocyte activation induced by agonists of TLR2, but the attribution of this effect to the HDL-associated S1P was not unequivocally demonstrated (Dueñas et al. 2008).

7 HDL-Induced Cell Signaling: Future Challenges and Opportunities

HDL constitutes a large and heterogenous macromolecule carrying more than 50 proteins, several amphipathic peptides, miRNAs, and hundreds of lipid species representing almost all known lipid classes. Several proteins and lipids identified in HDL—among others, apolipoprotein H, haptoglobin, and steroid hormones—are recognized receptor ligands and inductors of intracellular signaling events. However, the rigid evidence demonstrating that these proteins or lipids contribute to the HDL-induced signaling governing diverse functional responses to this lipoprotein has been not provided to date or is at best equivocal. For instance, several studies documented that HDL carries estradiol in the form of fatty acyl esters produced in a reaction catalyzed by LCAT. HDL-derived estradiol esters are internalized via SR-

BI-mediated mechanisms and subsequently hydrolyzed intracellularly to unfold their effects. However, the specific signaling events and/or physiological effects clearly attributable to the estradiol content in HDL particles have not been unequivocally defined. In one study, HDL-associated estradiol was found to stimulate eNOS and vasodilation in an SR-BI- and Akt-dependent manner, but these results were questioned by others (Gong et al. 2003; Nofer et al. 2004). In other study, dehydroepiandrosterone (DHEA) fatty acyl esters once incorporated in HDL were found to induce stronger vasodilatory response, and this effect appeared also to be mediated by SR-BI (Paatela et al. 2011). However, no clear signaling link between the HDL-associated DHEA and vasodilation has been established in this study. These two examples clearly show that the assignment of ligands present in HDL to the appending signaling machinery and to various functions exerted by these lipoproteins in physiology will be a major challenge in the future.

The composition of HDL particles is variable and influenced both by normal metabolism and in various states of pathology. For instance, HDL particles isolated from subjects suffering from acute or chronic inflammatory diseases were found to lose proteins and enzymes with established or presumed antiatherogenic function such as apoA-I or LCAT and to concomitantly acquire proinflammatory or prooxidative factors such as serum amyloid A (SAA), ceruloplasmin, lipoprotein-specific phospholipase A_2, or myeloperoxidase. Not surprisingly, such inflammatory HDL particles are severely impeded in their capacity to trigger certain signaling pathways, but at the same time may acquire new ligands and thereby alter their signaling properties. For instance, the relatively high content of apolipoprotein J (apoJ) in HDL particles obtained from healthy subjects endowed them with important antiatherogenic property to promote endothelial survival by stimulation of the expression of antiapoptotic protein Bcl-XL (Riwanto et al. 2013). By contrast, HDL particles isolated from patients with stable coronary heart disease (CHD) or acute coronary syndromes were found to be poor in apoJ, but instead enriched in apoC-I or apoC-III (McNeal et al. 2013; Riwanto et al. 2013). Furthermore, the increased content of these apolipoproteins turned HDL particles into potent apoptosis inducers in vascular smooth muscle and endothelial cells by virtue of stimulating the expression of proapoptotic protein tBid. In addition to CHD, dysfunctional HDL is also encountered in chronic conditions such as chronic kidney disease (CKD). Abnormal HDL composition characterized primarily by increased serum amyloid A and apoC-III content as well as compromised HDL functionality has been repeatedly reported in uremic patients (Holzer et al. 2011; Weichhart et al. 2012). In one recent study, SAA-enriched HDL encountered in CKD was found to induce production of a proinflammatory chemokine MCP1 in smooth muscle cells by interaction with formyl peptide receptor 2 (FPR2) (Tölle et al. 2012). In other study conducted in CKD patients, accumulation of symmetric dimethylarginine (SDMA) in HDL was found to turn this lipoprotein into a noxious particle reducing NO bioavailability, evoking endothelial dysfunction, and subsequently increasing arterial blood pressure. The deleterious effects of SDMA-modified HDL were attributed to the noncanonical activation of TLR2 (Speer et al. 2013). This exemplifies another challenge facing HDL research in the future, namely, identification of ligands, receptors, and signaling cascades attributable to HDL in various states of pathology.

Notwithstanding the physiological or pathological character of HDL-induced signal transduction, its elucidation may open new opportunities for the development of antiatherogenic, anti-inflammatory, or antithrombotic drugs. Several interventional studies conducted recently and aiming at cardiovascular risk reduction through simple elevation of plasma HDL cholesterol fall through. This epic failure emphasizes the necessity to develop more sophisticated strategies based on targeted exploitation of selected antiatherogenic or anti-inflammatory activities exerted by HDL particles. The proper understanding of HDL-induced signaling may help to specifically enhance beneficial physiological or to disable harmful pathological effects of HDL and thereby to design more efficacious therapeutic approaches for cardiovascular and other inflammatory diseases.

Open Access This chapter is distributed under the terms of the Creative Commons Attribution Noncommercial License, which permits any noncommercial use, distribution, and reproduction in any medium, provided the original author(s) and source are credited.

References

Al-Jarallah A, Trigatti BL (2010) A role for the scavenger receptor, class B type I in high density lipoprotein dependent activation of cellular signaling pathways. Biochim Biophys Acta 1801:1239–1248

Arakawa R, Yokoyama S (2002) Helical apolipoproteins stabilize ATP-binding cassette transporter A1 by protecting it from thiol protease-mediated degradation. J Biol Chem 277:22426–22429

Argraves KM, Gazzolo PJ, Groh EM et al (2008) High density lipoprotein-associated sphingosine 1-phosphate promotes endothelial barrier function. J Biol Chem 283:25074–25081

Argraves KM, Sethi AA, Gazzolo PJ et al (2011) S1P, dihydro-S1P and C24:1-ceramide levels in the HDL-containing fraction of serum inversely correlate with occurrence of ischemic heart disease. Lipids Health Dis 10:70

Assanasen C, Mineo C, Seetharam D et al (2005) Cholesterol binding, efflux, and a PDZ-interacting domain of scavenger receptor-BI mediate HDL-initiated signaling. J Clin Invest 115:969–977

Becker S, von Otte S, Robenek H et al (2011) Follicular fluid high-density lipoprotein-associated sphingosine 1-phosphate (S1P) promotes human granulosa lutein cell migration via S1P receptor type 3 and small G-protein RAC1. Biol Reprod 84:604–612

Bensinger SJ, Bradley MN, Joseph SB et al (2008) LXR signaling couples sterol metabolism to proliferation in the acquired immune response. Cell 134:97–111

Bodzioch M, Orso E, Klucken J et al (1999) The gene encoding ATP-binding cassette transporter 1 is mutated in Tangier disease. Nat Genet 22:347–351

Brooks-Wilson A, Marcil M, Clee SM et al (1999) Mutations in ABC1 in Tangier disease and familial high-density lipoprotein deficiency. Nat Genet 22:336–345

Cai L, Wang Z, Meyer JM et al (2012) Macrophage SR-BI regulates LPS-induced pro-inflammatory signaling in mice and isolated macrophages. J Lipid Res 53:1472–1481

Cavelier C, Ohnsorg PM, Rohrer L et al (2012) The β-chain of cell surface F(0)F(1) ATPase modulates apoA-I and HDL transcytosis through aortic endothelial cells. Arterioscler Thromb Vasc Biol 32:131–139

Chrisman TD, Perkins DT, Garbers DL (2003) Identification of a potent serum factor that causes desensitization of the receptor for C-Type natriuretic peptide. Cell Commun Signal 1:4

Christoffersen C, Obinata H, Kumaraswamy SB et al (2011) Endothelium-protective sphingosine-1-phosphate provided by HDL-associated apolipoprotein M. Proc Natl Acad Sci USA 108:9613–9618

Chroni A, Liu T, Fitzgerald ML, Freeman MW et al (2004) Cross-linking and lipid efflux properties of apoA-I mutants suggest direct association between apoA-I helices and ABCA1. Biochemistry 43:2126–2139

Dahm F, Nocito A, Bielawska A et al (2006) Distribution and dynamic changes of sphingolipids in blood in response to platelet activation. J Thromb Haemost 4:2704–2709

Damirin A, Tomura H, Komachi M et al (2007) Role of lipoprotein-associated lysophospholipids in migratory activity of coronary artery smooth muscle cells. Am J Physiol Heart Circ Physiol 292:H2513–H2522

Danilo C, Gutierrez-Pajares JL et al (2013) Scavenger receptor class B type I regulates cellular cholesterol metabolism and cell signaling associated with breast cancer development. Breast Cancer Res 15:R87

Deeg MA, Bowen RF, Oram JF et al (1997) High density lipoproteins stimulate mitogen-activated protein kinases in human skin fibroblasts. Arterioscler Thromb Vasc Biol 17:1667–1674

Drobnik W, Borsukova H, Böttcher A et al (2002) Apo AI/ABCA1-dependent and HDL3-mediated lipid efflux from compositionally distinct cholesterol-based microdomains. Traffic 3:268–278

Dueñas AI, Aceves M, Fernández-Pisonero I et al (2008) Selective attenuation of Toll-like receptor 2 signalling may explain the atheroprotective effect of sphingosine 1-phosphate. Cardiovasc Res 79:537–544

Fabre AC, Vantourout P, Champagne E et al (2006) Cell surface adenylate kinase activity regulates the F(1)-ATPase/P2Y (13)-mediated HDL endocytosis pathway on human hepatocytes. Cell Mol Life Sci 63:2829–2837

Feng H, Guo L, Wang D et al (2011) Deficiency of scavenger receptor BI leads to impaired lymphocyte homeostasis and autoimmune disorders in mice. Arterioscler Thromb Vasc Biol 31:2543–2551

Fitzgerald ML, Morris AL, Rhee JS et al (2002) Naturally occurring mutations in the largest extracellular loops of ABCA1 can disrupt its direct interaction with apolipoprotein A-I. J Biol Chem 277:33178–33187

Fitzgerald ML, Morris AL, Chroni A et al (2004) ABCA1 and amphipathic apolipoproteins form high-affinity molecular complexes required for cholesterol efflux. J Lipid Res 45:287–294

Francis GA, Knopp RH, Oram JF (1995) Defective removal of cellular cholesterol and phospholipids by apolipoprotein A-I in Tangier disease. J Clin Invest 96:78–87

Frias MA, James RW, Gerber-Wicht C et al (2009) Native and reconstituted HDL activate Stat3 in ventricular cardiomyocytes via ERK1/2: role of sphingosine-1-phosphate. Cardiovasc Res 82:313–323

Frias MA, Lang U, Gerber-Wicht C et al (2010) Native and reconstituted HDL protect cardiomyocytes from doxorubicin-induced apoptosis. Cardiovasc Res 85:118–126

Glomset JA (1968) The plasma lecithin: cholesterol acyltransferase reaction. J Lipid Res 9:155–167

Gong M, Wilson M, Kelly T et al (2003) HDL-associated estradiol stimulates endothelial NO synthase and vasodilation in an SR-BI-dependent manner. J Clin Invest 111:1579–1587

González-Díez M, Rodríguez C, Badimon L et al (2008) Prostacyclin induction by high-density lipoprotein (HDL) in vascular smooth muscle cells depends on sphingosine 1-phosphate receptors: effect of simvastatin. Thromb Haemost 100:119–126

Graham DL, Oram JF (1987) Identification and characterization of a high density lipoprotein-binding protein in cell membranes by ligand blotting. J Biol Chem 262:7439–7442

Guo L, Song Z, Li M, Li XA et al (2009) Scavenger receptor BI protects against septic death through Its role in modulating inflammatory response. J Biol Chem 284:19826–19834

Haidar B, Mott S, Boucher B et al (2001) Cellular cholesterol efflux is modulated by phospholipid-derived signaling molecules in familial HDL deficiency/Tangier disease fibroblasts. J Lipid Res 42:249–257

Haidar B, Denis M, Krimbou L et al (2002) cAMP induces ABCA1 phosphorylation activity and promotes cholesterol efflux from fibroblasts. J Lipid Res 43:2087–2094

Haidar B, Denis M, Marcil M et al (2004) Apolipoprotein A-I activates cellular cAMP signaling through the ABCA1 transporter. J Biol Chem 279:9963–9969

Hänel P, Andréani P, Gräler MH (2007) Erythrocytes store and release sphingosine 1-phosphate in blood. FASEB J 21:1202–1209

Hirano K, Matsuura F, Tsukamoto K et al (2000) Decreased expression of a member of the Rho GTPase family, Cdc42Hs, in cells from Tangier disease – the small G protein may play a role in cholesterol efflux. FEBS Lett 484:275–279

Hoekstra M, Van Berkel TJ, Van Eck M (2010) Scavenger receptor BI: a multi-purpose player in cholesterol and steroid metabolism. World J Gastroenterol 16:5916–5924

Holzer M, Birner-Gruenberger R, Stojakovic T et al (2011) Uremia alters HDL composition and function. J Am Soc Nephrol 22:1631–1641

Howard AD, Verghese PB, Arrese EL et al (2011) The β-subunit of ATP synthase is involved in cellular uptake and resecretion of apoA-I but does not control apoA-I-induced lipid efflux in adipocytes. Mol Cell Biochem 348:155–164

Hu YW, Ma X, Li XX et al (2009) Eicosapentaenoic acid reduces ABCA1 serine phosphorylation and impairs ABCA1-dependent cholesterol efflux through cyclic AMP/protein kinase A signaling pathway in THP-1 macrophage-derived foam cells. Atherosclerosis 204:e35–e43

Igarashi J, Miyoshi M, Hashimoto T et al (2007) Statins induce S1P1 receptors and enhance endothelial nitric oxide production in response to high-density lipoproteins. Br J Pharmacol 150:470–479

Imachi H, Murao K, Cao W et al (2003) Expression of human scavenger receptor B1 on and in human platelets. Arterioscler Thromb Vasc Biol 23:898–904

Iwamoto N, Lu R, Tanaka N, Abe-Dohmae S et al (2010) Calmodulin interacts with ATP binding cassette transporter A1 to protect from calpain-mediated degradation and upregulates high-density lipoprotein generation. Arterioscler Thromb Vasc Biol 30:1446–1452

Karuna R, Park R, Othman A et al (2011) Plasma levels of sphingosine-1-phosphate and apolipoprotein M in patients with monogenic disorders of HDL metabolism. Atherosclerosis 219:855–863

Karwatsky J, Ma L, Dong F et al (2010) Cholesterol efflux to apoA-I in ABCA1-expressing cells is regulated by Ca2+-dependent calcineurin signaling. J Lipid Res 51:1144–1156

Kimura T, Sato K, Kuwabara A et al (2001) Sphingosine 1-phosphate may be a major component of plasma lipoproteins responsible for the cytoprotective actions in human umbilical vein endothelial cells. J Biol Chem 276:31780–31785

Kimura T, Tomura H, Mogi C et al (2006) Role of scavenger receptor class B type I and sphingosine 1-phosphate receptors in high density lipoprotein-induced inhibition of adhesion molecule expression in endothelial cells. J Biol Chem 281:37457–37467

Kimura T, Mogi C, Tomura H et al (2008) Induction of scavenger receptor class B type I is critical for simvastatin enhancement of high-density lipoprotein-induced anti-inflammatory actions in endothelial cells. J Immunol 181:7332–7340

Kimura T, Tomura H, Sato K et al (2010) Mechanism and role of high density lipoprotein-induced activation of AMP-activated protein kinase in endothelial cells. J Biol Chem 285:4387–4397

Koseki M, Hirano K, Masuda D et al (2007) Increased lipid rafts and accelerated lipopolysaccharide-induced tumor necrosis factor-alpha secretion in Abca1-deficient macrophages. J Lipid Res 48:299–306

Krishna SM, Seto SW, Moxon JV et al (2012) Fenofibrate increases high-density lipoprotein and sphingosine 1 phosphate concentrations limiting abdominal aortic aneurysm progression in a mouse model. Am J Pathol 181:706–718

Landry YD, Denis M, Nandi S et al (2006) ATP-binding cassette transporter A1 expression disrupts raft membrane microdomains through its ATPase-related functions. J Biol Chem 281:36091–36101

Lee MH, Hammad SM, Semler AJ et al (2010) HDL3, but not HDL2, stimulates plasminogen activator inhibitor-1 release from adipocytes: the role of sphingosine-1-phosphate. J Lipid Res 51:2619–2628

Levkau B (2013) Cardiovascular effects of sphingosine-1-phosphate (S1P). Handb Exp Pharmacol 216:147–170

Li Q, Yokoyama S (1995) Independent regulation of cholesterol incorporation into free apolipoprotein-mediated cellular lipid efflux in rat vascular smooth muscle cells. J Biol Chem 270:26216–26223

Li Q, Tsujita M, Yokoyama S (1997) Selective down-regulation by protein kinase C inhibitors of apolipoprotein-mediated cellular cholesterol efflux in macrophages. Biochemistry 36:12045–12052

Liu D, Ji L, Tong X et al (2011) Human apolipoprotein A-I induces cyclooxygenase-2 expression and prostaglandin I-2 release in endothelial cells through ATP-binding cassette transporter A1. Am J Physiol Cell Physiol 301:C739–C748

Liu M, Seo J, Allegood J, Bi X et al (2013a) Hepatic ApoM overexpression stimulates formation of larger, ApoM/S1P-enriched plasma HDL. J Biol Chem 289:2801–2814

Liu XY, Lu Q, Ouyang XP et al (2013b) Apelin-13 increases expression of ATP-binding cassette transporter A1 via activating protein kinase C α signaling in THP-1 macrophage-derived foam cells. Atherosclerosis 226:398–407

Ma L, Dong F, Denis M et al (2011) Ht31, a protein kinase A anchoring inhibitor, induces robust cholesterol efflux and reverses macrophage foam cell formation through ATP-binding cassette transporter A1. J Biol Chem 286:3370–3378

Maceyka M, Harikumar KB, Milstien S et al (2012) Sphingosine-1-phosphate signaling and its role in disease. Trends Cell Biol 22:50–60

Malaval C, Laffargue M, Barbaras R et al (2009) RhoA/ROCK I signalling downstream of the P2Y13 ADP-receptor controls HDL endocytosis in human hepatocytes. Cell Signal 21:120–127

Malchinkhuu E, Sato K, Muraki T et al (2003) Assessment of the role of sphingosine 1-phosphate and its receptors in high-density lipoprotein-induced stimulation of astroglial cell function. Biochem J 370:817–827

Martinez LO, Agerholm-Larsen B, Wang N et al (2003a) Phosphorylation of a pest sequence in ABCA1 promotes calpain degradation and is reversed by ApoA-I. J Biol Chem 278:37368–37374

Martinez LO, Jacquet S, Esteve JP et al (2003b) Ectopic beta-chain of ATP synthase is an apolipoprotein A-I receptor in hepatic HDL endocytosis. Nature 421:75–79

Matsuo Y, Miura S, Kawamura A et al (2007) Newly developed reconstituted high-density lipoprotein containing sphingosine-1-phosphate induces endothelial tube formation. Atherosclerosis 194:159–168

McNeal CJ, Chatterjee S, Hou J et al (2013) Human HDL containing a novel apoC-I isoform induces smooth muscle cell apoptosis. Cardiovasc Res 98:83–93

Mendez AJ, Oram JF, Bierman EL (1991) Protein kinase C as a mediator of high density lipoprotein receptor-dependent efflux of intracellular cholesterol. J Biol Chem 266:10104–10111

Mineo C, Shaul PW (2012) Functions of scavenger receptor class B, type I in atherosclerosis. Curr Opin Lipidol 23:487–493

Mineo C, Yuhanna IS, Quon MJ et al (2003) High density lipoprotein induced endothelial nitric-oxide synthase activation is mediated by Akt and MAP kinases. J Biol Chem 278:9142–9149

Miura S, Fujino M, Matsuo Y et al (2003) High density lipoprotein-induced angiogenesis requires the activation of Ras/MAP kinase in human coronary artery endothelial cells. Arterioscler Thromb Vasc Biol 23:802–808

Mogilenko DA, Orlov SV, Trulioff AS et al (2012) Endogenous apolipoprotein A-I stabilizes ATP-binding cassette transporter A1 and modulates Toll-like receptor 4 signaling in human macrophages. FASEB J 26:2019–2030

Morel S, Frias MA, Rosker C et al (2012) The natural cardioprotective particle HDL modulates connexin43 gap junction channels. Cardiovasc Res 93:41–49

Murphy AJ, Woollard KJ, Hoang A et al (2008) High-density lipoprotein reduces the human monocyte inflammatory response. Arterioscler Thromb Vasc Biol 28:2071–2077

Murphy AJ, Woollard KJ, Suhartoyo A et al (2011) Neutrophil activation is attenuated by high-density lipoprotein and apolipoprotein A-I in in vitro and in vivo models of inflammation. Arterioscler Thromb Vasc Biol 31:1333–1341

Nofer JR, van Eck M (2011) HDL scavenger receptor class B type I and platelet function. Curr Opin Lipidol 22:277–282

Nofer J-R, Tepel M, Kehrel B et al (1996) High density lipoproteins enhance the Na^+/H^+ antiport in human platelets. Thromb Haemost 75:635–641

Nofer JR, Walter M, Kehrel B et al (1998) HDL3-mediated inhibition of thrombin-induced platelet aggregation and fibrinogen binding occurs via decreased production of phosphoinositide-derived second messengers 1,2-diacylglycerol and inositol 1,4,5-tris-phosphate. Arterioscler Thromb Vasc Biol 18:861–869

Nofer JR, Fobker M, Höbbel G et al (2000) Activation of phosphatidylinositol-specific phospholipase C by HDL-associated lysosphingolipid. Involvement in mitogenesis but not in cholesterol efflux. Biochemistry 39:15199–15207

Nofer JR, Junker R, Pulawski E et al (2001a) High density lipoproteins induce cell cycle entry in vascular smooth muscle cells via mitogen activated protein kinase-dependent pathway. Thromb Haemost 85:730–735

Nofer JR, Levkau B, Wolinska I et al (2001b) Suppression of endothelial cell apoptosis by high density lipoproteins (HDL) and HDL-associated lysosphingolipids. J Biol Chem 276:34480–34485

Nofer JR, Feuerborn R, Levkau B et al (2003) Involvement of Cdc42 signaling in apoA-I-induced cholesterol efflux. J Biol Chem 278:53055–53062

Nofer JR, van der Giet M, Tölle M et al (2004) HDL induces NO-dependent vasorelaxation via the lysophospholipid receptor S1P3. J Clin Invest 113:569–581

Nofer JR, Remaley AT, Feuerborn R et al (2006) Apolipoprotein A-I activates Cdc42 signaling through the ABCA1 transporter. J Lipid Res 47:794–803

Nofer JR, Brodde M, Korporaal SJ et al (2011) Native high density lipoproteins inhibit thrombin-induced platelet activation via scavenger receptor B1 (SR-B1). Atherosclerosis 215:374–382

Norata GD, Callegari E, Marchesi M et al (2005) High-density lipoproteins induce transforming growth factor-beta2 expression in endothelial cells. Circulation 111:2805–2811

Norata GD, Marchesi P, Pirillo A et al (2008) Long pentraxin 3, a key component of innate immunity, is modulated by high density lipoproteins in endothelial cells. Arterioscler Thromb Vasc Biol 28:925–931

Paatela H, Vihma V, Jauhiainen M et al (2011) Dehydroepiandrosterone fatty acyl esters in high density lipoprotein: interaction with human vascular endothelial cells and vascular responses ex vivo. Steroids 76:376–380

Radojkovic C, Genoux A, Pons V et al (2009) Stimulation of cell surface F1-ATPase activity by apolipoprotein A-I inhibits endothelial cell apoptosis and promotes proliferation. Arterioscler Thromb Vasc Biol 29:1125–1130

Riwanto M, Rohrer L, Roschitzki B et al (2013) Altered activation of endothelial anti- and proapoptotic pathways by high-density lipoprotein from patients with coronary artery disease: role of high-density lipoprotein-proteome remodeling. Circulation 127:891–904

Rust S, Rosier M, Funke H et al (1999) Tangier disease is caused by mutations in the gene encoding ATP-binding cassette transporter 1. Nat Genet 22:352–355

Sachinidis A, Kettenhofen R, Seewald S et al (1999) Evidence that lipoproteins are carriers of bioactive factors. Arterioscler Thromb Vasc Biol 19:2412–2421

Saddar S, Carriere V, Lee WR et al (2013) Scavenger receptor class B type I is a plasma membrane cholesterol sensor. Circ Res 112:140–151

Sano O, Kobayashi A, Nagao K et al (2007) Sphingomyelin-dependence of cholesterol efflux mediated by ABCG1. J Lipid Res 48:2377–2384

Sato K, Malchinkhuu E, Horiuchi Y et al (2007) HDL-like lipoproteins in cerebrospinal fluid affect neural cell activity through lipoprotein-associated sphingosine 1-phosphate. Biochem Biophys Res Commun 359:649–654

Sattler KJ, Elbasan S, Keul P et al (2010) Sphingosine 1-phosphate levels in plasma and HDL are altered in coronary artery disease. Basic Res Cardiol 105:821–832

See RH, Caday-Malcolm RA, Singaraja RR et al (2002) Protein kinase A site-specific phosphorylation regulates ATP-binding cassette A1 (ABCA1)-mediated phospholipid efflux. J Biol Chem 277:41835–41842

Seetharam D, Mineo C, Gormley AK et al (2006) High-density lipoprotein promotes endothelial cell migration and reendothelialization via scavenger receptor-B type I. Circ Res 98:63–72

Sekine Y, Suzuki K, Remaley AT (2011) HDL and sphingosine-1-phosphate activate stat3 in prostate cancer DU145 cells via ERK1/2 and S1P receptors, and promote cell migration and invasion. Prostate 71:690–699

Slotte JP, Oram JF, Bierman EL (1987) Binding of high density lipoproteins to cell receptors promotes translocation of cholesterol from intracellular membranes to the cell surface. J Biol Chem 262:12904–12907

Speer T, Rohrer L, Blyszczuk P et al (2013) Abnormal high-density lipoprotein induces endothelial dysfunction via activation of toll-like receptor-2. Immunity 38:754–768

Sun Y, Ishibashi M, Seimon T et al (2009) Free cholesterol accumulation in macrophage membranes activates Toll-like receptors and p38 mitogen-activated protein kinase and induces cathepsin K. Circ Res 104:455–465

Tamama K, Tomura H, Sato K et al (2005) High density lipoprotein inhibits migration of vascular smooth muscle cells through its sphingosine 1-phosphate component. Atherosclerosis 178:19–23

Tan JT, Prosser HC, Vanags LZ et al (2013) High-density lipoproteins augment hypoxia-induced angiogenesis via regulation of post-translational modulation of hypoxia-inducible factor 1α. FASEB J 28:206–217

Tang C, Vaughan AM, Oram JF (2004) Janus kinase 2 modulates the apolipoprotein interactions with ABCA1 required for removing cellular cholesterol. J Biol Chem 279:7622–7628

Tang C, Vaughan AM, Anantharamaiah GM et al (2006) Janus kinase 2 modulates the lipid-removing but not protein-stabilizing interactions of amphipathic helices with ABCA1. J Lipid Res 47:107–114

Tang C, Liu Y, Kessler PS, Vaughan AM, Oram JF (2009) The macrophage cholesterol exporter ABCA1 functions as an anti-inflammatory receptor. J Biol Chem 284:32336–32343

Tao R, Hoover HE, Honbo N et al (2010) High-density lipoprotein determines adult mouse cardiomyocyte fate after hypoxia-reoxygenation through lipoprotein-associated sphingosine 1-phosphate. Am J Physiol Heart Circ Physiol 298:H1022–H1028

Tatematsu S, Francis SA, Natarajan P et al (2013) Endothelial lipase is a critical determinant of high-density lipoprotein-stimulated sphingosine 1-phosphate-dependent signaling in vascular endothelium. Arterioscler Thromb Vasc Biol 33:1788–1794

Tölle M, Pawlak A, Schuchardt M et al (2008) HDL-associated lysosphingolipids inhibit NAD(P) H oxidase-dependent monocyte chemoattractant protein-1 production. Arterioscler Thromb Vasc Biol 28:1542–1548

Tölle M, Huang T, Schuchardt M et al (2012) High-density lipoprotein loses its anti-inflammatory capacity by accumulation of pro-inflammatory-serum amyloid A. Cardiovasc Res 94:154–162

Tsukamoto K, Hirano K, Tsujii K et al (2001) ATP-binding cassette transporter-1 induces rearrangement of actin cytoskeletons possibly through Cdc42/N-WASP. Biochem Biophys Res Commun 287:757–765

Tsukamoto K, Hirano K, Yamashita S et al (2002) Retarded intracellular lipid transport associated with reduced expression of Cdc42, a member of Rho-GTPases, in human aged skin fibroblasts: a possible function of Cdc42 in mediating intracellular lipid transport. Arterioscler Thromb Vasc Biol 22:1899–1904

Umemoto T, Han CY, Mitra P et al (2013) Apolipoprotein AI and high-density lipoprotein have anti-inflammatory effects on adipocytes via cholesterol transporters: ATP-binding cassette A-1, ATP-binding cassette G-1, and scavenger receptor B-1. Circ Res 112:1345–1354

Vaughan AM, Tang C, Oram JF (2009) ABCA1 mutants reveal an interdependency between lipid export function, apoA-I binding activity, and Janus kinase 2 activation. J Lipid Res 50:285–292

von Otte S, Paletta JR, Becker S et al (2006) Follicular fluid high density lipoprotein-associated sphingosine 1-phosphate is a novel mediator of ovarian angiogenesis. J Biol Chem 281:5398–5405

Walter M, Reinecke H, Nofer JR et al (1995) HDL3 stimulates multiple signaling pathways in human skin fibroblasts. Arterioscler Thromb Vasc Biol 15:1975–1986

Walter M, Reinecke H, Gerdes U et al (1996) Defective regulation of phosphatidylcholine-specific phospholipases C and D in a kindred with Tangier disease. Evidence for the involvement of phosphatidylcholine breakdown in HDL-mediated cholesterol efflux mechanisms. J Clin Invest 98:2315–2323

Wang N, Silver DL, Costet P, Tall AR (2000) Specific binding of ApoA-I, enhanced cholesterol efflux, and altered plasma membrane morphology in cells expressing ABC1. J Biol Chem 275:33053–33058

Wang N, Chen W, Linsel-Nitschke P et al (2003) A PEST sequence in ABCA1 regulates degradation by calpain protease and stabilization of ABCA1 by apoA-I. J Clin Invest 111:99–107

Wang N, Lan D, Chen W et al (2004) ATP-binding cassette transporters G1 and G4 mediate cellular cholesterol efflux to high-density lipoproteins. Proc Natl Acad Sci USA 101:9774–9779

Wang N, Ranalletta M, Matsuura F et al (2006) LXR-induced redistribution of ABCG1 to plasma membrane in macrophages enhances cholesterol mass efflux to HDL. Arterioscler Thromb Vasc Biol 26:1310–1316

Weichhart T, Kopecky C, Kubicek M et al (2012) Serum amyloid A in uremic HDL promotes inflammation. J Am Soc Nephrol 23:934–947

Westerterp M, Gourion-Arsiquaud S, Murphy AJ et al (2012) Regulation of hematopoietic stem and progenitor cell mobilization by cholesterol efflux pathways. Cell Stem Cell 11:195–206

Wilhelm AJ, Zabalawi M, Grayson JM et al (2009) Apolipoprotein A-I and its role in lymphocyte cholesterol homeostasis and autoimmunity. Arterioscler Thromb Vasc Biol 29:843–849

Wilkerson BA, Grass GD, Wing SB et al (2012) Sphingosine 1-phosphate (S1P) carrier-dependent regulation of endothelial barrier: high density lipoprotein (HDL)-S1P prolongs endothelial barrier enhancement as compared with albumin-S1P via effects on levels, trafficking, and signaling of S1P1. J Biol Chem 287:44645–44653

Xu J, Qian J, Xie X et al (2012) High density lipoprotein cholesterol promotes the proliferation of bone-derived mesenchymal stem cells via binding scavenger receptor-B type I and activation of PI3K/Akt, MAPK/ERK1/2 pathways. Mol Cell Biochem 371:55–64

Yamauchi Y, Hayashi M, Abe-Dohmae S et al (2003) Apolipoprotein A-I activates protein kinase C alpha signaling to phosphorylate and stabilize ATP binding cassette transporter A1 for the high density lipoprotein assembly. J Biol Chem 278:47890–47897

Yamauchi Y, Chang CC, Hayashi M et al (2004) Intracellular cholesterol mobilization involved in the ABCA1/apolipoprotein-mediated assembly of high density lipoprotein in fibroblasts. J Lipid Res 45:1943–1951

Yin K, Deng X, Mo ZC et al (2011) Tristetraprolin-dependent post-transcriptional regulation of inflammatory cytokine mRNA expression by apolipoprotein A-I: role of ATP-binding membrane cassette transporter A1 and signal transducer and activator of transcription 3. J Biol Chem 286:13834–13845

Yin K, Chen WJ, Zhou ZG et al (2012) Apolipoprotein A-I inhibits CD40 proinflammatory signaling via ATP-binding cassette transporter A1-mediated modulation of lipid raft in macrophages. J Atheroscler Thromb 19:823–836

Yuhanna IS, Zhu Y, Cox BE et al (2001) High-density lipoprotein binding to scavenger receptor-BI activates endothelial nitric oxide synthase. Nat Med 7:853–857

Yvan-Charvet L, Welch C, Pagler TA et al (2008) Increased inflammatory gene expression in ABC transporter-deficient macrophages: free cholesterol accumulation, increased signaling via toll-like receptors, and neutrophil infiltration of atherosclerotic lesions. Circulation 118:1837–1847

Yvan-Charvet L, Pagler TA, Seimon TA et al (2010a) ABCA1 and ABCG1 protect against oxidative stress-induced macrophage apoptosis during efferocytosis. Circ Res 106:1861–1869

Yvan-Charvet L, Pagler T, Gautier EL et al (2010b) ATP-binding cassette transporters and HDL suppress hematopoietic stem cell proliferation. Science 328:1689–1693

Zhang Z, Hirano K, Tsukamoto K et al (2005) Defective cholesterol efflux in Werner syndrome fibroblasts and its phenotypic correction by Cdc42, a RhoGTPase. Exp Gerontol 40:286–294

Zhang Q, Zhang Y, Feng H et al (2011) High density lipoprotein (HDL) promotes glucose uptake in adipocytes and glycogen synthesis in muscle cells. PLoS ONE 6:e23556

Zhang QH, Zu XY, Cao RX et al (2012) An involvement of SR-B1 mediated PI3K-Akt-eNOS signaling in HDL-induced cyclooxygenase 2 expression and prostacyclin production in endothelial cells. Biochem Biophys Res Commun 420:17–23

Zhu X, Lee JY, Timmins JM et al (2008a) Increased cellular free cholesterol in macrophage-specific Abca1 knock-out mice enhances pro-inflammatory response of macrophages. J Biol Chem 283:22930–22941

Zhu W, Saddar S, Seetharam D et al (2008b) The scavenger receptor class B type I adaptor protein PDZK1 maintains endothelial monolayer integrity. Circ Res 102:480–487

Part II
Pathology of HDL

Epidemiology: Disease Associations and Modulators of HDL-Related Biomarkers

Markku J. Savolainen

Contents

1	Protective Role of HDL: Evidence from Epidemiological Studies	261
2	HDL Cholesterol as a Risk Factor for Atherosclerosis and Its Complications	262
3	HDL Cholesterol as a Risk Factor for Other Diseases	262
4	Total HDL-C in Various Populations	263
5	Total HDL-C Modulated by Environmental Factors	264
6	HDL-C in Diseases and Conditions	266
7	High HDL Levels Do Not Add to the Protection	267
8	Effect of HDL on Stroke	267
9	Time Trends in Total HDL-C	268
10	Are There Other Biomarkers than the Total HDL-C?	268
11	HDL Fractions	268
12	HDL Particle Size	269
13	HDL Particle Number	269
14	HDL Lipids	269
15	HDL Apolipoproteins	270
16	HDL Proteomics	270
17	HDL Function	271
18	Pleiotropy	272
19	Future Approaches of Epidemiological Studies	272
References		274

Abstract

Epidemiological studies have shown an inverse association between high-density lipoprotein cholesterol (HDL-C) levels and risk of ischemic heart disease. In addition, a low level of HDL-C has been shown to be a risk factor for other diseases not related to atherosclerosis. However, recent studies have not

---

M.J. Savolainen (✉)
Department of Internal Medicine, Institute of Clinical Medicine, University of Oulu, Kajaanintie 50, P.O.Box 5000, 90014 Oulu, Finland
e-mail: markku.savolainen@oulu.fi

supported a causal effect of HDL-C in the development of atherosclerosis. Furthermore, new drugs markedly elevating HDL-C levels have been disappointing with respect to clinical endpoints. Earlier, most studies have focused almost exclusively on the total HDL-C without regard to the chemical composition or multiple subclasses of HDL particles. Recently, there have been efforts to dissect the HDL fraction into as many well-defined subfractions and individual molecules of HDL particles as possible. On the other hand, the focus is shifting from the structure and composition to the function of HDL particles. Biomarkers and mechanisms that could potentially explain the beneficial characteristics of HDL particles unrelated to their cholesterol content have been sought with sophisticated methods such as proteomics, lipidomics, metabonomics, and function studies including efflux capacity. These new approaches have been used in order to resolve the complex effects of diseases, conditions, environmental factors, and genes in relation to the protective role of HDL but high-throughput methods are still needed for large-scale epidemiological studies.

Keywords

High-density lipoproteins • Cholesterol • Atherosclerosis • Coronary heart disease • Apolipoproteins • Cholesterol efflux • Diabetes • Obesity • Cancer • Proteomics • Lipidomics • Metabonomics

Abbreviations

ApoA-I	Apolipoprotein A-I
ApoA-II	Apolipoprotein A-II
ApoC-III	Apolipoprotein C-III
CHD	Coronary heart disease
CETP	Cholesteryl ester transfer protein
CVD	Cardiovascular disease
EPA	Eicosapentaenoic acid
DHA	Docosahexaenoic acid
GI	Glycemic index
GL	Glycemic load
HDL	High-density lipoproteins
HDL-C	High-density lipoprotein cholesterol
HDL2	High-density lipoprotein fraction 2
LDL	Low-density lipoprotein
LDL-C	Low-density lipoprotein cholesterol
NMR	Nuclear magnetic resonance
PON1	Paraoxonase 1
S1P	Sphingosine-1-phosphate
T2DM	Type 2 diabetes mellitus

Low levels of high-density lipoprotein (HDL) particles in the plasma of patients with coronary heart disease (CHD) were observed already in the early 1950s (Barr et al. 1951; Nikkilä 1953). At that time the lipoprotein fraction was called alpha lipoprotein as electrophoresis methods were used. Twenty years later the same relationship was confirmed in large epidemiological studies using ultracentrifugation or precipitation methods. The new interest for HDL research was then stimulated by the reverse cholesterol transport hypothesis (Glomset et al. 1966) and a Lancet review written by Miller and Miller (1975). These authors proposed that low plasma HDL concentration accelerates the development of atherosclerosis by impairing the clearance of cholesterol from the arterial wall.

1 Protective Role of HDL: Evidence from Epidemiological Studies

The protective role of HDL has been well established in several epidemiological studies. An increment of 2.5 % corresponding to 1 mg/dl or 0.04 mmol/l is associated with a 2 and 3 % reduction in CHD risk in men and women, respectively (Gordon et al. 1989; Jacobs et al. 1990). So far, it is mainly the total cholesterol concentration in HDL particles that has been determined using standardized methods applicable also to routine clinical work. However, cholesterol represents only a small fraction, approximately 15 %, of the HDL particle mass. Furthermore, the proportion of free cholesterol to esterified cholesterol varies between lipoprotein particles.

The relationship between HDL-C and CHD is complex and HDL-C may not be an appropriate indicator of the impact of this lipoprotein fraction on cardiovascular risk. This has been demonstrated, e.g., by the fact that carotid intima media thickness is not increased in apoA-I(Milano) mutation carriers with very low levels of HDL-C (Sirtori et al. 2001). Moreover, recently, drugs that increase HDL cholesterol (HDL-C) level have failed to reduce cardiovascular risk, and Mendelian randomization studies have failed to show a causal relationship between HDL-C and cardiovascular diseases (van Capelleveen et al. 2013).

Despite this controversy, the hard evidence for low HDL-C level as a risk factor of atherosclerosis will be described in the first part of this presentation. It is noteworthy that the cholesterol concentration in HDL fraction is not necessarily associated with the antiatherogenic properties of HDL. Therefore, later in this chapter, other potential HDL-related biomarkers for the prevention of atherosclerosis by HDL will be presented. These include not only the other major lipid and apolipoprotein components of HDL but also minor bioactive lipid molecules residing in the HDL particles. Furthermore, the physicochemical characteristics of various subfractions of HDL as well as the large number of molecules circulating more or less firmly bound to HDL particles may contribute to the antiatherogenic potential of HDL via antioxidative and anti-inflammatory effects or cholesterol transport capacity.

2 HDL Cholesterol as a Risk Factor for Atherosclerosis and Its Complications

Until recently, the clinical evaluation of HDL as a risk factor has focused almost exclusively on the total HDL-C without regard to the chemical composition or multiple subclasses of HDL particles. A large body of epidemiological research has shown a solid inverse and independent relationship between HDL-C and the risk of cardiovascular disease (Toth et al. 2013).

Gofman et al. first reported an inverse association between HDL-C levels and the risk of ischemic heart disease (Gofman et al. 1966). This was shown in case–control studies from Framingham and Livermore cohorts with 10–12 years of follow-up. Later, the inverse relationship has been observed also in several larger studies in the USA (the Honolulu Heart Program, Rhoads et al. 1976, and the Framingham Heart Study, Gordon et al. 1977), in Norway (the Tromsø Heart Study, Miller et al. 1977), in Germany (the Prospective Cardiovascular Münster Study, Assmann et al. 1996), and in Israel (the Israeli Ischemic Heart Disease Study, Goldbourt et al. 1997). Recent meta-analyses have corroborated the relationship between HDL-C and atherosclerosis and its complications (Chirovsky et al. 2009; Boekholdt et al. 2013; Touboul et al. 2014). The association is independent of triglyceride levels and other risk factors (Goldbourt et al. 1997). Many CHD risk algorithms have included HDL-C as a factor to improve the prediction of CHD events (Cooper et al. 2005; Halcox et al. 2013; Hippisley-Cox et al. 2013; Tehrani et al. 2013).

Recently, the picture has become less clear. Mendelian randomization studies have not supported a causal effect of HDL-C in the atherosclerotic disease process (Voight et al. 2012; Holmes et al. 2014). Moreover, statin trials have shown that HDL-C is predictive among patients treated with statin even at low LDL levels (Barter et al. 2007), whereas it is not predictive among patients taking placebo (Ridker et al. 2010; Mora et al. 2012). Further research is needed to clarify the role of HDL-C as a risk factor. It is possible that other characteristics of HDL particles may be more important in this respect than the total cholesterol concentration.

In patients with CHD, the protective role of HDL-C is controversial (Silbernagel et al. 2013). Some studies have shown that low HDL-C is associated with atherosclerotic progression in myocardial infarction survivors (Johansson et al. 1991; Duffy et al. 2012; Liosis et al. 2013), provides additional prognostic value also in patients with acute coronary syndrome (Correia et al. 2009), and reduces the risk after coronary interventions (Sattler et al. 2009), whereas some studies have shown that HDL-C has no protective role in the secondary prevention of CHD after bypass operation (Angeloni et al. 2013).

3 HDL Cholesterol as a Risk Factor for Other Diseases

The low level of HDL-C has been shown to be a risk factor for other diseases not related to atherosclerosis. Recent reports have shown that HDL-C may be a biomarker for diseases like psoriasis (Holzer et al. 2012), rheumatoid arthritis

(Raterman et al. 2013), and liver fibrosis in hepatitis C patients (Gangadharan et al. 2012). Alterations in HDL composition have been observed also in hemodialysis patients (Mangé et al. 2012).

Special interest has been recently focused on HDL in cancer patients. Several epidemiological studies have shown that low HDL-C level may be a risk and/or prognostic factor of biliary tract cancers (Andreotti et al. 2008), prostate cancer (Mondul et al. 2011; Kotani et al. 2013), colon cancer (van Duijnhoven et al. 2011), breast cancer (Furberg et al. 2005, Melvin et al. 2012) and gastric cancer (Tamura et al. 2012), but not that of endometrial cancer (Cust et al. 2007; Esposito et al. 2014) or rectal cancer (van Duijnhoven et al. 2011). Confounding factors related to obesity and metabolic syndrome may be more strongly associated with the latter cancer types (Kotani et al. 2013).

The association with low HDL-C levels is shared among many types of cancer, and it is mainly linked to obesity and inflammation, suggesting a common pathway (Melvin et al. 2013; Vílchez et al. 2014). The mechanism of cancer protection is not known. Lipoprotein particles may carry cancerogenic molecules but at least one small study has shown that persistent organogenic pollutants in HDL particles are more associated with CVD than cancer (Ljunggren et al. 2014).

The potential importance of HDL particles in cancer protection has recently led to attempts to develop cancer treatments by HDL-mimetic synthetic nanoparticles (Zheng et al. 2013; Yang et al. 2013). These interesting approaches need to be followed closely in the future.

4 Total HDL-C in Various Populations

Racial differences. Several studies have shown differences in HDL-C between various ethnic groups (Thelle et al. 1982; Heiss et al. 1984; Haffner et al. 1986; Saha 1987). The differences may in part be due to genetic factors but the role of behavioral and anthropometric variables seem to affect the HDL-C (Thelle et al. 1982; Haffner et al. 1986). Environmental factors such as diet, smoking, and alcohol use may confound the differences observed between ethnic groups.

In the USA, the African-American population has higher HDL-C than Caucasians after adjustment for weight, smoking, and alcohol consumption than Caucasians (Heiss et al. 1984), while people originating from India have lower HDL-C than Europeans or many other populations (Saha 1987; de Munter et al. 2011). Differences in HDL-C and other lipoproteins between various ethnic groups can be observed already in children (van Vliet et al. 2011). Even at an early age, the HDL-C levels are confounded by other cardiovascular risk factors including blood pressure, overweight, and obesity.

Gender differences. Women have higher HDL-C levels than men (Heiss et al. 1980; Seidell et al. 1991), and significantly increased CHD risk is defined at levels below 50 mg/dl (1.29 mmol/l) and 40 mg/dl (1.03 mmol/l), respectively. The reason for the gender difference may mainly be sex hormones but the fat distribution seems to play a central role since adjustment for waist/thigh ratio almost totally removed the gender difference in HDL cholesterol (Seidell et al. 1991). In women HDL-C levels decline after menopause (Heiss et al. 1980) but the sex difference remains significant even in the seventh decade of life (Ostlund et al. 1990).

Age-related differences. Lipoprotein levels are very low at birth and then increase during childhood. HDL-C levels decrease in males during puberty and early adulthood and thereafter remain lower than those in women (Kreisberg and Kasim 1987; Walter 2009). HDL-C levels decline with age and low HDL cholesterol remains a powerful risk predictor into old age (Kreisberg and Kasim 1987; Walter 2009). A selection bias by HDL-lowering genetic variation may explain why HDL deficiency is rare among very old people (Kervinen et al. 1994; Baggio et al. 1998).

5   Total HDL-C Modulated by Environmental Factors

Alcohol consumption in moderation is associated with a reduced risk for cardiovascular diseases. Alcohol consumption increases the concentration of HDL-C, possibly secondary to an inhibition of CETP (cholesteryl ester transfer protein) (Savolainen et al. 1990; Hannuksela et al. 1992). A common polymorphism in the CETP gene (TaqIB2) modifies the relationship of alcohol intake with HDL-C suggesting a gene-environment interaction on the risk of CHD (Jensen et al. 2008). However, it has not been definitely shown whether the HDL-C elevation associated with alcohol consumption is cardioprotective.

The quantity of alcohol needed to increase HDL-C by 0.1 mmol/l (3.87 mg/dl) is about 30 g/d (Brien et al. 2011). Therefore, alcohol drinking cannot be recommended as a method to raise low HDL-C levels. The alcohol-induced increase in HDL-C occurs commonly without significant changes in other lipids although alcohol may increase triglyceride levels especially in subjects with elevated triglyceride concentration.

The death rate from CHD among moderate alcohol users is lower than in total abstainers or heavy drinkers (Rimm et al. 1999; Brien et al. 2011; Bergmann et al. 2013). Even very low alcohol consumption, e.g., a couple of drinks per week, seems to protect from CHD even though it does not have any significant effect on HDL-C. To explain this fact, it has been suggested that ethanol metabolism may produce specific bioactive lipids, e.g., phosphatidylethanol, that could serve as a memory molecule in the body (Liisanantti et al. 2004). If HDL particles of alcohol drinkers contain this bioactive lipid, it could circulate for several days even without daily alcohol drinking, and when HDL enters into the endothelial cell, it could then exert its positive effects on the vascular endothelium (Liisanantti and Savolainen 2005). However, the concentration of phosphatidylethanol in HDL may be in the low nanomolar range which makes the analysis challenging and may hamper its determination of epidemiological studies.

Smoking reduces HDL-C level and smoking cessation is associated with an increase in the plasma concentration of HDL-C (Maeda et al. 2003). The mechanism by which smoking reduces HDL-C is not known. It is noteworthy that in many cases smoking is associated with alcohol drinking and therefore smoking could attenuate the alcohol-induced increase in HDL-C.

Physical activity is associated with high HDL-C (Marti 1991). Low level of physical activity is very common in the developed countries, and therefore, increasing the level of exercise might be more beneficial for HDL-C than any other preventive measure.

Education and socioeconomic status. HDL-C levels increase with income and educational attainment after controlling diet, exercise, and other risk factors for elevated cholesterol (Muennig et al. 2007). The mechanisms are not clear. It has been suggested that stress differences by social class may play a role (Muennig et al. 2007). However, several lifestyle factors including leisure-time physical activity, smoking, alcohol drinking, and dietary habits correlate with the socioeconomic status, classified as the degree of educational level (Schröder et al. 2004).

Dietary carbohydrates affect the lipoprotein profile. The epidemiological studies can be divided into three categories. First, the type of carbohydrate modulates the impact of carbohydrates on plasma lipoproteins. The intake of refined carbohydrates has increased in Western societies, and they have more deleterious effects on abdominal obesity and consequently on insulin resistance and hepatic lipogenesis in comparison with complex carbohydrates or starches (Ma et al. 2012; Stanhope et al. 2009). The end result of high intake of refined carbohydrates is a low HDL-C level (Heiss et al. 1980; Sonestedt et al. 2012).

Second, trials focusing on dietary carbohydrate restriction have shown modulation of atherogenic dyslipidemia. The effect on HDL-C is modest but commonly greater than that on total cholesterol, and thus, the atherogenic burden is improved. Results from epidemiological studies are difficult to interpret due to differences in diets.

The third type of studies involves the replacement of carbohydrate with different fats in order to maintain isocaloric intake of macronutrients. A classic meta-analysis of 60 trials (Mensink et al. 2003) showed that replacement of 10 % of energy from carbohydrate with saturated fat, monounsaturated fat, and polyunsaturated fat increased HDL-C by 4.7, 3.4, and 2.8 mg/dl (0.12, 0.09, and 0.07 mmol/l), respectively. However, LDL cholesterol increased by 13 mg/dl (0.34 mmol/l) with saturated fat substitution and decreased by 3.3 mg/dl (0.09 mmol/l) with polyunsaturated fat substitution while the substitution with monounsaturated fat had no effect on LDL-C. Thus, the atherogenicity of the plasma lipoprotein profile improved with the unsaturated fat substitution of carbohydrate.

The effects of dietary carbohydrates on HDL-C and CHD risk have been analyzed also on the basis of their glycemic index (GI, the effect on blood glucose level) or glycemic load (GL, including carbohydrate content and intake of foods in addition to GI). High GL and GI were associated with significant increased risk of CVDs, specifically for women. Several cross-sectional studies have reported inverse associations of low GI and GL diets with HDL-C, but meta-analyses (Kelly et al. 2004; Goff et al. 2013) have not found any effect on HDL-C.

Fatty acids and especially omega-3 fatty acids have been the focus of many epidemiological studies. The most important omega-3 fatty acids in this respect are eicosapentaenoic acid (EPA) and docosahexaenoic acid (DHA) that increase HDL-C by 1–3 % (Balk et al. 2006).

6   HDL-C in Diseases and Conditions

Type 2 diabetes. Patients with type 2 diabetes may have various types of dyslipidemias that usually are accompanied with a low concentration of HDL-C (Chehade et al. 2013; Morgantini et al. 2014). Low HDL-C contributes to diabetes-related CHD risk more in women than in men, the diabetes-related hazard ratio for a major CHD event being 3 times higher in women after adjustment for other cardiovascular risk factors (Juutilainen et al. 2004). On the other hand, lipid composition of HDL particles is associated with the development of metabolic syndrome (Abbasi et al. 2013; Onat et al. 2013).

Dyslipidemias. Traditionally, dyslipidemia has been characterized by an elevation in plasma triglycerides and total or LDL cholesterol and reduction in HDL-C. This is commonly referred to as the atherogenic triad. LDL particles are small and dense (Mooradian 2009). Genetic studies have indicated linkage of apoB gene to peak LDL size and plasma triglycerides, HDL-C, and apoB levels.

Obesity. Overweight and obese subjects usually have low HDL-C concentrations (Seidell et al. 1991). The underlying cause may be insulin resistance that promotes free fatty acid flux to the liver, stimulates hepatic lipogenesis, and finally enhances the secretion of triglyceride-rich apoB-containing lipoproteins from the liver. The excess triglyceride-rich particles also enhance the CETP-mediated exchange of cholesteryl esters from HDL to apoB-containing lipoprotein particles and simultaneous transfer of triglycerides into HDL (Mann et al. 1991; Liinamaa et al. 1997). This may further reduce HDL-C levels.

Weight reduction is an important modulator of the lipoprotein profile and long-term weight reduction increases HDL-C levels especially in subjects with type 2 diabetes. It is noteworthy that the effect of weight reduction on HDL-C is different in the initial weight loss period compared with the weight maintenance phase. A meta-analysis has shown that HDL-C is increased by 0.35 mg/dl (0.009 mmol/l) per kilogram weight lost during the stable weight reduction. During active weight loss, however, HDL-C is reduced by 0.27 mg/dl (0.007 mmol/l) for every kilogram of weight lost (Browning et al. 2011). Theoretically, an increase in HDL-C of 2.45 mg/dl (0.06 mmol/l) could be expected during the stable weight maintenance stage after a weight reduction of 7 % in an individual with an initial weight of 100 kg. Furthermore, this would translate into 7.4 % reduction in CHD risk in women.

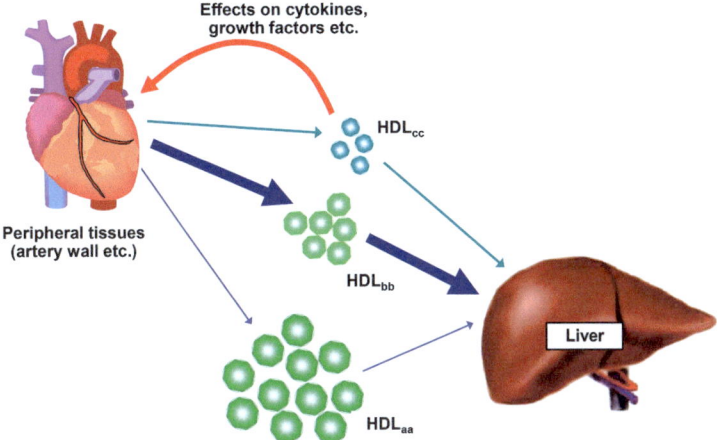

Fig. 1 Potential differences in the functions of HDL particles. HDL_{aa} depicts a theoretical subclass of particles comprising the bulk of plasma HDL-C concentration but without any significant role in reverse cholesterol transport or pleiotropic effects of HDL, whereas HDL_{bb} may be a smaller fraction but more active in reverse cholesterol transport. HDL_{cc} could be a small subclass without any significant contribution to the HDL-C quantity. These particles may, however, have many pleiotropic effects on cytokines, growth factors, etc. [Modified from Hannuksela et al. (2004)]

7 High HDL Levels Do Not Add to the Protection

While low levels of HDL-C are associated with increased CHD risk, high HDL-C levels are not uniformly atheroprotective. At higher concentrations the curve of CHD risk gradually tapers off as depicted in Fig. 1 which is based on the results from the Framingham Heart Study (Gordon et al. 1977), and there is no additional effect of higher HDL-C levels compared with average HDL-C concentrations.

8 Effect of HDL on Stroke

In contrast to the role of HDL-C as a major risk factor of CHD, the role of HDL-C in the pathogenesis of ischemic stroke is less clear. Epidemiological studies of carotid intima media thickness and stroke protection have provided conflicting results. Even in the studies reporting positive results, the effect of HDL-C on protection of stroke is modest (a 10 mg/dl or 0.38 mmol/l increase in HDL-C reduces stroke risk by 11–15 %) compared with its protective effect on CHD. The discordant results in prospective cohort studies as well as in case-control studies may be due to the heterogeneity of stroke, since dyslipidemia including low HDL-C levels may not be involved in the pathogenesis of some subtypes such as lacunar and cardioembolic strokes (Amarenco et al. 2008).

9 Time Trends in Total HDL-C

During the last decades, favorable trends in HDL-C levels have occurred in US adults and youths aged 6–19 years despite changes in physical activity, obesity, and diabetes (Carroll et al. 2012; Kit et al. 2012). Similar trends have been observed also in Europe and India (Muntoni et al. 2009; Gupta et al. 2012).

10 Are There Other Biomarkers than the Total HDL-C?

The cholesterol in HDL (i.e., HDL-C) may reflect the rate of reverse cholesterol transport from the peripheral tissues to liver. However, it may be that the bulk of HDL cholesterol does not necessarily represent the particles that are involved in the reverse cholesterol transport or any other potentially antiatherogenic activity of HDL (Fig. 1).

HDL particles comprise a heterogeneous fraction of lipoprotein which vary by the apolipoprotein and lipid content and consequently also in density, shape, size, and charge. Various subfractions can be analyzed, e.g., by ultracentrifugation (density), electron microscopy (shape), nondenaturing gel electrophoresis, gel filtration or nuclear magnetic resonance (NMR) spectroscopy (size), NMR or immunoaffinity (composition), and 2-dimensional electrophoresis (charge).

11 HDL Fractions

Analytic ultracentrifugation was first used to separate subclasses of HDL (DeLalla et al. 1954). Later, the sequential ultracentrifugation and various precipitation methods enabled the analysis of larger series in epidemiological studies.

Sequential ultracentrifugation methods have enabled separation of two major HDL subfractions, namely, HDL2 and HDL3, while the isolation of further subclasses has required more tedious methods such as gradient ultracentrifugation. Nuclear magnetic resonance (NMR) spectroscopy has simplified the differentiation into five subclasses. However, the nomenclature has varied and it is generally quite difficult to compare HDL subclasses separated by different methods. Recently, a unified nomenclature for a simplified differentiation into 5 subclasses according to particle size (very small, small, medium, large, and very large HDL) has been proposed (Rosenson et al. 2011).

Controversial results have been obtained in studies aimed at distinguishing cardiovascular differences between HDL subclasses separated by density (Superko et al. 2012; Pirillo et al. 2013). A majority of the studies, however, have found HDL2-C to be more predictive of CHD risk than total HDL-C or HDL3-C (Johansson et al. 1991; Drexel et al. 1992; Lamarche et al. 1997). A recent study with 29-year follow-up of the Gofman's Livermore cohort showed that HDL2 and HDL3 are independently related to CHD risk (Williams and Feldman 2011).

12 HDL Particle Size

Further characterization of HDL subfractions by nondenaturing polyacrylamide gels has identified three HDL3 and two HDL2 subclasses. Very large HDL particles (HDL2b subfraction with diameter between 9.7 and 12.9 nm) are strongly correlated with the total HDL-C concentration and most strongly inversely related to CHD risk in normotriglyceridemic subjects (Johansson et al. 1991). On the other hand, an increased concentration small HDL particles (HDL3b subfraction, 7.8–8.2 nm) is associated with an atherogenic lipoprotein profile characterized by low HDL2b levels, high plasma triglyceride concentration, and increased level of small, dense LDL particles (Berneis and Krauss 2002). Low concentration of HDL2b subclass has also been shown in patients with T2DM (Xian et al. 2009).

13 HDL Particle Number

The complexity of HDL metabolism leads to the formation of multiple HDL subpopulations with varying density, size, charge, and chemical composition. Two individuals with the same HDL-C concentration may have HDL particles of different size distribution, and consequently, the particle number is different. Recently, NMR methods have enabled high-throughput determination of HDL particle concentration. Plasma levels of large particles and low particle number are consistently associated with low CHD prevalence (Mora et al. 2012).

14 HDL Lipids

Some of the effects of HDL on cellular functions are mediated by sphingomyelin-1-phosphate (S1P) that is a bioactive lipid in HDL. Alterations in S1P content in HDL particles could explain the dysfunction of HDL since the HDL-associated S1P seems to be responsible for many of the pleiotropic effects of HDL by activating special S1P receptors. In epidemiological studies low levels of S1P in HDL has been associated with coronary heart disease (Argraves et al. 2011). Sphingomyelin content in HDL particles has been associated with coronary heart disease in postmenopausal women (Horter et al. 2002). Sphingomyelin has also been associated with kidney disease in patients with type 1 diabetes (Mäkinen et al. 2012).

15 HDL Apolipoproteins

The structure of HDL particles is very complex, apoA-I and apoA-II being the major protein components. ApoA-I accounts for about two-thirds of the protein content of HDL and is also functionally important since it is the acceptor in the efflux of phospholipids and free cholesterol from peripheral cells. The measurement of apolipoprotein levels is more expensive and time-consuming than that of lipid concentrations. To circumvent this, data from large epidemiological cohorts was recently used for the development of computer software that enables accurate estimation of apolipoprotein levels on the basis of lipid measurements (Raitakari et al. 2013).

High plasma levels of apoA-I are protective against atherosclerosis in some studies (Arsenault et al. 2011; Emerging Risk Factor Collaboration 2012). It has been proposed that apoA-I concentration could even be a better predictor of atherosclerosis development than HDL-C. A recent study reported that adjustment for apoA-I changes the direction of the association between HDL-C and the severity of atherosclerotic lesions as determined with coronary artery calcium score (Sung et al. 2013). Among patients treated with statin therapy, apoA-I levels are strongly associated with a reduced cardiovascular risk, even among those achieving very low LDL-C level (Boekholdt et al. 2013).

ApoA-II is the second major apolipoprotein in HDL particles. It is present in most but not all particles and the clinical significance of apoA-II-containing and apoA-II-free particles is controversial (Rosenson et al. 2011).

Some studies have reported enrichment with apoC-III in the HDL of patients with CHD (Vaisar et al. 2010; Kavo et al. 2012; Jensen et al. 2012). Interestingly, Jensen and coworkers found that HDL particles without apoC-III were inversely associated with the risk of CHD while apoC-III-containing HDL particles were directly associated with increased risk of CHD (Jensen et al. 2012). HDL with apoC-III comprised about 13 % of the total HDL-C.

16 HDL Proteomics

Several proteins circulate attached to HDL particles although most of them in much lower numbers than apolipoproteins mentioned above. The protein content of HDL particles has been characterized recently using various methods commonly described as proteomics. It has been shown that HDL2 fraction of CHD patients carries a distinct protein cargo (Vaisar et al. 2010; Gordon et al. 2010; Kavo et al. 2012; Alwaili et al. 2012).

Proteomics as a new promising approach for detecting biomarkers and mechanisms could potentially explain the beneficial characteristics of HDL particles unrelated to their cholesterol content. To this end, novel sophisticated methods have been recently developed (Gordon et al. 2010; Mazur et al. 2010;

Burillo et al. 2013; Hoofnagle et al. 2012; Mazur and Cardasis 2013; Riwanto et al. 2013). High-throughput methods are urgently needed for large-scale epidemiological studies.

17 HDL Function

Cholesterol efflux is the first step of reverse cholesterol transport. Excess cellular cholesterol from peripheral tissues is effluxed to extracellular HDL-based acceptor particles through the action of active transporters and passive diffusion. Recently, it was reported that the cholesterol efflux measured from cultured macrophages enriched with free cholesterol to apoB-depleted serum as cholesterol acceptor was inversely associated with the history of CHD independent of HDL-C levels (Khera et al. 2011). However, more recently, enhanced efflux has been associated with high cardiovascular risk (Li et al. 2013). Further studies are urgently needed to resolve the issue with conflicting results.

Antioxidative properties of HDL are mainly attributed to apoA-I although paraoxonase 1 (PON1) may also play an important role. Antioxidative activity of HDL subfractions increases with increment in density as follows: $HDL_{2b} < HDL_{2a} < HDL_{3a} < HDL_{3b} < HDL_{3c}$ (Kontush et al. 2003). The determination of the antioxidative capacity of HDL particles is technically challenging, and therefore, indirect approaches measuring arylesterase or paraoxonase activities have been used in larger cohorts.

Molecules carried by HDL particles. Recent studies have shown that short noncoding RNAs (microRNAs, miRNAs) are present in the circulation as a result of cellular damage or secretion (Laterza et al. 2009). miRNAs are ideal biomarkers since they are stable and their sequences can be easily amplified. Alterations in circulating miRNA profiles have been associated with cardiovascular risk factors such as hypertension, diabetes, and dyslipidemias as well as with cardiovascular diseases such as coronary heart disease, myocardial infarction, and heart failure (Fichtlscherer et al. 2011; de Rosa et al. 2011). However, this association has not been found in all studies (Wagner et al. 2013).

Several proteins are also attached to HDL particles. Paraoxonase 1 (PON1) is an HDL-associated enzyme that has been suggested to mediate many antiatherogenic and cardioprotective effects of HDL particles (Mackness et al. 2004; Aviram and Vaya 2013).

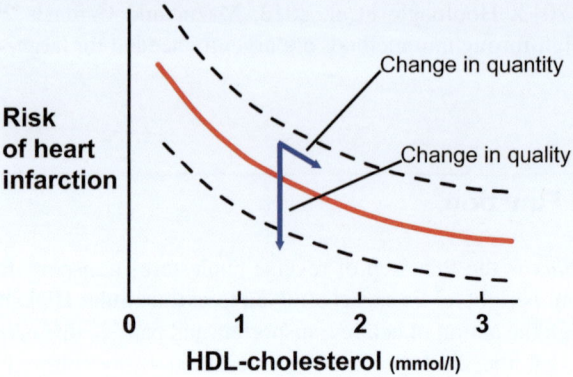

Fig. 2 Relation between HDL-C and risk of myocardial infarction. The *solid curve* shows the average risk level by HDL-C concentration. The risk of individuals, however, may vary substantially, and even at the same HDL-C level, the CHD risk could be anywhere between the *dotted lines*. Furthermore, the nearly *horizontal arrow* depicts how previous approaches have attempted to improve the atheroprotective capacity of HDL by increasing the quantity of HDL-C. Nowadays, several studies have suggested that it might be more important to change the quality of HDL particles as shown by the *arrow* down. Various biomarkers and function assays have been commonly used to determine the improved quality that ultimately could lead to lower risk of myocardial infarction

18 Pleiotropy

In addition to the antioxidative properties HDL particles have also other functions independent of the effects of HDL particles on cholesterol homeostasis. These include anti-inflammatory, anti-infective, antithrombotic, and endotoxin-neutralizing effects as well as effects on endothelial function. These pleiotropic effects have been recently reviewed (Annema and von Eckardstein 2013). The laboratory assays used for analyzing the pleiotropic effects of HDL are demanding and need to be standardized before they can be used in large epidemiological studies in order to improve the CHD risk assessment.

19 Future Approaches of Epidemiological Studies

New approaches have been used in order to resolve the complex effects of diseases, conditions, environmental factors, and genes in relation to the protective role of HDL. On one hand, there have been efforts to dissect the HDL fraction into as many well-defined subfractions and individual molecules of HDL particles as possible. On the other hand, as discussed above, the focus has been shifted from the structure and composition to the function of HDL particles (Fig. 2). Moreover, recent development in the field of "omics" (e.g., metabonomics, lipidomics, proteomics, etc.) has enabled the analysis of more holistic patterns on lipoproteins and subfractions and their relation to the risk of CHD (Ala-Korpela 2008; Rosenson et al. 2011; Würtz et al. 2012).

Dietary patterns have been used as a complementary approach to the traditional single-nutrient analysis (Randall et al. 1990; Bogl et al. 2013). Dietary patterns are, e.g., "fruit and vegetables," "meat," "sweets and desserts," "junk food," and "fish." The "junk food" pattern is characterized by higher intakes of energy-dense nutrient-poor foods, such as hamburger, pizza, French fries, salty snacks, and liquorices and lower intakes of porridge, rye bread, and fruit. The "junk food" pattern distinguishes from the commonly used definition of "Western pattern" in that "junk food" does not contain high amounts of meat, eggs, or high-fat dairy. The healthy diet pattern has been associated with higher levels of HDL-C, whereas the fast-food dietary pattern, high in saturated fat, did not have this effect (Hamer and Mishra 2010).

The idea of analyzing the whole diet through dietary patterns is based on the fact that people do not eat single purified nutrients or simple foods, but mixed meals consisting of several foods and nutrients at a time. The term "nutritional epidemiology" depicting food patterns was coined already in the 1970s (Krehl 1977) but has only recently used more frequently.

The Mediterranean diet refers to a dietary profile commonly available in the early 1960s in the Mediterranean regions and characterized by a high consumption of fruit, vegetables, legumes, and complex carbohydrates, with a moderate consumption of fish, and the consumption of olive oil as the main source of fats and a low-to-moderate amount of red wine during meals (Sofi et al. 2010). A recent meta-analysis of 50 studies summarized the impact of the Mediterranean diet on CHD risk factors (Kastorini et al. 2011). The diet is associated with a 3 % increase in HDL-C, but many other lipid and non-lipid risk factors such as hypertriglyceridemia, blood pressure, glucose, insulin resistance, and abdominal obesity are also affected.

> ### Conclusion
> HDL-C is a strong and independent predictor of major cardiovascular events in a wide range of populations, in men and women with or without preceding CHD. As molecular biology and related approaches have revealed new biomarkers or profiles in HDL structure and function, the epidemiological research is also moving from the classical determination of total HDL-C to structure-function analyses. Recent methodological breakthroughs in HDL structure analyses have enabled their use also in large cohorts. In the future, further investigations are urgently needed for validation and clinical applications of HDL function assays.

Open Access This chapter is distributed under the terms of the Creative Commons Attribution Noncommercial License, which permits any noncommercial use, distribution, and reproduction in any medium, provided the original author(s) and source are credited.

References

Abbasi A, Corpeleijn E, Gansevoort RT, Gans RO, Hillege HL, Stolk RP, Navis G, Bakker SJ, Dullaart RP (2013) Role of HDL cholesterol and estimates of HDL particle composition in future development of type 2 diabetes in the general population: the PREVEND study. J Clin Endocrinol Metab 98:E1352–E1359. doi:10.1210/jc.2013-1680

Ala-Korpela M (2008) Critical evaluation of 1H NMR metabonomics of serum as a methodology for disease risk assessment and diagnostics. Clin Chem Lab Med 46:27–42

Alwaili K, Bailey D, Awan Z, Bailey SD, Ruel I, Hafiane A, Krimbou L, Laboissiere S, Genest J (2012) The HDL proteome in acute coronary syndromes shifts to an inflammatory profile. Biochim Biophys Acta 1821:405–415. doi:10.1016/j.bbalip.2011.07.013

Amarenco P, Labreuche J, Touboul PJ (2008) High-density lipoprotein-cholesterol and risk of stroke and carotid atherosclerosis: a systematic review. Atherosclerosis 196:489–496

Andreotti G, Chen J, Gao YT, Rashid A, Chang SC, Shen MC, Wang BS, Han TQ, Zhang BH, Danforth KN, Althuis MD, Hsing AW (2008) Serum lipid levels and the risk of biliary tract cancers and biliary stones: a population-based study in China. Int J Cancer 122:2322–2329

Angeloni E, Paneni F, Landmesser U, Benedetto U, Melina G, Lüscher TF, Volpe M, Sinatra R, Cosentino F (2013) Lack of protective role of HDL-C in patients with coronary artery disease undergoing elective coronary artery bypass grafting. Eur Heart J 34:3557–3562. doi:10.1093/eurheartj/eht163

Annema W, von Eckardstein A (2013) High-density lipoproteins. Multifunctional but vulnerable protections from atherosclerosis. Circ J 77:2432–2448

Argraves KM, Sethi AA, Gazzolo PJ, Wilkerson BA, Remaley AT, Tybjaerg-Hansen A, Nordestgaard BG, Yeatts SD, Nicholas KS, Barth JL, Argraves WS (2011) S1P, dihydro-S1P and C24:1-ceramide levels in the HDL-containing fraction of serum inversely correlate with occurrence of ischemic heart disease. Lipids Health Dis 10:70. doi:10.1186/1476-511X-10-70

Arsenault BJ, Boekholdt SM, Kastelein JJ (2011) Lipid parameters for measuring risk of cardiovascular disease. Nat Rev Cardiol 8:197–206. doi:10.1038/nrcardio.2010.223

Assmann G, Schulte H, von Eckardstein A, Huang Y (1996) High-density lipoprotein cholesterol as a predictor of coronary heart disease risk. The PROCAM experience and pathophysiological implications for reverse cholesterol transport. Atherosclerosis 124(Suppl):S11–S20

Aviram M, Vaya J (2013) Paraoxonase 1 activities, regulation, and interactions with atherosclerotic lesion. Curr Opin Lipidol 24:339–344. doi:10.1097/MOL.0b013e32835ffcfd

Baggio G, Donazzan S, Monti D, Mari D, Martini S, Gabelli C, Dalla Vestra M, Previato L, Guido M, Pigozzo S, Cortella I, Crepaldi G, Franceschi C (1998) Lipoprotein(a) and lipoprotein profile in healthy centenarians: a reappraisal of vascular risk factors. FASEB J 12:433–437

Balk EM, Lichtenstein AH, Chung M, Kupelnick B, Chew P, Lau J (2006) Effects of omega-3 fatty acids on serum markers of cardiovascular disease risk: a systematic review. Atherosclerosis 189:19–30

Barr DP, Russ EM, Eder HA (1951) Protein-lipid relationships in human plasma. II. In atherosclerosis and related conditions. Am J Med 11:480–493

Barter P, Gotto AM, LaRosa JC, Maroni J, Szarek M, Grundy SM, Kastelein JJ, Bittner V, Fruchart JC (2007) Treating to new targets investigators. HDL cholesterol, very low levels of LDL cholesterol, and cardiovascular events. N Engl J Med 357:1301–1310

Bergmann MM, Rehm J, Klipstein-Grobusch K, Boeing H, Schütze M, Drogan D, Overvad K, Tjønneland A, Halkjær J, Fagherazzi G, Boutron-Ruault MC, Clavel-Chapelon F, Teucher B, Kaaks R, Trichopoulou A, Benetou V, Trichopoulos D, Palli D, Pala V, Tumino R, Vineis P, Beulens JW, Redondo ML, Duell EJ, Molina-Montes E, Navarro C, Barricarte A, Arriola L, Allen NE, Crowe FL, Khaw KT, Wareham N, Romaguera D, Wark PA, Romieu I, Nunes L, Riboli E, Ferrari P (2013) The association of pattern of lifetime alcohol use and cause of death in the European prospective investigation into cancer and nutrition (EPIC) study. Int J Epidemiol 42:1772–1790. doi:10.1093/ije/dyt154

Berneis KK, Krauss RM (2002) Metabolic origins and clinical significance of LDL heterogeneity. J Lipid Res 43:1363–1379

Boekholdt SM, Arsenault BJ, Hovingh GK, Mora S, Pedersen TR, Larosa JC, Welch KM, Amarenco P, Demicco DA, Tonkin AM, Sullivan DR, Kirby A, Colhoun HM, Hitman GA, Betteridge DJ, Durrington PN, Clearfield MB, Downs JR, Gotto AM Jr, Ridker PM, Kastelein JJ (2013) Levels and changes of HDL cholesterol and apolipoprotein A-I in relation to risk of cardiovascular events among statin-treated patients: a meta-analysis. Circulation 128:1504–1512

Bogl LH, Pietiläinen KH, Rissanen A, Kangas AJ, Soininen P, Rose RJ, Ala-Korpela M, Kaprio J (2013) Association between habitual dietary intake and lipoprotein subclass profile in healthy young adults. Nutr Metab Cardiovasc Dis 23:1071–1078. doi:10.1016/j.numecd.2012.11.007

Brien SE, Ronksley PE, Turner BJ, Mukamal KJ, Ghali WA (2011) Effect of alcohol consumption on biological markers associated with risk of coronary heart disease: systematic review and meta-analysis of interventional studies. BMJ 342:d636. doi:10.1136/bmj.d636

Browning JD, Baker JA, Rogers T, Davis J, Satapati S, Burgess SC (2011) Short-term weight loss and hepatic triglyceride reduction: evidence of a metabolic advantage with dietary carbohydrate restriction. Am J Clin Nutr 93:1048–1052. doi:10.3945/ajcn.110.007674

Burillo E, Vazquez J, Jorge I (2013) Quantitative proteomics analysis of high-density lipoproteins by stable 18O-isotope labeling. Methods Mol Biol 1000:139–156. doi:10.1007/978-1-62703-405-0_11

Carroll MD, Kit BK, Lacher DA, Shero ST, Mussolino ME (2012) Trends in lipids and lipoproteins in US adults, 1988–2010. JAMA 308:1545–1554. doi:10.1001/jama.2012.13260

Chehade JM, Gladysz M, Mooradian AD (2013) Dyslipidemia in type 2 diabetes: prevalence, pathophysiology, and management. Drugs 73:327–339. doi:10.1007/s40265-013-0023-5

Chirovsky DR, Fedirko V, Cui Y, Sazonov V, Barter P (2009) Prospective studies on the relationship between high-density lipoprotein cholesterol and cardiovascular risk: a systematic review. Eur J Cardiovasc Prev Rehabil 16:404–423. doi:10.1097/HJR.0b013e32832c8891

Cooper JA, Miller GJ, Humphries SE (2005) A comparison of the PROCAM and Framingham point-scoring systems for estimation of individual risk of coronary heart disease in the Second Northwick Park Heart Study. Atherosclerosis 181:93–100

Correia LC, Rocha MS, Esteves JP (2009) HDL-cholesterol level provides additional prognosis in acute coronary syndromes. Int J Cardiol 136:307–314. doi:10.1016/j.ijcard.2008.05.067

Cust AE, Kaaks R, Friedenreich C, Bonnet F, Laville M, Tjønneland A, Olsen A, Overvad K, Jakobsen MU, Chajès V, Clavel-Chapelon F, Boutron-Ruault MC, Linseisen J, Lukanova A, Boeing H, Pischon T, Trichopoulou A, Christina B, Trichopoulos D, Palli D, Berrino F, Panico S, Tumino R, Sacerdote C, Gram IT, Lund E, Quirós JR, Travier N, Martínez-García C, Larrañaga N, Chirlaque MD, Ardanaz E, Berglund G, Lundin E, Bueno-de-Mesquita HB, van Duijnhoven FJ, Peeters PH, Bingham S, Khaw KT, Allen N, Key T, Ferrari P, Rinaldi S, Slimani N, Riboli E (2007) Metabolic syndrome, plasma lipid, lipoprotein and glucose levels, and endometrial cancer risk in the European prospective investigation into cancer and nutrition (EPIC). Endocr Relat Cancer 14:755–767

de Munter JS, van Valkengoed IG, Stronks K, Agyemang C (2011) Total physical activity might not be a good measure in the relationship with HDL cholesterol and triglycerides in a multi-ethnic population: a cross-sectional study. Lipids Health Dis 10:223. doi:10.1186/1476-511X-10-223

de Rosa S, Fichtlscherer S, Lehmann R, Assmus B, Dimmeler S, Zeiher AM (2011) Transcoronary concentration gradients of circulating microRNAs. Circulation 124(18):1936–1944. doi:10.1161/CIRCULATIONAHA.111.037572

DeLalla OF, Elliott HA, Gofman JW (1954) Ultracentrifugal studies of high density serum lipoproteins in clinically healthy adults. Am J Physiol 179:333–337

Drexel H, Amann FW, Rentsch K, Neuenschwander C, Luethy A, Khan SI, Follath F (1992) Relation of the level of high-density lipoprotein subfractions to the presence and extent of coronary artery disease. Am J Cardiol 70:436–440

Duffy D, Holmes DN, Roe MT, Peterson ED (2012) The impact of high-density lipoprotein cholesterol levels on long-term outcomes after non-ST-elevation myocardial infarction. Am Heart J 163:705–713. doi:10.1016/j.ahj.2012.01.029

Emerging Risk Factors Collaboration, Di Angelantonio E, Gao P, Pennells L, Kaptoge S, Caslake M, Thompson A, Butterworth AS, Sarwar N, Wormser D, Saleheen D, Ballantyne CM, Psaty BM, Sundström J, Ridker PM, Nagel D, Gillum RF, Ford I, Ducimetiere P, Kiechl S, Koenig W, Dullaart RP, Assmann G, D'Agostino RB Sr, Dagenais GR, Cooper JA, Kromhout D, Onat A, Tipping RW, Gómez-de-la-Cámara A, Rosengren A, Sutherland SE, Gallacher J, Fowkes FG, Casiglia E, Hofman A, Salomaa V, Barrett-Connor E, Clarke R, Brunner E, Jukema JW, Simons LA, Sandhu M, Wareham NJ, Khaw KT, Kauhanen J, Salonen JT, Howard WJ, Nordestgaard BG, Wood AM, Thompson SG, Boekholdt SM, Sattar N, Packard C, Gudnason V, Danesh J (2012) Lipid-related markers and cardiovascular disease prediction. JAMA 307:2499–2506. doi:10.1001/jama.2012.6571

Esposito K, Chiodini P, Capuano A, Bellastella G, Maiorino MI, Giugliano D (2014) Metabolic syndrome and endometrial cancer: a meta-analysis. Endocrine 45:28–36. doi:10.1007/s12020-013-9973-3

Fichtlscherer S, Zeiher AM, Dimmeler S (2011) Circulating microRNAs: biomarkers or mediators of cardiovascular diseases? Arterioscler Thromb Vasc Biol 31:2383–2390. doi:10.1161/ATVBAHA.111.226696

Furberg AS, Jasienska G, Bjurstam N, Torjesen PA, Emaus A, Lipson SF, Ellison PT, Thune I (2005) Metabolic and hormonal profiles: HDL cholesterol as a plausible biomarker of breast cancer risk. The Norwegian EBBA study. Cancer Epidemiol Biomarkers Prev 14:33–40

Gangadharan B, Bapat M, Rossa J, Antrobus R, Chittenden D, Kampa B, Barnes E, Klenerman P, Dwek RA, Zitzmann N (2012) Discovery of novel biomarker candidates for liver fibrosis in hepatitis C patients: a preliminary study. PLoS ONE 7:e39603. doi:10.1371/journal.pone.0039603

Glomset JA, Janssen ET, Kennedy R, Dobbins J (1966) Role of plasma lecithin:cholesterol acyltransferase in the metabolism of high density lipoproteins. J Lipid Res 7:638–648

Goff LM, Cowland DE, Hooper L, Frost GS (2013) Low glycaemic index diets and blood lipids: a systematic review and meta-analysis of randomised controlled trials. Nutr Metab Cardiovasc Dis 23:1–10. doi:10.1016/j.numecd.2012.06.002

Gofman JW, Young W, Tandy R (1966) Ischemic heart disease, atherosclerosis, and longevity. Circulation 34:679–697

Goldbourt U, Yaari S, Medalie JH (1997) Isolated low HDL cholesterol as a risk factor for coronary heart disease mortality. A 21-year follow-up of 8000 men. Arterioscler Thromb Vasc Biol 17:107–113

Gordon T, Castelli WP, Hjortland MC, Kannel WB, Dawber TR (1977) High density lipoprotein as a protective factor against coronary heart disease. The Framingham study. Am J Med 62:707–714

Gordon DJ, Probstfield JL, Garrison RJ, Neaton JD, Castelli WP, Knoke JD, Jacobs DR Jr, Bangdiwala S, Tyroler HA (1989) High-density lipoprotein cholesterol and cardiovascular disease. Four prospective American studies. Circulation 79:8–15

Gordon SM, Deng J, Lu LJ, Davidson WS (2010) Proteomic characterization of human plasma high density lipoprotein fractionated by gel filtration chromatography. J Proteome Res 9:5239–5249. doi:10.1021/pr100520x

Gupta R, Guptha S, Gupta VP, Agrawal A, Gaur K, Deedwania PC (2012) Twenty-year trends in cardiovascular risk factors in India and influence of educational status. Eur J Prev Cardiol 19(6):1258–1271. doi:10.1177/1741826711424567

Haffner SM, Stern MP, Hazuda HP, Rosenthal M, Knapp JA (1986) The role of behavioral variables and fat patterning in explaining ethnic differences in serum lipids and lipoproteins. Am J Epidemiol 123:830–839

Halcox JP, Tubach F, Sazova O, Sweet S, Medina J, on behalf of the EURIKA steering committee (2013) Reclassification of European patients' cardiovascular risk using the updated systematic coronary risk evaluation algorithm. Eur J Prev Cardiol (in press)

Hamer M, Mishra GD (2010) Dietary patterns and cardiovascular risk markers in the UK. Low Income Diet and Nutrition Survey. Nutr Metab Cardiovasc Dis 20(7):491–497. doi:10.1016/j.numecd.2009.05.002

Hannuksela M, Marcel YL, Kesäniemi YA, Savolainen MJ (1992) Reduction in the concentration and activity of plasma cholesteryl ester transfer protein by alcohol. J Lipid Res 33:737–744

Hannuksela ML, Rämet ME, Nissinen AE, Liisanantti MK, Savolainen MJ (2004) Effects of ethanol on lipids and atherosclerosis. Pathophysiology 10:93–103

Heiss G, Johnson NJ, Reiland S, Davis CE, Tyroler HA (1980) The epidemiology of plasma high-density lipoprotein cholesterol levels. The lipid research clinics program prevalence study. Summary. Circulation 62(4 Pt 2):IV116–IV136

Heiss G, Schonfeld G, Johnson JL, Heyden S, Hames CG, Tyroler HA (1984) Black-white differences in plasma levels of apolipoproteins: the Evans County heart study. Am Heart J 108:807–814

Hippisley-Cox J, Coupland C, Brindle P (2013) Derivation and validation of QStroke score for predicting risk of ischaemic stroke in primary care and comparison with other risk scores: a prospective open cohort study. BMJ 346:f2573. doi:10.1136/bmj.f2573

Holmes MV, Asselbergs FW, Palmer TM, Drenos F, Lanktree MB, Nelson CP, Dale CE, Padmanabhan S, Finan C, Swerdlow DI, Tragante V, van Iperen EP, Sivapalaratnam S, Shah S, Elbers CC, Shah T, Engmann J, Giambartolomei C, White J, Zabaneh D, Sofat R, McLachlan S, Doevendans PA, Balmforth AJ, Hall AS, North KE, Almoguera B, Hoogeveen RC, Cushman M, Fornage M, Patel SR, Redline S, Siscovick DS, Tsai MY, Karczewski KJ, Hofker MH, Verschuren WM, Bots ML, van der Schouw YT, Melander O, Dominiczak AF, Morris R, Ben-Shlomo Y, Price J, Kumari M, Baumert J, Peters A, Thorand B, Koenig W, Gaunt TR, Humphries SE, Clarke R, Watkins H, Farrall M, Wilson JG, Rich SS, de Bakker PI, Lange LA, Davey Smith G, Reiner AP, Talmud PJ, Kivimäki M, Lawlor DA, Dudbridge F, Samani NJ, Keating BJ, Hingorani AD, Casás JP, on behalf of the UCLEB consortium (2014) Mendelian randomization of blood lipids for coronary heart disease. Eur Heart J (in press)

Holzer M, Wolf P, Curcic S, Birner-Gruenberger R, Weger W, Inzinger M, El-Gamal D, Wadsack C, Heinemann A, Marsche G (2012) Psoriasis alters HDL composition and cholesterol efflux capacity. J Lipid Res 53:1618–1624. doi:10.1194/jlr.M027367

Hoofnagle AN, Becker JO, Oda MN, Cavigiolio G, Mayer P, Vaisar T (2012) Multiple-reaction monitoring-mass spectrometric assays can accurately measure the relative protein abundance in complex mixtures. Clin Chem 58:777–781. doi:10.1373/clinchem.2011.173856

Horter MJ, Sondermann S, Reinecke H, Bogdanski J, Woltering A, Kerber S, Breithardt G, Assmann G, Von Eckardstein A (2002) Associations of HDL phospholipids and paraoxonase activity with coronary heart disease in postmenopausal women. Acta Physiol Scand 176:123–130

Jacobs DR Jr, Mebane IL, Bangdiwala SI, Criqui MH, Tyroler HA (1990) High density lipoprotein cholesterol as a predictor of cardiovascular disease mortality in men and women: the follow-up study of the lipid research clinics prevalence study. Am J Epidemiol 131:32–47

Jensen MK, Mukamal KJ, Overvad K, Rimm EB (2008) Alcohol consumption, TaqIB polymorphism of cholesteryl ester transfer protein, high-density lipoprotein cholesterol, and risk of coronary heart disease in men and women. Eur Heart J 29:104–112

Jensen MK, Rimm EB, Furtado JD, Sacks FM (2012) Apolipoprotein C-III as a potential modulator of the association between HDL-cholesterol and incident coronary heart disease. J Am Heart Assoc 1. pii:jah3-e000232. doi:10.1161/JAHA.111.000232

Johansson J, Carlson LA, Landou C, Hamsten A (1991) High density lipoproteins and coronary atherosclerosis. A strong inverse relation with the largest particles is confined to normotriglyceridemic patients. Arterioscler Thromb 11:174–182

Juutilainen A, Kortelainen S, Lehto S, Rönnemaa T, Pyörälä K, Laakso M (2004) Gender difference in the impact of type 2 diabetes on coronary heart disease risk. Diabetes Care 27:2898–2904

Kastorini CM, Milionis HJ, Ioannidi A, Kalantzi K, Nikolaou V, Vemmos KN, Goudevenos JA, Panagiotakos DB (2011) Adherence to the Mediterranean diet in relation to acute coronary syndrome or stroke nonfatal events: a comparative analysis of a case/case–control study. Am Heart J 162(4):717–724. doi:10.1016/j.ahj.2011.07.012

Kavo AE, Rallidis LS, Sakellaropoulos GC, Lehr S, Hartwig S, Eckel J, Bozatzi PI, Anastasiou-Nana M, Tsikrika P, Kypreos KE (2012) Qualitative characteristics of HDL in young patients of an acute myocardial infarction. Atherosclerosis 220:257–264. doi:10.1016/j.atherosclerosis.2011.10.017

Kelly S, Frost G, Whittaker V, Summerbell C (2004) Low glycaemic index diets for coronary heart disease. Cochrane Database Syst Rev 4, CD004467

Kervinen K, Savolainen MJ, Salokannel J, Hynninen A, Heikkinen J, Ehnholm C, Koistinen MJ, Kesäniemi YA (1994) Apolipoprotein E and B polymorphisms – longevity factors assessed in nonagenarians. Atherosclerosis 105:89–95

Khera AV, Cuchel M, de la Llera-Moya M, Rodrigues A, Burke MF, Jafri K, French BC, Phillips JA, Mucksavage ML, Wilensky RL, Mohler ER, Rothblat GH, Rader DJ (2011) Cholesterol efflux capacity, high-density lipoprotein function, and atherosclerosis. N Engl J Med 364:127–135

Kit BK, Carroll MD, Lacher DA, Sorlie PD, DeJesus JM, Ogden C (2012) Trends in serum lipids among US youths aged 6 to 19 years, 1988–2010. JAMA 308(6):591–600. doi:10.1001/jama.2012.9136

Kontush A, Chantepie S, Chapman MJ (2003) Small, dense HDL particles exert potent protection of atherogenic LDL against oxidative stress. Arterioscler Thromb Vasc Biol 23:1881–1888

Kotani K, Sekine Y, Ishikawa S, Ikpot IZ, Suzuki K, Remaley AT (2013) High-density lipoprotein and prostate cancer: an overview. J Epidemiol 23:313–319

Krehl WA (1977) The nutritional epidemiology of cardiovascular disease. Ann N Y Acad Sci 300:335–359

Kreisberg RA, Kasim S (1987) Cholesterol metabolism and aging. Am J Med 82(1B):54–60

Lamarche B, Moorjani S, Cantin B, Dagenais GR, Lupien PJ, Després JP (1997) Associations of HDL2 and HDL3 subfractions with ischemic heart disease in men. Prospective results from the Québec cardiovascular study. Arterioscler Thromb Vasc Biol 17:1098–1105

Laterza OF, Lim L, Garrett-Engele PW, Vlasakova K, Muniappa N, Tanaka WK, Johnson JM, Sina JF, Fare TL, Sistare FD, Glaab WE (2009) Plasma microRNAs as sensitive and specific biomarkers of tissue injury. Clin Chem 55:1977–1983. doi:10.1373/clinchem.2009.131797

Li XM, Tang WH, Mosior MK, Huang Y, Wu Y, Matter W, Gao V, Schmitt D, Didonato JA, Fisher EA, Smith JD, Hazen SL (2013) Paradoxical association of enhanced cholesterol efflux with increased incident cardiovascular risks. Arterioscler Thromb Vasc Biol 33:1696–1705. doi:10.1161/ATVBAHA.113.301373

Liinamaa MJ, Hannuksela ML, Kesäniemi YA, Savolainen MJ (1997) Altered transfer of cholesteryl esters and phospholipids in plasma from alcohol abusers. Arterioscler Thromb Vasc Biol 17:2940–2947

Liisanantti MK, Savolainen MJ (2005) Phosphatidylethanol in high density lipoproteins increases the vascular endothelial growth factor in smooth muscle cells. Atherosclerosis 180:263–269

Liisanantti MK, Hannuksela ML, Rämet ME, Savolainen MJ (2004) Lipoprotein-associated phosphatidylethanol increases the plasma concentration of vascular endothelial growth factor. Arterioscler Thromb Vasc Biol 24:1037–1042

Liosis S, Bauer T, Schiele R, Gohlke H, Gottwik M, Katus H, Sabin G, Zahn R, Schneider S, Rauch B, Senges J, Zeymer U (2013) Predictors of 1-year mortality in patients with contemporary guideline-adherent therapy after acute myocardial infarction: results from the OMEGA study. Clin Res Cardiol 102:671–677. doi:10.1007/s00392-013-0581-2

Ljunggren SA, Helmfrid I, Salihovic S, van Bavel B, Wingren G, Lindahl M, Karlsson H (2014) Persistent organic pollutants distribution in lipoprotein fractions in relation to cardiovascular disease and cancer. Environ Int 65:93–99. doi:10.1016/j.envint.2013.12.017

Ma XY, Liu JP, Song ZY (2012) Glycemic load, glycemic index and risk of cardiovascular diseases: meta-analyses of prospective studies. Atherosclerosis 223:491–496. doi:10.1016/j.atherosclerosis.2012.05.028

Mackness M, Durrington P, Mackness B (2004) Paraoxonase 1 activity, concentration and genotype in cardiovascular disease. Curr Opin Lipidol 15:399–404

Maeda K, Noguchi Y, Fukui T (2003) The effects of cessation from cigarette smoking on the lipid and lipoprotein profiles: a meta-analysis. Prev Med 37:283–290

Mäkinen VP, Tynkkynen T, Soininen P, Forsblom C, Peltola T, Kangas AJ, Groop PH, Ala-Korpela M (2012) Sphingomyelin is associated with kidney disease in type 1 diabetes (The FinnDiane study). Metabolomics 8:369–375

Mangé A, Goux A, Badiou S, Patrier L, Canaud B, Maudelonde T, Cristol JP, Solassol J (2012) HDL proteome in hemodialysis patients: a quantitative nanoflow liquid chromatography-tandem mass spectrometry approach. PLoS ONE 7:e34107. doi:10.1371/journal.pone.0034107

Mann CJ, Yen FT, Grant AM, Bihain BE (1991) Mechanism for plasma cholesteryl ester transfer in hyperlipidemia. J Clin Invest 88:2059–2066

Marti B (1991) Health effects of recreational running in women. Some epidemiological and preventive aspects. Sports Med 11:20–51

Mazur MT, Cardasis HL (2013) Quantitative analysis of apolipoproteins in human HDL by top-down differential mass spectrometry. Methods Mol Biol 1000:115–137. doi:10.1007/978-1-62703-405-0_10

Mazur MT, Cardasis HL, Spellman DS, Liaw A, Yates NA, Hendrickson RC (2010) Quantitative analysis of intact apolipoproteins in human HDL by top-down differential mass spectrometry. Proc Natl Acad Sci U S A 107:7728–7733. doi:10.1073/pnas.0910776107

Melvin JC, Seth D, Holmberg L, Garmo H, Hammar N, Jungner I, Walldius G, Lambe M, Wigertz A, Van Hemelrijck M (2012) Lipid profiles and risk of breast and ovarian cancer in the Swedish AMORIS study. Cancer Epidemiol Biomarkers Prev 21:1381–1384. doi:10.1158/1055-9965.EPI-12-0188

Melvin JC, Holmberg L, Rohrmann S, Loda M, Van Hemelrijck M (2013) Serum lipid profiles and cancer risk in the context of obesity: four meta-analyses. J Cancer Epidemiol 2013:823849. doi:10.1155/2013/823849

Mensink RP, Zock PL, Kester AD, Katan MB (2003) Effects of dietary fatty acids and carbohydrates on the ratio of serum total to HDL cholesterol and on serum lipids and apolipoproteins: a meta-analysis of 60 controlled trials. Am J Clin Nutr 77(5):1146–1155

Miller GJ, Miller NE (1975) Plasma-high-density-lipoprotein concentration and development of ischaemic heart-disease. Lancet 1:16–19

Miller NE, Thelle DS, Forde OH, Mjos OD (1977) The Tromsø heart-study. High-density lipoprotein and coronary heart-disease: a prospective case-control study. Lancet 8019:965–968

Mondul AM, Weinstein SJ, Virtamo J, Albanes D (2011) Serum total and HDL cholesterol and risk of prostate cancer. Cancer Causes Control 22:1545–1552. doi:10.1007/s10552-011-9831-7

Mooradian AD (2009) Dyslipidemia in type 2 diabetes mellitus. Nat Clin Pract Endocrinol Metab 5:150–159. doi:10.1038/ncpendmet1066

Mora S, Wenger NK, Demicco DA, Breazna A, Boekholdt SM, Arsenault BJ, Deedwania P, Kastelein JJ, Waters DD (2012) Determinants of residual risk in secondary prevention patients treated with high- versus low-dose statin therapy: the treating to new targets (TNT) study. Circulation 125:1979–1987. doi:10.1161/CIRCULATIONAHA.111.088591

Morgantini C, Meriwether D, Baldi S, Venturi E, Pinnola S, Wagner AC, Fogelman AM, Ferrannini E, Natali A, Reddy ST (2014) HDL lipid composition is profoundly altered in patients with type 2 diabetes and atherosclerotic vascular disease. Nutr Metab Cardiovasc Dis. pii:S0939-4753(14)00004-0. doi:10.1016/j.numecd.2013.12.011

Muennig P, Sohler N, Mahato B (2007) Socioeconomic status as an independent predictor of physiological biomarkers of cardiovascular disease: evidence from NHANES. Prev Med 45:35–40

Muntoni S, Atzori L, Mereu R, Manca A, Satta G, Gentilini A, Bianco P, Baule A, Baule GM, Muntoni S (2009) Risk factors for cardiovascular disease in Sardinia from 1978 to 2001: a comparative study with Italian mainland. Eur J Intern Med 20(4):373–377. doi:10.1016/j.ejim. 2008.10.007

Nikkilä EA (1953) Studies on the lipid-protein relationships in normal and pathological sera and the effect of heparin on serum lipoproteins. Scand J Clin Lab Invest 5(Suppl 8):1–101

Onat A, Altuğ Çakmak H, Can G, Yüksel M, Köroğlu B, Yüksel H (2013) Serum total and high-density lipoprotein phospholipids: independent predictive value for cardiometabolic risk. Clin Nutr. pii:S0261-5614(13)00279-3. doi:10.1016/j.clnu.2013.10.020

Ostlund RE Jr, Staten M, Kohrt WM, Schultz J, Malley M (1990) The ratio of waist-to-hip circumference, plasma insulin level, and glucose intolerance as independent predictors of the HDL2 cholesterol level in older adults. N Engl J Med 322(4):229–234

Pirillo A, Norata GD, Catapano AL (2013) High-density lipoprotein subfractions–what the clinicians need to know. Cardiology 124:116–125

Raitakari OT, Mäkinen VP, McQueen MJ, Niemi J, Juonala M, Jauhiainen M, Salomaa V, Hannuksela ML, Savolainen MJ, Kesäniemi YA, Kovanen PT, Sundvall J, Solakivi T, Loo BM, Marniemi J, Hernesniemi J, Lehtimäki T, Kähönen M, Peltonen M, Leiviskä J, Jula A, Anand SS, Miller R, Yusuf S, Viikari JS, Ala-Korpela M (2013) Computationally estimated apolipoproteins B and A1 in predicting cardiovascular risk. Atherosclerosis 226:245–251. doi:10.1016/j.atherosclerosis.2012.10.049

Randall E, Marshall JR, Graham S, Brasure J (1990) Patterns in food use and their associations with nutrient intakes. Am J Clin Nutr 52:739–745

Raterman HG, Levels H, Voskuyl AE, Lems WF, Dijkmans BA, Nurmohamed MT (2013) HDL protein composition alters from proatherogenic into less atherogenic and proinflammatory in rheumatoid arthritis patients responding to rituximab. Ann Rheum Dis 72:560–565. doi:10. 1136/annrheumdis-2011-201228

Rhoads GG, Gulbrandsen CL, Kagan A (1976) Serum lipoproteins and coronary heart disease in a population study of Hawaii Japanese men. N Engl J Med 294:293–298

Ridker PM, Genest J, Boekholdt SM, Libby P, Gotto AM, Nordestgaard BG, Mora S, MacFadyen JG, Glynn RJ, Kastelein JJ, JUPITER Trial Study Group (2010) HDL cholesterol and residual risk of first cardiovascular events after treatment with potent statin therapy: an analysis from the JUPITER trial. Lancet 376:333–339. doi:10.1016/S0140-6736(10)60713-1

Riwanto M, Rohrer L, Roschitzki B, Besler C, Mocharla P, Mueller M, Perisa D, Heinrich K, Altwegg L, von Eckardstein A, Lüscher TF, Landmesser U (2013) Altered activation of endothelial anti- and proapoptotic pathways by high-density lipoprotein from patients with coronary artery disease: role of high-density lipoprotein-proteome remodeling. Circulation 127:891–904. doi:10.1161/CIRCULATIONAHA.112.108753

Rosenson RS, Brewer HB Jr, Chapman MJ, Fazio S, Hussain MM, Kontush A, Krauss RM, Otvos JD, Remaley AT, Schaefer EJ (2011) HDL measures, particle heterogeneity, proposed nomenclature, and relation to atherosclerotic cardiovascular events. Clin Chem 57:392–410. doi:10. 1373/clinchem.2010.155333

Saha N (1987) Serum high density lipoprotein cholesterol, apolipoprotein A-I, A-II and B levels in Singapore ethnic groups. Atherosclerosis 68:117–121

Sattler KJ, Herrmann J, Yün S, Lehmann N, Wang Z, Heusch G, Sack S, Erbel R, Levkau B (2009) High high-density lipoprotein-cholesterol reduces risk and extent of percutaneous coronary intervention-related myocardial infarction and improves long-term outcome in patients undergoing elective percutaneous coronary intervention. Eur Heart J 30:1894–1902. doi:10. 1093/eurheartj/ehp183

Savolainen MJ, Hannuksela M, Seppänen S, Kervinen K, Kesäniemi YA (1990) Increased high-density lipoprotein cholesterol concentration in alcoholics is related to low cholesteryl ester transfer protein activity. Eur J Clin Invest 20:593–599

Schröder H, Rohlfs I, Schmelz EM, Marrugat J, REGICOR investigators (2004) Relationship of socioeconomic status with cardiovascular risk factors and lifestyle in a Mediterranean population. Eur J Nutr 43:77–85

Seidell JC, Cigolini M, Charzewska J, Ellsinger BM, Björntorp P, Hautvast JG, Szostak W (1991) Fat distribution and gender differences in serum lipids in men and women from four European communities. Atherosclerosis 87:203–210

Silbernagel G, Schöttker B, Appelbaum S, Scharnagl H, Kleber ME, Grammer TB, Ritsch A, Mons U, Holleczek B, Goliasch G, Niessner A, Boehm BO, Schnabel RB, Brenner H, Blankenberg S, Landmesser U, März W (2013) High-density lipoprotein cholesterol, coronary artery disease, and cardiovascular mortality. Eur Heart J 34:3563–3571. doi:10.1093/eurheartj/eht343

Sirtori CR, Calabresi L, Franceschini G, Baldassarre D, Amato M, Johansson J, Salvetti M, Monteduro C, Zulli R, Muiesan ML, Agabiti-Rosei E (2001) Cardiovascular status of carriers of the apolipoprotein A-I(Milano) mutant: the Limone sul Garda study. Circulation 103:1949–1954

Stanhope KL, Schwarz JM, Keim NL, Griffen SC, Bremer AA, Graham JL, Hatcher B, Cox CL, Dyachenko A, Zhang W, McGahan JP, Seibert A, Krauss RM, Chiu S, Schaefer EJ, Ai M, Otokozawa S, Nakajima K, Nakano T, Beysen C, Hellerstein MK, Berglund L, Havel PJ (2009) Consuming fructose-sweetened, not glucose-sweetened, beverages increases visceral adiposity and lipids and decreases insulin sensitivity in overweight/obese humans. J Clin Invest 119(5):1322–1334

Sofi F, Abbate R, Gensini GF, Casini A (2010) Accruing evidence on benefits of adherence to the Mediterranean diet on health: an updated systematic review and meta-analysis. Am J Clin Nutr 92:1189–1196. doi:10.3945/ajcn.2010.29673

Sonestedt E, Wirfält E, Wallström P, Gullberg B, Drake I, Hlebowicz J, Nordin Fredrikson G, Hedblad B, Nilsson J, Krauss RM, Orho-Melander M (2012) High disaccharide intake associates with atherogenic lipoprotein profile. Br J Nutr 107:1062–1069. doi:10.1017/S0007114511003783

Sung KC, Wild SH, Byrne CD (2013) Controlling for apolipoprotein A-I concentrations changes the inverse direction of the relationship between high HDL-C concentration and a measure of pre-clinical atherosclerosis. Atherosclerosis 231:181–186. doi:10.1016/j.atherosclerosis.2013.09.009

Superko HR, Pendyala L, Williams PT, Momary KM, King SB 3rd, Garrett BC (2012) High-density lipoprotein subclasses and their relationship to cardiovascular disease. J Clin Lipidol 6:496–523

Tamura T, Inagawa S, Hisakura K, Enomoto T, Ohkohchi N (2012) Evaluation of serum high-density lipoprotein cholesterol levels as a prognostic factor in gastric cancer patients. J Gastroenterol Hepatol 27:1635–1640. doi:10.1111/j.1440-1746.2012.07189.x

Tehrani DM, Gardin JM, Yanez D, Hirsch CH, Lloyd-Jones DM, Stein PK, Wong ND (2013) Impact of inflammatory biomarkers on relation of high density lipoprotein-cholesterol with incident coronary heart disease: cardiovascular Health Study. Atherosclerosis 231:246–251. doi:10.1016/j.atherosclerosis.2013.08.036

Thelle DS, Førde OH, Arnesen E (1982) Distribution of high-density lipoprotein cholesterol according to age, sex, and ethnic origin: cardiovascular disease study in Finnmark 1977. J Epidemiol Community Health 36:243–247

Toth PP, Barter PJ, Rosenson RS, Boden WE, Chapman MJ, Cuchel M, D'Agostino RB Sr, Davidson MH, Davidson WS, Heinecke JW, Karas RH, Kontush A, Krauss RM, Miller M, Rader DJ (2013) High-density lipoproteins: a consensus statement from the National Lipid Association. J Clin Lipidol 7:484–525. doi:10.1016/j.jacl.2013.08.001

Touboul PJ, Labreuche J, Bruckert E, Schargrodsky H, Prati P, Tosetto A, Hernandez-Hernandez-R, Woo KS, Silva H, Vicaut E, Amarenco P (2014) HDL-C, triglycerides and carotid IMT: a meta-analysis of 21,000 patients with automated edge detection IMT measurement. Atherosclerosis 232:65–71. doi:10.1016/j.atherosclerosis.2013.10.011

Vaisar T, Mayer P, Nilsson E, Zhao XQ, Knopp R, Prazen BJ (2010) HDL in humans with cardiovascular disease exhibits a proteomic signature. Clin Chim Acta 411:972–979. doi:10.1016/j.cca.2010.03.023

van Capelleveen JC, Bochem AE, Motazacker MM, Hovingh GK, Kastelein JJ (2013) Genetics of HDL-C: a causal link to atherosclerosis? Curr Atheroscler Rep 15:326. doi:10.1007/s11883-013-0326-8

van Duijnhoven FJ, Bueno-De-Mesquita HB, Calligaro M, Jenab M, Pischon T, Jansen EH, Frohlich J, Ayyobi A, Overvad K, Toft-Petersen AP, Tjønneland A, Hansen L, Boutron-Ruault MC, Clavel-Chapelon F, Cottet V, Palli D, Tagliabue G, Panico S, Tumino R, Vineis P, Kaaks R, Teucher B, Boeing H, Drogan D, Trichopoulou A, Lagiou P, Dilis V, Peeters PH, Siersema PD, Rodríguez L, González CA, Molina-Montes E, Dorronsoro M, Tormo MJ, Barricarte A, Palmqvist R, Hallmans G, Khaw KT, Tsilidis KK, Crowe FL, Chajes V, Fedirko V, Rinaldi S, Norat T, Riboli E (2011) Blood lipid and lipoprotein concentrations and colorectal cancer risk in the European prospective investigation into cancer and nutrition. Gut 60:1094–1102. doi:10.1136/gut.2010.225011

van Vliet M, Heymans MW, von Rosenstiel IA, Brandjes DP, Beijnen JH, Diamant M (2011) Cardiometabolic risk variables in overweight and obese children: a worldwide comparison. Cardiovasc Diabetol 10:106. doi:10.1186/1475-2840-10-106

Vílchez JA, Martínez-Ruiz A, Sancho-Rodríguez N, Martínez-Hernández P, Noguera-Velasco JA (2014) The real role of prediagnostic high-density lipoprotein cholesterol and the cancer risk: a concise review. Eur J Clin Invest 44:103–114. doi:10.1111/eci.12185

Voight BF, Peloso GM, Orho-Melander M, Frikke-Schmidt R, Barbalic M, Jensen MK, Hindy G, Hólm H, Ding EL, Johnson T, Schunkert H, Samani NJ, Clarke R, Hopewell JC, Thompson JF, Li M, Thorleifsson G, Newton-Cheh C, Musunuru K, Pirruccello JP, Saleheen D, Chen L, Stewart A, Schillert A, Thorsteinsdottir U, Thorgeirsson G, Anand S, Engert JC, Morgan T, Spertus J, Stoll M, Berger K, Martinelli N, Girelli D, McKeown PP, Patterson CC, Epstein SE, Devaney J, Burnett MS, Mooser V, Ripatti S, Surakka I, Nieminen MS, Sinisalo J, Lokki ML, Perola M, Havulinna A, de Faire U, Gigante B, Ingelsson E, Zeller T, Wild P, de Bakker PI, Klungel OH, Maitland-van der Zee AH, Peters BJ, de Boer A, Grobbee DE, Kamphuisen PW, Deneer VH, Elbers CC, Onland-Moret NC, Hofker MH, Wijmenga C, Verschuren WM, Boer JM, van der Schouw YT, Rasheed A, Frossard P, Demissie S, Willer C, Do R, Ordovas JM, Abecasis GR, Boehnke M, Mohlke KL, Daly MJ, Guiducci C, Burtt NP, Surti A, Gonzalez E, Purcell S, Gabriel S, Marrugat J, Peden J, Erdmann J, Diemert P, Willenborg C, König IR, Fischer M, Hengstenberg C, Ziegler A, Buysschaert I, Lambrechts D, Van de Werf F, Fox KA, El Mokhtari NE, Rubin D, Schrezenmeir J, Schreiber S, Schäfer A, Danesh J, Blankenberg S, Roberts R, McPherson R, Watkins H, Hall AS, Overvad K, Rimm E, Boerwinkle E, Tybjaerg-Hansen A, Cupples LA, Reilly MP, Melander O, Mannucci PM, Ardissino D, Siscovick D, Elosua R, Stefansson K, O'Donnell CJ, Salomaa V, Rader DJ, Peltonen L, Schwartz SM, Altshuler D, Kathiresan S (2012) Plasma HDL cholesterol and risk of myocardial infarction: a mendelian randomisation study. Lancet 380:572–580. doi:10.1016/S0140-6736(12)60312-2

Wagner J, Riwanto M, Besler C, Knau A, Fichtlscherer S, Röxe T, Zeiher AM, Landmesser U, Dimmeler S (2013) Characterization of levels and cellular transfer of circulating lipoprotein-bound microRNAs. Arterioscler Thromb Vasc Biol 33:1392–1400. doi:10.1161/ATVBAHA.112.300741

Walter M (2009) Interrelationships among HDL metabolism, aging, and atherosclerosis. Arterioscler Thromb Vasc Biol 29:1244–1250. doi:10.1161/ATVBAHA.108.181438

Williams PT, Feldman DE (2011) Prospective study of coronary heart disease vs. HDL2, HDL3, and other lipoproteins in Gofman's Livermore Cohort. Atherosclerosis 214:196–202. doi:10.1016/j.atherosclerosis.2010.10.024 (Epub 2010 Oct 23)

Würtz P, Raiko JR, Magnussen CG, Soininen P, Kangas AJ, Tynkkynen T, Thomson R, Laatikainen R, Savolainen MJ, Laurikka J, Kuukasjärvi P, Tarkka M, Karhunen PJ, Jula A, Viikari JS, Kähönen M, Lehtimäki T, Juonala M, Ala-Korpela M, Raitakari OT (2012) High-throughput quantification of circulating metabolites improves prediction of subclinical atherosclerosis. Eur Heart J 33:2307–2316. doi:10.1093/eurheartj/ehs020

Yang S, Damiano MG, Zhang H, Tripathy S, Luthi AJ, Rink JS, Ugolkov AV, Singh AT, Dave SS, Gordon LI, Thaxton CS (2013) Biomimetic, synthetic HDL nanostructures for lymphoma. Proc Natl Acad Sci U S A 110:2511–2516. doi:10.1073/pnas.1213657110

Zheng Y, Liu Y, Jin H, Pan S, Qian Y, Huang C, Zeng Y, Luo Q, Zeng M, Zhang Z (2013) Scavenger receptor B1 is a potential biomarker of human nasopharyngeal carcinoma and its growth is inhibited by HDL-mimetic nanoparticles. Theranostics 3:477–486. doi:10.7150/thno.6617

Beyond the Genetics of HDL: Why Is HDL Cholesterol Inversely Related to Cardiovascular Disease?

J.A. Kuivenhoven and A.K. Groen

Contents

1 General 286
2 Determinants of Plasma HDL Cholesterol Levels 288
   2.1 Established Primary Regulators of Plasma HDL Cholesterol 288
   2.2 Established Secondary Regulators of Plasma HDL Cholesterol 289
   2.3 Missing Heritability 290
3 Novel Insight into HDL Biology 291
   3.1 De Novo Synthesis of HDL and HDL Binding 292
      3.1.1 Bone Morphogenetic Protein-1 and Procollagen C-Proteinase Enhancer-2 292
      3.1.2 Apolipoprotein M 292
      3.1.3 CTP:Phosphocholine Cytidylyltransferase Alpha (CT Alpha) 293
      3.1.4 Apolipoprotein F 293
      3.1.5 Glucuronic Acid Epimerase 293
      3.1.6 Beta-Chain of ATP Synthase 293
   3.2 HDL Conversion and Remodeling 294
      3.2.1 Angptl Family of Proteins 294
      3.2.2 Tribbles Homolog 1 294
      3.2.3 Tetratricopeptide Repeat Domain/Glycogen-Targeting PP1 Subunit G(L) ... 295
      3.2.4 ppGalNAc-T2 295
      3.2.5 Glucokinase (Hexokinase 4) Regulator 296
References 297

J.A. Kuivenhoven (✉)
Department of Pediatrics, Section Molecular Genetics, University of Groningen, University Medical Center Groningen, Antonius Deusinglaan 1, 9713GZ Groningen, The Netherlands
e-mail: j.a.kuivenhoven@umcg.nl

A.K. Groen
Department of Pediatrics, University of Groningen, University Medical Center Groningen, Antonius Deusinglaan 1, 9713GZ Groningen, The Netherlands

© The Author(s) 2015
A. von Eckardstein, D. Kardassis (eds.), *High Density Lipoproteins*, Handbook of Experimental Pharmacology 224, DOI 10.1007/978-3-319-09665-0_8

Abstract

There is unequivocal evidence that high-density lipoprotein (HDL) cholesterol levels in plasma are inversely associated with the risk of cardiovascular disease (CVD). Studies of families with inherited HDL disorders and genetic association studies in general (and patient) population samples have identified a large number of factors that control HDL cholesterol levels. However, they have not resolved why HDL cholesterol and CVD are inversely related. A growing body of evidence from nongenetic studies shows that HDL in patients at increased risk of CVD has lost its protective properties and that increasing the cholesterol content of HDL does not result in the desired effects. Hopefully, these insights can help improve strategies to successfully intervene in HDL metabolism. It is clear that there is a need to revisit the HDL hypothesis in an unbiased manner. True insights into the molecular mechanisms that regulate plasma HDL cholesterol and triglycerides or control HDL function could provide the handholds that are needed to develop treatment for, e.g., type 2 diabetes and the metabolic syndrome. Especially genome-wide association studies have provided many candidate genes for such studies. In this review we have tried to cover the main molecular studies that have been produced over the past few years. It is clear that we are only at the very start of understanding how the newly identified factors may control HDL metabolism. In addition, the most recent findings underscore the intricate relations between HDL, triglyceride, and glucose metabolism indicating that these parameters need to be studied simultaneously.

Keywords

Gene • Dyslipidemia • Hyperalphalipoproteinemia • Hypoalphalipoproteinemia

Abbreviations

CVD	Cardiovascular disease
GWA	Genome-wide association
HDL	High-density lipoprotein
LDL	Low-density lipoprotein
VLDL	Very-low-density lipoprotein

1 General

Ever since plasma HDL cholesterol concentration was found to be inversely correlated with the risk of cardiovascular disease (CVD), there has been a strong interest in the biological mechanisms that can explain this correlation. From a large number of studies in humans, animals, and tissue culture, it has become clear that HDL exerts many beneficial functions with prominent roles in cellular cholesterol efflux and protection against inflammation. For recent overviews on this topic, see Luscher et al. (2014) and Rye and Barter (2014).

Paradoxically and not always recognized, however, is that rare inborn errors of HDL metabolism have illustrated that an almost complete loss of HDL or very high HDL cholesterol levels do not automatically translate in accelerated or protection from atherosclerosis, respectively. These observations may be related to the small numbers of patients available for studies or the absence/presence of concomitant established risk factors for CVD (such as increased LDL cholesterol, smoking, etc.) in these individuals. However, these findings by themselves indicate that the relation of HDL cholesterol with atherosclerosis is not as straightforward as for LDL cholesterol since in this case increases and decreases are always associated with increased and decreased risk, respectively.

Genetic approaches are frequently used to study whether changes in plasma HDL cholesterol concentration affect atherosclerosis. Such studies are conducted in families, larger patient population samples sharing large-impact mutations in HDL genes, as well as general population samples. Using candidate gene approaches, these studies mostly generated contrasting or confusing results [a short summary of these findings can be found in Chapman et al. (2011)]. Illustrative in this regard were investigations into variation at the locus encoding for the ATP-binding cassette transporter A1 (ABCA1). While a complete loss of ABCA1 function causes near HDL deficiency and often accelerated atherosclerosis in Tangier patients who are referred to the clinic, studies in general population samples indicated that *ABCA1* gene variation is not necessarily related with plasma HDL cholesterol concentration and risk of CVD (for review, see Frikke-Schmidt (2011)).

More recently, it has become possible to study the impact of whole-genome variation on complex diseases which has shed light on our understanding whether or not genes and their products are related to plasma lipid traits and the risk of CVD. In this regard, particularly Mendelian randomization studies showed that genetic variation associated with increased HDL cholesterol does not protect from atherosclerosis (Voight et al. 2012). In this case, it concerned a study of common variants in HDL candidate genes; however, more recently, it was also shown that low-frequency coding variants (frequencies between 0.1 and 2 %) with relatively large effects on HDL cholesterol and/or triglycerides were also not associated with risk for coronary heart disease (Peloso et al. 2014).

These and other studies have placed HDL cholesterol as a pharmaceutical target under heavy fire especially in the context of several large clinical trials that tested drugs which increased HDL cholesterol but did reduce atherosclerosis (for reviews on CETP inhibitors and the use of niacin: Ginsberg and Reyes-Soffer 2013; Rader and Degoma 2014). While there is still hope for HDL-related interventions as outlined in a recent review (van Capelleveen et al. 2014), it is clear that there is a need to revisit the mechanisms that have been put forward to explain the unequivocal relation between HDL cholesterol and risk of CVD in epidemiological studies.

To date, it is repeatedly been pointed out that a focus on the cholesterol content of HDL should maybe be replaced by a focus on the functions that are associated with this lipoprotein (Feig et al. 2014; Peloso et al. 2014; Luscher et al. 2014; Riwanto and Landmesser 2013), but unfortunately, HDL functionality studies have thus far not provided a solution to the problems encountered. So far it is not clear which of HDL properties should and could be targeted. There is evidence for a focus on HDL as an acceptor of cholesterol (Khera et al. 2011) but later studies were not able to confirm this (Li et al. 2013a, b). In other words, the association with CVD risk has only been firmly established for the concentration of cholesterol in HDL while at the same time there is little evidence of a causal relation between HDL and CVD (Vergeer et al. 2010). Thus, it is possible that HDL cholesterol is a proxy for an unknown correlating factor. A renewed focus on the clinical application of HDL-based strategies for certain indications *on the basis of functional properties of HDL but also on significant preclinical and clinical data*, as was recently suggested, may hopefully bring relief (Gordts et al. 2013).

In this light, the current chapter focuses on recent studies that have shed new light on how genes and/or their products affect HDL metabolism. The first section shortly describes established regulators of HDL metabolism as a general framework to help understand the new insights that are described in the second section.

2 Determinants of Plasma HDL Cholesterol Levels

Twin studies have indicated that genetic and environmental parameters equally contribute to the levels of cholesterol in HDL in the blood (Goode et al. 2007). This paragraph only shortly describes the major primary and secondary regulators of HDL cholesterol. It is important to underscore, however, that changes in HDL cholesterol without changes in other plasma lipid traits are very rare. They are mostly seen in the context of changes in plasma triglycerides. At the population level, genome-wide association (GWA) studies have recently confirmed that in many cases, genetic variation is associated with changes in more than one lipid trait (Teslovich et al. 2010; Willer et al. 2013).

2.1 Established Primary Regulators of Plasma HDL Cholesterol

For the de novo production of HDL, the small intestine and liver need to produce apolipoprotein (apo) A-I, ATP-binding cassette protein A1, and lecithin–cholesterol acyltransferase encoded by the *APOAI*, *ABCA1*, and *LCAT* genes, respectively. When the production of any of these proteins is attenuated (through functional large-impact mutations), it immediately translates into a reduction of HDL cholesterol in the circulation. Other established modulators of HDL are cholesteryl ester transfer protein (encoded by *CETP*) and scavenger receptor class B member 1 (SRB1, encoded by *SCARB1*). While CETP mediates the transfer of cholesterol ester from HDL to triglyceride-rich lipoproteins in exchange for triglycerides (thereby controlling the levels of cholesteryl ester in HDL), SRB1 mediates the selective cellular uptake of cholesteryl ester in the liver and steroidogenic organs. Figure 1 gives a schematic representation of how the above genes relate to HDL biology.

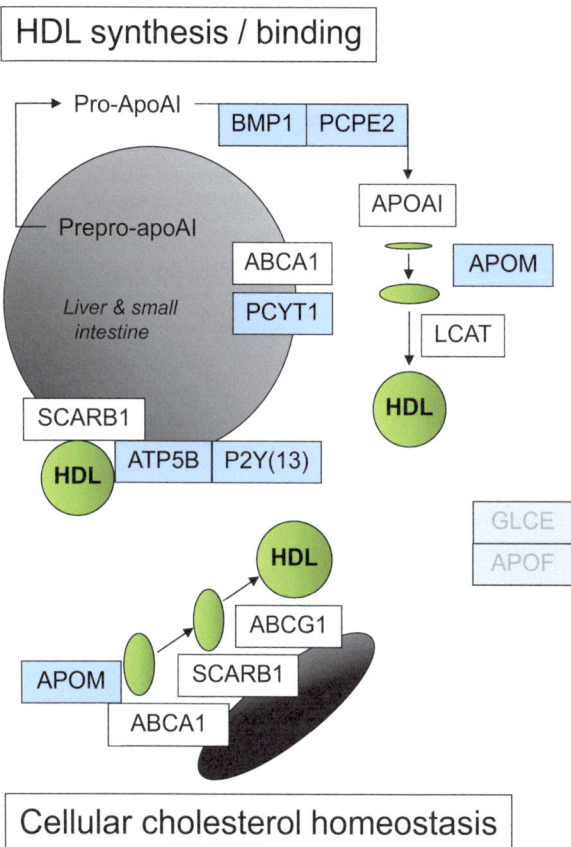

Fig. 1 A schematic presentation of factors that control HDL synthesis/binding and/or HDL-mediated cellular cholesterol homeostasis. The genes in the white boxes encode for key HDL proteins and enzymes, while the genes in the blue boxes encode for less established (new) factors that affect HDL metabolism. The roles of the genes in the *gray boxes* are addressed but solid evidence has not yet been provided. Abbreviations (proteins encoded by the gene names): *ABCA1* ATP-binding cassette A1, *ABCG1* ATP-binding cassette G1, *APOAI* apolipoprotein A-I, *APOF* apolipoprotein F, *APOM* apolipoprotein M, *BMP1* bone morphogenetic protein-1, *GLCE* glucuronic acid epimerase, *LCAT* lecithin–cholesterol acyltransferase, *PCPE2* procollagen C-proteinase enhancer 2, *PCYT1* CTP:phosphocholine cytidylyltransferase alpha, *SCARB1* encoding for scavenger receptor class B member 1

2.2 Established Secondary Regulators of Plasma HDL Cholesterol

There are a large number of other factors that affect HDL cholesterol concentration through modulating the lipolysis of plasma triglycerides [reviewed in Oldoni et al. (2014)]. Figure 2 only illustrates the major players. Most of these regulate the activity of lipoprotein lipase (LPL), the sole enzyme capable of breaking down triglycerides (packaged in chylomicrons and very-low-density lipoproteins

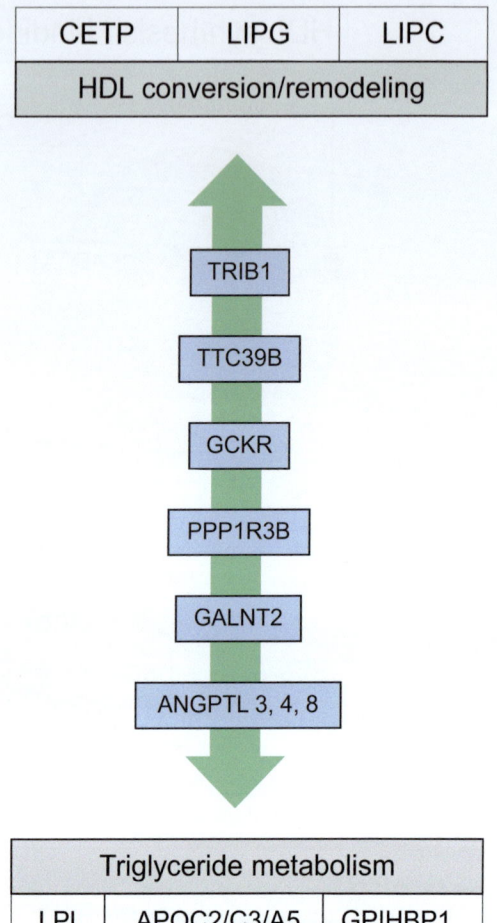

Fig. 2 A schematic presentation of factors that affect HDL through roles in the conversion and remodeling of HDL or through effects on triglyceride/glucose metabolism. The genes in the *white boxes* encode for established factors of HDL conversion/remodeling or plasma TG hydrolysis. The genes in the *blue boxes* encode for less established (new) factors. Abbreviations (proteins encoded by the gene names): *ANGPTL* angiopoietin-like protein, *TRIB1* tribbles homolog 1, *TTC39B* tetratricopeptide repeat domain 39B, *PPP1R3B* glycogen-targeting PP1 subunit G(L), *GALNT2* ppGalNAc-T2, *GCKR* glucokinase (hexokinase 4) regulator

(VLDL)) in the circulation. These include genes that encode for LPL's cofactor apoC-II (*APOCII*) as well as apoA-V (*APOAV*) and inhibitors of the LPL reaction (*APOCIII, ANGPTL3, ANGPTL4*). More recently, also GPI-anchored HDL-binding protein 1 (*GPIHBP1*) was shown to affect HDL concentration through ultimately its effect on LPL activity. In addition, hepatic lipase (encoded by *LIPC*) and endothelial lipase (*LIPG*) also exert marked effects on HDL cholesterol concentration mainly through modulating HDL phospholipids and HDL triglycerides, respectively.

2.3 Missing Heritability

So far, studies of over 40 genes have provided solid evidence that their products affect plasma HDL cholesterol concentration. With so many established genes, one may expect these accounting for the estimated 50 % heritability of this trait. This appears however not to be the case when GWA data are analyzed with the current statistical methods and datasets: The most recent of meta-analysis indicates that common genetic variation can only explain 12 % variation of HDL cholesterol levels while in this study, both variations in established loci as well as newly identified loci ($n = 46$) were taken into account (Willer et al. 2013). However, in these calculations, gene–gene and gene–environment interactions are not for. In addition, the estimated impact of gene variation on the phenotype is based on the presence and frequency of such variations and these are not constant factors over the genome.

From a different angle, candidate gene resequencing studies in individuals with very high or low HDL cholesterol (selected from the general population) have shown that apparent functional mutations are only found in a few percent of the cases (Cohen et al. 2004; Haase et al. 2012). Also in individuals that were referred to the clinic because of extreme levels of HDL cholesterol, resequencing studies of candidate genes only provided satisfying clues in a minority of the subjects studied (Candini et al. 2010; Holleboom et al. 2011a, b; Kiss et al. 2007). It may be noted, however, that most studies conducted thus far focused only on *APOAI, LCAT,* and *ABCA1* leaving ample room for large-effect variants in other genes. Another study focused on the origin of high HDL cholesterol levels through sequencing *CETP, LIPG,* and *GALNT,* showed an enrichment of rare coding and splicing mutations in 171 probands (Tietjen et al. 2012). A second study conducted a search for mutations in 197 lipid-related genes in 80 individuals with extreme HDL cholesterol phenotypes. The outcome was that multiple mutations in different genes combined could be responsible for extreme low or high HDL cholesterol levels (Motazacker et al. 2013). Although a polygenic origin of a complex trait like HDL cholesterol level appears logical, especially in view of similar studies for plasma triglycerides (Johansen et al. 2010) or LDL cholesterol (Talmud et al. 2013), this needs to be confirmed. Larger comprehensive resequencing efforts are warranted to study to what extent large-impact mutations in established and candidate genes can explain HDL cholesterol concentration in plasma and how such mutations relate to CVD risk.

3 Novel Insight into HDL Biology

As indicated above, GWA studies have identified many genomic regions that affect plasma lipid traits. One of the most famous papers in this regard published in 2010 (Teslovich et al. 2010) already listed 38 loci with HDL cholesterol as lead trait. Importantly, these and many other GWA studies rediscovered many of the established "HDL genes" which underscored the potential importance of the newly identified loci. The most comprehensive meta-analysis in this field of research to date lists an additional 46 loci that are associated with HDL cholesterol (Willer et al. 2013). This section tries to capture the novel molecular insights that have been published over the last few years. Many studies focused on genes that on the basis of their proximity to genetic markers in GWA studies were "annotated as HDL genes." In addition, other studies (not initiated through genetic insight) which produced interesting novel insight in HDL biology are shortly discussed.

3.1 De Novo Synthesis of HDL and HDL Binding

3.1.1 Bone Morphogenetic Protein-1 and Procollagen C-Proteinase Enhancer-2

In 2007, bone morphogenetic protein-1 encoded by *BMP-1* (aliases OI13, PCOLC, PCP, PCP2, TLD) was identified as the metalloproteinase that stimulates the conversion of newly secreted proapoA-I to its phospholipid-binding form (Chau et al. 2007). While these results were obtained through in vitro studies, early genetic association studies showed the possible involvement of procollagen C-proteinase enhancer-2 (encoded by *PCPE2* or *PCOLCE2*) in modulating HDL cholesterol levels (Hinds et al. 2004). It was subsequently shown that PCPE2 together with BMP1 and proapoAI forms a ternary complex and that it is involved in the regulation of apoA-I posttranslational processing (Zhu et al. 2009). In a later study, it was shown that *Pcpe2* KO mice have strongly increased plasma apoA-I and HDL cholesterol levels compared with wild-type littermates, regardless of gender or diet (Francone et al. 2011). Changes in HDL particle size and electrophoretic mobility observed in *Pcpe2* KO mice suggest that the presence of proapoA-I impairs the maturation of HDL. Although initially picked up by GWA studies, *PCPE2* is not listed in the subsequent respective meta-analyses (Teslovich et al. 2010; Willer et al. 2013), while *BMP1* is also not listed in the latter studies. So far, the role of *BMP1* or *PCPE2* in posttranslational processing in humans has not been reported which in the light of the larger GWA studies performed is now dependent on the identification of large-impact mutations in individuals with HDL disorders.

3.1.2 Apolipoprotein M

Originally identified as a protein associated with HDL in 1999 (Xu and Dahlback 1999), apoM was later implicated to play a role in the generation of prebeta-HDL and cholesterol efflux mediated by HDL (Wolfrum et al. 2005). These results have recently been supported by evidence in plasma of humans (Plomgaard et al. 2009) while others have found that human *APOM* gene variation affects HDL cholesterol (Park et al. 2013; Aung et al. 2013). Through carrying sphingosine-1-phosphate with effects on angiogenesis, endothelial cell migration, and inflammation, apoM (Christoffersen and Nielsen 2013) may be a link to many of the functions that are attributed to HDL. While there is thus substantial evidence for a role of apoM in human HDL metabolism, the gene has thus far not been picked up by GWA studies, while large-impact mutations in this gene have so far not been reported either.

3.1.3 CTP:Phosphocholine Cytidylyltransferase Alpha (CT Alpha)

Through an interest in the synthesis of phosphatidylcholine (the primary phospholipid in cellular membranes but also of HDL), the group of Dennis Vance studied a conditional *PCTY1* knockout mouse and showed in 2004 that hepatic CTP:phosphocholine cytidylyltransferase alpha reduces HDL cholesterol and apoA-I while it also affected VLDL (Jacobs et al. 2004). In 2008, the same investigators subsequently showed that in hepatocytes isolated from these mice, ABCA1 expression was reduced. Other findings included normalization of plasma HDL and VLDL in these mice after adenoviral delivery of CT alpha which clearly implicated its importance in HDL genesis (Jacobs et al. 2008).

3.1.4 Apolipoprotein F

In proteomic studies, apoF has been identified to be associated with HDL. This protein is also known as lipid transfer inhibitor protein which is able to inhibit CETP. Plasma apoF levels were found to be positively associated with HDL cholesterol in males but not in females (Morton et al. 2008). The Rader group subsequently showed that overexpression of apoF in mice reduced HDL cholesterol levels by accelerating clearance from the circulation (Lagor et al. 2009). A later study showed that a murine apoF knockout model had no substantial effect on plasma lipid concentrations, HDL size, lipid, or protein (Lagor et al. 2012) although a reduced ability to promote cholesterol efflux was observed. Until now, no other studies on apoF have been published.

3.1.5 Glucuronic Acid Epimerase

Through studies in Turkish families, a linkage peak with low HDL cholesterol was identified on chromosome 15 which was >20 cM wide including the gene encoding hepatic lipase (*LIPC*), which has important functions in HDL metabolism. However, the same investigators suspected that variations in *LIPC* might not fully explain this linkage peak and sought additional gene(s) that might contribute to the peak. This is how they identified *GLCE* encoding glucuronic acid epimerase, a heparan sulfate proteoglycan biosynthetic enzyme (Hodoglugil et al. 2011). So far, no other studies on the role of this enzyme in HDL metabolism have been conducted while the gene was not found associated with HDL cholesterol levels in the latest meta-analyses GWA study (Willer et al. 2013).

3.1.6 Beta-Chain of ATP Synthase

In 2003, beta-chain of ATP synthase (encoded by *ATPB5*), a principal protein complex of the mitochondrial inner membrane, was surprisingly identified as a high-affinity HDL receptor for apoA-I receptor with a role in hepatic HDL endocytosis (Martinez et al. 2003). A subsequent study demonstrated a major role of cytoskeleton reorganization in F(1)-ATPase/P2Y(13)-dependent HDL endocytosis under the control of the small GTPase RhoA and its effector ROCK I (Malaval et al. 2009). Others showed that binding of HDL to this receptor triggers the generation of ADP, which via the activation of the purinergic receptor P2Y13 stimulates the uptake and transport of HDL and initially lipid-free apoA-I by endothelial cells (Cavelier et al. 2012). So far, there is no genetic support for the involvement of beta-chain of ATP synthase and/or P2Y(13), but even mild mutations could have in the case of *ATPB5* lethal consequences. Most recently, it was reported that there may be a role for beta-chain of ATP synthase in regulating HDL cholesterol levels as mitochondrial inhibitor factor 1, which inhibits ATPB5 and can be measured in serum, is associated with HDL cholesterol levels (Genoux et al. 2013).

3.2 HDL Conversion and Remodeling

3.2.1 Angptl Family of Proteins

In the year 2000, *ANGPTL4* was identified as a peroxisome proliferator-activated receptor α response gene (Kersten et al. 2000) while *ANGPTL3* is since 2002 known for its effects on glucose and lipid metabolism through a hypolipidemic mouse strain (Koishi et al. 2002). Both Angptl3 and Angptl4 are known to inhibit LPL, thereby having an indirect effect on HDL cholesterol concentration in plasma, but Angptl3 may also exert direct effects on HDL remodeling through inhibition of endothelial lipase (Shimamura et al. 2007). In GWA studies published in 2008 and 2010, *ANGPTL3* was shown to be associated primarily with triglycerides (Kathiresan et al. 2008; Teslovich et al. 2010) while more recently it was shown that human *ANGPTL3* deficiency also causes marked reduction in HDL cholesterol (Musunuru et al. 2010). In 2010, *ANGPTL4* was reported to be associated with HDL cholesterol as primary trait (Teslovich et al. 2010) while later on *ANGPTL4* variants were found associated with lower triglycerides and elevated HDL cholesterol (Romeo et al. 2007). Another study showed that mutant alleles of *ANGPTL3* and *ANGPTL4* that were associated with low plasma triglyceride levels interfered either with the synthesis or secretion of the protein or with the ability of the *ANGPTL* protein to inhibit LPL (Romeo et al. 2009). More recently, *ANGPTL8* (also known as betatrophin) was shown to also affect HDL cholesterol through affecting the activity of Angpl3 (Quagliarini et al. 2012). In addition, low-frequency variants in *ANGPTL8* were shown to affect HDL cholesterol levels (Peloso et al. 2014). For a recent review on the current eight Angptl proteins, please see Santulli (2014).

3.2.2 Tribbles Homolog 1

Variation at the *TRIB1* gene locus has been reported to be associated with triglycerides (as lead trait), HDL cholesterol, total cholesterol, and LDL cholesterol (Kathiresan et al. 2008; Teslovich et al. 2010). It is one of the few genes subsequently studied in depth in mouse models, in vitro experiments, and further human genetic association studies. Viral-mediated overexpression in the liver of mice simultaneously reduced plasma triglycerides and cholesterol of all major lipoproteins (including HDL). On the other hand, *TRIB1* knockout mice showed elevated levels of triglycerides (Burkhardt et al. 2010) without statistically significant effects on HDL cholesterol. These effects were related to an on VLDL production but it is not known what mechanisms are responsible. Further epidemiological studies confirmed that *TRIB1* is associated with HDL cholesterol but interestingly without affecting apoA-I concentration (Varbo et al. 2011). More recently it was shown that Trib1 is important to adipose tissue maintenance and suppression of metabolic disorders by controlling the differentiation of tissue-resident M2-like macrophages (Satoh et al. 2013). In addition, a *TRIB1* single-nucleotide polymorphism was found associated with nonalcoholic fatty liver disease in humans while the same investigators showed that *TRIB1* expression affects hepatic lipogenesis and glycogenesis through multiple molecular interactions (including a molecular interaction with *MLXIPL or CHREBP*, a hepatic lipogenic master regulator) (Ishizuka et al. 2014). In addition to a role in lipid metabolism, there is a substantial literature on the role of tribbles in cancer (Liang et al. 2013).

3.2.3 Tetratricopeptide Repeat Domain/Glycogen-Targeting PP1 Subunit G(L)

TTC39B was discovered through GWA studies in 2009 (Kathiresan et al. 2009). When knocked down in mice (using a viral-mediated strategy), it was shown to increase HDL cholesterol levels (Teslovich et al. 2010). However, no studies have since revealed insight into how tetratricopeptide repeat domain 39B may affect lipid metabolism while another GWA study could not replicate the effect of *TTC39B* variation on plasma lipid levels (Dumitrescu et al. 2011).

PPP1R3B was also found to be associated with HDL cholesterol as primary trait (Teslovich et al. 2010), and in the same paper, viral-mediated overexpression was shown to reduce HDL cholesterol levels in mice. The gene encoding for glycogen-targeting PP1 (protein phosphatase 1) subunit G(L) has also been associated with type 2 diabetes and maturity-onset diabetes of the young but sequence variants at the *PPP1R3B* locus were not found to be related to diabetes in mostly Caucasian families (Dunn et al. 2006). Other studies proposed a role for this gene in inflammation (Dehghan et al. 2011), Alzheimer's disease (Kamboh et al. 2012), and hepatic steatosis (Palmer et al. 2013). As for *TTC39B*, there are no mechanistic studies that have revealed insight into how *PPP1R3B* gene products may affect lipid metabolism.

3.2.4 ppGalNAc-T2

Variation at the *GALNT2* locus was also shown to be associated with HDL cholesterol (Teslovich et al. 2010). Viral-mediated overexpression and knockdown were shown to decrease and increase HDL cholesterol levels in mice, respectively. In a later study a rare functional *GALNT2* variant, identified in two index cases with very high HDL cholesterol, was reported to affect HDL metabolism indirectly via an effect on the lipolysis of triglycerides (Holleboom et al. 2011a, b). This was suggested to be in part due to attenuated glycosylation of apoC-III which impaired its capacity to inhibit LPL in the catabolism of triglycerides. Other investigators reported that *GALNT2* regulates plasma lipid levels through the glycosylation of *ANGTPL3*, another inhibitor of the LPL reaction (Schjoldager et al. 2010). Recently, *ppGalNAc-T2* was identified as a novel regulator of insulin signaling (Marucci et al. 2013) while there appears to be a role for *GALNT2* in hypertension too (Pendergrass et al. 2013; McDonough et al. 2013).

3.2.5 Glucokinase (Hexokinase 4) Regulator

GCKR (also known as GKRP) encodes for a regulatory protein that inhibits glucokinase in the liver and pancreatic islet cells by binding non-covalently to form an inactive complex with the enzyme. This gene is considered a susceptibility gene candidate for a form of maturity-onset diabetes of the young. GCKR was picked up in both GWA studies with a focus on plasma lipids but also with type 2 diabetes (van de Bunt and Gloyn 2010). Varbo et al. showed that a *GCKR* SNP was associated with increased triglycerides, decreased HDL cholesterol, and remarkably increased apoA-I (Varbo et al. 2011). Several genetic association studies have shown that *GCKR* variants are associated with triglycerides, insulin resistance (Shen et al. 2013), type 2 diabetes fasting plasma glucose (Li et al. 2013a, b), hyperglycemia (Stancakova et al. 2012), and nonalcoholic fatty liver disease (Lin et al. 2014). In 2013, the crystal structure was resolved (Pautsch et al. 2013) while two potent small-molecule GK–GKRP disruptors were recently reported to normalize blood glucose levels in several rodent models of diabetes (Lloyd et al. 2013).

Conclusions and Perspectives

The harvest of screening the most recent published literature for novel HDL influencing genes is extensive. However, it also shows that there are only very few mechanistic clues on how "the new kids on the block" (primarily identified through GWA studies) affect HDL cholesterol or HDL function.

It is interesting that none of the genes encoding for the proteins and enzymes that have been discussed in the section on the de novo synthesis of HDL are listed in the current largest GWA studies. This may be surprising in light of the evidence for the roles of especially bone morphogenetic protein-1, procollagen C-proteinase enhancer 2, apoM, and CT alpha in HDL metabolism. It shows that GWA strategies, like any approach, come with limitations. Thus, prioritizing research interest on the basis of the top ranking GWA hits is not necessarily the most promising route to go.

In the section on (new) insights in HDL conversion and remodeling, it is clear that the factors discussed are only indirectly associated with HDL cholesterol levels through effects on triglyceride and/or glucose metabolism. It may be noted in this regard that HDL cholesterol was proposed to be the lead trait for *ANGPTL4, TTC39B, PPP1R3B, and GALNT2,* in GWA studies which highlights a need for more comprehensive studies with focus on the mechanistic relations between HDL cholesterol, triglycerides, and glucose metabolism. Such studies may need the use of genome-scale metabolic maps.

Finally, this review underlines an urgent need of mechanistic studies. But where to start? Answering this question is today's challenge as evidence-based prioritization tools and high-throughput functional studies are yet to come.

Acknowledgments I wish to apologize to colleagues whose work may not be cited in this review: It remains an attempt to cover the existing literature. This work is supported by Fondation LeDucq (Transatlantic Network, 2009–2014), the Netherlands CardioVascular Research Initiative (CVON2012-17; Genius), and the European Union (COST BM0904; Resolve: FP7-305707; TransCard: FP7-603091-2).

Open Access This chapter is distributed under the terms of the Creative Commons Attribution Noncommercial License, which permits any noncommercial use, distribution, and reproduction in any medium, provided the original author(s) and source are credited.

References

Aung LH et al (2013) Association of the apolipoprotein M gene polymorphisms and serum lipid levels. Mol Biol Rep 40(2):1843–1853
Burkhardt R et al (2010) Trib1 is a lipid- and myocardial infarction-associated gene that regulates hepatic lipogenesis and VLDL production in mice. J Clin Invest 120(12):4410–4414
Candini C et al (2010) Identification and characterization of novel loss of function mutations in ATP-binding cassette transporter A1 in patients with low plasma high-density lipoprotein cholesterol. Atherosclerosis 213(2):492–498
Cavelier C et al (2012) The beta-chain of cell surface F(0)F(1) ATPase modulates apoA-I and HDL transcytosis through aortic endothelial cells. Arterioscler Thromb Vasc Biol 32(1):131–139
Chapman MJ et al (2011) Triglyceride-rich lipoproteins and high-density lipoprotein cholesterol in patients at high risk of cardiovascular disease: evidence and guidance for management. Eur Heart J 32(11):1345–1361
Chau P et al (2007) Bone morphogenetic protein-1 (BMP-1) cleaves human proapolipoprotein A1 and regulates its activation for lipid binding. Biochemistry 46(28):8445–8450
Christoffersen C, Nielsen LB (2013) Apolipoprotein M: bridging HDL and endothelial function. Curr Opin Lipidol 24(4):295–300
Cohen JC et al (2004) Multiple rare alleles contribute to low plasma levels of HDL cholesterol. Science 305(5685):869–872
Dehghan A et al (2011) Meta-analysis of genome-wide association studies in >80,000 subjects identifies multiple loci for C-reactive protein levels. Circulation 123(7):731–738
Dumitrescu L et al (2011) Genetic determinants of lipid traits in diverse populations from the population architecture using genomics and epidemiology (PAGE) study. PLoS Genet 7(6):e1002138
Dunn JS et al (2006) Examination of PPP1R3B as a candidate gene for the type 2 diabetes and MODY loci on chromosome 8p23. Ann Hum Genet 70(Pt 5):587–593
Feig JE et al (2014) High-density lipoprotein and atherosclerosis regression: evidence from preclinical and clinical studies. Circ Res 114(1):205–213
Francone OL et al (2011) Disruption of the murine procollagen C-proteinase enhancer 2 gene causes accumulation of proapoA-I and increased HDL levels. J Lipid Res 52(11):1974–1983
Frikke-Schmidt R (2011) Genetic variation in ABCA1 and risk of cardiovascular disease. Atherosclerosis 218(2):281–282
Genoux A et al (2013) Serum IF1 concentration is independently associated to HDL levels and to coronary heart disease: the GENES study. J Lipid Res 54(9):2550–2558
Ginsberg HN, Reyes-Soffer G (2013) Niacin: a long history, but a questionable future. Curr Opin Lipidol 24(6):475–479
Goode EL et al (2007) Heritability of longitudinal measures of body mass index and lipid and lipoprotein levels in aging twins. Twin Res Hum Genet 10(5):703–711

Gordts SC et al (2013) Pleiotropic effects of HDL: towards new therapeutic areas for HDL-targeted interventions. Curr Mol Med 14(4):481–503

Haase CL et al (2012) Population-based resequencing of APOA1 in 10,330 individuals: spectrum of genetic variation, phenotype, and comparison with extreme phenotype approach. PLoS Genet 8(11):e1003063

Hinds DA et al (2004) Application of pooled genotyping to scan candidate regions for association with HDL cholesterol levels. Hum Genomics 1(6):421–434

Hodoglugil U et al (2011) Glucuronic acid epimerase is associated with plasma triglyceride and high-density lipoprotein cholesterol levels in Turks. Ann Hum Genet 75(3):398–417

Holleboom AG et al (2011a) Heterozygosity for a loss-of-function mutation in GALNT2 improves plasma triglyceride clearance in man. Cell Metab 14(6):811–818

Holleboom AG et al (2011b) High prevalence of mutations in LCAT in patients with low HDL cholesterol levels in The Netherlands: identification and characterization of eight novel mutations. Hum Mutat 32(11):1290–1298

Ishizuka Y et al (2014) TRIB1 down-regulates hepatic lipogenesis and glycogenesis via multiple molecular interactions. J Mol Endocrinol 52(2):145–158

Jacobs RL et al (2004) Targeted deletion of hepatic CTP:phosphocholine cytidylyltransferase alpha in mice decreases plasma high density and very low density lipoproteins. J Biol Chem 279(45):47402–47410

Jacobs RL et al (2008) Hepatic CTP:phosphocholine cytidylyltransferase-alpha is a critical predictor of plasma high density lipoprotein and very low density lipoprotein. J Biol Chem 283(4):2147–2155

Johansen CT et al (2010) Excess of rare variants in genes identified by genome-wide association study of hypertriglyceridemia. Nat Genet 42(8):684–687

Kamboh MI et al (2012) Genome-wide association study of Alzheimer's disease. Transl Psychiatry 2:e117

Kathiresan S et al (2008) Six new loci associated with blood low-density lipoprotein cholesterol, high-density lipoprotein cholesterol or triglycerides in humans. Nat Genet 40(2):189–197

Kathiresan S et al (2009) Common variants at 30 loci contribute to polygenic dyslipidemia. Nat Genet 41(1):56–65

Kersten S et al (2000) Characterization of the fasting-induced adipose factor FIAF, a novel peroxisome proliferator-activated receptor target gene. J Biol Chem 275(37):28488–28493

Khera AV et al (2011) Cholesterol efflux capacity, high-density lipoprotein function, and atherosclerosis. N Engl J Med 364(2):127–135

Kiss RS et al (2007) Genetic etiology of isolated low HDL syndrome: incidence and heterogeneity of efflux defects. Arterioscler Thromb Vasc Biol 27(5):1139–1145

Koishi R et al (2002) Angptl3 regulates lipid metabolism in mice. Nat Genet 30(2):151–157

Lagor WR et al (2009) Overexpression of apolipoprotein F reduces HDL cholesterol levels in vivo. Arterioscler Thromb Vasc Biol 29(1):40–46

Lagor WR et al (2012) The effects of apolipoprotein F deficiency on high density lipoprotein cholesterol metabolism in mice. PLoS ONE 7(2):e31616

Li H et al (2013a) Association of glucokinase regulatory protein polymorphism with type 2 diabetes and fasting plasma glucose: a meta-analysis. Mol Biol Rep 40(6):3935–3942

Li XM et al (2013b) Paradoxical association of enhanced cholesterol efflux with increased incident cardiovascular risks. Arterioscler Thromb Vasc Biol 33(7):1696–1705

Liang KL et al (2013) Tribbles in acute leukemia. Blood 121(21):4265–4270

Lin YC et al (2014) Genetic variants in GCKR and PNPLA3 confer susceptibility to nonalcoholic fatty liver disease in obese individuals. Am J Clin Nutr 99(4):869–874

Lloyd DJ et al (2013) Antidiabetic effects of glucokinase regulatory protein small-molecule disruptors. Nature 504(7480):437–440

Luscher TF et al (2014) High-density lipoprotein: vascular protective effects, dysfunction, and potential as therapeutic target. Circ Res 114(1):171–182

Malaval C et al (2009) RhoA/ROCK I signalling downstream of the P2Y13 ADP-receptor controls HDL endocytosis in human hepatocytes. Cell Signal 21(1):120–127

Martinez LO et al (2003) Ectopic beta-chain of ATP synthase is an apolipoprotein A-I receptor in hepatic HDL endocytosis. Nature 421(6918):75–79

Marucci A et al (2013) Role of GALNT2 in the modulation of ENPP1 expression, and insulin signaling and action: GALNT2: a novel modulator of insulin signaling. Biochim Biophys Acta 1833(6):1388–1395

McDonough CW et al (2013) Atenolol induced HDL-C change in the pharmacogenomic evaluation of antihypertensive responses (PEAR) study. PLoS ONE 8(10):e76984

Morton RE et al (2008) Lipid transfer inhibitor protein (apolipoprotein F) concentration in normolipidemic and hyperlipidemic subjects. J Lipid Res 49(1):127–135

Motazacker MM et al (2013) Evidence of a polygenic origin of extreme high-density lipoprotein cholesterol levels. Arterioscler Thromb Vasc Biol 33(7):1521–1528

Musunuru K et al (2010) Exome sequencing, ANGPTL3 mutations, and familial combined hypolipidemia. N Engl J Med 363(23):2220–2227

Oldoni F et al (2014) Mendelian disorders of high-density lipoprotein metabolism. Circ Res 114(1):124–142

Palmer ND et al (2013) Characterization of European ancestry nonalcoholic fatty liver disease-associated variants in individuals of African and Hispanic descent. Hepatology 58(3):966–975

Park YJ et al (2013) The APOM polymorphism as a novel risk factor for dyslipidaemia in rheumatoid arthritis: a possible shared link between disease susceptibility and dyslipidaemia. Clin Exp Rheumatol 31(2):180–188

Pautsch A et al (2013) Crystal structure of glucokinase regulatory protein. Biochemistry 52(20):3523–3531

Peloso GM et al (2014) Association of low-frequency and rare coding-sequence variants with blood lipids and coronary heart disease in 56,000 whites and blacks. Am J Hum Genet 94(2):223–232

Pendergrass SA et al (2013) Phenome-wide association study (PheWAS) for detection of pleiotropy within the population architecture using genomics and epidemiology (PAGE) network. PLoS Genet 9(1):e1003087

Plomgaard P et al (2009) Apolipoprotein M predicts pre-beta-HDL formation: studies in type 2 diabetic and nondiabetic subjects. J Intern Med 266(3):258–267

Quagliarini F et al (2012) Atypical angiopoietin-like protein that regulates ANGPTL3. Proc Natl Acad Sci USA 109(48):19751–19756

Rader DJ, Degoma EM (2014) Future of cholesteryl ester transfer protein inhibitors. Annu Rev Med 65:385–403

Riwanto M, Landmesser U (2013) High density lipoproteins and endothelial functions: mechanistic insights and alterations in cardiovascular disease. J Lipid Res 54(12):3227–3243

Romeo S et al (2007) Population-based resequencing of ANGPTL4 uncovers variations that reduce triglycerides and increase HDL. Nat Genet 39(4):513–516

Romeo S et al (2009) Rare loss-of-function mutations in ANGPTL family members contribute to plasma triglyceride levels in humans. J Clin Invest 119(1):70–79

Rye KA, Barter PJ (2014) Cardioprotective functions of HDLs. J Lipid Res 55(2):168–179

Santulli G (2014) Angiopoietin-like proteins: a comprehensive look. Front Endocrinol (Lausanne) 5:4

Satoh T et al (2013) Critical role of Trib1 in differentiation of tissue-resident M2-like macrophages. Nature 495(7442):524–528

Schjoldager KT et al (2010) O-glycosylation modulates proprotein convertase activation of angiopoietin-like protein 3: possible role of polypeptide GalNAc-transferase-2 in regulation of concentrations of plasma lipids. J Biol Chem 285(47):36293–36303

Shen Y et al (2013) GCKR variants increase triglycerides while protecting from insulin resistance in Chinese children. PLoS ONE 8(1):e55350

Shimamura M et al (2007) Angiopoietin-like protein3 regulates plasma HDL cholesterol through suppression of endothelial lipase. Arterioscler Thromb Vasc Biol 27(2):366–372

Stancakova A et al (2012) Hyperglycemia and a common variant of GCKR are associated with the levels of eight amino acids in 9,369 Finnish men. Diabetes 61(7):1895–1902

Talmud PJ et al (2013) Use of low-density lipoprotein cholesterol gene score to distinguish patients with polygenic and monogenic familial hypercholesterolaemia: a case-control study. Lancet 381(9874):1293–1301

Teslovich TM et al (2010) Biological, clinical and population relevance of 95 loci for blood lipids. Nature 466(7307):707–713

Tietjen I et al (2012) Segregation of LIPG, CETP, and GALNT2 mutations in Caucasian families with extremely high HDL cholesterol. PLoS ONE 7(8):e37437

van Capelleveen JC et al (2014) Novel therapies focused on the high-density lipoprotein particle. Circ Res 114(1):193–204

van de Bunt M, Gloyn AL (2010) From genetic association to molecular mechanism. Curr Diab Rep 10(6):452–466

Varbo A et al (2011) TRIB1 and GCKR polymorphisms, lipid levels, and risk of ischemic heart disease in the general population. Arterioscler Thromb Vasc Biol 31(2):451–457

Vergeer M et al (2010) The HDL hypothesis: does high-density lipoprotein protect from atherosclerosis? J Lipid Res 51(8):2058–2073

Voight BF et al (2012) Plasma HDL cholesterol and risk of myocardial infarction: a mendelian randomisation study. Lancet 380(9841):572–580

Willer CJ et al (2013) Discovery and refinement of loci associated with lipid levels. Nat Genet 45(11):1274–1283

Wolfrum C et al (2005) Apolipoprotein M is required for prebeta-HDL formation and cholesterol efflux to HDL and protects against atherosclerosis. Nat Med 11(4):418–422

Xu N, Dahlback B (1999) A novel human apolipoprotein (apoM). J Biol Chem 274(44):31286–31290

Zhu J et al (2009) Regulation of apoAI processing by procollagen C-proteinase enhancer-2 and bone morphogenetic protein-1. J Lipid Res 50(7):1330–1339

Mouse Models of Disturbed HDL Metabolism

Menno Hoekstra and Miranda Van Eck

Contents

1. Introduction 303
2. Apolipoprotein A-I 303
3. ATP-Binding Cassette Transporter A1 305
4. ATP-Binding Cassette Transporter G1 309
5. Lecithin-Cholesterol Acyltransferase 311
6. Phospholipid Transfer Protein 312
7. Scavenger Receptor BI 315
8. Insights from Intercrossing of the Different Knockout Mice 319
9. Conclusions from the Gene Knockout Mouse Studies 320
10. Cholesterol Ester Transfer Protein Transgenic Mice 321
References 325

Abstract

High-density lipoprotein (HDL) is considered to be an anti-atherogenic lipoprotein moiety. Generation of genetically modified (total body and tissue-specific knockout) mouse models has significantly contributed to our understanding of HDL function. Here we will review data from knockout mouse studies on the importance of HDL's major alipoprotein apoA-I, the ABC transporters A1 and G1, lecithin:cholesterol acyltransferase, phospholipid transfer protein, and scavenger receptor BI for HDL's metabolism and its protection against atherosclerosis in mice. The initial generation and maturation of HDL particles as well as the selective delivery of its cholesterol to the liver are essential parameters in the life cycle of HDL. Detrimental atherosclerosis effects observed in response to HDL deficiency in mice cannot be solely attributed to the low HDL levels per se, as

M. Hoekstra (✉) • M. Van Eck
Division of Biopharmaceutics, Gorlaeus Laboratories, Leiden Academic Centre for Drug Research, Leiden University, Einsteinweg 55, 2333CC Leiden, The Netherlands
e-mail: hoekstra@lacdr.leidenuniv.nl; m.eck@lacdr.leidenuniv.nl

© The Author(s) 2015
A. von Eckardstein, D. Kardassis (eds.), *High Density Lipoproteins*, Handbook of Experimental Pharmacology 224, DOI 10.1007/978-3-319-09665-0_9

the low HDL levels are in most models paralleled by changes in non-HDL-cholesterol levels. However, the cholesterol efflux function of HDL is of critical importance to overcome foam cell formation and the development of atherosclerotic lesions in mice. Although HDL is predominantly studied for its atheroprotective action, the mouse data also suggest an essential role for HDL as cholesterol donor for steroidogenic tissues, including the adrenals and ovaries. Furthermore, it appears that a relevant interaction exists between HDL-mediated cellular cholesterol efflux and the susceptibility to inflammation, which (1) provides strong support for the novel concept that inflammation and metabolism are intertwining biological processes and (2) identifies the efflux function of HDL as putative therapeutic target also in other inflammatory diseases than atherosclerosis.

Keywords
Knockout mice • High-density lipoprotein • Reverse cholesterol transport • Atherosclerosis • Steroidogenesis • Inflammation • ABC transporters • Scavenger receptor BI • Apolipoprotein • Lecithin-cholesterol acyltransferase • Phospholipid transfer protein

Abbreviations

ABCA1	ATP-binding cassette transporter A1
ABCG1	ATP-binding cassette transporter G1
apoA-I	Apolipoprotein A-I
apoA-II	Apolipoprotein A-II
apoA-IV	Apolipoprotein A-IV
apoB	Apolipoprotein B
apoE	Apolipoprotein E
CETP	Cholesterol ester transfer protein
HDL	High-density lipoprotein
LCAT	Lecithin-cholesterol acyltransferase
LDL	Low-density lipoprotein
LDLr	Low-density lipoprotein receptor
LPS	Lipopolysaccharide
PLTP	Phospholipid transfer protein
SR-BI	Scavenger receptor BI
VLDL	Very-low-density lipoprotein

1 Introduction

High levels of low-density lipoprotein (LDL) cholesterol are associated with an increased risk for cardiovascular disease. In contrast, plasma levels of cholesterol associated with high-density lipoprotein (HDL) are inversely correlated with the risk of cardiovascular disease (Gordon et al. 1977). Since standard treatment with statins to lower plasma levels of LDL cholesterol only reduces the risk of cardiovascular disease by ~30 % (Cholesterol Treatment Trialists' (CTT) Collaboration et al. 2010; Cholesterol Treatment Trialists' (CTT) Collaborators et al. 2012), raising plasma levels of HDL cholesterol has been considered a potential additional therapeutic strategy to overcome disease.

The term HDL refers to a class of alpha-migrating protein/lipid complexes that differ in size (5–12 nm), shape, and lipidation pattern. In order to be able to effectively modulate plasma HDL-cholesterol levels, it is of critical importance to understand the anti-atherogenic potential and metabolism of the different HDL particles. Due to the large biological/genetic variation and lifestyle of people, it is difficult to gain clear insight into the relation between the flux of cholesterol through the HDL pathway and the development of atherosclerosis—the primary underlying cause of cardiovascular disease—in the general human population. Small animal models, in particular mice, have been proven valuable tools to increase the understanding of the complexity of HDL metabolism and the consequences of interfering in specific pathways or atherosclerosis susceptibility. Here we will review data from several genetically modified (total body and tissue-specific knockout) mouse models to (1) show the importance of specific gene products in the life cycle of HDL and their contribution to HDL's primary function, reverse cholesterol transport, i.e., the flux of cholesterol from peripheral cells back to the liver for subsequent excretion, and (2) highlight the potential of modulation of HDL metabolism as an approach to lower atherosclerotic disease burden.

2 Apolipoprotein A-I

Apolipoprotein A-I (apoA-I) is produced by the liver (Zannis et al. 1983) and intestine (Gordon et al. 1982) and represents the primary apolipoprotein constituent of HDL particles (Scanu et al. 1969). Several mutations in the apoA-I gene have been causally linked to HDL deficiency in humans (Tilly-Kiesi et al. 1995; Leren et al. 1997; Matsunaga et al. 1999; Hovingh et al. 2004; Dastani et al. 2006; Wada et al. 2009; Berge and Leren 2010; Lee et al. 2013).

The groups of Nobuyo Maeda (Williamson et al. 1992) and Jan Breslow (Plump et al. 1997) both inactivated the apoA-I gene in mice using dedicated homologous recombination strategies, leading to the absence of detectable amounts of apoA-I protein in plasma. In accordance with an important function for apoA-I in the formation and stability of HDL particles, genetic disruption of apoA-I in mice is associated with an 83 % decrease in the level of HDL cholesterol, contributing to a 68 % decrease in plasma total cholesterol levels (Williamson et al. 1992). The limited amount of HDL cholesterol present in apoA-I knockout mice is carried by particles that are enriched in triglycerides (Plump et al. 1997). However, HDL particle size and shape are similar to that of wild-type controls. ApoA-I-deficient HDL particles display increased levels of other apolipoprotein subspecies, including apolipoprotein A-II (apoA-II), apolipoprotein E (apoE), and apolipoprotein A-IV (apoA-IV) (Li et al. 1993; Plump et al. 1997; Moore et al. 2003). Homozygous apoA-I deficiency in mice not only lowers plasma levels of HDL cholesterol, but also decreases levels of cholesterol associated with apoB-containing very-low-density lipoproteins (VLDL) (Plump et al. 1997). The cholesterol ester to unesterified cholesterol ratio is lower in plasma of apoA-I-deficient mice, which can be attributed to a reduced endogenous cholesterol esterification rate (Parks et al. 1995). Furthermore, the cholesterol ester fatty acid composition of plasma is markedly different between the different genotypes, with apoA-I knockout mice displaying a higher degree of fatty acid saturation as compared to wild-type mice (Parks et al. 1995). ApoA-I, in addition to its structural role, thus also serves as an essential activator of the enzyme lecithin-cholesterol acyltransferase (LCAT) that esterifies cholesterol.

ApoA-I deficiency in mice does, in general, not affect tissue cholesterol levels (Plump et al. 1997). In the liver, decreased delivery of cholesterol esters to hepatocytes is compensated by a concomitant decrease in bile acid formation as judged from a decrease in hepatic mRNA expression levels of cholesterol 7alpha-hydroxylase (CYP7A1) in apoA-I knockout mice (Plump et al. 1997). In contrast, the adrenals are not able to cope with the diminished plasma cholesterol flux due to the lack of apoA-I. Cortical cells located in the zona glomerulosa and zona fasciculata from the adrenals of wild-type mice are filled with large lipid droplets containing cholesterol esters, while adrenals from apoA-I-deficient mice completely lack cholesterol esters and display a parallel decrease in microvillar channel width (Plump et al. 1996). Importantly, apoA-I-deficient mice show a diminished adrenal glucocorticoid output in response to steroidogenic triggers (Plump et al. 1996), suggesting that a lack of apoA-I-containing HDL particles impairs normal adrenal function in mice. Although the cholesterol ester content of the testis is similar between apoA-I knockout and wild-type mice, interstitial and theca cells within the ovaries of apoA-I knockout mice are similarly deprived of cholesterol esters (Plump et al. 1996). Currently no evidence is present for an altered female fertility, despite the clear ovarian cholesterol depletion phenotype. Combined, these findings indicate that apoA-I-containing HDL particles act as important cholesterol donors for several, but not all, steroidogenic tissues.

Human apoA-I mutation carriers may develop premature atherosclerosis (Hovingh et al. 2004; Dastani et al. 2006). The formation of atherosclerotic lesions in normolipidemic mice can be induced by feeding them a cholic acid-containing diet enriched in cholesterol and fat that increases levels of pro-atherogenic apoB-containing lipoproteins (Paigen et al. 1987; Ishida et al. 1991). Strikingly, deletion of apoA-I in wild-type mice does not alter the incidence or extent of atherogenic diet-induced lesion formation (Li et al. 1993). This may be explained by the fact that, under cholic acid-containing diet feeding conditions, levels of anti-atherogenic

HDL as well as levels of pro-atherogenic VLDL/LDL are lower in apoA-I knockout mice as compared to wild-type controls (Li et al. 1993). To further determine the impact of apoA-I deficiency on atherosclerosis susceptibility, apoA-I knockout mice have been crossbred with genetically hyperlipidemic low-density lipoprotein receptor (LDLr) knockout mice that spontaneously develop atherosclerotic lesions on a chow diet, albeit at a slow rate (Ishibashi et al. 1994). Chow diet-fed apoA-I x LDLr double knockout (DKO) mice as compared to LDLr single knockout controls exhibit an increase in the plasma non-HDL-cholesterol to HDL-cholesterol ratio. This can be attributed to a marked increase in VLDL-cholesterol levels and a decrease in plasma HDL-cholesterol levels (Moore et al. 2003; Zabalawi et al. 2003). In accordance with the more atherogenic lipoprotein profile (higher VLDL cholesterol and lower HDL cholesterol), atherosclerotic lesion burden is increased in chow diet-fed apoA-I x LDLr DKO mice, as measured by en face analysis of the aorta, irrespective of the age of the mice (Moore et al. 2003). Feeding a Western-type high fat/high cholesterol diet induces a rapid progression of atherosclerotic lesion development in LDLr knockout mice (Ishibashi et al. 1994). In great contrast with the chow diet findings, the extent of atherosclerotic lesion formation was not significantly different between LDLr KO and apoA-I x LDLr DKO mice upon feeding an atherogenic diet containing 10 % saturated fat from palm oil and 0.1 % cholesterol for 16 weeks (Zabalawi et al. 2003). However, it should be noted that apoA-I x LDLr DKO mice show a much lower diet-induced increase in VLDL-cholesterol levels. Identical aortic cholesterol contents are thus found in the context of a ~70 % lower plasma total cholesterol level (Zabalawi et al. 2003). It therefore appears that ablation of apoA-I function and the concomitant decrease in plasma HDL-cholesterol levels do predispose for atherosclerotic lesion development in mice, but only after correcting for apoA-I genotype-associated changes in VLDL-cholesterol levels.

ApoA-I x LDLr DKO mice develop severe (fatal) skin lesions containing lipid-filled macrophages upon feeding the high cholesterol/high fat diet (Zabalawi et al. 2003). In addition, B-cells, T-cells, and dendritic cells within lymph nodes of apoA-I x LDLr DKO mice are enriched in cholesterol esters (Wilhelm et al. 2009). This is associated with an autoimmune phenotype characterized by relatively high autoantibody titers in plasma upon feeding an atherogenic high fat/high cholesterol diet (Wilhelm et al. 2009). ApoA-I-containing HDL particles thus also play an essential role in maintaining cholesterol homeostasis in the skin as well as in cells from the lymphoid system.

3 ATP-Binding Cassette Transporter A1

The cholesterol used to transform pre-beta1HDL into discoidal alpha-migrating HDL particles can be supplied through passive diffusion driven by a concentration gradient. However, it has become clear that energy-dependent flux of cholesterol and phospholipid across the membrane facilitated by members of the ATP-binding cassette (ABC) family of transporters contributes significantly to the lipidation of apoA-I.

Human carriers of a mutation in the full size ABC transporter ABCA1 gene suffer from familial HDL deficiency (Brooks-Wilson et al. 1999; Bodzioch et al. 1999; Hong et al. 2002). In accordance with a prominent role for ABCA1 in the control of HDL biogenesis, total body ABCA1-deficient mice are characterized by an almost complete lack of HDL cholesterol (>90 % decrease) in the context of similar levels of triglyceride-rich lipoproteins (Christiansen-Weber et al. 2000; Orsó et al. 2000; Calpe-Berdiel et al. 2005; Brunham et al. 2006). Mature alpha-migrating HDL particles are virtually absent in ABCA1 knockout mice. The remaining HDL cholesterol is carried by pre-beta HDL particles that are relatively enriched in triglycerides and exhibit an altered phospholipid species distribution (Orsó et al. 2000; Francone et al. 2003). ABCA1 knockout mice display a diminished LCAT activity (Francone et al. 2003), which highlights that lipidation of apoA-I via ABCA1 is required for LCAT activity and maturation of HDL.

Adrenocortical cells (Orsó et al. 2000) and testical Sertoli cells (Christiansen-Weber et al. 2000) of mice lacking ABCA1 show distinct accumulation of cholesterol esters. This can possibly be attributed to the observed increase in the activity of the cholesterol synthesis gene HMG-CoA reductase (Drobnik et al. 2001). However, ABCA1 may also act as a local regulator of steroidogenic tissue cholesterol homeostasis. In further support of a crucial role of ABCA1 in the control of normal steroidogenesis and development, pregnant ABCA1-deficient females exhibit lower plasma progesterone and estrogen levels (Christiansen-Weber et al. 2000). Furthermore, placentas of ABCA1 knockout mice often show malformations, which translates into a diminished birth and survival rate of ABCA1 knockout pups (Christiansen-Weber et al. 2000).

Livers and intestines of ABCA1 knockout mice are depleted of cholesterol esters (Orsó et al. 2000). Within the liver of wild-type mice, relatively high mRNA and protein expression levels of ABCA1 are found in hepatocytes and tissue macrophages (Kupffer cells) (Lawn et al. 2001; Hoekstra et al. 2003), while in the intestine ABCA1 expression seems to be mostly restricted to macrophages in the lamina propria of villi (Lawn et al. 2001). Dedicated gene targeting strategies have been employed to delineate the contribution of the different cell compartments to plasma HDL-cholesterol levels and total body cholesterol homeostasis. Conditional disruption of ABCA1 function specifically in parenchymal liver cells with the use of the Cre/Flox gene targeting system impairs the apoA-I-mediated efflux of cholesterol and phospholipid from hepatocytes without altering the efflux of cholesterol from macrophages to apoA-I (Timmins et al. 2005). Plasma HDL-cholesterol levels are ~80 % lower as a result of liver-specific ABCA1 deficiency (Timmins et al. 2005). Liver-specific deletion of ABCA1 is also associated with hypercatabolism of apoA-I by the kidneys resulting in an increased turnover of HDL particles and a marked decrease in steady-state plasma apoA-I levels (Timmins et al. 2005; Singaraja et al. 2006). Although no data on the effect of general ABCA1 deficiency in mice on apoA-I flux have been described, it can be assumed total body ABCA1 knockout mice similarly show an increased catabolism and clearance of apoA-I by the kidneys. ABCA1 deficiency in hepatocytes is associated with a rise in plasma levels and size of triglyceride-rich VLDL particles, which can be attributed to an increase in hepatic VLDL secretion and a parallel decrease in lipoprotein lipase (LPL)-mediated catabolism of triglycerides (Sahoo et al. 2004; Chung et al. 2010). ABCA1-mediated transport of cholesterol from hepatocytes to the plasma compartment thus not only controls HDL biogenesis, but also impacts on the synthesis and catabolism of non-HDL particles. Deletion of ABCA1 specifically in intestinal epithelial cells is associated with an accumulation of cholesterol esters in the intestine, a lower plasma apoA-I concentration, and a ~30 % decrease in plasma HDL-cholesterol levels (Brunham et al. 2006). In contrast to total body and hepatocyte-specific knockout mice, intestinal-specific ABCA1 knockout mice do not show major changes in non-HDL-cholesterol and plasma triglyceride levels (Brunham et al. 2006), which further highlights the role of hepatocyte ABCA1 in the control of plasma triglyceride levels. Importantly, mice genetically lacking ABCA1 in both the liver and intestine show a greater decrease in plasma HDL-cholesterol levels as compared to their single tissue knockout controls, almost reaching similarly low levels as those found in total body ABCA1 knockout mice (Brunham et al. 2006). Hepatocytes and the intestinal epithelium can thus be considered the primary cellular sources of the cholesterol that is used for lipidation of apoA-I and generation of HDL.

Loss-of-function mutations in the ABCA1 gene could not be associated with increased coronary heart disease in the general population (Frikke-Schmidt et al. 2008a). However, several studies have suggested that premature atherosclerosis and cardiovascular disease are a common finding in human carriers of functional mutations in the ABCA1 gene (Huang et al. 2001; Hong et al. 2002; Frikke-Schmidt et al. 2008b). In contrast, total body ABCA1 deficiency does not alter the susceptibility to atherosclerosis in hyperlipidemic apoE knockout and LDL receptor (LDLr) knockout mice fed either regular chow or an atherogenic high fat/high cholesterol diet (Aiello et al. 2002). However, deletion of ABCA1 function in the context of genetic hyperlipidemia is not only associated with HDL deficiency but also with a significant decrease in the plasma concentration of cholesterol carried by pro-atherogenic VLDL/LDL (Aiello et al. 2002). As is also the case for apoA-I, normal ABCA1 function in mice thus only protects against atherosclerosis after correcting for ABCA1 genotype-associated confounding differences in atherogenic lipoprotein levels. The atheroprotective effect of ABCA1 may be attributed to its general impact on plasma HDL-cholesterol levels or specifically due to its potential role in the efflux of cholesterol from macrophages. Lipid-filled foam cells can be found in all macrophage-rich tissues of hyperlipidemic total body ABCA1 knockout mice (Aiello et al. 2002), which already hints to an important in vivo role for ABCA1 in macrophage cholesterol efflux. The capacity of ABCA1 knockout mice to facilitate reverse cholesterol transport, the transport from peripheral cells—i.e., macrophages—back to the liver for subsequent excretion, has been validated by measuring the recovery in the plasma compartment and feces of radiolabeled cholesterol from intraperitoneal administered macrophages. In line with the hypothesis that ABCA1-mediated cholesterol efflux is an important determinant of reverse cholesterol transport, a lower amount of radiolabeled cholesterol is excreted under the condition of both total body and macrophage-specific ABCA1 deficiency (Calpe-Berdiel et al. 2005; Wang et al. 2007).

Aiello et al. (2002) and Van Eck et al. (2002) utilized bone marrow transplantation to explore the specific effect of ABCA1-dependent macrophage cholesterol efflux on atherosclerosis susceptibility. In accordance with a negligible impact of macrophage ABCA1 on total HDL biogenesis, bone marrow ABCA1 deficiency does not alter plasma HDL-cholesterol levels (Haghpassand et al. 2001; Aiello et al. 2002; Van Eck et al. 2002). Atherosclerotic lesion development in apoE knockout mice is markedly enhanced—at multiple sites—in response to disruption of ABCA1 expression in bone marrow (Aiello et al. 2002). Transplantation of ABCA1 knockout bone marrow into lethally irradiated LDLr knockout mice similarly stimulates atherosclerotic lesion development (Van Eck et al. 2002). ABCA1-mediated efflux of cholesterol from leukocytes is thus an important protective mechanism to inhibit macrophage foam cell formation and atherosclerosis in mice. Remarkably, macrophage-specific ABCA1 deficiency does not affect the extent of atherosclerosis in LDLr knockout mice (Brunham et al. 2009). It can therefore be concluded that ABCA1's function in macrophages does not confer protection against atherosclerosis, but that rather an efficient efflux of cholesterol from lymphocytes or other immune cells is essential to overcome disease.

Relative HDL deficiency in liver-specific ABCA1 knockout mice on a hyperlipidemic LDLr knockout background is not associated with any enhanced susceptibility for the initial development of macrophage-rich atherosclerotic lesions, while the development of more advanced (collagen-containing) plaques is inhibited (Bi et al. 2013). In accordance with a prominent role for macrophage ABCA1 in reverse cholesterol transport, no change is observed in the flux of cholesterol from macrophages to the feces in LDLr knockout mice lacking ABCA1 function only in hepatocytes (Bi et al. 2013). Strikingly, apoE knockout mice that contain hepatocyte-specific ABCA1 deficiency do show an increased susceptibility to atherosclerosis, despite lowered plasma VLDL/LDL-cholesterol levels (Brunham et al. 2009). A critical review of these data has suggested that the discrepancy in the atherosclerosis findings from liver-specific ABCA1 knockout mice may be attributed to differences in inflammatory status of the two hyperlipidemic mouse models (Van Eck and Van Berkel 2013), which makes proper interpretation of the findings difficult. However, due to the inconsistency of the aforementioned results, it seems clear that the decrease in plasma HDL-cholesterol levels due to deletion of ABCA1 function should not be considered the driving force for the increase in atherosclerosis in total body ABCA1 knockout mice.

A consistent decrease in plasma VLDL-cholesterol levels is observed in LDLr knockout mice that have been transplanted with ABCA1-deficient bone marrow upon feeding an atherogenic Western-type diet (Van Eck et al. 2002; Lammers et al. 2011, 2012). This finding suggests that modulation of the leukocyte cholesterol efflux rate can directly impact on the metabolism of VLDL particles in vivo. However, to date, the mechanism underlying this effect remains to be resolved. It is becoming more and more evident that immunology and metabolism are not just two separate disciplines, but that they intervene and impact each other at multiple levels (Mathis and Shoelson 2011). ABCA1-deficient macrophages display a higher sensitivity to the bacterial membrane component lipopolysaccharide (LPS) (Zhu et al. 2008; Yvan-Charvet et al. 2008) and myeloid cell-specific ABCA1 knockout mice are protected against bacterial infection (Zhu et al. 2012). Furthermore, ABCA1 deficiency in bone marrow is associated with an increase in blood leukocyte counts (Van Eck et al. 2002). Given the apparent link between ABCA1 and immune function in vivo, future research should be aimed at uncovering to what extent ABCA1 contributes to the interplay between inflammation and cholesterol metabolism.

4 ATP-Binding Cassette Transporter G1

ABCG1, formerly known as the ABC8/white gene, belongs to the family of ABC half transporters that need to dimerize to execute their function. ABCG1 expression is mostly restricted to macrophage-rich tissues, including liver, spleen, and lung (Klucken et al. 2000), where it acts as an intracellular sterol transporter (Tarling and Edwards 2011). ABCG1 mRNA expression can actually be used as a measure for the hepatic macrophage content in rodent livers, despite detectable levels of ABCG1 expression in hepatocytes (Hoekstra et al. 2003; Ye et al. 2008; Li et al. 2012). The capacity to efflux cholesterol to HDL is diminished in ABCG1-deficient macrophages, while the apoA-I-mediated cholesterol efflux rate is similar as compared to wild-type macrophages (Out et al. 2006). ABCG1 and ABCA1 thus interact with different HDL subspecies (mature HDL vs. pre-beta HDL) to facilitate their cholesterol efflux function. To date, no association between ABCG1 mutations and HDL-cholesterol levels has been described in humans, suggesting that ABCG1 does not significantly contribute to HDL biogenesis. In accordance, under basal chow diet feeding conditions, ABCG1 knockout mice do not exhibit a change in plasma total or HDL-cholesterol levels (Out et al. 2007; Wiersma et al. 2009).

Chow-fed ABCG1-deficient mice exhibit a severe pulmonary lipidosis phenotype, starting from the age of ~6 months, which is characterized by accumulation of phospholipids, cholesterol crystals, and inflammatory infiltrates in sub-pleural areas of the lung (Baldán et al. 2006a, 2008; Wojcik et al. 2008). The distinct lung phenotype in total body ABCG1 knockout mice can be specifically attributed to the lack of ABCG1 in bone marrow-derived cells, presumably macrophages. Lethally irradiated normolipidemic mice transplanted with ABCG1-deficient bone marrow also show inflammatory cell infiltrates in their lungs, while normalization of macrophage ABCG1 function prevents the occurrence of pulmonary lipidosis in total body ABCG1 knockout mice (Wojcik et al. 2008). Alveolar macrophages from ABCG1 knockout mice have a foamy appearance and display compensatory upregulation of the cholesterol efflux gene ABCA1 (Baldán et al. 2006a). In vitro cultured macrophages lacking ABCG1 are more susceptible to surfactant-induced cholesterol accumulation (Baldán et al. 2006a). ABCG1-mediated cholesterol efflux to HDL is thus essential for the protection against specifically surfactant-induced foam cell formation under normolipidemic conditions, since no accumulation of lipid has been noted in other macrophage-rich tissues in chow-fed ABCG1-deficient mice (Out et al. 2008a). The pulmonary lipidosis phenotype is exacerbated upon challenging the mice with a cholic acid-containing atherogenic diet (Out et al. 2007). Under atherogenic diet feeding conditions, ABCG1 knockout mice also show marked lipid accumulation in cells within and surrounding the germinal centers of the spleen (Out et al. 2007). This finding highlights that ABCG1 is not solely involved in maintaining cholesterol homeostasis locally within the lungs.

Variations in the human ABCG1 gene have been linked to an altered risk for atherosclerotic disease that cannot be explained by changes in lipoprotein levels (Furuyama et al. 2009). In accordance with a role for murine ABCG1 in the protection against atherosclerosis, a higher extent of lesion formation, in the context of similar plasma lipoprotein levels, is detected in ABCG1 knockout mice as compared to wild-type littermate controls upon feeding atherogenic diet (Out et al. 2007). The bone marrow transplantation technique has been applied to elucidate the specific contribution of macrophages in the atheroprotective effect of ABCG1. In line with the prominent role of macrophage ABCG1 in the prevention of pulmonary lipidosis, extensive lipid accumulation is observed also in lungs of hyperlipidemic mice transplanted with ABCG1 knockout bone marrow (Out et al. 2006; Baldán et al. 2006b). However, contrasting effects on atherosclerosis outcome have been noted upon bone marrow-specific ABCG1 deletion. Macrophage ABCG1 appears to protect against atherosclerosis in LDLr knockout mice as evident from the initial studies by Out et al. (2006). In marked contrast, similar studies by Baldán et al. in LDLr and apoE knockout mice (Baldán et al. 2006b) and Ranalletta et al. in LDLr-deficient mice (Ranalletta et al. 2006) have identified macrophage ABCG1 as being a pro-atherogenic factor. Importantly, following critical review of all the present data and additional bone marrow transplantation studies, Meurs et al. have been able to clarify the evident discrepancy in the findings regarding the role of macrophage ABCG1 in atherosclerosis. The effect of disruption of ABCG1 function in macrophages on atherogenesis appears to be highly dependent on the stage of lesion development (Meurs et al. 2012). Diminished ABCG1-mediated efflux of cholesterol to HDL stimulates the formation of foam cells and accelerates lesion development in initial plaques that primarily contain macrophages. During later stages of the disease, macrophage ABCG1 deficiency inhibits plaque progression as a result of activation of compensatory atheroprotective mechanisms, i.e., hypersecretion of apoE from macrophages (Ranalletta et al. 2006), and/or an increased apoptosis rate of lipid-laden macrophage foam cells (Baldán et al. 2006b).

5 Lecithin-Cholesterol Acyltransferase

Discoidal small HDL particles are able to carry a relatively limited load of cholesterol. In order for nascent HDL species to become fully enriched in cholesterol, the unesterified cholesterol acquired upon cellular efflux has to be esterified for subsequent storage in the core of more mature spherical HDL particles. The hepatocyte-derived enzyme lecithin-cholesterol acyltransferase (LCAT) is considered to be the sole mediator of cholesterol esterification in HDL in humans, since familial LCAT deficiency is characterized by a >95 % decrease in plasma HDL-cholesterol ester levels (Glomset et al. 1970). As such, homozygous carriers of loss-of-function mutations in LCAT present with HDL deficiency, while half of normal HDL-cholesterol values are observed in heterozygous carriers (Santamarina-Fojo et al. 2000; Hovingh et al. 2005). Two different strains of LCAT knockout mice have been generated through genetic deletion of respectively exon 1 or exons 2–5. In accordance with a major role for LCAT in the maturation of HDL, both types of LCAT-deficient mice show severe HDL-cholesterol deficiency (>90 % decreased; Ng et al. 1997; Sakai et al. 1997). This coincides with a significant decrease in plasma apoA-I levels (Ng et al. 1997), probably as a result of apoA-I hypercatabolism by the kidneys. The remaining plasma HDL cholesterol in LCAT knockout mice is contained in its unesterified form (Ng et al. 1997; Sakai et al. 1997).

Besides the clear HDL deficiency phenotype, LCAT knockout mice display significant changes in the metabolism of triglyceride-rich lipoproteins. Homozygous LCAT-deficient mice show hypertriglyceridemia and an increase in plasma VLDL-cholesterol levels on a chow diet (Sakai et al. 1997; Lambert et al. 2001; Ng et al. 2002). This phenotype is preserved when mice are fed a Western-type diet enriched in fat (Li et al. 2007). Furthermore, extensive triglyceride deposition can be observed within livers of LCAT knockout mice upon Western-type diet feeding (Li et al. 2007). Mechanistic studies in LCAT x LDLr double knockout mice have indicated that the hypertriglyceridemia can be attributed to an increase in secretion of triglyceride-rich VLDL particles and a decrease in the LPL-mediated lipolysis rate (Ng et al. 2004). As evidenced by lower fasting glucose and insulin levels, LCAT knockout mice exhibit an improved glucose tolerance and insulin sensitivity when crossbred onto the hyperlipidemic LDLr knockout background (Ng et al. 2004; Li et al. 2007). LCAT-mediated cholesterol ester formation thus not only contributes to the formation of mature HDL particles, but also (indirectly) impacts on fatty acid and glucose metabolism.

A genetic defect in LCAT does, in general, not affect the tissue cholesterol balance in mice (Ng et al. 1997). However, adrenals of LCAT knockout mice show gross morphological changes associated with neutral lipid depletion, i.e., they have a more red/brownish color instead of the normal white appearance after perfusion (Ng et al. 1997). Specifically adrenocortical cells within the zona fasciculata, but not zona glomerulosa, are deprived from neutral lipid stores (Hoekstra et al. 2013a). This results in an overall ~80 % decrease in adrenal cholesterol ester content (Ng et al. 1997). Probably as a compensatory response to overcome cholesterol insufficiency, marked increases in the expression of genes involved in the de novo synthesis and extracellular acquisition of cholesterol can be detected in LCAT knockout adrenals (Ng et al. 1997). Importantly, in support of the notion that HDL act as cholesterol donors for the synthesis of glucocorticoids by the adrenals, LCAT knockout mice display a 40–50 % reduction in the maximal glucocorticoid output (Hoekstra et al. 2013a). A minor decrease in tissue cholesterol content is also noted in ovaries, but not testis, in response to LCAT deficiency in mice (Tomimoto et al. 2001), which does not translate into apparent changes in female fertility.

LCAT knockout mice on a normolipidemic background are virtually protected against cholic acid containing diet-induced formation of atherosclerotic lesions (Lambert et al. 2001). Chow diet-fed LCAT x apoE double knockout mice also display a decrease in atherosclerotic lesion formation, as compared with apoE single knockout controls (Lambert et al. 2001; Ng et al. 2002). Ablation of LCAT function is associated with marked HDL deficiency under all hyperlipidemic conditions. Strikingly, mixed results on the effect of LCAT deficiency on atherosclerosis susceptibility, however, have been noted in hyperlipidemic mice upon feeding diets enriched in fat. LCAT x LDLr double knockout mice display a lower atherosclerotic lesion burden upon feeding a cholic acid-containing diet (Lambert et al. 2001). In contrast, LCAT deficiency stimulates the formation of atherosclerotic lesions feeding in LDLr knockout mice upon feeding a Western-type diet devoid of cholic acid, as judged by the aortic cholesterol content (Furbee et al. 2002; Lee et al. 2004). Furthermore, LCAT x apoE double knockout mice also exhibit an increase in their aortic cholesterol content upon feeding the Western-type high fat diet (Furbee et al. 2002). Importantly, lower plasma levels of non-HDL cholesterol are found in the hyperlipidemic LCAT knockout mice under cholic acid-containing diet feeding conditions (Lambert et al. 2001), while non-HDL-cholesterol levels are actually increased in response to LCAT deficiency upon Western-type diet feeding (Furbee et al. 2002; Lee et al. 2004). It thus seems that the diverse effects of LCAT deficiency on atherosclerosis outcome in the different mouse models cannot be attributed to the HDL deficiency, but are rather a consequence of the genotype-associated changes in plasma levels of apoB-containing lipoproteins (Kunnen and Van Eck 2012).

6   Phospholipid Transfer Protein

Transfer of cholesterol esters generated by LCAT to the core of the HDL particles and the subsequent maturation of the HDL particles require the action of phospholipid transfer protein (PLTP), which supplies phospholipids allowing surface expansion of the shell of the particles. The phospholipids are liberated during the lipolysis of triglycerides in the core of apoB-containing lipoproteins by the action of lipoprotein lipase (Tall et al. 1985). In addition, PLTP facilitates the fusion of HDL3 particles to enlarged particles (Lusa et al. 1996). During this process lipid-poor apoA-I particles are released that can act as substrate for ABCA1-mediated cholesterol efflux from macrophages. In humans, common variants of PLTP have been identified that are associated with alterations in serum HDL cholesterol and the accumulation of small HDL particles (Albers et al. 2012). Furthermore, PLTP expression and activity is regulated during several diseases, including sepsis, multiple sclerosis, cancer, and cardiovascular disease (Albers et al. 2012), but it is largely unknown if the regulation of PLTP is a causative factor or simply a consequence of the processes underlying the diseases. The generation of PLTP knockout mice by the group of Alan Tall 15 years ago has greatly contributed to the general understanding of the role of PLTP in HDL metabolism. PLTP was inactivated in mice by replacing exon 2 containing the translation initiation codon, the signal peptide, and the first 16 amino acids of mature PLTP with a neomycin-resistant gene (Jiang et al. 1999). The plasma transfer activity of the major phospholipids, including phosphatidylcholine, phosphatidylethanolamine, phosphatidylinositol, and sphingomyelin into HDL, was completely blocked, while also the transfer of free cholesterol was impaired in PLTP knockout mice. Deletion of PLTP in mice led to markedly decreased levels of HDL phospholipids and cholesterol (65–70 %), illustrating the importance of transfer of surface phospholipids from apoB-containing lipoproteins by PLTP for maintaining HDL levels (Jiang et al. 1999). Furthermore, similarly as described for LCAT knockout mice the decrease in HDL cholesterol coincided with a significant decrease in plasma apoA-I levels. Hepatocytes isolated from PLTP knockout mice synthesized normal amounts of apoA-I, albeit with reduced amounts of phosphatidylcholine (Siggins et al. 2007), indicating that the reduced plasma apoA-I levels were not due to impaired production. Qin et al. showed by in vivo turnover studies using autologous HDL that the reduced apoA-I levels are likely the consequence of increased catabolism of HDL in the PLTP knockout mice (Qin et al. 2000).

On regular chow diet, PLTP-deficient mice absorb less cholesterol in the intestine (Liu et al. 2007), while hepatic phospholipids are increased and triglycerides are reduced in mice lacking PLTP (Siggins et al. 2007).

Feeding the mice a high saturated fat diet, containing 20 % hydrogenated coconut oil and 0.15 % cholesterol, led to the accumulation of surface components of apoB-containing lipoproteins as evidenced by a massive increase in VLDL and LDL phospholipids and cholesterol in the absence of changes in apoB (Jiang et al. 1999). Subsequent studies showed that the animals accumulated phospholipid and free cholesterol-rich lamellar particles containing apoA-IV and apoE (Qin et al. 2000). The accumulation of the lamellar particles specifically in coconut oil-fed PLTP knockout mice has been attributed to markedly reduced removal of these particles via scavenger receptor BI (SR-BI) by parenchymal liver cells (Kawano et al. 2002) and impaired secretion of biliary cholesterol and phospholipid (Yeang et al. 2010). Interestingly, these particles did not accumulate in the circulation when the mice were fed regular chow diet or a Western diet, containing 20 % milk fat and 0.15 % cholesterol (Kawano et al. 2002). In contrast, while feeding a Western diet, containing milk fat and 0.15 % cholesterol, PLTP deficiency was associated with an attenuated diet-induced hypercholesterolemia due to twofold lower concentrations of cholesterol transported by apoB-containing lipoproteins and HDL. Hepatic and intestinal lipid levels were not affected under these conditions, but cholesterol absorption in the intestine was reduced (Shelly et al. 2008).

Both on chow and while feeding the milk fat Western diet, PLTP knockout mice display reduced systemic inflammation as evidenced by lower levels of interleukin 6 and reduced expression of intercellular adhesion molecule 1 (ICAM1) and vascular adhesion molecule 1 (VCAM1) in aorta, indicating that mice lacking PLTP might have a reduced atherosclerosis susceptibility (Shelly et al. 2008).

To study the effect of PLTP deficiency on atherosclerotic lesion development, PLTP knockout mice were crossbred with human apoB transgenic mice, apoE knockout mice, and LDLr knockout mice (Jiang et al. 2001). Both in the apoB transgenic and the apoE knockout background, PLTP deficiency led, on top of the reduction in HDL lipids, to substantially lower VLDL lipid levels. Interestingly, this effect on VLDL lipids was not found in the LDLr knockout background, indicating the involvement of the LDLr in the lowering of VLDL lipids. In line, apoB production was reduced in the apoB transgenic and apoE knockout background, but not in the LDLr knockout background (Jiang et al. 2001). The content of the antioxidant vitamin E was significantly increased in all three backgrounds in the absence of PLTP, while autoantibodies to oxidized LDL were largely decreased (Jiang et al. 2002). HDL isolated from PLTP knockout mice crossbred to both the apoB transgenic and the LDLr knockout background displayed improved anti-inflammatory properties and reduced the ability of LDL to induce monocyte chemotaxis (Yan et al. 2004). After 6 months of feeding apoB transgenic mice Western diet and after feeding apoE knockout mice chow for 3 months, atherosclerotic lesion size was reduced fivefold in mice lacking PLTP. In contrast, in LDLr knockout mice PLTP deficiency led to a twofold reduction in atherosclerotic lesion size after 8 weeks Western diet, whereas no significant effects were observed after 12 weeks Western diet feeding (Jiang et al. 2001). Thus, while PLTP deficiency can reduce early atherosclerotic lesion development in mice lacking the LDLr, the pro-atherogenic effects of PLTP should be primarily attributed to its effects on the production of apoB-containing lipoproteins in mice fed regular chow or a Western diet. Notably, when apoE knockout mice lacking PLTP were challenged with a coconut oil-enriched, high fat diet for 7 weeks plasma levels of free cholesterol were 23 % higher due to the accumulation of the lamellar free cholesterol and phospholipid-rich particles. Under these conditions no effects on atherosclerotic lesion development were found (Yeang et al. 2010).

PLTP is widely distributed with expression in placenta > pancreas > lung > kidney > heart > liver > skeletal muscle > brain (Day et al. 1994). In addition, PLTP is found in endothelial cells (Day et al. 1994) and in smooth muscle cell and macrophage foam cells in atherosclerotic lesions (Desrumaux et al. 2003; Laffitte et al. 2003; O'Brien et al. 2003). Interestingly, Vikstedt and colleagues showed that the expression of PLTP is sixfold higher in Kupffer cells, macrophages of the liver as compared to hepatocytes (Vikstedt et al. 2007). Mice with a hepatocyte-specific deletion of PLTP were generated by injection of PLTP-flox/flippase animals with adenovirus-associated virus expressing Cre-recombinase under control of the thyroxine binding globulin promoter (Yazdanyar et al. 2013). PLTP activity was reduced by approximately 25 % in these animals with a hepatocyte-specific deletion of PLTP, leading to a 20 % reduction of HDL cholesterol. In addition, non-HDL cholesterol was 29 % lower due to an impaired production of apoB-containing lipoproteins (Yazdanyar et al. 2013). Conversely, liver-specific expression of PLTP in a PLTP knockout background promoted the secretion of apoB-containing lipoproteins by the liver (Yazdanyar and Jiang 2012). The contribution of macrophage PLTP to plasma PLTP activity was determined by bone marrow transplantation studies in LDLr knockout mice. Selective deletion of PLTP in bone marrow-derived cells led to a decrease in plasma PLTP activity on chow and Western diet (Vikstedt et al. 2007). Vikstedt et al. showed that after 9 weeks Western diet feeding atherosclerotic lesion size was 29 % smaller in LDLr knockout mice transplanted with PLTP-deficient bone marrow, which coincided with decreased serum VLDL-cholesterol and phospholipid levels, while HDL phospholipid and apoA-I were increased (Vikstedt et al. 2007). In contrast, Valenta et al. also found lower levels of cholesterol in apoB-containing lipoproteins upon disruption of PLTP in bone marrow-derived cells of LDLr knockout mice, but surprisingly atherosclerotic lesion development was increased (Valenta et al. 2006). No effect on atherosclerosis was observed upon transplantation of PLTP-deficient bone marrow into LDLr knockout mice overexpressing human apoA-I (Valenta et al. 2006). The studies by Valenta et al. suggest that locally in the arterial wall PLTP produced by macrophages can be anti-atherogenic. A possible explanation is that PLTP production by macrophages promotes ABCA1-mediated cholesterol efflux, as evidenced by decreased efflux to apoA-I from PLTP-deficient macrophages (Lee-Rueckert et al. 2006). In agreement, macrophages isolated from PLTP knockout mice were shown to accumulate more cholesterol upon incubation with native or acetylated LDL (Ogier et al. 2007). Importantly, the higher levels of lipid-poor apoA-I in LDLr knockout mice overexpressing human apoA-I can overcome the pro-atherogenic effects of deletion of PLTP in macrophages, probably by stimulating the cholesterol efflux capacity of the macrophages.

In summary, the effects of PLTP on atherosclerotic lesion development are determined by a balance between systemic effects influencing the levels of anti-atherogenic HDL and pro-atherogenic apoB-containing lipoproteins and the antioxidant vitamin E and local effects of PLTP produced by macrophages in the arterial wall influencing macrophage apoE production and ABCA1-mediated cholesterol efflux.

7 Scavenger Receptor BI

During the final step of the reverse cholesterol transport process, mature HDL particles deliver their cholesterol load to the liver for subsequent excretion via the bile. The transmembrane glycoprotein scavenger receptor BI (SR-BI) is considered to be the sole mediator of selective uptake of cholesterol esters from HDL by hepatocytes in mice (Out et al. 2004). Several genetic variations at the SR-BI locus—either translating in an altered SR-BI protein expression or functionality—have been associated with a change in plasma HDL-cholesterol levels (Hsu et al. 2003; West et al. 2009; Vergeer et al. 2011; Brunham et al. 2011; Chadwick and Sahoo 2012), which highlights the crucial role for SR-BI in human HDL metabolism. To uncover the role of SR-BI in lipoprotein metabolism, the group of Monty Krieger developed a mouse model carrying a targeted mutation in the SR-BI gene that leads to complete absence of functional SR-BI protein in all tissues (Rigotti et al. 1997). In addition, a strain of SR-BIatt mice has been generated by the group of Dennis Huszar in which the SR-BI promoter is mutated, resulting in a lowered protein expression of SR-BI in liver, testis, and adrenals (Varban et al. 1998). Furthermore, hypomorphic SR-BI knockout mice have been constructed that do not suffer from complete SR-BI deficiency, but show a restricted expression pattern of the protein in liver, kidney, aorta, and steroidogenic tissues (Huby et al. 2006).

Total body SR-BI knockout mice are characterized by a diet-independent hypercholesterolemia, which can be primarily attributed to a marked rise in plasma HDL-cholesterol levels (Rigotti et al. 1997; Van Eck et al. 2003, 2008). The fractional catabolic rate, plasma decay, and subsequent uptake of HDL-cholesterol esters by the liver are diminished in SR-BI knockout mice (Out et al. 2004). SR-BIatt mice similarly show a diminished fractional catabolic rate and hepatic selective uptake of HDL-associated cholesterol, also leading to a higher steady-state level of HDL cholesterol in the plasma of these mice (Varban et al. 1998; Ji et al. 1999). In accordance with a role for hepatic SR-BI in the regulation of plasma HDL-cholesterol levels, both total body and hepatocyte-specific restriction of SR-BI expression in hypomorphic SR-BI knockout mice is also associated with an increase in plasma HDL-cholesterol levels (Huby et al. 2006). Although SR-BI is thought to primarily mediate the uptake HDL-cholesterol esters, in particular free cholesterol levels are higher in plasma in the different SR-BI-deficient mice (Van Eck et al. 2003; Huby et al. 2006; El Bouhassani et al. 2011). The increase in plasma free cholesterol to cholesterol ester ratio underlies anemia and thrombocytopenia phenotypes in SR-BI knockout mice (Holm et al. 2002; Meurs et al. 2005; Dole et al. 2008; Korporaal et al. 2011) and can be attributed to a reduction in the LCAT activity (El Bouhassani et al. 2011). As judged by the unchanged plasma apoA-I levels in total body SR-BI knockout mice, only the size and cholesterol content but not the absolute number of HDL particles is increased in response to SR-BI deficiency (Rigotti et al. 1997). The enlarged HDL particles observed in plasma of SR-BI knockout mice are highly enriched in apoE. In contrast, apoA-II cannot be detected in the HDL fraction of SR-BI-deficient mice, which suggests a generally altered apolipoprotein composition (Rigotti et al. 1997). In line with the notion that SR-BI facilitates HDL-mediated reverse cholesterol transport, the recovery of radiolabeled cholesterol in feces upon injection of either peritoneal or J774 macrophages is lower in SR-BI knockout mice as compared to their wild-type littermate controls, despite an increased appearance of radioactivity in the plasma (HDL) compartment (Zhang et al. 2005a; Zhao et al. 2011).

SR-BI knockout mice also display a significant increase in non-HDL-cholesterol levels (Rigotti et al. 1997; Van Eck et al. 2003, 2008). Hepatic VLDL secretion and LPL-mediated lipolysis rates are not affected by SR-BI deficiency (Van Eck et al. 2008). SR-BI has been shown to be directly involved in the selective uptake of cholesterol esters from LDL and VLDL particles by hepatocytes (Rhainds et al. 2003; Van Eck et al. 2008). It can therefore be anticipated that the increase in the level of cholesterol associated with apoB-containing lipoproteins in SR-BI knockout mice is directly resulting from a diminished clearance by the liver.

Two variations in the SR-BI gene have been linked to changes in fertility in humans (Yates et al. 2011). Without dietary addition of the lipid-lowering drug probucol, SR-BI knockout female mice are virtually sterile, with a ~90 % reduction in fertility and pup yield as compared to wild-type controls (Miettinen et al. 2001). The development of SR-BI knockout embryos is impaired in ex vivo cultures (Trigatti et al. 1999). Furthermore, developmental abnormalities are also observed in SR-BI knockout embryos in vivo (Santander et al. 2013). Although steroidogenic tissues, i.e., ovaries and adrenals, from SR-BI knockout mice exhibit an apparent decrease in their cholesterol ester stores (Rigotti et al. 1997; Trigatti et al. 1999), levels of the steroid hormone progesterone as well as the number of oocytes and cyclic activity are unaffected by murine SR-BI deficiency (Trigatti et al. 1999). The lack of pups generated from SR-BI-deficient breedings can thus rather be attributed to embryonic defects than to infertility of the mothers.

Human carriers of functional mutations in the SR-BI gene exhibit a decrease in their adrenal steroidogenesis rate (Vergeer et al. 2011). Similarly, depletion of adrenocortical cholesterol stores due to SR-BI deficiency translates into a significant change of the adrenal steroid function in mice. Basal glucocorticoid levels in plasma are similar between SR-BI knockout and wild-type mice (Hoekstra et al. 2008). However, total body SR-BI deficiency is associated with primary glucocorticoid insufficiency. Total body SR-BI knockout mice display increased plasma ACTH levels and are unable to increase their adrenal glucocorticoid output in response to established steroidogenic triggers (Cai et al. 2008; Hoekstra et al. 2008, 2009). It appears that SR-BI-mediated uptake of HDL-cholesterol esters by the adrenals is necessary for a proper steroid output, since the glucocorticoid function is also impaired in adrenocortical cell-specific SR-BI knockout mice (Hoekstra et al. 2013b). In further support, hepatocyte-specific SR-BI deficiency does not lower adrenal cholesterol levels or impair the adrenal corticosterone response to fasting (El Bouhassani et al. 2011).

No atherosclerotic lesion development is normally observed in normolipidemic wild-type mice upon feeding a regular chow diet or a high cholesterol high fat Western-type diet that does not contain cholic acid. No atherosclerotic plaques can also be detected in the aortic root of chow-fed SR-BI knockout mice (Zhao et al. 2011). In contrast, SR-BI knockout mice with increased HDL-cholesterol levels do readily develop macrophage-rich fatty streak lesions after 20 weeks of Western-type diet feeding (Van Eck et al. 2003). SR-BI knockout mice display an increase in non-HDL cholesterol under these feeding conditions (Van Eck et al. 2003). However, the measured levels of apoB-containing lipoproteins upon

Western-type diet feeding are still considered to be too low to stimulate the formation of atherosclerotic lesions. This suggests that other SR-BI-related mechanisms may underlie the susceptibility for Western-type diet-induced atherogenesis in SR-BI knockout mice. Follow-up studies in liver-specific hypomorphic SR-BI knockout mice have indicated that the SR-BI-mediated atheroprotection originates from both hepatic and peripheral cell sources (Huby et al. 2006; El Bouhassani et al. 2011). A similarly increased susceptibility for atherosclerosis upon ablation of total body SR-BI function has been noted in response to feeding a cholic acid-containing atherogenic diet (Zhang et al. 2005b). In accordance with a role for SR-BI in the clearance of apoB-containing lipoproteins, high fat diet-fed SR-BIatt x LDLr knockout mice exhibit higher plasma LDL-cholesterol levels and an increased extent of atherosclerosis as compared to LDLr knockout controls (Huszar et al. 2000). In contrast, total body SR-BI deficiency in high fat diet-fed LDLr knockout mice is associated with lower VLDL/LDL levels, but a similarly enhanced formation of atherosclerotic lesions in the context of a higher plasma HDL-cholesterol concentration (Covey et al. 2003). Loss of total body SR-BI function in chow diet-fed hyperlipidemic apoE knockout mice is associated with severe cardiac dysfunction, i.e., a reduced contractility and ejection fraction, and occlusion of the coronary arteries, leading to premature death of these animals at the age of 6–8 weeks (Braun et al. 2002). Occlusive coronary atherosclerosis and an increased extent of atherosclerosis in the aortic root are also seen upon genetic deletion of SR-BI in atherogenic diet-fed hypomorphic APOER61$^{h/h}$ mice (Zhang et al. 2005b). An optimal total body SR-BI function thus protects against atherosclerosis and myocardial infarction, through (1) its role in hepatic uptake of lipoprotein-derived cholesterol and (2) via its action extrahepatic/peripheral cells.

In rodents, high expression of SR-BI is not only found in liver and steroidogenic tissues (Acton et al. 1996), but also in tissue macrophages such as liver Kupffer cells (Fluiter et al. 1998; Malerød et al. 2002; Hoekstra et al. 2003), which suggests that SR-BI can also contribute to macrophage cholesterol homeostasis locally within atherosclerotic lesions. Studies using isolated SR-BI-deficient peritoneal macrophages have suggested that SR-BI may facilitate cholesterol efflux to mature HDL (Van Eck et al. 2004). However, macrophage SR-BI does not contribute to reverse cholesterol in vivo, based upon radiolabeled cholesterol recovery studies (Wang et al. 2007). Bone marrow transplantation studies have been executed to delineate the specific role of macrophage SR-BI in the protection against atherosclerosis. Transplantation of SR-BI knockout bone marrow into LDLr knockout mice stimulates atherogenesis without altering plasma lipoprotein cholesterol levels (Covey et al. 2003; Van Eck et al. 2004). Inactivation of macrophage SR-BI also promotes the development of atherosclerotic plaques in the context of unchanged lipid levels in apoE knockout mice (Zhang et al. 2003). In addition, the cardiac hypertrophy and dysfunction and coronary atherosclerosis development are attenuated upon restoration of macrophage SR-BI function in apoE x SR-BI double knockout mice (Pei et al. 2013). Strikingly, in marked contrast to the findings from hyperlipidemic mice, disruption of macrophage SR-BI expression in normolipidemic C57BL/6 mice results in a lower susceptibility for atherosclerosis upon feeding a cholic acid-containing atherogenic diet (Van Eck et al. 2004). C57Bl/6 mice fed a cholic acid-containing atherogenic diet develop small initial atherosclerotic lesions, suggesting that SR-BI might be pro-atherogenic in this early stage of lesion development. In line, early atherosclerotic lesion development in LDLr knockout mice challenged with Western diet for only 4 weeks was reduced upon disruption of SR-BI in bone marrow-derived cells. These combined studies suggest that macrophage SR-BI plays a dual role in foam cell formation and atherogenesis, which may be dependent on the stage of atherosclerotic lesion development.

8   Insights from Intercrossing of the Different Knockout Mice

Since some of the aforementioned proteins execute both different and overlapping functions in the formation of HDL particles and reverse cholesterol transport, it is conceivable that they may act synergistically and/or be able to compensate for each other's loss. To uncover interactions between specific gene products several double knockout mice have been generated via intercrossing. In line with a differential role in HDL formation and maturation, studies in mice lacking both ABCA1 and LCAT function have suggested that these two proteins act synergistically. Hepatic cholesterol accumulation is higher, while peripheral tissue cholesterol levels are lower in ABCA1 x LCAT double knockout mice as compared to the respective single knockout mice, despite an even higher HDL fractional catabolic rate (Hossain et al. 2009). In parallel, probucol-induced inhibition of ABCA1 function further lowers adrenal cholesterol levels in LCAT knockout mice (Tomimoto et al. 2001). The ABCA1-mediated formation of HDL particles rather than the SR-BI-mediated cholesterol uptake is the driving force in the maintenance of plasma HDL-cholesterol levels, as HDL-cholesterol levels in ABCA1 x SR-BI double knockout mice are still close to zero (Zhao et al. 2011). In contrast, ablation of either ABCA1 or SR-BI function is equally effective in lowering the rate of in vivo macrophage reverse cholesterol transport (Zhao et al. 2011). The mild pulmonary lipidosis phenotype of ABCA1 knockout mice is severely aggravated in response to deletion of SR-BI function, which suggests that macrophage SR-BI is able to partially compensate for the inability of ABCA1 to mediate cholesterol efflux (Zhao et al. 2011). In accordance, specific disruption of both ABCA1 and SR-BI in bone marrow-derived cells of LDLr knockout mice leads to an added increase in (tissue) macrophage foam cell formation and Western-type diet-induced atherosclerotic lesion development, in the context of lower plasma cholesterol levels, when compared to both single knockout bone marrow transplanted controls (Zhao et al. 2010). Importantly, ABCG1 seems to play a more essential role in the compensatory response to macrophage ABCA1 deficiency. Mass efflux of cholesterol from macrophages to both apoA-I and HDL is virtually absent as a result of combined deletion of ABCA1 and ABCG1 (Out et al. 2008a). Already on a chow diet ABCA1 x ABCG1 double knockout mice display marked accumulation of cholesterol not only within the lungs (reminiscent of the ABCG1 knockout phenotype), but also in primary and secondary lymphoid organs and macrophage-rich areas of the liver and intestine (Out et al. 2008a). Furthermore, the double deletion exacerbates high fat diet-induced lymphocytosis (Yvan-Charvet et al. 2008). Transplantation of bone marrow from ABCA1 x ABCG1 double knockout mice into LDLr knockout mice is associated with resistance to Western-type diet-induced hyperlipidemia (Out et al. 2008b). However, despite the rather low plasma cholesterol levels under Western-type diet feeding conditions, ABCA1 x ABCG1 double knockout bone marrow recipient mice show significant plaque development and exhibit distinct lipid deposition in macrophage-rich tissues (Out et al. 2008b). This highlights the necessity for these transporters to overcome macrophage foam cell formation and atherosclerosis. Subsequent bone marrow transplantations into mice with a heterozygous mutation in the LDLr gene (LDLr$^{+/-}$) have validated the increased susceptibility to macrophage cholesterol accumulation and atherosclerosis in response to the combined deficiency in macrophage ABCA1 and ABCG1 upon challenge with a cholate-containing diet (Yvan-Charvet et al. 2007). Furthermore, these studies have suggested that the combined deficiency of these ABC transporters in bone marrow also impacts on the inflammatory function and susceptibility to apoptosis of macrophages as well as the cyclic activity of hematopoietic stem cells (Yvan-Charvet et al. 2008, 2010a).

9 Conclusions from the Gene Knockout Mouse Studies

Generation of the different knockout mice has significantly contributed to our understanding of HDL's metabolism and function (see Fig. 1 for overview). It has become clear that the initial generation and maturation of HDL particles as well as the selective uptake of its cholesterol by the liver are important parameters in the life cycle of HDL. In line with the proposed atheroprotective function of HDL, murine HDL deficiency consistently predisposes to the development of atherosclerotic lesions. However, it should be noted that many of the detrimental atherosclerosis effects observed in response to HDL deficiency in mice cannot be solely attributed to the low HDL levels per se, as the low HDL levels are in most models paralleled by changes in non-HDL-cholesterol levels. Although in the human setting, low HDL-cholesterol levels predict increased cardiovascular risk at all plasma non-HDL-cholesterol levels, the mouse data firmly establish that HDL and non-HDL metabolism are heavily intertwined.

Based upon the bone marrow transplantation studies in hyperlipidemic mice, it appears that the cholesterol efflux function of HDL is of critical importance to overcome foam cell formation and the development of atherosclerotic lesions. We therefore consider it therapeutically highly relevant to increase the expression and/or activity of the cholesterol transporters in macrophages. In this regard, special focus should be given to macrophage ABCA1 as this transporter seems to be the driving force in the atheroprotection.

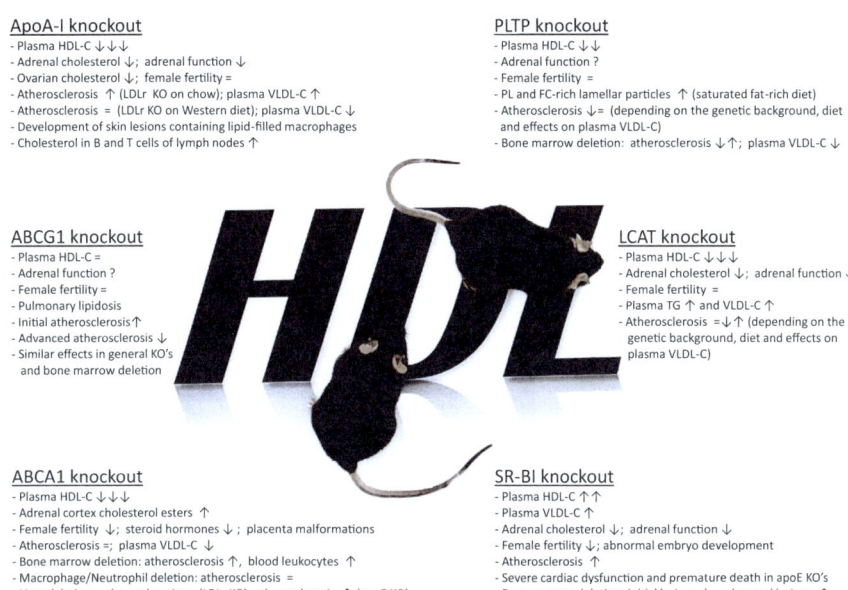

ApoA-I knockout
- Plasma HDL-C ↓↓↓
- Adrenal cholesterol ↓; adrenal function ↓
- Ovarian cholesterol ↓; female fertility =
- Atherosclerosis ↑ (LDLr KO on chow); plasma VLDL-C ↑
- Atherosclerosis = (LDLr KO on Western diet); plasma VLDL-C ↓
- Development of skin lesions containing lipid-filled macrophages
- Cholesterol in B and T cells of lymph nodes ↑

PLTP knockout
- Plasma HDL-C ↓↓
- Adrenal function ?
- Female fertility =
- PL and FC-rich lamellar particles ↑ (saturated fat-rich diet)
- Atherosclerosis ↓= (depending on the genetic background, diet and effects on plasma VLDL-C)
- Bone marrow deletion: atherosclerosis ↓↑; plasma VLDL-C ↓

ABCG1 knockout
- Plasma HDL-C =
- Adrenal function ?
- Female fertility =
- Pulmonary lipidosis
- Initial atherosclerosis ↑
- Advanced atherosclerosis ↓
- Similar effects in general KO's and bone marrow deletion

LCAT knockout
- Plasma HDL-C ↓↓↓
- Adrenal cholesterol ↓; adrenal function ↓
- Female fertility =
- Plasma TG ↑ and VLDL-C ↑
- Atherosclerosis =↓↑ (depending on the genetic background, diet and effects on plasma VLDL-C)

ABCA1 knockout
- Plasma HDL-C ↓↓↓
- Adrenal cortex cholesterol esters ↑
- Female fertility ↓; steroid hormones ↓ ; placenta malformations
- Atherosclerosis =; plasma VLDL-C ↓
- Bone marrow deletion: atherosclerosis ↑, blood leukocytes ↑
- Macrophage/Neutrophil deletion: atherosclerosis =
- Liver deletion: atherosclerosis = (LDLr KO), atherosclerosis ↑ (apoE KO)

SR-BI knockout
- Plasma HDL-C ↑↑
- Plasma VLDL-C ↑
- Adrenal cholesterol ↓; adrenal function ↓
- Female fertility ↓; abnormal embryo development
- Atherosclerosis ↑
- Severe cardiac dysfunction and premature death in apoE KO's
- Bone marrow deletion: initial lesions ↓, advanced lesions ↑

Fig. 1 Major phenotypic changes in mouse models of disturbed HDL metabolism. An overview is given of the effects of deletion of apolipoprotein A-I (apoA-I), the ABC transporters A1 (ABCA1) and G1 (ABCG1), lecithin-cholesterol acyltransferase (LCAT), phospholipid transfer protein (PLTP), and scavenger receptor BI (SR-BI) for HDL's metabolism, adrenal function, female fertility, and its protection against atherosclerosis in mice. = no effect, ↓ decreased, ↑ increased, C cholesterol, PL phospholipids, TG triglycerides, KO knockout

Although HDL is predominantly studied for its atheroprotective action, it can be appreciated that the mouse data suggest an essential role for HDL as cholesterol donor for steroidogenic tissues, including the adrenals and ovaries. Furthermore, it appears that a relevant interaction exists between HDL-mediated cellular cholesterol efflux and the susceptibility to inflammation, which (1) provides strong support for the novel concept that inflammation and metabolism are intertwining biological processes and (2) identifies the efflux function of HDL as putative therapeutic target also in other inflammatory diseases than atherosclerosis, including sepsis and diabetes.

10 Cholesterol Ester Transfer Protein Transgenic Mice

Despite the fact that a great deal regarding the function of specific gene products has been learned from the murine transgenic studies it should be acknowledged that one should take care when directly translating the data to the human situation. Humans, in contrast to mice, naturally express cholesterol ester transfer protein (CETP) that plays a major role in total body cholesterol homeostasis as it is able to exchange cholesterol esters and triglycerides between cholesterol-rich HDL and triglyceride-rich VLDL/LDL particles. In agreement with a major role for CETP in the modulation of plasma HDL-cholesterol levels, genetic CETP deficiency is associated with marked hyperalphalipoproteinemia, i.e., relatively high HDL-cholesterol levels, in humans (Rhyne et al. 2006; Calabresi et al. 2009). To overcome this apparent discrepancy between mice and humans in studies dealing with lipoprotein metabolism, the group of Alan Tall generated a mouse line (NFR-CETP) expressing the human CETP transgene under the control of its natural flanking regulatory elements (Jiang et al. 1992). Initial cloning studies (Drayna et al. 1987) indicated that CETP is primarily derived from liver, small intestine, adrenals, and spleen in humans. In accordance, relatively high CETP mRNA levels are found in liver, spleen, and small intestine of human NFR-CETP transgenic mice (Jiang et al. 1992). Of note, the Tall group also generated a transgenic mouse line expressing CETP under control of the methallothionein-I (MT) gene (Jiang et al. 1992). In MT-CETP transgenic mice, additional high expression of CETP can be detected in kidney, adipose tissue, and heart. Importantly, the expression of CETP in MT-CETP transgenic is not responsive to high cholesterol/high fat diet feeding, which contrasts the findings from NFR-CETP transgenic mice (Jiang et al. 1992) and species naturally expressing CETP, i.e., monkeys and rabbits (Quinet et al. 1990; Pape et al. 1991). Given the apparent limitations of MT-CETP transgenic mice, in our opinion the NFR-CETP mouse model should preferentially be used in studies aimed at translation to the human situation. Therefore, findings from NFR-CETP transgenic mice will only be reviewed in the next part.

Chow-fed CETP transgenic mice exhibit a ~15 % lower plasma total cholesterol level as compared to non-transgenic controls (Masucci-Magoulas et al. 1996; Harada et al. 2007). In contrast, plasma triglyceride levels are marginally increased in CETP carrying mice upon feeding chow diet (Zhou et al. 2006; Harada et al. 2007) due to a slightly diminished clearance of triglyceride-rich lipoproteins (Salerno et al. 2009). The decrease in total cholesterol in CETP transgenic mice can primarily be attributed to lower levels of cholesterol associated with HDL (Zhou et al. 2006). The fractional catabolic rate and uptake of HDL-cholesterol esters by the liver, adrenals, and adipose tissue are increased in CETP transgenic mice, which is associated with an increase in hepatic cholesterol stores (Harada et al. 2007). As a result, CETP transgenic mice exhibit a rise in the plasma total cholesterol over HDL-cholesterol ratio (Clee et al. 1997). The decrease in HDL-cholesterol levels cannot be attributed to a change in LCAT levels. Intriguingly, the presence of CETP actually enhances the saturation level of the LCAT-mediated cholesterol esterification reaction (Oliveira et al. 1997). This suggests that CETP generates particles that can serve as optimal substrates for LCAT.

Despite the marked increase in plasma HDL-cholesterol levels associated with complete CETP deficiency, variable effects on atherosclerosis and cardiovascular disease risk in response to genetic variations in the CETP gene have been noted in the human setting (Zhong et al. 1996; Agerholm-Larsen et al. 2000; Curb et al. 2004; Zheng et al. 2004; Rhyne et al. 2006). To provide clear insight into the contribution of CETP to atherosclerotic lesion development, NFR-CETP transgenic mice have been bred onto several different atherosclerosis-susceptible mouse backgrounds. CETP expression in heterozygous LDLr knockout mice on a cholic acid-containing atherogenic diet does not significantly alter plasma HDL- or non-HDL-cholesterol levels. As a result, CETP does not impact on the susceptibility to atherosclerosis in heterozygous LDLr knockout mice (Casquero et al. 2006). In contrast, CETP executes a clear pro-atherogenic effect in hyperlipidemic apoE knockout and homozygous LDLr knockout mice (Plump et al. 1999). CETP lowers plasma HDL-cholesterol levels in chow-fed apoE knockout mice without changing levels of cholesterol associated with non-HDL particles. In addition, triglyceride levels are increased in the plasma of CETP-expressing apoE knockout mice. CETP expression does not alter plasma HDL-cholesterol levels in LDLr knockout mice challenged with a Western-type high fat/high cholesterol diet. However, levels of cholesterol associated with pro-atherogenic apoB-containing lipoproteins are markedly increased in response to CETP expression in LDLr knockout mice. It thus appears that the increased atherosclerotic lesion development in apoE and LDLr knockout mice upon CETP expression can be directly related to an increased atherogenic index of the plasma compartment (higher non-HDL/HDL ratio) in both types of hyperlipidemic mice. In further support, a marked increase in plasma non-HDL-cholesterol levels and atherosclerotic lesion development—in the context of a marginal but significant decrease in HDL-cholesterol levels—was noted when we expressed CETP only in bone marrow-derived cells of LDLr knockout mice through bone marrow transplantation (Van Eck et al. 2007).

A clear drawback of using LDLr and apoE knockout mice for studies regarding lipoprotein metabolism is that they do not exhibit a normal clearance of apoE-containing VLDL/LDL particles. To determine the effect of CETP expression on lipoprotein metabolism and atherosclerosis development in a more relevant "humanized" setting, CETP transgenic mice have also been crossbred with hyperlipidemic apoE*3-Leiden mice. Transgenic apoE*3-Leiden mice express a mutation of the human apoE3 gene resulting in a slightly attenuated clearance of apoB-containing particles via the LDLr pathway (van den Maagdenberg et al. 1993). As a result, these mice (1) exhibit a lipoprotein cholesterol distribution profile that is highly similar to that found in humans (van den Maagdenberg et al. 1993; van Vlijmen et al. 1994), (2) are susceptible to atherosclerotic lesion development when fed a high fat/high cholesterol Western-type diet devoid of cholic acid (van Vlijmen et al. 1994), and (3) show a similar plasma lipid response as human subjects to several drug interventions (van der Hoogt et al. 2007; de Haan et al. 2008a; van der Hoorn et al. 2008). ApoE*3-Leiden x CETP double transgenic mice as compared to apoE*3-Leiden controls display a significant increase in plasma total cholesterol levels on both chow and under Western-type diet feeding conditions (Westerterp et al. 2006). Under these dietary conditions, VLDL-cholesterol levels are increased twofold, while HDL-cholesterol levels as well as plasma apoA-I levels are 25 % lower in apoE*3-Leiden x CETP transgenic mice (Westerterp et al. 2006). Importantly, apoE*3-Leiden x CETP transgenic mice exhibit increased atherosclerotic lesion development in the aortic root as compared to apoE*3-Leiden controls (Westerterp et al. 2006). Based upon these combined findings, CETP should clearly be considered a pro-atherogenic protein moiety.

In the last decade, therapeutic strategies have focused on inhibiting CETP activity through the use of a variety of CETP-binding drugs, i.e., torcetrapib, that form an inactive complex between CETP and HDL. Studies in apoE*3-Leiden x CETP transgenic mice have suggested that the HDL increase associated with torcetrapib treatment does not add to the beneficial impact of statin-induced lipid-lowering on atherosclerosis (de Haan et al. 2008b). Importantly, the ILLUMINATE and dal-OUTCOMES trials also failed to show the beneficial effect of CETP inhibition in statin-treated cardiovascular disease patients (Barter et al. 2007; Schwartz et al. 2012). Combination treatment with torcetrapib and statin increases the macrophage content in atherosclerotic lesions in apoE*3-Leiden x CETP transgenic mice as compared to statin treatment alone (de Haan et al. 2008b). A higher lesional macrophage content is generally associated with an increased susceptibility for plaque rupture and future cardiovascular events in the human situation (Redgrave et al. 2006; Marnane et al. 2014). An effect of CETP inhibition on plaque stability may therefore possibly have contributed to the negative outcome of the aforementioned torcetrapib and dalcetrapib trials.

Concluding Remark

The use of dedicated knockout and transgenic mouse models has significantly increased our knowledge on the different functions and metabolism of HDL. However, as evident from the recent failures of the CETP inhibition-based clinical trials, more research is clearly needed to better understand how we can use HDL as target in the treatment of patients at risk of cardiovascular disease and other immune-related disorders. The generation of "humanized" mouse models such as the apoE*3-Leiden x CETP transgenic mice will hopefully aid in the development of novel therapeutic approaches beneficially impacting on HDL functionality and atherosclerosis outcome also in the clinical setting.

Acknowledgments This work was supported by the European COST Action BM0904 "HDL–From Biological Understanding to Clinical Exploitation," the Netherlands Heart Foundation (Established Investigator Grant 2007 T056 to M. Van Eck and Senior Postdoc Grants 2008 T70 and 2012 T80 to M. Hoekstra), and VICI Grant 91813603 from the Netherlands Organization for Scientific Research.

Open Access This chapter is distributed under the terms of the Creative Commons Attribution Noncommercial License, which permits any noncommercial use, distribution, and reproduction in any medium, provided the original author(s) and source are credited.

References

Acton S, Rigotti A, Landschulz KT, Xu S, Hobbs HH, Krieger M (1996) Identification of scavenger receptor SR-BI as a high density lipoprotein receptor. Science 271:518–520

Agerholm-Larsen B, Tybjaerg-Hansen A, Schnohr P, Steffensen R, Nordestgaard BG (2000) Common cholesteryl ester transfer protein mutations, decreased HDL cholesterol, and possible decreased risk of ischemic heart disease: the Copenhagen City heart study. Circulation 102:2197–2203

Aiello RJ, Brees D, Bourassa PA, Royer L, Lindsey S, Coskran T, Haghpassand M, Francone OL (2002) Increased atherosclerosis in hyperlipidemic mice with inactivation of ABCA1 in macrophages. Arterioscler Thromb Vasc Biol 22:630–637

Albers JJ, Vuletic S, Cheung MC (2012) Role of plasma phospholipid transfer protein in lipid and lipoprotein metabolism. Biochim Biophys Acta 1821:345–357

Baigent C, Blackwell L, Emberson J, Holland LE, Reith C, Bhala N, Peto R, Barnes EH, Keech A, Simes J, Collins R, Cholesterol Treatment Trialists' (CTT) Collaboration (2010) Efficacy and safety of more intensive lowering of LDL cholesterol: a meta-analysis of data from 170,000 participants in 26 randomised trials. Lancet 376:1670–1681

Baldán A, Tarr P, Vales CS, Frank J, Shimotake TK, Hawgood S, Edwards PA (2006a) Deletion of the transmembrane transporter ABCG1 results in progressive pulmonary lipidosis. J Biol Chem 281:29401–29410

Baldán A, Pei L, Lee R, Tarr P, Tangirala RK, Weinstein MM, Frank J, Li AC, Tontonoz P, Edwards PA (2006b) Impaired development of atherosclerosis in hyperlipidemic Ldlr-/- and ApoE-/- mice transplanted with Abcg1-/- bone marrow. Arterioscler Thromb Vasc Biol 26:2301–2307

Baldán A, Gomes AV, Ping P, Edwards PA (2008) Loss of ABCG1 results in chronic pulmonary inflammation. J Immunol 180:3560–3568

Barter PJ, Caulfield M, Eriksson M, Grundy SM, Kastelein JJ, Komajda M, Lopez-Sendon J, Mosca L, Tardif JC, Waters DD, Shear CL, Revkin JH, Buhr KA, Fisher MR, Tall AR, Brewer B, Investigators ILLUMINATE (2007) Effects of torcetrapib in patients at high risk for coronary events. N Engl J Med 357:2109–2122

Berge KE, Leren TP (2010) Mutations in APOA-I and ABCA1 in Norwegians with low levels of HDL cholesterol. Clin Chim Acta 411:2019–2023

Bi X, Zhu X, Duong M, Boudyguina EY, Wilson MD, Gebre AK, Parks JS (2013) Liver ABCA1 deletion in LDLrKO mice does not impair macrophage reverse cholesterol transport or exacerbate atherogenesis. Arterioscler Thromb Vasc Biol 33:2288–2296

Bodzioch M, Orsó E, Klucken J, Langmann T, Böttcher A, Diederich W, Drobnik W, Barlage S, Büchler C, Porsch-Ozcürümez M, Kaminski WE, Hahmann HW, Oette K, Rothe G, Aslanidis C, Lackner KJ, Schmitz G (1999) The gene encoding ATP-binding cassette transporter 1 is mutated in Tangier disease. Nat Genet 22:347–351

Braun A, Trigatti BL, Post MJ, Sato K, Simons M, Edelberg JM, Rosenberg RD, Schrenzel M, Krieger M (2002) Loss of SR-BI expression leads to the early onset of occlusive atherosclerotic coronary artery disease, spontaneous myocardial infarctions, severe cardiac dysfunction, and premature death in apolipoprotein E-deficient mice. Circ Res 90:270–276

Brooks-Wilson A, Marcil M, Clee SM, Zhang LH, Roomp K, van Dam M, Yu L, Brewer C, Collins JA, Molhuizen HO, Loubser O, Ouelette BF, Fichter K, Ashbourne-Excoffon KJ, Sensen CW, Scherer S, Mott S, Denis M, Martindale D, Frohlich J, Morgan K, Koop B, Pimstone S, Kastelein JJ, Genest J Jr, Hayden MR (1999) Mutations in ABC1 in Tangier disease and familial high-density lipoprotein deficiency. Nat Genet 22:336–345

Brunham LR, Kruit JK, Iqbal J, Fievet C, Timmins JM, Pape TD, Coburn BA, Bissada N, Staels B, Groen AK, Hussain MM, Parks JS, Kuipers F, Hayden MR (2006) Intestinal ABCA1 directly contributes to HDL biogenesis in vivo. J Clin Invest 116:1052–1062

Brunham LR, Singaraja RR, Duong M, Timmins JM, Fievet C, Bissada N, Kang MH, Samra A, Fruchart JC, McManus B, Staels B, Parks JS, Hayden MR (2009) Tissue-specific roles of ABCA1 influence susceptibility to atherosclerosis. Arterioscler Thromb Vasc Biol 29:548–554

Brunham LR, Tietjen I, Bochem AE, Singaraja RR, Franchini PL, Radomski C, Mattice M, Legendre A, Hovingh GK, Kastelein JJ, Hayden MR (2011) Novel mutations in scavenger receptor BI associated with high HDL cholesterol in humans. Clin Genet 79:575–581

Cai L, Ji A, de Beer FC, Tannock LR, van der Westhuyzen DR (2008) SR-BI protects against endotoxemia in mice through its roles in glucocorticoid production and hepatic clearance. J Clin Invest 118:364–375

Calabresi L, Nilsson P, Pinotti E, Gomaraschi M, Favari E, Adorni MP, Bernini F, Sirtori CR, Calandra S, Franceschini G, Tarugi P (2009) A novel homozygous mutation in CETP gene as a cause of CETP deficiency in a Caucasian kindred. Atherosclerosis 205:506–511

Calpe-Berdiel L, Rotllan N, Palomer X, Ribas V, Blanco-Vaca F, Escolà-Gil JC (2005) Direct evidence in vivo of impaired macrophage-specific reverse cholesterol transport in ATP-binding cassette transporter A1-deficient mice. Biochim Biophys Acta 1738:6–9

Casquero AC, Berti JA, Salerno AG, Bighetti EJ, Cazita PM, Ketelhuth DF, Gidlund M, Oliveira HC (2006) Atherosclerosis is enhanced by testosterone deficiency and attenuated by CETP expression in transgenic mice. J Lipid Res 47:1526–1534

Chadwick AC, Sahoo D (2012) Functional characterization of newly-discovered mutations in human SR-BI. PLoS ONE 7:e45660

Christiansen-Weber TA, Voland JR, Wu Y, Ngo K, Roland BL, Nguyen S, Peterson PA, Fung-Leung WP (2000) Functional loss of ABCA1 in mice causes severe placental malformation, aberrant lipid distribution, and kidney glomerulonephritis as well as high-density lipoprotein cholesterol deficiency. Am J Pathol 157:1017–1029

Chung S, Timmins JM, Duong M, Degirolamo C, Rong S, Sawyer JK, Singaraja RR, Hayden MR, Maeda N, Rudel LL, Shelness GS, Parks JS (2010) Targeted deletion of hepatocyte ABCA1 leads to very low density lipoprotein triglyceride overproduction and low density lipoprotein hypercatabolism. J Biol Chem 285:12197–12209

Clee SM, Zhang H, Bissada N, Miao L, Ehrenborg E, Benlian P, Shen GX, Angel A, LeBoeuf RC, Hayden MR (1997) Relationship between lipoprotein lipase and high density lipoprotein cholesterol in mice: modulation by cholesteryl ester transfer protein and dietary status. J Lipid Res 38:2079–2089

Covey SD, Krieger M, Wang W, Penman M, Trigatti BL (2003) Scavenger receptor class B type I-mediated protection against atherosclerosis in LDL receptor-negative mice involves its expression in bone marrow-derived cells. Arterioscler Thromb Vasc Biol 23:1589–1594

Curb JD, Abbott RD, Rodriguez BL, Masaki K, Chen R, Sharp DS, Tall AR (2004) A prospective study of HDL-C and cholesteryl ester transfer protein gene mutations and the risk of coronary heart disease in the elderly. J Lipid Res 45:948–953

Dastani Z, Dangoisse C, Boucher B, Desbiens K, Krimbou L, Dufour R, Hegele RA, Pajukanta P, Engert JC, Genest J, Marcil M (2006) A novel nonsense apolipoprotein A-I mutation (apoA-I (E136X)) causes low HDL cholesterol in French Canadians. Atherosclerosis 185:127–136

Day JR, Albers JJ, Lofton-Day CE, Gilbert TL, Ching AF, Grant FJ, O'Hara PJ, Marcovina SM, Adolphson JL (1994) Complete cDNA encoding human phospholipid transfer protein from human endothelial cells. J Biol Chem 269:9388–9391

de Haan W, van der Hoogt CC, Westerterp M, Hoekstra M, Dallinga-Thie GM, Princen HM, Romijn JA, Jukema JW, Havekes LM, Rensen PC (2008a) Atorvastatin increases HDL cholesterol by reducing CETP expression in cholesterol-fed APOE*3-Leiden.CETP mice. Atherosclerosis 197:57–63

de Haan W, de Vries-van der Weij J, van der Hoorn JW, Gautier T, van der Hoogt CC, Westerterp M, Romijn JA, Jukema JW, Havekes LM, Princen HM, Rensen PC (2008b) Torcetrapib does not reduce atherosclerosis beyond atorvastatin and induces more proinflammatory lesions than atorvastatin. Circulation 117:2515–2522

Desrumaux CM, Mak PA, Boisvert WA, Masson D, Stupack D, Jauhiainen M, Ehnholm C, Curtiss LK (2003) Phospholipid transfer protein is present in human atherosclerotic lesions and is expressed by macrophages and foam cells. J Lipid Res 44:1453–1461

Dole VS, Matuskova J, Vasile E, Yesilaltay A, Bergmeier W, Bernimoulin M, Wagner DD, Krieger M (2008) Thrombocytopenia and platelet abnormalities in high-density lipoprotein receptor-deficient mice. Arterioscler Thromb Vasc Biol 28:1111–1116

Drayna D, Jarnagin AS, McLean J, Henzel W, Kohr W, Fielding C, Lawn R (1987) Cloning and sequencing of human cholesteryl ester transfer protein cDNA. Nature 327:632–634

Drobnik W, Lindenthal B, Lieser B, Ritter M, Christiansen Weber T, Liebisch G, Giesa U, Igel M, Borsukova H, Büchler C, Fung-Leung WP, Von Bergmann K, Schmitz G (2001) ATP-binding cassette transporter A1 (ABCA1) affects total body sterol metabolism. Gastroenterology 120:1203–1211

El Bouhassani M, Gilibert S, Moreau M, Saint-Charles F, Tréguier M, Poti F, Chapman MJ, Le Goff W, Lesnik P, Huby T (2011) Cholesteryl ester transfer protein expression partially attenuates the adverse effects of SR-BI receptor deficiency on cholesterol metabolism and atherosclerosis. J Biol Chem 286:17227–17238

Fluiter K, van der Westhuijzen DR, van Berkel TJ (1998) In vivo regulation of scavenger receptor BI and the selective uptake of high density lipoprotein cholesteryl esters in rat liver parenchymal and Kupffer cells. J Biol Chem 273:8434–8438

Francone OL, Subbaiah PV, van Tol A, Royer L, Haghpassand M (2003) Abnormal phospholipid composition impairs HDL biogenesis and maturation in mice lacking Abca1. Biochemistry 42:8569–8578

Frikke-Schmidt R, Nordestgaard BG, Stene MC, Sethi AA, Remaley AT, Schnohr P, Grande P, Tybjaerg-Hansen A (2008a) Association of loss-of-function mutations in the ABCA1 gene with high-density lipoprotein cholesterol levels and risk of ischemic heart disease. JAMA 299:2524–2532

Frikke-Schmidt R, Nordestgaard BG, Jensen GB, Steffensen R, Tybjaerg-Hansen A (2008b) Genetic variation in ABCA1 predicts ischemic heart disease in the general population. Arterioscler Thromb Vasc Biol 28:180–186

Furbee JW Jr, Sawyer JK, Parks JS (2002) Lecithin:cholesterol acyltransferase deficiency increases atherosclerosis in the low density lipoprotein receptor and apolipoprotein E knockout mice. J Biol Chem 277:3511–3519

Furuyama S, Uehara Y, Zhang B, Baba Y, Abe S, Iwamoto T, Miura S, Saku K (2009) Genotypic effect of ABCG1 gene promoter $-257T > G$ polymorphism on coronary artery disease severity in Japanese men. J Atheroscler Thromb 16:194–200

Glomset JA, Norum KR, King W (1970) Plasma lipoproteins in familial lecithin: cholesterol acyltransferase deficiency: lipid composition and reactivity in vitro. J Clin Invest 49:1827–1837

Gordon T, Castelli WP, Hjortland MC, Kannel WB, Dawber TR (1977) High density lipoprotein as a protective factor against coronary heart disease. The Framingham study. Am J Med 62:707–714

Gordon JI, Smith DP, Andy R, Alpers DH, Schonfeld G, Strauss AW (1982) The primary translation product of rat intestinal apolipoprotein A-I mRNA is an unusual preproprotein. J Biol Chem 257:971–978

Haghpassand M, Bourassa PA, Francone OL, Aiello RJ (2001) Monocyte/macrophage expression of ABCA1 has minimal contribution to plasma HDL levels. J Clin Invest 108:1315–1320

Harada LM, Amigo L, Cazita PM, Salerno AG, Rigotti AA, Quintão EC, Oliveira HC (2007) CETP expression enhances liver HDL-cholesteryl ester uptake but does not alter VLDL and biliary lipid secretion. Atherosclerosis 191:313–318

Hoekstra M, Kruijt JK, Van Eck M, Van Berkel TJ (2003) Specific gene expression of ATP-binding cassette transporters and nuclear hormone receptors in rat liver parenchymal, endothelial, and Kupffer cells. J Biol Chem 278:25448–25453

Hoekstra M, Meurs I, Koenders M, Out R, Hildebrand RB, Kruijt JK, Van Eck M, Van Berkel TJ (2008) Absence of HDL cholesteryl ester uptake in mice via SR-BI impairs an adequate adrenal glucocorticoid-mediated stress response to fasting. J Lipid Res 49:738–745

Hoekstra M, Ye D, Hildebrand RB, Zhao Y, Lammers B, Stitzinger M, Kuiper J, Van Berkel TJ, Van Eck M (2009) Scavenger receptor class B type I-mediated uptake of serum cholesterol is essential for optimal adrenal glucocorticoid production. J Lipid Res 50:1039–1046

Hoekstra M, Korporaal SJ, van der Sluis RJ, Hirsch-Reinshagen V, Bochem AE, Wellington CL, Van Berkel TJ, Kuivenhoven JA, Van Eck M (2013a) LCAT deficiency in mice is associated with a diminished adrenal glucocorticoid function. J Lipid Res 54:358–364

Hoekstra M, van der Sluis RJ, Van Eck M, Van Berkel TJ (2013b) Adrenal-specific scavenger receptor BI deficiency induces glucocorticoid insufficiency and lowers plasma very-low-density and low-density lipoprotein levels in mice. Arterioscler Thromb Vasc Biol 33:e39–e46

Holm TM, Braun A, Trigatti BL, Brugnara C, Sakamoto M, Krieger M, Andrews NC (2002) Failure of red blood cell maturation in mice with defects in the high-density lipoprotein receptor SR-BI. Blood 99:1817–1824

Hong SH, Rhyne J, Zeller K, Miller M (2002) ABCA1(Alabama): a novel variant associated with HDL deficiency and premature coronary artery disease. Atherosclerosis 164:245–250

Hossain MA, Tsujita M, Akita N, Kobayashi F, Yokoyama S (2009) Cholesterol homeostasis in ABCA1/LCAT double-deficient mouse. Biochim Biophys Acta 1791:1197–1205

Hovingh GK, Brownlie A, Bisoendial RJ, Dube MP, Levels JH, Petersen W, Dullaart RP, Stroes ES, Zwinderman AH, de Groot E, Hayden MR, Kuivenhoven JA, Kastelein JJ (2004) A novel apoA-I mutation (L178P) leads to endothelial dysfunction, increased arterial wall thickness, and premature coronary artery disease. J Am Coll Cardiol 44:1429–1435

Hovingh GK, Hutten BA, Holleboom AG, Petersen W, Rol P, Stalenhoef A, Zwinderman AH, de Groot E, Kastelein JJ, Kuivenhoven JA (2005) Compromised LCAT function is associated with increased atherosclerosis. Circulation 112:879–884

Hsu LA, Ko YL, Wu S, Teng MS, Peng TY, Chen CF, Chen CF, Lee YS (2003) Association between a novel 11-base pair deletion mutation in the promoter region of the scavenger receptor class B type I gene and plasma HDL cholesterol levels in Taiwanese Chinese. Arterioscler Thromb Vasc Biol 23:1869–1874

Huang W, Moriyama K, Koga T, Hua H, Ageta M, Kawabata S, Mawatari K, Imamura T, Eto T, Kawamura M, Teramoto T, Sasaki J (2001) Novel mutations in ABCA1 gene in Japanese patients with Tangier disease and familial high density lipoprotein deficiency with coronary heart disease. Biochim Biophys Acta 1537:71–78

Huby T, Doucet C, Dachet C, Ouzilleau B, Ueda Y, Afzal V, Rubin E, Chapman MJ, Lesnik P (2006) Knockdown expression and hepatic deficiency reveal an atheroprotective role for SR-BI in liver and peripheral tissues. J Clin Invest 116:2767–2776

Huszar D, Varban ML, Rinninger F, Feeley R, Arai T, Fairchild-Huntress V, Donovan MJ, Tall AR (2000) Increased LDL cholesterol and atherosclerosis in LDL receptor-deficient mice with attenuated expression of scavenger receptor B1. Arterioscler Thromb Vasc Biol 20:1068–1073

Ishibashi S, Goldstein JL, Brown MS, Herz J, Burns DK (1994) Massive xanthomatosis and atherosclerosis in cholesterol-fed low density lipoprotein receptor-negative mice. J Clin Invest 93:1885–1893

Ishida BY, Blanche PJ, Nichols AV, Yashar M, Paigen B (1991) Effects of atherogenic diet consumption on lipoproteins in mouse strains C57BL/6 and C3H. J Lipid Res 32:559–568

Ji Y, Wang N, Ramakrishnan R, Sehayek E, Huszar D, Breslow JL, Tall AR (1999) Hepatic scavenger receptor BI promotes rapid clearance of high density lipoprotein free cholesterol and its transport into bile. J Biol Chem 274:33398–33402

Jiang XC, Agellon LB, Walsh A, Breslow JL, Tall A (1992) Dietary cholesterol increases transcription of the human cholesteryl ester transfer protein gene in transgenic mice. Dependence on natural flanking sequences. J Clin Invest 90:1290–1295

Jiang XC, Bruce C, Mar J, Lin M, Ji Y, Francone OL, Tall AR (1999) Targeted mutation of plasma phospholipid transfer protein gene markedly reduces high-density lipoprotein levels. J Clin Invest 103:907–914

Jiang XC, Qin S, Qiao C, Kawano K, Lin M, Skold A, Xiao X, Tall AR (2001) Apolipoprotein B secretion and atherosclerosis are decreased in mice with phospholipid-transfer protein deficiency. Nat Med 7:847–852

Jiang XC, Tall AR, Qin S, Lin M, Schneider M, Lalanne F, Deckert V, Desrumaux C, Athias A, Witztum JL, Lagrost L (2002) Phospholipid transfer protein deficiency protects circulating lipoproteins from oxidation due to the enhanced accumulation of vitamin E. J Biol Chem 277:31850–31856

Kawano K, Qin S, Vieu C, Collet X, Jiang XC (2002) Role of hepatic lipase and scavenger receptor BI in clearing phospholipid/free cholesterol-rich lipoproteins in PLTP-deficient mice. Biochim Biophys Acta 1583:133–140

Klucken J, Büchler C, Orsó E, Kaminski WE, Porsch-Özcürümez M, Liebisch G, Kapinsky M, Diederich W, Drobnik W, Dean M, Allikmets R, Schmitz G (2000) ABCG1 (ABC8), the human homolog of the Drosophila white gene, is a regulator of macrophage cholesterol and phospholipid transport. Proc Natl Acad Sci USA 97:817–822

Korporaal SJ, Meurs I, Hauer AD, Hildebrand RB, Hoekstra M, Cate HT, Praticò D, Akkerman JW, Van Berkel TJ, Kuiper J, Van Eck M (2011) Deletion of the high-density lipoprotein receptor scavenger receptor BI in mice modulates thrombosis susceptibility and indirectly affects platelet function by elevation of plasma free cholesterol. Arterioscler Thromb Vasc Biol 31:34–42

Kunnen S, Van Eck M (2012) Lecithin:cholesterol acyltransferase: old friend or foe in atherosclerosis? J Lipid Res 53:1783–1799

Laffitte BA, Joseph SB, Chen M, Castrillo A, Repa J, Wilpitz D, Mangelsdorf D, Tontonoz P (2003) The phospholipid transfer protein gene is a liver X receptor target expressed by macrophages in atherosclerotic lesions. Mol Cell Biol 23:2182–2191

Lambert G, Sakai N, Vaisman BL, Neufeld EB, Marteyn B, Chan CC, Paigen B, Lupia E, Thomas A, Striker LJ, Blanchette-Mackie J, Csako G, Brady JN, Costello R, Striker GE, Remaley AT, Brewer HB Jr, Santamarina-Fojo S (2001) Analysis of glomerulosclerosis and atherosclerosis in lecithin cholesterol acyltransferase-deficient mice. J Biol Chem 276:15090–15098

Lammers B, Zhao Y, Hoekstra M, Hildebrand RB, Ye D, Meurs I, Van Berkel TJ, Van Eck M (2011) Augmented atherogenesis in LDL receptor deficient mice lacking both macrophage ABCA1 and ApoE. PLoS ONE 6:e26095

Lammers B, Zhao Y, Foks AC, Hildebrand RB, Kuiper J, Van Berkel TJ, Van Eck M (2012) Leukocyte ABCA1 remains atheroprotective in splenectomized LDL receptor knockout mice. PLoS ONE 7:e48080

Lawn RM, Wade DP, Couse TL, Wilcox JN (2001) Localization of human ATP-binding cassette transporter 1 (ABC1) in normal and atherosclerotic tissues. Arterioscler Thromb Vasc Biol 21:378–385

Lee RG, Kelley KL, Sawyer JK, Farese RV Jr, Parks JS, Rudel LL (2004) Plasma cholesteryl esters provided by lecithin: cholesterol acyltransferase and acyl-coenzyme a:cholesterol acyltransferase 2 have opposite atherosclerotic potential. Circ Res 95:998–1004

Lee EY, Klementowicz PT, Hegele RA, Asztalos BF, Schaefer EJ (2013) HDL deficiency due to a new insertion mutation (ApoA-INashua) and review of the literature. J Clin Lipidol 7:169–173

Lee-Rueckert M, Vikstedt R, Metso J, Ehnholm C, Kovanen PT, Jauhiainen M (2006) Absence of endogenous phospholipid transfer protein impairs ABCA1-dependent efflux of cholesterol from macrophage foam cells. J Lipid Res 47:1725–1732

Leren TP, Bakken KS, Daum U, Ose L, Berg K, Assmann G, von Eckardstein A (1997) Heterozygosity for apolipoprotein A-I(R160L)Oslo is associated with low levels of high density lipoprotein cholesterol and HDL-subclass LpA-I/A-II but normal levels of HDL-subclass LpA-I. J Lipid Res 38:121–131

Li H, Reddick RL, Maeda N (1993) Lack of apoA-I is not associated with increased susceptibility to atherosclerosis in mice. Arterioscler Thromb 13:1814–1821

Li L, Naples M, Song H, Yuan R, Ye F, Shafi S, Adeli K, Ng DS (2007) LCAT-null mice develop improved hepatic insulin sensitivity through altered regulation of transcription factors and suppressors of cytokine signaling. Am J Physiol Endocrinol Metab 293:E587–E594

Li Z, Wang Y, van der Sluis RJ, van der Hoorn JW, Princen HM, Van Eck M, Van Berkel TJ, Rensen PC, Hoekstra M (2012) Niacin reduces plasma CETP levels by diminishing liver macrophage content in CETP transgenic mice. Biochem Pharmacol 84:821–829

Liu R, Iqbal J, Yeang C, Wang DQ, Hussain MM, Jiang XC (2007) Phospholipid transfer protein-deficient mice absorb less cholesterol. Arterioscler Thromb Vasc Biol 27:2014–2021

Lusa S, Jauhiainen M, Metso J, Somerharju P, Ehnholm C (1996) The mechanism of human plasma phospholipid transfer protein-induced enlargement of high-density lipoprotein particles: evidence for particle fusion. Biochem J 313:275–282

Malerød L, Juvet K, Gjøen T, Berg T (2002) The expression of scavenger receptor class B, type I (SR-BI) and caveolin-1 in parenchymal and nonparenchymal liver cells. Cell Tissue Res 307:173–180

Marnane M, Prendeville S, McDonnell C, Noone I, Barry M, Crowe M, Mulligan N, Kelly PJ (2014) Plaque inflammation and unstable morphology are associated with early stroke recurrence in symptomatic carotid stenosis. Stroke 45:801–806

Masucci-Magoulas L, Plump A, Jiang XC, Walsh A, Breslow JL, Tall AR (1996) Profound induction of hepatic cholesteryl ester transfer protein transgene expression in apolipoprotein E and low density lipoprotein receptor gene knockout mice. A novel mechanism signals changes in plasma cholesterol levels. J Clin Invest 97:154–161

Mathis D, Shoelson SE (2011) Immunometabolism: an emerging frontier. Nat Rev Immunol 11:81

Matsunaga A, Sasaki J, Han H, Huang W, Kugi M, Koga T, Ichiki S, Shinkawa T, Arakawa K (1999) Compound heterozygosity for an apolipoprotein A1 gene promoter mutation and a structural nonsense mutation with apolipoprotein A1 deficiency. Arterioscler Thromb Vasc Biol 19:348–355

Meurs I, Hoekstra M, van Wanrooij EJ, Hildebrand RB, Kuiper J, Kuipers F, Hardeman MR, Van Berkel TJ, Van Eck M (2005) HDL cholesterol levels are an important factor for determining the lifespan of erythrocytes. Exp Hematol 33:1309–1319

Meurs I, Lammers B, Zhao Y, Out R, Hildebrand RB, Hoekstra M, Van Berkel TJ, Van Eck M (2012) The effect of ABCG1 deficiency on atherosclerotic lesion development in LDL receptor knockout mice depends on the stage of atherogenesis. Atherosclerosis 221:41–47

Miettinen HE, Rayburn H, Krieger M (2001) Abnormal lipoprotein metabolism and reversible female infertility in HDL receptor (SR-BI)-deficient mice. J Clin Invest 108:1717–1722

Mihaylova B, Emberson J, Blackwell L, Keech A, Simes J, Barnes EH, Voysey M, Gray A, Collins R, Baigent C, Cholesterol Treatment Trialists' (CTT) Collaborators (2012) The effects of lowering LDL cholesterol with statin therapy in people at low risk of vascular disease: meta-analysis of individual data from 27 randomised trials. Lancet 380:581–590

Moore RE, Kawashiri MA, Kitajima K, Secreto A, Millar JS, Pratico D, Rader DJ (2003) Apolipoprotein A-I deficiency results in markedly increased atherosclerosis in mice lacking the LDL receptor. Arterioscler Thromb Vasc Biol 23:1914–1920

Ng DS, Francone OL, Forte TM, Zhang J, Haghpassand M, Rubin EM (1997) Disruption of the murine lecithin: cholesterol acyltransferase gene causes impairment of adrenal lipid delivery and up-regulation of scavenger receptor class B type I. J Biol Chem 272:15777–15781

Ng DS, Maguire GF, Wylie J, Ravandi A, Xuan W, Ahmed Z, Eskandarian M, Kuksis A, Connelly PW (2002) Oxidative stress is markedly elevated in lecithin: cholesterol acyltransferase-deficient mice and is paradoxly reversed in the apolipoprotein E knockout background in association with a reduction in atherosclerosis. J Biol Chem 277:11715–11720

Ng DS, Xie C, Maguire GF, Zhu X, Ugwu F, Lam E, Connelly PW (2004) Hypertriglyceridemia in lecithin-cholesterol acyltransferase-deficient mice is associated with hepatic overproduction of triglycerides, increased lipogenesis, and improved glucose tolerance. J Biol Chem 279:7636–7642

O'Brien KD, Vuletic S, McDonald TO, Wolfbauer G, Lewis K, Tu AY, Marcovina S, Wight TN, Chait A, Albers JJ (2003) Cell-associated and extracellular phospholipid transfer protein in human coronary atherosclerosis. Circulation 108:270–274

Ogier N, Klein A, Deckert V, Athias A, Bessède G, Le Guern N, Lagrost L, Desrumaux C (2007) Cholesterol accumulation is increased in macrophages of phospholipid transfer protein-deficient mice: normalization by dietary alpha-tocopherol supplementation. Arterioscler Thromb Vasc Biol 27:2407–2412

Oliveira HC, Ma L, Milne R, Marcovina SM, Inazu A, Mabuchi H, Tall AR (1997) Cholesteryl ester transfer protein activity enhances plasma cholesteryl ester formation. Studies in CETP transgenic mice and human genetic CETP deficiency. Arterioscler Thromb Vasc Biol 17:1045–1052

Orsó E, Broccardo C, Kaminski WE, Böttcher A, Liebisch G, Drobnik W, Götz A, Chambenoit O, Diederich W, Langmann T, Spruss T, Luciani MF, Rothe G, Lackner KJ, Chimini G, Schmitz G (2000) Transport of lipids from golgi to plasma membrane is defective in tangier disease patients and Abc1-deficient mice. Nat Genet 24:192–196

Out R, Hoekstra M, Spijkers JA, Kruijt JK, van Eck M, Bos IS, Twisk J, Van Berkel TJ (2004) Scavenger receptor class B type I is solely responsible for the selective uptake of cholesteryl esters from HDL by the liver and the adrenals in mice. J Lipid Res 45:2088–2095

Out R, Hoekstra M, Hildebrand RB, Kruit JK, Meurs I, Li Z, Kuipers F, Van Berkel TJ, Van Eck M (2006) Macrophage ABCG1 deletion disrupts lipid homeostasis in alveolar macrophages and moderately influences atherosclerotic lesion development in LDL receptor-deficient mice. Arterioscler Thromb Vasc Biol 26:2295–2300

Out R, Hoekstra M, Meurs I, de Vos P, Kuiper J, Van Eck M, Van Berkel TJ (2007) Total body ABCG1 expression protects against early atherosclerotic lesion development in mice. Arterioscler Thromb Vasc Biol 27:594–599

Out R, Jessup W, Le Goff W, Hoekstra M, Gelissen IC, Zhao Y, Kritharides L, Chimini G, Kuiper J, Chapman MJ, Huby T, Van Berkel TJ, Van Eck M (2008a) Coexistence of foam cells and hypocholesterolemia in mice lacking the ABC transporters A1 and G1. Circ Res 102:113–120

Out R, Hoekstra M, Habets K, Meurs I, de Waard V, Hildebrand RB, Wang Y, Chimini G, Kuiper J, Van Berkel TJ, Van Eck M (2008b) Combined deletion of macrophage ABCA1 and ABCG1 leads to massive lipid accumulation in tissue macrophages and distinct atherosclerosis at relatively low plasma cholesterol levels. Arterioscler Thromb Vasc Biol 28:258–264

Paigen B, Morrow A, Holmes PA, Mitchell D, Williams RA (1987) Quantitative assessment of atherosclerotic lesions in mice. Atherosclerosis 68:231–240

Pape ME, Rehberg EF, Marotti KR, Melchior GW (1991) Molecular cloning, sequence, and expression of cynomolgus monkey cholesteryl ester transfer protein. Inverse correlation between hepatic cholesteryl ester transfer protein mRNA levels and plasma high density lipoprotein levels. Arterioscler Thromb 11:1759–1771

Parks JS, Li H, Gebre AK, Smith TL, Maeda N (1995) Effect of apolipoprotein A-I deficiency on lecithin: cholesterol acyltransferase activation in mouse plasma. J Lipid Res 36:349–355

Pei Y, Chen X, Aboutouk D, Fuller MT, Dadoo O, Yu P, White EJ, Igdoura SA, Trigatti BL (2013) SR-BI in bone marrow derived cells protects mice from diet induced coronary artery atherosclerosis and myocardial infarction. PLoS ONE 8:e72492

Plump AS, Erickson SK, Weng W, Partin JS, Breslow JL, Williams DL (1996) Apolipoprotein A-I is required for cholesteryl ester accumulation in steroidogenic cells and for normal adrenal steroid production. J Clin Invest 97:2660–2671

Plump AS, Azrolan N, Odaka H, Wu L, Jiang X, Tall A, Eisenberg S, Breslow JL (1997) ApoA-I knockout mice: characterization of HDL metabolism in homozygotes and identification of a post-RNA mechanism of apoA-I up-regulation in heterozygotes. J Lipid Res 38:1033–1047

Plump AS, Masucci-Magoulas L, Bruce C, Bisgaier CL, Breslow JL, Tall AR (1999) Increased atherosclerosis in ApoE and LDL receptor gene knock-out mice as a result of human cholesteryl ester transfer protein transgene expression. Arterioscler Thromb Vasc Biol 19:1105–1110

Qin S, Kawano K, Bruce C, Lin M, Bisgaier C, Tall AR, Jiang X (2000) Phospholipid transfer protein gene knock-out mice have low high density lipoprotein levels, due to hypercatabolism, and accumulate apoA-IV-rich lamellar lipoproteins. J Lipid Res 41:269–276

Quinet EM, Agellon LB, Kroon PA, Marcel YL, Lee YC, Whitlock ME, Tall AR (1990) Atherogenic diet increases cholesteryl ester transfer protein messenger RNA levels in rabbit liver. J Clin Invest 85:357–363

Ranalletta M, Wang N, Han S, Yvan-Charvet L, Welch C, Tall AR (2006) Decreased atherosclerosis in low-density lipoprotein receptor knockout mice transplanted with Abcg1-/- bone marrow. Arterioscler Thromb Vasc Biol 26:2308–2315

Redgrave JN, Lovett JK, Gallagher PJ, Rothwell PM (2006) Histological assessment of 526 symptomatic carotid plaques in relation to the nature and timing of ischemic symptoms: the Oxford plaque study. Circulation 113:2320–2328

Rhainds D, Brodeur M, Lapointe J, Charpentier D, Falstrault L, Brissette L (2003) The role of human and mouse hepatic scavenger receptor class B type I (SR-BI) in the selective uptake of low-density lipoprotein-cholesteryl esters. Biochemistry 42:7527–7538

Rhyne J, Ryan MJ, White C, Chimonas T, Miller M (2006) The two novel CETP mutations Gln87X and Gln165X in a compound heterozygous state are associated with marked hyperalphalipoproteinemia and absence of significant coronary artery disease. J Mol Med (Berl) 84:647–650

Rigotti A, Trigatti BL, Penman M, Rayburn H, Herz J, Krieger M (1997) A targeted mutation in the murine gene encoding the high density lipoprotein (HDL) receptor scavenger receptor class B type I reveals its key role in HDL metabolism. Proc Natl Acad Sci USA 94:12610–12615

Sahoo D, Trischuk TC, Chan T, Drover VA, Ho S, Chimini G, Agellon LB, Agnihotri R, Francis GA, Lehner R (2004) ABCA1-dependent lipid efflux to apolipoprotein A-I mediates HDL particle formation and decreases VLDL secretion from murine hepatocytes. J Lipid Res 45:1122–1131

Sakai N, Vaisman BL, Koch CA, Hoyt RF Jr, Meyn SM, Talley GD, Paiz JA, Brewer HB Jr, Santamarina-Fojo S (1997) Targeted disruption of the mouse lecithin: cholesterol acyltransferase (LCAT) gene. Generation of a new animal model for human LCAT deficiency. J Biol Chem 272:7506–7510

Salerno AG, Patrício PR, Berti JA, Oliveira HC (2009) Cholesteryl ester transfer protein (CETP) increases postprandial triglyceridaemia and delays triacylglycerol plasma clearance in transgenic mice. Biochem J 419:629–634

Santamarina-Fojo S, Lambert G, Hoeg JM, Brewer HB Jr (2000) Lecithin-cholesterol acyltransferase: role in lipoprotein metabolism, reverse cholesterol transport and atherosclerosis. Curr Opin Lipidol 11:267–275

Santander NG, Contreras-Duarte S, Awad MF, Lizama C, Passalacqua I, Rigotti A, Busso D (2013) Developmental abnormalities in mouse embryos lacking the HDL receptor SR-BI. Hum Mol Genet 22:1086–1096

Scanu A, Toth J, Edelstein C, Koga S, Stiller E (1969) Fractionation of human serum high density lipoprotein in urea solutions. Evidence for polypeptide heterogeneity. Biochemistry 8:3309–3316

Schwartz GG, Olsson AG, Abt M, Ballantyne CM, Barter PJ, Brumm J, Chaitman BR, Holme IM, Kallend D, Leiter LA, Leitersdorf E, McMurray JJ, Mundl H, Nicholls SJ, Shah PK, Tardif JC, Wright RS, dal-OUTCOMES Investigators (2012) Effects of dalcetrapib in patients with a recent acute coronary syndrome. N Engl J Med 367:2089–2099

Shelly L, Royer L, Sand T, Jensen H, Luo Y (2008) Phospholipid transfer protein deficiency ameliorates diet-induced hypercholesterolemia and inflammation in mice. J Lipid Res 49:773–781

Siggins S, Bykov I, Hermansson M, Somerharju P, Lindros K, Miettinen TA, Jauhiainen M, Olkkonen VM, Ehnholm C (2007) Altered hepatic lipid status and apolipoprotein A-I metabolism in mice lacking phospholipid transfer protein. Atherosclerosis 190:114–123

Singaraja RR, Stahmer B, Brundert M, Merkel M, Heeren J, Bissada N, Kang M, Timmins JM, Ramakrishnan R, Parks JS, Hayden MR, Rinninger F (2006) Hepatic ATP-binding cassette transporter A1 is a key molecule in high-density lipoprotein cholesteryl ester metabolism in mice. Arterioscler Thromb Vasc Biol 26:1821–1827

Tall AR, Krumholz S, Olivecrona T, Deckelbaum RJ (1985) Plasma phospholipid transfer protein enhances transfer and exchange of phospholipids between very low density lipoproteins and high density lipoproteins during lipolysis. J Lipid Res 26:842–851

Tarling EJ, Edwards PA (2011) ATP binding cassette transporter G1 (ABCG1) is an intracellular sterol transporter. Proc Natl Acad Sci USA 108:19719–19724

Tilly-Kiesi M, Zhang Q, Ehnholm S, Kahri J, Lahdenperä S, Ehnholm C, Taskinen MR (1995) ApoA-IHelsinki (Lys107–>0) associated with reduced HDL cholesterol and LpA-I:A-II deficiency. Arterioscler Thromb Vasc Biol 15:1294–1306

Timmins JM, Lee JY, Boudyguina E, Kluckman KD, Brunham LR, Mulya A, Gebre AK, Coutinho JM, Colvin PL, Smith TL, Hayden MR, Maeda N, Parks JS (2005) Targeted inactivation of hepatic Abca1 causes profound hypoalphalipoproteinemia and kidney hypercatabolism of apoA-I. J Clin Invest 115:1333–1342

Tomimoto S, Tsujita M, Okazaki M, Usui S, Tada T, Fukutomi T, Ito S, Itoh M, Yokoyama S (2001) Effect of probucol in lecithin-cholesterol acyltransferase-deficient mice: inhibition of 2 independent cellular cholesterol-releasing pathways in vivo. Arterioscler Thromb Vasc Biol 21:394–400

Trigatti B, Rayburn H, Viñals M, Braun A, Miettinen H, Penman M, Hertz M, Schrenzel M, Amigo L, Rigotti A, Krieger M (1999) Influence of the high density lipoprotein receptor SR-BI on reproductive and cardiovascular pathophysiology. Proc Natl Acad Sci USA 96:9322–9327

Valenta DT, Ogier N, Bradshaw G, Black AS, Bonnet DJ, Lagrost L, Curtiss LK, Desrumaux CM (2006) Atheroprotective potential of macrophage-derived phospholipid transfer protein in low-density lipoprotein receptor-deficient mice is overcome by apolipoprotein AI overexpression. Arterioscler Thromb Vasc Biol 26:1572–1578

van den Maagdenberg AM, Hofker MH, Krimpenfort PJ, de Bruijn I, van Vlijmen B, van der Boom H, Havekes LM, Frants RR (1993) Transgenic mice carrying the apolipoprotein E3-Leiden gene exhibit hyperlipoproteinemia. J Biol Chem 268:10540–10545

van der Hoogt CC, de Haan W, Westerterp M, Hoekstra M, Dallinga-Thie GM, Romijn JA, Princen HM, Jukema JW, Havekes LM, Rensen PC (2007) Fenofibrate increases HDL-cholesterol by reducing cholesteryl ester transfer protein expression. J Lipid Res 48:1763–1771

van der Hoorn JW, de Haan W, Berbée JF, Havekes LM, Jukema JW, Rensen PC, Princen HM (2008) Niacin increases HDL by reducing hepatic expression and plasma levels of cholesteryl ester transfer protein in APOE*3Leiden.CETP mice. Arterioscler Thromb Vasc Biol 28:2016–2022

Van Eck M, Van Berkel TJ (2013) ATP-binding cassette transporter A1 in lipoprotein metabolism and atherosclerosis: a new piece of the complex puzzle. Arterioscler Thromb Vasc Biol 33:2281–2283

Van Eck M, Bos IS, Kaminski WE, Orsó E, Rothe G, Twisk J, Böttcher A, Van Amersfoort ES, Christiansen-Weber TA, Fung-Leung WP, Van Berkel TJ, Schmitz G (2002) Leukocyte ABCA1 controls susceptibility to atherosclerosis and macrophage recruitment into tissues. Proc Natl Acad Sci USA 99:6298–6303

Van Eck M, Twisk J, Hoekstra M, Van Rij BT, Van der Lans CA, Bos IS, Kruijt JK, Kuipers F, Van Berkel TJ (2003) Differential effects of scavenger receptor BI deficiency on lipid metabolism in cells of the arterial wall and in the liver. J Biol Chem 278:23699–23705

Van Eck M, Bos IS, Hildebrand RB, Van Rij BT, Van Berkel TJ (2004) Dual role for scavenger receptor class B, type I on bone marrow-derived cells in atherosclerotic lesion development. Am J Pathol 165:785–794

Van Eck M, Ye D, Hildebrand RB, Kar Kruijt J, de Haan W, Hoekstra M, Rensen PC, Ehnholm C, Jauhiainen M, Van Berkel TJ (2007) Important role for bone marrow-derived cholesteryl ester transfer protein in lipoprotein cholesterol redistribution and atherosclerotic lesion development in LDL receptor knockout mice. Circ Res 100:678–685

Van Eck M, Hoekstra M, Out R, Bos IS, Kruijt JK, Hildebrand RB, Van Berkel TJ (2008) Scavenger receptor BI facilitates the metabolism of VLDL lipoproteins in vivo. J Lipid Res 49:136–146

van Vlijmen BJ, van den Maagdenberg AM, Gijbels MJ, van der Boom H, HogenEsch H, Frants RR, Hofker MH, Havekes LM (1994) Diet-induced hyperlipoproteinemia and atherosclerosis in apolipoprotein E3-Leiden transgenic mice. J Clin Invest 93:1403–1410

Varban ML, Rinninger F, Wang N, Fairchild-Huntress V, Dunmore JH, Fang Q, Gosselin ML, Dixon KL, Deeds JD, Acton SL, Tall AR, Huszar D (1998) Targeted mutation reveals a central role for SR-BI in hepatic selective uptake of high density lipoprotein cholesterol. Proc Natl Acad Sci USA 95:4619–4624

Vergeer M, Korporaal SJ, Franssen R, Meurs I, Out R, Hovingh GK, Hoekstra M, Sierts JA, Dallinga-Thie GM, Motazacker MM, Holleboom AG, Van Berkel TJ, Kastelein JJ, Van Eck M, Kuivenhoven JA (2011) Genetic variant of the scavenger receptor BI in humans. N Engl J Med 364:136–1345

Vikstedt R, Ye D, Metso J, Hildebrand RB, Van Berkel TJ, Ehnholm C, Jauhiainen M, Van Eck M (2007) Macrophage phospholipid transfer protein contributes significantly to total plasma phospholipid transfer activity and its deficiency leads to diminished atherosclerotic lesion development. Arterioscler Thromb Vasc Biol 27:578–586

Wada M, Iso T, Asztalos BF, Takama N, Nakajima T, Seta Y, Kaneko K, Taniguchi Y, Kobayashi H, Nakajima K, Schaefer EJ, Kurabayashi M (2009) Marked high density lipoprotein deficiency due to apolipoprotein A-I Tomioka (codon 138 deletion). Atherosclerosis 207:157–161

Wang X, Collins HL, Ranalletta M, Fuki IV, Billheimer JT, Rothblat GH, Tall AR, Rader DJ (2007) Macrophage ABCA1 and ABCG1, but not SR-BI, promote macrophage reverse cholesterol transport in vivo. J Clin Invest 117:2216–2224

West M, Greason E, Kolmakova A, Jahangiri A, Asztalos B, Pollin TI, Rodriguez A (2009) Scavenger receptor class B type I protein as an independent predictor of high-density lipoprotein cholesterol levels in subjects with hyperalphalipoproteinemia. J Clin Endocrinol Metab 94:1451–1457

Westerterp M, van der Hoogt CC, de Haan W, Offerman EH, Dallinga-Thie GM, Jukema JW, Havekes LM, Rensen PC (2006) Cholesteryl ester transfer protein decreases high-density lipoprotein and severely aggravates atherosclerosis in APOE*3-Leiden mice. Arterioscler Thromb Vasc Biol 26:2552–2559

Wiersma H, Nijstad N, de Boer JF, Out R, Hogewerf W, Van Berkel TJ, Kuipers F, Tietge UJ (2009) Lack of Abcg1 results in decreased plasma HDL cholesterol levels and increased biliary cholesterol secretion in mice fed a high cholesterol diet. Atherosclerosis 206:141–147

Wilhelm AJ, Zabalawi M, Grayson JM, Weant AE, Major AS, Owen J, Bharadwaj M, Walzem R, Chan L, Oka K, Thomas MJ, Sorci-Thomas MG (2009) Apolipoprotein A-I and its role in lymphocyte cholesterol homeostasis and autoimmunity. Arterioscler Thromb Vasc Biol 29:843–849

Williamson R, Lee D, Hagaman J, Maeda N (1992) Marked reduction of high density lipoprotein cholesterol in mice genetically modified to lack apolipoprotein A-I. Proc Natl Acad Sci USA 89:7134–7138

Wojcik AJ, Skaflen MD, Srinivasan S, Hedrick CC (2008) A critical role for ABCG1 in macrophage inflammation and lung homeostasis. J Immunol 180:4273–4282

Yan D, Navab M, Bruce C, Fogelman AM, Jiang XC (2004) PLTP deficiency improves the anti-inflammatory properties of HDL and reduces the ability of LDL to induce monocyte chemotactic activity. J Lipid Res 45:1852–1858

Yates M, Kolmakova A, Zhao Y, Rodriguez A (2011) Clinical impact of scavenger receptor class B type I gene polymorphisms on human female fertility. Hum Reprod 26:1910–1916

Yazdanyar A, Jiang XC (2012) Liver phospholipid transfer protein (PLTP) expression with a PLTP-null background promotes very low-density lipoprotein production in mice. Hepatology 56:576–584

Yazdanyar A, Quan W, Jin W, Jiang XC (2013) Liver-specific phospholipid transfer protein deficiency reduces high-density lipoprotein and non-high-density lipoprotein production in mice. Arterioscler Thromb Vasc Biol 33:2058–2064

Ye D, Hoekstra M, Out R, Meurs I, Kruijt JK, Hildebrand RB, Van Berkel TJ, Van Eck M (2008) Hepatic cell-specific ATP-binding cassette (ABC) transporter profiling identifies putative novel candidates for lipid homeostasis in mice. Atherosclerosis 196:650–658

Yeang C, Qin S, Chen K, Wang DQ, Jiang XC (2010) Diet-induced lipid accumulation in phospholipid transfer protein-deficient mice: its atherogenicity and potential mechanism. J Lipid Res 51:2993–3002

Yvan-Charvet L, Ranalletta M, Wang N, Han S, Terasaka N, Li R, Welch C, Tall AR (2007) Combined deficiency of ABCA1 and ABCG1 promotes foam cell accumulation and accelerates atherosclerosis in mice. J Clin Invest 117:3900–3908

Yvan-Charvet L, Welch C, Pagler TA, Ranalletta M, Lamkanfi M, Han S, Ishibashi M, Li R, Wang N, Tall AR (2008) Increased inflammatory gene expression in ABC transporter-deficient macrophages: free cholesterol accumulation, increased signaling via toll-like receptors, and neutrophil infiltration of atherosclerotic lesions. Circulation 118:1837–1847

Yvan-Charvet L, Pagler TA, Seimon TA, Thorp E, Welch CL, Witztum JL, Tabas I, Tall AR (2010a) ABCA1 and ABCG1 protect against oxidative stress-induced macrophage apoptosis during efferocytosis. Circ Res 106:1861–1869

Yvan-Charvet L, Pagler T, Gautier EL, Avagyan S, Siry RL, Han S, Welch CL, Wang N, Randolph GJ, Snoeck HW, Tall AR (2010b) ATP-binding cassette transporters and HDL suppress hematopoietic stem cell proliferation. Science 328:1689–1693

Zabalawi M, Bhat S, Loughlin T, Thomas MJ, Alexander E, Cline M, Bullock B, Willingham M, Sorci-Thomas MG (2003) Induction of fatal inflammation in LDL receptor and ApoA-I double-knockout mice fed dietary fat and cholesterol. Am J Pathol 163:1201–1213

Zannis VI, Karathanasis SK, Keutmann HT, Goldberger G, Breslow JL (1983) Intracellular and extracellular processing of human apolipoprotein A-I: secreted apolipoprotein A-I isoprotein 2 is a propeptide. Proc Natl Acad Sci USA 80:2574–2578

Zhang W, Yancey PG, Su YR, Babaev VR, Zhang Y, Fazio S, Linton MF (2003) Inactivation of macrophage scavenger receptor class B type I promotes atherosclerotic lesion development in apolipoprotein E-deficient mice. Circulation 108:2258–2263

Zhang Y, Da Silva JR, Reilly M, Billheimer JT, Rothblat GH, Rader DJ (2005a) Hepatic expression of scavenger receptor class B type I (SR-BI) is a positive regulator of macrophage reverse cholesterol transport in vivo. J Clin Invest 115:2870–2874

Zhang S, Picard MH, Vasile E, Zhu Y, Raffai RL, Weisgraber KH, Krieger M (2005b) Diet-induced occlusive coronary atherosclerosis, myocardial infarction, cardiac dysfunction, and premature death in scavenger receptor class B type I-deficient, hypomorphic apolipoprotein ER61 mice. Circulation 111:3457–3464

Zhao Y, Pennings M, Hildebrand RB, Ye D, Calpe-Berdiel L, Out R, Kjerrulf M, Hurt-Camejo E, Groen AK, Hoekstra M, Jessup W, Chimini G, Van Berkel TJ, Van Eck M (2010) Enhanced foam cell formation, atherosclerotic lesion development, and inflammation by combined deletion of ABCA1 and SR-BI in Bone marrow-derived cells in LDL receptor knockout mice on western-type diet. Circ Res 107:e20–e31

Zhao Y, Pennings M, Vrins CL, Calpe-Berdiel L, Hoekstra M, Kruijt JK, Ottenhoff R, Hildebrand RB, van der Sluis R, Jessup W, Le Goff W, Chapman MJ, Huby T, Groen AK, Van Berkel TJ,

Van Eck M (2011) Hypocholesterolemia, foam cell accumulation, but no atherosclerosis in mice lacking ABC-transporter A1 and scavenger receptor BI. Atherosclerosis 218:314–322

Zheng KQ, Zhang SZ, Zhang L, Huang DJ, Liao LC, Hou YP (2004) A novel missense mutation (L296 Q) in cholesteryl ester transfer protein gene related to coronary heart disease. Acta Biochim Biophys Sin (Shanghai) 36:33–36

Zhong S, Sharp DS, Grove JS, Bruce C, Yano K, Curb JD, Tall AR (1996) Increased coronary heart disease in Japanese-American men with mutation in the cholesteryl ester transfer protein gene despite increased HDL levels. J Clin Invest 97(12):2917–2923

Zhou H, Li Z, Silver DL, Jiang XC (2006) Cholesteryl ester transfer protein (CETP) expression enhances HDL cholesteryl ester liver delivery, which is independent of scavenger receptor BI, LDL receptor related protein and possibly LDL receptor. Biochim Biophys Acta 1761:1482–1488

Zhu X, Lee JY, Timmins JM, Brown JM, Boudyguina E, Mulya A, Gebre AK, Willingham MC, Hiltbold EM, Mishra N, Maeda N, Parks JS (2008) Increased cellular free cholesterol in macrophage-specific Abca1 knock-out mice enhances pro-inflammatory response of macrophages. J Biol Chem 283:22930–22941

Zhu X, Westcott MM, Bi X, Liu M, Gowdy KM, Seo J, Cao Q, Gebre AK, Fessler MB, Hiltbold EM, Parks JS (2012) Myeloid cell-specific ABCA1 deletion protects mice from bacterial infection. Circ Res 111:1398–1409

Dysfunctional HDL: From Structure-Function-Relationships to Biomarkers

Meliana Riwanto, Lucia Rohrer, Arnold von Eckardstein, and Ulf Landmesser

Contents

1	Introduction	340
2	HDL and Reverse Cholesterol Transport	341
	2.1 Mechanisms Under Physiological Conditions	341
	2.2 Alterations of the Cholesterol Efflux Capacity of HDL in Cardiovascular Disease	342
3	Effects of HDL on LDL Oxidation	343
	3.1 Mechanisms Under Physiological Conditions	343
	3.2 Impairment of the Anti-Oxidative Effects of HDL in Patients After Surgery and With Cardiovascular Disease	345
4	Effects of HDL on Endothelial Nitric Oxide Bioavailability	346
	4.1 Mechanisms Under Physiological Conditions	346
	4.2 Impaired HDL Capacity to Stimulate NO Production in Patients with Cardiovascular Disease	347
5	Endothelial Anti-Inflammatory Effects of HDL	348
	5.1 Mechanisms Under Physiological Conditions	348
	5.2 Impaired Endothelial Anti-Inflammatory Effects of HDL in Patients with CAD, Diabetes, or Chronic Kidney Dysfunction	350
6	Effects of HDL on Endothelial Cell Apoptotic Pathways	351
	6.1 Mechanisms Under Physiological Conditions	351
	6.2 Impairment of the Endothelial Anti-Apoptotic Effects of HDL in Patients with Cardiovascular Disease	352
7	HDL Protein Cargo and Prognostic Biomarkers	353
References		355

M. Riwanto • U. Landmesser (✉)
Cardiology, University Heart Center, University Hospital Zurich and Center of Molecular Cardiology, University of Zurich, Rämistrasse 100, CH 8091 Zurich, Switzerland
e-mail: ulf.landmesser@usz.ch

L. Rohrer • A. von Eckardstein
Institute of Clinical Chemistry, University Hospital Zurich, Zurich, Switzerland

© The Author(s) 2015
A. von Eckardstein, D. Kardassis (eds.), *High Density Lipoproteins*, Handbook of Experimental Pharmacology 224, DOI 10.1007/978-3-319-09665-0_10

Abstract

Reduced plasma levels of HDL-C are associated with an increased risk of CAD and myocardial infarction, as shown in various prospective population studies. However, recent clinical trials on lipid-modifying drugs that increase plasma levels of HDL-C have not shown significant clinical benefit. Notably, in some recent clinical studies, there is no clear association of higher HDL-C levels with a reduced risk of cardiovascular events observed in patients with existing CAD. These observations have prompted researchers to shift from a cholesterol-centric view of HDL towards assessing the function and composition of HDL particles.

Of importance, experimental and translational studies have further demonstrated various potential antiatherogenic effects of HDL. HDL has been proposed to promote macrophage reverse cholesterol transport and to protect endothelial cell functions by prevention of oxidation of LDL and its adverse endothelial effects. Furthermore, HDL from healthy subjects can directly stimulate endothelial cell production of nitric oxide and exert anti-inflammatory and antiapoptotic effects. Of note, increasing evidence suggests that the vascular effects of HDL can be highly heterogeneous and HDL may lose important anti-atherosclerotic properties and turn dysfunctional in patients with chronic inflammatory disorders. A greater understanding of mechanisms of action of HDL and its altered vascular effects is therefore critical within the context of HDL-targeted therapies.

Keywords

HDL dysfunction • Endothelial function • Atherosclerosis • CAD

Abbreviations

ABCA1	ATP-binding cassette transporter A1
ABCG1	ATP-binding cassette transporter G1
ACS	Acute coronary syndrome
apoA-I	Apolipoprotein A1
apoA-II	Apolipoprotein A2
apoB	Apolipoprotein B
apoA-IV	Apolipoprotein A-IV
apoC-III	Apolipoprotein C-III
apoE	Apolipoprotein E
apoJ	Apolipoprotein J
Bad	Bcl-2-associated death promoter
Bcl-xL	B-cell lymphoma-extra large
Bid	BH3 interacting domain death agonist

CAD	Coronary artery disease
caspase	Cysteine-aspartic proteases
CETP	Cholesteryl ester transfer protein
CHF	Chronic heart failure
CKD	Chronic kidney disease
eNOS	Endothelial nitric oxide synthase
HDL	High-density lipoprotein
HDL-C	High-density lipoprotein cholesterol
HPLC	High-performance liquid chromatography
ICAM-1	Intercellular Adhesion Molecule 1
LCAT	Lecithin cholesterol acyltransferase
MS/MS	Tandem mass spectrometry
LDL	Low density lipoprotein
L-NAME	L-NG-Nitroarginine Methyl Ester
LOX-1	Lectin-type oxidized LDL receptor 1
LpPLA2	Lipoprotein-associated phospholipase A2
MALDI-TOF-MS	Matrix-assisted laser desorption/ionization-time-of-flight mass spectrometry
MAPK	Mitogen-activated protein kinase
MCP	Monocyte chemoattractant protein
MDA	Malondialdehyde
MI	Myocardial infarction
MPO	Myeloperoxidase
NF-kB	Nuclear factor kappa-light-chain-enhancer of activated B cells
NO	Nitric oxide
oxLDL	Oxidized low density lipoprotein
PAF-AH	Platelet-activating factor acetylhydrolase
PI3K	Phosphoinositide-3-kinase
PLTP	Phospholipid transport protein
PON1	Paraoxonase 1
POPC	1-palmitoyl-2-oleoyl-phosphatidylcholine
RCT	Reverse cholesterol transport
rHDL	Reconstituted high-density lipoprotein
S1P	Sphingosine-1-phosphate
SAA	Serum amyloid A
SNP	Single nucleotide polymorphism
sPLA$_2$-IIa	Secretory phospholipase A$_2$-IIa
SR-BI	Scavenger receptor type B1
TNF-α	Tumor necrosis factor-alpha
VCAM-1	Vascular cell adhesion molecule 1
VLDL	Very low density lipoprotein

1 Introduction

It was shown in multiple large prospective studies of cardiovascular risk factors that reduced plasma levels of HDL-cholesterol (HDL-C) are associated with an increased risk of coronary artery disease (CAD) (Castelli et al. 1986; Cullen et al. 1997; Di Angelantonio et al. 2009; Gordon et al. 1977; Sharrett et al. 2001). Multiple biological functions of HDL have been identified, whereby HDL may exert antiatherogenic effects (Annema and von Eckardstein 2013; Barter et al. 2004; Mineo et al. 2006; Rader 2006; Riwanto and Landmesser 2013), e.g., HDL from healthy subjects has been shown to directly promote endothelial antiapoptotic, anti-inflammatory, and antithrombotic effects (Mineo et al. 2006; Nofer et al. 2004; Rye and Barter 2008; Tall et al. 2008; Yuhanna et al. 2001). Accordingly, interventions to improve HDL-C levels and/or HDL function are being intensely evaluated as a potential therapeutic strategy to reduce cardiovascular risk. However, increasing evidence suggests that the endothelial and vascular effects of HDL are highly heterogeneous and vasoprotective properties of HDL are impaired in patients with diabetes, CAD, or chronic kidney dysfunction (Besler et al. 2011; Khera et al. 2011; Riwanto et al. 2013; Sorrentino et al. 2010).

Intriguingly, several recent clinical trials testing the effects of HDL-C-raising therapies have failed to demonstrate cardiovascular risk reduction in patients with CAD. The Investigation of Lipid Level Management to Understand its Impact in Atherosclerotic Events (ILLUMINATE) trial testing the impact of the CETP inhibitor torcetrapib on clinical outcome showed increased risk of mortality and morbidity in patients at high risk for coronary events, despite a substantial increase of HDL-C levels (Barter et al. 2007). Dalcetrapib, another CETP inhibitor, modestly increased HDL-C levels, but the phase 3 trial dal-OUTCOMES study was terminated before completion to a lack of efficacy (Schwartz et al. 2012). More recently, the HPS2-THRIVE trial results showed that adding extended-release niacin/laropiprant, another HDL-cholesterol-raising agent, to statins did not reduce the risk of cardiovascular event (Haynes et al. 2013). Taken together, these observations strongly suggest that plasma HDL-C levels per se are not an optimal therapeutic target.

Of note, accumulating evidence suggests that the vascular effects of HDL can be highly heterogeneous and may turn dysfunctional. HDL loses potential anti-atherosclerotic properties in patients with chronic inflammatory disorders, such as the antiphospholipid syndrome (Charakida et al. 2009), systemic lupus erythematosus and rheumatoid arthritis (McMahon et al. 2006), scleroderma (Weihrauch et al. 2007), metabolic syndrome (de Souza et al. 2008), diabetes (Persegol et al. 2006; Sorrentino et al. 2010), and CAD (Ansell et al. 2003; Besler et al. 2011; Riwanto et al. 2013). In a study of 189 patients with chronic kidney disease on hemodialysis, an impaired anti-inflammatory capacity of HDL was correlated with a poor clinical outcome (Kalantar-Zadeh et al. 2007). Furthermore, HDL isolated from subjects with type 1 or type 2 diabetes mellitus or abdominal obesity had reduced capacity to reverse the inhibition of aortic ring endothelium-dependent relaxation by oxLDL as compared to HDL from healthy control subjects

(Persegol et al. 2006, 2007). Importantly, the heterogeneity of the vascular effects of HDL may be attributed to changes in the HDL-associated proteome and lipidome, i.e., changes in the amount and type of proteins and lipids bound to the HDL particle and also posttranslational modifications. In particular, HDL is known to be susceptible to modification in vitro by a variety of oxidants, such as metal ions, peroxyl and hydroxyl radicals, aldehydes, various myeloperoxidase (MPO)-generated oxidants, lipoxygenase, phospholipase A2, elastase, nonenzymatic glycation, and homocysteinylation (Ferretti et al. 2006). These oxidative modifications may contribute to the generation of dysfunctional proinflammatory HDL.

2   HDL and Reverse Cholesterol Transport

2.1   Mechanisms Under Physiological Conditions

One of the antiatherogenic effects of HDL has been attributed to its function in macrophage reverse cholesterol transport (RCT), i.e., the removal of excess cholesterol from lipid-laden macrophage foam cells in the atherosclerotic plaque and its transport to the liver for excretion in the bile (Rader 2006). The first step of macrophage RCT involves the hydrolysis of cytoplasmic cholesteryl ester into free cholesterol followed by the efflux of the free cholesterol to mature HDL or extracellular lipid-poor apoA-I. Macrophage cholesterol efflux is mediated by active transport systems, which include the ATP-binding cassette transporters ABCA1 and ABCG1 (Rader 2006; Tall et al. 2008). Studies in macrophages from ABCA1-knockout or ABCA1-overexpressing mice have shown that ABCA1 primarily mediates the cholesterol efflux to lipid-poor apoA-I (Bortnick et al. 2000; Haghpassand et al. 2001). In contrast, ABCG1 largely mediates cholesterol efflux from macrophages to mature HDL (Tall et al. 2008).

A study by Wang et al. quantitatively assessed the roles of SR-BI, ABCA1, and ABCG1 in macrophage RCT in mice in vivo (Wang et al. 2007a). Using primary macrophages lacking SR-BI, the authors demonstrated that SR-BI did not promote macrophage RCT in vivo after intraperitoneal injection. In contrast, both ABCA1 and ABCG1 contributed to macrophage RCT in vivo. The study also demonstrated that transplantation of bone marrow from ABCA1/ABCG1-deficient mice accelerated atherosclerotic lesion formation in LDL receptor-deficient mice (Wang et al. 2007a). Interestingly, Yvan-Charvet et al. also demonstrated that SR-BI failed to stimulate net cholesterol efflux from HEK293 cells to plasma HDL and inhibited ABCG1-mediated cholesterol efflux, which was at least in part due to the increased uptake of HDL cholesteryl esters into these cells (Yvan-Charvet et al. 2008).

HDL-associated cholesterol is subsequently esterified by LCAT which transfers a fatty acyl residue from phospholipids to the 3-beta-hydroxy group of cholesterol. The final steps of RCT involve the uptake of HDL-C by the liver and its excretion in the bile. Cholesterol esters and free cholesterol in HDL can either be directly taken up by the liver via SR-BI or transferred to apoB containing lipoproteins in a process mediated by CETP (Cuchel et al. 2009). In the latter pathway, cholesterol esters are taken up by the liver via the LDL receptor and then hydrolyzed to free cholesterol which can be excreted directly into the bile or following conversion to bile acid.

2.2 Alterations of the Cholesterol Efflux Capacity of HDL in Cardiovascular Disease

A case-control study by Khera et al. showed that the cholesterol efflux capacity of apoB-depleted serum was inversely related to carotid IMT in healthy volunteers and to the likelihood of angiographic CAD, even after adjustment for HDL-C and apoA-I levels (Khera et al. 2011). In a separate study, enhanced cholesterol efflux activity from ABCA1-stimulated macrophages was associated with reduced risk of prevalent CAD in unadjusted models in two case-control cohorts: an angiographic cohort comprising stable subjects undergoing elective diagnostic coronary angiography and an outpatient cohort (Li et al. 2013). However, the inverse risk relationship remained significant after adjustment for traditional CAD risk factors only within the outpatient cohort (Li et al. 2013). Surprisingly, higher cholesterol efflux activity was associated with an increase in prospective 3-year risk of myocardial infarction/stroke and major adverse cardiovascular events. The authors further observed that the HDL fraction ($1.063 < d < 1.21$) contained only a minority (≈ 40 %) of [(14)C] cholesterol released, with the majority found within the lipoprotein particle-depleted fraction, where ≈ 60 % was recovered after apoA-I immunoprecipitation. This is in agreement with previous findings on plasmas from patients with apoA-I deficiency, such as Tangier disease and LCAT deficiency, whereby the cholesterol efflux capacity amounted to at least 40 and 20 % of total and apoB-depleted plasmas from healthy controls, respectively (von Eckardstein et al. 1995a). However, in these studies the considerable residual cholesterol efflux capacity of HDL-free or HDL-poor plasma was explained by the abundance of apoA-I-free HDL particles containing apoA-IV, apoA-II, or apoE rather than by albumin (Huang et al. 1994, 1995; von Eckardstein et al. 1995a, b).

Importantly, impaired cholesterol efflux capacity of HDL has been associated with structural changes of the HDL components. Two groups independently demonstrated that oxidative modification of apoA-I, particularly by oxidation of methionine, tyrosine, or tryptophan residues through the myeloperoxidase pathway, dramatically reduced the ability of apoA-I to promote cholesterol efflux through the ABCA1 pathway (Bergt et al. 2004; Huang et al. 2014; Shao et al. 2006, 2010). Both groups also demonstrated that apoA-I in patients with cardiovascular disease was oxidatively modified and defective in promoting cholesterol efflux from macrophages via ABCA1 (Bergt et al. 2004; Shao et al. 2012; Zheng et al. 2004).

Of note, the cholesterol efflux capacity as determined by the assay protocol of Khera et al. which is widely used after its prominent publications is potentially confounded by several HDL-independent factors: the 4-h long incubation with cells raises the possibility that serum components deregulate the activity of ABCA1 and

ABCG1. For example, free fatty acids and cytokines are known to alter the expression of ABCA1 and ABCG1 on both the transcriptional and posttranslational level. The use of serum instead of plasma implies the activation of thrombin and other proteases which are known to degrade prebeta-HDL (Eckardstein 2012).

3 Effects of HDL on LDL Oxidation

3.1 Mechanisms Under Physiological Conditions

The antioxidative properties of HDL were first reported by Bowry (Bowry et al. 1992). It has been suggested that HDL limits the formation of oxidized LDL (oxLDL) and oxidation of LDL has been proposed for decades to be a key event in atherogenesis (Heinecke 1998; Witztum and Steinberg 1991). LDL accumulates in the subendothelial space where it is oxidized by different pathways such as lipoxygenase, myeloperoxidase, or NADPH oxidase pathways (Diaz et al. 1997). OxLDL induces the expression of various proinflammatory cytokines, chemokines, and adhesion molecules (Navab et al. 1996). In addition, it promotes monocyte chemotaxis, their differentiation into macrophages, and the subsequent oxLDL uptake that convert them into foam cells, a hallmark of atherosclerotic plaques (Chisolm et al. 1999).

HDL is a major carrier of lipid oxidation products and specifically apoA-I, the main protein constituent of HDL, binds to and removes lipid hydroperoxides of LDL in vitro and in vivo (Navab et al. 2000b). In addition, apoA-I may directly reduce cholesteryl ester hydroperoxides and phosphatidylcholine hydroperoxides by oxidation of specific Met residues in the protein (Garner et al. 1998). Interestingly, the resulting oxidized HDL is more rapidly and selectively removed by hepatocytes than native HDL (Garner et al. 1998). Treatment of human artery wall cells with apoA-I but not apoA-II, or by addition of native HDL, apoA-I peptide mimetics, or paraoxonase, prevents the cells from oxidizing LDL in vitro (Navab et al. 2000a). The antioxidative function of apoA-I was corroborated by the finding that recombinant HDL containing only apoA-I and 1-palmitoyl-2-oleoyl-phosphatidylcholine (POPC) was as effective as native HDL in preventing LDL oxidation. In addition, in vivo studies have demonstrated that apoA-I can act as an antioxidative, anti-inflammatory, and anti-atherosclerotic agent (Nicholls et al. 2005a, b; Paszty et al. 1994). This further underlines a key antioxidant role for the HDL major protein apoA-I (Zerrad-Saadi et al. 2009).

Several other HDL-associated apolipoproteins have also been shown to contribute to its antioxidant effects. ApoE has been demonstrated to show isoform-dependent antioxidant activity (Miyata and Smith 1996). ApoE2 stimulates endothelial NO release and exerts anti-inflammatory effects (Sacre et al. 2003). In contrast, apoE4 has been suggested to be proinflammatory (Ophir et al. 2005). Another HDL-associated protein, apoA-IV, showed anti-atherosclerotic, anti-inflammatory, and antioxidant actions in vivo (Ostos et al. 2001; Recalde et al. 2004; Vowinkel et al. 2004). In addition, apoJ, also called clusterin, has been reported to block LDL oxidation by artery wall cells (Navab et al. 1997).

Furthermore, ApoA-II-enriched HDL isolated from transgenic human apoA-II mice protected VLDL from oxidation more efficiently than control HDL (Boisfer et al. 2002). In contrast, in other studies, dyslipidemic mice overexpressing human apoA-II showed accelerated atherosclerosis and reduced antioxidative activity of HDL, which may be attributed to the displacement of apoA-I and PON1 by apoA-II in HDL particles (Ribas et al. 2004; Rotllan et al. 2005). Notably, in apparently healthy subjects from the prospective EPIC-Norfolk (European Prospective Investigation into Cancer and Nutrition-Norfolk) cohort, apoA-II was found to be associated with a decreased risk of future CAD in a nested case-control study (Birjmohun et al. 2007).

Importantly, HDL carries antioxidant enzymes that have the capacity to prevent lipid oxidation or degrade lipid hydroperoxides such as serum paraoxonase/arylesterase 1 (PON1), lecithin/cholesterol acyltransferase (LCAT), and platelet-activating factor acetylhydrolase (PAF-AH). Specifically PON1 has been suggested to be an important regulator of the potential antiatherogenic capacity of HDL (Shih et al. 1998; Tward et al. 2002). It is secreted from the liver and associates with HDL in the plasma. Many studies revealed that PON1 degrades oxidized proinflammatory lipids, measured as a reduction in lipid peroxides (Mackness et al. 1991; Shih et al. 2000; Watson et al. 1995). Higher PON1 activity has been reported to be associated with a lower incidence of major cardiovascular events in human. Conversely, reduced activity of PON1 has been associated with pathological conditions such as diabetes, chronic renal failure rheumatoid arthritis, and various dementia (Soran et al. 2009).

LCAT is another HDL-associated enzyme with antioxidative properties. The antioxidant role of LCAT is thought to be its capacity to hydrolyze oxidized acyl chains from phosphatidylcholine-based oxidized phospholipids and oxidized free fatty acids (Goyal et al. 1997; Subramanian et al. 1999). Genetics and biochemical characterizations of HDLs from patients with low HDL disorders demonstrated that HDL-LCAT activity was reduced in all LCAT mutation carriers as well as in patients with HDL deficiency due to mutation in APOA1 or ABCA1 (Soran et al. 2009; von Eckardstein et al. 1995a). In vivo study in mice deficient for LDL receptor and leptin revealed that LCAT overexpression decreased autoantibodies to oxLDL (Mertens et al. 2003). Besides LCAT and PON1, PAF-AH, another HDL-associated enzyme, nowadays also known as lipoprotein-associated phospholipase A2 (LpPLA2), is able to hydrolyze oxidized phospholipids (Marathe et al. 2003; Noto et al. 2003). The local expression of PAF-AH in arteries of non-hyperlipidemic rabbits reduced the accumulation of oxidatively modified LDL without changing plasma levels of PAF-AH and reduced the expression of endothelial cell adhesion molecules (Arakawa et al. 2005). PAF-AH deficiency caused by a missense mutation in the gene has been indicated to be an independent risk factor for CAD in Japanese men (McIntyre et al. 2009). In contrast to this both activity and mass concentration of LpPLA2 in total serum has been associated with increased rather than decreased cardiovascular risk (Lp et al. 2010). However, recently HDL-associated LpPLA2 was found to be inversely related to risk of coronary events in stable CAD patients (Rallidis et al. 2012). Hence, the lipoprotein distribution of LpPLA2 may determine the pro- or antiatherogenic role of PAF-AH/LpPLA2. Furthermore, plasma levels of PAF-AH are also shown to be an independent risk marker of CAD (Garza et al. 2007). However, a study by Holleboom et al. showed that the reduction in LCAT and PAF-AH activites due to LCAT mutations was not associated with increased plasma lipid peroxidation (Holleboom et al. 2012).

3.2 Impairment of the Anti-Oxidative Effects of HDL in Patients After Surgery and With Cardiovascular Disease

In the acute phase response to surgery, van Lenten et al. demonstrated that the serum amyloid A (SAA) levels in HDL was increased and the activities of PON1 and PAF-AH were reduced (Van Lenten et al. 1995). Concomitant with these changes the anti-inflammatory capacity of HDL was lost in both humans and rabbits (Van Lenten et al. 1995). Isolated HDL from patients before and immediately after surgery were compared for the effects of HDL on LDL-induced monocyte transmigration and lipid hydroperoxide formation (Van Lenten et al. 1995). Prior to surgery, HDL completely inhibited the LDL-induced increase in monocyte transmigration and lipid hydroperoxide formation. In marked contrast, the "acute phase" HDL obtained from the same patients 2–3 days after surgery revealed a significant LDL-induced monocyte transmigration and was less effective in inhibiting lipid hydroperoxide formation. In other words, HDL in the same patient had been transformed from anti-inflammatory towards proinflammatory particles (Van Lenten et al. 1995). Similarly findings were reported in mice infected with influenza A or *Chlamydia pneumonia*. HDL isolated from virus-infected wild-type mice had impaired capacity to block LDL oxidation as well as LDL-induced monocyte chemotactic activity in human artery wall cell co-cultures (Van Lenten et al. 2001). The mice infected with *C. pneumoniae* had decreased PON1 activity and reduced HDL capacity to prevent LDL oxidation at days 2 and 3 postinfection (Campbell et al. 2010). Of note, studies by Navab et al. showed that HDL isolated from patients with CAD failed to prevent LDL oxidation by human artery wall cells (Navab et al. 2000a) and had impaired capacity to inhibit LDL-induced monocyte chemotactiy activity (Navab et al. 2001b). Similar results were obtained with HDL isolated from mice genetically predisposed to diet-induced atherosclerosis; it became proinflammatory when the mice were fed an atherogenic diet (Navab et al. 2001a). A subsequent small study by Ansell et al. suggested that the capacity of HDL to alter LDL-induced monocyte chemotactic activity in patients with CAD was somewhat improved after 6 weeks of simvastatin therapy (Ansell et al. 2003). However, HDL from patients with CAD on statin therapy remained proinflammatory when compared to HDL from healthy subjects, despite a significant improvement in plasma lipid levels.

Interestingly, impaired antioxidant properties and increased oxidized fatty acids content were found in HDL from patients with type 2 diabetes (Morgantini et al. 2011). The authors speculated that the elevated levels of oxidized fatty acids in HDL from these patients may account for the impaired antioxidant properties of the lipoprotein (Morgantini et al. 2011). Moreover, in a different study, the ability of HDL to prevent LDL oxidation was found to be reduced in patients with acute coronary syndrome but not in patients with stable coronary artery syndrome (Patel et al. 2011). The antioxidative and cholesterol efflux capacities of HDL were also found to be reduced in ischemic cardiomyopathy (Patel et al. 2013).

4 Effects of HDL on Endothelial Nitric Oxide Bioavailability

4.1 Mechanisms Under Physiological Conditions

Endothelial nitric oxide plays a crucial role in the regulation of vascular tone, platelet aggregation, and angiogenesis. Endothelial nitric oxide synthase (eNOS)-derived nitric oxide (NO) has been shown to exert a variety of atheroprotective effects in the vasculature, such as anti-inflammatory and antithrombotic effects (Landmesser et al. 2004). Consequently, reduced endothelial NO bioavailability has been proposed to promote initiation and progression of atherosclerosis (Landmesser et al. 2004).

It was shown that HDL can directly stimulate eNOS-mediated NO production (Yuhanna et al. 2001). There have been several studies that consistently demonstrated the capacity of HDL to modulate eNOS expression and to stimulate endothelial NO production in vitro and in vivo (Besler et al. 2011; Kuvin et al. 2002; Mineo et al. 2003; Nofer et al. 2004; Ramet et al. 2003; Sorrentino et al. 2010). Furthermore, administration of reconstituted HDL (rHDL) has been shown to improve endothelial function in patients with hypercholesterolemia or in subjects with isolated low HDL due to heterozygous loss-of-function mutations in the ABCA1 gene locus (Bisoendial et al. 2003; Spieker et al. 2002). Several mechanisms have been proposed to account for the endothelial NO-stimulating capacity of HDL. Early studies have demonstrated that HDL prevents oxLDL-mediated eNOS displacements from caveolae and restores enzyme stimulation (Uittenbogaard et al. 2000). Yuhanna et al. demonstrated that HDL binding to endothelial SR-BI is required for eNOS activation (Yuhanna et al. 2001). This binding activates the phosphatidylinositol-3-kinase/Akt signaling pathway and the MAP kinase/extracellular signal-regulated kinase pathway (Mineo et al. 2003). Activation of endothelial Akt by HDL stimulates phosphorylation of eNOS at serine residue 1177 (Mineo et al. 2003; Nofer et al. 2004), which is known to be an important regulatory mechanism leading to eNOS activation (Dimmeler et al. 1999).

There are many different components of HDL known to play a role in the endothelial NO-stimulating capacity. The potential interaction of apoA-I with eNOS in cultured endothelial cells has been previously reported (Ramet et al. 2003). However, lipid-free apoA-I failed to activate eNOS despite being the ligand for SR-BI, suggesting that other HDL components may be important or are required to support the conformation of apoA-I to allow its interaction with SR-BI and to stimulate eNOS (de Beer et al. 2001). In isolated endothelial cell plasma membranes, anti-apoA-I antibody blocks eNOS activation by HDL in vitro (Yuhanna et al. 2001). In contrast anti-apoA-II antibody further enhances eNOS stimulation by HDL (Yuhanna et al. 2001). Furthermore, lysophospholipids may play a role in eNOS activation as demonstrated in several studies. A study by Nofer et al. suggested that HDL-associated sphingolipids such as sphingosylphosphorylcholine, sphingosine-1-phosphate, and lysosulfatide caused eNOS-dependent relaxation of precontracted aortic rings from mice by binding to the lysophospholipid receptor S1P3 expressed in endothelial cells (Nofer et al. 2004). However, the vasodilatory response of HDL was not completely abolished in S1P3-deficient mice (Nofer et al. 2004).

Recently, we demonstrated that the HDL-associated enzyme PON1 is participating in HDL's capacity to stimulate endothelial NO production and to exert NO-dependent endothelial atheroprotective effects (Besler et al. 2011). Inhibition of PON1 in HDL from healthy subjects impaired the capacity of HDL to stimulate endothelial NO production in human aortic endothelial cells. In addition HDL isolated from PON1-deficient mice failed to stimulate NO production in mouse aortic endothelial cells (Besler et al. 2011). Furthermore, inhibition of eNOS-mediated NO production by the pharmacological inhibitor L-NAME prevented the inhibitory effects of HDL from healthy subjects on nuclear factor κB (NF-κB) activity, vascular cell adhesion molecule (VCAM)-1 expression, and endothelial monocyte adhesion. These clearly indicate that the capacity of HDL to stimulate endothelial NO production is important for these endothelial anti-inflammatory effects of HDL (Besler et al. 2011).

4.2 Impaired HDL Capacity to Stimulate NO Production in Patients with Cardiovascular Disease

We and others have recently shown that HDL isolated from patients with stable CAD, diabetes, chronic kidney diseases (CKD), or acute coronary artery syndrome (ACS) displays altered endothelial effects when compared to HDL from healthy subjects. Of note, HDL from patients with diabetes in contrast to HDL from healthy subjects failed to stimulate endothelial cell NO production and to promote endothelial repair in a carotid artery injury model in mice (Sorrentino et al. 2010). Moreover, HDL from patients with either stable CAD or an ACS, in contrast to HDL from age- and gender-matched healthy subjects, inhibited endothelial cell NO production and lost the capacity to limit endothelial inflammatory activation as well as to promote endothelial repair in vivo (Besler et al. 2011). Interestingly, the capacity to stimulate endothelial NO production could be improved upon exercise training (Adams et al. 2013).

HDL from patients with CAD and chronic heart failure (CHF) showed an elevated malondialdehyde (MDA) content as compared to HDL from healthy subjects, which may contribute to the impaired endothelial NO production (Besler et al. 2011). The MDA-lysine adducts in HDL was determined to act via lectin-type oxidized LDL receptor 1 (LOX-1) activation of protein kinase C-βII, which blocks Akt-activating phosphorylation at Ser473 and eNOS-activating phosphorylation at Ser1177. Furthermore, inactivation of PON1 in HDL from healthy subjects resulted in attenuated NO production, in greater protein kinase C-βII activation, decreased activating eNOS-Ser1177 phosphorylation, and increased inactivating eNOS-Thr495 phosphorylation. Furthermore, HDL isolated from PON1-deficient mice failed to stimulate endothelial cell NO production (Besler et al. 2011). These observations likely suggest that alterations of HDL-associated PON1 may severely affect endothelial effects of HDL.

Very recently, we have also shown that MPO and PON1 reciprocally modulate each other's function in vivo (Huang et al. 2013). MPO may promote site-specific oxidative modification of PON1, therefore limiting its activity. In addition, HDL isolated from patients with an ACS carries enhanced chlorotyrosine content, site-specific PON1 methionine oxidation, and reduced PON1 activity (Huang et al. 2013).

Several groups have reported an inverse relationship between PON1 serum activity and cardiovascular events (Bhattacharyya et al. 2008; Regieli et al. 2009). Interestingly, analysis of SNPs for PON1 identified in genome wide association studies did not reveal a significant association between SNPs, associated with mildly reduced PON1 activity, and the risk of cardiovascular events (Tang et al. 2012). The interpretation of PON1 activity is difficult because it is not known to which extent the paraoxonase and arylesterase activities represent biologically relevant functions. Furthermore, we and others observed important post-translational modifications of the proteins, which could explain alterations of biological properties of the enzymes (Aviram et al. 1999; Besler et al. 2011; Huang et al. 2013).

5 Endothelial Anti-Inflammatory Effects of HDL

5.1 Mechanisms Under Physiological Conditions

Atherosclerosis is a chronic inflammatory disorder. The cellular inflammatory response is triggered by modified LDL accumulated in the subendothelium and the induction of proinflammatory cytokines and adhesion protein expression. As a consequence monocytes/macrophages infiltrate and accumulate in the arterial wall and express scavenger receptors and toll-like receptors to sequester cholesterol. The cholesterol-loaded macrophages together with T lymphocytes produce a wide array of proinflammatory cytokines and accelerate plague progression (Hansson 2005).

In vitro, HDL has been shown to inhibit the expression of monocyte chemoattractant protein (MCP)-1, an important proinflammatory chemokine in endothelial cells (Mackness et al. 2004; Navab et al. 1991). Furthermore, the potential anti-inflammatory effects of HDL have been demonstrated by several in vivo studies. Infusion of native HDL as well as reconstituted HDL containing apoA-I or apoA-I Milano suppressed cytokine and chemokine expression in animal models of inflammation (Calabresi et al. 1997; Cockerill et al. 1995). Reduced VCAM-1 expression and decreased monocyte/macrophage infiltration were reported after carotid artery cuff injury in apoE-deficient mice (Dimayuga et al. 1999). In the streptozotocin-induced diabetic cardiomyopathy rat model, human apoA-I gene transfer increased HDL-C plasma levels and blocked the diabetes-induced myocardial mRNA expression of VCAM-1 and ICAM-1 (Van Linthout et al. 2008). In contrast, overexpression of human apoA-I in apoE-deficient mice did not result in any change of endothelial VCAM-1 expression and monocyte adherence in early atherosclerotic lesions at the aortic branch sites, despite overall reduction in aortic atherosclerotic lesion formation (Dansky et al. 1999). These studies support the concept that the anti-inflammatory capacity of HDL is heterogeneous, and is dependent on the pathophysiological conditions. In vitro studies showed variations in the capacity of HDL isolated from different human subjects to reduce TNF-α stimulated endothelial VCAM-1 expression (Ashby et al. 1998; Van Lenten et al. 1995). In human studies, the administration of reconstituted HDL increased the anti-inflammatory capacity of HDL from patients with type-2 diabetes (Patel et al. 2009).

One of the proposed mechanisms for the anti-inflammatory effect of HDL is the removal of cholesterol from cell membranes of macrophages and also endothelial cells. The cholesterol depletion and consequently the lipid rafts disruption may downregulate signaling pathways and interfere with the antigen presentation or the expression of toll-like receptor (Anderson et al. 2000; Norata and Catapano 2012; Wang et al. 2012). Additional studies suggest that HDL and apoA-I are able to inhibit the capacity of antigen-presenting cells to stimulate T-cell activation. This inhibition was attributed to cholesterol efflux through ABCA1, an ATP-binding transporter (Tang et al. 2009; Zhu et al. 2010).

The anti-inflammatory capacity of HDL has been mainly attributed to apoA-I, the major protein constituent of HDL. In an in vivo study, apoA-I infusion in rabbits in acute vascular inflammation model reduced neutrophil infiltration and endothelial cell inflammatory activation (Puranik et al. 2008). In addition, apoA-1 mimetic peptides have been shown to reduce vascular inflammation in type I diabetic rats and improve insulin sensitivity in obese mice (Peterson et al. 2007, 2008). It has also been demonstrated that treatment with lipid-free apoA-I and rHDL treatment reduced the expression of chemokines and chemokine receptors in vitro and in vivo by modulating NF-κB and peroxisome proliferator-activated receptor γ (Bursill et al. 2010). Interestingly, apoA-I has also been shown to inhibit palmitate-induced NF-κB activation by reducing Toll-like receptor-4 recruitment into lipid rafts (Cheng et al. 2012). More recently, De Nardo et al. using a systems biology approach identified the transcription factor ATF3 as an HDL-inducible gene to suppress Toll-like receptor-induced proinflammatory cytokines. Mice deficient in ATF3 and apoE revealed enhanced atherosclerotic lesions, and administration of rHDL into apoE-deficient mice resulted in ATF3 upregulation. The authors used rHDL in their study, suggesting that apoA-I is a key mechanistic player, and of note, the effects seem to be unrelated to cholesterol transport (De Nardo et al. 2014).

The lipid moiety of HDL has also been proposed to be important for the anti-inflammatory effects of HDL. In vitro studies using reconstituted HDL containing only apoA-I with a few phospholipid molecules suggested that inhibitory effects of HDL on endothelial cell adhesion molecule expression are also, at least in part, dependent on HDL-associated phospholipid species (Baker et al. 2000). The inhibition of cytokine-induced expression of VCAM-1 by reconstituted HDL varied when different phosphatidylcholine species were used. This suggests that the lipid composition of HDL influences its anti-inflammatory capacity and is likely an important determinant of HDL functionality (Baker et al. 2000; Barter et al. 2004). In addition there are multiple reports suggesting the involvement of sphingosine-1-phosphate (S1P), a biologically active sphingolipid that plays key functions in the immune, inflammatory, and cardiovascular systems. S1P carried by HDL is believed to regulate arterial tone, vascular permeability, and tissue perfusion. During inflammation it induces endothelial adhesion molecules and recruits inflammatory cells, and furthermore it activates a negative feedback loop that consecutively decreases vascular leakage by improving the endothelial barrier function and preventing cytokine-induced leukocyte adhesion (Garcia et al. 2001; Lucke and Levkau 2010; Takeya et al. 2003).

5.2 Impaired Endothelial Anti-Inflammatory Effects of HDL in Patients with CAD, Diabetes, or Chronic Kidney Dysfunction

HDL isolated from patients with inflammatory disease such as CAD, diabetes, or chronic kidney disease display reduced eNOS activation and impaired endothelial repair capacity in a carotid artery injury mouse model. There have been various mechanisms proposed to account for the impaired endothelial anti-inflammatory effects of HDL. Reduced HDL apoA-I levels in inflammatory states has been related to accelerated HDL catabolism and apoA-I substitution in HDL particles by serum amyloid A (SAA) (Esteve et al. 2005; Khovidhunkit et al. 2004). In acute phase, SAA is able to replace apoA-I in HDL, resulting in reduced plasma levels of apoA-I (Parks and Rudel 1985). ApoA-I can be completely replaced by SAA in rabbits and mice, in a subset of small, dense HDL particles and therefore functioning as a structural apolipoprotein (Cabana et al. 1999). HDL isolated from patients with chronic kidney disease is enriched in SAA, which correlated with its reduced anti-inflammatory capacity to inhibit monocyte chemoattractant protein-1 formation in vascular smooth muscle cells (Tolle et al. 2012).

Moreover, HDL carries potential cardioprotective molecules that are significantly altered in compositions in patients with CAD or ACS (Riwanto et al. 2013; Vaisar et al. 2007). Oxidative modifications of HDL, in particular the amino acid residues in apoA-I, have been shown to contribute to the generation of dysfunctional HDL (Bergt et al. 2004; Shao et al. 2006, 2012; Zheng et al. 2004). In vitro study has demonstrated that MPO-catalyzed oxidative modification of HDL or apoA-I converts HDL into a proinflammatory particle which promotes NF-κB activation and endothelial VCAM-1 expression (Undurti et al. 2009). Furthermore, glycation of HDL and apoA-I, a process that is known to occur in diabetes in vivo (Curtiss and Witztum 1985), has been shown to impair the anti-inflammatory capacity of HDL (Nobecourt et al. 2010). Infusion of glycated lipid-free apoA-I failed to decrease adhesion molecule expression following vascular injury (Nobecourt et al. 2010). Moreover, glycation of HDL has also been shown to inhibit the HDL capacity to inhibit oxLDL-induced monocyte adhesion to human aortic endothelial cells in vitro (Hedrick et al. 2000).

6 Effects of HDL on Endothelial Cell Apoptotic Pathways

6.1 Mechanisms Under Physiological Conditions

Endothelial cell dysfunction and injury has been associated with the pathogenesis of atherosclerosis (Landmesser et al. 2004; Nabel and Braunwald 2012; Ross 1999). Studies in pigs have shown that atherosclerotic lesion-prone regions are characterized by a high endothelial cell turnover (Caplan and Schwartz 1973), which has been suggested to be due to increased endothelial cell apoptosis (Caplan and Schwartz 1973). Several studies have indicated an important contribution of endothelial cell apoptosis to the pathophysiology of CAD (Burke et al. 1997; Rossig et al. 2001). Furthermore, it is evident that apoptosis of endothelial cells and smooth muscle cells is detrimental to plaque stability (Bombeli et al. 1997; Kockx and Herman 2000). In the normal arterial wall, endothelial cells may protect themselves against apoptosis using NO-dependent mechanisms. The situation is completely different in atherosclerotic plaques, in a high oxidative stress environment where macrophages produce high amounts of nitric oxide or peroxynitrite that could induce apoptotic cell death (Kockx and Herman 2000). In the aorta the capacity of HDL to quench endothelial cell apoptosis may therefore represent an antiatherogenic property of HDL (de Souza et al. 2010; Nofer et al. 2001; Suc et al. 1997; Sugano et al. 2000).

HDL has been shown to attenuate endothelial cell apoptosis induced by different stimuli such as TNF-α, oxLDL, and growth factor deprivation (de Souza et al. 2010; Nofer et al. 2001; Suc et al. 1997; Sugano et al. 2000). It has been suggested that HDL may inhibit both death-receptor and mitochondrial-mediated apoptotic pathways. Importantly, both the protein and lipid moieties have been reported to contribute to the antiapoptotic capacity of HDL. ApoA-I has been shown to inhibit endothelial cell apoptosis induced by oxLDL, VLDL, and TNF-α (Speidel et al. 1990; Suc et al. 1997; Sugano et al. 2000). In a more detailed study, HDL subpopulations enriched with apoA-I account for approximately 70 % of the antiapoptotic activity of HDL in a cell culture model with human microvascular endothelial cells that were treated with mildly oxidized LDL. Using rHDL consisting of apoA-I, cholesterol and phospholipids potently inhibited oxLDL-induced apoptosis in these cells (de Souza et al. 2010). This suggests that apoA-I plays an important role for the antiapoptotic capacity of HDL in oxLDL-stimulated endothelial cells.

HDL-associated lysosphingolipids have been identified to inhibit endothelial cell apoptosis induced by growth factor deprivation (Kimura et al. 2001, 2003; Nofer et al. 2001). The antiapoptotic capacity of the lipid moiety of HDL was further supported by the findings that the ratio of sphingosine-1-phosphate and sphingomyelin in small dense HDL3 particles was increased and correlated positively with the characteristics of these HDL subpopulations to inhibit endothelial cell apoptosis (Kontush et al. 2007).

Several mechanisms have been suggested for the endothelial antiapoptotic effects of HDL, depending on the stimulus of apoptosis. Endothelial cell death triggered by oxLDL causes a delayed but sustained increase in intracellular calcium which is shown to be prevented by HDL (Suc et al. 1997). HDL inhibits tumor necrosis factor-α-induced endothelial cell death by association as well as suppression of caspase-3, which is a key factor of all primary apoptotic pathways (Sugano et al. 2000). Nofer et al. suggested that sphingosine-1-phosphate (S1P) carried by HDL prevents the growth factor deprived activation of the intrinsic pathway to cell death (Nofer et al. 2001). Of note, lysophospholipids have been shown to enhance endothelial cell survival, but the effects were inhibited by S1P receptor knockdown, as well as by the presence of phosphoinositide 3 (PI3) kinase and Erk pathway antagonists (Kimura et al. 2003).

In addition, HDL activates Akt and causes phosphorylation of the Akt target Bcl-2-associated death promoter (BAD), preventing it from binding to the antiapoptotic protein Bcl-xL (Nofer et al. 2001). HDL also causes phosphoinositide 3 (PI3) kinase-mediated upregulation of the antiapoptotic Bcl-2 family protein Bcl-xL expression (Riwanto et al. 2013). Interestingly, HDL retained its antiapoptotic activity after knockdown of eNOS using specific RNA interference or pharmacological inhibition using L-NAME (Riwanto et al. 2013), indicating that HDL may exert its antiapoptotic activity independently of eNOS activation.

6.2 Impairment of the Endothelial Anti-Apoptotic Effects of HDL in Patients with Cardiovascular Disease

Recently, we reported that HDL isolated from patients with stable CAD or ACS, in contrast to HDL from healthy subjects, failed to inhibit endothelial cell apoptosis in vitro and in apoE-deficient-mice in vivo. Using proteomic analysis of HDL, we observed reduced clusterin and increased apo C-III content in HDL isolated from patients with CAD. Both, supplementing HDL from healthy subjects with apoC-III or pretreatment of healthy HDL with an antibody against clusterin, lead to the activation of proapoptotic signaling pathways in endothelial cells (Riwanto et al. 2013). HDL isolated from these patients stimulated endothelial proapoptotic pathways, in particular p38-MAPK-mediated activation of the proapoptotic Bcl-2-protein tBid. Our study further suggests that differences in the proteome of HDL from patients with CAD, in particular reduced HDL-associated clusterin and increased HDL-associated apoC-III, play an important role for altered activation of endothelial anti- and proapoptotic signaling pathways (Riwanto et al. 2013). In line with this, a nested case-control study found that cholesterol levels in apoC-III-containing HDL particles show a positive association with cardiovascular risk, whereas both total HDL-C and cholesterol in apoC-III-free HDL particles show the expected inverse associations (Jensen et al. 2012). In addition, Undurti et al. demonstrated that MPO-catalyzed oxidation of HDL impaired its capacity to inhibit endothelial apoptosis in vitro (Undurti et al. 2009).

7   HDL Protein Cargo and Prognostic Biomarkers

As highlighted above, alterations in HDL functions are likely, at least in part, attributed to changes in the composition of HDL particles. HDL particles are highly heterogenous in their structure and compositions. They undergo continuous remodeling through apolipoprotein exchanges with other circulating lipoproteins and tissues (Lund-Katz et al. 2003; Saito et al. 2004).

An early study has shown that acute phase response causes significant changes in the HDL-associated proteins and apolipoprotein composition, i.e., enrichment in C-reactive protein (CRP), secretory phospholipase A_2-IIa (sPLA$_2$-IIa), serum amyloid A (SAA), and cholesterol ester transfer protein (CETP) (Coetzee et al. 1986). It has been suggested that such remodeling may compromise the antiatherogenic functions of HDL (Jahangiri et al. 2009). These observations have prompted explorative studies into HDL composition. Evaluations of the HDL proteome have been reported with one- or two-dimensional electrophoresis combined with high performance liquid chromatography (HPLC) and tandem mass spectrometry (MS/MS) or matrix-assisted laser desorption/ionization-time flight mass spectrometry (MALDI-TOF-MS) (Heller et al. 2007; Karlsson et al. 2005). More recent studies have used shotgun proteomics, through label-free MS techniques, such as spectral counting and XIC inspection, to analyze HDL protein compositions (Riwanto et al. 2013; Vaisar et al. 2007). Overall, the identification techniques currently available for the analysis of the HDL proteome are semi-quantitative.

Given the complexity of HDL and its multitude of biological functions, it is conceivable to expect that sets of functionally associated proteins can provide information about their participation in the spectrum of atheroprotective actions attributed to HDL. This has further led to studies comparing the HDL proteome of healthy subjects with that of particles from patients with dyslipidemias or CAD (Green et al. 2008; Heller et al. 2007; Riwanto et al. 2013; Vaisar et al. 2007). In the study by Vaisar et al., HDL was shown to carry apolipoproteins and proteins with functions in lipid metabolism, the acute phase response, complement regulation, and blood coagulation (Vaisar et al. 2007). The same group further compared the proteome of HDL3 from 6 CAD patients before and after 1-year treatment with combined statin/niacin therapy (Green et al. 2008). In this study, HDL3 of CAD patients is significantly enriched in apoE and apoCII and carry less apoJ and phospholipid transport protein (PLTP) as compared to HDL from control subjects. Treatment with niacin/statin decreases HDL3 apoE and raises apoJ and PLTP levels

(Green et al. 2008). In our recent study, we have observed that reduced clusterin and increased apoC-III content in HDL isolated from patients with CAD lead to activation of proapoptotic signaling pathways in endothelial cells (Riwanto et al. 2013).

Of note, oxidative modifications of HDL-associated proteins have also been demonstrated to contribute to the functional impairment of HDL. In particular, MPO-mediated oxidation of apoA-I has been linked to the generation of dysfunctional forms of HDL (Huang et al. 2014; Shao et al. 2010; Undurti et al. 2009; Wang et al. 2007b; Zheng et al. 2004). Recently, we have shown that MPO and PON1 interact and reciprocally modulate each other's function in vivo, i.e., PON1 partially inhibits MPO activity, while MPO oxidizes and inactivates PON1 (Huang et al. 2013). Furthermore, HDL isolated from patients with an ACS carries enhanced chlorotyrosine content, site-specific PON1 methionine oxidation, and reduced PON1 activity (Huang et al. 2013).

Taken together, these observations indicate that the HDL proteomic composition may be altered in different disease states which likely have an impact on HDL function. Unraveling the complexities of the HDL through proteomics studies may therefore allow for better understanding of the HDL composition and function. However, proteomics studies of HDL have been limited in size and number due to the labor-intensive nature of the approach. Furthermore, given the diverse concentration range of the proteins in the HDL particles, some proteomics approaches may fail to reflect the actual amount or alteration of low abundant proteins due to signal suppression in the presence of highly abundant proteins. Targeted proteomics using selected reaction monitoring is a powerful tandem mass spectrometry method that can be used to monitor target peptides within a complex protein digest (Picotti and Aebersold 2012; Picotti et al. 2010). The approach will enable absolute quantification of the individual protein contained within HDL particles.

Conclusion and Perspectives

Over the past two decades, many studies have emerged showing various antiatherogenic effects of HDL, including the capacity of HDL to mediate reverse cholesterol transport, and antioxidative, anti-inflammatory, and antiapoptotic effects (Riwanto and Landmesser 2013). Growing evidence has shown that HDL particles are highly heterogeneous and the vasoprotective effects of HDL are altered in patients with CAD, diabetes, and chronic kidney dysfunction, i.e., patients with a high cardiovascular risk profile. Of note, these findings have more recently been supported by observations in cohort studies, suggesting that in patients with advanced CAD, higher plasma levels of HDL-C are no longer associated with reduced risk of cardiovascular events (Angeloni et al. 2013; Schwartz et al. 2012). Importantly, the increase in HDL-C observed with some lipid-modifying drugs has not been uniformly associated with clinical benefit (Barter et al. 2007; Schwartz et al. 2012). These findings have clearly indicated that the association between HDL and cardiovascular disease is far more complex than previously thought and is likely mediated by different HDL functional properties, independent of the cholesterol component of the HDL particle (Heinecke 2011).

It is increasingly evident that the functional heterogeneity of HDL may be attributed to the complexity of the HDL particles, containing proteins and lipids that can be modified or altered in their composition. Of note, rapid advances in the MS proteome technology have allowed tremendous leaps in characterizing the HDL proteome. Studies investigating the lipidome of HDL are still somewhat limited by the available technologies. Importantly, several studies have associated changes in specific HDL protein either due to altered composition or modification to the loss of function of HDL (Besler et al. 2011; Huang et al. 2013; Riwanto et al. 2013; Shao et al. 2006, 2012; Zheng et al. 2004). Nevertheless, it remains to be determined if the alterations are secondary to changes that occur during the disease progression or if the alteration has indeed a causal effect on disease etiology. In this context it is also important to emphasize the problem of confounding between different diseases: for example, many diabetic patients have (sub)clinical chronic kidney disease and/or CAD. As a consequence one must also expect considerable overlap of HDL modifications which were initially identified in patients with either diabetes, CAD, or CKD. It is hence rather unlikely that each disease has its specific footprint of HDL dysfunction and alterations in proteome and lipidome.

Taken together, it is becoming clear that plasma HDL-C levels are not an appropriate marker of vascular effects of high-density lipoproteins and therefore do not represent a reliable therapeutic target. Importantly, HDL-targeted treatment approaches need to take into account the altered vascular effects of HDL in cardiovascular disease. Therefore, targeting HDL-mediated antiatherogenic mechanisms, rather than plasma HDL-C levels, may represent a more promising and interesting therapeutic target. Studies looking into the specific mechanisms leading to a loss of antiatherogenic effects of HDL are of particular importance, in order to develop therapeutic approaches that aim to restore the vasoprotective properties of HDL.

Open Access This chapter is distributed under the terms of the Creative Commons Attribution Noncommercial License, which permits any noncommercial use, distribution, and reproduction in any medium, provided the original author(s) and source are credited.

References

Adams V, Besler C, Fischer T, Riwanto M, Noack F, Hollriegel R, Oberbach A, Jehmlich N, Volker U, Winzer EB, Lenk K, Hambrecht R, Schuler G, Linke A, Landmesser U, Erbs S (2013) Exercise training in patients with chronic heart failure promotes restoration of high-density lipoprotein functional properties. Circ Res 113:1345–1355

Anderson HA, Hiltbold EM, Roche PA (2000) Concentration of MHC class II molecules in lipid rafts facilitates antigen presentation. Nat Immunol 1:156–162

Angeloni E, Paneni F, Landmesser U, Benedetto U, Melina G, Luscher TF, Volpe M, Sinatra R, Cosentino F (2013) Lack of protective role of HDL-C in patients with coronary artery disease undergoing elective coronary artery bypass grafting. Eur Heart J 34(46):3557–3562

Annema W, von Eckardstein A (2013) High-density lipoproteins. Multifunctional but vulnerable protections from atherosclerosis. Circ J 77:2432–2448

Ansell BJ, Navab M, Hama S, Kamranpour N, Fonarow G, Hough G, Rahmani S, Mottahedeh R, Dave R, Reddy ST, Fogelman AM (2003) Inflammatory/antiinflammatory properties of high-density lipoprotein distinguish patients from control subjects better than high-density lipoprotein cholesterol levels and are favorably affected by simvastatin treatment. Circulation 108:2751–2756

Arakawa H, Qian JY, Baatar D, Karasawa K, Asada Y, Sasaguri Y, Miller ER, Witztum JL, Ueno H (2005) Local expression of platelet-activating factor-acetylhydrolase reduces accumulation of oxidized lipoproteins and inhibits inflammation, shear stress-induced thrombosis, and neointima formation in balloon-injured carotid arteries in nonhyperlipidemic rabbits. Circulation 111:3302–3309

Ashby DT, Rye KA, Clay MA, Vadas MA, Gamble JR, Barter PJ (1998) Factors influencing the ability of HDL to inhibit expression of vascular cell adhesion molecule-1 in endothelial cells. Arterioscler Thromb Vasc Biol 18:1450–1455

Aviram M, Rosenblat M, Billecke S, Erogul J, Sorenson R, Bisgaier CL, Newton RS, La Du B (1999) Human serum paraoxonase (PON 1) is inactivated by oxidized low density lipoprotein and preserved by antioxidants. Free Radic Biol Med 26:892–904

Baker PW, Rye KA, Gamble JR, Vadas MA, Barter PJ (2000) Phospholipid composition of reconstituted high density lipoproteins influences their ability to inhibit endothelial cell adhesion molecule expression. J Lipid Res 41:1261–1267

Barter PJ, Nicholls S, Rye KA, Anantharamaiah GM, Navab M, Fogelman AM (2004) Antiinflammatory properties of HDL. Circ Res 95:764–772

Barter PJ, Caulfield M, Eriksson M, Grundy SM, Kastelein JJ, Komajda M, Lopez-Sendon J, Mosca L, Tardif JC, Waters DD, Shear CL, Revkin JH, Buhr KA, Fisher MR, Tall AR, Brewer B (2007) Effects of torcetrapib in patients at high risk for coronary events. N Engl J Med 357:2109–2122

Bergt C, Pennathur S, Fu X, Byun J, O'Brien K, McDonald TO, Singh P, Anantharamaiah GM, Chait A, Brunzell J, Geary RL, Oram JF, Heinecke JW (2004) The myeloperoxidase product hypochlorous acid oxidizes HDL in the human artery wall and impairs ABCA1-dependent cholesterol transport. Proc Natl Acad Sci USA 101:13032–13037

Besler C, Heinrich K, Rohrer L, Doerries C, Riwanto M, Shih DM, Chroni A, Yonekawa K, Stein S, Schaefer N, Mueller M, Akhmedov A, Daniil G, Manes C, Templin C, Wyss C, Maier W, Tanner FC, Matter CM, Corti R, Furlong C, Lusis AJ, von Eckardstein A, Fogelman AM, Luscher TF, Landmesser U (2011) Mechanisms underlying adverse effects of HDL on eNOS-activating pathways in patients with coronary artery disease. J Clin Invest 121:2693–2708

Bhattacharyya T, Nicholls SJ, Topol EJ, Zhang R, Yang X, Schmitt D, Fu X, Shao M, Brennan DM, Ellis SG, Brennan ML, Allayee H, Lusis AJ, Hazen SL (2008) Relationship of paraoxonase 1 (PON1) gene polymorphisms and functional activity with systemic oxidative stress and cardiovascular risk. JAMA 299:1265–1276

Birjmohun RS, Dallinga-Thie GM, Kuivenhoven JA, Stroes ES, Otvos JD, Wareham NJ, Luben R, Kastelein JJ, Khaw KT, Boekholdt SM (2007) Apolipoprotein A-II is inversely associated with risk of future coronary artery disease. Circulation 116:2029–2035

Bisoendial RJ, Hovingh GK, Levels JH, Lerch PG, Andresen I, Hayden MR, Kastelein JJ, Stroes ES (2003) Restoration of endothelial function by increasing high-density lipoprotein in subjects with isolated low high-density lipoprotein. Circulation 107:2944–2948

Boisfer E, Stengel D, Pastier D, Laplaud PM, Dousset N, Ninio E, Kalopissis AD (2002) Antioxidant properties of HDL in transgenic mice overexpressing human apolipoprotein A-II. J Lipid Res 43:732–741

Bombeli T, Karsan A, Tait JF, Harlan JM (1997) Apoptotic vascular endothelial cells become procoagulant. Blood 89:2429–2442

Bortnick AE, Rothblat GH, Stoudt G, Hoppe KL, Royer LJ, McNeish J, Francone OL (2000) The correlation of ATP-binding cassette 1 mRNA levels with cholesterol efflux from various cell lines. J Biol Chem 275:28634–28640

Bowry VW, Stanley KK, Stocker R (1992) High density lipoprotein is the major carrier of lipid hydroperoxides in human blood plasma from fasting donors. Proc Natl Acad Sci USA 89:10316–10320

Burke AP, Farb A, Malcom GT, Liang YH, Smialek J, Virmani R (1997) Coronary risk factors and plaque morphology in men with coronary disease who died suddenly. N Engl J Med 336:1276–1282

Bursill CA, Castro ML, Beattie DT, Nakhla S, van der Vorst E, Heather AK, Barter PJ, Rye KA (2010) High-density lipoproteins suppress chemokines and chemokine receptors in vitro and in vivo. Arterioscler Thromb Vasc Biol 30:1773–1778

Cabana VG, Reardon CA, Wei B, Lukens JR, Getz GS (1999) SAA-only HDL formed during the acute phase response in apoA-I+/+and apoA-I-/- mice. J Lipid Res 40:1090–1103

Calabresi L, Franceschini G, Sirtori CR, De Palma A, Saresella M, Ferrante P, Taramelli D (1997) Inhibition of VCAM-1 expression in endothelial cells by reconstituted high density lipoproteins. Biochem Biophys Res Commun 238:61–65

Campbell LA, Yaraei K, Van Lenten B, Chait A, Blessing E, Kuo CC, Nosaka T, Ricks J, Rosenfeld ME (2010) The acute phase reactant response to respiratory infection with Chlamydia pneumoniae: implications for the pathogenesis of atherosclerosis. Microbes Infect 12:598–606

Caplan BA, Schwartz CJ (1973) Increased endothelial cell turnover in areas of in vivo Evans Blue uptake in the pig aorta. Atherosclerosis 17:401–417

Castelli WP, Garrison RJ, Wilson PW, Abbott RD, Kalousdian S, Kannel WB (1986) Incidence of coronary heart disease and lipoprotein cholesterol levels. The Framingham study. JAMA 256:2835–2838

Charakida M, Besler C, Batuca JR, Sangle S, Marques S, Sousa M, Wang G, Tousoulis D, Delgado Alves J, Loukogeorgakis SP, Mackworth-Young C, D'Cruz D, Luscher T, Landmesser U, Deanfield JE (2009) Vascular abnormalities, paraoxonase activity, and dysfunctional HDL in primary antiphospholipid syndrome. JAMA 302:1210–1217

Cheng AM, Handa P, Tateya S, Schwartz J, Tang C, Mitra P, Oram JF, Chait A, Kim F (2012) Apolipoprotein A-I attenuates palmitate-mediated NF-kappaB activation by reducing Toll-like receptor-4 recruitment into lipid rafts. PLoS ONE 7:e33917

Chisolm GM 3rd, Hazen SL, Fox PL, Cathcart MK (1999) The oxidation of lipoproteins by monocytes-macrophages. Biochemical and biological mechanisms. J Biol Chem 274:25959–25962

Cockerill GW, Rye KA, Gamble JR, Vadas MA, Barter PJ (1995) High-density lipoproteins inhibit cytokine-induced expression of endothelial cell adhesion molecules. Arterioscler Thromb Vasc Biol 15:1987–1994

Coetzee GA, Strachan AF, van der Westhuyzen DR, Hoppe HC, Jeenah MS, de Beer FC (1986) Serum amyloid A-containing human high density lipoprotein 3. Density, size, and apolipoprotein composition. J Biol Chem 261:9644–9651

Cuchel M, Lund-Katz S, de la Llera-Moya M, Millar JS, Chang D, Fuki I, Rothblat GH, Phillips MC, Rader DJ (2009) Pathways by which reconstituted high-density lipoprotein mobilizes free cholesterol from whole body and from macrophages. Arterioscler Thromb Vasc Biol 30:526–532

Cullen P, Schulte H, Assmann G (1997) The Munster heart study (PROCAM): total mortality in middle-aged men is increased at low total and LDL cholesterol concentrations in smokers but not in nonsmokers. Circulation 96:2128–2136

Curtiss LK, Witztum JL (1985) Plasma apolipoproteins AI, AII, B, CI, and E are glucosylated in hyperglycemic diabetic subjects. Diabetes 34:452–461

Dansky HM, Charlton SA, Barlow CB, Tamminen M, Smith JD, Frank JS, Breslow JL (1999) Apo A-I inhibits foam cell formation in Apo E-deficient mice after monocyte adherence to endothelium. J Clin Invest 104:31–39

de Beer MC, Durbin DM, Cai L, Jonas A, de Beer FC, van der Westhuyzen DR (2001) Apolipoprotein A-I conformation markedly influences HDL interaction with scavenger receptor BI. J Lipid Res 42:309–313

De Nardo D, Labzin LI, Kono H, Seki R, Schmidt SV, Beyer M, Xu D, Zimmer S, Lahrmann C, Schildberg FA, Vogelhuber J, Kraut M, Ulas T, Kerksiek A, Krebs W, Bode N, Grebe A, Fitzgerald ML, Hernandez NJ, Williams BR, Knolle P, Kneilling M, Rocken M, Lutjohann D, Wright SD, Schultze JL, Latz E (2014) High-density lipoprotein mediates anti-inflammatory reprogramming of macrophages via the transcriptional regulator ATF3. Nat Immunol 15:152–160

de Souza JA, Vindis C, Hansel B, Negre-Salvayre A, Therond P, Serrano CV Jr, Chantepie S, Salvayre R, Bruckert E, Chapman MJ, Kontush A (2008) Metabolic syndrome features small, apolipoprotein A-I-poor, triglyceride-rich HDL3 particles with defective anti-apoptotic activity. Atherosclerosis 197:84–94

de Souza JA, Vindis C, Negre-Salvayre A, Rye KA, Couturier M, Therond P, Chantepie S, Salvayre R, Chapman MJ, Kontush A (2010) Small, dense HDL 3 particles attenuate apoptosis in endothelial cells: pivotal role of apolipoprotein A-I. J Cell Mol Med 14:608–620

Di Angelantonio E, Sarwar N, Perry P, Kaptoge S, Ray KK, Thompson A, Wood AM, Lewington S, Sattar N, Packard CJ, Collins R, Thompson SG, Danesh J (2009) Major lipids, apolipoproteins, and risk of vascular disease. JAMA 302:1993–2000

Diaz MN, Frei B, Vita JA, Keaney JF Jr (1997) Antioxidants and atherosclerotic heart disease. N Engl J Med 337:408–416

Dimayuga P, Zhu J, Oguchi S, Chyu KY, Xu XO, Yano J, Shah PK, Nilsson J, Cercek B (1999) Reconstituted HDL containing human apolipoprotein A-1 reduces VCAM-1 expression and neointima formation following periadventitial cuff-induced carotid injury in apoE null mice. Biochem Biophys Res Commun 264:465–468

Dimmeler S, Fleming I, Fisslthaler B, Hermann C, Busse R, Zeiher AM (1999) Activation of nitric oxide synthase in endothelial cells by Akt-dependent phosphorylation. Nature 399:601–605

Eckardstein A (2012) Tachometer for reverse cholesterol transport? J Am Heart Assoc 1:e003723

Esteve E, Ricart W, Fernandez-Real JM (2005) Dyslipidemia and inflammation: an evolutionary conserved mechanism. Clin Nutr 24:16–31

Ferretti G, Bacchetti T, Negre-Salvayre A, Salvayre R, Dousset N, Curatola G (2006) Structural modifications of HDL and functional consequences. Atherosclerosis 184:1–7

Garcia JG, Liu F, Verin AD, Birukova A, Dechert MA, Gerthoffer WT, Bamberg JR, English D (2001) Sphingosine 1-phosphate promotes endothelial cell barrier integrity by Edg-dependent cytoskeletal rearrangement. J Clin Invest 108:689–701

Garner B, Waldeck AR, Witting PK, Rye KA, Stocker R (1998) Oxidation of high density lipoproteins. II. Evidence for direct reduction of lipid hydroperoxides by methionine residues of apolipoproteins AI and AII. J Biol Chem 273:6088–6095

Garza CA, Montori VM, McConnell JP, Somers VK, Kullo IJ, Lopez-Jimenez F (2007) Association between lipoprotein-associated phospholipase A2 and cardiovascular disease: a systematic review. Mayo Clin Proc 82:159–165

Gordon T, Castelli WP, Hjortland MC, Kannel WB, Dawber TR (1977) High density lipoprotein as a protective factor against coronary heart disease. The Framingham study. Am J Med 62:707–714

Goyal J, Wang K, Liu M, Subbaiah PV (1997) Novel function of lecithin-cholesterol acyltransferase. Hydrolysis of oxidized polar phospholipids generated during lipoprotein oxidation. J Biol Chem 272:16231–16239

Green PS, Vaisar T, Pennathur S, Kulstad JJ, Moore AB, Marcovina S, Brunzell J, Knopp RH, Zhao XQ, Heinecke JW (2008) Combined statin and niacin therapy remodels the high-density lipoprotein proteome. Circulation 118:1259–1267

Haghpassand M, Bourassa PA, Francone OL, Aiello RJ (2001) Monocyte/macrophage expression of ABCA1 has minimal contribution to plasma HDL levels. J Clin Invest 108:1315–1320

Hansson GK (2005) Inflammation, atherosclerosis, and coronary artery disease. N Engl J Med 352:1685–1695

Haynes R, Jiang L, Hopewell JC, Li J, Chen F, Parish S, Landray MJ, Collins R, Armitage J, Collins R, Armitage J, Baigent C, Chen Z, Landray M, Chen Y, Jiang L, Pedersen TR, Landray M, Bowman L, Chen F, Hill M, Haynes R, Knott C, Rahimi K, Tobert J, Sleight P, Simpson D, Parish S, Baxter A, Lay M, Bray C, Wincott E, Leijenhorst G, Skattebol A, Moen G, Mitchel Y, Kuznetsova O, Macmahon S, Kjekshus J, Hill C, Lam TH, Sandercock P, Peto R, Hopewell JC (2013) HPS2-THRIVE randomized placebo-controlled trial in 25 673 high-risk patients of ER niacin/laropiprant: trial design, pre-specified muscle and liver outcomes, and reasons for stopping study treatment. Eur Heart J 34:1279–1291

Hedrick CC, Thorpe SR, Fu MX, Harper CM, Yoo J, Kim SM, Wong H, Peters AL (2000) Glycation impairs high-density lipoprotein function. Diabetologia 43:312–320

Heinecke JW (1998) Oxidants and antioxidants in the pathogenesis of atherosclerosis: implications for the oxidized low density lipoprotein hypothesis. Atherosclerosis 141:1–15

Heinecke J (2011) HDL and cardiovascular-disease risk–time for a new approach? N Engl J Med 364:170–171

Heller M, Schlappritzi E, Stalder D, Nuoffer JM, Haeberli A (2007) Compositional protein analysis of high density lipoproteins in hypercholesterolemia by shotgun LC-MS/MS and probabilistic peptide scoring. Mol Cell Proteomics 6:1059–1072

Holleboom AG, Daniil G, Fu X, Zhang R, Hovingh GK, Schimmel AW, Kastelein JJ, Stroes ES, Witztum JL, Hutten BA, Tsimikas S, Hazen SL, Chroni A, Kuivenhoven JA (2012) Lipid oxidation in carriers of lecithin:cholesterol acyltransferase gene mutations. Arterioscler Thromb Vasc Biol 32:3066–3075

Huang Y, von Eckardstein A, Wu S, Maeda N, Assmann G (1994) A plasma lipoprotein containing only apolipoprotein E and with gamma mobility on electrophoresis releases cholesterol from cells. Proc Natl Acad Sci USA 91:1834–1838

Huang Y, von Eckardstein A, Wu S, Assmann G (1995) Cholesterol efflux, cholesterol esterification, and cholesteryl ester transfer by LpA-I and LpA-I/A-II in native plasma. Arterioscler Thromb Vasc Biol 15:1412–1418

Huang Y, Wu Z, Riwanto M, Gao S, Levison BS, Gu X, Fu X, Wagner MA, Besler C, Gerstenecker G, Zhang R, Li XM, DiDonato AJ, Gogonea V, Tang WH, Smith JD, Plow EF, Fox PL, Shih DM, Lusis AJ, Fisher EA, DiDonato JA, Landmesser U, Hazen SL (2013) Myeloperoxidase, paraoxonase-1, and HDL form a functional ternary complex. J Clin Invest 123:3815–3828

Huang Y, DiDonato JA, Levison BS, Schmitt D, Li L, Wu Y, Buffa J, Kim T, Gerstenecker GS, Gu X, Kadiyala CS, Wang Z, Culley MK, Hazen JE, Didonato AJ, Fu X, Berisha SZ, Peng D, Nguyen TT, Liang S, Chuang CC, Cho L, Plow EF, Fox PL, Gogonea V, Tang WH, Parks JS, Fisher EA, Smith JD, Hazen SL (2014) An abundant dysfunctional apolipoprotein A1 in human atheroma. Nat Med 20:193–203

Jahangiri A, de Beer MC, Noffsinger V, Tannock LR, Ramaiah C, Webb NR, van der Westhuyzen DR, de Beer FC (2009) HDL remodeling during the acute phase response. Arterioscler Thromb Vasc Biol 29:261–267

Jensen MK, Rimm EB, Furtado JD, Sacks FM (2012) Apolipoprotein C-III as a potential modulator of the association between HDL-cholesterol and incident coronary heart disease. J Am Heart Assoc 1

Kalantar-Zadeh K, Kopple JD, Kamranpour N, Fogelman AM, Navab M (2007) HDL-inflammatory index correlates with poor outcome in hemodialysis patients. Kidney Int 72:1149–1156

Karlsson H, Leanderson P, Tagesson C, Lindahl M (2005) Lipoproteomics II: mapping of proteins in high-density lipoprotein using two-dimensional gel electrophoresis and mass spectrometry. Proteomics 5:1431–1445

Khera AV, Cuchel M, de la Llera-Moya M, Rodrigues A, Burke MF, Jafri K, French BC, Phillips JA, Mucksavage ML, Wilensky RL, Mohler ER, Rothblat GH, Rader DJ (2011) Cholesterol efflux capacity, high-density lipoprotein function, and atherosclerosis. N Engl J Med 364:127–135

Khovidhunkit W, Kim MS, Memon RA, Shigenaga JK, Moser AH, Feingold KR, Grunfeld C (2004) Effects of infection and inflammation on lipid and lipoprotein metabolism: mechanisms and consequences to the host. J Lipid Res 45:1169–1196

Kimura T, Sato K, Kuwabara A, Tomura H, Ishiwara M, Kobayashi I, Ui M, Okajima F (2001) Sphingosine 1-phosphate may be a major component of plasma lipoproteins responsible for the cytoprotective actions in human umbilical vein endothelial cells. J Biol Chem 276:31780–31785

Kimura T, Sato K, Malchinkhuu E, Tomura H, Tamama K, Kuwabara A, Murakami M, Okajima F (2003) High-density lipoprotein stimulates endothelial cell migration and survival through sphingosine 1-phosphate and its receptors. Arterioscler Thromb Vasc Biol 23:1283–1288

Kockx MM, Herman AG (2000) Apoptosis in atherosclerosis: beneficial or detrimental? Cardiovasc Res 45:736–746

Kontush A, Therond P, Zerrad A, Couturier M, Negre-Salvayre A, de Souza JA, Chantepie S, Chapman MJ (2007) Preferential sphingosine-1-phosphate enrichment and sphingomyelin depletion are key features of small dense HDL3 particles: relevance to antiapoptotic and antioxidative activities. Arterioscler Thromb Vasc Biol 27:1843–1849

Kuvin JT, Ramet ME, Patel AR, Pandian NG, Mendelsohn ME, Karas RH (2002) A novel mechanism for the beneficial vascular effects of high-density lipoprotein cholesterol: enhanced vasorelaxation and increased endothelial nitric oxide synthase expression. Am Heart J 144:165–172

Landmesser U, Hornig B, Drexler H (2004) Endothelial function: a critical determinant in atherosclerosis? Circulation 109:II27–II33

Li XM, Tang WH, Mosior MK, Huang Y, Wu Y, Matter W, Gao V, Schmitt D, Didonato JA, Fisher EA, Smith JD, Hazen SL (2013) Paradoxical association of enhanced cholesterol efflux with increased incident cardiovascular risks. Arterioscler Thromb Vasc Biol 33:1696–1705

Lp PLASC, Thompson A, Gao P, Orfei L, Watson S, Di Angelantonio E, Kaptoge S, Ballantyne C, Cannon CP, Criqui M, Cushman M, Hofman A, Packard C, Thompson SG, Collins R, Danesh J (2010) Lipoprotein-associated phospholipase A(2) and risk of coronary disease, stroke, and mortality: collaborative analysis of 32 prospective studies. Lancet 375:1536–1544

Lucke S, Levkau B (2010) Endothelial functions of sphingosine-1-phosphate. Cell Physiol Biochem 26:87–96

Lund-Katz S, Liu L, Thuahnai ST, Phillips MC (2003) High density lipoprotein structure. Front Biosci 8:d1044–d1054

Mackness MI, Arrol S, Durrington PN (1991) Paraoxonase prevents accumulation of lipoperoxides in low-density lipoprotein. FEBS Lett 286:152–154

Mackness B, Hine D, Liu Y, Mastorikou M, Mackness M (2004) Paraoxonase-1 inhibits oxidised LDL-induced MCP-1 production by endothelial cells. Biochem Biophys Res Commun 318:680–683

Marathe GK, Zimmerman GA, McIntyre TM (2003) Platelet-activating factor acetylhydrolase, and not paraoxonase-1, is the oxidized phospholipid hydrolase of high density lipoprotein particles. J Biol Chem 278:3937–3947

McIntyre TM, Prescott SM, Stafforini DM (2009) The emerging roles of PAF acetylhydrolase. J Lipid Res 50(Suppl):S255–S259

McMahon M, Grossman J, FitzGerald J, Dahlin-Lee E, Wallace DJ, Thong BY, Badsha H, Kalunian K, Charles C, Navab M, Fogelman AM, Hahn BH (2006) Proinflammatory high-density lipoprotein as a biomarker for atherosclerosis in patients with systemic lupus erythematosus and rheumatoid arthritis. Arthritis Rheum 54:2541–2549

Mertens A, Verhamme P, Bielicki JK, Phillips MC, Quarck R, Verreth W, Stengel D, Ninio E, Navab M, Mackness B, Mackness M, Holvoet P (2003) Increased low-density lipoprotein oxidation and impaired high-density lipoprotein antioxidant defense are associated with increased macrophage homing and atherosclerosis in dyslipidemic obese mice: LCAT gene transfer decreases atherosclerosis. Circulation 107:1640–1646

Mineo C, Yuhanna IS, Quon MJ, Shaul PW (2003) High density lipoprotein-induced endothelial nitric-oxide synthase activation is mediated by Akt and MAP kinases. J Biol Chem 278:9142–9149

Mineo C, Deguchi H, Griffin JH, Shaul PW (2006) Endothelial and antithrombotic actions of HDL. Circ Res 98:1352–1364

Miyata M, Smith JD (1996) Apolipoprotein E allele-specific antioxidant activity and effects on cytotoxicity by oxidative insults and beta-amyloid peptides. Nat Genet 14:55–61

Morgantini C, Natali A, Boldrini B, Imaizumi S, Navab M, Fogelman AM, Ferrannini E, Reddy ST (2011) Anti-inflammatory and antioxidant properties of HDLs are impaired in type 2 diabetes. Diabetes 60:2617–2623

Nabel EG, Braunwald E (2012) A tale of coronary artery disease and myocardial infarction. N Engl J Med 366:54–63

Navab M, Imes SS, Hama SY, Hough GP, Ross LA, Bork RW, Valente AJ, Berliner JA, Drinkwater DC, Laks H et al (1991) Monocyte transmigration induced by modification of low density lipoprotein in cocultures of human aortic wall cells is due to induction of monocyte chemotactic protein 1 synthesis and is abolished by high density lipoprotein. J Clin Invest 88:2039–2046

Navab M, Berliner JA, Watson AD, Hama SY, Territo MC, Lusis AJ, Shih DM, Van Lenten BJ, Frank JS, Demer LL, Edwards PA, Fogelman AM (1996) The Yin and Yang of oxidation in the development of the fatty streak. A review based on the 1994 George Lyman Duff Memorial Lecture. Arterioscler Thromb Vasc Biol 16:831–842

Navab M, Hama-Levy S, Van Lenten BJ, Fonarow GC, Cardinez CJ, Castellani LW, Brennan ML, Lusis AJ, Fogelman AM, La Du BN (1997) Mildly oxidized LDL induces an increased apolipoprotein J/paraoxonase ratio. J Clin Invest 99:2005–2019

Navab M, Hama SY, Anantharamaiah GM, Hassan K, Hough GP, Watson AD, Reddy ST, Sevanian A, Fonarow GC, Fogelman AM (2000a) Normal high density lipoprotein inhibits three steps in the formation of mildly oxidized low density lipoprotein: steps 2 and 3. J Lipid Res 41:1495–1508

Navab M, Hama SY, Cooke CJ, Anantharamaiah GM, Chaddha M, Jin L, Subbanagounder G, Faull KF, Reddy ST, Miller NE, Fogelman AM (2000b) Normal high density lipoprotein inhibits three steps in the formation of mildly oxidized low density lipoprotein: step 1. J Lipid Res 41:1481–1494

Navab M, Berliner JA, Subbanagounder G, Hama S, Lusis AJ, Castellani LW, Reddy S, Shih D, Shi W, Watson AD, Van Lenten BJ, Vora D, Fogelman AM (2001a) HDL and the inflammatory response induced by LDL-derived oxidized phospholipids. Arterioscler Thromb Vasc Biol 21:481–488

Navab M, Hama SY, Hough GP, Subbanagounder G, Reddy ST, Fogelman AM (2001b) A cell-free assay for detecting HDL that is dysfunctional in preventing the formation of or inactivating oxidized phospholipids. J Lipid Res 42:1308–1317

Nicholls SJ, Cutri B, Worthley SG, Kee P, Rye KA, Bao S, Barter PJ (2005a) Impact of short-term administration of high-density lipoproteins and atorvastatin on atherosclerosis in rabbits. Arterioscler Thromb Vasc Biol 25:2416–2421

Nicholls SJ, Dusting GJ, Cutri B, Bao S, Drummond GR, Rye KA, Barter PJ (2005b) Reconstituted high-density lipoproteins inhibit the acute pro-oxidant and proinflammatory vascular changes induced by a periarterial collar in normocholesterolemic rabbits. Circulation 111:1543–1550

Nobecourt E, Tabet F, Lambert G, Puranik R, Bao S, Yan L, Davies MJ, Brown BE, Jenkins AJ, Dusting GJ, Bonnet DJ, Curtiss LK, Barter PJ, Rye KA (2010) Nonenzymatic glycation impairs the antiinflammatory properties of apolipoprotein A-I. Arterioscler Thromb Vasc Biol 30:766–772

Nofer JR, Levkau B, Wolinska I, Junker R, Fobker M, von Eckardstein A, Seedorf U, Assmann G (2001) Suppression of endothelial cell apoptosis by high density lipoproteins (HDL) and HDL-associated lysosphingolipids. J Biol Chem 276:34480–34485

Nofer JR, van der Giet M, Tolle M, Wolinska I, von Wnuck LK, Baba HA, Tietge UJ, Godecke A, Ishii I, Kleuser B, Schafers M, Fobker M, Zidek W, Assmann G, Chun J, Levkau B (2004) HDL induces NO-dependent vasorelaxation via the lysophospholipid receptor S1P3. J Clin Invest 113:569–581

Norata GD, Catapano AL (2012) HDL and adaptive immunity: a tale of lipid rafts. Atherosclerosis 225:34–35

Noto H, Hara M, Karasawa K, Iso ON, Satoh H, Togo M, Hashimoto Y, Yamada Y, Kosaka T, Kawamura M, Kimura S, Tsukamoto K (2003) Human plasma platelet-activating factor acetylhydrolase binds to all the murine lipoproteins, conferring protection against oxidative stress. Arterioscler Thromb Vasc Biol 23:829–835

Ophir G, Amariglio N, Jacob-Hirsch J, Elkon R, Rechavi G, Michaelson DM (2005) Apolipoprotein E4 enhances brain inflammation by modulation of the NF-kappaB signaling cascade. Neurobiol Dis 20:709–718

Ostos MA, Conconi M, Vergnes L, Baroukh N, Ribalta J, Girona J, Caillaud JM, Ochoa A, Zakin MM (2001) Antioxidative and antiatherosclerotic effects of human apolipoprotein A-IV in apolipoprotein E-deficient mice. Arterioscler Thromb Vasc Biol 21:1023–1028

Parks JS, Rudel LL (1985) Alteration of high density lipoprotein subfraction distribution with induction of serum amyloid A protein (SAA) in the nonhuman primate. J Lipid Res 26:82–91

Paszty C, Maeda N, Verstuyft J, Rubin EM (1994) Apolipoprotein AI transgene corrects apolipoprotein E deficiency-induced atherosclerosis in mice. J Clin Invest 94:899–903

Patel S, Drew BG, Nakhla S, Duffy SJ, Murphy AJ, Barter PJ, Rye KA, Chin-Dusting J, Hoang A, Sviridov D, Celermajer DS, Kingwell BA (2009) Reconstituted high-density lipoprotein increases plasma high-density lipoprotein anti-inflammatory properties and cholesterol efflux capacity in patients with type 2 diabetes. J Am Coll Cardiol 53:962–971

Patel PJ, Khera AV, Jafri K, Wilensky RL, Rader DJ (2011) The anti-oxidative capacity of high-density lipoprotein is reduced in acute coronary syndrome but not in stable coronary artery disease. J Am Coll Cardiol 58:2068–2075

Patel PJ, Khera AV, Wilensky RL, Rader DJ (2013) Anti-oxidative and cholesterol efflux capacities of high-density lipoprotein are reduced in ischaemic cardiomyopathy. Eur J Heart Fail 15(11):1215–1219

Persegol L, Verges B, Foissac M, Gambert P, Duvillard L (2006) Inability of HDL from type 2 diabetic patients to counteract the inhibitory effect of oxidised LDL on endothelium-dependent vasorelaxation. Diabetologia 49:1380–1386

Persegol L, Verges B, Gambert P, Duvillard L (2007) Inability of HDL from abdominally obese subjects to counteract the inhibitory effect of oxidized LDL on vasorelaxation. J Lipid Res 48:1396–1401

Peterson SJ, Husney D, Kruger AL, Olszanecki R, Ricci F, Rodella LF, Stacchiotti A, Rezzani R, McClung JA, Aronow WS, Ikehara S, Abraham NG (2007) Long-term treatment with the apolipoprotein A1 mimetic peptide increases antioxidants and vascular repair in type I diabetic rats. J Pharmacol Exp Ther 322:514–520

Peterson SJ, Drummond G, Kim DH, Li M, Kruger AL, Ikehara S, Abraham NG (2008) L-4 F treatment reduces adiposity, increases adiponectin levels, and improves insulin sensitivity in obese mice. J Lipid Res 49:1658–1669

Picotti P, Aebersold R (2012) Selected reaction monitoring-based proteomics: workflows, potential, pitfalls and future directions. Nat Methods 9:555–566

Picotti P, Rinner O, Stallmach R, Dautel F, Farrah T, Domon B, Wenschuh H, Aebersold R (2010) High-throughput generation of selected reaction-monitoring assays for proteins and proteomes. Nat Methods 7:43–46

Puranik R, Bao S, Nobecourt E, Nicholls SJ, Dusting GJ, Barter PJ, Celermajer DS, Rye KA (2008) Low dose apolipoprotein A-I rescues carotid arteries from inflammation in vivo. Atherosclerosis 196:240–247

Rader DJ (2006) Molecular regulation of HDL metabolism and function: implications for novel therapies. J Clin Invest 116:3090–3100

Rallidis LS, Tellis CC, Lekakis J, Rizos I, Varounis C, Charalampopoulos A, Zolindaki M, Dagres N, Anastasiou-Nana M, Tselepis AD (2012) Lipoprotein-associated phospholipase A (2) bound on high-density lipoprotein is associated with lower risk for cardiac death in stable coronary artery disease patients: a 3-year follow-up. J Am Coll Cardiol 60:2053–2060

Ramet ME, Ramet M, Lu Q, Nickerson M, Savolainen MJ, Malzone A, Karas RH (2003) High-density lipoprotein increases the abundance of eNOS protein in human vascular endothelial cells by increasing its half-life. J Am Coll Cardiol 41:2288–2297

Recalde D, Ostos MA, Badell E, Garcia-Otin AL, Pidoux J, Castro G, Zakin MM, Scott-Algara D (2004) Human apolipoprotein A-IV reduces secretion of proinflammatory cytokines and atherosclerotic effects of a chronic infection mimicked by lipopolysaccharide. Arterioscler Thromb Vasc Biol 24:756–761

Regieli JJ, Jukema JW, Doevendans PA, Zwinderman AH, Kastelein JJ, Grobbee DE, van der Graaf Y (2009) Paraoxonase variants relate to 10-year risk in coronary artery disease: impact of a high-density lipoprotein-bound antioxidant in secondary prevention. J Am Coll Cardiol 54:1238–1245

Ribas V, Sanchez-Quesada JL, Anton R, Camacho M, Julve J, Escola-Gil JC, Vila L, Ordonez-Llanos J, Blanco-Vaca F (2004) Human apolipoprotein A-II enrichment displaces paraoxonase from HDL and impairs its antioxidant properties: a new mechanism linking HDL protein composition and antiatherogenic potential. Circ Res 95:789–797

Riwanto M, Landmesser U (2013) High density lipoproteins and endothelial functions: mechanistic insights and alterations in cardiovascular disease. J Lipid Res 54:3227–3243

Riwanto M, Rohrer L, Roschitzki B, Besler C, Mocharla P, Mueller M, Perisa D, Heinrich K, Altwegg L, von Eckardstein A, Luscher TF, Landmesser U (2013) Altered activation of endothelial anti- and pro-apoptotic pathways by high-density lipoprotein from patients with coronary artery disease: role of HDL-proteome remodeling. Circulation 127(8):891–904

Ross R (1999) Atherosclerosis – an inflammatory disease. N Engl J Med 340:115–126

Rossig L, Dimmeler S, Zeiher AM (2001) Apoptosis in the vascular wall and atherosclerosis. Basic Res Cardiol 96:11–22

Rotllan N, Ribas V, Calpe-Berdiel L, Martin-Campos JM, Blanco-Vaca F, Escola-Gil JC (2005) Overexpression of human apolipoprotein A-II in transgenic mice does not impair macrophage-specific reverse cholesterol transport in vivo. Arterioscler Thromb Vasc Biol 25:e128–e132

Rye KA, Barter PJ (2008) Antiinflammatory actions of HDL: a new insight. Arterioscler Thromb Vasc Biol 28:1890–1891

Sacre SM, Stannard AK, Owen JS (2003) Apolipoprotein E (apoE) isoforms differentially induce nitric oxide production in endothelial cells. FEBS Lett 540:181–187

Saito H, Lund-Katz S, Phillips MC (2004) Contributions of domain structure and lipid interaction to the functionality of exchangeable human apolipoproteins. Prog Lipid Res 43:350–380

Schwartz GG, Olsson AG, Abt M, Ballantyne CM, Barter PJ, Brumm J, Chaitman BR, Holme IM, Kallend D, Leiter LA, Leitersdorf E, McMurray JJ, Mundl H, Nicholls SJ, Shah PK, Tardif JC, Wright RS (2012) Effects of dalcetrapib in patients with a recent acute coronary syndrome. N Engl J Med 367:2089–2099

Shao B, Oda MN, Bergt C, Fu X, Green PS, Brot N, Oram JF, Heinecke JW (2006) Myeloperoxidase impairs ABCA1-dependent cholesterol efflux through methionine oxidation and site-specific tyrosine chlorination of apolipoprotein A-I. J Biol Chem 281:9001–9004

Shao B, Tang C, Heinecke JW, Oram JF (2010) Oxidation of apolipoprotein A-I by myeloperoxidase impairs the initial interactions with ABCA1 required for signaling and cholesterol export. J Lipid Res 51:1849–1858

Shao B, Pennathur S, Heinecke JW (2012) Myeloperoxidase targets apolipoprotein A-I, the major high density lipoprotein protein, for site-specific oxidation in human atherosclerotic lesions. J Biol Chem 287:6375–6386

Sharrett AR, Ballantyne CM, Coady SA, Heiss G, Sorlie PD, Catellier D, Patsch W (2001) Coronary heart disease prediction from lipoprotein cholesterol levels, triglycerides, lipoprotein(a), apolipoproteins A-I and B, and HDL density subfractions: the atherosclerosis risk in communities (ARIC) study. Circulation 104:1108–1113

Shih DM, Gu L, Xia YR, Navab M, Li WF, Hama S, Castellani LW, Furlong CE, Costa LG, Fogelman AM, Lusis AJ (1998) Mice lacking serum paraoxonase are susceptible to organophosphate toxicity and atherosclerosis. Nature 394:284–287

Shih DM, Xia YR, Wang XP, Miller E, Castellani LW, Subbanagounder G, Cheroutre H, Faull KF, Berliner JA, Witztum JL, Lusis AJ (2000) Combined serum paraoxonase knockout/apolipoprotein E knockout mice exhibit increased lipoprotein oxidation and atherosclerosis. J Biol Chem 275:17527–17535

Soran H, Younis NN, Charlton-Menys V, Durrington P (2009) Variation in paraoxonase-1 activity and atherosclerosis. Curr Opin Lipidol 20:265–274

Sorrentino SA, Besler C, Rohrer L, Meyer M, Heinrich K, Bahlmann FH, Mueller M, Horvath T, Doerries C, Heinemann M, Flemmer S, Markowski A, Manes C, Bahr MJ, Haller H, von Eckardstein A, Drexler H, Landmesser U (2010) Endothelial-vasoprotective effects of high-density lipoprotein are impaired in patients with type 2 diabetes mellitus but are improved after extended-release niacin therapy. Circulation 121:110–122

Speidel MT, Booyse FM, Abrams A, Moore MA, Chung BH (1990) Lipolyzed hypertriglyceridemic serum and triglyceride-rich lipoprotein cause lipid accumulation in and are cytotoxic to cultured human endothelial cells. High density lipoproteins inhibit this cytotoxicity. Thromb Res 58:251–264

Spieker LE, Sudano I, Hurlimann D, Lerch PG, Lang MG, Binggeli C, Corti R, Ruschitzka F, Luscher TF, Noll G (2002) High-density lipoprotein restores endothelial function in hypercholesterolemic men. Circulation 105:1399–1402

Subramanian VS, Goyal J, Miwa M, Sugatami J, Akiyama M, Liu M, Subbaiah PV (1999) Role of lecithin-cholesterol acyltransferase in the metabolism of oxidized phospholipids in plasma: studies with platelet-activating factor-acetyl hydrolase-deficient plasma. Biochim Biophys Acta 1439:95–109

Suc I, Escargueil-Blanc I, Troly M, Salvayre R, Negre-Salvayre A (1997) HDL and ApoA prevent cell death of endothelial cells induced by oxidized LDL. Arterioscler Thromb Vasc Biol 17:2158–2166

Sugano M, Tsuchida K, Makino N (2000) High-density lipoproteins protect endothelial cells from tumor necrosis factor-alpha-induced apoptosis. Biochem Biophys Res Commun 272:872–876

Takeya H, Gabazza EC, Aoki S, Ueno H, Suzuki K (2003) Synergistic effect of sphingosine 1-phosphate on thrombin-induced tissue factor expression in endothelial cells. Blood 102:1693–1700

Tall AR, Yvan-Charvet L, Terasaka N, Pagler T, Wang N (2008) HDL, ABC transporters, and cholesterol efflux: implications for the treatment of atherosclerosis. Cell Metab 7:365–375

Tang C, Liu Y, Kessler PS, Vaughan AM, Oram JF (2009) The macrophage cholesterol exporter ABCA1 functions as an anti-inflammatory receptor. J Biol Chem 284:32336–32343

Tang WH, Hartiala J, Fan Y, Wu Y, Stewart AF, Erdmann J, Kathiresan S, Roberts R, McPherson R, Allayee H, Hazen SL (2012) Clinical and genetic association of serum paraoxonase and arylesterase activities with cardiovascular risk. Arterioscler Thromb Vasc Biol 32:2803–2812

Tolle M, Huang T, Schuchardt M, Jankowski V, Prufer N, Jankowski J, Tietge UJ, Zidek W, van der Giet M (2012) High-density lipoprotein loses its anti-inflammatory capacity by accumulation of pro-inflammatory-serum amyloid A. Cardiovasc Res 94:154–162

Tward A, Xia YR, Wang XP, Shi YS, Park C, Castellani LW, Lusis AJ, Shih DM (2002) Decreased atherosclerotic lesion formation in human serum paraoxonase transgenic mice. Circulation 106:484–490

Uittenbogaard A, Shaul PW, Yuhanna IS, Blair A, Smart EJ (2000) High density lipoprotein prevents oxidized low density lipoprotein-induced inhibition of endothelial nitric-oxide synthase localization and activation in caveolae. J Biol Chem 275:11278–11283

Undurti A, Huang Y, Lupica JA, Smith JD, DiDonato JA, Hazen SL (2009) Modification of high density lipoprotein by myeloperoxidase generates a pro-inflammatory particle. J Biol Chem 284:30825–30835

Vaisar T, Pennathur S, Green PS, Gharib SA, Hoofnagle AN, Cheung MC, Byun J, Vuletic S, Kassim S, Singh P, Chea H, Knopp RH, Brunzell J, Geary R, Chait A, Zhao XQ, Elkon K, Marcovina S, Ridker P, Oram JF, Heinecke JW (2007) Shotgun proteomics implicates protease inhibition and complement activation in the antiinflammatory properties of HDL. J Clin Invest 117:746–756

Van Lenten BJ, Hama SY, de Beer FC, Stafforini DM, McIntyre TM, Prescott SM, La Du BN, Fogelman AM, Navab M (1995) Anti-inflammatory HDL becomes pro-inflammatory during the acute phase response. Loss of protective effect of HDL against LDL oxidation in aortic wall cell cocultures. J Clin Invest 96:2758–2767

Van Lenten BJ, Wagner AC, Nayak DP, Hama S, Navab M, Fogelman AM (2001) High-density lipoprotein loses its anti-inflammatory properties during acute influenza a infection. Circulation 103:2283–2288

Van Linthout S, Spillmann F, Riad A, Trimpert C, Lievens J, Meloni M, Escher F, Filenberg E, Demir O, Li J, Shakibaei M, Schimke I, Staudt A, Felix SB, Schultheiss HP, De Geest B, Tschope C (2008) Human apolipoprotein A-I gene transfer reduces the development of experimental diabetic cardiomyopathy. Circulation 117:1563–1573

von Eckardstein A, Huang Y, Wu S, Funke H, Noseda G, Assmann G (1995a) Reverse cholesterol transport in plasma of patients with different forms of familial HDL deficiency. Arterioscler Thromb Vasc Biol 15:691–703

von Eckardstein A, Huang Y, Wu S, Sarmadi AS, Schwarz S, Steinmetz A, Assmann G (1995b) Lipoproteins containing apolipoprotein A-IV but not apolipoprotein A-I take up and esterify cell-derived cholesterol in plasma. Arterioscler Thromb Vasc Biol 15:1755–1763

Vowinkel T, Mori M, Krieglstein CF, Russell J, Saijo F, Bharwani S, Turnage RH, Davidson WS, Tso P, Granger DN, Kalogeris TJ (2004) Apolipoprotein A-IV inhibits experimental colitis. J Clin Invest 114:260–269

Wang X, Collins HL, Ranalletta M, Fuki IV, Billheimer JT, Rothblat GH, Tall AR, Rader DJ (2007a) Macrophage ABCA1 and ABCG1, but not SR-BI, promote macrophage reverse cholesterol transport in vivo. J Clin Invest 117:2216–2224

Wang Z, Nicholls SJ, Rodriguez ER, Kummu O, Horkko S, Barnard J, Reynolds WF, Topol EJ, DiDonato JA, Hazen SL (2007b) Protein carbamylation links inflammation, smoking, uremia and atherogenesis. Nat Med 13:1176–1184

Wang SH, Yuan SG, Peng DQ, Zhao SP (2012) HDL and ApoA-I inhibit antigen presentation-mediated T cell activation by disrupting lipid rafts in antigen presenting cells. Atherosclerosis 225:105–114

Watson AD, Berliner JA, Hama SY, La Du BN, Faull KF, Fogelman AM, Navab M (1995) Protective effect of high density lipoprotein associated paraoxonase. Inhibition of the biological activity of minimally oxidized low density lipoprotein. J Clin Invest 96:2882–2891

Weihrauch D, Xu H, Shi Y, Wang J, Brien J, Jones DW, Kaul S, Komorowski RA, Csuka ME, Oldham KT, Pritchard KA (2007) Effects of D-4 F on vasodilation, oxidative stress, angiostatin, myocardial inflammation, and angiogenic potential in tight-skin mice. Am J Physiol Heart Circ Physiol 293:H1432–H1441

Witztum JL, Steinberg D (1991) Role of oxidized low density lipoprotein in atherogenesis. J Clin Invest 88:1785–1792

Yuhanna IS, Zhu Y, Cox BE, Hahner LD, Osborne-Lawrence S, Lu P, Marcel YL, Anderson RG, Mendelsohn ME, Hobbs HH, Shaul PW (2001) High-density lipoprotein binding to scavenger receptor-BI activates endothelial nitric oxide synthase. Nat Med 7:853–857

Yvan-Charvet L, Pagler TA, Wang N, Senokuchi T, Brundert M, Li H, Rinninger F, Tall AR (2008) SR-BI inhibits ABCG1-stimulated net cholesterol efflux from cells to plasma HDL. J Lipid Res 49:107–114

Zerrad-Saadi A, Therond P, Chantepie S, Couturier M, Rye KA, Chapman MJ, Kontush A (2009) HDL3-mediated inactivation of LDL-associated phospholipid hydroperoxides is determined by the redox status of apolipoprotein A-I and HDL particle surface lipid rigidity: relevance to inflammation and atherogenesis. Arterioscler Thromb Vasc Biol 29:2169–2175

Zheng L, Nukuna B, Brennan ML, Sun M, Goormastic M, Settle M, Schmitt D, Fu X, Thomson L, Fox PL, Ischiropoulos H, Smith JD, Kinter M, Hazen SL (2004) Apolipoprotein A-I is a selective target for myeloperoxidase-catalyzed oxidation and functional impairment in subjects with cardiovascular disease. J Clin Invest 114:529–541

Zhu X, Owen JS, Wilson MD, Li H, Griffiths GL, Thomas MJ, Hiltbold EM, Fessler MB, Parks JS (2010) Macrophage ABCA1 reduces MyD88-dependent toll-like receptor trafficking to lipid rafts by reduction of lipid raft cholesterol. J Lipid Res 51:3196–3206

Part III

Possible Indications and Target Mechanisms of HDL Therapy

HDL and Atherothrombotic Vascular Disease

Wijtske Annema, Arnold von Eckardstein, and Petri T. Kovanen

Contents

1	Introduction	371
2	Interactions of HDLs with the Endothelium	373
	2.1 Preservation of Endothelial Integrity	373
	2.2 Preservation of Endothelial Function	375
3	Anti-inflammatory Effects	376
	3.1 Suppression of Myelopoiesis	376
	3.2 Suppression of Monocyte Extravasation	377
	3.3 Interference with Macrophage Differentiation and Activation	379
4	Effects of HDLs on Vascular Lipid and Lipoprotein Homeostasis	379
	4.1 From LDL Retention to Atheroma Formation	379
	4.1.1 Initial Lipid Accumulation in Atherosclerosis-Prone Arterial Intima: A Sign of LDL Retention	379
	4.1.2 Formation of a Fatty Streak	381
	4.1.3 Development of an Atheroma	382
	4.2 Inhibition of LDL Modification by HDLs	383
	4.3 Macrophage Cholesterol Efflux and Reverse Cholesterol Transport	385
	4.3.1 Transendothelial HDL Transport	385
	4.3.2 Cholesterol Efflux from Macrophages	386
	4.3.3 Exit from the Arterial Wall	387
	4.3.4 Delivery of Cholesterol to the Liver and Intestine	388

W. Annema (✉)
Institute of Clinical Chemistry, University Hospital Zurich, Zurich, Switzerland
e-mail: wijtske.annema@usz.ch

A. von Eckardstein
Institute of Clinical Chemistry, University Hospital Zurich, Zurich, Switzerland

Center of integrative Human Physiology, University of Zurich, Zurich, Switzerland
e-mail: arnold.voneckardstein@usz.ch

P.T. Kovanen
Wihuri Research Institute, Helsinki, Finland
e-mail: petri.kovanen@wri.fi

© The Author(s) 2015
A. von Eckardstein, D. Kardassis (eds.), *High Density Lipoproteins*, Handbook of Experimental Pharmacology 224, DOI 10.1007/978-3-319-09665-0_11

5	Atherothrombosis	388
	5.1 Effects of HDLs on Smooth Muscle Cells	390
	5.2 Antithrombotic Effects of HDLs	391
	5.2.1 HDLs and Platelets	391
	5.2.2 HDLs and Coagulation	392
6	Myocardial Injury	393
References		394

Abstract

High-density lipoproteins (HDLs) exert many beneficial effects which may help to protect against the development or progression of atherosclerosis or even facilitate lesion regression. These activities include promoting cellular cholesterol efflux, protecting low-density lipoproteins (LDLs) from modification, preserving endothelial function, as well as anti-inflammatory and antithrombotic effects. However, questions remain about the relative importance of these activities for atheroprotection. Furthermore, the many molecules (both lipids and proteins) associated with HDLs exert both distinct and overlapping activities, which may be compromised by inflammatory conditions, resulting in either loss of function or even gain of dysfunction. This complexity of HDL functionality has so far precluded elucidation of distinct structure–function relationships for HDL or its components. A better understanding of HDL metabolism and structure–function relationships is therefore crucial to exploit HDLs and its associated components and cellular pathways as potential targets for anti-atherosclerotic therapies and diagnostic markers.

Keywords

Atherosclerosis • Cholesterol efflux • Endothelial function • Foam cells • High-density lipoproteins • Inflammation • Macrophages • Oxidation • Thrombosis • Smooth muscle cells

List of Abbreviations

AAPH	2,2′-Azobis(2-amidinopropane) dihydrochloride
ApoA-I	Apolipoprotein A-I
ABCA1	ATP-binding cassette transporter A1
ABCG1	ATP-binding cassette transporter G1
ATF3	Activating transcription factor 3
β-ATPase	β-Chain of F0F1 ATPase
BSEP	Bile salt export pump
CETP	Cholesteryl ester transfer protein
CYP7A1	Cholesterol 7α-hydroxylase
eNOS	Endothelial nitric oxide synthase
EPC	Endothelial progenitor cell
HDL	High-density lipoprotein
HO-1	Heme oxygenase-1

IL-1β	Interleukin-1β
LCAT	Lecithin-cholesterol acyltransferase
LDL	Low-density lipoprotein
Lp-PLA$_2$	Lipoprotein-associated phospholipase A$_2$
MAP	Mitogen-activated protein
MCP-1	Monocyte chemotactic protein-1
MDA	Malondialdehyde
MMP	Matrix metalloproteinase
NF-κβ	Nuclear factor-κβ
NO	Nitric oxide
PAF-AH	Platelet-activating factor acetylhydrolase
PDZK1	PDZ domain-containing protein 1
PI3K	Phosphatidylinositol 3-kinase
PON1	Paraoxonase 1
rHDL	Reconstituted high-density lipoprotein
ROS	Reactive oxygen species
S1P	Sphingosine-1-phosphate
SMC	Smooth muscle cell
SR-BI	Scavenger receptor BI
STAT3	Signal transducer and activator of transcription 3
TCFA	Thin-cap fibroatheroma
TLR	Toll-like receptor
TNFα	Tumor necrosis factor α
VCAM-1	Vascular cell adhesion molecule 1
VSMC	Vascular smooth muscle cell

1 Introduction

Atherothrombotic diseases, such as coronary heart disease, stroke, and peripheral arterial disease, are the world's leading cause of death. The first-line therapy for prevention and treatment of coronary heart disease is low-density lipoprotein (LDL) lowering by statin therapy. Although statin drugs are very effective in reducing plasma LDL cholesterol concentrations, the risk of major cardiovascular events is only decreased by 15–40 % (Baigent et al. 2005; Ridker et al. 2010). Additional therapeutic approaches are needed to minimize this considerable remaining cardiovascular risk in statin-treated patients. It has been postulated that part of this residual risk may be explained by low circulating levels of high-density lipoproteins (HDLs). Clinical and large population-based epidemiological studies have consistently shown that low HDL cholesterol is independently associated with an increased risk of atherothrombotic diseases (Di Angelantonio et al. 2009; Barter et al. 2007a). However, this long-standing inverse relationship between HDL and atherosclerotic cardiovascular disease has been challenged by recent findings.

Indeed, high plasma concentrations of HDL cholesterol might not be beneficial for certain subgroups of patients. Thus, in high-risk postinfarction patients characterized by high total cholesterol and C-reactive protein, elevated levels of HDL cholesterol were paradoxically found to be related to poor clinical outcome (Corsetti et al. 2006). Similarly, the inverse relationship between HDL cholesterol levels and major cardiovascular events was found lost in patients with established coronary artery disease undergoing diagnostic angiography or coronary artery bypass grafting (Angeloni et al. 2013; Silbernagel et al. 2013). Moreover, several large randomized clinical trials assessing the effects of new therapies targeting HDL metabolism, such as cholesteryl ester transfer protein (CETP) inhibitors and niacin, reported disappointing results. The Investigation of Lipid Level Management to Understand its Impact in Atherosclerotic Events (ILLUMINATE) trial revealed that the combination of atorvastatin with the CETP inhibitor torcetrapib increased HDL cholesterol levels by 72.1 % relative to atorvastatin alone (Barter et al. 2007b). This raise in HDL cholesterol upon torcetrapib administration did, however, not lower the risk of the major cardiovascular outcome (Barter et al. 2007b). On the contrary, patients in the atorvastatin/torcetrapib group unexpectedly experienced significantly more adverse events (Barter et al. 2007b). More recently, another phase 3 clinical trial involving patients with a recent acute coronary syndrome, named dal-OUTCOMES, was designed to evaluate the effects of the CETP inhibitor dalcetrapib (Schwartz et al. 2012). Addition of dalcetrapib to statin therapy raised HDL cholesterol levels by 31–40 % but failed to reduce the risk of recurrent cardiovascular events (Schwartz et al. 2012). Niacin is another effective agent for raising HDL cholesterol and also lowering LDL cholesterol, triglycerides, and lipoprotein(a). Data from the Atherothrombosis Intervention in Metabolic Syndrome with Low HDL/High Triglycerides: Impact on Global Health Outcomes (AIM-HIGH) trial, which included patients with established cardiovascular disease, showed that adding extended-release niacin to statin therapy was not effective in reducing the risk of cardiovascular events during a follow-up of 36 months despite a 25 % increase in HDL cholesterol (Boden et al. 2011).

Genetic studies also questioned a role for HDLs in atherothrombotic diseases. Lifelong reductions of plasma HDL cholesterol levels due to heterozygosity for a loss-of-function mutation in *ABCA1* did not affect the risk for ischemic heart disease (Frikke-Schmidt et al. 2008). Likewise, genetic variation in the apolipoprotein A-I (apoA-I) gene, *APOA1*, affecting apoA-I and HDL cholesterol levels did not predict ischemic heart disease or myocardial infarction in the general population (Haase et al. 2010). A more recent Mendelian randomization analysis demonstrated that there is no significant association between risk of myocardial infarction and a loss-of-function SNP in the gene coding for endothelial lipase (LIPG Asn396Ser), which results in elevated levels of HDL cholesterol (Voight et al. 2012). Hence, there is ongoing discussion whether HDLs are causally involved in the development of atherothrombotic diseases or are merely an innocent bystander. On the other hand, there is considerable evidence from in vitro and in vivo studies that HDL particles isolated from human plasma as well as artificially reconstituted HDLs (rHDLs) exert multiple atheroprotective effects (Annema and von Eckardstein 2013).

The above-outlined controversies have shifted the interest of the cardiovascular research field to HDL functionality, meaning that the functional quality of the particles might be more important than the absolute amount of HDL cholesterol. Currently, the available HDL-targeted therapies for atherothrombotic diseases raise plasma HDL cholesterol levels per se. Nonetheless, plasma concentrations of HDL cholesterol do not necessarily reflect the functional capacity of HDL particles. In addition, there is evidence that the anti-atherogenic functions of HDLs are impaired in patients with cardiovascular disease (Annema and von Eckardstein 2013), and these dysfunctional HDLs might even promote atherothrombotic diseases. Therefore, a better understanding of the molecular mechanisms behind the protective functions of HDLs in the context of atherothrombotic diseases is fundamental for successful exploitation of HDLs for the treatment of coronary heart disease.

2 Interactions of HDLs with the Endothelium

2.1 Preservation of Endothelial Integrity

Perturbations in the function of endothelial cells lining the lumen of blood vessels are an important hallmark of the earliest stages of the atherosclerotic process. HDLs may improve or help reverse endothelial dysfunction by several endothelial-vasoprotective means. Several in vitro studies indicate that HDLs are able to counteract endothelial cell death. Treatment of endothelial cells with HDLs caused a marked attenuation of apoptosis in response to pro-atherogenic signals such as oxidized LDL (Suc et al. 1997; de Souza et al. 2010; Kimura et al. 2001), tumor necrosis factor α (TNFα) (Sugano et al. 2000; Riwanto et al. 2013), and growth factor deprivation (Kimura et al. 2001; Nofer et al. 2001b; Riwanto et al. 2013). Although the underlying mechanism has not been fully elucidated, it has been suggested that HDLs regulate endothelial cell survival by interfering with critical steps in apoptotic pathways. Several studies reported that HDLs block the activation of the mitochondrial pathway of apoptosis by preventing depolarization of the mitochondrial transmembrane and subsequent mitochondrial release of cytochrome c and apoptosis-inducing factor into the cytoplasm (Nofer et al. 2001b; de Souza et al. 2010). As a consequence, HDLs inhibit the cytochrome c-initiated caspase activation cascade (Sugano et al. 2000; Nofer et al. 2001b; de Souza et al. 2010). Moreover, levels of the apoptosis-inducing protein truncated Bid are reduced in endothelial cells cultured in the presence of HDLs, whereas endothelial expression of the antiapoptotic protein Bcl-xL is enhanced (Riwanto et al. 2013). The capacity of HDLs to attenuate apoptosis in endothelial cells requires induction of phosphatidylinositol 3-kinase (PI3K)/Akt/endothelial nitric oxide synthase (eNOS) signaling (Nofer et al. 2001b; Riwanto et al. 2013) and was attributed to apoA-I (Radojkovic et al. 2009) as well as HDL-associated lysophospholipids (Kimura et al. 2001; Nofer et al. 2001b). More recently, a lower clusterin (or apoJ) and increased apoC-III content in the HDLs of patients with stable coronary artery disease or acute coronary syndrome was found to be associated with reduced antiapoptotic activities in human endothelial cells (Riwanto et al. 2013).

Disruption of vascular endothelium integrity is another central event promoting atherosclerotic plaque formation. Sphingosine-1-phosphate (S1P), a bioactive sphingolipid metabolite, has been identified as an important endothelial barrier function-enhancing agonist (Garcia et al. 2001). In plasma S1P preferentially associates with HDL particles via binding to apoM (Christoffersen et al. 2011). In vitro studies provided evidence that delivery of S1P carried by apoM-containing HDLs to S1P1 receptors on endothelial cells leads to activation of Akt and eNOS, which in turn promotes endothelial barrier integrity (Argraves et al. 2008; Wilkerson et al. 2012). Interestingly, the lung vascular barrier function, as determined by Evan's blue extravasation, is decreased in mice that lack S1P in their HDL fraction due to deficiency of apoM (Christoffersen et al. 2011). These data underline a significant role of HDL-associated S1P in endothelial barrier protection in vivo. In this context it is important to highlight the submicromolar concentration of S1P in HDLs which amounts to only 5–10 % of HDL particle concentration but is in the range of the affinity constants of the G-protein-coupled S1P receptors (Annema and von Eckardstein 2013; Karuna et al. 2011). Moreover, erythrocytes release S1P in response to HDLs (Lucke and Levkau 2010) so that in vivo the local concentration of S1P may be considerably higher than in isolated HDLs.

Acceleration of reendothelialization after vascular injury is another mechanism proposed to contribute to the positive influence of HDLs on endothelial function. In this respect, HDLs might be specifically important in driving endothelial cell migration during repair. Incubation of cultured endothelial cells with HDLs stimulated cell migration in response to wounding (Seetharam et al. 2006). Moreover, hepatic expression of human apoA-I rescued the impaired carotid artery reendothelialization following perivascular electric injury in apoA-I knockout mice (Seetharam et al. 2006). The promotion of endothelial cell migration by HDLs has been demonstrated to involve scavenger receptor BI (SR-BI)-mediated activation of Rac-directed formation of actin-based lamellipodia (Seetharam et al. 2006). An SR-BI-dependent mechanism for the promigratory actions of HDL was confirmed by experiments showing that the capacity of HDLs to stimulate endothelial cell migration was reduced by transfection of endothelial cells with siRNA specific to PDZ domain-containing protein 1 (PDZK1), an SR-BI adaptor protein that controls steady-state SR-BI levels, and that the reendothelialization capacity of injured vessels is lost in mice in which the *pdzk1* gene has been disrupted (Zhu et al. 2008). Other studies suggested that HDL-induced endothelial cell migration is mediated by stimulation of Akt/eNOS phosphorylation through HDL-bound S1P in an S1P1-dependent manner and that this process is facilitated by endothelial lipase (Tatematsu et al. 2013).

Besides endothelial cell migration, also recruitment and incorporation of bone marrow-derived endothelial progenitor cells (EPCs) to sites of vessel injury contributes to the repair of the damaged vessel wall. HDLs activate the PI3K/Akt pathway in EPCs resulting in enhanced EPC proliferation, migration, and survival (Sumi et al. 2007; Petoumenos et al. 2009; Zhang et al. 2010). Moreover, preincubation of EPCs with HDL augmented their adhesive capacity to endothelial cells and potentiated their transendothelial migration (Petoumenos et al. 2009). Mice that were injected with rHDLs had an increased number of circulating EPCs and showed improved reendothelialization after carotid artery injury (Petoumenos et al. 2009). Additional mouse studies indicated that infusion of rHDLs enhanced bone marrow progenitor cell mobilization, neovascularization, and functional recovery after hind limb ischemia in wild-type mice (Sumi et al. 2007). In contrast, no beneficial effects of rHDL treatment on ischemia-induced angiogenesis were observed in eNOS knockout mice (Sumi et al. 2007), suggesting an essential role for eNOS in mediating HDL-stimulated EPC recruitment and the subsequent functional effects on the ischemic limb. There is some evidence that HDLs also affect circulating EPCs in humans. In patients with coronary artery disease, higher HDL cholesterol levels were significantly correlated with higher circulating EPC numbers (Petoumenos et al. 2009). These findings are corroborated by data from a study in patients with type 2 diabetes mellitus which found that rHDL infusion at a dose of 80 mg/kg body weight markedly increased circulating EPC levels compared to baseline (van Oostrom et al. 2007).

2.2 Preservation of Endothelial Function

Other reports on HDL functionality in endothelial cells established that HDLs have direct effects on vascular endothelium-relaxing properties. HDLs induce endothelium-derived vasorelaxation in aortic segments ex vivo in a dose-dependent manner (Nofer et al. 2004). Mechanistically, binding of HDL to SR-BI and the lysophospholipid receptor S1P3 on endothelial cells leads to PI3K-stimulated activation of Akt and mitogen-activated protein (MAP) kinase pathways, and Akt in turn phosphorylates eNOS at Ser-1177 to increase endothelial nitric oxide (NO) production and ultimately endothelium-dependent vasorelaxation (Nofer et al. 2004; Mineo et al. 2003; Yuhanna et al. 2001). The stimulatory effect of HDLs on NO-dependent vasorelaxation could be mimicked by three HDL-associated lysophospholipids, i.e., sphingosylphosphorylcholine, S1P, and lysosulfatide (Nofer et al. 2004). In addition to eNOS activation, modulation of Akt and MAP signaling by HDLs has been reported to delay the degradation of eNOS mRNA in endothelial cells, thereby upregulating eNOS protein levels (Ramet et al. 2003). Dietary oxysterols, in particular 7-oxysterols, impair endothelial vasorelaxant function by disrupting the active dimeric form of eNOS and by inducing production of reactive oxygen species (ROS) (Terasaka et al. 2010). HDLs can counteract the negative effects of 7-oxysterols on endothelial eNOS function by promoting efflux of 7-oxysterols from endothelial cells via ATP-binding cassette transporter G1 (ABCG1) (Terasaka et al. 2010). The same researchers revealed that HDL-mediated removal of cholesterol and oxysterols from endothelial cells through ABCG1 could reverse the cholesterol loading-induced inactivation of eNOS resulting from its increased interaction with caveolin (Terasaka et al. 2010). Similar to NO, there is evidence to suggest that the release of the endothelium-derived vasodilator prostacyclin can be influenced by HDLs. In human umbilical vein endothelial cells, HDLs elevated cyclooxygenase-2 protein expression and thereby induced prostacyclin synthesis via a pathway that involved PKC, ERK1/2, and sphingosine kinase SphK-2 (Xiong et al. 2014). The human relevance of these findings is underscored by studies demonstrating that infusion of rHDLs in hypercholesterolemic patients led to normalization of endothelium-dependent vasodilation in response to acetylcholine (Spieker et al. 2002). Moreover, heterozygosity for a loss-of-function mutation in *ABCA1* has been reported to be associated with low levels of HDL cholesterol as well as an impaired endothelium-dependent forearm vasodilatory response, a phenotype which could be restored by intravenous administration of apoA-I/phosphatidylcholine discs (Bisoendial et al. 2003).

3 Anti-inflammatory Effects

3.1 Suppression of Myelopoiesis

Vascular inflammation, one of the main forces driving the formation of atherosclerotic plaques, is accompanied by the rolling, firm adhesion, and subsequent migration of monocytes across the vascular endothelium. A first critical determinant for monocyte transmigration is the number of monocytes present in the circulation. HDL-mediated cholesterol efflux pathways appear to play a prime role in the proliferation of myeloid cells in the bone marrow, cell mobilization from the bone marrow, and subsequent myeloid differentiation, at least in mice. Mice that lack both ABCA1 and ABCG1 have increased numbers of myeloid progenitor cells in bone marrow, blood, spleen, and liver, suggestive of aberrant myeloproliferation, myeloid progenitor cell mobilization, and extramedullary hematopoiesis (Yvan-Charvet et al. 2010; Westerterp et al. 2012). As a consequence monocyte counts in blood are elevated in ABCA1 x ABCG1 double knockout mice (Yvan-Charvet et al. 2010). In addition, the myeloproliferative phenotype and the enhanced mobilization of hematopoietic progenitor cells associated with bone marrow deficiency of ABCA1 and ABCG1 were reversed in mice carrying a human apoA-I transgene, and in high cholesterol diet-fed LDL receptor knockout mice, this was associated with less severe atherosclerosis (Yvan-Charvet et al. 2010; Westerterp et al. 2012). Correspondingly, raising HDL cholesterol levels in other murine models of myeloproliferative disorders also efficiently reduced the appearance of hematopoietic progenitor cells in the circulation (Westerterp et al. 2012). Direct proof for a link between HDL and myeloproliferation was provided by experiments showing that incubation of myeloid progenitor cells with HDLs ex vivo suppressed cell expansion (Yvan-Charvet et al. 2010). Taken together, these findings indicate that HDLs may beneficially impact the number of circulating monocytes. In humans there is some evidence for a correlation between HDL cholesterol levels and monocyte numbers in patients with familial hypercholesterolemia (Tolani et al. 2013). However, only few and inconsistent data on the association between monocyte counts and risk of atherosclerotic vascular disease in humans has been reported. In addition, it will be important to consider the heterogeneity of monocytes for any association with coronary heart disease (Zawada et al. 2012).

3.2 Suppression of Monocyte Extravasation

The arrest of circulating monocytes on inflamed endothelial monolayers is triggered by chemokines and their corresponding receptors, which both have been demonstrated to be subject to regulation by HDLs. Infusions of lipid-free apoA-I in apoE-deficient mice that were fed a high-fat diet reduced the relative abundance of cells stained positive for the chemokine receptors CCR2 and CX3CR1 in atherosclerotic plaques in the aortic sinus and decreased the circulating levels of the chemokines CCL2 and CCL5 (Bursill et al. 2010). Moreover, in vitro rHDLs were found to downregulate cell surface expression of chemokine receptors on monocytes and to render human monocytes and endothelial cells partly resistant to phytohemagglutinin- or cytokine-induced expression of chemokines (Bursill et al. 2010; Spirig et al. 2013). Although it was shown that HDLs counterbalance the upregulation of chemokines in response to pro-inflammatory cytokines by reducing nuclear factor-$\kappa\beta$ (NF-$\kappa\beta$) pathway activation, how HDLs modulate monocyte expression of chemokine receptors on the molecular level remains to be established (Bursill et al. 2010). In addition to these effects on monocytes and endothelial cells, HDLs were also able to limit the production of CCL2 in vascular smooth muscle cells (VSMCs) exposed to thrombin (Tolle et al. 2008). The study identified HDL as a negative regulator of thrombin-induced activation of Rac1-dependent NADPH oxidases and subsequent ROS generation. This process is mediated by binding of HDL to SR-BI on VSMCs, and moreover lysosphingolipids present in HDL particles can act on local S1P3 receptor to inhibit CCL2 expression (Tolle et al. 2008).

Once monocytes are attracted by chemokines to inflamed endothelium, monocyte–endothelial cell interactions are mediated by several cell adhesion molecules. Experiments in vitro provided compelling evidence that cytokine-induced adhesion molecule expression on the endothelial cell surface is diminished in the presence of either native HDLs (Cockerill et al. 1995; Ashby et al. 1998) or rHDLs (Baker et al. 1999; Clay et al. 2001). Notably, the effects of HDLs on endothelial expression of adhesion molecules not only occurred in cultured cells but also in the in vivo setting under inflammatory conditions. For example, a single injection of discoidal rHDL particles in pigs prevented upregulation of E-selectin expression in response to intradermal interleukin-1α (Cockerill et al. 2001). Moreover, daily treatment of normocholesterolemic rabbits with rHDLs or lipid-free apoA-I markedly attenuated periarterial collar-induced endothelial expression of vascular cell adhesion molecule 1 (VCAM-1) and intercellular adhesion molecule 1 as well as concomitant neutrophil infiltration into the vessel wall and generation of ROS (Nicholls et al. 2005). The positive effects of lipid-free apoA-I on pro-inflammatory vascular changes were still observed when administered as a single low dose 3 h after insertion of the non-occlusive periarterial collar (Puranik et al. 2008), indicating that infusion of apoA-I may also be able to rescue preestablished vascular inflammation. In patients suffering from claudication, treatment with rHDLs led to a significant reduction in the percentage of VCAM-1 positive cells in femoral artery lesions (Shaw et al. 2008). Diminution of adhesion molecule expression on endothelial cells by HDLs may involve the activation of multiple receptors and signaling cascades. There are data supporting that HDLs alter adhesion molecule expression as a consequence of perturbation of cellular sphingolipid metabolism and the NF-κβ signaling pathway. HDL3 was found to impede TNFα-induced upregulation of VCAM-1 and E-selectin expression in human umbilical vein endothelial cells by antagonizing activation of sphingosine kinase and subsequent intracellular formation of S1P (Xia et al. 1999). Furthermore, it has been documented that in cultured endothelial cells, incubation with HDLs prevented NF-κβ activation, movement of p65 to the nucleus, and expression of NF-κβ target genes in response to TNFα (Xia et al. 1999; Park et al. 2003; Kimura et al. 2006; McGrath et al. 2009). The basal molecular mechanisms of the beneficial effects of HDL on endothelial inflammation are dependent on an increase in the expression of 3β-hydroxysteroid-Δ24 reductase that results in activation of PI3K and downstream activation of eNOS and heme oxygenase-1 (HO-1) (McGrath et al. 2009; Wu et al. 2013). Specific knockdown of SR-BI or S1P1 receptor by treatment of human endothelial cells with small interfering RNAs against SR-BI and PDZK1 or S1P1, respectively, markedly reduced the ability of HDLs to inhibit VCAM-1 expression and NF-κβ activation (Kimura et al. 2006; Wu et al. 2013), suggesting that both receptors may be critical for the anti-inflammatory endothelial actions of HDLs. Corresponding with the possible involvement of the S1P1 receptor, the inhibitory activity of HDLs on the expression of endothelial cell adhesion molecules has been proposed to be a function of biologically active lysosphingolipids carried by HDL particles (Nofer et al. 2003; Kimura et al. 2006). Besides the lipid component, the anti-inflammatory properties of HDLs could also be related to the microRNA content of HDL particles. Recently, it has been reported that transfer of microRNA-223 by HDLs to human aortic endothelial cells significantly repressed expression of intracellular adhesion molecule 1 (Tabet et al. 2014).

Monocytes bind to adhesion molecules on activated endothelial cells through monocytic cell surface receptors, such as CD11b. Ex vivo incubation of stimulated human peripheral blood monocytes with apoA-I in the lipid-free or HDL-associated state caused a reduction in cell surface expression of CD11b (Murphy et al. 2008). These results might be directly translatable to the human in vivo situation, as evidenced by the observation that infusion of rHDLs in patients with atherosclerosis disease in the femoral artery led to a significantly lower expression of CD11b on circulating monocytes (Shaw et al. 2008). The fact that cellular depletion of cholesterol by cyclodextrin mimicked the protective effects of apoA-I on the expression of monocyte CD11b and that inhibition of CD11b expression is lost in the presence of an antibody against ABCA1 or in monocytes derived from Tangier disease patients is indicative of a role for ABCA1-mediated cholesterol efflux in the mechanism of monocyte deactivation by apoA-I (Murphy et al. 2008).

3.3 Interference with Macrophage Differentiation and Activation

Monocytes that have entered the arterial intima subsequently differentiate into resident macrophages of at least two major subtypes that differentially produce a variety of pro- and anti-inflammatory mediators. In the subendothelial space, the macrophages become activated and comprise key targets for the anti-inflammatory action of HDLs. Incubation with HDLs primed isolated murine bone marrow macrophages into alternative M2 macrophages with anti-inflammatory features (Sanson et al. 2013). Moreover, HDLs were potent inhibitors of interferon-gamma-mediated expression of pro-inflammatory M1 markers in isolated mouse macrophages, and, interestingly, experiments performed with macrophages from STAT6-deficient mice demonstrated that STAT6 plays an essential role in the HDL-mediated shift of the macrophage phenotype to M2 (Sanson et al. 2013). However, other investigators could not replicate these observations, when using primary human monocyte-derived macrophages (Colin et al. 2014). A recent study provided novel mechanistic insights into the anti-inflammatory effects of HDLs in macrophages. Thus, administration of native HDL, rHDL, or apoA-I to mice challenged with a toll-like receptor (TLR) agonist protected from liver damage and reduced serum levels of pro-inflammatory cytokines, which was supposed to be due the ability of HDLs to reduce cytokine production in activated macrophages (De Nardo et al. 2014). The anti-inflammatory action of HDL particles on macrophages was explained by their ability to induce the expression of activating transcription factor 3 (ATF3), which functions as a repressor of pro-inflammatory target genes (De Nardo et al. 2014).

During the past decades, a considerable amount of data has been published revealing that natural regulatory T cells, which dampen inflammatory responses, may help decrease the atherosclerotic disease burden. The findings of a recent murine study imply that apoA-I could promote the expansion of regulatory T cells. Subcutaneous injection of human lipid-free apoA-I in LDL receptor/apoA-I double knockout mice fed an atherogenic diet restored the regulatory T-cell population in the lymph nodes (Wilhelm et al. 2010). In these animals the raise in the number of regulatory T cells in the lymph nodes was accompanied by a reversal of the inflammatory and autoimmune phenotype.

4 Effects of HDLs on Vascular Lipid and Lipoprotein Homeostasis

4.1 From LDL Retention to Atheroma Formation

4.1.1 Initial Lipid Accumulation in Atherosclerosis-Prone Arterial Intima: A Sign of LDL Retention

The intima is disproportionally thick at the outer curvatures of the proximal branches of coronary and carotid arteries, where turbulent flow conditions prevail and the pulse wave amplitude is relatively high (Wentzel et al. 2012). The thickening is a physiological response to these conditions and ensues when the vascular smooth muscle cells (VSMCs) divide and secrete various components of the extracellular matrix at the stressed sites. One major factor explaining the relationship between the thickness of the intima and its susceptibility to atherosclerotic lesion development is the lack of a capillary delivery system and lack of a lymphatic drainage system in this tissue (Hulten and Levin 2009; Eliska et al. 2006). As a consequence, the inflowing apoB100-containing lipoprotein particles (notably LDL and also VLDL remnants) are only slowly conveyed by the intimal fluid from the subendothelial superficial layer into the deeper intimal layers and ultimately to the medial layer of the arterial wall where lymphatic capillaries are present. Indeed, in cholesterol-fed rabbits, intimal LDL concentration initially increases in atherosclerosis-susceptible sites, where their intimal residence time of the LDL particles is long, the transit time being calculated to be hours (Schwenke and Carew 1989a, b).

Since progressive accumulation of LDL cholesterol is a key feature of atherogenesis, we can deduce that some intimal LDL particles never leave the site. The constant inflow without compensatory outflow of LDL particles results in a dramatic increase in their concentration in the intimal fluid to reach values equaling or even exceeding (by twofold) those in the corresponding human blood plasma (Smith 1990). Thus, the concentration of LDL in the arterial intimal fluid is 10–20 times higher than in the extracellular fluids of other extrahepatic tissues, in which its concentration is about 1/10th of that in the corresponding plasma, so rendering the intima a truly "hypercholesterolemic" site.

During their lengthy passage across the thick avascular tissue, the LDL particles tend to bind to the subendothelial extracellular proteoglycan matrix, a phenomenon known as retention of LDL (Williams and Tabas 1995). The proteoglycans form a tight network that is negatively charged. To this network the LDL particles bind via ionic interactions, when positively charged lysine and arginine residues located in the proteoglycan binding site of apoB100 interact with the negatively charged sulfate groups of the glycosaminoglycan components of the proteoglycans (Camejo et al. 1998). The prolonged residence time of LDL particles in the intima, whether the particles are in solution or trapped in the matrix, enhances their exposure to a variety of local LDL-modifying factors. Thus, during their long extracellular residence time, the LDL particles are modified, for example, by oxygen radicals secreted by the resident cells of the intima, notably by endothelial cells, and also by SMCs. It is of particular relevance that the endothelial cells become activated to secrete pro-oxidative compounds at sites with turbulent flow, i.e., at the atherogenesis-prone sites of the arterial tree (Nigro et al. 2011). Moreover, the subendothelially located LDL particles are not protected by the powerful antioxidant systems present in the circulating blood plasma, and they also gradually become depleted of their own antioxidant molecules capable of protecting them from free radical attack and oxidation (Esterbauer et al. 1991). In addition to the oxidative processes, various hydrolytic (both proteolytic and lipolytic) enzymes may also modify the LDL particles in the intima (Tabas 1999). Importantly, the oxidative and hydrolytic changes of LDL particles render them unstable, trigger their aggregation and fusion into lipid droplets, and markedly strengthen their binding to arterial proteoglycans (Oorni et al. 2000). These in vitro observations have their close counterparts in various animal models, and also in the human carotid arteries, in which tiny proteoglycan-, collagen-, and elastin-associated lipid deposits, detectable only by aid of special electron microscopic techniques, are located in the subendothelial extracellular space even before any microscopic signs of atherosclerosis are visible (Nievelstein-Post et al. 1994; Tamminen et al. 1999; Pasquinelli et al. 1989).

4.1.2 Formation of a Fatty Streak

As stated above, atherosclerosis typically develops at certain predilection sites as a response to the hemodynamic stress and other types of injury affecting the endothelial cells. Importantly, in such stressed areas with disturbed flow conditions, endothelial inflammatory signaling pathways become activated and a local pro-inflammatory environment is created (Nigro et al. 2011). We can consider this local scenario as a singular starting condition to the development of atherosclerosis. Thus, the flow-mediated preconditioning of the intima renders it a suitable soil for the retention, modification, and accumulation of LDL particles. Because the subendothelially located modified LDL particles and the products released from them are pro-inflammatory, they are likely to have profound local cell-activating effects and to worsen the inflammation by inducing local generation of various LDL-modifying enzymes and agents and so creating a positive feedback loop (Pentikainen et al. 2000; Tabas et al. 2007). Moreover, synthesis and secretion of pro-inflammatory cytokines and chemokines, such as the monocyte chemotactic protein-1 (MCP-1; CCL2), are induced (Libby 2002; Zernecke and Weber 2010). Indeed, monocytes preferentially adhere to the endothelium covering inflamed tissue sites, i.e., regarding atherogenesis, they tend to enter the intimal areas containing oxidized or otherwise modified LDL particles (Schmitt et al. 2014a, b).

The extracellularly located modified LDL particles, whether free floating in the intimal fluid or whether matrix bound, are ingested by intimal macrophages and, to some extent, also by intimal SMCs (Witztum 2005; Wang et al. 1996). On their surfaces, the macrophages express scavenger receptors, which recognize oxidized and otherwise modified LDL particles (Krieger and Herz 1994). The macrophages may also phagocytose modified LDL particles, provided the particles are in aggregate form or have been converted into droplets via fusion. Since uptake of LDL cholesterol by macrophages is not negatively feedback regulated by the inflowing cholesterol, uptake of LDL cholesterol, be it scavenger receptor dependent or mediated by other mechanisms, may continue until the cells are filled with cholesterol (Brown and Goldstein 1983). Of great interest, also unmodified native LDLs can be ingested by macrophages at a rate which is linear to their concentration in the extracellular fluid (Kruth 2011). Since the concentration of LDL in the intimal fluid is exceptionally high (see above), such micropinocytotic uptake of LDLs by macrophages may actually be of great significance in the atherosclerosis-prone arterial intima.

In the macrophages, the excess of intracellular cholesterol is esterified with fatty acids and stored in their cytoplasm as cholesteryl ester droplets (Brown and Goldstein 1983). Macrophages full of such droplets look foamy under the microscope and are called foam cells. When numerous foam cells form clusters, they have yellow appearance visible to the naked eye and the site appears as a fatty streak (Stary et al. 1994). Importantly, when fatty streaks develop in human carotid arteries, the subendothelial perifibrous lipid droplets disappear (Pasquinelli et al. 1989). This switch conceivably reflects uptake of the perifibrous droplets by the subendothelial macrophages. Thus, in a fatty streak the LDL-derived cholesterol is located mostly intracellularly. Taken together, an important role of subendothelial macrophages in the arterial intima is to clear unmodified or modified LDL particles from their surrounding and to store their cholesterol as cholesteryl esters. Indeed, foam cells in the arterial intima are the most defining feature of atherosclerosis.

HDLs directly bind unesterified cholesterol. As the consequence, HDL can mobilize unesterified cholesterol both from cells and extracellular locations (Freeman et al. 2014). Regarding cholesteryl esters, the extracellularly stored esters, present either in the core of the matrix-bound LDL particles or in the core of the perifibrous lipid droplets derived from modified LDL particles, are likely to fully resist transfer to HDLs. In sharp contrast, intracellular cholesterol stored as cholesteryl esters in the cytoplasmic lipid droplets of macrophage foam cells is sensitive to transfer to HDLs (Rosenson et al. 2012). Thus, regarding the return of LDL cholesterol from the arterial intima back to the circulating blood, it appears to be of fundamental importance that the LDL particles are first taken up and metabolized by intimal macrophages. Accordingly, this macrophage-dependent preparative step appears to be mandatory for the initiation of reverse cholesterol transport in the arterial intima. However, since the macrophages accumulate LDL-derived cholesterol, an imbalance between cholesterol influx and cholesterol efflux exists in the atherosclerosis-prone arterial intima. One reason for the relative inability of HDLs to effectively remove cholesterol from the intimal macrophages is their functional modification including proteolysis, oxidation, and lipolysis to various extents (Lee-Rueckert and Kovanen 2011; DiDonato et al. 2013).

4.1.3 Development of an Atheroma

Conversion of a fatty streak into an atheroma is the next phase in atherogenesis (Stary et al. 1995). It is characterized by extracellular accumulation of cholesterol in the deep layers of the intima, i.e., below the superficial foam cell layers. However, in sharp contrast to the superficial layers of the intima, where atherogenesis initiates and where macrophages swiftly arrive and efficiently remove the lipids so preventing their extracellular accumulation, in the deep layers, the macrophages arrive late (Nakashima et al. 2008). Accordingly, once formed, the deep extracellular lipids are not rapidly scavenged by macrophages. Rather, the continuous formation of lipid droplets without their concomitant removal leads to their progressive deposition, which ultimately leads to the appearance of "lipid lakes". The lakes consist of huge numbers of such droplets, and together the lakes form a lipid core (Guyton and Klemp 1996). The presence of a lipid core is the hallmark of an atheroma.

During the process of lipid core formation, macrophages start to migrate from the superficial layer into the deep intimal area where they ingest the extracellular lipid droplets in the periphery of the lipid core and themselves become converted into foam cells. Since the deep intimal region is distant from the circulating blood, it easily becomes hypoxic (Hulten and Levin 2009). Although survival mechanisms are activated in macrophages trapped in such hypoxic areas (Ramkhelawon et al. 2013), promotion of lipid accumulation, ATP depletion, and immobilization of the macrophage foam cells is likely to ultimately lead to their death. Of note, death of a foam cell is bound to liberation of its cytoplasmic cholesteryl ester droplets into the extracellular space, where the droplets mix with the preexisting extracellular lipid droplets. Moreover, the nonviable macrophages (foam cells) fail to effectively remove the cellular debris derived from the dying or dead foam cells (Thorp et al. 2011), so leaving the formed debris deposited among the lipids of the growing core. A lipid core also containing remnants derived from dead cells is called a "necrotic lipid core".

In summary, regarding cholesterol metabolism in the arterial intima, the superficial and the deep layers appear to be fundamentally different. Mechanistically, in the superficial compartment of the arterial intima, viable macrophages ingest and store LDL cholesterol as cholesteryl esters, and transfer it to HDLs for ultimate return to the circulation, continual hydrolysis of the cytoplasmic cholesteryl esters being the critical factor for the transfer of cholesterol from LDL to HDL. In contrast, in the deep intimal compartment, macrophages only inefficiently remove LDL cholesterol, and, provided a macrophage is converted into a foam cell, it may die and release its cholesteryl ester cargo into the extracellular space, where the cholesteryl esters are not available for removal by HDLs. Thus, in the absence of an efficient transfer of cholesterol from LDL to HDL, the necrotic lipid core grows progressively and may convert an atheroma into a clinically significant atherosclerotic lesion.

4.2   Inhibition of LDL Modification by HDLs

Resident macrophages in the subendothelial space take up oxidized LDLs via scavenger receptors and become macrophage foam cells, the hallmark cell type of atherosclerotic lesions. HDLs have been proven to help to reduce the formation of lipid-laden macrophage foam cells by rendering LDL particles resistant to oxidation. Preexposure of LDL particles to apoA-I or mature HDL particles completely prevented the oxidative modification of LDLs by cultured human aortic endothelial cells and SMCs (Navab et al. 2000). Moreover, the human artery wall cells were no longer able to oxidize LDLs isolated from mice and humans that received a single treatment with human apoA-I. Likewise, HDLs have been shown to possess potent antioxidative activity in chemical in vitro models for oxidative modifications of

LDLs, using, for example, copper ions (Parthasarathy et al. 1990) or the more mild oxidizing agent 2,2′-azobis(2-amidinopropane) dihydrochloride (AAPH) (Kontush et al. 2003). At least part of the ability of HDLs to inhibit LDL oxidation was attributed to the capacity of apoA-I to remove seeding molecules present in freshly isolated LDLs (Navab et al. 2000). Notably, the reduction of lipid hydroperoxides to their corresponding hydroxides by HDLs was associated with the selective oxidation of Met residues in apoA-I to their Met sulfoxide form (Garner et al. 1998), suggesting that Met residues in apoA-I may act as endogenous antioxidants within the HDL particle.

In addition to apoA-I, antioxidant properties of HDLs are conferred by paraoxonase 1 (PON1), an antioxidant enzyme transported in the circulation by HDL particles. The inhibitory effect of HDLs on LDL oxidation by human artery wall cells or copper could be mimicked by incubation of LDLs with purified PON1 (Navab et al. 2000; Liu et al. 2008). After treatment of oxidized phospholipids with PON1, oxygenated polyunsaturated fatty acids were no longer detectable (Watson et al. 1995a), indicating that PON1 was able to destroy these multi-oxygenated biologically active molecules which are abundantly formed during the atherogenic oxidation of LDL particles. Studies in transgenic and knockout mice on the role of PON1 in the pathogenesis of atherosclerosis have supported the protective effects of PON1 on LDL oxidation. The lack of PON1 in the HDL fraction of PON1 knockout mice resulted in an inability of these HDL particles to prevent the formation of oxidized LDLs, and accordingly *pon1*-null mice were more susceptible to atherosclerosis than their wild-type littermates (Shih et al. 1998). Expression of human PON1 in transgenic mice led to the opposite phenotype. On the atherosclerosis-prone *apoE*$^{-/-}$ background, mice transgenic for human PON1 developed significantly less atherosclerotic plaques in the aorta than the control mice. Moreover, in vitro PON1-enriched HDLs from PON1 transgenic/apoE knockout were more potent inhibitors of copper-catalyzed oxidation of LDL particles (Tward et al. 2002).

Another antioxidant enzyme that has been related to the cardioprotective antioxidative function of HDL particles is lipoprotein-associated phospholipase A_2 (Lp-PLA$_2$), also known as platelet-activating factor acetylhydrolase (PAF-AH). In human plasma a small proportion of the circulating Lp-PLA$_2$ enzyme activity is found in HDL particles (Tellis and Tselepis 2009). Inhibition of the enzymatic activity of Lp-PLA$_2$ rendered HDLs unable to prevent monocyte binding induced by oxidized LDLs in endothelial cells, and supplementation of the dysfunctional HDL particles with purified Lp-PLA$_2$ restored the normal protective function (Watson et al. 1995b). In mice, adenovirus-mediated expression of human Lp-PLA$_2$ leads to a reduction in the autoantibody titers against oxidized LDLs and malondialdehyde (MDA)-modified LDLs (Quarck et al. 2001; Noto et al. 2003). Despite the fact that the antioxidative potential of HDL-bound Lp-PLA$_2$ has been well documented, several lines of evidence suggest that the presence of Lp-PLA$_2$ in LDL particles may actually stimulate atherogenesis. The pro-atherogenic actions of Lp-PLA$_2$ in LDLs are thought to arise from oxidized free fatty acids and lysophospholipids, two inflammatory mediators which are released after the hydrolysis of oxidized phospholipids by Lp-PLA$_2$ (Tellis and Tselepis 2009). LDLs are the preferred lipoprotein carriers for Lp-PLA$_2$ in the circulation, which would explain why previous prospective cohort studies paradoxically linked high plasma Lp-PLA$_2$ levels with adverse cardiovascular outcomes (Brilakis et al. 2005; Koenig et al. 2004; Sabatine et al. 2007). In support of the hypothesis that Lp-PLA$_2$ is atheroprotective only when incorporated in HDL particles, Rallidis et al. found that elevated total plasma Lp-PLA$_2$ mass and activity increased the risk for future cardiac death in patients with stable coronary artery disease, while Lp-PLA$_2$ mass and activity in apoB-depleted plasma showed a significant association with future cardiac death in the opposite direction (Rallidis et al. 2012).

The protective effect of HDLs on LDL oxidation has been tentatively ascribed to the antioxidant activities of lecithin-cholesterol acyltransferase (LCAT). Similar to HDLs, purified LCAT has been shown to diminish the formation of both lipid hydroperoxides and conjugated dienes in copper-oxidized LDLs (Vohl et al. 1999). Moreover, adenovirus-mediated transfer of the human LCAT gene into mice with combined leptin and LDL receptor deficiency decreased circulating levels of autoantibodies binding to MDA-LDL by 40 % and accordingly limited the accumulation of oxidized LDLs in the arterial wall (Mertens et al. 2003).

4.3 Macrophage Cholesterol Efflux and Reverse Cholesterol Transport

The uptake of modified LDLs in arterial macrophages leads to the cellular accumulation of cholesterol and oxidized lipids. This process causes the macrophages to transform into lipid-laden macrophage foam cells which can no longer escape the arterial wall (Potteaux et al. 2011). The excessive cellular cholesterol in macrophage foam cells can be efficiently removed by the potent cholesterol acceptors lipid-free apoA-I and HDLs to be transported back to the liver by a process termed reverse cholesterol transport.

4.3.1 Transendothelial HDL Transport

In order to reach these resident macrophage foam cells in the innermost layer of the arterial wall, apoA-I and HDLs first need to passage across the endothelial cell layer. There is cell culture evidence that lipid-free apoA-I binds to specific saturable high-affinity binding sites on aortic endothelial cells, which is followed by internalization, transcytosis, and delivery of lipidated apoA-I to the basolateral side (Rohrer et al. 2006). A similar phenomenon was observed when cultivated aortic endothelial cells were treated with mature HDLs (Rohrer et al. 2009). However, the mechanisms involved in regulating this endothelial transport of apoA-I and HDLs are not fully understood. The ABCA1 transporter is thought to play major role in the apical-to-basolateral transport of apoA-I in aortic endothelial cells (Cavelier et al. 2006). Induction of ABCA1 expression by a combination of oxysterol and 9-cis-retinoic acid increased the binding and internalization of apoA-I in endothelial cells, whereas knockdown of endothelial ABCA1 by RNA interference reduced the cellular uptake and transcytosis of apoA-I (Cavelier et al. 2006). Conversely, siRNA-mediated gene silencing experiments revealed that the cell surface binding and transport of HDLs through aortic endothelial cells are highly dependent on SR-BI and ABCG1, but not on ABCA1 (Rohrer et al. 2009). Subsequent research has demonstrated expression of the ectopic β-chain of F0F1 ATPase (β-ATPase) on the endothelial cell surface (Cavelier et al. 2012). The results of this study further suggest that the binding of lipid-free apoA-I to endothelial cell surface β-ATPase facilitates the uptake and transport of lipidated apoA-I and mature HDLs by enhancing β-ATPase-mediated hydrolysis of extracellular ATP into ADP and inducing consecutive activation of the purinergic P2Y12 receptor (Cavelier et al. 2012). More recently, transendothelial transport of HDL particles was found to be modulated by endothelial lipase both through its enzymatic lipolytic activity and its ability to bridge the binding of lipoproteins to the endothelial surface (Robert et al. 2013). Initiation of an atherosclerosis-related inflammatory phenotype after stimulation of vascular endothelial cells with interleukin-6 significantly increased the binding, cell association, and transport of HDLs, which was linked mechanistically to the increase in endothelial lipase expression in response to interleukin-6 (Robert et al. 2013).

4.3.2 Cholesterol Efflux from Macrophages

Once apoA-I and HDLs have reached macrophage foam cells in the atheromatous vessel wall, macrophage cholesterol efflux can be elicited via several different pathways. It is well documented that lipid-free apoA-I and pre-beta HDL particles are able to remove cholesterol and phospholipids from macrophage foam cells via the ABCA1 transporter (Wang et al. 2000; Wang and Tall 2003). It has been postulated that apoA-I-dependent cholesterol efflux from cells is mediated either by a direct protein–protein interaction between apoA-I and ABCA1 or indirectly by ABCA1-induced changes in the membrane cholesterol distribution (Wang and Tall 2003). The ability of ABCA1 to mobilize cellular lipids to apoA-I is essential for the initial lipidation of apoA-I. The ABCG1 transporter is responsible for a major part of the macrophage cholesterol efflux towards mature HDL particles (Wang et al. 2004). Since cellular removal of cholesterol via ABCG1 was not accompanied by specific binding of HDLs to the plasma membrane (Wang et al. 2004), the molecular basis of efflux of cholesterol mediated by ABCG1, like ABCA1, is not fully understood and warrants further investigation. Another cell surface receptor involved in macrophage cholesterol efflux elicited by mature HDLs is SR-BI. Binding of HDL to macrophage SR-BI facilitates a bidirectional exchange of unesterified cholesterol and other lipids between the cell membrane and HDL acceptor particles according to the cholesterol concentration gradient (de La Llera-Moya et al. 2001). This cholesterol concentration gradient between cells and HDL not only allows cholesterol transfer via SR-BI but also passive diffusion of cholesterol molecules to nearby HDL particles by receptor-independent processes (von Eckardstein et al. 2001; Yancey et al. 2003). The HDL-associated enzyme LCAT catalyzes the esterification of cholesterol in HDL particles (Calabresi and Franceschini 2010). The generated cholesteryl esters leave the particle surface and form the core of the maturing HDL particle; hence, a local gradient of unesterified cholesterol is created that maintains a continuous flow of unesterified cholesterol from macrophages towards HDLs (Calabresi and Franceschini 2010).

4.3.3 Exit from the Arterial Wall

There is little reported data on the route used by HDLs to leave the arterial intima. Recent lines of evidence support a role for the lymphatic vasculature in the transport of HDLs from the interstitial space back to the bloodstream. Following injection of fluorescently labeled HDLs in the mouse footpad, HDLs were subsequently found in the draining lymph node and its afferent and efferent lymphatic vessels (Lim et al. 2013). Moreover, the movement of cholesterol from macrophages in the footpad to the lymph and plasma was substantially reduced in mice with surgical interruption of the afferent lymphatic vessels or in Chy mutant mice that lack dermal lymphatic capillaries (Lim et al. 2013; Martel et al. 2013). The importance of the lymphatic system in the reentry of cholesterol originating from macrophages via HDLs in the circulation was further confirmed by additional experiments showing that the appearance of macrophage-derived cholesterol in the plasma was impaired after implantation of [^{3}H]-cholesterol-labeled macrophages in the tail skin of apoA-I transgenic mice with microsurgical ablation of the major lymphatic conduits in the tail (Martel et al. 2013). To provide more direct proof for the involvement of the lymphatic vasculature in the transport of cholesterol from the atherosclerotic plaque, aortic segments with advanced atherosclerotic lesions were loaded with D6-cholesterol and transplanted in apoE knockout mice (Martel et al. 2013). Inhibition of regrowth of a functional lymphatic vasculature in the aortic donor wall by an antibody blocking the function of vascular endothelial growth factor receptor 3 markedly suppressed the apoE-induced removal of D6-cholesterol from the transplanted aorta (Martel et al. 2013), hence emphasizing the pivotal role of the lymphatic system for the egress of cholesterol from the aortic wall. The entry of HDL particles into lymphatic vessels appears to be primarily dependent on SR-BI. Experiments in cultured cells demonstrate that the internalization and transcytosis of HDL particles by lymphatic endothelial cells are diminished by an SR-BI blocking antibody or a selective SR-BI inhibitor (Lim et al. 2013). In agreement, lymphatic transport of HDLs is compromised in SR-BI-null mice and in mice treated with an antibody against SR-BI (Lim et al. 2013). Although not formally tested, it is also possible that, in order to reach the circulation again, HDL particles undergo transcytosis through the luminal endothelial cell barrier or transfer cholesterol to endothelial cells for delivery to circulating HDLs. The importance of a well-regulated homeostasis between entry and exit of HDL particles into and from the arterial wall, respectively, is indicated by the enrichment of modified and dysfunctional HDLs and apoA-I in atherosclerotic lesions (DiDonato et al. 2013, 2014; Huang et al. 2014).

4.3.4 Delivery of Cholesterol to the Liver and Intestine

After having left the arterial wall and entered the circulation, the HDL particles transport the cholesterol either directly or indirectly via apoB-containing lipoproteins to the liver for either secretion into bile or for de novo assembly of lipoproteins. The first direct way by which HDL can deliver cholesterol to the liver is via SR-BI, a cell surface receptor that binds HDLs and mediates the selective uptake of HDL-associated cholesteryl esters and the subsequent resecretion of cholesteryl ester-poor HDLs (Acton et al. 1996). Secondly, HDL particles can be taken up from the circulation by the liver via HDL holoparticle endocytosis (i.e., uptake of the whole HDL particle). The binding of apoA-I to the ectopic β-chain of F0F1 ATPase on hepatocytes triggers extracellular production of ADP, which in turn acts on the P2Y13 receptor to stimulate HDL holoparticle endocytosis (Martinez et al. 2003; Jacquet et al. 2005). In humans, the major fraction of the cholesteryl esters in HDL particles are shuttled via the action of CETP to apoB-containing lipoproteins (Charles and Kane 2012). These cholesteryl esters originating from HDLs are internalized as part of apoB-containing lipoproteins by hepatic LDL receptors. Following hepatic uptake, HDL-derived cholesterol is, at least in part, targeted for fecal excretion via the biliary route (Lewis and Rader 2005; Nijstad et al. 2011). Hepatic cholesterol can either be secreted directly in the free form into the bile by the ABCG5/G8 heterodimer or alternatively be converted by cholesterol 7α-hydroxylase (CYP7A1) and other enzymes into bile acids for hepatic excretion via the bile salt export pump (BSEP) (Dikkers and Tietge 2010). Secretion into bile enables cholesterol to move from the hepatocyte to the intestinal lumen for final excretion via the stool. This major HDL-driven pathway to eliminate atherogenic cholesterol from the body is also known as reverse cholesterol transport (Annema and Tietge 2012). Although reverse cholesterol transport from plaque macrophages to feces only represents a very small proportion of the total reverse cholesterol transport flux from peripheral tissues, it is key component of the reverse transport potentially capable of initiating and maintaining the regression of atherosclerosis (Lee-Rueckert et al. 2013).

5 Atherothrombosis

The most important mechanism of acute coronary syndromes, i.e., unstable angina, acute myocardial infarction, and sudden cardiac death, is rupture or erosion of a coronary atheroma (Libby 2013b). The risk of atheromatous rupture or erosion appears to critically depend on both the cellular and extracellular compositions of the atheroma. The superficial layer of an atheroma separating a necrotic lipid core from the circulating blood is called the "fibrous cap". Although a cap usually contains foam cells, its critical components are VSMCs, which synthesize and secrete collagen and other components of the extracellular matrix. It is current understanding that the cap is thinning as the necrotic lipid core is growing. Whether the thinning is a response to the enlargement of the necrotic core, or whether thinning is an independent process and just allows the core to expand, has not been established.

As expected, a lesion with a thick fibrous cap and a small lipid core is stable, whereas a lesion with a thin fibrous cap and a large lipid core is unstable and prone to rupture. Indeed, it is now well recognized that acute coronary syndromes most commonly result from rupture of a thin-cap fibroatheroma (TCFA), which is characterized by a large necrotic core with an overlying thin fibrous cap measuring less than 65 μm (Sakakura et al. 2013; Fujii et al. 2013).

Provided a stable lesion with a thick fibrous cap causes significant stenosis of the arterial lumen, and turbulent flow conditions are created, some endothelial cells may detach, that is, the plaque surface becomes eroded (Libby 1995; Farb et al. 1996). Apoptosis of endothelial cells also leads to endothelial denudation, i.e., to plaque erosion when, for example, subendothelial mast cells degrade the basement membrane of endothelial cells (Durand et al. 2004; Mayranpaa et al. 2006). The exposed subendothelial structures are prothrombotic and a platelet-rich thrombus is formed locally. Since the activated platelets secrete platelet-derived growth factor which stimulates SMCs to divide and generate extracellular matrix, a neointima is formed at the site of erosion. This is one mechanism by which the fibrotic cap of a stenosing plaque becomes thicker and the plaque grows into the lumen and so aggravates the local stenosis.

When an unstable TCFA ruptures, the flowing blood becomes exposed to the necrotic lipid core. The various components of the core, notably tissue factor released by macrophages, apoptotic bodies, remnants of dead cells, and extracellularly located lipids, all are contributing to the strong thrombogenic response usually associated with a plaque rupture (Fuster et al. 2005; Corti et al. 2003). Thus, the platelet-rich arterial thrombus growing at the site of plaque rupture easily becomes occlusive, even if the plaque only had caused a mild stenosis which was hemodynamically nonsignificant.

In view of the importance of the thickness of collagen-rich matrix of the fibrous cap in determining plaque stability and instability, it is essential to understand the factors that may regulate its thickness. Regarding such regulation, two related observations made in advanced human atherosclerotic plaques are of prime importance. First, irrespective of the structure of an atherosclerotic plaque, the actual site of atheromatous ulceration with ensuing local formation of a platelet-rich thrombus is characterized by an inflammatory process (van der Wal et al. 1994). Thus, atheromatous coronary and carotid ulcerations have invariably shown the presence of inflammatory cell infiltrates, composed of macrophages, T cells, and mast cells (Kovanen et al. 1995; Bui et al. 2009). When compared with the evolving atherosclerotic plaques, which contain inflammatory cell infiltrates, composed of various subsets of macrophages and T cells and also of mast cells and neutrophils (Woollard 2013), the vulnerable plaques appear to be equally, if not even more immunologically, active. The second critical aspect of a vulnerable atherosclerotic plaque is the fact that SMCs are dying in their fibrous caps (Bennett 1999).

Fundamentally, the various activities of SMCs tend to thicken the cap, while those of the inflammatory cells tend to oppose them. The net production of matrix components depends on the number of the matrix-producing SMCs and the ability of the individual cells to produce such components. The number of SMCs depends on the balance of cell replication and cell death. In human atherosclerotic plaques, the infiltrating lymphocytes are of the Th1 subpopulation of CD4+ cells and the pro-inflammatory Th1-derived cytokines, notably interferon-gamma, dominate (Hansson 2005). This cytokine is able to inhibit the capacity of SMCs to produce collagen (Libby 1995) and together with TNFα and interleukin-1β (IL-1β) contribute to their apoptotic death. Thus, activation of Th1 cells, together with IL-1-β-producing macrophages and TNFα-producing mast cells within the cap of an atheroma, could lower the local rate of collagen production by lowering the number of SMCs and by inhibiting the rate of synthesis in those SMCs that survive (Kaartinen et al. 1996). Importantly, the fibrous caps often contain cholesterol crystals, which have the potential to activate the NLRP3 inflammasome in macrophages and so trigger release of IL-1β from the macrophages (Rajamaki et al. 2010; Duewell et al. 2010). Since it is unlikely that HDLs are capable of removing crystalline cholesterol from the intima, the crystals, once formed, may permanently stimulate cap macrophages to secrete IL-1β and so contribute to the ongoing cap thinning.

The various components of the extracellular matrix are degraded by members of the superfamily of matrix metalloproteinases (MMPs) primarily produced by the macrophages of the caps (Newby 2005). For the MMPs to become active, they need to be proteolytically activated. Among the proteolytic enzymes capable of activating plaque MMPs are plasmin, chymase, and tryptase, the two latter ones being neutral proteases secreted by activated mast cells in atherosclerotic lesions (Johnson et al. 1998). Once activated, interstitial collagenase, MMP-1, can degrade the otherwise protease-resistant fibrillar collagen fibers into fragments, and another member of the family, MMP-9, can break collagen fragments further. Importantly, stromelysin, MMP-3, has a broad spectrum and can degrade other components of the extracellular matrix as well, including proteoglycans and elastin (Libby 2013a). Cathepsins are another group of enzymes capable of effectively degrading the various components of the fibrous cap (Fonovic and Turk 2014). Thus, the presence of inflammatory cells in the caps of advanced plaques tends to render the plaque unstable and prone to rupture by both decreasing the production and increasing the degradation of the extracellular matrix of the fibrous cap.

5.1 Effects of HDLs on Smooth Muscle Cells

VSCMs can indirectly influence the development of atherosclerosis by the secretion of growth factors and vasoactive agents, and the protective effects of HDLs on atherosclerotic lesion formation might, at least partly, be related to their impact on the secretory function of VSMCs. Exposure of aortic SMCs to mildly oxidized LDLs enhanced the production and release of platelet-derived growth factor and basic fibroblast growth factor, while addition of HDLs abolished the stimulatory effect of oxidized LDLs on growth factor secretion by these VSMCs (Cucina et al. 1998, 2006). Conversely, treatment of rabbit SMCs in culture with HDLs potentiated the synthesis of the vasodilator prostacyclin via a mechanism involving MAP kinase

kinase signaling and increased cyclooxygenase-2 expression (Vinals et al. 1997, 1999). VSMCs are also considered to protect the atherosclerotic plaque against destabilization and rupture. In this respect, HDLs regulate both the proliferation and migration of VSMCs. Nofer and colleagues showed that HDLs accelerate proliferation of VSMCs by driving the cell cycle forward in the G1 phase via cyclin D1-induced phosphorylation of pRb (Nofer et al. 2001a). Elevated cyclin D1 expression in VSMCs in response to treatment with HDLs in turn relies on activation of the Raf/MEK/ERK signal transduction pathway (Nofer et al. 2001a). On the other hand, the platelet-derived growth factor-mediated migration of VCMCs is limited by HDLs through the interaction of HDL-bound S1P with the S1P2 receptor (Tamama et al. 2005). Finally, HDLs were found to suppress inflammatory responses in VSMCs. Oxidized LDLs, via formation of ROS, induced NF-κβ activation in cultured VSMCs, while their pretreatment with HDLs inhibited the intracellular rise in ROS and the subsequent upregulation of NF-κβ elicited by the oxidized LDLs (Robbesyn et al. 2003).

5.2 Antithrombotic Effects of HDLs

5.2.1 HDLs and Platelets

The majority of acute coronary syndromes are caused by the rupture of an atherosclerotic plaque in a coronary artery and subsequent formation of an occlusive coronary thrombus. There appears to be a relationship between plasma HDLs and the thrombogenic potential of blood, as evidenced by the fact that the total thrombus area in an ex vivo flow chamber model for coronary thrombosis was lower in human subjects with higher HDL cholesterol levels (Naqvi et al. 1999). HDLs have several favorable effects on platelets that might explain its interference with thrombus formation in the coronary vasculature. First, it has been reported that HDLs can block the activation and aggregation of platelets. In a small clinical trail, 13 patients with type 2 diabetes mellitus were randomized to a single infusion of rHDLs (CSL-111) or placebo (Calkin et al. 2009). Study results established that a 40 % increase in HDL cholesterol levels in patients with type 2 diabetes mellitus was associated with a 50–75 % reduction in platelet aggregation (Calkin et al. 2009). A similar decrease in the aggregation of platelets was observed when isolated human platelets were incubated with rHDLs or native HDLs in vitro (Calkin et al. 2009; Nofer et al. 1998; Badrnya et al. 2013). Native HDLs and rHDLs not only modulate platelet aggregation but also prevent platelet degranulation and spreading as well as adhesion of platelets to fibrinogen (Calkin et al. 2009; Nofer et al. 1998; Badrnya et al. 2013). In addition, rHDLs limited the incorporation of platelets in growing thrombi under flow conditions in vitro (Calkin et al. 2009). The ability of HDLs to block platelet aggregation was proposed to be due to a reduction in the availability of the second messengers 1,2-diacylglycerol and inositol 1,4,5-trisphosphate, a decreased mobilization of intracellular calcium, activation of protein kinase C, and elevated Na^+/H^+ exchange (Nofer et al. 1996; 1998). The role of SR-BI for the anti-aggregatory effect of HDLs is controversial.

One study reported that inhibition of platelet reactivity by HDLs results from cholesterol depletion in lipid rafts but that this is independent of ABCG1 and SR-BI (Calkin et al. 2009). In contrast, Brodde et al. found that SR-BI is the receptor responsible for binding of HDL to platelets and that HDLs are no longer able to attenuate thrombin-induced adhesion of platelets to fibrinogen when using platelets isolated from SR-BI knockout mice (Brodde et al. 2011). The inhibitory activity of HDLs on platelets is mimicked by soybean phosphocholine (Calkin et al. 2009) and phosphatidylserine-containing liposomes (Brodde et al. 2011), underlining the importance of the phospholipid component in this antithrombotic function of HDLs. As a final point, HDLs may regulate the platelet coagulant activity by reducing the synthesis of platelet-activating factor in endothelial cells (Sugatani et al. 1996).

Another step of the arterial atherothrombotic process in which HDLs may interfere is the production of platelets from bone marrow megakaryocyte progenitor cells. In vitro experiments have revealed that rHDLs promote the removal of cholesterol from megakaryocyte progenitor cells in an ABCG4-dependent fashion, thereby decreasing cellular proliferation (Murphy et al. 2013). Consistent with these observations, intravenous administration of rHDLs resulted in a decrease in the amount of bone marrow megakaryocyte progenitor cells and concomitantly reduced circulating platelets in $Ldlr^{-/-}$ mice transplanted with wild-type bone marrow, while rHDLs lost their effects on bone marrow megakaryocyte progenitor cells and platelet numbers when injected into $Ldlr^{-/-}$ mice that were transplanted with ABCG4-deficient bone marrow (Murphy et al. 2013). A small placebo-controlled clinical study demonstrated that a single injection of rHDLs significantly reduced platelet counts in patients with peripheral vascular disease (Murphy et al. 2013). The importance of deregulation of cholesterol homeostasis in megakaryocyte progenitor cells for atherothrombosis was supported by evidence that selective disruption of ABCG4 in bone marrow cells of LDL receptor knockout mice accelerated the progression of atherosclerotic lesion formation and was associated with marked thrombocytosis and an increased tendency to develop arterial thrombosis (Murphy et al. 2013).

5.2.2 HDLs and Coagulation

Various publications have described an impact of HDLs on the extrinsic coagulation pathway that might result in a decreased tendency for thrombosis. Inhibition of the catalytic activity of the tissue factor–factor VIIA complex by HDLs has been shown to impede the activation of the procoagulant factor X (Carson 1981). Besides direct inactivation of tissue factor and factor Va, HDLs are able to modulate the tissue factor-activated coagulation cascade by downregulating tissue factor expression in thrombin-stimulated endothelial cells through suppression of RhoA and induction of PI3K (Viswambharan et al. 2004). In addition, incorporation of anionic phospholipids into apoA-I-containing rHDL particles resulted in the loss of their procoagulant properties, revealing that, by scavenging anionic phospholipids, apoA-I may control blood coagulation (Oslakovic et al. 2009). HDLs have also been shown to promote the ability of the anticoagulant activated protein C and its cofactor protein S to cause inactivation of factor Va through Arg^{306} cleavage, and in normal healthy subjects, plasma levels of apoA-I were positively correlated with the anticoagulant potency of activated protein C and protein S in a prothrombin-time clotting assay (Griffin et al. 1999). Of note, these findings were later questioned by another study revealing that the previously observed capacity of the HDL fraction to boost the activated protein C system was not an actual property of HDLs, but instead was due to co-isolation of anionic phospholipid membranes with the HDL fraction during ultracentrifugation (Oslakovic et al. 2010).

6 Myocardial Injury

The main pathophysiological manifestation of coronary heart disease is myocardial damage due to ischemia–reperfusion. It has been postulated that HDLs can provide protection against myocardial injury following ischemia–reperfusion. In an in vivo murine ischemia–reperfusion model, it was found that the infarcted area was decreased when mice were treated 30 min before ischemia–reperfusion with either human HDLs or S1P (Theilmeier et al. 2006). In this regard, HDL or S1P injection mitigated the inflammatory response to ischemia–reperfusion as manifested by a lower accumulation of leukocytes in the infarcted area (Theilmeier et al. 2006). Furthermore, cardiomyocytes were better able to resist ischemia–reperfusion-induced apoptosis in the presence of HDLs or S1P (Theilmeier et al. 2006), potentially due to a higher tolerance of HDL-treated cardiomyocytes to hypoxia-induced opening of the mitochondrial permeability transition pore (Frias et al. 2013). S1P3 was identified as the receptor responsible for the beneficial effect of HDLs and S1P on reperfusion injury (Theilmeier et al. 2006). The finding that HDLs ameliorate myocardial ischemia–reperfusion damage has subsequently been confirmed and extended by others. Ex vivo exposure of isolated hearts from wild-type mice to HDLs provided dose-dependent cardioprotection from reperfusion injury (Frias et al. 2013). On the other hand, HDL-mediated protection of isolated hearts or cardiomyocytes against hypoxia was lost in mice deficient in tumor necrosis factor and in mice with cardiomyocyte-restricted knockout of signal transducer and activator of transcription 3 (STAT3) (Frias et al. 2013), suggesting a role for the survivor activating factor enhancement pathway which is initiated by tumor necrosis factor and involves activation of STAT3. In another study, HDL cholesterol levels were raised in LDL receptor-deficient mice using adenovirus-mediated gene transfer of human apoA-I (Gordts et al. 2013). After induction of myocardial infarction by permanent ligation of the left anterior descending coronary artery, mice that were injected with an adenovirus expressing human apoA-I exhibited a higher survival rate, reduced infarct expansion, attenuated dilatation of the left ventricle, as well as improved systolic and diastolic cardiac function when compared with controls (Gordts et al. 2013).

Conclusions

HDL particles exert many effects which may help either to protect arteries from the development, progression, and complication of atherosclerosis or even facilitate repair and regression of lesions. Nevertheless, HDL has not yet been successfully exploited for preventive or curative therapy of cardiovascular diseases. One potential reason for this shortfall is the structural and functional complexity of HDL particles which is further increased in several inflammatory conditions including CAD itself and which is not discerned by the biomarker HDL cholesterol. Moreover, the relative importance of the many physiological and pathological activities of normal and dysfunctional HDL, respectively, for the pathogenesis of atherosclerosis is unknown. The elucidation of structure–function relationships of HDL-associated molecules is essential to exploit HDL for the development of anti-atherogenic drugs as well as of diagnostic biomarkers for the identification, personalized treatment stratification, and monitoring of patients at increased cardiovascular risk. In addition to the metabolism of HDL or its anti-atherogenic molecules, also the cellular pathways positively regulated by HDL are interesting targets for anti-atherosclerotic therapies.

Open Access This chapter is distributed under the terms of the Creative Commons Attribution Noncommercial License, which permits any noncommercial use, distribution, and reproduction in any medium, provided the original author(s) and source are credited.

References

Acton S, Rigotti A, Landschulz KT et al (1996) Identification of scavenger receptor SR-BI as a high density lipoprotein receptor. Science 271(5248):518–520

Angeloni E, Paneni F, Landmesser U et al (2013) Lack of protective role of HDL-C in patients with coronary artery disease undergoing elective coronary artery bypass grafting. Eur Heart J 34(46):3557–3562

Annema W, Tietge UJ (2012) Regulation of reverse cholesterol transport - a comprehensive appraisal of available animal studies. Nutr Metab (Lond) 9(1):25

Annema W, von Eckardstein A (2013) High-density lipoproteins. Multifunctional but vulnerable protections from atherosclerosis. Circ J 77(10):2432–2448

Argraves KM, Gazzolo PJ, Groh EM et al (2008) High density lipoprotein-associated sphingosine 1-phosphate promotes endothelial barrier function. J Biol Chem 283(36):25074–25081

Ashby DT, Rye KA, Clay MA et al (1998) Factors influencing the ability of HDL to inhibit expression of vascular cell adhesion molecule-1 in endothelial cells. Arterioscler Thromb Vasc Biol 18(9):1450–1455

Badrnya S, Assinger A, Volf I (2013) Native high density lipoproteins (HDL) interfere with platelet activation induced by oxidized low density lipoproteins (OxLDL). Int J Mol Sci 14(5):10107–10121

Baigent C, Keech A, Kearney PM et al (2005) Efficacy and safety of cholesterol-lowering treatment: prospective meta-analysis of data from 90,056 participants in 14 randomised trials of statins. Lancet 366(9493):1267–1278

Baker PW, Rye KA, Gamble JR et al (1999) Ability of reconstituted high density lipoproteins to inhibit cytokine-induced expression of vascular cell adhesion molecule-1 in human umbilical vein endothelial cells. J Lipid Res 40(2):345–353

Barter P, Gotto AM, LaRosa JC et al (2007a) HDL cholesterol, very low levels of LDL cholesterol, and cardiovascular events. N Engl J Med 357(13):1301–1310

Barter PJ, Caulfield M, Eriksson M et al (2007b) Effects of torcetrapib in patients at high risk for coronary events. N Engl J Med 357(21):2109–2122

Bennett MR (1999) Apoptosis of vascular smooth muscle cells in vascular remodelling and atherosclerotic plaque rupture. Cardiovasc Res 41(2):361–368

Bisoendial RJ, Hovingh GK, Levels JH et al (2003) Restoration of endothelial function by increasing high-density lipoprotein in subjects with isolated low high-density lipoprotein. Circulation 107(23):2944–2948

Boden WE, Probstfield JL, Anderson T et al (2011) Niacin in patients with low HDL cholesterol levels receiving intensive statin therapy. N Engl J Med 365(24):2255–2267

Brilakis ES, McConnell JP, Lennon RJ et al (2005) Association of lipoprotein-associated phospholipase A2 levels with coronary artery disease risk factors, angiographic coronary artery disease, and major adverse events at follow-up. Eur Heart J 26(2):137–144

Brodde MF, Korporaal SJ, Herminghaus G et al (2011) Native high-density lipoproteins inhibit platelet activation via scavenger receptor BI: role of negatively charged phospholipids. Atherosclerosis 215(2):374–382

Brown MS, Goldstein JL (1983) Lipoprotein metabolism in the macrophage: implications for cholesterol deposition in atherosclerosis. Annu Rev Biochem 52:223–261

Bui QT, Prempeh M, Wilensky RL (2009) Atherosclerotic plaque development. Int J Biochem Cell Biol 41(11):2109–2113

Bursill CA, Castro ML, Beattie DT et al (2010) High-density lipoproteins suppress chemokines and chemokine receptors in vitro and in vivo. Arterioscler Thromb Vasc Biol 30(9):1773–1778

Calabresi L, Franceschini G (2010) Lecithin:cholesterol acyltransferase, high-density lipoproteins, and atheroprotection in humans. Trends Cardiovasc Med 20(2):50–53

Calkin AC, Drew BG, Ono A et al (2009) Reconstituted high-density lipoprotein attenuates platelet function in individuals with type 2 diabetes mellitus by promoting cholesterol efflux. Circulation 120(21):2095–2104

Camejo G, Hurt-Camejo E, Wiklund O et al (1998) Association of apo B lipoproteins with arterial proteoglycans: pathological significance and molecular basis. Atherosclerosis 139(2):205–222

Carson SD (1981) Plasma high density lipoproteins inhibit the activation of coagulation factor X by factor VIIa and tissue factor. FEBS Lett 132(1):37–40

Cavelier C, Rohrer L, von Eckardstein A (2006) ATP-Binding cassette transporter A1 modulates apolipoprotein A-I transcytosis through aortic endothelial cells. Circ Res 99(10):1060–1066

Cavelier C, Ohnsorg PM, Rohrer L et al (2012) The beta-chain of cell surface F(0)F(1) ATPase modulates apoA-I and HDL transcytosis through aortic endothelial cells. Arterioscler Thromb Vasc Biol 32(1):131–139

Charles MA, Kane JP (2012) New molecular insights into CETP structure and function: a review. J Lipid Res 53(8):1451–1458

Christoffersen C, Obinata H, Kumaraswamy SB et al (2011) Endothelium-protective sphingosine-1-phosphate provided by HDL-associated apolipoprotein M. Proc Natl Acad Sci USA 108 (23):9613–9618

Clay MA, Pyle DH, Rye KA et al (2001) Time sequence of the inhibition of endothelial adhesion molecule expression by reconstituted high density lipoproteins. Atherosclerosis 157(1):23–29

Cockerill GW, Rye KA, Gamble JR et al (1995) High-density lipoproteins inhibit cytokine-induced expression of endothelial cell adhesion molecules. Arterioscler Thromb Vasc Biol 15(11):1987–1994

Cockerill GW, Huehns TY, Weerasinghe A et al (2001) Elevation of plasma high-density lipoprotein concentration reduces interleukin-1-induced expression of E-selectin in an in vivo model of acute inflammation. Circulation 103(1):108–112

Colin S, Fanchon M, Belloy L et al (2014) HDL does not influence the polarization of human monocytes toward an alternative phenotype. Int J Cardiol 172(1):179–184

Corsetti JP, Zareba W, Moss AJ et al (2006) Elevated HDL is a risk factor for recurrent coronary events in a subgroup of non-diabetic postinfarction patients with hypercholesterolemia and inflammation. Atherosclerosis 187(1):191–197

Corti R, Fuster V, Badimon JJ (2003) Pathogenetic concepts of acute coronary syndromes. J Am Coll Cardiol 41(4 Suppl S):7S–14S

Cucina A, Pagliei S, Borrelli V et al (1998) Oxidised LDL (OxLDL) induces production of platelet derived growth factor AA (PDGF AA) from aortic smooth muscle cells. Eur J Vasc Endovasc Surg 16(3):197–202

Cucina A, Scavo MP, Muzzioli L et al (2006) High density lipoproteins downregulate basic fibroblast growth factor production and release in minimally oxidated-LDL treated smooth muscle cells. Atherosclerosis 189(2):303–309

de La Llera-Moya M, Connelly MA, Drazul D et al (2001) Scavenger receptor class B type I affects cholesterol homeostasis by magnifying cholesterol flux between cells and HDL. J Lipid Res 42(12):1969–1978

De Nardo D, Labzin LI, Kono H et al (2014) High-density lipoprotein mediates anti-inflammatory reprogramming of macrophages via the transcriptional regulator ATF3. Nat Immunol 15(2): 152–160

de Souza JA, Vindis C, Negre-Salvayre A et al (2010) Small, dense HDL 3 particles attenuate apoptosis in endothelial cells: pivotal role of apolipoprotein A-I. J Cell Mol Med 14(3): 608–620

Di Angelantonio E, Sarwar N, Perry P et al (2009) Major lipids, apolipoproteins, and risk of vascular disease. JAMA 302(18):1993–2000

DiDonato JA, Huang Y, Aulak KS et al (2013) Function and distribution of apolipoprotein A1 in the artery wall are markedly distinct from those in plasma. Circulation 128(15):1644–1655

DiDonato JA, Aulak K, Huang Y et al (2014) Site-specific nitration of apolipoprotein A-I at tyrosine 166 is both abundant within human atherosclerotic plaque and dysfunctional. J Biol Chem 289(15):10276–10292

Dikkers A, Tietge UJ (2010) Biliary cholesterol secretion: more than a simple ABC. World J Gastroenterol 16(47):5936–5945

Duewell P, Kono H, Rayner KJ et al (2010) NLRP3 inflammasomes are required for atherogenesis and activated by cholesterol crystals. Nature 464(7293):1357–1361

Durand E, Scoazec A, Lafont A et al (2004) In vivo induction of endothelial apoptosis leads to vessel thrombosis and endothelial denudation: a clue to the understanding of the mechanisms of thrombotic plaque erosion. Circulation 109(21):2503–2506

Eliska O, Eliskova M, Miller AJ (2006) The absence of lymphatics in normal and atherosclerotic coronary arteries in man: a morphologic study. Lymphology 39(2):76–83

Esterbauer H, Puhl H, Dieber-Rotheneder M et al (1991) Effect of antioxidants on oxidative modification of LDL. Ann Med 23(5):573–581

Farb A, Burke AP, Tang AL et al (1996) Coronary plaque erosion without rupture into a lipid core. A frequent cause of coronary thrombosis in sudden coronary death. Circulation 93(7): 1354–1363

Fonovic M, Turk B (2014) Cysteine cathepsins and extracellular matrix degradation. Biochim Biophys Acta 1840(8):2560–2570

Freeman SR, Jin X, Anzinger JJ et al (2014) ABCG1-mediated generation of extracellular cholesterol microdomains. J Lipid Res 55(1):115–127

Frias MA, Pedretti S, Hacking D et al (2013) HDL protects against ischemia reperfusion injury by preserving mitochondrial integrity. Atherosclerosis 228(1):110–116

Frikke-Schmidt R, Nordestgaard BG, Stene MC et al (2008) Association of loss-of-function mutations in the ABCA1 gene with high-density lipoprotein cholesterol levels and risk of ischemic heart disease. JAMA 299(21):2524–2532

Fujii K, Hao H, Ohyanagi M et al (2013) Intracoronary imaging for detecting vulnerable plaque. Circ J 77(3):588–595

Fuster V, Moreno PR, Fayad ZA et al (2005) Atherothrombosis and high-risk plaque: part I: evolving concepts. J Am Coll Cardiol 46(6):937–954

Garcia JG, Liu F, Verin AD et al (2001) Sphingosine 1-phosphate promotes endothelial cell barrier integrity by Edg-dependent cytoskeletal rearrangement. J Clin Invest 108(5):689–701

Garner B, Waldeck AR, Witting PK et al (1998) Oxidation of high density lipoproteins. II. Evidence for direct reduction of lipid hydroperoxides by methionine residues of apolipoproteins AI and AII. J Biol Chem 273(11):6088–6095

Gordts SC, Muthuramu I, Nefyodova E et al (2013) Beneficial effects of selective HDL-raising gene transfer on survival, cardiac remodelling and cardiac function after myocardial infarction in mice. Gene Ther 20(11):1053–1061

Griffin JH, Kojima K, Banka CL et al (1999) High-density lipoprotein enhancement of anticoagulant activities of plasma protein S and activated protein C. J Clin Invest 103(2):219–227

Guyton JR, Klemp KF (1996) Development of the lipid-rich core in human atherosclerosis. Arterioscler Thromb Vasc Biol 16(1):4–11

Haase CL, Tybjaerg-Hansen A, Grande P et al (2010) Genetically elevated apolipoprotein A-I, high-density lipoprotein cholesterol levels, and risk of ischemic heart disease. J Clin Endocrinol Metab 95(12):E500–E510

Hansson GK (2005) Inflammation, atherosclerosis, and coronary artery disease. N Engl J Med 352(16):1685–1695

Huang Y, DiDonato JA, Levison BS et al (2014) An abundant dysfunctional apolipoprotein A1 in human atheroma. Nat Med 20(2):193–203

Hulten LM, Levin M (2009) The role of hypoxia in atherosclerosis. Curr Opin Lipidol 20(5): 409–414

Jacquet S, Malaval C, Martinez LO et al (2005) The nucleotide receptor P2Y13 is a key regulator of hepatic high-density lipoprotein (HDL) endocytosis. Cell Mol Life Sci 62(21):2508–2515

Johnson JL, Jackson CL, Angelini GD et al (1998) Activation of matrix-degrading metalloproteinases by mast cell proteases in atherosclerotic plaques. Arterioscler Thromb Vasc Biol 18(11):1707–1715

Kaartinen M, Penttila A, Kovanen PT (1996) Mast cells in rupture-prone areas of human coronary atheromas produce and store TNF-alpha. Circulation 94(11):2787–2792

Karuna R, Park R, Othman A et al (2011) Plasma levels of sphingosine-1-phosphate and apolipoprotein M in patients with monogenic disorders of HDL metabolism. Atherosclerosis 219(2): 855–863

Kimura T, Sato K, Kuwabara A et al (2001) Sphingosine 1-phosphate may be a major component of plasma lipoproteins responsible for the cytoprotective actions in human umbilical vein endothelial cells. J Biol Chem 276(34):31780–31785

Kimura T, Tomura H, Mogi C et al (2006) Role of scavenger receptor class B type I and sphingosine 1-phosphate receptors in high density lipoprotein-induced inhibition of adhesion molecule expression in endothelial cells. J Biol Chem 281(49):37457–37467

Koenig W, Khuseyinova N, Lowel H et al (2004) Lipoprotein-associated phospholipase A2 adds to risk prediction of incident coronary events by C-reactive protein in apparently healthy middle-aged men from the general population: results from the 14-year follow-up of a large cohort from southern Germany. Circulation 110(14):1903–1908

Kontush A, Chantepie S, Chapman MJ (2003) Small, dense HDL particles exert potent protection of atherogenic LDL against oxidative stress. Arterioscler Thromb Vasc Biol 23(10):1881–1888

Kovanen PT, Kaartinen M, Paavonen T (1995) Infiltrates of activated mast cells at the site of coronary atheromatous erosion or rupture in myocardial infarction. Circulation 92(5): 1084–1088

Krieger M, Herz J (1994) Structures and functions of multiligand lipoprotein receptors: macrophage scavenger receptors and LDL receptor-related protein (LRP). Annu Rev Biochem 63: 601–637

Kruth HS (2011) Receptor-independent fluid-phase pinocytosis mechanisms for induction of foam cell formation with native low-density lipoprotein particles. Curr Opin Lipidol 22(5):386–393

Lee-Rueckert M, Kovanen PT (2011) Extracellular modifications of HDL in vivo and the emerging concept of proteolytic inactivation of prebeta-HDL. Curr Opin Lipidol 22(5):394–402

Lee-Rueckert M, Blanco-Vaca F, Kovanen PT et al (2013) The role of the gut in reverse cholesterol transport–focus on the enterocyte. Prog Lipid Res 52(3):317–328

Lewis GF, Rader DJ (2005) New insights into the regulation of HDL metabolism and reverse cholesterol transport. Circ Res 96(12):1221–1232

Libby P (1995) Molecular bases of the acute coronary syndromes. Circulation 91(11):2844–2850

Libby P (2002) Inflammation in atherosclerosis. Nature 420(6917):868–874

Libby P (2013a) Collagenases and cracks in the plaque. J Clin Invest 123(8):3201–3203

Libby P (2013b) Mechanisms of acute coronary syndromes and their implications for therapy. N Engl J Med 368(21):2004–2013

Lim HY, Thiam CH, Yeo KP et al (2013) Lymphatic vessels are essential for the removal of cholesterol from peripheral tissues by SR-BI-mediated transport of HDL. Cell Metab 17(5):671–684

Liu Y, Mackness B, Mackness M (2008) Comparison of the ability of paraoxonases 1 and 3 to attenuate the in vitro oxidation of low-density lipoprotein and reduce macrophage oxidative stress. Free Radic Biol Med 45(6):743–748

Lucke S, Levkau B (2010) Endothelial functions of sphingosine-1-phosphate. Cell Physiol Biochem 26(1):87–96

Martel C, Li W, Fulp B et al (2013) Lymphatic vasculature mediates macrophage reverse cholesterol transport in mice. J Clin Invest 123(4):1571–1579

Martinez LO, Jacquet S, Esteve JP et al (2003) Ectopic beta-chain of ATP synthase is an apolipoprotein A-I receptor in hepatic HDL endocytosis. Nature 421(6918):75–79

Mayranpaa MI, Heikkila HM, Lindstedt KA et al (2006) Desquamation of human coronary artery endothelium by human mast cell proteases: implications for plaque erosion. Coron Artery Dis 17(7):611–621

McGrath KC, Li XH, Puranik R et al (2009) Role of 3beta-hydroxysteroid-delta 24 reductase in mediating antiinflammatory effects of high-density lipoproteins in endothelial cells. Arterioscler Thromb Vasc Biol 29(6):877–882

Mertens A, Verhamme P, Bielicki JK et al (2003) Increased low-density lipoprotein oxidation and impaired high-density lipoprotein antioxidant defense are associated with increased macrophage homing and atherosclerosis in dyslipidemic obese mice: LCAT gene transfer decreases atherosclerosis. Circulation 107(12):1640–1646

Mineo C, Yuhanna IS, Quon MJ et al (2003) High density lipoprotein-induced endothelial nitric-oxide synthase activation is mediated by Akt and MAP kinases. J Biol Chem 278(11):9142–9149

Murphy AJ, Woollard KJ, Hoang A et al (2008) High-density lipoprotein reduces the human monocyte inflammatory response. Arterioscler Thromb Vasc Biol 28(11):2071–2077

Murphy AJ, Bijl N, Yvan-Charvet L et al (2013) Cholesterol efflux in megakaryocyte progenitors suppresses platelet production and thrombocytosis. Nat Med 19(5):586–594

Nakashima Y, Wight TN, Sueishi K (2008) Early atherosclerosis in humans: role of diffuse intimal thickening and extracellular matrix proteoglycans. Cardiovasc Res 79(1):14–23

Naqvi TZ, Shah PK, Ivey PA et al (1999) Evidence that high-density lipoprotein cholesterol is an independent predictor of acute platelet-dependent thrombus formation. Am J Cardiol 84(9):1011–1017

Navab M, Hama SY, Cooke CJ et al (2000) Normal high density lipoprotein inhibits three steps in the formation of mildly oxidized low density lipoprotein: step 1. J Lipid Res 41(9):1481–1494

Newby AC (2005) Dual role of matrix metalloproteinases (matrixins) in intimal thickening and atherosclerotic plaque rupture. Physiol Rev 85(1):1–31

Nicholls SJ, Dusting GJ, Cutri B et al (2005) Reconstituted high-density lipoproteins inhibit the acute pro-oxidant and proinflammatory vascular changes induced by a periarterial collar in normocholesterolemic rabbits. Circulation 111(12):1543–1550

Nievelstein-Post P, Mottino G, Fogelman A et al (1994) An ultrastructural study of lipoprotein accumulation in cardiac valves of the rabbit. Arterioscler Thromb 14(7):1151–1161

Nigro P, Abe J, Berk BC (2011) Flow shear stress and atherosclerosis: a matter of site specificity. Antioxid Redox Signal 15(5):1405–1414

Nijstad N, Gautier T, Briand F et al (2011) Biliary sterol secretion is required for functional in vivo reverse cholesterol transport in mice. Gastroenterology 140(3):1043–1051

Nofer JR, Tepel M, Kehrel B et al (1996) High density lipoproteins enhance the Na^+/H^+ antiport in human platelets. Thromb Haemost 75(4):635–641

Nofer JR, Walter M, Kehrel B et al (1998) HDL3-mediated inhibition of thrombin-induced platelet aggregation and fibrinogen binding occurs via decreased production of phosphoinositide-derived second messengers 1,2-diacylglycerol and inositol 1,4,5-tris-phosphate. Arterioscler Thromb Vasc Biol 18(6):861–869

Nofer JR, Junker R, Pulawski E et al (2001a) High density lipoproteins induce cell cycle entry in vascular smooth muscle cells via mitogen activated protein kinase-dependent pathway. Thromb Haemost 85(4):730–735

Nofer JR, Levkau B, Wolinska I et al (2001b) Suppression of endothelial cell apoptosis by high density lipoproteins (HDL) and HDL-associated lysosphingolipids. J Biol Chem 276(37): 34480–34485

Nofer JR, Geigenmuller S, Gopfert C et al (2003) High density lipoprotein-associated lysosphingolipids reduce E-selectin expression in human endothelial cells. Biochem Biophys Res Commun 310(1):98–103

Nofer JR, van der Giet M, Tolle M et al (2004) HDL induces NO-dependent vasorelaxation via the lysophospholipid receptor S1P3. J Clin Invest 113(4):569–581

Noto H, Hara M, Karasawa K et al (2003) Human plasma platelet-activating factor acetylhydrolase binds to all the murine lipoproteins, conferring protection against oxidative stress. Arterioscler Thromb Vasc Biol 23(5):829–835

Oorni K, Pentikainen MO, Ala-Korpela M et al (2000) Aggregation, fusion, and vesicle formation of modified low density lipoprotein particles: molecular mechanisms and effects on matrix interactions. J Lipid Res 41(11):1703–1714

Oslakovic C, Krisinger MJ, Andersson A et al (2009) Anionic phospholipids lose their procoagulant properties when incorporated into high density lipoproteins. J Biol Chem 284 (9):5896–5904

Oslakovic C, Norstrom E, Dahlback B (2010) Reevaluation of the role of HDL in the anticoagulant activated protein C system in humans. J Clin Invest 120(5):1396–1399

Park SH, Park JH, Kang JS et al (2003) Involvement of transcription factors in plasma HDL protection against TNF-alpha-induced vascular cell adhesion molecule-1 expression. Int J Biochem Cell Biol 35(2):168–182

Parthasarathy S, Barnett J, Fong LG (1990) High-density lipoprotein inhibits the oxidative modification of low-density lipoprotein. Biochim Biophys Acta 1044(2):275–283

Pasquinelli G, Preda P, Vici M et al (1989) Electron microscopy of lipid deposits in human atherosclerosis. Scanning Microsc 3(4):1151–1159

Pentikainen MO, Oorni K, Ala-Korpela M et al (2000) Modified LDL - trigger of atherosclerosis and inflammation in the arterial intima. J Intern Med 247(3):359–370

Petoumenos V, Nickenig G, Werner N (2009) High-density lipoprotein exerts vasculoprotection via endothelial progenitor cells. J Cell Mol Med 13(11–12):4623–4635

Potteaux S, Gautier EL, Hutchison SB et al (2011) Suppressed monocyte recruitment drives macrophage removal from atherosclerotic plaques of Apoe-/- mice during disease regression. J Clin Invest 121(5):2025–2036

Puranik R, Bao S, Nobecourt E et al (2008) Low dose apolipoprotein A-I rescues carotid arteries from inflammation in vivo. Atherosclerosis 196(1):240–247

Quarck R, De Geest B, Stengel D et al (2001) Adenovirus-mediated gene transfer of human platelet-activating factor-acetylhydrolase prevents injury-induced neointima formation and reduces spontaneous atherosclerosis in apolipoprotein E-deficient mice. Circulation 103 (20):2495–2500

Radojkovic C, Genoux A, Pons V et al (2009) Stimulation of cell surface F1-ATPase activity by apolipoprotein A-I inhibits endothelial cell apoptosis and promotes proliferation. Arterioscler Thromb Vasc Biol 29(7):1125–1130

Rajamaki K, Lappalainen J, Oorni K et al (2010) Cholesterol crystals activate the NLRP3 inflammasome in human macrophages: a novel link between cholesterol metabolism and inflammation. PLoS ONE 5(7):e11765

Rallidis LS, Tellis CC, Lekakis J et al (2012) Lipoprotein-associated phospholipase A(2) bound on high-density lipoprotein is associated with lower risk for cardiac death in stable coronary artery disease patients: a 3-year follow-up. J Am Coll Cardiol 60(20):2053–2060

Ramet ME, Ramet M, Lu Q et al (2003) High-density lipoprotein increases the abundance of eNOS protein in human vascular endothelial cells by increasing its half-life. J Am Coll Cardiol 41(12):2288–2297

Ramkhelawon B, Yang Y, van Gils JM et al (2013) Hypoxia induces netrin-1 and Unc5b in atherosclerotic plaques: mechanism for macrophage retention and survival. Arterioscler Thromb Vasc Biol 33(6):1180–1188

Ridker PM, Genest J, Boekholdt SM et al (2010) HDL cholesterol and residual risk of first cardiovascular events after treatment with potent statin therapy: an analysis from the JUPITER trial. Lancet 376(9738):333–339

Riwanto M, Rohrer L, Roschitzki B et al (2013) Altered activation of endothelial anti- and proapoptotic pathways by high-density lipoprotein from patients with coronary artery disease: role of high-density lipoprotein-proteome remodeling. Circulation 127(8):891–904

Robbesyn F, Garcia V, Auge N et al (2003) HDL counterbalance the proinflammatory effect of oxidized LDL by inhibiting intracellular reactive oxygen species rise, proteasome activation, and subsequent NF-kappaB activation in smooth muscle cells. FASEB J 17(6):743–745

Robert J, Lehner M, Frank S et al (2013) Interleukin 6 stimulates endothelial binding and transport of high-density lipoprotein through induction of endothelial lipase. Arterioscler Thromb Vasc Biol 33(12):2699–2706

Rohrer L, Cavelier C, Fuchs S et al (2006) Binding, internalization and transport of apolipoprotein A-I by vascular endothelial cells. Biochim Biophys Acta 1761(2):186–194

Rohrer L, Ohnsorg PM, Lehner M et al (2009) High-density lipoprotein transport through aortic endothelial cells involves scavenger receptor BI and ATP-binding cassette transporter G1. Circ Res 104(10):1142–1150

Rosenson RS, Brewer HB Jr, Davidson WS et al (2012) Cholesterol efflux and atheroprotection: advancing the concept of reverse cholesterol transport. Circulation 125(15):1905–1919

Sabatine MS, Morrow DA, O'Donoghue M et al (2007) Prognostic utility of lipoprotein-associated phospholipase A2 for cardiovascular outcomes in patients with stable coronary artery disease. Arterioscler Thromb Vasc Biol 27(11):2463–2469

Sakakura K, Nakano M, Otsuka F et al (2013) Pathophysiology of atherosclerosis plaque progression. Heart Lung Circ 22(6):399–411

Sanson M, Distel E, Fisher EA (2013) HDL induces the expression of the M2 macrophage markers arginase 1 and Fizz-1 in a STAT6-dependent process. PLoS ONE 8(8):e74676

Schmitt MM, Fraemohs L, Hackeng TM et al (2014a) Atherogenic mononuclear cell recruitment is facilitated by oxidized lipoprotein-induced endothelial junctional adhesion molecule-A redistribution. Atherosclerosis 234(2):254–264

Schmitt MM, Megens RT, Zernecke A et al (2014b) Endothelial junctional adhesion molecule-a guides monocytes into flow-dependent predilection sites of atherosclerosis. Circulation 129 (1):66–76

Schwartz GG, Olsson AG, Abt M et al (2012) Effects of dalcetrapib in patients with a recent acute coronary syndrome. N Engl J Med 367(22):2089–2099

Schwenke DC, Carew TE (1989a) Initiation of atherosclerotic lesions in cholesterol-fed rabbits. I. Focal increases in arterial LDL concentration precede development of fatty streak lesions. Arteriosclerosis 9(6):895–907

Schwenke DC, Carew TE (1989b) Initiation of atherosclerotic lesions in cholesterol-fed rabbits. II. Selective retention of LDL vs. selective increases in LDL permeability in susceptible sites of arteries. Arteriosclerosis 9(6):908–918

Seetharam D, Mineo C, Gormley AK et al (2006) High-density lipoprotein promotes endothelial cell migration and reendothelialization via scavenger receptor-B type I. Circ Res 98(1):63–72

Shaw JA, Bobik A, Murphy A et al (2008) Infusion of reconstituted high-density lipoprotein leads to acute changes in human atherosclerotic plaque. Circ Res 103(10):1084–1091

Shih DM, Gu L, Xia YR et al (1998) Mice lacking serum paraoxonase are susceptible to organophosphate toxicity and atherosclerosis. Nature 394(6690):284–287

Silbernagel G, Schottker B, Appelbaum S et al (2013) High-density lipoprotein cholesterol, coronary artery disease, and cardiovascular mortality. Eur Heart J 34(46):3563–3571

Smith EB (1990) Transport, interactions and retention of plasma proteins in the intima: the barrier function of the internal elastic lamina. Eur Heart J 11(Suppl E):72–81

Spieker LE, Sudano I, Hurlimann D et al (2002) High-density lipoprotein restores endothelial function in hypercholesterolemic men. Circulation 105(12):1399–1402

Spirig R, Schaub A, Kropf A et al (2013) Reconstituted high-density lipoprotein modulates activation of human leukocytes. PLoS ONE 8(8):e71235

Stary HC, Chandler AB, Glagov S et al (1994) A definition of initial, fatty streak, and intermediate lesions of atherosclerosis. A report from the committee on vascular lesions of the council on arteriosclerosis, American heart association. Arterioscler Thromb 14(5):840–856

Stary HC, Chandler AB, Dinsmore RE et al (1995) A definition of advanced types of atherosclerotic lesions and a histological classification of atherosclerosis. A report from the committee on vascular lesions of the council on arteriosclerosis, American heart association. Circulation 92(5):1355–1374

Suc I, Escargueil-Blanc I, Troly M et al (1997) HDL and ApoA prevent cell death of endothelial cells induced by oxidized LDL. Arterioscler Thromb Vasc Biol 17(10):2158–2166

Sugano M, Tsuchida K, Makino N (2000) High-density lipoproteins protect endothelial cells from tumor necrosis factor-alpha-induced apoptosis. Biochem Biophys Res Commun 272(3):872–876

Sugatani J, Miwa M, Komiyama Y et al (1996) High-density lipoprotein inhibits the synthesis of platelet-activating factor in human vascular endothelial cells. J Lipid Mediat Cell Signal 13(1):73–88

Sumi M, Sata M, Miura S et al (2007) Reconstituted high-density lipoprotein stimulates differentiation of endothelial progenitor cells and enhances ischemia-induced angiogenesis. Arterioscler Thromb Vasc Biol 27(4):813–818

Tabas I (1999) Nonoxidative modifications of lipoproteins in atherogenesis. Annu Rev Nutr 19:123–139

Tabas I, Williams KJ, Boren J (2007) Subendothelial lipoprotein retention as the initiating process in atherosclerosis: update and therapeutic implications. Circulation 116(16):1832–1844

Tabet F, Vickers KC, Cuesta Torres LF et al (2014) HDL-transferred microRNA-223 regulates ICAM-1 expression in endothelial cells. Nat Commun 5:3292

Tamama K, Tomura H, Sato K et al (2005) High-density lipoprotein inhibits migration of vascular smooth muscle cells through its sphingosine 1-phosphate component. Atherosclerosis 178(1):19–23

Tamminen M, Mottino G, Qiao JH et al (1999) Ultrastructure of early lipid accumulation in ApoE-deficient mice. Arterioscler Thromb Vasc Biol 19(4):847–853

Tatematsu S, Francis SA, Natarajan P et al (2013) Endothelial lipase is a critical determinant of high-density lipoprotein-stimulated sphingosine 1-phosphate-dependent signaling in vascular endothelium. Arterioscler Thromb Vasc Biol 33(8):1788–1794

Tellis CC, Tselepis AD (2009) The role of lipoprotein-associated phospholipase A2 in atherosclerosis may depend on its lipoprotein carrier in plasma. Biochim Biophys Acta 1791(5):327–338

Terasaka N, Westerterp M, Koetsveld J et al (2010) ATP-binding cassette transporter G1 and high-density lipoprotein promote endothelial NO synthesis through a decrease in the interaction of caveolin-1 and endothelial NO synthase. Arterioscler Thromb Vasc Biol 30(11):2219–2225

Theilmeier G, Schmidt C, Herrmann J et al (2006) High-density lipoproteins and their constituent, sphingosine-1-phosphate, directly protect the heart against ischemia/reperfusion injury in vivo via the S1P3 lysophospholipid receptor. Circulation 114(13):1403–1409

Thorp E, Subramanian M, Tabas I (2011) The role of macrophages and dendritic cells in the clearance of apoptotic cells in advanced atherosclerosis. Eur J Immunol 41(9):2515–2518

Tolani S, Pagler TA, Murphy AJ et al (2013) Hypercholesterolemia and reduced HDL-C promote hematopoietic stem cell proliferation and monocytosis: studies in mice and FH children. Atherosclerosis 229(1):79–85

Tolle M, Pawlak A, Schuchardt M et al (2008) HDL-associated lysosphingolipids inhibit NAD(P) H oxidase-dependent monocyte chemoattractant protein-1 production. Arterioscler Thromb Vasc Biol 28(8):1542–1548

Tward A, Xia YR, Wang XP et al (2002) Decreased atherosclerotic lesion formation in human serum paraoxonase transgenic mice. Circulation 106(4):484–490

van der Wal AC, Becker AE, van der Loos CM et al (1994) Site of intimal rupture or erosion of thrombosed coronary atherosclerotic plaques is characterized by an inflammatory process irrespective of the dominant plaque morphology. Circulation 89(1):36–44

van Oostrom O, Nieuwdorp M, Westerweel PE et al (2007) Reconstituted HDL increases circulating endothelial progenitor cells in patients with type 2 diabetes. Arterioscler Thromb Vasc Biol 27(8):1864–1865

Vinals M, Martinez-Gonzalez J, Badimon JJ et al (1997) HDL-induced prostacyclin release in smooth muscle cells is dependent on cyclooxygenase-2 (Cox-2). Arterioscler Thromb Vasc Biol 17(12):3481–3488

Vinals M, Martinez-Gonzalez J, Badimon L (1999) Regulatory effects of HDL on smooth muscle cell prostacyclin release. Arterioscler Thromb Vasc Biol 19(10):2405–2411

Viswambharan H, Ming XF, Zhu S et al (2004) Reconstituted high-density lipoprotein inhibits thrombin-induced endothelial tissue factor expression through inhibition of RhoA and stimulation of phosphatidylinositol 3-kinase but not Akt/endothelial nitric oxide synthase. Circ Res 94(7):918–925

Vohl MC, Neville TA, Kumarathasan R et al (1999) A novel lecithin-cholesterol acyltransferase antioxidant activity prevents the formation of oxidized lipids during lipoprotein oxidation. Biochemistry 38(19):5976–5981

Voight BF, Peloso GM, Orho-Melander M et al (2012) Plasma HDL cholesterol and risk of myocardial infarction: a mendelian randomisation study. Lancet 380(9841):572–580

von Eckardstein A, Nofer JR, Assmann G (2001) High density lipoproteins and arteriosclerosis. Role of cholesterol efflux and reverse cholesterol transport. Arterioscler Thromb Vasc Biol 21 (1):13–27

Wang N, Tall AR (2003) Regulation and mechanisms of ATP-binding cassette transporter A1-mediated cellular cholesterol efflux. Arterioscler Thromb Vasc Biol 23(7):1178–1184

Wang Y, Lindstedt KA, Kovanen PT (1996) Phagocytosis of mast cell granule remnant-bound LDL by smooth muscle cells of synthetic phenotype: a scavenger receptor-mediated process that effectively stimulates cytoplasmic cholesteryl ester synthesis. J Lipid Res 37(10): 2155–2166

Wang N, Silver DL, Costet P et al (2000) Specific binding of ApoA-I, enhanced cholesterol efflux, and altered plasma membrane morphology in cells expressing ABC1. J Biol Chem 275(42): 33053–33058

Wang N, Lan D, Chen W et al (2004) ATP-binding cassette transporters G1 and G4 mediate cellular cholesterol efflux to high-density lipoproteins. Proc Natl Acad Sci USA 101(26): 9774–9779

Watson AD, Berliner JA, Hama SY et al (1995a) Protective effect of high density lipoprotein associated paraoxonase. Inhibition of the biological activity of minimally oxidized low density lipoprotein. J Clin Invest 96(6):2882–2891

Watson AD, Navab M, Hama SY et al (1995b) Effect of platelet activating factor-acetylhydrolase on the formation and action of minimally oxidized low density lipoprotein. J Clin Invest 95 (2):774–782

Wentzel JJ, Chatzizisis YS, Gijsen FJ et al (2012) Endothelial shear stress in the evolution of coronary atherosclerotic plaque and vascular remodelling: current understanding and remaining questions. Cardiovasc Res 96(2):234–243

Westerterp M, Gourion-Arsiquaud S, Murphy AJ et al (2012) Regulation of hematopoietic stem and progenitor cell mobilization by cholesterol efflux pathways. Cell Stem Cell 11(2):195–206

Wilhelm AJ, Zabalawi M, Owen JS et al (2010) Apolipoprotein A-I modulates regulatory T cells in autoimmune LDLr-/-, ApoA-I-/- mice. J Biol Chem 285(46):36158–36169

Wilkerson BA, Grass GD, Wing SB et al (2012) Sphingosine 1-phosphate (S1P) carrier-dependent regulation of endothelial barrier: high density lipoprotein (HDL)-S1P prolongs endothelial barrier enhancement as compared with albumin-S1P via effects on levels, trafficking, and signaling of S1P1. J Biol Chem 287(53):44645–44653

Williams KJ, Tabas I (1995) The response-to-retention hypothesis of early atherogenesis. Arterioscler Thromb Vasc Biol 15(5):551–561

Witztum JL (2005) You are right too! J Clin Invest 115(8):2072–2075

Woollard KJ (2013) Immunological aspects of atherosclerosis. Clin Sci (Lond) 125(5):221–235

Wu BJ, Chen K, Shrestha S et al (2013) High-density lipoproteins inhibit vascular endothelial inflammation by increasing 3beta-hydroxysteroid-Delta24 reductase expression and inducing heme oxygenase-1. Circ Res 112(2):278–288

Xia P, Vadas MA, Rye KA et al (1999) High density lipoproteins (HDL) interrupt the sphingosine kinase signaling pathway. A possible mechanism for protection against atherosclerosis by HDL. J Biol Chem 274(46):33143–33147

Xiong SL, Liu X, Yi GH (2014) High-density lipoprotein induces cyclooxygenase-2 expression and prostaglandin I-2 release in endothelial cells through sphingosine kinase-2. Mol Cell Biochem 389(1–2):197–207

Yancey PG, Bortnick AE, Kellner-Weibel G et al (2003) Importance of different pathways of cellular cholesterol efflux. Arterioscler Thromb Vasc Biol 23(5):712–719

Yuhanna IS, Zhu Y, Cox BE et al (2001) High-density lipoprotein binding to scavenger receptor-BI activates endothelial nitric oxide synthase. Nat Med 7(7):853–857

Yvan-Charvet L, Pagler T, Gautier EL et al (2010) ATP-binding cassette transporters and HDL suppress hematopoietic stem cell proliferation. Science 328(5986):1689–1693

Zawada AM, Rogacev KS, Schirmer SH et al (2012) Monocyte heterogeneity in human cardiovascular disease. Immunobiology 217(12):1273–1284

Zernecke A, Weber C (2010) Chemokines in the vascular inflammatory response of atherosclerosis. Cardiovasc Res 86(2):192–201

Zhang Q, Yin H, Liu P et al (2010) Essential role of HDL on endothelial progenitor cell proliferation with PI3K/Akt/cyclin D1 as the signal pathway. Exp Biol Med (Maywood) 235 (9):1082–1092

Zhu W, Saddar S, Seetharam D et al (2008) The scavenger receptor class B type I adaptor protein PDZK1 maintains endothelial monolayer integrity. Circ Res 102(4):480–487

HDLs, Diabetes, and Metabolic Syndrome

Peter Vollenweider, Arnold von Eckardstein, and Christian Widmann

Contents

1 Type 2 Diabetes Mellitus and the Metabolic Syndrome 406
2 Low HDL Cholesterol Levels in Diabetes and Metabolic Syndrome 408
   2.1 Insulin Resistance, Hypertriglyceridemia, and Decreased HDL Cholesterol Levels 408
   2.2 FFAs and Decreased HDL Levels 410
   2.3 HDL Metabolism, Subclasses, and Catabolism 410
3 Does HDL Play a Causal Role in the Pathogenesis of Diabetes? 411
   3.1 Effects of HDL on Obesity 412
   3.2 Effects of HDL on Insulin Sensitivity and Glucose Utilization by Skeletal Muscle 413
   3.3 The Beneficial Effects of HDL in Pancreatic Beta Cells 414
      3.3.1 Beneficial Effects of HDL on Insulin Secretion 414
      3.3.2 Effects of HDL on Pancreatic Beta Cell Survival and ER Stress 415
Conclusion 416
References 417

Abstract

The prevalence of type 2 diabetes mellitus and of the metabolic syndrome is rising worldwide and reaching epidemic proportions. These pathologies are associated with significant morbidity and mortality, in particular with an excess of cardiovascular deaths. Type 2 diabetes mellitus and the cluster of pathologies

---

P. Vollenweider
Department of Internal Medicine, University Hospital Center (CHUV) and University of Lausanne, Lausanne, Switzerland

A. von Eckardstein
Institute of Clinical Chemistry, University Hospital Zurich, Zurich, Switzerland

C. Widmann (✉)
Department of Physiology, University of Lausanne, Bugnon 7, 1005 Lausanne, Switzerland
e-mail: christian.widmann@unil.ch

including insulin resistance, central obesity, high blood pressure, and hypertriglyceridemia that constitute the metabolic syndrome are associated with low levels of HDL cholesterol and the presence of dysfunctional HDLs. We here review the epidemiological evidence and the potential underlying mechanisms of this association. We first discuss the well-established association of type 2 diabetes mellitus and insulin resistance with alterations of lipid metabolism and how these alterations may lead to low levels of HDL cholesterol and the occurrence of dysfunctional HDLs. We then present and discuss the evidence showing that HDL modulates insulin sensitivity, insulin-independent glucose uptake, insulin secretion, and beta cell survival. A dysfunction in these actions could play a direct role in the pathogenesis of type 2 diabetes mellitus.

Keywords
HDLs • Diabetes • Metabolic syndrome

1 Type 2 Diabetes Mellitus and the Metabolic Syndrome

Diabetes mellitus (DM) is characterized by increased fasting and/or postprandial glucose levels. The disease had been known for many centuries (Laios et al. 2012). Its etiology[1] however started to be unraveled in the late nineteenth century/early twentieth century. During this period, it was eventually discovered that a compound in pancreatic extracts (later identified to be insulin) could cure young diabetic patients (Banting et al. 1922). From the early work of H. P. Himsworth in 1936, a concept emerged that diabetes could be divided into what he termed "insulin sensitive" and "insulin insensitive" [the pioneering work of Himsworth is nicely portrayed in a historical perspective written by Gerald Reaven (Reaven 2005)]. But it is only once a reliable radioimmunoassay became available to measure circulating insulin levels (Bray 1996) that it was realized that some patients suffered from a disease associated with low to undetectable insulin levels and prone to ketoacidosis (type 1 DM [T1DM]), while others, the vast majority, had high circulating insulin levels despite increased glucose levels (type 2 DM [T2DM]) (Inzucchi 2012). Patients with T2DM are often overweight and obese, and their diabetes develops later in life. These patients are characterized as having insulin resistance (IR), i.e., a decreased response of insulin-sensitive tissues to the effects of the hormone, in particular its ability to induce glucose uptake. Insulin resistance is usually associated with compensatory hyperinsulinemia in an attempt to maintain normoglycemia. Insulin resistance is a hallmark of T2DM as well as of overweight and obesity. Insulin resistance precedes the development of T2DM, and it is once that the beta cell can no longer compensate to the insulin resistance that the full-blown disease with hyperglycemia develops (Tabak et al. 2009). Gerald

---

[1] The study of causes or origins (of a disease).

Table 1 Definitions of the metabolic syndrome

	NCEP ATP III	WHO (1998)	IDF (2005)
Criteria	Three of the following	Presence of DM, IGF, IGT, or insulin resistance **and** two of the following	Central obesity (ethnicity specificity) **and** two of the following
Anthropometric data	Waist circumference >102 cm (men) >88 cm (women)	WHR > 0.9 (men) WHR > 0.85 (women) and/or BMI > 30 kg/m2	
Lipids	TG ≥ 1.7 mmol/l and/or HDL-C < 1.03 mmol/l (men) or < 1.29 mmol/l (women) or treatment for these specific lipid abnormalities	TG ≥ 1.7 mmol/l and/or HDL-C < 0.9 mmol/l (men) or <1.0 mmol/l (women)	TG ≥ 1.7 mmol/l and/or HDL-C < 1.03 mmol/l (men) or < 1.29 mmol/l (women) or treatment for these specific lipid abnormalities
Blood pressure	≥130/85 mmHg or treatment	≥140/90 mmHg	≥130/85 mmHg r treatment
Blood glucose	FPG ≥ 5.6 mmol/l or treatment		FPG ≥ 5.6 mmol/l or treatment
Others		Urinary albumin excretion ≥ 20 μg/min or albumin/creatinine ratio ≥ 30 mg/g	

BMI body mass index, *DM* diabetes mellitus, *FPG* fasting plasma glucose, *HDL-C* high-density lipoprotein cholesterol, *IDF* International Diabetes Federation, *IFG* impaired fasting glucose, *IGT* impaired glucose tolerance, *NCEP* National Cholesterol Education Program, *TG* triglycerides, *WHO* World Health Organization, *WHR* waist-hip ratio

Reaven, in his seminal Banting Lecture in 1988 (Reaven 1988), summarized his work that put insulin resistance not only at the center of diabetes research but proposed that insulin resistance may directly contribute to the increased risk of coronary artery disease associated with T2DM or obesity. In this chapter, he suggested that IR and hyperinsulinemia were, together with increased VLDL/triglyceride levels, low HDL cholesterol levels, and hypertension, directly associated with and responsible for the development of diabetes. He termed "syndrome[2] X" the clustering of these features. Syndrome X was also named "insulin resistance syndrome" but is now generally called "metabolic syndrome." This syndrome corresponds to a cluster of increased abdominal obesity (a good proxy of insulin resistance), increased blood glucose levels, high blood pressure and elevated triglycerides, and low HDL cholesterol levels. Several definitions of the

---

[2] A group of symptoms that collectively indicate or characterize a disease.

metabolic syndrome have been proposed with the National Cholesterol Education Program Adult Treatment Panel III (NCEP ATP III) definition being the one most widely used. The World Health Organization (WHO) and the International Diabetes Federation (IDF) have suggested other definitions. Table 1 summarizes these various definitions. These current definitions, albeit arbitrary in the cutoffs that were chosen for the different variables included, are intended to provide a useful tool for clinicians to identify individuals at increased risk for the development of T2DM, atherosclerotic cardiovascular disease, and cardiovascular death. T2DM and the metabolic syndrome are characterized not only by insulin resistance but also by a dyslipidemia that includes high triglyceride and low HDL cholesterol levels. With the presence of small dense LDLs in T2DM, these lipid abnormalities substantially contribute to the increased cardiovascular morbidity and mortality associated with these conditions (Grundy 2012). Several epidemiological studies have clearly shown that these lipid abnormalities are present before the development of T2DM. Furthermore, longitudinal studies have provided evidence that these abnormalities represent an independent risk factor for T2DM (von Eckardstein et al. 2000). Hence, the concept has emerged that lipid abnormalities associated with IR and the metabolic syndrome are not only bystanders of the metabolic dysfunctions but contribute directly to the progression to type DM.

2 Low HDL Cholesterol Levels in Diabetes and Metabolic Syndrome

T2DM and the metabolic syndrome are characterized by the presence of an atherogenic dyslipidemia typically including low HDL cholesterol, hypertriglyceridemia, and increased small dense LDL blood concentrations. Individuals with low HDL cholesterol are often insulin resistant and at increased risk of developing T2DM (Li et al. 2014; von Eckardstein and Sibler 2011). T2DM, the metabolic syndrome, and IR are characterized by hyperinsulinemia and increased free fatty acids (FFA). Both insulin resistance and FFAs appear to contribute to the pathogenesis associated with decreased HDL levels (Rohrer et al. 2004).

2.1 Insulin Resistance, Hypertriglyceridemia, and Decreased HDL Cholesterol Levels

Insulin has pleiotropic effects on lipid metabolism, in particular on the metabolism of HDLs and the TG-rich VLDLs. In rats, insulin can acutely decrease the hepatic production of ApoB100 (Chirieac et al. 2000), the apolipoprotein required for VLDL assembly, by favoring ApoB100 degradation. In the Hep2G liver cell line, insulin also decreases the levels of microsomal TG transfer protein that is essential for VLDL synthesis and secretion (Lin et al. 1995). Insulin could therefore negatively affect the hepatic production of VLDL. However, the situation is more complex than this because there are two VLDL subtypes that are differentially regulated by insulin, the larger and more buoyant VLDL1 particles and the smaller and denser VLDL2 particles. VLDL2 particles are intermediates in the synthesis of VLDL1 in the liver, but they can also be secreted directly by this organ. Additionally, VLDL2 can be generated in the circulation from VLDL1 through lipoprotein lipase activity (Taskinen 2003). Insulin does not appear to affect the hepatic production of VLDL2 (at least not in humans, see below) but seems to only inhibit hepatic VLDL1 production by inhibiting the conversion of VLDL2 into VLDL1 (Taskinen 2003). VLDL1 particles are seen by some authors as "liver-derived" chylomicron-like particles (Taskinen 2003). It makes sense therefore that insulin inhibits their production following meals in a situation where chylomicrons are produced by the intestine. In contrast, VLDL1 particle production could be favored in the fasting state. While insulin inhibits VLDL1 production, it favors the hepatic production of VLDL2 particles because insulin can increase the activity of the sterol regulatory element-binding protein (SREBP) 1c transcription factor that induces the expression of enzymes required for fatty acid and TG synthesis, such as fatty acid synthase (FAS) and acetyl-CoA carboxylase (ACC) (Horton et al. 2002). However, insulin was not found to modulate VLDL2 blood levels in humans (Malmstrom et al. 1997a, b). The physiological importance of the regulation of fatty acid and TG synthesis by insulin on VLDL production is therefore unclear. In type 2 diabetic individuals (i.e., in the presence of an insulin-resistant state), insulin is no longer able to inhibit VLDL1 production (Malmstrom et al. 1997a), and this can thus favor overall VLDL hepatic secretion. This presumably contributes to the hypertriglyceridemia seen in type 2 diabetic and metabolic syndrome patients (Brown and Goldstein 2008). In addition, hyperinsulinemia resulting from insulin resistance can enhance SREBP1c transcription and thereby microRNA33b (miR33b) that is encoded by an intron of the SREBP1c gene. Both SREBP1c and miR33b can drive VLDL production and favor hypertriglyceridemia by stimulating fatty acid and triglyceride synthesis (Rayner et al. 2011).

A high production of VLDLs can decrease the stability of HDL particles by increasing the transfer of TG from TG-rich particles to HDLs via the activity of the cholesteryl ester transfer protein (CETP) enzyme (Rashid et al. 2003). This is one possible mechanism linking high TG blood levels with decreased HDL cholesterol levels. Insulin participates in the clearance of TG from TG-rich lipoproteins by stimulating the secretion of lipoprotein lipase (LPL) in certain adipose tissues (e.g., subcutaneous depot) (Fried et al. 1993). Consequently, insulin resistance would lead to increased circulating TG-rich particles and augmented TG loading of HDLs. Additionally, reduced TG hydrolysis in VLDLs and chylomicrons will decrease the release of surface remnants which contribute to the maturation of HDL (Eisenberg 1984; Rashid et al. 2003). These two effects combined—decreased surface remnant production and enhanced CETP activity—favor the reduction in HDL cholesterol levels via the mechanism mentioned above (i.e., decreased HDL stability due TG enrichment of the particle).

Insulin has the potential to directly affect HDL production by the liver, either in a positive or in a negative manner. In the Hep2G liver cell line, insulin can stimulate the expression of apolipoprotein A-I (ApoA-I) (Murao et al. 1998), the most abundant apolipoprotein in HDLs (Eisenberg 1984; Rye and Barter 2012), raising the possibility that hepatic insulin resistance contributes to decreased production of HDLs. The relevance of this effect has however not been validated in vivo. In contrast, insulin may also negatively impact on hepatic HDL production by enhancing miR33 transcription. Indeed, miR33 suppresses the production of the adenosine triphosphate-binding cassette transporter A1 (ABCA1) at the posttranscriptional level and thereby compromises phospholipid and cholesterol efflux needed for HDL formation (Besler et al. 2012; Rader 2006; Rayner et al. 2011; Tang and Oram 2009).

2.2   FFAs and Decreased HDL Levels

In vitro, FFAs decrease the expression of SREBP1c by inhibiting the transcriptional activity of liver X receptor (LXR) (Ou et al. 2001; Yoshikawa et al. 2002). This represents a negative feedback loop ensuring that the production of VLDLs occurs when there is a need to deliver TGs to peripheral tissues (muscles, adipose tissue, etc.). Reduction in LXR activity, and consequently VLDL production, would be expected to lead to increased HDL cholesterol blood levels as a result of increased HDL stability. However, in the liver, unsaturated FFAs inhibit the expression of ABCA1 (Uehara et al. 2002). Moreover, free fatty acids stimulate the proteasomal degradation of ABCA1 (Wang and Oram 2005, 2007). So even though FFAs can lower the production of hepatic triglycerides, which will favor HDL stability, FFAs, at the same time, directly inhibit the generation of HDL particles.

2.3   HDL Metabolism, Subclasses, and Catabolism

Nascent HDLs are the precursors of the mature HDL_3 and HDL_2 particles. They are generated from lipid-free ApoA-I that acquires phospholipids and cholesterol extracellularly. ABCA1-mediated lipid efflux from cells is important for this initial lipidation. Further acquisition of lipids and subsequent LCAT-driven cholesterol esterification leads to the formation of mature HDL_3 particles. These particles can absorb more phospholipids and free cholesterol by a mechanism involving ABCG1. Altogether, this transforms the HDL_3 particles into larger and less dense HDL_2 particles. Like ABCA1, this transporter is regulated by LXR and hence suppressed by unsaturated fatty acids at the transcriptional level (Uehara et al. 2007) so that the production of larger HDL_2 is compromised. In addition, the conversion of HDL_2 into HDL_3 is enhanced in diabetic dyslipidemia by increased activities of CETP and hepatic lipase: First, the transfer rate of cholesteryl esters from HDL to VLDL in normal individuals depends on the concentration of CE acceptors, that is, VLDL particles, but not on CETP activity (Mann et al. 1991). However, in individuals with hypertriglyceridemia, cholesteryl ester transfer from HDL to VLDL can be augmented by increasing CETP plasma activity (Mann et al. 1991). Second, HL and endothelial lipase can convert HDL_2 particles back to HDL_3 particles that begin a new cycle of maturation or are further degraded by the liver or cleared through the kidneys (Borggreve et al. 2003; Ginsberg 1998; Rader 2006). HL activity is increased in T2DM (Tan et al. 1999) and contributes to the production of smaller and denser HDL particles (i.e., HDL_2 to HDL_3) that are then catabolized as indicated above.

3 Does HDL Play a Causal Role in the Pathogenesis of Diabetes?

Several epidemiological studies have shown the association between low plasma concentrations of HDL-C and increased risks not only of CHD but also T2DM (Di et al. 2009; von Eckardstein et al. 2000).

We have seen that some of the metabolic perturbations (hyperglycemia, hypertriglyceridemia, etc.) associated with diabetes (and the metabolic syndrome) are mechanistically linked to decreased HDL cholesterol levels in blood. Traditionally, a reduced plasmatic level of HDL cholesterol has been interpreted to be the result of such metabolic perturbations, which are already present in the prediabetic state, and hence an innocent bystander of T2DM. However, several more recent studies suggest that HDLs help to maintain euglycemia via both insulin-dependent and insulin-independent pathways (Drew et al. 2012; von Eckardstein and Sibler 2011). The infusion of artificially reconstituted HDL (rHDL) was found to improve glycemia in patients with T2DM (Drew et al. 2009). In a post hoc analysis of data from diabetic participants in the ILLUMINATE trial, the addition of the CETP inhibitor torcetrapib to atorvastatin was found to increase HDL-C and to improve glycemic control (Barter et al. 2011). Several in vitro and animal experiments provided evidence that HDL improves the function and survival of beta cells as well as glucose-lowering metabolism in the muscle, liver, and adipose tissue (Drew et al. 2012; von Eckardstein and Sibler 2011; von Eckardstein and Widmann 2014). In mice, the absence of ApoA-I leads to hyperinsulinemia and hyperglycemia as well as altered responses to insulin tolerance tests. Vice versa, ApoA-I transgenic mice exhibit lower fasting glucose levels and improved glucose tolerance test compared with wild-type mice (Han et al. 2007; Lehti et al. 2013). These data point to a modulating role of HDLs in maintaining normal glucose homeostasis. Moreover, genetically increased HDL cholesterol and ApoA-I levels in mice prevented high-fat diet-induced glucose homeostasis impairment (Lehti et al. 2013).

Although it remains to be determined whether this is related to differences in skeletal muscle insulin resistance, adipose tissue mass, or differences in beta cell function, these data obtained in mice indicate that ApoA-I plays a modulatory role in the regulation of glucose homeostasis. This notion is also supported by studies in humans displaying altered HDL metabolism due to polymorphisms in the ApoA-I and ABCA1 genes that have been associated with increased risk of diabetes and impaired glucose uptake (Morcillo et al. 2005; Villarreal-Molina et al. 2008). Heterozygotes for rare mutations in the ABCA1 gene that profoundly decrease HDL-C levels were found to be less glucose tolerant than their unaffected relatives (Vergeer et al. 2010). However, it should also be noted that larger population studies did not find any association between heterozygosity for ABCA1 mutations and risk of diabetes (Schou et al. 2012).

3.1 Effects of HDL on Obesity

Augmented adipose tissue content is the hallmark of obesity. There is increasing evidence that HDL or ApoA-I regulate adipose tissue content. In humans, polymorphisms within the ApoA-I gene have been associated with increased risk of developing obesity and hypertension in a Brazilian cohort (Chen et al. 2009). More evidence for a potential role of HDL cholesterol on adipose tissue content regulation and obesity stems from genetic mouse models either lacking or overexpressing ApoA-I or treated with ApoA-I mimetic peptides. For example, ApoA-I-deficient mice have an increased fat content (Han et al. 2007). On the other hand, mice overexpressing ApoA-I or treated with the D-4F ApoA-I mimetic peptide[3] (Navab et al. 2005; Van Lenten et al. 2009) showed decreased white adipose tissue mass gain when fed a high-fat diet for 3 months (Ruan et al. 2011). ApoA-I transgenic and D-4F-treated mice had slightly improved insulin sensitivity when fed a high-fat diet compared to wild-type mice. This was associated with increased energy expenditure as measured in metabolic cages and related to increased expression of uncoupling protein 1 (UCP-1) in brown adipose tissue and AMPK phosphorylation (Ruan et al. 2011). Consistent with these results is the observation that administration of D-4F and L-4F in high-fat diet-fed mice reduced weight gain and improved insulin sensitivity when compared with age-matched vehicle-treated mice (Peterson et al. 2009). Taken together, these observations implicate a potential antiobesity effect of ApoA-I. In another study, ApoA-I gene transfer in the liver of mice increased adiponectin plasma levels and potently attenuated the rise of triglycerides and free fatty acids induced by lipopolysaccharide administration (Van Linthout et al. 2010). In vitro, incubation of 3 T3-L1 adipocytes with HDLs induced adiponectin expression in a phosphatidylinositol-3-kinase-dependent manner (Van Linthout et al. 2010). This set of experiments showed that ApoA-I and HDLs affect adipose tissue metabolism and adiponectin expression. ApoA-I and HDL have been shown to have anti-inflammatory properties in adipocytes similarly to what has been observed in macrophages and endothelial cells. It was found that both HDL and ApoA-I inhibit palmitate-mediated gene induction of pro-inflammatory chemotactic factor (e.g., monocyte chemotactic protein-1, serum amyloid A3) in cultured adipocytes (Umemoto et al. 2013). Furthermore, there was a reduction in chemotactic factor

---

[3] This peptide, albeit lacking any sequence homology to ApoA-I, forms a class A amphipatic helix similar to those found in ApoA-I. D-4F favors HDL formation and has anti-inflammatory and anti-atherogenic properties. The "D" in D-4F indicates that the peptide was synthesized with nonnatural D-amino acids (if synthesized with the natural L-amino acids, the peptide is called L-4F). D-amino peptides are very stable in biological fluids as they are not efficiently recognized by proteases (Chorev and Goodman 1995; Fischer 2003; Michod et al. 2009).

expression and macrophage accumulation in ApoA-I transgenic mice fed a high-fat diet compared to similarly treated control mice (Umemoto et al. 2013). Because macrophage accumulation and adipose tissue inflammation are strongly correlated with insulin resistance, these observations provide potential implications for the prevention and management of insulin resistance.

The molecular mechanisms involved in the antiobesity effects of ApoA-I are poorly understood. A recent suggestion is that ApoA-I and HDL might regulate autophagy in adipocytes. Decreased autophagy in pre-adipocytes has been associated with a preferential differentiation into "brown-like" adipocytes and improved energy expenditure, with reduced generation of white adipose tissue that serves mostly as an energy storage tissue (Wang and Peng 2012). Another potential mechanism is the direct capture of cholesterol from adipocytes by HDLs or ApoA-I (von Eckardstein and Sibler 2011; Zhang et al. 2010).

3.2 Effects of HDL on Insulin Sensitivity and Glucose Utilization by Skeletal Muscle

There is increasing evidence that HDLs regulate glucose uptake in insulin-sensitive tissues. In a set of elegant experiments in patients with T2 DM, Drew et al. showed that reconstituted HDL (rHDL) infusion over 4 h induced larger decreases of plasma glucose levels than placebo. This effect was associated with increased insulin secretion (Drew et al. 2009). In the C2C12 myocyte cell line, ApoA-I increased glucose uptake in an insulin-independent manner and also potentiated insulin- and adiponectin-induced glucose uptake (Han et al. 2007). ApoA-I effects on glucose uptake were dependent on the endocytosis of the protein and subsequent phosphorylation of AMPK and acetyl-coenzyme A carboxylase (ACC). In agreement with these observations, ApoA-I also increased glucose uptake in isolated mouse soleus muscle. Furthermore, in ApoA-I-deficient mice, AMPK phosphorylation status was decreased in skeletal muscle as well as in liver (Han et al. 2007). These data suggest that in vivo, ApoA-I levels are able to regulate AMPK phosphorylation. In ApoA-I$^{-/-}$ mice, expression of liver enzymes regulating gluconeogenesis (e.g., phosphoenolpyruvate carboxykinase, glucose-6-phosphatase) was also increased (Han et al. 2007), suggesting that hepatic glucose output is increased in animals lacking the main HDL apolipoprotein.

HDLs were also found to directly enhance glucose oxidation by increasing glycolysis and mitochondrial respiration rate in C2C12 muscle cells (Lehti et al. 2013). ATP synthesis was blunted in mitochondria isolated from the gastrocnemius muscle of ApoA-I knockout mice. Endurance capacity of ApoA-I knockout mice was reduced upon exercise exhaustion testing. Conversely, mice transgenically overexpressing human ApoA-I exhibited increased lactate levels, reduced fat mass, and protection against age-induced decline of endurance capacity. Circulating levels of fibroblast growth factor 21, a factor potentially involved in mitochondrial respiratory chain activity and inhibition of white adipose lipolysis, were significantly reduced in ApoA-I transgenic mice (Lehti et al. 2013).

3.3 The Beneficial Effects of HDL in Pancreatic Beta Cells

HDLs also exert potentially antidiabetogenic functions on pancreatic beta cells by potently inhibiting stress-induced cell death and possibly enhancing glucose-stimulated insulin secretion.

3.3.1 Beneficial Effects of HDL on Insulin Secretion

In vivo and in vitro evidence suggests that HDLs support the insulin secretory capacity of beta cells (Drew et al. 2012). Infusion of reconstituted HDLs in type 2 diabetes patients increased their HOMA-B index, an indirect measurement of pancreatic beta cell function (Drew et al. 2009). Also, the increase of HDL cholesterol levels in healthy volunteers treated for 2 weeks with a CETP inhibitor was found to increase postprandial insulin and C-peptide plasma levels (Siebel et al. 2013). The plasma of the CETP inhibited volunteers but not the CETP inhibitor itself also showed an increased capacity to enhance glucose-stimulated insulin secretion from MIN6 cells.

In vitro data on the effects of HDL or ApoA-I on insulin secretion are however controversial. Fryirs and collaborators showed that stimulation of the Min6 insulinoma cell line for 1 h with lipid-free recombinant ApoA-I, ApoA-II, or discoidal reconstituted HDLs dose-dependently increased both basal and glucose-stimulated insulin secretion (Fryirs et al. 2010). By RNA interference, the authors showed that the stimulatory effect of lipid-free ApoA-I and reconstituted HDLs on insulin secretion depended on ABCA1 and ABCG1, respectively (Fryirs et al. 2010). By contrast, other experiments in Min6 cells as well as islets of mice and humans did not find conclusive evidence that HDL enhances insulin production or basal and glucose-stimulated insulin secretion (Abderrahmani et al. 2007; Roehrich et al. 2003; Rutti et al. 2009). Therefore and considering the supra-physiological concentrations of lipid-free ApoA-I used by Fryirs et al., it appears uncertain of whether HDLs have a direct effect on the insulin secretory capacity of beta cells.

However, there is good evidence that ABCA1 and ABCG1 modulate insulin secretion from pancreatic beta cells. Mice with a targeted knockout of ABCA1 in pancreatic beta cells and crossbred with hypercholesterolemic LDL-receptor knockout mice were found to be less glucose tolerant than LDL-receptor knockout only mice (Brunham et al. 2007). The beta cell-specific ABCA1 knockout mice also showed reduced insulin secretion in response to glucose administration. Islets isolated from these mice showed altered cholesterol homeostasis and impaired insulin secretion in vitro (Brunham et al. 2007). In vitro, beta cells lacking ABCA1 showed impaired depolarization-induced insulin granule release, disturbances in membrane micro-domain organization, and alteration in Golgi and insulin granule morphology. Acute cholesterol depletion rescued the exocytotic defect in beta cells lacking ABCA1, suggesting that elevated islet cholesterol accumulation directly impairs granule fusion and insulin secretion (Kruit et al. 2011). Posttranscriptional suppression of ABCA1 in beta cells by adenoviral overexpression of miR33a and miR145 also led to increased cholesterol levels and to decreased glucose-stimulated insulin secretion (Kang et al. 2013; Wijesekara et al. 2012). This compromised insulin secretion was again rescued by cholesterol depletion. Inhibition of miR33a expression in apolipoprotein E knockout islets and ABCA1 overexpression in beta cell-specific ABCA1 knockout islets rescued normal insulin secretion and reduced islet cholesterol (Wijesekara et al. 2012). ABCG1 knockout mice also show glucose intolerance due to reduced insulin secretion (Sturek et al. 2010). In contrast to ABCA1 knockout, this defect was rescued by the addition of cholesterol rather than by depletion of cholesterol (Sturek et al. 2010). These studies show that a delicate balance of cholesterol concentrations between different subcellular compartments must be achieved to allow optimal beta cell functionality. As the consequence of the nonredundant roles of ABCA1 and ABCG1 in beta cell activity, the combined inactivation of ABCA1 and ABCG1 increased intracellular cholesterol accumulation and induced inflammation in beta cells and aggravated the diabetic phenotype found in the single knockout animals (Kruit et al. 2012).

Taken together, these findings indicate that cholesterol homeostasis and its regulation by ABCA1 and ABCG1 are critical for the secretory function of beta cells. However, the human genetic data mentioned before do not unequivocally support the findings made in the genetic mouse models. On the one hand, decreased glucose-induced insulin secretion has been reported in ABCA1-deficient patients with Tangier disease or heterozygous carriers of ABCA1 mutations (Koseki et al. 2009; Vergeer et al. 2010). On the other hand, mutations in ABCA1 have not been associated with increased risk of diabetes (Schou et al. 2012).

3.3.2 Effects of HDL on Pancreatic Beta Cell Survival and ER Stress

Several studies have shown that HDLs are very efficient in inhibiting apoptosis of beta cells induced by a variety of stimuli including inflammatory cytokines, free fatty acids (e.g., palmitate), thapsigargin, tunicamycin, or protein overexpression (Abderrahmani et al. 2007; Petremand et al. 2012; Roehrich et al. 2003; Rutti et al. 2009). Many of these stimuli induce endoplasmic reticulum (ER) stress. As ER stress has been proposed to be a driving parameter in beta cell dysfunction and death in the course of diabetes development (Eizirik et al. 2008; Oyadomari et al. 2002; Volchuk and Ron 2010), the capacity of HDLs to protect beta cells from ER stressors could be one mechanism underlying their potential ability to prevent T2DM.

Pancreatic beta cells are professional secretory cells. Half of the proteins produced by these cells are insulin. These cells have therefore developed an extensive ER network to fulfill their physiological function (Like and Chick 1970; Marsh et al. 2001). In response to the development of insulin resistance, triggered by obesity, for example, more insulin needs to be secreted by beta cells to compensate for insulin resistance and to achieve proper glycemic control. Such a sustained demand in insulin production can eventually cause ER dysfunction leading to chronic ER stress. This may promote beta cell dysfunction and ultimately beta cell death (Eizirik et al. 2008; Eizirik and Cnop 2010; Volchuk and Ron 2010). There is a wealth of evidence demonstrating that HDLs activate antiapoptotic responses in beta cells (Abderrahmani et al. 2007; Cnop et al. 2002; Petremand et al. 2009; Rutti et al. 2009). HDLs are particularly potent in inhibiting beta cell death induced by ER stressors (Petremand et al. 2009; Rutti et al. 2009). Recently, it was shown that HDLs maintain, despite the presence of ER stressors, a proper ER morphology and functionality in terms of the capacity of the ER to fold proteins and to allow protein trafficking out of the ER. Maintenance of the functionality of the ER was shown to be required for the beta cell protective activity of HDLs (Petremand et al. 2012). The ER stressors used included palmitate that is thought to mimic pro-diabetogenic conditions found in individuals with impaired glucose tolerance that are at risk of developing diabetes (Biden et al. 2004; Maedler et al. 2001; Shimabukuro et al. 1998). While this work identified a cellular mechanism mediating the beneficial effect of HDLs on beta cells against pro-diabetogenic factors, the underlying molecular mechanisms remain largely unresolved.

The agonist for the potentially antidiabetic activity of HDL on beta cell apoptosis has not yet been identified. The available information is mostly of negative nature. In beta cells, in contrast to endothelial cells, the SR-BI receptor is not required for HDL-mediated antiapoptotic responses (Petremand et al. 2012; Rutti et al. 2009). There is also no evidence that Akt, which plays a pivotal role in mediating the antiapoptotic activity of HDL in endothelial cells, participates in the protective function of HDLs in beta cells (Puyal et al. 2013). It appears therefore that the molecular mechanisms underlying the antiapoptotic property of HDLs in beta cells differ from those employed in endothelial cells. These mechanisms remain unknown (von Eckardstein and Widmann 2014).

Conceivably, individual components of HDLs may mediate their protective functions in beta cells. Actually, there is evidence that this could be the case. In primary islets, IL1β-induced beta cell apoptosis can be inhibited by either the delipidated HDL protein or the deproteinated lipid moieties of HDL as well as by ApoA-I and by S1P (Rutti et al. 2009). This is a good indication that proteins and lipids found in HDL particles carry pro-survival activities. However, pancreatic islets do not only contain beta cells (70–80 %) but also alpha cells (15–20 %), delta cells, PP cells, and immigrating blood-borne cells such as macrophages. Therefore, it is not clear whether ApoA-I and S1P exert their antiapoptotic effects on beta cells directly or indirectly, for example, by interfering with the release of proapoptotic signals from non-beta cells. To distinguish direct from indirect effects and to unravel the molecular basis for the antiapoptotic effects, it will be essential to substitute the complex model of pancreatic islets with purified beta cells or surrogate beta cell lines.

Conclusion

HDL possibly enhances glucose-stimulated insulin secretion (Drew et al. 2012; Fryirs et al. 2010; von Eckardstein and Sibler 2011) and reverses the proapoptotic effects of native and oxidized LDLs, IL1β, or thapsigargin (Abderrahmani et al. 2007; Petremand et al. 2012; Roehrich et al. 2003; Rutti et al. 2009). In addition, HDL and ApoA-I promote glucose uptake and activate AMP-kinase in primary human skeletal muscle cells and differentiated adipocytes by an insulin-independent way (Drew et al. 2009; Han et al. 2007; Zhang et al. 2011) and enhance oxidative metabolism in skeletal muscle through phosphorylation of acetyl-CoA carboxylase (Drew et al. 2009). Unfortunately, the identity of the agonists and their cognate receptors as well as the elicited signaling pathways and downstream targets of the antidiabetogenic HDL activities are yet poorly resolved. Understanding these pathways is a prerequisite for the development of drugs that stimulate or mimic the antidiabetic effects of HDLs and thereby help to lower the risk, for example, in overweight patients, to manifest T2DM.

Open Access This chapter is distributed under the terms of the Creative Commons Attribution Noncommercial License, which permits any noncommercial use, distribution, and reproduction in any medium, provided the original author(s) and source are credited.

References

Abderrahmani A, Niederhauser G, Favre D, Abdelli S, Ferdaoussi M, Yang JY, Regazzi R, Widmann C, Waeber G (2007) Human high-density lipoprotein particles prevent activation of the JNK pathway induced by human oxidised low-density lipoprotein particles in pancreatic b cells. Diabetologia 50:1304–1314

Banting FG, Best CH, Collip JB, Campbell WR, Fletcher AA (1922) Pancreatic extracts in the treatment of diabetes mellitus. Can Med Assoc J 12:141–146

Barter PJ, Rye KA, Tardif JC, Waters DD, Boekholdt SM, Breazna A, Kastelein JJ (2011) Effect of torcetrapib on glucose, insulin, and hemoglobin A1c in subjects in the Investigation of Lipid Level Management to Understand its Impact in Atherosclerotic Events (ILLUMINATE) trial. Circulation 124:555–562

Besler C, Luscher TF, Landmesser U (2012) Molecular mechanisms of vascular effects of High-density lipoprotein: alterations in cardiovascular disease. EMBO Mol Med 4:251–268

Biden TJ, Robinson D, Cordery D, Hughes WE, Busch AK (2004) Chronic effects of fatty acids on pancreatic b-cell function: new insights from functional genomics. Diabetes 53(Suppl 1): S159–S165

Borggreve SE, De VR, Dullaart RP (2003) Alterations in high-density lipoprotein metabolism and reverse cholesterol transport in insulin resistance and type 2 diabetes mellitus: role of lipolytic enzymes, lecithin:cholesterol acyltransferase and lipid transfer proteins. Eur J Clin Invest 33:1051–1069

Bray GA (1996) Methods and obesity research: the radioimmunoassay of insulin. Obes Res 4:579–582

Brown MS, Goldstein JL (2008) Selective versus total insulin resistance: a pathogenic paradox. Cell Metab 7:95–96

Brunham LR, Kruit JK, Pape TD, Timmins JM, Reuwer AQ, Vasanji Z, Marsh BJ, Rodrigues B, Johnson JD, Parks JS, Verchere CB, Hayden MR (2007) b-cell ABCA1 influences insulin secretion, glucose homeostasis and response to thiazolidinedione treatment. Nat Med 13:340–347

Chen ES, Mazzotti DR, Furuya TK, Cendoroglo MS, Ramos LR, Araujo LQ, Burbano RR, de Arruda Cardoso SM (2009) Apolipoprotein A1 gene polymorphisms as risk factors for hypertension and obesity. Clin Exp Med 9:319–325

Chirieac DV, Chirieac LR, Corsetti JP, Cianci J, Sparks CE, Sparks JD (2000) Glucose-stimulated insulin secretion suppresses hepatic triglyceride-rich lipoprotein and apoB production. Am J Physiol Endocrinol Metab 279:E1003–E1011

Chorev M, Goodman M (1995) Recent developments in retro peptides and proteins – an ongoing topochemical exploration. Trends Biotechnol 13:438–445

Cnop M, Hannaert JC, Grupping AY, Pipeleers DG (2002) Low density lipoprotein can cause death of islet beta-cells by its cellular uptake and oxidative modification. Endocrinology 143:3449–3453

Di AE, Sarwar N, Perry P, Kaptoge S, Ray KK, Thompson A, Wood AM, Lewington S, Sattar N, Packard CJ, Collins R, Thompson SG, Danesh J (2009) Major lipids, apolipoproteins, and risk of vascular disease. JAMA 302:1993–2000

Drew BG, Duffy SJ, Formosa MF, Natoli AK, Henstridge DC, Penfold SA, Thomas WG, Mukhamedova N, de Court FJM, Yap FY, Kaye DM, van Hall G, Febbraio MA, Kemp BE, Sviridov D, Steinberg GR, Kingwell BA (2009) High-density lipoprotein modulates glucose metabolism in patients with type 2 diabetes mellitus. Circulation 119:2103–2111

Drew BG, Rye KA, Duffy SJ, Barter P, Kingwell BA (2012) The emerging role of HDL in glucose metabolism. Nat Rev Endocrinol 8:237–245

Eisenberg S (1984) High density lipoprotein metabolism. J Lipid Res 25:1017–1058

Eizirik DL, Cnop M (2010) ER stress in pancreatic beta cells: the thin red line between adaptation and failure. Sci Signal 3:e7

Eizirik DL, Cardozo AK, Cnop M (2008) The role for endoplasmic reticulum stress in diabetes mellitus. Endocr Rev 29:42–61

Fischer PM (2003) The design, synthesis and application of stereochemical and directional peptide isomers: a critical review. Curr Protein Pept Sci 4:339–356

Fried SK, Russell CD, Grauso NL, Brolin RE (1993) Lipoprotein lipase regulation by insulin and glucocorticoid in subcutaneous and omental adipose tissues of obese women and men. J Clin Invest 92:2191–2198

Fryirs MA, Barter PJ, Appavoo M, Tuch BE, Tabet F, Heather AK, Rye KA (2010) Effects of high-density lipoproteins on pancreatic beta-cell insulin secretion. Arterioscler Thromb Vasc Biol 30:1642–1648

Ginsberg HN (1998) Lipoprotein physiology. Endocrinol Metab Clin North Am 27:503–519

Grundy SM (2012) Pre-diabetes, metabolic syndrome, and cardiovascular risk. J Am Coll Cardiol 59:635–643

Han R, Lai R, Ding Q, Wang Z, Luo X, Zhang Y, Cui G, He J, Liu W, Chen Y (2007) Apolipoprotein A-I stimulates AMP-activated protein kinase and improves glucose metabolism. Diabetologia 50:1960–1968

Horton JD, Goldstein JL, Brown MS (2002) SREBPs: activators of the complete program of cholesterol and fatty acid synthesis in the liver. J Clin Invest 109:1125–1131

Inzucchi SE (2012) Clinical practice. Diagnosis of diabetes. N Engl J Med 367:542–550

Kang MH, Zhang LH, Wijesekara N, de Haan W, Butland S, Bhattacharjee A, Hayden MR (2013) Regulation of ABCA1 protein expression and function in hepatic and pancreatic islet cells by miR-145. Arterioscler Thromb Vasc Biol 33:2724–2732

Koseki M, Matsuyama A, Nakatani K, Inagaki M, Nakaoka H, Kawase R, Yuasa-Kawase M, Tsubakio-Yamamoto K, Masuda D, Sandoval JC, Ohama T, Nakagawa-Toyama Y, Matsuura F, Nishida M, Ishigami M, Hirano K, Sakane N, Kumon Y, Suehiro T, Nakamura T, Shimomura I, Yamashita S (2009) Impaired insulin secretion in four Tangier disease patients with ABCA1 mutations. J Atheroscler Thromb 16:292–296

Kruit JK, Wijesekara N, Fox JE, Dai XQ, Brunham LR, Searle GJ, Morgan GP, Costin AJ, Tang R, Bhattacharjee A, Johnson JD, Light PE, Marsh BJ, Macdonald PE, Verchere CB, Hayden MR (2011) Islet cholesterol accumulation due to loss of ABCA1 leads to impaired exocytosis of insulin granules. Diabetes 60:3186–3196

Kruit JK, Wijesekara N, Westwell-Roper C, Vanmierlo T, de Haan W, Bhattacharjee A, Tang R, Wellington CL, LutJohann D, Johnson JD, Brunham LR, Verchere CB, Hayden MR (2012) Loss of both ABCA1 and ABCG1 results in increased disturbances in islet sterol homeostasis, inflammation, and impaired beta-cell function. Diabetes 61:659–664

Laios K, Karamanou M, Saridaki Z, Androutsos G (2012) Aretaeus of Cappadocia and the first description of diabetes. Hormones (Athens) 11:109–113

Lehti M, Donelan E, Abplanalp W, Al-Massadi O, Habegger KM, Weber J, Ress C, Mansfeld J, Somvanshi S, Trivedi C, Keuper M, Ograjsek T, Striese C, Cucuruz S, Pfluger PT, Krishna R, Gordon SM, Silva RA, Luquet S, Castel J, Martinez S, D'Alessio D, Davidson WS, Hofmann SM (2013) High-density lipoprotein maintains skeletal muscle function by modulating cellular respiration in mice. Circulation 128:2364–2371

Li N, Fu J, Koonen DP, Kuivenhoven JA, Snieder H, Hofker MH (2014) Are hypertriglyceridemia and low HDL causal factors in the development of insulin resistance? Atherosclerosis 233:130–138

Like AA, Chick WL (1970) Studies in the diabetic mutant mouse. II. Electron microscopy of pancreatic islets. Diabetologia 6:216–242

Lin MC, Gordon D, Wetterau JR (1995) Microsomal triglyceride transfer protein (MTP) regulation in HepG2 cells: insulin negatively regulates MTP gene expression. J Lipid Res 36:1073–1081

Maedler K, Spinas GA, Dyntar D, Moritz W, Kaiser N, Donath MY (2001) Distinct effects of saturated and monounsaturated fatty acids on beta-cell turnover and function. Diabetes 50:69–76

Malmstrom R, Packard CJ, Caslake M, Bedford D, Stewart P, Yki-Jarvinen H, Shepherd J, Taskinen MR (1997a) Defective regulation of triglyceride metabolism by insulin in the liver in NIDDM. Diabetologia 40:454–462

Malmstrom R, Packard CJ, Watson TD, Rannikko S, Caslake M, Bedford D, Stewart P, Yki-Jarvinen H, Shepherd J, Taskinen MR (1997b) Metabolic basis of hypotriglyceridemic effects of insulin in normal men. Arterioscler Thromb Vasc Biol 17:1454–1464

Mann CJ, Yen FT, Grant AM, Bihain BE (1991) Mechanism of plasma cholesteryl ester transfer in hypertriglyceridemia. J Clin Invest 88:2059–2066

Marsh BJ, Mastronarde DN, Buttle KF, Howell KE, McIntosh JR (2001) Organellar relationships in the Golgi region of the pancreatic beta cell line, HIT-T15, visualized by high resolution electron tomography. Proc Natl Acad Sci USA 98:2399–2406

Michod D, Annibaldi A, Schaefer S, Dapples C, Rochat B, Widmann C (2009) Effect of RasGAP N2 fragment-derived peptide on tumor growth in mice. J Natl Cancer Inst 101:828–832

Morcillo S, Cardona F, Rojo-Martinez G, Esteva I, Ruiz-de-Adana MS, Tinahones F, Gomez-Zumaquero JM, Soriguer F (2005) Association between MspI polymorphism of the APO AI gene and type 2 diabetes mellitus. Diabet Med 22:782–788

Murao K, Wada Y, Nakamura T, Taylor AH, Mooradian AD, Wong NC (1998) Effects of glucose and insulin on rat apolipoprotein A-I gene expression. J Biol Chem 273:18959–18965

Navab M, Anantharamaiah GM, Reddy ST, Hama S, Hough G, Grijalva VR, Yu N, Ansell BJ, Datta G, Garber DW, Fogelman AM (2005) Apolipoprotein A-I mimetic peptides. Arterioscler Thromb Vasc Biol 25:1325–1331

Ou J, Tu H, Shan B, Luk A, DeBose-Boyd RA, Bashmakov Y, Goldstein JL, Brown MS (2001) Unsaturated fatty acids inhibit transcription of the sterol regulatory element-binding protein-1c (SREBP-1c) gene by antagonizing ligand-dependent activation of the LXR. Proc Natl Acad Sci USA 98:6027–6032

Oyadomari S, Araki E, Mori M (2002) Endoplasmic reticulum stress-mediated apoptosis in pancreatic beta-cells. Apoptosis 7:335–345

Peterson SJ, Kim DH, Li M, Positano V, Vanella L, Rodella LF, Piccolomini F, Puri N, Gastaldelli A, Kusmic C, L'Abbate A, Abraham NG (2009) The L-4F mimetic peptide prevents insulin resistance through increased levels of HO-1, pAMPK, and pAKT in obese mice. J Lipid Res 50:1293–1304

Petremand J, Bulat N, Butty AC, Poussin C, Rutti S, Au K, Ghosh S, Mooser V, Thorens B, Yang J-Y, Widmann C, Waeber G (2009) Involvement of 4E-BP1 in the protection induced by HDLs on pancreatic beta cells. Mol Endocrinol 23:1572–1586

Petremand J, Puyal J, Chatton JY, Duprez J, Allagnat F, Frias M, James RW, Waeber G, Jonas JC, Widmann C (2012) HDLs protect pancreatic beta-cells against ER stress by restoring protein folding and trafficking. Diabetes 61:1100–1111

Puyal J, Petremand J, Dubuis G, Rummel C, Widmann C (2013) HDLs protect the MIN6 insulinoma cell line against tunicamycin-induced apoptosis without inhibiting ER stress and without restoring ER functionality. Mol Cell Endocrinol 381:291–301

Rader DJ (2006) Molecular regulation of HDL metabolism and function: implications for novel therapies. J Clin Invest 116:3090–3100

Rashid S, Watanabe T, Sakaue T, Lewis GF (2003) Mechanisms of HDL lowering in insulin resistant, hypertriglyceridemic states: the combined effect of HDL triglyceride enrichment and elevated hepatic lipase activity. Clin Biochem 36:421–429

Rayner KJ, Esau CC, Hussain FN, McDaniel AL, Marshall SM, van Gils JM, Ray TD, Sheedy FJ, Goedeke L, Liu X, Khatsenko OG, Kaimal V, Lees CJ, Fernandez-Hernando C, Fisher EA, Temel RE, Moore KJ (2011) Inhibition of miR-33a/b in non-human primates raises plasma HDL and lowers VLDL triglycerides. Nature 478:404–407

Reaven GM (1988) Banting lecture 1988. Role of insulin resistance in human disease. Diabetes 37:1595–1607

Reaven GM (2005) Why Syndrome X? From Harold Himsworth to the insulin resistance syndrome. Cell Metab 1:9–14

Roehrich ME, Mooser V, Lenain V, Herz J, Nimpf J, Azhar S, Bideau M, Capponi A, Nicod P, Haefliger JA, Waeber G (2003) Insulin-secreting b-cell dysfunction induced by human lipoproteins. J Biol Chem 278:18368–18375

Rohrer L, Hersberger M, von Eckardstein A (2004) High density lipoproteins in the intersection of diabetes mellitus, inflammation and cardiovascular disease. Curr Opin Lipidol 15:269–278

Ruan X, Li Z, Zhang Y, Yang L, Pan Y, Wang Z, Feng GS, Chen Y (2011) Apolipoprotein A-I possesses an anti-obesity effect associated with increase of energy expenditure and up-regulation of UCP1 in brown fat. J Cell Mol Med 15:763–772

Rutti S, Ehses JA, Sibler RA, Prazak R, Rohrer L, Georgopoulos S, Meier DT, Niclauss N, Berney T, Donath MY, von Eckardstein A (2009) Low and high-density lipoproteins modulate function, apoptosis and proliferation of primary human and murine pancreatic beta cells. Endocrinology 150:4521–4530

Rye KA, Barter PJ (2012) Predictive value of different HDL particles for the protection against or risk of coronary heart disease. Biochim Biophys Acta 1821:473–480

Schou J, Tybjaerg-Hansen A, Moller HJ, Nordestgaard BG, Frikke-Schmidt R (2012) ABC transporter genes and risk of type 2 diabetes: a study of 40,000 individuals from the general population. Diabetes Care 35:2600–2606

Shimabukuro M, Zhou YT, Levi M, Unger RH (1998) Fatty acid-induced b cell apoptosis: a link between obesity and diabetes. Proc Natl Acad Sci USA 95:2498–2502

Siebel AL, Natoli AK, Yap FY, Carey AL, Reddy-Luthmoodoo M, Sviridov D, Weber CI, Meneses-Lorente G, Maugeais C, Forbes JM, Kingwell BA (2013) Effects of high-density lipoprotein elevation with cholesteryl ester transfer protein inhibition on insulin secretion. Circ Res 113:167–175

Sturek JM, Castle JD, Trace AP, Page LC, Castle AM, Evans-Molina C, Parks JS, Mirmira RG, Hedrick CC (2010) An intracellular role for ABCG1-mediated cholesterol transport in the regulated secretory pathway of mouse pancreatic beta cells. J Clin Invest 120:2575–2589

Tabak AG, Jokela M, Akbaraly TN, Brunner EJ, Kivimaki M, Witte DR (2009) Trajectories of glycaemia, insulin sensitivity, and insulin secretion before diagnosis of type 2 diabetes: an analysis from the Whitehall II study. Lancet 373:2215–2221

Tan KC, Shiu SW, Chu BY (1999) Roles of hepatic lipase and cholesteryl ester transfer protein in determining low density lipoprotein subfraction distribution in Chinese patients with non-insulin-dependent diabetes mellitus. Atherosclerosis 145:273–278

Tang C, Oram JF (2009) The cell cholesterol exporter ABCA1 as a protector from cardiovascular disease and diabetes. Biochim Biophys Acta 1791:563–572

Taskinen MR (2003) Diabetic dyslipidaemia: from basic research to clinical practice. Diabetologia 46:733–749

Uehara Y, Engel T, Li Z, Goepfert C, Rust S, Zhou X, Langer C, Schachtrup C, Wiekowski J, Lorkowski S, Assmann G, von Eckardstein A (2002) Polyunsaturated fatty acids and acetoacetate downregulate the expression of the ATP-binding cassette transporter A1. Diabetes 51:2922–2928

Uehara Y, Miura S, von Eckardstein A, Abe S, Fujii A, Matsuo Y, Rust S, Lorkowski S, Assmann G, Yamada T, Saku K (2007) Unsaturated fatty acids suppress the expression of the ATP-binding cassette transporter G1 (ABCG1) and ABCA1 genes via an LXR/RXR responsive element. Atherosclerosis 191:11–21

Umemoto T, Han CY, Mitra P, Averill MM, Tang C, Goodspeed L, Omer M, Subramanian S, Wang S, Den Hartigh LJ, Wei H, Kim EJ, Kim J, O'Brien KD, Chait A (2013) Apolipoprotein AI and high-density lipoprotein have anti-inflammatory effects on adipocytes via cholesterol transporters: ATP-binding cassette A-1, ATP-binding cassette G-1, and scavenger receptor B-1. Circ Res 112:1345–1354

Van Lenten BJ, Wagner AC, Anantharamaiah GM, Navab M, Reddy ST, Buga GM, Fogelman AM (2009) Apolipoprotein A-I mimetic peptides. Curr Atheroscler Rep 11:52–57

Van Linthout S, Foryst-Ludwig A, Spillmann F, Peng J, Feng Y, Meloni M, Van Craeyveld E, Kintscher U, Schultheiss HP, De GB, Tschope C (2010) Impact of HDL on adipose tissue metabolism and adiponectin expression. Atherosclerosis 210:438–444

Vergeer M, Brunham LR, Koetsveld J, Kruit JK, Verchere CB, Kastelein JJ, Hayden MR, Stroes ES (2010) Carriers of loss-of-function mutations in ABCA1 display pancreatic beta-cell dysfunction. Diabetes Care 33:869–874

Villarreal-Molina MT, Flores-Dorantes MT, Arellano-Campos O, Villalobos-Comparan M, Rodriguez-Cruz M, Miliar-Garcia A, Huertas-Vazquez A, Menjivar M, Romero-Hidalgo S, Wacher NH, Tusie-Luna MT, Cruz M, Aguilar-Salinas CA, Canizales-Quinteros S (2008) Association of the ATP-binding cassette transporter A1 R230C variant with early-onset type 2 diabetes in a Mexican population. Diabetes 57:509–513

Volchuk A, Ron D (2010) The endoplasmic reticulum stress response in the pancreatic beta-cell. Diabetes Obes Metab 12(Suppl 2):48–57

von Eckardstein A, Sibler RA (2011) Possible contributions of lipoproteins and cholesterol to the pathogenesis of diabetes mellitus type 2. Curr Opin Lipidol 22:26–32

von Eckardstein A, Widmann C (2014) HDL, beta cells and diabetes. Cardiovasc Res 103(3):384–394

von Eckardstein A, Schulte H, Assmann G (2000) Risk for diabetes mellitus in middle-aged Caucasian male participants of the PROCAM study: implications for the definition of impaired fasting glucose by the American Diabetes Association. J Clin Endocrinol Metab 85:3101–3108

Wang Y, Oram JF (2005) Unsaturated fatty acids phosphorylate and destabilize ABCA1 through a phospholipase D2 pathway. J Biol Chem 280:35896–35903

Wang Y, Oram JF (2007) Unsaturated fatty acids phosphorylate and destabilize ABCA1 through a protein kinase Cd pathway. J Lipid Res 48:1062–1068

Wang S, Peng D (2012) Regulation of adipocyte autophagy – the potential anti-obesity mechanism of high density lipoprotein and ApolipoproteinA-I. Lipids Health Dis 11:131

Wijesekara N, Zhang LH, Kang MH, Abraham T, Bhattacharjee A, Warnock GL, Verchere CB, Hayden MR (2012) miR-33a modulates ABCA1 expression, cholesterol accumulation, and insulin secretion in pancreatic islets. Diabetes 61:653–658

Yoshikawa T, Shimano H, Yahagi N, Ide T, Amemiya-Kudo M, Matsuzaka T, Nakakuki M, Tomita S, Okazaki H, Tamura Y, Iizuka Y, Ohashi K, Takahashi A, Sone H, Osuga JJ, Gotoda T, Ishibashi S, Yamada N (2002) Polyunsaturated fatty acids suppress sterol regulatory element-binding protein 1c promoter activity by inhibition of liver X receptor (LXR) binding to LXR response elements. J Biol Chem 277:1705–1711

Zhang Y, McGillicuddy FC, Hinkle CC, O'Neill S, Glick JM, Rothblat GH, Reilly MP (2010) Adipocyte modulation of high-density lipoprotein cholesterol. Circulation 121:1347–1355

Zhang Q, Zhang Y, Feng H, Guo R, Jin L, Wan R, Wang L, Chen C, Li S (2011) High density lipoprotein (HDL) promotes glucose uptake in adipocytes and glycogen synthesis in muscle cells. PLoS ONE 6:e23556

High-Density Lipoprotein: Structural and Functional Changes Under Uremic Conditions and the Therapeutic Consequences

Mirjam Schuchardt, Markus Tölle, and Markus van der Giet

Contents

1 Introduction 425
2 Chronic Kidney Disease: Epidemiology and Pathophysiology 426
3 Dyslipidemia, Lipid-Modulating Therapy, and Cardiovascular Risk in CKD Patients .. 428
   3.1 Dyslipidemia in CKD Patients 428
   3.2 Lipid-Modulating Therapy in CKD Patients 429
   3.3 Protective Effects of HDL on the Kidney 434
4 Structural and Functional Modifications of HDL in CKD Patients 436
   4.1 Dysregulation of Proteins in HDL Metabolism 436
   4.2 Changes of HDL Apolipoproteins (Modifications and Levels) 438
   4.3 Loss of Protective Proteins or Lipids 440
   4.4 Increase of Molecules within HDL with a Fatal Function in the Vascular Wall ... 441
5 Possibility of Functional Restoration of HDL 442
6 Laboratory Tests to Measure HDL Function 443
Conclusion and Perspective 444
References 444

Abstract

High-density lipoprotein (HDL) has attracted interest as a therapeutic target in cardiovascular diseases in recent years. Although many functional mechanisms of the vascular protective effects of HDL have been identified, increasing the HDL plasma level has not been successful in all patient cohorts with increased cardiovascular risk. The composition of the HDL particle is very complex and

---

M. Schuchardt • M. Tölle
Department of Nephrology, Campus Benjamin Franklin, Charité – Universitaetsmedizin Berlin, Charité Centrum 13, Hindenburgdamm 30, 12203 Berlin, Germany

M. van der Giet (✉)
Department of Nephrology, Charité – Campus Benjamin Franklin, Hindenburgdamm 30, 12203 Berlin, Germany
e-mail: Markus.vanderGiet@charite.de

includes diverse lipids and proteins that can be modified in disease conditions. In patients with chronic kidney disease (CKD), the accumulation of uremic toxins, high oxidative stress, and chronic micro-inflammatory conditions contribute to changes in the HDL composition and may also account for protein/lipid modifications. These conditions are associated with a decreased protective function of HDL. Therefore, the HDL quantity and the functional quality of the particle must be considered.

This review summarizes the current knowledge of dyslipidemia in CKD patients, the effects of lipid-modulating therapy, and the structural modifications of HDL that are associated with dysfunction.

Keywords
Chronic kidney disease • Dyslipidemia • High-density lipoprotein • Uremia • Uremic toxin

Abbreviations

ABCA1	ATP-binding cassette subfamily A
ADMA	Asymmetric dimethylarginine
AGE	Advanced glycated end product
AIM-HIGH	Atherothrombosis intervention in metabolic syndrome with low HDL/high triglycerides: impact on global health outcomes
ALERT	Assessment of LEscol in Renal Transplantation
apo	Apolipoprotein
AURORA	A study to evaluate the use of rosuvastatin in subjects on regular hemodialysis: an assessment of survival and cardiovascular events
CETP	Cholesterol ester transfer protein
CKD	Chronic kidney disease
cLDL	Carbamylated low-density lipoprotein
4D	Deutsche Diabetes Dialyse Studie
DEFINE	Determining the Efficacy and Tolerability of CETP inhibition with Anacetrapib
DMA	Dimethylarginine
eGFR	Estimated glomerular filtration rate
EPIC	European Prospective Investigation into Cancer and Nutrition
ESRD	end-stage renal disease
GFR	Glomerular filtration rate
HD	Hemodialysis
HDL	High-density lipoprotein
HDL-C	High-density lipoprotein cholesterol
HMG-CoA	3-hydroxy-3-methylglutaryl-coenzyme A
HPS2-THRIVE	Treatment of HDL to Reduce the Incidence of Vascular Events
IDL	Intermediate-density lipoprotein

ILLUMINATE	Investigation of Lipid Level Management to Understand Its Impact in Atherosclerotic Events
LCAT	Lecithin-cholesterol acyltransferase
LDL	Low-density lipoprotein
LDL-C	Low-density lipoprotein cholesterol
Lp(a)	Lipoprotein a
Lp-PLA	Lipoprotein-associated phospholipase
LPL	Lipoprotein lipase
MESA	Multi-Ethnic Study of Atherosclerosis
MI	Myocardial infarction
MPO	Myeloperoxidase
NO	Nitric oxide
n3-PUFA	n3-polyunsaturated fatty acids
ox-apo	Oxidized apolipoprotein
PAF-AH	Platelet-activating factor acetylhydrolase
PD	Peritoneal dialysis
PLA	Phospholipase
PON	Paraoxonase
PPAR	Peroxisome proliferator-activated receptor
PREVEND-IT	Prevention of Renal and Vascular End-Stage Disease Intervention Trial
RBP4	Retinol-binding protein 4
RCT	Reverse cholesterol transport
RECORD	Rosiglitazone evaluated for cardiovascular outcomes in oral agent combination therapy for type 2 diabetes
REVEAL	Randomized evaluation of the effects of anacetrapib through lipid-modification
ROS	Reactive oxygen species
SAA	Serum amyloid A
SDMA	Symmetric dimethylarginine
SHARP	Study of Heart and Renal Protection
S1P	Sphingosine-1-phosphate
sPLA2	Secretory phospholipase type 2
SR-BI	Scavenger receptor BI
VA-HIT	Veterans' Affairs High-Density Lipoprotein Intervention Trial
VLDL	Very low-density lipoprotein

1 Introduction

Plasma lipoproteins are composed of non-covalent aggregates of different lipids and proteins. They transport water-insoluble substances in the blood by building micelle-like structures. These structures are classified by their density into the following groups: very low-density lipoprotein (VLDL), intermediate-density lipoprotein (IDL), low-density lipoprotein (LDL), and high-density lipoprotein (HDL).

In previous years, several clinical studies have found that HDL cholesterol (HDL-C) plasma levels were inversely correlated with cardiovascular risk, whereas high LDL-C levels were related to increased cardiovascular mortality (Di Angelantonio et al. 2009; Gordon et al. 1977; Mori et al. 1999). Initially, due to the crucial role of HDL in reverse cholesterol transport (RCT), HDL was associated with a decreased cardiovascular risk. Therefore, the theory of the "good HDL" and "bad LDL" emerged. Further studies, primarily in vitro or animal studies, identified the pleiotropic cardiovascular protective effects of HDL in addition to its RCT function: HDL has anti-atherosclerotic, anti-inflammatory, antioxidative, antithrombotic, and endothelial-protective properties (Navab et al. 2011). Accordingly, there is growing interest in the study of HDL metabolism and the cellular and the molecular signaling pathways involved in its vascular protective effects.

Therefore, enhancing HDL plasma levels has been a principal approach for the reduction of cardiovascular risks in different patient cohorts. Pharmacologically active substances, such as cholesterol ester transfer protein (CETP) inhibitors, fibrates, and niacin, have been tested in large clinical trials to evaluate the occurrence of cardiovascular endpoints (Longenecker et al. 2005; van Capelleveen et al. 2014). However, raising HDL-C plasma levels did not protect against cardiovascular events in all patient cohorts tested. In addition, despite a significant HDL-C increase in patients, large clinical trials were terminated because of a lack of positive effects or, in some cases, an increase in the rate of cardiovascular events.

These findings indicate that a better understanding of lipoprotein modifications in disease conditions is necessary to establish possible indications and target mechanisms in the therapy of cardiovascular diseases. Patients with chronic kidney disease (CKD) suffer from a dramatic increase in cardiovascular morbidity and mortality. The disease condition is associated with dyslipidemia and/or modifications in lipoprotein composition and function (Keane et al. 2013; Vaziri 2006).

This review summarizes current knowledge regarding the structural and functional changes of HDL in the context of renal dysfunction. A brief overview of CKD epidemiology and pathophysiology is presented, followed by aspects of dyslipidemia and its therapy in CKD patients. Subsequently, CKD-dependent structural and functional changes of HDL are summarized.

2 Chronic Kidney Disease: Epidemiology and Pathophysiology

Kidney disease is defined as an abnormality of kidney structure or function. The criteria for abnormalities include: albuminuria, urine sediment abnormalities, electrolyte and other abnormalities due to tubular disorders, abnormalities detected by

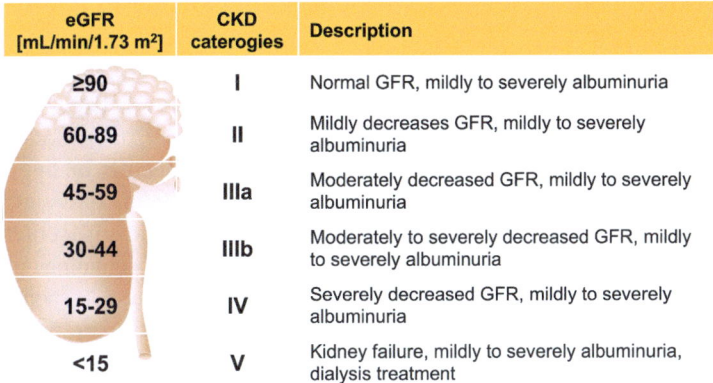

Fig. 1 Categories of chronic renal failure. CKD categories were divided into subgroups according to KDIGO 2013

histology or imaging, and history of kidney transplantation (Levey et al. 2005). Kidney disease can occur abruptly, and it can either resolve or become chronic. CKD is defined as kidney damage that persists for more than 3 months (Levey et al. 2005). CKD is one of the major medical concerns associated with premature morbidity and mortality, especially due to cardiovascular complications. The epidemic of CKD is globally driven by demographic aging, as well as an increase in other risk factors leading to CKD (e.g., diabetes mellitus, hypertension, and obesity).

CKD is divided into different stages, which are summarized in Fig. 1, based on the glomerular filtration rate (GFR) (Levey et al. 2005). CKD patients have a high occurrence of cardiovascular disease. In recent years, several studies have implicated an inverse correlation of kidney function and cardiovascular mortality (Tonelli et al. 2006). Thereby, the cardiovascular phenotype of the CKD population is heterogeneous, and the cardiovascular risk depends on the CKD stage. In addition to traditional risk factors, such as hypertension, diabetes mellitus, and smoking, nontraditional risk factors also contribute to cardiovascular diseases. In CKD patients, the nontraditional risk factors of anemia and enhanced oxidative stress or uremic toxins play a major role in cardiovascular disease progression (Mizobuchi et al. 2009). Under normal conditions, uremic toxins are cleared by the kidneys, and many of these toxins have been recently identified (Duranton et al. 2012). Uremic toxins are classified as water-soluble molecules with low molecular weight (e.g., uric acid), middle molecules (e.g., parathyroid hormone), and protein-bound toxins (e.g., hippuric acid) (Duranton et al. 2012). A number of different uremic toxins are elusive, and biological functions have not been identified for all toxins. Clinical studies have shown that cardiovascular morbidity and mortality in CKD patients are associated with uremic toxin accumulation, which leads to a progression of vascular alterations. Recently, Moradi et al. summarized the influence of uremic toxins on vascular cells and pathological pathways in the vascular wall (Moradi et al. 2013). The activation of leukocytes;

endothelial damage, for example, by the disruption of glycocalyx and production of reactive oxygen species (ROS); the proliferating effects on smooth muscle cells; and platelet activation are only some of the described effects (Moradi et al. 2013). In addition, uremic toxins influence lipoprotein modifications. Urea induces the formation of carbamylated LDL (cLDL) (Apostolov et al. 2010). Some uremic toxins are lipid-bound and may occur in a different subfraction of lipoproteins. Additionally, HDL function may be affected by the accumulation of uremic toxins.

Nonetheless, further studies are necessary to characterize the impact of different compounds on cardiovascular outcome and the precise signaling pathways involved. The therapeutic goal should be to remove the solutes associated with the highest cardiovascular risk to minimize fatal cardiovascular outcomes in patients with CKD. To date, cardiovascular morbidity in patients with CKD remains high despite the utility of dialysis and renal transplantation (Foley et al. 1998).

3 Dyslipidemia, Lipid-Modulating Therapy, and Cardiovascular Risk in CKD Patients

In the general population, plasma lipid levels correlate with the level of cardiovascular risk (Di Angelantonio et al. 2009). Increased LDL-C levels have been associated with cardiovascular mortality. LDL-C is a primary risk biomarker for cardiovascular disease in the general population; however, it loses its prognostic association in CKD patients (Kilpatrick et al. 2007), especially for patients with end-stage renal disease (ESRD) (Baigent et al. 2011; Kovesdy et al. 2007). In addition, traditional risk factors, such as hypercholesterolemia, hypertension, and obesity, no longer appear to be related to the level of cardiovascular risk in ESRD patients. One explanation might be the accumulation of nontraditional risk factors in CKD patients (Appel 2004; Coresh et al. 1998; Muntner et al. 2004). Presumably, the complexity of traditional and nontraditional risk factors determines patient outcome. An alternative explanation may be that patient outcome is determined by the numerous quantitative and qualitative lipid abnormalities in triglycerides, phospholipids, and lipoproteins that are observed in CKD patient cohorts (Keane et al. 2013; Vaziri 2006).

3.1 Dyslipidemia in CKD Patients

Several studies have demonstrated a characteristic switch in the lipid phenotype at different stages of CKD. Data from the "Multi-Ethnic Study of Atherosclerosis" (MESA) have indicated an elevation of triglyceride-rich lipoproteins (Lamprea-Montealegre et al. 2013). Triglyceride-rich VLDL or other apoB-rich lipoproteins accumulate in patients with ESRD as a consequence of increased triglyceride levels (Vaziri and Norris 2011). The VLDL particle increase depends on the CKD stage and is aggravated by a decline in kidney function (Lamprea-Montealegre et al. 2013). Hypertriglyceridemia is caused by reduced lipoprotein lipase (LPL) levels (Vaziri 2006; Vaziri et al. 2012) and results in the limited delivery of triglyceride-rich fuel lipoproteins to adipocytes and myocytes (Vaziri et al. 2012). The enrichment of triglyceride-containing lipoprotein particles results in a higher susceptibility to oxidative modifications. These particles are highly pro-inflammatory (Vaziri 2013). In addition, other apoB-rich lipoproteins, such as small dense LDL or IDL-C, are commonly increased in patients with CKD (Saland and Ginsberg 2007; Vaziri 2006). The reduced clearance rate of LDL particles contributes to the increase in LDL levels (Ikewaki 2013).

Lipoprotein a (Lp(a)), which consists of apoA and apoB and is similar to LDL-C, is increased in CKD patients (Longenecker et al. 2005; Muntner et al. 2004). Lp (a) serves as an independent biomarker that predicts cardiovascular complications (Jacobson 2013; Thompson and Seed 2013).

Dyslipidemia is further enhanced by hyperparathyroidism, which frequently occurs in CKD patients (Moe and Sprague 2012). Normal lipid metabolism depends on proper parathyroid function, and hyperparathyroidism tends to result in hypertriglyceridemia (Liang et al. 1998) as well as in a deficiency in lipid metabolism enzymes, such as hepatic lipase and LPL (Klin et al. 1996). The enrichment of apoB-containing lipoproteins is correlated with decreased HDL-C levels (Saland and Ginsberg 2007; Vaziri 2006).

The lipid profile is altered in CKD patients compared with healthy subjects and differs depending on the stage of CKD progression (e.g., stage I to V, pre/post-dialysis, and time on dialysis) and the dialysis procedure (e.g., hemodialysis (HD) or peritoneal dialysis (PD)). As a consequence, the lipid profiles of the CKD population are heterogeneous. Extrinsic factors, including drug administration for kidney disease and associated disorders (hypertension, diabetes mellitus, and malnutrition), contribute to dyslipidemia in this patient cohort.

To overcome the complex dyslipidemia in CKD patients, diverse clinical trials have addressed the influence of lipid-modulating therapies on cardiovascular outcomes. The results of these therapies in the general population suggest that the incidence of cardiovascular events in CKD patients is expected to decrease (van Capelleveen et al. 2014).

3.2 Lipid-Modulating Therapy in CKD Patients

Currently, drugs that treat dyslipidemia are in use or under investigation. Statins, which primarily lower the LDL-C level; peroxisome proliferator-activated receptor (PPAR) agonists; inhibitors of lipid metabolism (CETP, niacin); and apoAI mimetics have been evaluated in preclinical and/or clinical studies. The currently established lipidemic drugs have been reviewed elsewhere; therefore, this article summarizes aspects of therapy related to CKD. Although lipid-modulating therapies have had a positive impact on the cardiovascular outcome in patients with normal kidney function, the effect in CKD patients has been disappointing. These observations have originated predominantly from statin trials.

Statin treatment leads to a reduction in LDL-C through the inhibition of 3-hydroxy-3-methylglutaryl-coenzyme A (HMG-CoA) reductase, the initial enzyme in endogenous cholesterol biosynthesis. In several clinical trials, it has been clearly demonstrated that for patients without renal disease, a reduction of LDL-C upon statin therapy is associated with a decreased cardiovascular mortality rate (The Long-Term Intervention with Pravastatin in Ischaemic Disease (LIPID) Study Group 1998; Scandinavian Simvastatin Survival Study Group 1994; Baigent et al. 2005; Collins et al. 2003; LaRosa et al. 2005; Sever et al. 2003). Due to the substantially high cardiovascular risk in CKD patients, the effect of statin therapy was also evaluated in three large clinical trials (>1,000 patients included): the "Deutsche Diabetes Dialyse Studie" (4D) (Wanner et al. 2005), "A Study to Evaluate the Use of Rosuvastatin in Subjects on Regular Hemodialysis: An Assessment of Survival and Cardiovascular Events" (AURORA) (Fellstrom et al. 2009a), and "Study of Heart and Renal Protection" (SHARP) (Baigent et al. 2011). The "Prevention of Renal and Vascular End-Stage Disease Intervention Trial" (PREVEND-IT) (Asselbergs et al. 2004; Brouwers et al. 2011) has also addressed the effect of statin therapy in a smaller cohort (<1,000 patients) but included a longer follow-up period. The "Assessment of LEscol in Renal Transplantation" (ALERT) (Holdaas et al. 2003) trial investigated cardiovascular outcomes in renal transplant recipients. The 4D study consisted of a randomized control trial that examined Western HD patients with diabetes (Wanner et al. 2005). No benefits in the primary endpoints were observed in a follow-up period of approximately 4 years (Wanner et al. 2005) (Table 1). A subsequent subgroup analysis of the 4D study cohort showed a reduction in cardiovascular disease after statin treatment in patients with LDL-C >145 mg/dL (Marz et al. 2011). The 4D trial with 1,225 patients was followed by the larger AURORA trial, which enrolled 2,776 HD patients with and without diabetes mellitus. Nonetheless, the reduction in LDL-C that was observed did not have an effect on cardiovascular outcomes (Fellstrom et al. 2009a). Another large randomized trial, the SHARP trial, enrolled HD patients and CKD patients in stages III to V: (Baigent et al. 2011). The combination therapy of simvastatin and ezetimibe lowered the LDL-C level but did not reduce the overall vascular mortality (Baigent et al. 2011). However, the risk for atherosclerotic events was reduced in a wide range of CKD patients (Baigent et al. 2011). The longest follow-up period, up to 9.5 years, was in the PREVEND-IT trial (Asselbergs et al. 2004; Brouwers et al. 2011). Here, patients with microalbuminuria were enrolled and treated with pravastatin and fosinopril. In the follow-up period of 46 months, pravastatin treatment had no effect on cardiovascular endpoints (Asselbergs et al. 2004). The extended follow-up period of 9.5 years showed that elevated urinary albumin excretion was associated with increased cardiovascular morbidity and mortality (Brouwers et al. 2011).

Thus, the large clinical trials examining statin therapy in CKD patients failed to show a significant reduction in overall cardiovascular mortality for patients undergoing dialysis treatment (Table 1). Based on the data from the US Renal Data System, the majority of cardiovascular deaths in CKD patients are due to chronic heart failure and are not influenced by statin treatment (Collins et al. 2013).

Table 1 CKD patients in clinical trials on lipid-modulating therapy

	Study	Follow-up (years)	Drug	Patients	Outcome	References
Statins	4D	3.9	Atorvastatin (20 mg daily)	1,255 patients, DM type II on HD	Decreased rate of all cardiac events, increased rate of fatal stroke	Wanner et al. (2005)
	AURORA	3.2	Rosuvastatin (10 mg daily)	2,776 patients, HD with and without DM	No reduction of cardiovascular mortality, nonfatal MI, nonfatal stroke	Fellstrom et al. (2009b), Holdaas et al. (2011)
	PREVEND-IT	9.5	Pravastatin (40 mg daily), fosinopril (20 mg daily)	864 patients with microalbuminuria	No effect on cardiovascular events	Asselbergs et al. (2004), Brouwers et al. (2011)
	SHARP	4.9	Ezetimibe (10 mg daily) in combination with simvastatin (20 mg daily)	9,270 CKD patients (stage III to V, including dialysis)	17 % reduction in major atherosclerotic events, 15 % reduction in major vascular events	Baigent et al. (2011)
	ALERT	5–6	Fluvastatin	2,102 renal transplant recipients with stable graft function	Fewer cardiac deaths and nonfatal MIs, no reduction in mortality	Holdaas et al. (2003)
Fibrates and fatty acids	VA-HIT	5–7	Gemfibrozil (1,200 mg daily)	1,046 CKD patients with creatinine clearance <75 ml/min	No reduction of overall mortality, decreased cardiovascular events and nonfatal MIs	Rubins et al. (1993), Tonelli et al. (2004)
	OPACH	2	n3-PUFA	717 HD patients	No reduction of cardiovascular events or death	Svensson et al. (2006)

For full study names, please see text

CKD chronic kidney disease, *HD* hemodialysis, *DM* diabetes mellitus, *n3-PUFA* n3-polyunsaturated fatty acids, *MI* myocardial infarction

Therefore, other uremia-related pathways and/or the previously existing vascular damage may contribute to the increase in cardiovascular risk. A trial that investigated the effect of statin therapy in renal transplant recipients observed similar effects. In the ALERT trial (Holdaas et al. 2003), statin therapy reduced the LDL-C plasma concentration in treated patients (Holdaas et al. 2003). Although fewer cardiac deaths and nonfatal MIs were observed, no significant reduction in the primary endpoint was achieved (Holdaas et al. 2003).

Based on the findings of the current clinical trials with CKD patients, the therapies tested had either no real impact on cardiovascular outcome or resulted in fatal effects (Kilpatrick et al. 2007). Interestingly, despite a robust reduction in the LDL-C level, only negative or neutral effects on acute cardiovascular events were measured (Baigent et al. 2011; Fellstrom et al. 2009b; Wanner et al. 2005). Effective prevention of cardiovascular events was only documented in patients with mild to moderate CKD (Baigent et al. 2011).

In addition to statins, which primarily lower LDL-C plasma levels, other lipid-modulating drugs are currently used in non-renal-insufficient patients. Here, different targets in HDL metabolism have been identified to overcome dyslipidemia. Initial studies addressed the effect of enzyme inhibition on lipid metabolism, such as CETP inhibition. PPAR agonists, niacin, and apoAI mimetics are other therapeutic options that have been tested in clinical and/or preclinical studies. However, these studies often excluded patients with CKD from recruitment. Therefore, further studies are necessary to determine the effects of these therapies on patients with renal insufficiency.

CETP is responsible for the shuttling of cholesteryl esters between HDL and apoB-containing lipoprotein particles. Inhibition of this enzyme results in a dramatic increase in HDL levels. Torcetrapib was the first CETP inhibitor examined in a large clinical trial, "Investigation of Lipid Level Management to Understand Its Impact in Atherosclerotic Events" (ILLUMINATE) which included approximately 15,000 patients (Barter et al. 2007). The treatment increased the HDL concentration by more than 50 % (Barter et al. 2007). However, the study was prematurely terminated because an increased risk of cardiovascular events was observed (e.g., increase in arterial blood pressure) (Barter et al. 2007). In the ILLUMINATE trial, patients with severe CKD were not enrolled. The mean eGFR in patients included in the study was 79.5 ml/min/1.73 m^2 with creatinine concentration of 1 mg/dL (Barter et al. 2007). In subsequent trials with other CETP inhibitors, which attempted to overcome the off-target effects of torcetrapib, CKD patients were excluded. The "Determining the Efficacy and Tolerability of CETP Inhibition with Anacetrapib" (DEFINE) study (Cannon et al. 2010) investigated anacetrapib, and a key exclusion criteria in this study was an eGFR <30 ml/min/1.73 m^2 or severe renal impairment (Cannon et al. 2010). In the dal-OUTCOMES trial with dalcetrapib (Schwartz et al. 2009, 2012), the exclusion criteria was a creatinine level of >2.2 mg/dL (Schwartz et al. 2009). For the "Randomized Evaluation of the Effects of Anacetrapib Through Lipid-Modification" (REVEAL) study, patient recruitment is ongoing, and the study is expected to be completed in 2017 (Gutstein et al. 2012). On the basis of the current CETP inhibitor studies, the effects of CETP inhibition on cardiovascular mortality cannot be determined, and patients with severe CKD were excluded from all studies.

Niacin treatment most effectively raises HDL-C levels and decreases triglyceride, LDL-C, and Lp(a) levels. The mechanism underlying the influence of niacin on HDL metabolism is unknown (Linsel-Nitschke and Tall 2005). The beneficial effects on lipoprotein levels and the corresponding cardiovascular outcome have been described in several small-scale clinical trials. Unfortunately, the large-scale trials AIM-HIGH (AIM-HIGH Investigators 2011; Boden et al. 2011) and HPS2-THRIVE (2013), in which patients were treated with niacin in combination with a statin, were prematurely terminated because of a high prevalence of side effects. In both studies, no patients with severe CKD were included. In AIM-HIGH, the mean eGFR was 82.8 ml/min/1.72 m^2 (2011), and in HPS2-THRIVE, severe renal insufficiency was an exclusion criteria for patient recruitment (2013).

The family of PPAR transcription factors is involved in the regulation of fatty acid metabolism and influences lipid levels. The family consists of three members: PPAR-α, PPAR-γ, and PPAR-δ. PPAR-γ agonists, such as rosiglitazone and pioglitazone, are high-affinity agonists at the receptor site (Linsel-Nitschke and Tall 2005), and their effects on cardiovascular outcome have been tested. However, only a minimal benefit or an increase in heart failure rates was observed in a study population with diabetes in the "Rosiglitazone Evaluated for Cardiovascular Outcomes in Oral Agent Combination Therapy for Type 2 Diabetes" (RECORD) trial (Home et al. 2009). The mean serum creatinine level in the study cohort was approximately 62 μmol/L (Home et al. 2009). The activation of PPAR-α via fibrates, which are weak agonists of this receptor (Linsel-Nitschke and Tall 2005), results in reductions in LDL-C and triglycerides, whereas HDL-C plasma levels are moderately increased (Linsel-Nitschke and Tall 2005; Sahebkar et al. 2014). The "Veterans' Affairs High-Density Lipoprotein Intervention Trial" (VA-HIT) (Tonelli et al. 2004) investigated the effects of the fibrate gemfibrozil in a cohort of individuals with CKD. Gemfibrozil treatment did not reduce overall mortality, but the rate of major cardiovascular events and nonfatal myocardial infarctions (MIs) was significantly reduced in patients with mild to moderate CKD (Tonelli et al. 2004). Currently, there are no synthetic ligands for PPAR-δ in clinical use (Sahebkar et al. 2014).

In the secondary prevention of cardiovascular disease, there is evidence that n3-polyunsaturated fatty acids (n3-PUFAs) are effective (Studer et al. 2005). A study with HD patients failed to show a reduction in cardiovascular events or death after treatment with an n3-PUFA (Svensson et al. 2006). However, a significant reduction of MIs was observed in this patient cohort (Svensson et al. 2006).

Additional pharmacotherapeutic strategies are currently under development. apoAI-based drugs and reconstituted/engineered HDL particles have been tested in animal and human studies. Recently, van Capelleveen et al. summarized current knowledge of these novel therapeutics (van Capelleveen et al. 2014). The human trials were completed with a limited patient number (<200 patients) and a short follow-up period (several weeks). However, the clinical impact of these novel drugs is highly interesting and must be evaluated in larger clinical trials. Several trials are currently under way (van Capelleveen et al. 2014); however, the high treatment costs may limit their use in routine clinical practice. After the promising results in animal studies, treatment with apoAI mimetics did not improve biomarkers that were selected to describe HDL function (Watson et al. 2011).

To date, the benefit of apoAI-based therapy for CKD patients remains elusive.

The studies described indicate that with a decline in renal function, lipid-modulating therapies lose their capacity to prevent the occurrence of cardiovascular nonfatal and fatal events. Consequently, the lipid compositional and corresponding functional changes that occur with declining renal function are not fully understood. In Table 1, the different therapeutic approaches in patients with variable renal function and the effects on cardiovascular outcome are summarized.

Recent guidelines for the management of lipid disorders in patients with chronic renal failure recommend lipid-lowering therapy only in patients with CKD who are not undergoing renal replacement therapy (National Kidney Foundation 2003; Tonelli and Wanner 2013). The guidelines clearly indicate our inadequate understanding of how to address the complex lipid changes in CKD patients.

3.3 Protective Effects of HDL on the Kidney

Based on the current evidence, the causal association between the HDL level, HDL function, and cardiovascular outcome in CKD patients is not fully understood. Dysregulated HDL metabolism might be caused by reduced kidney function. Abnormalities in lipid metabolism enzymes and transport processes have been described (Hirano 2013; Pahl et al. 2009; Vaziri and Norris 2011). The changes involve enrichment of free cholesterol, triglycerides, and fatty acids as well as the depletion of cholesterol esters within HDL. However, there is evidence that HDL might be an independent protective factor for kidney function. Epidemiologic data from a multi-study cohort demonstrated an association between a low HDL-C level and reduced kidney function (Odden et al. 2013). Additional trials addressed the question of whether HDL-C is associated with the progression of CKD. A recent study, which included 3,303 patients with CKD stages III to V and a median follow-up period of 2.8 years, supports the hypothesis that dyslipidemia is independently associated with rapid renal progression (Chen et al. 2013). Two meta-analyses suggest that statin therapy has a benefit on GFR and inhibits GFR decline (Fried et al. 2001; Sandhu et al. 2006). However, other studies found no impact on renal outcome with statin treatment (Atthobari et al. 2006; Rahman et al. 2008). Focusing on HDL-C, Baragetti and coworkers observed an association between low HDL-C and earlier entry in dialysis or doubling of the plasma creatinine in a patient cohort of 176 subjects with mild to moderate kidney dysfunction (Baragetti et al. 2013). Moreover, a cross-sectional analysis of 4,925 patients with normal kidney function strengthened the association between HDL-C and eGFR (Wang et al. 2013). Confounders (e.g., age, blood pressure, and lipid parameters) influenced the relationship (Wang et al. 2013). Furthermore, malnutrition and hypoalbuminemia,

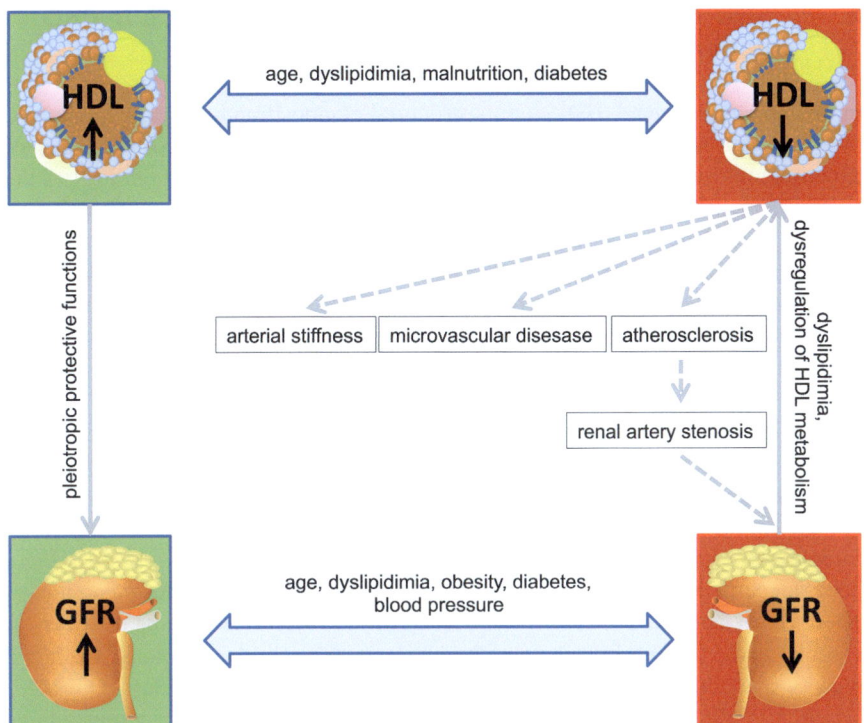

Fig. 2 Interaction of HDL level/function with the kidney. HDL has a pleiotropic protective function. A decrease in HDL level/function has an impact on the vascular alterations associated with kidney damage. The decline in kidney function induces/strengthens the decrease in HDL level/function. For further information and references, refer to the text

which are frequently present in patients with CKD (Kaysen 2009; Peev et al. 2014), affected the HDL level and functionality (Khovidhunkit et al. 2004).

Further studies are necessary to determine whether HDL-targeted therapies are beneficial for renal outcome. The difference in the HDL quantity and functional quality in CKD patients must be considered. The results of clinical trials, such as ILLUMINATE (Barter et al. 2007) and HPS2-THRIVE (2013), suggest that increasing HDL-C plasma levels is not an optimal therapeutic target. Furthermore, as shown in the "European Prospective Investigation into Cancer and Nutrition" (EPIC) study and the MESA study, intima/media thickness and cardiovascular events had a stronger association with HDL particle number compared with HDL cholesterol levels (Arsenault et al. 2009; El Harchaoui et al. 2009; Mackey et al. 2012). Figure 2 summarizes the potential protective effects of HDL on kidney function.

Additional studies that address the structure-dependent functions of the heterogeneous HDL particles are necessary to assess the benefit of HDL-increasing therapy for different disease conditions.

4 Structural and Functional Modifications of HDL in CKD Patients

In the last several years, HDL and its structural composition in disease conditions, especially cardiovascular, metabolic, and renal diseases, have been a focus of experimental research. Compositional changes of the proteome, alterations of the lipid moiety, and posttranslational modifications of HDL isolated from these patient cohorts contribute to a lower protective function of the lipoprotein particle (Annema and von Eckardstein 2013). The following section summarizes the current knowledge of the HDL changes identified in patients with reduced renal function. It is important to keep in mind that the investigations were performed in a heterogeneous group of patients with different stages of renal failure. Therefore, it is often difficult to define the role of uremia on the functionality of HDL.

The HDL particle contains hundreds of different lipids and a multitude of proteins that form a lipoprotein with different shapes, sizes, and densities throughout its metabolism (Camont et al. 2011; Shah et al. 2013). Because of its amphiphilic character and the presence of specific binding proteins, HDL serves as cargo for vitamins, hormones, toxins, and microRNA (Vickers and Remaley 2014). There is evidence of biological activity for many of the lipid and protein components. These components can bind to specific receptors and activate signaling pathways in vascular cells, thereby influencing atherogenesis, thrombosis, apoptosis, oxidative reactions, endothelial properties, and inflammatory reactions (Calabresi et al. 2003; Navab et al. 1991; Nofer et al. 2004; Watson et al. 1995). The functions of the associated proteins and lipids in the vascular wall have been characterized in vitro and in animal studies, as well as in clinical trials. It has been shown that compositional changes are related to functionality and are influenced by uremic toxins. For example, (1) the dysregulation of proteins in the HDL metabolism affects HDL plasma level and cholesterol clearance. Furthermore, (2) changes in HDL apolipoprotein composition occur, and finally (3) the loss of proteins with a protective function or (4) the accumulation of substances with a fatal function within HDL influences its vascular protective properties. The compositional changes may not only result in a decrease in function but may also lead to increased dysfunction (Fig. 3), and both are associated with an increased cardiovascular risk profile.

4.1 Dysregulation of Proteins in HDL Metabolism

As discussed in the preceding paragraph, CKD is associated with a decreased HDL-C plasma concentration and dyslipidemia. The uremic condition is responsible for the disrupted protein metabolism (Vaziri 2006) due to uremic-induced liver damage (Yeung et al. 2013). Many proteins important in HDL metabolism are predominantly produced within the liver, and their production is dysregulated in uremia. For proper HDL maturation and metabolism, enzymes such as LCAT and CETP are responsible. Furthermore, diverse receptors on intestinal and peripheral cells, as well as hepatic cells, are necessary for cholesterol transport via HDL to the

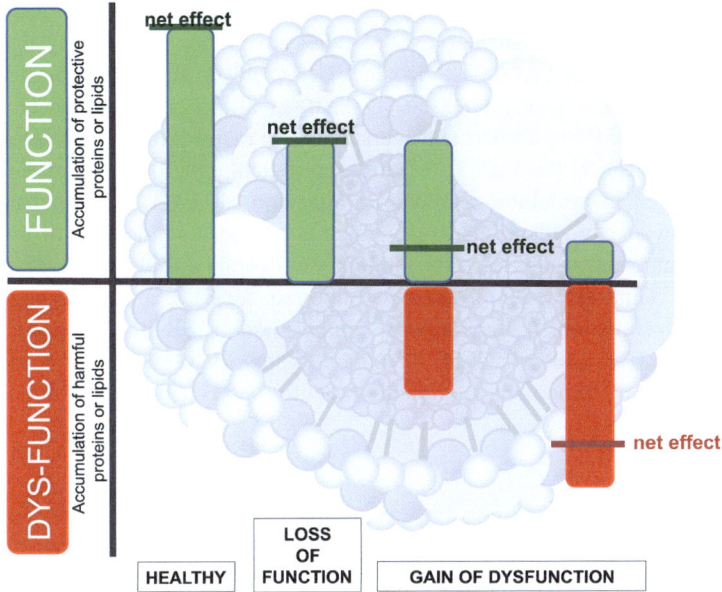

Fig. 3 Functional vs. dysfunctional HDL particles. Depending on the substances that accumulate within the HDL, the particle may exert functional or dysfunctional properties

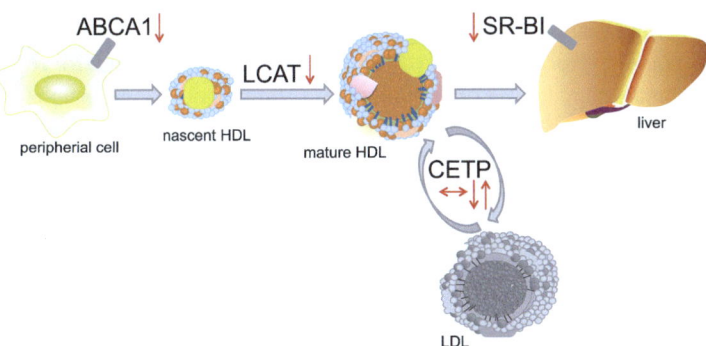

Fig. 4 Dysregulation in HDL metabolism under uremic conditions. ↓ decrease, ↑ increase, ↔ no effect. For references, refer to the text

liver for biliary excretion (Fig. 4). There is evidence that the expression of enzymes and receptors in HDL metabolism is altered in patients who suffer from kidney disease. To date, most studies have been completed with CKD stage V patients dependent on dialysis (ESRD), and less information on enzyme expression has been obtained from patients prior to dialysis. LCAT is important for HDL maturation. In ESRD patients, the LCAT enzyme concentration and activity were reduced (Miida et al. 2003; Moradi et al. 2009; Pahl et al. 2009; Shoji et al. 1992; Tolle et al. 2012).

The association between the plasma levels of cholesterol transport proteins, such as CETP and CKD, remains unclear. Some studies have failed to show any significant changes in the CETP protein level or activity (Pahl et al. 2009), whereas other studies have identified an increased (Dullaart et al. 1993) or decreased (Miida et al. 2003) CETP level. These controversial data may result from the heterogeneity of CKD patients and the analysis of different patient subgroups: increased levels were identified in proteinuric patients (Dullaart et al. 1993), and unchanged or decreased levels were identified in HD patients with different lengths of minimum dialysis treatment (minimum of 3 months vs. 1 year) (Miida et al. 2003; Pahl et al. 2009). The expression of receptors important for RCT appears to be affected by kidney dysfunction. In nephrectomy animal models, SR-BI (Liang and Vaziri 1999) and ABCA1 (Zuo et al. 2009) receptor expression is decreased. A reduction of these receptors in humans would subsequently be associated with altered HDL metabolism in CKD patients.

4.2 Changes of HDL Apolipoproteins (Modifications and Levels)

HDL contains different apolipoproteins. Under uremic conditions, the composition of the proteins within HDL changes: some protein levels decrease, whereas other protein levels increase. The main apolipoprotein of HDL under normal physiological conditions is apoAI. The presence of altered apoAI levels in CKD patients remains controversial. Some studies have identified decreased apoAI levels in dialysis patients (Holzer et al. 2011a; Moradi et al. 2009; Vaziri et al. 1999, 2009), whereas other studies did not identify a significant change in apoAI concentration (Shoji et al. 1992; Tolle et al. 2012). While apoAII and apoCI levels decreased in CKD patients (Holzer et al. 2011a), apoCII (Holzer et al. 2011a; Weichhart et al. 2012) and apoCIII (Holzer et al. 2011a) increased. ApoAIV was not identified in the HDL obtained from healthy subjects, but was detected in dialysis patients via mass spectrometry analysis (Holzer et al. 2011a). No significant difference was found in apoD and apoE levels, whereas apoM levels decreased (Holzer et al. 2011a). Table 2 summarizes the main apolipoprotein changes in CKD patients. Currently, most identified changes are based on studies with HD patients. The discrepancy in apolipoprotein levels may be a result of the heterogeneous patient population (e.g., CKD stage, time on dialysis, medication, or secondary disease), different HDL isolation protocols (e.g., one-step vs. multistep ultracentrifugation, chromatography, or differentiation between HDL2/3), and different apolipoprotein detection methods (e.g., enzyme-linked immunosorbent assay vs. mass spectrometry).

In addition to changes in the apolipoprotein levels, modifications of the proteins within HDL occurred, which were closely connected to altered functionality. To date, oxidative modifications, carbamylation, and glycosylation have been detected. These effects may be a consequence of the high level of ROS in patients with renal failure.

Table 2 Differences in apolipoprotein composition within the HDL in CKD patients compared with healthy control subjects (↑ increase, ↓decrease, ↔ no change vs. healthy control)

Apolipoprotein	Change in the HDL in CKD patients	References
apoAI	↓↔	Holzer et al. (2011a), Moradi et al. (2009), Shoji et al. (1992), Tolle et al. (2012), Vaziri et al. (1999), Vaziri et al. (2009)
apoAII	↓	Holzer et al. (2011a)
apoAIV	↑	Holzer et al. (2011a)
apoCI	↓	Holzer et al. (2011a)
apoCII	↑	Holzer et al. (2011a), Weichhart et al. (2012)
apoCIII	↑	Holzer et al. (2011a)
apoD	↔	Holzer et al. (2011a)
apoE	↔	Holzer et al. (2011a)
apoM	↓	Holzer et al. (2011a)

Patients with CKD suffer from increased reactive oxygen stress (Sung et al. 2013; Tucker et al. 2013). The pro-oxidant state is multifactorial but is related to high amounts of uremic toxins. Uremic toxins are directly involved in the oxidative response, e.g., phenyl acetic acid is a strong inducer of ROS (Schmidt et al. 2008), and indoxyl sulfate has a dual role that consists of pro-oxidant properties in the uremic condition and antioxidative properties under normal physiological conditions (Miyamoto et al. 2011). The increase in highly reactive radicals contributes to the oxidation of proteins and lipids. For example, the oxidation of LDL is known to occur in CKD patients (Ribeiro et al. 2012; Samouilidou et al. 2012). Furthermore, oxidative modifications of apoAI have been identified (Nicholls et al. 2005; Undurti et al. 2009; Zheng et al. 2004). ApoAI oxidation is dependent on the binding activity between apoAI and its receptors ABCA1 (Zheng et al. 2004) and SR-BI (Undurti et al. 2009). Therefore, ox-apoAI impairs HDL metabolism. In addition, an association between oxidative modifications of HDL and a higher risk for cardiovascular events in dialysis patients has been identified (Honda et al. 2012).

Another reactive compound that leads to protein modification is reactive cyanate, which induces the carbamylation of proteins. Cyanates emerge from the degradation of urea or via myeloperoxidase (MPO) at the sites of inflammation (Holzer et al. 2012; Sirpal 2009). CKD patients are prone to carbamylated proteins because urea levels are increased in the uremic condition, and these patients experience increased inflammation. MPO associates with HDL (Nicholls et al. 2005); thus, HDL proteins in addition to LDL proteins (Apostolov et al. 2005) become targets for carbamylation. Carbamylation impairs HDL function by decreasing the cholesterol efflux capacity of HDL from macrophages (Hadfield et al. 2013; Holzer et al. 2011b).

CKD patients are also prone to glycated protein modifications caused by insulin resistance, which is a condition that is frequently observed with reduced renal function (DeFronzo et al. 1981; Kobayashi et al. 2005). Insulin resistance appears to be associated with a high burden of oxidative stress. The reduction of oxidative stress by treatment with a superoxide dismutase/catalase mimetic reduces not only ROS production but also insulin resistance in a CKD mouse model (D'Apolito et al. 2010). The modified proteins are advanced glycated end products (AGEs), which are usually found at high levels in CKD patients. Glycated apoAI is likely an AGE (Lapolla et al. 2008). It has been speculated that these modified apoAIs have reduced RCT activity.

Additional proteins are also associated with HDL under physiological and pathophysiological conditions. Many proteins within HDL have been identified using a proteomic approach. Some proteins are increased, whereby other proteins are decreased in HD patients (Holzer et al. 2011a; Weichhart et al. 2012). However, an association between all of the proteins identified by proteome analysis and HDL functionality/dysfunctionality in the vessel wall could not be identified. Retinol-binding protein 4 (RBP4) was detected in the HDL from CKD patients, whereas it was not found within the HDL of healthy controls (Holzer et al. 2011a). Plasma levels of RBP4 were increased in dialysis patients (Frey et al. 2008), and the increase was dependent upon the CKD stage (Henze et al. 2010). Furthermore, surfactant protein B and α-1-microglobulin/bikunin precursor proteins were significantly increased in ESRD patients (Weichhart et al. 2012). Transthyretin was not identified within the HDL from healthy controls, but was detectable in the HDL from dialysis patients (Holzer et al. 2011a). An overview of the identified proteins was recently presented (Holzer et al. 2011a; Weichhart et al. 2012).

4.3 Loss of Protective Proteins or Lipids

Oxidative stress participates in the pathogenesis and progression of CKD and was increased in CKD patients (Mimic-Oka et al. 1999; Puchades et al. 2013). This results from an imbalance of pro-oxidative and antioxidative signals that is favored by uremic toxins (Mimic-Oka et al. 1999). Navab et al. postulated that HDL itself can become pro-inflammatory (Navab et al. 2006). In this condition, serum lipoproteins are prone to oxidative modification. The enzymes linked to HDL that have an antioxidative capacity are thought to facilitate oxidative protection. A higher pro-inflammatory index of HDL in CKD patients resulted in a higher adjusted death hazard ratio (Kalantar-Zadeh et al. 2007). Furthermore, it was shown that the in vitro antioxidative function of HDL was impaired in CKD patients (Moradi et al. 2009). The primary antioxidative HDL-linked enzymes are PON and glutathione peroxidase, and of these two enzymes, PON is the most studied one. The PON family consists of three members: PON1, PON2, and PON3. In contrast to PON2, PON1 and PON3 are secreted in the blood and are associated within HDL (Macharia et al. 2012). PON1 remains the best-studied enzyme of this family in cardiovascular disease and kidney dysfunction. In CKD patients, proteomic analysis did not indicate a significant difference in PON1 compared with healthy individuals (Holzer et al. 2011a), whereas its enzyme activity was reduced in CKD patients (Dantoine et al. 1998; Kennedy et al. 2013;

Moradi et al. 2009). The reduction negatively correlated with the CKD stage (Dantoine et al. 1998). For glutathione peroxidase, decreased concentration and activity were identified in dialysis patients (Moradi et al. 2009).

The loss of this antioxidative capacity combined with an increased level of pro-oxidative molecules (see below) may, at least in part, account for the increased oxidative stress in CKD patients.

A component that contributes to several of the protective functions of HDL is sphingosine-1-phosphate (S1P) (Nofer et al. 2004, 2007; Schuchardt et al. 2011; Theilmeier et al. 2006). HDL-associated apoM binds S1P (Christoffersen et al. 2011). In CKD patients, apoM levels are decreased (Holzer et al. 2011a), which may lead to higher levels of S1P free from HDL. There are several indications that non-HDL-bound S1P signaling differs compared with HDL-bound S1P (Schuchardt et al. 2011).

4.4 Increase of Molecules within HDL with a Fatal Function in the Vascular Wall

In vitro and in vivo studies and/or clinical trials have identified proteins that accumulate within HDL. Interestingly, the albumin content in HDL from uremic patients was elevated, and apolipoprotein displacement may occur (Holzer et al. 2011a).

The acute phase protein serum amyloid A (SAA), which is secreted by the liver during inflammation, is primarily transported by HDL within the blood (Uhlar and Whitehead 1999). In uremic patients, elevated SAA (Holzer et al. 2011a; Tolle et al. 2012; Weichhart et al. 2012; Zimmermann et al. 1999) is a sign of a chronic inflammatory status. This is closely related to an increased cardiovascular risk in humans (Zimmermann et al. 1999). In vitro studies support these findings, as the accumulation of SAA in HDL was correlated with a reduced anti-inflammatory capacity of HDL and a pro-inflammatory potential (Tolle et al. 2012; Weichhart et al. 2012). Furthermore, the cholesterol efflux capacity was decreased when the SAA level dramatically increased during acute sepsis (Annema et al. 2010). In addition, SAA accumulation results in the replacement of apoAI and influences HDL remodeling and metabolism. Nonetheless, the replacement was not identified in all patient cohorts (Tolle et al. 2012), which may be because of the different experimental settings. A displacement of other protective proteins during the acute phase response by SAA was also observed for PON1 and PAF-AH (Van Lenten et al. 1995). Thus, SAA enrichment within HDL affects the anti-inflammatory response. Recently, a pro-inflammatory response of SAA-rich HDL in CKD patients was observed (Tolle et al. 2012; Weichhart et al. 2012). The effects of SAA accumulation on the cholesterol efflux function of HDL remain controversial. In some experiments, a reduction of the cholesterol efflux capacity of HDL was observed if it was enriched with SAA (Artl et al. 2000; Marsche et al. 2007), whereas other studies found normal efflux capacity even with SAA-rich HDL (Banka et al. 1995; van der Westhuyzen et al. 2005).

Furthermore, phospholipases (PLA) associated with cardiovascular mortality are elevated in uremic patients. For example, lipoprotein-associated PLA (Lp-PLA) 2 was increased in uremic patients (Holzer et al. 2011a). The secreted PLA 2 (sPLA2) concentration and activity were also higher in ESRD patients. This elevation contributes to excessive oxidative stress in these patients (van der Giet et al. 2010a).

Dimethylarginines (DMAs), such as asymmetric dimethylarginine (ADMA) and its structural isomer symmetric DMA (SDMA), have been correlated with cardiovascular risk factors. As uremic toxins, they have been associated with cardiovascular outcomes and renal dysfunction (Duranton et al. 2012; Kielstein et al. 2006). DMAs originate from protein proteolysis and are mainly excreted in the urine. DMAs influence nitric oxide (NO) synthesis and negatively influence its vascular protective effects (Bode-Boger et al. 2006). The accumulation of SDMA in HDL from CKD patients and its association with decreased NO have recently been demonstrated (Speer et al. 2013).

MPO is another protein associated with the pathogenesis of cardiovascular disease because of its pro-oxidative and carbamylating potential (Nicholls et al. 2005; Undurti et al. 2009; Zheng et al. 2004). As discussed earlier, MPO associates with HDL (Nicholls et al. 2005) and contributes to reduced RCT in inflammatory diseases (Annema et al. 2010; Zheng et al. 2004). Furthermore, MPO-derived oxidative products modify apoAI (Hadfield et al. 2013). The modifications induced by MPO result in a pro-inflammatory HDL particle (Undurti et al. 2009). In patients with diabetes mellitus type 2, MPO activity was increased in HDL (Sorrentino et al. 2010). Interestingly, in a CKD cohort, the MPO plasma level decreased with advancing renal failure (Madhusudhana Rao et al. 2011). Further studies are necessary to determine the role of MPO in HDL dysfunction in a CKD cohort.

The HDL from ESRD patients on HD was less effective at accepting cholesterol from macrophages compared with the HDL from healthy subjects (Holzer et al. 2011a; Yamamoto et al. 2012). The reverse cholesterol transport may be affected in uremic patients.

5   Possibility of Functional Restoration of HDL

HDL function has been a recent focus of experimental research. With respect to weak effects on cardiovascular outcome after an increase of HDL quantity in CKD patients, HDL function capacity appears important for the design of new therapeutic approaches. Therefore, the question arose whether structural modifications and associated dysfunctions under uremic conditions are reversible. The best approach to test this hypothesis in the case of renal function is to study patients after successful renal transplantation. A limitation of this model is that these patients often require immunosuppressive drugs, which may affect lipid metabolism and function (Badiou et al. 2009). Some investigations have attempted to observe changes in HDL function after renal transplantation. Dantoine and coworkers reported that the PON enzyme activity in patients after kidney transplantation was comparable to that in control subjects, whereas in dialysis patients, it was decreased (Dantoine et al. 1998). Furthermore, the anti-inflammatory capacity of HDL from ESRD patients compared with healthy controls increased after successful renal transplantation (van der Giet et al. 2010b). The observed effect appears to be related to a decreased SAA level in transplant recipients. Other indicators of the reversibility of HDL dysfunction are based on studies with chronic heart failure patients. In this cohort, exercise training, which is an accepted intervention strategy to decrease cardiovascular risk in heart failure patients, resulted in increased HDL. The authors demonstrated that changes in HDL function induced by exercise training correlated with improved endothelial function (Adams et al. 2013).

According to these data it has been suggested that a restoration of HDL is possible. The lipid metabolism and the composition of different lipoprotein particles are primarily influenced by the metabolic condition. To date, little information is available regarding whether the CKD stage before transplantation is dependent on the HDL functionality after kidney transplantation or whether a point-of-no-return exists for functional restoration, vessel damage, and cardiovascular risk.

6   Laboratory Tests to Measure HDL Function

According to previous clinical and experimental trials, it has become clear that HDL particle modifications in disease conditions are associated with reduced HDL function; specific factors that determine HDL functionality remain unclear. It is known that various structural features are associated with HDL function, which can be measured by in vitro studies. A limitation arises in the difficulty of comparing experimental results. The standardization of the experimental design for HDL isolation, separation of proteins/lipids, sample preparation for proteomic approaches, and detection methods for components within HDL would help to overcome certain limitations that may result in discrepancies between findings. In addition, in vitro functionality assays are very complex and further hamper comparability. The complexity and cost preclude these tests from routine clinical laboratory analyses, thus limiting the validation of experimental findings of functionality in large clinical trials. Standardization, validation, and optimization for high throughput must first be established.

Recently, an overview of the laboratory tests that measure HDL subclasses (shape, density) and several HDL functions was described (Eren et al. 2012). There is a growing need to identify an optimal biomarker that describes HDL functionality. The development of reproducible, standardized, and validated methods to assess HDL function for routine use is of substantial interest. The knowledge regarding individual HDL function can help identify patients who may benefit from HDL-C increasing therapy or patients with a normal HDL-C level but at particularly high risk for cardiovascular events.

> **Conclusion and Perspective**
> HDL is a plasma lipoprotein with many pleiotropic protective functions in the vascular wall, including anti-atherosclerotic properties. Emerging evidence from clinical and laboratory studies indicates that HDL-C plasma levels in humans do not adequately represent HDL function. The pleiotropic protective effects of HDL depend on its composition, which is influenced by pathophysiological conditions. With the decline of renal function, HDL modifications occur. Evidence suggests that HDL composition, rather than plasma level, may be an important determinant for its pleiotropic protective function in the vascular wall. The goal is to identify robust biomarkers that describe HDL functionality and are measurable in validated, standardized assays that can be used routinely. Improvement of HDL functionality may serve as an interesting therapeutic target in the future for populations beyond CKD patients.

Acknowledgements The authors thank Jaqueline Herrmann for proofreading the manuscript.

The research projects of the authors are supported by the EFSD/Boehringer Ingelheim award (M.vdG), the Hans und Gertie Fischer-Stiftung (M.T), and the Peter und Traudl Engelhorn Stiftung scholarship (M.S). M.vdG is a member of the European Cooperation in Science and Technology (COST) Action "HDLnet" (BM904).

Conflict of Interest Statement None declared.

Open Access This chapter is distributed under the terms of the Creative Commons Attribution Noncommercial License, which permits any noncommercial use, distribution, and reproduction in any medium, provided the original author(s) and source are credited.

References

Adams V, Besler C, Fischer T, Riwanto M, Noack F, Hollriegel R, Oberbach A, Jehmlich N, Volker U, Winzer EB, Lenk K, Hambrecht R, Schuler G, Linke A, Landmesser U, Erbs S (2013) Exercise training in patients with chronic heart failure promotes restoration of high-density lipoprotein functional properties. Circ Res 113(12):1345–1355

AIM-HIGH Investigators (2011) The role of niacin in raising high-density lipoprotein cholesterol to reduce cardiovascular events in patients with atherosclerotic cardiovascular disease and optimally treated low-density lipoprotein cholesterol: baseline characteristics of study participants. The atherothrombosis intervention in metabolic syndrome with low HDL/high triglycerides: impact on global health outcomes (AIM-HIGH) trial. Am Heart J 161(3):538–543

Annema W, von Eckardstein A (2013) High-density lipoproteins. Multifunctional but vulnerable protections from atherosclerosis. Circ J 77(10):2432–2448

Annema W, Nijstad N, Tolle M, de Boer JF, Buijs RV, Heeringa P, van der Giet M, Tietge UJ (2010) Myeloperoxidase and serum amyloid A contribute to impaired in vivo reverse cholesterol transport during the acute phase response but not group IIA secretory phospholipase A(2). J Lipid Res 51(4):743–754

Apostolov EO, Shah SV, Ok E, Basnakian AG (2005) Quantification of carbamylated LDL in human sera by a new sandwich ELISA. Clin Chem 51(4):719–728

Apostolov EO, Ray D, Savenka AV, Shah SV, Basnakian AG (2010) Chronic uremia stimulates LDL carbamylation and atherosclerosis. J Am Soc Nephrol 21(11):1852–1857

Appel LJ (2004) Beyond (or back to) traditional risk factors: preventing cardiovascular disease in patients with chronic kidney disease. Ann Intern Med 140(1):60–61

Arsenault BJ, Lemieux I, Despres JP, Gagnon P, Wareham NJ, Stroes ES, Kastelein JJ, Khaw KT, Boekholdt SM (2009) HDL particle size and the risk of coronary heart disease in apparently healthy men and women: the EPIC-Norfolk prospective population study. Atherosclerosis 206 (1):276–281

Artl A, Marsche G, Lestavel S, Sattler W, Malle E (2000) Role of serum amyloid A during metabolism of acute-phase HDL by macrophages. Arterioscler Thromb Vasc Biol 20 (3):763–772

Asselbergs FW, Diercks GF, Hillege HL, van Boven AJ, Janssen WM, Voors AA, de Zeeuw D, de Jong PE, van Veldhuisen DJ, van Gilst WH (2004) Effects of fosinopril and pravastatin on cardiovascular events in subjects with microalbuminuria. Circulation 110(18):2809–2816

Atthobari J, Brantsma AH, Gansevoort RT, Visser ST, Asselbergs FW, van Gilst WH, de Jong PE, de Jong-van den Berg LT (2006) The effect of statins on urinary albumin excretion and glomerular filtration rate: results from both a randomized clinical trial and an observational cohort study. Nephrol Dial Transplant 21(11):3106–3114

Badiou S, Cristol JP, Mourad G (2009) Dyslipidemia following kidney transplantation: diagnosis and treatment. Curr Diab Rep 9(4):305–311

Baigent C, Keech A, Kearney PM, Blackwell L, Buck G, Pollicino C, Kirby A, Sourjina T, Peto R, Collins R, Simes R (2005) Efficacy and safety of cholesterol-lowering treatment: prospective meta-analysis of data from 90,056 participants in 14 randomised trials of statins. Lancet 366 (9493):1267–1278

Baigent C, Landray MJ, Reith C, Emberson J, Wheeler DC, Tomson C, Wanner C, Krane V, Cass A, Craig J, Neal B, Jiang L, Hooi LS, Levin A, Agodoa L, Gaziano M, Kasiske B, Walker R, Massy ZA, Feldt-Rasmussen B, Krairittichai U, Ophascharoensuk V, Fellstrom B, Holdaas H, Tesar V, Wiecek A, Grobbee D, de Zeeuw D, Gronhagen-Riska C, Dasgupta T, Lewis D, Herrington W, Mafham M, Majoni W, Wallendszus K, Grimm R, Pedersen T, Tobert J, Armitage J, Baxter A, Bray C, Chen Y, Chen Z, Hill M, Knott C, Parish S, Simpson D, Sleight P, Young A, Collins R (2011) The effects of lowering LDL cholesterol with simvastatin plus ezetimibe in patients with chronic kidney disease (Study of Heart and Renal Protection): a randomised placebo-controlled trial. Lancet 377(9784):2181–2192

Banka CL, Yuan T, de Beer MC, Kindy M, Curtiss LK, de Beer FC (1995) Serum amyloid A (SAA): influence on HDL-mediated cellular cholesterol efflux. J Lipid Res 36(5):1058–1065

Baragetti A, Norata GD, Sarcina C, Rastelli F, Grigore L, Garlaschelli K, Uboldi P, Baragetti I, Pozzi C, Catapano AL (2013) High density lipoprotein cholesterol levels are an independent predictor of the progression of chronic kidney disease. J Intern Med 274(3):252–262

Barter PJ, Caulfield M, Eriksson M, Grundy SM, Kastelein JJ, Komajda M, Lopez-Sendon J, Mosca L, Tardif JC, Waters DD, Shear CL, Revkin JH, Buhr KA, Fisher MR, Tall AR, Brewer B (2007) Effects of torcetrapib in patients at high risk for coronary events. N Engl J Med 357 (21):2109–2122

Bode-Boger SM, Scalera F, Kielstein JT, Martens-Lobenhoffer J, Breithardt G, Fobker M, Reinecke H (2006) Symmetrical dimethylarginine: a new combined parameter for renal function and extent of coronary artery disease. J Am Soc Nephrol 17(4):1128–1134

Boden WE, Probstfield JL, Anderson T, Chaitman BR, Desvignes-Nickens P, Koprowicz K, McBride R, Teo K, Weintraub W (2011) Niacin in patients with low HDL cholesterol levels receiving intensive statin therapy. N Engl J Med 365(24):2255–2267

Brouwers FP, Asselbergs FW, Hillege HL, de Boer RA, Gansevoort RT, van Veldhuisen DJ, van Gilst WH (2011) Long-term effects of fosinopril and pravastatin on cardiovascular events in subjects with microalbuminuria: ten years of follow-up of Prevention of Renal and Vascular End-stage Disease Intervention Trial (PREVEND IT). Am Heart J 161(6):1171–1178

Calabresi L, Gomaraschi M, Franceschini G (2003) Endothelial protection by high-density lipoproteins: from bench to bedside. Arterioscler Thromb Vasc Biol 23(10):1724–1731

Camont L, Chapman MJ, Kontush A (2011) Biological activities of HDL subpopulations and their relevance to cardiovascular disease. Trends Mol Med 17(10):594–603

Cannon CP, Shah S, Dansky HM, Davidson M, Brinton EA, Gotto AM, Stepanavage M, Liu SX, Gibbons P, Ashraf TB, Zafarino J, Mitchel Y, Barter P (2010) Safety of anacetrapib in patients with or at high risk for coronary heart disease. N Engl J Med 363(25):2406–2415

Chen SC, Hung CC, Kuo MC, Lee JJ, Chiu YW, Chang JM, Hwang SJ, Chen HC (2013) Association of dyslipidemia with renal outcomes in chronic kidney disease. PLoS ONE 8(2): e55643

Christoffersen C, Obinata H, Kumaraswamy SB, Galvani S, Ahnstrom J, Sevvana M, Egerer-Sieber C, Muller YA, Hla T, Nielsen LB, Dahlback B (2011) Endothelium-protective sphingosine-1-phosphate provided by HDL-associated apolipoprotein M. Proc Natl Acad Sci USA 108(23):9613–9618

Collins R, Armitage J, Parish S, Sleigh P, Peto R (2003) MRC/BHF Heart Protection Study of cholesterol-lowering with simvastatin in 5963 people with diabetes: a randomised placebo-controlled trial. Lancet 361(9374):2005–2016

Collins AJ, Foley RN, Herzog C, Chavers B, Gilbertson D, Ishani A, Johansen K, Kasiske B, Kutner N, Liu J, St Peter W, Ding S, Guo H, Kats A, Lamb K, Li S, Roberts T, Skeans M, Snyder J, Solid C, Thompson B, Weinhandl E, Xiong H, Yusuf A, Zaun D, Arko C, Chen SC, Daniels F, Ebben J, Frazier E, Hanzlik C, Johnson R, Sheets D, Wang X, Forrest B, Constantini E, Everson S, Eggers P, Agodoa L (2013) US renal data system 2012 annual data report. Am J Kidney Dis 61(1 Suppl 1):A7, e1–e476

Coresh J, Longenecker JC, Miller ER 3rd, Young HJ, Klag MJ (1998) Epidemiology of cardiovascular risk factors in chronic renal disease. J Am Soc Nephrol 9(12 Suppl):S24–S30

Dantoine TF, Debord J, Charmes JP, Merle L, Marquet P, Lachatre G, Leroux-Robert C (1998) Decrease of serum paraoxonase activity in chronic renal failure. J Am Soc Nephrol 9 (11):2082–2088

D'Apolito M, Du X, Zong H, Catucci A, Maiuri L, Trivisano T, Pettoello-Mantovani M, Campanozzi A, Raia V, Pessin JE, Brownlee M, Giardino I (2010) Urea-induced ROS generation causes insulin resistance in mice with chronic renal failure. J Clin Invest 120 (1):203–213

DeFronzo RA, Alvestrand A, Smith D, Hendler R, Hendler E, Wahren J (1981) Insulin resistance in uremia. J Clin Investig 67(2):563–568

Di Angelantonio E, Sarwar N, Perry P, Kaptoge S, Ray KK, Thompson A, Wood AM, Lewington S, Sattar N, Packard CJ, Collins R, Thompson SG, Danesh J (2009) Major lipids, apolipoproteins, and risk of vascular disease. JAMA 302(18):1993–2000

Dullaart RP, Gansevoort RT, Dikkeschei BD, de Zeeuw D, de Jong PE, Van Tol A (1993) Role of elevated lecithin: cholesterol acyltransferase and cholesteryl ester transfer protein activities in abnormal lipoproteins from proteinuric patients. Kidney Int 44(1):91–97

Duranton F, Cohen G, De Smet R, Rodriguez M, Jankowski J, Vanholder R, Argiles A (2012) Normal and pathologic concentrations of uremic toxins. J Am Soc Nephrol 23(7):1258–1270

El Harchaoui K, Arsenault BJ, Franssen R, Despres JP, Hovingh GK, Stroes ES, Otvos JD, Wareham NJ, Kastelein JJ, Khaw KT, Boekholdt SM (2009) High-density lipoprotein particle size and concentration and coronary risk. Ann Intern Med 150(2):84–93

Eren E, Yilmaz N, Aydin O (2012) High density lipoprotein and it's dysfunction. Open Biochem J 6:78–93

Fellstrom B, Holdaas H, Jardine AG, Svensson MK, Gottlow M, Schmieder RE, Zannad F (2009a) Cardiovascular disease in patients with renal disease: the role of statins. Curr Med Res Opin 25 (1):271–285

Fellstrom BC, Jardine AG, Schmieder RE, Holdaas H, Bannister K, Beutler J, Chae DW, Chevaile A, Cobbe SM, Gronhagen-Riska C, De Lima JJ, Lins R, Mayer G, McMahon AW, Parving HH, Remuzzi G, Samuelsson O, Sonkodi S, Sci D, Suleymanlar G, Tsakiris D,

Tesar V, Todorov V, Wiecek A, Wuthrich RP, Gottlow M, Johnsson E, Zannad F (2009b) Rosuvastatin and cardiovascular events in patients undergoing hemodialysis. N Engl J Med 360(14):1395–1407

Foley RN, Parfrey PS, Sarnak MJ (1998) Epidemiology of cardiovascular disease in chronic renal disease. J Am Soc Nephrol 9(12 Suppl):S16–S23

Frey SK, Nagl B, Henze A, Raila J, Schlosser B, Berg T, Tepel M, Zidek W, Weickert MO, Pfeiffer AF, Schweigert FJ (2008) Isoforms of retinol binding protein 4 (RBP4) are increased in chronic diseases of the kidney but not of the liver. Lipids Health Dis 7:29

Fried LF, Orchard TJ, Kasiske BL (2001) Effect of lipid reduction on the progression of renal disease: a meta-analysis. Kidney Int 59(1):260–269

Gordon T, Castelli WP, Hjortland MC, Kannel WB, Dawber TR (1977) High density lipoprotein as a protective factor against coronary heart disease. The Framingham study. Am J Med 62 (5):707–714

Gutstein DE, Krishna R, Johns D, Surks HK, Dansky HM, Shah S, Mitchel YB, Arena J, Wagner JA (2012) Anacetrapib, a novel CETP inhibitor: pursuing a new approach to cardiovascular risk reduction. Clin Pharmacol Ther 91(1):109–122

Hadfield KA, Pattison DI, Brown BE, Hou L, Rye KA, Davies MJ, Hawkins CL (2013) Myeloperoxidase-derived oxidants modify apolipoprotein A-I and generate dysfunctional high-density lipoproteins: comparison of hypothiocyanous acid (HOSCN) with hypochlorous acid (HOCl). Biochem J 449(2):531–542

Henze A, Frey SK, Raila J, Scholze A, Spranger J, Weickert MO, Tepel M, Zidek W, Schweigert FJ (2010) Alterations of retinol-binding protein 4 species in patients with different stages of chronic kidney disease and their relation to lipid parameters. Biochem Biophys Res Commun 393(1):79–83

Hirano T (2013) Abnormal lipoprotein metabolism in diabetic nephropathy. Clin Exp Nephrol 1–4

Holdaas H, Fellstrom B, Jardine AG, Holme I, Nyberg G, Fauchald P, Gronhagen-Riska C, Madsen S, Neumayer HH, Cole E, Maes B, Ambuhl P, Olsson AG, Hartmann A, Solbu DO, Pedersen TR (2003) Effect of fluvastatin on cardiac outcomes in renal transplant recipients: a multicentre, randomised, placebo-controlled trial. Lancet 361(9374):2024–2031

Holdaas H, Holme I, Schmieder RE, Jardine AG, Zannad F, Norby GE, Fellstrom BC (2011) Rosuvastatin in diabetic hemodialysis patients. J Am Soc Nephrol 22(7):1335–1341

Holzer M, Birner-Gruenberger R, Stojakovic T, El-Gamal D, Binder V, Wadsack C, Heinemann A, Marsche G (2011a) Uremia alters HDL composition and function. J Am Soc Nephrol 22(9):1631–1641

Holzer M, Gauster M, Pfeifer T, Wadsack C, Fauler G, Stiegler P, Koefeler H, Beubler E, Schuligoi R, Heinemann A, Marsche G (2011b) Protein carbamylation renders high-density lipoprotein dysfunctional. Antioxid Redox Signal 14(12):2337–2346

Holzer M, Zangger K, El-Gamal D, Binder V, Curcic S, Konya V, Schuligoi R, Heinemann A, Marsche G (2012) Myeloperoxidase-derived chlorinating species induce protein carbamylation through decomposition of thiocyanate and urea: novel pathways generating dysfunctional high-density lipoprotein. Antioxid Redox Signal 17(8):1043–1052

Home PD, Pocock SJ, Beck-Nielsen H, Curtis PS, Gomis R, Hanefeld M, Jones NP, Komajda M, McMurray JJ (2009) Rosiglitazone evaluated for cardiovascular outcomes in oral agent combination therapy for type 2 diabetes (RECORD): a multicentre, randomised, open-label trial. Lancet 373(9681):2125–2135

Honda H, Ueda M, Kojima S, Mashiba S, Michihata T, Takahashi K, Shishido K, Akizawa T (2012) Oxidized high-density lipoprotein as a risk factor for cardiovascular events in prevalent hemodialysis patients. Atherosclerosis 220(2):493–501

HPS2-THRIVE Collaborative Group (2013) HPS2-THRIVE randomized placebo-controlled trial in 25 673 high-risk patients of ER niacin/laropiprant: trial design, pre-specified muscle and liver outcomes, and reasons for stopping study treatment. Eur Heart J 34(17):1279–1291

Ikewaki K (2013) In vivo kinetic studies to further understand pathogenesis of abnormal lipoprotein metabolism in chronic kidney disease. Clin Exp Nephrol 1–4

Jacobson TA (2013) Lipoprotein(a), cardiovascular disease, and contemporary management. Mayo Clin Proc 88(11):1294–1311

Kalantar-Zadeh K, Kopple JD, Kamranpour N, Fogelman AM, Navab M (2007) HDL-inflammatory index correlates with poor outcome in hemodialysis patients. Kidney Int 72(9):1149–1156

Kaysen GA (2009) Biochemistry and biomarkers of inflamed patients: why look, what to assess. Clin J Am Soc Nephrol 4(Suppl 1):S56–S63

Keane WF, Tomassini JE, Neff DR (2013) Lipid abnormalities in patients with chronic kidney disease: implications for the pathophysiology of atherosclerosis. J Atheroscler Thromb 20 (2):123–133

Kennedy DJ, Wilson Tang WH, Fan Y, Wu Y, Mann S, Pepoy M, Hazen SL (2013) Diminished antioxidant activity of high-density lipoprotein-associated proteins in chronic kidney disease. J Am Heart Assoc 2(2):e000104

Khovidhunkit W, Kim MS, Memon RA, Shigenaga JK, Moser AH, Feingold KR, Grunfeld C (2004) Effects of infection and inflammation on lipid and lipoprotein metabolism: mechanisms and consequences to the host. J Lipid Res 45(7):1169–1196

Kielstein JT, Salpeter SR, Bode-Boeger SM, Cooke JP, Fliser D (2006) Symmetric dimethylarginine (SDMA) as endogenous marker of renal function–a meta-analysis. Nephrol Dial Transplant 21(9):2446–2451

Kilpatrick RD, McAllister CJ, Kovesdy CP, Derose SF, Kopple JD, Kalantar-Zadeh K (2007) Association between serum lipids and survival in hemodialysis patients and impact of race. J Am Soc Nephrol 18(1):293–303

Klin M, Smogorzewski M, Ni Z, Zhang G, Massry SG (1996) Abnormalities in hepatic lipase in chronic renal failure: role of excess parathyroid hormone. J Clin Invest 97(10):2167–2173

Kobayashi S, Maesato K, Moriya H, Ohtake T, Ikeda T (2005) Insulin resistance in patients with chronic kidney disease. Am J Kidney Dis 45(2):275–280

Kovesdy CP, Anderson JE, Kalantar-Zadeh K (2007) Inverse association between lipid levels and mortality in men with chronic kidney disease who are not yet on dialysis: effects of case mix and the malnutrition-inflammation-cachexia syndrome. J Am Soc Nephrol 18(1):304–311

Lamprea-Montealegre JA, McClelland RL, Astor BC, Matsushita K, Shlipak M, de Boer IH, Szklo M (2013) Chronic kidney disease, plasma lipoproteins, and coronary artery calcium incidence: the Multi-Ethnic Study of Atherosclerosis. Arterioscler Thromb Vasc Biol 33(3):652–658

Lapolla A, Brioschi M, Banfi C, Tremoli E, Cosma C, Bonfante L, Cristoni S, Seraglia R, Traldi P (2008) Nonenzymatically glycated lipoprotein ApoA-I in plasma of diabetic and nephropathic patients. Ann N Y Acad Sci 1126:295–299

LaRosa JC, Grundy SM, Waters DD, Shear C, Barter P, Fruchart JC, Gotto AM, Greten H, Kastelein JJ, Shepherd J, Wenger NK (2005) Intensive lipid lowering with atorvastatin in patients with stable coronary disease. N Engl J Med 352(14):1425–1435

Levey AS, Eckardt KU, Tsukamoto Y, Levin A, Coresh J, Rossert J, De Zeeuw D, Hostetter TH, Lameire N, Eknoyan G (2005) Definition and classification of chronic kidney disease: a position statement from Kidney Disease: Improving Global Outcomes (KDIGO). Kidney Int 67(6):2089–2100

Liang K, Vaziri ND (1999) Down-regulation of hepatic high-density lipoprotein receptor, SR-B1, in nephrotic syndrome. Kidney Int 56(2):621–626

Liang K, Oveisi F, Vaziri ND (1998) Role of secondary hyperparathyroidism in the genesis of hypertriglyceridemia and VLDL receptor deficiency in chronic renal failure. Kidney Int 53 (3):626–630

Linsel-Nitschke P, Tall AR (2005) HDL as a target in the treatment of atherosclerotic cardiovascular disease. Nat Rev Drug Discov 4(3):193–205

Longenecker JC, Klag MJ, Marcovina SM, Liu YM, Jaar BG, Powe NR, Fink NE, Levey AS, Coresh J (2005) High lipoprotein(a) levels and small apolipoprotein(a) size prospectively predict cardiovascular events in dialysis patients. J Am Soc Nephrol 16(6):1794–1802

Macharia M, Hassan MS, Blackhurst D, Erasmus RT, Matsha TE (2012) The growing importance of PON1 in cardiovascular health: a review. J Cardiovasc Med 13(7):443–453

Mackey RH, Greenland P, Goff DC Jr, Lloyd-Jones D, Sibley CT, Mora S (2012) High-density lipoprotein cholesterol and particle concentrations, carotid atherosclerosis, and coronary events: MESA (multi-ethnic study of atherosclerosis). J Am Coll Cardiol 60(6):508–516

Madhusudhana Rao A, Anand U, Anand CV (2011) Myeloperoxidase in chronic kidney disease. Indian J Clin Biochem 26(1):28–31

Marsche G, Frank S, Raynes JG, Kozarsky KF, Sattler W, Malle E (2007) The lipidation status of acute-phase protein serum amyloid A determines cholesterol mobilization via scavenger receptor class B, type I. Biochem J 402(1):117–124

Marz W, Genser B, Drechsler C, Krane V, Grammer TB, Ritz E, Stojakovic T, Scharnagl H, Winkler K, Holme I, Holdaas H, Wanner C (2011) Atorvastatin and low-density lipoprotein cholesterol in type 2 diabetes mellitus patients on hemodialysis. Clin J Am Soc Nephrol 6 (6):1316–1325

Miida T, Miyazaki O, Hanyu O, Nakamura Y, Hirayama S, Narita I, Gejyo F, Ei I, Tasaki K, Kohda Y, Ohta T, Yata S, Fukamachi I, Okada M (2003) LCAT-dependent conversion of prebeta1-HDL into alpha-migrating HDL is severely delayed in hemodialysis patients. J Am Soc Nephrol 14(3):732–738

Mimic-Oka J, Simic T, Djukanovic L, Reljic Z, Davicevic Z (1999) Alteration in plasma antioxidant capacity in various degrees of chronic renal failure. Clin Nephrol 51(4):233–241

Miyamoto Y, Watanabe H, Otagiri M, Maruyama T (2011) New insight into the redox properties of uremic solute indoxyl sulfate as a pro- and anti-oxidant. Ther Apher Dial 15(2):129–131

Mizobuchi M, Towler D, Slatopolsky E (2009) Vascular calcification: the killer of patients with chronic kidney disease. J Am Soc Nephrol 20(7):1453–1464

Moe SM, Sprague SM (2012) Chronic kidney disease—mineral bone disorder. In: Taal MW (ed) Brenner and Rector's the kidney, 9th edn. Elsevier, Philadelphia, pp 2021–2058

Moradi H, Pahl MV, Elahimehr R, Vaziri ND (2009) Impaired antioxidant activity of high-density lipoprotein in chronic kidney disease. Transl Res 153(2):77–85

Moradi H, Sica DA, Kalantar-Zadeh K (2013) Cardiovascular burden associated with uremic toxins in patients with chronic kidney disease. Am J Nephrol 38(2):136–148

Mori K, Shioi A, Jono S, Nishizawa Y, Morii H (1999) Dexamethasone enhances In vitro vascular calcification by promoting osteoblastic differentiation of vascular smooth muscle cells. Arterioscler Thromb Vasc Biol 19(9):2112–2118

Muntner P, Hamm LL, Kusek JW, Chen J, Whelton PK, He J (2004) The prevalence of nontraditional risk factors for coronary heart disease in patients with chronic kidney disease. Ann Intern Med 140(1):9–17

National Kidney Foundation (2003) I. Introduction. Am J Kidney Dis 41:S11–S21

Navab M, Imes SS, Hama SY, Hough GP, Ross LA, Bork RW, Valente AJ, Berliner JA, Drinkwater DC, Laks H et al (1991) Monocyte transmigration induced by modification of low density lipoprotein in cocultures of human aortic wall cells is due to induction of monocyte chemotactic protein 1 synthesis and is abolished by high density lipoprotein. J Clin Invest 88 (6):2039–2046

Navab M, Anantharamaiah GM, Reddy ST, Van Lenten BJ, Ansell BJ, Fogelman AM (2006) Mechanisms of disease: proatherogenic HDL–an evolving field. Nat Clin Pract Endocrinol Metab 2(9):504–511

Navab M, Reddy ST, Van Lenten BJ, Fogelman AM (2011) HDL and cardiovascular disease: atherogenic and atheroprotective mechanisms. Nat Rev Cardiol 8(4):222–232

Nicholls SJ, Zheng L, Hazen SL (2005) Formation of dysfunctional high-density lipoprotein by myeloperoxidase. Trends Cardiovasc Med 15(6):212–219

Nofer JR, van der Giet M, Tolle M, Wolinska I, von Wnuck LK, Baba HA, Tietge UJ, Godecke A, Ishii I, Kleuser B, Schafers M, Fobker M, Zidek W, Assmann G, Chun J, Levkau B (2004) HDL induces NO-dependent vasorelaxation via the lysophospholipid receptor S1P3. J Clin Invest 113(4):569–581

Nofer JR, Bot M, Brodde M, Taylor PJ, Salm P, Brinkmann V, van Berkel T, Assmann G, Biessen EA (2007) FTY720, a synthetic sphingosine 1 phosphate analogue, inhibits development of atherosclerosis in low-density lipoprotein receptor-deficient mice. Circulation 115(4):501–508

Odden MC, Tager IB, Gansevoort RT, Bakker SJ, Fried LF, Newman AB, Katz R, Satterfield S, Harris TB, Sarnak MJ, Siscovick D, Shlipak MG (2013) Hypertension and low HDL cholesterol were associated with reduced kidney function across the age spectrum: a collaborative study. Ann Epidemiol 23(3):106–111

Pahl MV, Ni Z, Sepassi L, Moradi H, Vaziri ND (2009) Plasma phospholipid transfer protein, cholesteryl ester transfer protein and lecithin:cholesterol acyltransferase in end-stage renal disease (ESRD). Nephrol Dial Transplant 24(8):2541–2546

Peev V, Nayer A, Contreras G (2014) Dyslipidemia, malnutrition, inflammation, cardiovascular disease and mortality in chronic kidney disease. Curr Opin Lipidol 25(1):54–60

Puchades MJ, Saez G, Munoz MC, Gonzalez M, Torregrosa I, Juan I, Miguel A (2013) Study of oxidative stress in patients with advanced renal disease and undergoing either hemodialysis or peritoneal dialysis. Clin Nephrol 80(3):177–186

Rahman M, Baimbridge C, Davis BR, Barzilay J, Basile JN, Henriquez MA, Huml A, Kopyt N, Louis GT, Pressel SL, Rosendorff C, Sastrasinh S, Stanford C (2008) Progression of kidney disease in moderately hypercholesterolemic, hypertensive patients randomized to pravastatin versus usual care: a report from the antihypertensive and lipid-lowering treatment to prevent heart attack trial (ALLHAT). Am J Kidney Dis 52(3):412–424

Ribeiro S, Faria Mdo S, Silva G, Nascimento H, Rocha-Pereira P, Miranda V, Vieira E, Santos R, Mendonca D, Quintanilha A, Costa E, Belo L, Santos-Silva A (2012) Oxidized low-density lipoprotein and lipoprotein(a) levels in chronic kidney disease patients under hemodialysis: influence of adiponectin and of a polymorphism in the apolipoprotein(a) gene. Hemodial Int 16(4):481–490

Rubins HB, Robins SJ, Iwane MK, Boden WE, Elam MB, Fye CL, Gordon DJ, Schaefer EJ, Schectman G, Wittes JT (1993) Rationale and design of the Department of Veterans Affairs High-Density Lipoprotein Cholesterol Intervention Trial (HIT) for secondary prevention of coronary artery disease in men with low high-density lipoprotein cholesterol and desirable low-density lipoprotein cholesterol. Am J Cardiol 71(3):45–52

Sahebkar A, Chew GT, Watts GF (2014) New peroxisome proliferator-activated receptor agonists: potential treatments for atherogenic dyslipidemia and non-alcoholic fatty liver disease. Expert Opin Pharmacother 15(4):493–503

Saland JM, Ginsberg HN (2007) Lipoprotein metabolism in chronic renal insufficiency. Pediatr Nephrol 22(8):1095–1112

Samouilidou EC, Karpouza AP, Kostopoulos V, Bakirtzi T, Pantelias K, Petras D, Tzanatou-Exarchou H, Grapsa EJ (2012) Lipid abnormalities and oxidized LDL in chronic kidney disease patients on hemodialysis and peritoneal dialysis. Ren Fail 34(2):160–164

Sandhu S, Wiebe N, Fried LF, Tonelli M (2006) Statins for improving renal outcomes: a meta-analysis. J Am Soc Nephrol 17(7):2006–2016

Scandinavian Simvastatin Survival Study Group (1994) Randomised trial of cholesterol lowering in 4444 patients with coronary heart disease: the Scandinavian Simvastatin Survival Study (4S). Lancet 344(8934):1383–1389

Schmidt S, Westhoff TH, Krauser P, Zidek W, van der Giet M (2008) The uraemic toxin phenylacetic acid increases the formation of reactive oxygen species in vascular smooth muscle cells. Nephrol Dial Transplant 23(1):65–71

Schuchardt M, Tolle M, Prufer J, van der Giet M (2011) Pharmacological relevance and potential of sphingosine 1-phosphate in the vascular system. Br J Pharmacol 163(6):1140–1162

Schwartz GG, Olsson AG, Ballantyne CM, Barter PJ, Holme IM, Kallend D, Leiter LA, Leitersdorf E, McMurray JJ, Shah PK, Tardif JC, Chaitman BR, Duttlinger-Maddux R, Mathieson J (2009) Rationale and design of the dal-OUTCOMES trial: efficacy and safety of dalcetrapib in patients with recent acute coronary syndrome. Am Heart J 158(6):896–901 e893

Schwartz GG, Olsson AG, Abt M, Ballantyne CM, Barter PJ, Brumm J, Chaitman BR, Holme IM, Kallend D, Leiter LA, Leitersdorf E, McMurray JJ, Mundl H, Nicholls SJ, Shah PK, Tardif JC, Wright RS (2012) Effects of dalcetrapib in patients with a recent acute coronary syndrome. N Engl J Med 367(22):2089–2099

Sever PS, Dahlof B, Poulter NR, Wedel H, Beevers G, Caulfield M, Collins R, Kjeldsen SE, Kristinsson A, McInnes GT, Mehlsen J, Nieminen M, O'Brien E, Ostergren J (2003) Prevention of coronary and stroke events with atorvastatin in hypertensive patients who have average or lower-than-average cholesterol concentrations, in the Anglo-Scandinavian Cardiac Outcomes Trial--Lipid Lowering Arm (ASCOT-LLA): a multicentre randomised controlled trial. Lancet 361(9364):1149–1158

Shah AS, Tan L, Long JL, Davidson WS (2013) Proteomic diversity of high density lipoproteins: our emerging understanding of its importance in lipid transport and beyond. J Lipid Res 54 (10):2575–2585

Shoji T, Nishizawa Y, Nishitani H, Yamakawa M, Morii H (1992) Impaired metabolism of high density lipoprotein in uremic patients. Kidney Int 41(6):1653–1661

Sirpal S (2009) Myeloperoxidase-mediated lipoprotein carbamylation as a mechanistic pathway for atherosclerotic vascular disease. Clin Sci (Lond) 116(9):681–695

Sorrentino SA, Besler C, Rohrer L, Meyer M, Heinrich K, Bahlmann FH, Mueller M, Horvath T, Doerries C, Heinemann M, Flemmer S, Markowski A, Manes C, Bahr MJ, Haller H, von Eckardstein A, Drexler H, Landmesser U (2010) Endothelial-vasoprotective effects of high-density lipoprotein are impaired in patients with type 2 diabetes mellitus but are improved after extended-release niacin therapy. Circulation 121(1):110–122

Speer T, Rohrer L, Blyszczuk P, Shroff R, Kuschnerus K, Krankel N, Kania G, Zewinger S, Akhmedov A, Shi Y, Martin T, Perisa D, Winnik S, Muller MF, Sester U, Wernicke G, Jung A, Gutteck U, Eriksson U, Geisel J, Deanfield J, von Eckardstein A, Luscher TF, Fliser D, Bahlmann FH, Landmesser U (2013) Abnormal high-density lipoprotein induces endothelial dysfunction via activation of Toll-like receptor-2. Immunity 38(4):754–768

Studer M, Briel M, Leimenstoll B, Glass TR, Bucher HC (2005) Effect of different antilipidemic agents and diets on mortality: a systematic review. Arch Intern Med 165(7):725–730

Sung CC, Hsu YC, Chen CC, Lin YF, Wu CC (2013) Oxidative stress and nucleic acid oxidation in patients with chronic kidney disease. Oxid Med Cell Longev 2013:301982

Svensson M, Schmidt EB, Jorgensen KA, Christensen JH (2006) N-3 fatty acids as secondary prevention against cardiovascular events in patients who undergo chronic hemodialysis: a randomized, placebo-controlled intervention trial. Clin J Am Soc Nephrol 1(4):780–786

The long-term intervention with pravastatin in ischaemic disease (LIPID) study group (1998) Prevention of cardiovascular events and death with pravastatin in patients with coronary heart disease and a broad range of initial cholesterol levels. N Engl J Med 339(19):1349–1357

Theilmeier G, Schmidt C, Herrmann J, Keul P, Schafers M, Herrgott I, Mersmann J, Larmann J, Hermann S, Stypmann J, Schober O, Hildebrand R, Schulz R, Heusch G, Haude M, von Wnuck LK, Herzog C, Schmitz M, Erbel R, Chun J, Levkau B (2006) High-density lipoproteins and their constituent, sphingosine-1-phosphate, directly protect the heart against ischemia/reperfusion injury in vivo via the S1P3 lysophospholipid receptor. Circulation 114(13):1403–1409

Thompson GR, Seed M (2013) Lipoprotein(a): the underestimated cardiovascular risk factor. Heart 100(7):534–535

Tolle M, Huang T, Schuchardt M, Jankowski V, Prufer N, Jankowski J, Tietge UJ, Zidek W, van der Giet M (2012) High-density lipoprotein loses its anti-inflammatory capacity by accumulation of pro-inflammatory-serum amyloid A. Cardiovasc Res 94(1):154–162

Tonelli M, Wanner C (2013) Lipid management in chronic kidney disease: synopsis of the kidney disease: improving global outcomes 2013 clinical practice guideline. Ann Intern Med 160 (3):182–189

Tonelli M, Collins D, Robins S, Bloomfield H, Curhan GC (2004) Gemfibrozil for secondary prevention of cardiovascular events in mild to moderate chronic renal insufficiency. Kidney Int 66(3):1123–1130

Tonelli M, Wiebe N, Culleton B, House A, Rabbat C, Fok M, McAlister F, Garg AX (2006) Chronic kidney disease and mortality risk: a systematic review. J Am Soc Nephrol 17(7):2034–2047

Tucker PS, Dalbo VJ, Han T, Kingsley MI (2013) Clinical and research markers of oxidative stress in chronic kidney disease. Biomarkers 18(2):103–115

Uhlar CM, Whitehead AS (1999) Serum amyloid A, the major vertebrate acute-phase reactant. Eur J Biochem 265(2):501–523

Undurti A, Huang Y, Lupica JA, Smith JD, DiDonato JA, Hazen SL (2009) Modification of high density lipoprotein by myeloperoxidase generates a pro-inflammatory particle. J Biol Chem 284(45):30825–30835

van Capelleveen JC, Brewer HB, Kastelein JJ, Hovingh GK (2014) Novel therapies focused on the high-density lipoprotein particle. Circ Res 114(1):193–204

van der Giet M, Tölle M, Pratico D, Lufft V, Schuchardt M, Hörl M, Zidek W, Tietge UF (2010a) Increased type IIA secretory phospholipase A2 expression contributes to oxidative stress in end-stage renal disease. J Mol Med 88(1):75–83

van der Giet M, Schuchardt M, Huang T, Prüfer N, Prüfer J, Tölle M (2010b) Dysfunctional HDL from patients with end-stage renal disease recover after successful renal transplantation. Am J Transplant 10(Abstract #50):53

van der Westhuyzen DR, Cai L, de Beer MC, de Beer FC (2005) Serum amyloid A promotes cholesterol efflux mediated by scavenger receptor B-I. J Biol Chem 280(43):35890–35895

Van Lenten BJ, Hama SY, de Beer FC, Stafforini DM, McIntyre TM, Prescott SM, La Du BN, Fogelman AM, Navab M (1995) Anti-inflammatory HDL becomes pro-inflammatory during the acute phase response. Loss of protective effect of HDL against LDL oxidation in aortic wall cell cocultures. J Clin Invest 96(6):2758–2767

Vaziri ND (2006) Dyslipidemia of chronic renal failure: the nature, mechanisms, and potential consequences. Am J Physiol 290(2):F262–F272

Vaziri ND (2013) Role of dyslipidemia in impairment of energy metabolism, oxidative stress, inflammation and cardiovascular disease in chronic kidney disease. Clin Exp Nephrol 1–4

Vaziri ND, Norris K (2011) Lipid disorders and their relevance to outcomes in chronic kidney disease. Blood Purif 31(1–3):189–196

Vaziri ND, Deng G, Liang K (1999) Hepatic HDL receptor, SR-B1 and Apo A-I expression in chronic renal failure. Nephrol Dial Transplant 14(6):1462–1466

Vaziri ND, Moradi H, Pahl MV, Fogelman AM, Navab M (2009) In vitro stimulation of HDL anti-inflammatory activity and inhibition of LDL pro-inflammatory activity in the plasma of patients with end-stage renal disease by an apoA-1 mimetic peptide. Kidney Int 76(4):437–444

Vaziri ND, Yuan J, Ni Z, Nicholas SB, Norris KC (2012) Lipoprotein lipase deficiency in chronic kidney disease is accompanied by down-regulation of endothelial GPIHBP1 expression. Clin Exp Nephrol 16(2):238–243

Vickers KC, Remaley AT (2014) HDL and cholesterol: life after the divorce? J Lipid Res 55(1):4–12

Wang F, Zheng J, Ye P, Luo L, Bai Y, Xu R, Sheng L, Xiao T, Wu H (2013) Association of high-density lipoprotein cholesterol with the estimated glomerular filtration rate in a community-based population. PLoS ONE 8(11):e79738

Wanner C, Krane V, Marz W, Olschewski M, Mann JF, Ruf G, Ritz E (2005) Atorvastatin in patients with type 2 diabetes mellitus undergoing hemodialysis. N Engl J Med 353(3):238–248

Watson AD, Berliner JA, Hama SY, La Du BN, Faull KF, Fogelman AM, Navab M (1995) Protective effect of high density lipoprotein associated paraoxonase. Inhibition of the biological activity of minimally oxidized low density lipoprotein. J Clin Invest 96(6):2882–2891

Watson CE, Weissbach N, Kjems L, Ayalasomayajula S, Zhang Y, Chang I, Navab M, Hama S, Hough G, Reddy ST, Soffer D, Rader DJ, Fogelman AM, Schecter A (2011) Treatment of patients with cardiovascular disease with L-4 F, an apo-A1 mimetic, did not improve select biomarkers of HDL function. J Lipid Res 52(2):361–373

Weichhart T, Kopecky C, Kubicek M, Haidinger M, Doller D, Katholnig K, Suarna C, Eller P, Tolle M, Gerner C, Zlabinger GJ, van der Giet M, Horl WH, Stocker R, Saemann MD (2012) Serum amyloid A in uremic HDL promotes inflammation. J Am Soc Nephrol 23(5):934–947

Yamamoto S, Yancey PG, Ikizler TA, Jerome WG, Kaseda R, Cox B, Bian A, Shintani A, Fogo AB, Linton MF, Fazio S, Kon V (2012) Dysfunctional high-density lipoprotein in patients on chronic hemodialysis. J Am Coll Cardiol 60(23):2372–2379

Yeung CK, Shen DD, Thummel KE, Himmelfarb J (2013) Effects of chronic kidney disease and uremia on hepatic drug metabolism and transport. Kidney Int 85:522–528

Zheng L, Nukuna B, Brennan ML, Sun M, Goormastic M, Settle M, Schmitt D, Fu X, Thomson L, Fox PL, Ischiropoulos H, Smith JD, Kinter M, Hazen SL (2004) Apolipoprotein A-I is a selective target for myeloperoxidase-catalyzed oxidation and functional impairment in subjects with cardiovascular disease. J Clin Invest 114(4):529–541

Zimmermann J, Herrlinger S, Pruy A, Metzger T, Wanner C (1999) Inflammation enhances cardiovascular risk and mortality in hemodialysis patients. Kidney Int 55(2):648–658

Zuo Y, Yancey P, Castro I, Khan WN, Motojima M, Ichikawa I, Fogo AB, Linton MF, Fazio S, Kon V (2009) Renal dysfunction potentiates foam cell formation by repressing ABCA1. Arterioscler Thromb Vasc Biol 29(9):1277–1282

Impact of Systemic Inflammation and Autoimmune Diseases on apoA-I and HDL Plasma Levels and Functions

Fabrizio Montecucco, Elda Favari, Giuseppe Danilo Norata, Nicoletta Ronda, Jerzy-Roch Nofer, and Nicolas Vuilleumier

Contents

1	Introduction	457
2	HDL/apoA-I Structure and Function	458
3	HDL and apoA-I as Members of the Innate Immune System	459
4	Relationship Between Lipid Raft Modulation and Lymphocyte Function	460
5	Modulation of Sphingosine-1-Phosphate (S1P)/S1P-Receptor Axis and Lymphocyte Function	462
6	Inflammatory Diseases	464

F. Montecucco
Division of Laboratory Medicine, Department of Genetics and Laboratory Medicine, Geneva University Hospitals, 4 rue Gabrielle Perret-Gentil, 1211 Geneva, Switzerland

Department of Human Protein Science, Geneva, Switzerland

Division of Cardiology, Geneva University Hospital, Geneva, Switzerland

E. Favari • N. Ronda
Department of Pharmacy, University of Parma, Parma, Italy

G.D. Norata
Department of Pharmacological and Biomolecular Sciences, Università degli Studi di Milano, Milan, Italy

Center for the Study of Atherosclerosis, Società Italiana Studio Aterosclerosi, Ospedale Bassini, Cinisello Balsamo, Italy

The Blizard Institute, Centre for Diabetes, Barts and The London School of Medicine & Dentistry, Queen Mary University, London, UK

J.-R. Nofer
Center for Laboratory Medicine, University Hospital Münster, Münster, Germany

N. Vuilleumier (✉)
Division of Laboratory Medicine, Department of Genetics and Laboratory Medicine, Geneva University Hospitals, 4 rue Gabrielle Perret-Gentil, 1211 Geneva, Switzerland

Department of Human Protein Science, Geneva, Switzerland
e-mail: nicolas.vuilleumier@hcuge.ch

© The Author(s) 2015
A. von Eckardstein, D. Kardassis (eds.), *High Density Lipoproteins*, Handbook of Experimental Pharmacology 224, DOI 10.1007/978-3-319-09665-0_14

6.1	Inflammatory Bowel Diseases	464
6.2	Vasculitis	464
6.3	Psoriasis	466
7	Autoimmune Diseases: The Role of Autoantibodies	467
7.1	Systemic Lupus Erythematosus (SLE)	467
7.2	Rheumatoid Arthritis (RA)	468
8	Impact of Anti-inflammatory Treatments	469
Conclusions		472
References		472

Abstract

The cholesterol of high-density lipoproteins (HDLs) and its major proteic component, apoA-I, have been widely investigated as potential predictors of acute cardiovascular (CV) events. In particular, HDL cholesterol levels were shown to be inversely and independently associated with the risk of acute CV diseases in different patient populations, including autoimmune and chronic inflammatory disorders. Some relevant and direct anti-inflammatory activities of HDL have been also recently identified targeting both immune and vascular cell subsets. These studies recently highlighted the improvement of HDL function (instead of circulating levels) as a promising treatment strategy to reduce inflammation and associated CV risk in several diseases, such as systemic lupus erythematosus and rheumatoid arthritis. In these diseases, anti-inflammatory treatments targeting HDL function might improve both disease activity and CV risk. In this narrative review, we will focus on the pathophysiological relevance of HDL and apoA-I levels/functions in different acute and chronic inflammatory pathophysiological conditions.

Keywords

Systemic inflammation • Autoimmune disease • apoA-I • HDL • Innate immunity

Abbreviations

ABCA	ATP-binding cassette transporters
ABCG1	ATP-binding cassette G1
APC	Antigen-presenting cells
apoA-I	Apolipoprotein A-1
APR	Acute-phase response
BD	Behçet's disease
CAD	Coronary artery disease
CD	Crohn's disease
CVD	Cardiovascular diseases
EBV	Epstein-Barr virus
HDL-C	High-density lipoprotein cholesterol
hs-CRP	High-sensitivity C-reactive protein

IBD	Inflammatory bowel diseases
KD	Kawasaki disease
LPL	Lipoprotein lipase
Lp-PLA2	Lipoprotein-specific phospholipase A2
LPS	Lipopolysaccharide
LTA	Lipoteichoic acid
MPO	Myeloperoxidase
piHDL	Proinflammatory HDL
PON1	Paraoxonase 1
RA	Rheumatoid arthritis
SAA	Serum amyloid A
SLE	Systemic lupus erythematosus
SMAD	Small mothers against decapentaplegic
SR-BI	Scavenger receptor BI
Th1	T helper type 1
TGF-β	Transforming growth factor beta
UC	Ulcerative colitis

1 Introduction

Since the seminal publication by Miller and coworkers in 1975, HDL-C has consistently been shown to be inversely and independently associated with the risk of acute CVD (such as myocardial infarction) (Miller and Miller 1975; Castelli et al. 1986; Di Angelantonio et al. 2009) and became one of the most commonly measured biomarkers (Expert Panel 2001). Nevertheless, the recent failures of HDL-C-raising therapies to reduce CV complications and atherosclerosis (Davidson 2012) as well as Mendelian randomization studies to demonstrate associations of genetically determined HDL-C levels with altered CV risk (Voight et al. 2012; Shah et al. 2013; Haase et al. 2012) raised the question whether HDL is an innocent bystander or a mediator of atherogenesis and CVD.

Among the important progresses in our understanding about HDL physiopathology, many studies stress the importance of HDL functions rather than HDL-C blood levels. This in turn could be of key importance to understand the negative outcome of both interventional trials and Mendelian randomization studies which did not take HDL functionality into account. Indeed, there is a growing body of evidence indicating that both acute and chronic inflammatory conditions induce posttranslational modifications of HDL, impairing both its lipid homoeostasis-regulating and anti-inflammatory properties, and even turn HDL into a proinflammatory molecule (Navab et al. 2011).

Those observations are shifting the attention from HDL-C levels to HDL function, emphasizing the importance of taking into account the clinical situations in which HDL-C levels are measured. Therefore, the precise knowledge of the different chronic or acute conditions susceptible to affect both the quantity and quality of HDL is likely to be paramount in order to understand the actual controversy about the causal role of HDL in CVD. Furthermore, because many of the functional properties of HDL can be recapitulated by apolipoprotein A-1 (apoA-I) (Phillips 2013), we will review the different acute and chronic inflammatory pathophysiological conditions reported to affect both HDL and apoA-I levels and functions, as potential modulators of the innate immune system.

2 HDL/apoA-I Structure and Function

Since HDL structure is predominantly affected by the presence of apoA-I (Phillips 2013), both molecules will be presented in a joint manner. Depending on the lipid state of apolipoproteins, HDL molecules are heterogeneous in shape, size, and density. The predominant HDL species are spherical HDL, consisting of a core of cholesteryl ester (CE) and triacylglycerol, encapsulated by a monolayer of phospholipids, unesterified cholesterol, and different lipoproteins. The latter is mostly represented by apoA-I which constitutes the major protein fraction of HDL and represents up to 80 % of HDL mass (Heinecke 2009). The second more abundant lipoprotein of HDL is apoA-II, followed by less abundant proteins including apoCs, D, E, J, and M; several enzymes such as lecithin-cholesterol acyltransferase (LCAT), serum paraoxonase 1 (PON1), and platelet-activating factor acetylhydrolase (PAF-AH); and sphingosine-1-phosphate (S1P) (see chapter "Structure of HDL: Particle Subclasses and Molecular Components" for more details).

The atheroprotective role of HDL on the cardiovascular system has been attributed to the pleiotropic effects of HDL, including reverse cholesterol transport, vasodilatation, antithrombotic, anticoagulant, and anti-inflammatory effects (Gordon and Davidson 2012). Reflecting those versatile properties, mass spectrometry analyses revealed that HDL encompasses up to 80 different proteins. Two-thirds of them are either acute-phase proteins, proteases, antioxidants, antithrombotic enzymes, or proteins involved in complement regulation, and only one-third are dedicated to lipid transport (Heinecke 2009).

On top of being the major protein fraction of HDL and a limiting factor for HDL formation, apoA-I per se executes many of the HDL-related properties, ranging from reverse cholesterol efflux and LPS and LTA scavenging to the inhibition of different proinflammatory, pro-oxidant, and prothrombotic pathways (Gu et al. 2000; Thuahnai et al. 2003; Yuhanna et al. 2001; Vuilleumier et al. 2013). Thus, HDL and apoA-I appear to contribute to host defense against many biological and chemical hazards. Nevertheless, chronic or acute inflammation induces major changes in HDL-C levels and functions, compromising and perverting the protective activities of HDL to harm.

Infection and inflammation generate an acute-phase response (APR), leading to important changes affecting the metabolism of different lipids and lipoproteins and thereby resulting in increased plasma levels of triglycerides and very-low-density lipoprotein (VLDL) cholesterol (reviewed in Khovidhunkit et al. 2004). Concomitantly, the APR also induces major changes in HDL functions since in this context apoA-I is replaced by other acute-phase proteins, such as SAA protein, ceruloplasmin, and haptoglobin (Navab et al. 2011), with SAA representing up to 87 % of HDL proteins and transforming HDL into a proinflammatory molecule (Van Lenten et al. 1995). Furthermore, recent evidence suggests that APR can also induce specific posttranslational modifications, such as chlorination, nitration, and carbamylation of amino acids by myeloperoxidase (MPO), oxidation by reactive carbonyls, as well as glycation, which compromise HDL and apoA-I functions, perverting them to harm (Vuilleumier et al. 2013).

3   HDL and apoA-I as Members of the Innate Immune System

The innate immune system represents the first line of defense against infectious agents and modified self-antigens (modified/oxidized lipids and apoptotic cells) consisting of a cellular and a humoral part. Cellular components are represented mostly by antigen-presenting cells (APCs), such as macrophages and dendritic cells, whereas the humoral counterpart is mostly represented by the pentraxin family and the complement system. Being extensively conserved throughout the evolution, and existing in early evolutionary species not affected by CVD (Babin et al. 1997), HDLs are believed to be part of the humoral innate immune system, helping mammals to fight against invading pathogens. Indeed, thanks to the presence of different proteins on HDL molecules, such as apoA-I, apoL-1, and haptoglobin-related proteins, HDLs are known to behave as antimicrobial agents protecting mammals against different parasites and bacteria, such as *Trypanosoma brucei, Escherichia coli, and Klebsiella pneumoniae* (Shiflett et al. 2005; Beck et al. 2013) (see chapter "HDL in Infectious Diseases and Sepsis").

Furthermore, on top of interfering directly with invading pathogens, HDL and apoA-I also modulate the APR-induced activation of the innate immune system by both neutralizing major bacterial membrane components, such as LPS of Gram-negative bacteria and LTA of Gram-positive bacteria, and modulating the proinflammatory signaling at the level of innate receptors, such as Toll-like receptors (TLRs), scavenger receptors (SRs), and Nod-like receptors (NLRs). Mathison and colleagues demonstrated for the first time in 1979 that HDL could reduce LPS toxicity in vivo (Mathison and Ulevitch 1979). This observation was subsequently inferred to the ability of HDL to sequester LPS, preventing the latter to elicit a proinflammatory response through TLR4/CD14 complex interaction (Levine et al. 1993). Later on, this scavenging effect of HDL was attributed to a specific apoA-I region on the N-terminal segment of apoA-I (aa: 52–74) (Wang et al. 2008). More recently, apoA-I was shown to have a similar ability to bind to LTA and to neutralize its proinflammatory effect (Jiao and Wu 2008). Interestingly, this effect was much weaker with HDL, suggesting that apoA-I is the key effector of

LTA scavenging (Grunfeld et al. 1999). The integrity of innate immune receptor signaling is largely dependent on their localization into lipid rafts, as well as on the integrity of the latter (Triantafilou et al. 2002). In this respect, apoA-I has been shown to deplete cholesterol from lipid rafts, leading to a decrease in TLR4 functionality, followed by an inhibition of LPS-induced inflammatory responses (Triantafilou et al. 2002). In the same line of thought, on top of altering the lipid raft composition, apoA-I has also been shown to impede TLR4 transport into lipid rafts, thereby preventing its ability to promote an efficient proinflammatory response (Smythies et al. 2010; Cheng et al. 2012). These findings indicate that apoA-I and HDL, albeit to a lesser extent, can effectively modulate TLR activity at the preceptor, receptor, and post-receptor levels (De Nardo et al. 2014).

4   Relationship Between Lipid Raft Modulation and Lymphocyte Function

Adaptive immune response relies on the activation of lymphocytes and the expansion of specific subsets in response to antigens. B lymphocytes represent an essential component of the humoral adaptive response through the synthesis of immunoglobulins, while T lymphocytes play a central role in cell-mediated immunity, and indeed the different T phenotypes orchestrate the immune response through different mechanisms. Two main aspects suggest a link between lipoproteins and adaptive immunity: first, the key receptors of B and T cells (BCR and TCR, respectively) are located in the lipid rafts, and their activity is modulated upon changes in the lipid raft composition and structure (Gupta and DeFranco 2007; Kabouridis and Jury 2008; Norata et al. 2012); second, B- and T-cell trafficking and T-cell subset differentiation are controlled also by lysosphingolipids, mainly S1P (Mandala et al. 2002; Liu et al. 2010). Lipid rafts, which concentrate specific proteins, thus limiting their ability to freely diffuse over the plasma membrane, act as platforms, bringing together molecules essential for the activation of immune cells (immunological synapse), but also separating such molecules when the conditions for activation are not appropriate (Ehrenstein et al. 2005). Lipid rafts compartmentalize key signaling molecules during the different stages of B-cell activation including BCR-initiated signal transduction, endocytosis of BCR-antigen complexes, loading of antigenic peptides onto MHC class II molecules, MHC-II-associated antigen presentation to T cells, and receipt of helper signals via the CD40 receptor (Gupta and DeFranco 2007). Critical regulators of BCR signaling lose their association with membrane rafts in disease conditions; for instance, the LMP2A gene product of the EBV constitutively resides in membrane raft of EBV-transformed human B cells and blocks the entry of ligand-clustered BCRs and BCR translocation (Longnecker and Miller 1996). Furthermore, an alteration of Lyn, an accessory protein of BCR signaling, has been reported in patients with systemic lupus erythematosus (SLE) (Flores-Borja et al. 2005). It is therefore reasonable to speculate that HDL, by removing cholesterol from lipid rafts, could affect B-cell function. While a direct evidence for HDL effects on B-cell function is still lacking, several reports indicate that HDL and apoA-I stimulate cholesterol efflux from cells, leading to cholesterol depletion and disruption of lipid rafts, which induces profound functional changes (Smythies et al. 2010) in macrophages and also affects antigen presentation and TCR signaling (Gruaz et al. 2010; Norata and Catapano 2012).

The response of T lymphocytes to antigen is orchestrated by a number of molecules that cluster in lipid rafts. TCR complex integrity is vital for the induction of optimal and efficient immune responses (Baniyash 2004). In immune-mediated disorders, such as rheumatoid arthritis (RA) and SLE, and in chronic infectious diseases, T cells are dysfunctional with a characteristic loss of expression of the TCRζ chain (also called CD249), a key component of the TCR complex that couples surface antigen recognition with intracellular signal transduction (Baniyash 2004). Also other molecules associated with the TCR signaling are reduced or altered in autoimmune disorders, such as lymphocyte-specific protein tyrosine kinase (Lck) (Jury et al. 2006). T-cell immunological synapses are altered in circulating T cells from patients with coronary artery disease (CAD), and increased memory T-cell subsets were observed in particular in CAD patients with increased inflammatory markers (Ammirati et al. 2012a, b). Molecular mechanisms of how reduced levels of blood lipids can affect lipid rafts in immune-mediated disorders still remain to be addressed. However, atorvastatin reversed many of the signaling defects characteristic of T cells from patients with SLE (Jury et al. 2006). The possibility that atorvastatin targets lipid raft-associated signaling abnormalities in autoreactive T cells has been proposed as the rationale for its use in the therapy of autoimmune disease (Jury et al. 2006). This is further supported by a large observational study that demonstrated an association between persistence with statin therapy and reduced risk of developing RA (Chodick et al. 2010). More recently, Yvan-Charvet et al. reported that two key proteins involved in HDL cholesterol efflux such as ATP-binding cassette transporters ABCA1 and ABCG1 play a key role in hematopoietic stem and multipotential progenitor cell proliferation, thus further linking cholesterol efflux and lipid raft modulation to immune cell function (Yvan-Charvet et al. 2010a, b). Finally, a key role for apoA-I in controlling cholesterol-associated lymphocyte activation and proliferation in peripheral lymph nodes was observed in animal models (Wilhelm et al. 2009). The prevalence of classical CD14++/CD16- but not of intermediate CD14++/CD16+ monocytes in hypoalphalipoproteinemia should also be taken into account as could impact on the different polarization on APC cells (Sala et al. 2013). More recently, Wang et al. (2012) demonstrated that HDL and apoA-I-induced cholesterol depletion and consequent disruption of plasma membrane lipid rafts in APCs inhibit their capacity to stimulate T-cell activation. This mechanism is highly dependent on the reduction of MHC class II molecules present on the cell surface following ABCA1 activation and on cholesterol efflux supporting a role for HDL in controlling also lymphocyte-mediated responses.

5 Modulation of Sphinghosine-1-Phosphate (S1P)/ S1P-Receptor Axis and Lymphocyte Function

Among the lipids that concentrate in lipid rafts, sphingolipids represent a major class that is metabolized to generate ceramide and subsequently sphingosine that in turn could be phosphorylated by sphingosine kinase (SPHK) (expressed mainly in platelets and in other peripheral blood cells) to generate S1P (Scanu and Edelstein 2008). Free or albumin-bound S1P is more susceptible to degradation than S1P bound to HDL (Yatomi 2008), which suggests that the latter might have a role in determining the uptake, cellular degradation, and systemic function. S1P carried by HDL positively correlates with HDL cholesterol, apoA-I, and apoA-II levels; furthermore, S1P is enriched in small dense HDL3 (Scanu and Edelstein 2008). S1P signals through five known G protein-coupled receptors (S1P1–S1P5). Over the last few years, it become apparent that S1P and the key enzymes SPHKs play a central role in the pathogenesis of several inflammatory disorders, including rheumatoid arthritis, asthma, and atherosclerosis by modulating macrophage function through the control of apoptosis as well as cell trafficking (Weigert et al. 2009). Many of these effects might depend on the activation of different S1P receptors. The activation of the lysosphingolipid receptor-PI3K/Akt axis by sphingosine-1-phosphate or other S1P mimetics is responsible for the induction by HDL of several genes involved in the immune response including the long pentraxin PTX3 or the transforming growth factor beta 2 (Norata et al. 2005, 2008). The S1P/S1P-receptor axis also plays a key role in lymphocyte function. The activation of S1P receptors and the consequent downstream signaling facilitate the egress of T cells from lymphoid organs (Mandala et al. 2002; Matloubian et al. 2004) and play a role in the lineage determination of peripheral T cells (Liu et al. 2010). S1P inhibits the differentiation of forkhead box P3 (FoxP3)$^+$ regulatory T cells (T_{regs}) while promoting the development of T helper type 1 (Th1) in a reciprocal manner (Liu et al. 2010). S1P receptor antagonizes TGF-β receptor function through an inhibitory effect on SMAD-3 activities to control the dichotomy between these two T-cell lineages (Liu et al. 2010). In animal models, apoA-I reduces inflammation in LDL receptor (−/−), apoA-I(−/−) mice by augmenting the effectiveness of the lymph nodes' T_{reg} response, with an increase in T_{reg} and a decrease in the percentage of effector/effector memory T cells (Wilhelm et al. 2010). While marked changes of T_{reg} number/function (two- to threefold difference) have been associated with atheroprotective functions in animal models (Ait-Oufella et al. 2006; Mor et al. 2007), in humans the correlation between T_{reg}, immune, and cardiometabolic disorders is less clear. We found no association between circulating CD3 + CD4 + CD25highCD127low T_{reg} levels and the extent or progression of human atherosclerotic disease at carotid and coronary sites (Ammirati et al. 2010). In a series of immune-mediated diseases such as rheumatoid arthritis, increased T_{reg} levels in synovial fluid of inflamed joints were observed (Mottonen et al. 2005; Bacchetta et al. 2007). Furthermore, Liu et al. observed relatively high T_{reg}-cell levels in patients with type 1 diabetes mellitus (Liu et al. 2006), and pathogenic T cells have been shown to have a paradoxical protective effect in murine autoimmune diabetes by boosting T_{regs} (Grinberg-Bleyer et al. 2010). Of note, excessive IL-6 or TNF-alpha production was associated with increased T_{reg} levels (Fujimoto et al. 2011; Bilate and Lafaille 2010). This evidence suggests the possibility that T_{reg} numbers could increase during some stages of disease as an attempt to regulate effector-cell activity. Unexpectedly, in the general population, we observed an inverse relation between HDL-C and T_{reg} count (Ammirati et al. 2010); whether this finding suggests that HDL could influence the polarization of lymphocyte subsets or could be a bystander of the T-cell status remains to be addressed. Recent evidence on the role of S1P in immune surveillance and the discovery of regulatory mechanisms in S1P-mediated immune trafficking has prompted extensive investigation in the field of S1P-receptor pharmacology. Fingolimod (FTY720), an S1P-receptor modulator, prevents lymphocyte egress from lymph nodes and modulates lymphocyte differentiation (Chi 2011). Initially, fingolimod was used as immunosuppressant in solid organ transplantation (Tedesco-Silva et al. 2006), while MS has been the first disease in which fingolimod was tested. In MS the myelin sheaths around the axons of the brain and spinal cord are damaged by inflammatory processes, leading to demyelination. Current therapeutic approaches are focused on the suppression of the immune system and on the blockage of T-cell blood–brain barrier transmigration into the brain parenchyma. Treatment with fingolimod was effective in reducing the disability progression on a large cohort of patients with relapsing MS (Cohen et al. 2010; Kappos et al. 2010). Of note, MS patients during the phase of clinical remission showed increased levels of HDL and total cholesterol levels (Salemi et al. 2010), whereas in immune-mediated disorders with increased markers of systemic inflammation, HDL levels are often decreased. Fingolimod was assessed also in diseases with an immunological component, such as atherosclerosis (Nofer et al. 2007; Keul et al. 2007). In LDL receptor-deficient mice, fingolimod inhibits atherosclerosis by modulating lymphocyte and macrophage function. In this study plasma lipids remained unchanged during the course of fingolimod treatment, whereas fingolimod lowered blood lymphocyte count (Nofer et al. 2007). As S1P levels are increased in many inflammatory conditions, such as in asthma and autoimmunity, the exact mechanism by which S1PR agonists could modulate its function is debated (Rivera et al. 2008). Interestingly, fingolimod is highly active in inducing the internalization, ubiquitination, and subsequent degradation of S1PR1 (Graler and Goetzl 2004) which suggests that its inhibitory action on immune cell trafficking might be through receptor downregulation. It is therefore crucial to understand the effect of HDL contained S1P on receptor expression and activity in immune disorders.

In spite of the presence of a number of experimental and clinical observations suggesting a relation between HDL and innate immunity, several questions remain to be addressed. Are the altered HDL-C levels a consequence of the atherogenic process, a cause of increased atherosclerosis observed in immune disorders, or independently related to the latter? How is HDL function altered in these diseases? Does raising HDL-C improve the outcome of immune disorders? As HDLs are a reservoir for several biologically active substances that may impact the immune system (Norata et al. 2012), how does the fine-tuning of lipid and protein exchange among lipoproteins affect HDL-related immune functions? Is there a specific HDL subfraction that is relevant? Addressing these aspects will be critical to understand the connection between HDL and the immune response.

6 Inflammatory Diseases

6.1 Inflammatory Bowel Diseases

The inflammatory bowel diseases (IBD) Crohn's disease (CD) and ulcerative colitis (UC) are chronic inflammatory diseases whose pathogenesis is not completely understood (Williams et al. 2012) but seems to be linked to autoimmune phenomena in a genetically prone background, following an abnormal immune response to colonic bacteria, and facilitated by a Western-type diet. IBD are associated with marked atherosclerosis and increased cardiovascular risk. Infection and chronic inflammation impair and alter lipoprotein metabolism and cause a variety of changes in plasma concentrations of lipids and lipoproteins (Borba et al. 2006; de Carvalho et al. 2008). An increase in inflammatory cytokines may result in a decrease in LPL enzyme activity, leading to a characteristic lipoprotein profile with decreased HDL-C levels as seen in patients with IBD that have high circulatory levels of inflammatory cytokines (Williams et al. 2012; Sappati Biyyani et al. 2010).

CD is a chronic inflammatory bowel disease that can affect any region of the gastrointestinal tract (Grand et al. 1995) with increased chronic inflammatory cell infiltrates in the mucosal lesions. The excessive local production of soluble mediators from activated monocytes and polymorphonuclear leukocytes has been implicated in mediating the tissue injury (Weiss 1989). Important among these mediators are oxygen free radicals. The chronic gut inflammation promotes an imbalance between oxidant and antioxidant mechanisms at the tissue level (Buffinton and Doe 1995) and may even compromise circulating antioxidant concentrations. A chronic inflammatory state is a risk factor for accelerated atherogenesis (van Leuven et al. 2007). A recent exploratory analysis demonstrated that CD is associated with an acceleration of the atherosclerotic process, as illustrated by an increased carotid intima-media thickness (IMT) in CD patients compared to healthy controls. In addition, CD patients were characterized during an inflammatory exacerbation by profoundly decreased levels of HDL combined with biochemical changes of the HDL particles, such as association with serum amyloid A, suggesting that early detection of atherosclerosis and subsequent cardiovascular prevention in patients with CD might be warranted (Romanato et al. 2009).

6.2 Vasculitis

Behçet's disease (BD) is a systemic vasculitis, most common in the Mediterranean area and in Asia, which can involve nearly every organ system and results from the interplay between infectious agent exposure and genetic factors. High production of cytokines, T and B lymphocyte activation, autoantibody production, and hypercoagulable/prothrombotic state are all characteristics of BD. Evidence for accelerated atherosclerosis in BD has been observed, but the relationship between cardiovascular risk factors and accelerated atherosclerosis in patients with BD is still controversial (Messedi et al. 2011). A study published by Messedi and colleagues demonstrated that in BD patients, HDL concentration and their subfraction levels are decreased. The same study reported that the percentage of HDL2 subpopulation was also decreased and HDL3 subfraction was significantly higher. The LDL-C/HDL-C ratio and CRP level were increased, and HDL and its subfractions were correlated with CRP and TG levels, suggesting that all these parameters may be considered as important predictors of cardiovascular events in BD patients (Messedi et al. 2011). This study confirms what seen in some previous studies in which, compared to control subjects, BD patients were characterized by reduced levels of HDL (Cimen et al. 2012; Musabak et al. 2005; Orem et al. 2002).

Kawasaki disease (KD) is an acute vasculitis that predominantly occurs in infancy and early childhood. It is commonly thought that KD results from the exposure of a genetically predisposed individual to an as-yet unidentified, possibly infectious environmental trigger. Coronary artery aneurysms or ectasia develops in approximately 15–25 % of affected children (Dhillon et al. 1996; Newburger et al. 1991; Cheung et al. 2004). There is increasing evidence to suggest that children with a history of KD might be predisposed to premature atherosclerosis and a significant association between carotid IMT and systemic arterial stiffness in children after KD has been demonstrated (Cheung et al. 2007). This syndrome is associated with significant abnormalities in lipid profile. In one of the first studies on this subject, it was shown that in the earliest days of illness, mean plasma concentrations of total cholesterol and HDL cholesterol are profoundly depressed, whereas mean triglyceride concentration is very high. Total cholesterol values rapidly return to normal and remain stable for more than 3 months after the onset of illness. HDL concentration recovered more slowly after illness onset, and mean HDL was significantly lower than expected more than 3 years after illness onset (Newburger et al. 1991). The persistence of low HDL for many years suggests a more lasting effect of KD on endothelial function, perhaps attributable to diminished activity of lipoprotein lipase. This enzyme resides on the capillary walls of most tissues and functions at the luminal surface of the vascular endothelium (Eckel 1989). This observation is confirmed, at least in part, in later studies on the lipid derangement in KD (Cheung et al. 2004; Chiang et al. 1997; Cabana et al. 1997). The results showed that during the acute phase, the concentrations of plasma HDL-C, apoA-I, and apoA-II were significantly reduced and the reduction of HDL was mainly related to the lowering of esterified and unesterified cholesterol in HDL2 (Chiang et al. 1997). In parallel, another study demonstrated that the lipid changes involved not only HDL-C concentration but also HDL composition. The authors showed that children with KD have extremely low serum HDL-C and apoA-I levels at the time of the acute illness and that serum amyloid A (SAA) is present in the acute stage and is associated mainly with HDL particles of HDL3 density (Cabana et al. 1997). Moreover, a more recent study demonstrated a significant induction of MCP-1, CCR2, and iNOS expression in THP-1 macrophages in vitro by the serum of children with a history of KD, showing that this induction correlated positively with serum high-sensitivity C-reactive protein (hs-CRP) and LDL and negatively with HDL-C levels (Cheung et al. 2005).

6.3 Psoriasis

Psoriasis is a chronic inflammatory skin disease associated with arthritis in up to 40 % of cases. Genetic and environmental factors contribute to the activation of lymphocytes, particularly Th1 and Th17, monocytes/macrophages, and dendritic cells and to the high cytokine content typical of psoriatic lesions. Psoriasis is associated with increased incidence of stroke and CVD (Yu et al. 2012) and with increased mean carotid IMT (Troitzsch et al. 2012). The mechanism behind such associations is still unknown. It is possible that the chronic inflammatory environment and high cytokine production typical of psoriasis induce abnormal HDL particle composition (El Harchaoui et al. 2009; McGillicuddy et al. 2009). A similar chronic inflammatory environment is observed in psoriasis (Davidovici et al. 2010) and contributes to the increased incidence of aortic inflammation, stroke, and myocardial infarction seen in this patient population (Mehta et al. 2011; Gelfand et al. 2006).

Lipoprotein profiling by NMR spectroscopy showed a reduction of HDL in psoriatic patients compared to controls, with an atherogenic profile with respect to HDL particles. In particular, high HDL-C concentration and large HDL size were associated with less aortic inflammation, while small HDL particles were more prevalent in cases with strong inflammation. Such association persisted following adjustment for CV risk factors, suggesting that HDL particle characteristics may play an important role in psoriatic vascular inflammation and CVD (Yu et al. 2012). HDL reduction in psoriatic patients has also been described to be associated to a reduction of serum cholesterol efflux capacity compared to control subjects (Mehta et al. 2012). In this study, psoriasis activity was inversely associated with HDL efflux capacity in a way depending on HDL particle size. Holzer and colleagues demonstrated that the protein composition of HDL is markedly altered in patients with psoriasis. ApoA-I and apoM levels are decreased, whereas the levels of several acute-phase proteins such as SAA, prothrombin, α-2-HS-glycoprotein, and α-1-acid glycoprotein 1 are increased (Holzer et al. 2012). Additionally, also the lipid composition of HDL from patients with psoriasis was altered, with a decrease in total cholesterol, cholesterol ester, free cholesterol, phosphatidylcholine, and sphingomyelin. These investigators confirmed that HDL from patients with psoriasis was less efficient in promoting cholesterol efflux from macrophages and that this defect in HDL function correlated with the severity of psoriasis. Surprisingly, the antioxidant properties of HDL were similar in control and psoriatic HDL and PON activity was not altered. However, Lp-PLA2 activity was increased and correlated with disease activity (Holzer et al. 2012). This observation does not completely agree with a report showing that the PON1 55 M allele is a risk factor for psoriasis.

Carriers of this allele have high levels of apoB and Lp(a) and a high apoB/apoA-I ratio, indicating that oxidative stress, impairment of the antioxidant system, and abnormal lipid metabolism may play a role in the pathogenesis and progression of psoriasis and its related complications (Asefi et al. 2012). Moreover, the paper by Holzer and coworkers demonstrates that a relatively mild chronic inflammatory state can similarly result in dysfunctional HDL, leading to decreased cholesterol efflux from macrophages (Holzer et al. 2012).

7 Autoimmune Diseases: The Role of Autoantibodies

7.1 Systemic Lupus Erythematosus (SLE)

SLE is a systemic autoimmune disease of multifactorial origin, in which genetic and environmental factors induce innate and acquired immunity derangement, with type I interferon (INF) production, T- and B-cell dysregulation, autoantibody production, and finally multiple organ damage. SLE is associated with accelerated atherosclerosis and markedly increased cardiovascular risk (CVR) (Shoenfeld et al. 2005). As in other autoimmune disorders, CVR cannot be fully explained by traditional risk factors, and various specific immune and inflammatory mechanisms have been demonstrated or proposed. The modifications of HDL level and function occurring in SLE have been indicated as possible factors contributing to cardiovascular damage.

Decreased circulating HDL-C levels have been often reported, although not unanimously, in SLE patients with active disease (de Carvalho et al. 2008; Hahn et al. 2008; Kiss et al. 2007), but the actual relevance of this finding with respect to CVR has not been clarified so far. The most important pro-atherogenic modifications seem to be those relative to composition and function of HDL. HDLs in SLE have impaired anti-inflammatory and antioxidant properties. In SLE, the so-called proinflammatory HDL (piHDL) can be detected in as many as 45 % of patients, correlating with increased oxidized LDL formation, carotid plaque, and carotid intima-media thickness (cIMT) (McMahon et al. 2011). piHDL are characterized by decreased content in the protective proteins paraoxonase (PON1) and apoA-I and by markedly increased levels of the pro-oxidant SAA (Hahn et al. 2008). These HDLs can be defined as proinflammatory because they actually enhance the oxidation of LDL and therefore monocyte attraction/activation, anti-oxLDL antibody production, and immune complex formation (Hahn et al. 2008; Teixeira et al. 2012; Carbone et al. 2013; Vuilleumier et al. 2014). It has been shown that the reduced activity of PON1 in some cases may be due also to the action of anti-apoA-I and anti-HDL antibodies (O'Neill et al. 2010; Batuca et al. 2007). Anti-apoA-I antibodies, first described in SLE patients but detected also in patients with acute coronary syndrome, have shown independent association and predictive value with respect to cardiovascular events in various patient populations (Carbone et al. 2013). Proposed mechanisms for such association include the ability of anti-apoA-I antibodies to induce HDL dysfunction, scavenger receptor B1 (SR-BI) function impairment in endothelial cells, neutrophil infiltration and matrix-metalloproteinase 9 production in plaques, activation of NFkB via TLR2/CD14 complex interaction, and cytokine release (Carbone et al. 2013). Interestingly, anti-apoA-I and anti-HDL antibodies have been shown to correlate also with disease activity in SLE patients (O'Neill et al. 2010).

In addition, HDL dysfunction in SLE involves their capacity to promote cell cholesterol efflux (Ronda et al. 2014). In particular, in SLE patients HDL cholesterol efflux capacity (CEC) is reduced with respect to the ABCG1 and ABCA1 transporter pathways, while the SR-BI-mediated CEC is unchanged. In addition the correlation between SR-BI-mediated CEC and HDL-C levels was stronger in SLE plasmas as compared to control plasma. This pattern is consistent with a possible reduction/dysfunction of the small HDL populations (Favari et al. 2009) and a shift to larger HDL, typical acceptors of cholesterol effluxed by SR-BI. Indeed, HDLs of SLE patients were found increased in size (Hua et al. 2009; Juárez-Rojas et al. 2008). The impaired ABCA1- and ABCG1-mediated CEC in SLE patients may have a great impact because cholesterol efflux not only opposes lipid deposition in vessels but is also crucial for the modulation of macrophage, endothelial, and T-cell inflammatory functions (Prosser et al. 2012; Yvan-Charvet et al. 2010a, b).

7.2 Rheumatoid Arthritis (RA)

RA is a systemic autoimmune disease characterized by lymphocyte activation, autoantibody production, high serum and tissue cytokine levels, and strong inflammation of synovia and vessels; similarly to SLE, it is associated with high CV risk (Shoenfeld et al. 2005). Circulating levels of HDL-C are often reduced, especially in active disease, and generally return to normal values during drug-induced remission; however, as such modifications are mirrored by those relative to LDL cholesterol and total cholesterol, the actual clinical significance of HDL-C variation is not clear. Complex and specific mechanisms underlie lipid metabolism derangement in RA, in which cardiovascular risk inversely correlates with circulating LDL-C levels (the so-called RA lipid paradox) (Myasoedova et al. 2011). As in SLE, the clinical relevance of HDL-C lowering (when present) in RA patients has yet to be clarified.

Composition and functional characteristics of HDL are altered in RA. piHDL levels are increased in RA patients as compared to healthy controls (Hahn et al. 2008) and correlate positively with disease activity (Charles-Schoeman et al. 2009). Profound modifications of HDL composition in patients with active RA have been described, particularly with respect to protein content. These changes include increased amount of serum amyloid A (SAA), apoJ, fibrinogen, and haptoglobin and reduced PON1 (Charles-Schoeman et al. 2009; Watanabe et al. 2012). As in SLE, anti-apoA-I autoantibodies are detectable in the serum of RA patients. In particular, in this population anti-apoA-I IgG levels have been shown to predict acute cardiovascular events and improve the prognostic power of the Framingham Risk Score (Carbone et al. 2013).

Finally, functional impairment of HDL as cholesterol acceptors for cell cholesterol efflux has been reported in RA (Ronda et al. 2014; Charles-Schoeman et al. 2012), independent of serum HDL-C levels and with a pattern of modifications differing from that found in SLE patients with respect to single cholesterol transporters (Ronda et al. 2014). In particular, ABCG1-mediated CEC is impaired in RA patients independently of HDL-C levels but inversely correlating with disease activity (Ronda et al. 2014). On the one hand, such correlation may reflect the impact of inflammation and autoimmunity on HDL function, but on the other hand, it may indicate the adverse effect of reduced ABCG1-mediated CEC on vessel inflammation and immune reaction promotion. In fact, by promoting cholesterol and 7-ketocholesterol release, ABCG1 promotes anti-inflammatory phenotypes of macrophages and endothelial cells (Terasaka et al. 2007; Hassan et al. 2006; O'Connell et al. 2004).

Modifications of HDL occurring in SLE and RA compromise cholesterol efflux-promoting, anti-inflammatory, and antioxidant properties of HDL and thereby may have an important role in the progression of both atherosclerosis and autoimmune as well as inflammatory phenomena typical for these diseases.

8   Impact of Anti-inflammatory Treatments

HDL exerts potent and multifactorial anti-inflammatory effects, which makes it an attractive target for pharmacological intervention. As extensively discussed in this chapter, several molecular principles exploited by HDL particles to interfere with innate or acquired immunity might be well utilized for designing new drugs effectively combating atherosclerosis and/or autoimmune diseases. By contrast, little efforts have been devoted to assess the influence of medication traditionally used in acute or chronic inflammatory diseases on HDL-C levels and HDL function. Most information regarding the interrelationship between HDL and anti-inflammatory therapy stems from interventional studies designed to examine the impact of drugs on the primary disease rather than on lipid metabolism. Despite this limited approach, currently available data strongly suggest that anti-inflammatory medication may exert positive adjuvant effects both on HDL quantity and its functionality.

It is well known that HDL is a negative acute-phase reactant and that HDL-C levels decline—sometimes dramatically—at the onset of acute inflammation and infection. For instance, sepsis, extensive surgery, or viral infections are commonly associated with low HDL-C levels that are significantly and rapidly increased following spontaneous or treatment-related reductions in disease activity (Marik 2006; Akgun et al. 1998; Marchesi et al. 2005). In addition, several chronic autoimmune disorders including SLE, RA, Kawasaki disease, and Behçet's disease as well as periodontal disease are accompanied by decreased HDL levels in plasma (Haas and Mooradian 2011). Hence, the administration of nonsteroidal anti-inflammatory drugs (NSAIDs), which represent the most often prescribed class of anti-inflammatory drugs, might be expected to produce HDL-C elevations by the simple reversal of the acute-phase reaction. However, the effect of NSAIDs on the quantity and quality of HDL remains controversial. In animal studies rising effects of aspirin on ABCA1 expression and cholesterol efflux as well as on lecithin-cholesterol acyltransferase (LCAT) activity and plasma HDL-C levels were occasionally observed (Sethi et al. 2011; Jafarnejad et al. 2008; Viñals et al. 2005). However, these effects apparently did not translate into increased HDL generation, as similar plasma HDL-C levels were noted in a small prospective study in aspirin- and placebo-treated groups (Eritsland et al. 1989). In one study, the administration of ibuprofen was found to slightly elevate HDL-C levels both in smokers and nonsmokers, whereas in another study treatment with naproxen failed to show any effect (Zapolska-Downar et al. 2000; Young et al. 1995). The influence of selective cyclooxygenase-2 inhibitors such as celecoxib or rofecoxib on HDL-C plasma levels has not been reported to date. In a more recent investigation, atreleuton (VIA-2291)—the first clinically tested 5-lipooxygenase inhibitor—was found to decrease leukotriene B4 production and CRP levels in blood and to concomitantly reduce coronary plaque burden, but these favorable effects were observed in the absence of any changes in plasma lipid profile including HDL-C (Tardif et al. 2010).

In a major contrast to nonsteroidal antiphlogistics, corticosteroids were repeatedly reported to increase plasma HDL-C levels in patients with chronic autoimmune diseases including RA, psoriatic arthritis, and SLE as well as in a non-autoimmune chronic inflammation (sarcoidosis) (Ettinger et al. 1987; Salazar et al. 2002; Boers et al. 2003; Sarkissian et al. 2007; Peters et al. 2007; Georgiadis et al. 2008; García-Gómez et al. 2008). Furthermore, the addition of corticosteroids to disease-modifying antirheumatic drugs (DMARDs, such as methotrexate or sulfasalazine), which on their own increase HDL-C in plasma, significantly potentiated their effects on HDL-C levels (Boers et al. 2003). In one study, corticosteroids were also found to favorably affect HDL subfractions by preferentially increasing the amount of HDL2-C over HDL3-C (García-Gómez et al. 2008). The beneficial effects exerted by corticosteroids on HDL quantity may appear at the first glimpse counterintuitive, since protracted therapy with these compounds is known to enhance insulin resistance and to produce a prediabetic state, which is almost obligatorily accompanied by low HDL-C. Actually, endogenous hypercortisolism (Cushing disease) has been related both to decreased HDL-C levels and to increased cardiovascular risk in several studies (Faggiano et al. 2003). However, the relationship between steroid use and cardiovascular risk is complicated by the fact that these drugs tend to be used more often in patients with severe or intractable chronic disease; in such patients, the anti-inflammatory HDL-C elevating effect of corticosteroids may outweigh the HDL-C decreasing effect related to aggravation of insulin resistance. The mechanisms underlying modulatory effects of corticosteroids on HDL metabolism remain obscure. These compounds were found to increase the activity of lipoprotein lipase (LPL) and to decrease the activity of hepatic triglyceride lipase, which are both critically involved in the generation of HDL precursors and in HDL particle remodeling (Ewart et al. 1997; Dolinsky et al. 2004). In addition, corticosteroids may stimulate the production of nascent

HDL as they enhance apoA-I gene expression in hepatocytes (Hargrove et al. 1999). Finally, corticosteroids were found to inhibit the activity of cholesteryl ester transfer protein (CETP), which may well explain their modulatory effect on the HDL subfraction composition (Georgiadis et al. 2006).

Studies involving biological pharmaceuticals, which selectively interfere with proinflammatory signaling pathways, add further evidence underscoring the beneficial effects of anti-inflammatory therapies on HDL quantity in plasma. The greatest amount of research into the effects of biologicals on lipids has been performed with TNF-alpha antagonists, in particular, with infliximab. In general, these studies demonstrated a consistent action of this drug in patients with RA, ankylosing spondylitis, and inflammatory bowel disease characterized by increases in total cholesterol mostly due to the elevation of HDL-C (Choy and Sattar 2009; Pollono et al. 2010; Mathieu et al. 2010; Koutroubakis et al. 2009; Parmentier-Decrucq et al. 2009). Similar effects on HDL-C were observed in studies utilizing two other TNF-alpha antagonists, adalimumab and, most recently, golimumab, whereas the application of etanercept—a TNF-alpha receptor antagonist—produced less consistent results (Pollono et al. 2010; Stagakis et al. 2012; Navarro-Millán et al. 2013; Lestre et al. 2011; Kirkham et al. 2014). Positive effects on plasma lipid profile encompassing the elevation of HDL-C levels were also observed in patients with RA undergoing IL-6 receptor blockade with tocilizumab or CD20 signaling blockade with rituximab (Pollono et al. 2010; Kawashiri et al. 2011; Kerekes et al. 2009). In a major contrast, canakinumab—a compound neutralizing IL-1beta—failed to change plasma lipid profile in a large randomized study involving 556 men and women at high cardiovascular risk, albeit the treatment led to significant reductions in acute-phase proteins such as CRP and fibrinogen (Ridker et al. 2012). The molecular mechanisms underlying the beneficial effects of biological therapies on HDL-C levels are poorly understood. While it cannot be entirely excluded that HDL-C increases are at least partly related to the retardation of acute-phase reaction brought about by the inhibition of selected proinflammatory signaling pathways, it seems more likely that treatment with TNF-alpha or IL-6 antagonists leads to derepression of the APOA1 gene, the activity of which is known to be downregulated in hepatocytes exposed to proinflammatory cytokines (Haas and Mooradian 2011).

In addition to the reduction of HDL quantity, acute-phase reaction was demonstrated to profoundly affect HDL composition. HDL particles isolated from subjects suffering from acute or chronic inflammatory diseases were found to lose proteins and enzymes with established or presumed antiatherogenic function such as apoA-I, LCAT, or PON1 and to concomitantly acquire proinflammatory or pro-oxidative factors such as SAA, ceruloplasmin, Lp-PLA2, or MPO. Such inflammatory HDL particles are severely impeded in their ability to exert several antiatherogenic functions including initiation of reversed cholesterol transport, inhibition of pro-oxidative processes, inhibition of leukocyte migration and recruitment into arterial wall, or inhibition of thrombocyte activation. The picture emerging from few recent studies suggests that anti-inflammatory medications may not only elevate HDL quantity in plasma but also help to restore its proper composition and thereby the antiatherogenic functionality. For instance, treatment of RA patients with adalimumab, etanercept, tocilizumab, or rituximab decreased and increased the content of SAA and apoA-I in HDL particles, respectively (Raterman et al. 2013; McInnes et al. 2014; Jamnitski et al. 2013). In addition, prolonged therapy with methotrexate was found to improve the capacity of HDL to inhibit both LDL oxidation and leukocyte migration in patients with RA (Charles-Schoeman et al. 2009). Increased levels of HDL-associated PON1 and improved anti-oxidative capacity were seen in patients with psoriasis treated with etanercept (Bacchetti et al. 2013).

Conclusions

HDL particles have been shown to play a protective and an anti-inflammatory role in autoimmune and inflammatory disorders. Their pathophysiological relevance directly implies the regulation of both immune and vascular cell functions that influence the common inflammatory processes underlying disease progression and the associated CV risk. Numerous studies have demonstrated that inflammatory disorders increase the risk of CVD and that this increase cannot be totally accounted for by traditional risk factors. Alterations in the quantity, composition, and function of HDL may contribute to the promotion of atherosclerosis process. In this prospective the assessment of HDL function, evaluated as the capacity to promote cell cholesterol efflux, may offer a better prediction of CVD than classical HDL-C levels. In addition, HDL function impairment, which involves their anti-inflammatory and antioxidant properties as well as their ability to interact with cellular cholesterol transporters, may have an important role in accelerating atherosclerosis but also autoimmune and inflammatory mechanisms typical for these diseases. Therefore, the improvement of HDL function (instead of HDL-C levels) represents an interesting therapeutic strategy to reduce inflammation and associated CV risk in several immune diseases, such as systemic lupus erythematosus and rheumatoid arthritis. We believe that selective treatments improving HDL function or reducing the adverse modifications of HDL structure might be of pathophysiological relevance. Both basic and clinical studies are needed to validate this promising therapeutic issue in a near future.

Open Access This chapter is distributed under the terms of the Creative Commons Attribution Noncommercial License, which permits any noncommercial use, distribution, and reproduction in any medium, provided the original author(s) and source are credited.

References

Ait-Oufella H, Salomon BL, Potteaux S, Robertson AK, Gourdy P, Zoll J, Merval R, Esposito B, Cohen JL, Fisson S, Flavell RA, Hansson GK, Klatzmann D, Tedgui A, Mallat Z (2006) Natural regulatory T cells control the development of atherosclerosis in mice. Nat Med 12:178–180

Akgun S, Ertel NH, Mosenthal A, Oser W (1998) Postsurgical reduction of serum lipoproteins: interleukin-6 and the acute-phase response. J Lab Clin Med 131:103–108

Ammirati E, Cianflone D, Banfi M, Vecchio V, Palini A, De Metrio M, Marenzi G, Panciroli C, Tumminello G, Anzuini A, Palloshi A, Grigore L, Garlaschelli K, Tramontana S, Tavano D, Airoldi F, Manfredi AA, Catapano AL, Norata GD (2010) Circulating CD4+CD25hiCD127lo regulatory t-cell levels do not reflect the extent or severity of carotid and coronary atherosclerosis. Arterioscler Thromb Vasc Biol 30:1832–1841

Ammirati E, Cianflone D, Vecchio V, Banfi M, Vermi AC, De Metrio M, Grigore L, Pellegatta F, Pirillo A, Garlaschelli K, Manfredi AA, Catapano AL, Maseri A, Palini AG, Norata GD (2012a) Effector memory T cells are associated with atherosclerosis in humans and animal models. J Am Heart Assoc 1:27–41

Ammirati E, Monaco C, Norata GD (2012b) Antigen-dependent and antigen-independent pathways modulate CD4+CD28null T-cells during atherosclerosis. Circ Res 111:e48–e49

Asefi M, Vaisi-Raygani A, Bahrehmand F, Kiani A, Rahimi Z, Nomani H, Ebrahimi A, Tavilani H, Pourmotabbed T (2012) Paraoxonase 1 (PON1) 55 polymorphism, lipid profiles and psoriasis. Br J Dermatol 167:1279–1286

Babin PJ, Thisse C, Durliat M, Andre M, Akimenko MA, Thisse B (1997) Both apolipoprotein E and A-I genes are present in a nonmammalian vertebrate and are highly expressed during embryonic development. Proc Natl Acad Sci USA 94:8622–8627

Bacchetta R, Gambineri E, Roncarolo MG (2007) Role of regulatory T cells and FOXP3 in human diseases. J Allergy Clin Immunol 120:227–235

Bacchetti T, Campanati A, Ferretti G, Simonetti O, Liberati G, Offidani AM (2013) Oxidative stress and psoriasis: the effect of antitumour necrosis factor-α inhibitor treatment. Br J Dermatol 168:984–989

Baniyash M (2004) TCR zeta-chain downregulation: curtailing an excessive inflammatory immune response. Nat Rev Immunol 4:675–687

Batuca JR, Ames PR, Isenberg DA, Alves JD (2007) Antibodies toward high-density lipoprotein components inhibit paraoxonase activity in patients with systemic lupus erythematosus. Ann N Y Acad Sci 1108:137–146

Beck WH, Adams CP, Biglang-Awa IM, Patel AB, Vincent H, Haas-Stapleton EJ, Weers PM (2013) Apolipoprotein A-I binding to anionic vesicles and lipopolysaccharides: role for lysine residues in antimicrobial properties. Biochim Biophys Acta 1828:1503–1510

Bilate AM, Lafaille JJ (2010) Can TNF-alpha boost regulatory T cells? J Clin Invest 120:4190–4192

Boers M, Nurmohamed MT, Doelman CJ, Lard LR, Verhoeven AC, Voskuyl AE, Huizinga TW, van de Stadt RJ, Dijkmans BA, van der Linden S (2003) Influence of glucocorticoids and disease activity on total and high density lipoprotein cholesterol in patients with rheumatoid arthritis. Ann Rheum Dis 62:842–845

Borba EF, Carvalho JF, Bonfa E (2006) Mechanisms of dyslipoproteinemias in systemic lupus erythematosus. Clin Dev Immunol 13:203–208

Buffinton GD, Doe WF (1995) Depleted mucosal antioxidant defences in inflammatory bowel disease. Free Radic Biol Med 19:911–918

Cabana VG, Gidding SS, Getz GS, Chapman J, Shulman ST (1997) Serum amyloid A and high density lipoprotein participate in the acute phase response of Kawasaki disease. Pediatr Res 42:651–655

Carbone F, Nencioni A, Mach F, Vuilleumier N, Montecucco F (2013) Evidence on the pathogenic role of auto-antibodies in acute cardiovascular diseases. Thromb Haemost 109:854–868

Castelli WP, Garrison RJ, Wilson PW, Abbott RD, Kalousdian S, Kannel WB (1986) Incidence of coronary heart disease and lipoprotein cholesterol levels. The Framingham study. JAMA 256:2835–2838

Charles-Schoeman C, Watanabe J, Lee YY, Furst DE, Amjadi S, Elashoff D, Park G, McMahon M, Paulus HE, Fogelman AM, Reddy ST (2009) Abnormal function of high-density lipoprotein is associated with poor disease control and an altered protein cargo in rheumatoid arthritis. Arthritis Rheum 60:2870–2879

Charles-Schoeman C, Lee YY, Grijalva V, Amjadi S, FitzGerald J, Ranganath VK, Taylor M, McMahon M, Paulus HE, Reddy ST (2012) Cholesterol efflux by high density lipoproteins is impaired in patients with active rheumatoid arthritis. Ann Rheum Dis 71:1157–1162

Cheng AM, Handa P, Tateya S, Schwartz J, Tang C, Mitra P, Oram JF, Chait A, Kim F (2012) Apolipoprotein A-I attenuates palmitate-mediated NF-κB activation by reducing Toll-like receptor-4 recruitment into lipid rafts. PLoS ONE 7:e33917

Cheung YF, Yung TC, Tam SC, Ho MH, Chau AK (2004) Novel and traditional cardiovascular risk factors in children after Kawasaki disease: implications for premature atherosclerosis. J Am Coll Cardiol 43:120–124

Cheung YF, O K, Tam SC, Siow YL (2005) Induction of MCP1, CCR2, and iNOS expression in THP-1 macrophages by serum of children late after Kawasaki disease. Pediatr Res 58:1306–1310

Cheung YF, Wong SJ, Ho MH (2007) Relationship between carotid intima-media thickness and arterial stiffness in children after Kawasaki disease. Arch Dis Child 92:43–47

Chi H (2011) Sphingosine-1-phosphate and immune regulation: trafficking and beyond. Trends Pharmacol Sci 32:16–24

Chiang AN, Hwang B, Shaw GC, Lee BC, Lu JH, Meng CC, Chou P (1997) Changes in plasma levels of lipids and lipoprotein composition in patients with Kawasaki disease. Clin Chim Acta 260:15–26

Chodick G, Amital H, Shalem Y, Kokia E, Heymann AD, Porath A, Shalev V (2010) Persistence with statins and onset of rheumatoid arthritis: a population-based cohort study. PLoS Med 7:e1000336

Choy E, Sattar N (2009) Interpreting lipid levels in the context of high-grade inflammatory states with a focus on rheumatoid arthritis: a challenge to conventional cardiovascular risk actions. Ann Rheum Dis 68:460–469

Cimen F, Yildirmak ST, Ergen A, Cakmak M, Dogan S, Yenice N, Sezgin F (2012) Serum lipid, lipoprotein and oxidatively modified low density lipoprotein levels in active or inactive patients with Behcet's disease. Indian J Dermatol 57:97–101

Cohen JA, Barkhof F, Comi G, Hartung HP, Khatri BO, Montalban X, Pelletier J, Capra R, Gallo P, Izquierdo G, Tiel-Wilck K, de Vera A, Jin J, Stites T, Wu S, Aradhye S, Kappos L, TRANSFORMS Study Group (2010) Oral fingolimod or intramuscular interferon for relapsing multiple sclerosis. N Engl J Med 362:402–415

Davidovici BB, Sattar N, Prinz J, Puig L, Emery P, Barker JN, van de Kerkhof P, Stahle M, Nestle FO, Girolomoni G, Krueger JG (2010) Psoriasis and systemic inflammatory diseases: potential mechanistic links between skin disease and co-morbid conditions. J Invest Dermatol 130:1785–1796

Davidson MH (2012) HDL and CETP inhibition: will this DEFINE the future? Curr Treat Options Cardiovasc Med 14:384–390

de Carvalho JF, Bonfá E, Borba EF (2008) Systemic lupus erythematosus and "lupus dyslipoproteinemia". Autoimmun Rev 7:246–250

De Nardo D, Labzin LI, Kono H, Seki R, Schmidt SV, Beyer M, Xu D, Zimmer S, Lahrmann C, Schildberg FA, Vogelhuber J, Kraut M, Ulas T, Kerksiek A, Krebs W, Bode N, Grebe A, Fitzgerald ML, Hernandez NJ, Williams BR, Knolle P, Kneilling M, Röcken M, Lütjohann D, Wright SD, Schultze JL, Latz E (2014) High-density lipoprotein mediates anti-inflammatory reprogramming of macrophages via the transcriptional regulator ATF3. Nat Immunol 15:152–160

Dhillon R, Clarkson P, Donald AE, Powe AJ, Nash M, Novelli V, Dillon MJ, Deanfield JE (1996) Endothelial dysfunction late after Kawasaki disease. Circulation 94:2103–2106

Di Angelantonio E, Sarwar N, Perry P, Kaptoge S, Ray KK, Thompson A, Wood AM, Lewington S, Sattar N, Packard CJ, Collins R, Thompson SG, Danesh J, Emerging Risk

Factors Collaboration (2009) Major lipids, apolipoproteins, and risk of vascular disease. JAMA 302:1993–2000

Dolinsky VW, Douglas DN, Lehner R, Vance DE (2004) Regulation of the enzymes of hepatic microsomal triacylglycerol lipolysis and re-esterification by the glucocorticoid dexamethasone. Biochem J 378:967–974

Eckel RH (1989) Lipoprotein lipase. A multifunctional enzyme relevant to common metabolic diseases. N Engl J Med 320:1060–1068

Ehrenstein MR, Jury EC, Mauri C (2005) Statins for atherosclerosis–as good as it gets? N Engl J Med 352:73–75

El Harchaoui K, Arsenault BJ, Franssen R, Despres JP, Hovingh GK, Stroes ES, Otvos JD, Wareham NJ, Kastelein JJ, Khaw KT, Boekholdt SM (2009) High-density lipoprotein particle size and concentration and coronary risk. Ann Intern Med 150:84–93

Eritsland J, Arnesen H, Smith P, Seljeflot I, Dahl K (1989) Effects of highly concentrated omega-3 polyunsaturated fatty acids and acetylsalicylic acid, alone and combined, on bleeding time and serum lipid profile. J Oslo City Hosp 39:97–101

Ettinger WH, Klinefelter HF, Kwiterovitch PO (1987) Effect of short-term, low-dose corticosteroids on plasma lipoprotein lipids. Atherosclerosis 63:167–172

Ewart HS, Carroll R, Severson DL (1997) Lipoprotein lipase activity in rat cardiomyocytes is stimulated by insulin and dexamethasone. Biochem J 327:439–442

Expert Panel on Detection Evaluation, and Treatment of High Blood Cholesterol in Adults (2001) Executive summary of the third report of the national cholesterol education program (NCEP) expert panel on detection, evaluation, and treatment of high blood cholesterol in adults (Adult Treatment Panel III). JAMA 285:2486–2497

Faggiano A, Pivonello R, Spiezia S, De Martino MC, Filippella M, Di Somma C, Lombardi G, Colao A (2003) Cardiovascular risk factors and common carotid artery calibre and stiffness in patients with Cushing's disease during active disease and 1 year after disease remission. J Clin Endocrinol Metab 88:2527–2533

Favari E, Calabresi L, Adorni MP, Jessup W, Simonelli S, Franceschini G, Bernini F (2009) Small discoidal pre-beta1 HDL particles are efficient acceptors of cell cholesterol via ABCA1 and ABCG1. Biochemistry 48:11067–11074

Flores-Borja F, Kabouridis PS, Jury EC, Isenberg DA, Mageed RA (2005) Decreased Lyn expression and translocation to lipid raft signaling domains in B lymphocytes from patients with systemic lupus erythematosus. Arthritis Rheum 52:3955–3965

Fujimoto M, Nakano M, Terabe F, Kawahata H, Ohkawara T, Han Y, Ripley B, Serada S, Nishikawa T, Kimura A, Nomura S, Kishimoto T, Naka T (2011) The influence of excessive IL-6 production in vivo on the development and function of Foxp3+ regulatory T cells. J Immunol 186:32–40

García-Gómez C, Nolla JM, Valverde J, Narváez J, Corbella E, Pintó X (2008) High HDL-cholesterol in women with rheumatoid arthritis on low-dose glucocorticoid therapy. Eur J Clin Invest 38:686–692

Gelfand JM, Neimann AL, Shin DB, Wang X, Margolis DJ, Troxel AB (2006) Risk of myocardial infarction in patients with psoriasis. JAMA 296:1735–1741

Georgiadis AN, Papavasiliou EC, Lourida ES, Alamanos Y, Kostara C, Tselepis AD (2006) Atherogenic lipid profile is a feature characteristic of patients with early rheumatoid arthritis: effect of early treatment-a prospective, controlled study. Arthritis Res Ther 8:R82

Georgiadis AN, Voulgari PV, Argyropoulou MI, Alamanos Y, Elisaf M, Tselepis AD, Drosos AA (2008) Early treatment reduces the cardiovascular risk factors in newly diagnosed rheumatoid arthritis patients. Semin Arthritis Rheum 38:13–19

Gordon SM, Davidson WS (2012) Apolipoprotein A-I mimetics and high-density lipoprotein function. Curr Opin Endocrinol Diabetes Obes 19:109–114

Graler MH, Goetzl EJ (2004) The immunosuppressant FTY720 down-regulates sphingosine 1-phosphate G-protein-coupled receptors. FASEB J 18:551–553

Grand RJ, Ramakrishna J, Calenda KA (1995) Inflammatory bowel disease in the pediatric patient. Gastroenterol Clin North Am 24:613–632

Grinberg-Bleyer Y, Saadoun D, Baeyens A, Billiard F, Goldstein JD, Grégoire S, Martin GH, Elhage R, Derian N, Carpentier W, Marodon G, Klatzmann D, Piaggio E, Salomon BL (2010) Pathogenic T cells have a paradoxical protective effect in murine autoimmune diabetes by boosting Tregs. J Clin Invest 120:4558–4568

Gruaz L, Delucinge-Vivier C, Descombes P, Dayer JM, Burger D (2010) Blockade of T cell contact-activation of human monocytes by high-density lipoproteins reveals a new pattern of cytokine and inflammatory genes. PLoS ONE 5:e9418

Grunfeld C, Marshall M, Shigenaga JK, Moser AH, Tobias P, Feingold KR (1999) Lipoproteins inhibit macrophage activation by lipoteichoic acid. J Lipid Res 40:245–252

Gu X, Lawrence R, Krieger M (2000) Dissociation of the high density lipoprotein and low density lipoprotein binding activities of murine scavenger receptor class B type I (mSR-BI) using retrovirus library-based activity dissection. J Biol Chem 275:9120–9130

Gupta N, DeFranco AL (2007) Lipid rafts and B cell signaling. Semin Cell Dev Biol 18:616–626

Haas MJ, Mooradian AD (2011) Inflammation, high-density lipoprotein and cardiovascular dysfunction. Curr Opin Infect Dis 24:265–272

Haase CL, Tybjærg-Hansen A, Qayyum AA, Schou J, Nordestgaard BG, Frikke-Schmidt RLCAT (2012) HDL cholesterol and ischemic cardiovascular disease: a Mendelian randomization study of HDL cholesterol in 54,500 individuals. J Clin Endocrinol Metab 97:E248–E256

Hahn BH, Grossman J, Ansell BJ, Skaggs BJ, McMahon M (2008) Altered lipoprotein metabolism in chronic inflammatory states: proinflammatory high-density lipoprotein and accelerated atherosclerosis in systemic lupus erythematosus and rheumatoid arthritis. Arthritis Res Ther 10:213

Hargrove GM, Junco A, Wong NC (1999) Hormonal regulation of apolipoprotein AI. J Mol Endocrinol 22:103–111

Hassan HH, Denis M, Krimbou L, Marcil M, Genest J (2006) Cellular cholesterol homeostasis in vascular endothelial cells. Can J Cardiol 22:35B–40B

Heinecke JW (2009) The HDL, proteome: a marker–and perhaps mediator–of coronary artery disease. J Lipid Res 50(Suppl):S167–S171

Holzer M, Wolf P, Curcic S, Birner-Gruenberger R, Weger W, Inzinger M, El-Gamal D, Wadsack C, Heinemann A, Marsche G (2012) Psoriasis alters HDL composition and cholesterol efflux capacity. J Lipid Res 53:1618–1624

Hua X, Su J, Svenungsson E, Hurt-Camejo E, Jensen-Urstad K, Angelin B, Båvenholm P, Frostegård J (2009) Dyslipidaemia and lipoprotein pattern in systemic lupus erythematosus (SLE) and SLE-related cardiovascular disease. Scand J Rheumatol 38:184–189

Jafarnejad A, Bathaie SZ, Nakhjavani M, Hassan MZ (2008) Investigation of the mechanisms involved in the high-dose and long-term acetyl salicylic acid therapy of type I diabetic rats. J Pharmacol Exp Ther 324:850–857

Jamnitski A, Levels JH, van den Oever IA, Nurmohamed MT (2013) High-density lipoprotein profiling changes in patients with rheumatoid arthritis treated with tumor necrosis factor inhibitors: a cohort study. J Rheumatol 40:825–830

Jiao YL, Wu MP (2008) Apolipoprotein A-I diminishes acute lung injury and sepsis in mice induced by lipoteichoic acid. Cytokine 43:83–87

Juárez-Rojas J, Medina-Urrutia A, Posadas-Sánchez R, Jorge-Galarza E, Mendoza-Pérez E, Caracas-Portilla N, Cardoso-Saldaña G, Muñoz-Gallegos G, Posadas-Romero C (2008) High-density lipoproteins are abnormal in young women with uncomplicated systemic lupus erythematosus. Lupus 17:981–987

Jury EC, Isenberg DA, Mauri C, Ehrenstein MR (2006) Atorvastatin restores Lck expression and lipid raft-associated signaling in T cells from patients with systemic lupus erythematosus. J Immunol 177:7416–7422

Kabouridis PS, Jury EC (2008) Lipid rafts and T-lymphocyte function: implications for autoimmunity. FEBS Lett 582:3711–3718

Kappos L, Radue EW, O'Connor P, Polman C, Hohlfeld R, Calabresi P, Selmaj K, Agoropoulou C, Leyk M, Zhang-Auberson L, Burtin P, FREEDOMS Study Group (2010) A placebo-controlled trial of oral fingolimod in relapsing multiple sclerosis. N Engl J Med 362:387–401

Kawashiri SY, Kawakami A, Yamasaki S, Imazato T, Iwamoto N, Fujikawa K, Aramaki T, Tamai M, Nakamura H, Ida H, Origuchi T, Ueki Y, Eguchi K (2011) Effects of the anti-interleukin-6 receptor antibody, tocilizumab, on serum lipid levels in patients with rheumatoid arthritis. Rheumatol Int 31:451–456

Kerekes G, Soltész P, Dér H, Veres K, Szabó Z, Végvári A, Szegedi G, Shoenfeld Y, Szekanecz Z (2009) Effects of rituximab treatment on endothelial dysfunction, carotid atherosclerosis, and lipid profile in rheumatoid arthritis. Clin Rheumatol 28:705–710

Keul P, Tölle M, Lucke S, von Wnuck LK, Heusch G, Schuchardt M, van der Giet M, Levkau B (2007) The sphingosine-1-phosphate analogue FTY720 reduces atherosclerosis in apolipoprotein E-deficient mice. Arterioscler Thromb Vasc Biol 27:607–613

Khovidhunkit W, Kim MS, Memon RA, Shigenaga JK, Moser AH, Feingold KR, Grunfeld C (2004) Effects of infection and inflammation on lipid and lipoprotein metabolism: mechanisms and consequences to the host. J Lipid Res 45:1169–1196

Kirkham BW, Wasko MC, Hsia EC, Fleischmann RM, Genovese MC, Matteson EL, Liu H, Rahman MU (2014) Effects of golimumab, an anti-tumour necrosis factor-α human monoclonal antibody, on lipids and markers of inflammation. Ann Rheum Dis 73:161–169

Kiss E, Seres I, Tarr T, Kocsis Z, Szegedi G, Paragh G (2007) Reduced paraoxonase1 activity is a risk for atherosclerosis in patients with systemic lupus erythematosus. Ann N Y Acad Sci 1108:83–91

Koutroubakis IE, Oustamanolakis P, Malliaraki N, Karmiris K, Chalkiadakis I, Ganotakis E, Karkavitsas N, Kouroumalis EA (2009) Effects of tumor necrosis factor alpha inhibition with infliximab on lipid levels and insulin resistance in patients with inflammatory bowel disease. Eur J Gastroenterol Hepatol 21:283–288

Lestre S, Diamantino F, Veloso L, Fidalgo A, Ferreira A (2011) Effects of etanercept treatment on lipid profile in patients with moderate-to-severe chronic plaque psoriasis: a retrospective cohort study. Eur J Dermatol 21:916–920

Levine DM, Parker TS, Donnelly TM, Walsh A, Rubin AL (1993) In vivo protection against endotoxin by plasma high density lipoprotein. Proc Natl Acad Sci USA 90:12040–12044

Liu W, Putnam AL, Xu-Yu Z, Szot GL, Lee MR, Zhu S, Gottlieb PA, Kapranov P, Gingeras TR, Fazekas de St Groth B, Clayberger C, Soper DM, Ziegler SF, Bluestone JA (2006) CD127 expression inversely correlates with FoxP3 and suppressive function of human CD4+ T reg cells. J Exp Med 203:1701–1711

Liu G, Yang K, Burns S, Shrestha S, Chi H (2010) The S1P(1)-mTOR axis directs the reciprocal differentiation of T(H)1 and T(reg) cells. Nat Immunol 11:1047–1056

Longnecker R, Miller CL (1996) Regulation of Epstein-Barr virus latency by latent membrane protein 2. Trends Microbiol 4:38–42

Mandala S, Hajdu R, Bergstrom J, Quackenbush E, Xie J, Milligan J, Thornton R, Shei GJ, Card D, Keohane C, Rosenbach M, Hale J, Lynch CL, Rupprecht K, Parsons W, Rosen H (2002) Alteration of lymphocyte trafficking by sphingosine-1-phosphate receptor agonists. Science 296:346–349

Marchesi S, Lupattelli G, Lombardini R, Sensini A, Siepi D, Mannarino M, Vaudo G, Mannarino E (2005) Acute inflammatory state during influenza infection and endothelial function. Atherosclerosis 178:345–350

Marik PE (2006) Dyslipidemia in the critically ill. Crit Care Clin 22:151–159

Mathieu S, Dubost JJ, Tournadre A, Malochet-Guinamand S, Ristori JM, Soubrier M (2010) Effects of 14 weeks of TNF alpha blockade treatment on lipid profile in ankylosing spondylitis. Joint Bone Spine 77:50–52

Mathison JC, Ulevitch RJ (1979) The clearance, tissue distribution, and cellular localization of intravenously injected lipopolysaccharide in rabbits. J Immunol 123:2133–2143

Matloubian M, Lo CG, Cinamon G, Lesneski MJ, Xu Y, Brinkmann V, Allende ML, Proia RL, Cyster JG (2004) Lymphocyte egress from thymus and peripheral lymphoid organs is dependent on S1P receptor 1. Nature 427:355–360

McGillicuddy FC, de la Llera MM, Hinkle CC, Joshi MR, Chiquoine EH, Billheimer JT, Rothblat GH, Reilly MP (2009) Inflammation impairs reverse cholesterol transport in vivo. Circulation 119:1135–1145

McInnes IB, Thompson L, Giles JT, Bathon JM, Salmon JE, Beaulieu AD, Codding CE, Carlson TH, Delles C, Lee JS, Sattar N (2014) Effect of interleukin-6 receptor blockade on surrogates of vascular risk in rheumatoid arthritis: MEASURE, a randomised, placebo-controlled study. Ann Rheum Dis. doi:10.1136/annrheumdis-2013-204345 (in press)

McMahon M, Skaggs BJ, Sahakian L, Grossman J, Fitzgerald J, Ragavendra N, Charles-Schoeman C, Chernishof M, Gorn A, Witztum JL, Wong WK, Weisman M, Wallace DJ, La Cava A, Hahn BH (2011) High plasma leptin levels confer increased risk of atherosclerosis in women with systemic lupus erythematosus, and are associated with inflammatory oxidised lipids. Ann Rheum Dis 70:1619–1624

Mehta NN, Yu Y, Saboury B, Foroughi N, Krishnamoorthy P, Raper A, Baer A, Antigua J, Van Voorhees AS, Torigian DA, Alavi A, Gelfand JM (2011) Systemic and vascular inflammation in patients with moderate to severe psoriasis as measured by [18F]-fluorodeoxyglucose positron emission tomography-computed tomography (FDG-PET/CT): a pilot study. Arch Dermatol 147:1031–1039

Mehta NN, Li R, Krishnamoorthy P, Yu Y, Farver W, Rodrigues A, Raper A, Wilcox M, Baer A, DerOhannesian S, Wolfe M, Reilly MP, Rader DJ, Van Voorhees A, Gelfand JM (2012) Abnormal lipoprotein particles and cholesterol efflux capacity in patients with psoriasis. Atherosclerosis 224:218–221

Messedi M, Jamoussi K, Frigui M, Laporte F, Turki M, Chaabouni K, Mnif E, Jaloulli M, Kaddour N, Bahloul Z, Ayedi F (2011) Atherogenic lipid profile in Behcet's disease: evidence of alteration of HDL subclasses. Arch Med Res 42:211–218

Miller GJ, Miller NE (1975) Plasma-high-density-lipoprotein concentration and development of ischaemic heart-disease. Lancet 1:16–19

Mor A, Planer D, Luboshits G, Afek A, Metzger S, Chajek-Shaul T, Keren G, George J (2007) Role of naturally occurring $CD4^+$ $CD25^+$ regulatory T cells in experimental atherosclerosis. Arterioscler Thromb Vasc Biol 27:893–900

Mottonen M, Heikkinen J, Mustonen L, Isomaki P, Luukkainen R, Lassila O (2005) $CD4^+$ $CD25^+$ T cells with the phenotypic and functional characteristics of regulatory T cells are enriched in the synovial fluid of patients with rheumatoid arthritis. Clin Exp Immunol 140:360–367

Musabak U, Baylan O, Cetin T, Yesilova Z, Sengul A, Saglam K, Inal A, Kocar IH (2005) Lipid profile and anticardiolipin antibodies in Behcet's disease. Arch Med Res 36:387–392

Myasoedova E, Crowson CS, Kremers HM, Roger VL, Fitz-Gibbon PD, Therneau TM, Gabriel SE (2011) Lipid paradox in rheumatoid arthritis: the impact of serum lipid measures and systemic inflammation on the risk of cardiovascular disease. Ann Rheum Dis 70:482–487

Navab M, Reddy ST, Van Lenten BJ, Fogelman AM (2011) HDL and cardiovascular disease: atherogenic and atheroprotective mechanisms. Nat Rev Cardiol 8:222–232

Navarro-Millán I, Charles-Schoeman C, Yang S, Bathon JM, Bridges SL Jr, Chen L, Cofield SS, Dell'Italia LJ, Moreland LW, O'Dell JR, Paulus HE, Curtis JR (2013) Changes in lipoproteins associated with methotrexate or combination therapy in early rheumatoid arthritis: results from the treatment of early rheumatoid arthritis trial. Arthritis Rheum 65:1430–1438

Newburger JW, Burns JC, Beiser AS, Loscalzo J (1991) Altered lipid profile after Kawasaki syndrome. Circulation 84:625–631

Nofer JR, Bot M, Brodde M, Taylor PJ, Salm P, Brinkmann V, van Berkel T, Assmann G, Biessen EA (2007) FTY720, a synthetic sphingosine 1 phosphate analogue, inhibits development of atherosclerosis in low-density lipoprotein receptor-deficient mice. Circulation 115:501–508

Norata GD, Catapano AL (2012) HDL and adaptive immunity: a tale of lipid rafts. Atherosclerosis 225:34–35

Norata GD, Callegari E, Marchesi M, Chiesa G, Eriksson P, Catapano AL (2005) High-density lipoproteins induce transforming growth factor-beta2 expression in endothelial cells. Circulation 111:2805–2811

Norata GD, Marchesi P, Pirillo A, Uboldi P, Chiesa G, Maina V, Garlanda C, Mantovani A, Catapano AL (2008) Long pentraxin 3, a key component of innate immunity, is modulated by high-density lipoproteins in endothelial cells. Arterioscler Thromb Vasc Biol 28:925–931

Norata GD, Pirillo A, Ammirati E, Catapano AL (2012) Emerging role of high density lipoproteins as a player in the immune system. Atherosclerosis 220:11–21

O'Connell BJ, Denis M, Genest J (2004) Cellular physiology of cholesterol efflux in vascular endothelial cells. Circulation 110:2881–2888

O'Neill SG, Giles I, Lambrianides A, Manson J, D'Cruz D, Schrieber L, March LM, Latchman DS, Isenberg DA, Rahman A (2010) Antibodies to apolipoprotein A-I, high-density lipoprotein, and C-reactive protein are associated with disease activity in patients with systemic lupus erythematosus. Arthritis Rheum 62:845–854

Orem A, Yandi YE, Vanizor B, Cimsit G, Uydu HA, Malkoc M (2002) The evaluation of autoantibodies against oxidatively modified low-density lipoprotein (LDL), susceptibility of LDL to oxidation, serum lipids and lipid hydroperoxide levels, total antioxidant status, antioxidant enzyme activities, and endothelial dysfunction in patients with Behcet's disease. Clin Biochem 35:217–224

Parmentier-Decrucq E, Duhamel A, Ernst O, Fermont C, Louvet A, Vernier-Massouille G, Cortot A, Colombel JF, Desreumaux P, Peyrin-Biroulet L (2009) Effects of infliximab therapy on abdominal fat and metabolic profile in patients with Crohn's disease. Inflamm Bowel Dis 15:1476–1484

Peters MJ, Vis M, van Halm VP, Wolbink GJ, Voskuyl AE, Lems WF, Dijkmans BA, Twisk JW, de Koning MH, van de Stadt RJ, Nurmohamed MT (2007) Changes in lipid profile during infliximab and corticosteroid treatment in rheumatoid arthritis. Ann Rheum Dis 66:958–961

Phillips MC (2013) New insights into the determination of HDL structure by apolipoproteins: thematic review series: high density lipoprotein structure, function, and metabolism. J Lipid Res 54:2034–2048

Pollono EN, Lopez-Olivo MA, Lopez JA, Suarez-Almazor ME (2010) A systematic review of the effect of TNF-alpha antagonists on lipid profiles in patients with rheumatoid arthritis. Clin Rheumatol 29:947–955

Prosser HC, Ng MK, Bursill CA (2012) The role of cholesterol efflux in mechanisms of endothelial protection by HDL. Curr Opin Lipidol 23:182–189

Raterman HG, Levels H, Voskuyl AE, Lems WF, Dijkmans BA, Nurmohamed MT (2013) HDL protein composition alters from proatherogenic into less atherogenic and proinflammatory in rheumatoid arthritis patients responding to rituximab. Ann Rheum Dis 72:560–565

Ridker PM, Howard CP, Walter V, Everett B, Libby P, Hensen J, Thuren T, CANTOS Pilot Investigative Group (2012) Effects of interleukin-1β inhibition with canakinumab on hemoglobin A1c, lipids, C-reactive protein, interleukin-6, and fibrinogen: a phase IIb randomized, placebo-controlled trial. Circulation 126:2739–2748

Rivera J, Proia RL, Olivera A (2008) The alliance of sphingosine-1-phosphate and its receptors in immunity. Nat Rev Immunol 8:753–763

Romanato G, Scarpa M, Angriman I, Faggian D, Ruffolo C, Marin R, Zambon S, Basato S, Zanoni S, Filosa T, Pilon F, Manzato E (2009) Plasma lipids and inflammation in active inflammatory bowel diseases. Aliment Pharmacol Ther 29:298–307

Ronda N, Favari E, Borghi MO, Ingegnoli F, Gerosa M, Chighizola C, Zimetti F, Adorni MP, Bernini F, Meroni PL (2014) Impaired serum cholesterol efflux capacity in rheumatoid arthritis and systemic lupus erythematosus. Ann Rheum Dis 73(3):609–615. doi:10.1136/annrheumdis-2012-202914

Sala F, Cutuli L, Grigore L, Pirillo A, Chiesa G, Catapano AL, Norata GD (2013) Prevalence of classical $CD14^{++}/CD16^{-}$ but not of intermediate $CD14^{++}/CD16^{+}$ monocytes in hypoalphalipoproteinemia. Int J Cardiol 168:2886–2889

Salazar A, Mana J, Pinto X, Argimon JM, Hurtado I, Pujol R (2002) Corticosteroid therapy increases HDL-cholesterol concentrations in patients with active sarcoidosis and hypoalphalipoproteinemia. Clin Chim Acta 320:59–64

Salemi G, Gueli MC, Vitale F, Battaglieri F, Guglielmini E, Ragonese P, Trentacosti A, Massenti MF, Savettieri G, Bono A (2010) Blood lipids, homocysteine, stress factors, and vitamins in clinically stable multiple sclerosis patients. Lipids Health Dis 9:19

Sappati Biyyani RS, Putka BS, Mullen KD (2010) Dyslipidemia and lipoprotein profiles in patients with inflammatory bowel disease. J Clin Lipidol 4:478–482

Sarkissian T, Beyene J, Feldman B, McCrindle B, Silverman ED (2007) Longitudinal examination of lipid profiles in pediatric systemic lupus erythematosus. Arthritis Rheum 56:631–638

Scanu AM, Edelstein C (2008) HDL: bridging past and present with a look at the future. FASEB J 22:4044–4054

Sethi A, Parmar HS, Kumar A (2011) The effect of aspirin on atherogenic diet-induced diabetes mellitus. Basic Clin Pharmacol Toxicol 108:371–377

Shah S, Casas JP, Drenos F, Whittaker J, Deanfield J, Swerdlow DI, Holmes MV, Kivimaki M, Langenberg C, Wareham N, Gertow K, Sennblad B, Strawbridge RJ, Baldassarre D, Veglia F, Tremoli E, Gigante B, de Faire U, Kumari M, Talmud PJ, Hamsten A, Humphries SE, Hingorani AD (2013) Causal relevance of blood lipid fractions in the development of carotid atherosclerosis: Mendelian randomization analysis. Circ Cardiovasc Genet 6:63–72

Shiflett AM, Bishop JR, Pahwa A, Hajduk SL (2005) Human high density lipoproteins are platforms for the assembly of multi-component innate immune complexes. J Biol Chem 280:32578–32585

Shoenfeld Y, Gerli R, Doria A, Matsuura E, Cerinic MM, Ronda N, Jara LJ, Abu-Shakra M, Meroni PL, Sherer Y (2005) Accelerated atherosclerosis in autoimmune rheumatic diseases. Circulation 112:3337–3347

Smythies LE, White CR, Maheshwari A, Palgunachari MN, Anantharamaiah GM, Chaddha M, Kurundkar AR, Datta G (2010) Apolipoprotein A-I mimetic 4F alters the function of human monocyte-derived macrophages. Am J Physiol Cell Physiol 298:C1538–C1548

Stagakis I, Bertsias G, Karvounaris S, Kavousanaki M, Virla D, Raptopoulou A, Kardassis D, Boumpas DT, Sidiropoulos PI (2012) Anti-tumor necrosis factor therapy improves insulin resistance, beta cell function and insulin signaling in active rheumatoid arthritis patients with high insulin resistance. Arthritis Res Ther 14:R141

Tardif JC, L'allier PL, Ibrahim R, Grégoire JC, Nozza A, Cossette M, Kouz S, Lavoie MA, Paquin J, Brotz TM, Taub R, Pressacco J (2010) Treatment with 5-lipoxygenase inhibitor VIA-2291 (Atreleuton) in patients with recent acute coronary syndrome. Circ Cardiovasc Imaging 3:298–307

Tedesco-Silva H, Pescovitz MD, Cibrik D, Rees MA, Mulgaonkar S, Kahan BD, Gugliuzza KK, Rajagopalan PR, Esmeraldo Rde M, Lord H, Salvadori M, Slade JM, FTY720 Study Group (2006) Randomized controlled trial of FTY720 versus MMF in de novo renal transplantation. Transplantation 82:1689–1697

Teixeira PC, Cutler P, Vuilleumier N (2012) Autoantibodies to apolipoprotein A-1 in cardiovascular diseases: current perspectives. Clin Dev Immunol 2012:868251. doi:10.1155/2012/868251

Terasaka N, Wang N, Yvan-Charvet L, Tall AR (2007) High-density lipoprotein protects macrophages from oxidized low-density lipoprotein-induced apoptosis by promoting efflux of 7-ketocholesterol via ABCG1. Proc Natl Acad Sci USA 104:15093–15098

Thuahnai ST, Lund-Katz S, Anantharamaiah GM, Williams DL, Phillips MC (2003) A quantitative analysis of apolipoprotein binding to SR-BI: multiple binding sites for lipid-free and lipid-associated apolipoproteins. J Lipid Res 44:1132–1142

Triantafilou M, Miyake K, Golenbock DT, Triantafilou K (2002) Mediators of innate immune recognition of bacteria concentrate in lipid rafts and facilitate lipopolysaccharide-induced cell activation. J Cell Sci 115:2603–2611

Troitzsch P, Paulista Markus MR, Dorr M, Felix SB, Junger M, Schminke U, Schmidt CO, Volzke H, Baumeister SE, Arnold A (2012) Psoriasis is associated with increased intima-media thickness–the Study of Health in Pomerania (SHIP). Atherosclerosis 225:486–490

Van Lenten BJ, Hama SY, de Beer FC, Stafforini DM, McIntyre TM, Prescott SM, La Du BN, Fogelman AM, Navab M (1995) Anti-inflammatory HDL becomes pro-inflammatory during the acute phase response. Loss of protective effect of HDL against LDL oxidation in aortic wall cell cocultures. J Clin Invest 96:2758–2767

van Leuven SI, Hezemans R, Levels JH, Snoek S, Stokkers PC, Hovingh GK, Kastelein JJ, Stroes ES, de Groot E, Hommes DW (2007) Enhanced atherogenesis and altered high density lipoprotein in patients with Crohn's disease. J Lipid Res 48:2640–2646

Viñals M, Bermúdez I, Llaverias G, Alegret M, Sanchez RM, Vázquez-Carrera M, Laguna JC (2005) Aspirin increases CD36, SR-BI, and ABCA1 expression in human THP-1 macrophages. Cardiovasc Res 66:141–149

Voight BF, Peloso GM, Orho-Melander M, Frikke-Schmidt R, Barbalic M, Jensen MK, Hindy G, Hólm H, Ding EL, Johnson T, Schunkert H, Samani NJ, Clarke R, Hopewell JC, Thompson JF, Li M, Thorleifsson G, Newton-Cheh C, Musunuru K, Pirruccello JP, Saleheen D, Chen L, Stewart A, Schillert A, Thorsteinsdottir U, Thorgeirsson G, Anand S, Engert JC, Morgan T, Spertus J, Stoll M, Berger K, Martinelli N, Girelli D, McKeown PP, Patterson CC, Epstein SE, Devaney J, Burnett MS, Mooser V, Ripatti S, Surakka I, Nieminen MS, Sinisalo J, Lokki ML, Perola M, Havulinna A, de Faire U, Gigante B, Ingelsson E, Zeller T, Wild P, de Bakker PI, Klungel OH, Maitland-van der Zee Zee AH, Peters BJ, de Boer A, Grobbee DE, Kamphuisen PW, Deneer VH, Elbers CC, Onland-Moret NC, Hofker MH, Wijmenga C, Verschuren WM, Boer JM, van der Schouw YT, Rasheed A, Frossard P, Demissie S, Willer C, Do R, Ordovas JM, Abecasis GR, Boehnke M, Mohlke KL, Daly MJ, Guiducci C, Burtt NP, Surti A, Gonzalez E, Purcell S, Gabriel S, Marrugat J, Peden J, Erdmann J, Diemert P, Willenborg C, König IR, Fischer M, Hengstenberg C, Ziegler A, Buysschaert I, Lambrechts D, Van de Werf F, Fox KA, El Mokhtari NE, Rubin D, Schrezenmeir J, Schreiber S, Schäfer A, Danesh J, Blankenberg S, Roberts R, McPherson R, Watkins H, Hall AS, Overvad K, Rimm E, Boerwinkle E, Tybjaerg-Hansen A, Cupples LA, Reilly MP, Melander O, Mannucci PM, Ardissino D, Siscovick D, Elosua R, Stefansson K, O'Donnell CJ, Salomaa V, Rader DJ, Peltonen L, Schwartz SM, Altshuler D, Kathiresan S (2012) Plasma HDL cholesterol and risk of myocardial infarction: a mendelian randomisation study. Lancet 380:572–580

Vuilleumier N, Dayer JM, von Eckardstein A, Roux-Lombard P (2013) Pro- or anti-inflammatory role of apolipoprotein A-1 in high-density lipoproteins? Swiss Med Wkly 143:w13781

Vuilleumier N, Montecucco F, Hartley O (2014) Autoantibodies to apolipoprotein A-1 as a biomarker of cardiovascular autoimmunity. World J Cardiol 6(5):314–326. doi:10.4330/wjc.v6.i5.314

Wang Y, Zhu X, Wu G, Shen L, Chen B (2008) Effect of lipid-bound apoA-I cysteine mutants on lipopolysaccharide-induced endotoxemia in mice. J Lipid Res 49:1640–1645

Wang SH, Yuan SG, Peng DQ, Zhao SP (2012) HDL and ApoA-I inhibit antigen presentation-mediated T cell activation by disrupting lipid rafts in antigen presenting cells. Atherosclerosis 225:105–114

Watanabe J, Charles-Schoeman C, Miao Y, Elashoff D, Lee YY, Katselis G, Lee TD, Reddy ST (2012) Proteomic profiling following immunoaffinity capture of high-density lipoprotein: association of acute-phase proteins and complement factors with proinflammatory high-density lipoprotein in rheumatoid arthritis. Arthritis Rheum 64:1828–1837

Weigert A, Weis N, Brune B (2009) Regulation of macrophage function by sphingosine-1-phosphate. Immunobiology 214:748–760

Weiss SJ (1989) Tissue destruction by neutrophils. N Engl J Med 320:365–376

Wilhelm AJ, Zabalawi M, Grayson JM, Weant AE, Major AS, Owen J, Bharadwaj M, Walzem R, Chan L, Oka K, Thomas MJ, Sorci-Thomas MG (2009) Apolipoprotein A-I and its role in lymphocyte cholesterol homeostasis and autoimmunity. Arterioscler Thromb Vasc Biol 29:843–849

Wilhelm AJ, Zabalawi M, Owen JS, Shah D, Grayson JM, Major AS, Bhat S, Gibbs DP Jr, Thomas MJ, Sorci-Thomas MG (2010) Apolipoprotein A-I modulates regulatory T cells in autoimmune LDLr-/-, ApoA-I-/- mice. J Biol Chem 285:36158–36169

Williams HR, Willsmore JD, Cox IJ, Walker DG, Cobbold JF, Taylor-Robinson SD, Orchard TR (2012) Serum metabolic profiling in inflammatory bowel disease. Dig Dis Sci 57:2157–2165

Yatomi Y (2008) Plasma sphingosine 1-phosphate metabolism and analysis. Biochim Biophys Acta 1780:606–611

Young D, Peterson C, Basch C, Halladay SC (1995) Effects of naproxen and nabumetone on serum cholesterol levels in patients with osteoarthritis. Clin Ther 17:231–240

Yu Y, Sheth N, Krishnamoorthy P, Saboury B, Raper A, Baer A, Ochotony R, Doveikis J, Derohannessian S, Voorhees AS, Torigian DA, Alavi A, Gelfand JM, Mehta NN (2012) Aortic vascular inflammation in psoriasis is associated with HDL particle size and concentration: a pilot study. Am J Cardiovasc Dis 2:285–292

Yuhanna IS, Zhu Y, Cox BE, Hahner LD, Osborne-Lawrence S, Lu P, Marcel YL, Anderson RG, Mendelsohn ME, Hobbs HH, Shaul PW (2001) High-density lipoprotein binding to scavenger receptor-BI activates endothelial nitric oxide synthase. Nat Med 7:853–857

Yvan-Charvet L, Pagler T, Gautier EL, Avagyan S, Siry RL, Han S, Welch CL, Wang N, Randolph GJ, Snoeck HW, Tall AR (2010a) ATP-binding cassette transporters and HDL suppress hematopoietic stem cell proliferation. Science 328:1689–1693

Yvan-Charvet L, Wang N, Tall AR (2010b) Role of HDL, ABCA1, and ABCG1 transporters in cholesterol efflux and immune responses. Arterioscler Thromb Vasc Biol 30:139–143

Zapolska-Downar D, Naruszewicz M, Zapolski-Downar A, Markiewski M, Bukowska H, Millo B (2000) Ibuprofen inhibits adhesiveness of monocytes to endothelium and reduces cellular oxidative stress in smokers and non-smokers. Eur J Clin Invest 30:1002–1010

HDL in Infectious Diseases and Sepsis

Angela Pirillo, Alberico Luigi Catapano, and Giuseppe Danilo Norata

Contents

1	Introduction	485
2	HDL and Bacterial Infections	486
	2.1　Interaction of HDL with LPS and Gram-Negative Bacteria	488
	2.2　Interaction of HDL with LTA and Gram-Positive Bacteria	490
	2.3　HDL and Mycobacteria	491
	2.4　General Innate Host Defense Mechanisms Exerted by HDL After Bacterial Infection	491
3	HDL and Parasitic Infections	494
4	HDL and Viral Infections	495
	4.1　HIV Infection	495
	4.2　HCV Infection	497
References		500

Abstract

During infection significant alterations in lipid metabolism and lipoprotein composition occur. Triglyceride and VLDL cholesterol levels increase, while reduced HDL cholesterol (HDL-C) and LDL cholesterol (LDL-C) levels are observed. More importantly, endotoxemia modulates HDL composition and size: phospholipids are reduced as well as apolipoprotein (apo) A-I, while

---

A. Pirillo • A.L. Catapano
SISA Center for the Study of Atherosclerosis, Bassini Hospital, Cinisello Balsamo, Italy

IRCCS, Multimedica, Milan, Italy

G.D. Norata (✉)
SISA Center for the Study of Atherosclerosis, Bassini Hospital, Cinisello Balsamo, Italy

Department of Pharmacological Sciences, Università degli Studi di Milano, Milan, Italy

Department of Pharmacological and Biomolecular Sciences, University of Milan,
Via Balzaretti 9, 20133 Milan, Italy
e-mail: Danilo.Norata@unimi.it

© The Author(s) 2015
A. von Eckardstein, D. Kardassis (eds.), *High Density Lipoproteins*, Handbook of Experimental Pharmacology 224, DOI 10.1007/978-3-319-09665-0_15

serum amyloid A (SAA) and secretory phospholipase A2 (sPLA2) dramatically increase, and, although the total HDL particle number does not change, a significant decrease in the number of small- and medium-size particles is observed. Low HDL-C levels inversely correlate with the severity of septic disease and associate with an exaggerated systemic inflammatory response. HDL, as well as other plasma lipoproteins, can bind and neutralize Gram-negative bacterial lipopolysaccharide (LPS) and Gram-positive bacterial lipoteichoic acid (LTA), thus favoring the clearance of these products. HDLs are emerging also as a relevant player during parasitic infections, and a specific component of HDL, namely, apoL-1, confers innate immunity against trypanosome by favoring lysosomal swelling which kills the parasite. During virus infections, proteins associated with the modulation of cholesterol bioavailability in the lipid rafts such as ABCA1 and SR-BI have been shown to favor virus entry into the cells. Pharmacological studies support the benefit of recombinant HDL or apoA-I mimetics during bacterial infection, while apoL-1–nanobody complexes were tested for trypanosome infection. Finally, SR-BI antagonism represents a novel and forefront approach interfering with hepatitis C virus entry which is currently tested in clinical studies. From the coming years, we have to expect new and compelling observations further linking HDL to innate immunity and infections.

Keywords
HDL • Infections • Bacteria • Parasites • Virus

Abbreviations

ABCA-1	ATP-binding cassette transporter 1
ApoA-I	Apolipoprotein A-I
ApoE	Apolipoprotein E
ApoL-1	Apolipoprotein L-1
ApoM	Apolipoprotein M
IL-1β	Interleukin-1β
IL-6	Interleukin-6
HCV	Hepatitis C virus
HIV	Human immunodeficiency virus
HDL-C	HDL cholesterol
LBP	LPS-binding protein
LCAT	Lecithin-cholesterol acyltransferase
LDL-C	LDL cholesterol
LPS	Gram-negative bacterial lipopolysaccharide
LTA	Gram-positive bacterial lipoteichoic acid
PAF-AH	Platelet-activating factor acetylhydrolase

PON1	Serum paraoxonase
PTX3	Pentraxin 3
rHDL	Reconstituted HDL
SAA	Serum amyloid A
S1P	Sphingosine-1-phosphate
sPLA2	Secretory phospholipase A2
SR-BI	Scavenger receptor class B member 1
TG	Triglycerides
TLF-1	Trypanosome lytic factor-1
TLRs	Toll-like receptors
TNF-α	Tumor necrosis factor-α

1   Introduction

HDLs are heterogeneous particles generated by the continuous remodeling by lipolytic enzymes and lipid transporters and by lipid and apolipoprotein exchange with other circulating lipoproteins and tissues (Kontush and Chapman 2006). Mature HDL particles have a hydrophobic core containing cholesteryl esters and triglycerides, while proteins are embedded in a lipid monolayer composed mainly of phospholipids and free cholesterol. HDLs contain two main proteins, apolipoprotein A-I (apoA-I) and apoA-II, but several other minor apoproteins as well as enzymes such as lecithin-cholesterol acyltransferase (LCAT), serum paraoxonase (PON1), and platelet-activating factor acetylhydrolase (PAF-AH) are associated with HDL particles (Navab et al. 2004). HDLs possess several biological functions (Pirillo et al. 2013), but the role of HDL in innate immunity has emerged in the 1970s with the first observation associating HDL cholesterol (HDL-C) plasma levels to protection against sepsis. In the coming years, it has emerged that the ability of HDL to modulate cholesterol bioavailability in the lipid rafts, membrane microdomains enriched in glycosphingolipids and cholesterol, is evolutionary conserved and affects the properties of cells involved in the innate and adaptive immune response, tuning inflammatory response and antigen presentation functions in macrophages as well as activation of B and T cells. In the context of infections, HDL and their components have been linked with protection toward Gram-negative and Gram-positive bacteria and parasites, while the role during virus infection is debated (Fig. 1). Furthermore, HDLs influence humoral innate immunity by tuning the activation of the complement system and the expression of pentraxin 3 (PTX3). HDLs are critical not only during sepsis but also in other bacterial, parasitic, and viral infection. The aim of this chapter is to discuss the relevant findings on the link between HDL and immune response, shedding a new light on the role of these lipoproteins during sepsis and infectious disease.

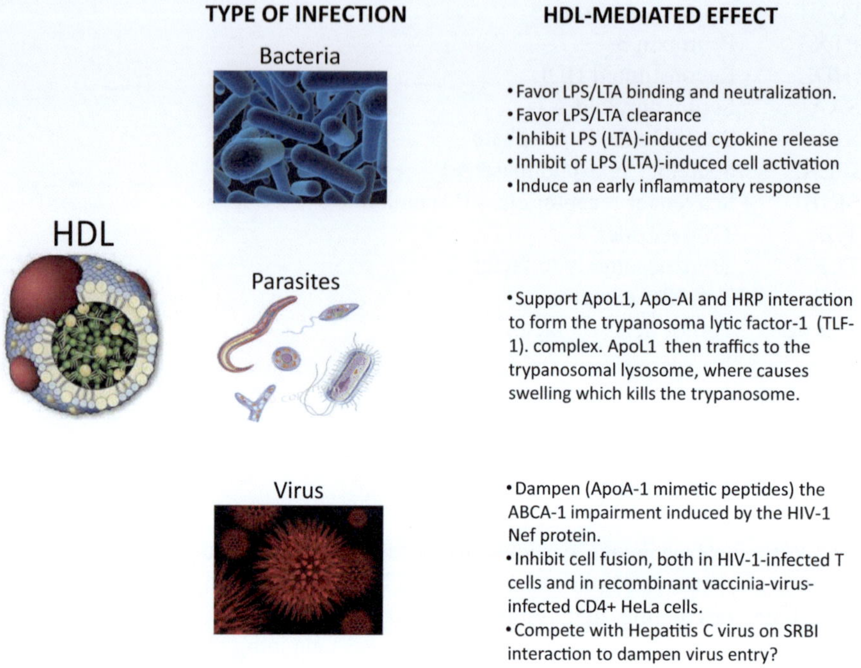

TYPE OF INFECTION	HDL-MEDIATED EFFECT
Bacteria	• Favor LPS/LTA binding and neutralization. • Favor LPS/LTA clearance • Inhibit LPS (LTA)-induced cytokine release • Inhibit of LPS (LTA)-induced cell activation • Induce an early inflammatory response
Parasites	• Support ApoL1, Apo-AI and HRP interaction to form the trypanosoma lytic factor-1 (TLF-1). complex. ApoL1 then traffics to the trypanosomal lysosome, where causes swelling which kills the trypanosome.
Virus	• Dampen (ApoA-1 mimetic peptides) the ABCA-1 impairment induced by the HIV-1 Nef protein. • Inhibit cell fusion, both in HIV-1-infected T cells and in recombinant vaccinia-virus-infected CD4+ HeLa cells. • Compete with Hepatitis C virus on SRBI interaction to dampen virus entry?

Fig. 1 HDL and infections

2   HDL and Bacterial Infections

Increasing observations suggest that persistent low-grade inflammation is associated with the pathogenesis of severe chronic diseases such as atherosclerosis, diabetes, and other aging-related neurological diseases. Low levels of circulating Gram-negative bacterial endotoxin lipopolysaccharide (LPS) appear to sustain a non-resolving low-grade inflammation. As a consequence, low-grade endotoxinemia may skew host immune environment into a mild non-resolving pro-inflammatory state, which eventually leads to the pathogenesis and progression of inflammatory diseases.

During infection, significant changes in the lipid metabolism are observed. At first, plasma levels of lipid and lipoproteins may change: triglyceride (TG) and VLDL cholesterol levels increase due to several mechanisms, including reduction of TG hydrolysis, LPS- and pro-inflammatory cytokines-induced de novo free fatty acid production, and TG synthesis in the liver and reduction of lipoprotein lipase activity thus resulting in reduced VLDL clearance and increased TG levels (Wendel et al. 2007). In addition, the increase in free fatty acids induces insulin resistance, thus contributing to increased glucose levels during systemic inflammation. On the other hand, HDL-C and LDL-C levels decrease during sepsis, and a low plasma

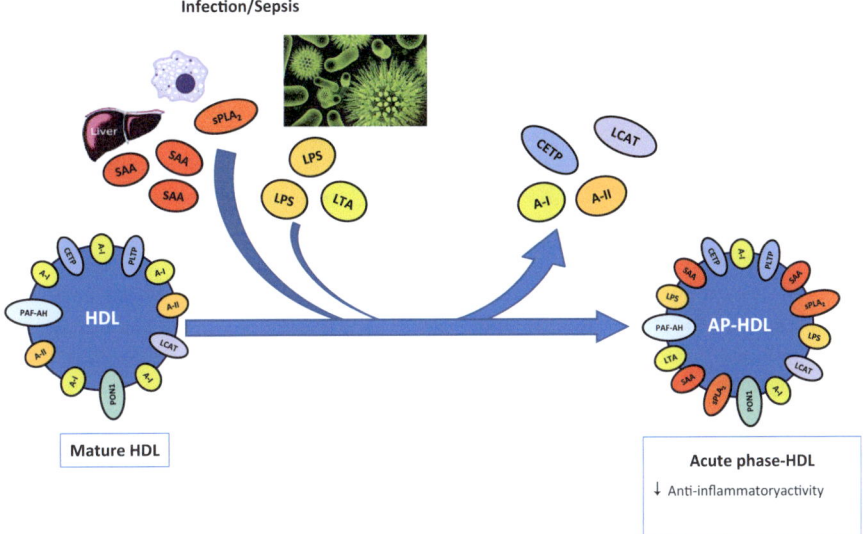

Fig. 2 Acute-phase HDL

HDL-C level (associated with a low plasma apoA-I level) is a poor prognostic factor for severe sepsis, as it is associated with increased mortality and adverse clinical outcomes (Chien et al. 2005); more importantly, significant alterations in lipoprotein composition are observed, and increased levels of acute-phase proteins, including serum amyloid A (SAA) and secretory phospholipase A2 (sPLA2), may contribute to decreased HDL-C levels, by replacing some structural and functional HDL components (Fig. 2).

Endotoxinemia also modulates HDL composition and size: phospholipids are reduced while SAA dramatically increases and apoA-I decreases, and, although the total HDL particle number does not change, a significant decrease in the numbers of small- and medium-size particles is observed (de la Llera et al. 2012). The apoA-I content is reduced due to the rapid association of SAA, which displaces apoA-I and becomes the main protein of acute-phase HDL (Coetzee et al. 1986; Khovidhunkit et al. 2004); the content of other proteins associated with HDL (PON1, PAF-AH) is altered, resulting in reduced antioxidant properties of HDL (Feingold et al. 1998) and increased content of pro-atherogenic lipids (Cao et al. 1998; Memon et al. 1999). Also the lipid composition of HDL is altered during the acute-phase response (Khovidhunkit et al. 2004). Endotoxemia induces the increase of some enzymes involved in HDL remodeling, including endothelial lipase (Badellino et al. 2008) and secretory phospholipase A2 (de la Llera et al. 2012), and the decrease of other, such as CETP and LCAT (de la Llera et al. 2012; Wendel et al. 2007). Altogether these changes result in the loss of functional properties of HDL (Banka et al. 1995; de la Llera et al. 2012; McGillicuddy et al. 2009).

2.1 Interaction of HDL with LPS and Gram-Negative Bacteria

Sepsis is a major cause of death in hospitalized patients. Mortality is mainly due to the cytotoxic actions of lipid components of the bacterial outer membrane Lipopolysaccharide (LPS) is the toxic component of endotoxin in the outer membrane of Gram-negative bacteria. Lipoteichoic acid (LTA) is a heat-stable component of the cell membrane and wall of most Gram-positive bacteria that shares structural similarities with LPS and induces cytokine cascades alike LPS (Grunfeld et al. 1999).

LPS, the major pathogenic factor in Gram-negative sepsis, is an essential component of the bacterial cell wall, and it is not toxic when incorporated into the membrane; after release in the blood following bacterial cell reproduction, lysis, or death, lipid A, the most essential part of LPS, induces an inflammatory response (Van Amersfoort et al. 2003). This is mediated by pro-inflammatory cytokines released primarily from monocytes/macrophages and neutrophils, such as tumor necrosis factor-α (TNF-α), interleukin-1β (IL-1β), and interleukin-6 (IL-6) (Levine et al. 1993). LPS is recognized mainly by toll-like receptor 4 (TLR4) in cooperation with other proteins including MD-2, CD14, and LPS-binding protein (LBP) (Jerala 2007). LBP catalyzes the transfer of LPS to CD14, thus enhancing LPS-induced cell activation (Van Amersfoort et al. 2003). To prevent exaggerated responses to LPS, the host has developed several control mechanisms that include inhibitory LPS-binding proteins and plasma lipoproteins (Van Amersfoort et al. 2003).

In patients with severe sepsis, HDL-C decreases rapidly, and SAA is the major protein present in HDL (45 %) at the start of sepsis and is slowly replaced by apoA-I during recovery (van Leeuwen et al. 2003). Low HDL-C levels inversely correlate with the severity of septic disease and associate with an exaggerated systemic inflammatory response (Wendel et al. 2007), although it is difficult to establish whether changes in plasma lipoproteins simply reflect the severity of disease or they can directly modify the host response to inflammation. Also in healthy subjects low HDL levels are associated with increased inflammatory response on endotoxin challenge compared to subjects with normal or high HDL levels (Birjmohun et al. 2007), without differences in the HDL proteome (Levels et al. 2011). These observations indicate a positive role of HDL in the protection against sepsis.

Several mechanisms are involved in HDL-mediated protection. HDL, as well as other plasma lipoproteins (LDL, TG-rich lipoproteins), can bind and neutralize Gram-negative bacterial LPS as well as Gram-positive bacterial lipoteichoic acid (LTA) (Grunfeld et al. 1999; Murch et al. 2007). ApoA-I knockout mice, which lack HDL, exhibit decreased LPS neutralization in the serum compared with serum from control mice (Guo et al. 2013); overexpression of apoA-I moderately improves survival compared to controls, suggesting that HDL elevation may protect against septic death. In endotoxemic rats, in which LPS has been infused after HDL administration, HDLs attenuate LPS-induced organ damage accompanied by lower TNF-α and nitric oxide production (Lee et al. 2007). In addition to its role in LPS neutralization, HDLs exert its protection against sepsis also by promoting LPS clearance; in fact, almost all LPS exists in the complex LPS–HDL in the blood (Ulevitch et al. 1979, 1981), the HDL receptor SR-BI binds and mediates LPS uptake, and HDLs promote SR-BI-mediated LPS uptake (Vishnyakova et al. 2003).

Administration of reconstituted HDL (rHDL) efficiently inhibits the LPS-induced cytokine release from the whole-blood system in vitro (Parker et al. 1995); however, the first study that demonstrated the LPS-neutralizing ability of rHDL in humans was performed by intravenous infusion of rHDL before induction of endotoxemia in healthy volunteers (Pajkrt et al. 1996). rHDL significantly reduced endotoxemia-induced inflammatory response, as it reduced clinical symptoms, reduced inflammatory cytokine (TNF-α, IL-6, IL-8) production, and attenuated LPS-induced leukocyte activation, in part due to the downregulation of the main LPS receptor monocyte-bound CD14 (Pajkrt et al. 1996). In LPS-challenged macrophages, HDLs selectively inhibit the activation of type I IFN response genes (Suzuki et al. 2010), which play a critical role in the antiviral response of cells, although emerging evidence also implicates this response in host defense during bacterial infection. This inhibitory effect of HDL does not require LPS binding to lipoproteins (Suzuki et al. 2010). HDLs (and apoA-I) attenuate also LPS-induced neutrophil activation (Murphy et al. 2011). More recently Di Nardo et al. (2014) have shown that HDLs promote the expression of ATF3 in macrophages, a transcriptional regulator which inhibits TLR2 expression. Of note the protective effects of HDL against TLR-induced inflammation were shown to be fully dependent on ATF3 in vitro and in vivo.

Another mechanism by which HDLs exert a protection against sepsis is by inducing an early inflammatory response to Gram-negative bacteria, thus helping to maintain a sensitive host response to LPS; HDLs exert this effect by suppressing the inhibitory activity of LBP (Thompson and Kitchens 2006). In contrast to native HDL, recombinant HDL did not increase cell response at early time points and are strongly inhibitory of cell response; this different effect is due to the composition of rHDL, which contain only apoA-I and PC and are optimized for LPS binding and neutralization (Thompson and Kitchens 2006).

Indeed apoA-I is a major HDL component that plays a central role in the anti-inflammatory functions of this lipoprotein class and exhibits protective effects against sepsis. ApoA-I in fact can directly inactivate bacterial endotoxin by protein–protein interaction (Emancipator et al. 1992), being the C-terminal half of apoA-I the main domain responsible for LPS neutralization (Henning et al. 2011), but also inhibits LPS-induced cytokine release from human monocytes (Flegel et al. 1993); in addition, apoA-I reduces TNF-α levels during LPS challenge in rats and increases the survival rates (Humphries et al. 2006), suggesting that apoA-I might inhibit LPS binding to macrophages thus inhibiting the production of inflammatory cytokines that are related to sepsis. In mice, the overexpression of apoA-I (that results in an increased serum level of both apoA-I and HDL) attenuates LPS-induced acute injury in lung and kidney (Li et al. 2008); LPS-induced pro-inflammatory cytokines decrease, as well as CD14 expression in liver and lung, resulting in a protective effect against systemic inflammation and multiple organ damage (Li et al. 2008). Similar to what reported in mice, subjects with low plasma

HDL levels (hypoalphalipoproteinemia) present an increased prevalence of classical CD14++/CD16− but not of intermediate CD14++/CD16+ monocytes, further linking HDL- to LPS-mediated responses (Sala et al. 2013). Septic HDLs are almost depleted of apoC-I (Barlage et al. 2001). ApoC-I contains a consensus LPS-binding motif and is able to enhance the biological response to LPS thus reducing mortality in mice with Gram-negative-induced sepsis (Berbee et al. 2006). In fact, apoC-I binds to LPS and prevents its clearance by the liver and spleen, resulting in the stimulation of the LPS-induced pro-inflammatory response and protection against septic death (Berbee et al. 2006). These findings suggest that when LPS is released into the plasma following bacteria proliferation in the blood, apoC-I binds to LPS and presents it to responsive cells such as macrophages, leading to a rapid and enhanced production of pro-inflammatory cytokines, which are essential for effective eradication of the bacterial infection.

2.2 Interaction of HDL with LTA and Gram-Positive Bacteria

The effects of HDL on components of the bacterial outer membrane are not restricted to Gram-negative bacteria but also involve cell wall components of Gram-positive bacteria such as LTA which is able to induce an inflammatory response similar to that induced by LPS. Indeed it induces a massive production of mediators of inflammation that may result in systemic inflammatory response syndrome, septic shock, and multiorgan damage (Bhakdi et al. 1991; De Kimpe et al. 1995). LTA may also trigger disturbances of lipid metabolism, interfering with both lipoprotein production and lipoprotein clearance (Grunfeld et al. 1999).

All lipoproteins can bind LTA, but the majority of LTA is found in the HDL fraction (Levels et al. 2003), suggesting that HDLs have the highest affinity. In contrast to LPS, HDLs (as well as other lipoproteins) alone do not inhibit the cytokine production induced by LTA, but require the presence of lipoprotein-depleted plasma (Grunfeld et al. 1999), suggesting that lipoproteins contain cofactors in sufficient amounts to facilitate LPS binding to lipoproteins but not LTA binding. LBP is a plasma component (normally bound to HDL, but it can be found in lipoprotein-depleted plasma following ultracentrifugation) which enhances the activation of macrophages by LPS in the absence of lipoproteins and facilitates LPS binding to lipoproteins; LBP can also bind LTA (Tobias et al. 1989) and allows HDL to efficiently inactivate LTA (Grunfeld et al. 1999). The inhibition of LBP with neutralizing antibodies significantly decreases (53 %) the ability of lipoprotein-depleted plasma to facilitate LTA inactivation, but also suggests the presence of other plasma factors playing a role in HDL inactivation of LTA (Grunfeld et al. 1999).

ApoA-I has been shown to bind directly LTA in vitro and to attenuate LTA-induced NF-kB activation (Jiao and Wu 2008); apoA-I dose-dependently inhibits L-929 cell death induced by LTA-activated macrophages, and lipoprotein-depleted plasma strengthened this effect of apoA-I (Jiao and Wu 2008). In mice, apoA-I attenuates LTA-induced acute lung injury and significantly inhibits LTA-induced pro-inflammatory cytokine production (Jiao and Wu 2008).

These findings suggest that apoA-I can inhibit LTA activation by multiple mechanisms, by direct binding to LTA, and by interfering with LTA-mediated inflammatory response.

2.3 HDL and Mycobacteria

HDLs were reported to be protective also toward intracellular bacteria such as mycobacteria (Cruz et al. 2008). Intracellular pathogens survive by evading the host immune system and accessing host metabolic pathways to obtain nutrients for their growth. *Mycobacterium leprae*, the causative agent of leprosy, is thought to be the mycobacterium most dependent on host metabolic pathways, including host-derived lipids. Although fatty acids and phospholipids accumulate in the lesions of individuals with the lepromatous (also known as disseminated) form of human leprosy (L-lep), the origin and significance of these lipids remains unclear. Host-derived oxidized phospholipids were detected in macrophages within L-lep lesions, and one specific oxidized phospholipid, 1-palmitoyl-2-(5,6-epoxyisoprostane E2)-sn-glycero-3phosphorylcholine (PEIPC), accumulates in macrophages infected with live mycobacteria (Cruz et al. 2008). Mycobacterial infection and host-derived oxidized phospholipids both inhibited innate immune responses, and this inhibition was reversed by the addition of normal HDL, a scavenger of oxidized phospholipids, but not by HDL from patients with L-lep (Cruz et al. 2008). The accumulation of host-derived oxidized phospholipids in L-lep lesions is strikingly similar to observations in atherosclerosis, which suggests that the link between host lipid metabolism and innate immunity could contribute to the pathogenesis of both microbial infection and metabolic disease.

2.4 General Innate Host Defense Mechanisms Exerted by HDL After Bacterial Infection

In addition to limiting LPS or LTA responses during infection, HDLs exert additional functions during the innate immune response. In normal plasma, about 5 % of HDL particles contain apolipoprotein M (apoM) which binds sphingosine-1-phosphate (S1P), an important bioactive lipid mediator known to be associated with HDL (Christoffersen et al. 2011). The inhibition of apoM production results in the decrease of HDL-C levels and changes in HDL size, subclass profile, and functions (Wolfrum et al. 2005); apoM is a negative acute-phase protein that decreases during infection and inflammation (Feingold et al. 2008); in patients with severe sepsis and systemic inflammatory response syndrome (SIRS), a leading cause of mortality in non-coronary intensive care units, apoM plasma levels decrease dramatically suggesting a reduction of the vasculoprotective effects of apoM and its ligand S1P, with a strong correlation between apoM decrease and the severity of disease (Kumaraswamy et al. 2012). It is still unclear whether apoM and S1P levels may have prognostic value and whether changes in apoM levels contribute to the pathogenesis of SIRS and septic shock.

Another key player during the immune response regulated by HDL is the long pentraxin 3 (PTX3). This protein belongs, together with the C-reactive protein (CRP) and other acute-phase proteins, to the pentraxin superfamily: soluble, multi-functional, pattern recognition proteins. Pentraxins share a common C-terminal pentraxin domain, which in the case of PTX3 is coupled to an unrelated long N-terminal domain (Bonacina et al. 2013). PTX3, which is the prototypic long pentraxin, was identified in the early 1990s, as a molecule rapidly induced by IL-1 in endothelial cells (ECs) or by TNF-α in both ECs and fibroblasts (Breviario et al. 1992; Lee et al. 1993). The protein presents a high degree of conservation from mouse to human (82 % identical and 92 % conserved amino acids) and is induced in a variety of somatic and innate immunity cells by primary inflammatory stimuli (Garlanda et al. 2005). PTX3 is a key player of the humoral arm of the innate immunity, and its physiological functions are associated with the recognition and binding to different ligands, including microbial moieties, complement components, and P-selectin. This protein plays also a key role in cardiovascular diseases including atherosclerosis (Norata et al. 2009, 2010). Similarly to short pentraxins, PTX3 recognizes the highly conserved pathogen-associated molecular patterns (PAMPs) expressed by microorganisms (Iwasaki and Medzhitov 2010) and binds a number of bacteria, fungi, and viruses. A specific binding has been observed to conidia of *Aspergillus fumigatus* (Garlanda et al. 2002), *Paracoccidioides brasiliensis*, and zymosan (Diniz et al. 2004), to selected Gram-positive and Gram-negative bacteria (Bozza et al. 2006; Garlanda et al. 2002; Jeannin et al. 2005), and finally to some viral strains, including human and murine cyto-megalovirus and influenza virus type A (IVA) (Bozza et al. 2006; Reading et al. 2008). Both short pentraxin and PTX3 bind apoptotic cells and facilitate their clearance (Doni et al. 2012; Jaillon et al. 2009). Surface bound CRP activates the classical pathway of complement through interaction with C1q, thus leading to cell elimination (Nauta et al. 2003). Cell-bound PTX3 might favor the clearance of apoptotic cells (Jaillon et al. 2009; Poon et al. 2010) by enhancing the deposition of both C1q and C3 on cell surfaces (Nauta et al. 2003). On the contrary, when in the fluid phase, PTX3 interacts with C1q and dampens the deposition on apoptotic cells and the resulting phagocytosis by dendritic cells and phagocytes (Baruah et al. 2006; Gershov et al. 2000; Rovere et al. 2000; van Rossum et al. 2004). In addition to PTX3, C1q recognizes and binds to ficolin-2 and mannose-binding lectin (MBL), thus modulating the classical and the lectin pathways of complement activation (Bottazzi et al. 1997). The best described and characterized ligand of PTX3 is the first component of the classical complement system C1q (Bottazzi et al. 1997; Nauta et al. 2003); PTX3 interacts with the globular head of the protein (Roumenina et al. 2006) thus resulting in the activation of the classical complement cascade only when C1q is plastic immobilized, a situation that mimics C1q bound to a microbial surface. Anti-inflammatory molecules were shown to modulate PTX3 expression. Under inflammatory conditions, glucocorticoid hormones (GCs) induce and enhance the protein expression in fibroblasts but not in myeloid cells (Doni et al. 2008). Also HDLs, which possess a series of vascular protective activities, induce PTX3 expression in endothelial cells (Norata et al. 2008). The latter mechanism requires the activation of the PI3K/Akt pathway through G-coupled lysosphingolipid receptors and is mimicked by sphingosine-1-phosphate and others S1P mimetics (Norata et al. 2008), physiologically present in HDL and responsible for some of the activities linking HDL to the immunoinflammatory response (Norata et al. 2005, 2012). In vivo, an increased expression of PTX3 mRNA was detected in the aorta of transgenic mice overexpressing human apoA-I, compared to apoA-I knockout mice, and plasma levels of PTX3 are significantly increased in C57BL/6 mice injected with HDL (Norata et al. 2008). These data suggest that some of the beneficial effects in immunity of HDL may result also from the modulation of molecules that act as sensors of the immunoinflammatory balance.

In summary, all the observations showing that the increase of HDL is associated with an attenuation of LPS-induced inflammatory response (Levine et al. 1993; van Leeuwen et al. 2003) strongly favor the hypothesis that raising plasma HDL may represent a therapeutic approach in the treatment of sepsis and its complications. The picture however is more complicated, and not only HDL quantity but also HDL quality/composition is critical. For instance, the increase in sepsis-related mortality of the ILLUMINATE trial observed in the torcetrapib (a cholesteryl ester transfer protein inhibitor, increasing HDL-C levels) arm (Barter et al. 2007) was unlikely due to a direct effect of torcetrapib on LBP or bactericidal/permeability increasing protein function nor to inhibition of an interaction of CETP with LPS (Clark et al. 2010). It is rather possible that changes in plasma lipoprotein composition despite increased HDL levels, or the known off-target effects of torcetrapib, such as aldosterone elevation, could have aggravated the effects of sepsis (Clark et al. 2010). Apolipoprotein mimetic peptides represent an emerging area of HDL therapy; the most effective apoA-I mimetic peptide is 4F, which has been shown to improve HDL quality/function (Sherman et al. 2010; White et al. 2009). 4F mimics also anti-inflammatory properties of HDL: in vitro, 4F inhibits the expression of pro-inflammatory mediators in LPS-treated cells by directly binding to LPS, thus resulting in the inhibition of LPS binding to LBP (Gupta et al. 2005). In endotoxemic rats, the administration of 4F after LPS injection results in the attenuation of acute lung injury and increased survival, probably due to the preservation of circulating HDL-C and the downregulation of inflammatory pathways (Kwon et al. 2012); 4F also prevents defects in vascular functions and is associated with a decrease in plasma endotoxin activity in rats (Dai et al. 2010) and improved cardiac performance in LPS-treated rats (Datta et al. 2011). These observations indicate that, by scavenging LPS, 4F may prevent LPS-induced release of pro-inflammatory cytokines and changes in HDL composition resulting in an effective reduction of clinical complications associated with sepsis. Although promising, future studies are warranted to translate these findings into the clinical setting.

3 HDL and Parasitic Infections

The connection between HDL and parasitic infections mainly relates on the ability of specific apolipoproteins, which circulate as part of the HDL3 complex, to limit *Trypanosoma brucei* or *Leishmania* infection. The most relevant apolipoprotein in this context is apoL-1 which was discovered in 1997 (Duchateau et al. 1997) and was shown to be a part of a primate-specific complex named trypanosome lytic factor-1 (TLF-1) that also contains apoA-I and haptoglobin-related protein as the main protein components. TLF-1 has lytic activity toward African *Trypanosoma brucei brucei* and renders humans and most other primates resistant to infection with this parasite causing endemic infections of African cattle. The *Trypanosoma* species *brucei rhodesiense* and *brucei gambiense*, however, are resistant to TLF-1 and cause sleeping sickness in humans. Sleeping sickness is fatal when untreated and thus is an important health problem in many African countries. The trypanosome lytic activity has been associated with apoL-1 (Vanhamme et al. 2003); the apoL-1-containing complex is taken up by *T. b. brucei* via a receptor that binds hemoglobin and haptoglobin-related protein (Hb-Hpr). ApoL-1 traffics to the trypanosomal lysosome, where the acidic pH causes a conformational change, leading to activation of anion channel function in the apoL-1 N terminus. Lysosomal swelling kills the trypanosome (Perez-Morga et al. 2005). Thus, apoL-1 confers innate immunity against this parasite (Wheeler 2010). Over time, *T. b. brucei* developed a virulence factor called SRA that can inactivate apoL-1 protein, although the cellular location of this interaction is unknown. These SRA-expressing trypanosomes evolved into *T. b. rhodesiense*, the etiologic agent that causes acute African sleeping sickness (Wheeler 2010). This discovery has now been used to engineer a potential fusion protein drug for treating sleeping sickness caused by *brucei rhodesiense* in human (Baral et al. 2006). The active component in the novel protein drug candidate is a recombinant apoL-1 where the C-terminal SRA-binding region of wild-type apoL-1 has been deleted. This truncated apoL-1 alone probably has low biological activity when injected into plasma due to competition with endogenous apoL-1 (~6 mg/l) for uptake by trypanosomes. To circumvent this problem and effectively target recombinant apoL-1 to trypanosomes, the SRA-resistant apoL-1 without the C-terminal sequences was fused with a fragment of a high-affinity camel antibody (designated nanobody) specifically recognizing conserved epitopes of variant surface glycoprotein on trypanosomes (Baral et al. 2006). In vitro, the recombinant apoL-1–nanobody specifically bound to trypanosomes and was capable of lysing them in vivo. When mice infected with *Trypanosoma brucei brucei* were treated with the apoL-1–nanobody fusion protein, the parasites were promptly cleared from the circulation.

Another trypanosomatida, *Leishmania*, is also targeted by apoL-1 (Samanovic et al. 2009). It is certainly possible that apoL-1 has broad innate immunity properties, shaping the relative frequencies of the *APOL1* alleles. Of note the chronic kidney disease (CKD)-associated G1 and G2 variants (Parsa et al. 2013) encode forms of apoL-1 that evade SRA and remain active against *T. b. rhodesiense*. This, and/or other biological effects, may have conferred a selective advantage to G1 and G2 heterozygotes, causing a selective sweep. The contribution of apoL-1 to CKD could also be related to the HDL functionality (Baragetti et al. 2013) under specific immunopathological conditions such as those observed during CKD. Hence, HDL could represent the bridge between apoL-1 and CKD and should be taken into consideration when exploring the contribution of apoL-1 to the disease. The delineation of how primates in late evolution have exploited HDL to fight parasite infections highlights the need of investigations on other potential roles of plasma HDL beyond those in lipid metabolism and reverse cholesterol transport.

4   HDL and Viral Infections

Changes in plasma HDL-C levels have been reported to occur also during infection with viruses including human immunodeficiency virus (HIV) and hepatitis C virus (HCV). Furthermore, proteins related to HDL life cycle, such as SR-BI, were shown to play a key role in HCV infection. So far HDLs were proposed to increase virus infection and inhibit virus neutralizing antibodies; however, recent findings are challenging previous data and proposing a more complicated picture based on the ability of virus to take advantage of HDL–lipid transfer activity in host cells. Most of the available evidences linking HDL to viral infection are available for HIV and HCV.

4.1   HIV Infection

HIV infects and depletes CD4 lymphocytes, resulting in immunodeficiency and a slowly progressive disease. HIV is associated with dyslipidemia, namely, hypocholesterolemia, low levels of LDL, and hypertriglyceridemia (Riddler et al. 2003; Shor-Posner et al. 1993). HIV infection is commonly associated also with hypoalphalipoproteinemia; however, it is unclear whether virion replication plays a causative role in these changes. Some data suggest that hypoalphalipoproteinemia in patients with HIV is likely to be secondary to HIV infection itself (Rose et al. 2006). Systemic inflammation has been shown to lower the antioxidant and anti-inflammatory activity by transforming HDL to a pro-oxidant, pro-inflammatory acute-phase HDL (Kelesidis et al. 2013; Norata et al. 2006). A small pilot study of HIV-1-infected individuals with suppressed viremia on combination antiretroviral therapy showed that oxidative stress and inflammation in HIV-1 are associated with a marked reduction of HDL antioxidant–anti-inflammatory activities. In vitro, these abnormalities were significantly improved by treatment with the apoA-1 mimetic peptide 4F (Kelesidis et al. 2011). HIV infection is associated with modified HDL metabolism redirecting cholesterol to the apoB-containing lipoproteins and likely reducing the functionality of reverse cholesterol transport (Rose et al. 2008). Of note, the HIV-1 Nef protein can impair ABCA1 cholesterol efflux from macrophages, thus supporting atherosclerosis. This viral inhibition of efflux was correlated with a direct interaction between ABCA1 and Nef (Fitzgerald et al. 2010; Mujawar et al. 2006). More recently it was shown that

Nef downregulates ABCA1 function by a posttranslational mechanism that stimulates ABCA1 degradation but does not require the ability of Nef to bind ABCA1 (Mujawar et al. 2010).

Not all data are however concordant on this, and although HDL cholesterol and preβ1-HDL were significantly lower in all HIV-infected groups ($p < 0.05$), mean levels of apoA-I and the ability of plasma to promote cholesterol efflux were similar in treatment-naïve HIV-infected patients or in HIV-infected patients on long therapy break. Of note a positive correlation between apoA-I and levels of CD4+ cells was also observed ($r^2 = 0.3, p < 0.001$) (Rose et al. 2008). Furthermore apoA-I, the major protein component of high-density lipoprotein, and its amphipathic peptide analogue were found to inhibit cell fusion, both in HIV-1-infected T cells and in recombinant vaccinia-virus-infected CD4+ HeLa cells expressing HIV envelope protein on their surfaces (Srinivas et al. 1990). The amphipathic peptides inhibited the infectivity of HIV-1. The inhibitory effects were manifest when the virus, but not cells, was pretreated with the peptides. Also, a reduction in HIV-induced cell killing was observed when virus-infected cell cultures were maintained in the presence of amphipathic peptides. These results have potential implications for HIV biology and therapy (Srinivas et al. 1990).

An aspect debated is how much the HIV infection and/or treatment contribute to the changes in HDL-C levels. With highly active antiretroviral therapy (HAART) intervention, mortality due to HIV was greatly reduced (Madamanchi et al. 2002). However, there have been several reports of increases in cardiovascular complications in patients with HIV. It is now established that some HAART regimens cause severe dyslipidemia, characterized by high levels of TC and LDL-C, hypertriglyceridemia, and hypoalphalipoproteinemia (Riddler et al. 2003). This clearly pro-atherogenic lipoprotein profile is associated with a rise in the incidence of CAD (Depairon et al. 2001). The rate of inflammation predicts changes in HDL-C and apoA-I following the initiation of antiretroviral therapy and indeed in a subgroup of participants not taking ART at study entry who were randomized in the Strategies for Management of Antiretroviral Therapy (SMART) to immediately initiate ART ("VS group") or to defer it ("DC group"); HDL-C and ApoA-I levels increased among VS participants ($n = 128$) after starting ART compared to DC. The effect of starting ART on changes in HDL-C and apoA-I was greater for those with higher versus lower baseline levels of IL-6 or hsCRP indicating that the activation of inflammatory pathways could contribute to HIV-associated changes in HDL (Baker et al. 2011). Also non-nucleoside reverse transcriptase inhibitors (NNRTI), such as nevirapine (NVP), were shown to increase apoA-I production, which contributes to the HDL-C increase after introduction of NVP-containing regimens. In view of the potent anti-atherogenic effects of apoA-I, the observed increase was suggested to contribute to the favorable cardiovascular profile of NVP (Franssen et al. 2009). Also efavirenz, another NNRTI antiretroviral treatment, was associated with HDL particles with a better antioxidant function, i.e., with a higher PON-1 activity. The PON-1 activity of black patients is higher than that found in whites regardless of treatment suggesting that ethnicity should be taken into consideration when studying drug effects on PON-1 activity (Pereira et al. 2009).

Overall the available evidences suggest that HIV infection could be associated with modified HDL metabolism redirecting cholesterol to the apoB-containing lipoproteins and likely reducing the functionality of reverse cholesterol transport and promote atherosclerosis. Additional pro-atherogenic mechanisms could be associated with a decrease in the anti-inflammatory properties of HDL.

4.2 HCV Infection

HCV is a major cause of liver cirrhosis and hepatocellular carcinoma. Viral entry is required for initiation, spread, and maintenance of infection and thus is a promising target for antiviral therapy. HCV exists in heterogenous forms in human serum and may be associated with VLDL, LDL, and HDL which can shield the virus from neutralizing antibodies targeting the HCV envelope glycoproteins (Agnello et al. 1999; Hijikata et al. 1993; Nielsen et al. 2006; Thomssen et al. 1992). HCV binding and entry into hepatocytes is a complex process involving the viral envelope glycoproteins E1 and E2, as well as several host factors, including highly sulfated heparan sulfate, CD81, the low-density lipoprotein receptor, claudin-1, occludin, and receptor tyrosine kinases (Lupberger et al. 2011; Zeisel et al. 2011).

Also SR-BI which binds a variety of lipoproteins and mediates the selective uptake of HDL cholesterol ester (CE) as well as bidirectional free cholesterol transport at the cell membrane emerged as a critical receptor affecting HCV entry. SR-BI directly binds HCV E2 (Bartosch et al. 2003; Evans et al. 2007; Scarselli et al. 2002), but virus-associated lipoproteins, including apoB containing, also contribute to host cell binding and uptake (Dao Thi et al. 2012; Maillard et al. 2006). Moreover, physiological SR-BI ligands modulate HCV infection (Bartosch et al. 2005; Voisset et al. 2005; von Hahn et al. 2006), suggesting the existence of a complex interplay between lipoproteins (not only HDL), SR-BI and HCV envelope glycoproteins for HCV entry. Earlier studies using small molecule inhibitors indicated a role for SR-BI lipid transfer function in HCV infection and HDL-mediated entry enhancement (Bartosch et al. 2003; Dreux et al. 2009; Syder et al. 2011; Voisset et al. 2005). A human anti-SR-BI mAb has been reported to inhibit HDL binding, to interfere with cholesterol efflux, and to decrease cell culture-derived HCV (HCVcc) entry during attachment steps without having a relevant impact on SR-BI-mediated post-binding steps (Catanese et al. 2007, 2010). However, SR-BI mediates the uptake of HDL-C in a two-step process including HDL binding and subsequent transfer of CE into the cell without internalization of HDL; a novel emerging hypothesis suggests that the interference with SR-BI lipid transfer function may be relevant for both initiation of HCV infection and viral dissemination independently of HDL function (Zahid et al. 2013).

Indeed SR-BI has also been demonstrated to mediate post-binding events during HCV entry (Haberstroh et al. 2008; Syder et al. 2011; Zeisel et al. 2007). HCV–SR-BI interaction during post-binding steps occurs at similar time points as the HCV utilization of CD81 and claudin-1 suggesting that HCV entry may be mediated through the formation of co-receptor complexes (Harris et al. 2008; Krieger et al. 2010; Zeisel et al. 2007). Also the SR-BI partner PDZK1 was shown to facilitate hepatitis C virus entry (Eyre et al. 2010).

These data suggest that SR-BI plays a multifunctional role during HCV entry at both binding and post-binding steps (Catanese et al. 2010). Furthermore the HCV post-binding function of human SR-BI can be dissociated from its E2-binding function. Murine SR-BI does not bind E2 although it is capable of promoting HCV entry (Catanese et al. 2010; Ploss et al. 2009), and SR-BI, although able to directly bind E2 and virus-associated lipoproteins, could play additional functions during HCV infection (Bartosch et al. 2003; Dreux et al. 2009; Zeisel et al. 2007).

Although the addition of HDL enhances the efficiency of HCVcc infection, anti-SR-BI antibodies and SR-BI-specific siRNA efficiently inhibited HCV infection independently of lipoproteins (Zeisel et al. 2007). In this context, the post-binding activity of SR-BI is of key relevance for cell-free HCV infection as well as cell-to-cell transmission and by using antibodies which do not inhibit HDL binding to SR-BI; it was observed that post-binding function of SR-BI appears to be unrelated to HDL interaction but to be directly linked to its lipid transfer function (Zahid et al. 2013). So far small molecules and mAbs targeting SR-BI and interfering with HCV infection have been described (Bartosch et al. 2003; Catanese et al. 2007; Syder et al. 2011). A codon-optimized version of this mAb has been demonstrated to prevent HCV spread in vivo (Meuleman et al. 2012), and ITX5061, a SR-BI inhibitor, is in clinical development as HCV entry inhibitor (phase I, http://clinicaltrials.gov/ct2/show/NCT01560468?term=ITX+5061&rank=3).

Despite this promising approach, some open questions remain. First, it has been shown in vitro that apoA-I is required for HCV production (Mancone et al. 2011) and that serum amyloid A has antiviral activities against HCV which are reduced when HDL are co-incubated with SAA (Lavie et al. 2006). Given the changes between apoA-I and SAA occurring in HDL during the acute phase (Norata et al. 2012), it is still unknown whether this mechanism could be the consequence of viral infection or may represent part of the immune response which HCV learned to escape. Second, as the inhibition of SR-BI represents one of the most promising targets for HCV infection, the potential side effects on the impairment of HDL function should be carefully evaluated. The new data showing that HCV infection does not require receptor-E2–HDL interactions, coupled with the observation that HCV entry and dissemination can be inhibited without blocking HDL–SR-BI binding (Zahid et al. 2013), open a novel perspective for the design of entry inhibitors interfering specifically with the proviral function of SR-BI.

Conclusion

HDLs are emerging as a relevant player in both innate and adaptive immunity (Norata et al. 2011, 2012; Sala et al. 2012). HDL activities rely not only on the ability to modulate cholesterol availability in immune cells but also on the role of specific molecules shuttled by HDL. During infections and acute conditions, HDL-C levels decrease very rapidly, and HDL particles undergo changes that dramatically alter their composition and function. Whether this is the consequence of a humoral innate response aimed at scavenging lipid bacterial

Table 1 HDL-related therapies and infections

	Compound	References
Endotoxemia/ LPS scavenging	Recombinant HDL	Parker et al. (1995), Pajkrt et al. (1996)
	ApoA-I mimetics	Gupta et al. (2005)
Parasites infections Trypanosoma Leishmania	ApoL-1– nanobody	Baral et al. (2006), Samanovic et al. (2009)
Hepatitis C virus entry	SR-BI monoclonal antibodies	Bartosch et al. (2003), Catanese et al. (2007), Meuleman et al. (2012), Syder et al. (2011)

products such as LPS from the circulation and driving them into the liver for catabolism and elimination is still debated. Experimental evidence in genetically manipulated animal models suggests, however, that alterations in HDL structure/composition are associated with poor prognosis following endotoxemia or sepsis, further supporting a protective role for HDL. The same is true for some parasitic infections, where the key player appears to be a specific and minor apolipoprotein of HDL–apoL-1. For viral infections, the landscape is more complicated; SR-BI was clearly indicated as a player favoring virus entry; however, it is not clear whether viruses, such as HCV, evolved to take advantage of the HDL–SR-BI interaction to entry liver cells as in vitro studies suggest or whether in vivo HDL can compete for the interaction between HCV and SR-BI. Further studies are warranted in this context. Despite this, proteins related to HDL physiology represent already a target in clinical development for infections, and SR-BI antagonism represents a novel and forefront approach to interfere with hepatitis C virus entry.

All the observations showing that the increase of HDL-C is associated with an attenuation of LPS-induced inflammatory responses strongly favor the hypothesis that raising plasma HDL may represent a therapeutic approach in the treatment of sepsis and its complications (Table 1). HDL-related therapies are of great interest also in the context of parasites and virus infections (Table 1). As of beginning of 2014, the research in the field of HDL and immunity is in its infancy compared to the body of data available for HDL and atherosclerosis. From the coming years, we have to expect new and compelling observations further linking HDL to innate immunity.

Open Access This chapter is distributed under the terms of the Creative Commons Attribution Noncommercial License, which permits any noncommercial use, distribution, and reproduction in any medium, provided the original author(s) and source are credited.

References

Agnello V, Abel G, Elfahal M, Knight GB, Zhang QX (1999) Hepatitis C virus and other flaviviridae viruses enter cells via low density lipoprotein receptor. Proc Natl Acad Sci U S A 96:12766–12771

Badellino KO, Wolfe ML, Reilly MP, Rader DJ (2008) Endothelial lipase is increased in vivo by inflammation in humans. Circulation 117:678–685. doi:10.1161/CIRCULATIONAHA.107. 707349

Baker JV, Neuhaus J, Duprez D, Cooper DA, Hoy J, Kuller L, Lampe FC, Liappis A, Friis-Moller N, Otvos J, Paton NI, Tracy R, Neaton JD, Group ISS (2011) Inflammation predicts changes in high-density lipoprotein particles and apolipoprotein A1 following initiation of antiretroviral therapy. AIDS 25:2133–2142. doi:10.1097/QAD.0b013e32834be088

Banka CL, Yuan T, de Beer MC, Kindy M, Curtiss LK, de Beer FC (1995) Serum amyloid A (SAA): influence on HDL-mediated cellular cholesterol efflux. J Lipid Res 36:1058–1065

Baragetti A, Norata GD, Sarcina C, Rastelli F, Grigore L, Garlaschelli K, Uboldi P, Baragetti I, Pozzi C, Catapano AL (2013) High density lipoprotein cholesterol levels are an independent predictor of the progression of chronic kidney disease. J Intern Med 274:252–262. doi:10.1111/joim.12081

Baral TN, Magez S, Stijlemans B, Conrath K, Vanhollebeke B, Pays E, Muyldermans S, De Baetselier P (2006) Experimental therapy of African trypanosomiasis with a nanobody-conjugated human trypanolytic factor. Nat Med 12:580–584. doi:10.1038/nm1395

Barlage S, Frohlich D, Bottcher A, Jauhiainen M, Muller HP, Noetzel F, Rothe G, Schutt C, Linke RP, Lackner KJ, Ehnholm C, Schmitz G (2001) ApoE-containing high density lipoproteins and phospholipid transfer protein activity increase in patients with a systemic inflammatory response. J Lipid Res 42:281–290

Barter PJ, Caulfield M, Eriksson M, Grundy SM, Kastelein JJ, Komajda M, Lopez-Sendon J, Mosca L, Tardif JC, Waters DD, Shear CL, Revkin JH, Buhr KA, Fisher MR, Tall AR, Brewer B (2007) Effects of torcetrapib in patients at high risk for coronary events. N Engl J Med 357:2109–2122. doi:10.1056/NEJMoa0706628

Bartosch B, Vitelli A, Granier C, Goujon C, Dubuisson J, Pascale S, Scarselli E, Cortese R, Nicosia A, Cosset FL (2003) Cell entry of hepatitis C virus requires a set of co-receptors that include the CD81 tetraspanin and the SR-B1 scavenger receptor. J Biol Chem 278:41624–41630. doi:10.1074/jbc.M305289200

Bartosch B, Verney G, Dreux M, Donot P, Morice Y, Penin F, Pawlotsky JM, Lavillette D, Cosset FL (2005) An interplay between hypervariable region 1 of the hepatitis C virus E2 glycoprotein, the scavenger receptor BI, and high-density lipoprotein promotes both enhancement of infection and protection against neutralizing antibodies. J Virol 79:8217–8229. doi:10.1128/JVI.79.13.8217-8229.2005

Baruah P, Dumitriu IE, Peri G, Russo V, Mantovani A, Manfredi AA, Rovere-Querini P (2006) The tissue pentraxin PTX3 limits C1q-mediated complement activation and phagocytosis of apoptotic cells by dendritic cells. J Leukoc Biol 80:87–95. doi:10.1189/jlb.0805445

Berbee JF, van der Hoogt CC, Kleemann R, Schippers EF, Kitchens RL, van Dissel JT, Bakker-Woudenberg IA, Havekes LM, Rensen PC (2006) Apolipoprotein CI stimulates the response to lipopolysaccharide and reduces mortality in gram-negative sepsis. Faseb J 20:2162–2164

Bhakdi S, Klonisch T, Nuber P, Fischer W (1991) Stimulation of monokine production by lipoteichoic acids. Infect Immun 59:4614–4620

Birjmohun RS, van Leuven SI, Levels JH, van't Veer C, Kuivenhoven JA, Meijers JC, Levi M, Kastelein JJ, van der Poll T, Stroes ES (2007) High-density lipoprotein attenuates inflammation and coagulation response on endotoxin challenge in humans. Arterioscler Thromb Vasc Biol 27:1153–1158. doi:10.1161/ATVBAHA.106.136325

Bonacina F, Baragetti A, Catapano AL, Norata GD (2013) Long pentraxin 3: experimental and clinical relevance in cardiovascular diseases. Mediators Inflamm 2013:725102. doi:10.1155/2013/725102

Bottazzi B, Vouret-Craviari V, Bastone A, De Gioia L, Matteucci C, Peri G, Spreafico F, Pausa M, D'Ettorre C, Gianazza E, Tagliabue A, Salmona M, Tedesco F, Introna M, Mantovani A (1997) Multimer formation and ligand recognition by the long pentraxin PTX3. Similarities and differences with the short pentraxins C-reactive protein and serum amyloid P component. J Biol Chem 272:32817–32823

Bozza S, Bistoni F, Gaziano R, Pitzurra L, Zelante T, Bonifazi P, Perruccio K, Bellocchio S, Neri M, Iorio AM, Salvatori G, De Santis R, Calvitti M, Doni A, Garlanda C, Mantovani A, Romani L (2006) Pentraxin 3 protects from MCMV infection and reactivation through TLR sensing pathways leading to IRF3 activation. Blood 108:3387–3396. doi:10.1182/blood-2006-03-009266

Breviario F, d'Aniello EM, Golay J, Peri G, Bottazzi B, Bairoch A, Saccone S, Marzella R, Predazzi V, Rocchi M et al (1992) Interleukin-1-inducible genes in endothelial cells. Cloning of a new gene related to C-reactive protein and serum amyloid P component. J Biol Chem 267:22190–22197

Cao Y, Stafforini DM, Zimmerman GA, McIntyre TM, Prescott SM (1998) Expression of plasma platelet-activating factor acetylhydrolase is transcriptionally regulated by mediators of inflammation. J Biol Chem 273:4012–4020

Catanese MT, Graziani R, von Hahn T, Moreau M, Huby T, Paonessa G, Santini C, Luzzago A, Rice CM, Cortese R, Vitelli A, Nicosia A (2007) High-avidity monoclonal antibodies against the human scavenger class B type I receptor efficiently block hepatitis C virus infection in the presence of high-density lipoprotein. J Virol 81:8063–8071. doi:10.1128/JVI.00193-07

Catanese MT, Ansuini H, Graziani R, Huby T, Moreau M, Ball JK, Paonessa G, Rice CM, Cortese R, Vitelli A, Nicosia A (2010) Role of scavenger receptor class B type I in hepatitis C virus entry: kinetics and molecular determinants. J Virol 84:34–43. doi:10.1128/JVI.02199-08

Chien JY, Jerng JS, Yu CJ, Yang PC (2005) Low serum level of high-density lipoprotein cholesterol is a poor prognostic factor for severe sepsis. Crit Care Med 33:1688–1693

Christoffersen C, Obinata H, Kumaraswamy SB, Galvani S, Ahnstrom J, Sevvana M, Egerer-Sieber C, Muller YA, Hla T, Nielsen LB, Dahlback B (2011) Endothelium-protective sphingosine-1-phosphate provided by HDL-associated apolipoprotein M. Proc Natl Acad Sci USA 108:9613–9618. doi:10.1073/pnas.1103187108

Clark RW, Cunningham D, Cong Y, Subashi TA, Tkalcevic GT, Lloyd DB, Boyd JG, Chrunyk BA, Karam GA, Qiu X, Wang IK, Francone OL (2010) Assessment of cholesteryl ester transfer protein inhibitors for interaction with proteins involved in the immune response to infection. J Lipid Res 51:967–974. doi:10.1194/jlr.M002295

Coetzee GA, Strachan AF, van der Westhuyzen DR, Hoppe HC, Jeenah MS, de Beer FC (1986) Serum amyloid A-containing human high density lipoprotein 3. Density, size, and apolipoprotein composition. J Biol Chem 261:9644–9651

Cruz D, Watson AD, Miller CS, Montoya D, Ochoa MT, Sieling PA, Gutierrez MA, Navab M, Reddy ST, Witztum JL, Fogelman AM, Rea TH, Eisenberg D, Berliner J, Modlin RL (2008) Host-derived oxidized phospholipids and HDL regulate innate immunity in human leprosy. J Clin Invest 118:2917–2928. doi:10.1172/JCI34189

Dai L, Datta G, Zhang Z, Gupta H, Patel R, Honavar J, Modi S, Wyss JM, Palgunachari M, Anantharamaiah GM, White CR (2010) The apolipoprotein A-I mimetic peptide 4F prevents defects in vascular function in endotoxemic rats. J Lipid Res 51:2695–2705. doi:10.1194/jlr.M008086

Dao Thi VL, Granier C, Zeisel MB, Guerin M, Mancip J, Granio O, Penin F, Lavillette D, Bartenschlager R, Baumert TF, Cosset FL, Dreux M (2012) Characterization of hepatitis C virus particle subpopulations reveals multiple usage of the scavenger receptor BI for entry steps. J Biol Chem 287:31242–31257. doi:10.1074/jbc.M112.365924

Datta G, Gupta H, Zhang Z, Mayakonda P, Anantharamaiah GM, White CR (2011) HDL mimetic peptide administration improves left ventricular filling and cardiac output in lipopolysaccharide-treated rats. J Clin Exp Cardiolog 2:pii: 1000172. doi:10.4172/2155-9880.1000172

De Kimpe SJ, Kengatharan M, Thiemermann C, Vane JR (1995) The cell wall components peptidoglycan and lipoteichoic acid from Staphylococcus aureus act in synergy to cause shock and multiple organ failure. Proc Natl Acad Sci USA 92:10359–10363

de la Llera MM, McGillicuddy FC, Hinkle CC, Byrne M, Joshi MR, Nguyen V, Tabita-Martinez J, Wolfe ML, Badellino K, Pruscino L, Mehta NN, Asztalos BF, Reilly MP (2012) Inflammation modulates human HDL composition and function in vivo. Atherosclerosis 222:390–394. doi:10.1016/j.atherosclerosis.2012.02.032/S0021-9150(12)00146-3

Depairon M, Chessex S, Sudre P, Rodondi N, Doser N, Chave JP, Riesen W, Nicod P, Darioli R, Telenti A, Mooser V, Swiss HIVCS (2001) Premature atherosclerosis in HIV-infected individuals – focus on protease inhibitor therapy. AIDS 15:329–334

Diniz SN, Nomizo R, Cisalpino PS, Teixeira MM, Brown GD, Mantovani A, Gordon S, Reis LF, Dias AA (2004) PTX3 function as an opsonin for the dectin-1-dependent internalization of zymosan by macrophages. J Leukoc Biol 75:649–656. doi:10.1189/jlb.0803371

Doni A, Mantovani G, Porta C, Tuckermann J, Reichardt HM, Kleiman A, Sironi M, Rubino L, Pasqualini F, Nebuloni M, Signorini S, Peri G, Sica A, Beck-Peccoz P, Bottazzi B, Mantovani A (2008) Cell-specific regulation of PTX3 by glucocorticoid hormones in hematopoietic and nonhematopoietic cells. J Biol Chem 283:29983–29992. doi:10.1074/jbc.M805631200

Doni A, Garlanda C, Bottazzi B, Meri S, Garred P, Mantovani A (2012) Interactions of the humoral pattern recognition molecule PTX3 with the complement system. Immunobiology 217:1122–1128. doi:10.1016/j.imbio.2012.07.004/S0171-2985(12)00168-4

Dreux M, Dao Thi VL, Fresquet J, Guerin M, Julia Z, Verney G, Durantel D, Zoulim F, Lavillette D, Cosset FL, Bartosch B (2009) Receptor complementation and mutagenesis reveal SR-BI as an essential HCV entry factor and functionally imply its intra- and extra-cellular domains. PLoS Pathog 5:e1000310. doi:10.1371/journal.ppat.1000310

Duchateau PN, Pullinger CR, Orellana RE, Kunitake ST, Naya-Vigne J, O'Connor PM, Malloy MJ, Kane JP (1997) Apolipoprotein L, a new human high density lipoprotein apolipoprotein expressed by the pancreas. Identification, cloning, characterization, and plasma distribution of apolipoprotein L. J Biol Chem 272:25576–25582

Emancipator K, Csako G, Elin RJ (1992) In vitro inactivation of bacterial endotoxin by human lipoproteins and apolipoproteins. Infect Immun 60:596–601

Evans MJ, von Hahn T, Tscherne DM, Syder AJ, Panis M, Wolk B, Hatziioannou T, McKeating JA, Bieniasz PD, Rice CM (2007) Claudin-1 is a hepatitis C virus co-receptor required for a late step in entry. Nature 446:801–805. doi:10.1038/nature05654

Eyre NS, Drummer HE, Beard MR (2010) The SR-BI partner PDZK1 facilitates hepatitis C virus entry. PLoS Pathog 6:e1001130. doi:10.1371/journal.ppat.1001130

Feingold KR, Memon RA, Moser AH, Grunfeld C (1998) Paraoxonase activity in the serum and hepatic mRNA levels decrease during the acute phase response. Atherosclerosis 139:307–315

Feingold KR, Shigenaga JK, Chui LG, Moser A, Khovidhunkit W, Grunfeld C (2008) Infection and inflammation decrease apolipoprotein M expression. Atherosclerosis 199:19–26. doi:10.1016/j.atherosclerosis.2007.10.007/S0021-9150(07)00660-0

Fitzgerald ML, Mujawar Z, Tamehiro N (2010) ABC transporters, atherosclerosis and inflammation. Atherosclerosis 211:361–370. doi:10.1016/j.atherosclerosis.2010.01.011

Flegel WA, Baumstark MW, Weinstock C, Berg A, Northoff H (1993) Prevention of endotoxin-induced monokine release by human low- and high-density lipoproteins and by apolipoprotein A-I. Infect Immun 61:5140–5146

Franssen R, Sankatsing RR, Hassink E, Hutten B, Ackermans MT, Brinkman K, Oesterholt R, Arenas-Pinto A, Storfer SP, Kastelein JJ, Sauerwein HP, Reiss P, Stroes ES (2009) Nevirapine increases high-density lipoprotein cholesterol concentration by stimulation of apolipoprotein A-I production. Arterioscler Thromb Vasc Biol 29:1336–1341. doi:10.1161/ATVBAHA.109.192088

Garlanda C, Hirsch E, Bozza S, Salustri A, De Acetis M, Nota R, Maccagno A, Riva F, Bottazzi B, Peri G, Doni A, Vago L, Botto M, De Santis R, Carminati P, Siracusa G, Altruda F, Vecchi A, Romani L, Mantovani A (2002) Non-redundant role of the long pentraxin PTX3 in anti-fungal innate immune response. Nature 420:182–186

Garlanda C, Bottazzi B, Bastone A, Mantovani A (2005) Pentraxins at the crossroads between innate immunity, inflammation, matrix deposition, and female fertility. Annu Rev Immunol 23:337–366

Gershov D, Kim S, Brot N, Elkon KB (2000) C-Reactive protein binds to apoptotic cells, protects the cells from assembly of the terminal complement components, and sustains an antiinflammatory innate immune response: implications for systemic autoimmunity. J Exp Med 192:1353–1364

Grunfeld C, Marshall M, Shigenaga JK, Moser AH, Tobias P, Feingold KR (1999) Lipoproteins inhibit macrophage activation by lipoteichoic acid. J Lipid Res 40:245–252

Guo L, Ai J, Zheng Z, Howatt DA, Daugherty A, Huang B, Li XA (2013) High density lipoprotein protects against polymicrobe-induced sepsis in mice. J Biol Chem 288:17947–17953. doi:10.1074/jbc.M112.442699

Gupta H, Dai L, Datta G, Garber DW, Grenett H, Li Y, Mishra V, Palgunachari MN, Handattu S, Gianturco SH, Bradley WA, Anantharamaiah GM, White CR (2005) Inhibition of lipopolysaccharide-induced inflammatory responses by an apolipoprotein AI mimetic peptide. Circ Res 97:236–243

Haberstroh A, Schnober EK, Zeisel MB, Carolla P, Barth H, Blum HE, Cosset FL, Koutsoudakis G, Bartenschlager R, Union A, Depla E, Owsianka A, Patel AH, Schuster C, Stoll-Keller F, Doffoel M, Dreux M, Baumert TF (2008) Neutralizing host responses in hepatitis C virus infection target viral entry at postbinding steps and membrane fusion. Gastroenterology 135(1719–1728):e1. doi:10.1053/j.gastro.2008.07.018

Harris HJ, Farquhar MJ, Mee CJ, Davis C, Reynolds GM, Jennings A, Hu K, Yuan F, Deng H, Hubscher SG, Han JH, Balfe P, McKeating JA (2008) CD81 and claudin 1 coreceptor association: role in hepatitis C virus entry. J Virol 82:5007–5020. doi:10.1128/JVI.02286-07

Henning MF, Herlax V, Bakas L (2011) Contribution of the C-terminal end of apolipoprotein AI to neutralization of lipopolysaccharide endotoxic effect. Innate Immun 17:327–337

Hijikata M, Shimizu YK, Kato H, Iwamoto A, Shih JW, Alter HJ, Purcell RH, Yoshikura H (1993) Equilibrium centrifugation studies of hepatitis C virus: evidence for circulating immune complexes. J Virol 67:1953–1958

Humphries SE, Whittall RA, Hubbart CS, Maplebeck S, Cooper JA, Soutar AK, Naoumova R, Thompson GR, Seed M, Durrington PN, Miller JP, Betteridge DJ, Neil HA (2006) Genetic causes of familial hypercholesterolaemia in patients in the UK: relation to plasma lipid levels and coronary heart disease risk. J Med Genet 43:943–949

Iwasaki A, Medzhitov R (2010) Regulation of adaptive immunity by the innate immune system. Science 327:291–295. doi:10.1126/science.1183021/327/5963/291

Jaillon S, Jeannin P, Hamon Y, Fremaux I, Doni A, Bottazzi B, Blanchard S, Subra JF, Chevailler A, Mantovani A, Delneste Y (2009) Endogenous PTX3 translocates at the membrane of late apoptotic human neutrophils and is involved in their engulfment by macrophages. Cell Death Differ 16:465–474. doi:10.1038/cdd.2008.173

Jeannin P, Bottazzi B, Sironi M, Doni A, Rusnati M, Presta M, Maina V, Magistrelli G, Haeuw JF, Hoeffel G, Thieblemont N, Corvaia N, Garlanda C, Delneste Y, Mantovani A (2005) Complexity and complementarity of outer membrane protein A recognition by cellular and humoral innate immunity receptors. Immunity 22:551–560. doi:10.1016/j.immuni.2005.03.008/S1074-7613(05)00103-2

Jerala R (2007) Structural biology of the LPS recognition. Int J Med Microbiol 297:353–363. doi:10.1016/j.ijmm.2007.04.001/S1438-4221(07)00068-9

Jiao YL, Wu MP (2008) Apolipoprotein A-I diminishes acute lung injury and sepsis in mice induced by lipoteichoic acid. Cytokine 43:83–87. doi:10.1016/j.cyto.2008.04.002/S1043-4666(08)00099-9

Kelesidis T, Yang OO, Currier JS, Navab K, Fogelman AM, Navab M (2011) HIV-1 infected patients with suppressed plasma viremia on treatment have pro-inflammatory HDL. Lipids Health Dis 10:35. doi:10.1186/1476-511X-10-35

Kelesidis T, Yang OO, Kendall MA, Hodis HN, Currier JS (2013) Dysfunctional HDL and progression of atherosclerosis in HIV-1-infected and -uninfected adults. Lipids Health Dis 12:23. doi:10.1186/1476-511X-12-23

Khovidhunkit W, Kim MS, Memon RA, Shigenaga JK, Moser AH, Feingold KR, Grunfeld C (2004) Effects of infection and inflammation on lipid and lipoprotein metabolism: mechanisms and consequences to the host. J Lipid Res 45:1169–1196

Kontush A, Chapman MJ (2006) Functionally defective high-density lipoprotein: a new therapeutic target at the crossroads of dyslipidemia, inflammation, and atherosclerosis. Pharmacol Rev 58:342–374

Krieger SE, Zeisel MB, Davis C, Thumann C, Harris HJ, Schnober EK, Mee C, Soulier E, Royer C, Lambotin M, Grunert F, Dao Thi VL, Dreux M, Cosset FL, McKeating JA, Schuster C, Baumert TF (2010) Inhibition of hepatitis C virus infection by anti-claudin-1 antibodies is mediated by neutralization of E2-CD81-claudin-1 associations. Hepatology 51:1144–1157. doi:10.1002/hep.23445

Kumaraswamy SB, Linder A, Akesson P, Dahlback B (2012) Decreased plasma concentrations of apolipoprotein M in sepsis and systemic inflammatory response syndromes. Crit Care 16:R60. doi:10.1186/cc11305/cc11305

Kwon WY, Suh GJ, Kim KS, Kwak YH, Kim K (2012) 4F, apolipoprotein AI mimetic peptide, attenuates acute lung injury and improves survival in endotoxemic rats. J Trauma Acute Care Surg 72:1576–1583. doi:10.1097/TA.0b013e3182493ab4/01586154-201206000-00023

Lavie M, Voisset C, Vu-Dac N, Zurawski V, Duverlie G, Wychowski C, Dubuisson J (2006) Serum amyloid A has antiviral activity against hepatitis C virus by inhibiting virus entry in a cell culture system. Hepatology 44:1626–1634. doi:10.1002/hep.21406

Lee GW, Lee TH, Vilcek J (1993) TSG-14, a tumor necrosis factor- and IL-1-inducible protein, is a novel member of the pentaxin family of acute phase proteins. J Immunol 150:1804–1812

Lee RP, Lin NT, Chao YF, Lin CC, Harn HJ, Chen HI (2007) High-density lipoprotein prevents organ damage in endotoxemia. Res Nurs Health 30:250–260. doi:10.1002/nur.20187

Levels JH, Abraham PR, van Barreveld EP, Meijers JC, van Deventer SJ (2003) Distribution and kinetics of lipoprotein-bound lipoteichoic acid. Infect Immun 71:3280–3284

Levels JH, Geurts P, Karlsson H, Maree R, Ljunggren S, Fornander L, Wehenkel L, Lindahl M, Stroes ES, Kuivenhoven JA, Meijers JC (2011) High-density lipoprotein proteome dynamics in human endotoxemia. Proteome Sci 9:34. doi:10.1186/1477-5956-9-34

Levine DM, Parker TS, Donnelly TM, Walsh A, Rubin AL (1993) In vivo protection against endotoxin by plasma high density lipoprotein. Proc Natl Acad Sci USA 90:12040–12044

Li Y, Dong JB, Wu MP (2008) Human ApoA-I overexpression diminishes LPS-induced systemic inflammation and multiple organ damage in mice. Eur J Pharmacol 590:417–422. doi:10.1016/j.ejphar.2008.06.047/S0014-2999(08)00656-0

Lupberger J, Zeisel MB, Xiao F, Thumann C, Fofana I, Zona L, Davis C, Mee CJ, Turek M, Gorke S, Royer C, Fischer B, Zahid MN, Lavillette D, Fresquet J, Cosset FL, Rothenberg SM, Pietschmann T, Patel AH, Pessaux P, Doffoel M, Raffelsberger W, Poch O, McKeating JA, Brino L, Baumert TF (2011) EGFR and EphA2 are host factors for hepatitis C virus entry and possible targets for antiviral therapy. Nat Med 17:589–595. doi:10.1038/nm.2341

Madamanchi NR, Patterson C, Runge MS (2002) HIV therapies and atherosclerosis: answers or questions? Arterioscler Thromb Vasc Biol 22:1758–1760

Maillard P, Huby T, Andreo U, Moreau M, Chapman J, Budkowska A (2006) The interaction of natural hepatitis C virus with human scavenger receptor SR-BI/Cla1 is mediated by ApoB-containing lipoproteins. FASEB J 20:735–737. doi:10.1096/fj.05-4728fje

Mancone C, Steindler C, Santangelo L, Simonte G, Vlassi C, Longo MA, D'Offizi G, Di Giacomo C, Pucillo LP, Amicone L, Tripodi M, Alonzi T (2011) Hepatitis C virus production requires apolipoprotein A-I and affects its association with nascent low-density lipoproteins. Gut 60:378–386. doi:10.1136/gut.2010.211292

McGillicuddy FC, de la Llera MM, Hinkle CC, Joshi MR, Chiquoine EH, Billheimer JT, Rothblat GH, Reilly MP (2009) Inflammation impairs reverse cholesterol transport in vivo. Circulation 119:1135–1145

Memon RA, Fuller J, Moser AH, Feingold KR, Grunfeld C (1999) In vivo regulation of plasma platelet-activating factor acetylhydrolase during the acute phase response. Am J Physiol 277: R94–R103

Meuleman P, Catanese MT, Verhoye L, Desombere I, Farhoudi A, Jones CT, Sheahan T, Grzyb K, Cortese R, Rice CM, Leroux-Roels G, Nicosia A (2012) A human monoclonal antibody targeting scavenger receptor class B type I precludes hepatitis C virus infection and viral spread in vitro and in vivo. Hepatology 55:364–372. doi:10.1002/hep.24692

Mujawar Z, Rose H, Morrow MP, Pushkarsky T, Dubrovsky L, Mukhamedova N, Fu Y, Dart A, Orenstein JM, Bobryshev YV, Bukrinsky M, Sviridov D (2006) Human immunodeficiency virus impairs reverse cholesterol transport from macrophages. PLoS Biol 4:e365. doi:10.1371/journal.pbio.0040365

Mujawar Z, Tamehiro N, Grant A, Sviridov D, Bukrinsky M, Fitzgerald ML (2010) Mutation of the ATP cassette binding transporter A1 (ABCA1) C-terminus disrupts HIV-1 Nef binding but does not block the Nef enhancement of ABCA1 protein degradation. Biochemistry 49:8338–8349. doi:10.1021/bi100466q

Murch O, Collin M, Hinds CJ, Thiemermann C (2007) Lipoproteins in inflammation and sepsis. I. Basic science. Intensive Care Med 33:13–24

Murphy AJ, Woollard KJ, Suhartoyo A, Stirzaker RA, Shaw J, Sviridov D, Chin-Dusting JP (2011) Neutrophil activation is attenuated by high-density lipoprotein and apolipoprotein a-I in in vitro and in vivo models of inflammation. Arterioscler Thromb Vasc Biol 31:1333–1341

Nauta AJ, Bottazzi B, Mantovani A, Salvatori G, Kishore U, Schwaeble WJ, Gingras AR, Tzima S, Vivanco F, Egido J, Tijsma O, Hack EC, Daha MR, Roos A (2003) Biochemical and functional characterization of the interaction between pentraxin 3 and C1q. Eur J Immunol 33:465–473. doi:10.1002/immu.200310022

Navab M, Ananthramaiah GM, Reddy ST, Van Lenten BJ, Ansell BJ, Fonarow GC, Vahabzadeh K, Hama S, Hough G, Kamranpour N, Berliner JA, Lusis AJ, Fogelman AM (2004) The oxidation hypothesis of atherogenesis: the role of oxidized phospholipids and HDL. J Lipid Res 45:993–1007

Nielsen SU, Bassendine MF, Burt AD, Martin C, Pumeechockchai W, Toms GL (2006) Association between hepatitis C virus and very-low-density lipoprotein (VLDL)/LDL analyzed in iodixanol density gradients. J Virol 80:2418–2428. doi:10.1128/JVI.80.5.2418-2428.2006

Norata GD, Callegari E, Marchesi M, Chiesa G, Eriksson P, Catapano AL (2005) High-density lipoproteins induce transforming growth factor-beta2 expression in endothelial cells. Circulation 111:2805–2811

Norata GD, Pirillo A, Catapano AL (2006) Modified HDL: biological and physiopathological consequences. Nutr Metab Cardiovasc Dis 16(5):371–386

Norata GD, Marchesi P, Pirillo A, Uboldi P, Chiesa G, Maina V, Garlanda C, Mantovani A, Catapano AL (2008) Long pentraxin 3, a key component of innate immunity, is modulated by high-density lipoproteins in endothelial cells. Arterioscler Thromb Vasc Biol 28:925–931. doi:10.1161/ATVBAHA.107.160606

Norata GD, Marchesi P, Pulakazhi Venu VK, Pasqualini F, Anselmo A, Moalli F, Pizzitola I, Garlanda C, Mantovani A, Catapano AL (2009) Deficiency of the long pentraxin PTX3 promotes vascular inflammation and atherosclerosis. Circulation 120:699–708. doi:10.1161/CIRCULATIONAHA.108.806547

Norata GD, Garlanda C, Catapano AL (2010) The long pentraxin PTX3: a modulator of the immunoinflammatory response in atherosclerosis and cardiovascular diseases. Trends Cardiovasc Med 20:35–40. doi:10.1016/j.tcm.2010.03.005/S1050-1738(10)00036-8

Norata GD, Pirillo A, Catapano AL (2011) HDLs, immunity, and atherosclerosis. Curr Opin Lipidol 22:410–416. doi:10.1097/MOL.0b013e32834adac3

Norata GD, Pirillo A, Ammirati E, Catapano AL (2012) Emerging role of high density lipoproteins as a player in the immune system. Atherosclerosis 220:11–21. doi:10.1016/j.atherosclerosis. 2011.06.045

Pajkrt D, Doran JE, Koster F, Lerch PG, Arnet B, van der Poll T, ten Cate JW, van Deventer SJ (1996) Antiinflammatory effects of reconstituted high-density lipoprotein during human endotoxemia. J Exp Med 184:1601–1608

Parker TS, Levine DM, Chang JC, Laxer J, Coffin CC, Rubin AL (1995) Reconstituted high-density lipoprotein neutralizes gram-negative bacterial lipopolysaccharides in human whole blood. Infect Immun 63:253–258

Parsa A, Kao WH, Xie D, Astor BC, Li M, Hsu CY, Feldman HI, Parekh RS, Kusek JW, Greene TH, Fink JC, Anderson AH, Choi MJ, Wright JT Jr, Lash JP, Freedman BI, Ojo A, Winkler CA, Raj DS, Kopp JB, He J, Jensvold NG, Tao K, Lipkowitz MS, Appel LJ, AASK Study Investigators, CRIC Study Investigators (2013) APOL1 risk variants, race, and progression of chronic kidney disease. N Engl J Med 369(23):2183–2196. doi:10.1056/NEJMoa1310345

Pereira SA, Batuca JR, Caixas U, Branco T, Delgado-Alves J, Germano I, Lampreia F, Monteiro EC (2009) Effect of efavirenz on high-density lipoprotein antioxidant properties in HIV-infected patients. Br J Clin Pharmacol 68:891–897. doi:10.1111/j.1365-2125.2009. 03535.x

Perez-Morga D, Vanhollebeke B, Paturiaux-Hanocq F, Nolan DP, Lins L, Homble F, Vanhamme L, Tebabi P, Pays A, Poelvoorde P, Jacquet A, Brasseur R, Pays E (2005) Apolipoprotein L-I promotes trypanosome lysis by forming pores in lysosomal membranes. Science 309:469–472. doi:10.1126/science.1114566

Pirillo A, Norata GD, Catapano AL (2013) Treating high density lipoprotein cholesterol (HDL-C): quantity versus quality. Curr Pharm Des 19:3841–3857

Ploss A, Evans MJ, Gaysinskaya VA, Panis M, You H, de Jong YP, Rice CM (2009) Human occludin is a hepatitis C virus entry factor required for infection of mouse cells. Nature 457:882–886. doi:10.1038/nature07684

Poon IK, Hulett MD, Parish CR (2010) Molecular mechanisms of late apoptotic/necrotic cell clearance. Cell Death Differ 17:381–397. doi:10.1038/cdd.2009.195

Reading PC, Bozza S, Gilbertson B, Tate M, Moretti S, Job ER, Crouch EC, Brooks AG, Brown LE, Bottazzi B, Romani L, Mantovani A (2008) Antiviral activity of the long chain pentraxin PTX3 against influenza viruses. J Immunol 180:3391–3398

Riddler SA, Smit E, Cole SR, Li R, Chmiel JS, Dobs A, Palella F, Visscher B, Evans R, Kingsley LA (2003) Impact of HIV infection and HAART on serum lipids in men. JAMA 289:2978–2982. doi:10.1001/jama.289.22.2978

Rose H, Woolley I, Hoy J, Dart A, Bryant B, Mijch A, Sviridov D (2006) HIV infection and high-density lipoprotein: the effect of the disease vs the effect of treatment. Metabolism 55:90–95. doi:10.1016/j.metabol.2005.07.012

Rose H, Hoy J, Woolley I, Tchoua U, Bukrinsky M, Dart A, Sviridov D (2008) HIV infection and high density lipoprotein metabolism. Atherosclerosis 199:79–86. doi:10.1016/j.atherosclerosis.2007.10.018

Roumenina LT, Ruseva MM, Zlatarova A, Ghai R, Kolev M, Olova N, Gadjeva M, Agrawal A, Bottazzi B, Mantovani A, Reid KB, Kishore U, Kojouharova MS (2006) Interaction of C1q with IgG1, C-reactive protein and pentraxin 3: mutational studies using recombinant globular head modules of human C1q A, B, and C chains. Biochemistry 45:4093–4104. doi:10.1021/bi052646f

Rovere P, Peri G, Fazzini F, Bottazzi B, Doni A, Bondanza A, Zimmermann VS, Garlanda C, Fascio U, Sabbadini MG, Rugarli C, Mantovani A, Manfredi AA (2000) The long pentraxin PTX3 binds to apoptotic cells and regulates their clearance by antigen-presenting dendritic cells. Blood 96:4300–4306

Sala F, Catapano AL, Norata GD (2012) High density lipoproteins and atherosclerosis: emerging aspects. J Geriatr Cardiol 9:401–407. doi:10.3724/SP.J.1263.2011.12282

Sala F, Cutuli L, Grigore L, Pirillo A, Chiesa G, Catapano AL, Norata GD (2013) Prevalence of classical $CD14^{++}/CD16^{-}$ but not of intermediate $CD14^{++}/CD16^{+}$ monocytes in hypoalphalipoproteinemia. Int J Cardiol 168:2886–2889. doi:10.1016/j.ijcard.2013.03.103

Samanovic M, Molina-Portela MP, Chessler AD, Burleigh BA, Raper J (2009) Trypanosome lytic factor, an antimicrobial high-density lipoprotein, ameliorates Leishmania infection. PLoS Pathog 5:e1000276. doi:10.1371/journal.ppat.1000276

Scarselli E, Ansuini H, Cerino R, Roccasecca RM, Acali S, Filocamo G, Traboni C, Nicosia A, Cortese R, Vitelli A (2002) The human scavenger receptor class B type I is a novel candidate receptor for the hepatitis C virus. EMBO J 21:5017–5025

Sherman CB, Peterson SJ, Frishman WH (2010) Apolipoprotein A-I mimetic peptides: a potential new therapy for the prevention of atherosclerosis. Cardiol Rev 18:141–147. doi:10.1097/CRD. 0b013e3181c4b508

Shor-Posner G, Basit A, Lu Y, Cabrejos C, Chang J, Fletcher M, Mantero-Atienza E, Baum MK (1993) Hypocholesterolemia is associated with immune dysfunction in early human immunodeficiency virus-1 infection. Am J Med 94:515–519

Srinivas RV, Birkedal B, Owens RJ, Anantharamaiah GM, Segrest JP, Compans RW (1990) Antiviral effects of apolipoprotein A-I and its synthetic amphipathic peptide analogs. Virology 176:48–57

Suzuki M, Pritchard DK, Becker L, Hoofnagle AN, Tanimura N, Bammler TK, Beyer RP, Bumgarner R, Vaisar T, de Beer MC, de Beer FC, Miyake K, Oram JF, Heinecke JW (2010) High-density lipoprotein suppresses the type I interferon response, a family of potent antiviral immunoregulators, in macrophages challenged with lipopolysaccharide. Circulation 122:1919–1927

Syder AJ, Lee H, Zeisel MB, Grove J, Soulier E, Macdonald J, Chow S, Chang J, Baumert TF, McKeating JA, McKelvy J, Wong-Staal F (2011) Small molecule scavenger receptor BI antagonists are potent HCV entry inhibitors. J Hepatol 54:48–55. doi:10.1016/j.jhep.2010.06.024

Thompson PA, Kitchens RL (2006) Native high-density lipoprotein augments monocyte responses to lipopolysaccharide (LPS) by suppressing the inhibitory activity of LPS-binding protein. J Immunol 177:4880–4887. doi:10.4049/jimmunol.177.7.4880

Thomssen R, Bonk S, Propfe C, Heermann KH, Kochel HG, Uy A (1992) Association of hepatitis C virus in human sera with beta-lipoprotein. Med Microbiol Immunol 181:293–300

Tobias PS, Soldau K, Ulevitch RJ (1989) Identification of a lipid A binding site in the acute phase reactant lipopolysaccharide binding protein. J Biol Chem 264:10867–10871

Ulevitch RJ, Johnston AR, Weinstein DB (1979) New function for high density lipoproteins. Their participation in intravascular reactions of bacterial lipopolysaccharides. J Clin Invest 64:1516–1524. doi:10.1172/JCI109610

Ulevitch RJ, Johnston AR, Weinstein DB (1981) New function for high density lipoproteins. Isolation and characterization of a bacterial lipopolysaccharide-high density lipoprotein complex formed in rabbit plasma. J Clin Invest 67:827–837. doi:10.1172/JCI110100

Van Amersfoort ES, Van Berkel TJ, Kuiper J (2003) Receptors, mediators, and mechanisms involved in bacterial sepsis and septic shock. Clin Microbiol Rev 16:379–414

van Leeuwen HJ, Heezius EC, Dallinga GM, van Strijp JA, Verhoef J, van Kessel KP (2003) Lipoprotein metabolism in patients with severe sepsis. Crit Care Med 31:1359–1366

van Rossum AP, Fazzini F, Limburg PC, Manfredi AA, Rovere-Querini P, Mantovani A, Kallenberg CG (2004) The prototypic tissue pentraxin PTX3, in contrast to the short pentraxin serum amyloid P, inhibits phagocytosis of late apoptotic neutrophils by macrophages. Arthritis Rheum 50:2667–2674. doi:10.1002/art.20370

Vanhamme L, Paturiaux-Hanocq F, Poelvoorde P, Nolan DP, Lins L, Van Den Abbeele J, Pays A, Tebabi P, Van Xong H, Jacquet A, Moguilevsky N, Dieu M, Kane JP, De Baetselier P, Brasseur R, Pays E (2003) Apolipoprotein L-I is the trypanosome lytic factor of human serum. Nature 422:83–87. doi:10.1038/nature01461

Vishnyakova TG, Bocharov AV, Baranova IN, Chen Z, Remaley AT, Csako G, Eggerman TL, Patterson AP (2003) Binding and internalization of lipopolysaccharide by Cla-1, a human orthologue of rodent scavenger receptor B1. J Biol Chem 278:22771–22780. doi:10.1074/jbc. M211032200

Voisset C, Callens N, Blanchard E, Op De Beeck A, Dubuisson J, Vu-Dac N (2005) High density lipoproteins facilitate hepatitis C virus entry through the scavenger receptor class B type I. J Biol Chem 280:7793–7799. doi:10.1074/jbc.M411600200

von Hahn T, Lindenbach BD, Boullier A, Quehenberger O, Paulson M, Rice CM, McKeating JA (2006) Oxidized low-density lipoprotein inhibits hepatitis C virus cell entry in human hepatoma cells. Hepatology 43:932–942. doi:10.1002/hep.21139

Wendel M, Paul R, Heller AR (2007) Lipoproteins in inflammation and sepsis. II. Clinical aspects. Intensive Care Med 33:25–35. doi:10.1007/s00134-006-0433-x

Wheeler RJ (2010) The trypanolytic factor-mechanism, impacts and applications. Trends Parasitol 26:457–464. doi:10.1016/j.pt.2010.05.005

White CR, Datta G, Mochon P, Zhang Z, Kelly O, Curcio C, Parks D, Palgunachari M, Handattu S, Gupta H, Garber DW, Anantharamaiah GM (2009) Vasculoprotective effects of apolipoprotein mimetic peptides: an evolving paradigm in HDL therapy. Vasc Dis Prev 6:122–130. doi:10. 2174/1567270000906010122

Wolfrum C, Poy MN, Stoffel M (2005) Apolipoprotein M is required for prebeta-HDL formation and cholesterol efflux to HDL and protects against atherosclerosis. Nat Med 11:418–422. doi:10.1038/nm1211

Zahid MN, Turek M, Xiao F, Thi VL, Guerin M, Fofana I, Bachellier P, Thompson J, Delang L, Neyts J, Bankwitz D, Pietschmann T, Dreux M, Cosset FL, Grunert F, Baumert TF, Zeisel MB (2013) The postbinding activity of scavenger receptor class B type I mediates initiation of hepatitis C virus infection and viral dissemination. Hepatology 57:492–504. doi:10.1002/hep. 26097

Zeisel MB, Koutsoudakis G, Schnober EK, Haberstroh A, Blum HE, Cosset FL, Wakita T, Jaeck D, Doffoel M, Royer C, Soulier E, Schvoerer E, Schuster C, Stoll-Keller F, Bartenschlager R, Pietschmann T, Barth H, Baumert TF (2007) Scavenger receptor class B type I is a key host factor for hepatitis C virus infection required for an entry step closely linked to CD81. Hepatology 46:1722–1731. doi:10.1002/hep.21994

Zeisel MB, Fofana I, Fafi-Kremer S, Baumert TF (2011) Hepatitis C virus entry into hepatocytes: molecular mechanisms and targets for antiviral therapies. J Hepatol 54:566–576. doi:10.1016/j. jhep.2010.10.014

High-Density Lipoproteins in Stroke

Olivier Meilhac

Contents

1   Introduction 510
2   Pleiotropic Effects of HDLs 511
3   HDL Potential Effects on the Blood–Brain Barrier 511
    3.1   Reconstituted HDLs and HDL Mimetics 512
4   Part 1: Pleiotropic Effects of HDLs—In Vitro and In Vivo Data 512
    4.1   HDLs and Nitric Oxide (NO) 512
    4.2   HDLs and Sphingosine 1-Phosphate 513
    4.3   Antioxidative Stress 514
    4.4   Anti-inflammatory and Antiprotease Properties 514
    4.5   Endothelial Cell Integrity, EPCs and Antiapoptotic Action 515
    4.6   Antithrombotic Actions 516
5   HDLs in the Cerebral Compartment 517
    5.1   Lipoproteins in the Brain 517
    5.2   Plasma HDLs in the Brain 518
6   Part 2: HDLs and Acute Stroke (In Vivo Data) 518
    6.1   HDLs and Acute Stroke: Experimental Models 518
References 520

Abstract

Besides their well-documented function of reverse transport of cholesterol, high-density lipoproteins (HDLs) display pleiotropic effects due to their antioxidant, antithrombotic, anti-inflammatory and antiapoptotic properties that may play a

---

O. Meilhac (✉)
Inserm U1148, Paris, France

Université de La Réunion, Saint-Denis, Ile de la Réunion, France

CHU de La Réunion, Saint-Denis, Ile de la Réunion, France

Laboratory for Vascular Translational Research, Inserm U1148,
46 Rue Henri Huchard, 75018 Paris, France
e-mail: olivier.meilhac@inserm.fr

major protective role in acute stroke, in particular by limiting the deleterious effects of ischaemia on the blood–brain barrier (BBB) and on the parenchymal cerebral compartment. HDLs may also modulate leukocyte and platelet activation, which may also represent an important target that would justify the use of HDL-based therapy in acute stroke. In this review, we will present an update of all the recent findings in HDL biology that could support a potential clinical use of HDL therapy in ischaemic stroke.

Keywords
HDL • Ischaemic stroke • Endothelium • Animal models

1   Introduction

Stroke is usually divided according to the type of cerebrovascular lesion into ischaemic and haemorrhagic stroke. Ischaemia is the leading cause of stroke, due to obstruction within a blood vessel supplying blood to the brain. It accounts for about 80–90 % of all stroke cases. Haemorrhagic stroke occurs when a weakened vessel ruptures due to an aneurysm or arteriovenous malformations. Uncontrolled hypertension is often associated with haemorrhagic stroke.

Numerous epidemiological studies demonstrate that HDLs are a strong independent negative predictor of vascular events (Miller and Miller 1975; Nofer et al. 2002), supporting HDL-raising therapies. The SPARCL study (Stroke Prevention by Aggressive Reduction in Cholesterol Levels) was designed to assess the effects of atorvastatin 80 mg/day in patients who previously experienced a stroke or transient ischaemic attack, but without known coronary heart disease (Amarenco et al. 2003). In this study, a negative correlation between HDL-C levels and stroke recurrence was found among these 4,731 patients, but only when ischaemic stroke was considered. Only baseline HDL-C and LDL/HDL ratio were associated with a good outcome in the ischaemic stroke subgroup (Amarenco et al. 2009). Each 13.7 mg/dL (0.35 mmol/L) increment in HDL-C was associated with a 13 % reduction in the risk of ischaemic stroke (Amarenco et al. 2008). Interestingly, decreased LDL levels and high HDL levels did not have any effect on haemorrhagic stroke. In a different study, low HDL-C levels have been also associated with a worse outcome after ischaemic stroke. Of 489 patients treated with rtPA for IS, the low concentration of HDL-C was also associated with a poor prognosis at 3 months (Makihara et al. 2012). More recently, another study in patients with atherosclerotic ischaemic stroke reported that low levels of HDL-C ($\leq$35 mg/dL) at admission were associated with higher stroke severity and poor clinical outcome during follow-up (Yeh et al. 2013).

Data from studies investigating the effect of HDL subfractions on vascular prognosis are conflicting, mainly due to the variety of techniques used to assess HDL particle size (Camont et al. 2011; Krauss 2010). Only the inverse relationship between large HDLs (alpha1) and vascular risk was found in several studies

(Asztalos et al. 2004; Schaefer and Asztalos 2007; Asztalos and Schaefer 2003). Two large cohorts comprising a total of more than 160,000 people used the Mendelian randomisation single nucleotide polymorphism associated with HDL-C to assess a causative role of HDLs in preventing myocardial infarction (Voight et al. 2012). The authors found no significant relationship and therefore concluded that HDL-C was not causal in vascular disease but merely a risk marker. Another interpretation could be that HDL-C, i.e. cholesterol associated with HDL, is not an appropriate prognostic marker or therapeutic target. Given the pleiotropic functions of HDLs (see below), in particular their protective action on the endothelium (Tran-Dinh et al. 2013), their functionality should be evaluated in addition to the concentration of HDL cholesterol.

The long-term beneficial effect of HDLs has been largely attributed to their role in reverse cholesterol transport and atherosclerotic burden regression. However, short-term HDL elevation has also been shown to be beneficial: in patients with acute coronary syndromes, each 1 mg/dL increment of HDL-C during the course of a 16-week treatment with atorvastatin resulted in a 1.4 % risk reduction for recurrent adverse events (Olsson et al. 2005).

2   Pleiotropic Effects of HDLs

HDLs represent a heterogeneous class of lipoproteins as far as size, shape and composition are concerned. Whereas their principal function is to transport cholesterol from peripheral tissues to the liver, additional functions have been attributed to HDLs, which may depend on their composition in lipids, proteins, vitamins or antioxidant molecules. In this review, we will focus on pleiotropic effects of HDLs other than reverse transport of cholesterol that can impact on stroke at different levels and particularly on ischaemic stroke.

3   HDL Potential Effects on the Blood–Brain Barrier

The blood–brain barrier (BBB) is composed of endothelial cells interconnected by means of tight junctions lining the microvasculature of the central nervous system (CNS), pericytes, astrocytes, end feet and neuron processes. The first barrier separating the blood and cerebral compartments is represented by endothelial cells, which play a pivotal role in regulating the traffic of molecules and blood cells. Many protective properties of HDLs have been documented in endothelial cells (Tran-Dinh et al. 2013) and are developed below, including vasodilatation, antioxidant, anti-inflammatory and antiapoptotic effects. However, most of the in vitro results have been obtained using HUVECs (human umbilical vein endothelial cells) or BAEC (bovine aortic endothelial cells), which do not present the same characteristics as endothelial cells composing the BBB, in terms of permeability and establishment of tight junctions. In this review, we will summarise the different effects attributed to HDLs that could support their use as therapeutic tools in acute stroke or in secondary prevention. Effects of reconstituted HDLs, HDL mimetics and HDLs isolated from plasma will be presented.

3.1 Reconstituted HDLs and HDL Mimetics

Apo A-1 is the major protein of HDLs, able to recruit lipids and organise HDL particles. Reconstituted HDLs consist of an in vitro combination of apo A-I and phospholipids, producing disc-shaped particles resembling nascent HDL. Apo A-I may be either purified from human plasma or produced by recombinant technology. Carriers of the apo A-I Milano mutation, characterised by the replacement of an arginine by a cysteine at position 173, have reduced plasma levels of apo A-I and HDL-C without increased cardiovascular disease (Sirtori et al. 2001). This observation has led to a strong interest in using apo A-I Milano peptides or proteins as potential therapeutic agent to treat cardiovascular disease (Nissen et al. 2003). Intravenous injection of apo A-I Milano in humans was shown to reduce the atheromatous plaque volume in both murine models and in humans.

However, controversial data have been published as to whether apo A-I Milano is more effective than apo A-I in reverse transporting cholesterol. In mice, gene transfer of wild-type human apo A-I and human apo A-I Milano inhibited the progression of native atherosclerosis and allograft vasculopathy to a similar extent. An equivalent increase in endothelial progenitor cell (EPC) number/function and endothelial regeneration in allografts was also observed (Feng et al. 2009). The use of either native or mutated apo A-I may however be beneficial for endothelial protection and repair in pathological situations such as ischaemic stroke.

The lipid-binding activity of apo A-1 was attempted to be mimicked by synthetic peptides containing the class A amphipathic helix modified by phenylalanine residues on the hydrophobic face. Improvement of the apo A-1 mimetic peptide stability was reached by using D-amino acids instead of L-amino acids which are more readily degraded in plasma. For example, D4F was shown to reduce atherosclerotic lesions in apo E-null mice (Navab et al. 2005) and to improve the anti-inflammatory properties of HDL in humans (Bloedon et al. 2008). Very few data are currently available on the potential beneficial effects, in stroke, of reconstituted HDLs containing wild-type or mutated apo A-I.

4 Part 1: Pleiotropic Effects of HDLs—In Vitro and In Vivo Data

4.1 HDLs and Nitric Oxide (NO)

Hypertension represents a major risk factor for stroke, and polymorphisms governing endothelial NO synthase (NOS3) gene expression/function have been reported to increase the risk of stroke (Berger et al. 2007; Howard et al. 2005). A recent meta-analysis indicates that NOS3 gene 4b/a, T-786C and G894T polymorphism could be associated with IS (Wang et al. 2013). NO may participate in neuroprotection in cerebral ischaemia by its vasodilatory effects but also by increasing erythrocyte membrane fluidity which could limit the microviscosity (Tsuda et al. 2000). In experimental stroke, administration of NO appears to reduce the infarct volume in both permanent and transient models (Willmot et al. 2005). The results of ongoing clinical studies using NO donors are needed to validate experimental data (Sare et al. 2009). Several lines of evidence suggest that HDLs increase NO bioavailability, in particular via the induction of NOS3 [increased expression and activity (Mineo et al. 2006)]. NO exerts many beneficial effects on the microvasculature during reperfusion including vasodilatation, reflow and decreased permeability (Schulz et al. 2004). The results of the trial "Efficacy of Nitric Oxide in Stroke" are still not available. This study should give more insight into the effects of the glyceryl trinitrate patch (transdermal diffusion of NO) on the outcome post-stroke (ENOS Trial Investigators, 2006).

4.2 HDLs and Sphingosine 1-Phosphate

Sphingosine 1-Phosphate (S1P) is an important lipid component identified in HDLs (Kimura et al. 2001) that may account for NO-mediated vasodilatory effects of HDLs (Nofer et al. 2004). S1P is secreted by activated platelets (Yatomi et al. 1995) but mainly transported in plasma by HDL particles (54 %) (Argraves et al. 2008).

HDLs, and in particular S1P, were demonstrated to directly protect the heart against ischaemia/reperfusion in a mouse model (Theilmeier et al. 2006). The authors report that this protection could be attributed to inhibition of neutrophil recruitment and cardiomyocyte apoptosis in the infarcted area by a mechanism involving NO and S1P signalling.

S1P has been shown to promote endothelial barrier function in cultured pulmonary endothelial cells (Garcia et al. 2001) by enhancing tight junction formation [in HUVECs (Lee et al. 2006)] and cortical actin assembly (Garcia et al. 2001), resulting in a decreased permeability. HDL-associated S1P was also reported to promote endothelial motility, a process of potential importance in case of vascular injury, via Gi-coupled S1P receptors and the Akt signalling pathway (in HUVECs (Argraves et al. 2008)).

In a model of transient middle cerebral artery occlusion in rats, an agonist or the S1P receptor 1 was shown to be neuroprotective, after intraperitoneal injection at the time of reperfusion (Hasegawa et al. 2010). Most studies using this agonist (fingolimod or FTY720) report reduced infarct volumes and improved functional outcomes, as summarised in a recent meta-analysis (Liu et al. 2013). This protective effect could be due to a reduced cerebral lymphocyte infiltration (Rolland et al. 2013). In a thromboembolic model, FTY720 was reported to reduce rtPA-associated haemorrhagic transformations (Campos et al. 2013). Since HDL particles are the major transporter of S1P, it could be expected that at least part of their beneficial effect on the BBB could be attributed to this sphingolipid.

4.3 Antioxidative Stress

In stroke, a plethora of studies have shown that oxidative stress is associated with ischaemia and reperfusion and still more so after rtPA treatment. Many experimental studies report that antioxidant treatment displays neuroprotective effects (Margaill et al. 2005). However, hitherto, no concluding clinical study could provide evidence of a potential benefit of antioxidant therapy in stroke (Amaro and Chamorro 2011).

Antiatherogenic functions of HDLs have been related, at least in part, to their antioxidant properties. HDLs contain lipid-soluble vitamins, antioxidants and enzymes such as paraoxonase 1 (PON1), platelet-activating factor acetylhydrolase (PAF-AH) and glutathione phospholipid peroxidase (Florentin et al. 2008). Apo A-I and apo A-II also display antioxidant properties. This antioxidant arsenal confers to HDLs the capacity to limit LDL oxidation and to scavenge lipid hydroperoxides (Navab et al. 2000; Negre-Salvayre et al. 2006). In a rat model of renal ischaemia/reperfusion, rHDLs were shown to reduce the severity of acute ischaemic renal failure associated with decreased malondialdehyde, suggesting attenuation of lipid peroxidation subsequent to oxidative stress (Thiemermann et al. 2003).

4.4 Anti-inflammatory and Antiprotease Properties

Acute stroke is characterised by an activated endothelium favouring the recruitment of leukocytes. In particular, neutrophil proteases such as elastase and matrix metalloprotease-9 may induce BBB breakdown and produce deleterious effects on the parenchymal compartment (Stowe et al. 2009). Proinflammatory cytokines such as tumour necrosis factor-alpha (TNF-alpha), interleukin-1 (IL-1) and IL-6 are increased in plasma of patients with acute stroke (Tuttolomondo et al. 2008).

HDLs may exert anti-inflammatory effects on the endothelium but also on leukocytes. In endothelial cells, HDLs (both native and rHDLs) inhibited TNFα and IL-1 induction of leukocyte adhesion molecules VCAM-1, ICAM-1 and E-selectin but had no effect on the expression of platelet endothelial cell adhesion molecule (PECAM) (Cockerill et al. 1995). More recently, McGrath and colleagues reported that the modulatory effects of rHDL on endothelial cells stimulated TNF-alpha and were mediated by DHCR24, an antioxidant enzyme involved in cholesterol biosynthesis (McGrath et al. 2009). Reconstituted HDLs were also reported to limit PMN adhesion to endothelial cells stimulated by TNFα and LPS (Moudry et al. 1997). These effects were later confirmed in vitro using both native and rHDL on HUVECs but also in vivo, in a rat model of haemorrhagic shock (Cockerill et al. 2001). In monocytes, HDL reduced activation of CD11b induced by PMA leading to decreased adhesion to an endothelial cell monolayer, monocytic spreading under shear flow, and transmigration. This process was mimicked by apo A-1 and reported to be ABCA1-dependent (Murphy et al. 2008). More recently, the same group reported that plasma HDLs were potent inhibitors of neutrophil activation both in vitro and in vivo, using mice models of inflammation (Murphy et al. 2011). In patients with peripheral vascular disease treated by rHDL, they showed a decreased expression of CD11b by neutrophils 5–7 days post-injection versus saline-injected patients. HDLs can also neutralise circulating inflammatory molecules such as C-reactive protein (Wadham et al. 2004) and LPS (Wurfel et al. 1994).

In addition, proteomic studies have shown that HDLs may transport different antiproteases (Karlsson et al. 2005; Vaisar et al. 2007) and in particular alpha-1 antitrypsin (AAT), the natural inhibitor of elastase (Ortiz-Munoz et al. 2009). HDLs displayed antielastase activity and protected vascular cells against elastase-induced apoptosis. Using an in vitro model of BBB (Weksler et al. 2005), we have recently reported that HDLs could limit the deleterious, elastase-mediated role of activated neutrophils under oxygen-glucose deprivation conditions leading to BBB disruption (Bao Dang et al. 2013). In stroke, inhibition of neutrophil elastase may be a therapeutic target, as shown by using specific inhibitors (Ikegame et al. 2010).

4.5 Endothelial Cell Integrity, EPCs and Antiapoptotic Action

Apoptosis is an important mechanism involved in BBB breakdown and associated cerebral damage. In vivo imaging of apoptotic cells using annexin V was reported to be correlated with BBB permeability in patients with acute stroke (Lorberboym et al. 2006). Prevention of apoptosis of all cells composing the neurovascular unit is therefore of major importance to reduce deleterious effects of ischaemia. Plasma HDL-C and apo A-I were reported to prevent apoptosis of endothelial cells induced by mildly oxidised LDL independently of paraoxonase activity (Suc et al. 1997). In this in vitro model (bovine and human endothelial cell lines), HDL binding was specific (receptor-mediated) and HDLs blocked the intracellular calcium increase preceding apoptosis (Escargueil-Blanc et al. 1997). It was suggested that the Apo A-I moiety mediates this cytoprotective effect rather than SP1 (de Souza et al. 2010).

Hypoxia has been reported to induce an autophagic process in endothelial cells (Zhang et al. 2011). Autophagy can be regarded as an adaptive response of the cell to deleterious environmental conditions that could delay apoptosis but could also represent an early step of the apoptotic process. HDLs were shown to inhibit endoplasmic reticulum stress and autophagic response induced by oxidised LDLs in endothelial cells (Muller et al. 2011), suggesting that they impact on very early events of the apoptotic cascade. In the brain, autophagy was significantly increased in the cortex immediately following experimental subarachnoid haemorrhage (SAH) (Lee et al. 2009). During SAH, the BBB is disrupted subsequent to endothelial cell death by apoptosis. In this model, inhibition of apoptosis was shown to significantly reduce the formation of cerebral oedema and associated mortality (Yan et al. 2011). HDLs could thus modulate apoptosis of BBB endothelial cells by preventing autophagy.

Endothelial cell progenitor (EPC) therapy has been envisaged as a potential therapy in stroke [for review, see (Rouhl et al. 2008)]. EPCs can repair damaged vessels and form new ones that could promote recovery after ischaemic injury. In a mouse model of transient middle cerebral artery occlusion (tMCAO), systemic delivery of EPCs limited brain damage associated with ischaemic injury (i.e. improved neurovascular repair and long-term neurobehavioral outcomes) (Fan et al. 2010). In rats, EPCs injected 24 h after tMCAO were shown to reach the injured area and improved functional recovery, potentially attributable to antiapoptotic factors secreted by EPCs (Moubarik et al. 2011). In humans, erythropoietin therapy significantly increased the number of circulating EPCs and improved a 90-day major adverse neurological effect (Yip et al. 2011). Different studies report that reconstituted HDLs may increase circulating EPC number (Tso et al. 2006) and their differentiation from mononuclear cells as well as their angiogenic capacity (Sumi et al. 2007). This was shown in mouse models of endothelial injury in response to LPS and hind limb ischaemia but also in humans after injection of rHDLs in type 2 diabetic patients (van Oostrom et al. 2007). This increase in circulating EPCs was significant at day 7 post-injection. In hypercholesterolemic subjects, HDL concentration was linked to EPC number and function (Rossi et al. 2010). Low EPC number was also reported to be an independent risk factor for endothelial dysfunction in these patients. Petoumenos et al. reported a correlation between circulating EPCs and HDL concentrations in patients with coronary artery disease (Petoumenos et al. 2009). In addition, they have shown that recombinant HDLs improved the function of circulating EPCs in a model of endothelial denudation, which could be one explanation for the vasculoprotective actions of HDLs.

4.6 Antithrombotic Actions

In stroke, haemostasis disorders are associated with dysfunction of vascular endothelium, or with abnormalities of or interference with anticoagulant proteins (i.e. protein C, protein S and antithrombin III) (Coull and Clark 1993). However, all factors favouring clot formation may increase the risk of ischaemic stroke, involving all cellular components (endothelium, erythrocytes, platelets and leukocytes).

HDLs may modulate haemostasis by impacting on platelet, red blood cell (RBC) and endothelial functions (Mineo et al. 2006). For example, thrombin-induced endothelial tissue factor expression in vitro was shown to be downregulated by rHDLs (Viswambharan et al. 2004). Since tissue factor induction on endothelial cells is regulated by NO production, HDLs may inhibit its production in response to endotoxins or cytokines, by increasing the synthesis of NO (Yang and Loscalzo 2000).

Apo A-I, its amphipathic peptide analogues and HDLs were also shown to protect erythrocytes against the generation of procoagulant activity (Epand et al. 1994). HDL_2 subfraction was reported to be inversely correlated with erythrocyte aggregation in hypercholesterolemic patients (Razavian et al. 1994). A multiple regression analysis demonstrated that this association was independent of fibrinogen, the major determinant of erythrocyte aggregation. The mechanisms by which HDL_2 may prevent RBC aggregability have not been elucidated, but this effect is relevant in small-sized arteries where increased blood viscosity may trigger clot formation. Apo A-I, its amphipathic peptide analogues and HDLs were also shown to protect erythrocytes against the generation of procoagulant activity (Epand et al. 1994). HDLs may inhibit phospholipid flip-flop and the associated procoagulant activity. In healthy volunteers receiving low doses of LPS, injection of rHDLs reduced collagen-induced platelet aggregation and modified the procoagulant state associated with endotoxemia (Pajkrt et al. 1997). More recently, reconstituted HDL infusion in patients with type 2 diabetes induced a 50–75 % attenuation of platelet aggregation in response to ADP and collagen (Calkin et al. 2009). In vitro, this effect was shown to be dose-dependent and remained after removal of rHDLs. Ex vivo, under blood flow conditions, rHDL limited thrombus formation on a matrix of collagen. This effect was attributed to the phospholipid moiety of rHDLs, since it was not reproduced by apo A-I, whereas phosphatidylcholine reached the same level of reduction of aggregation as intact rHDLs. However, in this study, plasma HDLs did not display similar in vitro effects on platelet aggregation. The authors suggest that this may be due to limited efflux of cholesterol from platelet membranes induced by native HDLs.

5 HDLs in the Cerebral Compartment

5.1 Lipoproteins in the Brain

Various studies have described the capacity of the brain to synthesise lipoproteins that are different from plasma lipoproteins (for review, see Wang and Eckel 2014). Lipoproteins found in the cerebrospinal fluid (CSF) mainly originate from astrocytes and are characterised by a small size (resembling the size and density of HDLs) and contain mainly apo E (Ladu et al. 2000; Koch et al. 2001). Apo E, J, D and A-I are the main apolipoproteins synthesised in the brain, in an age-dependent manner for apo J and E (Elliott et al. 2010). Apo E and more particularly apo J are thought to be involved in the clearance of amyloid β-peptide by astrocytes and microglial cells (Mulder et al. 2014; Calero et al. 2000). Most of lipoproteins isolated from the CSF are associated with amyloid β-peptide and may participate in its polymerisation, transport and clearance (Ladu et al. 2000). The BBB requires a local homeostasis of cholesterol. Whereas astrocytes are thought to be the major lipoprotein factory, neurons may also participate in the regulation of their synthesis and redistribution in the brain (Pfrieger and Ungerer 2011). After cortical spreading depression (a stimulus that provides long-lasting ischaemic tolerance), apo J expression is increased, suggesting that this apolipoprotein may participate in neuroprotection subsequently to an ischaemic episode (Wiggins et al. 2003). In a model of middle cerebral artery occlusion, apo J-deficient mice displayed a worse structural restoration in the vicinity of the infarct scar as compared to WT mice (Imhof et al. 2006).

5.2 Plasma HDLs in the Brain

Under physiological conditions, it is not clear whether HDLs reach the cerebral parenchymal compartment. Based on in vitro results using bovine cerebral endothelial cells, HDLs have been suggested to cross the BBB via paracellular transport (de Vries et al. 1995). More recent data describe an active process of transcytosis involving receptors for HDLs at the surface of endothelial cells. In particular, SRB1 seems to be involved in HDL uptake whereas ABCA1 (ATP-binding cassette transporter A1) is responsible for internalisation of apo A-1 alone (for review, see von Eckardstein and Rohrer 2009). However, most studies on transendothelial transport of HDLs were performed on aortic endothelial cells that may differ from endothelial cells of the BBB. Few studies are available on HDL actions on neurons and astrocytes. Ferretti et al. reported that HDLs reduced oxidative stress and cell death induced by copper ions in rat astrocytes (Ferretti et al. 2003). In rat type I astrocytes and glioma cells, HDLs stimulated DNA synthesis and expression of fibroblast growth factor-2, a potent neurotrophic factor, which was associated with the activation of proliferative intracellular signalling (Malchinkhuu et al. 2003). It is likely that in pathological conditions (haemorrhage or transient increase of BBB permeability), HDLs have an improved access to the cerebral parenchymal compartment and then act on both neurons and astrocytes (Lapergue et al. 2010, 2013).

6 Part 2: HDLs and Acute Stroke (In Vivo Data)

Under acute stroke conditions, the lipid profile has been reported to be modified (Santos-Silva et al. 2002). These authors have shown that patients with diagnosed ischaemic stroke (CT imaging, $n = 21$) had lower plasma concentrations of both HDL and apo A-I than healthy subjects with no history of cardiovascular events and normal haematologic values ($n = 29$). Blood was sampled after a 12-h fasting period. In these patients, leukocyte count was increased relative to controls, but the concentration of granulocytes accounted for most of this increase (monocytes were moderately augmented whereas lymphocyte count was decreased in stroke patients). More importantly, markers of neutrophil activation, including lactoferrin and elastase, were strongly increased in plasma of stroke patients when compared to controls.

6.1 HDLs and Acute Stroke: Experimental Models

Few studies have attempted to use HDL therapy in acute stroke. Based on the observation that low HDL cholesterol levels were associated with cerebrovascular events (Sacco et al. 2001; Wannamethee et al. 2000), Paterno et al. showed that pretreatment with reconstituted HDL reduced neuronal damage in two experimental models of stroke in rats (Paterno et al. 2004). They showed that high doses of rHDLs (120 mg/kg) infused 2 h before the onset of stroke reduced the brain necrotic area by 61 and 76 %, respectively, in an excitotoxic (NMDA) and a transient MCAO model of stroke in rats. More recently, we have shown that plasma HDLs (10 mg/kg), injected immediately or at 3 or 5 h after stroke, also reduced cerebral infarct volume by 74, 68 and 70 %, respectively, in a rat model of thromboembolic stroke (Lapergue et al. 2010). This was associated with a reduced BBB breakdown and decreased neutrophil recruitment in the infarct area. In this model, injection of fluorescent-labelled HDLs showed a staining of endothelial cells (but also glial cells), suggesting that their primary protective effect was on the BBB. Thus, HDLs may be a powerful neuroprotective tool for the treatment of cerebrovascular diseases by preventing BBB breakdown. We have confirmed these in vivo results by a study using a model of BBB in vitro (Bao Dang et al. 2013).

In humans, recombinant tissue plasminogen activator (rtPA) is the only effective fibrinolytic treatment in ischaemic stroke. The disruption of the BBB is involved in oedema and haemorrhagic transformation following tPA treatment. We assessed whether HDL-rtPA combined treatment could improve the safety and efficacy of tPA in experimental stroke. We showed that HDL injection decreased tPA-induced haemorrhagic transformation in two models of focal middle cerebral artery occlusion (MCAO) (embolic and 4-h monofilament MCAO). Both the blood–brain barrier in vitro model and in vivo results support the vasculoprotective action of HDLs on BBB under ischaemic conditions. Finally, HDLs do not interfere with the fibrinolytic activity of rtPA (Lapergue et al. 2013).

Conclusion

In addition to reverse transport of cholesterol, HDL particles display different protective effects that could support their use in the acute phase of stroke. Different neuroprotective drugs proven to be effective in animal models have failed to translate into clinical settings (Xu and Pan 2013). Early treatments that could be used without interfering with fibrinolytic treatments should be considered as a good option to limit the deleterious effects of ischaemic stroke. HDLs may represent a good candidate, particularly with respect to their protective effects on the endothelium in ischaemia/reperfusion conditions. As discussed by Navab et al., HDL mimetic peptides should be more appropriate for use in chronic settings since they can be administrated orally or by subcutaneous route, whereas reconstituted HDLs may be more suitable for acute treatment by intravenous injection (Navab et al. 2010). Whether apo A-I combined with phospholipids are sufficient to prevent the effects of ischaemia is questionable. It is likely that enrichment of reconstituted HDLs with protective molecules such as antioxidants or antiproteases should provide an optimal beneficial effect as observed when using plasma HDLs isolated from healthy subjects in different animal models of stroke.

Acknowledgements OM's group at Inserm is supported by the Fondation de France, the Fondation Coeur et Artères and the Agence Nationale de la Recherche (ANR JCJC 2010).

Open Access This chapter is distributed under the terms of the Creative Commons Attribution Noncommercial License, which permits any noncommercial use, distribution, and reproduction in any medium, provided the original author(s) and source are credited.

References

Amarenco P, Bogousslavsky J, Callahan AS, Goldstein L, Hennerici M, Sillsen H, Welch MA, Zivin J (2003) Design and baseline characteristics of the stroke prevention by aggressive reduction in cholesterol levels (SPARCL) study. Cerebrovasc Dis 16(4):389–395

Amarenco P, Labreuche J, Touboul PJ (2008) High-density lipoprotein-cholesterol and risk of stroke and carotid atherosclerosis: a systematic review. Atherosclerosis 196(2):489–496. doi:10.1016/j.atherosclerosis.2007.07.033

Amarenco P, Goldstein LB, Messig M, O'Neill BJ, Callahan A 3rd, Sillesen H, Hennerici MG, Zivin JA, Welch KM, Investigators S (2009) Relative and cumulative effects of lipid and blood pressure control in the stroke prevention by aggressive reduction in cholesterol levels trial. Stroke 40(7):2486–2492. doi:10.1161/STROKEAHA.108.546735

Amaro S, Chamorro A (2011) Translational stroke research of the combination of thrombolysis and antioxidant therapy. Stroke 42(5):1495–1499. doi:10.1161/STROKEAHA.111.615039

Argraves KM, Gazzolo PJ, Groh EM, Wilkerson BA, Matsuura BS, Twal WO, Hammad SM, Argraves WS (2008) High density lipoprotein-associated sphingosine 1-phosphate promotes endothelial barrier function. J Biol Chem 283(36):25074–25081

Asztalos BF, Schaefer EJ (2003) High-density lipoprotein subpopulations in pathologic conditions. Am J Cardiol 91(7A):12E–17E

Asztalos BF, Cupples LA, Demissie S, Horvath KV, Cox CE, Batista MC, Schaefer EJ (2004) High-density lipoprotein subpopulation profile and coronary heart disease prevalence in male participants of the Framingham Offspring Study. Arterioscler Thromb Vasc Biol 24 (11):2181–2187. doi:10.1161/01.ATV.0000146325.93749.a8

Bao Dang Q, Lapergue B, Tran-Dinh A, Diallo D, Moreno JA, Mazighi M, Romero IA, Weksler B, Michel JB, Amarenco P, Meilhac O (2013) High-density lipoproteins limit neutrophil-induced damage to the blood-brain barrier in vitro. J Cereb Blood Flow Metab 33(4):575–582. doi:10. 1038/jcbfm.2012.206

Berger K, Stogbauer F, Stoll M, Wellmann J, Huge A, Cheng S, Kessler C, John U, Assmann G, Ringelstein EB, Funke H (2007) The glu298asp polymorphism in the nitric oxide synthase 3 gene is associated with the risk of ischemic stroke in two large independent case-control studies. Hum Genet 121(2):169–178. doi:10.1007/s00439-006-0302-2

Bloedon LT, Dunbar R, Duffy D, Pinell-Salles P, Norris R, DeGroot BJ, Movva R, Navab M, Fogelman AM, Rader DJ (2008) Safety, pharmacokinetics, and pharmacodynamics of oral apoA-I mimetic peptide D-4F in high-risk cardiovascular patients. J Lipid Res 49 (6):1344–1352

Calero M, Rostagno A, Matsubara E, Zlokovic B, Frangione B, Ghiso J (2000) Apolipoprotein J (clusterin) and Alzheimer's disease. Microsc Res Tech 50(4):305–315. doi:10.1002/1097-0029 (20000815)50:4<305::AID-JEMT10>3.0.CO;2-L

Calkin AC, Drew BG, Ono A, Duffy SJ, Gordon MV, Schoenwaelder SM, Sviridov D, Cooper ME, Kingwell BA, Jackson SP (2009) Reconstituted high-density lipoprotein attenuates platelet function in individuals with type 2 diabetes mellitus by promoting cholesterol efflux. Circulation 120(21):2095–2104. doi:10.1161/CIRCULATIONAHA.109.870709

Camont L, Chapman MJ, Kontush A (2011) Biological activities of HDL subpopulations and their relevance to cardiovascular disease. Trends Mol Med 17(10):594–603. doi:10.1016/j.molmed. 2011.05.013

Campos F, Qin T, Castillo J, Seo JH, Arai K, Lo EH, Waeber C (2013) Fingolimod reduces hemorrhagic transformation associated with delayed tissue plasminogen activator treatment in a mouse thromboembolic model. Stroke 44(2):505–511. doi:10.1161/STROKEAHA.112. 679043

Cockerill GW, Rye KA, Gamble JR, Vadas MA, Barter PJ (1995) High-density lipoproteins inhibit cytokine-induced expression of endothelial cell adhesion molecules. Arterioscler Thromb Vasc Biol 15(11):1987–1994

Cockerill GW, McDonald MC, Mota-Filipe H, Cuzzocrea S, Miller NE, Thiemermann C (2001) High density lipoproteins reduce organ injury and organ dysfunction in a rat model of hemorrhagic shock. Faseb J 15(11):1941–1952

Coull BM, Clark WM (1993) Abnormalities of hemostasis in ischemic stroke. Med Clin North Am 77(1):77–94

de Souza JA, Vindis C, Negre-Salvayre A, Rye KA, Couturier M, Therond P, Chantepie S, Salvayre R, Chapman MJ, Kontush A (2010) Small, dense HDL 3 particles attenuate apoptosis in endothelial cells: pivotal role of apolipoprotein A-I. J Cell Mol Med 14(3):608–620. doi:10.1111/j.1582-4934.2009.00713.x/JCMM713

de Vries HE, Breedveld B, Kuiper J, de Boer AG, Van Berkel TJ, Breimer DD (1995) High-density lipoprotein and cerebral endothelial cells in vitro: interactions and transport. Biochem Pharmacol 50(2):271–273. doi:10.1016/0006-2952(95)00127-L

Elliott DA, Weickert CS, Garner B (2010) Apolipoproteins in the brain: implications for neurological and psychiatric disorders. Clin Lipidol 51(4):555–573. doi:10.2217/CLP.10.37

ENOS Trial Investigators ET (2006) Glyceryl trinitrate vs. control, and continuing vs. stopping temporarily prior antihypertensive therapy, in acute stroke: rationale and design of the efficacy of nitric oxide in stroke (ENOS) trial (ISRCTN99414122). Int J Stroke 1(4):245–249. doi:10.1111/j.1747-4949.2006.00059.x

Epand RM, Stafford A, Leon B, Lock PE, Tytler EM, Segrest JP, Anantharamaiah GM (1994) HDL and apolipoprotein A-I protect erythrocytes against the generation of procoagulant activity. Arterioscler Thromb 14(11):1775–1783

Escargueil-Blanc I, Meilhac O, Pieraggi MT, Arnal JF, Salvayre R, Negre-Salvayre A (1997) Oxidized LDLs induce massive apoptosis of cultured human endothelial cells through a calcium-dependent pathway. Prevention by aurintricarboxylic acid. Arterioscler Thromb Vasc Biol 17(2):331–339

Fan Y, Shen F, Frenzel T, Zhu W, Ye J, Liu J, Chen Y, Su H, Young WL, Yang GY (2010) Endothelial progenitor cell transplantation improves long-term stroke outcome in mice. Ann Neurol 67(4):488–497. doi:10.1002/ana.21919

Feng Y, Van Craeyveld E, Jacobs F, Lievens J, Snoeys J, De Geest B (2009) Wild-type apo A-I and apo A-I(Milano) gene transfer reduce native and transplant arteriosclerosis to a similar extent. J Mol Med 87(3):287–297

Ferretti G, Bacchetti T, Moroni C, Vignini A, Curatola G (2003) Protective effect of human HDL against Cu(2+)-induced oxidation of astrocytes. J Trace Elem Med Biol 17(Suppl 1):25–30

Florentin M, Liberopoulos EN, Wierzbicki AS, Mikhailidis DP (2008) Multiple actions of high-density lipoprotein. Curr Opin Cardiol 23(4):370–378. doi:10.1097/HCO.0b013e3283043806/00001573-200807000-0001

Garcia JG, Liu F, Verin AD, Birukova A, Dechert MA, Gerthoffer WT, Bamberg JR, English D (2001) Sphingosine 1-phosphate promotes endothelial cell barrier integrity by Edg-dependent cytoskeletal rearrangement. J Clin Invest 108(5):689–701

Hasegawa Y, Suzuki H, Sozen T, Rolland W, Zhang JH (2010) Activation of sphingosine 1-phosphate receptor-1 by FTY720 is neuroprotective after ischemic stroke in rats. Stroke 41 (2):368–374. doi:10.1161/STROKEAHA.109.568899

Howard TD, Giles WH, Xu J, Wozniak MA, Malarcher AM, Lange LA, Macko RF, Basehore MJ, Meyers DA, Cole JW, Kittner SJ (2005) Promoter polymorphisms in the nitric oxide synthase 3 gene are associated with ischemic stroke susceptibility in young black women. Stroke 36 (9):1848–1851. doi:10.1161/01.STR.0000177978.97428.53

Ikegame Y, Yamashita K, Hayashi S, Yoshimura S, Nakashima S, Iwama T (2010) Neutrophil elastase inhibitor prevents ischemic brain damage via reduction of vasogenic edema. Hypertens Res 33(7):703–707. doi:10.1038/hr.2010.58

Imhof A, Charnay Y, Vallet PG, Aronow B, Kovari E, French LE, Bouras C, Giannakopoulos P (2006) Sustained astrocytic clusterin expression improves remodeling after brain ischemia. Neurobiol Dis 22(2):274–283. doi:10.1016/j.nbd.2005.11.009

Karlsson H, Leanderson P, Tagesson C, Lindahl M (2005) Lipoproteomics II: mapping of proteins in high-density lipoprotein using two-dimensional gel electrophoresis and mass spectrometry. Proteomics 5(5):1431–1445. doi:10.1002/pmic.200401010

Kimura T, Sato K, Kuwabara A, Tomura H, Ishiwara M, Kobayashi I, Ui M, Okajima F (2001) Sphingosine 1-phosphate may be a major component of plasma lipoproteins responsible for the cytoprotective actions in human umbilical vein endothelial cells. J Biol Chem 276 (34):31780–31785. doi:10.1074/jbc.M104353200

Koch S, Donarski N, Goetze K, Kreckel M, Stuerenburg HJ, Buhmann C, Beisiegel U (2001) Characterization of four lipoprotein classes in human cerebrospinal fluid. J Lipid Res 42 (7):1143–1151

Krauss RM (2010) Lipoprotein subfractions and cardiovascular disease risk. Curr Opin Lipidol 21 (4):305–311. doi:10.1097/MOL.0b013e32833b7756

Ladu MJ, Reardon C, Van Eldik L, Fagan AM, Bu G, Holtzman D, Getz GS (2000) Lipoproteins in the central nervous system. Ann N Y Acad Sci 903:167–175

Lapergue B, Moreno JA, Dang BQ, Coutard M, Delbosc S, Raphaeli G, Auge N, Klein I, Mazighi M, Michel JB, Amarenco P, Meilhac O (2010) Protective effect of high-density lipoprotein-based therapy in a model of embolic stroke. Stroke 41(7):1536–1542. doi:10.1161/STROKEAHA.110.581512

Lapergue B, Dang BQ, Desilles JP, Ortiz-Munoz G, Delbosc S, Loyau S, Louedec L, Couraud PO, Mazighi M, Michel JB, Meilhac O, Amarenco P (2013) High-density lipoprotein-based therapy reduces the hemorrhagic complications associated with tissue plasminogen activator treatment in experimental stroke. Stroke 44(3):699–707. doi:10.1161/STROKEAHA.112.667832

Lee JF, Zeng Q, Ozaki H, Wang L, Hand AR, Hla T, Wang E, Lee MJ (2006) Dual roles of tight junction-associated protein, zonula occludens-1, in sphingosine 1-phosphate-mediated endothelial chemotaxis and barrier integrity. J Biol Chem 281(39):29190–29200

Lee JY, He Y, Sagher O, Keep R, Hua Y, Xi G (2009) Activated autophagy pathway in experimental subarachnoid hemorrhage. Brain Res 1287:126–135. doi:10.1016/j.brainres.2009.06.028/S0006-8993(09)01185-8

Liu J, Zhang C, Tao W, Liu M (2013) Systematic review and meta-analysis of the efficacy of sphingosine-1-phosphate (S1P) receptor agonist FTY720 (fingolimod) in animal models of stroke. Int J Neurosci 123(3):163–169. doi:10.3109/00207454.2012.749255

Lorberboym M, Blankenberg FG, Sadeh M, Lampl Y (2006) In vivo imaging of apoptosis in patients with acute stroke: correlation with blood-brain barrier permeability. Brain Res 1103 (1):13–19. doi:10.1016/j.brainres.2006.05.073/S0006-8993(06)01491-0

Makihara N, Okada Y, Koga M, Shiokawa Y, Nakagawara J, Furui E, Kimura K, Yamagami H, Hasegawa Y, Kario K, Okuda S, Naganuma M, Toyoda K (2012) Effect of serum lipid levels on stroke outcome after rt-PA therapy: SAMURAI rt-PA registry. Cerebrovasc Dis 33 (3):240–247. doi:10.1159/000334664

Malchinkhuu E, Sato K, Muraki T, Ishikawa K, Kuwabara A, Okajima F (2003) Assessment of the role of sphingosine 1-phosphate and its receptors in high-density lipoprotein-induced stimulation of astroglial cell function. Biochem J 370(Pt 3):817–827. doi:10.1042/BJ20020867

Margaill I, Plotkine M, Lerouet D (2005) Antioxidant strategies in the treatment of stroke. Free Radic Biol Med 39(4):429–443. doi:10.1016/j.freeradbiomed.2005.05.003/S0891-5849(05) 00266-2

McGrath KC, Li XH, Puranik R, Liong EC, Tan JT, Dy VM, DiBartolo BA, Barter PJ, Rye KA, Heather AK (2009) Role of 3beta-hydroxysteroid-delta 24 reductase in mediating antiinflammatory effects of high-density lipoproteins in endothelial cells. Arterioscler Thromb Vasc Biol 29(6):877–882. doi:10.1161/ATVBAHA.109.184663

Miller GJ, Miller NE (1975) Plasma-high-density-lipoprotein concentration and development of ischaemic heart-disease. Lancet 1(7897):16–19

Mineo C, Deguchi H, Griffin JH, Shaul PW (2006) Endothelial and antithrombotic actions of HDL. Circ Res 98(11):1352–1364

Moubarik C, Guillet B, Youssef B, Codaccioni JL, Piercecchi MD, Sabatier F, Lionel P, Dou L, Foucault-Bertaud A, Velly L, Dignat-George F, Pisano P (2011) Transplanted late outgrowth endothelial progenitor cells as cell therapy product for stroke. Stem Cell Rev 7(1):208–220. doi:10.1007/s12015-010-9157-y

Moudry R, Spycher MO, Doran JE (1997) Reconstituted high density lipoprotein modulates adherence of polymorphonuclear leukocytes to human endothelial cells. Shock 7(3):175–181

Mulder SD, Nielsen HM, Blankenstein MA, Eikelenboom P, Veerhuis R (2014) Apolipoproteins E and J interfere with amyloid-beta uptake by primary human astrocytes and microglia in vitro. Glia. doi:10.1002/glia.22619

Muller C, Salvayre R, Negre-Salvayre A, Vindis C (2011) HDLs inhibit endoplasmic reticulum stress and autophagic response induced by oxidized LDLs. Cell Death Differ 18(5):817–828. doi:10.1038/cdd2010149

Murphy AJ, Woollard KJ, Hoang A, Mukhamedova N, Stirzaker RA, McCormick SP, Remaley AT, Sviridov D, Chin-Dusting J (2008) High-density lipoprotein reduces the human monocyte inflammatory response. Arterioscler Thromb Vasc Biol 28(11):2071–2077. doi:10.1161/ATVBAHA.108.168690

Murphy AJ, Woollard KJ, Suhartoyo A, Stirzaker RA, Shaw J, Sviridov D, Chin-Dusting JP (2011) Neutrophil activation is attenuated by high-density lipoprotein and apolipoprotein A-I in vitro and in vivo models of inflammation. Arterioscler Thromb Vasc Biol 31 (6):1333–1341. doi:10.1161/ATVBAHA.111.226258

Navab M, Hama SY, Cooke CJ, Anantharamaiah GM, Chaddha M, Jin L, Subbanagounder G, Faull KF, Reddy ST, Miller NE, Fogelman AM (2000) Normal high density lipoprotein inhibits three steps in the formation of mildly oxidized low density lipoprotein: step 1. J Lipid Res 41 (9):1481–1494

Navab M, Anantharamaiah GM, Hama S, Hough G, Reddy ST, Frank JS, Garber DW, Handattu S, Fogelman AM (2005) D-4F and statins synergize to render HDL antiinflammatory in mice and monkeys and cause lesion regression in old apolipoprotein E-null mice. Arterioscler Thromb Vasc Biol 25(7):1426–1432

Navab M, Shechter I, Anantharamaiah GM, Reddy ST, Van Lenten BJ, Fogelman AM (2010) Structure and function of HDL mimetics. Arterioscler Thromb Vasc Biol 30(2):164–168

Negre-Salvayre A, Dousset N, Ferretti G, Bacchetti T, Curatola G, Salvayre R (2006) Antioxidant and cytoprotective properties of high-density lipoproteins in vascular cells. Free Radic Biol Med 41(7):1031–1040. doi:10.1016/j.freeradbiomed.2006.07.006/S0891-5849(06)00430-8

Nissen SE, Tsunoda T, Tuzcu EM, Schoenhagen P, Cooper CJ, Yasin M, Eaton GM, Lauer MA, Sheldon WS, Grines CL, Halpern S, Crowe T, Blankenship JC, Kerensky R (2003) Effect of recombinant ApoA-I Milano on coronary atherosclerosis in patients with acute coronary syndromes: a randomized controlled trial. JAMA 290(17):2292–2300

Nofer JR, Kehrel B, Fobker M, Levkau B, Assmann G, von Eckardstein A (2002) HDL and arteriosclerosis: beyond reverse cholesterol transport. Atherosclerosis 161(1):1–16

Nofer JR, van der Giet M, Tolle M, Wolinska I, von Wnuck LK, Baba HA, Tietge UJ, Godecke A, Ishii I, Kleuser B, Schafers M, Fobker M, Zidek W, Assmann G, Chun J, Levkau B (2004) HDL induces NO-dependent vasorelaxation via the lysophospholipid receptor S1P3. J Clin Invest 113(4):569–581

Olsson AG, Schwartz GG, Szarek M, Sasiela WJ, Ezekowitz MD, Ganz P, Oliver MF, Waters D, Zeiher A (2005) High-density lipoprotein, but not low-density lipoprotein cholesterol levels influence short-term prognosis after acute coronary syndrome: results from the MIRACL trial. Eur Heart J 26(9):890–896. doi:10.1093/eurheartj/ehi186

Ortiz-Munoz G, Houard X, Martin-Ventura JL, Ishida BY, Loyau S, Rossignol P, Moreno JA, Kane JP, Chalkley RJ, Burlingame AL, Michel JB, Meilhac O (2009) HDL antielastase activity prevents smooth muscle cell anoikis, a potential new antiatherogenic property. Faseb J 23 (9):3129–3139. doi:10.1096/fj.08-127928

Pajkrt D, Lerch PG, van der Poll T, Levi M, Illi M, Doran JE, Arnet B, van den Ende A, ten Cate JW, van Deventer SJ (1997) Differential effects of reconstituted high-density lipoprotein on coagulation, fibrinolysis and platelet activation during human endotoxemia. Thromb Haemost 77(2):303–307

Paterno R, Ruocco A, Postiglione A, Hubsch A, Andresen I, Lang MG (2004) Reconstituted high-density lipoprotein exhibits neuroprotection in two rat models of stroke. Cerebrovasc Dis 17 (2–3):204–211

Petoumenos V, Nickenig G, Werner N (2009) High-density lipoprotein exerts vasculoprotection via endothelial progenitor cells. J Cell Mol Med 13(11–12):4623–4635

Pfrieger FW, Ungerer N (2011) Cholesterol metabolism in neurons and astrocytes. Prog Lipid Res 50(4):357–371. doi:10.1016/j.plipres.2011.06.002

Razavian SM, Atger V, Giral P, Cambillau M, Del-Pino M, Simon AC, Moatti N, Levenson J (1994) Influence of HDL subfractions on erythrocyte aggregation in hypercholesterolemic me. PCVMETRA group. Arterioscler Thromb 14(3):361–366

Rolland WB, Lekic T, Krafft PR, Hasegawa Y, Altay O, Hartman R, Ostrowski R, Manaenko A, Tang J, Zhang JH (2013) Fingolimod reduces cerebral lymphocyte infiltration in experimental models of rodent intracerebral hemorrhage. Exp Neurol 241:45–55. doi:10.1016/j.expneurol. 2012.12.009

Rossi F, Bertone C, Montanile F, Miglietta F, Lubrano C, Gandini L, Santiemma V (2010) HDL cholesterol is a strong determinant of endothelial progenitor cells in hypercholesterolemic subjects. Microvasc Res 80(2):274–279. doi:10.1016/j.mvr.2010.05.003/S0026-2862(10) 00091-9

Rouhl RP, van Oostenbrugge RJ, Damoiseaux J, Tervaert JW, Lodder J (2008) Endothelial progenitor cell research in stroke: a potential shift in pathophysiological and therapeutical concepts. Stroke 39(7):2158–2165. doi:10.1161/STROKEAHA.107.507251

Sacco RL, Benson RT, Kargman DE, Boden-Albala B, Tuck C, Lin IF, Cheng JF, Paik MC, Shea S, Berglund L (2001) High-density lipoprotein cholesterol and ischemic stroke in the elderly: the Northern Manhattan Stroke Study. JAMA 285(21):2729–2735

Santos-Silva A, Rebelo I, Castro E, Belo L, Catarino C, Monteiro I, Almeida MD, Quintanilha A (2002) Erythrocyte damage and leukocyte activation in ischemic stroke. Clin Chim Acta 320 (1–2):29–35

Sare GM, Gray LJ, Wardlaw J, Chen C, Bath PM (2009) Is lowering blood pressure hazardous in patients with significant ipsilateral carotid stenosis and acute ischaemic stroke? Interim assessment in the 'Efficacy of Nitric Oxide in Stroke' trial. Blood Press Monit 14(1):20–25. doi:10.1097/MBP.0b013e32831e30bd

Schaefer EJ, Asztalos BF (2007) Increasing high-density lipoprotein cholesterol, inhibition of cholesteryl ester transfer protein, and heart disease risk reduction. Am J Cardiol 100(11 A): n25–n31. doi:10.1016/j.amjcard.2007.08.010

Schulz R, Kelm M, Heusch G (2004) Nitric oxide in myocardial ischemia/reperfusion injury. Cardiovasc Res 61(3):402–413

Sirtori CR, Calabresi L, Franceschini G, Baldassarre D, Amato M, Johansson J, Salvetti M, Monteduro C, Zulli R, Muiesan ML, Agabiti-Rosei E (2001) Cardiovascular status of carriers of the apolipoprotein A-I(Milano) mutant: the Limone sul Garda study. Circulation 103 (15):1949–1954

Stowe AM, Adair-Kirk TL, Gonzales ER, Perez RS, Shah AR, Park TS, Gidday JM (2009) Neutrophil elastase and neurovascular injury following focal stroke and reperfusion. Neurobiol Dis 35(1):82–90. doi:10.1016/j.nbd.2009.04.006/S0969-9961(09)00083-7

Suc I, Escargueil-Blanc I, Troly M, Salvayre R, Negre-Salvayre A (1997) HDL and ApoA prevent cell death of endothelial cells induced by oxidized LDL. Arterioscler Thromb Vasc Biol 17 (10):2158–2166

Sumi M, Sata M, Miura S, Rye KA, Toya N, Kanaoka Y, Yanaga K, Ohki T, Saku K, Nagai R (2007) Reconstituted high-density lipoprotein stimulates differentiation of endothelial progenitor cells and enhances ischemia-induced angiogenesis. Arterioscler Thromb Vasc Biol 27 (4):813–818. doi:10.1161/01.ATV.0000259299.38843.64

Theilmeier G, Schmidt C, Herrmann J, Keul P, Schafers M, Herrgott I, Mersmann J, Larmann J, Hermann S, Stypmann J, Schober O, Hildebrand R, Schulz R, Heusch G, Haude M, von Wnuck LK, Herzog C, Schmitz M, Erbel R, Chun J, Levkau B (2006) High-density lipoproteins and their constituent, sphingosine-1-phosphate, directly protect the heart against ischemia/reperfusion injury in vivo via the S1P3 lysophospholipid receptor. Circulation 114(13):1403–1409

Thiemermann C, Patel NS, Kvale EO, Cockerill GW, Brown PA, Stewart KN, Cuzzocrea S, Britti D, Mota-Filipe H, Chatterjee PK (2003) High density lipoprotein (HDL) reduces renal ischemia/reperfusion injury. J Am Soc Nephrol 14(7):1833–1843

Tran-Dinh A, Diallo D, Delbosc S, Varela-Perez LM, Dang QB, Lapergue B, Burillo E, Michel JB, Levoye A, Martin-Ventura JL, Meilhac O (2013) HDL and endothelial protection. Br J Pharmacol 169(3):493–511. doi:10.1111/bph.12174

Tso C, Martinic G, Fan WH, Rogers C, Rye KA, Barter PJ (2006) High-density lipoproteins enhance progenitor-mediated endothelium repair in mice. Arterioscler Thromb Vasc Biol 26(5):1144–1149. doi:10.1161/01.ATV.0000216600.37436.cf

Tsuda K, Kimura K, Nishio I, Masuyama Y (2000) Nitric oxide improves membrane fluidity of erythrocytes in essential hypertension: an electron paramagnetic resonance investigation. Biochem Biophys Res Commun 275(3):946–954. doi:10.1006/bbrc.2000.3408(S0006-291X(00)93408-9)

Tuttolomondo A, Di Raimondo D, di Sciacca R, Pinto A, Licata G (2008) Inflammatory cytokines in acute ischemic stroke. Curr Pharm Des 14(33):3574–3589

Vaisar T, Pennathur S, Green PS, Gharib SA, Hoofnagle AN, Cheung MC, Byun J, Vuletic S, Kassim S, Singh P, Chea H, Knopp RH, Brunzell J, Geary R, Chait A, Zhao XQ, Elkon K, Marcovina S, Ridker P, Oram JF, Heinecke JW (2007) Shotgun proteomics implicates protease inhibition and complement activation in the antiinflammatory properties of HDL. J Clin Invest 117(3):746–756. doi:10.1172/JCI26206

van Oostrom O, Nieuwdorp M, Westerweel PE, Hoefer IE, Basser R, Stroes ES, Verhaar MC (2007) Reconstituted HDL increases circulating endothelial progenitor cells in patients with type 2 diabetes. Arterioscler Thromb Vasc Biol 27(8):1864–1865. doi:10.1161/ATVBAHA.107.143875

Viswambharan H, Ming XF, Zhu S, Hubsch A, Lerch P, Vergeres G, Rusconi S, Yang Z (2004) Reconstituted high-density lipoprotein inhibits thrombin-induced endothelial tissue factor expression through inhibition of RhoA and stimulation of phosphatidylinositol 3-kinase but not Akt/endothelial nitric oxide synthase. Circ Res 94(7):918–925. doi:10.1161/01.RES.0000124302.20396.B7

Voight BF, Peloso GM, Orho-Melander M, Frikke-Schmidt R, Barbalic M, Jensen MK, Hindy G, Holm H, Ding EL, Johnson T, Schunkert H, Samani NJ, Clarke R, Hopewell JC, Thompson JF, Li M, Thorleifsson G, Newton-Cheh C, Musunuru K, Pirruccello JP, Saleheen D, Chen L, Stewart A, Schillert A, Thorsteinsdottir U, Thorgeirsson G, Anand S, Engert JC, Morgan T, Spertus J, Stoll M, Berger K, Martinelli N, Girelli D, McKeown PP, Patterson CC, Epstein SE, Devaney J, Burnett MS, Mooser V, Ripatti S, Surakka I, Nieminen MS, Sinisalo J, Lokki ML, Perola M, Havulinna A, de Faire U, Gigante B, Ingelsson E, Zeller T, Wild P, de Bakker PI, Klungel OH, Maitland-van der Zee AH, Peters BJ, de Boer A, Grobbee DE, Kamphuisen PW, Deneer VH, Elbers CC, Onland-Moret NC, Hofker MH, Wijmenga C, Verschuren WM, Boer JM, van der Schouw YT, Rasheed A, Frossard P, Demissie S, Willer C, Do R, Ordovas JM, Abecasis GR, Boehnke M, Mohlke KL, Daly MJ, Guiducci C, Burtt NP, Surti A, Gonzalez E, Purcell S, Gabriel S, Marrugat J, Peden J, Erdmann J, Diemert P, Willenborg C, Konig IR, Fischer M, Hengstenberg C, Ziegler A, Buysschaert I, Lambrechts D, Van de Werf F, Fox KA, El Mokhtari NE, Rubin D, Schrezenmeir J, Schreiber S, Schafer A, Danesh J, Blankenberg S, Roberts R, McPherson R, Watkins H, Hall AS, Overvad K, Rimm E, Boerwinkle E, Tybjaerg-Hansen A, Cupples LA, Reilly MP, Melander O, Mannucci PM, Ardissino D, Siscovick D, Elosua R, Stefansson K, O'Donnell CJ, Salomaa V, Rader DJ, Peltonen L, Schwartz SM, Altshuler D, Kathiresan S (2012) Plasma HDL cholesterol and risk of myocardial infarction: a mendelian randomisation study. Lancet 380(9841):572–580. doi:10.1016/S0140-6736(12)60312-2 von Eckardstein A, Rohrer L (2009) Transendothelial lipoprotein transport and regulation of endothelial permeability and integrity by lipoproteins. Curr Opin Lipidol 20(3):197–205. doi:10.1097/MOL.0b013e32832afd63

Wadham C, Albanese N, Roberts J, Wang L, Bagley CJ, Gamble JR, Rye KA, Barter PJ, Vadas MA, Xia P (2004) High-density lipoproteins neutralize C-reactive protein proinflammatory activity. Circulation 109(17):2116–2122. doi:10.1161/01.CIR.0000127419.45975.26

Wang H, Eckel RH (2014) What are lipoproteins doing in the brain? Trends Endocrinol Metab 25 (1):8–14. doi:10.1016/j.tem.2013.10.003

Wang M, Jiang X, Wu W, Zhang D (2013) Endothelial NO synthase gene polymorphisms and risk of ischemic stroke in Asian population: a meta-analysis. PLoS ONE 8(3):e60472. doi:10.1371/journal.pone.0060472

Wannamethee SG, Shaper AG, Ebrahim S (2000) HDL-cholesterol, total cholesterol, and the risk of stroke in middle-aged British men. Stroke 31(8):1882–1888

Weksler BB, Subileau EA, Perriere N, Charneau P, Holloway K, Leveque M, Tricoire-Leignel H, Nicotra A, Bourdoulous S, Turowski P, Male DK, Roux F, Greenwood J, Romero IA, Couraud PO (2005) Blood-brain barrier-specific properties of a human adult brain endothelial cell line. Faseb J 19(13):1872–1874. doi:10.1096/fj.04-3458fje

Wiggins AK, Shen PJ, Gundlach AL (2003) Delayed, but prolonged increases in astrocytic clusterin (ApoJ) mRNA expression following acute cortical spreading depression in the rat: evidence for a role of clusterin in ischemic tolerance. Brain Res Mol Brain Res 114(1):20–30

Willmot M, Gray L, Gibson C, Murphy S, Bath PM (2005) A systematic review of nitric oxide donors and L-arginine in experimental stroke; effects on infarct size and cerebral blood flow. Nitric Oxide 12(3):141–149. doi:10.1016/j.niox.2005.01.003/S1089-8603(05)00004-2

Wurfel MM, Kunitake ST, Lichenstein H, Kane JP, Wright SD (1994) Lipopolysaccharide (LPS)-binding protein is carried on lipoproteins and acts as a cofactor in the neutralization of LPS. J Exp Med 180(3):1025–1035

Xu SY, Pan SY (2013) The failure of animal models of neuroprotection in acute ischemic stroke to translate to clinical efficacy. Med Sci Monit Basic Res 19:37–45

Yan J, Li L, Khatibi NH, Yang L, Wang K, Zhang W, Martin RD, Han J, Zhang J, Zhou C (2011) Blood-brain barrier disruption following subarachnoid hemorrhage may be facilitated through PUMA induction of endothelial cell apoptosis from the endoplasmic reticulum. Exp Neurol 230(2):240–247. doi:10.1016/j.expneurol.2011.04.022/S0014-4886(11)00153-1

Yang Y, Loscalzo J (2000) Regulation of tissue factor expression in human microvascular endothelial cells by nitric oxide. Circulation 101(18):2144–2148

Yatomi Y, Ruan F, Hakomori S, Igarashi Y (1995) Sphingosine-1-phosphate: a platelet-activating sphingolipid released from agonist-stimulated human platelets. Blood 86(1):193–202

Yeh PS, Yang CM, Lin SH, Wang WM, Chen PS, Chao TH, Lin HJ, Lin KC, Chang CY, Cheng TJ, Li YH (2013) Low levels of high-density lipoprotein cholesterol in patients with atherosclerotic stroke: a prospective cohort study. Atherosclerosis 228(2):472–477. doi:10.1016/j.atherosclerosis.2013.03.015

Yip HK, Tsai TH, Lin HS, Chen SF, Sun CK, Leu S, Yuen CM, Tan TY, Lan MY, Liou CW, Lu CH, Chang WN (2011) Effect of erythropoietin on level of circulating endothelial progenitor cells and outcome in patients after acute ischemic stroke. Crit Care 15(1):R40. doi:10.1186/cc10002

Zhang R, Zhu F, Ren J, Huang L, Liu P, Wu G (2011) Beclin1/PI3K-mediated autophagy prevents hypoxia-induced apoptosis in EAhy926 cell line. Cancer Biother Radiopharm 26(3):335–343. doi:10.1089/cbr.2010.0814

Therapeutic Potential of HDL in Cardioprotection and Tissue Repair

Sophie Van Linthout, Miguel Frias, Neha Singh, and Bart De Geest

Contents

1	Introduction	529
2	In Vitro Effects of HDL on Cardiomyocytes	530
	2.1 Effects of HDL in Rat Neonatal Cardiomyocytes	531
	2.2 Adult Mouse Cardiomyocytes	531
	2.3 Apo A-I and Sphingosine-1-Phosphate Mediate the Cytoprotective Effects of HDL	532
3	HDL Confer Protection Against Ischaemia Reperfusion Injury	533
	3.1 IRI in the Isolated Heart Model	534
	3.2 IRI in In Vivo Models	534
	3.3 Apo A-I and S1P Confer Cardioprotective Effects on HDL	535
4	Human *Apo A-I* Gene Transfer Attenuates Diabetic Cardiomyopathy	536
	4.1 Human Apo A-I Gene Transfer Influences Metabolic Parameters in Streptozotocin-Induced Diabetes Mellitus	538
	4.2 Human Apo A-I Gene Transfer Attenuates Diabetes-Associated Oxidative Stress, Cardiac Fibrosis, and Endothelial Dysfunction	538
	4.3 Human Apo A-I Gene Transfer Reduces Diabetes-Induced Cardiac Inflammation	540
	4.4 Human Apo A-I Gene Transfer Reduces Diabetes-Associated Cardiac Apoptosis and Improves the Cardiac Endothelial Integrity	542
5	HDL and Tissue Repair: Modulation of EPC Biology via SR-BI	543
6	Development of Topical HDL Therapy for Cutaneous Wound Healing	547

S. Van Linthout
Charité-University-Medicine Berlin, Campus Virchow, Berlin-Brandenburg Center for Regenerative Therapy (BCRT), Berlin, Germany

M. Frias
Laboratoire des Lipides, Hôpitaux Universitaires de Genève, Geneva, Switzerland

N. Singh • B. De Geest (✉)
Molecular and Vascular Biology, Department of Cardiovascular Sciences, Center for Molecular and Vascular Biology, Katholieke Universiteit Leuven, Campus Gasthuisberg, Herestraat 49 bus 911, 3000 Leuven, Belgium
e-mail: bart.degeest@med.kuleuven.be

© The Author(s) 2015
A. von Eckardstein, D. Kardassis (eds.), *High Density Lipoproteins*, Handbook of Experimental Pharmacology 224, DOI 10.1007/978-3-319-09665-0_17

7 Beneficial Effects of Selective HDL-Raising Gene Transfer on Cardiac Remodelling
   and Cardiac Function After Myocardial Infarction in Mice 549
Conclusions 552
References 554

Abstract

Epidemiological studies support a strong association between high-density lipoprotein (HDL) cholesterol levels and heart failure incidence. Experimental evidence from different angles supports the view that low HDL is unlikely an innocent bystander in the development of heart failure. HDL exerts direct cardioprotective effects, which are mediated via its interactions with the myocardium and more specifically with cardiomyocytes. HDL may improve cardiac function in several ways. Firstly, HDL may protect the heart against ischaemia/reperfusion injury resulting in a reduction of infarct size and thus in myocardial salvage. Secondly, HDL can improve cardiac function in the absence of ischaemic heart disease as illustrated by beneficial effects conferred by these lipoproteins in diabetic cardiomyopathy. Thirdly, HDL may improve cardiac function by reducing infarct expansion and by attenuating ventricular remodelling post-myocardial infarction. These different mechanisms are substantiated by in vitro, ex vivo, and in vivo intervention studies that applied treatment with native HDL, treatment with reconstituted HDL, or human *apo A-I* gene transfer. The effect of human *apo A-I* gene transfer on infarct expansion and ventricular remodelling post-myocardial infarction illustrates the beneficial effects of HDL on tissue repair. The role of HDL in tissue repair is further underpinned by the potent effects of these lipoproteins on endothelial progenitor cell number, function, and incorporation, which may in particular be relevant under conditions of high endothelial cell turnover. Furthermore, topical HDL therapy enhances cutaneous wound healing in different models. In conclusion, the development of HDL-targeted interventions in these strategically chosen therapeutic areas is supported by a strong clinical rationale and significant preclinical data.

Keywords

High-density lipoproteins • Apolipoprotein A-I • Heart failure • Ventricular remodelling • Drug development • Ischaemia/reperfusion injury • Tissue repair • Wound healing • Diabetic ulcer

Abbreviations

Apo	Apolipoprotein
AT1R	Angiotensin II receptor, type 1
BOEC	Blood outgrowth endothelial cells

CAV	Cardiac allograft vasculopathy
ECFCs	Endothelial colony-forming cells
eNOS	Endothelial nitric oxide synthase
EPC	Endothelial progenitor cell
ERK	Extracellular signal-regulated kinase
HDL	High-density lipoprotein
ICAM	Intercellular adhesion molecule
IDL	Intermediate-density lipoprotein
HFpEF	Heart failure with preserved ejection fraction
HFrEF	Heart failure with reduced ejection fraction
IL	Interleukin
IRI	Ischaemia/reperfusion injury
JNK	c-Jun N-terminal kinase
LDL	Low-density lipoprotein
L-NAME	L-NG-nitroarginine methyl ester
NADPH	Nicotinamide adenine dinucleotide phosphate
NO	Nitric oxide
p38 MAPK	p38 mitogen-activated protein kinase
PI3K	Phosphoinositide 3-kinase
S1P	Sphingosine-1-phosphate
STAT3	Signal transducer and activator of transcription 3
SR-BI	Scavenger receptor class B, type I
TBARS	Thiobarbituric acid reactive substances
TLR4	Toll-like receptor 4
TNF-α	Tumour necrosis factor-α
VCAM	Vascular cell adhesion molecule
VEGF	Vascular endothelial growth factor
VLDL	Very-low-density lipoprotein

1   Introduction

Heart failure is the pathophysiological state in which cardiac dysfunction is responsible for failure of the heart to pump blood at a rate commensurate with the requirements of metabolising tissues. This clinical syndrome is characterised by symptoms and signs of increased tissue and organ water and of decreased tissue and organ perfusion. Heart failure is a growing public health problem, the leading cause of hospitalisation and a major cause of mortality (Rathi and Deedwania 2012). The prevalence of heart failure can be estimated at 2 % in the Western world, and the incidence approaches 5–10 per 1,000 persons per year (Mosterd and Hoes 2007). However, the prevalence and incidence progressively increase with age. The prevalence of heart failure is 7 % in the age group 75–84 years and over 10 % in those older than 85 years (Mosterd et al. 1999). Projections show that by 2030, the prevalence of heart failure will increase 25 % from 2013 estimates (Go et al. 2013). The 5-year age-adjusted mortality rates after onset of heart failure are 50 % in men and 46 % in women (Roger et al. 2004).

Epidemiological studies support a strong association between high-density lipoprotein (HDL) cholesterol levels and heart failure incidence. In Framingham Heart Study participants free of coronary heart disease at baseline, low HDL cholesterol levels were independently associated with heart failure incidence after adjustment for interim myocardial infarction and clinical covariates (Velagaleti et al. 2009). These seminal prospective data are in agreement with earlier cross-sectional studies showing that post-infarct ejection fraction is lower in patients with low HDL cholesterol levels (Kempen et al. 1987; Wang et al. 1998). Low HDL cholesterol levels and low levels of apolipoprotein (apo) A-I, the main apo of HDL, indicate an unfavourable prognosis in patients with heart failure independent of the aetiology (Iwaoka et al. 2007; Mehra et al. 2009). An intriguing observation is that apo A-I is expressed in the human heart and in the heart of human *apo A-I* transgenic mice (Baroukh et al. 2004; Lowes et al. 2006).

The observed relationship between HDL and heart failure in epidemiological studies might be due to residual confounding. Low HDL may be an integrated biomarker of adverse metabolic processes including abnormal metabolism of triglyceride-rich lipoproteins, insulin resistance, and ongoing tissue inflammation. Crosstalk between inflammatory processes and metabolic dysregulation may be a critical player in the pathogenesis of heart failure (Palomer et al. 2013). Here, we will present experimental evidence from different angles, which support the view that low HDL is unlikely an innocent bystander in the development of heart failure. As will be discussed in the next sections, HDL exerts direct cardioprotective effects, which are mediated via their interactions with the myocardium and more specifically with cardiomyocytes (Sect. 2). HDL may improve cardiac function in several ways. Firstly, HDL may protect the heart against ischaemia/reperfusion injury resulting in a reduction of infarct size and thus in myocardial salvage (Sect. 3). Secondly, HDL can improve cardiac function in the absence of ischaemic heart disease as illustrated by its beneficial effects in diabetic cardiomyopathy (Sect. 4). Thirdly, HDL may improve cardiac function by reducing infarct expansion and by attenuating ventricular remodelling (Sect. 7). Since this latter effect is an example of the role of HDL in tissue repair, the reparative functions of HDL will be further dealt with in Sect. 5 on HDL and endothelial progenitor cells (EPCs) and in Sect. 6 on topical HDL therapy for cutaneous wound healing.

2   In Vitro Effects of HDL on Cardiomyocytes

HDL may exert an indirect protective influence on the heart derived from their ability to limit atherosclerotic plaque formation, thus preventing myocardial ischaemia and loss of myocardial tissue in acute coronary syndromes. Data are now accumulating to indicate that HDL have a more immediate and direct effect on cardiomyocytes. Discussion of these direct effects in this section will provide a more complete picture of the cardioprotective role of these lipoproteins and will also raise questions about the use of the HDL complex itself as a therapeutic agent.

Although HDL is found in plasma, they are also present in the interstitial space in amounts that correspond to approximately 25 % of their plasma concentration (Parini et al. 2006). It is therefore physiologically relevant to evaluate their direct impact on cells. The direct impact of HDL on cardiomyocytes has mainly been studied in two in vitro models: (1) rat neonatal cardiomyocytes and (2) adult mouse cardiomyocytes. These models allow the investigation of the specific actions of HDL on cardiomyocytes independent of the systemic parameters and of endothelial cells whose activity is regulated by HDL. Although studies investigating the direct role of HDL on cardiomyocytes are relatively limited, they all show a beneficial impact of HDL. These protective effects have been attributed to the activation of intracellular signalling cascades involved in pro-survival cell fate.

2.1 Effects of HDL in Rat Neonatal Cardiomyocytes

In cultured rat neonatal cardiomyocytes, HDL treatment inhibits the cytotoxic effects induced by oxidative stress from glucose and growth factor depletion or doxorubicin incubation. This protective effect has mainly been evaluated by its capacity to counteract the proapoptotic signals such as caspase 3 activation and DNA fragmentation (Frias et al. 2010; Theilmeier et al. 2006). HDL incubation induces the phosphorylation of phosphoinositide 3-kinase (PI3K)/Akt, extracellular signal-regulated kinase (ERK) 1/2, p38 mitogen-activated protein kinase (p38 MAPK), and the transcription factor signal transducer and activator of transcription 3 (STAT3) (Frias et al. 2009). Although the precise mechanism of action has not been completely elucidated, experiments using specific inhibitors have defined their role in counteracting apoptotic signals induced by doxorubicin. These studies suggest that the cardioprotective effect involves the activation of the pro-survival proteins ERK1/2 and STAT3, but not p38 MAPK. HDL also induces the phosphorylation of connexin 43 (Morel et al. 2012). This phosphorylation leads to a reduction in gap junction permeability, which may limit the spread of mediators implicated in death pathways. Connexin 43 phosphorylation requires the activation of protein kinase C (Morel et al. 2012). Although there is not yet direct proof of the implication of connexin 43 in the protective effects of HDL, connexin 43 has been shown to be implicated in ischaemic conditioning (Schulz et al. 2007; Schulz and Heusch 2004).

2.2 Adult Mouse Cardiomyocytes

In mouse adult cardiomyocytes, the protective actions of HDL have been demonstrated using a model of hypoxia and reperfusion. Incubation of cardiomyocytes with HDL before the hypoxia period (preconditioning) improved cell survival (Frias et al. 2013). This protection involves the preservation of mitochondrial integrity, as HDL treatment inhibits mitochondrial permeability transition pore opening. Similarly to the results obtained in neonatal cardiomyocytes, HDL induces that activation of intracellular signalling pathways such as PI3K/Akt, ERK1/2, and STAT3, which play a role in this protection. This protective action of HDL is significantly inhibited in cardiomyocytes from cardiomyocyte-restricted STAT3 knockout mice (Frias et al. 2013). Similarly, cardiomyocytes treated with HDL after hypoxia (at reperfusion) show improved cell survival compared to non-treated cells. In this context, the effects of HDL were attributed to PI3K/Akt and ERK1/2 (Tao et al. 2010). The possible role of STAT3 has not been investigated in this reperfusion protocol.

Taken together, HDL-induced activation of several target proteins in cardiomyocytes appears to be cardioprotective. These beneficial effects cannot be attributed to its role in cholesterol transport.

2.3 Apo A-I and Sphingosine-1-Phosphate Mediate the Cytoprotective Effects of HDL

As mentioned above, HDL can modulate the activation of signalling pathways in cardiac cells. HDL is a complex particle, which is composed of numerous proteins and bioactive lipids. Among these constituents, apo A-I and sphingosine-1-phosphate (S1P) are the two major effectors that have been involved in the activation of signalling pathways. Activation occurs via the binding of the HDL particle to scavenger receptor class B, type I (SR-BI), and of S1P to S1P receptors. S1P binds to five (S1P1–S1P5) high-affinity G protein-coupled receptors generating multiple downstream signals. Of the different S1P receptor subtypes, only S1P1, S1P2, and S1P3 receptors are expressed in the heart (Means and Brown 2009). S1P has been shown to be cytoprotective and to confer cardioprotection. For a more detailed discussion of the cardioprotective role of S1P, the interested reader is referred to recent reviews (Karliner 2013; Sattler and Levkau 2009).

In vitro, most of the beneficial effects of HDL mentioned above have been attributed to S1P. Data from experiments using pharmacological antagonists specific for S1P1, S1P2, and S1P3 receptors demonstrate an inhibition of the HDL actions. For example, several experiments suggest that S1P3 is involved in PI3K/Akt signalling and that S1P2 and/or S1P1 might be involved in ERK1/2 and STAT3 signalling (Frias et al. 2009; Tao et al. 2010). S1P2 was shown to play a key role in the HDL-induced protection against the apoptotic effects of doxorubicin (Frias et al. 2010). In the protection against hypoxia reperfusion injury, a predominant, mediatory role was indicated for the S1P1 and S1P3 receptors, via ERK1/2 and PI3K/Akt activation. No role was attributed to S1P2, but neither was it investigated (Tao et al. 2010). The evidence that reinforces the cytoprotective role of S1P comes from experiments, where addition of S1P to reconstituted HDL (rHDL) containing apo A-I improved the protection against the apoptotic action of doxorubicin (Frias et al. 2010).

It should be acknowledged that S1P cannot freely circulate in plasma and that the major plasma lipoprotein source of this lipid is HDL. Therefore, it is not surprising that the effects of HDL can be mediated by S1P. However, conflicting data suggest S1P-independent effects of apo A-I. One explanation proposed initially by Nofer and colleagues (2004) is that SR-BI can anchor the HDL particle at the cell membrane and allow S1P to interact with its specific receptors. In this context, Means and colleagues (Means et al. 2008) have shown that S1P1 is present in lipid raft structures from adult mouse cardiomyocytes. Lipid rafts are known to harbour SR-BI (Babitt et al. 1997). Although increasing evidence suggests a major role of S1P, the mechanism of signal transduction from HDL particle to cell is still under intensive investigation, and a possible role of SR-BI cannot be eliminated.

Crosstalk between intracellular signalling pathways increases the complexity of the regulation of HDL-induced cascades and impedes identification of the specific role of HDL constituents. Nevertheless, it is important to keep in mind that such complexity can result in differing and conflicting data. Further investigations will be necessary to elucidate the precise mechanism of action of HDL in cardiomyocytes in vitro.

3 HDL Confer Protection Against Ischaemia Reperfusion Injury

Whereas the previous section was dedicated to HDL-induced signalling in cardiomyocytes and to cytoprotective effects of HDL in vitro, the role of HDL in this section is extended to cardiac ischaemia/reperfusion injury (IRI).

Ischaemic heart disease is the leading cause of death in the world. After myocardial infarction, early reperfusion is the most effective strategy to limit cell death and subsequent complications that can lead to heart failure. However, re-establishing perfusion is paradoxically detrimental to the heart, giving rise to IRI. This phenomenon is termed lethal reperfusion injury and is defined as a myocardial injury caused by the restoration of coronary flow after an ischaemic episode. This injury culminates in the death of cardiomyocytes that were viable immediately before the myocardial reperfusion. Combating IRI is a notable focus of attempts to improve treatment strategies.

Experimental animal models of IRI are adequate to investigate the precise mechanisms of action and the potential targets of therapeutic drugs. Most investigations used the isolated rodent heart (ex vivo model) or transient ligature of the left anterior descending coronary artery (in vivo model). All studies that evaluated the effects of HDL on the heart submitted to the IRI protocol are unanimous and demonstrate a beneficial effect on cardiac function. Among these studies, a few demonstrated similar effects using rHDL, composed of apo A-I and phospholipids (palmitoyl-oleoyl-phosphatidylcholine). Some studies used rHDL containing the apo A-I Milano mutant.

3.1 IRI in the Isolated Heart Model

An early study using HDL perfusion in isolated hearts suggested that HDL significantly reduced post-ischaemic arrhythmias (Mochizuki et al. 1991). In contrast, perfusion with low-density lipoproteins (LDL) had no effects indicating HDL specificity. The mechanism was not defined, but it was hypothesised that the lipoprotein stabilised bioactive arachidonic acid metabolites (Mochizuki et al. 1991). These antiarrhythmogenic effects were later confirmed in vivo when treatment with rHDL decreased the incidence of post-ischaemic arrhythmias (Imaizumi et al. 2008). Surprisingly, treatment with apo A-I or phospholipids did not significantly reduce these arrhythmogenic responses. The effect of rHDL was mediated via the production of nitric oxide (NO), through an Akt/ERK/NO pathway in endothelial cells (Imaizumi et al. 2008). Unfortunately, the direct impact on cardiomyocytes was not investigated.

The direct cardioprotective impact of HDL on IRI has been studied by Calabresi and colleagues (Calabresi et al. 2003). They reported that treatment with rHDL improved post-ischaemic functional recovery and decreased creatine kinase release in the coronary effluent. Lipid-free apo A-I or phosphatidylcholine liposomes were not effective in protecting the heart from IRI. rHDL caused a dose-dependent reduction of ischaemia-induced cardiac tumour necrosis factor-α (TNF-α) expression and content, which correlated with the improved functional recovery (Calabresi et al. 2003). In agreement with this prior study, rHDL improved cardiac function (Rossoni et al. 2004) and reduced infarct size compared to non-treated animals (Gu et al. 2007). Recent studies showed that treatment with HDL significantly reduced the infarct size in rodents submitted to IRI compared to non-treated IRI animals (Frias et al. 2013; Morel et al. 2012). The protective impact of HDL or rHDL did not depend on the timing of injection. Indeed, the treatment at the moment of reperfusion was also effective, but to a lesser extent than the treatment before the ischaemic period.

The precise mechanism of HDL-induced cardioprotection is under intensive investigation, and the activation of intracellular signalling pathways plays an important role in such cardioprotection. In this context, the protective role of STAT3, previously shown in cardiomyocytes, has been confirmed in the isolated heart model. The reduction of the infarct size induced by HDL was significantly inhibited in STAT3 knockout mice (Frias et al. 2013). Mitochondria play an important role in cell survival and may be the key player in cardioprotection. Interestingly, HDL and rHDL containing apo A-I Milano have been shown to preserve mitochondrial integrity (Frias et al. 2013; Marchesi et al. 2008), consistent with a direct impact of HDL and rHDL on cardiac cells.

3.2 IRI in In Vivo Models

The data obtained ex vivo were extended to in vivo models. The first in vivo data on cardiac function after HDL injection demonstrated that HDL injection stimulates myocardial perfusion. This effect was mediated via the activation of endothelial nitric oxide synthase (eNOS) (Nofer et al. 2004). Although this study did not investigate IRI, it underlines the potential response of the heart to HDL in vivo. The effect of HDL was extended to an in vivo transient ligature of the left anterior descending coronary artery model (Theilmeier et al. 2006). One single injection of native HDL before ischaemia reduced infarct size, apoptotic cell death, and neutrophil infiltration. These protective effects were also significantly inhibited in the presence of the eNOS inhibitor L-NG-nitroarginine methyl ester (L-NAME) and were partially inhibited in S1P receptor subtype 3 KO mice (Theilmeier et al. 2006). Taken together, these results suggest that the HDL action is mediated via NO production and is dependent on the S1P content of HDL.

Similarly to the results obtained in the isolated heart, the protective role of rHDL containing apo A-I Milano during ischaemia or at reperfusion significantly reduced the infarct size induced by IRI in rabbits (Marchesi et al. 2004). Protective effects of apo A-I were also demonstrated (Gu et al. 2007). Injection of apo A-I 10 min before reperfusion significantly improved cardiac function, which was associated with a reduction of myocardial TNF-α and interleukin (IL)-6 levels. Histological analysis of cardiac tissue from rats treated with apo A-I showed a decrease in intercellular adhesion molecule (ICAM) expression and neutrophil infiltration compared to non-treated animals (Gu et al. 2007). It should be acknowledged that apo A-I and rHDL can absorb phospholipids, including S1P, in plasma. This could modulate their impact on IRI.

3.3 Apo A-I and S1P Confer Cardioprotective Effects on HDL

Based on all published reports, it is not easy to delineate the precise role of S1P or apo A-I in the cardioprotective effects of HDL. Both constituents have been investigated individually, and both have been shown to protect against IRI individually. However, their precise impact on the actions of HDL has not always been investigated in the same study or in the same model. Interestingly, levels of HDL-associated S1P were diminished in patients with coronary artery disease (Sattler et al. 2010). Very recent data demonstrate a key role of apolipoprotein M (apoM) in S1P binding to HDL particles (Christoffersen et al. 2011). Indeed, the amount of S1P contained in HDL was very low in apoM KO and higher in apoM overexpressing mice. Unfortunately, IRI in these mice has not yet been investigated.

With respect to S1P, given that its plasma concentration is many times higher than its affinity for S1P receptors, it has been suggested that HDL might play the role of neutralising and limiting the actions of circulating S1P (Murata et al. 2000). The impact of S1P association with HDL on the function of HDL still remains to be elucidated. It is probable that in the HDL particle, both are necessary to obtain the maximal positive response.

The mechanisms of HDL-induced cardioprotection against IRI are complex and cannot be attributed to the role of HDL in reverse cholesterol transport. More precise knowledge on the role of different HDL constituents in intracellular

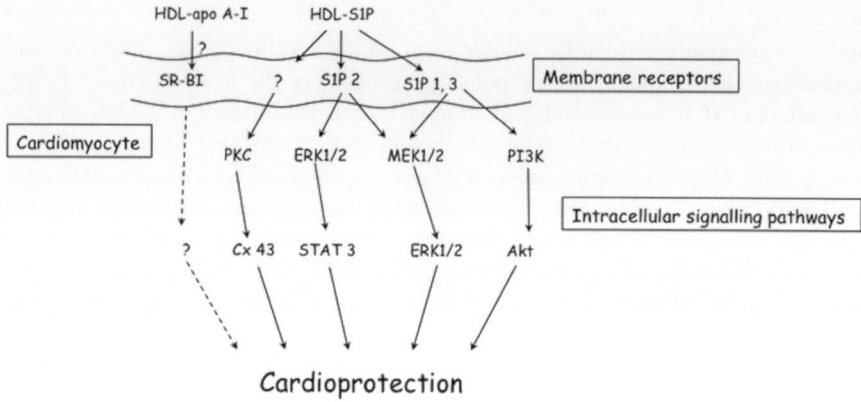

Fig. 1 Proposed protective intracellular signalling pathways induced by HDL in cardiomyocytes. HDL induces the activation of intracellular signalling pathways in cardiomyocytes. This activation leads to cardioprotection. Most of the activated signalling pathways induced by HDL involve its S1P constituent (HDL-S1P) and the participation of S1P receptor subtypes. Further investigations are necessary to elucidate the precise role of SR-BI in cardioprotection induced by HDL. The arrows indicate intracellular signalling as described in cardiomyocytes. Cx 43; connexin 43; PKC protein kinase C; MEK1/2: ERK (MAPK) kinase 1/2

signalling and cardioprotection would help to design a therapeutic compound, which replicates all the beneficial aspects of HDL in the heart. Figure 1 illustrates conceptually the direct actions of HDL on cardiomyocytes.

4 Human *Apo A-I* Gene Transfer Attenuates Diabetic Cardiomyopathy

Diabetic cardiomyopathy is a cardiac disorder, which takes place in the absence of coronary artery disease and hypertension, and is characterised by impaired left ventricular function due to cardiac inflammation, oxidative stress, cardiomyocyte apoptosis, cardiac perivascular and interstitial fibrosis, intramyocardial microangiopathy, endothelial dysfunction, disturbed cardiac substrate metabolism, cardiomyocyte hypertrophy, abnormal intracellular Ca^{2+} handling, and impaired functionality of cardiac stem cells. This cardiac entity has first been recognised by Rubler et al. (1972), and its existence has been confirmed during the last three decades via epidemiological, clinical, and experimental studies. Hyperglycaemia, hyperinsulinaemia, and dyslipidaemia each contribute to the pathogenesis of diabetic cardiomyopathy via triggering cellular signalling and subsequent specific alterations in cardiac structure. However, hyperglycaemia-induced oxidative stress has been proposed to be the key determinant in the development of diabetic cardiomyopathy. Left ventricular diastolic dysfunction represents the earliest preclinical manifestation of diabetic cardiomyopathy, preceding systolic dysfunction, which can progress to symptomatic heart failure (Cosson and Kevorkian 2003; Freire et al. 2007; Raev 1994).

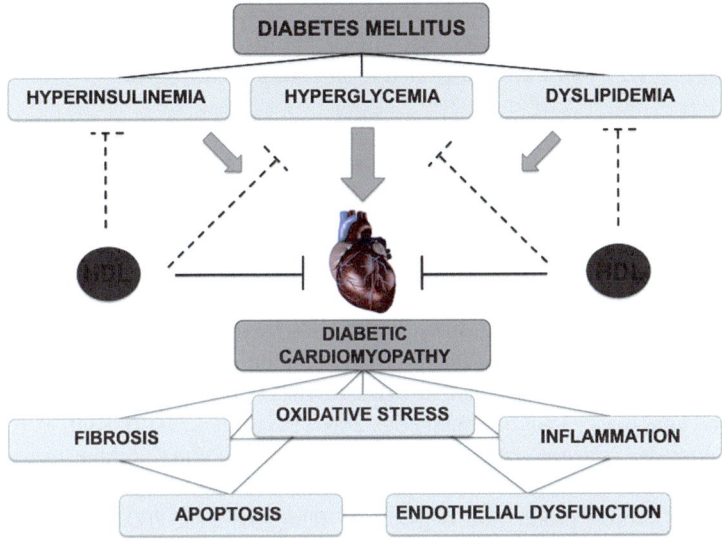

Fig. 2 HDL reduces the development of diabetic cardiomyopathy. Type 2 diabetes mellitus is associated with hyperglycaemia, dyslipidaemia, and hyperinsulinaemia. Hyperglycaemia is the main trigger that contributes to the development of diabetic cardiomyopathy via the induction of oxidative stress. Hallmarks of diabetic cardiomyopathy are, besides cardiac oxidative stress, cardiac fibrosis, inflammation, apoptosis, and endothelial dysfunction, which reciprocally affect one another (interconnected *grey lines*). HDLs have antidiabetic features (*dotted lines*): they reduce hyperglycaemia and dyslipidaemia via—among other mechanisms—their capacity to decrease pancreatic β-cell apoptosis and adipocyte lipolysis, respectively, and they increase insulin sensitivity via the induction of adiponectin, abrogating the development of hyperinsulinaemia. Besides interference with the triggers that lead to diabetic cardiomyopathy, HDLs reduce the features of diabetic cardiomyopathy via their pleiotropic effects, including their anti-oxidative, antifibrotic, anti-inflammatory, antiapoptotic, and endothelial-protective effects (*full lines*)

HDLs have antidiabetic properties. They reduce hyperglycaemia via their capacity to decrease pancreatic β-cell apoptosis (Abderrahmani et al. 2007; Fryirs et al. 2010; Rutti et al. 2009) and to stimulate glucose disposal in skeletal muscle (Han et al. 2007). They decrease dyslipidaemia via lowering adipocyte lipolysis (Van Linthout et al. 2010a) and induce insulin sensitivity (Berg et al. 2001) via upregulating the expression of adiponectin (Van Linthout et al. 2010a), which abrogates the development of hyperinsulinaemia. Consequently, increased HDL may, from a theoretical point of view, antagonise the main triggers of diabetic cardiomyopathy.

The main hallmarks of diabetic cardiomyopathy, including cardiac oxidative stress, interstitial inflammation, apoptosis, fibrosis, and endothelial dysfunction, on the one hand and the pleiotropic effects of HDL on the other hand further support the hypothesis that an increase in HDL may reduce the development of diabetic cardiomyopathy (Fig. 2). Via a human *apo A-I* gene transfer strategy, this hypothesis has been evaluated in an experimental rat model of streptozotocin-induced diabetes mellitus associated with severe hyperglycaemia and an LDL to HDL ratio of 1:1 (Van Linthout et al. 2007; Young et al. 1988).

4.1 Human Apo A-I Gene Transfer Influences Metabolic Parameters in Streptozotocin-Induced Diabetes Mellitus

Apo A-I transfer resulted in a 60 % increase in HDL cholesterol, which was paralleled with a significant decline in very-low-density lipoprotein (VLDL) cholesterol, intermediate-density lipoprotein (IDL) cholesterol, and triglycerides, whereas LDL cholesterol was unaffected (Van Linthout et al. 2008). Despite the well-described antidiabetic effects of HDL involving their ability to reduce pancreatic β-cell apoptosis (Abderrahmani et al. 2007; Fryirs et al. 2010; Rutti et al. 2009), *apo A-I* transfer did not reduce blood glucose levels (Van Linthout et al. 2008), potentially due to the severity of the streptozotocin model, which is associated with a remaining insulin production below 1 % (Hughes et al. 2001). The decline in triglycerides and in the triglyceride-rich lipoproteins VLDL and IDL (Sztalryd and Kraemer 1995) after *apo A-I* transfer suggests an HDL-mediated decrease in lipolysis in adipose tissue, leading to less free fatty acids in the circulation, less triglyceride synthesis in the liver, and subsequent less VLDL and IDL synthesis (Tunaru et al. 2003). This hypothesis is corroborated by the ability of HDL to reduce the expression of hormone-sensitive lipase (Van Linthout et al. 2010a), the rate-limiting enzyme of adipocyte lipolysis in abdominal fat (Sztalryd and Kraemer 1995), and to increase the phosphorylation of the PI3K downstream target Akt in abdominal fat (Van Linthout et al. 2010a). Akt is involved in the anti-lipolytic and lipogenic effects of insulin in adipose tissue (Whiteman et al. 2002) and in the regulation of the expression of the adipokine adiponectin (Cong et al. 2007; Pereira and Draznin 2005), which is known to improve insulin sensitivity under diabetes (Peterson et al. 2008). However, an involvement of HDL in the hepatic expression of genes involved in triglyceride metabolism may not be excluded. The decrease in cardiac glycogen content (Van Linthout et al. 2008) suggests that *apo A-I* transfer in streptozotocin rats partly restored glucose metabolism as an energy source in the heart instead of nearly exclusive reliance on fatty acid metabolism for production of ATP (Kota et al. 2011).

4.2 Human Apo A-I Gene Transfer Attenuates Diabetes-Associated Oxidative Stress, Cardiac Fibrosis, and Endothelial Dysfunction

Hyperglycaemia induces oxidative stress via creating a disbalance between the generation of reactive oxygen species and their resolution by antioxidant enzymes, like superoxide dismutases, which convert O_2^- anions into molecular oxygen and hydrogen peroxide (Nishikawa et al. 2000). Reactive oxygen species initiate diverse stress-signalling pathways including ERK, c-Jun N-terminal kinase (JNK), and p38 MAPK, alter cellular proteins, and induce lipid peroxidation. In diabetic cardiomyocytes, reactive oxygen species are predominantly generated by mitochondria, due to mitochondrial oxidation of fatty acids (Kota et al. 2011), and by nicotinamide adenine dinucleotide phosphate (NADPH) oxidases (Li et al. 2010b). The pathological importance of p38 MAPK in the diabetic heart follows from the observation that p38 MAPK inhibition reduces cardiac inflammation and improves left ventricular dysfunction in streptozotocin-induced diabetic mice (Westermann et al. 2006). Besides decreasing the activity of p38 MAPK, *apo A-I* transfer reduced cardiac oxidative stress via upregulating the expression of the diabetes-downregulated mitochondrial superoxide dismutase-2 and extracellular-superoxide dismutase (Van Linthout et al. 2008). Both forms of superoxide dismutase are important for the heart as outlined in studies whereby cardiac overexpression of superoxide dismutase-2 protected the mitochondrial respiratory function and blocked apoptosis induction (Shen et al. 2006; Suzuki et al. 2002) and overexpression of extracellular-superoxide dismutase decreased macrophage infiltration and fibrosis and improved left ventricular dysfunction (Dewald et al. 2003). These studies postulate that the decrease in cardiac fibrosis after *apo A-I* transfer in streptozotocin-diabetic rats can be partly explained by the reduction in oxidative stress and inflammation, including downregulated expression of pro-fibrotic cytokines (inflammatory fibrosis) (Tschope et al. 2005), as well as by the decrease in cardiac apoptosis (*cfr. supra*) and subsequent replacement fibrosis, resulting in improved left ventricular function. The anti-inflammatory and antifibrotic effects of HDL are supported by findings in *apo A-I* knockout mice, which are associated with increased inflammatory cells and collagen deposition in the lung (Wang et al. 2010), and from observations with the mimetic apo A-I peptide 4F demonstrating an L-4F-mediated decreased endothelial cell matrix production (Ou et al. 2005). Since cardiac NADPH oxidases play a predominant role in the development of diabetic cardiomyopathy (Guo et al. 2007; Wold et al. 2006) and in diabetic cardiac remodelling (Li et al. 2010b), decreased cardiac NADPH oxidase activity following *apo A-I* transfer (Van Linthout et al. 2009) may be a critical mediator of reduced cardiac oxidative stress and subsequent fibrosis in the myocardium.

Systemically, *apo A-I* transfer decreased the oxidative stress in rats with streptozotocin-induced diabetes as indicated by the decline in serum levels of thiobarbituric acid reactive substances (TBARS) (Van Linthout et al. 2008), a marker of lipid peroxidation (Tschope et al. 2005). This finding is corroborated by Mackness et al. (1991), who reported that HDLs decrease the formation of TBARS on oxidised LDL. This is likely mediated by an increase in the activity of paraoxonase or platelet-activating factor-acetyl hydrolase, 2 enzymes with antioxidative features, which are associated with HDL and which are known to be induced after *apo A-I* gene transfer (De Geest et al. 2000).

In the vasculature of diabetic rats, *apo A-I* transfer decreased oxidative stress as indicated by the reduction in diabetes-enhanced NADPH oxidase activity and eNOS uncoupling (Van Linthout et al. 2009), a phenomenon occurring when eNOS produces O_2^- rather than NO. Uncoupling of eNOS is, besides NADPH oxidases (Cai 2005), an important source of reactive oxygen species in diseased, including diabetic, blood vessels (Hink et al. 2001). Consistent with the demonstrated role of the angiotensin II receptor, type 1 (AT1R) in mediating increased NADPD oxidase activity, and eNOS uncoupling in diabetes (Oak and Cai 2007), the downregulation in diabetes-induced AT1R (Hodroj et al. 2007; Nyby et al. 2007) after *apo A-I* transfer (Van Linthout et al. 2009) was postulated to be the predominant mediator of reduced NADPH oxidase activity and eNOS uncoupling. This hypothesis is further supported by in vitro findings showing that the HDL-mediated reduction in AT1R expression in human aortic endothelial cells was associated with a decline in hyperglycaemia-induced oxidative stress and a reduced responsiveness to angiotensin II (Van Linthout et al. 2009). These observations underline the finding of Tolle et al. (2008), who showed that HDL reduce NADPH oxidase-dependent reactive oxygen species generation via inhibition of the activation of Rac1, which is a downstream AT1R-dependent mediator of angiotensin II (Ohtsu et al. 2006). The exact mechanism by which HDL affect AT1R regulation under diabetes mellitus requires further fundamental studies. However, since oxidised LDL (Li et al. 2000) and reactive oxygen species (Gragasin et al. 2003) play a role in the induction of the AT1R in human aortic endothelial cells, one may speculate that HDL via intrinsic anti-oxidative features (cfr. *supra*, via paraoxonase and platelet-activating factor-acetyl hydrolase) may contribute to the downregulation of the AT1R under diabetes mellitus, which results in less NADPH oxidase activity and reactive oxygen formation and in turn may decrease AT1R expression. Concomitant with the reduced vascular oxidative stress, including decreased eNOS uncoupling, *apo A-I* transfer in streptozotocin rats resulted in an enhanced NO bioavailability (Chalupsky and Cai 2005) and consequently in a decrease in endothelial dysfunction (Van Linthout et al. 2009), which is another hallmark of diabetic cardiomyopathy.

4.3 Human Apo A-I Gene Transfer Reduces Diabetes-Induced Cardiac Inflammation

In agreement with the direct anti-inflammatory properties of HDL (Cockerill et al. 2001; Hyka et al. 2001), *apo A-I* transfer decreased the diabetes-induced intramyocardial inflammation (Westermann et al. 2007b, c) as indicated by the reduction in *ICAM-1, vascular cell adhesion molecule (VCAM)-1,* and *TNF-α* mRNA expression (Van Linthout et al. 2008) and VCAM-1 protein expression (Van Linthout et al. 2010b). Downregulation of VCAM-1/ICAM-1 expression suppresses monocyte-endothelial cell adhesion and subsequent transendothelial migration of inflammatory cells. This in turn reduced local expression of cytokines like TNF-α and consequently may attenuate the potentiation of the intramyocardial inflammatory reaction. Hyperglycaemia (Li et al. 2010a; Piga et al. 2007), oxidised LDL (Lee et al. 2010) (Hodgkinson et al. 2008; Peterson et al. 2007), and increased angiotensin II (Alvarez et al. 2004; Rius et al. 2010) under diabetes mellitus upregulate the expression of VCAM-1/ICAM-1. Since HDL did not affect blood glucose levels, the reduction in adhesion molecule expression was not an indirect consequence of an HDL-mediated antidiabetic effect. However, the reduced levels of TBARS, which are also retrieved on oxidised LDL, in diabetic rats treated with *apo A-I* transfer compared to diabetic rats treated with control vector suggest that *apo A-I* transfer decreases oxidised LDL and subsequent cardiac inflammation. Oxidised LDLs are agonists of Toll-like receptor 4 (TLR4), which is expressed on the cell surface of cardiac cells, including cardiomyocytes, smooth muscle cells, and endothelial cells. A role for TLR4 in the development of diabetic cardiomyopathy has recently been suggested (Zhang et al. 2010). The documented HDL-mediated reduction in endothelial TLR4 expression and subsequent decrease in NF-κB activity (Van Linthout et al. 2011) suggest that the reduction in cardiac inflammation after *apo A-I* transfer is further partly mediated via a decrease in endothelial TLR4 expression, limiting oxidised LDL-endothelial TLR4 interactions and subsequent activation of NF-κB. Furthermore, the activation of the AT1R, whose expression is upregulated under diabetes mellitus (Cohen 1993; Nyby et al. 2007), contributes to the development of diabetic cardiomyopathy. After all, AT1R antagonism under diabetes mellitus improves endothelial function (Cheetham et al. 2000) and reduces cardiac inflammation and fibrosis, resulting in an improvement in cardiac function (Westermann et al. 2007a). The observations that *apo A-I* transfer reduces aortic AT1R expression in streptozotocin-induced diabetic rats and that HDLs decrease the hyperglycaemia-induced AT1R expression in endothelial cells (Van Linthout et al. 2009) suggest that *apo A-I* transfer reduces cardiac endothelial AT1R and subsequent angiotensin II responsiveness including the induction of VCAM-1/ICAM-1 expression and subsequent monocyte-endothelial cell adhesion (Alvarez et al. 2004; Rius et al. 2010). Downregulation of the expression of the anti-inflammatory adipokine adiponectin (Ouchi et al. 2000, 2001; Ouedraogo et al. 2007) as observed in diabetic patients and in streptozotocin rats (Thule et al. 2006) has been implicated in chronic inflammatory phenotype in these subjects/rats. The positive correlation between apo A-I/HDL plasma levels and adiponectin concentrations (Verges et al. 2006) and the finding that *apo A-I* transfer increases the expression of adiponectin in an experimental model of extreme inflammation (Van Linthout et al. 2010a) further suggest that the effects of HDL on adiponectin expression may contribute to its anti-inflammatory effects and to the attenuation of cardiac inflammation in rats with streptozotocin-induced diabetes. Finally, since apo A-I induces regulatory T cells (Wilhelm et al. 2010), which protect against the pro-inflammatory status of endothelial cells (He et al. 2010), an induction in regulatory T cells following *apo A-I* transfer may also partly account for the reduced cardiac inflammation in diabetic rats.

Since inflammation triggers fibrosis (Ismahil et al. 2013; Savvatis et al. 2012; Westermann et al. 2011), the less pronounced diabetes-induced cardiac fibrosis (*cfr. infra*) following *apo A-I* transfer might partly be explained by the decrease in cardiac inflammation. In this context, the influence of a systemic HDL-mediated immunomodulation on cardiac inflammation and fibrosis needs further investigation.

4.4 Human Apo A-I Gene Transfer Reduces Diabetes-Associated Cardiac Apoptosis and Improves the Cardiac Endothelial Integrity

Cardiac apoptosis is another hallmark of diabetic cardiomyopathy. The incidence of cardiac apoptosis is higher in diabetic patients (Frustaci et al. 2000) and in diabetic animals (Cai et al. 2002) compared to non-diabetic controls and is directly linked to hyperglycaemia-induced oxidative stress (Cai et al. 2002). Mitochondria play an important role in oxidative stress-induced apoptosis. Caspase 3 and caspase 7 are essential mediators in the mitochondrial processes of apoptosis (Lakhani et al. 2006). In accordance with the reduction in oxidative stress, including the upregulation of cardiac mitochondrial superoxide dismutase-2 expression, *apo A-I* transfer decreased the induced caspase 3/7 activity in streptozotocin rats (Van Linthout et al. 2008). Concomitantly, *apo A-I* transfer raised the ratio of the antiapoptotic Bcl-2, a 'guardian' against mitochondrial initiation of caspase activation (Susin et al. 1996), towards the proapoptotic Bax. This ratio is a marker of increased cardiomyocyte survival probability (Condorelli et al. 1999). Furthermore, *apo A-I* transfer normalised the diabetes-reduced phosphorylation of the pro-survival protein kinase B Akt (Montanari et al. 2005; Uchiyama et al. 2004) and of its effector eNOS to levels found in non-diabetic hearts (Van Linthout et al. 2008). This finding was further corroborated by experiments in cardiomyocytes. HDL supplementation on cardiomyocytes in the presence of in vitro hyperglycaemia reduced apoptosis in a PI3K- and NO-dependent manner (Van Linthout et al. 2008). The antiapoptotic effects of *apo A-I* transfer were reflected at ultrastructural level by a reduced number of cardiomyocytes with swollen mitochondria and apoptotic bodies and a more intact endothelium and basement membrane (Van Linthout et al. 2008). These findings and the presence of activated Akt in cardiomyocytes, as well as in cardiac endothelial cells documented via immunofluorescence staining, underscore the importance of Akt activation after *apo A-I* transfer in the reduction of cardiomyocyte apoptosis as well as in the improvement in endothelial integrity. The ameliorated integrity of the endothelium suggests a potential restoration of the microvascular homeostasis, which has been demonstrated to reduce cardiomyocyte apoptosis and to result in the recovery of cardiac function in diabetic cardiomyopathy (Yoon et al. 2005). The significance of the cardiac endothelium on the contractile state and Ca^{2+} handling of subjacent cardiomyocytes (Brutsaert 2003; Nishida et al. 1993; Ramaciotti et al. 1992) suggests that part of the HDL-mediated improvement in left ventricular function in rats with streptozotocin-induced diabetes was indirectly due to their protective effect on the cardiac endothelium. This hypothesis is supported by the improvement in endothelial function found after *apo A-I* transfer in streptozotocin-induced diabetic rats (Van Linthout et al. 2009). In addition, HDL supplementation on isolated cardiomyocytes directly improved their contractility under hyperglycaemia-induced stress in a PI3K- and NO-dependent manner (Van Linthout et al. 2008).

Besides the direct HDL-mediated cardiomyocyte-protective effects and the endothelial-protective features, the decrease in cardiac apoptosis after *apo A-I* transfer in diabetic rats might be attributed to several other in parallel HDL-mediated triggered processes: (1) HDL may act as biological buffers capable of rapidly removing active proapoptotic TNF-α from the heart; (2) they have the potential to increase (Rossoni et al. 2004), stabilise, and activate prostaglandins (Aoyama et al. 1990). This enhanced prostanoid availability/activity may contribute to the HDL-mediated cardioprotection, by acting directly on cardiomyocytes (Zacharowski et al. 1999) and/or by inhibiting cardiac TNF-α production (Shinomiya et al. 2001); (3) HDL may reduce the expression of TLR 4 (Van Linthout et al. 2011), which has recently been shown to play an important role in cardiac apoptosis in diabetic cardiomyopathy.

In conclusion, *apo A-I* transfer attenuates the development of experimental diabetic cardiomyopathy via its anti-oxidative, anti-inflammatory, antifibrotic, and antiapoptotic actions (Fig. 2). Beyond beneficial vascular/cardioprotective long-term effects, direct myocardial effects of HDL may contribute to the improvement of cardiac function under severe streptozotocin-induced stress. Further studies are required to investigate the potential of *apo A-I* transfer to ameliorate established diabetic cardiomyopathy, especially in the context of type 2 diabetes mellitus, and to improve the function of HDL, which is impaired under diabetic conditions (Hedrick et al. 2000; Persegol et al. 2006; Sorrentino et al. 2009; Zheng et al. 2004). Finally, also the impact of the apo A-I/HDL-induced immunomodulatory effects on diabetes-associated endothelial dysfunction and cardiac fibrosis needs further clarification.

5   HDL and Tissue Repair: Modulation of EPC Biology via SR-BI

The term EPC is most often used in a very broad sense. There is no consensus definition of EPCs, which is a major source of confusion in the literature. EPCs encompass different categories of cells with different phenotypic and functional properties that affect neovascularisation and reendothelialisation. In a narrow, literally correct, and unambiguous sense, EPCs are immature precursor cells capable of differentiating into mature endothelial cells in vivo (Hagensen et al. 2010). Thus, an EPC sensu stricto corresponds to an in vivo category and is defined from a functional point of view. However, there are very significant methodological challenges to detect incorporation of circulating progenitor cells in the endothelium (Hagensen et al. 2010). Incorporation of EPCs is in general a relatively rare event (Feng et al. 2008, 2009a, b; Hagensen et al. 2010). High-resolution multichannel sequential confocal scanning microscopy provides a platform for colocalisation analysis with high specificity, which is required in *lege artis* cell tracking studies (De Geest 2009). Alternatively, EPCs may act in a paracrine way by releasing angiogenic factors and proteases to stimulate sprouting of local vessels or promote reendothelialisation indirectly (Rehman et al. 2003; Takakura et al. 2000; Urbich

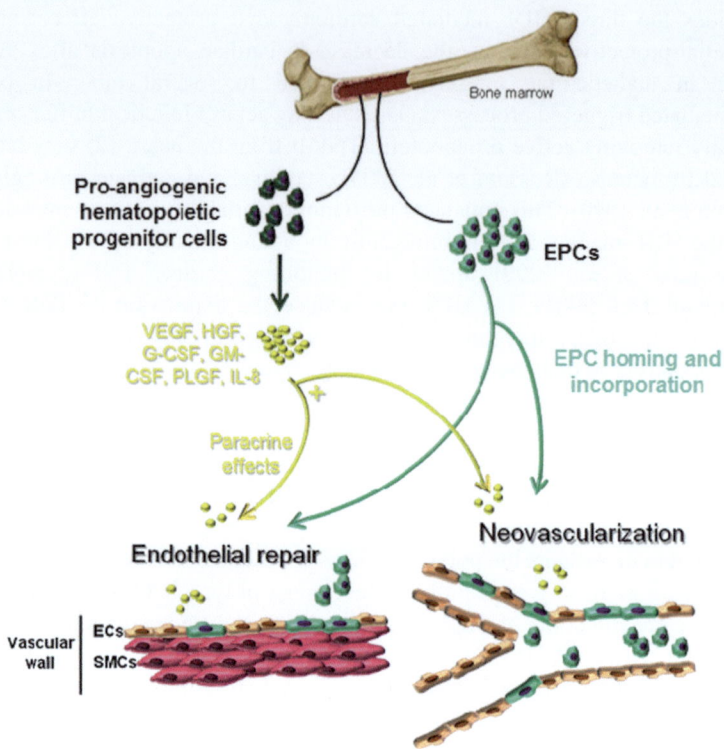

Fig. 3 Role of bone marrow-derived proangiogenic haematopoietic progenitor cells and EPCs sensu stricto in endothelial repair and neovascularisation. Proangiogenic haematopoietic progenitor cells, the in vivo equivalent of 'early EPCs', contribute to vascularisation and endothelial repair indirectly by secreting paracrine factors such as VEGF, hepatocyte growth factor (HGF), granulocyte colony-stimulating factor (G-CSF), granulocyte-macrophage colony-stimulating factor (GM-CSF), placental growth factor (PLGF), and IL-8. These factors stimulate local endothelial cell proliferation. On the other hand, EPCs sensu stricto enhance neovascularisation and endothelial repair directly by incorporation into the endothelium. EPCs sensu stricto can be considered to constitute the in vivo equivalent of ECFCs or BOECs ('late outgrowth EPCs'). *ECs* endothelial cells; *SMCs* smooth muscle cells

et al. 2005). To make a distinction with EPCs sensu stricto, these cells may be designated as proangiogenic haematopoietic progenitor cells (Fig. 3).

Different protocols for short-term culture of blood mononuclear cells on fibronectin (with or without gelatin)-coated plates have been established by Vasa et al. (2001) and Hill et al. (2003). These cells are stimulated by in vitro culture conditions to mimic many features of endothelial cells (Hirschi et al. 2008). However, the putative EPCs identified by these assays do not give rise to a lineage of endothelial progeny, but cultured cells in these assays consist of monocytes or a population of haematopoietic cells with monocyte-macrophage or T-cell lineage commitment. Nevertheless, these so-called early EPCs or early outgrowth EPCs may regulate the angiogenic response in a paracrine way (Hirschi et al. 2008). In contrast, endothelial colony-forming cells (ECFCs) (also called 'late outgrowth EPCs' or 'blood outgrowth endothelial cells' (BOECs)) are immature precursor cells capable of differentiating into mature endothelial cells in vivo. They are derived from long-term culture of adherent peripheral blood mononuclear cells in endothelial conditions ('late outgrowth EPCs') and consist of colonies that display a cobblestone morphology (Ingram et al. 2004). ECFCs express cell surface antigens like primary endothelium, clonally propagate, can be replated into secondary and tertiary ECFCs, form capillary-like structures in vitro, and become endothelial cells in vivo (Ingram et al. 2004). ECFCs display postnatal vasculogenic activity upon transplantation in a matrix scaffold (Richardson and Yoder 2011). ECFCs could be considered to be similar to EPCs sensu stricto: they have a potential for postnatal vasculogenesis and endothelial repair at sites of endothelial damage. However, one should always consider that cell categories that are defined based on in vitro cell culture conditions have always a rather artificial character.

HDLs enhance EPC-mediated repair (Feng et al. 2008, 2009a, b; Tso et al. 2006). The effect of HDL on EPC number, function, and incorporation occurs via SR-BI (Feng et al. 2009b). Since HDLs have been demonstrated to increase vascular endothelial growth factor (VEGF) production by early EPCs in vitro, HDLs may improve reendothelialisation by its effects on these circulating proangiogenic cells (Feng et al. 2011). On the other hand, HDLs also improve ECFC function in vitro, and these effects are dependent on signalling via SR-BI, ERKs, and NO and on increased $\beta 1$ integrin expression (Feng et al. 2011). Taken together, HDL exerts potent effects on different functional categories of EPCs via SR-BI.

The endothelial-protective and endothelial-reparative properties of HDL may be pathophysiologically and clinically relevant under conditions of high endothelial turnover. In healthy individuals, the endothelial layer is renewed at a low replication rate of 0–1 % per day (Erdbruegger et al. 2006). In pathological conditions of arteriosclerosis, endothelial turnover may be higher. Arteriosclerosis is a broad term that encompasses all diseases that lead to arterial hardening, including native atherosclerosis, post-injury neointima formation, cardiac allograft vasculopathy (CAV), and vein graft atherosclerosis. Areas of low shear stress in human arteries have an increased rate of endothelial turnover (Tricot et al. 2000). Augmented endothelial turnover is also observed in atherosclerosis-prone areas in $apo\ E^{-/-}$ mice (Foteinos et al. 2008). However, the highest degree of endothelial turnover is likely observed after arterial injury, in allografts (Rahmani et al. 2006), and in vein grafts. Although there is significant evidence that HDLs exert beneficial effects in models of angioplasty- or injury-induced arteriosclerosis (Ameli et al. 1994; De Geest et al. 1997; Shah and Amin 1992), we will focus here on allograft vasculopathy and vein graft atherosclerosis.

Orthotopic heart transplantation is a well-established therapy for selected patients with end-stage congestive heart failure. The long-term success of heart transplantation is limited by CAV. EPC-mediated repair may inhibit the progression of CAV. Although CAV is primarily the result of chronic rejection, non-immunological factors are modifiers of CAV progression. In the 'response to injury' concept of CAV, vascular lesions are considered to be the result of cumulative endothelial injury both by alloimmune responses and by non-alloimmune insults (Vassalli et al. 2003). T-cell alloimmunity, antibody-mediated immune attack, and non-immune factors induce endothelial cell death or endothelial dysfunction. Since the endothelium regulates vascular tone, inflammation, smooth muscle cell proliferation, and thrombosis (Behrendt and Ganz 2002), restoration of endothelial integrity and function is pivotal to attenuate the development of CAV (Pinney and Mancini 2004). Markers of endothelial injury have been shown to discriminate between CAV-positive and CAV-negative heart transplant recipients (Singh et al. 2012), which is consistent with a postulated pivotal role of endothelial injury in the pathogenesis of CAV.

Current strategies to prevent or treat CAV are clearly insufficient. Disruption of the endothelial lining due to endothelial cell apoptosis can be restored by proliferation of adjacent endothelial cells but also by incorporation of circulating bone marrow-derived EPCs (Friedrich et al. 2006; Walter et al. 2002; Werner et al. 2002, 2003). Hu et al. (2003) have demonstrated that EPCs contribute to endothelial regeneration in murine allografts. Moreover, increased HDL induced by hepatocyte-directed human *apo A-I* gene transfer in C57BL/6 *apo E*$^{-/-}$ mice enhanced the incorporation of bone marrow-derived EPCs into the injured endothelium of allografts from donor BALB/c mice and reduced neointima formation in these allografts (Feng et al. 2008). In subsequent experiments using BALB/c allografts in wild-type C57BL/6 mice, it was demonstrated that increased EPC incorporation and attenuation of allograft vasculopathy induced by hepatocyte-directed human *apo A-I* gene transfer are strictly dependent on SR-BI expression in bone marrow and bone marrow-derived cells. This conclusion was reached based on experiments in mice transplanted with wild-type bone marrow or alternatively with SR-BI$^{-/-}$ bone marrow. These data suggest that the effect of HDL on EPC incorporation and/or paracrine effects of EPCs are critical for the attenuation of allograft vasculopathy in this model. Nevertheless, the implications of these data for clinical CAV are unknown since murine experiments are performed in the absence of immunosuppressive therapies.

Surgical revascularisation using autologous vein grafts is limited by vein graft failure, which is frequently due to an aggressive form of atherosclerosis (Campeau et al. 1984). Vein graft failure is observed in 30–50 % of coronary bypass grafts within 10 years (Fitzgibbon et al. 1996). Hypercholesterolaemia is a risk factor for vein graft failure (Goldman et al. 2004), which may be related to adverse effects on endothelial integrity and function (Raja et al. 2004). Previously, it has been demonstrated that endothelial regeneration is retarded, and neointima formation is accelerated in vein grafts in hypercholesterolaemic *apo E*$^{-/-}$ mice compared to normocholesterolaemic controls (Dietrich et al. 2000; Xu et al. 2003). Topical HDL therapy reduced vein graft atherosclerosis in *apo E*$^{-/-}$ mice (Feng et al. 2011). Topical HDL application involves formulation of HDL in 20 % pluronic F-127 gel (pH 7.2) (Feng et al. 2011). Pluronic F-127 is a biocompatible and non-toxic substance and is characterised by thermoreversible gel formation at temperatures above 21 °C (Hu et al. 1999). One of the main advantages of topical HDL therapy is that the 'distribution volume' of the therapeutic agent is small compared to systemic administration. In addition, the extracellular protein concentration of HDL is 300–400 µg/ml (Parini et al. 2006; Sloop et al. 1987). This is approximately 20 % of the plasma protein concentration. Therefore, therapeutic effects of topical HDL can be expected at HDL concentrations that are substantially lower than the plasma concentration. Topical HDL administration on the adventitial side of vein grafts improved vein graft patency and function (Feng et al. 2011). Caval veins of C57BL/6 *apo E*$^{-/-}$ mice were grafted to the right carotid arteries of recipient C57BL/6 *TIE2-LacZ/apo E*$^{-/-}$ mice. HDL in 20 % pluronic F-127 gel was applied on the adventitial side of vein grafts. Topical HDL application reduced intimal area by 55 % ($p < 0.001$) at day 28 compared to control mice. Blood flow quantified by micro-magnetic resonance imaging at day 28 was 2.8-fold ($p < 0.0001$) higher in grafts of topical HDL-treated mice than in control mice. Topical HDL potently reduced intimal inflammation and resulted in enhanced endothelial regeneration as evidenced by an increase in the number of CD31-positive endothelial cells. As stated supra, HDL potently enhanced migration and adhesion of ECFCs in vitro, and these effects were dependent on signalling via SR-BI, ERK, and NO and on increased β1-integrin expression. Correspondingly, the number of CD31 β-galactosidase double-positive cells, reflecting incorporated circulating progenitor cells, was 3.9-fold ($p < 0.01$) higher in grafts of HDL-treated mice than in control grafts. Importantly, the effects of topical HDL therapy on vein graft atherosclerosis were similar compared to the effects of systemically increased HDL cholesterol after human *apo A-I* gene transfer. Taken together, topical HDL application is a new paradigm of HDL therapy.

Whereas this section was focused on HDL- and EPC-mediated endothelial repair, the next two sections will deal with HDL and tissue repair. In these therapeutic areas, the effect of HDL on EPCs may be an important mediator of the beneficial effects of HDL-targeted therapies.

6 Development of Topical HDL Therapy for Cutaneous Wound Healing

Wound healing results from complex interactions between extracellular matrix molecules, soluble mediators, resident skin cells, and infiltrating leukocytes as well as infiltrating EPCs. Rather artificially, wound healing can be divided in an inflammation phase, a phase of tissue formation with accumulation of granulation tissue and reepithelialisation, and finally a phase of tissue remodelling (Diegelmann and Evans 2004).

Granulation tissue contains tissue macrophages, fibroblasts, numerous new vessels, and extracellular matrix molecules. Macrophages stimulate fibroplasia and angiogenesis by secretion of various growth factors. Fibroblasts deposit and remodel the extracellular matrix of wounds. Blood vessels are needed to provide oxygen and nutrients, whereas extracellular matrix provides a conduit for cell migration and cell ingrowth. Formation of granulation tissue starts a few days after injury and is mainly triggered by growth factors such as platelet-derived growth factor and transforming growth factor-β, which stimulate fibroblast proliferation and migration (Singer and Clark 1999). Fibroblasts start to deposit and remodel extracellular matrix, which initially consists of fibronectin and hyaluronan but is later on replaced by proteoglycans and type I and III collagens (Toriseva and Kahari 2009). The formation of new blood vessels is a critical component of to granulation tissue formation. Angiogenesis is initially induced by tissue destruction and hypoxia and is subsequently stimulated by various molecules such as basic fibroblast growth factor, VEGF, and transforming growth factor-β, secreted by macrophages, keratinocytes, and endothelial cells. A role of EPCs in granulation tissue formation is directly suggested by experiments demonstrating that cell therapy with EPCs enhances wound healing in mice (Suh et al. 2005). The release of various growth factors such as VEGF and platelet-derived growth factor by EPCs appears to enhance wound healing (Suh et al. 2005). Reepithelialisation is the term used to describe the appearance of a new epithelial layer on top of a healing skin wound. It is dependent on the formation of a provisional wound bed matrix and involves the migration and proliferation of keratinocytes, the differentiation and stratification of new epithelium, and finally the reformation of the basement membrane.

Typical chronic cutaneous wounds are diabetic ulcers, ischaemic ulcers, and pressure ulcers. Diabetic ulcers occur in patients with type 1 and type 2 diabetes, whereas ischaemic ulcers are observed in patients with peripheral arterial diseases. Pressure ulcers occur in patients with paralysis or in other conditions that inhibit movement of body parts. Pressure ulcers or decubitus wounds typically occur on sacrum, shoulder blades, and heels. Several factors contribute to deficient wound healing in patients with diabetes: deficient growth factor production (Galkowska et al. 2006), impaired neovascularisation (Galiano et al. 2004b), attenuated keratinocyte and fibroblast proliferation and migration (Gibran et al. 2002), and altered balance between extracellular matrix accumulation and remodelling of the extracellular matrix by matrix metalloproteinases (Lobmann et al. 2002). Foot ulcers occur in a diabetic population with a prevalence of 5 % and lifetime incidence of 15 % (Abbott et al. 2002; Muller et al. 2002). Even with *lege artis* treatment, amputation is required in 14–24 % of cases. Every 30 s, a lower limb is lost due to diabetes. Approximately 85 % of these amputations are preceded by an ulcer. Mortality is up to 50 % in the first 3 years after amputation. Various studies show that costs to treat diabetic foot ulcers are extremely high, especially if hospitalisation is required (Matricali et al. 2007).

HDL may beneficially affect wound healing by accelerating resolution of inflammation, by enhancing granulation tissue formation involving increased EPC incorporation and increased paracrine effects of EPCs, and by accelerating reepithelialisation. Keratinocytes express SR-BI, and SR-BI expression is upregulated in dividing keratinocytes (Tsuruoka et al. 2002). Whereas in humans healing is primarily the result of reepithelialisation and granulation tissue formation, wound healing in mice occurs predominantly by wound contraction (Greenhalgh 2003). In the excisional wound healing model (Galiano et al. 2004a), a circular full-thickness wound is applied on the back of each mouse. Subsequently, a silicone splint is fixed around the wound with nylon sutures to counteract wound contraction. Consequently, wound healing in this model occurs by granulation tissue formation and reepithelialisation from the border. Granulation tissue formation and reepithelialisation were significantly delayed in C57BL/6 apo $E^{-/-}$ mice compared to C57BL/6 mice (Gordts et al. 2014). Topical administration of HDL (protein concentration 800 μg/ml; volume 80 μl) formulated in 20 % pluronic F-127 gel (pH 7.2) every 2 days on wounds in C57BL/6 apo $E^{-/-}$ mice significantly improved granulation tissue formation and reepithelialisation. Wound healing in mice treated with topical HDL therapy was very similar compared to C57BL/6 mice. Topical gel without HDL did not enhance wound healing. Improved wound healing induced by topical HDL therapy was also observed in C57BL/6 mice with streptozotocin-induced diabetes mellitus and in male type 2 diabetic leptin receptor-deficient mice (unpublished data).

Further preclinical studies are required to evaluate the robustness of this strategy in other models of delayed wound healing. However, murine models of diabetic wound healing differ in many respects to clinical diabetic wound healing. They do not take into account that there is a significant degree of heterogeneity of ulcers in patients with diabetes. Three categories can be discerned: 'purely neurotrophic' diabetic ulcers, 'purely ischaemic' ulcers in patients with diabetes, and poorly healing ulcers in diabetic patients with microangiopathy. The 'purely neurotrophic' diabetic ulcers heal well after off-loading. The 'purely ischaemic' ulcers should be treated with revascularisation strategies. Typically, diabetic ulcers in patients with more pronounced microangiopathy and macroangiopathy are characterised by defects in granulation tissue formation and constitute a potential therapeutic area for HDL-targeted therapies. However, no murine model can adequately mimic these chronic diabetic ulcers. Decubitus wounds constitute another target for topical HDL therapy but lack of adequate animal models hinders preclinical development.

7   Beneficial Effects of Selective HDL-Raising Gene Transfer on Cardiac Remodelling and Cardiac Function After Myocardial Infarction in Mice

Loss of myocardial tissue following acute myocardial infarction results in a decreased systolic ejection and increased left ventricular end-diastolic volume and pressure. The Frank-Starling mechanism, implying that an increased end-diastolic volume results in an increased pressure developed during systole, may help to restore cardiac output. However, the concomitant increased wall stress may induce regional hypertrophy in the non-infarcted segment, whereas in the infarcted area expansion and thinning may occur. Experimental animal studies show that the infarcted ventricle hypertrophies and that the degree of hypertrophy is dependent on the infarct size (Anversa and Sonnenblick 1990). Taken together, architectural remodelling is characterised by the formation of a discrete collagen scar, ventricular dilatation, and ventricular hypertrophy. This process may continue for weeks or months until the distending forces are counterbalanced by the tensile strength of the collagen scar. Remodelling post-myocardial infarction is complex since it involves the infarct area, the infarct border zone, and the non-infarcted myocardium (Sutton and Sharpe 2000). Following myocardial infarction, dyskinetic bulging may occur in infarct area. The myocardial fibres contiguous to this segment in the infarct border zone become exposed to a more pronounced increase in wall stress because of the more prominent change in the radius of curvature induced by the regional dilatation. Cardiomyocytes in the border zone of the infarct become larger than cardiomyocytes in the remote area of the ventricle. This is consistent with the view that the infarct-induced stress on the chamber walls is an important determinant of cardiomyocyte hypertrophy (Cohn 1993).

Post-infarct remodelling occurs in the setting of volume overload, since the stretched and dilated infarcted tissue increases the left ventricular volume. An increased ventricular volume not only implies increased preload (passive ventricular wall stress at the end of diastole) but also increased afterload (total myocardial wall stress during systolic ejection). Afterload is increased since the systolic radius is increased. Since both systolic and diastolic wall stress are increased, remodelling and hypertrophy post-myocardial infarction are characterised by mixed features of pressure overload and volume overload.

The myocardium consists of 3 integrated components: cardiomyocytes, extracellular matrix, and capillary microcirculation. All 3 components are involved in the remodelling process. The role of the extracellular matrix is distinct in the early phase of remodelling (within 72 h) and the late phase (beyond 72 h). Neutrophil infiltration of matrix metalloproteinases induces degradation of intermyocyte collagen struts and cardiomyocyte slippage. This leads to infarct expansion characterised by the disproportionate thinning and dilatation of the infarct segment (Erlebacher et al. 1984). Infarct expansion predisposes to myocardial rupture and congestive heart failure (Eaton et al. 1979; Erlebacher et al. 1982; Jugdutt and Michorowski 1987; Schuster and Bulkley 1979). Increased wall stress as a result of infarct expansion leads to mechanical stretch-elicited local angiotensin II release and activation of a fetal gene programme (Sutton and Sharpe 2000). Local angiotensin II release promotes cardiomyocyte hypertrophy. In later stages of remodelling, interstitial fibrosis is induced. Transforming growth factor-β1 transforms fibroblasts into myofibroblasts and induces activation of tissue inhibitors of metalloproteinases and production of type I and type III collagen. The resultant interstitial fibrosis negatively affects the diastolic properties of the heart.

Post-infarct ejection fraction is lower in patients with low HDL cholesterol levels (Kempen et al. 1987; Wang et al. 1998). Although this could be due to differences in atherosclerosis or in the microvasculature or could be related to a decrease in infarct size due to beneficial effects of HDL in IRI, the possibility that HDL beneficially affects infarct expansion and ventricular remodelling should be considered. This can be investigated in models of permanent ligation of, e.g. the left anterior descending coronary artery. An attenuation of post-infarct left ventricular remodelling and improved infarct healing was induced by infusion of rHDL once a week for 4 weeks in rats following ligation of the proximal left coronary artery

(Kiya et al. 2009). Interestingly, rHDL-treated rats also showed an increase of phosphorylation of ERK1/2 in the left ventricular tissue, but not of p38 MAPK or JNK (Kiya et al. 2009). Gordts et al. (2013) have recently shown that human *apo A-I* gene transfer in C57BL/6 LDL receptor-deficient mice increases survival, decreases infarct expansion, attenuates left ventricular dilatation, and improves cardiac function following permanent ligation of the left anterior descending coronary artery. Gene transfer in C57BL/6 LDL receptor-deficient mice was performed with the E1E3E4-deleted adenoviral vector AdA-I, inducing hepatocyte-specific expression of human apo A-I, or with the control vector Adnull. A ligation of the left anterior descending coronary artery was performed 2 weeks after transfer or saline injection. Permanent ligation of the left anterior descending coronary artery excludes salutary coronary effects of HDL on myocardial salvage as observed in models of IRI (Theilmeier et al. 2006). Consequently, the infarct area and the area at risk were nearly identical at 24 h after ligation implying that the initial increase in loading conditions was not different between human *apo A-I* gene transfer myocardial infarction mice and control myocardial infarction mice. HDL cholesterol levels were persistently 1.5 times ($p < 0.0001$) higher in AdA-I mice compared to controls. Survival was increased ($p < 0.01$) in AdA-I myocardial infarction mice compared to control myocardial infarction mice during the 28-day follow-up period (hazard ratio for mortality 0.42; 95 % CI 0.24–0.76). Longitudinal morphometric analysis demonstrated attenuated infarct expansion (reduced infarct length and increased infarct thickness of the infarct in AdA-I myocardial infarction mice) and inhibition of left ventricular dilatation in AdA-I myocardial infarction mice compared to controls. AdA-I transfer exerted immunomodulatory effects and increased neovascularisation in the infarct zone. Increased HDL after AdA-I transfer significantly improved systolic and diastolic cardiac function post-MI and led to a preservation of peripheral blood pressure.

The effects of AdA-I transfer on remodelling have significant consequences for the progressive development of heart failure. Ventricular dilatation increases the loading conditions of the heart. After all, 'preload' at the organ level corresponds to passive ventricular wall stress at the end of diastole (preload$_{LV}$ = (EDP$_{LV}$) (EDR$_{LV}$)/2WT$_{LV}$ where EDP is end-diastolic pressure, EDR is end-diastolic radius, and WT is wall thickness). 'Afterload' at the organ level reflects myocardial wall stress during systolic ejection (afterload$_{LV}$ = (SP$_{LV}$) (SR$_{LV}$)/2WT$_{LV}$ where SP is systolic pressure, SR is systolic radius, and WT is wall thickness). Thus, the attenuation of infarct expansion and ventricular dilatation induced by increased HDL is critical for long-term preservation of left ventricular function. Although an increase of end-diastolic volume may be physiologically beneficial in terms of the Frank–Starling mechanism, the increase in afterload in a larger ventricle is clearly not beneficial. The increase in afterload corresponds at the cellular level to an increase in tension that each muscle fibre must develop, and correspondingly, oxygen consumption increases. In addition, ventricular hypertrophy may be associated with an imbalance between the vascular and cardiomyocyte compartment in the myocardium (Shimizu et al. 2010; Tirziu et al. 2007). Increased oxygen consumption together with a decrease in relative vascularity may lead to tissue hypoxia, cardiomyocyte dysfunction, and late cardiomyocyte death. There was a trend for a higher capillary density and a higher relative vascularity at day 28 in AdA-I myocardial infarction mice compared to control myocardial infarction mice. Taken together, ventricular dilatation initially produced by an external stimulus (ligation of the left anterior descending coronary artery) induces a vicious cycle where an increase of loading conditions leads to an intrinsic pathological heart muscle (cardiomyopathy) characterised by cardiomyocyte death and dysfunction (reduced intrinsic myocardial contractility and relaxation), collagen deposition, and progressive dilatation of the heart. Although no follow-up was performed in the current study after 28 days, we speculate that differences in ventricular function between control MI mice and human *apo A-I* gene transfer MI mice will be more marked after longer follow-up.

Neovascularisation may play a role in scar formation and scar tissue remodelling (Wang et al. 2005). Increased neovascularisation was observed in the infarcts of AdA-I myocardial infarction mice at day 28 after gene transfer. HDL exerts potent effects on the endothelium. These effects occur via enhanced endothelial survival (Nofer et al. 2001), endothelial cell migration (Seetharam et al. 2006), and EPC-mediated repair (Feng et al. 2009a, b; Tso et al. 2006). Beneficial effects on EPCs may have contributed to improved infarct healing.

Taken together, attenuation of remodelling following human *apo A-I* gene transfer may be considered to constitute the cardiac equivalent of enhanced wound healing and tissue repair induced by HDL.

Conclusions

One of the pitfalls in the field of HDL-targeted interventions is the lack of selective HDL-raising drugs. The development of gene transfer technologies with a sufficiently high therapeutic index may pave the road for a selective and effective HDL-targeted therapeutic intervention (Jacobs et al. 2012; Van Craeyveld et al. 2010). Nathwani et al. (2011) performed a landmark study that demonstrated for the first time long-term expression of a transgene product at therapeutic levels following systemic hepatocyte-directed gene transfer in humans. In this study (Nathwani et al. 2011), haemophilia B was successfully treated with an adeno-associated viral serotype 8 human FIX expressing vector. These seminal data highlight that hepatocyte-directed adeno-associated viral serotype 8 gene transfer may become a clinical reality in the next decades. The recent positive assessment by the European Medicines Agency's Committee for Medicinal Products for Human Use of the marketing authorisation for the first gene transfer product in Europe (alipogene tiparvovec (Glybera®)) highlights an important paradigm shift by regulatory agencies, as well as the biotechnological entrepreneurs and investors. Nevertheless, there is no short-term perspective for widespread clinical use of gene transfer technologies.

Three modalities discussed in this review are selective HDL-targeted therapies: infusion of rHDL, human *apo A-I* gene transfer, and topical HDL therapy. The main strategic questions with regard to HDL-targeted therapies are related to the choice of therapeutic areas in which a real clinical difference could be made. These questions are inextricably linked to the current state of the art of evidence-based medicine. Since a significant part of this review is focused on heart failure, we will illustrate this point on the basis of a very significant dichotomous classification of these patients.

Among patients with hospitalised heart failure, approximately 50 % have heart failure with reduced ejection fraction (HFrEF), and 50 % have heart failure with preserved ejection fraction (HFpEF) with the proportion of patients with HFpEF increasing with time (Liu et al. 2013; Vaduganathan and Fonarow 2013). The cutoff value for preserved ejection fraction is 0.5. Most clinical heart failure trials have been focused on patients with HFrEF. Inhibition of the renin-angiotensin-aldosterone and sympathetic nervous systems improves survival and decreases hospitalisations in patients with HFrEF (Reed et al. 2014). In contrast to these significant advances in treatment and reduction in mortality in patients with HFrEF, drug strategies with strong evidence in HFrEF have proved unsuccessful in HFpEF, and the mortality in patients with HFpEF has remained unchanged (Liu et al. 2013). HFpEF is a complex clinical syndrome that is characterised by classical heart failure symptoms with increased left ventricular filling pressures and preserved left ventricular ejection fraction. Increased end-diastolic ventricular stiffness is observed in HFpEF, but the exact mechanisms that induce HFpEF are unknown. This heart failure subtype disproportionately affects women and the elderly and is commonly associated with other cardiovascular comorbidities, such as hypertension and diabetes. Most therapeutic gains can likely be made in the field HFpEF. Interestingly, human *apo A-I* gene transfer has been shown to improve diastolic function in C57BL/6 LDL receptor-deficient mice (Gordts et al. 2012). The same strategic questions should also be raised in relation to the role of HDL in prevention and treatment of coronary artery disease. One might speculate that perspectives for an HDL-targeted therapy in patients with stable coronary artery disease are limited, whereas beneficial coronary and myocardial effects in the setting of acute coronary syndromes may still constitute a window of opportunity.

Conflict of Interest None.

Acknowledgments Dr. Sophie Van Linthout received grant support from the European Foundation for the Study of Diabetes (EFSD). Research in the labs of Dr. Frias was supported by grants from the Fondation Gustave et Simone Prévot, Wolfermann Nageli Stiftung, Novartis Consumable Foundation, Jubiläumsstiftung, and Fondation pour la luttre contre le cancer et investigations médico-biologiques. Dr. Bart De Geest received grant support of the FWO-Vlaanderen (grants G.0529.10N and G0A3114N) and of the KU Leuven (grant OT/13/090). Dr. Frias and Dr. De Geest participated in and received financial support from COST Action BM0904 'HDL—from biological understanding to clinical exploitation'.

Open Access This chapter is distributed under the terms of the Creative Commons Attribution Noncommercial License, which permits any noncommercial use, distribution, and reproduction in any medium, provided the original author(s) and source are credited.

References

Abbott CA, Carrington AL, Ashe H, Bath S, Every LC, Griffiths J, Hann AW, Hussein A, Jackson N, Johnson KE, Ryder CH, Torkington R, Van Ross ER, Whalley AM, Widdows P, Williamson S, Boulton AJ (2002) The North-West Diabetes Foot Care Study: incidence of, and risk factors for, new diabetic foot ulceration in a community-based patient cohort. Diabet Med 19:377–384

Abderrahmani A, Niederhauser G, Favre D, Abdelli S, Ferdaoussi M, Yang JY, Regazzi R, Widmann C, Waeber G (2007) Human high-density lipoprotein particles prevent activation of the JNK pathway induced by human oxidised low-density lipoprotein particles in pancreatic beta cells. Diabetologia 50:1304–1314

Alvarez A, Cerda-Nicolas M, Naim Abu Nabah Y, Mata M, Issekutz AC, Panes J, Lobb RR, Sanz MJ (2004) Direct evidence of leukocyte adhesion in arterioles by angiotensin II. Blood 104:402–408

Ameli S, Hultgardh-Nilsson A, Cercek B, Shah PK, Forrester JS, Ageland H, Nilsson J (1994) Recombinant apolipoprotein A-I Milano reduces intimal thickening after balloon injury in hypercholesterolemic rabbits. Circulation 90:1935–1941

Anversa P, Sonnenblick EH (1990) Ischemic cardiomyopathy: pathophysiologic mechanisms. Prog Cardiovasc Dis 33:49–70

Aoyama T, Yui Y, Morishita H, Kawai C (1990) Prostaglandin I2 half-life regulated by high density lipoprotein is decreased in acute myocardial infarction and unstable angina pectoris. Circulation 81:1784–1791

Babitt J, Trigatti B, Rigotti A, Smart EJ, Anderson RG, Xu S, Krieger M (1997) Murine SR-BI, a high density lipoprotein receptor that mediates selective lipid uptake, is N-glycosylated and fatty acylated and colocalizes with plasma membrane caveolae. J Biol Chem 272:13242–13249

Baroukh N, Lopez CE, Saleh MC, Recalde D, Vergnes L, Ostos MA, Fiette L, Fruchart JC, Castro G, Zakin MM, Ochoa A (2004) Expression and secretion of human apolipoprotein A-I in the heart. FEBS Lett 557:39–44

Behrendt D, Ganz P (2002) Endothelial function. From vascular biology to clinical applications. Am J Cardiol 90:40L–48L

Berg AH, Combs TP, Du X, Brownlee M, Scherer PE (2001) The adipocyte-secreted protein Acrp30 enhances hepatic insulin action. Nat Med 7:947–953

Brutsaert DL (2003) Cardiac endothelial-myocardial signaling: its role in cardiac growth, contractile performance, and rhythmicity. Physiol Rev 83:59–115

Cai H (2005) NAD(P)H oxidase-dependent self-propagation of hydrogen peroxide and vascular disease. Circ Res 96:818–822

Cai L, Li W, Wang G, Guo L, Jiang Y, Kang YJ (2002) Hyperglycemia-induced apoptosis in mouse myocardium: mitochondrial cytochrome C-mediated caspase-3 activation pathway. Diabetes 51:1938–1948

Calabresi L, Rossoni G, Gomaraschi M, Sisto F, Berti F, Franceschini G (2003) High-density lipoproteins protect isolated rat hearts from ischemia-reperfusion injury by reducing cardiac tumor necrosis factor-alpha content and enhancing prostaglandin release. Circ Res 92:330–337

Campeau L, Enjalbert M, Lesperance J, Bourassa MG, Kwiterovich P Jr, Wacholder S, Sniderman A (1984) The relation of risk factors to the development of atherosclerosis in saphenous-vein bypass grafts and the progression of disease in the native circulation. A study 10 years after aortocoronary bypass surgery. N Engl J Med 311:1329–1332

Chalupsky K, Cai H (2005) Endothelial dihydrofolate reductase: critical for nitric oxide bioavailability and role in angiotensin II uncoupling of endothelial nitric oxide synthase. Proc Natl Acad Sci USA 102:9056–9061

Cheetham C, Collis J, O'Driscoll G, Stanton K, Taylor R, Green D (2000) Losartan, an angiotensin type 1 receptor antagonist, improves endothelial function in non-insulin-dependent diabetes. J Am Coll Cardiol 36:1461–1466

Christoffersen C, Obinata H, Kumaraswamy SB, Galvani S, Ahnstrom J, Sevvana M, Egerer-Sieber C, Muller YA, Hla T, Nielsen LB, Dahlback B (2011) Endothelium-protective sphingosine-1-phosphate provided by HDL-associated apolipoprotein M. Proc Natl Acad Sci USA 108:9613–9618

Cockerill GW, Huehns TY, Weerasinghe A, Stocker C, Lerch PG, Miller NE, Haskard DO (2001) Elevation of plasma high-density lipoprotein concentration reduces interleukin-1-induced expression of E-selectin in an in vivo model of acute inflammation. Circulation 103:108–112

Cohen RA (1993) Dysfunction of vascular endothelium in diabetes mellitus. Circulation 87:V67–V76

Cohn JN (1993) Post-MI remodeling. Clin Cardiol 16:II21–II24

Condorelli G, Morisco C, Stassi G, Notte A, Farina F, Sgaramella G, de Rienzo A, Roncarati R, Trimarco B, Lembo G (1999) Increased cardiomyocyte apoptosis and changes in proapoptotic and antiapoptotic genes bax and bcl-2 during left ventricular adaptations to chronic pressure overload in the rat. Circulation 99:3071–3078

Cong L, Chen K, Li J, Gao P, Li Q, Mi S, Wu X, Zhao AZ (2007) Regulation of adiponectin and leptin secretion and expression by insulin through a PI3K-PDE3B dependent mechanism in rat primary adipocytes. Biochem J 403:519–525

Cosson S, Kevorkian JP (2003) Left ventricular diastolic dysfunction: an early sign of diabetic cardiomyopathy? Diabetes Metab 29:455–466

De Geest B (2009) The origin of intimal smooth muscle cells: are we on a steady road back to the past? Cardiovasc Res 81:7–8

De Geest B, Zhao Z, Collen D, Holvoet P (1997) Effects of adenovirus-mediated human apo A-I gene transfer on neointima formation after endothelial denudation in apo E-deficient mice. Circulation 96:4349–4356

De Geest B, Stengel D, Landeloos M, Lox M, Le Gat L, Collen D, Holvoet P, Ninio E (2000) Effect of overexpression of human apo A-I in C57BL/6 and C57BL/6 apo E-deficient mice on 2 lipoprotein-associated enzymes, platelet-activating factor acetylhydrolase and paraoxonase. Comparison of adenovirus-mediated human apo A-I gene transfer and human apo A-I transgenesis. Arterioscler Thromb Vasc Biol 20:E68–E75

Dewald O, Frangogiannis NG, Zoerlein M, Duerr GD, Klemm C, Knuefermann P, Taffet G, Michael LH, Crapo JD, Welz A, Entman ML (2003) Development of murine ischemic cardiomyopathy is associated with a transient inflammatory reaction and depends on reactive oxygen species. Proc Natl Acad Sci USA 100:2700–2705

Diegelmann RF, Evans MC (2004) Wound healing: an overview of acute, fibrotic and delayed healing. Front Biosci 9:283–289

Dietrich H, Hu Y, Zou Y, Huemer U, Metzler B, Li C, Mayr M, Xu Q (2000) Rapid development of vein graft atheroma in ApoE-deficient mice. Am J Pathol 157:659–669

Eaton LW, Weiss JL, Bulkley BH, Garrison JB, Weisfeldt ML (1979) Regional cardiac dilatation after acute myocardial infarction: recognition by two-dimensional echocardiography. N Engl J Med 300:57–62

Erdbruegger U, Haubitz M, Woywodt A (2006) Circulating endothelial cells: a novel marker of endothelial damage. Clin Chim Acta 373:17–26

Erlebacher JA, Weiss JL, Eaton LW, Kallman C, Weisfeldt ML, Bulkley BH (1982) Late effects of acute infarct dilation on heart size: a two dimensional echocardiographic study. Am J Cardiol 49:1120–1126

Erlebacher JA, Weiss JL, Weisfeldt ML, Bulkley BH (1984) Early dilation of the infarcted segment in acute transmural myocardial infarction: role of infarct expansion in acute left ventricular enlargement. J Am Coll Cardiol 4:201–208

Feng Y, Jacobs F, Van Craeyveld E, Brunaud C, Snoeys J, Tjwa M, Van Linthout S, De Geest B (2008) Human ApoA-I transfer attenuates transplant arteriosclerosis via enhanced incorporation of bone marrow-derived endothelial progenitor cells. Arterioscler Thromb Vasc Biol 28:278–283

Feng Y, Van Craeyveld E, Jacobs F, Lievens J, Snoeys J, De Geest B (2009a) Wild-type apo A-I and apo A-I(Milano) gene transfer reduce native and transplant arteriosclerosis to a similar extent. J Mol Med 87:287–297

Feng Y, van Eck M, Van Craeyveld E, Jacobs F, Carlier V, Van Linthout S, Erdel M, Tjwa M, De Geest B (2009b) Critical role of scavenger receptor-BI-expressing bone marrow-derived endothelial progenitor cells in the attenuation of allograft vasculopathy after human apo A-I transfer. Blood 113:755–764

Feng Y, Gordts SC, Chen F, Hu Y, Van Craeyveld E, Jacobs F, Carlier V, Zhang Z, Xu Q, Ni Y, De Geest B (2011) Topical HDL administration reduces vein graft atherosclerosis in apo E deficient mice. Atherosclerosis 214:271–278

Fitzgibbon GM, Kafka HP, Leach AJ, Keon WJ, Hooper GD, Burton JR (1996) Coronary bypass graft fate and patient outcome: angiographic follow-up of 5,065 grafts related to survival and reoperation in 1,388 patients during 25 years. J Am Coll Cardiol 28:616–626

Foteinos G, Hu Y, Xiao Q, Metzler B, Xu Q (2008) Rapid endothelial turnover in atherosclerosis-prone areas coincides with stem cell repair in apolipoprotein E-deficient mice. Circulation 117:1856–1863

Freire CM, Moura AL, Barbosa Mde M, Machado LJ, Nogueira AI, Ribeiro-Oliveira A Jr (2007) Left ventricle diastolic dysfunction in diabetes: an update. Arq Bras Endocrinol Metabol 51:168–175

Frias MA, James RW, Gerber-Wicht C, Lang U (2009) Native and reconstituted HDL activate Stat3 in ventricular cardiomyocytes via ERK1/2: role of sphingosine-1-phosphate. Cardiovasc Res 82:313–323

Frias MA, Lang U, Gerber-Wicht C, James RW (2010) Native and reconstituted HDL protect cardiomyocytes from doxorubicin-induced apoptosis. Cardiovasc Res 85:118–126

Frias MA, Pedretti S, Hacking D, Somers S, Lacerda L, Opie LH, James RW, Lecour S (2013) HDL protects against ischemia reperfusion injury by preserving mitochondrial integrity. Atherosclerosis 228:110–116

Friedrich EB, Walenta K, Scharlau J, Nickenig G, Werner N (2006) CD34-/CD133+/VEGFR-2+ endothelial progenitor cell subpopulation with potent vasoregenerative capacities. Circ Res 98: e20–e25

Frustaci A, Kajstura J, Chimenti C, Jakoniuk I, Leri A, Maseri A, Nadal-Ginard B, Anversa P (2000) Myocardial cell death in human diabetes. Circ Res 87:1123–1132

Fryirs MA, Barter PJ, Appavoo M, Tuch BE, Tabet F, Heather AK, Rye KA (2010) Effects of high-density lipoproteins on pancreatic beta-cell insulin secretion. Arterioscler Thromb Vasc Biol 30:1642–1648

Galiano RD, Jt M, Dobryansky M, Levine JP, Gurtner GC (2004a) Quantitative and reproducible murine model of excisional wound healing. Wound Repair Regen 12:485–492

Galiano RD, Tepper OM, Pelo CR, Bhatt KA, Callaghan M, Bastidas N, Bunting S, Steinmetz HG, Gurtner GC (2004b) Topical vascular endothelial growth factor accelerates diabetic wound healing through increased angiogenesis and by mobilizing and recruiting bone marrow-derived cells. Am J Pathol 164:1935–1947

Galkowska H, Wojewodzka U, Olszewski WL (2006) Chemokines, cytokines, and growth factors in keratinocytes and dermal endothelial cells in the margin of chronic diabetic foot ulcers. Wound Repair Regen 14:558–565

Gibran NS, Jang YC, Isik FF, Greenhalgh DG, Muffley LA, Underwood RA, Usui ML, Larsen J, Smith DG, Bunnett N, Ansel JC, Olerud JE (2002) Diminished neuropeptide levels contribute to the impaired cutaneous healing response associated with diabetes mellitus. J Surg Res 108:122–128

Go AS, Mozaffarian D, Roger VL, Benjamin EJ, Berry JD, Borden WB, Bravata DM, Dai S, Ford ES, Fox CS, Franco S, Fullerton HJ, Gillespie C, Hailpern SM, Heit JA, Howard VJ, Huffman MD, Kissela BM, Kittner SJ, Lackland DT, Lichtman JH, Lisabeth LD, Magid D, Marcus GM, Marelli A, Matchar DB, McGuire DK, Mohler ER, Moy CS, Mussolino ME, Nichol G, Paynter NP, Schreiner PJ, Sorlie PD, Stein J, Turan TN, Virani SS, Wong ND, Woo D, Turner MB

(2013) Heart disease and stroke statistics–2013 update: a report from the American Heart Association. Circulation 127:e6–e245

Goldman S, Zadina K, Moritz T, Ovitt T, Sethi G, Copeland JG, Thottapurathu L, Krasnicka B, Ellis N, Anderson RJ, Henderson W (2004) Long-term patency of saphenous vein and left internal mammary artery grafts after coronary artery bypass surgery: results from a Department of Veterans Affairs Cooperative Study. J Am Coll Cardiol 44:2149–2156

Gordts SC, Van Craeyveld E, Muthuramu I, Singh N, Jacobs F, De Geest B (2012) Lipid lowering and HDL raising gene transfer increase endothelial progenitor cells, enhance myocardial vascularity, and improve diastolic function. PLoS ONE 7:e46849

Gordts SC, Muthuramu I, Nefyodova E, Jacobs F, Van Craeyveld E, De Geest B (2013) Beneficial effects of selective HDL raising gene transfer on survival, cardiac remodelling, and cardiac function after myocardial infarction in mice. Gene Ther 20:1053–1061

Gordts SC, Muthuramu I, Amin R, Jacobs F, De Geest B (2014) The impact of lipoproteins on wound healing: topical HDL therapy corrects delayed wound healing in apolipoprotein E deficient mice. Pharmaceuticals 7(4):419–432

Gragasin FS, Xu Y, Arenas IA, Kainth N, Davidge ST (2003) Estrogen reduces angiotensin II-induced nitric oxide synthase and NAD(P)H oxidase expression in endothelial cells. Arterioscler Thromb Vasc Biol 23:38–44

Greenhalgh DG (2003) Wound healing and diabetes mellitus. Clin Plast Surg 30:37–45

Gu SS, Shi N, Wu MP (2007) The protective effect of ApolipoproteinA-I on myocardial ischemia-reperfusion injury in rats. Life Sci 81:702–709

Guo Z, Xia Z, Jiang J, McNeill JH (2007) Downregulation of NADPH oxidase, antioxidant enzymes, and inflammatory markers in the heart of streptozotocin-induced diabetic rats by N-acetyl-L-cysteine. Am J Physiol Heart Circ Physiol 292:H1728–H1736

Hagensen MK, Shim J, Thim T, Falk E, Bentzon JF (2010) Circulating endothelial progenitor cells do not contribute to plaque endothelium in murine atherosclerosis. Circulation 121:898–905

Han R, Lai R, Ding Q, Wang Z, Luo X, Zhang Y, Cui G, He J, Liu W, Chen Y (2007) Apolipoprotein A-I stimulates AMP-activated protein kinase and improves glucose metabolism. Diabetologia 50:1960–1968

He S, Li M, Ma X, Lin J, Li D (2010) CD4$^+$CD25$^+$Foxp3$^+$ regulatory T cells protect the proinflammatory activation of human umbilical vein endothelial cells. Arterioscler Thromb Vasc Biol 30:2621–2630

Hedrick CC, Thorpe SR, Fu MX, Harper CM, Yoo J, Kim SM, Wong H, Peters AL (2000) Glycation impairs high-density lipoprotein function. Diabetologia 43:312–320

Hill JM, Zalos G, Halcox JP, Schenke WH, Waclawiw MA, Quyyumi AA, Finkel T (2003) Circulating endothelial progenitor cells, vascular function, and cardiovascular risk. N Engl J Med 348:593–600

Hink U, Li H, Mollnau H, Oelze M, Matheis E, Hartmann M, Skatchkov M, Thaiss F, Stahl RA, Warnholtz A, Meinertz T, Griendling K, Harrison DG, Forstermann U, Munzel T (2001) Mechanisms underlying endothelial dysfunction in diabetes mellitus. Circ Res 88:E14–E22

Hirschi KK, Ingram DA, Yoder MC (2008) Assessing identity, phenotype, and fate of endothelial progenitor cells. Arterioscler Thromb Vasc Biol 28:1584–1595

Hodgkinson CP, Laxton RC, Patel K, Ye S (2008) Advanced glycation end-product of low density lipoprotein activates the toll-like 4 receptor pathway implications for diabetic atherosclerosis. Arterioscler Thromb Vasc Biol 28:2275–2281

Hodroj W, Legedz L, Foudi N, Cerutti C, Bourdillon MC, Feugier P, Beylot M, Randon J, Bricca G (2007) Increased insulin-stimulated expression of arterial angiotensinogen and angiotensin type 1 receptor in patients with type 2 diabetes mellitus and atheroma. Arterioscler Thromb Vasc Biol 27:525–531

Hu Y, Zou Y, Dietrich H, Wick G, Xu Q (1999) Inhibition of neointima hyperplasia of mouse vein grafts by locally applied suramin. Circulation 100:861–868

Hu Y, Davison F, Zhang Z, Xu Q (2003) Endothelial replacement and angiogenesis in arteriosclerotic lesions of allografts are contributed by circulating progenitor cells. Circulation 108:3122–3127

Hughes SJ, Powis SH, Press M (2001) Surviving native beta-cells determine outcome of syngeneic intraportal islet transplantation. Cell Transplant 10:145–151

Hyka N, Dayer JM, Modoux C, Kohno T, Edwards CK 3rd, Roux-Lombard P, Burger D (2001) Apolipoprotein A-I inhibits the production of interleukin-1beta and tumor necrosis factor-alpha by blocking contact-mediated activation of monocytes by T lymphocytes. Blood 97:2381–2389

Imaizumi S, Miura S, Nakamura K, Kiya Y, Uehara Y, Zhang B, Matsuo Y, Urata H, Ideishi M, Rye KA, Sata M, Saku K (2008) Antiarrhythmogenic effect of reconstituted high-density lipoprotein against ischemia/reperfusion in rats. J Am Coll Cardiol 51:1604–1612

Ingram DA, Mead LE, Tanaka H, Meade V, Fenoglio A, Mortell K, Pollok K, Ferkowicz MJ, Gilley D, Yoder MC (2004) Identification of a novel hierarchy of endothelial progenitor cells using human peripheral and umbilical cord blood. Blood 104:2752–2760

Ismahil MA, Hamid T, Bansal SS, Patel B, Kingery JR, Prabhu SD (2013) Remodeling of the mononuclear phagocyte network underlies chronic inflammation and disease progression in heart failure: critical importance of the cardiosplenic axis. Circ Res 114(2):266–282

Iwaoka M, Obata JE, Abe M, Nakamura T, Kitta Y, Kodama Y, Kawabata K, Takano H, Fujioka D, Saito Y, Kobayashi T, Hasebe H, Kugiyama K (2007) Association of low serum levels of apolipoprotein A-I with adverse outcomes in patients with nonischemic heart failure. J Card Fail 13:247–253

Jacobs F, Gordts S, Muthuramu I, De Geest B (2012) The liver as a target organ for gene therapy: state of the art, challenges, and future perspectives. Pharmaceuticals 5:1372–1392

Jugdutt BI, Michorowski BL (1987) Role of infarct expansion in rupture of the ventricular septum after acute myocardial infarction: a two-dimensional echocardiographic study. Clin Cardiol 10:641–652

Karliner JS (2013) Sphingosine kinase and sphingosine 1-phosphate in the heart: a decade of progress. Biochim Biophys Acta 1831:203–212

Kempen HJ, van Gent CM, Buytenhek R, Buis B (1987) Association of cholesterol concentrations in low-density lipoprotein, high-density lipoprotein, and high-density lipoprotein subfractions, and of apolipoproteins AI and AII, with coronary stenosis and left ventricular function. J Lab Clin Med 109:19–26

Kiya Y, Miura S, Imaizumi S, Uehara Y, Matsuo Y, Abe S, Jimi S, Urata H, Rye KA, Saku K (2009) Reconstituted high-density lipoprotein attenuates postinfarction left ventricular remodeling in rats. Atherosclerosis 203:137–144

Kota SK, Kota SK, Jammula S, Panda S, Modi KD (2011) Effect of diabetes on alteration of metabolism in cardiac myocytes: therapeutic implications. Diabetes Technol Ther 13:1155–1160

Lakhani SA, Masud A, Kuida K, Porter GA Jr, Booth CJ, Mehal WZ, Inayat I, Flavell RA (2006) Caspases 3 and 7: key mediators of mitochondrial events of apoptosis. Science 311:847–851

Lee WJ, Ou HC, Hsu WC, Chou MM, Tseng JJ, Hsu SL, Tsai KL, Sheu WH (2010) Ellagic acid inhibits oxidized LDL-mediated LOX-1 expression, ROS generation, and inflammation in human endothelial cells. J Vasc Surg 52:1290–1300

Li D, Saldeen T, Romeo F, Mehta JL (2000) Oxidized LDL upregulates angiotensin II type 1 receptor expression in cultured human coronary artery endothelial cells: the potential role of transcription factor NF-kappaB. Circulation 102:1970–1976

Li J, Jin HB, Sun YM, Su Y, Wang LF (2010a) KB-R7943 inhibits high glucose-induced endothelial ICAM-1 expression and monocyte-endothelial adhesion. Biochem Biophys Res Commun 392:516–519

Li J, Zhu H, Shen E, Wan L, Arnold JM, Peng T (2010b) Deficiency of rac1 blocks NADPH oxidase activation, inhibits endoplasmic reticulum stress, and reduces myocardial remodeling in a mouse model of type 1 diabetes. Diabetes 59:2033–2042

Liu Y, Haddad T, Dwivedi G (2013) Heart failure with preserved ejection fraction: current understanding and emerging concepts. Curr Opin Cardiol 28:187–196

Lobmann R, Ambrosch A, Schultz G, Waldmann K, Schiweck S, Lehnert H (2002) Expression of matrix-metalloproteinases and their inhibitors in the wounds of diabetic and non-diabetic patients. Diabetologia 45:1011–1016

Lowes BD, Zolty R, Minobe WA, Robertson AD, Leach S, Hunter L, Bristow MR (2006) Serial gene expression profiling in the intact human heart. J Heart Lung Transplant 25:579–588

Mackness MI, Arrol S, Durrington PN (1991) Paraoxonase prevents accumulation of lipoperoxides in low-density lipoprotein. FEBS Lett 286:152–154

Marchesi M, Booth EA, Davis T, Bisgaier CL, Lucchesi BR (2004) Apolipoprotein A-IMilano and 1-palmitoyl-2-oleoyl phosphatidylcholine complex (ETC-216) protects the in vivo rabbit heart from regional ischemia-reperfusion injury. J Pharmacol Exp Ther 311:1023–1031

Marchesi M, Booth EA, Rossoni G, Garcia RA, Hill KR, Sirtori CR, Bisgaier CL, Lucchesi BR (2008) Apolipoprotein A-IMilano/POPC complex attenuates post-ischemic ventricular dysfunction in the isolated rabbit heart. Atherosclerosis 197:572–578

Matricali GA, Dereymaeker G, Muls E, Flour M, Mathieu C (2007) Economic aspects of diabetic foot care in a multidisciplinary setting: a review. Diabetes Metab Res Rev 23:339–347

Means CK, Brown JH (2009) Sphingosine-1-phosphate receptor signalling in the heart. Cardiovasc Res 82:193–200

Means CK, Miyamoto S, Chun J, Brown JH (2008) S1P1 receptor localization confers selectivity for Gi-mediated cAMP and contractile responses. J Biol Chem 283:11954–11963

Mehra MR, Uber PA, Lavie CJ, Milani RV, Park MH, Ventura HO (2009) High-density lipoprotein cholesterol levels and prognosis in advanced heart failure. J Heart Lung Transplant 28:876–880

Mochizuki S, Okumura M, Tanaka F, Sato T, Kagami A, Tada N, Nagano M (1991) Ischemia-reperfusion arrhythmias and lipids: effect of human high- and low-density lipoproteins on reperfusion arrhythmias. Cardiovasc Drugs Ther 5(Suppl 2):269–276

Montanari D, Yin H, Dobrzynski E, Agata J, Yoshida H, Chao J, Chao L (2005) Kallikrein gene delivery improves serum glucose and lipid profiles and cardiac function in streptozotocin-induced diabetic rats. Diabetes 54:1573–1580

Morel S, Frias MA, Rosker C, James RW, Rohr S, Kwak BR (2012) The natural cardioprotective particle HDL modulates connexin43 gap junction channels. Cardiovasc Res 93:41–49

Mosterd A, Hoes AW (2007) Clinical epidemiology of heart failure. Heart 93:1137–1146

Mosterd A, Hoes AW, de Bruyne MC, Deckers JW, Linker DT, Hofman A, Grobbee DE (1999) Prevalence of heart failure and left ventricular dysfunction in the general population. The Rotterdam study. Eur Heart J 20:447–455

Muller IS, de Grauw WJ, van Gerwen WH, Bartelink ML, van Den Hoogen HJ, Rutten GE (2002) Foot ulceration and lower limb amputation in type 2 diabetic patients in dutch primary health care. Diabetes Care 25:570–574

Murata N, Sato K, Kon J, Tomura H, Yanagita M, Kuwabara A, Ui M, Okajima F (2000) Interaction of sphingosine 1-phosphate with plasma components, including lipoproteins, regulates the lipid receptor-mediated actions. Biochem J 352(Pt 3):809–815

Nathwani AC, Tuddenham EG, Rangarajan S, Rosales C, McIntosh J, Linch DC, Chowdary P, Riddell A, Pie AJ, Harrington C, O'Beirne J, Smith K, Pasi J, Glader B, Rustagi P, Ng CY, Kay MA, Zhou J, Spence Y, Morton CL, Allay J, Coleman J, Sleep S, Cunningham JM, Srivastava D, Basner-Tschakarjan E, Mingozzi F, High KA, Gray JT, Reiss UM, Nienhuis AW, Davidoff AM (2011) Adenovirus-associated virus vector-mediated gene transfer in hemophilia B. N Engl J Med 365:2357–2365

Nishida M, Springhorn JP, Kelly RA, Smith TW (1993) Cell-cell signaling between adult rat ventricular myocytes and cardiac microvascular endothelial cells in heterotypic primary culture. J Clin Invest 91:1934–1941

Nishikawa T, Edelstein D, Du XL, Yamagishi S, Matsumura T, Kaneda Y, Yorek MA, Beebe D, Oates PJ, Hammes HP, Giardino I, Brownlee M (2000) Normalizing mitochondrial superoxide production blocks three pathways of hyperglycaemic damage. Nature 404:787–790

Nofer JR, Levkau B, Wolinska I, Junker R, Fobker M, von Eckardstein A, Seedorf U, Assmann G (2001) Suppression of endothelial cell apoptosis by high density lipoproteins (HDL) and HDL-associated lysosphingolipids. J Biol Chem 276:34480–34485

Nofer JR, van der Giet M, Tolle M, Wolinska I, von Wnuck LK, Baba HA, Tietge UJ, Godecke A, Ishii I, Kleuser B, Schafers M, Fobker M, Zidek W, Assmann G, Chun J, Levkau B (2004) HDL induces NO-dependent vasorelaxation via the lysophospholipid receptor S1P3. J Clin Invest 113:569–581

Nyby MD, Abedi K, Smutko V, Eslami P, Tuck ML (2007) Vascular Angiotensin type 1 receptor expression is associated with vascular dysfunction, oxidative stress and inflammation in fructose-fed rats. Hypertens Res 30:451–457

Oak JH, Cai H (2007) Attenuation of angiotensin II signaling recouples eNOS and inhibits nonendothelial NOX activity in diabetic mice. Diabetes 56:118–126

Ohtsu H, Suzuki H, Nakashima H, Dhobale S, Frank G, Motley E, Eguchi S (2006) Angiotensin II signal transduction through small GTP-binding proteins: mechanism and significance in vascular smooth muscle cells. Hypertension 48:534–540

Ou J, Wang J, Xu H, Ou Z, Sorci-Thomas MG, Jones DW, Signorino P, Densmore JC, Kaul S, Oldham KT, Pritchard KA Jr (2005) Effects of D-4F on vasodilation and vessel wall thickness in hypercholesterolemic LDL receptor-null and LDL receptor/apolipoprotein A-I double-knockout mice on Western diet. Circ Res 97:1190–1197

Ouchi N, Kihara S, Arita Y, Okamoto Y, Maeda K, Kuriyama H, Hotta K, Nishida M, Takahashi M, Muraguchi M, Ohmoto Y, Nakamura T, Yamashita S, Funahashi T, Matsuzawa Y (2000) Adiponectin, an adipocyte-derived plasma protein, inhibits endothelial NF-kappaB signaling through a cAMP-dependent pathway. Circulation 102:1296–1301

Ouchi N, Kihara S, Arita Y, Nishida M, Matsuyama A, Okamoto Y, Ishigami M, Kuriyama H, Kishida K, Nishizawa H, Hotta K, Muraguchi M, Ohmoto Y, Yamashita S, Funahashi T, Matsuzawa Y (2001) Adipocyte-derived plasma protein, adiponectin, suppresses lipid accumulation and class A scavenger receptor expression in human monocyte-derived macrophages. Circulation 103:1057–1063

Ouedraogo R, Gong Y, Berzins B, Wu X, Mahadev K, Hough K, Chan L, Goldstein BJ, Scalia R (2007) Adiponectin deficiency increases leukocyte-endothelium interactions via upregulation of endothelial cell adhesion molecules in vivo. J Clin Invest 117:1718–1726

Palomer X, Salvado L, Barroso E, Vazquez-Carrera M (2013) An overview of the crosstalk between inflammatory processes and metabolic dysregulation during diabetic cardiomyopathy. Int J Cardiol 168:3160–3172

Parini P, Johansson L, Broijersen A, Angelin B, Rudling M (2006) Lipoprotein profiles in plasma and interstitial fluid analyzed with an automated gel-filtration system. Eur J Clin Invest 36:98–104

Pereira RI, Draznin B (2005) Inhibition of the phosphatidylinositol 3′-kinase signaling pathway leads to decreased insulin-stimulated adiponectin secretion from 3T3-L1 adipocytes. Metabolism 54:1636–1643

Persegol L, Verges B, Foissac M, Gambert P, Duvillard L (2006) Inability of HDL from type 2 diabetic patients to counteract the inhibitory effect of oxidised LDL on endothelium-dependent vasorelaxation. Diabetologia 49:1380–1386

Peterson SJ, Husney D, Kruger AL, Olszanecki R, Ricci F, Rodella LF, Stacchiotti A, Rezzani R, McClung JA, Aronow WS, Ikehara S, Abraham NG (2007) Long-term treatment with the apolipoprotein A1 mimetic peptide increases antioxidants and vascular repair in type I diabetic rats. J Pharmacol Exp Ther 322:514–520

Peterson SJ, Drummond G, Kim DH, Li M, Kruger AL, Ikehara S, Abraham NG (2008) L-4F treatment reduces adiposity, increases adiponectin levels, and improves insulin sensitivity in obese mice. J Lipid Res 49:1658–1669

Piga R, Naito Y, Kokura S, Handa O, Yoshikawa T (2007) Short-term high glucose exposure induces monocyte-endothelial cells adhesion and transmigration by increasing VCAM-1 and MCP-1 expression in human aortic endothelial cells. Atherosclerosis 193:328–334

Pinney SP, Mancini D (2004) Cardiac allograft vasculopathy: advances in understanding its pathophysiology, prevention, and treatment. Curr Opin Cardiol 19:170–176

Raev DC (1994) Which left ventricular function is impaired earlier in the evolution of diabetic cardiomyopathy? An echocardiographic study of young type I diabetic patients. Diabetes Care 17:633–639

Rahmani M, Cruz RP, Granville DJ, McManus BM (2006) Allograft vasculopathy versus atherosclerosis. Circ Res 99:801–815

Raja SG, Haider Z, Ahmad M, Zaman H (2004) Saphenous vein grafts: to use or not to use? Heart Lung Circ 13:403–409

Ramaciotti C, Sharkey A, McClellan G, Winegrad S (1992) Endothelial cells regulate cardiac contractility. Proc Natl Acad Sci USA 89:4033–4036

Rathi S, Deedwania PC (2012) The epidemiology and pathophysiology of heart failure. Med Clin North Am 96:881–890

Reed BN, Rodgers JE, Sueta CA (2014) Polypharmacy in heart failure: drugs to use and avoid. Heart Fail Clin 10(4):577–590

Rehman J, Li J, Orschell CM, March KL (2003) Peripheral blood "endothelial progenitor cells" are derived from monocyte/macrophages and secrete angiogenic growth factors. Circulation 107:1164–1169

Richardson MR, Yoder MC (2011) Endothelial progenitor cells: quo vadis? J Mol Cell Cardiol 50 (2):266–272

Rius C, Abu-Taha M, Hermenegildo C, Piqueras L, Cerda-Nicolas JM, Issekutz AC, Estan L, Cortijo J, Morcillo EJ, Orallo F, Sanz MJ (2010) Trans- but not cis-resveratrol impairs angiotensin-II-mediated vascular inflammation through inhibition of NF-kappaB activation and peroxisome proliferator-activated receptor-gamma upregulation. J Immunol 185:3718–3727

Roger VL, Weston SA, Redfield MM, Hellermann-Homan JP, Killian J, Yawn BP, Jacobsen SJ (2004) Trends in heart failure incidence and survival in a community-based population. JAMA 292:344–350

Rossoni G, Gomaraschi M, Berti F, Sirtori CR, Franceschini G, Calabresi L (2004) Synthetic high-density lipoproteins exert cardioprotective effects in myocardial ischemia/reperfusion injury. J Pharmacol Exp Ther 308:79–84

Rubler S, Dlugash J, Yuceoglu YZ, Kumral T, Branwood AW, Grishman A (1972) New type of cardiomyopathy associated with diabetic glomerulosclerosis. Am J Cardiol 30:595–602

Rutti S, Ehses JA, Sibler RA, Prazak R, Rohrer L, Georgopoulos S, Meier DT, Niclauss N, Berney T, Donath MY, von Eckardstein A (2009) Low- and high-density lipoproteins modulate function, apoptosis, and proliferation of primary human and murine pancreatic beta-cells. Endocrinology 150:4521–4530

Sattler K, Levkau B (2009) Sphingosine-1-phosphate as a mediator of high-density lipoprotein effects in cardiovascular protection. Cardiovasc Res 82:201–211

Sattler KJ, Elbasan S, Keul P, Elter-Schulz M, Bode C, Graler MH, Brocker-Preuss M, Budde T, Erbel R, Heusch G, Levkau B (2010) Sphingosine 1-phosphate levels in plasma and HDL are altered in coronary artery disease. Basic Res Cardiol 105:821–832

Savvatis K, van Linthout S, Miteva K, Pappritz K, Westermann D, Schefold JC, Fusch G, Weithauser A, Rauch U, Becher PM, Klingel K, Ringe J, Kurtz A, Schultheiss HP, Tschope C (2012) Mesenchymal stromal cells but not cardiac fibroblasts exert beneficial systemic immunomodulatory effects in experimental myocarditis. PLoS ONE 7:e41047

Schulz R, Heusch G (2004) Connexin 43 and ischemic preconditioning. Cardiovasc Res 62:335–344

Schulz R, Boengler K, Totzeck A, Luo Y, Garcia-Dorado D, Heusch G (2007) Connexin 43 in ischemic pre- and postconditioning. Heart Fail Rev 12:261–266

Schuster EH, Bulkley BH (1979) Expansion of transmural myocardial infarction: a pathophysiologic factor in cardiac rupture. Circulation 60:1532–1538

Seetharam D, Mineo C, Gormley AK, Gibson LL, Vongpatanasin W, Chambliss KL, Hahner LD, Cummings ML, Kitchens RL, Marcel YL, Rader DJ, Shaul PW (2006) High-density lipoprotein promotes endothelial cell migration and reendothelialization via scavenger receptor-B type I. Circ Res 98:63–72

Shah PK, Amin J (1992) Low high density lipoprotein level is associated with increased restenosis rate after coronary angioplasty. Circulation 85:1279–1285

Shen X, Zheng S, Metreveli NS, Epstein PN (2006) Protection of cardiac mitochondria by overexpression of MnSOD reduces diabetic cardiomyopathy. Diabetes 55:798–805

Shimizu I, Minamino T, Toko H, Okada S, Ikeda H, Yasuda N, Tateno K, Moriya J, Yokoyama M, Nojima A, Koh GY, Akazawa H, Shiojima I, Kahn CR, Abel ED, Komuro I (2010) Excessive cardiac insulin signaling exacerbates systolic dysfunction induced by pressure overload in rodents. J Clin Invest 120:1506–1514

Shinomiya S, Naraba H, Ueno A, Utsunomiya I, Maruyama T, Ohuchida S, Ushikubi F, Yuki K, Narumiya S, Sugimoto Y, Ichikawa A, Oh-ishi S (2001) Regulation of TNFalpha and interleukin-10 production by prostaglandins I(2) and E(2): studies with prostaglandin receptor-deficient mice and prostaglandin E-receptor subtype-selective synthetic agonists. Biochem Pharmacol 61:1153–1160

Singer AJ, Clark RA (1999) Cutaneous wound healing. N Engl J Med 341:738–746

Singh N, Van Craeyveld E, Tjwa M, Ciarka A, Emmerechts J, Droogne W, Gordts SC, Carlier V, Jacobs F, Fieuws S, Vanhaecke J, Van Cleemput J, De Geest B (2012) Circulating apoptotic endothelial cells and apoptotic endothelial microparticles independently predict the presence of cardiac allograft vasculopathy. J Am Coll Cardiol 60:324–331

Sloop CH, Dory L, Roheim PS (1987) Interstitial fluid lipoproteins. J Lipid Res 28:225–237

Sorrentino SA, Besler C, Rohrer L, Meyer M, Heinrich K, Bahlmann FH, Mueller M, Horvath T, Doerries C, Heinemann M, Flemmer S, Markowski A, Manes C, Bahr MJ, Haller H, von Eckardstein A, Drexler H, Landmesser U (2009) Endothelial-vasoprotective effects of high-density lipoprotein are impaired in patients with type 2 diabetes mellitus but are improved after extended-release niacin therapy. Circulation 121(1):110–122

Suh W, Kim KL, Kim JM, Shin IS, Lee YS, Lee JY, Jang HS, Lee JS, Byun J, Choi JH, Jeon ES, Kim DK (2005) Transplantation of endothelial progenitor cells accelerates dermal wound healing with increased recruitment of monocytes/macrophages and neovascularization. Stem Cells 23:1571–1578

Susin SA, Zamzami N, Castedo M, Hirsch T, Marchetti P, Macho A, Daugas E, Geuskens M, Kroemer G (1996) Bcl-2 inhibits the mitochondrial release of an apoptogenic protease. J Exp Med 184:1331–1341

Sutton MG, Sharpe N (2000) Left ventricular remodeling after myocardial infarction: pathophysiology and therapy. Circulation 101:2981–2988

Suzuki K, Murtuza B, Sammut IA, Latif N, Jayakumar J, Smolenski RT, Kaneda Y, Sawa Y, Matsuda H, Yacoub MH (2002) Heat shock protein 72 enhances manganese superoxide dismutase activity during myocardial ischemia-reperfusion injury, associated with mitochondrial protection and apoptosis reduction. Circulation 106:I270–I276

Sztalryd C, Kraemer FB (1995) Regulation of hormone-sensitive lipase in streptozotocin-induced diabetic rats. Metabolism 44:1391–1396

Takakura N, Watanabe T, Suenobu S, Yamada Y, Noda T, Ito Y, Satake M, Suda T (2000) A role for hematopoietic stem cells in promoting angiogenesis. Cell 102:199–209

Tao R, Hoover HE, Honbo N, Kalinowski M, Alano CC, Karliner JS, Raffai R (2010) High-density lipoprotein determines adult mouse cardiomyocyte fate after hypoxia-reoxygenation through lipoprotein-associated sphingosine 1-phosphate. Am J Physiol Heart Circ Physiol 298:H1022–H1028

Theilmeier G, Schmidt C, Herrmann J, Keul P, Schafers M, Herrgott I, Mersmann J, Larmann J, Hermann S, Stypmann J, Schober O, Hildebrand R, Schulz R, Heusch G, Haude M, von Wnuck

LK, Herzog C, Schmitz M, Erbel R, Chun J, Levkau B (2006) High-density lipoproteins and their constituent, sphingosine-1-phosphate, directly protect the heart against ischemia/reperfusion injury in vivo via the S1P3 lysophospholipid receptor. Circulation 114:1403–1409

Thule PM, Campbell AG, Kleinhenz DJ, Olson DE, Boutwell JJ, Sutliff RL, Hart CM (2006) Hepatic insulin gene therapy prevents deterioration of vascular function and improves adipocytokine profile in STZ-diabetic rats. Am J Physiol Endocrinol Metab 290:E114–E122

Tirziu D, Chorianopoulos E, Moodie KL, Palac RT, Zhuang ZW, Tjwa M, Roncal C, Eriksson U, Fu Q, Elfenbein A, Hall AE, Carmeliet P, Moons L, Simons M (2007) Myocardial hypertrophy in the absence of external stimuli is induced by angiogenesis in mice. J Clin Invest 117:3188–3197

Tolle M, Pawlak A, Schuchardt M, Kawamura A, Tietge UJ, Lorkowski S, Keul P, Assmann G, Chun J, Levkau B, van der Giet M, Nofer JR (2008) HDL-associated lysosphingolipids inhibit NAD(P)H oxidase-dependent monocyte chemoattractant protein-1 production. Arterioscler Thromb Vasc Biol 28:1542–1548

Toriseva M, Kahari VM (2009) Proteinases in cutaneous wound healing. Cell Mol Life Sci 66:203–224

Tricot O, Mallat Z, Heymes C, Belmin J, Leseche G, Tedgui A (2000) Relation between endothelial cell apoptosis and blood flow direction in human atherosclerotic plaques. Circulation 101:2450–2453

Tschope C, Walther T, Escher F, Spillmann F, Du J, Altmann C, Schimke I, Bader M, Sanchez-Ferrer CF, Schultheiss HP, Noutsias M (2005) Transgenic activation of the kallikrein-kinin system inhibits intramyocardial inflammation, endothelial dysfunction and oxidative stress in experimental diabetic cardiomyopathy. Faseb J 19:2057–2059

Tso C, Martinic G, Fan WH, Rogers C, Rye KA, Barter PJ (2006) High-density lipoproteins enhance progenitor-mediated endothelium repair in mice. Arterioscler Thromb Vasc Biol 26:1144–1149

Tsuruoka H, Khovidhunkit W, Brown BE, Fluhr JW, Elias PM, Feingold KR (2002) Scavenger receptor class B type I is expressed in cultured keratinocytes and epidermis. Regulation in response to changes in cholesterol homeostasis and barrier requirements. J Biol Chem 277:2916–2922

Tunaru S, Kero J, Schaub A, Wufka C, Blaukat A, Pfeffer K, Offermanns S (2003) PUMA-G and HM74 are receptors for nicotinic acid and mediate its anti-lipolytic effect. Nat Med 9:352–355

Uchiyama T, Engelman RM, Maulik N, Das DK (2004) Role of Akt signaling in mitochondrial survival pathway triggered by hypoxic preconditioning. Circulation 109:3042–3049

Urbich C, Aicher A, Heeschen C, Dernbach E, Hofmann WK, Zeiher AM, Dimmeler S (2005) Soluble factors released by endothelial progenitor cells promote migration of endothelial cells and cardiac resident progenitor cells. J Mol Cell Cardiol 39:733–742

Vaduganathan M, Fonarow GC (2013) Epidemiology of hospitalized heart failure: differences and similarities between patients with reduced versus preserved ejection fraction. Heart Fail Clin 9:271–276, v

Van Craeyveld E, Gordts S, Jacobs F, De Geest B (2010) Gene therapy to improve high-density lipoprotein metabolism and function. Curr Pharm Des 16:1531–1544

Van Linthout S, Riad A, Dhayat N, Spillmann F, Du J, Dhayat S, Westermann D, Hilfiker-Kleiner D, Noutsias M, Laufs U, Schultheiss HP, Tschope C (2007) Anti-inflammatory effects of atorvastatin improve left ventricular function in experimental diabetic cardiomyopathy. Diabetologia 50:1977–1986

Van Linthout S, Spillmann F, Riad A, Trimpert C, Lievens J, Meloni M, Escher F, Filenberg E, Demir O, Li J, Shakibaei M, Schimke I, Staudt A, Felix SB, Schultheiss HP, De Geest B, Tschope C (2008) Human apolipoprotein A-I gene transfer reduces the development of experimental diabetic cardiomyopathy. Circulation 117:1563–1573

Van Linthout S, Spillmann F, Lorenz M, Meloni M, Jacobs F, Egorova M, Stangl V, De Geest B, Schultheiss HP, Tschope C (2009) Vascular-protective effects of high-density lipoprotein include the downregulation of the angiotensin II type 1 receptor. Hypertension 53:682–687

Van Linthout S, Foryst-Ludwig A, Spillmann F, Peng J, Feng Y, Meloni M, Van Craeyveld E, Kintscher U, Schultheiss HP, De Geest B, Tschope C (2010a) Impact of HDL on adipose tissue metabolism and adiponectin expression. Atherosclerosis 210:438–444

Van Linthout S, Spillmann F, Schultheiss HP, Tschope C (2010b) High-density lipoprotein at the interface of type 2 diabetes mellitus and cardiovascular disorders. Curr Pharm Des 16:1504–1516

Van Linthout S, Spillmann F, Graiani G, Miteva K, Peng J, Van Craeyveld E, Meloni M, Tolle M, Escher F, Subasiguller A, Doehner W, Quaini F, De Geest B, Schultheiss HP, Tschope C (2011) Down-regulation of endothelial TLR4 signalling after apo A-I gene transfer contributes to improved survival in an experimental model of lipopolysaccharide-induced inflammation. J Mol Med (Berl) 89:151–160

Vasa M, Fichtlscherer S, Adler K, Aicher A, Martin H, Zeiher AM, Dimmeler S (2001) Increase in circulating endothelial progenitor cells by statin therapy in patients with stable coronary artery disease. Circulation 103:2885–2890

Vassalli G, Gallino A, Weis M, von Scheidt W, Kappenberger L, von Segesser LK, Goy JJ (2003) Alloimmunity and nonimmunologic risk factors in cardiac allograft vasculopathy. Eur Heart J 24:1180–1188

Velagaleti RS, Massaro J, Vasan RS, Robins SJ, Kannel WB, Levy D (2009) Relations of lipid concentrations to heart failure incidence: the Framingham Heart Study. Circulation 120:2345–2351

Verges B, Petit JM, Duvillard L, Dautin G, Florentin E, Galland F, Gambert P (2006) Adiponectin is an important determinant of apoA-I catabolism. Arterioscler Thromb Vasc Biol 26:1364–1369

Walter DH, Rittig K, Bahlmann FH, Kirchmair R, Silver M, Murayama T, Nishimura H, Losordo DW, Asahara T, Isner JM (2002) Statin therapy accelerates reendothelialization: a novel effect involving mobilization and incorporation of bone marrow-derived endothelial progenitor cells. Circulation 105:3017–3024

Wang TD, Wu CC, Chen WJ, Lee CM, Chen MF, Liau CS, Sung FC, Lee YT (1998) Dyslipidemias have a detrimental effect on left ventricular systolic function in patients with a first acute myocardial infarction. Am J Cardiol 81:531–537

Wang B, Ansari R, Sun Y, Postlethwaite AE, Weber KT, Kiani MF (2005) The scar neovasculature after myocardial infarction in rats. Am J Physiol Heart Circ Physiol 289: H108–H113

Wang W, Xu H, Shi Y, Nandedkar S, Zhang H, Gao H, Feroah T, Weihrauch D, Schulte ML, Jones DW, Jarzembowski J, Sorci-Thomas M, Pritchard KA Jr (2010) Genetic deletion of apolipoprotein A-I increases airway hyperresponsiveness, inflammation, and collagen deposition in the lung. J Lipid Res 51:2560–2570

Werner N, Priller J, Laufs U, Endres M, Bohm M, Dirnagl U, Nickenig G (2002) Bone marrow-derived progenitor cells modulate vascular reendothelialization and neointimal formation: effect of 3-hydroxy-3-methylglutaryl coenzyme a reductase inhibition. Arterioscler Thromb Vasc Biol 22:1567–1572

Werner N, Junk S, Laufs U, Link A, Walenta K, Bohm M, Nickenig G (2003) Intravenous transfusion of endothelial progenitor cells reduces neointima formation after vascular injury. Circ Res 93:e17–e24

Westermann D, Rutschow S, Van Linthout S, Linderer A, Bucker-Gartner C, Sobirey M, Riad A, Pauschinger M, Schultheiss HP, Tschope C (2006) Inhibition of p38 mitogen-activated protein kinase attenuates left ventricular dysfunction by mediating pro-inflammatory cardiac cytokine levels in a mouse model of diabetes mellitus. Diabetologia 49:2507–2513

Westermann D, Rutschow S, Jager S, Linderer A, Anker S, Riad A, Unger T, Schultheiss HP, Pauschinger M, Tschope C (2007a) Contributions of inflammation and cardiac matrix metalloproteinase activity to cardiac failure in diabetic cardiomyopathy: the role of angiotensin type 1 receptor antagonism. Diabetes 56:641–646

Westermann D, Van Linthout S, Dhayat S, Dhayat N, Escher F, Bucker-Gartner C, Spillmann F, Noutsias M, Riad A, Schultheiss HP, Tschope C (2007b) Cardioprotective and anti-inflammatory effects of interleukin converting enzyme inhibition in experimental diabetic cardiomyopathy. Diabetes 56:1834–1841

Westermann D, Van Linthout S, Dhayat S, Dhayat N, Schmidt A, Noutsias M, Song XY, Spillmann F, Riad A, Schultheiss HP, Tschope C (2007c) Tumor necrosis factor-alpha antagonism protects from myocardial inflammation and fibrosis in experimental diabetic cardiomyopathy. Basic Res Cardiol 102:500–507

Westermann D, Lindner D, Kasner M, Zietsch C, Savvatis K, Escher F, von Schlippenbach J, Skurk C, Steendijk P, Riad A, Poller W, Schultheiss HP, Tschope C (2011) Cardiac inflammation contributes to changes in the extracellular matrix in patients with heart failure and normal ejection fraction. Circ Heart Fail 4:44–52

Whiteman EL, Cho H, Birnbaum MJ (2002) Role of Akt/protein kinase B in metabolism. Trends Endocrinol Metab 13:444–451

Wilhelm AJ, Zabalawi M, Owen JS, Shah D, Grayson JM, Major AS, Bhat S, Gibbs DP Jr, Thomas MJ, Sorci-Thomas MG (2010) Apolipoprotein A-I modulates regulatory T cells in autoimmune LDLr-/-, ApoA-I-/- mice. J Biol Chem 285(46):36158–36169

Wold LE, Ceylan-Isik AF, Fang CX, Yang X, Li SY, Sreejayan N, Privratsky JR, Ren J (2006) Metallothionein alleviates cardiac dysfunction in streptozotocin-induced diabetes: role of $Ca2^+$ cycling proteins, NADPH oxidase, poly(ADP-Ribose) polymerase and myosin heavy chain isozyme. Free Radic Biol Med 40:1419–1429

Xu Q, Zhang Z, Davison F, Hu Y (2003) Circulating progenitor cells regenerate endothelium of vein graft atherosclerosis, which is diminished in ApoE-deficient mice. Circ Res 93:e76–e86

Yoon YS, Uchida S, Masuo O, Cejna M, Park JS, Gwon HC, Kirchmair R, Bahlman F, Walter D, Curry C, Hanley A, Isner JM, Losordo DW (2005) Progressive attenuation of myocardial vascular endothelial growth factor expression is a seminal event in diabetic cardiomyopathy: restoration of microvascular homeostasis and recovery of cardiac function in diabetic cardiomyopathy after replenishment of local vascular endothelial growth factor. Circulation 111:2073–2085

Young NL, Lopez DR, McNamara DJ (1988) Contributions of absorbed dietary cholesterol and cholesterol synthesized in small intestine to hypercholesterolemia in diabetic rats. Diabetes 37:1151–1156

Zacharowski K, Olbrich A, Piper J, Hafner G, Kondo K, Thiemermann C (1999) Selective activation of the prostanoid EP(3) receptor reduces myocardial infarct size in rodents. Arterioscler Thromb Vasc Biol 19:2141–2147

Zhang Y, Peng T, Zhu H, Zheng X, Zhang X, Jiang N, Cheng X, Lai X, Shunnar A, Singh M, Riordan N, Bogin V, Tong N, Min WP (2010) Prevention of hyperglycemia-induced myocardial apoptosis by gene silencing of Toll-like receptor-4. J Transl Med 8:133

Zheng L, Nukuna B, Brennan ML, Sun M, Goormastic M, Settle M, Schmitt D, Fu X, Thomson L, Fox PL, Ischiropoulos H, Smith JD, Kinter M, Hazen SL (2004) Apolipoprotein A-I is a selective target for myeloperoxidase-catalyzed oxidation and functional impairment in subjects with cardiovascular disease. J Clin Invest 114:529–541

Part IV

Treatments for Dyslipidemias and Dysfunction of HDL

HDL and Lifestyle Interventions

Joan Carles Escolà-Gil, Josep Julve, Bruce A. Griffin, Dilys Freeman, and Francisco Blanco-Vaca

Contents

1	HDL and Diet	571
	1.1 Effects of Substituting Dietary Saturated Fatty Acids	571
	1.2 Dietary MUFA	572
	1.3 n-3 Polyunsaturated Fatty Acids	572
	1.4 Carbohydrate and Extrinsic Sugars	573
	1.5 Effects of Dietary Fatty Acids and Cholesterol on HDL Function	573
2	Weight Loss	574
	2.1 Impact of Adiposity on HDL Concentration and Function	574
	2.2 Effect of Weight-Loss Therapies on HDL Concentration and Function	575
3	Effects of Regular Aerobic Exercise	576
	3.1 Effects of Exercise on HDL	576
	3.2 Effects of Exercise on Prevention and Treatment of Cardiometabolic Risk	577
4	Smoking Cessation	578
	4.1 Smoking and HDL-C Concentration	579
	4.2 Potential Mechanisms for the Smoking-Related Reduction in HDL-C	579
	4.3 Smoking Cessation Intervention to Increase Plasma HDL Concentration	581

J.C. Escolà-Gil • J. Julve • F. Blanco-Vaca (✉)
Institut d'Investigació Biomèdica (IIB) Sant Pau, Barcelona, Spain

CIBER de Diabetes y Enfermedades Metabólicas Asociadas (CIBERDEM), Barcelona, Spain

Departament de Bioquímica i Biologia Molecular, Universitat Autònoma de Barcelona, Barcelona, Spain
e-mail: fblancova@santpau.cat

B.A. Griffin
Department of Nutritional Sciences, University of Surrey, Surrey, UK

D. Freeman
Institute of Cardiovascular and Medical Sciences, University of Glasgow, Glasgow, UK

5	Effects of Alcohol on HDL and Cardiovascular Risk	582
	5.1 Alcohol Intake and HDL-C	582
	5.2 Effects of Alcohol on HDL Atheroprotective Functions	583
References		584

Abstract

The main lifestyle interventions to modify serum HDL cholesterol include physical exercise, weight loss with either caloric restriction or specific dietary approaches, and smoking cessation. Moderate alcohol consumption can be permitted in some cases. However, as these interventions exert multiple effects, it is often difficult to discern which is responsible for improvement in HDL outcomes. It is particularly noteworthy that recent data questions the use of HDL cholesterol as a risk factor and therapeutic target since randomised interventions and Mendelian randomisation studies failed to provide evidence for such an approach. Therefore, these current data should be considered when reading and interpreting this review. Further studies are needed to document the effect of lifestyle changes on HDL structure–function and health.

Keywords

Alcohol • Exercise • HDL • Fat • Smoking • Weight

Abbreviations

ABC	Adenosine triphosphate-binding cassette transporter
Apo	Apolipoprotein
CVD	Cardiovascular disease
CETP	Cholesteryl ester transfer protein
HL	Hepatic lipase
HDL-C	High-density lipoprotein cholesterol
LCAT	Lecithin–cholesterol acyltransferase
LDL-C	Low-density lipoprotein cholesterol
LPL	Lipoprotein lipase
Lp-PLA2	Lipoprotein-associated phospholipase A2
MUFA	Monounsaturated fatty acids
PLTP	Phospholipid transfer protein
PON1	Paraoxonase 1
PUFA	Polyunsaturated fatty acids
RCT	Reverse cholesterol transport
SFA	Saturated fatty acids
SR-BI	Scavenger receptor class-BI
TAG	Triacylglycerol

1 HDL and Diet

The landmark cross-cultural and migration studies of Ancel Keys in the 1950s were the first to identify associations between serum HDL cholesterol (HDL-C) and diet and lifestyle factors in free-living populations (Keys 1980a). These early studies laid the foundation for the development of the diet–heart hypothesis, by revealing links between the incidence of coronary heart disease, raised serum cholesterol and energy derived from saturated fat. They also helped to establish diet as a determinant of low HDL in obesity (Keys 1980b). It has been established that certain dietary saturated fatty acids (SFA), with a chain length of between 12 and 16 carbons, and dietary cholesterol can increase both serum low-density lipoprotein cholesterol (LDL-C) and, somewhat paradoxically with respect to cardiovascular disease (CVD) risk, HDL-C (Grundy and Denke 1990). Conversely, the replacement of SFA with either polyunsaturated (PUFA) or monounsaturated fatty acids (MUFA) has been shown to reduce serum HDL-C. Moreover, the replacement of dietary fat with carbohydrate has also been shown to produce significant reductions in serum HDL-C (Katan et al. 1997; Stanhope et al. 2013). Heightened awareness of the risk associated with dietary sugars in the early 1970s (Yudkin 1972) has resurfaced recently, in part, through escalation of obesity-related cardiometabolic risk and increased understanding of how extrinsic sugars influence HDL by adversely affecting the metabolism of triacylglycerol (TAG) (Lustig 2010). The history of diet and HDL has evolved in the era of the LDL-lowering statins, which are relatively ineffective in raising HDL-C, and leave behind what is often described as untreated 'residual risk' (Belsey et al. 2008). This situation has created a need for alternative HDL-targeted drugs and also increased awareness of the importance of diet as a modifier of HDL structure and function.

1.1 Effects of Substituting Dietary Saturated Fatty Acids

Serum HDL-C has been shown to be influenced by both the amount and quality of dietary fatty acids and carbohydrate. In the past, the relative effects of dietary fatty acids on HDL-C have been described in absolute terms, that is, SFA tends to raise HDL-C, whilst PUFA, *trans* fatty acids and carbohydrate, all tend to reduce HDL-C, with MUFA being relatively neutral (Grundy and Denke 1990). However, in reality the absolute effect of fatty acids, or any macronutrient, on HDL-C cannot be measured, as its addition or replacement may be counter-affected by whatever fatty acid or carbohydrate takes its place to maintain a feasible diet. This phenomenon of substitution limits the ability to interpret the impact of dietary macronutrients on HDL and CVD risk to 'relative', rather than 'absolute' effects. Meta-analyses of intervention studies have provided strong evidence to show that the iso-energetic replacement of SFA with PUFA, MUFA and carbohydrate decreases HDL-C, with carbohydrate exerting the greatest impact in lowering both HDL-C and the total cholesterol–HDL-C ratio (Mensink et al. 2003). In contrast, the iso-energetic replacement of carbohydrate with all dietary fatty acids, other than *trans* fatty acids, tends to raise HDL-C (Micha and Mozaffarian 2010). Estimates for the relative magnitude of change in the concentration of serum HDL-C in response to these dietary substitutions are directly proportional to the amount of energy being exchanged. An iso-energetic exchange of 5 % energy from carbohydrate to SFA, MUFA and PUFA is associated with increases in HDL-C of 0.05, 0.04 and 0.03 mmol/L, respectively, with *trans* fatty acids reducing HDL by 0.02 mmol/L. In addition, individual SFA is known to exert differential effects on serum HDL-C, so that replacing 5 % energy as carbohydrate with lauric, myristic, palmitic and stearic, increases HDL-C by 0.13, 0.09, 0.05 and <0.01 mmol/L, respectively. These associations between dietary fatty acids and serum HDL-C are statistically robust and have been used to formulate dietary guidelines. However, because they rely heavily upon data from measures of dietary intake, as estimated from dietary recall, food frequency questionnaires or diet diaries, this can limit their value in predicting the biological effects of complex foods on HDL (Astrup et al. 2011). In other words, foods are not single nutrients, but complex mixtures of nutrients within a food matrix, all of which interact together to produce a biological effect. Because a food is rich in one particular fatty acid, it does not mean that consumption of this food will result in an effect on HDL that is typical of that fatty acid.

1.2 Dietary MUFA

Interest in the potential health benefits of dietary MUFA originated from the use of olive oil in the Mediterranean diet. Whilst substitution of SFA with MUFA is less effective in lowering LDL cholesterol than n-6 PUFA, there is evidence to suggest that this substitution may be more effective in preventing the decrease in HDL-C that accompanies the removal of SFA (Schwingshackl and Hoffmann 2012). This finding is supported by data to show that MUFA is relatively less effective in stimulating cholesteryl ester transfer protein (CETP), and thus the remodelling and increased clearance of HDL from serum, than either SFA or n-6 PUFA (Groener et al. 1991; Lagrost et al. 1999). This finding is in accord with data from a recent study which concluded that dietary MUFA reduced the catabolic rate of the principal apolipoprotein (apo) in HDL, apo A-I (Labonte et al. 2013).

1.3 n-3 Polyunsaturated Fatty Acids

The principal essential fatty acid of the n-3 series, α-linolenic acid (18:2) is the most abundant fatty acid on earth, but is consumed in significantly less quantity by humans than linoleic acid (18:2 n-6) (National Diet and Nutrition Survey. Department of Health 2011). When fed in physiologically relevant amounts, α-linolenic acid has been shown to be equivalent to linoleic acid as a substitute for SFA in lowering LDL cholesterol (Harris 1997). However, even though a recent meta-analysis indicated a benefit of α-linolenic acid intake on CVD risk (Pan et al. 2012), human interventions with α-linolenic acid-enriched diets have shown variable effects on serum HDL-C (Harper et al. 2006; Goyens and Mensink 2006; Kaul et al. 2008; Griffin et al. 2006). The longer-chain derivatives of α-linolenic acid, chiefly eicosapentaenoic and docosahexaenoic acids (EPA, DHA), which in humans are mainly obtained directly from oily fish, exert only a moderate elevating effect on HDL-C (Harris 1989). This is perhaps surprising given the potent TAG-lowering effect of these long-chain fatty acids.

1.4 Carbohydrate and Extrinsic Sugars

The replacement of dietary fat with carbohydrate in low-fat, high-carbohydrate diets has long been associated with a reduction in serum HDL-C that may be linked to the carbohydrate-induced increase in TAG (Katan et al. 1997). The latter is known to promote lipid exchanges between HDL- and TAG-rich lipoproteins that remodel HDL into smaller and denser particles with an increased catabolic rate and thus reduced residence time in serum. Adverse effects of carbohydrate on HDL-C in the longer term may also be mediated through increased body weight and the accumulation of body fat (Stanhope et al. 2013). Whilst the reduction in HDL-C has been attributed to diets with a high glycaemic index (Frost et al. 1999), there is now little doubt that the extrinsic sugars, sucrose and fructose make a major contribution to this effect (Lustig 2010). The findings of a recent meta-analysis which concluded that a very high intake of fructose (>100 g/d) increases serum LDL cholesterol, but has no significant effects on HDL-C, are somewhat surprising in both respects (Zhang et al. 2013) and in contrast to the outcome of dietary interventions with beverages sweetened with fructose. One example of the latter showed marked increases in cardiometabolic risk factors, including significant reductions in serum HDL-C, relative to glucose-sweetened beverages (Stanhope et al. 2009). A vital question to be answered is whether populations are consuming amounts of extrinsic sugars which are sufficient to elicit these adverse changes in HDL in the long term. Dietary intakes in the United Kingdom (National Diet and Nutrition Survey. Department of Health 2011) indicate that the upper 2.5th percentile of the population may be approaching intakes of extrinsic sugars which have the potential to induce adverse effects on metabolism and increase body weight. The overconsumption of sweetened beverages in adolescents is of particular concern in promoting premature obesity and lowering of HDL-C, as shown in a recent Australian study (Ambrosini et al. 2013).

1.5 Effects of Dietary Fatty Acids and Cholesterol on HDL Function

HDL-C is a surrogate marker of HDL particle size and number and may convey little or no information about the anti-atherosclerotic properties of HDL in the process of reverse cholesterol. There is evidence to suggest that paradoxical increase in HDL-C induced by dietary SFA and cholesterol in mice (Escola-Gil et al. 2011) and in egg-fed humans (Andersen et al. 2013) results in a beneficial increase in cholesterol efflux capacity. Beneficial effects of dietary fatty acids on cholesterol efflux capacity have also been described for EPA and DHA supplementation in hamsters (Kasbi Chadli et al. 2013) and MUFA-rich extra-virgin olive oil consumption in humans (Helal et al. 2013). Conversely, several intervention studies in humans showed no effect on cholesterol efflux capacity of replacing of SFA with either PUFA (Kralova Lesna et al. 2008) or carbohydrate (De Vries et al. 2005), or differences between diets enriched with *trans* fatty acids (8.3 % energy), SFA (13.2 % energy) and PUFA (14.6 %) elicited by either total plasma HDL or HDL subfractions (Buonacorso et al. 2007).

In conclusion, evidence from meta-analyses to support the relative effects of dietary fatty acids on HDL-C may be statistically incontrovertible, but they do not necessarily translate directly to the effects of complex foods and diets on HDL-C and CVD risk. Reduction in HDL-C induced by the overconsumption of dietary extrinsic sugars in sugar-sweetened beverages may have major implications for cardiometabolic health, especially in adolescents. Finally, evidence for the effects of dietary components on the anti-atherogenic, functional properties of HDL is inconclusive and warrants further study.

2 Weight Loss

Obesity, defined as an excess of body fat, is a major health problem in Western societies (Tchernof and Despres 2013; Bays et al. 2013). Excess body fat is also an independent risk factor for CVD (Tchernof and Despres 2013). Excessive adipose tissue accumulation, but more importantly its distribution in the body, particularly visceral obesity, has been identified as a main correlate of cardiometabolic disorders including not only alterations in glucose tolerance and insulin sensitivity but also in blood lipids, including HDL-C.

2.1 Impact of Adiposity on HDL Concentration and Function

Abnormally low plasma HDL-C concentrations accompany hypertriacylglycerolaemia and are a common trait in many obese subjects that contributes to the development of atherosclerosis (Bays et al. 2013; Tchernof and Despres 2013). In obesity, the low plasma HDL-C concentration has been mainly attributed to increased HDL clearance, in part, due to enhanced uptake of HDL by adipocytes (Wang and Peng 2011; Rashid and Genest 2007). Plasma concentrations of larger HDL particles are decreased in patients with high cardiovascular risk (Pirillo et al. 2013) and, similarly, appear to be low in overweight and obese subjects (Tian et al. 2006, 2011). Moreover, a switch towards a higher proportion of smaller HDL particles in relation to larger HDL has been reported in the plasma of obese subjects with ectopic fat and dysfunctional adipose tissue (Tchernof and Despres

2013). Obesity is commonly associated with increased oxidative stress (Wang and Peng 2011), and, in this context, there may be impairment of HDL functions, including reverse cholesterol transport and protection against lipoprotein oxidation. In this regard, the ability of HDL to mediate cholesterol efflux from cells, which is considered the first step of reverse cholesterol transport (RCT) (Escola-Gil et al. 2009), is reduced in the obese (Sasahara et al. 1998; Vazquez et al. 2012). In addition, although the protection against LDL oxidation conferred by HDL has not been assessed to date, the antioxidant action of apoA-I has been reported to be attenuated in scenarios of increased oxidative stress (Kontush and Chapman 2010), such as that present in obesity (Wang and Peng 2011). Other HDL-associated proteins such as paraoxonase 1 (PON1) and lipoprotein-associated phospholipase A2 (Lp-PLA2), which are also main contributors to the antioxidant properties of HDL (Kontush and Chapman 2010), have been found to be altered in obesity (Seres et al. 2010; da Silva et al. 2013).

2.2 Effect of Weight-Loss Therapies on HDL Concentration and Function

Weight loss has been widely thought to be an effective means of achieving a substantial reduction in cardiometabolic risk (Vest et al. 2012). Obesity-associated metabolic disorders can, in theory, be treated by lifestyle modifications including reduced energy intake and/or increased regular, moderate physical activity (Sacks et al. 2009; Bays et al. 2013). A further strategy to achieve excess weight loss in severely obese patients is bariatric surgery (Bays et al. 2013). In particular, the Roux-en-Y gastric bypass (RYGBP) has been reported to induce a sustained weight reduction without relapse in morbidly obese patients. In contrast, the metabolic results of lifestyle interventions are variable and insufficient to achieve sustained weight loss (Vest et al. 2012; Bays et al. 2013), whilst surgery has been shown to stably improve the lipid profile in most randomised, prospective trials (Bays et al. 2013; Vest et al. 2012). However, in both approaches, only modest effects on plasma HDL-C concentrations were reported in most recent systematic reviews and meta-analyses of stabilised weight-loss clinical trials which examined the effects of different dietary (Bays et al. 2013; Chapman et al. 2011; Dattilo and Kris-Etherton 1992; Hu et al. 2012; Sacks et al. 2009; Schwingshackl and Hoffmann 2013; Singh et al. 2007; Poobalan et al. 2004) or bariatric surgery programmes (Bays et al. 2013; Vest et al. 2012; Poobalan et al. 2004). There are few reports directly assessing the effect of weight reduction on HDL remodelling and its relationship, if any, with its anti-atherogenic properties. In this regard, recent studies showed that weight loss exerts beneficial changes in HDL subpopulations in the obese (Aron-Wisnewsky et al. 2011; Asztalos et al. 2010), leading to a switch towards predominantly larger HDL particles and increased circulating plasma HDL mass despite no significant impact on plasma HDL-C concentrations. Interestingly, in one of these studies, an elevation in HDL mass was strongly associated with an increased capacity of sera from postoperative patients to mediate cholesterol efflux via adenosine triphosphate-binding cassette transporter (ABC) G1 and scavenger receptor class-BI (SR-BI) pathways, relative to baseline values (Aron-Wisnewsky et al. 2011), thus suggesting that the removal of excess cholesterol from cells by HDL particles might be enhanced after surgery. Excess weight loss also improves oxidative stress (Uzun et al. 2004; Gletsu-Miller et al. 2009; Rector et al. 2007). However, whilst the impact of weight loss on HDL's antioxidant properties in severely obese patients undergoing bariatric surgery has been reported in observational studies, it has not yet been examined in detail (Gletsu-Miller et al. 2009). For instance, increased PON1 activity levels have been found in serum of patients undergoing gastric surgery (Uzun et al. 2004), whereas the Lp-PLA2 mass concentration did not change in another study (Hanusch-Enserer et al. 2009). Given that the distribution of Lp-PLA2 activity between serum lipoproteins is still uncertain, and evidence to suggest that Lp-PLA2 may be anti-atherogenic when associated with HDL (Kontush and Chapman 2010), it is reasonable to speculate that the increase in plasma concentration of larger HDL particles following weight loss might lead to a proportional increase in anti-atherogenic HDL-associated Lp-PLA2.

Taken together, HDL changes may, in part, account for the significant improvement in the cardiovascular risk observed in obese patients after weight reduction. Thus, in summary, weight loss has long-term beneficial effects on plasma lipids in the obese. Regarding the plasma concentration of HDL-C, the impact of lifestyle interventions appears to be variable and modest in response to different weight-loss treatments. However, data on the effect of weight-reducing lifestyle interventions on the functional properties of HDL are scant. Bariatric surgery is becoming a frequent approach for reducing adiposity and improving the lipid profile in the severely obese. The excess weight loss following intervention may provide an interesting scenario for assessing the potential favourable changes in HDL remodelling and functionality.

3 Effects of Regular Aerobic Exercise

Sedentary behaviour is a major health problem that is associated with increased cardiometabolic risk. Regular aerobic physical activity results in improved exercise performance which depends on an increased ability to utilise oxygen to derive energy for work. These effects usually require exercise intensities ranging from 40 to 85 % of maximal oxygen consumption (VO_2 max) or heart rate reserve, with higher exercise intensities being necessary for higher levels of initial fitness, and vice versa (Chapman et al. 2011).

3.1 Effects of Exercise on HDL

Regular aerobic physical activity has a positive effect on many of the established risk factors for CVD by causing a long-lasting reduction in TAGs by up to 20 % and increasing HDL-C by up to 10 %. It also increases the size of LDL and HDL particles. All these changes are considered potentially protective against atherosclerosis (Kraus et al. 2002; Tall 2002). Given the close interrelation between plasma TAG-rich particles and HDL, changes in HDL-C after regular aerobic exercise are probably dependent on an increased HDL synthesis induced by raised TAG lipolysis (Tall 2002; Olchawa et al. 2004) and, perhaps, also on a reduced HDL catabolism secondary to decreased hepatic lipase (HL) activity (Bergeron et al. 2001).

The effect of physical activity on TAG levels varies depending on baseline TAGs and exercise characteristics such as intensity, caloric expenditure and duration (Chapman et al. 2011; Miller et al. 2011; Kraus et al. 2002; Kodama et al. 2007). The reduction in plasma TAGs of 20 % induced by daily exercise was similar to that obtained by caloric restriction in nonobese subjects (Fontana et al. 2007). Daily exercise blocked, at least in the short term, the increase in TAGs induced by a high-carbohydrate diet in healthy postmenopausal women (Koutsari et al. 2001). Even with a variable adherence to diet, exercise did have a significant effect on plasma TAG and HDL-C levels (Huffman et al. 2012). In healthy, non-smoking men, the increase in HDL-C was dependent on exercise intensity (for instance, miles run per week) (Kokkinos et al. 1995; Tambalis et al. 2009; Duncan et al. 2005), thus suggesting a dose–response relationship. Between 1,200 and 2,200 Kcal of energy expenditure per week (15–20 miles/week of brisk walking or jogging) has been estimated to be the threshold for obtaining a significant change in plasma TAGs and HDL-C (Durstine et al. 2002). The TAG decrease was greater in individuals with fasting TAGs >1.69 mmol/L than in those with TAGs <1.69 mmol/L (Couillard et al. 2001). HDL-C increased, particularly in men with TAGs >1.69 mmol/L and low HDL-C, compared with those with low HDL-C and TAGs <1.69 mmol/L, and was associated with a reduction in abdominal adipose tissue (Couillard et al. 2001). This improvement may have been associated with decreased insulin resistance, adiposity and TAG levels, at least in part, through increased fatty acid oxidation by the muscle under the control of Rev-erb-alpha (Tall 2002; Woldt et al. 2013). Whilst data regarding the changes in HDL functionality induced by regular aerobic training are to date scant and inconsistent, there is evidence to implicate beneficial changes in some functions of HDL (Iborra et al. 2008; Kazeminasab et al. 2013; Meissner et al. 2010, 2011).

3.2 Effects of Exercise on Prevention and Treatment of Cardiometabolic Risk

Obviously, exercise exerts many different effects at a molecular level, rendering it very difficult to attribute the final effects on disease prevention or treatment to one effect, including HDL-C.

Several prospective cohort trials have shown a reduction in the risk of type 2 diabetes in individuals participating in physical activity of moderate intensity compared with being sedentary (Wing 2010; Uusitupa et al. 2009). This difference was also found in people who walked regularly (typically >2.5 h of brisk walking) (reviewed by Jeon et al. 2007).

Regular aerobic exercise is widely considered an important tool for primary and secondary cardiovascular prevention since it is thought to reduce the risk of fatal and nonfatal events in the general population (Chapman et al. 2011; Heran et al. 2011). Although the volume of moderate-intensity physical activity able to provide a reduction in cardiovascular mortality was initially considered to range from 2.5 to 5 h/week, similar results appear to be obtainable with 1–1.5 h/week of vigorous-intensity or a combination of vigorous- and moderate-intensity exercises (Wen et al. 2011; Sattelmair et al. 2011). However, a systematic review of exercise-based rehabilitation found a decrease in total and cardiovascular mortality but no difference in the number of new myocardial infarctions (Heran et al. 2011). Further, the Look AHEAD Trial which had, as a goal, to achieve and maintain a weight loss of at least 7 % by both reducing caloric intake and increasing physical activity in type 2 diabetic patients found no reduction in cardiovascular events in a 10-year follow-up (Wing et al. 2013). Similar negative results were obtained by the Finnish Diabetes Prevention Study after a 10-year lifestyle intervention (diet + exercise) in middle-aged, overweight people with impaired glucose tolerance (Uusitupa et al. 2009). To interpret these discrepant results, one has to consider differences in patient disease (type 2 diabetes versus established coronary artery disease) versus the general population, follow-up time and patient intervention type(s) (Sattelmair et al. 2011; Wen et al. 2011; Heran et al. 2011; Uusitupa et al. 2009).

Exercise improved the components of the metabolic syndrome in affected patients (Pattyn et al. 2013). In patients with established type 2 diabetes mellitus included in the Look AHEAD Trial, increased physical activity had beneficial effects on glycaemic control and cardiovascular risk factors, including HDL-C and TAGs (Wing 2010). However, in other studies, aerobic and/or resistance training was shown to improve glycaemic control, but did not change HDL-C or TAGs in adults with type 2 diabetes (Sigal et al. 2007). There were major differences in the number of patients and follow-up time between studies that should be considered when interpreting these different results (Wing 2010; Sigal et al. 2007). Furthermore, one study included a hypo-caloric diet together with exercise (Wing 2010), whereas the other only included exercise (Sigal et al. 2007).

In summary, there appears to be a consensus that regular aerobic exercise raises HDL-C in a way closely related to plasma TAG reduction. Regular aerobic exercise may prevent type 2 diabetes and CVD, probably with more intensity in primary than in secondary prevention. In particular, aerobic exercise does not seem to prevent CVD in patients with established type 2 diabetes. The role of exercise-induced changes in HDL in these outcomes is largely unknown.

4   Smoking Cessation

Smoking is a well-documented risk factor for CVD that is amenable to intervention. A potential mechanism for the atherogenic effect of cigarette smoke is via plasma HDL. Most research to date has looked at smoking effects on HDL-C concentration rather than HDL structure or function. One issue that is particularly difficult to address in observational smoking research is the presence of confounding lifestyle behaviours that may cluster with smoking behaviour (such as low physical activity, poor diet and increased alcohol consumption). Thus, comparison of the properties of HDL isolated from smokers and non-smokers does not allow direct causal relationships to be determined, and there is a need for more direct experimental approaches to dissect the biological effects of smoking and to identify specific mediating factors (i.e. the combustible products of tobacco such as nicotine, carbon monoxide and other gaseous products and free radicals). To further complicate the issue, smoking has effects on TAG metabolism, via increased sympathetic drive, insulin resistance or both, leading to increased plasma TAG concentrations and consequent remodelling effects of plasma HDL resulting in higher plasma turnover [reviewed in Freeman and Packard (1995)].

4.1   Smoking and HDL-C Concentration

A meta-analysis carried out almost 25 years ago indicated that smoking has a strong independent effect on plasma HDL-C levels with smokers having on average 6 % lower HDL-C concentrations compared to non-smokers (Craig et al. 1989). A more recent meta-analysis, which compared within individual differences before and after stopping smoking, indicated that the absolute HDL-C concentration difference was between 0.06 and 0.11 mmol/L (Forey et al. 2013). There also appears to be a larger smoking effect in women than men (Freeman and Packard 1995). HDL-C concentrations rise after stopping smoking and fall on restarting, and the magnitude of the effect on HDL-C is related to the number of cigarettes smoked (Fortmann et al. 1986; Moffatt et al. 1995; Stubbe et al. 1982; Tuomilehto et al. 1986). There are no long-lasting effects of smoking on HDL-C concentration after cessation (Forey et al. 2013), and there is no association between number of years stopped and plasma HDL-C (Wilson et al. 1983). Passive smoking is also associated with lower plasma HDL (Moffatt et al. 1995; Neufeld et al. 1997), and a reduction in HDL-C has been observed acutely only 6 h after exposure to environmental tobacco smoke (Moffatt et al. 2004). The reduction in plasma HDL-C levels appears to be limited to a 0.15 mmol/L reduction in the larger HDL_2 particles, independent of confounders, as determined by analytical ultracentrifugation (Freeman et al. 1993). This has been confirmed by others (Moriguchi et al. 1991; Shennan et al. 1985; Moffatt et al. 2004), although one study attributed the acute effects of smoking to the smaller HDL_3 fraction (Gnasso et al. 1984). Plasma apoA-I levels are also 6 % lower in smokers (Craig et al. 1989), and in some studies apoA-II has also been shown to be reduced in smokers (Berg et al. 1979; Haffner et al. 1985) and ex-smokers (Richard et al. 1997).

4.2   Potential Mechanisms for the Smoking-Related Reduction in HDL-C

Because of the intrinsic link between plasma TAG and HDL metabolism, demonstrated by their negative association in populations, and the biological effects of smoking on plasma TAG metabolism (Freeman and Packard 1995), it is likely that changes in plasma HDL-C concentration could be an indirect effect of smoking-related increases in plasma TAG concentration. Indeed, much of the effect of smoking on plasma HDL is lost after statistical correction for changes in plasma TAGs (Freeman et al. 1993; Phillips et al. 1981). Cigarette smoking affects the activities of plasma enzymes involved in regulating HDL size and turnover: lecithin–cholesterol acyltransferase (LCAT) (Freeman et al. 1998; Haffner et al. 1985) and lipoprotein lipase (LPL) (Freeman et al. 1998; Elkeles et al. 1983) activities are reduced, whilst HL (Moriguchi et al. 1991) and CETP (Dullaart et al. 1994) activities are increased. These changes result in a shift in the size distribution of HDL into smaller particles which have increased clearance from the plasma compartment (Brinton et al. 1994). However, there is a residual effect of smoking after correction for changes in plasma TAG, suggesting TAG-independent effects of smoking on plasma HDL concentration also exist. These TAG-independent effects appear to be more important in men than women (Freeman et al. 1993).

There is evidence for structural/compositional changes in HDL brought about by smoking. Ex vivo experiments, in which human plasma was acutely exposed to cigarette smoke, resulted in cross-linking of apoA-I and apoA-II (McCall et al. 1994) that may impair activation of LCAT. Similarly, chemically cross-linked HDL has an increased clearance in rodents (Senault et al. 1990). Smoking is associated with a reduction in the HDL content of Lp-PLA2, an enzyme thought to play an anti-atherogenic role in HDL (Tselepis et al. 2009). Early studies showed that chronic inhalation of cigarette smoke in pigeons inhibits liver HDL uptake (Mulligan et al. 1983). Recent data have shown that a by-product of cigarette smoking, benzo(a)pyrene, inhibited apoA-I synthesis in HepG2 cells, via activation of the aryl hydrocarbon nuclear steroid receptor, whilst nicotine had no effect (Naem et al. 2012). Modelling of the monocyte transcriptome in smokers compared to non-smokers identified *SLC39A8* to be on a causal pathway between smoking and plaque formation (Verdugo et al. 2013). This gene is known to be associated with the cellular uptake of cadmium from tobacco and was negatively associated with HDL cholesterol levels in this study.

There are very few studies on the effects of smoking on HDL function. HDL isolated from smokers showed reduced ability to induce cholesterol efflux from macrophages, possibly via apoA-I-mediated effects (Kralova Lesna et al. 2012). Ex vivo cigarette smoke treated HDL, which resulted in an increased conjugated diene and denatured apoA-I content, reduced the efflux capacity of HDL to a level similar to that of copper-oxidised HDL (Ueyama et al. 1998). Co-incubation with superoxide dismutase prevented approximately half of the impairment and reduced the level of conjugated dienes, but not the apoA-I denaturation. There is also some evidence that plasma thiocyanates found in high levels in smokers could cause HDL oxidation and reduced apoA-I cholesterol efflux ability (Hadfield et al. 2013). Smoking is associated with reduced activity and concentration of PON1, an HDL-associated antioxidant enzyme, effects which are reversed after smoking cessation (James et al. 2000). The mechanistic data for the effects of smoking on HDL function are far from comprehensive and merit further investigation.

4.3 Smoking Cessation Intervention to Increase Plasma HDL Concentration

The majority of smoking cessation studies are small, and most, but not all, have been shown to result in a rise in HDL concentration. A meta-analysis of 27 studies incorporating over 6,000 subjects indicated that HDL-C increased by 0.10 mmol/L after smoking cessation, whilst plasma total cholesterol, LDL-C and TAG did not change (Maeda et al. 2003).

A number of lifestyle changes may also occur when smokers cease smoking, and at least one study has shown that the rise in HDL after smoking cessation is not independent of the change in diet (Quensel et al. 1989b). However, refuting this, a recent large randomised, double-blind controlled trial was carried out in over 1,500 smokers smoking an average of 21 cigarettes per day (Gepner et al. 2011). Patients were randomised to one of six treatments: nicotine lozenge, nicotine patch, sustained-release bupropion, nicotine patch plus nicotine lozenge, sustained-release bupropion plus nicotine lozenge, or placebo. Of the 923 participants who returned after 1 year intervention, 36 % who had stopped smoking showed a significantly greater rise in HDL-C than those who did not, despite gaining an average of 4 kg more weight than those who continued to smoke. The effects of smoking cessation in this study were particularly evident in women.

Transdermal nicotine replacement therapy is a commonly used approach to support smoking cessation. A small study was carried out to compare the effects of transdermal nicotine patches, used as part of a smoking cessation intervention, on plasma HDL-C levels (Moffatt et al. 2000). Subjects who used transdermal nicotine patches over 35 days showed no improvement in HDL-C concentration. However, if the use of patches was then stopped, HDL-C concentrations rose to normal, non-smoking levels over the next 42 days. Others have observed contrasting effects in a larger study where smoking cessation, accompanied by the use of higher-dose nicotine transdermal patches, resulted in an immediate increase in HDL, whereas in contrast, low-dose nicotine did not (Allen et al. 1994). These latter findings are somewhat contradictory to data indicating that nicotine administration to non-smokers, via chewing gum, does not affect HDL-C levels (Quensel et al. 1989a). The potential effects of nicotine patches during smoking cessation on plasma HDL have yet to be clarified.

In summary, smoking leads to a reduction in plasma HDL-C, HDL2-C, apo A-I and probably apoA-II concentrations. The effects of smoking on HDL are dose dependent and reversed upon smoking cessation. Much of the effect of smoking may be TAG dependent where increased plasma TAG concentration leads to remodelling of HDL to a smaller particle size which is more rapidly cleared from the circulation. TAG-independent effects have not yet been thoroughly investigated but include modification of apoA-I, reductions in HDL antioxidant enzyme activity and reduced ability of HDL to promote cholesterol efflux. The effect of nicotine aids, used to support smoking cessation interventions, on HDL-C concentration and function is unclear.

5 Effects of Alcohol on HDL and Cardiovascular Risk

Multiple studies have established a known J-shaped relationship between alcohol intake and cardiovascular risk including coronary heart disease and ischaemic stroke (Krenz and Korthuis 2012). The term 'drink' is imprecise; however, the amount of alcohol in one drink is similar for wine, spirits or beer. Thus, a 100 ml glass of table wine at 13 % alcohol, 35 ml of distilled spirits at 40 % alcohol and 300 ml of beer at 5 % alcohol all contain around 10–12 g of pure ethyl alcohol. The general consensus is that men consuming two standard alcoholic drinks per day and women consuming half that intake appear to have a lower cardiovascular event rate than persons abstaining from alcohol (Krenz and Korthuis 2012). Estimates of the risk reduction associated with moderate alcohol intake drinkers compared with those abstaining from alcohol range from 25 to 30 % (Krenz and Korthuis 2012).

5.1 Alcohol Intake and HDL-C

Several plausible mechanisms have been proposed to explain the positive moderate alcohol intake-mediated effects on CVDs: reduced thrombogenic and coagulation factors (platelet aggregation and fibrinogen levels), low plasma concentrations of inflammation markers (C-reactive protein, interleukin-6 and adiponectin), lowered blood pressure, increased insulin sensitivity and lipoprotein profile via a lowering of LDL-C and increase in HDL-C (Klatsky 2010). However, this alcohol–CVD relationship is nonlinear, and excessive alcohol consumption has adverse effects on hypertension and other pro-inflammatory factors, despite increasing HDL-C (Foerster et al. 2009).

Some reports have suggested that wine carries a lower risk of mortality than beer or spirits and that this may be related to the ability of nonalcoholic phenolic compounds to inhibit LDL oxidation and its pro-inflammatory effects (Frankel et al. 1993; Gronbaek et al. 2000; Klatsky et al. 2003). However, other studies have failed to identify an additional advantage associated with red wine (Di Castelnuovo et al. 2002; Mukamal et al. 2003), thereby suggesting that ethyl alcohol is the main factor for the cardioprotective effect.

A recent meta-analysis demonstrated that alcohol exerts favourable effects on several cardiovascular biomarkers (higher HDL-C, apoA-I and adiponectin levels and lower fibrinogen levels) which seems to be independent of the alcoholic beverage type (Brien et al. 2011). This systematic review found that 1–2 drinks per day increased HDL-C by 0.10 mmol/L (Brien et al. 2011). However, the contribution of alcohol-induced HDL-C-raising effects to cardiovascular risk remains largely unknown. Early studies indicated that the atheroprotective effects were mediated, largely, by the increase in HDL-C, since the addition of HDL-C to the multivariate model attenuated the inverse association between alcohol intake and myocardial infarction (Gaziano et al. 1993). Nevertheless, one recent report describing a large Norwegian cohort with extensive control for confounding factors showed that HDL-C levels were not on the causal pathway connecting alcohol to the lower risk of death from coronary heart disease (Magnus et al. 2011). These findings raise the question of which specific HDL subpopulation particles are affected by alcohol intake and their significance for coronary heart disease protection.

5.2 Effects of Alcohol on HDL Atheroprotective Functions

Studies on the effect of alcohol consumption on HDL particle size are inconsistent. In some studies, moderate alcohol consumption increased small HDL subfractions, mainly HDL3 (Haskell et al. 1984; Gardner et al. 2000; Nishiwaki et al. 1994), and lipid-poor preβ-HDL particles (Beulens et al. 2004). These effects on HDL size could be related to the ability of alcohol to stimulate LPL activity (Nishiwaki et al. 1994). However, other studies on moderate drinkers found increased concentrations of both HDL2 and HDL3 (Gaziano et al. 1993; Clevidence et al. 1995; De Oliveira et al. 2000), and human kinetic studies demonstrated that moderate alcohol consumption resulted in dose-dependent increases in HDL-C, apoA-I and apoA-II through an increase in the HDL apolipoprotein transport rate, but without altering their catabolic rate and LPL activity (De Oliveira et al. 2000). These changes might explain the ability of HDL from moderate alcohol drinkers to enhance macrophage cholesterol efflux (Beulens et al. 2004). However, cholesterol efflux was also enhanced in heavy drinkers who usually show larger HDL (Makela et al. 2008); these changes were associated with decreased CETP activity and increased phospholipid transfer protein (PLTP) activity (Makela et al. 2008). It should be noted that these macrophage cholesterol efflux assays only quantified the first-step RCT pathway without addressing the efficiency of the rest of the RCT steps or the other HDL cardioprotective functions. Nevertheless, moderate alcohol intake was not associated with changes in the activities of other HDL remodelling factors, such as LCAT or CETP (Nishiwaki et al. 1994; Riemens et al. 1997; Sierksma et al. 2004), and, furthermore, the ability of HDL to deliver cholesterol into the liver cells seemed to be decreased in both moderate and heavy drinkers (Rao et al. 2000; Marmillot et al. 2007).

Beyond the RCT pathway, moderate alcohol consumption increased serum PON1 (van der Gaag et al. 1999; Sierksma et al. 2002; Rao et al. 2003), along with liver PON1 mRNA levels (Rao et al. 2003), although this effect was not found in one study with moderate red wine drinkers (Sarandol et al. 2003). In contrast, heavy alcohol drinking produced the opposite effects (Rao et al. 2003). Furthermore, alcohol promotes the conversion of phosphatidylcholine into phosphatidylethanol, and it was reported that HDL-associated phosphatidylethanol increased endothelial secretion of the vascular endothelial growth factor (Liisanantti and Savolainen 2009), which could partly explain the positive effect of alcohol on angiogenesis induction.

In summary, the inverse relationship between moderate and regular alcohol intake and cardiovascular risk as well as the ability of alcohol to increase HDL-C is well documented in the literature. Although some studies reported favourable effects of alcohol on HDL cardioprotective functions, it is unclear how many of the alcohol-mediated cardioprotective effects were mediated by the HDL increase. Finally, the American Heart Association recommendation that alcoholic beverage consumption should be limited to no more than 2 drinks per day for men and 1 drink per day for women and should be consumed with meals should be taken into account. Alcohol should not be considered a therapeutic option for cardiovascular risk reduction (Lichtenstein et al. 2006).

Acknowledgements This work was partly funded by European Cooperation in Science and Technology (COST) action BM0904, Ministerio de Sanidad y Consumo, Instituto de Salud Carlos III, FIS 10-00277 and CP13-00070 (to J.J.), FIS 11-0176 (to F.V-B.) and FIS 12-00291 (to J.C.E-G). CIBER de Diabetes y Enfermedades Metabólicas Asociadas is an Instituto de Salud Carlos III Project.

Open Access This chapter is distributed under the terms of the Creative Commons Attribution Noncommercial License, which permits any noncommercial use, distribution, and reproduction in any medium, provided the original author(s) and source are credited.

References

Allen SS, Hatsukami D, Gorsline J (1994) Cholesterol changes in smoking cessation using the transdermal nicotine system. Transdermal nicotine study group. Prev Med 23(2):190–196
Ambrosini GL, Oddy WH, Huang RC, Mori TA, Beilin LJ, Jebb SA (2013) Prospective associations between sugar-sweetened beverage intakes and cardiometabolic risk factors in adolescents. Am J Clin Nutr 98(2):327–334
Andersen CJ, Blesso CN, Lee J, Barona J, Shah D, Thomas MJ, Fernandez ML (2013) Egg consumption modulates HDL lipid composition and increases the cholesterol-accepting capacity of serum in metabolic syndrome. Lipids 48(6):557–567
Aron-Wisnewsky J, Julia Z, Poitou C, Bouillot JL, Basdevant A, Chapman MJ, Clement K, Guerin M (2011) Effect of bariatric surgery-induced weight loss on SR-BI-, ABCG1-, and ABCA1-mediated cellular cholesterol efflux in obese women. J Clin Endocrinol Metab 96(4):1151–1159
Astrup A, Dyerberg J, Elwood P, Hermansen K, Hu FB, Jakobsen MU, Kok FJ, Krauss RM, Lecerf JM, LeGrand P, Nestel P, Riserus U, Sanders T, Sinclair A, Stender S, Tholstrup T, Willett WC (2011) The role of reducing intakes of saturated fat in the prevention of cardiovascular disease: where does the evidence stand in 2010? Am J Clin Nutr 93(4):684–688
Asztalos BF, Swarbrick MM, Schaefer EJ, Dallal GE, Horvath KV, Ai M, Stanhope KL, Austrheim-Smith I, Wolfe BM, Ali M, Havel PJ (2010) Effects of weight loss, induced by gastric bypass surgery, on HDL remodeling in obese women. J Lipid Res 51(8):2405–2412
Bays HE, Toth PP, Kris-Etherton PM, Abate N, Aronne LJ, Brown WV, Gonzalez-Campoy JM, Jones SR, Kumar R, La Forge R, Samuel VT (2013) Obesity, adiposity, and dyslipidemia: a consensus statement from the National Lipid Association. J Clin Lipidol 7(4):304–383
Belsey J, de Lusignan S, Chan T, van Vlymen J, Hague N (2008) Abnormal lipids in high-risk patients achieving cholesterol targets: a cross-sectional study of routinely collected UK general practice data. Curr Med Res Opin 24(9):2551–2560
Berg K, Borresen AL, Dahlen G (1979) Effect of smoking on serum levels of HDL apoproteins. Atherosclerosis 34(3):339–343
Bergeron J, Couillard C, Despres JP, Gagnon J, Leon AS, Rao DC, Skinner JS, Wilmore JH, Bouchard C (2001) Race differences in the response of postheparin plasma lipoprotein lipase

and hepatic lipase activities to endurance exercise training in men: results from the HERITAGE Family Study. Atherosclerosis 159(2):399–406

Beulens JW, Sierksma A, van Tol A, Fournier N, van Gent T, Paul JL, Hendriks HF (2004) Moderate alcohol consumption increases cholesterol efflux mediated by ABCA1. J Lipid Res 45(9):1716–1723

Brien SE, Ronksley PE, Turner BJ, Mukamal KJ, Ghali WA (2011) Effect of alcohol consumption on biological markers associated with risk of coronary heart disease: systematic review and meta-analysis of interventional studies. BMJ 342:d636

Brinton EA, Eisenberg S, Breslow JL (1994) Human HDL cholesterol levels are determined by apoA-I fractional catabolic rate, which correlates inversely with estimates of HDL particle size. Effects of gender, hepatic and lipoprotein lipases, triglyceride and insulin levels, and body fat distribution. Arterioscler Thromb 14(5):707–720

Buonacorso V, Nakandakare ER, Nunes VS, Passarelli M, Quintao EC, Lottenberg AM (2007) Macrophage cholesterol efflux elicited by human total plasma and by HDL subfractions is not affected by different types of dietary fatty acids. Am J Clin Nutr 86(5):1270–1277

Chapman MJ, Ginsberg HN, Amarenco P, Andreotti F, Boren J, Catapano AL, Descamps OS, Fisher E, Kovanen PT, Kuivenhoven JA, Lesnik P, Masana L, Nordestgaard BG, Ray KK, Reiner Z, Taskinen MR, Tokgozoglu L, Tybjaerg-Hansen A, Watts GF (2011) Triglyceride-rich lipoproteins and high-density lipoprotein cholesterol in patients at high risk of cardiovascular disease: evidence and guidance for management. Eur Heart J 32(11):1345–1361

Clevidence BA, Reichman ME, Judd JT, Muesing RA, Schatzkin A, Schaefer EJ, Li Z, Jenner J, Brown CC, Sunkin M et al (1995) Effects of alcohol consumption on lipoproteins of premenopausal women. A controlled diet study. Arterioscler Thromb Vasc Biol 15(2):179–184

Couillard C, Despres JP, Lamarche B, Bergeron J, Gagnon J, Leon AS, Rao DC, Skinner JS, Wilmore JH, Bouchard C (2001) Effects of endurance exercise training on plasma HDL cholesterol levels depend on levels of triglycerides: evidence from men of the Health, Risk Factors, Exercise Training and Genetics (HERITAGE) Family Study. Arterioscler Thromb Vasc Biol 21(7):1226–1232

Craig WY, Palomaki GE, Haddow JE (1989) Cigarette smoking and serum lipid and lipoprotein concentrations: an analysis of published data. BMJ 298(6676):784–788

da Silva IT, Timm Ade S, Damasceno NR (2013) Influence of obesity and cardiometabolic makers on lipoprotein-associated phospholipase A2 (Lp-PLA2) activity in adolescents: the healthy young cross-sectional study. Lipids Health Dis 12:19

Dattilo AM, Kris-Etherton PM (1992) Effects of weight reduction on blood lipids and lipoproteins: a meta-analysis. Am J Clin Nutr 56(2):320–328

De Oliveira ESER, Foster D, McGee Harper M, Seidman CE, Smith JD, Breslow JL, Brinton EA (2000) Alcohol consumption raises HDL cholesterol levels by increasing the transport rate of apolipoproteins A-I and A-II. Circulation 102(19):2347–2352

De Vries R, Beusekamp BJ, Kerstens MN, Groen AK, Van Tol A, Dullaart RP (2005) A low-saturated-fat, low-cholesterol diet decreases plasma CETP activity and pre beta-HDL formation but does not affect cellular cholesterol efflux to plasma from type 1 diabetic patients. Scand J Clin Lab Invest 65(8):729–737

Di Castelnuovo A, Rotondo S, Iacoviello L, Donati MB, De Gaetano G (2002) Meta-analysis of wine and beer consumption in relation to vascular risk. Circulation 105(24):2836–2844

Dullaart RP, Hoogenberg K, Dikkeschei BD, van Tol A (1994) Higher plasma lipid transfer protein activities and unfavorable lipoprotein changes in cigarette-smoking men. Arterioscler Thromb 14(10):1581–1585

Duncan GE, Anton SD, Sydeman SJ, Newton RL Jr, Corsica JA, Durning PE, Ketterson TU, Martin AD, Limacher MC, Perri MG (2005) Prescribing exercise at varied levels of intensity and frequency: a randomized trial. Arch Intern Med 165(20):2362–2369

Durstine JL, Grandjean PW, Cox CA, Thompson PD (2002) Lipids, lipoproteins, and exercise. J Cardiopulm Rehabil 22(6):385–398

Elkeles RS, Khan SR, Chowdhury V, Swallow MB (1983) Effects of smoking on oral fat tolerance and high density lipoprotein cholesterol. Clin Sci (Lond) 65(6):669–672

Escola-Gil JC, Rotllan N, Julve J, Blanco-Vaca F (2009) In vivo macrophage-specific RCT and antioxidant and antiinflammatory HDL activity measurements: new tools for predicting HDL atheroprotection. Atherosclerosis 206(2):321–327

Escola-Gil JC, Llaverias G, Julve J, Jauhiainen M, Mendez-Gonzalez J, Blanco-Vaca F (2011) The cholesterol content of Western diets plays a major role in the paradoxical increase in high-density lipoprotein cholesterol and upregulates the macrophage reverse cholesterol transport pathway. Arterioscler Thromb Vasc Biol 31(11):2493–2499

Foerster M, Marques-Vidal P, Gmel G, Daeppen JB, Cornuz J, Hayoz D, Pecoud A, Mooser V, Waeber G, Vollenweider P, Paccaud F, Rodondi N (2009) Alcohol drinking and cardiovascular risk in a population with high mean alcohol consumption. Am J Cardiol 103(3):361–368

Fontana L, Villareal DT, Weiss EP, Racette SB, Steger-May K, Klein S, Holloszy JO (2007) Calorie restriction or exercise: effects on coronary heart disease risk factors. A randomized, controlled trial. Am J Physiol Endocrinol Metab 293(1):E197–E202

Forey BA, Fry JS, Lee PN, Thornton AJ, Coombs KJ (2013) The effect of quitting smoking on HDL-cholesterol – a review based on within-subject changes. Biomark Res 1(1):26

Fortmann SP, Haskell WL, Williams PT (1986) Changes in plasma high density lipoprotein cholesterol after changes in cigarette use. Am J Epidemiol 124(4):706–710

Frankel EN, Kanner J, German JB, Parks E, Kinsella JE (1993) Inhibition of oxidation of human low-density lipoprotein by phenolic substances in red wine. Lancet 341(8843):454–457

Freeman DJ, Packard CJ (1995) Smoking and plasma lipoprotein metabolism. Clin Sci (Lond) 89 (4):333–342

Freeman DJ, Griffin BA, Murray E, Lindsay GM, Gaffney D, Packard CJ, Shepherd J (1993) Smoking and plasma lipoproteins in man: effects on low density lipoprotein cholesterol levels and high density lipoprotein subfraction distribution. Eur J Clin Invest 23(10):630–640

Freeman DJ, Caslake MJ, Griffin BA, Hinnie J, Tan CE, Watson TD, Packard CJ, Shepherd J (1998) The effect of smoking on post-heparin lipoprotein and hepatic lipase, cholesteryl ester transfer protein and lecithin:cholesterol acyl transferase activities in human plasma. Eur J Clin Invest 28(7):584–591

Frost G, Leeds AA, Dore CJ, Madeiros S, Brading S, Dornhorst A (1999) Glycaemic index as a determinant of serum HDL-cholesterol concentration. Lancet 353(9158):1045–1048

Gardner CD, Tribble DL, Young DR, Ahn D, Fortmann SP (2000) Associations of HDL, HDL(2), and HDL(3) cholesterol and apolipoproteins A-I and B with lifestyle factors in healthy women and men: the Stanford Five City Project. Prev Med 31(4):346–356

Gaziano JM, Buring JE, Breslow JL, Goldhaber SZ, Rosner B, VanDenburgh M, Willett W, Hennekens CH (1993) Moderate alcohol intake, increased levels of high-density lipoprotein and its subfractions, and decreased risk of myocardial infarction. N Engl J Med 329 (25):1829–1834

Gepner AD, Piper ME, Johnson HM, Fiore MC, Baker TB, Stein JH (2011) Effects of smoking and smoking cessation on lipids and lipoproteins: outcomes from a randomized clinical trial. Am Heart J 161(1):145–151

Gletsu-Miller N, Hansen JM, Jones DP, Go YM, Torres WE, Ziegler TR, Lin E (2009) Loss of total and visceral adipose tissue mass predicts decreases in oxidative stress after weight-loss surgery. Obesity (Silver Spring) 17(3):439–446

Gnasso A, Haberbosch W, Schettler G, Schmitz G, Augustin J (1984) Acute influence of smoking on plasma lipoproteins. Klin Wochenschr 62(Suppl 2):36–42

Goyens PL, Mensink RP (2006) Effects of alpha-linolenic acid versus those of EPA/DHA on cardiovascular risk markers in healthy elderly subjects. Eur J Clin Nutr 60(8):978–984

Griffin MD, Sanders TA, Davies IG, Morgan LM, Millward DJ, Lewis F, Slaughter S, Cooper JA, Miller GJ, Griffin BA (2006) Effects of altering the ratio of dietary n-6 to n-3 fatty acids on insulin sensitivity, lipoprotein size, and postprandial lipemia in men and postmenopausal women aged 45–70 y: the OPTILIP Study. Am J Clin Nutr 84(6):1290–1298

Groener JE, van Ramshorst EM, Katan MB, Mensink RP, van Tol A (1991) Diet-induced alteration in the activity of plasma lipid transfer protein in normolipidemic human subjects. Atherosclerosis 87(2–3):221–226

Gronbaek M, Becker U, Johansen D, Gottschau A, Schnohr P, Hein HO, Jensen G, Sorensen TI (2000) Type of alcohol consumed and mortality from all causes, coronary heart disease, and cancer. Ann Intern Med 133(6):411–419

Grundy SM, Denke MA (1990) Dietary influences on serum lipids and lipoproteins. J Lipid Res 31 (7):1149–1172

Hadfield KA, Pattison DI, Brown BE, Hou L, Rye KA, Davies MJ, Hawkins CL (2013) Myeloperoxidase-derived oxidants modify apolipoprotein A-I and generate dysfunctional high-density lipoproteins: comparison of hypothiocyanous acid (HOSCN) with hypochlorous acid (HOCl). Biochem J 449(2):531–542

Haffner SM, Applebaum-Bowden D, Wahl PW, Hoover JJ, Warnick GR, Albers JJ, Hazzard WR (1985) Epidemiological correlates of high density lipoprotein subfractions, apolipoproteins A-I, A-II, and D, and lecithin cholesterol acyltransferase. Effects of smoking, alcohol, and adiposity. Arteriosclerosis 5(2):169–177

Hanusch-Enserer U, Zorn G, Wojta J, Kopp CW, Prager R, Koenig W, Schillinger M, Roden M, Huber K (2009) Non-conventional markers of atherosclerosis before and after gastric banding surgery. Eur Heart J 30(12):1516–1524

Harper CR, Edwards MC, Jacobson TA (2006) Flaxseed oil supplementation does not affect plasma lipoprotein concentration or particle size in human subjects. J Nutr 136(11):2844–2848

Harris WS (1989) Fish oils and plasma lipid and lipoprotein metabolism in humans: a critical review. J Lipid Res 30(6):785–807

Harris WS (1997) n-3 fatty acids and serum lipoproteins: human studies. Am J Clin Nutr 65 (5 Suppl):1645S–1654S

Haskell WL, Camargo C Jr, Williams PT, Vranizan KM, Krauss RM, Lindgren FT, Wood PD (1984) The effect of cessation and resumption of moderate alcohol intake on serum high-density-lipoprotein subfractions. A controlled study. N Engl J Med 310(13):805–810

Helal O, Berrougui H, Loued S, Khalil A (2013) Extra-virgin olive oil consumption improves the capacity of HDL to mediate cholesterol efflux and increases ABCA1 and ABCG1 expression in human macrophages. Br J Nutr 109(10):1844–1855

Heran BS, Chen JM, Ebrahim S, Moxham T, Oldridge N, Rees K, Thompson DR, Taylor RS (2011) Exercise-based cardiac rehabilitation for coronary heart disease. Cochrane Database Syst Rev 7, CD001800

Hu T, Mills KT, Yao L, Demanelis K, Eloustaz M, Yancy WS Jr, Kelly TN, He J, Bazzano LA (2012) Effects of low-carbohydrate diets versus low-fat diets on metabolic risk factors: a meta-analysis of randomized controlled clinical trials. Am J Epidemiol 176(Suppl 7):S44–S54

Huffman KM, Hawk VH, Henes ST, Ocampo CI, Orenduff MC, Slentz CA, Johnson JL, Houmard JA, Samsa GP, Kraus WE, Bales CW (2012) Exercise effects on lipids in persons with varying dietary patterns-does diet matter if they exercise? Responses in Studies of a Targeted Risk Reduction Intervention through Defined Exercise I. Am Heart J 164(1):117–124

Iborra RT, Ribeiro IC, Neves MQ, Charf AM, Lottenberg SA, Negrao CE, Nakandakare ER, Passarelli M (2008) Aerobic exercise training improves the role of high-density lipoprotein antioxidant and reduces plasma lipid peroxidation in type 2 diabetes mellitus. Scand J Med Sci Sports 18(6):742–750

James RW, Leviev I, Righetti A (2000) Smoking is associated with reduced serum paraoxonase activity and concentration in patients with coronary artery disease. Circulation 101 (19):2252–2257

Jeon CY, Lokken RP, Hu FB, van Dam RM (2007) Physical activity of moderate intensity and risk of type 2 diabetes: a systematic review. Diabetes Care 30(3):744–752

Kasbi Chadli F, Nazih H, Krempf M, Nguyen P, Ouguerram K (2013) Omega 3 fatty acids promote macrophage reverse cholesterol transport in hamster fed high fat diet. PLoS ONE 8 (4):e61109

Katan MB, Grundy SM, Willett WC (1997) Should a low-fat, high-carbohydrate diet be recommended for everyone? Beyond low-fat diets. N Engl J Med 337(8):563–566, discussion 566-567

Kaul N, Kreml R, Austria JA, Richard MN, Edel AL, Dibrov E, Hirono S, Zettler ME, Pierce GN (2008) A comparison of fish oil, flaxseed oil and hempseed oil supplementation on selected parameters of cardiovascular health in healthy volunteers. J Am Coll Nutr 27(1):51–58

Kazeminasab F, Marandi M, Ghaedi K, Esfarjani F, Moshtaghian J (2013) Endurance training enhances LXRalpha gene expression in Wistar male rats. Eur J Appl Physiol 113(9):2285–2290

Keys A (1980a) Alpha lipoprotein (HDL) cholesterol in the serum and the risk of coronary heart disease and death. Lancet 2(8195 pt 1):603–606

Keys A (1980b) W. O. Atwater memorial lecture: overweight, obesity, coronary heart disease and mortality. Nutr Rev 38(9):297–307

Klatsky AL (2010) Alcohol and cardiovascular health. Physiol Behav 100(1):76–81

Klatsky AL, Friedman GD, Armstrong MA, Kipp H (2003) Wine, liquor, beer, and mortality. Am J Epidemiol 158(6):585–595

Kodama S, Tanaka S, Saito K, Shu M, Sone Y, Onitake F, Suzuki E, Shimano H, Yamamoto S, Kondo K, Ohashi Y, Yamada N, Sone H (2007) Effect of aerobic exercise training on serum levels of high-density lipoprotein cholesterol: a meta-analysis. Arch Intern Med 167(10):999–1008

Kokkinos PF, Holland JC, Narayan P, Colleran JA, Dotson CO, Papademetriou V (1995) Miles run per week and high-density lipoprotein cholesterol levels in healthy, middle-aged men. A dose-response relationship. Arch Intern Med 155(4):415–420

Kontush A, Chapman MJ (2010) Antiatherogenic function of HDL particle subpopulations: focus on antioxidative activities. Curr Opin Lipidol 21(4):312–318

Koutsari C, Karpe F, Humphreys SM, Frayn KN, Hardman AE (2001) Exercise prevents the accumulation of triglyceride-rich lipoproteins and their remnants seen when changing to a high-carbohydrate diet. Arterioscler Thromb Vasc Biol 21(9):1520–1525

Kralova Lesna I, Suchanek P, Kovar J, Stavek P, Poledne R (2008) Replacement of dietary saturated FAs by PUFAs in diet and reverse cholesterol transport. J Lipid Res 49(11):2414–2418

Kralova Lesna I, Poledne R, Pagacova L, Stavek P, Pitha J (2012) HDL and apolipoprotein A1 concentrations as markers of cholesterol efflux in middle-aged women: interaction with smoking. Neuro Endocrinol Lett 33(Suppl 2):38–42

Kraus WE, Houmard JA, Duscha BD, Knetzger KJ, Wharton MB, McCartney JS, Bales CW, Henes S, Samsa GP, Otvos JD, Kulkarni KR, Slentz CA (2002) Effects of the amount and intensity of exercise on plasma lipoproteins. N Engl J Med 347(19):1483–1492

Krenz M, Korthuis RJ (2012) Moderate ethanol ingestion and cardiovascular protection: from epidemiologic associations to cellular mechanisms. J Mol Cell Cardiol 52(1):93–104

Labonte ME, Jenkins DJ, Lewis GF, Chiavaroli L, Wong JM, Kendall CW, Hogue JC, Couture P, Lamarche B (2013) Adding MUFA to a dietary portfolio of cholesterol-lowering foods reduces apoAI fractional catabolic rate in subjects with dyslipidaemia. Br J Nutr 110(3):426–436

Lagrost L, Mensink RP, Guyard-Dangremont V, Temme EH, Desrumaux C, Athias A, Hornstra G, Gambert P (1999) Variations in serum cholesteryl ester transfer and phospholipid transfer activities in healthy women and men consuming diets enriched in lauric, palmitic or oleic acids. Atherosclerosis 142(2):395–402

Lichtenstein AH, Appel LJ, Brands M, Carnethon M, Daniels S, Franch HA, Franklin B, Kris-Etherton P, Harris WS, Howard B, Karanja N, Lefevre M, Rudel L, Sacks F, Van Horn L, Winston M, Wylie-Rosett J (2006) Diet and lifestyle recommendations revision 2006: a scientific statement from the American Heart Association Nutrition Committee. Circulation 114(1):82–96

Liisanantti MK, Savolainen MJ (2009) Phosphatidylethanol mediates its effects on the vascular endothelial growth factor via HDL receptor in endothelial cells. Alcohol Clin Exp Res 33(2):283–288

Lustig RH (2010) Fructose: metabolic, hedonic, and societal parallels with ethanol. J Am Diet Assoc 110(9):1307–1321

Maeda K, Noguchi Y, Fukui T (2003) The effects of cessation from cigarette smoking on the lipid and lipoprotein profiles: a meta-analysis. Prev Med 37(4):283–290

Magnus P, Bakke E, Hoff DA, Hoiseth G, Graff-Iversen S, Knudsen GP, Myhre R, Normann PT, Naess O, Tambs K, Thelle DS, Morland J (2011) Controlling for high-density lipoprotein cholesterol does not affect the magnitude of the relationship between alcohol and coronary heart disease. Circulation 124(21):2296–2302

Makela SM, Jauhiainen M, Ala-Korpela M, Metso J, Lehto TM, Savolainen MJ, Hannuksela ML (2008) HDL2 of heavy alcohol drinkers enhances cholesterol efflux from raw macrophages via phospholipid-rich HDL 2b particles. Alcohol Clin Exp Res 32(6):991–1000

Marmillot P, Munoz J, Patel S, Garige M, Rosse RB, Lakshman MR (2007) Long-term ethanol consumption impairs reverse cholesterol transport function of high-density lipoproteins by depleting high-density lipoprotein sphingomyelin both in rats and in humans. Metabolism 56 (7):947–953

McCall MR, van den Berg JJ, Kuypers FA, Tribble DL, Krauss RM, Knoff LJ, Forte TM (1994) Modification of LCAT activity and HDL structure. New links between cigarette smoke and coronary heart disease risk. Arterioscler Thromb 14(2):248–253

Meissner M, Nijstad N, Kuipers F, Tietge UJ (2010) Voluntary exercise increases cholesterol efflux but not macrophage reverse cholesterol transport in vivo in mice. Nutr Metab (Lond) 7:54

Meissner M, Lombardo E, Havinga R, Tietge UJ, Kuipers F, Groen AK (2011) Voluntary wheel running increases bile acid as well as cholesterol excretion and decreases atherosclerosis in hypercholesterolemic mice. Atherosclerosis 218(2):323–329

Mensink RP, Zock PL, Kester AD, Katan MB (2003) Effects of dietary fatty acids and carbohydrates on the ratio of serum total to HDL cholesterol and on serum lipids and apolipoproteins: a meta-analysis of 60 controlled trials. Am J Clin Nutr 77(5):1146–1155

Micha R, Mozaffarian D (2010) Saturated fat and cardiometabolic risk factors, coronary heart disease, stroke, and diabetes: a fresh look at the evidence. Lipids 45(10):893–905

Miller M, Stone NJ, Ballantyne C, Bittner V, Criqui MH, Ginsberg HN, Goldberg AC, Howard WJ, Jacobson MS, Kris-Etherton PM, Lennie TA, Levi M, Mazzone T, Pennathur S (2011) Triglycerides and cardiovascular disease: a scientific statement from the American Heart Association. Circulation 123(20):2292–2333

Moffatt RJ, Stamford BA, Biggerstaff KD (1995) Influence of worksite environmental tobacco smoke on serum lipoprotein profiles of female nonsmokers. Metabolism 44(12):1536–1539

Moffatt RJ, Biggerstaff KD, Stamford BA (2000) Effects of the transdermal nicotine patch on normalization of HDL-C and its subfractions. Prev Med 31(2 Pt 1):148–152

Moffatt RJ, Chelland SA, Pecott DL, Stamford BA (2004) Acute exposure to environmental tobacco smoke reduces HDL-C and HDL2-C. Prev Med 38(5):637–641

Moriguchi EH, Fusegawa Y, Tamachi H, Goto Y (1991) Effects of smoking on HDL subfractions in myocardial infarction patients: effects on lecithin-cholesterol acyltransferase and hepatic lipase. Clin Chim Acta 195(3):139–143

Mukamal KJ, Conigrave KM, Mittleman MA, Camargo CA Jr, Stampfer MJ, Willett WC, Rimm EB (2003) Roles of drinking pattern and type of alcohol consumed in coronary heart disease in men. N Engl J Med 348(2):109–118

Mulligan JJ, Cluette JE, Kew RR, Stack DJ, Hojnacki JL (1983) Cigarette smoking impairs hepatic uptake of high density lipoproteins. Biochem Biophys Res Commun 112(3):843–850

Naem E, Alcalde R, Gladysz M, Mesliniene S, Jaimungal S, Sheikh-Ali M, Haas MJ, Wong NC, Mooradian AD (2012) Inhibition of apolipoprotein A-I gene by the aryl hydrocarbon receptor: a potential mechanism for smoking-associated hypoalphalipoproteinemia. Life Sci 91 (1–2):64–69

National Diet and Nutrition Survey (2011) Headline results from Years 1 and 2 (combined) of the rolling programme 2008/9–2009/10. http://www.dh.gov.uk/prod_consum_dh/groups/dh_digitalassets/documents/digitalasset/dh_128550.pdf.

Neufeld EJ, Mietus-Snyder M, Beiser AS, Baker AL, Newburger JW (1997) Passive cigarette smoking and reduced HDL cholesterol levels in children with high-risk lipid profiles. Circulation 96(5):1403–1407

Nishiwaki M, Ishikawa T, Ito T, Shige H, Tomiyasu K, Nakajima K, Kondo K, Hashimoto H, Saitoh K, Manabe M et al (1994) Effects of alcohol on lipoprotein lipase, hepatic lipase, cholesteryl ester transfer protein, and lecithin:cholesterol acyltransferase in high-density lipoprotein cholesterol elevation. Atherosclerosis 111(1):99–109

Olchawa B, Kingwell BA, Hoang A, Schneider L, Miyazaki O, Nestel P, Sviridov D (2004) Physical fitness and reverse cholesterol transport. Arterioscler Thromb Vasc Biol 24(6):1087–1091

Pan A, Chen M, Chowdhury R, Wu JH, Sun Q, Campos H, Mozaffarian D, Hu FB (2012) alpha-Linolenic acid and risk of cardiovascular disease: a systematic review and meta-analysis. Am J Clin Nutr 96(6):1262–1273

Pattyn N, Cornelissen VA, Eshghi SR, Vanhees L (2013) The effect of exercise on the cardiovascular risk factors constituting the metabolic syndrome: a meta-analysis of controlled trials. Sports Med 43(2):121–133

Phillips NR, Havel RJ, Kane JP (1981) Levels and interrelationships of serum and lipoprotein cholesterol and triglycerides. Association with adiposity and the consumption of ethanol, tobacco, and beverages containing caffeine. Arteriosclerosis 1(1):13–24

Pirillo A, Norata GD, Catapano AL (2013) High-density lipoprotein subfractions–what the clinicians need to know. Cardiology 124(2):116–125

Poobalan A, Aucott L, Smith WC, Avenell A, Jung R, Broom J, Grant AM (2004) Effects of weight loss in overweight/obese individuals and long-term lipid outcomes–a systematic review. Obes Rev 5(1):43–50

Quensel M, Agardh CD, Nilsson-Ehle P (1989a) Nicotine does not affect plasma lipoprotein concentrations in healthy men. Scand J Clin Lab Invest 49(2):149–153

Quensel M, Soderstrom A, Agardh CD, Nilsson-Ehle P (1989b) High density lipoprotein concentrations after cessation of smoking: the importance of alterations in diet. Atherosclerosis 75(2–3):189–193

Rao MN, Liu QH, Marmillot P, Seeff LB, Strader DB, Lakshman MR (2000) High-density lipoproteins from human alcoholics exhibit impaired reverse cholesterol transport function. Metabolism 49(11):1406–1410

Rao MN, Marmillot P, Gong M, Palmer DA, Seeff LB, Strader DB, Lakshman MR (2003) Light, but not heavy alcohol drinking, stimulates paraoxonase by upregulating liver mRNA in rats and humans. Metabolism 52(10):1287–1294

Rashid S, Genest J (2007) Effect of obesity on high-density lipoprotein metabolism. Obesity (Silver Spring) 15(12):2875–2888

Rector RS, Warner SO, Liu Y, Hinton PS, Sun GY, Cox RH, Stump CS, Laughlin MH, Dellsperger KC, Thomas TR (2007) Exercise and diet induced weight loss improves measures of oxidative stress and insulin sensitivity in adults with characteristics of the metabolic syndrome. Am J Physiol Endocrinol Metab 293(2):E500–E506

Richard F, Marecaux N, Dallongeville J, Devienne M, Tiem N, Fruchart JC, Fantino M, Zylberberg G, Amouyel P (1997) Effect of smoking cessation on lipoprotein A-I and lipoprotein A-I:A-II levels. Metabolism 46(6):711–715

Riemens SC, van Tol A, Hoogenberg K, van Gent T, Scheek LM, Sluiter WJ, Dullaart RP (1997) Higher high density lipoprotein cholesterol associated with moderate alcohol consumption is not related to altered plasma lecithin:cholesterol acyltransferase and lipid transfer protein activity levels. Clin Chim Acta 258(1):105–115

Sacks FM, Bray GA, Carey VJ, Smith SR, Ryan DH, Anton SD, McManus K, Champagne CM, Bishop LM, Laranjo N, Leboff MS, Rood JC, de Jonge L, Greenway FL, Loria CM, Obarzanek E, Williamson DA (2009) Comparison of weight-loss diets with different compositions of fat, protein, and carbohydrates. N Engl J Med 360(9):859–873

Sarandol E, Serdar Z, Dirican M, Safak O (2003) Effects of red wine consumption on serum paraoxonase/arylesterase activities and on lipoprotein oxidizability in healthy-men. J Nutr Biochem 14(9):507–512

Sasahara T, Nestel P, Fidge N, Sviridov D (1998) Cholesterol transport between cells and high density lipoprotein subfractions from obese and lean subjects. J Lipid Res 39(3):544–554

Sattelmair J, Pertman J, Ding EL, Kohl HW 3rd, Haskell W, Lee IM (2011) Dose response between physical activity and risk of coronary heart disease: a meta-analysis. Circulation 124(7):789–795

Schwingshackl L, Hoffmann G (2012) Monounsaturated fatty acids and risk of cardiovascular disease: synopsis of the evidence available from systematic reviews and meta-analyses. Nutrients 4(12):1989–2007

Schwingshackl L, Hoffmann G (2013) Long-term effects of low-fat diets either low or high in protein on cardiovascular and metabolic risk factors: a systematic review and meta-analysis. Nutr J 12:48

Senault C, Mahlberg FH, Renaud G, Girard-Globa A, Chacko GK (1990) Effect of apoprotein cross-linking on the metabolism of human HDL3 in rat. Biochim Biophys Acta 1046(1):81–88

Seres I, Bajnok L, Harangi M, Sztanek F, Koncsos P, Paragh G (2010) Alteration of PON1 activity in adult and childhood obesity and its relation to adipokine levels. Adv Exp Med Biol 660:129–142

Shennan NM, Seed M, Wynn V (1985) Variation in serum lipid and lipoprotein levels associated with changes in smoking behaviour in non-obese Caucasian males. Atherosclerosis 58 (1–3):17–25

Sierksma A, van der Gaag MS, van Tol A, James RW, Hendriks HF (2002) Kinetics of HDL cholesterol and paraoxonase activity in moderate alcohol consumers. Alcohol Clin Exp Res 26 (9):1430–1435

Sierksma A, Vermunt SH, Lankhuizen IM, van der Gaag MS, Scheek LM, Grobbee DE, van Tol A, Hendriks HF (2004) Effect of moderate alcohol consumption on parameters of reverse cholesterol transport in postmenopausal women. Alcohol Clin Exp Res 28(4):662–666

Sigal RJ, Kenny GP, Boule NG, Wells GA, Prud'homme D, Fortier M, Reid RD, Tulloch H, Coyle D, Phillips P, Jennings A, Jaffey J (2007) Effects of aerobic training, resistance training, or both on glycemic control in type 2 diabetes: a randomized trial. Ann Intern Med 147 (6):357–369

Singh IM, Shishehbor MH, Ansell BJ (2007) High-density lipoprotein as a therapeutic target: a systematic review. JAMA 298(7):786–798

Stanhope KL, Schwarz JM, Keim NL, Griffen SC, Bremer AA, Graham JL, Hatcher B, Cox CL, Dyachenko A, Zhang W, McGahan JP, Seibert A, Krauss RM, Chiu S, Schaefer EJ, Ai M, Otokozawa S, Nakajima K, Nakano T, Beysen C, Hellerstein MK, Berglund L, Havel PJ (2009) Consuming fructose-sweetened, not glucose-sweetened, beverages increases visceral adiposity and lipids and decreases insulin sensitivity in overweight/obese humans. J Clin Invest 119 (5):1322–1334

Stanhope KL, Schwarz JM, Havel PJ (2013) Adverse metabolic effects of dietary fructose: results from the recent epidemiological, clinical, and mechanistic studies. Curr Opin Lipidol 24 (3):198–206

Stubbe I, Eskilsson J, Nilsson-Ehle P (1982) High-density lipoprotein concentrations increase after stopping smoking. BMJ 284(6328):1511–1513

Tall AR (2002) Exercise to reduce cardiovascular risk – how much is enough? N Engl J Med 347 (19):1522–1524

Tambalis K, Panagiotakos DB, Kavouras SA, Sidossis LS (2009) Responses of blood lipids to aerobic, resistance, and combined aerobic with resistance exercise training: a systematic review of current evidence. Angiology 60(5):614–632

Tchernof A, Despres JP (2013) Pathophysiology of human visceral obesity: an update. Physiol Rev 93(1):359–404

Tian L, Jia L, Mingde F, Tian Y, Xu Y, Tian H, Yang Y (2006) Alterations of high density lipoprotein subclasses in obese subjects. Lipids 41(8):789–796

Tian L, Xu Y, Fu M, Peng T, Liu Y, Long S (2011) The impact of plasma triglyceride and apolipoproteins concentrations on high-density lipoprotein subclasses distribution. Lipids Health Dis 10:17

Tselepis AD, Panagiotakos DB, Pitsavos C, Tellis CC, Chrysohoou C, Stefanadis C (2009) Smoking induces lipoprotein-associated phospholipase A2 in cardiovascular disease free adults: the ATTICA Study. Atherosclerosis 206(1):303–308

Tuomilehto J, Tanskanen A, Salonen JT, Nissinen A, Koskela K (1986) Effects of smoking and stopping smoking on serum high-density lipoprotein cholesterol levels in a representative population sample. Prev Med 15(1):35–45

Ueyama K, Yokode M, Arai H, Nagano Y, Li ZX, Cho M, Kita T (1998) Cholesterol efflux effect of high density lipoprotein is impaired by whole cigarette smoke extracts through lipid peroxidation. Free Radic Biol Med 24(1):182–190

Uusitupa M, Peltonen M, Lindstrom J, Aunola S, Ilanne-Parikka P, Keinanen-Kiukaanniemi S, Valle TT, Eriksson JG, Tuomilehto J (2009) Ten-year mortality and cardiovascular morbidity in the Finnish Diabetes Prevention Study – secondary analysis of the randomized trial. PLoS ONE 4(5):e5656

Uzun H, Zengin K, Taskin M, Aydin S, Simsek G, Dariyerli N (2004) Changes in leptin, plasminogen activator factor and oxidative stress in morbidly obese patients following open and laparoscopic Swedish adjustable gastric banding. Obes Surg 14(5):659–665

van der Gaag MS, van Tol A, Scheek LM, James RW, Urgert R, Schaafsma G, Hendriks HF (1999) Daily moderate alcohol consumption increases serum paraoxonase activity; a diet-controlled, randomised intervention study in middle-aged men. Atherosclerosis 147(2):405–410

Vazquez E, Sethi AA, Freeman L, Zalos G, Chaudhry H, Haser E, Aicher BO, Aponte A, Gucek M, Kato GJ, Waclawiw MA, Remaley AT, Cannon RO 3rd (2012) High-density lipoprotein cholesterol efflux, nitration of apolipoprotein A-I, and endothelial function in obese women. Am J Cardiol 109(4):527–532

Verdugo RA, Zeller T, Rotival M, Wild PS, Munzel T, Lackner KJ, Weidmann H, Ninio E, Tregouet DA, Cambien F, Blankenberg S, Tiret L (2013) Graphical modeling of gene expression in monocytes suggests molecular mechanisms explaining increased atherosclerosis in smokers. PLoS ONE 8(1):e50888

Vest AR, Heneghan HM, Agarwal S, Schauer PR, Young JB (2012) Bariatric surgery and cardiovascular outcomes: a systematic review. Heart 98(24):1763–1777

Wang H, Peng DQ (2011) New insights into the mechanism of low high-density lipoprotein cholesterol in obesity. Lipids Health Dis 10:176

Wen CP, Wai JP, Tsai MK, Yang YC, Cheng TY, Lee MC, Chan HT, Tsao CK, Tsai SP, Wu X (2011) Minimum amount of physical activity for reduced mortality and extended life expectancy: a prospective cohort study. Lancet 378(9798):1244–1253

Wilson PW, Garrison RJ, Abbott RD, Castelli WP (1983) Factors associated with lipoprotein cholesterol levels. The Framingham study. Arteriosclerosis 3(3):273–281

Wing RR (2010) Long-term effects of a lifestyle intervention on weight and cardiovascular risk factors in individuals with type 2 diabetes mellitus: four-year results of the Look AHEAD trial. Arch Intern Med 170(17):1566–1575

Wing RR, Bolin P, Brancati FL, Bray GA, Clark JM, Coday M, Crow RS, Curtis JM, Egan CM, Espeland MA, Evans M, Foreyt JP, Ghazarian S, Gregg EW, Harrison B, Hazuda HP, Hill JO, Horton ES, Hubbard VS, Jakicic JM, Jeffery RW, Johnson KC, Kahn SE, Kitabchi AE, Knowler WC, Lewis CE, Maschak-Carey BJ, Montez MG, Murillo A, Nathan DM, Patricio J, Peters A, Pi-Sunyer X, Pownall H, Reboussin D, Regensteiner JG, Rickman AD, Ryan DH, Safford M, Wadden TA, Wagenknecht LE, West DS, Williamson DF, Yanovski SZ (2013) Cardiovascular effects of intensive lifestyle intervention in type 2 diabetes. N Engl J Med 369(2):145–154

Woldt E, Sebti Y, Solt LA, Duhem C, Lancel S, Eeckhoute J, Hesselink MK, Paquet C, Delhaye S, Shin Y, Kamenecka TM, Schaart G, Lefebvre P, Neviere R, Burris TP, Schrauwen P, Staels B, Duez H (2013) Rev-erb-alpha modulates skeletal muscle oxidative capacity by regulating mitochondrial biogenesis and autophagy. Nat Med 19(8):1039–1046

Yudkin J (1972) Sugar and disease. Nature 239(5369):197–199

Zhang YH, An T, Zhang RC, Zhou Q, Huang Y, Zhang J (2013) Very high fructose intake increases serum LDL-cholesterol and total cholesterol: a meta-analysis of controlled feeding trials. J Nutr 143(9):1391–1398

Effects of Established Hypolipidemic Drugs on HDL Concentration, Subclass Distribution, and Function

Monica Gomaraschi, Maria Pia Adorni, Maciej Banach, Franco Bernini, Guido Franceschini, and Laura Calabresi

Contents

1	Effects on Plasma HDL-C Concentration	595
	1.1 Statins	595
	1.2 Fibrates	595
	1.3 Niacin	596
2	Effects on HDL Subclass Distribution	596
	2.1 Statins	596
	2.2 Fibrates	597
	2.3 Niacin	597
3	Effects on Cholesterol Efflux Capacity	597
	3.1 Statins	598
	3.2 Fibrates	599
	3.3 Niacin	599
4	Effect on HDL Ability to Preserve Endothelial Cell Homeostasis	600
	4.1 Statins	601
	4.2 Fibrates	602
	4.3 Niacin	602
5	Effect on HDL Antioxidant Properties	603
	5.1 Statins	604
	5.2 Fibrates	604
	5.3 Niacin	605
References		607

M. Gomaraschi • G. Franceschini • L. Calabresi (✉)
Center E. Grossi Paoletti, Department of Pharmacological and Biomolecular Sciences, University of Milano, Via Balzaretti 9, 20133 Milan, Italy
e-mail: monica.gomaraschi@unimi.it; guido.franceschini@unimi.it; laura.calabresi@unimi.it

M.P. Adorni • F. Bernini
Department of Pharmacy, University of Parma, Parma, Italy
e-mail: mariapia.adorni@unipr.it; fbernini@unipr.it

M. Banach
Medical University of Lodz, Łódź, Poland
e-mail: maciejbanach@aol.co.uk

© The Author(s) 2015
A. von Eckardstein, D. Kardassis (eds.), *High Density Lipoproteins*, Handbook of Experimental Pharmacology 224, DOI 10.1007/978-3-319-09665-0_19

Abstract

The knowledge of an inverse relationship between plasma high-density lipoprotein cholesterol (HDL-C) concentrations and rates of cardiovascular disease has led to the concept that increasing plasma HDL-C levels would be protective against cardiovascular events. Therapeutic interventions presently available to correct the plasma lipid profile have not been designed to specifically act on HDL, but have modest to moderate effects on plasma HDL-C concentrations. Statins, the first-line lipid-lowering drug therapy in primary and secondary cardiovascular prevention, have quite modest effects on plasma HDL-C concentrations (2–10 %). Fibrates, primarily used to reduce plasma triglyceride levels, also moderately increase HDL-C levels (5–15 %). Niacin is the most potent available drug in increasing HDL-C levels (up to 30 %), but its use is limited by side effects, especially flushing.

The present chapter reviews the effects of established hypolipidemic drugs (statins, fibrates, and niacin) on plasma HDL-C levels and HDL subclass distribution, and on HDL functions, including cholesterol efflux capacity, endothelial protection, and antioxidant properties.

Keywords

HDL • HDL subclasses • Cell cholesterol efflux • Endothelial cells • Statins • Fibrates • Niacin

Abbreviations

ABCA1	ATP-binding cassette A1
CETP	Cholesteryl ester transfer protein
CEC	Cholesterol efflux capacity
CRP	Reactive protein
eNOS	Endothelial nitric oxide synthase
EPC	Endothelial progenitor cells
FMD	Flow-mediated vasodilation
GSPx	Glutathione selenoperoxidase
HDL-C	HDL cholesterol
LDL-C	LDL cholesterol
LCAT	Lecithin:cholesterol acyltransferase
LOOHs	Lipoprotein lipid hydroperoxides
LpA-I	HDL containing apoA-I only
LpA-I:A-II	HDL containing apoA-I and apoA-II
Lp-PLA2	Lipoprotein-associated phospholipase A2
NO	Nitric oxide
PGI$_2$	Prostacyclin
PON1	Paraoxonase 1

ROS   Reactive oxygen species
SR-BI   Scavenger receptor class B type I
S1P   Sphingolipid sphingosine-1-phosphate
T2DM   Type 2 diabetes
TNFα   Tumor necrosis factor alpha

1 Effects on Plasma HDL-C Concentration

1.1 Statins

Statins are inhibitors of the hydroxymethylglutaryl coenzyme A (HMG-CoA) reductase enzyme, a key enzyme in cholesterol synthesis. In addition to effectively reducing low-density lipoprotein cholesterol (LDL-C), all statins modestly increase HDL-C levels across the range of doses used, with important differences between different statins (Jones et al. 2003). The increase in HDL-C is dose dependent with some statins, e.g., simvastatin and rosuvastatin, but not with others. In the case of atorvastatin, the highest increase in HDL-C levels is observed at the lowest dose used. Statin-induced changes in HDL-C are normally paralleled by an increase in apoA-I levels (Barter et al. 2010). Interestingly, variations in LDL-C and in HDL-C levels are independent from each other (Barter et al. 2010). The mechanism by which statins increase plasma HDL-C and apoA-I levels remains unknown, and it is apparently unrelated to the inhibition of the HMG-CoA reductase enzyme. The ability of statins to upregulate hepatic ABCA1 gene expression (Tamehiro et al. 2007), as well as the described statin-induced inhibition of cholesteryl ester transfer protein (CETP) (Guerin et al. 2000; Kassai et al. 2007) can contribute to explain the effect of statin on HDL-C levels.

1.2 Fibrates

Fibrates (fibric acid derivates), synthetic ligands for PPAR-α, significantly reduce plasma triglyceride levels (30–50 %) and moderately increase HDL-C concentrations (5–15 %), with no major differences among available molecules (Khoury and Goldberg 2011). Activation of PPAR-α leads to β-oxidation of free fatty acids in the liver reducing VLDL synthesis and secretion. Furthermore, the expression of the gene coding for lipoprotein lipase is increased and apolipoprotein C-III expression in the liver is decreased, resulting in increased hydrolysis of triglyceride-rich lipoproteins. HDL-C raises results from the increased expression of apoA-II and, at a much lesser extent, apoA-I (Yetukuri et al. 2011), as well as to the enhanced cholesterol efflux via the induction of ABCA1 in the liver (Berger et al. 2005). In addition, animal studies have shown that fenofibrate reduces CETP hepatic expression (van der Hoogt et al. 2007).

1.3 Niacin

Niacin (nicotinic acid), a GPR109a agonist used clinically for more than 50 years to lower cholesterol and triglycerides, increases HDL-C levels up to 30 % (Vega and Grundy 1994). All the different niacin formulations (regular and extended release, ER) are equally effective in increasing HDL-C levels, whereas acipimox (a nicotinic acid analog) only induces a 7 % increase (Birjmohun et al. 2005). Niacin side-effect profile, including flushing, gastrointestinal upset, and liver function testing abnormalities, has, however, limited its use, especially in warmer Mediterranean countries (Birjmohun et al. 2005). Coadministration of the prostaglandin receptor antagonist laropiprant significantly reduces but does not eliminate niacin-induced skin symptoms (Yadav et al. 2012). It is not quite clear how niacin leads to an increase in HDL-C levels, and various theories have been proposed (Soudijn et al. 2007). These include the ability of niacin to inhibit CETP (Hernandez et al. 2007), to induce ABCA1 expression (Rubic et al. 2004), and to reduce HDL catabolism (Bodor and Offermanns 2008).

2 Effects on HDL Subclass Distribution

Plasma HDL is highly heterogeneous and can be separated in several subclasses according to density, size, shape, and lipid and protein composition (Calabresi et al. 2010). According to density, two HDL subclasses can be identified, the less dense HDL2 and the more dense HDL3 (De Lalla and Gofman 1954). HDL2 and HDL3 could be further fractionated in five distinct subclasses on the basis of particle size (from large to small: HDL2b, HDL2a, HDL3a, HDL3b, and HDL3c) (Nichols et al. 1986). According to surface charge, HDL could be separated into α-HDL and preβ-HDL. The separation by charge and size, using a two-dimensional (2-D) gel electrophoresis, allow the identification of up to 12 distinct apoA-I-containing HDL subclasses (Asztalos and Schaefer 2003). Finally, according to the protein component, HDL can be separated into particles containing only apoA-I (LpA-I) or containing also apoA-II (LpA-I:A-II).

2.1 Statins

The reported effects of statins on HDL subclass distribution are not consistent, likely depending on the technique used for the analysis. Some studies showed no effect of statins on HDL subclasses (Alaupovic et al. 1994; Bard et al. 1989; Franceschini et al. 2007; Tomas et al. 2000), and on HDL particle size distribution (Franceschini et al. 1989; Homma et al. 1995). On the contrary, studies evaluating HDL subclasses by 2-D gel electrophoresis showed that all statins, although with different potency, increase the levels of the large HDL particles and decrease the levels of small particles (Asztalos et al. 2002a, b, 2007).

2.2 Fibrates

A number of studies have shown that treatment with fibrates causes a shift of HDL from large to small particles in low HDL (Franceschini et al. 2007; Otvos et al. 2006), hypertriglyceridemic (Miida et al. 2000; Sasaki et al. 2002), and hypercholesterolemic (Franceschini et al. 1989) patients. These effects could be explained by the fibrate-induced increase in hepatic lipase activity (Miida et al. 2000; Patsch et al. 1984). In addition, bezafibrate has been shown to increase small discoidal preβ1-HDL particles in hypertriglyceridemic patients (Miida et al. 2000). The analysis of HDL subclasses by 2-D electrophoresis in the large cohort of the VA-HIT trial confirmed that gemfibrozil treatment is associated with a decrease in large (α-1 and α-2) HDL and with an increase in the small, lipid-poor HDL particles (Asztalos et al. 2008).

2.3 Niacin

Sheperd et al. were the first to demonstrate in the late 1970s that niacin (3 g/day) in healthy volunteers significantly increases the large HDL2 particles, associated with a parallel reduction of the small HDL3 (Shepherd et al. 1979). Several studies have confirmed this observation in patients with different types of dyslipidemia treated with 2, 3, or 4 g/day of regular or ER niacin (Johansson and Carlson 1990; Knopp et al. 1998; Morgan et al. 2003; Wahlberg et al. 1990) and in patients with coronary artery disease (Kuvin et al. 2006). A very recent study showed that also 1 g/day of ER niacin increases large HDL particles in dyslipidemic patients, with a concomitant significant increase in mean HDL size (Franceschini et al. 2013). The analysis of HDL subclasses by 2-D gel electrophoresis recently confirmed the observation that niacin promotes the maturation of HDL into large particles (Lamon-Fava et al. 2008), likely explained by a reduction in CETP activity. In addition, Sakai et al. reported that niacin selectively increases LpA-I particles in patients with low HDL-C, due to a prolongation of apoA-I residence time (Sakai et al. 2001).

3 Effects on Cholesterol Efflux Capacity

The most relevant antiatherogenic function of HDL is the ability to promote reverse cholesterol transport, the physiological process by which excess cholesterol is removed from peripheral tissues and transported to the liver for excretion (Cuchel and Rader 2006). The first and limiting step of this process is the efflux of cellular cholesterol from lipid-laden macrophages of the arterial wall (Lewis and Rader 2005). Multiple mechanisms for efflux of cholesterol exist in macrophages (Adorni et al. 2007). The release of free cholesterol via passive diffusion occurs in all cell types and is accelerated when the scavenger receptor class B type I (SR-BI) is present on the cell plasma membrane. Both passive diffusion and SR-BI-mediated efflux occur to phospholipid-containing acceptors (i.e., HDL and lipidated apolipoproteins). ATP-binding cassette A1 (ABCA1) belongs to the ATP-binding cassette transporter family and promotes unidirectional efflux of membrane cholesterol and phospholipids to lipid-poor apolipoproteins. Macrophages may also release cholesterol via the ABCG1 pathway, another ABC transporter that mediates net mass efflux of cellular cholesterol to mature HDL (Favari et al. 2009). HDL cholesterol efflux capacity (CEC) from macrophages can be measured in vitro and provides a reliable measure of HDL functionality (Khera et al. 2011).

3.1 Statins

The impact of statin therapy on HDL ability to promote cholesterol efflux has been investigated in a number of studies. Khera et al. measured HDL CEC in patients treated with 10 and 80 mg atorvastatin or 40 mg pravastatin for 16 weeks and reported no effects of statin therapy on HDL CEC from macrophages expressing all the known cholesterol efflux pathways (ABCA1, ABCG1, SR-BI, and passive diffusion) (Khera et al. 2011). As opposed to these results, Guerin et al. have shown that atorvastatin therapy for 12 weeks (10 and 40 mg daily) could increase SR-BI-dependent CEC of plasma from patients with type IIB hyperlipidemia, in a dose-dependent manner (Guerin et al. 2002). The effect of simvastatin on HDL's ability to promote cholesterol efflux was assessed in a randomized, double-blind, parallel group trial carried out in dyslipidemic patients with low HDL-C levels receiving either fenofibrate (160 mg/day) or simvastatin (40 mg/day) for 8 weeks (Franceschini et al. 2007). Simvastatin led to a small, significant increase in the capacity of plasma to promote SR-BI-mediated cholesterol efflux, likely related to the slight increase in HDL-C plasma levels that occurred after simvastatin treatment. The latter results are in contrast with data obtained by de Vries et al. who demonstrated that HDL from moderately hypercholesterolemic type 1 diabetic patients after 10, 20, and 40 mg simvastatin treatment does not increase their ability to promote cellular cholesterol efflux via SR-BI (De Vries et al. 2005). The effect of simvastatin on HDL functionality was further investigated recently in a small number of diabetic patients treated with simvastatin 40 mg/day or bezafibrate 400 mg/day, alone or in combination. CEC of apoB-depleted plasma samples from cholesterol-loaded macrophages, a cellular model which expresses simultaneously all the pathways of known relevance in cholesterol efflux, was increased by 14 % in response to simvastatin compared to placebo (Triolo et al. 2013a). Sviridov et al. showed that treatment with 40 mg/day rosuvastatin in overweight subjects with defined metabolic syndrome did not significantly modify HDL capacity to promote cholesterol efflux from LXR-activated human macrophages (Sviridov et al. 2008). Finally, a recent study showed a moderate ability of pitavastatin to enhance HDL CEC from macrophage foam cells in dyslipidemic subjects (Miyamoto-Sasaki et al. 2013). Overall, the impact of statin therapy on HDL CEC does not appear to be substantial, in agreement with the modest effect of statins on HDL-C concentration and on HDL subclass distribution.

3.2 Fibrates

The effect of fibrates on HDL cholesterol efflux capacity was assessed by Guerin et al. showing a significant ciprofibrate-mediated 13 % elevation in the capacity of plasma from type IIB hyperlipidemic subjects to mediate cholesterol efflux from SR-BI-expressing Fu5AH hepatoma cells (Guerin et al. 2003). The study by Franceschini et al., comparing fenofibrate and simvastatin in dyslipidemic patients (Franceschini et al. 2007), provided evidence that fenofibrate increases the plasma capacity to promote ABCA1-mediated efflux with no changes in SR-BI efflux. These observed results might be explained by a shift of HDL from large to small particles, the preferential cholesterol acceptors for ABCA1 (Adorni et al. 2007). This result was not confirmed in a more recent paper by the same group (Franceschini et al. 2013), likely because of the different HDL particle profile in the two investigated populations of patients. In the first study (Guerin et al. 2002), patients displayed a greater content of large HDL particles, whereas in patients with dyslipidemia of the last study (Franceschini et al. 2013), the small, ABCA1-interacting HDL was the predominant fraction, and no further increase in this HDL population was observed after fenofibrate therapy, consistent with the lack of changes in ABCA1-mediated CEC. Whether fenofibrate therapy modulate the ability of plasma or HDL to facilitate cholesterol efflux from macrophages has been investigated also in a subset of the FIELD study showing that cholesterol efflux values from macrophage foam cells to HDL and plasma in the fenofibrate group (200 mg/day for 5 years) were comparable to those of the placebo group (Maranghi et al. 2011). Finally, in a small study carried out in diabetic patients, it was shown that bezafibrate increases CEC of apoB-depleted plasma samples from cholesterol-loaded THP-1 macrophages (Triolo et al. 2013a). Overall, studies on the ability of fibrates to modulate HDL cholesterol efflux capacity seem not consistent, likely due to the different cell models used in the different studies and also to the different types of patients.

3.3 Niacin

The impact of niacin on HDL cholesterol efflux capacity was investigated in a placebo-controlled study carried out by Khera et al. who measured HDL CEC before and after ER niacin therapy in a small number of patients with carotid atherosclerosis (Khera et al. 2013). The study showed that niacin treatment led to favorable changes in patients' lipid profiles without significantly improving CEC of HDL from macrophages. In the study by Franceschini et al., the effect of ER niacin therapy on plasma CEC was evaluated in a population of dyslipidemic subjects (Franceschini et al. 2013). The tendency of ER niacin in enhancing passive diffusion, SR-BI-, and ABCG1-mediated CEC observed in this study is in line with the results of a previous investigation, which demonstrated that niacin increased both HDL-C levels and CEC from human macrophages (Yvan-Charvet et al. 2010). In this study the increase in CEC achieved statistical significance, possibly because of a greater increase in plasma HDL-C levels compared with that observed in study by

Franceschini et al., because passive diffusion, SR-BI-, and partially ABCG1-mediated CECs are known to be dependent on HDL-C concentrations. Consistently, a previous investigation by Morgan et al. reported that ER niacin therapy in patients with a history of primary dyslipoproteinemia had a beneficial effect on SR-BI-mediated efflux and that it was related to the change in level of HDL-C (Morgan et al. 2007). Another study evaluated specific cholesterol efflux pathway-mediated CECs (ABCA1, SR-BI, and ABCG1) of serum from men with HDL deficiency before and after treatment with niacin, but no changes were reported in all the considered efflux pathways (Alrasadi et al. 2008). Taken together, available data on the effect of niacin on the HDL-mediated ability to promote cell cholesterol efflux show modest effects on the pathways involving large HDL, in agreement with the changes observed on HDL particles.

4 Effect on HDL Ability to Preserve Endothelial Cell Homeostasis

Several in vitro and in vivo evidences indicate that the anti-atherosclerotic activity of HDL is also due to their ability to preserve endothelial cell homeostasis (Calabresi et al. 2003). HDL is able to promote vasorelaxation, to inhibit the production of cell adhesion and proinflammatory molecules, and to favor the integrity and repair of the endothelial layer, thus preventing and correcting the main features of endothelial dysfunction typical of the atherosclerotic process. HDL can modulate vascular tone by promoting the release of vasoactive molecules like nitric oxide (NO) and prostacyclin (PGI_2) which also exert powerful antithrombotic effects. HDL is able to increase the protein abundance and to induce the activation of endothelial nitric oxide synthase (eNOS) (Kuvin et al. 2002; Yuhanna et al. 2001). This latter effect is driven by the interaction of HDL with the scavenger receptor SR-BI and by the subsequent activation of the PI3K/Akt signaling pathway leading to eNOS phosphorylation; in addition, the sphingolipid sphingosine-1-phosphate (S1P) carried by HDL can also mediate eNOS activation by binding with its receptor S1P3 (Nofer et al. 2004). PGI2 is a metabolite of arachidonate and HDL was shown to increase its production by different mechanisms. HDL can provide endothelial cells with the substrate arachidonate or promote its release from cellular phospholipids by activating calcium-sensitive membrane-bound phospholipases (Calabresi et al. 2003). Another feature of endothelial dysfunction is the increased production of molecules favoring blood cell adhesion to the endothelial layer and their consequent extravasation and activation. Studies on endothelial cells have shown that HDL are able to inhibit the cytokine-induced expression of cell adhesion molecules (Calabresi et al. 1997; Cockerill et al. 1995), but the precise mechanism responsible for this effect has not been fully elucidated to date (Barter et al. 2002). Different studies have shown an HDL-mediated inhibition of sphingosine kinase activity and NF-kB nuclear translocation and an increased expression of heme-oxygenase 1 mediated by SR-BI (McGrath et al. 2009; Xia et al. 1999). HDL was also shown to inhibit the production of proinflammatory cytokines, as interleukin-6 and chemokines (Gomaraschi et al. 2005). Finally, HDL was shown to promote endothelial cell migration and proliferation, to inhibit cell apoptosis through different mechanisms (Mineo and Shaul 2012), and to promote endothelial progenitor cells (EPC) differentiation and survival (Mineo and Shaul 2012; Petoumenos et al. 2009).

In certain pathological conditions HDL may lose their ability to protect the endothelium and even become dysfunctional. In particular, conditions associated with both acute and chronic inflammatory processes, such as ischemic vascular events, infections, type 2 diabetes (T2DM), obesity, metabolic syndrome, autoimmune disorders, and chronic kidney disease, have been associated with HDL displaying impaired endothelial protective activities (Riwanto and Landmesser 2013). Indeed, HDL isolated from these patients failed to promote NO production, to exert antioxidant and anti-inflammatory effects, and to favor endothelial integrity in vitro (Riwanto and Landmesser 2013).

The measurement of vasodilation in response to changes in forearm blood flow (flow-mediated vasodilation, FMD) is currently considered the best method to evaluate endothelial function in vivo (Deanfield et al. 2007) and is commonly used to assess in vivo vascular condition and to evaluate the effects of pharmacological treatments. In addition, plasma levels of the soluble forms of adhesion molecules and of proinflammatory cytokines can be used as in vivo markers of endothelial inflammatory activation. Recently, circulating EPC number and migratory capacity emerged as novel biomarkers of endothelial repair (Petoumenos et al. 2009).

4.1 Statins

Several studies have shown that statins are able to improve FMD, to reduce the plasma levels of inflammatory markers, and to increase EPC number in dyslipidemic subjects (Antonopoulos et al. 2012; Liu et al. 2012; Reriani et al. 2011). These effects are likely not related to statin-induced changes in HDL. In vitro experiments have indeed shown that statin-mediated improvement of endothelial function is due to a direct effect of the drug on endothelial cells. Incubation of endothelial cells with different statins promotes a significant increase expression of SR-BI and S1P receptors and of eNOS, with a consequent increased production of NO after incubation with HDL (Igarashi et al. 2007; Kimura et al. 2008). A recent study performed on subjects with isolated low HDL-C showed that FMD was significantly improved after 4 weeks treatment with pravastatin, and multiple regression analysis revealed that changes in HDL-C and EPC number were independent predictors of FMD increase (Higashi et al. 2010). Despite the limitations due to the small number of analyzed subjects, this latter study suggests a potential role of statins on HDL capacity to maintain endothelial homeostasis.

4.2 Fibrates

Different fibrates have been tested for their ability to modulate endothelial function in dyslipidemic patients. Fenofibrate was shown to improve FMD and to decrease inflammatory biomarkers, as C-reactive protein (CRP), tumor necrosis factor alpha (TNFα), interleukin-1, and soluble CD40 ligand, in patients with combined hyperlipidemia (Malik et al. 2001; Wang et al. 2003). FMD improvement was inversely related to baseline HDL-C levels and indeed fenofibrate significantly increased FMD only in the subgroup with low HDL-C and not in subjects with normal HDL-C, in which a statin was instead effective (Wang et al. 2003). Fenofibrate, alone or in combination with a statin, was also effective in modulating FMD and decreasing plasma levels of CRP and TNFα in hypertriglyceridemic subjects (Koh et al. 2004, 2005), and a significant decrease of plasma levels of sCAMs was observed in subjects with low HDL-C levels (Calabresi et al. 2002); in this latter study, the decrease of plasma sCAMs was inversely related to the fenofibrate-induced increase of HDL-C, while no correlation was found with triglyceride reduction (Calabresi et al. 2002).

Recently, particular attention has been devoted to the effects of fibrates in subjects with T2DM. Four studies consistently demonstrated that different fibrates, i.e., ciprofibrate, gemfibrozil, fenofibrate, and bezafibrate, were all able to significantly improve FMD in T2DM, in a way that was not dependent on changes in lipid parameters, including HDL-C (Avogaro et al. 2001; Evans et al. 2000; Ghani et al. 2013; Triolo et al. 2013b). Interestingly, after 12 weeks of treatment with ciprofibrate, T2DM subjects displayed improved FMD measured in fasting conditions and in the postprandial phase (Evans et al. 2000). Ciprofibrate treatment also markedly modified lipid metabolism: the fasting triglyceride content of all lipoproteins was reduced and the increase during the postprandial phase blunted; the improvement of fasting FMD was correlated with the decreased triglyceride content in VLDL and HDL (but not with changes of HDL-C), while in the postprandial phase changes of FMD were correlated with the reduced lipoprotein TG content (Evans et al. 2000).

4.3 Niacin

The effect of niacin treatment on endothelial function has been assessed in a variety of clinical conditions, including established coronary artery disease, T2DM, metabolic syndrome, and low HDL-C. Overall, when niacin was used as monotherapy, the studies showed an increase in FMD and a reduction in inflammatory biomarkers but failed in demonstrating a correlation with the niacin-mediated increase of HDL-C levels. In 2002, Kuvin et al. demonstrated that a 3-month treatment with niacin improved FMD in patients with CAD and low HDL-C and that this improvement could be related to the HDL ability to increase eNOS protein abundance in vitro (Kuvin et al. 2002). Some years later, the same authors also showed that in statin treated patients with established CAD, the addition of niacin caused a reduction of the plasma levels of inflammatory biomarkers as CRP and lipoprotein-associated phospholipase A2 (Kuvin et al. 2006); unfortunately, likely due to the small number of participants, correlation analyses between changes of endothelial function parameters and the increase of HDL-C or other lipid variables were not performed (Kuvin et al. 2006). Later on, the ability of niacin to improve FMD and reduce inflammatory biomarkers as CRP was confirmed in subjects with isolated low HDL-C (Benjo et al. 2006), with the metabolic syndrome (Thoenes et al. 2007), and in patients after myocardial infarction (Bregar et al. 2013); the relationship between increase of HDL-C and changes of FMD was analyzed only in the latter case, and no correlation was found with any lipid parameter. The absence of correlation between niacin-mediated changes of HDL-C and endothelial function could be explained by the direct antioxidant, anti-inflammatory, and anti-adhesive activities of niacin which are independent from its role on lipid metabolism (Yadav et al. 2012). However, a recent study clearly showed that HDL isolated from subjects with T2DM after 3 months of treatment with niacin displayed an improved ability to stimulate NO production, to reduce reactive oxygen species, and to promote EPC-mediated endothelial repair in vitro when compared to HDL isolated at baseline (Sorrentino et al. 2010), further strengthening the concept that plasma HDL-C and its change may not be a reliable indicator of HDL function. Recent studies analyzed the effect of niacin treatment in patients with established CAD on top of the existing statin therapy; two of them failed to detect any significant improvement of FMD, while in one case FMD was significantly increased only in the subgroup of patients with a low baseline HDL-C level (Lee et al. 2009; Philpott et al. 2013; Warnholtz et al. 2009).

5 Effect on HDL Antioxidant Properties

ApoA-I plays a major role in HDL-mediated protection from oxidative damage (Rosenson et al. 2013). Many studies have indicated that the antioxidant function of HDL also depends upon such enzymes as lipoprotein-associated phospholipase A2 (Lp-PLA2), paraoxonase 1 (PON1), lecithin-cholesterol acyltransferase (LCAT), and glutathione selenoperoxidase (GSPx) (Florentin et al. 2008; Otocka-Kmiecik et al. 2012; Podrez 2010). GSPx can reduce lipoprotein lipid hydroperoxides (LOOHs) to the corresponding hydroxides and thereby detoxify them (Rosenson et al. 2013). Besides apoA-I, other apolipoproteins such as apoE, apoJ, apoA-II, and apoA-IV also display antioxidant properties but to a less degree than apoA-I (Otocka-Kmiecik et al. 2012). An antioxidant mechanism of HDL may also result from their ability to accept phospholipid-containing hydroperoxides (PLOOHs) and other lipid peroxidation products from oxidized LDL (Stremler et al. 1991) and their subsequent reduction by redox-active methionine residues of apoA-I, with the formation of redox-inactive phospholipid hydroxides (Kontush and Chapman 2010). In addition, HDL functions as an antioxidant enzyme by hydrolyzing oxidized phospholipids, such as F2-isoprostanes, formed during the oxidative modification of LDL. Antioxidant properties differ for the different HDL subclasses

(Kontush and Chapman 2010). Activities of HDL-associated enzymes—LCAT, PON1, and platelet-activating factor acetyl hydrolase (PAF-AH)—are elevated in small, dense HDL3c (Davidson et al. 2009), which thus seem to have a potent role in protecting LDL from oxidation (Podrez 2010). It is worth emphasizing that small, dense HDL3 particles are also more resistant to oxidative modification compared with large, light HDL2 (Shuhei et al. 2010).

5.1 Statins

The role of statins in improving plasma oxidative status has been reported in a number of studies, although this effect was not always related to drug-induced changes in HDL. Different statins have been shown to significantly enhance the antioxidant activity of PON1. Atorvastatin has a positive impact on plasma total antioxidant status and PON1 activity by reducing plasma susceptibility to lipid peroxidation induced by free radicals (Fuhrman et al. 2002; Harangi et al. 2004, 2009; Nagila et al. 2009). Pitavastatin was also shown to increase the promoter activity and protein expression of PON1 in vitro via p44/42 mitogen-activated protein kinase-mediated phosphorylation of sterol-regulatory-element-binding protein-2 and binding of Sp1 to PON1 DNA (Arii et al. 2009; Yamashita et al. 2010). The improvement of antioxidant properties of HDL, via increased PON1 activity, was also observed after simvastatin and fluvastatin therapy (Bergheanu et al. 2007; Mirdamadi et al. 2008; Muacevic-Katanec et al. 2007). Atorvastatin also suppresses the enhanced cellular uptake of oxidized LDL by macrophages and reduces expression of cellular scavenger receptors (Fuhrman et al. 2002; Otocka-Kmiecik et al. 2012). Similar results were observed in the study by Kassai et al. which showed a significant increase in serum PON1 specific activity, PON/HDL ratio, and LCAT activity, as well as significant reduction of oxidized LDL and CETP activities after atorvastatin therapy (Kassai et al. 2007). Finally, statin treatment enhances the defective antioxidative activity of HDL3 in patients with acute coronary syndrome by increasing the activity of HDL-associated enzymes (Otocka-Kmiecik et al. 2012).

5.2 Fibrates

Tsimihodimos et al. reported that fenofibrate induces redistribution of Lp-PLA2 from apoB-containing lipoproteins to HDL in dyslipidemic patients (Tsimihodimos et al. 2003). This effect enhanced the anti-inflammatory and antioxidant potential of the enzyme. In a recent study, Vazanna et al. examined whether HDL levels are related to in vivo oxidative stress and platelet activation in patients with coronary heart disease treated with fenofibrate (Vazzana et al. 2013). The authors measured urinary 8-iso-prostaglandin (PG) F2α and 11-dehydrothromboxane (TX) B2—in vivo markers of oxidative stress and platelet activation, respectively. They observed significantly higher urinary 8-iso-PGF2α and 11-dehydro-TXB2 in patients with low HDL than in individuals with higher HDL levels, as well as an inverse relation between the markers of oxidation and HDL concentrations. Fenofibrate treatment significantly reduced the 2 eicosanoids in healthy subjects, in parallel with an HDL increase (Vazzana et al. 2013). Tkac et al. showed a significant reduction of circulating conjugated dienes, a nonsignificant decrease in the production of malondialdehyde, and an increase in the GPx activity in patients with combined hyperlipidemia treated with fenofibrate (Tkac et al. 2006). Similar results were obtained in patients with visceral obesity and dyslipidemia treated with micronized fenofibrate (Broncel et al. 2006). Few studies have evaluated the effect of fibrates on PON1 activity. Paragh et al. evaluated this effect in a group of coronary patients with type IIb hyperlipidemia; after 3 months of treatment, HDL-C and apoA-I levels increased significantly, and PON1 specific activity also increased significantly (Paragh et al. 2003). Very similar results were obtained by Phuntuwate et al., who evaluated the effect of fenofibrate on PON1 levels, as well as the effects of the PON1 polymorphisms on lipid and PON1 responses to the drug, in dyslipidemic patients with low HDL (Phuntuwate et al. 2008). A significant increases in PON1 concentration and activity was observed after treatment, and a significant positive correlation between changes in HDL-C and changes in PON1 concentration/activity was detected (Phuntuwate et al. 2008). The response of lipid parameters to fenofibrate was independent of PON1 polymorphisms; however, PON1 Q192R and T-108C polymorphisms significantly affected the increase in PON1 activity and concentration (Phuntuwate et al. 2008). An increase of PON1 activity, which was significantly correlated with the changes in HDL-C, was also observed in patients with combined hyperlipidemia (Yesilbursa et al. 2005). On the contrary, Dullart et al. did not observe such increase after short-term (8 weeks) administration of simvastatin and bezafibrate, even when combined in diabetic patients (Dullaart et al. 2009). It needs to be also emphasized that besides the above mentioned positive HDL associated antioxidant properties of fibrates, they also elevate plasma total homocysteine, which is known to induce oxidative stress and endothelial dysfunction. Among others, a subgroup analysis of the FIELD study showed that fenofibrate treatment was associated with an increase of homocysteine levels, with no changes of serum PON-1 mass (Maranghi et al. 2011).

5.3 Niacin

There is still very limited data on the potential role of niacin therapy on HDL-related antioxidant properties. Recent studies indicate that niacin increases vascular endothelial cell redox state, resulting in the inhibition of oxidative stress (Kamanna and Kashyap 2008). These effects were investigated by Ganji et al. in vitro, who showed that in cultured endothelial cells niacin increases NADPH levels and reduces GSH (Ganji et al. 2009). In the same study, niacin also inhibited angiotensin II-induced reactive oxygen species (ROS) production (Ganji et al. 2009). These findings indicate for the first time that niacin inhibits vascular inflammation by decreasing endothelial ROS production and subsequent

LDL oxidation (Ganji et al. 2009). Kaplon et al. have recently tested the hypothesis that vascular endothelial function and oxidative stress are related to dietary niacin intake among healthy middle-aged and older adults (Kaplon et al. 2014). They observed that plasma ox-LDL was inversely related to niacin intake. Moreover, in endothelial cells sampled from the brachial artery of a subgroup patients, dietary niacin intake was inversely related to nitrotyrosine, a marker of oxidative damage, leading to the hypothesis that higher dietary niacin intake is associated with greater vascular endothelial function related to lower systemic and vascular oxidative stress (Kaplon et al. 2014). The antioxidant properties of niacin were also observed patients with hypercholesterolemia and low HDL-C treated with niacin for 12 weeks (Hamoud et al. 2013). Subjects with low HDL-C levels exhibited higher oxidative stress compared with subjects with normal HDL-C levels. Niacin treatment in hypercholesterolemic patients caused a significant increase in HDL-C and apoA-I levels and a decrease in TG levels. Niacin also significantly reduced oxidative stress, as measured by a significant decrease in the serum content of thiobarbituric acid reactive substances and lipid peroxides and increase of PON activity, compared with the levels before treatment (Hamoud et al. 2013).

Conclusions and Perspectives

Currently, optimal statin treatment manages to achieve a risk reduction for cardiovascular mortality of approximately 25–35 %, mainly due to a reduction of LDL-C. This leaves an important residual risk that needs to be considered for the optimal management of patients at risk of cardiovascular disease. Raising HDL-C was seen as a viable and promising way to further reduce the risk of cardiovascular mortality. However, things look to be difficult with HDL (Table 1).

Although all fibrates have been shown to increase HDL-C significantly, their beneficial effect on cardiovascular and all-cause mortality remains controversial (Saha et al. 2007). The early VA-HIT and BIP secondary prevention trials have shown that treating CHD patients with low plasma HDL-C levels and especially

Table 1 Effects of established hypolipidemic drugs on HDL concentration and function

	Statins	Fibrates	Niacin
HDL-C	↑ 2–10 %	↑ 5–15 %	↑ 15–30 %
HDL subclasses	↑ large HDL particles	↑ small HDL particles	↑ large HDL particles
Cell cholesterol efflux	↑ up to 14 % (mainly SR-BI mediated)	↑ up to 21 % (mainly ABCA1 mediated)	↑ up to 25 % (mainly SR-BI mediated)
Endothelial cell homeostasis	↑ FMD[a] ↑ HDL-induced NO production[b]	↑ FMD[a] ↓ markers of inflammation	↑ FMD ↑ EPC-mediated endothelial repair ↑ HDL-induced NO production
Antioxidant activity	↑ PON1 activity	↑ PON1 activity ↑ HDL-bound LpPLA$_2$	↑ PON1 activity

[a]Not necessarily related to HDL changes
[b]In vitro data

those with moderate triglyceride elevations and some degree of insulin resistance (Rubins et al. 2002; The Bezafibrate Infarction Prevention (BIP) Study Group 2000) with a fibrate leads to significant reductions in the risk of a recurrent cardiovascular event, part of the effect being mediated by the fibrate-induced increase in HDL-C (Robins et al. 2001). In the FIELD study in people with type 2 diabetes mellitus, fenofibrate induced a highly significant reduction of cardiovascular events in patients with low HDL-C levels, versus a minimal effectiveness of the drug in patients with higher HDL-C (The FIELD study investigators 2005). The more recent ACCORD trial shows no benefits of adding fibrate therapy to statin on cardiovascular events except possibly in patients with residual atherogenic dyslipidemia (Ginsberg et al. 2010).

Earlier studies with niacin suggested a favorable effect of the drug on cardiovascular mortality (Canner et al. 1986), but the AIM-HIGH trial, which compared the combination niacin/simvastatin with simvastatin alone, was stopped for lack of efficacy, as niacin on the top of statin, despite increasing HDL-C levels, failed to determine an incremental clinical benefit in patients with LDL-C values at target (Boden et al. 2011). The more recent and much larger HPS2-THRIVE trial, also designed to assess the effect of adding niacin to a statin therapy in patients with established cardiovascular disease, also gave disappointing results (HPS2-THRIVE Collaborative Group 2013).

Novel HDL-raising drugs targeting new players in the HDL system, and more potent than the available drugs, are urgently needed. The promising CETP inhibitors, despite substantially increasing HDL-C levels, did not give so far positive results on cardiovascular risk (Barter et al. 2007; Schwartz et al. 2013). Different is the case of the HDL mimetics, which appears to be more promising, at least from results on surrogate end points (Nissen et al. 2003; Tardif et al. 2007), awaiting morbidity and mortality studies.

Open Access This chapter is distributed under the terms of the Creative Commons Attribution Noncommercial License, which permits any noncommercial use, distribution, and reproduction in any medium, provided the original author(s) and source are credited.

References

Adorni MP, Zimetti F, Billheimer JT, Wang N, Rader DJ, Phillips MC, Rothblat GH (2007) The roles of different pathways in the release of cholesterol from macrophages. J Lipid Res 48:2453–2462

Alaupovic P, Hodis HN, Knight Gibson C, Mack WJ, LaBree L, Cashin Hemphill L, Corder CN, Kramsch DM, Blankenhorn DH (1994) Effects of lovastatin on ApoA- and ApoB-containing lipoproteins families in a subpopulation of patients participating in the monitored atherosclerosis regression study (MARS). Arterioscler Thromb 14:1906–1913

Alrasadi K, Awan Z, Alwaili K, Ruel I, Hafiane A, Krimbou L, Genest J (2008) Comparison of treatment of severe high-density lipoprotein cholesterol deficiency in men with daily atorvastatin (20 mg) versus fenofibrate (200 mg) versus extended-release niacin (2 g). Am J Cardiol 102:1341–1347

Antonopoulos AS, Margaritis M, Lee R, Channon K, Antoniades C (2012) Statins as anti-inflammatory agents in atherogenesis: molecular mechanisms and lessons from the recent clinical trials. Curr Pharm Des 18:1519–1530

Arii K, Suehiro T, Ota K, Ikeda Y, Kumon Y, Osaki F, Inoue M, Inada S, Ogami N, Takata H, Hashimoto K, Terada Y (2009) Pitavastatin induces PON1 expression through p44/42 mitogen-activated protein kinase signaling cascade in Huh7 cells. Atherosclerosis 202:439–445

Asztalos BF, Schaefer EJ (2003) High-density lipoprotein subpopulations in pathologic conditions. Am J Cardiol 91:12E–17E

Asztalos BF, Horvath K, McNamara J, Roheim P, Rubinstein J, Schaefer E (2002a) Comparing the effects of five different statins on the HDL subpopulation profiles of coronary heart disease patients. Atherosclerosis 164:361–369

Asztalos BF, Horvath KV, McNamara JR, Roheim PS, Rubinstein JJ, Schaefer EJ (2002b) Effects of atorvastatin on the HDL subpopulation profile of coronary heart disease patients. J Lipid Res 43:1701–1707

Asztalos BF, Le MF, Dallal GE, Stein E, Jones PH, Horvath KV, McTaggart F, Schaefer EJ (2007) Comparison of the effects of high doses of rosuvastatin versus atorvastatin on the subpopulations of high-density lipoproteins. Am J Cardiol 99:681–685

Asztalos BF, Collins D, Horvath KV, Bloomfield HE, Robins SJ, Schaefer EJ (2008) Relation of gemfibrozil treatment and high-density lipoprotein subpopulation profile with cardiovascular events in the veterans affairs high-density lipoprotein intervention trial. Metabolism 57:77–83

Avogaro A, Miola M, Favaro A, Gottardo L, Pacini G, Manzato E, Zambon S, Sacerdoti D, de Kreutzenberg S, Piliego T, Tiengo A, Del Prato S (2001) Gemfibrozil improves insulin sensitivity and flow-mediated vasodilatation in type 2 diabetic patients. Eur J Clin Invest 31:603–609

Bard JM, Luc G, Douste-Blazy P, Drouin P, Ziegler O, Jacotot B, Dachet C, de Gennes JL, Fruchart JC (1989) Effect of simvastatin on plasma lipids, apolipoproteins and lipoprotein particles in patients with primary hypercholesterolaemia. Eur J Clin Pharmacol 37:545–550

Barter PJ, Baker PW, Rye KA (2002) Effect of high-density lipoproteins on the expression of adhesion molecules in endothelial cells. Curr Opin Lipidol 13:285–288

Barter PJ, Caulfield M, Eriksson M, Grundy SM, Kastelein JJ, Komajda M, Lopez-Sendon J, Mosca L, Tardif JC, Waters DD, Shear CL, Revkin JH, Buhr KA, Fisher MR, Tall AR, Brewer B (2007) Effects of torcetrapib in patients at high risk for coronary events. N Engl J Med 357:2109–2122

Barter PJ, Brandrup-Wognsen G, Palmer MK, Nicholls SJ (2010) Effect of statins on HDL-C: a complex process unrelated to changes in LDL-C: analysis of the VOYAGER Database. J Lipid Res 51:1546–1553

Benjo AM, Maranhao RC, Coimbra SR, Andrade AC, Favarato D, Molina MS, Brandizzi LI, da Luz PL (2006) Accumulation of chylomicron remnants and impaired vascular reactivity occur in subjects with isolated low HDL cholesterol: effects of niacin treatment. Atherosclerosis 187:116–122

Berger JP, Akiyama TE, Meinke PT (2005) PPARs: therapeutic targets for metabolic disease. Trends Pharmacol Sci 26:244–251

Bergheanu SC, Van Tol A, Dallinga-Thie GM, Liem A, Dunselman PH, Van der Bom JG, Jukema JW (2007) Effect of rosuvastatin versus atorvastatin treatment on paraoxonase-1 activity in men with established cardiovascular disease and a low HDL-cholesterol. Curr Med Res Opin 23:2235–2240

Birjmohun RS, Hutten BA, Kastelein JJ, Stroes ES (2005) Efficacy and safety of high-density lipoprotein cholesterol-increasing compounds. A meta-analysis of randomized controlled trials. J Am Coll Cardiol 45:185–197

Boden WE, Probstfield JL, Anderson T, Chaitman BR, Desvignes-Nickens P, Koprowicz K, McBride R, Teo K, Weintraub W (2011) Niacin in patients with low HDL cholesterol levels receiving intensive statin therapy. N Engl J Med 365:2255–2267

Bodor ET, Offermanns S (2008) Nicotinic acid: an old drug with a promising future. Br J Pharmacol 153(Suppl 1):S68–S75

Bregar U, Jug B, Keber I, Cevc M, Sebestjen M (2013) Extended-release niacin/laropiprant improves endothelial function in patients after myocardial infarction. Heart Vessel 29 (3):313–319
Broncel M, Cieslak D, Koter-Michalak M, Duchnowicz P, Mackiewicz K, Chojnowska-Jezierska J (2006) The anti-inflammatory and antioxidants effects of micronized fenofibrate in patients with visceral obesity and dyslipidemia. Pol Merkur Lekarski 20:547–550
Calabresi L, Franceschini G, Sirtori CR, de Palma A, Saresella M, Ferrante P, Taramelli D (1997) Inhibition of VCAM-1 expression in endothelial cells by reconstituted high density lipoproteins. Biochem Biophys Res Commun 238:61–65
Calabresi L, Gomaraschi M, Villa B, Omoboni L, Dmitrieff C, Franceschini G (2002) Elevated soluble cellular adhesion molecules in subjects with low HDL-cholesterol. Arterioscler Thromb Vasc Biol 22:656–661
Calabresi L, Gomaraschi M, Franceschini G (2003) Endothelial protection by high-density lipoproteins: from bench to bedside. Arterioscler Thromb Vasc Biol 23:1724–1731
Calabresi L, Gomaraschi M, Franceschini G (2010) High-density lipoprotein quantity or quality for cardiovascular prevention? Curr Pharm Des 16:1494–1503
Canner PL, Berge KG, Wenger NK, Stamler J, Friedman L, Prineas RJ, Friedewald WT (1986) Fifteen year mortality in coronary drug project patients: longterm benefit with niacin. J Am Coll Cardiol 8:1245–1255
Cockerill GW, Rye KA, Gamble JR, Vadas MA, Barter PJ (1995) High-density lipoproteins inhibit cytokine-induced expression of endothelial cell adhesion molecules. Arterioscler Thromb Vasc Biol 15:1987–1994
Cuchel M, Rader DJ (2006) Macrophage reverse cholesterol transport: key to the regression of atherosclerosis? Circulation 113:2548–2555
Davidson WS, Silva RA, Chantepie S, Lagor WR, Chapman MJ, Kontush A (2009) Proteomic analysis of defined HDL subpopulations reveals particle-specific protein clusters: relevance to antioxidative function. Arterioscler Thromb Vasc Biol 29:870–876
De Lalla OF, Gofman JW (1954) Ultracentrifugal analysis of serum lipoproteins. Methods Biochem Anal 1:459–478
De Vries R, Kerstens MN, Sluiter WJ, Groen AK, Van TA, Dullaart RP (2005) Cellular cholesterol efflux to plasma from moderately hypercholesterolaemic type 1 diabetic patients is enhanced, and is unaffected by simvastatin treatment. Diabetologia 48:1105–1113
Deanfield JE, Halcox JP, Rabelink TJ (2007) Endothelial function and dysfunction: testing and clinical relevance. Circulation 115:1285–1295
Dullaart RP, De VR, Voorbij HA, Sluiter WJ, Van TA (2009) Serum paraoxonase-I activity is unaffected by short-term administration of simvastatin, bezafibrate, and their combination in type 2 diabetes mellitus. Eur J Clin Invest 39:200–203
Evans M, Anderson RA, Graham J, Ellis GR, Morris K, Davies S, Jackson SK, Lewis MJ, Frenneaux MP, Rees A (2000) Ciprofibrate therapy improves endothelial function and reduces postprandial lipemia and oxidative stress in type 2 diabetes mellitus. Circulation 101:1773–1779
Favari E, Calabresi L, Adorni MP, Jessup W, Simonelli S, Franceschini G, Bernini F (2009) Small discoidal pre-beta1 HDL particles are efficient acceptors of cell cholesterol via ABCA1 and ABCG1. Biochemistry 48:11067–11074
Florentin M, Liberopoulos EN, Wierzbicki AS, Mikhailidis DP (2008) Multiple actions of high-density lipoprotein. Curr Opin Cardiol 23:370–378
Franceschini G, Sirtori M, Vaccarino V, Gianfranceschi G, Chiesa G, Sirtori CR (1989) Plasma lipoprotein changes after treatment with pravastatin and gemfibrozil in patients with familial hypercholesterolemia. J Lab Clin Med 114:250–259
Franceschini G, Calabresi L, Colombo C, Favari E, Bernini F, Sirtori CR (2007) Effects of fenofibrate and simvastatin on HDL-related biomarkers in low-HDL patients. Atherosclerosis 195:385–391
Franceschini G, Favari E, Calabresi L, Simonelli S, Bondioli A, Adorni MP, Zimetti F, Gomaraschi M, Coutant K, Rossomanno S, Niesor EJ, Bernini F, Benghozi R (2013) Differential effects of fenofibrate and extended-release niacin on high-density lipoprotein particle

size distribution and cholesterol efflux capacity in dyslipidemic patients. J Clin Lipidol 7:414–422

Fuhrman B, Koren L, Volkova N, Keidar S, Hayek T, Aviram M (2002) Atorvastatin therapy in hypercholesterolemic patients suppresses cellular uptake of oxidized-LDL by differentiating monocytes. Atherosclerosis 164:179–185

Ganji SH, Qin S, Zhang L, Kamanna VS, Kashyap ML (2009) Niacin inhibits vascular oxidative stress, redox-sensitive genes, and monocyte adhesion to human aortic endothelial cells. Atherosclerosis 202:68–75

Ghani RA, Bin YI, Wahab NA, Zainudin S, Mustafa N, Sukor N, Wan Mohamud WN, Kadir KA, Kamaruddin NA (2013) The influence of fenofibrate on lipid profile, endothelial dysfunction, and inflammatory markers in type 2 diabetes mellitus patients with typical and mixed dyslipidemia. J Clin Lipidol 7:446–453

Ginsberg HN, Elam MB, Lovato LC, Crouse JR III, Leiter LA, Linz P, Friedewald WT, Buse JB, Gerstein HC, Probstfield J, Grimm RH, Ismail-Beigi F, Bigger JT, Goff DC Jr, Cushman WC, Simons-Morton DG, Byington RP (2010) Effects of combination lipid therapy in type 2 diabetes mellitus. N Engl J Med 362:1563–1574

Gomaraschi M, Basilico N, Sisto F, Taramelli D, Eligini S, Colli S, Sirtori CR, Franceschini G, Calabresi L (2005) High-density lipoproteins attenuate interleukin-6 production in endothelial cells exposed to pro-inflammatory stimuli. Biochim Biophys Acta 1736:136–143

Guerin M, Lassel TS, Le Goff W, Farnier M, Chapman MJ (2000) Action of atorvastatin in combined hyperlipidemia : preferential reduction of cholesteryl ester transfer from HDL to VLDL1 particles. Arterioscler Thromb Vasc Biol 20:189–197

Guerin M, Egger P, Soudant C, Le Goff W, van Tol A, Dupuis R, Chapman MJ (2002) Dose-dependent action of atorvastatin in type IIB hyperlipidemia: preferential and progressive reduction of atherogenic apoB-containing lipoprotein subclasses (VLDL-2, IDL, small dense LDL) and stimulation of cellular cholesterol efflux. Atherosclerosis 163:287–296

Guerin M, Le Goff W, Frisdal E, Schneider S, Milosavljevic D, Bruckert E, Chapman MJ (2003) Action of ciprofibrate in type IIb hyperlipoproteinemia: modulation of the atherogenic lipoprotein phenotype and stimulation of high-density lipoprotein-mediated cellular cholesterol efflux. J Clin Endocrinol Metab 88:3738–3746

Hamoud S, Kaplan M, Meilin E, Hassan A, Torgovicky R, Cohen R, Hayek T (2013) Niacin administration significantly reduces oxidative stress in patients with hypercholesterolemia and low levels of high-density lipoprotein cholesterol. Am J Med Sci 345:195–199

Harangi M, Seres I, Varga Z, Emri G, Szilvassy Z, Paragh G, Remenyik E (2004) Atorvastatin effect on high-density lipoprotein-associated paraoxonase activity and oxidative DNA damage. Eur J Clin Pharmacol 60:685–691

Harangi M, Mirdamadi HZ, Seres I, Sztanek F, Molnar M, Kassai A, Derdak Z, Illyes L, Paragh G (2009) Atorvastatin effect on the distribution of high-density lipoprotein subfractions and human paraoxonase activity. Transl Res 153:190–198

Hernandez M, Wright SD, Cai TQ (2007) Critical role of cholesterol ester transfer protein in nicotinic acid-mediated HDL elevation in mice. Biochem Biophys Res Commun 355:1075–1080

Higashi Y, Matsuoka H, Umei H, Sugano R, Fujii Y, Soga J, Kihara Y, Chayama K, Imaizumi T (2010) Endothelial function in subjects with isolated low HDL cholesterol: role of nitric oxide and circulating progenitor cells. Am J Physiol Endocrinol Metab 298:E202–E209

Homma Y, Ozawa H, Kobayashi T, Yamaguchi H, Sakane H, Nakamura H (1995) Effects of simvastatin on plasma lipoprotein subfractions, cholesterol esterification rate, and cholesteryl ester transfer protein in type II hyperlipoproteinemia. Atherosclerosis 114:223–234

HPS2-THRIVE Collaborative Group (2013) HPS2-THRIVE randomized placebo-controlled trial in 25 673 high-risk patients of ER niacin/laropiprant: trial design, pre-specified muscle and liver outcomes, and reasons for stopping study treatment. Eur Heart J 34:1279–1291

Igarashi J, Miyoshi M, Hashimoto T, Kubota Y, Kosaka H (2007) Statins induce S1P(1) receptors and enhance endothelial nitric oxide production in response to high-density lipoproteins. Br J Pharmacol 150:470–479

Johansson J, Carlson LA (1990) The effects of nicotinic acid treatment on high density lipoprotein particle size subclass levels in hyperlipidaemic subjects. Atherosclerosis 83:207–216

Jones PH, Davidson MH, Stein EA, Bays HE, McKenney JM, Miller E, Cain VA, Blasetto JW (2003) Comparison of the efficacy and safety of rosuvastatin versus atorvastatin, simvastatin, and pravastatin across doses (STELLAR* Trial). Am J Cardiol 92:152–160

Kamanna VS, Kashyap ML (2008) Mechanism of action of niacin. Am J Cardiol 101:20B–26B

Kaplon RE, Gano LB, Seals DR (2014) Vascular endothelial function and oxidative stress are related to dietary niacin intake among healthy middle-aged and older adults. J Appl Physiol (1985) 116:156–163

Kassai A, Illyes L, Mirdamadi HZ, Seres I, Kalmar T, Audikovszky M, Paragh G (2007) The effect of atorvastatin therapy on lecithin:cholesterol acyltransferase, cholesteryl ester transfer protein and the antioxidant paraoxonase. Clin Biochem 40:1–5

Khera AV, Cuchel M, De LL-M, Rodrigues A, Burke MF, Jafri K, French BC, Phillips JA, Mucksavage ML, Wilensky RL, Mohler ER, Rothblat GH, Rader DJ (2011) Cholesterol efflux capacity, high-density lipoprotein function, and atherosclerosis. N Engl J Med 364:127–135

Khera AV, Patel PJ, Reilly MP, Rader DJ (2013) The addition of niacin to statin therapy improves high-density lipoprotein cholesterol levels but not metrics of functionality. J Am Coll Cardiol 62:1909–1910

Khoury N, Goldberg AC (2011) The use of fibric acid derivatives in cardiovascular prevention. Curr Treat Option Cardiovasc Med 13:335–342

Kimura T, Mogi C, Tomura H, Kuwabara A, Im DS, Sato K, Kurose H, Murakami M, Okajima F (2008) Induction of scavenger receptor class B type I is critical for simvastatin enhancement of high-density lipoprotein-induced anti-inflammatory actions in endothelial cells. J Immunol 181:7332–7340

Knopp RH, Alagona P, Davidson M, Goldberg AC, Kafonek SD, Kashyap M, Sprecher D, Superko HR, Jenkins S, Marcovina S (1998) Equivalent efficacy of a time-release form of niacin (Niaspan) given once-a-night versus plain niacin in the management of hyperlipidemia. Metabolism 47:1097–1104

Koh KK, Yeal AJ, Hwan HS, Kyu JD, Sik KH, Cheon LK, Kyun SE, Sakuma I (2004) Effects of fenofibrate on lipoproteins, vasomotor function, and serological markers of inflammation, plaque stabilization, and hemostasis. Atherosclerosis 174:379–383

Koh KK, Quon MJ, Han SH, Chung WJ, Ahn JY, Seo YH, Choi IS, Shin EK (2005) Additive beneficial effects of fenofibrate combined with atorvastatin in the treatment of combined hyperlipidemia. J Am Coll Cardiol 45:1649–1653

Kontush A, Chapman MJ (2010) Antiatherogenic function of HDL particle subpopulations: focus on antioxidative activities. Curr Opin Lipidol 21:312–318

Kuvin JT, Ramet ME, Patel AR, Pandian NG, Mendelsohn ME, Karas RH (2002) A novel mechanism for the beneficial vascular effects of high-density lipoprotein cholesterol: enhanced vasorelaxation and increased endothelial nitric oxide synthase expression. Am Heart J 144:165–172

Kuvin JT, Dave DM, Sliney KA, Mooney P, Patel AR, Kimmelstiel CD, Karas RH (2006) Effects of extended-release niacin on lipoprotein particle size, distribution, and inflammatory markers in patients with coronary artery disease. Am J Cardiol 98:743–745

Lamon-Fava S, Diffenderfer MR, Barrett PH, Buchsbaum A, Nyaku M, Horvath KV, Asztalos BF, Otokozawa S, Ai M, Matthan NR, Lichtenstein AH, Dolnikowski GG, Schaefer EJ (2008) Extended-release niacin alters the metabolism of plasma apolipoprotein (Apo) A-I and ApoB-containing lipoproteins. Arterioscler Thromb Vasc Biol 28:1672–1678

Lee JM, Robson MD, Yu LM, Shirodaria CC, Cunnington C, Kylintireas I, Digby JE, Bannister T, Handa A, Wiesmann F, Durrington PN, Channon KM, Neubauer S, Choudhury RP (2009) Effects of high-dose modified-release nicotinic acid on atherosclerosis and vascular function a randomized, placebo-controlled, magnetic resonance imaging study. J Am Coll Cardiol 54:1787–1794

Lewis GF, Rader DJ (2005) New insights into the regulation of HDL metabolism and reverse cholesterol transport. Circ Res 96:1221–1232

Liu Y, Wei J, Hu S, Hu L (2012) Beneficial effects of statins on endothelial progenitor cells. Am J Med Sci 344:220–226

Malik J, Melenovsky V, Wichterle D, Haas T, Simek J, Ceska R, Hradec J (2001) Both fenofibrate and atorvastatin improve vascular reactivity in combined hyperlipidaemia (fenofibrate versus atorvastatin trial – FAT). Cardiovasc Res 52:290–298

Maranghi M, Hiukka A, Badeau R, Sundvall J, Jauhiainen M, Taskinen MR (2011) Macrophage cholesterol efflux to plasma and HDL in subjects with low and high homocysteine levels: a FIELD substudy. Atherosclerosis 219:259–265

McGrath KC, Li XH, Puranik R, Liong EC, Tan JT, Dy VM, DiBartolo BA, Barter PJ, Rye KA, Heather AK (2009) Role of 3beta-hydroxysteroid-Delta 24 reductase in mediating antiinflammatory effects of high-density lipoproteins in endothelial cells. Arterioscler Thromb Vasc Biol 29:877–882

Miida T, Sakai K, Ozaki K, Nakamura Y, Yamaguchi T, Tsuda T, Kashiwa T, Murakami T, Inano K, Okada M (2000) Bezafibrate increases Prebeta1-HDL at the expense of HDL(2b) in hypertriglyceridemia. Arterioscler Thromb Vasc Biol 20:2428–2433

Mineo C, Shaul PW (2012) Novel biological functions of high-density lipoprotein cholesterol. Circ Res 111:1079–1090

Mirdamadi HZ, Sztanek F, Derdak Z, Seres I, Harangi M, Paragh G (2008) The human paraoxonase-1 phenotype modifies the effect of statins on paraoxonase activity and lipid parameters. Br J Clin Pharmacol 66:366–374

Miyamoto-Sasaki M, Yasuda T, Monguchi T, Nakajima H, Mori K, Toh R, Ishida T, Hirata K (2013) Pitavastatin increases HDL particles functionally preserved with cholesterol efflux capacity and antioxidative actions in dyslipidemic patients. J Atheroscler Thromb 20:708–716

Morgan JM, Capuzzi DM, Baksh RI, Intenzo C, Carey CM, Reese D, Walker K (2003) Effects of extended-release niacin on lipoprotein subclass distribution. Am J Cardiol 91:1432–1436

Morgan JM, De LL-M, Capuzzi DM (2007) Effects of niacin and Niaspan on HDL lipoprotein cellular SR-BI-mediated cholesterol efflux. J Clin Lipidol 1:614–619

Muacevic-Katanec D, Bradamante V, Poljicanin T, Reiner Z, Babic Z, Simeon-Rudolf V, Katanec D (2007) Clinical study on the effect of simvastatin on paraoxonase activity. Arzneimittelforschung 57:647–653

Nagila A, Permpongpaiboon T, Tantrarongroj S, Porapakkham P, Chinwattana K, Deakin S, Porntadavity S (2009) Effect of atorvastatin on paraoxonase1 (PON1) and oxidative status. Pharmacol Rep 61:892–898

Nichols AV, Krauss RM, Musliner TA (1986) Nondenaturing polyacrylamide gradient gel electrophoresis. Methods Enzymol 128:417–431

Nissen SE, Tsunoda T, Tuzcu EM, Schoenhagen P, Cooper CJ, Yasin M, Eaton GM, Lauer MA, Sheldon WS, Grines CL, Halpern S, Crowe T, Blankenship JC, Kerensky R (2003) Effect of recombinant ApoA-I Milano on coronary atherosclerosis in patients with acute coronary syndromes: a randomized controlled trial. JAMA 290:2292–2300

Nofer JR, van der Giet M, Tolle M, Wolinska I, von Wnuck LK, Baba HA, Tietge UJ, Godecke A, Ishii I, Kleuser B, Schafers M, Fobker M, Zidek W, Assmann G, Chun J, Levkau B (2004) HDL induces NO-dependent vasorelaxation via the lysophospholipid receptor S1P3. J Clin Invest 113:569–581

Otocka-Kmiecik A, Mikhailidis DP, Nicholls SJ, Davidson M, Rysz J, Banach M (2012) Dysfunctional HDL: a novel important diagnostic and therapeutic target in cardiovascular disease? Prog Lipid Res 51:314–324

Otvos JD, Collins D, Freedman DS, Shalaurova I, Schaefer EJ, McNamara JR, Bloomfield HE, Robins SJ (2006) Low-density lipoprotein and high-density lipoprotein particle subclasses predict coronary events and are favorably changed by gemfibrozil therapy in the veterans affairs high-density lipoprotein intervention trial. Circulation 113:1556–1563

Paragh G, Seres I, Harangi M, Balogh Z, Illyes L, Boda J, Szilvassy Z, Kovacs P (2003) The effect of micronised fenofibrate on paraoxonase activity in patients with coronary heart disease. Diabetes Metab 29:613–618

Patsch JR, Prasad S, Gotto AM Jr, Bengtsson-Olivecrona G (1984) Postprandial lipemia. A key for the conversion of high density lipoprotein$_2$ into high density lipoprotein$_3$ by hepatic lipase. J Clin Invest 74:2017–2023

Petoumenos V, Nickenig G, Werner N (2009) High-density lipoprotein exerts vasculoprotection via endothelial progenitor cells. J Cell Mol Med 13:4623–4635

Philpott AC, Hubacek J, Sun YC, Hillard D, Anderson TJ (2013) Niacin improves lipid profile but not endothelial function in patients with coronary artery disease on high dose statin therapy. Atherosclerosis 226:453–458

Phuntuwate W, Suthisisang C, Koanantakul B, Chaloeiphap P, Mackness B, Mackness M (2008) Effect of fenofibrate therapy on paraoxonase1 status in patients with low HDL-C levels. Atherosclerosis 196:122–128

Podrez EA (2010) Anti-oxidant properties of high-density lipoprotein and atherosclerosis. Clin Exp Pharmacol Physiol 37:719–725

Reriani MK, Dunlay SM, Gupta B, West CP, Rihal CS, Lerman LO, Lerman A (2011) Effects of statins on coronary and peripheral endothelial function in humans: a systematic review and meta-analysis of randomized controlled trials. Eur J Cardiovasc Prev Rehabil 18:704–716

Riwanto M, Landmesser U (2013) High density lipoproteins and endothelial functions: mechanistic insights and alterations in cardiovascular disease. J Lipid Res 54:3227–3243

Robins SJ, Collins D, Wittes JT, Papademetriou V, Deedwania PC, Schaefer EJ, McNamara JR, Kashyap ML, Hershman JM, Wexler LF, Rubins HB (2001) Relation of gemfibrozil treatment and lipid levels with major coronary events: VA-HIT: a randomized controlled trial. JAMA 285:1585–1591

Rosenson RS, Brewer HB Jr, Ansell B, Barter P, Chapman MJ, Heinecke JW, Kontush A, Tall AR, Webb NR (2013) Translation of high-density lipoprotein function into clinical practice: current prospects and future challenges. Circulation 128:1256–1267

Rubic T, Trottmann M, Lorenz RL (2004) Stimulation of CD36 and the key effector of reverse cholesterol transport ATP-binding cassette A1 in monocytoid cells by niacin. Biochem Pharmacol 67:411–419

Rubins HB, Robins SJ, Collins D, Nelson DB, Elam MB, Schaefer EJ, Faas FH, Anderson JW (2002) Diabetes, plasma insulin, and cardiovascular disease: subgroup analysis from the department of veterans affairs high-density lipoprotein intervention trial (VA-HIT). Arch Intern Med 162:2597–2604

Saha SA, Kizhakepunnur LG, Bahekar A, Arora RR (2007) The role of fibrates in the prevention of cardiovascular disease–a pooled meta-analysis of long-term randomized placebo-controlled clinical trials. Am Heart J 154:943–953

Sakai T, Kamanna VS, Kashyap ML (2001) Niacin, but not gemfibrozil, selectively increases LP-AI, a cardioprotective subfraction of HDL, in patients with low HDL cholesterol. Arterioscler Thromb Vasc Biol 21:1783–1789

Sasaki J, Yamamoto K, Ageta M (2002) Effects of fenofibrate on high-density lipoprotein particle size in patients with hyperlipidemia: a randomized, double-blind, placebo-controlled, multicenter, crossover study. Clin Ther 24:1614–1626

Schwartz GG, Olsson AG, Barter PJ (2013) Dalcetrapib in patients with an acute coronary syndrome. N Engl J Med 368:869–870

Shepherd J, Packard CJ, Patsch JR, Gotto AM Jr, Taunton OD (1979) Effects of nicotinic acid therapy on plasma high density lipoprotein subfraction distribution and composition on apolipoprotein A metabolism. J Clin Invest 63:858–867

Shuhei N, Soderlund S, Jauhiainen M, Taskinen MR (2010) Effect of HDL composition and particle size on the resistance of HDL to the oxidation. Lipids Health Dis 9:104

Sorrentino SA, Besler C, Rohrer L, Meyer M, Heinrich K, Bahlmann FH, Mueller M, Horvath T, Doerries C, Heinemann M, Flemmer S, Markowski A, Manes C, Bahr MJ, Haller H, Von EA,

Drexler H, Landmesser U (2010) Endothelial-vasoprotective effects of high-density lipoprotein are impaired in patients with type 2 diabetes mellitus but are improved after extended-release niacin therapy. Circulation 121:110–122

Soudijn W, van Wijngaarden I, Ijzerman AP (2007) Nicotinic acid receptor subtypes and their ligands. Med Res Rev 27:417–433

Stremler KE, Stafforini DM, Prescott SM, McIntyre TM (1991) Human plasma platelet-activating factor acetylhydrolase. Oxidatively fragmented phospholipids as substrates. J Biol Chem 266:11095–11103

Sviridov D, Hoang A, Ooi E, Watts G, Barrett PH, Nestel P (2008) Indices of reverse cholesterol transport in subjects with metabolic syndrome after treatment with rosuvastatin. Atherosclerosis 197:732–739

Tamehiro N, Shigemoto-Mogami Y, Kakeya T, Okuhira K, Suzuki K, Sato R, Nagao T, Nishimaki-Mogami T (2007) Sterol regulatory element-binding protein-2- and liver X receptor-driven dual promoter regulation of hepatic ABC transporter A1 gene expression: mechanism underlying the unique response to cellular cholesterol status. J Biol Chem 282:21090–21099

Tardif JC, Gregoire J, L'Allier PL, Ibrahim R, Lesperance J, Heinonen TM, Kouz S, Berry C, Basser R, Lavoie MA, Guertin MC, Rodes-Cabau J (2007) Effects of reconstituted high-density lipoprotein infusions on coronary atherosclerosis: a randomized controlled trial. JAMA 297:1675–1682

The Bezafibrate Infarction Prevention (BIP) Study Group (2000) Secondary prevention by raising HDL cholesterol and reducing triglycerides in patients with coronary artery disease: the Bezafibrate Infarction Prevention (BIP) study. Circulation 102:21–27

The FIELD study investigators (2005) Effects of long-term fenofibrate therapy on cardiovascular events in 9795 people with type 2 diabetes mellitus (the FIELD study): randomised controlled trial. Lancet 366:1849–1861

Thoenes M, Oguchi A, Nagamia S, Vaccari CS, Hammoud R, Umpierrez GE, Khan BV (2007) The effects of extended-release niacin on carotid intimal media thickness, endothelial function and inflammatory markers in patients with the metabolic syndrome. Int J Clin Pract 61:1942–1948

Tkac I, Molcanyiova A, Javorsky M, Kozarova M (2006) Fenofibrate treatment reduces circulating conjugated diene level and increases glutathione peroxidase activity. Pharmacol Res 53:261–264

Tomas M, Senti M, Garcia-Faria F, Vila J, Torrents A, Covas M, Marrugat J (2000) Effect of simvastatin therapy on paraoxonase activity and related lipoproteins in familial hypercholesterolemic patients. Arterioscler Thromb Vasc Biol 20:2113–2119

Triolo M, Annema W, de Boer JF, Tietge UJ, Dullaart RP (2013a) Simvastatin and bezafibrate increase cholesterol efflux in men with type 2 diabetes. Eur J Clin Invest 44(3):240–248

Triolo M, Kwakernaak AJ, Perton FG, De Vries R, Dallinga-Thie GM, Dullaart RP (2013b) Low normal thyroid function enhances plasma cholesteryl ester transfer in type 2 diabetes mellitus. Atherosclerosis 228:466–471

Tsimihodimos V, Kakafika A, Tambaki AP, Bairaktari E, Chapman MJ, Elisaf M, Tselepis AD (2003) Fenofibrate induces HDL-associated PAF-AH but attenuates enzyme activity associated with apoB-containing lipoproteins. J Lipid Res 44:927–934

van der Hoogt CC, de Haan W, Westerterp M, Hoekstra M, Dallinga-Thie GM, Romijn JA, Princen HM, Jukema JW, Havekes LM, Rensen PC (2007) Fenofibrate increases HDL-cholesterol by reducing cholesteryl ester transfer protein expression. J Lipid Res 48:1763–1771

Vazzana N, Ganci A, Cefalu AB, Lattanzio S, Noto D, Santoro N, Saggini R, Puccetti L, Averna M, Davi G (2013) Enhanced lipid peroxidation and platelet activation as potential contributors to increased cardiovascular risk in the low-HDL phenotype. J Am Heart Assoc 2:e000063

Vega GL, Grundy SM (1994) Lipoprotein responses to treatment with lovastatin, gemfibrozil, and nicotinic acid in normolipidemic patients with hypoalphalipoproteinemia. Arch Intern Med 154:73–82

Wahlberg G, Walldius G, Olsson AG, Kirstein P (1990) Effects of nicotinic acid on serum cholesterol concentrations of high density lipoprotein subfractions HDL_2 and HDL_3 in hyperlipoproteinaemia. J Intern Med 228:151–157

Wang TD, Chen WJ, Lin JW, Cheng CC, Chen MF, Lee YT (2003) Efficacy of fenofibrate and simvastatin on endothelial function and inflammatory markers in patients with combined hyperlipidemia: relations with baseline lipid profiles. Atherosclerosis 170:315–323

Warnholtz A, Wild P, Ostad MA, Elsner V, Stieber F, Schinzel R, Walter U, Peetz D, Lackner K, Blankenberg S, Munzel T (2009) Effects of oral niacin on endothelial dysfunction in patients with coronary artery disease: results of the randomized, double-blind, placebo-controlled INEF study. Atherosclerosis 204:216–221

Xia P, Vadas MA, Rye KA, Barter PJ, Gamble JR (1999) High density lipoproteins (HDL) interrupt the sphingosine kinase signaling pathway. A possible mechanism for protection against atherosclerosis by HDL. J Biol Chem 274:33143–33147

Yadav R, France M, Younis N, Hama S, Ammori BJ, Kwok S, Soran H (2012) Extended-release niacin with laropiprant : a review on efficacy, clinical effectiveness and safety. Expert Opin Pharmacother 13:1345–1362

Yamashita S, Tsubakio-Yamamoto K, Ohama T, Nakagawa-Toyama Y, Nishida M (2010) Molecular mechanisms of HDL-cholesterol elevation by statins and its effects on HDL functions. J Atheroscler Thromb 17:436–451

Yesilbursa D, Serdar A, Saltan Y, Serdar Z, Heper Y, Guclu S, Cordan J (2005) The effect of fenofibrate on serum paraoxonase activity and inflammatory markers in patients with combined hyperlipidemia. Kardiol Pol 62:526–530

Yetukuri L, Huopaniemi I, Koivuniemi A, Maranghi M, Hiukka A, Nygren H, Kaski S, Taskinen MR, Vattulainen I, Jauhiainen M, Oresic M (2011) High density lipoprotein structural changes and drug response in lipidomic profiles following the long-term fenofibrate therapy in the FIELD substudy. PLoS ONE 6:e23589

Yuhanna IS, Zhu Y, Cox BE, Hahner LD, Osborne-Lawrence S, Lu P, Marcel YL, Anderson RG, Mendelsohn ME, Hobbs HH, Shaul PW (2001) High-density lipoprotein binding to scavenger receptor-BI activates endothelial nitric oxide synthase. Nat Med 7:853–857

Yvan-Charvet L, Kling J, Pagler T, Li H, Hubbard B, Fisher T, Sparrow CP, Taggart AK, Tall AR (2010) Cholesterol efflux potential and antiinflammatory properties of high-density lipoprotein after treatment with niacin or anacetrapib. Arterioscler Thromb Vasc Biol 30:1430–1438

Emerging Small Molecule Drugs

Sophie Colin, Giulia Chinetti-Gbaguidi, Jan A. Kuivenhoven, and Bart Staels

Contents

1	Introduction	618
2	Cholesteryl Ester Transfer Protein Inhibitors	619
	2.1 Biological Mechanisms	619
	2.2 Current State	619
	2.3 Future Perspectives	620
3	Novel PPAR Agonists	621
	3.1 Biological Mechanisms	621
	3.2 Current State	622
	3.3 Future Perspectives	622
4	Novel LXR Agonists	623
	4.1 Biological Mechanisms	623
	4.2 Current State	624
	4.3 Future Perspectives	624
5	RVX-208	625
	5.1 Biological Mechanisms	625
	5.2 Current State	625
References		626

S. Colin • G. Chinetti-Gbaguidi • B. Staels (✉)
Université Lille 2, F-59000 Lille, France

Inserm, U1011, F-59000 Lille, France

Institut Pasteur de Lille, F-59019 Lille, France

European Genomic Institute for Diabetes (EGID), FR 3508, F-59000 Lille, France
e-mail: bart.staels@pasteur-lille.fr

J.A. Kuivenhoven
Section Molecular Genetics, Department of Pediatrics, University Medical Center Groningen, University of Groningen, 9713 AV Groningen, The Netherlands

Abstract

Dyslipidaemia is a major risk factor for cardiovascular diseases. Pharmacological lowering of LDL-C levels using statins reduces cardiovascular risk. However, a substantial residual risk persists especially in patients with type 2 diabetes mellitus. Because of the inverse association observed in epidemiological studies of HDL-C with the risk for cardiovascular diseases, novel therapeutic strategies to raise HDL-C levels or improve HDL functionality are developed as complementary therapy for cardiovascular diseases. However, until now most therapies targeting HDL-C levels failed in clinical trials because of side effects or absence of clinical benefits. This chapter will highlight the emerging small molecules currently developed and tested in clinical trials to pharmacologically modulate HDL-C and functionality including new CETP inhibitors (anacetrapib, evacetrapib), novel PPAR agonists (K-877, CER-002, DSP-8658, INT131 and GFT505), LXR agonists (ATI-111, LXR-623, XL-652) and RVX-208.

Keywords

HDL therapy • CETP inhibitors • PPAR • LXR • RVX-208

Abbreviations

ABCA1	ATP-binding cassette transporter A1
ABCG1	ATP-binding cassette transporter G1
ACC	Acetyl-CoA carboxylase
Apo	apolipoprotein
CETP	Cholesteryl ester transfer protein
CVD	Cardiovascular disease
FAS	Fatty acid synthase
HDL	High-density lipoprotein
LDL	Low-density lipoprotein
LXR	Liver X receptor
MTTP	Microsomal triglyceride transfer protein
PPAR	Peroxisome proliferator-activated receptor
PCSK9	Proprotein convertase subtilisin/kexin type 9
SCD-1	Stearoyl-CoA desaturase-1
SREBP-1	Sterol regulatory element-binding protein

1 Introduction

Dyslipidaemia is a major risk factor for cardiovascular diseases, a main cause of morbidity and mortality worldwide, with 17.3 million deaths per year (Laslett et al. 2012). LDL-C-lowering therapy, especially with statins, has shown to be an efficient approach to reduce cardiovascular risk on average by 25–35 %. Although lowering LDL-C with statins has beneficial effects and reduces cardiovascular events, significant numbers of residual cardiovascular events remain in high-risk patients, prompting the search for alternative complementary approaches. Among these strategies, new agents combined with statins, such as proprotein convertase subtilisin/kexin type 9 (PCSK9) inhibitors, microsomal triglyceride transfer protein (MTTP) inhibitors or ezetimibe can provide additional lowering LDL-C effects.

Because HDL-C levels are inversely correlated with cardiovascular risk (Gordon et al. 1977), raising HDL-C levels has spawned high hopes as additional therapy for cardiovascular diseases. However, so far none of the pharmacological interventions aimed at raising HDL-C levels has yielded convincing results with respect to reduction of cardiovascular risk.

In this chapter we will focus on these emerging small molecule drugs in development, including cholesteryl ester transfer protein (CETP) inhibitors, novel peroxisome proliferator-activated receptor (PPAR) agonists, liver X receptor (LXR) agonists and RVX-208. For each drug, the biological mechanisms of these molecules, an overview of the current state of the clinical trials and the future perspectives will be provided.

2 Cholesteryl Ester Transfer Protein Inhibitors

2.1 Biological Mechanisms

The cholesteryl ester transfer protein (CETP) promotes the transfer of triglycerides from apoB-containing lipoproteins (LDL, IDL and VLDL) to HDL-C in exchange for cholesteryl esters. Interest in CETP inhibitor development came from studies in families with CETP deficiency with hyperalphalipoproteinaemia CETP deficiency (Inazu 1990) and epidemiological studies showing that CETP gene variants are associated with increased HDL-C levels and a lower risk of coronary heart disease events (Curb et al. 2004).

2.2 Current State

Four CETP inhibitors have been developed in humans: torcetrapib, anacetrapib, dalcetrapib and evacetrapib. The first CETP inhibitor designed by Pfizer and tested in phase III clinical trials was torcetrapib. The initial results of the ILLUMINATE (Investigation of Lipid Level Management to Understand its Impact in Atherosclerotic Events) trial were encouraging, with a 72 % increase in HDL-C and a 25 % decrease in LDL-C in patients with high CVD risk treated with torcetrapib on top of atorvastatin (Kastelein et al. 2007). However, all studies were interrupted because of an increased risk of cardiovascular events and an excess of mortality upon torcetrapib usage, possibly due to an increase in aldosterone level and blood pressure (Barter et al. 2007). Further analyses have demonstrated that the effect on blood pressure was independent of CETP inhibition. Indeed, torcetrapib increases blood pressure in mice that do not express CETP (Forrest et al. 2008). Furthermore, genetic association studies in 58,948 subjects with polymorphism in CETP gene report that CETP genotype was not associated with systolic nor diastolic blood pressure (Sofat et al. 2010). These results suggest that toxicity upon torcetrapib treatment could be CETP independent.

Even though the dal-OUTCOMES trial with dalcetrapib (Roche) showed an increase of HDL-C by 30 %, the results showed futility and trials were halted due to the absence of obvious benefit (Schwartz et al. 2012). Furthermore, the effect of dalcetrapib on vessel wall structure and vascular inflammation was investigated after a 2-year treatment in the dal-PLAQUE trial (Fayad et al. 2011). No significant benefit was reported in dalcetrapib-treated patients, but in patients with low HDL-C at baseline, beneficial effects on endothelial function were observed. However, in the dal-VESSEL study, designed to validate these results, the beneficial effects were not confirmed (Lüscher et al. 2012). Nevertheless, despite these failures, no vascular toxicity and blood pressure increase were observed on top of dalcetrapib treatment (Lüscher et al. 2012). Based on these results, two other more potent CETP inhibitors, anacetrapib and evacetrapib, have been developed and are still in phase III clinical trials.

2.3 Future Perspectives

Anacetrapib is a potent CETP inhibitor developed by Merck. In a first clinical trial, anacetrapib increased HDL-C by 138 % and reduced LDL-C by 40 % in patients with coronary artery disease or at high risk for coronary heart disease on statin therapy. This clinical trial with the acronym DEFINE (Determining the EFficacy and Tolerability of CETP Inhibition with Anacetrapib), demonstrated the safety of anacetrapib and absence of significant changes in blood pressure, aldosterone and electrolyte levels (Cannon et al. 2010). Currently, a phase III clinical trial is ongoing, acronym REVEAL (Randomized EValuation of the Effects of Anacetrapib Through Lipid-modification), which will determine whether lipid modification on anacetrapib therapy reduces the risk of coronary death, myocardial infarction (MI) or coronary revascularisation in patients with circulatory problems and low LDL-C on statin therapy (ClinicalTrials.gov identifier: NCT01252953). The results of this trial are expected at the beginning of 2017.

Evacetrapib, designed by Eli Lilly & Company, is the fourth member of the CETP inhibitor class that is tested in clinical trials. A phase II clinical trial evaluated the efficacy of evacetrapib as monotherapy or in combination with the most prescribed statins in patients with either hypercholesterolaemia or low HDL-C levels. As monotherapy, evacetrapib (30 mg, 100 mg or 500 mg/day) dose-dependently reduced LDL-C from 14 to 36 % and increased HDL-C from 54 to 129 %. Although the decrease of LDL-C was higher in combination with statins (49 % vs. 24 %), the increase of HDL-C was not stronger when compared to evacetrapib monotherapy (Nicholls et al. 2011). No adverse effects were observed in this trial, with no changes in blood pressure or aldosterone levels. Thus evacetrapib appears to be well-tolerated (Nicholls et al. 2011). The benefits of evacetrapib in combination with statins will be determined in a phase III randomised outcome trial, acronym ACCELERATE (Assessment of Clinical Effects of Cholesteryl Ester Transfer Protein Inhibition with Evacetrapib in Patients at a High-Risk for Vascular Outcomes) (Estimated end date: January 2016; ClinicalTrials.gov identifier: NCT01687998).

3 Novel PPAR Agonists

3.1 Biological Mechanisms

Peroxisome proliferator-activated receptors (PPARs) are a nuclear receptor subfamily with three members, PPARα, PPARγ and PPARβ/δ, encoded by distinct genes. The three isoforms display distinct patterns of expression with PPARα being highly expressed in liver, kidney, heart, muscle and brown adipose tissue, PPARγ is most abundant in adipose tissue, whereas PPARβ/δ is ubiquitously expressed (Lefebvre et al. 2006). Upon heterodimerisation with the retinoic X receptor, PPARs bind to PPAR response elements (PPRE) located in the promoters of their target genes and thus exert negative or positive control on their transcription.

The role of PPARα was initially studied in the liver where it enhances fatty acid oxidation, regulates gluconeogenesis through an increase of pyruvate dehydrogenase kinase 4 expression and ketone body production in response to the fasting state. Through its natural ligands, such as long-chain unsaturated fatty acids, arachidonic acid derivatives and oxidised phospholipids, PPARα regulates also some genes involved in lipid and lipoprotein metabolism. The increase of apoAV and lipoprotein lipase expression by PPARα activation associated with the reduction of apoCIII expression contributes to reduced plasma triglyceride levels in humans. Moreover, plasma HDL cholesterol levels increase as a result of the stimulation of two major HDL-associated apolipoproteins, apoAI and apoAII, by PPARα (Staels et al. 1998). In addition to these hepatic effects, PPARα activation enhances reverse cholesterol transport related to the increase of ATP-binding cassette transporter A1 (ABCA1) in macrophages (Chinetti et al. 2001) and exerts many pleiotropic effects on vascular remodelling and inflammatory responses.

PPARγ is activated by natural ligands such as polyunsaturated fatty acids, 15-deoxy-Δ12,14-prostaglandin J2 and oxidised fatty acids. In addition, pharmacological PPARγ agonists, the thiazolidinediones, are used as insulin sensitisers. Furthermore, PPARγ is the major regulator of adipogenesis (Tontonoz and Spiegelman 2008). PPARγ regulates the expression of adipokines such as adiponectin (Yu et al. 2002). In addition to this adipogenic effect, PPARγ displays anti-inflammatory actions and promotes the polarisation of monocytes towards alternative M2 macrophages (Bouhlel et al. 2007).

PPARβ/δ is activated by long-chain unsaturated fatty acids, and several synthetic ligands have been designed including L-165041, GW501516 and GW0742. However, no PPARβ/δ agonists are in clinical use yet. PPARβ/δ is highly expressed in skeletal muscle where it increases the expression of fatty acid oxidation-related genes. This nuclear receptor also improves lipid metabolism by reducing triglycerides and LDL-C levels and by increasing HDL-C levels. Moreover, PPARβ/δ activation increases insulin sensitivity (Oliver et al. 2001).

Overall, PPARs are involved in the control of lipid lipoprotein and glucose metabolism as well as in the inflammatory response (Lefebvre et al. 2006).

3.2 Current State

Among the pharmacologically used PPARα ligands are the fibrates, which are currently clinically used as hypolipidaemic drugs. Clinical benefits of fibrates were reported in primary and secondary intervention trials and reviewed in a meta-analysis (Jun et al. 2010). In line with their capacity to increase HDL-C and reduce triglycerides and LDL-C (Jun et al. 2010), their effects on the incidence of coronary heart disease appear most pronounced in patients with high triglycerides (>200 mg/dL) and/or low HDL-C (<40 mg/dL), although this has not yet been formally proven in a dedicated trial in type 2 diabetes patients (Suh et al. 2012; Keech et al. 2005).

Each PPAR isoform displays specific roles, with PPARα controlling lipid metabolism, whereas PPARδ improves also glucose metabolism. An interesting strategy consisted in the development of dual agonists with synergistic effects on different PPAR isoforms and minimal side effects (Rosenson et al. 2012). In this context, Roche developed a dual PPAR α/γ agonist, aleglitazar. Despite promising results in the synchrony study, with an improvement of lipid and glucose parameters in type 2 diabetes patients (Henry et al. 2009), the AleCardio phase III trial failed due to adverse effects on heart failure. Subsequently, all trials with aleglitazar were stopped (press report: http://www.roche.com/media/media_releases/med-cor-2013-07-10.htm).

3.3 Future Perspectives

Some limitations of fibrate therapy are their relatively weak activity on PPARα and their efficacy that depends on the targeted population (Staels 2010). To address these issues, several highly selective and potent PPAR agonists were developed (Fruchart 2013). The KOWA Company is developing K-877, a potent PPARα agonist. In a comparative clinical trial (International Clinical Trials identifier: JPRN-JapicCTI-121764), patients with hypertriglyceridaemia and low HDL-C were treated with K-877 or fenofibrate for 12 weeks. The first results showed that the increase of HDL-C is stronger by K-877 than by fenofibrate. In addition, none of the adverse effects induced by fenofibrate, such as increased levels of serum homocysteine and creatinine, were observed in K-877-treated patients (http://www.kenes.com/eas2012/abstracts/pdf/525.pdf). This molecule is currently in phase II in the USA and EU (International Clinical Trials identifier: EUCTR2013-001517-32-SE) and in phase III clinical trials for atherosclerotic dyslipidaemia in Japan (International Clinical Trials identifier: JPRN-JapicCTI-132067).

A HDL-inducer developed by Cerenis Therapeutics is a specific PPARδ agonist, CER-002. The phase I clinical trial demonstrated that CER-002 is well-tolerated without major adverse effects (press report: http://www.drugs.com/clinical_trials/ cerenis-therapeutics-announces-successful-completion-phase-clinical-trial-cer-002-cardiovascular-4272.html). Another selective PPARδ agonist, HPP593, is currently tested in healthy subjects, and its effect on LDL and HDL cholesterol as well as triglycerides will be determined (press report: http://www.ttpharma.com/ TherapeuticAreas/MetabolicDisorders/Dyslipidemia/HPP593/tabid/118/Default.aspx).

The synthetic, non-thiazolidinedione PPARγ compound, INT131, improves glucose tolerance in rodent models of diabetes to a similar extent as rosiglitazone. However, no major adverse events, such as weight gain, haemodilution or plasma volume increase, were observed in INT131-treated rats compared to rosiglitazone-treated rats (Motani et al. 2009). In a randomised, double-blind study, type 2 diabetes patients were treated 4 weeks with 1 or 10 mg of INT131. Consistent with the in vitro data, INT131 displayed a glucose-lowering activity and increased HDL-C levels without changing other lipid parameters (Dunn et al. 2011). This molecule is currently tested in a comparative clinical study in type 2 diabetes patients treated for 24 weeks with INT131 or pioglitazone (ClinicalTrials.gov Identifier: NCT00631007).

In addition, a phase I clinical trial is ongoing to evaluate safety, tolerability and pharmacokinetic behaviour of a new PPAR α/γ modulator, DSP-8658, in type 2 diabetes mellitus and healthy subjects (ClinicalTrials.gov Identifier: NCT01042106). DSP-8658 is a non-thiazolidinedione compound which exhibits potent anti-hyperglycaemic effects, reduces plasma triglycerides and increases HDL-C levels with less side effects on, e.g., body weight gain (press report: www.ds-pharma.com/ir/library/presentation/pdf).

Finally, the new dual PPAR α/δ agonist GFT505, developed by Genfit, reduced plasma triglycerides and increased HDL-C levels in abdominally obese patients with either dyslipidaemia or prediabetes (Cariou et al. 2011). A phase IIb clinical trial is ongoing to evaluate the efficacy of GFT505 in patients with non-alcoholic steatohepatitis (ClinicalTrials.gov identifier: NCT01694849), an unmet clinical need because of the continuous increasing incidence of fatty liver disease due to abdominal obesity (Tailleux et al. 2012).

These novel selective PPAR agonists, K-877, CER-002, DSP-8658, INT131 and GFT505 appear promising drugs to treat the cardiovascular risk associated with metabolic syndrome and type 2 diabetes.

4 Novel LXR Agonists

4.1 Biological Mechanisms

Liver X receptors (LXRs) are nuclear receptors. There are two isoforms, LXRα and LXRβ, with LXRα mainly expressed in the liver, intestine, kidney and spleen, whereas LXRβ is ubiquitously expressed (Repa and Mangelsdorf 2000). Oxysterols and other cholesterol metabolites are natural ligands for LXRs, and several synthetic ligands were also developed (T0901317, GW3965). After their binding, LXRs modulate the expression of genes involved in cholesterol metabolism and transport and glucose metabolism (Schultz et al. 2000; Laffitte et al. 2003). Activation of LXR in macrophages induces ABCA1, ABCG1 and apoE expression which promotes cholesterol efflux and reverses cholesterol transport (Sabol et al. 2005; Venkateswaran et al. 2000; Laffitte et al. 2001). Besides their effects on lipid metabolism, LXRs display anti-inflammatory properties (Joseph et al. 2003) and improve glucose tolerance. In contrast to their beneficial effects, LXR activation induces fatty acid synthesis (de novo lipogenesis) related to a modulation of the hepatic expression of sterol regulatory element-binding protein (SREBP-1), stearoyl-CoA desaturase-1 (SCD-1), fatty acid synthase (FAS) and acetyl-CoA carboxylase (ACC) (Schultz et al. 2000). This upregulation of hepatic SREBP-1c expression may contribute to the elevation of plasma triglycerides.

4.2 Current State

T0901317 and GW3965, the most studied agonists, have been extensively described to exert beneficial effects in preclinical animal models of cardiovascular diseases, neurodegenerative diseases and inflammation (Terasaka et al. 2003; Joseph et al. 2002, 2003; Zelcer et al. 2007). However, LXR ligands have not yet been tested in clinical trials because of their adverse effects such as an increase of hepatic lipogenesis, hypertriglyceridaemia and hepatosteatosis (Calkin and Tontonoz 2012). These lipogenic effects have been assigned to LXRα which is highly expressed in the liver (Lehrke et al. 2005; Bradley et al. 2007). Therapeutic strategies are now focusing on the development of selective LXR modulators, LXRβ-specific agonists and/or ligands which act selectively in specific tissues in order to maintain positive effects on cholesterol metabolism and minimise the lipid side effects.

LXR-623 (WAY-252623) is a novel synthetic ligand with higher potency for LXRβ, which induces plaque regression in combination with statins in a rabbit model of atherosclerosis (Giannarelli et al. 2012). LXR-623 entered in a phase I trial (NCT00366522) to test tolerance and safety in humans (Katz et al. 2009). However, its development was interrupted because of adverse effects in the central nervous system with potential induction of psychiatric disorders.

4.3 Future Perspectives

A novel synthetic, steroidal LXR ligand, ATI-111, has been developed. This molecule is most potent on LXRα with modest effects on LXRβ. The higher efficiency on LXRα allows its utilisation at lower concentrations than T0901317, which probably reduces cytotoxicity and also adverse effects such as hypertriglyceridaemia. To determine whether ATI-111 does not provoke hypertriglyceridaemia, mice were treated with 5 mg.kg^{-1}.day^{-1} of ATI-111 for 8 weeks. Interestingly, a decrease of plasma triglyceride levels and VLDL cholesterol was observed in ATI-111-treated mice (Peng et al. 2011). In addition, ATI-111 exhibits anti-inflammatory properties with a decrease of LPS-induced inflammatory gene expression. Furthermore, ATI-111 reduced atherosclerotic lesions in LDL-receptor-deficient mice (Peng et al. 2011). Altogether, accumulating proofs from in vitro and in vivo animal studies show beneficial effects of ATI-111 on atherosclerosis with anti-inflammatory effects, a reduction of hypertriglyceridaemia and a consequential decrease of atherosclerotic lesions. Further molecular investigations are necessary to assess the full potential of ATI-111 in clinical trials.

A phase I clinical trial with XL-652 (XL-014), a novel LXR ligand, is currently ongoing to evaluate its safety (www.exelixis.com/pipeline/xl652).

5 RVX-208

5.1 Biological Mechanisms

An alternative strategy to raise the serum level of HDL-C is to increase one of the major HDL proteins, apolipoprotein AI (apoAI). In this context, after a screening assay in HepG2 cells to identify molecules inducing apoAI, Resverlogix Corporation has selected and developed an oral quinazoline molecule, RVX-208 (RVX-000222). This molecule, a derivative of resveratrol, increases hepatic apoAI production (Bailey et al. 2010). Thus RVX-208 is a small molecule for potential treatment of cardiovascular diseases.

5.2 Current State

First in vitro experiments have shown that RVX-208 increases apoAI mRNA, protein and the release of apoAI in the medium of HepG2 cells (Bailey et al. 2010). To test its efficiency in vivo, male monkeys were orally treated with RVX-208 for 63 days. This in vivo study in nonhuman primates demonstrated that RVX-208 increases in a dose-dependent manner serum levels of apoAI and pre-β-HDL-C (Bailey et al. 2010). No adverse effects were observed in monkeys. In humans, 7 days of treatment increased preβ-HDL, apoAI and cholesterol efflux (Bailey et al. 2010). The efficacy and safety of RVX-208 was furthermore investigated in a phase II randomised trial including 299 statin-treated patients with coronary artery diseases (ASSERT study: NTC01058018). Twice daily administration of RVX-208 (50, 100 and 150 mg) was well tolerated; however, a transient elevation in transaminase levels was found in some RVX-208-treated patients. Furthermore, only a modest increase of HDL-C level (3.2–8.3 %) was observed (Nicholls et al. 2011). In a next phase IIb clinical study, the ASSURE (ApoAI Synthesis Stimulation and Intravascular Ultrasound for Coronary Atheroma Regression Evaluation) trial, the effect of 26 weeks treatment with RVX-208

(100 mg) was determined on the progression of coronary atherosclerosis in 323 patients with symptomatic coronary artery disease and low HDL-C levels. The primary end point, change in atheroma volume determined by intravascular ultrasound (IVUS), was not met. No significant change in plaque regression was observed between placebo and RVX-208 group. The question is whether this lack of beneficial effect on plaque regression is due to the weak efficacy of RVX-208 or the impossibility to confer beneficial effects on top of statins, since 84 % of patients were on statin therapy. Furthermore, the increases of apoAI and HDL-C levels did not differ from placebo, whereas transaminases were again found to be elevated. Thus, so far, no clinical benefit has been demonstrated with RVX-208 (press report: communication Nicholls SJ www.clinicaltrialresults.org/Nicholls_ASSURE).

Conclusion

To reduce residual cardiovascular risk persisting after statin therapy, novel therapeutic strategies based on raising HDL-C are currently under investigation. However, the utility of increasing HDL-C is not yet established, and pharmaceutical manipulation of HDL-C appears to be less efficient than lowering LDL-C to reduce cardiovascular risk. Moreover, the failure of several clinical trials with the first members of the CETP inhibitor class, torcetrapib and dalcetrapib, and the lack of beneficial effects of an increase of HDL-C and the reduction of cardiovascular events raises doubts about the relevance of the "HDL-C hypothesis". These disappointing results highlight the complexity of HDL metabolism in contrast to that of LDL.

Many biological activities of HDL, such as antioxidant, anti-inflammatory and antiapoptotic properties, are mediated by different HDL subclasses, and solely increasing HDL-C may not enhance these functions. Moreover, plasma HDL contains heterogeneous particle subpopulations whose composition, metabolism and functionality differ depending on the metabolic status (Besler et al. 2011). Thus far, the larger randomised placebo controlled phase III clinical trials have shown that an increase in HDL-C does not benefit the patient, and thus, future strategies should aim at improving HDL function.

Open Access This chapter is distributed under the terms of the Creative Commons Attribution Noncommercial License, which permits any noncommercial use, distribution, and reproduction in any medium, provided the original author(s) and source are credited.

References

Bailey D, Jahagirdar R, Gordon A, Hafiane A, Campbell S, Chatur S, Wagner GS, Hansen HC, Chiacchia FS, Johansson J, Krimbou L, Wong NCW, Genest J (2010) RVX-208: a small molecule that increases apolipoprotein AI and high-density lipoprotein cholesterol in vitro and in vivo. J Am Coll Cardiol 55:2580–2589

Barter PJ, Caulfield M, Eriksson M, Grundy SM, Kastelein JJP, Komajda M, Lopez-Sendon J, Mosca L, Tardif J, Waters DD, Shear CL, Revkin JH, Buhr KA, Fisher MR, Tall AR, Brewer B

(2007) Effects of Torcetrapib in patients at high risk for coronary events. N Engl J Med 357:2109–2122

Besler C, Heinrich K, Rohrer L, Doerries C, Riwanto M, Shih DM, Chroni A, Yonekawa K, Stein S, Schaefer N, Mueller M, Akhmedov A, Daniil G, Manes C, Templin C, Wyss C, Maier W, Tanner FC, Matter CM, Corti R, Furlong C, Lusis AJ, von Eckardstein A, Fogelman AM, Lüscher TF, Landmesser U (2011) Mechanisms underlying adverse effects of HDL on enos-activating pathways in patients with coronary artery disease. J Clin Invest 121:2693–2708

Bouhlel MA, Derudas B, Rigamonti E, Dièvart R, Brozek J, Haulon S, Zawadzki C, Jude B, Torpier G, Marx N, Staels B, Chinetti-Gbaguidi G (2007) PPARgamma activation primes human monocytes into alternative m2 macrophages with anti-inflammatory properties. Cell Metab 6:137–143

Bradley MN, Hong C, Chen M, Joseph SB, Wilpitz DC, Wang X, Lusis AJ, Collins A, Hseuh WA, Collins JL, Tangirala RK, Tontonoz P (2007) Ligand activation of LXR beta reverses atherosclerosis and cellular cholesterol overload in mice lacking LXR alpha and apoE. J Clin Invest 117:2337–2346

Calkin AC, Tontonoz P (2012) Transcriptional integration of metabolism by the nuclear sterol-activated receptors LXR and FXR. Nat Rev Mol Cell Biol 13:213–224

Cannon CP, Shah S, Dansky HM, Davidson M, Brinton EA, Gotto AM, Stepanavage M, Liu SX, Gibbons P, Ashraf TB, Zafarino J, Mitchel Y, Barter P (2010) Safety of Anacetrapib in patients with or at high risk for coronary heart disease. N Engl J Med 363:2406–2415

Cariou B, Zaïr Y, Staels B, Bruckert E (2011) Effects of the new dual PPAR α/δ agonist GFT505 on lipid and glucose homeostasis in abdominally obese patients with combined dyslipidemia or impaired glucose metabolism. Diabetes Care 34:2008–2014

Chinetti G, Lestavel S, Bocher V, Remaley AT, Neve B, Torra IP, Teissier E, Minnich A, Jaye M, Duverger N, Brewer HB, Fruchart JC, Clavey V, Staels B (2001) PPAR-alpha and PPAR-gamma activators induce cholesterol removal from human macrophage foam cells through stimulation of the ABCA1 pathway. Nat Med 7:53–58

Curb JD, Abbott RD, Rodriguez BL, Masaki K, Chen R, Sharp DS, Tall AR (2004) A prospective study of HDL-C and cholesteryl ester transfer protein gene mutations and the risk of coronary heart disease in the elderly. J Lipid Res 45:948–953

Dunn FL, Higgins LS, Fredrickson J, DePaoli AM (2011) Selective modulation of PPARγ activity can lower plasma glucose without typical thiazolidinedione side-effects in patients with type 2 diabetes. J Diabetes Complicat 25:151–158

Fayad ZA, Mani V, Woodward M, Kallend D, Abt M, Burgess T, Fuster V, Ballantyne CM, Stein EA, Tardif J, Rudd JHF, Farkouh ME, Tawakol A (2011) Safety and efficacy of Dalcetrapib on atherosclerotic disease using novel non-invasive multimodality imaging (dal-plaque): a randomised clinical trial. Lancet 378:1547–1559

Forrest MJ, Bloomfield D, Briscoe RJ, Brown PN, Cumiskey AM, Ehrhart J, Hershey JC, Keller WJ, Ma X, McPherson HE, Messina E, Peterson LB, Sharif-Rodriguez W, Siegl PK, Sinclair PJ, Sparrow CP, Stevenson AS, Sun SY, Tsai C, Vargas H, Walker M 3rd, West SH, White V, Woltmann RF (2008) Torcetrapib-induced blood pressure elevation is independent of cetp inhibition and is accompanied by increased circulating levels of aldosterone. Br J Pharmacol 154:1465–1473

Fruchart J (2013) Selective peroxisome proliferator-activated receptor α modulators (SPPARMα): the next generation of peroxisome proliferator-activated receptor α-agonists. Cardiovasc Diabetol 12:82

Giannarelli C, Cimmino G, Connolly TM, Ibanez B, Ruiz JMG, Alique M, Zafar MU, Fuster V, Feuerstein G, Badimon JJ (2012) Synergistic effect of Liver X Receptor activation and simvastatin on plaque regression and stabilization: an magnetic resonance imaging study in a model of advanced atherosclerosis. Eur Heart J 33:264–273

Gordon T, Castelli WP, Hjortland MC, Kannel WB, Dawber TR (1977) High density lipoprotein as a protective factor against coronary heart disease the framingham study. Am J Med 62:707–714

Henry RR, Lincoff AM, Mudaliar S, Rabbia M, Chognot C, Herz M (2009) Effect of the dual peroxisome proliferator-activated receptor-alpha/gamma agonist aleglitazar on risk of cardiovascular disease in patients with type 2 diabetes (synchrony): a phase II, randomised, dose-ranging study. Lancet 374:126–135

http://www.drugs.com/clinical_trials/cerenis-therapeutics-announces-successful-completion-phase-clinical-trial-cer-002

Inazu A, Brown ML, Hesler CB, Agellon LB, Koizumi J, Takata K, Maruhama Y, Mabuchi H, Tall AR (1990) Increased high-density lipoprotein levels caused by a common cholesteryl-ester transfer protein gene mutation. N Engl J Med 323:1234–1238

Joseph SB, McKilligin E, Pei L, Watson MA, Collins AR, Laffitte BA, Chen M, Noh G, Goodman J, Hagger GN, Tran J, Tippin TK, Wang X, Lusis AJ, Hsueh WA, Law RE, Collins JL, Willson TM, Tontonoz P (2002) Synthetic LXR ligand inhibits the development of atherosclerosis in mice. Proc Natl Acad Sci USA 99:7604–7609

Joseph SB, Castrillo A, Laffitte BA, Mangelsdorf DJ, Tontonoz P (2003) Reciprocal regulation of inflammation and lipid metabolism by Liver X Receptors. Nat Med 9:213–219

Jun M, Foote C, Lv J, Neal B, Patel A, Nicholls SJ, Grobbee DE, Cass A, Chalmers J, Perkovic V (2010) Effects of fibrates on cardiovascular outcomes: a systematic review and meta-analysis. Lancet 375:1875–1884

Kastelein JJP, van Leuven SI, Burgess L, Evans GW, Kuivenhoven JA, Barter PJ, Revkin JH, Grobbee DE, Riley WA, Shear CL, Duggan WT, Bots ML (2007) Effect of torcetrapib on carotid atherosclerosis in familial hypercholesterolemia. N Engl J Med 356:1620–1630

Katz A, Udata C, Ott E, Hickey L, Burczynski ME, Burghart P, Vesterqvist O, Meng X (2009) Safety, pharmacokinetics, and pharmacodynamics of single doses of LXR-623, a novel Liver X-Receptor agonist, in healthy participants. J Clin Pharmacol 49:643–649

Keech A, Simes RJ, Barter P, Best J, Scott R, Taskinen MR, Forder P, Pillai A, Davis T, Glasziou P, Drury P, Kesäniemi YA, Sullivan D, Hunt D, Colman P, D'Emden M, Whiting M, Ehnholm C, Laakso M (2005) Effects of long-term fenofibrate therapy on cardiovascular events in 9795 people with type 2 diabetes mellitus (the field study): randomised controlled trial. Lancet 366:1849–1861

Laffitte BA, Repa JJ, Joseph SB, Wilpitz DC, Kast HR, Mangelsdorf DJ, Tontonoz P (2001) LXRs control lipid-inducible expression of the apolipoprotein E gene in macrophages and adipocytes. Proc Natl Acad Sci USA 98:507–512

Laffitte BA, Chao LC, Li J, Walczak R, Hummasti S, Joseph SB, Castrillo A, Wilpitz DC, Mangelsdorf DJ, Collins JL, Saez E, Tontonoz P (2003) Activation of Liver X Receptor improves glucose tolerance through coordinate regulation of glucose metabolism in liver and adipose tissue. Proc Natl Acad Sci USA 100:5419–5424

Laslett LJ, Alagona PJ, Clark BA, Drozda JPJ, Saldivar F, Wilson SR, Poe C, Hart M (2012) The worldwide environment of cardiovascular disease: prevalence, diagnosis, therapy, and policy issues: a report from the american college of cardiology. J Am Coll Cardiol 60:S1–S49

Lefebvre P, Chinetti G, Fruchart J, Staels B (2006) Sorting out the roles of PPAR alpha in energy metabolism and vascular homeostasis. J Clin Invest 116:571–580

Lehrke M, Lebherz C, Millington SC, Guan H, Millar J, Rader DJ, Wilson JM, Lazar MA (2005) Diet-dependent cardiovascular lipid metabolism controlled by hepatic LXRalpha. Cell Metab 1:297–308

Lüscher TF, Taddei S, Kaski J, Jukema JW, Kallend D, Münzel T, Kastelein JJP, Deanfield JE (2012) Vascular effects and safety of Dalcetrapib in patients with or at risk of coronary heart disease: the dal-vessel randomized clinical trial. Eur Heart J 33:857–865

Motani A, Wang Z, Weiszmann J, McGee LR, Lee G, Liu Q, Staunton J, Fang Z, Fuentes H, Lindstrom M, Liu J, Biermann DHT, Jaen J, Walker NPC, Learned RM, Chen J, Li Y (2009) INT131: a selective modulator of PPAR gamma. J Mol Biol 386:1301–1311

Nicholls SJ, Brewer HB, Kastelein JJP, Krueger KA, Wang M, Shao M, Hu B, McErlean E, Nissen SE (2011) Effects of the cetp inhibitor Evacetrapib administered as monotherapy or in combination with statins on HDL and LDL cholesterol: a randomized controlled trial. JAMA 306:2099–2109

Oliver WRJ, Shenk JL, Snaith MR, Russell CS, Plunket KD, Bodkin NL, Lewis MC, Winegar DA, Sznaidman ML, Lambert MH, Xu HE, Sternbach DD, Kliewer SA, Hansen BC, Willson TM (2001) A selective peroxisome proliferator-activated receptor delta agonist promotes reverse cholesterol transport. Proc Natl Acad Sci USA 98:5306–5311

Peng D, Hiipakka RA, Xie J, Dai Q, Kokontis JM, Reardon CA, Getz GS, Liao S (2011) A novel potent synthetic steroidal Liver X Receptor agonist lowers plasma cholesterol and triglycerides and reduces atherosclerosis in LDLR(-/-) mice. Br J Pharmacol 162:1792–1804

Repa JJ, Mangelsdorf DJ (2000) The role of orphan nuclear receptors in the regulation of cholesterol homeostasis. Annu Rev Cell Dev Biol 16:459–481

Roche Group Communications (2013) Roche halts investigation of aleglitazar following regular safety review of phase III trial. http://www.roche.com/media/media_releases/med-cor-2013-07-10.html

Rosenson RS, Wright RS, Farkouh M, Plutzky J (2012) Modulating peroxisome proliferator-activated receptors for therapeutic benefit? Biology, clinical experience, and future prospects. Am Heart J 164:672–680

Sabol SL, Brewer HBJ, Santamarina-Fojo S (2005) The human ABCG1 gene: identification of LXR response elements that modulate expression in macrophages and liver. J Lipid Res 46:2151–2167

Schultz JR, Tu H, Luk A, Repa JJ, Medina JC, Li L, Schwendner S, Wang S, Thoolen M, Mangelsdorf DJ, Lustig KD, Shan B (2000) Role of LXRs in control of lipogenesis. Genes Dev 14:2831–2838

Schwartz GG, Olsson AG, Abt M, Ballantyne CM, Barter PJ, Brumm J, Chaitman BR, Holme IM, Kallend D, Leiter LA, Leitersdorf E, McMurray JJV, Mundl H, Nicholls SJ, Shah PK, Tardif J, Wright RS (2012) Effects of Dalcetrapib in patients with a recent acute coronary syndrome. N Engl J Med 367:2089–2099

Sofat R, Hingorani AD, Smeeth L, Humphries SE, Talmud PJ, Cooper J, Shah T, Sandhu MS, Ricketts SL, Boekholdt SM, Wareham N, Khaw KT, Kumari M, Kivimaki M, Marmot M, Asselbergs FW, van der Harst P, Dullaart RP, Navis G, van Veldhuisen DJ, Van Gilst WH, Thompson JF, McCaskie P, Palmer LJ, Arca M, Quagliarini F, Gaudio C, Cambien F, Nicaud V, Poirer O, Gudnason V, Isaacs A, Witteman JC, van Duijn CM, Pencina M, Vasan RS, D'Agostino RB Sr, Ordovas J, Li TY, Kakko S, Kauma H, Savolainen MJ, Kesäniemi YA, Sandhofer A, Paulweber B, Sorli JV, Goto A, Yokoyama S, Okumura K, Horne BD, Packard C, Freeman D, Ford I, Sattar N, McCormack V, Lawlor DA, Ebrahim S, Smith GD, Kastelein JJ, Deanfield J, Casas JP (2010) Separating the mechanism-based and off-target actions of cholesteryl ester transfer protein inhibitors with cetp gene polymorphisms. Circulation 121:52–62

Staels B (2010) Fibrates in CVD: a step towards personalised medicine. Lancet 375:1847–1848

Staels B, Dallongeville J, Auwerx J, Schoonjans K, Leitersdorf E, Fruchart JC (1998) Mechanism of action of fibrates on lipid and lipoprotein metabolism. Circulation 98:2088–2093

Suh HS, Hay JW, Johnson KA, Doctor JN (2012) Comparative effectiveness of statin plus fibrate combination therapy and statin monotherapy in patients with type 2 diabetes: use of propensity-score and instrumental variable methods to adjust for treatment-selection bias. Pharmacoepidemiol Drug Saf 21:470–484

Tailleux A, Wouters K, Staels B (2012) Roles of PPARs in NAFLD: potential therapeutic targets. Biochim Biophys Acta 1821:809–818

Terasaka N, Hiroshima A, Koieyama T, Ubukata N, Morikawa Y, Nakai D, Inaba T (2003) T-0901317, a synthetic Liver X Receptor ligand, inhibits development of atherosclerosis in LDL receptor-deficient mice. FEBS Lett 536:6–11

Tontonoz P, Spiegelman BM (2008) Fat and beyond: the diverse biology of PPARgamma. Annu Rev Biochem 77:289–312

Venkateswaran A, Laffitte BA, Joseph SB, Mak PA, Wilpitz DC, Edwards PA, Tontonoz P (2000) Control of cellular cholesterol efflux by the nuclear oxysterol receptor LXR alpha. Proc Natl Acad Sci USA 97:12097–12102
www.clinicaltrialresults.org/Nicholls_ASSURE
www.ds-pharma.com/ir/library/presentation/pdf
www.ttpharma.com/therapeuticAreas/MetabolicDisorders/Dyslipidemia/HPP593
Yu JG, Javorschi S, Hevener AL, Kruszynska YT, Norman RA, Sinha M, Olefsky JM (2002) The effect of thiazolidinediones on plasma adiponectin levels in normal, obese, and type 2 diabetic subjects. Diabetes 51:2968–2974
Zelcer N, Khanlou N, Clare R, Jiang Q, Reed-Geaghan EG, Landreth GE, Vinters HV, Tontonoz P (2007) Attenuation of neuroinflammation and alzheimer's disease pathology by Liver X Receptors. Proc Natl Acad Sci USA 104:10601–10606

ApoA-I Mimetics

R.M. Stoekenbroek, E.S. Stroes, and G.K. Hovingh

Contents

1 Introduction 633
2 ApoA-I Mimetic Peptides 634
   2.1 4F 635
   2.2 6F 635
   2.3 FX-5A 636
   2.4 ATI-5261 637
   2.5 ETC-642 637
3 ApoA-I-Based Infusion Therapy 638
   3.1 HDL-VHDL Infusions 638
   3.2 Purified ApoA-I Infusions 638
   3.3 ApoA-I$_{Milano}$ Infusions 639
   3.4 CSL-111 and CSL-112 Infusions 640
   3.5 CER-001 Infusions 641
   3.6 Modified ApoA-I 641
4 Selective Delipidation 642
Conclusion 642
References 643

Abstract

A wealth of evidence indicates that plasma levels of high-density lipoprotein cholesterol (HDL-C) are inversely related to the risk of cardiovascular disease (CVD). Consequently, HDL-C has been considered a target for therapy in order to reduce the residual CVD burden that remains significant, even after application of current state-of-the-art medical interventions. In recent years, however, a number of clinical trials of therapeutic strategies that increase HDL-C levels

---

R.M. Stoekenbroek • E.S. Stroes • G.K. Hovingh (✉)
Department of Vascular Medicine, Academic Medical Center, Po box 22660, 1100 DD Amsterdam, The Netherlands
e-mail: g.k.hovingh@amc.uva.nl

© The Author(s) 2015
A. von Eckardstein, D. Kardassis (eds.), *High Density Lipoproteins*, Handbook of Experimental Pharmacology 224, DOI 10.1007/978-3-319-09665-0_21

failed to show the anticipated beneficial effect on CVD outcomes. As a result, attention has begun to shift toward strategies to improve HDL functionality, rather than levels of HDL-C per se. ApoA-I, the major protein component of HDL, is considered to play an important role in many of the antiatherogenic functions of HDL, most notably reverse cholesterol transport (RCT), and several therapies have been developed to mimic apoA-I function, including administration of apoA-I, mutated variants of apoA-I, and apoA-I mimetic peptides. Based on the potential anti-inflammatory effects, apoA-I mimetics hold promise not only as anti-atherosclerotic therapy but also in other therapeutic areas.

Keywords
ApoA-I mimetics • ApoA-I analogues • ApoA-I infusion

Abbreviations

ACS	Acute coronary syndrome
ABCA1	ATP-binding cassette transporter A1
ABCG1	ATP-binding cassette transporter G1
ApoA-I	Apolipoprotein A-1
ApoB	Apolipoprotein B
ApoE	Apolipoprotein E
CVD	Cardiovascular disease
FH	Familial hypercholesterolemia
FPHA	Familial primary hypoalphalipoproteinemia
HCAEC	Human coronary artery endothelial cells
HDL-C	High-density lipoprotein cholesterol
HDL-sdl	HDL selectively delipidated
HDL-VHDL	High-density lipoprotein-very high-density lipoprotein
IVUS	Intravascular ultrasound
LDL-C	Low-density lipoprotein cholesterol
LDL-R	Low-density lipoprotein receptor
LPA	Lysophosphatidic acid
RCT	Reverse cholesterol transport
rHDL	Reconstituted HDL
ROS	Radical oxygen species
sdLDL	Small dense LDL
WD	Western diet

1 Introduction

The beneficial consequences of lowering low-density lipoprotein cholesterol (LDL-C) on cardiovascular (CVD) events have been unequivocally shown in numerous intervention studies. A plasma LDL-C reduction of 1.0 mmol/L has been shown to reduce the risk of major cardiovascular events by approximately 20 %, irrespective of baseline cholesterol or risk. Intensive LDL-C lowering is therefore advocated in most guidelines (Baigent et al. 2010). However, despite the efficiency of established therapies, the residual burden of disease remains substantial (Roger et al. 2012). Novel targets for therapy are therefore eagerly awaited in order to decrease the residual CVD risk.

A large number of epidemiological studies have shown that levels of high-density lipoprotein cholesterol (HDL-C) are inversely associated with CVD risk. In fact, it has been calculated from these studies that a 1 mg/dL (0.03 mmol/L) increase in HDL-C would translate into a 2–3 % reduction of risk for subsequent coronary events (Gordon et al. 1989). Moreover, levels of HDL-C have been considered a stronger predictive factor of incident coronary heart disease than levels of LDL-C (Gordon et al. 1977). Even among patients who attain low levels of LDL-C while receiving LDL-C lowering therapy, levels of HDL-C remain predictive for subsequent CVD events (Barter et al. 2007; Jafri et al. 2010).

These epidemiological data do not infer a causal relation between levels of HDL-C and CVD risk. However, a number of antiatherogenic mechanisms have been ascribed to HDL. HDL has been shown to play a pivotal role in reverse cholesterol transport (RCT), a pathway by which cholesterol is transported from peripheral cells (e.g., macrophages within the vessel wall) to the liver for biliary excretion (Fielding and Fielding 1995). In addition, HDL has been shown to possess antioxidant, anti-inflammatory (Barter et al. 2004), antithrombotic (Mineo et al. 2006), and antiapoptotic properties (Suc et al. 1997).

HDL has been considered a target for therapy to lower CVD risk, based on both these epidemiological and biological arguments of atheroprotection. In recent years, a number of large clinical trial programs investigating the efficacy of drugs with an established effect on HDL-C levels have been terminated because of the inability to induce improvement in clinical outcomes. In addition, treatment-induced changes in HDL-C were not associated with CVD risk after adjusting for LDL-C (Briel et al. 2009). Moreover, a number of common genetic variants in genes coding for proteins involved in HDL metabolism have been shown to alter levels of HDL-C without showing the anticipated effect on CVD risk (Haase et al. 2011; Voight et al. 2012). These findings challenge the concept that levels of circulating HDL-C are causally related to atherosclerosis, and assessment of HDL functionality has therefore been proposed to better reflect the therapeutic potential of therapies targeting HDL (deGoma et al. 2008).

Although levels of HDL-C alone may thus be a poor target for therapies, apolipoprotein A-1 (apoA-I) itself could in fact represent a promising target. ApoA-I is the major protein component of the HDL particle and considered to play a pivotal role in many of the antiatherogenic properties attributed to HDL. The role of apoA-I in the protection against atherosclerosis has been shown in a number of rodent studies. In low-density lipoprotein receptor (LDL-R) null mice, apoA-I deficiency was shown to result in increased atherosclerosis (Moore et al. 2003). Increased atherosclerotic lesion development was also seen in apoA-I knockout mice expressing human apolipoprotein B (apoB) when fed a Western diet (WD) (Voyiaziakis et al. 1998). Additional proof of the antiatherogenic role of apoA-I was derived from animal models of human apoA-I overexpression and in mice treated with apoA-I infusions, in which it was consistently shown that apoA-I provides protection against atherosclerotic lesion formation in proatherogenic animal models (Duverger et al. 1996; Miyazaki et al. 1995; Rubin et al. 1991). Moreover, mutations in the gene coding for apoA-I have been shown to alter CVD risk in humans. Most of the carriers of mutations in the apoA-I gene are characterized by low levels of HDL-C and an increased risk of CVD (Hovingh et al. 2004). ApoA-I is widely accepted as an attractive target for therapy based on the consistent data derived in animal and human studies. Several therapies have been developed that mimic apoA-I function, including administration of full length apoA-I, mutated variants of apoA-I, and apoA-I mimetic peptides.

2 ApoA-I Mimetic Peptides

ApoA-I comprises a total of 243 amino acids and its secondary structure resembles 10 amphipathic α-helices that are crucial for its efficient interaction with lipids. Recently, there has been increasing interest in the application of peptides that resemble the amphipathic helices in apoA-I as therapeutic agents.

Anantharamaiah et al. synthesized the first apoA-I mimetic peptide, 18A, comprising 18 amino acids. Subsequently, several modifications to 18A have been made to create peptides that more closely mimic apoA-I in its antiatherogenic functions. For example, blocking the ends of 18A with an amide group and an acetyl group, thereby creating a peptide called 2 F, increased its helicity and efficiency in inducing cholesterol efflux (Venkatachalapathi et al. 1993; Yancey et al. 1995). In addition, tandem peptides composed of more than one amphipathic helix were shown to have a superior lipid affinity and ability to induce cholesterol efflux from macrophages compared to peptides that contain only one helix (Anantharamaiah et al. 1985; Garber et al. 1992; Wool et al. 2008). D'Souza et al. investigated the effect on cholesterol efflux and anti-inflammatory and antioxidant properties of 22 different bihelical apoA-I mimetic peptides. None of the compounds was superior in all antiatherogenic functions, and each of the examined antiatherogenic functions was shown to be primarily affected by specific structural features (D'Souza et al. 2010). These results indicate that apoA-I mimetic peptides more closely resembling apoA-I do not necessarily improve all various anti-atherosclerotic functions. Moreover, combining several apoA-I mimetic peptides, each mimicking different structural aspects of apoA-I, may prove to be a valuable strategy to mimic the various anti-atherosclerotic properties of apoA-I.

2.1   4F

Despite potent anti-inflammatory and antioxidant effects in vitro and in experimental models, the promise of 4 F in humans has not been a success to date (Navab et al. 2002, 2004). Two trials of 4 F in patients with CHD or at high risk of CVD assessed the effect of orally and parenterally administered 4 F on the HDL inflammatory index, a measure of HDL-mediated protection against LDL-induced monocyte chemotaxis, providing conflicting results (Bloedon et al. 2008; Watson et al. 2011). The development of 4 F by Novartis has subsequently been discontinued. Although a number of differences between the studies may partially account for the observed discrepancy, the primary outcome measure "HDL inflammatory index" has been proven to be largely irreproducible in the majority of laboratories. In fact, it has been difficult to identify an efficacy parameter for apoA-I mimetic peptides in human trials altogether, and in lack of standardization, all trials using HDL quality as readout have been disappointing.

2.2   6F

A disadvantage of many apoA-I mimetic peptides is that they require end blocking to be effective, which precludes its synthesis by living organisms. Large-scale production of the peptides for therapeutic use would therefore be too expensive. In a search for apoA-I mimetic peptides that do not require end blocking, 6 F yielded promising results.

L-6 F has been shown to possess antioxidant and anti-inflammatory properties in several mice models (Chattopadhyay et al. 2013; Navab et al. 2013). Given that the small intestine may be an important site of action for apoA-I mimetic peptides, the observed reduction in plasma levels of the proinflammatory and proatherogenic lipid lysophosphatidic acid (LPA) following administration of L-6 F is of particular interest. Feeding LDL-R knockout mice a Western diet does result in changes in intestinal gene expression and proinflammatory and hypercholesterolemic changes in serum. Similar changes were observed when mice were fed unsaturated LPA. Administration of L-6 F reduced intestinal LPA levels and prevented the Western diet-mediated proatherogenic changes in intestinal gene expression. In addition, when L-6 F was administered to mice on chow supplemented with LPA, intestinal levels of unsaturated LPA were significantly reduced, and this decrease correlated with reduced systemic inflammation and hypercholesterolemia (Chattopadhyay et al. 2013; Navab et al. 2013). Thus, reducing levels of intestinal LPA may be an important mechanism of action for L-6 F.

To enable the large-scale production of the peptide, transgenic, 6 F expressing tomatoes have been developed. Feeding LDL-R knockout mice a Western diet with and without transgenic tomatoes expressing 6 F improved an array of biomarkers of inflammation and antioxidant status and significantly reduced aortic lesion area compared to controls following 13 weeks of treatment (Chattopadhyay et al. 2013).

2.3 FX-5A

The peptide 5A is synthesized by replacing 5 amino acids of the apoA-I mimetic peptide 37pA with alanine residues, thereby decreasing its lipid affinity. Although 37pA effectively induces cholesterol efflux, its high lipid affinity is associated with cytotoxicity because of adverse effects on the integrity of the plasma membrane (Remaley et al. 2003). As a result of the amino acid substitutions, 5A is less cytotoxic and induces cholesterol efflux more specifically through ATP-binding cassette transporter A1 (ABCA1) (Sethi et al. 2008).

The 5A-phospholipid complex has been shown to induce cholesterol efflux both through ABCA1 and ATP-binding cassette transporter G1 (ABCG1). Administration of a single intravenous dose of a 5A-phospholipid complex was shown to enhance reverse cholesterol transport, as evidenced by a flux of cholesterol from peripheral cells to plasma, an increased amount of cholesterol and phospholipids in the HDL fraction, and an enhanced fecal sterol and bile acid secretion (Amar et al. 2010). In addition, 5A has been shown to possess anti-inflammatory and antioxidant properties (Tabet et al. 2010). To mimic acute inflammation, a nonocclusive collar was placed around the carotid artery in rabbits following infusion with 5A-phospholipid. Treatment with 5A-phospholipid significantly reduced the expression of proinflammatory endothelial adhesion molecules and neutrophil infiltration compared to placebo, and this effect of 5A infusion was similar to the effect observed after infusing lipid-free apoA-I or recombinant HDL. In the same rabbit model, it was shown that 5A-phospholipid reduced the production of radical oxygen species (ROS), such as O_2^- (Tabet et al. 2010).

The ability of 5A-phospholipid to inhibit the formation of atherosclerosis in vivo has been demonstrated in apolipoprotein E (apoE) knockout mice. Treatment with 30 mg/kg of intravenous 5A-phospholipid three times per week for 13 weeks resulted in 30 % reduction in aortic plaque area in both young apoE knockout mice and older apoE knockout mice with established lesions. The efficiency of the complex to inhibit plaque formation was shown to be dependent on the constitution of the phospholipid component. Reconstituting 5A with a sphingomyelin-containing combination of phospholipids enhanced its anti-atherosclerotic properties. When apoE knockout mice were treated with this complex using the same dosing regimen as used in the previously described experiments, there was a superior 54 % reduction in aortic lesions compared to controls.

In addition to its potent anti-atherosclerotic effects, 5A has been shown to reduce airway inflammation in murine models of asthma (Yao et al. 2011). Thus, the therapeutic potential of 5A extends beyond atherosclerotic cardiovascular disease.

As of yet, 5A has not been tested in humans. However, the first clinical trial is planned in the near future.

2.4 ATI-5261

Bielicki et al. developed ATI-5261, a peptide consisting of 36 amino acids forming a single amphipathic helix with high aqueous solubility that induced ABCA1-mediated cholesterol efflux with similar efficiency as apoA-I (Bielicki et al. 2010). LDL-R knockout mice received a high-fat Western diet for 13 weeks while concomitantly receiving daily intraperitoneal injections of ATI-5261 during the last 6 weeks. In another model, ApoE knockout mice received a Western diet for 18 weeks, followed by a chow diet with concomitant intraperitoneal injections of ATI-5261 every other day for another 6 weeks. The aortic lesion area and plaque lipid content were significantly reduced in mice treated ATI-5261 compared to mice receiving placebo. The decrease in atherosclerosis was accompanied by increased fecal sterol excretion, which is indicative of increased RCT upon ATI-5261 treatment.

2.5 ETC-642

ETC-642 is a complex of a 22-amino acid peptide that forms an amphipathic helix and phospholipids. The complex has been shown to exert multiple favorable effects on LDL and HDL particles (Di Bartolo et al. 2011b). Following 12 weeks of treatment in rabbits, a shift in LDL subfractions toward less negatively charged particles was observed, indicating a reduction in proinflammatory oxidized LDL. Moreover, a reduction in the particularly atherogenic small dense LDL (sdLDL) subfraction was noticed. Shortly after infusion of ETC-642, there was also a shift in HDL subfractions toward the antiatherogenic pre-β fraction. Moreover, ETC-642 was shown to be a potent inducer of cholesterol efflux from human macrophages in in vitro assays. It has also been shown in vivo to increase the cholesterol content in the HDL fraction, which may indicate increased reverse cholesterol transport (Di Bartolo et al. 2011a; Iwata et al. 2011). The anti-inflammatory effects of ETC-642 have been demonstrated in rabbit models of acute and chronic inflammation (Di Bartolo et al. 2011a, b). ETC-642 reduced endothelial adhesion molecule expression both in collared carotid arteries and in the aorta of cholesterol-fed rabbits, and this reduction was similar to effect observed in animals treated with reconstituted HDL. ETC-642 has been shown to induce an anti-inflammatory effect, as HDL isolated from ETC-642-treated rabbits decreased TNF-α-induced expression of NF-kB and endothelial adhesion molecules in human coronary artery endothelial cells (HCAECs) (Di Bartolo et al. 2011b). In addition, ETC-642 inhibited TNF-α-induced monocyte adhesion in HCAECs (Di Bartolo et al. 2011a).

The efficiency of ETC-642 as an anti-atherosclerotic agent has been demonstrated using intravascular ultrasound (IVUS) in hyperlipidemic rabbits that were treated with either low or high dose (15 or 50 mg/kg, respectively) or placebo two times per week for 12 weeks. Treatment with high-dose ETC-642 significantly inhibited plaque formation compared to controls (Iwata et al. 2011).

3 ApoA-I-Based Infusion Therapy

Lipid-poor pre-β HDL is the main acceptor of cholesterol from peripheral cells, including macrophages in the subendothelial vessel wall, by means of the interaction between apoA-I and the ABCA1 (Kontush and Chapman 2006). Largely based on the findings that apoA-I overexpression and apoA-I infusion in animal models do inhibit plaque formation, it is anticipated that infusion of apoA-I-containing particles, such as lipid-poor pre-β HDL, has direct beneficial effects on atherosclerosis.

3.1 HDL-VHDL Infusions

The atheroprotective effect of infusions of apoA-I-containing particles was first shown in the 1980s by Badimon et al. Cholesterol-fed rabbits were treated with 8 weekly infusions of homologous high-density lipoprotein-very high-density lipoprotein (HDL-VHDL) obtained by ultracentrifugation. Following the 8-week treatment period, a significant reduction in the aortic surface area covered by atherosclerotic-like lipid-rich lesions was found in HDL-VHDL-treated animals, compared to controls (37.9 % vs. 14.9 %). The reduced formation of aortic lesions was accompanied by a reduced total lipid deposition in the vascular wall and liver, whereas no significant differences in plasma lipid profiles were noted between both groups (Badimon et al. 1989).

It was subsequently shown that HDL-VHDL infusions did not only inhibit plaque formation, but also reduced the extent of preexisting lesions in a model where rabbits were fed an atherogenic diet for 60 days to induce atherosclerotic lesions (group 1). Two groups were treated with an additional 30 days of cholesterol-rich diet, with or without weekly HDL-VHDL infusions during the last 30 days (groups 2 and 3). Treatment with HDL-VHDL significantly reduced the aortic surface area covered by fatty streaks (17.8 % in treated rabbits vs. 34 and 38.8 % in control groups) as well as aortic lipid accumulation (Badimon et al. 1990).

3.2 Purified ApoA-I Infusions

The initial studies performed by Badimon et al. showed that infusions of HDL-VHDL did result in protection against atherosclerosis. Miyazaki et al. subsequently tested the hypothesis that infusions of purified apoA-I would have similar effects. In their first experiment, rabbits were fed an atherogenic diet for 90 days. During the last 30 days, half of the rabbits received weekly injections of purified apoA-I. Treatment with purified apoA-I resulted in a significant reduction in the proportion of the aortic surface area covered by fatty streaks, to a similar extent to treatment with HDL-VHDL (46.0 % vs. 23.9 %). In a second experiment, Miyazaki et al. studied whether purified apoA-I infusions induced regression of established plaques. Rabbits were fed an atherogenic diet for 105 days. Group 1 was sacrificed on day 105. Groups 2, 3, and 4 were fed a chow diet for an additional 60 days. Group 3 received infusions of purified apoA-I every other day, and group 4 received weekly infusions of a higher dose, while rabbits in group 2 served as untreated controls. The aortic surface area covered by atherosclerotic lesions progressed from day 105 to day 165 despite chow diet (50.0 % vs. 86.2 % for groups 1 and 2, respectively). Compared to the controls in group 2, both groups 3 and 4 showed less aortic surface area covered by atherosclerosis (86.2 % vs. 70.2 % and 65.7 %, respectively). However, apoA-I infusions did not induce plaque regression (Miyazaki et al. 1995).

3.3 ApoA-I$_{Milano}$ Infusions

Carriers of the apoA-I$_{Milano}$ mutation are characterized by reduced levels of HDL-C without the anticipated increased risk of CVD (Franceschini et al. 1980). Compared to native apoA-I, apoA-I$_{Milano}$ has been shown to induce more ABCA1-mediated cholesterol efflux, and, in addition, apoA-I$_{Milano}$ has been shown to exert superior anti-inflammatory and plaque stabilizing properties (Ibanez et al. 2012). Ameli et al. showed that infusion of recombinant apoA-I$_{Milano}$ effectively reduced the formation of intimal lesions in cholesterol-fed rabbits following balloon-induced vascular injury. Twenty rabbits on an atherogenic diet underwent balloon injury of the femoral and iliac arteries. Eight animals received injections with a complex of apoA-I$_{Milano}$ and phospholipids on alternating days during a 10-day period, starting 5 days prior to the balloon injury. Eight other animals received only the phospholipid carrier, and 4 animals served as controls; they did not receive any treatment. Infusions of apoA-I$_{Milano}$ were found to significantly inhibit intimal thickening following balloon injury (0.49 mm^2 vs. 1.14 mm^2 and 1.69 mm^2) and to reduce intimal macrophage content. This effect was present while plasma cholesterol levels remained similar among the groups (Ameli et al. 1994). Subsequent animal studies have confirmed the ability of infusion of recombinant apoA-I$_{Milano}$ to inhibit plaque progression and reduce plaque area of established lesions in models of arterial injury and in apoE-deficient mice (Chiesa et al. 2002; Shah et al. 1998, 2001; Soma et al. 1995). A reduction in plaque lipid and macrophage content has been demonstrated after only 48 h of a single infusion (Shah et al. 2001).

In the "ApoA-I Milano Trial," 57 patients with an acute coronary syndrome (ACS) were randomized to 5 weekly infusions of either a high (45 mg/kg) or a low dose (15 mg/kg) of a complex of recombinant apoA-I$_{Milano}$ with phospholipid carriers (ETC-216) or placebo, and the treatment began within 2 weeks after ACS. Percentage and total atheroma volume and plaque thickness were significantly reduced in patients treated with apoA-I$_{Milano}$ compared to baseline as assessed by IVUS. The infusions with ETC-216 were generally well tolerated. The small pilot study lacked power to detect dose effects and differences between the treatment and placebo groups (Nissen et al. 2003). Additional analysis revealed that treatment with apoA-I$_{Milano}$ had a beneficial effect on arterial wall remodeling

(Nicholls et al. 2006). In conclusion, regression of atherosclerosis was demonstrated following short-term infusions, and the results of larger studies with longer follow-up and clinically relevant end points are eagerly awaited.

3.4 CSL-111 and CSL-112 Infusions

Two subsequent studies investigated the safety and efficacy of infusion of reconstituted HDL (rHDL), consisting of native apoA-I and phospholipids, on coronary atherosclerosis. CSL-111 is composed of human apoA-I and soybean phosphatidylcholine and thereby resembles native HDL. In the "Effect of rHDL on Atherosclerosis-Safety and Efficacy" (ERASE) trial, a total of 183 patients was randomized to 4 weekly infusions of either 40 or 80 mg/kg of CSL-111 or placebo, and this treatment was started within 2 weeks after an ACS. An IVUS and a quantitative coronary angiography were obtained prior to randomization and 2 weeks after the last infusion. Administration of high-dose CSL-111 was associated with a high incidence of transaminase elevations, which led to early discontinuation of the 80 mg/kg study arm. Treatment with 4 weekly infusions of 40 mg/kg of CSL-111 significantly reduced atheroma volume compared to baseline, to a similar extent as seen in the ApoA-I Milano Trial. However, the differences in atheroma volume between the study groups did not reach statistical significance. Nonetheless, the authors conclude that CSL-111 may exert a favorable treatment effect, as treatment improved plaque characterization index on IVUS and coronary score on quantitative angiography (Tardif et al. 2007).

Further development of CSL-111 was discontinued after the observation that the therapy induced a high incidence of liver function abnormalities, possibly related to the high cholate content of the CSL-111 particle.

CSL-112, a second-generation compound with no liver toxicity (Krause and Remaley 2013), has been tested in two phase 1 studies comprising a total of 93 healthy individuals. In the single-dose study, subjects received doses of up to 135 mg/kg, and in the multiple-dose study, 36 subjects were divided into four groups. Groups 1 and 2 received 4 weekly infusions of either low- or high-dose CSL-112, group 3 received a low dose of CSL-112 twice weekly for 4 weeks, and group 4 received placebo. Infusion of CSL-112 resulted in a dose-dependent increase in apoA-I, which remained above baseline levels for 3 days. A rapid 36-fold increase in pre-β HDL levels was noted, which was accompanied by increased ABCA1-mediated cholesterol efflux (maximum increase 270 % compared to baseline). Importantly, CSL-112 did not cause clinically relevant elevations of liver function parameters (Easton et al. 2013). A phase 2a study of CSL-112 in patients with stable CVD was recently completed, and the results are eagerly awaited.

3.5 CER-001 Infusions

CER-001, an engineered rHDL particle comprising recombinant human wild-type apoA-I and diphosphatidylglycerol and sphingomyelin, was designed to mimic nascent pre-β HDL. CER-001 infusions were shown to reduce atherosclerotic plaque size and aortic macrophage and lipid content in atherosclerosis-prone mice (Goffinet et al. 2012). A phase 1 randomized crossover study evaluated dose ranges from 0.25 to 45 mg/kg in 32 dyslipidemic volunteers. CER-001 infusions were well tolerated and did not show any safety issue at any dose; in particular there were no liver enzyme elevations. Following administration of a single infusion of low doses of CER-001, plasma total cholesterol and cholesterol in the HDL fraction were significantly elevated (almost 700 % increase in the HDL fraction at the 45 mg/kg dose), which may indicate mobilization of cholesterol from peripheral tissues (Keyserling et al. 2011). Cholesterol esterification was also observed, consistent with LCAT activation. Three phase 2 trials are currently investigating the effects of CER-001 on atherosclerotic plaques. The "Can HDL Infusions Significantly Quicken Atherosclerosis Regression" (CHI-SQUARE) trial is designed to investigate whether CER-001 infusions induce plaque regression in patients with ACS. In total, 507 patients have been randomized to 6 weekly infusions of three ascending doses of CER-001, or placebo. The primary outcome measure is the absolute change in plaque volume, as determined by IVUS, from baseline to 3 weeks after the last infusion (NCT01201837). In parallel, two open-label trials are currently evaluating the effects of CER-001 on plaque volume. In the Modifying Orphan Disease Evaluation (MODE) trial (NCT01412034), a total of 23 patients with homozygous familial hypercholesterolemia (FH) is treated for 6 months with biweekly infusions with CER-001. The primary outcome measure in this trial is the percent change in carotid mean vessel wall area as assessed by 3Tesla MRI. A similar regimen is followed in the investigator-driven SAMBA trial, conducted in patients with very low or absent HDL due to genetic defects—familial primary hypoalphalipoproteinemia (FPHA) (EudraCT number 2011-006188-23). Preliminary results presented at the 2013 PACE Snapshot session (ESC, Amsterdam) indicate that CER-001 increases RCT and may effectively reduce aortic atheroma volume.

These studies will provide valuable information about the clinical utility of CER-001 infusions in patients at increased CVD risk.

3.6 Modified ApoA-I

The rapid renal clearance of lipid-poor apoA-I may limit the therapeutic potential of apoA-I-based therapies. TripA-I represents a trimeric apoA-I analogue that functionally resembles wild-type apoA-I, with a prolonged half-life in plasma due to its increased size (Graversen et al. 2008; Ohnsorg et al. 2011). Intravenous treatment with tripA-I decreased the progression of atherosclerotic lesion size in LDL-R knockout mice on an atherogenic diet, although tripA-I did not affect aortic atherosclerotic lesion size in uremic apoE knockout mice (Graversen et al. 2008; Pedersen et al. 2009).

4 Selective Delipidation

Given that cholesterol-poor HDL particles are particularly effective in inducing reverse cholesterol transport, a procedure has been developed that selectively delipidates large, cholesterol-rich HDL particles, thereby creating small, lipid-poor HDL particles similar to small α- and pre-β-like HDL (HDL selectively delipidated or HDL-sdl). The technique involves mixing plasma with two delipidation reagents, followed by bulk separation by gravity and subsequent passage through a charcoal column to remove residual reagents (Sacks et al. 2009). Potential benefits of this technique are the fact that autologous particles are reinfused, which is anticipated to limit potential toxicity, and that the reinfused pre-β HDL particles are deemed particularly effective in promoting cholesterol efflux. Sacks et al. showed that 12 weekly infusions of selectively delipidated HDL reduced diet-induced atherosclerosis in monkeys, as assessed using IVUS. Delipidated small HDL particles were entirely converted into large α-HDL particles after infusion. Combined with the finding that selectively delipidated plasma induced a seven-fold increase in ABCA1-mediated cholesterol efflux, this may indicate enhanced reverse cholesterol transport (Sacks et al. 2009). Waksman et al. subsequently randomized 28 patients undergoing cardiac catheterization for ACS to 7 weekly infusions with either selectively delipidated HDL or control plasma to evaluate the safety of the Lipid Sciences Plasma Delipidation System-2 (LS PDS-2) and to explore the potential effects on atheroma volume. The treatment was well tolerated by all patients. On average, there was a 28-fold increase in pre-β-like HDL following delipidation compared to baseline. IVUS was obtained during cardiac catheterization for the ACS and after 2 weeks of the last treatment. A trend toward regression of total atheroma volume was found in patients treated with selective delipidation, while a small increase in atheroma volume was observed in the control group. Although the sample size was too small to detect significant effects on atheroma volume, the trend toward plaque regression can be considered as a beneficial effect. Additional studies in larger cohorts of patients will provide a more definite answer to the question whether delipidation is a potential benefit for patients suffering from an ACS (Waksman et al. 2010).

Conclusion
The development of apoA-I analogues has been driven by the observation that apoA-I plays a pivotal role in the antiatherogenic capacity of HDL particles.

A large number of apoA-I analogue proteins have been studied in vitro and some did show beneficial effects. However, "the proof of the pudding will be in the eating:" well-designed clinical trials in patients are indispensable to address the role of apoA-I mimetics and alike in atherosclerosis regression.

The clinical potential of rHDL particles reaches beyond the treatment of atherosclerosis. For example, rHDL particles labeled with contrast-generating agents, such as gadolinium or ^{125}I, can be used as natural nanoparticles in medical imaging to visualize natural targets such as macrophages in atherosclerotic plaques or other targets by attaching targeting molecules (Ryan 2010; Stanley 2014). In addition, rHDL can be used as a vector for drug transport by incorporating lipophilic drugs into rHDL particles, which may decrease drug toxicity and increase bioavailability (Stanley 2014).

Open Access This chapter is distributed under the terms of the Creative Commons Attribution Noncommercial License, which permits any noncommercial use, distribution, and reproduction in any medium, provided the original author(s) and source are credited.

References

Amar MJ, D'Souza W, Turner S, Demosky S, Sviridov D, Stonik J, Luchoomun J, Voogt J, Hellerstein M, Sviridov D, Remaley AT (2010) 5A apolipoprotein mimetic peptide promotes cholesterol efflux and reduces atherosclerosis in mice. J Pharmacol Exp Ther 334:634–641. doi:10.1124/jpet.110.167890

Ameli S, Hultgardh-Nilsson A, Cercek B, Shah PK, Forrester JS, Ageland H, Nilsson J (1994) Recombinant apolipoprotein A-I Milano reduces intimal thickening after balloon injury in hypercholesterolemic rabbits. Circulation 90:1935–1941

Anantharamaiah GM, Jones JL, Brouillette CG, Schmidt CF, Chung BH, Hughes TA, Bhown AS, Segrest JP (1985) Studies of synthetic peptide analogs of the amphipathic helix. Structure of complexes with dimyristoyl phosphatidylcholine. J Biol Chem 260:10248–10255

Badimon JJ, Badimon L, Galvez A, Dische R, Fuster V (1989) High density lipoprotein plasma fractions inhibit aortic fatty streaks in cholesterol-fed rabbits. Lab Invest 60:455–461

Badimon JJ, Badimon L, Fuster V (1990) Regression of atherosclerotic lesions by high density lipoprotein plasma fraction in the cholesterol-fed rabbit. J Clin Invest 85:1234–1241. doi:10.1172/jci114558

Baigent C, Blackwell L, Emberson J, Holland LE, Reith C, Bhala N, Peto R, Barnes EH, Keech A, Simes J, Collins R (2010) Efficacy and safety of more intensive lowering of LDL cholesterol: a meta-analysis of data from 170,000 participants in 26 randomised trials. Lancet 376:1670–1681. doi:10.1016/s0140-6736(10)61350-5

Barter PJ, Nicholls S, Rye KA, Anantharamaiah GM, Navab M, Fogelman AM (2004) Antiinflammatory properties of HDL. Circ Res 95:764–772. doi:10.1161/01.res.0000146094.59640.13

Barter P, Gotto AM, LaRosa JC, Maroni J, Szarek M, Grundy SM, Kastelein JJ, Bittner V, Fruchart JC (2007) HDL cholesterol, very low levels of LDL cholesterol, and cardiovascular events. N Engl J Med 357:1301–1310. doi:10.1056/NEJMoa064278

Bielicki JK, Zhang H, Cortez Y, Zheng Y, Narayanaswami V, Patel A, Johansson J, Azhar S (2010) A new HDL mimetic peptide that stimulates cellular cholesterol efflux with high efficiency greatly reduces atherosclerosis in mice. J Lipid Res 51:1496–1503. doi:10.1194/jlr.M003665

Bloedon LT, Dunbar R, Duffy D, Pinell-Salles P, Norris R, DeGroot BJ, Movva R, Navab M, Fogelman AM, Rader DJ (2008) Safety, pharmacokinetics, and pharmacodynamics of oral apoA-I mimetic peptide D-4F in high-risk cardiovascular patients. J Lipid Res 49:1344–1352. doi:10.1194/jlr.P800003-JLR200

Briel M, Ferreira-Gonzalez I, You JJ, Karanicolas PJ, Akl EA, Wu P, Blechacz B, Bassler D, Wei X, Sharman A, Whitt I, Alves da Silva S, Khalid Z, Nordmann AJ, Zhou Q, Walter SD,

Vale N, Bhatnagar N, O'Regan C, Mills EJ, Bucher HC, Montori VM, Guyatt GH (2009) Association between change in high density lipoprotein cholesterol and cardiovascular disease morbidity and mortality: systematic review and meta-regression analysis. BMJ 338:b92. doi:10.1136/bmj.b92

Chattopadhyay A, Navab M, Hough G, Gao F, Meriwether D, Grijalva V, Springstead JR, Palgnachari MN, Namiri-Kalantari R, Su F, Van Lenten BJ, Wagner AC, Anantharamaiah GM, Farias-Eisner R, Reddy ST, Fogelman AM (2013) A novel approach to oral apoA-I mimetic therapy. J Lipid Res 54:995–1010. doi:10.1194/jlr.M033555

Chiesa G, Monteggia E, Marchesi M, Lorenzon P, Laucello M, Lorusso V, Di Mario C, Karvouni E, Newton RS, Bisgaier CL, Franceschini G, Sirtori CR (2002) Recombinant apolipoprotein A-I(Milano) infusion into rabbit carotid artery rapidly removes lipid from fatty streaks. Circ Res 90:974–980

deGoma EM, deGoma RL, Rader DJ (2008) Beyond high-density lipoprotein cholesterol levels evaluating high-density lipoprotein function as influenced by novel therapeutic approaches. J Am Coll Cardiol 51:2199–2211. doi:10.1016/j.jacc.2008.03.016

Di Bartolo BA, Nicholls SJ, Bao S, Rye KA, Heather AK, Barter PJ, Bursill C (2011a) The apolipoprotein A-I mimetic peptide ETC-642 exhibits anti-inflammatory properties that are comparable to high density lipoproteins. Atherosclerosis 217:395–400. doi:10.1016/j.atherosclerosis.2011.04.001

Di Bartolo BA, Vanags LZ, Tan JT, Bao S, Rye KA, Barter PJ, Bursill CA (2011b) The apolipoprotein A-I mimetic peptide, ETC-642, reduces chronic vascular inflammation in the rabbit. Lipids Health Dis 10:224. doi:10.1186/1476-511x-10-224

D'Souza W, Stonik JA, Murphy A, Demosky SJ, Sethi AA, Moore XL, Chin-Dusting J, Remaley AT, Sviridov D (2010) Structure/function relationships of apolipoprotein a-I mimetic peptides: implications for antiatherogenic activities of high-density lipoprotein. Circ Res 107:217–227. doi:10.1161/circresaha.110.216507

Duverger N, Kruth H, Emmanuel F, Caillaud JM, Viglietta C, Castro G, Tailleux A, Fievet C, Fruchart JC, Houdebine LM, Denefle P (1996) Inhibition of atherosclerosis development in cholesterol-fed human apolipoprotein A-I-transgenic rabbits. Circulation 94:713–717

Easton R, Gille A, D'Andrea D, Davis R, Wright SD, Shear C (2013) A multiple ascending dose study of CSL112, an infused formulation of ApoA-I. J Clin Pharmacol. doi:10.1002/jcph.194

Fielding CJ, Fielding PE (1995) Molecular physiology of reverse cholesterol transport. J Lipid Res 36:211–228

Franceschini G, Sirtori CR, Capurso A 2nd, Weisgraber KH, Mahley RW (1980) A-IMilano apoprotein. Decreased high density lipoprotein cholesterol levels with significant lipoprotein modifications and without clinical atherosclerosis in an Italian family. J Clin Invest 66:892–900. doi:10.1172/jci109956

Garber DW, Venkatachalapathi YV, Gupta KB, Ibdah J, Phillips MC, Hazelrig JB, Segrest JP, Anantharamaiah GM (1992) Turnover of synthetic class A amphipathic peptide analogues of exchangeable apolipoproteins in rats. Correlation with physical properties. Arterioscler Thromb 12:886–894

Goffinet M, Tardy C, Bluteau A, Boubekeur N, Baron R, Keyserling C, Barbaras R, Lalwani N, Dasseux J (2012) Anti-atherosclerotic effect of CER-001, an engineered HDL-mimetic, in the high-fat diet-fed LDLr knock-out mouse. Circulation 126, A18667

Gordon T, Castelli WP, Hjortland MC, Kannel WB, Dawber TR (1977) High density lipoprotein as a protective factor against coronary heart disease. The Framingham Study. Am J Med 62:707–714

Gordon DJ, Probstfield JL, Garrison RJ, Neaton JD, Castelli WP, Knoke JD, Jacobs DR Jr, Bangdiwala S, Tyroler HA (1989) High-density lipoprotein cholesterol and cardiovascular disease. Four prospective American studies. Circulation 79:8–15

Graversen JH, Laurberg JM, Andersen MH, Falk E, Nieland J, Christensen J, Etzerodt M, Thøgersen HC, Moestrup SK (2008) Trimerization of apolipoprotein A-I retards plasma clearance and preserves antiatherosclerotic properties. J Cardiovasc Pharmacol 51:170–177. doi:10.1097/FJC.0b013e31815ed0b9

Haase CL, Frikke-Schmidt R, Nordestgaard BG, Kateifides AK, Kardassis D, Nielsen LB, Andersen CB, Kober L, Johnsen AH, Grande P, Zannis VI, Tybjaerg-Hansen A (2011) Mutation in APOA1 predicts increased risk of ischaemic heart disease and total mortality without low HDL cholesterol levels. J Intern Med 270:136–146. doi:10.1111/j.1365-2796.2011.02381.x

Hovingh GK, Brownlie A, Bisoendial RJ, Dube MP, Levels JH, Petersen W, Dullaart RP, Stroes ES, Zwinderman AH, de Groot E, Hayden MR, Kuivenhoven JA, Kastelein JJ (2004) A novel apoA-I mutation (L178P) leads to endothelial dysfunction, increased arterial wall thickness, and premature coronary artery disease. J Am Coll Cardiol 44:1429–1435. doi:10.1016/j.jacc.2004.06.070

Ibanez B, Giannarelli C, Cimmino G, Santos-Gallego CG, Alique M, Pinero A, Vilahur G, Fuster V, Badimon L, Badimon JJ (2012) Recombinant HDL(Milano) exerts greater anti-inflammatory and plaque stabilizing properties than HDL(wild-type). Atherosclerosis 220:72–77. doi:10.1016/j.atherosclerosis.2011.10.006

Iwata A, Miura S, Zhang B, Imaizumi S, Uehara Y, Shiomi M, Saku K (2011) Antiatherogenic effects of newly developed apolipoprotein A-I mimetic peptide/phospholipid complexes against aortic plaque burden in Watanabe-heritable hyperlipidemic rabbits. Atherosclerosis 218:300–307. doi:10.1016/j.atherosclerosis.2011.05.029

Jafri H, Alsheikh-Ali AA, Karas RH (2010) Meta-analysis: statin therapy does not alter the association between low levels of high-density lipoprotein cholesterol and increased cardiovascular risk. Ann Intern Med 153:800–808. doi:10.7326/0003-4819-153-12-201012210-00006

Keyserling CH, Hunt TL, Klepp HM, Scott RA, Barbaras R, Schwendeman A, Lalwani N, Dasseux J-L (2011) CER-001, a synthetic HDL-mimetic, safely mobilizes cholesterol in healthy dyslipidemic volunteers. Circulation 124, A15525

Kontush A, Chapman MJ (2006) Functionally defective high-density lipoprotein: a new therapeutic target at the crossroads of dyslipidemia, inflammation, and atherosclerosis. Pharmacol Rev 58:342–374. doi:10.1124/pr.58.3.1

Krause BR, Remaley AT (2013) Reconstituted HDL for the acute treatment of acute coronary syndrome. Curr Opin Lipidol 24:480–486. doi:10.1097/mol.0000000000000020

Mineo C, Deguchi H, Griffin JH, Shaul PW (2006) Endothelial and antithrombotic actions of HDL. Circ Res 98:1352–1364. doi:10.1161/01.res.0000225982.01988.93

Miyazaki A, Sakuma S, Morikawa W, Takiue T, Miake F, Terano T, Sakai M, Hakamata H, Sakamoto Y, Natio M et al (1995) Intravenous injection of rabbit apolipoprotein A-I inhibits the progression of atherosclerosis in cholesterol-fed rabbits. Arterioscler Thromb Vasc Biol 15:1882–1888

Moore RE, Kawashiri MA, Kitajima K, Secreto A, Millar JS, Pratico D, Rader DJ (2003) Apolipoprotein A-I deficiency results in markedly increased atherosclerosis in mice lacking the LDL receptor. Arterioscler Thromb Vasc Biol 23:1914–1920. doi:10.1161/01.atv.0000092328.66882.f5

Navab M, Anantharamaiah GM, Hama S, Garber DW, Chaddha M, Hough G, Lallone R, Fogelman AM (2002) Oral administration of an Apo A-I mimetic Peptide synthesized from D-amino acids dramatically reduces atherosclerosis in mice independent of plasma cholesterol. Circulation 105:290–292

Navab M, Anantharamaiah GM, Reddy ST, Hama S, Hough G, Grijalva VR, Wagner AC, Frank JS, Datta G, Garber D, Fogelman AM (2004) Oral D-4F causes formation of pre-beta high-density lipoprotein and improves high-density lipoprotein-mediated cholesterol efflux and reverse cholesterol transport from macrophages in apolipoprotein E-null mice. Circulation 109:3215–3220. doi:10.1161/01.cir.0000134275.90823.87

Navab M, Hough G, Buga GM, Su F, Wagner AC, Meriwether D, Chattopadhyay A, Gao F, Grijalva V, Danciger JS, Van Lenten BJ, Org E, Lusis AJ, Pan C, Anantharamaiah GM, Farias-Eisner R, Smyth SS, Reddy ST, Fogelman AM (2013) Transgenic 6F tomatoes act on the small intestine to prevent systemic inflammation and dyslipidemia caused by Western diet and intestinally derived lysophosphatidic acid. J Lipid Res 54:3403–3418. doi:10.1194/jlr.M042051

Nicholls SJ, Tuzcu EM, Sipahi I, Schoenhagen P, Crowe T, Kapadia S, Nissen SE (2006) Relationship between atheroma regression and change in lumen size after infusion of apolipoprotein A-I Milano. J Am Coll Cardiol 47:992–997. doi:10.1016/j.jacc.2005.11.040

Nissen SE, Tsunoda T, Tuzcu EM, Schoenhagen P, Cooper CJ, Yasin M, Eaton GM, Lauer MA, Sheldon WS, Grines CL, Halpern S, Crowe T, Blankenship JC, Kerensky R (2003) Effect of recombinant ApoA-I Milano on coronary atherosclerosis in patients with acute coronary syndromes: a randomized controlled trial. JAMA 290:2292–2300. doi:10.1001/jama.290.17.2292

Ohnsorg PM, Mary JL, Rohrer L, Pech M, Fingerle J, von Eckardstein A (2011) Trimerized apolipoprotein A-I (TripA) forms lipoproteins, activates lecithin: cholesterol acyltransferase, elicits lipid efflux, and is transported through aortic endothelial cells. Biochim Biophys Acta 1811:1115–1123. doi:10.1016/j.bbalip.2011.09.001

Pedersen TX, Bro S, Andersen MH, Etzerodt M, Jauhiainen M, Moestrup S, Nielsen LB (2009) Effect of treatment with human apolipoprotein A-I on atherosclerosis in uremic apolipoprotein-E deficient mice. Atherosclerosis 202:372–381. doi:10.1016/j.atherosclerosis.2008.04.041

Remaley AT, Thomas F, Stonik JA, Demosky SJ, Bark SE, Neufeld EB, Bocharov AV, Vishnyakova TG, Patterson AP, Eggerman TL, Santamarina-Fojo S, Brewer HB (2003) Synthetic amphipathic helical peptides promote lipid efflux from cells by an ABCA1-dependent and an ABCA1-independent pathway. J Lipid Res 44:828–836. doi:10.1194/jlr.M200475-JLR200

Roger VL, Go AS, Lloyd-Jones DM, Benjamin EJ, Berry JD, Borden WB, Bravata DM, Dai S, Ford ES, Fox CS, Fullerton HJ, Gillespie C, Hailpern SM, Heit JA, Howard VJ, Kissela BM, Kittner SJ, Lackland DT, Lichtman JH, Lisabeth LD, Makuc DM, Marcus GM, Marelli A, Matchar DB, Moy CS, Mozaffarian D, Mussolino ME, Nichol G, Paynter NP, Soliman EZ, Sorlie PD, Sotoodehnia N, Turan TN, Virani SS, Wong ND, Woo D, Turner MB (2012) Heart disease and stroke statistics–2012 update: a report from the American Heart Association. Circulation 125:e2–e220. doi:10.1161/CIR.0b013e31823ac046

Rubin EM, Krauss RM, Spangler EA, Verstuyft JG, Clift SM (1991) Inhibition of early atherogenesis in transgenic mice by human apolipoprotein AI. Nature 353:265–267. doi:10.1038/353265a0

Ryan RO (2010) Nanobiotechnology applications of reconstituted high density lipoprotein. J Nanobiotechnol 8:28. doi:10.1186/1477-3155-8-28

Sacks FM, Rudel LL, Conner A, Akeefe H, Kostner G, Baki T, Rothblat G, de la Llera-Moya M, Asztalos B, Perlman T, Zheng C, Alaupovic P, Maltais JA, Brewer HB (2009) Selective delipidation of plasma HDL enhances reverse cholesterol transport in vivo. J Lipid Res 50:894–907. doi:10.1194/jlr.M800622-JLR200

Sethi AA, Stonik JA, Thomas F, Demosky SJ, Amar M, Neufeld E, Brewer HB, Davidson WS, D'Souza W, Sviridov D, Remaley AT (2008) Asymmetry in the lipid affinity of bihelical amphipathic peptides. A structural determinant for the specificity of ABCA1-dependent cholesterol efflux by peptides. J Biol Chem 283:32273–32282. doi:10.1074/jbc.M804461200

Shah PK, Nilsson J, Kaul S, Fishbein MC, Ageland H, Hamsten A, Johansson J, Karpe F, Cercek B (1998) Effects of recombinant apolipoprotein A-I(Milano) on aortic atherosclerosis in apolipoprotein E-deficient mice. Circulation 97:780–785

Shah PK, Yano J, Reyes O, Chyu KY, Kaul S, Bisgaier CL, Drake S, Cercek B (2001) High-dose recombinant apolipoprotein A-I(milano) mobilizes tissue cholesterol and rapidly reduces plaque lipid and macrophage content in apolipoprotein e-deficient mice. Potential implications for acute plaque stabilization. Circulation 103:3047–3050

Soma MR, Donetti E, Parolini C, Sirtori CR, Fumagalli R, Franceschini G (1995) Recombinant apolipoprotein A-IMilano dimer inhibits carotid intimal thickening induced by perivascular manipulation in rabbits. Circ Res 76:405–411

Stanley S (2014) Biological nanoparticles and their influence on organisms. Curr Opin Biotechnol 28:69–74. doi:10.1016/j.copbio.2013.11.014

Suc I, Escargueil-Blanc I, Troly M, Salvayre R, Negre-Salvayre A (1997) HDL and ApoA prevent cell death of endothelial cells induced by oxidized LDL. Arterioscler Thromb Vasc Biol 17:2158–2166

Tabet F, Remaley AT, Segaliny AI, Millet J, Yan L, Nakhla S, Barter PJ, Rye KA, Lambert G (2010) The 5A apolipoprotein A-I mimetic peptide displays antiinflammatory and antioxidant properties in vivo and in vitro. Arterioscler Thromb Vasc Biol 30:246–252. doi:10.1161/atvbaha.109.200196

Tardif JC, Gregoire J, L'Allier PL, Ibrahim R, Lesperance J, Heinonen TM, Kouz S, Berry C, Basser R, Lavoie MA, Guertin MC, Rodes-Cabau J (2007) Effects of reconstituted high-density lipoprotein infusions on coronary atherosclerosis: a randomized controlled trial. JAMA 297:1675–1682. doi:10.1001/jama.297.15.jpc70004

Venkatachalapathi YV, Phillips MC, Epand RM, Epand RF, Tytler EM, Segrest JP, Anantharamaiah GM (1993) Effect of end group blockage on the properties of a class A amphipathic helical peptide. Proteins 15:349–359. doi:10.1002/prot.340150403

Voight BF, Peloso GM, Orho-Melander M, Frikke-Schmidt R, Barbalic M, Jensen MK, Hindy G, Holm H, Ding EL, Johnson T, Schunkert H, Samani NJ, Clarke R, Hopewell JC, Thompson JF, Li M, Thorleifsson G, Newton-Cheh C, Musunuru K, Pirruccello JP, Saleheen D, Chen L, Stewart A, Schillert A, Thorsteinsdottir U, Thorgeirsson G, Anand S, Engert JC, Morgan T, Spertus J, Stoll M, Berger K, Martinelli N, Girelli D, McKeown PP, Patterson CC, Epstein SE, Devaney J, Burnett MS, Mooser V, Ripatti S, Surakka I, Nieminen MS, Sinisalo J, Lokki ML, Perola M, Havulinna A, de Faire U, Gigante B, Ingelsson E, Zeller T, Wild P, de Bakker PI, Klungel OH, Maitland-van der Zee AH, Peters BJ, de Boer A, Grobbee DE, Kamphuisen PW, Deneer VH, Elbers CC, Onland-Moret NC, Hofker MH, Wijmenga C, Verschuren WM, Boer JM, van der Schouw YT, Rasheed A, Frossard P, Demissie S, Willer C, Do R, Ordovas JM, Abecasis GR, Boehnke M, Mohlke KL, Daly MJ, Guiducci C, Burtt NP, Surti A, Gonzalez E, Purcell S, Gabriel S, Marrugat J, Peden J, Erdmann J, Diemert P, Willenborg C, Konig IR, Fischer M, Hengstenberg C, Ziegler A, Buysschaert I, Lambrechts D, Van de Werf F, Fox KA, El Mokhtari NE, Rubin D, Schrezenmeir J, Schreiber S et al (2012) Plasma HDL cholesterol and risk of myocardial infarction: a mendelian randomisation study. Lancet 380:572–580. doi:10.1016/s0140-6736(12)60312-2

Voyiaziakis E, Goldberg IJ, Plump AS, Rubin EM, Breslow JL, Huang LS (1998) ApoA-I deficiency causes both hypertriglyceridemia and increased atherosclerosis in human apoB transgenic mice. J Lipid Res 39:313–321

Waksman R, Torguson R, Kent KM, Pichard AD, Suddath WO, Satler LF, Martin BD, Perlman TJ, Maltais JA, Weissman NJ, Fitzgerald PJ, Brewer HB Jr (2010) A first-in-man, randomized, placebo-controlled study to evaluate the safety and feasibility of autologous delipidated high-density lipoprotein plasma infusions in patients with acute coronary syndrome. J Am Coll Cardiol 55:2727–2735. doi:10.1016/j.jacc.2009.12.067

Watson CE, Weissbach N, Kjems L, Ayalasomayajula S, Zhang Y, Chang I, Navab M, Hama S, Hough G, Reddy ST, Soffer D, Rader DJ, Fogelman AM, Schecter A (2011) Treatment of patients with cardiovascular disease with L-4F, an apo-A1 mimetic, did not improve select biomarkers of HDL function. J Lipid Res 52:361–373. doi:10.1194/jlr.M011098

Wool GD, Reardon CA, Getz GS (2008) Apolipoprotein A-I mimetic peptide helix number and helix linker influence potentially anti-atherogenic properties. J Lipid Res 49:1268–1283. doi:10.1194/jlr.M700552-JLR200

Yancey PG, Bielicki JK, Johnson WJ, Lund-Katz S, Palgunachari MN, Anantharamaiah GM, Segrest JP, Phillips MC, Rothblat GH (1995) Efflux of cellular cholesterol and phospholipid to lipid-free apolipoproteins and class A amphipathic peptides. Biochemistry 34:7955–7965

Yao X, Dai C, Fredriksson K, Dagur PK, McCoy JP, Qu X, Yu ZX, Keeran KJ, Zywicke GJ, Amar MJ, Remaley AT, Levine SJ (2011) 5A, an apolipoprotein A-I mimetic peptide, attenuates the induction of house dust mite-induced asthma. J Immunol 186:576–583. doi:10.4049/jimmunol.1001534

Antisense Oligonucleotides, microRNAs, and Antibodies

Alberto Dávalos and Angeliki Chroni

Contents

1	Antisense Oligonucleotides	651
	1.1 Making Sense of Antisense	651
	1.2 Therapeutic Antisense Oligonucleotides to Treat Dyslipidemia	653
	1.3 Therapeutic Antisense Oligonucleotides for the Increase of HDL-Cholesterol Levels and Improvement of HDL Function	654
2	miRNAs	656
	2.1 miRNA-Based Therapy	657
	2.2 Micromanaging Cholesterol Efflux, RCT, HDL Levels, and HDL Function	661
3	Antibodies	666
	3.1 LDL-Cholesterol Lowering Approaches: Proprotein Convertase Subtilisin/Kexin Type 9 Blocking Antibodies	666
	3.2 Approaches to Antibody Therapy for the Increase of HDL-Cholesterol Levels ...	669
	3.3 Effect of Antibodies Used for the Treatment of Chronic Inflammatory Diseases on HDL Antiatherogenic Functions	671
	3.4 Vaccines Against Atherosclerosis	672
Conclusions		675
References		675

Abstract

The specificity of Watson–Crick base pairing and the development of several chemical modifications to oligonucleotides have enabled the development of novel drug classes for the treatment of different human diseases. This review

---

A. Dávalos (✉)
Laboratory of Disorders of Lipid Metabolism and Molecular Nutrition, Madrid Institute for Advanced Studies (IMDEA)-Food, Ctra. de Cantoblanco 8, 28049 Madrid, Spain
e-mail: alberto.davalos@imdea.org

A. Chroni (✉)
Institute of Biosciences and Applications, National Center for Scientific Research Demokritos, Patriarchou Grigoriou and Neapoleos, Agia Paraskevi, 15310 Athens, Greece
e-mail: achroni@bio.demokritos.gr

focuses on promising results of recent preclinical or clinical studies on targeting HDL metabolism and function by antisense oligonucleotides and miRNA-based therapies. Although many hurdles regarding basic mechanism of action, delivery, specificity, and toxicity need to be overcome, promising results from recent clinical trials and recent approval of these types of therapy to treat dyslipidemia suggest that the treatment of HDL dysfunction will benefit from these unique clinical opportunities. Moreover, an overview of monoclonal antibodies (mAbs) developed for the treatment of dyslipidemia and cardiovascular disease and currently being tested in clinical studies is provided. Initial studies have shown that these compounds are generally safe and well tolerated, but ongoing large clinical studies will assess their long-term safety and efficacy.

Keywords

Antisense • Oligonucleotides • microRNAs • Antibodies • High-density lipoproteins • Cholesterol efflux • LDL-C reduction • HDL-C increase • HDL antiatherogenic function improvement • Atherosclerotic lesion reduction

Abbreviations

AAV	Adeno-associated virus
Ago2	Protein Argonaute-2
Apo	Apolipoprotein
ASOs	Antisense oligonucleotides
ceRNA	Competing endogenous RNA
CRP	C-reactive protein
CETP	Cholesteryl ester transfer protein
CHD	Coronary heart disease
EL	Endothelial lipase
EMA	European Medicines Agency
FDA	US Food and Drug Administration
HDL-C	HDL cholesterol
HSPs	Heat shock proteins
LDL-C	LDL cholesterol
LDLR	LDL receptor
LNA	Locked nucleic acid
miRNA	microRNA
mAbs	Monoclonal antibodies
MDA	Malondialdehyde
MI	Myocardial infarction
oxLDL	Oxidized LDL
PCSK9	Proprotein convertase subtilisin/kexin type 9
VEGFR2	Vascular endothelial growth factor receptor 2

1 Antisense Oligonucleotides

Milestone discoveries of the specificity of Watson–Crick base pairing and of RNA interference and the importance of different RNAs in the genomic regulation of living organisms have led to the emergence of different RNA-based therapeutics that hold the promise for the silencing of "drug-able" targets. Although several classes of RNA-derived therapeutics have reached clinical trials, many hurdles and challenges have reduced the number of RNA drugs that reach the market. The recent approvals of an antisense oligonucleotide (mipomersen) by FDA and a gene therapy (Glybera) by EMA, as well as several ongoing clinical trials, envision to this type of therapeutics with real future successes. Antisense oligonucleotides (ASOs), ribozymes, aptamers, miRNAs, and siRNAs are only some examples of the wide range of RNA-based therapeutics.

1.1 Making Sense of Antisense

The concept underlying antisense technology is simple and straightforward. Making it simple, during the first step of protein synthesis, DNA is transcribed into the complementary sequence of a messenger RNA (mRNA) molecule. The sequences of these "sense" molecules are translated into amino acids to form proteins. The use of an oligonucleotide sequence tailored complementary, by virtue of Watson–Crick base pair formation—an "antisense" strand—recognizes the specific sequence on the mRNA strand, thus preventing it from being translated into a protein and blocking the function of a gene, as it was first shown by Paul C. Zamecnik (Zamecnik and Stephenson 1978). This conceptually simple approach has given rise to a wide variety of oligonucleotide chemistries, and a new class of therapeutic compounds called antisense drugs continues to expand.

A milestone in this therapeutic arena was the FDA approval of fomivirsen (marked as Vitravene) as the first antisense oligonucleotide-based therapeutic for the treatment of a human disease in 1998, namely, for the treatment for cytomegalovirus retinitis. After numerous clinical trials of therapeutic oligonucleotides against several types of diseases (Sehgal et al. 2013), it was not until 2013 (15 years later) that the FDA approved the second antisense oligonucleotide-based therapeutic, mipomersen (marked as Kynamro), for the treatment of homozygous familial hypercholesterolemia by the inhibition of apolipoprotein B.

ASOs, both RNA and DNA molecules, are unstable in vivo due to the large amount of nucleases in plasma or cells. Moreover, problems of immune activation delivery, specificity, and other hurdles have delayed the jump from clinical trials to the market. For this, significant effort has been expanded to develop nuclease-resistant oligonucleotides with reduced immunogenicity as well as improved pharmacokinetic and pharmacodynamic properties. Many chemical modifications (Fig. 1) have been successfully developed to overcome all these hurdles.

One of the earliest chemical modification for therapeutic oligonucleotides was the phosphorothioate backbone modification (Marcus-Sekura et al. 1987; Campbell

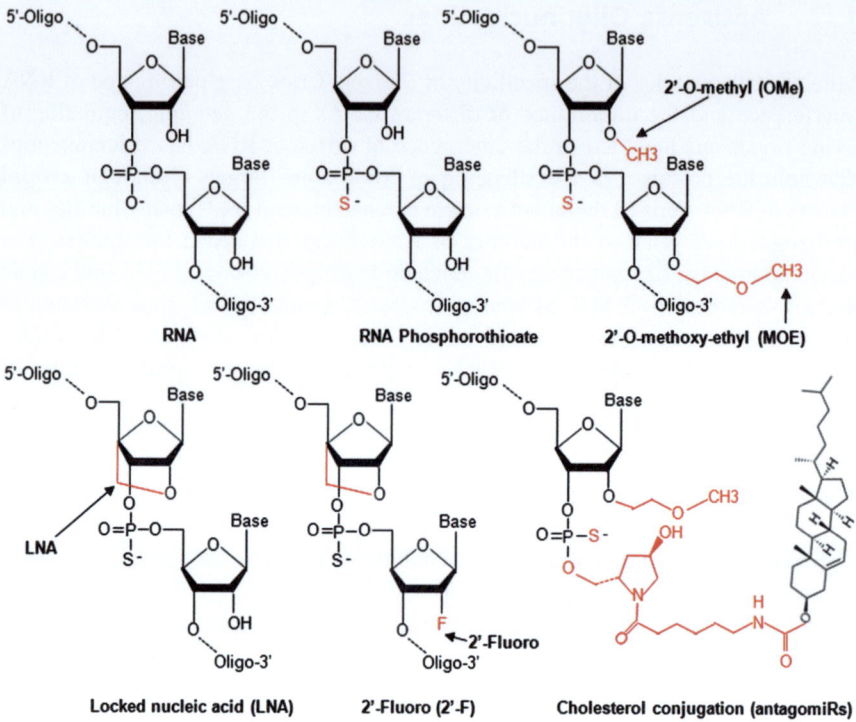

Fig. 1 Examples of chemical modifications used in antisense oligonucleotides and miRNA therapy

et al. 1990). The replacement of the nonbridging oxygens by sulfur dramatically changed the biological properties of the oligonucleotide, making them excellent candidates for antisense application: The modified oligonucleotides display increased resistance to nucleolytic degradation, increased affinity for plasma proteins, and thus reduced clearance. They are also highly soluble and elicit RNase H activity which mediate the cleavage of the target mRNA (Bennett and Swayze 2010).

Sugar modification—mostly chemical substitution at the 2′hydroxyl group—has conferred better drug-like properties to these molecules and have differentiated the various therapeutic strategies followed by pharmaceutical companies. 2′–O-methylation (2′-OMe) enhances both binding affinity and nuclease resistance, but reduces off-target effects (Yoo et al. 2004; Prakash et al. 2005). 2′-fluoro (2′-F) modification also increases the binding affinity for the target driven by the electronegative substituent at this position (Monia et al. 1993; Bennett and Swayze 2010). The 2′O-methoxyethyl (2′-MOE) modification increases binding affinity and resistance to nucleases as well (Geary et al. 2001; Yu et al. 2004). These kinds of modifications allow the oligonucleotides to adopt the most energy-favorable conformation, thus improving their pharmacological properties, and have enabled several MOE-modified drugs to enter clinical trials (Bennett and Swayze 2010). Another revolutionary sugar modification is the locked nucleic acids (LNA). In the LNA, the ribose moiety is modified with an extra bridge connecting the $2'$ oxygen and $4'$ carbon (Fig. 1) (Kumar et al. 1998). The bridge "locks" the ribose in the $3'$-endo (Northern) conformation of sugar, which is often found in the A-form duplexes. This chemical characteristic increases hybridization, potency, and nuclease resistance but also toxicity in some cases (Swayze et al. 2007). Several LNA-modified drugs have also entered clinical trials (Sehgal et al. 2013). Several other chemical modifications to the sugar moiety have been reported in the literature; however not much of them are on clinical trials yet.

An alternative interesting strategy that improves the cellular uptake, in vivo stability, and pharmacokinetic properties is the conjugation of the oligonucleotides to different ligands, for example, to certain types of cell-permeable/penetrating peptides (Oehlke et al. 2002) or carbohydrates (Zatsepin and Oretskaya 2004). Moreover several other conjugation methods have been experimentally developed including nanostructures, liposomes, bile acid, flavin, poly(ethylene glycol), and others (Karinaga et al. 2006; Singh et al. 2010; Gonzalez-Carmona et al. 2013). However, for most of them, several challenges remain to be addressed before their preclinical and clinical development (Lee et al. 2013a, b). Cholesterol conjugation of oligonucleotides has been reported to work for multiple antisense mechanisms. Cholesterol conjugation enhances the cellular uptake, particularly hepatic, and increases in vivo stability (Holasova et al. 2005; Krutzfeldt et al. 2005, 2007). Even when it is not free of unwanted side effects, as many other antisense chemistries, cholesterol conjugation is a promising strategy to develop drug therapies.

1.2 Therapeutic Antisense Oligonucleotides to Treat Dyslipidemia

After less than 7 years of clinical trials (Kastelein et al. 2006; Raal et al. 2010), the first antisense oligonucleotide to treat dyslipidemia reached the market in 2013. Mipomersen sodium (marketed as Kynamro™; ISIS Pharmaceutical) is an ASO inhibitor of apolipoprotein B-100 synthesis. It is indicated as an adjunct to lipid-lowering medications to reduce LDL cholesterol, apolipoprotein B (apoB), total cholesterol, and non-HDL cholesterol in patients with homozygous familial hypercholesterolemia. Other antisense therapies for the treatment of dyslipidemia are in clinical or preclinical studies, namely, the antisense drug to reduce apolipoprotein C-III (ISIS-APOCIIIRx) (Graham et al. 2013), intended to lower triglyceride production in patients with familial chylomicronemia or severe high hypertriglyceridemia, the antisense drug to reduce apolipoprotein(a) LP(a) in the liver (ISIS-APO(a)Rx) or the antisense drug to reduce angiopoietin-like 3 protein (ISIS-ANGPTL3Rx). An ASO against the hepatic microsomal triglyceride transfer protein (MTP) has also been evaluated, but even when it consistently reduced the hepatic VLDL/triglyceride secretion, it led to hepatic triglyceride accumulation and biomarkers of hepatotoxicity relative to apoB ASO, due in part to enhanced expression of peroxisome proliferator activated receptor γ target genes and the inability to reduce hepatic fatty acid synthesis (Lee et al. 2013a).

Thanks to genetic studies, mutations in the proprotein convertase subtilisin/kexin type 9 (PCSK9) were originally found to cause autosomal dominant hypercholesterolemia (Abifadel et al. 2003), which was further validated by loss-of-function mutations (Cohen et al. 2005) that lead to reduced cholesterol levels and reduced coronary heart disease (Cohen et al. 2006). PCSK9 encodes NARC-1 (neural apoptosis-regulated convertase), a human subtilase that is highly expressed in the liver and the intestine and circulates in plasma. PCSK9 binds to the LDL receptor and promotes its degradation in the endosomal/lysosomal pathway, thereby reducing LDL uptake from the circulation and increasing plasma cholesterol levels. After the premature Phase I trial termination of the PCSK9 phosphorothioate LNA RNase H antisense inhibitor SPC5001 (Santaris Pharma) and the phosphorothioate 2′ MOE RNase H PCSK9 antisense inhibitor (BMS-844421; BMS/ISIS), probably due to side effects (van Poelgeest et al. 2013), only ALN-PCS siRNA (Alnylam Pharmaceutical) entered Phase II clinical studies to target PCSK9 for the treatment of hypercholesterolemia. However, this strategy competes with monoclonal antibodies against PCSK9 that have already entered Phase III clinical trials (see below).

1.3 Therapeutic Antisense Oligonucleotides for the Increase of HDL-Cholesterol Levels and Improvement of HDL Function

HDL-cholesterol levels and function can be modified directly or indirectly by several pathways, some of which have been targeted by ASOs.

1.3.1 Cholesteryl Ester Transfer Protein

One of the first ASO experimentally approached to target HDL levels was directed against CETP. Sugano's lab first demonstrated that a single injection of the ASO, coupled with the complex asialoglycoprotein-poly-L-lysine, into cholesterol-fed rabbits, reduced CETP activity and increased plasma HDL-cholesterol levels (Sugano and Makino 1996). This effect was due to reduced liver CETP mRNA levels, which was accompanied by a reduction in LDL- and/or VLDL-cholesterol levels. In a longer study, 8-week treatment with the same molecule (30 μg/kg twice a week) reduced both CETP mass and atherosclerosis in cholesterol-fed rabbits (Sugano et al. 1998). While triglyceride levels did not change, LDL- and VLDL-cholesterol levels were significantly decreased by the ASO treatment (Sugano et al. 1998). Despite these promising preclinical antiatherogenic findings, the controversial clinical development of CETP inhibitors (Barter et al. 2007; Schaefer 2013) increased the caution for the future development of this type of therapy. In this context, recent preliminary finding suggests that inhibition of CETP by ASOs may differ from CETP inhibition by small-molecule inhibitors (Bell et al. 2013). Indeed, the 20-mer phosphorothioate ASO containing 2′-O-methoxyethyl (2′MOE) targeted to human CETP (ISIS Pharmaceutical) did not only reduce CETP activity and increase HDL-C levels but also enhance macrophage reverse cholesterol transport and reduce the accumulation of aortic cholesterol in a CETP transgenic $LDLR^{-/-}$ mice (Bell et al. 2013). This finding together with a previous study regarding the lack of association of genetic inhibition of CETP (Johannsen et al. 2012) and possible side effects previously reported for torcetrapib suggests that not all inhibitors of CETP are equal. Thus inhibition of CETP still holds promise as a beneficial therapeutic target, but as for other drugs, this needs to be experimentally and clinically validated.

1.3.2 Endothelial Lipase

Endothelial lipase plays an important role in HDL metabolism (Kuusi et al. 1980; Voight et al. 2012), and it has been suggested that its inhibition may improve cardioprotection (Singaraja et al. 2013). In a preliminary study, a 20-mer ASO containing $2'$-O-(methoxy)-ethyl ($2'$MOE) modifications on the first five and last five bases (ISIS Pharmaceutical) to target the rabbit endothelial lipase was tested in rabbits for 6 weeks (Zhang et al. 2012b). Even though the experimental protocol did not show a clear increase in HDL-cholesterol levels, the cholesterol content of large HDL (>12.1 nm) was increased (Zhang et al. 2012b). Whether other ASO chemistries may increase the impact on HDL-C levels and function is not known, but this deserves further investigation.

1.3.3 ACAT2

The sterol O-acyltransferase 2, encoded by the SOAT2 gene and originally named ACAT2 (referred here as ACAT2), is a membrane-bound enzyme, with an acyl-CoA cholesterol acyltransferase activity, localized in the endoplasmic reticulum. SOAT2/ACAT2 catalyzes the synthesis of cholesteryl esters from long-chain fatty acyl-CoA and cholesterol and is involved in cholesterol absorption and the secretion of cholesteryl esters into apoB-containing lipoproteins. ACAT2 is expressed exclusively in lipoprotein-producing cells, the enterocytes and hepatocytes (Anderson et al. 1998). While both hepatic and intestinal deletion of ACAT2 improves atherogenic hyperlipidemia and limits hepatic cholesteryl ester accumulation (Zhang et al. 2012a), it has been proposed that specific tissue ACAT inhibition would be beneficial for atheroprotection (Nissen et al. 2006; Brown et al. 2008). However, ACAT inhibition is not free of controversies in clinical development (Nissen et al. 2006). A 20-mer antisense phosphorothioate oligonucleotide containing 2-0-methoxyethyl groups at positions 1–5 and 15–20 was originally found to reduce hepatic ACAT2 levels and mediate protection against diet-induced hypercholesterolemia and aortic cholesteryl ester deposition (Bell et al. 2006). Interestingly, in mice this antisense oligonucleotide (ISIS Pharmaceutical) therapy (25 mg/kg biweekly for 8 weeks) promoted fecal neutral sterol excretion without altering biliary sterol secretion (Brown et al. 2008). This potentially important finding indicates that the antisense oligonucleotide promotes non-biliary fecal sterol loss and thus reverses cholesterol transport enhancement. Pharmacological inhibition of liver ACAT2 by using ASOs has also been shown to reduce cholesterol-associated hepatic steatosis (Alger et al. 2010) which may explain the hypertriglyceridemia observed in mice lacking ACAT2, probably by enhancing hepatic TG mobilization. In overall, ASO treatment against hepatic ACAT2 has uncovered other novel benefits distinct from that of HDL function and increased its therapeutic potential.

Although we are expecting the results of several ongoing clinical trials with ASOs for different pathologies (Sehgal et al. 2013), the recent approval for commercialization of Kynamro will really increase our interest to follow this therapeutic arena. Even when the long-term toxicity effects and other forms of delivery need to be evaluated, the promising preclinical results on ASOs to treat HDL levels and function make them an interesting alternative to small-molecule inhibitors. After all, opening new possible avenues to treat HDL dysfunction using ASOs makes more sense than simply awaiting for the discovery of potential small-molecule inhibitors.

2 miRNAs

Mature microRNAs (miRNAs) are single-stranded, ~21–23 nucleotide (nt) long, and noncoding RNAs that directly bind, via Watson–Crick base pairing, to sequences commonly located within the 3′untranslated region (3′-UTR) of target mRNAs. This interaction inhibits the translation and/or degradation of mRNAs (Guo et al. 2010; Krol et al. 2010). However, certain miRNAs can interact with other target mRNA regions including the 5′UTR, coding region, or intron–exon junction and even increase rather than decrease target mRNA expression (Vasudevan et al. 2007; Orom et al. 2008; Tay et al. 2008; Schnall-Levin et al. 2010). RNA sequencing studies have identified ~2,000 miRNAs in our human genome which are predicted to regulate ~ a third of our genes. The binding of the "seed" sequence (nucleotides 2–8 at the 5′ end of the mature miRNA) is critical for target selection (Bartel 2009). However, other regions of the miRNA can bind to the target mRNA and, therefore, almost 60 % of seed interactions are noncanonical (Helwak et al. 2013). Many miRNAs are evolutionary conserved among different species. While some of them are ubiquitously expressed, certain miRNAs are highly expressed or even restricted (Lagos-Quintana et al. 2002; Small and Olson 2011) to certain cell types and can only target their mRNA target if they are co-expressed in the same tissue at the same time.

Based on short sequences ("seed"), computational methods and previous validation studies have revealed that a single miRNA can target hundreds of genes with either multiple related or different functions in different physiological/pathological processes or tissues. Likewise, a single mRNA may have different miRNA binding sites, allowing a coordinated regulation by different miRNAs. While the primary role of miRNAs seems to be the "fine-tuning" of gene expression (Flynt and Lai 2008), the appearance of this complex RNA-based regulatory network suggests that miRNAs have probably evolved as buffers against deleterious variation in gene-expression programs. Even when a single miRNA exerts modest effects on many target mRNAs, the additive effect of coordinated regulation of a large suite of transcripts that govern the same biological process is believed to result in strong phenotypic outputs (Mendell and Olson 2012). These basic principles of miRNA mode of function are the basis for a novel and revolutionary type of therapeutics, called the miRNA-based therapy. However, as *the devil is in the details*, the high redundancy among related and non-related miRNAs in the regulation of gene expression described above reduces the importance of a particular miRNA under conditions of normal cellular homeostasis. Nevertheless, since under conditions of stress, the function of miRNAs becomes especially pronounced (Mendell and Olson 2012), the modulation of miRNA function may represent a real alternative to the conventional one-target drug therapy. However, many hurdles need to be overcome as other novel players have entered the equation, including pseudo genes (Poliseno et al. 2010), long noncoding RNAs (lncRNAs) (Cesana et al. 2011), and circular RNAs (circRNAs) that contain miRNA binding sites. As competing endogenous RNAs (ceRNAs), they may sequester miRNAs and prevent them from binding to their mRNA targets (Salmena et al. 2011).

2.1   miRNA-Based Therapy

miRNAs as potential therapeutics have received special attention from the scientific and clinical audience primarily because of their "promiscuous" mode of action and the multifactorial nature of most modern metabolic diseases. Moreover, previous antisense technology and gene therapy approaches, some of them already in market, have catalyzed the efforts to develop therapies to modulate miRNA levels in vivo. Although certain questions regarding their biological function, regulation, and delivery still remain to be answered, the simultaneous modulation of different components of a complex disease pathway by an miRNA offers a unique and alternative opportunity to treat disease in a manner that is completely different from our conventional classical one-target-directed drugs. Eventually this feature may also enable to bypass tissue insensitivity or drug resistance, characterizing classical one-target-directed drugs.

Different pharmacological tools have been developed to target miRNA pathways (van Rooij et al. 2008, 2012; van Rooij and Olson 2012) (Fig. 2). As miRNAs are generally inhibitors of gene expression, the use of therapies to increase or block gene expression will result in a decreased or derepression of their mRNA targets, respectively. Based on these opposite approaches we can classify the therapeutic application of miRNAs into two strategies. The first strategy involves an miRNA "gain of function" phenotype, also called "inhibitors," and aims to inhibit the function of miRNAs. Several approaches can be utilized for this purpose, including (1) small-molecule inhibitors directed to regulate miRNA expression, (2) miRNA masking due to molecules complementary to the 3′-UTR of the target miRNA, resulting in competitive inhibition of the downstream target effects, (3) miRNA sponges that utilize oligonucleotide constructs with multiple complementary miRNA binding sites to the target miRNA, and (4) antisense

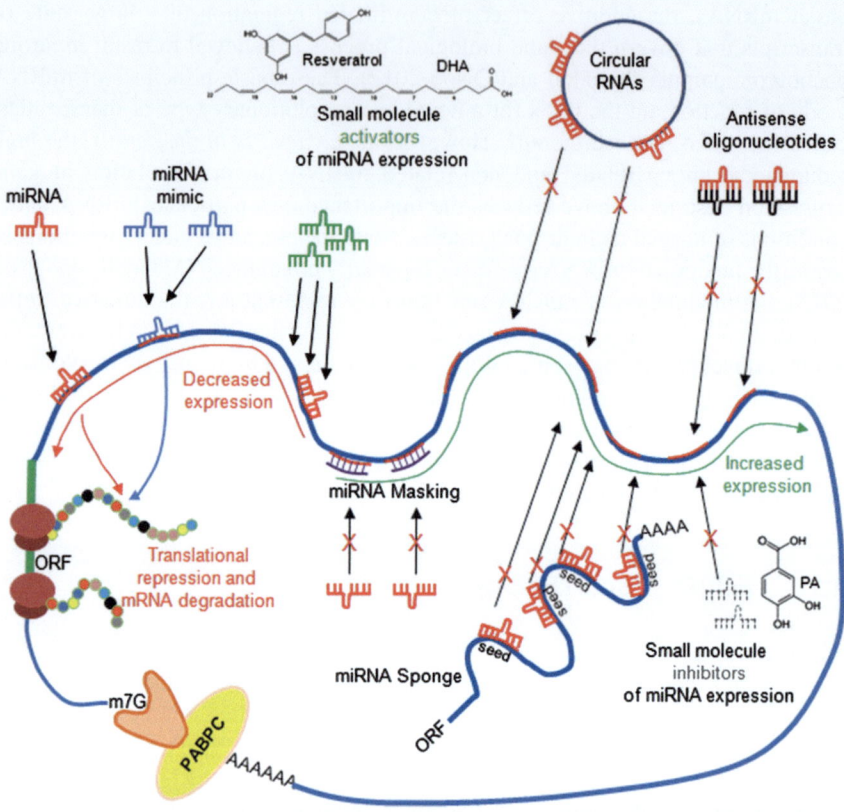

Fig. 2 miRNA-based therapy. Endogenous miRNAs bind to a complementary sequence generally localized within the 3′UTR of target genes and repress the synthesis of the corresponding protein or degrade the mRNA target. For miRNA replacement therapy, an exogenous miRNA mimic is delivered systemically to exert a repression of their target genes. Small-molecule activators of miRNA expression can also be used for this purpose. For endogenous miRNA inhibition, and thus derepression of their target genes, several approaches can be used. Small-molecule inhibitors can be directed to repress an miRNA expression. miRNA masking employs molecules complementary to the 3′-UTR of the target miRNA, resulting in competitive inhibition of the downstream target effects. miRNA sponges use oligonucleotide constructs with multiple complementary miRNA binding sites to the target miRNA, thereby preventing them from binding to their target mRNAs. Antisense oligonucleotides, also known as miRNA antagonists, inhibitors, or anti-miRs, complementary bind to a target miRNA inducing either duplex formation or miRNA degradation. Novel approaches can arise from recent discovery of other noncoding RNAs that regulate miRNA activity. That is, *circular* RNAs can sequester a large amount of miRNAs acting as competitive inhibitors for miRNA binding, thereby preventing the mRNA repression of the target miRNA. *PA* protocatechuic acid, *DHA* docosahexaenoic acid

oligonucleotides, also known as miRNA antagonists or inhibitors, such as anti-miRs, locked nucleic acids (LNA), or antagomiRs that by complementarity bind to miRNAs inducing either duplex formation or miRNA degradation (Davalos and Suarez 2013). The second strategy involves an miRNA "loss-of-function" phenotype, also called "mimic," and aims to enhance the function of miRNAs. The approaches that can be utilized for this strategy include (1) small-molecule activators or inductors of miRNA expression and (2) miRNA mimics, which as exogenous miRNAs aim to repress the function of their mRNA targets. They are also called "miRNA replacement therapy."

2.1.1 Therapeutic miRNA Mimics or miRNA Replacement Therapy

In principle, delivery of miRNA mimics as pharmacological therapy could be used in situations in which a reduction in miRNA levels is responsible for the development of a pathological state, such as those produced in the human rare Mendelian disorders or certain types of cancer, where regions containing miRNAs are deleted (Calin et al. 2002). Genetic mutation in either miRNA seed region or other miRNA regions that results in a reduced functional miRNA with a significant reduction of mRNA targeting required for normal function (Mencia et al. 2009; Ryan et al. 2010) could also benefit from these therapies. The use of miRNA mimics for therapy has been really challenging and their development has been catalyzed by gene therapy. Gene therapy was first conceptualized in 1972 (Friedmann and Roblin 1972) as an approach to deliver a gene or alter the expression of a gene in order to replace a mutated gene or deliver a therapeutic functional gene using a vector to treat a disease. Since then, the FDA has approved hundreds of clinical trials during the last 20 years using different approaches, for different diseases, with some promising results. However, it was not until 2012 that the European Medicines Agency (EMA) has approved the first gene therapy drug, alipogene tiparvovec (marked as Glybera)—for the treatment for lipoprotein lipase deficiency—and the first of its kind in the western society. It uses a viral vector, the adeno-associated virus serotype 1 (AAV1), to deliver a copy of the human lipoprotein lipase gene. As proof of concept, many preclinical data generated in animal models suggest that pharmacological delivery of miRNA mimics is feasible and current strategies to deliver miRNA mimics are promising. Indeed, a synthetic version of miR-34a (MRX34, Mirna Therapeutics), delivered using a liposomal delivery formulation, was the first miRNA to advance into a human Phase 1 clinical trial for cancer (clinicaltrials.gov number NCT01829971).

Different experimental strategies to deliver miRNA mimics have been tested. Synthetic miRNA or pre-miRNA duplexes, normally modified for better stability and cellular uptake, have been incorporated into different delivery systems, including lipid nanoparticles with surface receptor ligands or other components to increase tissue/cell specificity (Wiggins et al. 2010; Trang et al. 2011; Piao et al. 2012). Adeno-associated viruses (AAV) (Miyazaki et al. 2012) are another interesting alternative. Certain tissue specificity due to the natural tropism of different AAV serotypes (Zincarelli et al. 2008) could be achieved. Viral-based vectors, including adenoviruses and lentiviruses (Chistiakov et al. 2012; Langlois et al. 2012), consist another well-studied delivery method.

There are still questions regarding the biological function of miRNAs, particularly those related to extracellular miRNAs, intercellular communication by miRNAs, and their presence in numerous biological fluids that need to be addressed. As miRNAs can circulate in the blood or different biological fluids in microvesicles, exosomes, Ago2-containing complexes, or HDL (Arroyo et al. 2011; Vickers et al. 2011; Chen et al. 2012), opportunities will probably arise for therapeutically exploiting the physiologic forms of miRNA delivery (Davalos and Fernandez-Hernando 2013). The basic function of miRNAs, which is to target different mRNAs of different biological pathways, raises the possibility of unintended off-target effects (van Rooij et al. 2008). Several hurdles need to be solved including the delivery issues, as uptake of a miRNA by tissues that normally do not express them will result in the repression of their targets that could ultimately cause side effects. Moreover, the overexpression of a particular miRNA, even in its specific target cell, could modify either its own secretion or the secretion of other miRNAs that could target a different cell/tissue type causing unwanted side effects.

2.1.2 Therapeutic miRNA Inhibition or Anti-miR Therapy

In contrast to miRNA replacement therapy, miRNA inhibitors as therapy have benefits compared to existing antisense technology in the market. Over the last decade, several miRNAs have been characterized, and it was found that their induction or overexpression plays a causal role in a disease or directly contributes to it. Thus, pharmacological inhibition of miRNA activity in vivo has been achieved through the use of different chemically modified single-stranded reverse complement oligonucleotides known as antisense oligonucleotide (Fig. 1). Antisense oligonucleotides (ASOs) complementary to the mature miRNA sequence, "antagomiRs," were the first miRNA inhibitors in mammals (Krutzfeldt et al. 2005). Since then different chemical modifications were performed to ASOs in order to modify their pharmacological, pharmacokinetic, and pharmacodynamic properties: cholesterol, conjugated via a $2'$-O-methyl ($2'$-O-Me) linkage in the $3'$end, to increase cellular uptake and stability; phosphorothioate linkage to increase stability and reduce clearance by promoting plasma protein binding; $2'$-O-methyl ($2'$-O-methyl)-modified ribose sugar to protect from endonuclease activity (Krutzfeldt et al. 2005, 2007); $2',4'$-constrained $2'$-O-ethyl(cET)-modified nucleotides to improve potency and stability (Seth et al. 2010; Pallan et al. 2012); $2'$-O-methoxyethyl ($2'$-MOE) and $2'$-fluoro ($2'$-F) modifications to improve in vivo efficacy (Davis et al. 2009); and the $2'$-fluoro/methoxyethyl ($2'$-F/MOE) modified with phosphorothioate backbone-modified anti-miR technology which has been shown to be efficacious in nonhuman primates (Rayner et al. 2011a). Lastly, locked nucleic acid (LNA) gives promising properties in order to be used as miRNA therapy. As LNA anti-miR has high-binding affinity and increased selectivity to complementary RNA, the sequence length can be reduced. LNA also increases the duplex's melting temperature and stability in biological systems (Vester and Wengel 2004; Elmen et al. 2005; Veedu and Wengel 2010). In preclinical studies LNA-modified anti-miR technology has been widely shown to be efficacious in nonhuman primates (Elmen et al. 2008; Lanford et al. 2010). Moreover, it was the first anti-miR therapy to show efficacy in human trials (clinicaltrials.gov number NCT01200420) (Janssen et al. 2013). A phosphorothioate backbone tiny 8-mer LNA-modified anti-miRs for in vivo use (Obad et al. 2011) has also been developed particularly for reducing the activity of entire miRNA families that share a common seed region.

For now, anti-miR therapy is administered parenterally. Although miRNA inhibitors are generally water soluble, their size and charge prevent them to be absorbed by the intestine, thus becoming bad candidates for oral therapy. Their long-lasting effects shown in different studies (Krutzfeldt et al. 2005; Elmen et al. 2008; Lanford et al. 2010; Obad et al. 2011; Rayner et al. 2011a, b; van Rooij and Olson 2012) suggest their potential use for chronic rather than acute disease. Whereas under normal unstressed conditions miRNAs only slightly change protein expression (Selbach et al. 2008), pharmacological inhibition under pathological stress conditions may become relevant (Mendell and Olson 2012). Even when specific toxicity associated with the inhibition of a particular miRNA has not been clearly reported, as for other LNA-containing ASO therapies, they might not be free of potential off-target effects (Swayze et al. 2007). Thus, their evaluation might be challenging and should be done in long-lasting studies. Moreover, other miRNAs and mRNAs (independent of miRNA mediated) modified by the stress conditions and other regulatory mechanism exerted by ceRNA, lncRNAs, and circRNAs will greatly influence the pharmacodynamics of every particular anti-miR chemistry.

Although there are many aspects of anti-miR biology that need to be addressed, this therapeutic approach successfully led up to the first miRNA-based clinical trials for the treatment of hepatitis C virus infection by targeting miR-122 with an LNA-anti-miR (miravirsen or SPC3649; Santaris Pharma, Denmark) with very promising results (Janssen et al. 2013). Thus, the biological interest in controlling miRNAs level therapeutically anticipates the further development of this new class of drugs.

2.2 Micromanaging Cholesterol Efflux, RCT, HDL Levels, and HDL Function

Several miRNAs have been investigated for their potential use as therapeutics for different aspects of HDL function. Although still in preclinical studies, the pharmacological inhibition of the miRNA-33a/b is leading this aspect of research. These and other miRNAs directly or indirectly related to cellular cholesterol efflux, RCT, and HDL function that could potentially be used as pharmacological therapy will be discussed below.

2.2.1 miR-33a/b

The genomic localization of this family of miRNAs within the introns of the master regulators of lipid and cholesterol metabolism, the SREBPs-, has catalyzed the discovery of this miRNA as major player in HDL function and cholesterol efflux (Horton et al. 2002; Horie et al. 2010; Marquart et al. 2010; Najafi-Shoushtari et al. 2010; Rayner et al. 2010). Modulation of miR-33a/b levels in preclinical studies resulted in changes in cellular cholesterol efflux (Najafi-Shoushtari et al. 2010; Rayner et al. 2010; Davalos et al. 2011). In vivo modulation of miR-33a/b either by target disruption of the gene, LNA anti-miR, or viral delivery of sense and antisense oligonucleotides significantly alters circulating HDL-C and reverse cholesterol transport (Horie et al. 2010; Marquart et al. 2010; Najafi-Shoushtari et al. 2010; Rayner et al. 2010), which is consistent with the regulation of its targets ABCA1, ABCG1, and NPC1. miR-33 deficiency reduces the progression of atherosclerotic plaque (Horie et al. 2012). Likewise the antisense inhibition of this miRNA for 4 weeks in LDLR$^{-/-}$ mice led to the regression of atherosclerosis by enhancing *Abca1* expression and cholesterol removal in plaque macrophages, reducing the size and inflammatory gene expression of plaques, and increasing markers of plaque stability (Rayner et al. 2011b). Even when there is conflicting results in this aspect (Marquart et al. 2013), there is still therapeutic potential in inhibiting this miRNA family for atherosclerotic cardiovascular disease. The dramatic increase of SREBP1c (host of miR-33b)—in insulin resistance states—which contributes to both increased levels of plasma triglycerides and low HDL levels (Brown et al. 2010) also suggests the therapeutic use of anti-miR-33 for metabolic syndrome. Indeed, as proof of concept, in African green monkeys fed with a high carbohydrate diet, the inhibition of miR-33b for 12 weeks reduced hepatic expression of *Abca1*, increased the function of HDL evaluated as macrophage cholesterol efflux, raised plasma HDL levels, and reduced VLDL triglyceride levels (Rayner et al. 2011a).

Thus, the therapeutic potential of anti-miR-33 is not only based on *Abca1*, cholesterol efflux, and RCT, but miR-33a/b also controls the expression of important genes involved in fatty acid β-oxidation, insulin signaling, lipid metabolism, and biliary transporters (Allen et al. 2012; Horie et al. 2013) (Gerin et al. 2010; Davalos et al. 2011; Rayner et al. 2011a, b), including carnitine palmitoyltransferase 1A (*Cpt1a*), the carnitine O-octanoyltransferase (*Crot*), the mitochondrial beta hydroxyacyl-CoA dehydrogenase/3-ketoacyl-CoA thiolase/enoyl-CoA hydratase (*Hadhb*), sirtuin 6 (*Sirt6*), 5'-AMP-activated protein kinase catalytic subunit alpha-1 (PRKAA1 gene, *Ampkα*), insulin receptor substrate 2 (*Irs2*), SREBP-1, and the biliary transporters ABCB11 and ATP8B1. Although specific toxicity associated to miR-33 inhibition has not been reported, its safety should be carefully evaluated as other targets related to cell proliferation, cell cycle, and inflammation, including cyclin-dependent kinase 6 (*Cdk6*), cyclin D1 (*Ccnd1*), the tumor suppressor p53, and the nuclear receptor coregulator receptor interacting protein 140 (*Rip140*) have also been described (Herrera-Merchan et al. 2010; Ho et al. 2011; Cirera-Salinas et al. 2012).

Different anti-miR chemistries were tested for inhibiting miR-33 family members including LNA-antisense oligonucleotide (Najafi-Shoushtari et al. 2010) and 2'F/MOE-modified phosphorothioate backbone-modified ASO (Rayner et al. 2011a, b) (Regulus Therapeutics). Interestingly, the miR-33 family has special structural characteristics, not common in most mammalian miRNAs. They have a repetitive sequence similar to that of seed (UGCAUUG) between nucleotides 13 and 19 apart from their native seed sequence between nucleotides 2 and 8 at the 5' end of the mature miRNA. This could be benefited by the use of phosphorothioate backbone tiny 8-mer LNA-modified anti-miRs chemistry. Indeed recent promising results suggest the efficacy and safety of an 8-mer LNA anti-miR against miR-33 family during a 108-day treatment in a nonhuman primate metabolic disease model (Rottiers et al. 2013) (Santaris Pharma). Which anti-miR chemistry will have the best pharmacologic and safety profile for human use is not known, but pharmaceutical industries (Santaris Pharma and Regulus Therapeutics) are intensively researching on this topic and anti-miRs will probably soon enter clinical trials.

2.2.2 miR-758 and miR-106b

Like miR-33a/b, miR-758 and miR-106b target ABCA1. The expression of miR-758 is somehow mediated by high cholesterol levels and regulates cellular cholesterol efflux by directly targeting the 3′UTR of *Abca1*. As the relative expression of miR-758 is particularly elevated in the brain (Ramirez et al. 2011), it seems that it regulates other important proteins involved in several neurological functions including SLC38A1, IGF1, NTM, XTXBP1, and EPHA1. Although our understanding of the role of miR-758 under physiological and pathological conditions needs to be enhanced first, the development of appropriate anti-miR chemistries for targeting the brain miR-758 still remains to be dealt with, including bypassing the blood–brain barrier and delivery to specific cell types. Also in neuronal cells, miR-106b was found to directly target the 3′UTR of *Abca1* as having a perfect 8-mer and several supplementary pairing sites in mammals (Bartel 2009; Kim et al. 2012). The miR-106b not only reduces cholesterol efflux to apoA-I but also increases amyloid β (Aβ) peptide secretion and clearance (Kim et al. 2012). The production and/or aggregation of Aβ peptide is believed to play a central role in the pathogenesis of Alzheimer disease (AD). While the final effect might be directly linked to miR-106b effects on *Abca1* rather than other target genes in neuronal cells (Kim et al. 2012), we should not discard other indirect effects as this miRNA also targets various other proteins related to cell proliferation and differentiation (Brett et al. 2011). Moreover, the amyloid precursor protein (APP) is also a target of miR-106b (Hebert et al. 2009). The final phenotype of the inhibition of the neuronal miR-106b is not known, but in the context of cholesterol efflux in the CNS, miR-106b could be an interesting target for the regulation of neuronal cholesterol excess.

2.2.3 miR-26 and miR-144

The nuclear liver X receptors (LXRs) control distinct aspects of cholesterol homeostasis at the transcriptional level including uptake (IDOL) or efflux (ABCA1, ABCG5, ABCG8). Induction of LXR by using agonists revealed the repression of the miR-26 and the induction of the miR-144 family. LXR activation increased the expression of *Abca1* and ADP-ribosylation factor-like7 (*Arl7*), both of which participate in apoA-I-dependent cholesterol efflux (Engel et al. 2004). The miR-26-a-1 expression is also regulated by LXR (Sun et al. 2012) and directly targets the 3′UTR of *Abca1* and *Arl7*. Thus the inhibition of this miRNA, in principle, should enhance cholesterol efflux and RCT as LXR activation does. By contrast, LXR activation induces the expression of miR-144 (Ramirez et al. 2013). Interestingly, miR-144 is not only activated by LXR but also by the nuclear receptor farnesoid X receptor (FXR) (Vickers and Rader 2013). FXR is highly expressed in the liver and the intestine. It controls the hepatic sterol and bile acid levels through transcriptional regulation of lipid-associated and bile acid genes. The miR-144 directly targets the 3′UTR of ABCA1, thus reducing ABCA1 protein levels and cholesterol efflux, but not ABCA1 mRNA in all models tested (de Aguiar Vallim et al. 2013; Ramirez et al. 2013; Vickers and Rader 2013). In vivo therapeutic inhibition of miR-144 by using either 2′-fluoro/2′-methoxyethyl, phosphorothioate backbone-modified anti-miRs (Regulus Therapeutics) 5 mg/kg biweekly treatments (intraperitoneal injections) for 4 weeks (de Aguiar Vallim et al. 2013) or mirVana inhibitors (7 mg/kg) coupled with In vivo fectamine (Invitrogen) for intravenous injections twice every 3 days (Ramirez et al. 2013) increased both hepatic ABCA1 protein expression and HDL-C levels in mice. It is important to note that the hepatic effect of miR-144 might differ from that of miR-33a/b (Vickers and Rader 2013). Activation of FXR will induce both the scavenger receptor B1 (SCARB1) and miR-144, thereby increasing the uptake of plasma HDL cholesterol and reducing both ABCA1 protein levels and cholesterol efflux to lipid-poor apoA-I. This would lead to increased biliary excretion of cholesterol via ABCG5/ABCG8 rather than resecretion of cholesterol via ABCA1 to preβ-HDL and the formation of HDL (de Aguiar Vallim et al. 2013). The final therapeutic outcome by modulating miR-144 levels in vivo and other questions regarding safety issues still need to be experimentally tested, as other miR-144 targets are directly related to cancer proliferation (Guo et al. 2013; Zhang et al. 2013).

2.2.4 miR-10b, miR-128-2, miR-145

Several other miRNAs have been described to regulate ABCA1, ABCG1, and other genes related to cholesterol efflux. In the context of small-molecule activators to either induce or repress miRNAs expression, polyphenols and fatty acids are emerging as possible candidates to exert part of their biological effects by this mechanism (Visioli et al. 2012; Tome-Carneiro et al. 2013). Anthocyanidins are pigmented polyphenols found in different vegetables, fruits, as well as common beverages including grape and berry juice and red wine. Protocatechuic acid (PCA) was found to be an intestinal microbiota metabolite of Cyanidin-3-O-glucoside (Cy-3-G), a major anthocyanidin. Interestingly the antiatherogenic effect of PCA was recently found to be mediated through miR-10b (Wang et al. 2012). Indeed, PCA increases macrophage cholesterol efflux through the repression of miR-10b. The miR-10b directly represses *Abca1* and *Abcg1* and negatively regulates cholesterol efflux from lipid-loaded macrophages (Wang et al. 2012). Although several genes involved in cancer progression are targets of the oncogenic miR-10b (Gabriely et al. 2011; Tsukerman et al. 2012), its inhibition by either anti-miR chemistries or dietary intervention with anthocyanidins may be an interesting pharmacological approach to increase cholesterol efflux and RCT. The use of other pharmacological approaches, including small natural dietary compounds, is still under intense investigation (Visioli and Davalos 2011) and provides an attractive alternative to the use of ASOs or miRNA mimic technology.

miR-145 was described as a major regulator of smooth muscle fate by targeting a network of transcription factors, including Klf4, myocardin, and Elk-1 which regulate the quiescent versus proliferative phenotype of smooth muscle cells (Cordes et al. 2009). miR-145 also regulate ABCA1 expression and function. In pancreatic beta cells, its inhibition improves glucose-stimulated insulin secretion. Inhibition of miR-145b has been shown to increase ABCA1 expression, promoting HDL biogenesis in the liver and improving glucose-stimulated insulin secretion in islets (Kang et al. 2013). The miR-128-2 was described to be frequently downregulated in breast cancer, and its overexpression impeded several oncogenic traits of mammary carcinoma cells (Qian et al. 2012). ABCA1, ABCG1, and RXRα are direct targets of miR-128-2, and its inhibition induces cholesterol efflux (Adlakha et al. 2013). Although we lack in vivo evidence of their pharmacological modulation and HDL function, recent evidence of association of cholesterol levels and cancer (Nelson et al. 2013) warrants further investigation of these miRNAs.

2.2.5 Other miRNA Related to Cholesterol Efflux and Cholesterol Homeostasis

Several other miRNAs have been described to indirectly regulate different aspects of cholesterol efflux, RCT, and cholesterol metabolism, but their real physiological contribution to HDL function is not well understood. Several miRNAs have been described that regulate different targets in autophagy (Xu et al. 2012), the cell catabolism process by which unnecessary or dysfunctional cellular components are degraded through the lysosomal machinery. Lipid droplet cholesteryl ester hydrolysis is being recognized as an important step in cholesterol efflux (Ouimet et al. 2011); thus miRNAs that target key pathways in lipid-loaded macrophage autophagy and/or cholesterol ester hydrolases might be interesting targets to promote cholesterol efflux (Davalos and Fernandez-Hernando 2013). Caveolin, the major protein coat of caveolae, has also been proposed to contribute to cellular cholesterol efflux (Truong et al. 2010; Kuo et al. 2011). Even when there is increasing evidence of several miRNAs including miR-103, miR-107, miR-133a, miR-192, miR-802, and others that target caveolin (Nohata et al. 2011; Trajkovski et al. 2011), their contribution to cholesterol efflux and RCT remains unknown. miR-125a and miR-455 were found to repress the lipoprotein-supported steroidogenesis by targeting SR-BI (Hu et al. 2012). miR-185, miR-96, and miR-223 were also found to target SR-BI and repress HDL-cholesterol uptake (Wang et al. 2013). Even though there is lack of evidence for any effect on HDL metabolism and in vivo pharmacological modulation for any of those miRNAs, the major role of SR-BI in HDL metabolism warrants further research on this topic.

Although the potential of LNA anti-miR-122-based therapy (Miravirsen, Santaris Pharma) is fascinating, much more needs to be elucidated about miRNA biology and miRNA regulatory networks in human diseases before we can introduce such research into clinical care. Moreover, new opportunities for therapeutic interventions by exploiting the different physiologic forms of miRNA delivery in biological fluids (associated with microvesicles, exosomes, Ago2-containing complexes, or HDL particles) will arise. In addition, in the postgenomic and

"RNA world" era, new types and new roles of noncoding RNAs continue to emerge, suggesting that there is much yet to be discovered in this therapy arena.

3 Antibodies

The use of monoclonal antibodies (mAbs) for the treatment of various diseases, including cancers and autoimmune diseases, has been established for at least 15 years. The specificity of mAbs to the target antigen offers clear benefit for their use over conventional pharmacotherapy. Today mAbs are being developed for the treatment of dyslipidemia and cardiovascular disease. An overview of approaches to antibody therapy for the decrease of LDL cholesterol, increase of HDL cholesterol, treatment of HDL dysfunction, and reduction of cardiovascular events is provided below. In addition, approaches of active immunization to modulate atherosclerosis with promising results in preclinical studies are discussed.

3.1 LDL-Cholesterol Lowering Approaches: Proprotein Convertase Subtilisin/Kexin Type 9 Blocking Antibodies

Statins reduce LDL-cholesterol (LDL-C) levels by increasing the hepatic uptake of LDL through inhibiting HMG CoA reductase and subsequently cholesterol biosynthesis. A meta-analysis of data from 90,056 participants in 14 randomized trials showed that statin therapy can safely reduce the 5-year incidence of major coronary events, coronary revascularization, and stroke by about one fifth per mmol/L reduction in LDL-C (Baigent et al. 2005). In addition a newer meta-analysis of data from 170,000 participants in 26 randomized trials of statins showed that each 1 mmol/L LDL-cholesterol reduction reduces the risk of occlusive vascular events by about a fifth, irrespective of baseline cholesterol concentration, which implies that a 2–3 mmol/L reduction would reduce risk by about 40–50 % (Baigent et al. 2010). However, a significant proportion of patients treated with statins fail to achieve the recommended levels of LDL-C (Catapano 2009). Furthermore, even when LDL-C is reduced at the recommended levels by statins, there is a residual 50–60 % risk, and therefore, new targets for therapeutic intervention need to be developed.

Loss-of-function mutations in proprotein convertase subtilisin/kexin type 9 (PCSK9) gene result in low levels of LDL-C and protect against coronary heart disease. These observations have made PCSK9 one of the most intensively investigated novel targets to treat hypercholesterolemia (Cohen et al. 2005, 2006). PCSK9, a protein mainly expressed in the liver and intestine, is present in human plasma (Lambert et al. 2012). PCSK9 binds to LDL receptors (LDLRs) and thereby targets the internalized receptor to lysosomal degradation and thus limits recycling of receptor to the plasma membrane for LDL uptake (Zhang et al. 2007). Statins have been shown to increase PCSK9 expression, an effect that blunts the LDL-C lowering effectiveness of statins (Mayne et al. 2008). Therefore, the inhibition of interaction between PCSK9 and LDLR is expected to increase the LDLRs that are available in the plasma membrane of hepatocytes and as a consequence reduce plasma LDL-C. In addition, blocking the interactions between PCSK9 and LDLR may increase the lipid-lowering efficacy of statins.

Various approaches for decreasing PCSK9 levels or blocking PCSK9/LDLR interactions are being explored. By far the most advanced approach aims to inhibit PCSK9/LDLR interactions by the use of monoclonal antibodies targeting PCSK9 (Catapano and Papadopoulos 2013; Kramer 2013). Currently, clinical or preclinical trials for at least 13 different anti-PCSK9 antibodies are being conducted, with two compounds having entered Phase 3 of clinical development (Catapano and Papadopoulos 2013; Kramer 2013) (Table 1).

Preclinical studies in rodents and nonhuman primates showed that several anti-PCSK9 antibodies increased hepatic LDLR protein levels and reduced plasma LDL-C levels up to 80 % (Chan et al. 2009; Gusarova et al. 2012; Liang et al. 2012; Chaparro-Riggers et al. 2012; Zhang et al. 2012c; Ni et al. 2011). Reported results from Phase 1 studies in humans using REGN727/SAR236553 (alirocumab), AMG 145 (evolocumab), and PF-04950615 (RN316, bococizumab) anti-PCSK9 antibodies showed that the treatments are generally well tolerated and significantly reduce plasma LDL-C levels in healthy subjects or hypercholesterolemic patients, both as monotherapy and when added to statin treatment (Stein et al. 2012b; Dias et al. 2012; Gumbiner et al. 2012a, b). Phase 2 trial results have also been reported for alirocumab and evolocumab anti-PCSK9 antibodies. When alirocumab was administered subcutaneously at doses ranging from 50 to 150 mg every 2 weeks or 200–400 mg every 4 weeks to patients with primary hypercholesterolemia on top of ongoing stable atorvastatin therapy (10, 20, 40, 80 mg/day), additional reductions in LDL-C, than that accomplished with atorvastatin alone, were observed (McKenney et al. 2012; Roth et al. 2012). The reduction of LDL-C levels was found to be similar irrespective of statin dose, indicating that the coadministration of alirocumab with atorvastatin may provide benefit to patients that fail to achieve their LDL-C target using high-dose statins or are intolerant to high-dose statins (McKenney et al. 2012; Roth et al. 2012). Great reductions in LDL-C levels were also obtained when evolocumab was administered at doses ranging from 70 to 140 mg every 2 weeks or 280–420 mg every 4 weeks to patients with hypercholesterolemia, either as a monotherapy or in combination with a stable dose of statin with or without ezetimibe therapy, or to statin-intolerant patients due to muscle-related side effects on ezetimibe therapy (Giugliano et al. 2012; Koren et al. 2012; Sullivan et al. 2012). In other studies of patients with heterozygous familial hypercholesterolemia and elevated LDL-C on intensive statin use, with or without ezetimibe therapy, the administration of alirocumab or evolocumab resulted in substantial further LDL-C reduction (Stein et al. 2012a; Raal et al. 2012). In addition to their capacity to reduce LDL-C levels, alirocumab and evolocumab were shown to reduce lipoprotein(α) levels in patients with hypercholesterolemia receiving statin therapy (McKenney et al. 2012; Desai et al. 2013) or patients with heterozygous familial hypercholesterolemia on statins, with or without ezetimibe therapy (Stein et al. 2012a; Raal et al. 2012).

Table 1 PCSK9 blocking antibodies in clinical and preclinical development

Drug candidate	Company	Development phase	Literature or ClinicalTrials.gov Identifier
REGN727/ SAR236553 (alirocumab)	Sanofi/Regeneron Pharmaceuticals	Phase 3	NCT01709513 NCT01709500 NCT01730053 NCT01926782 NCT01644474 NCT01644188 NCT01663402 NCT01507831 NCT01623115 NCT01617655 NCT01644175 NCT01730040 NCT01954394
AMG 145 (evolocumab)	Amgen	Phase 3	NCT01652703 NCT01849497 NCT01813422 NCT01763918 NCT01516879 NCT01763866 NCT01763905 NCT01763827 NCT01764633 NCT01854918
PF-04950615 (RN316, bococizumab)	Pfizer	Phase 2	NCT01342211, NCT01592240, NCT01350141
LY-3015014	Eli Lilly and Company	Phase 2	NCT01890967
LGT209	Novartis	Phase 1	NCT01859455
PF-05335810 (RN317)	Pfizer	Phase 1	NCT01720537
J16	Pfizer	Preclinical	Liang et al. (2012)
J17	Pfizer	Preclinical	Chaparro-Riggers et al. (2012)
1B20	Merck	Preclinical	Zhang et al. (2012c)
1D05-IgG2	Merck	Preclinical	Ni et al. (2011)
LGT210	Novartis	Preclinical	Kramer (2013)
LGT211	Novartis	Preclinical	Kramer (2013)
ALD-306	Alder Biopharmaceuticals	Preclinical	Kramer (2013)

Alirocumab and evolocumab are currently further tested in 13 and 10 Phase 3 clinical trials, respectively (Table 1), for long-term efficacy and safety, either as monotherapy or on top of other lipid-modifying therapies and in various patient populations (e.g., subjects with primary hypercholesterolemia or mixed dyslipidemia, with high cardiovascular risk and with hyperlipidemia or mixed dyslipidemia, with heterozygous familial hypercholesterolemia, with clinically evident cardiovascular disease, with a 10-year Framingham risk score of 10 % or less, undergoing coronary catheterization, who recently experienced an acute coronary syndrome, with statin intolerance). Among these Phase 3 clinical studies, two studies, which are conducted in numerous study centers across the United States, Canada, Western and Eastern Europe, South America, Australia, Africa, and Asia, will assess the effect of candidate drugs on the occurrence of cardiovascular events for up to 64 months in large-sized patient groups. Specifically, the effect of alirocumab (ODYSSEY Outcomes, NCT01663402) on the occurrence of cardiovascular events (composite endpoint of coronary heart disease (CHD) death, nonfatal myocardial infarction (MI), fatal and nonfatal ischemic stroke, unstable angina requiring hospitalization) is evaluated in 18,000 patients who have experienced an acute coronary syndrome event 4–16 weeks prior to randomization and are treated with evidence-based medical and dietary management of dyslipidemia (time frame: up to month 64). Another objective of the study is the evaluation of the effect of alirocumab on secondary endpoints (any CHD event, major CHD event, any cardiovascular event, composite of all-cause mortality/nonfatal MI/nonfatal ischemic stroke, all-cause mortality), as well as on blood lipids and lipoprotein levels (time frame: up to month 64). In addition, the long-term safety and tolerability of alirocumab will be evaluated. In another large clinical trial (FOURIER, NCT01764633), the effect of evolocumab used in combination with statin therapy on additional LDL-C reduction and risk of cardiovascular death, MI, hospitalization for unstable angina, stroke, or coronary revascularization is evaluated in 22,500 patients with clinically evident cardiovascular disease. The primary endpoint is the time to cardiovascular death, MI, hospitalization for unstable angina, stroke, or coronary revascularization, whichever occurs first (time frame: 5 years). Another objective of the study is the evaluation of the effect of evolocumab on secondary endpoints, such as time to death by any cause, cardiovascular death, hospitalization for worsening heart failure, ischemic fatal or nonfatal stroke, and transient ischemic attack, whichever occurs first (time frame: 5 years).

3.2 Approaches to Antibody Therapy for the Increase of HDL-Cholesterol Levels

Numerous clinical and epidemiological studies have demonstrated an inverse association between HDL-cholesterol (HDL-C) levels and the risk of cardiovascular disease (Gordon et al. 1977; Assmann et al. 1996). Furthermore, HDL exerts a series of antiatherogenic properties (Navab et al. 2011). Thus, raising of HDL-C levels is expected to translate into a reduction of cardiovascular events and has led to serious efforts to develop new therapies that can increase the concentration of HDL-C. Therapeutic strategies using antibody-based blocking of proteins of the

HDL metabolism pathway that result in increase of HDL-C levels are discussed below.

3.2.1 Cholesteryl Ester Transfer Protein

Cholesteryl ester transfer protein (CETP) promotes net mass transfer of cholesteryl esters from HDL to other plasma lipoprotein fractions (Barter and Rye 2012). Therefore, inhibition of CETP can increase the concentration of HDL-C and CETP inhibitors have been capable to increase HDL-C levels in preliminary clinical trials, while clinical outcome trials are ongoing (Barter and Rye 2012; Landmesser et al. 2012). The properties and effects of different CETP small-molecule inhibitors that were observed in clinical trials are being described by Staels et al. in chapter "Emerging Small-Molecule Drugs."

An alternative approach involves the blocking of CETP action with an antibody, named CETi-1 (developed by AVANT Immunotherapeutics), raised against a dimerized synthetic peptide, including residues 461–476 of human CETP and T cell epitope of tetanus toxoid (residues 830–843), and formulated with aluminum-containing adjuvants (Davidson et al. 2003). In a Phase II study, CETi-1 was shown to be safe, and >90 % of treated patients with the highest dose of vaccine showed 1 year after vaccination with CETi-1 an immune response with an increase of HDL-C by 8 % (Komori 2004). However, this trial failed to meet the primary endpoint of increasing plasma HDL-C concentrations in the vaccine-treated groups as compared to the placebo group, due to the low titers of antihuman CETP antibody achieved in a number of the vaccinated subjects, and CETi-1 is no longer in development (Komori 2004).

A few years ago, in order to improve the efficacy of the CETi-1 vaccine, AVANT Immunotherapeutics researchers examined in mice and rabbits the immunogenicity of CETi-1 with the coadministration of the investigational TLR9 agonist VaxImmune^{TM} (CPG 7909) as an adjuvant (Thomas et al. 2009). In parallel, they studied the immunogenicity of another anti-CETP antibody, the PADRE-CETP, raised against a monomeric peptide, in which a PADRE T cell epitope (aK-Cha-VAAWTLKAa) replaces the TT(830–843) T cell epitope of CETi-1, with or without the coadministration of VaxImmune^{TM}. The studies showed that PADRE T cell epitope is more potent than the TT(830–843) epitope in providing help for the anti-CETP antibody response and that the coadministration of VaxImmune^{TM} with either vaccine increased immunogenicity as measured by antibody response (Thomas et al. 2009). However, there is no information for the initiation of clinical trials using these new vaccination approaches up to date. Another recent CETP vaccination approach involves the ATH-03 anti-CETP antibody (developed by Affiris AG), using a small peptide fragment of the CETP protein acting as a B cell epitope (Kramer 2013), which has entered Phase 1 trials to assess its safety and immunogenicity (ClinicalTrials.gov Identifier: NCT01284582).

3.2.2 Endothelial Lipase

Endothelial lipase (EL) is a phospholipase that participates in HDL metabolism and regulates HDL-C levels in humans and mice (Yasuda et al. 2010; Annema and Tietge 2011). Early studies in which EL was inhibited in wild-type, hepatic lipase-deficient, and human apolipoprotein (apo) A-I transgenic mice by intravenous infusion of a polyclonal inhibitory anti-mouse-EL antibody resulted in a 25–60 % increase in HDL-C levels in three mouse models, while triglyceride and non-HDL-cholesterol levels were not changed (Jin et al. 2003). In human apoA-I transgenic mice, apoA-I levels were also increased and the HDL phospholipid turnover was retarded (Jin et al. 2003). Based on this and other studies in mice lacking EL activity, as well as on studies in humans expressing loss-of-function EL variants, the inhibition of EL in humans would be expected to raise plasma HDL-C levels (Brown et al. 2009; Ishida et al. 2003; Ma et al. 2003; Edmondson et al. 2009). Although the effect of EL inhibition in the reduction of atherosclerotic cardiovascular disease risk has not been proven (Yasuda et al. 2010; Annema and Tietge 2011), EL remains a potential target for pharmacological inhibition, possibly by antibodies against EL, as a novel strategy to raise HDL-C and reduce the risk of cardiovascular disease.

3.3 Effect of Antibodies Used for the Treatment of Chronic Inflammatory Diseases on HDL Antiatherogenic Functions

Patients with chronic inflammatory rheumatic diseases, such as rheumatoid arthritis and systemic lupus erythematosus, have increased risk for cardiovascular disease morbidity and mortality (Onat and Direskeneli 2012; Farragher and Bruce 2006; Popa et al. 2012). Various studies have shown that during the course of these chronic inflammatory conditions, the levels as well as the antiatherogenic properties of HDL are affected (Onat and Direskeneli 2012; Popa et al. 2012). Specifically, patients with rheumatoid arthritis and systemic lupus erythematosus were found to have proinflammatory HDL (McMahon et al. 2006; Charles-Schoeman et al. 2009). In addition, cholesterol efflux capacity of HDL was impaired in rheumatoid arthritis patients with high disease activity and was correlated with systemic inflammation and HDL's antioxidant capacity (Charles-Schoeman et al. 2012). Recommendations for the treatment of rheumatic diseases propose a tight control of the inflammatory process which probably will favorably impact the risk of cardiovascular disease. New therapeutics include antibodies designed to block inflammatory proteins or cells that are produced in abundance during the disease, such as TNF-α, IL-6, or B cells (Onat and Direskeneli 2012; Popa et al. 2012). Anti-TNF therapy of rheumatoid arthritis patients with monoclonal antibody infliximab was shown to increase plasma paraoxonase-1 activity and to improve HDL antioxidative capacity, an effect that was sustained 6 months after anti-TNF therapy has been initiated (Popa et al. 2009). In addition, another recent study in rheumatoid arthritis patients treated with rituximab, a B cell depleting monoclonal antibody against the protein CD20, which is primarily found on the surface of B lymphocytes, showed beneficial changes in HDL composition (Raterman et al. 2013). Specifically, during 6 months of treatment with rituximab, HDL-associated serum amyloid A decreased in patients with good response to the therapy, rendering HDL from proatherogenic to less proatherogenic (Raterman et al. 2013). Future large-scale studies are needed to establish the value of monitoring HDL function during antibody therapy, as well as the impact of this therapy on cardiovascular disease risk of rheumatic disease patients and possibly other patients with coronary artery disease who remain at high vascular risk despite contemporary prevention strategies.

3.4 Vaccines Against Atherosclerosis

A large body of evidence has shown that atherosclerosis is a multifactorial, multiphase disease characterized by chronic inflammation and altered immune response. Therefore, approaches of active immunization have been developed to modulate atherosclerosis with promising results in preclinical studies. Many studies have reported reduced atherosclerosis in animal models after immunization using as antigens LDL (native or modified), apoB100 peptides, heat shock proteins, and other proteins or phospholipids associated with the initiation and progression of the atherosclerotic plaque.

Immunization with homologous LDL, oxLDL (copper-oxidized LDL), or MDA-LDL (LDL modified by malondialdehyde (MDA), an epitope of oxLDL) generated high titers of antibodies and reduced atherogenesis development in hypercholesterolemic rabbits, LDLR-deficient rabbits, or apoE-deficient mice (Palinski et al. 1995; Ameli et al. 1996; George et al. 1998; Zhou et al. 2001). Similar results were achieved using homologous plaque homogenates (containing immunogen(s) sharing epitopes on MDA-LDL, MDA-VLDL, and oxidized cardiolipin) as the antigen (Zhou et al. 2001). The generation of antibodies against oxidation epitopes in LDL has been proposed to inhibit the binding and uptake of oxLDL by macrophage scavenger receptors CD36 and SR-BI and therefore to reduce the formation of foam cells (Steinberg and Witztum 2010). In addition, induction of oral tolerance to oxLDL can induce a significant increase in CD4 +CD25+Foxp3+ T_{regs} in spleen and mesenteric lymph nodes, and these cells specifically respond to oxLDL with increased TGF-β production and significant attenuation of the initiation and progression of atherogenesis in LDLR-deficient mice (van Puijvelde et al. 2006). However, since atherosclerosis immune responses can been triggered against autoantigens, such as anti-oxLDL, all the efforts toward the development of a successful vaccine against oxLDL should result in the restoration of tolerance against autoantibodies and balance of pro- and antiatherogenic immune responses (Samson et al. 2012).

Screening of a library of 302 polypeptides covering the complete sequence of apoB-100, the major protein component in LDL, in their native state and after MDA modification, using pooled plasma derived from healthy control subjects, resulted in the identification of more than 100 different human antibodies reacting against

MDA-modified apoB100 sequences (Fredrikson et al. 2003a). Immunization with apoB100 peptide sequences, against which high levels of IgG and IgM antibodies are present in healthy human controls, was found to reduce atherosclerosis in apoE-deficient mice by about 60 % (Fredrikson et al. 2003b). Studies on the mechanisms underlying active immunization using a specific apoB100-related peptide, namely, the p210, indicated that the atheroprotective effect of immunization with this peptide is mediated via the activation of CD8+T cells (Chyu et al. 2012) and CD4+ CD25+Foxp3+ T_{regs} (Wigren et al. 2011).

Other approaches to reduce atherosclerosis involve the immunization with antigens other than LDL or apoB100-related peptides. Immunization of hypercholesterolemic rabbits with protein-free liposomes containing dimyristoyl phosphatidylcholine, dimyristoyl phosphatidylglycerol, cholesterol (71 %), and lipid A from *Salmonella* Minnesota R595 as adjuvant induced the generation of anticholesterol antibodies and reduced the diet-induced hypercholesterolemia and plaque formation (Alving et al. 1996). In another study, pneumococcal vaccination was also found to decrease the atherosclerotic lesion formation (Binder et al. 2003). The use of *Streptococcus pneumoniae* as an antigen was based on the finding that during the progression of atherosclerosis in apoE-deficient mice, autoantibodies against epitopes of oxLDL, and more specifically against the oxidized phospholipids, share complete genetic and structural identity with antibodies from the classic anti-phosphorylcholine B cell clone, T15, which protect against common infectious pathogens, including pneumococci (Shaw et al. 2000). The reduction of atherosclerosis in LDLR-deficient mice after immunization with *S. pneumoniae* suggested molecular mimicry between epitopes of oxLDL and *S. pneumonia* (Binder et al. 2003). In another recent study, it was found that T cell hybridomas from oxLDL-immunized human apoB100 transgenic mice responded against native LDL and purified apoB100, but not against oxLDL and expressed a single T cell receptor variable beta chain, TRBV31. Immunization of double human apoB100 transgenic and LDLR-deficient mice with a TRBV31-derived peptide generated anti-TRBV31 antibodies that blocked T cell recognition of apoB100 and reduced atherosclerosis (Hermansson et al. 2010).

In the last decade, various therapeutic approaches for reducing atherosclerosis by immunization were based on heat shock proteins (HSPs). HSPs, which are named according to their molecular weight, are a highly conserved group of proteins that have been implicated in atherogenesis (Kilic and Mandal 2012; Grundtman et al. 2011). Intracellularly, HSPs act as molecular chaperones and assist in the folding of misfolded proteins, but extracellularly, HSPs can promote immune responses. Immunization of rabbit or mouse atherosclerotic models with HSP65 resulted in conflicting results, with some studies showing induction of atherosclerotic lesions (Xu et al. 1992; George et al. 1999; Zhang et al. 2012d) and others reduction of atherosclerosis (Maron et al. 2002; Harats et al. 2002; Long et al. 2012; Klingenberg et al. 2012). Vaccination of high-cholesterol-fed atherosclerotic rabbits with a recombinant HSP65-CETP fusion protein resulted in production of more protective IL-10 and less adverse IFN-γ, reduction of total cholesterol and LDL-C levels, as well as decrease of the area of aortic lesions

(Jun et al. 2012). In addition, the immunization of apoE-deficient or double apoB48- and LDLR-deficient mice with chimeric proteins containing apoB-100 and/or hHSP60 peptides resulted in the reduction of atherosclerotic lesions (Li et al. 2011; Lu et al. 2010).

Another target for the treatment of atherosclerosis by immunization involves a specific cell population, the vascular endothelial growth factor receptor 2 (VEGFR2)-overexpressing cells. Highly VEGFR2 expressing cells include proliferating endothelial cells that are involved in angiogenesis, a process that may be associated with the initiation and progression of atherosclerosis (Shi et al. 1998; Li et al. 2002; Moulton et al. 2003). Vaccination of apoE- or LDLR-deficient mice against VEGFR2 by an orally administered DNA vaccine, comprising a plasmid encoding murine VEGFR2 carried by live attenuated *Salmonella typhimurium*, resulted in a marked induction of CD8+ cytotoxic T cells specific for VEGFR2 and attenuated the initiation and progression of atherosclerosis (Hauer et al. 2007). Oral administration of another DNA vaccine, one against CD99, a membrane protein expressed on vascular endothelium overlying atherosclerotic plaques, also generated antigen-specific cytotoxic CD8+ T cells and reduced atherosclerosis in LDLR-deficient mice (van Wanrooij et al. 2008).

In a recent study of a potential treatment option for atherosclerosis, immunization of apoE-deficient mice with a virus-like particle-based vaccine against IL-1α, a cytokine exerting proinflammatory functions and implicated in the development of atherosclerosis (Kamari et al. 2007), reduced both the inflammatory reaction in the plaque and plaque progression (Tissot et al. 2013). A current Phase 3 clinical study that aims to test the inflammatory hypothesis of atherothrombosis has been designed to evaluate whether Canakinumab can reduce rates of recurrent MI, stroke, and cardiovascular death among stable patients with coronary artery disease who remain at high vascular risk due to persistent elevations of high-sensitivity C-reactive protein (CRP) (>2 mg/L) despite contemporary secondary prevention strategies (Ridker et al. 2011). Canakinumab is a human monoclonal antibody that selectively neutralizes IL-1β, a cytokine that exerts proinflammatory effects and has been implicated in the pathogenesis of atherothrombosis (Fearon and Fearon 2008). The study which is named Canakinumab Anti-inflammatory Thrombosis Outcomes Study (CANTOS, ClinicalTrials.gov Identifier: NCT01327846) will randomly allocate 17,200 patients to either placebo or to Canakinumab at doses of 50, 150, or 300 mg every 3 months, administered subcutaneously, and all participants will be followed up over an estimated period of up to 4 years (Ridker et al. 2011). As part of the CANTOS study, a multinational Phase IIb randomized, placebo-controlled trial was conducted to evaluate the effects of Canakinumab agent on hemoglobin A1c, glucose, lipids, CRP, IL-6, and fibrinogen among 556 men and women with diabetes mellitus and high cardiovascular risk (Ridker et al. 2012). The patients were randomly allocated to subcutaneous placebo or to subcutaneous Canakinumab at doses of 5, 15, 50, or 150 mg monthly and followed over 4 months. No effects were seen for hemoglobin A1c, glucose, insulin, LDL-C, HDL-C, or non-HDL-cholesterol levels. By contrast, there were significant reductions in CRP, IL-6, and fibrinogen levels in both women and men. The CANTOS Investigative Group members support that these data provide a strong basis for the use of Canakinumab in humans as an approach to test whether therapeutic targeting of inflammation can reduce the risk of recurrent events in patients on current standard of care treatment after MI (Ridker et al. 2012).

Conclusions

Antisense oligonucleotide or gene therapy has been discussed for more than 30 years. The approval for commercial use of the first antisense oligonucleotide in 1998 by the FDA has catalyzed the development of these types of therapy for several diseases. Moreover, recent approvals of mipomersen or the gene therapy drug Glyvera have opened avenues for the use of these novel therapeutic approaches to treat lipid metabolism disorders. While miRNA and mAbs therapy to treat dyslipidemia have benefitted from previous existing technology, the recent clinical trials to target miR-122 through anti-miR technology and to target PCSK9 through mAbs technology will probably encourage the development of novel therapies to modulate HDL metabolism and function. mAbs are currently being investigated for their capacity to decrease LDL-C levels, increase HDL-C levels, and reduce the risk of cardiovascular events in clinical studies. Initial studies showed that these compounds are generally safe and well tolerated. Ongoing, large clinical studies will assess the long-term safety as well as the efficacy of these new antibody-based drugs in a wide range of dyslipidemic and cardiovascular disease patients. Even though the complex biological structure and functions of HDL are probably delaying the benefits of having a drug using any of these technologies in clinical care, ongoing preclinical or clinical studies using antisense oligonucleotides, miRNAs, or mAbs envision a bright future for the HDL field.

Acknowledgments Research by the groups of Drs Alberto Dávalos and Angeliki Chroni is supported by COST Action BM0904. Dr Angeliki Chroni's research is also supported by grants of the General Secretariat of Research and Technology of Greece (Grant Synergasia 09SYN-12-897) and the Ministry of Education of Greece (Grant Thalis MIS 377286). Dr Alberto Dávalos' research is also supported by a grant from the Spanish Instituto de Salud Carlos III (FIS, PI11/00315).

Open Access This chapter is distributed under the terms of the Creative Commons Attribution Noncommercial License, which permits any noncommercial use, distribution, and reproduction in any medium, provided the original author(s) and source are credited.

References

Abifadel M, Varret M, Rabes JP, Allard D, Ouguerram K, Devillers M, Cruaud C, Benjannet S, Wickham L, Erlich D, Derre A, Villeger L, Farnier M, Beucler I, Bruckert E, Chambaz J, Chanu B, Lecerf JM, Luc G, Moulin P, Weissenbach J, Prat A, Krempf M, Junien C, Seidah NG, Boileau C (2003) Mutations in PCSK9 cause autosomal dominant hypercholesterolemia. Nat Genet 34:154–156

Adlakha YK, Khanna S, Singh R, Singh VP, Agrawal A, Saini N (2013) Pro-apoptotic miRNA-128-2 modulates ABCA1, ABCG1 and RXRalpha expression and cholesterol homeostasis. Cell Death Dis 4:e780

Alger HM, Brown JM, Sawyer JK, Kelley KL, Shah R, Wilson MD, Willingham MC, Rudel LL (2010) Inhibition of acyl-coenzyme A:cholesterol acyltransferase 2 (ACAT2) prevents dietary cholesterol-associated steatosis by enhancing hepatic triglyceride mobilization. J Biol Chem 285:14267–14274

Allen RM, Marquart TJ, Albert CJ, Suchy FJ, Wang DQ, Ananthanarayanan M, Ford DA, Baldan A (2012) miR-33 controls the expression of biliary transporters, and mediates statin- and diet-induced hepatotoxicity. EMBO Mol Med 4:882–895

Alving CR, Swartz GM Jr, Wassef NM, Ribas JL, Herderick EE, Virmani R, Kolodgie FD, Matyas GR, Cornhill JF (1996) Immunization with cholesterol-rich liposomes induces anti-cholesterol antibodies and reduces diet-induced hypercholesterolemia and plaque formation. J Lab Clin Med 127:40–49

Ameli S, Hultgardh-Nilsson A, Regnstrom J, Calara F, Yano J, Cercek B, Shah PK, Nilsson J (1996) Effect of immunization with homologous LDL and oxidized LDL on early atherosclerosis in hypercholesterolemic rabbits. Arterioscler Thromb Vasc Biol 16:1074–1079

Anderson RA, Joyce C, Davis M, Reagan JW, Clark M, Shelness GS, Rudel LL (1998) Identification of a form of acyl-CoA:cholesterol acyltransferase specific to liver and intestine in nonhuman primates. J Biol Chem 273:26747–26754

Annema W, Tietge UJ (2011) Role of hepatic lipase and endothelial lipase in high-density lipoprotein-mediated reverse cholesterol transport. Curr Atheroscler Rep 13:257–265

Arroyo JD, Chevillet JR, Kroh EM, Ruf IK, Pritchard CC, Gibson DF, Mitchell PS, Bennett CF, Pogosova-Agadjanyan EL, Stirewalt DL, Tait JF, Tewari M (2011) Argonaute2 complexes carry a population of circulating microRNAs independent of vesicles in human plasma. Proc Natl Acad Sci USA 108:5003–5008

Assmann G, Schulte H, von Eckardstein A, Huang Y (1996) High-density lipoprotein cholesterol as a predictor of coronary heart disease risk. The PROCAM experience and pathophysiological implications for reverse cholesterol transport. Atherosclerosis 124(Suppl):S11–S20

Baigent C, Keech A, Kearney PM, Blackwell L, Buck G, Pollicino C, Kirby A, Sourjina T, Peto R, Collins R, Simes R (2005) Efficacy and safety of cholesterol-lowering treatment: prospective meta-analysis of data from 90,056 participants in 14 randomised trials of statins. Lancet 366:1267–1278

Baigent C, Blackwell L, Emberson J, Holland LE, Reith C, Bhala N, Peto R, Barnes EH, Keech A, Simes J, Collins R (2010) Efficacy and safety of more intensive lowering of LDL cholesterol: a meta-analysis of data from 170,000 participants in 26 randomised trials. Lancet 376:1670–1681

Bartel DP (2009) MicroRNAs: target recognition and regulatory functions. Cell 136:215–233

Barter PJ, Rye KA (2012) Cholesteryl ester transfer protein inhibition as a strategy to reduce cardiovascular risk. J Lipid Res 53:1755–1766

Barter PJ, Caulfield M, Eriksson M, Grundy SM, Kastelein JJ, Komajda M, Lopez-Sendon J, Mosca L, Tardif JC, Waters DD, Shear CL, Revkin JH, Buhr KA, Fisher MR, Tall AR, Brewer B (2007) Effects of torcetrapib in patients at high risk for coronary events. N Engl J Med 357:2109–2122

Bell TA 3rd, Brown JM, Graham MJ, Lemonidis KM, Crooke RM, Rudel LL (2006) Liver-specific inhibition of acyl-coenzyme a:cholesterol acyltransferase 2 with antisense oligonucleotides limits atherosclerosis development in apolipoprotein B100-only low-density lipoprotein receptor$^{-/-}$ mice. Arterioscler Thromb Vasc Biol 26:1814–1820

Bell TA, Graham MJ, Lee RG, Mullick AE, Fu W, Norris D, Crooke RM (2013) Antisense oligonucleotide inhibition of cholesteryl ester transfer protein enhances RCT in hyperlipidemic, CETP transgenic, LDLr$^{-/-}$ mice. J Lipid Res 54:2647–2657

Bennett CF, Swayze EE (2010) RNA targeting therapeutics: molecular mechanisms of antisense oligonucleotides as a therapeutic platform. Annu Rev Pharmacol Toxicol 50:259–293

Binder CJ, Horkko S, Dewan A, Chang MK, Kieu EP, Goodyear CS, Shaw PX, Palinski W, Witztum JL, Silverman GJ (2003) Pneumococcal vaccination decreases atherosclerotic lesion formation: molecular mimicry between Streptococcus pneumoniae and oxidized LDL. Nat Med 9:736–743

Brett JO, Renault VM, Rafalski VA, Webb AE, Brunet A (2011) The microRNA cluster miR-106b~25 regulates adult neural stem/progenitor cell proliferation and neuronal differentiation. Aging (Albany NY) 3:108–124

Brown JM, Bell TA 3rd, Alger HM, Sawyer JK, Smith TL, Kelley K, Shah R, Wilson MD, Davis MA, Lee RG, Graham MJ, Crooke RM, Rudel LL (2008) Targeted depletion of hepatic ACAT2-driven cholesterol esterification reveals a non-biliary route for fecal neutral sterol loss. J Biol Chem 283:10522–10534

Brown RJ, Edmondson AC, Griffon N, Hill TB, Fuki IV, Badellino KO, Li M, Wolfe ML, Reilly MP, Rader DJ (2009) A naturally occurring variant of endothelial lipase associated with elevated HDL exhibits impaired synthesis. J Lipid Res 50:1910–1916

Brown MS, Ye J, Goldstein JL (2010) Medicine. HDL miR-ed down by SREBP introns. Science 328:1495–1496

Calin GA, Dumitru CD, Shimizu M, Bichi R, Zupo S, Noch E, Aldler H, Rattan S, Keating M, Rai K, Rassenti L, Kipps T, Negrini M, Bullrich F, Croce CM (2002) Frequent deletions and down-regulation of micro- RNA genes miR15 and miR16 at 13q14 in chronic lymphocytic leukemia. Proc Natl Acad Sci USA 99:15524–15529

Campbell JM, Bacon TA, Wickstrom E (1990) Oligodeoxynucleoside phosphorothioate stability in subcellular extracts, culture media, sera and cerebrospinal fluid. J Biochem Biophys Methods 20:259–267

Catapano AL (2009) Perspectives on low-density lipoprotein cholesterol goal achievement. Curr Med Res Opin 25:431–447

Catapano AL, Papadopoulos N (2013) The safety of therapeutic monoclonal antibodies: implications for cardiovascular disease and targeting the PCSK9 pathway. Atherosclerosis 228:18–28

Cesana M, Cacchiarelli D, Legnini I, Santini T, Sthandier O, Chinappi M, Tramontano A, Bozzoni I (2011) A long noncoding RNA controls muscle differentiation by functioning as a competing endogenous RNA. Cell 147:358–369

Chan JC, Piper DE, Cao Q, Liu D, King C, Wang W, Tang J, Liu Q, Higbee J, Xia Z, Di Y, Shetterly S, Arimura Z, Salomonis H, Romanow WG, Thibault ST, Zhang R, Cao P, Yang XP, Yu T, Lu M, Retter MW, Kwon G, Henne K, Pan O, Tsai MM, Fuchslocher B, Yang E, Zhou L, Lee KJ, Daris M, Sheng J, Wang Y, Shen WD, Yeh WC, Emery M, Walker NP, Shan B, Schwarz M, Jackson SM (2009) A proprotein convertase subtilisin/kexin type 9 neutralizing antibody reduces serum cholesterol in mice and nonhuman primates. Proc Natl Acad Sci USA 106:9820–9825

Chaparro-Riggers J, Liang H, DeVay RM, Bai L, Sutton JE, Chen W, Geng T, Lindquist K, Casas MG, Boustany LM, Brown CL, Chabot J, Gomes B, Garzone P, Rossi A, Strop P, Shelton D, Pons J, Rajpal A (2012) Increasing serum half-life and extending cholesterol lowering in vivo by engineering antibody with pH-sensitive binding to PCSK9. J Biol Chem 287:11090–11097

Charles-Schoeman C, Watanabe J, Lee YY, Furst DE, Amjadi S, Elashoff D, Park G, McMahon M, Paulus HE, Fogelman AM, Reddy ST (2009) Abnormal function of high-density lipoprotein is associated with poor disease control and an altered protein cargo in rheumatoid arthritis. Arthritis Rheum 60:2870–2879

Charles-Schoeman C, Lee YY, Grijalva V, Amjadi S, FitzGerald J, Ranganath VK, Taylor M, McMahon M, Paulus HE, Reddy ST (2012) Cholesterol efflux by high density lipoproteins is impaired in patients with active rheumatoid arthritis. Ann Rheum Dis 71:1157–1162

Chen X, Liang H, Zhang J, Zen K, Zhang CY (2012) Secreted microRNAs: a new form of intercellular communication. Trends Cell Biol 22:125–132

Chistiakov DA, Sobenin IA, Orekhov AN (2012) Strategies to deliver microRNAs as potential therapeutics in the treatment of cardiovascular pathology. Drug Deliv 19:392–405

Chyu KY, Zhao X, Dimayuga PC, Zhou J, Li X, Yano J, Lio WM, Chan LF, Kirzner J, Trinidad P, Cercek B, Shah PK (2012) CD8+ T cells mediate the athero-protective effect of immunization with an ApoB-100 peptide. PLoS ONE 7:e30780

Cirera-Salinas D, Pauta M, Allen RM, Salerno AG, Ramirez CM, Chamorro-Jorganes A, Wanschel AC, Lasuncion MA, Morales-Ruiz M, Suarez Y, Baldan A, Esplugues E, Fernandez-Hernando C (2012) Mir-33 regulates cell proliferation and cell cycle progression. Cell Cycle 11:922–933

Cohen J, Pertsemlidis A, Kotowski IK, Graham R, Garcia CK, Hobbs HH (2005) Low LDL cholesterol in individuals of African descent resulting from frequent nonsense mutations in PCSK9. Nat Genet 37:161–165

Cohen JC, Boerwinkle E, Mosley TH, Hobbs HH (2006) Sequence variations in PCSK9, low LDL, and protection against coronary heart disease. N Engl J Med 354:1264–1272

Cordes KR, Sheehy NT, White MP, Berry EC, Morton SU, Muth AN, Lee TH, Miano JM, Ivey KN, Srivastava D (2009) miR-145 and miR-143 regulate smooth muscle cell fate and plasticity. Nature 460:705–710

Davalos A, Fernandez-Hernando C (2013) From evolution to revolution: miRNAs as pharmacological targets for modulating cholesterol efflux and reverse cholesterol transport. Pharmacol Res 75:60–72

Davalos A, Suarez Y (2013) MiRNA-based therapy: from bench to bedside. Pharmacol Res 75:1–2

Davalos A, Goedeke L, Smibert P, Ramirez CM, Warrier NP, Andreo U, Cirera-Salinas D, Rayner K, Suresh U, Pastor-Pareja JC, Esplugues E, Fisher EA, Penalva LO, Moore KJ, Suarez Y, Lai EC, Fernandez-Hernando C (2011) miR-33a/b contribute to the regulation of fatty acid metabolism and insulin signaling. Proc Natl Acad Sci USA 108:9232–9237

Davidson MH, Maki K, Umporowicz D, Wheeler A, Rittershaus C, Ryan U (2003) The safety and immunogenicity of a CETP vaccine in healthy adults. Atherosclerosis 169:113–120

Davis S, Propp S, Freier SM, Jones LE, Serra MJ, Kinberger G, Bhat B, Swayze EE, Bennett CF, Esau C (2009) Potent inhibition of microRNA in vivo without degradation. Nucleic Acids Res 37:70–77

de Aguiar Vallim TQ, Tarling EJ, Kim T, Civelek M, Baldan A, Esau C, Edwards PA (2013) MicroRNA-144 regulates hepatic ATP binding cassette transporter A1 and plasma high-density lipoprotein after activation of the nuclear receptor farnesoid X receptor. Circ Res 112:1602–1612

Desai NR, Kohli P, Giugliano RP, O'Donoghue ML, Somaratne R, Zhou J, Hoffman EB, Huang F, Rogers WJ, Wasserman SM, Scott R, Sabatine MS (2013) AMG145, a monoclonal antibody against proprotein convertase subtilisin kexin type 9, significantly reduces lipoprotein(a) in hypercholesterolemic patients receiving statin therapy: an analysis from the LDL-C assessment with proprotein convertase subtilisin kexin type 9 monoclonal antibody inhibition combined with statin therapy (LAPLACE)-thrombolysis in myocardial infarction (TIMI) 57 trial. Circulation 128:962–969

Dias CS, Shaywitz AJ, Wasserman SM, Smith BP, Gao B, Stolman DS, Crispino CP, Smirnakis KV, Emery MG, Colbert A, Gibbs JP, Retter MW, Cooke BP, Uy ST, Matson M, Stein EA (2012) Effects of AMG 145 on low-density lipoprotein cholesterol levels: results from 2 randomized, double-blind, placebo-controlled, ascending-dose phase 1 studies in healthy volunteers and hypercholesterolemic subjects on statins. J Am Coll Cardiol 60:1888–1898

Edmondson AC, Brown RJ, Kathiresan S, Cupples LA, Demissie S, Manning AK, Jensen MK, Rimm EB, Wang J, Rodrigues A, Bamba V, Khetarpal SA, Wolfe ML, Derohannessian S, Li M, Reilly MP, Aberle J, Evans D, Hegele RA, Rader DJ (2009) Loss-of-function variants in endothelial lipase are a cause of elevated HDL cholesterol in humans. J Clin Invest 119:1042–1050

Elmen J, Thonberg H, Ljungberg K, Frieden M, Westergaard M, Xu Y, Wahren B, Liang Z, Orum H, Koch T, Wahlestedt C (2005) Locked nucleic acid (LNA) mediated improvements in siRNA stability and functionality. Nucleic Acids Res 33:439–447

Elmen J, Lindow M, Schutz S, Lawrence M, Petri A, Obad S, Lindholm M, Hedtjarn M, Hansen HF, Berger U, Gullans S, Kearney P, Sarnow P, Straarup EM, Kauppinen S (2008) LNA-mediated microRNA silencing in non-human primates. Nature 452:896–899

Engel T, Lueken A, Bode G, Hobohm U, Lorkowski S, Schlueter B, Rust S, Cullen P, Pech M, Assmann G, Seedorf U (2004) ADP-ribosylation factor (ARF)-like 7 (ARL7) is induced by cholesterol loading and participates in apolipoprotein AI-dependent cholesterol export. FEBS Lett 566:241–246

Farragher TM, Bruce IN (2006) Cardiovascular risk in inflammatory rheumatic diseases: loose ends and common threads. J Rheumatol 33:2105–2107

Fearon WF, Fearon DT (2008) Inflammation and cardiovascular disease: role of the interleukin-1 receptor antagonist. Circulation 117:2577–2579

Flynt AS, Lai EC (2008) Biological principles of microRNA-mediated regulation: shared themes amid diversity. Nat Rev Genet 9:831–842

Fredrikson GN, Hedblad B, Berglund G, Alm R, Ares M, Cercek B, Chyu KY, Shah PK, Nilsson J (2003a) Identification of immune responses against aldehyde-modified peptide sequences in apoB associated with cardiovascular disease. Arterioscler Thromb Vasc Biol 23:872–878

Fredrikson GN, Soderberg I, Lindholm M, Dimayuga P, Chyu KY, Shah PK, Nilsson J (2003b) Inhibition of atherosclerosis in apoE-null mice by immunization with apoB-100 peptide sequences. Arterioscler Thromb Vasc Biol 23:879–884

Friedmann T, Roblin R (1972) Gene therapy for human genetic disease? Science 175:949–955

Gabriely G, Teplyuk NM, Krichevsky AM (2011) Context effect: microRNA-10b in cancer cell proliferation, spread and death. Autophagy 7:1384–1386

Geary RS, Watanabe TA, Truong L, Freier S, Lesnik EA, Sioufi NB, Sasmor H, Manoharan M, Levin AA (2001) Pharmacokinetic properties of 2′-O-(2-methoxyethyl)-modified oligonucleotide analogs in rats. J Pharmacol Exp Ther 296:890–897

George J, Afek A, Gilburd B, Levkovitz H, Shaish A, Goldberg I, Kopolovic Y, Wick G, Shoenfeld Y, Harats D (1998) Hyperimmunization of apo-E-deficient mice with homologous malondialdehyde low-density lipoprotein suppresses early atherogenesis. Atherosclerosis 138:147–152

George J, Shoenfeld Y, Afek A, Gilburd B, Keren P, Shaish A, Kopolovic J, Wick G, Harats D (1999) Enhanced fatty streak formation in C57BL/6 J mice by immunization with heat shock protein-65. Arterioscler Thromb Vasc Biol 19:505–510

Gerin I, Clerbaux LA, Haumont O, Lanthier N, Das AK, Burant CF, Leclercq IA, MacDougald OA, Bommer GT (2010) Expression of miR-33 from an SREBP2 intron inhibits cholesterol export and fatty acid oxidation. J Biol Chem 285:33652–33661

Giugliano RP, Desai NR, Kohli P, Rogers WJ, Somaratne R, Huang F, Liu T, Mohanavelu S, Hoffman EB, McDonald ST, Abrahamsen TE, Wasserman SM, Scott R, Sabatine MS (2012) Efficacy, safety, and tolerability of a monoclonal antibody to proprotein convertase subtilisin/kexin type 9 in combination with a statin in patients with hypercholesterolaemia (LAPLACE-TIMI 57): a randomised, placebo-controlled, dose-ranging, phase 2 study. Lancet 380:2007–2017

Gonzalez-Carmona MA, Quasdorff M, Vogt A, Tamke A, Yildiz Y, Hoffmann P, Lehmann T, Bartenschlager R, Engels JW, Kullak-Ublick GA, Sauerbruch T, Caselmann WH (2013) Inhibition of hepatitis C virus RNA translation by antisense bile acid conjugated phosphorothioate modified oligodeoxynucleotides (ODN). Antiviral Res 97:49–59

Gordon T, Castelli WP, Hjortland MC, Kannel WB, Dawber TR (1977) High density lipoprotein as a protective factor against coronary heart disease. The Framingham study. Am J Med 62:707–714

Graham MJ, Lee RG, Bell TA 3rd, Fu W, Mullick AE, Alexander VJ, Singleton W, Viney N, Geary R, Su J, Baker BF, Burkey J, Crooke ST, Crooke RM (2013) Antisense oligonucleotide inhibition of apolipoprotein C-III reduces plasma triglycerides in rodents, nonhuman primates, and humans. Circ Res 112:1479–1490

Grundtman C, Kreutmayer SB, Almanzar G, Wick MC, Wick G (2011) Heat shock protein 60 and immune inflammatory responses in atherosclerosis. Arterioscler Thromb Vasc Biol 31:960–968

Gumbiner B, Udata C, Joh T, Liang H, Wan H, Shelton D, Forgues P, Billote S, Pons J, Baum CM, Garzone PD (2012a) The effects of multiple dose administration of RN316 (PF-04950615), a humanized IgG2a monoclonal antibody binding proprotein convertase subtilisin kexin type 9, in hypercholesterolemic subjects. Circulation 126, A13524

Gumbiner B, Udata C, Joh T, Liang H, Wan H, Shelton D, Forgues P, Billote S, Pons J, Baum CM, Garzone PD (2012b) The Effects of Single Dose Administration of RN316 (PF-04950615), a Humanized IgG2a Monoclonal Antibody Binding Proprotein Convertase Subtilisin Kexin Type 9, in Hypercholesterolemic Subjects Treated with and without Atorvastatin. Circulation 126, A13322

Guo H, Ingolia NT, Weissman JS, Bartel DP (2010) Mammalian microRNAs predominantly act to decrease target mRNA levels. Nature 466:835–840

Guo Y, Ying L, Tian Y, Yang P, Zhu Y, Wang Z, Qiu F, Lin J (2013) miR-144 downregulation increases bladder cancer cell proliferation by targeting EZH2 and regulating Wnt signaling. FEBS J 280:4531–4538

Gusarova V, Howard VG, Okamoto H, Koehler-Stec EM, Papadopoulos N, Murphy AJ, Yancopoulos GD, Stahl N, Sleeman MW (2012) Reduction of LDL cholesterol by a monoclonal antibody to PCSK9 in rodents and nonhuman primates. Clin Lipidol 7:737–743

Harats D, Yacov N, Gilburd B, Shoenfeld Y, George J (2002) Oral tolerance with heat shock protein 65 attenuates Mycobacterium tuberculosis-induced and high-fat-diet-driven atherosclerotic lesions. J Am Coll Cardiol 40:1333–1338

Hauer AD, van Puijvelde GH, Peterse N, de Vos P, van Weel V, van Wanrooij EJ, Biessen EA, Quax PH, Niethammer AG, Reisfeld RA, van Berkel TJ, Kuiper J (2007) Vaccination against VEGFR2 attenuates initiation and progression of atherosclerosis. Arterioscler Thromb Vasc Biol 27:2050–2057

Hebert SS, Horre K, Nicolai L, Bergmans B, Papadopoulou AS, Delacourte A, De Strooper B (2009) MicroRNA regulation of Alzheimer's Amyloid precursor protein expression. Neurobiol Dis 33:422–428

Helwak A, Kudla G, Dudnakova T, Tollervey D (2013) Mapping the human miRNA interactome by CLASH reveals frequent noncanonical binding. Cell 153:654–665

Hermansson A, Ketelhuth DF, Strodthoff D, Wurm M, Hansson EM, Nicoletti A, Paulsson-Berne G, Hansson GK (2010) Inhibition of T cell response to native low-density lipoprotein reduces atherosclerosis. J Exp Med 207:1081–1093

Herrera-Merchan A, Cerrato C, Luengo G, Dominguez O, Piris MA, Serrano M, Gonzalez S (2010) miR-33-mediated downregulation of p53 controls hematopoietic stem cell self-renewal. Cell Cycle 9:3277–3285

Ho PC, Chang KC, Chuang YS, Wei LN (2011) Cholesterol regulation of receptor-interacting protein 140 via microRNA-33 in inflammatory cytokine production. FASEB J 25:1758–1766

Holasova S, Mojzisek M, Buncek M, Vokurkova D, Radilova H, Safarova M, Cervinka M, Haluza R (2005) Cholesterol conjugated oligonucleotide and LNA: a comparison of cellular and nuclear uptake by Hep2 cells enhanced by streptolysin-O. Mol Cell Biochem 276:61–69

Horie T, Ono K, Horiguchi M, Nishi H, Nakamura T, Nagao K, Kinoshita M, Kuwabara Y, Marusawa H, Iwanaga Y, Hasegawa K, Yokode M, Kimura T, Kita T (2010) MicroRNA-33 encoded by an intron of sterol regulatory element-binding protein 2 (Srebp2) regulates HDL in vivo. Proc Natl Acad Sci USA 107:17321–17326

Horie T, Baba O, Kuwabara Y, Chujo Y, Watanabe S, Kinoshita M, Horiguchi M, Nakamura T, Chonabayashi K, Hishizawa M, Hasegawa K, Kume N, Yokode M, Kita T, Kimura T, Ono K (2012) MicroRNA-33 deficiency reduces the progression of atherosclerotic plaque in ApoE(−/−) mice. J Am Heart Assoc 1:e003376

Horie T, Nishino T, Baba O, Kuwabara Y, Nakao T, Nishiga M, Usami S, Izuhara M, Sowa N, Yahagi N, Shimano H, Matsumura S, Inoue K, Marusawa H, Nakamura T, Hasegawa K,

Kume N, Yokode M, Kita T, Kimura T, Ono K (2013) MicroRNA-33 regulates sterol regulatory element-binding protein 1 expression in mice. Nat Commun 4:2883

Horton JD, Goldstein JL, Brown MS (2002) SREBPs: activators of the complete program of cholesterol and fatty acid synthesis in the liver. J Clin Invest 109:1125–1131

Hu Z, Shen WJ, Kraemer FB, Azhar S (2012) MicroRNAs 125a and 455 repress lipoprotein-supported steroidogenesis by targeting scavenger receptor class B type I in steroidogenic cells. Mol Cell Biol 32:5035–5045

Ishida T, Choi S, Kundu RK, Hirata K, Rubin EM, Cooper AD, Quertermous T (2003) Endothelial lipase is a major determinant of HDL level. J Clin Invest 111:347–355

Janssen HL, Reesink HW, Lawitz EJ, Zeuzem S, Rodriguez-Torres M, Patel K, van der Meer AJ, Patick AK, Chen A, Zhou Y, Persson R, King BD, Kauppinen S, Levin AA, Hodges MR (2013) Treatment of HCV infection by targeting microRNA. N Engl J Med 368:1685–1694

Jin W, Millar JS, Broedl U, Glick JM, Rader DJ (2003) Inhibition of endothelial lipase causes increased HDL cholesterol levels in vivo. J Clin Invest 111:357–362

Johannsen TH, Frikke-Schmidt R, Schou J, Nordestgaard BG, Tybjaerg-Hansen A (2012) Genetic inhibition of CETP, ischemic vascular disease and mortality, and possible adverse effects. J Am Coll Cardiol 60:2041–2048

Jun L, Jie L, Dongping Y, Xin Y, Taiming L, Rongyue C, Jie W, Jingjing L (2012) Effects of nasal immunization of multi-target preventive vaccines on atherosclerosis. Vaccine 30:1029–1037

Kamari Y, Werman-Venkert R, Shaish A, Werman A, Harari A, Gonen A, Voronov E, Grosskopf I, Sharabi Y, Grossman E, Iwakura Y, Dinarello CA, Apte RN, Harats D (2007) Differential role and tissue specificity of interleukin-1alpha gene expression in atherogenesis and lipid metabolism. Atherosclerosis 195:31–38

Kang MH, Zhang LH, Wijesekara N, de Haan W, Butland S, Bhattacharjee A, Hayden MR (2013) Regulation of ABCA1 protein expression and function in hepatic and pancreatic islet cells by miR-145. Arterioscler Thromb Vasc Biol 33:2724–2732

Karinaga R, Anada T, Minari J, Mizu M, Koumoto K, Fukuda J, Nakazawa K, Hasegawa T, Numata M, Shinkai S, Sakurai K (2006) Galactose-PEG dual conjugation of beta-(1–>3)-D-glucan schizophyllan for antisense oligonucleotides delivery to enhance the cellular uptake. Biomaterials 27:1626–1635

Kastelein JJ, Wedel MK, Baker BF, Su J, Bradley JD, Yu RZ, Chuang E, Graham MJ, Crooke RM (2006) Potent reduction of apolipoprotein B and low-density lipoprotein cholesterol by short-term administration of an antisense inhibitor of apolipoprotein B. Circulation 114:1729–1735

Kilic A, Mandal K (2012) Heat shock proteins: pathogenic role in atherosclerosis and potential therapeutic implications. Autoimmune Dis 2012:502813

Kim J, Yoon H, Ramirez CM, Lee SM, Hoe HS, Fernandez-Hernando C (2012) MiR-106b impairs cholesterol efflux and increases Abeta levels by repressing ABCA1 expression. Exp Neurol 235:476–483

Klingenberg R, Ketelhuth DF, Strodthoff D, Gregori S, Hansson GK (2012) Subcutaneous immunization with heat shock protein-65 reduces atherosclerosis in Apoe(-)/(-) mice. Immunobiology 217:540–547

Komori T (2004) CETi-1. AVANT. Curr Opin Investig Drugs 5:334–338

Koren MJ, Scott R, Kim JB, Knusel B, Liu T, Lei L, Bolognese M, Wasserman SM (2012) Efficacy, safety, and tolerability of a monoclonal antibody to proprotein convertase subtilisin/kexin type 9 as monotherapy in patients with hypercholesterolaemia (MENDEL): a randomised, double-blind, placebo-controlled, phase 2 study. Lancet 380:1995–2006

Kramer W (2013) Novel drug approaches in development for the treatment of lipid disorders. Exp Clin Endocrinol Diabetes 121:567–580

Krol J, Loedige I, Filipowicz W (2010) The widespread regulation of microRNA biogenesis, function and decay. Nat Rev Genet 11:597–610

Krutzfeldt J, Rajewsky N, Braich R, Rajeev KG, Tuschl T, Manoharan M, Stoffel M (2005) Silencing of microRNAs in vivo with 'antagomirs'. Nature 438:685–689

Krutzfeldt J, Kuwajima S, Braich R, Rajeev KG, Pena J, Tuschl T, Manoharan M, Stoffel M (2007) Specificity, duplex degradation and subcellular localization of antagomirs. Nucleic Acids Res 35:2885–2892

Kumar R, Singh SK, Koshkin AA, Rajwanshi VK, Meldgaard M, Wengel J (1998) The first analogues of LNA (locked nucleic acids): phosphorothioate-LNA and 2′-thio-LNA. Bioorg Med Chem Lett 8:2219–2222

Kuo CY, Lin YC, Yang JJ, Yang VC (2011) Interaction abolishment between mutant caveolin-1 (Delta62-100) and ABCA1 reduces HDL-mediated cellular cholesterol efflux. Biochem Biophys Res Commun 414:337–343

Kuusi T, Saarinen P, Nikkila EA (1980) Evidence for the role of hepatic endothelial lipase in the metabolism of plasma high density lipoprotein2 in man. Atherosclerosis 36:589–593

Lagos-Quintana M, Rauhut R, Yalcin A, Meyer J, Lendeckel W, Tuschl T (2002) Identification of tissue-specific microRNAs from mouse. Curr Biol 12:735–739

Lambert G, Sjouke B, Choque B, Kastelein JJ, Hovingh GK (2012) The PCSK9 decade. J Lipid Res 53:2515–2524

Landmesser U, von Eckardstein A, Kastelein J, Deanfield J, Luscher TF (2012) Increasing high-density lipoprotein cholesterol by cholesteryl ester transfer protein-inhibition: a rocky road and lessons learned? The early demise of the dal-HEART programme. Eur Heart J 33:1712–1715

Lanford RE, Hildebrandt-Eriksen ES, Petri A, Persson R, Lindow M, Munk ME, Kauppinen S, Orum H (2010) Therapeutic silencing of microRNA-122 in primates with chronic hepatitis C virus infection. Science 327:198–201

Langlois RA, Shapiro JS, Pham AM, tenOever BR (2012) In vivo delivery of cytoplasmic RNA virus-derived miRNAs. Mol Ther 20:367–375

Lee RG, Fu W, Graham MJ, Mullick AE, Sipe D, Gattis D, Bell TA, Booten S, Crooke RM (2013a) Comparison of the pharmacological profiles of murine antisense oligonucleotides targeting apolipoprotein B and microsomal triglyceride transfer protein. J Lipid Res 54:602–614

Lee SH, Castagner B, Leroux JC (2013b) Is there a future for cell-penetrating peptides in oligonucleotide delivery? Eur J Pharm Biopharm 85:5–11

Li Y, Wang MN, Li H, King KD, Bassi R, Sun H, Santiago A, Hooper AT, Bohlen P, Hicklin DJ (2002) Active immunization against the vascular endothelial growth factor receptor flk1 inhibits tumor angiogenesis and metastasis. J Exp Med 195:1575–1584

Li J, Zhao X, Zhang S, Wang S, Du P, Qi G (2011) ApoB-100 and HSP60 peptides exert a synergetic role in inhibiting early atherosclerosis in immunized ApoE-null mice. Protein Pept Lett 18:733–740

Liang H, Chaparro-Riggers J, Strop P, Geng T, Sutton JE, Tsai D, Bai L, Abdiche Y, Dilley J, Yu J, Wu S, Chin SM, Lee NA, Rossi A, Lin JC, Rajpal A, Pons J, Shelton DL (2012) Proprotein convertase substilisin/kexin type 9 antagonism reduces low-density lipoprotein cholesterol in statin-treated hypercholesterolemic nonhuman primates. J Pharmacol Exp Ther 340:228–236

Long J, Lin J, Yang X, Yuan D, Wu J, Li T, Cao R, Liu J (2012) Nasal immunization with different forms of heat shock protein-65 reduced high-cholesterol-diet-driven rabbit atherosclerosis. Int Immunopharmacol 13:82–87

Lu X, Chen D, Endresz V, Xia M, Faludi I, Burian K, Szabo A, Csanadi A, Miczak A, Gonczol E, Kakkar V (2010) Immunization with a combination of ApoB and HSP60 epitopes significantly reduces early atherosclerotic lesion in Apobtm2SgyLdlrtm1Her/J mice. Atherosclerosis 212:472–480

Ma K, Cilingiroglu M, Otvos JD, Ballantyne CM, Marian AJ, Chan L (2003) Endothelial lipase is a major genetic determinant for high-density lipoprotein concentration, structure, and metabolism. Proc Natl Acad Sci USA 100:2748–2753

Marcus-Sekura CJ, Woerner AM, Shinozuka K, Zon G, Quinnan GV (1987) Comparative inhibition of chloramphenicol acetyltransferase gene expression by antisense oligonucleotide analogues having alkyl phosphotriester, methylphosphonate and phosphorothioate linkages. Nucleic Acids Res 15:5749–5763

Maron R, Sukhova G, Faria AM, Hoffmann E, Mach F, Libby P, Weiner HL (2002) Mucosal administration of heat shock protein-65 decreases atherosclerosis and inflammation in aortic arch of low-density lipoprotein receptor-deficient mice. Circulation 106:1708–1715

Marquart TJ, Allen RM, Ory DS, Baldan A (2010) miR-33 links SREBP-2 induction to repression of sterol transporters. Proc Natl Acad Sci USA 107:12228–12232

Marquart TJ, Wu J, Lusis AJ, Baldan A (2013) Anti-miR-33 therapy does not alter the progression of atherosclerosis in low-density lipoprotein receptor-deficient mice. Arterioscler Thromb Vasc Biol 33:455–458

Mayne J, Dewpura T, Raymond A, Cousins M, Chaplin A, Lahey KA, Lahaye SA, Mbikay M, Ooi TC, Chretien M (2008) Plasma PCSK9 levels are significantly modified by statins and fibrates in humans. Lipids Health Dis 7:22

McKenney JM, Koren MJ, Kereiakes DJ, Hanotin C, Ferrand AC, Stein EA (2012) Safety and efficacy of a monoclonal antibody to proprotein convertase subtilisin/kexin type 9 serine protease, SAR236553/REGN727, in patients with primary hypercholesterolemia receiving ongoing stable atorvastatin therapy. J Am Coll Cardiol 59:2344–2353

McMahon M, Grossman J, FitzGerald J, Dahlin-Lee E, Wallace DJ, Thong BY, Badsha H, Kalunian K, Charles C, Navab M, Fogelman AM, Hahn BH (2006) Proinflammatory high-density lipoprotein as a biomarker for atherosclerosis in patients with systemic lupus erythematosus and rheumatoid arthritis. Arthritis Rheum 54:2541–2549

Mencia A, Modamio-Hoybjor S, Redshaw N, Morin M, Mayo-Merino F, Olavarrieta L, Aguirre LA, del Castillo I, Steel KP, Dalmay T, Moreno F, Moreno-Pelayo MA (2009) Mutations in the seed region of human miR-96 are responsible for nonsyndromic progressive hearing loss. Nat Genet 41:609–613

Mendell JT, Olson EN (2012) MicroRNAs in stress signaling and human disease. Cell 148:1172–1187

Miyazaki Y, Adachi H, Katsuno M, Minamiyama M, Jiang YM, Huang Z, Doi H, Matsumoto S, Kondo N, Iida M, Tohnai G, Tanaka F, Muramatsu S, Sobue G (2012) Viral delivery of miR-196a ameliorates the SBMA phenotype via the silencing of CELF2. Nat Med 18:1136–1141

Monia BP, Lesnik EA, Gonzalez C, Lima WF, McGee D, Guinosso CJ, Kawasaki AM, Cook PD, Freier SM (1993) Evaluation of 2′-modified oligonucleotides containing 2′-deoxy gaps as antisense inhibitors of gene expression. J Biol Chem 268:14514–14522

Moulton KS, Vakili K, Zurakowski D, Soliman M, Butterfield C, Sylvin E, Lo KM, Gillies S, Javaherian K, Folkman J (2003) Inhibition of plaque neovascularization reduces macrophage accumulation and progression of advanced atherosclerosis. Proc Natl Acad Sci USA 100:4736–4741

Najafi-Shoushtari SH, Kristo F, Li Y, Shioda T, Cohen DE, Gerszten RE, Naar AM (2010) MicroRNA-33 and the SREBP host genes cooperate to control cholesterol homeostasis. Science 328:1566–1569

Navab M, Reddy ST, Van Lenten BJ, Fogelman AM (2011) HDL and cardiovascular disease: atherogenic and atheroprotective mechanisms. Nat Rev Cardiol 8:222–232

Nelson ER, Wardell SE, Jasper JS, Park S, Suchindran S, Howe MK, Carver NJ, Pillai RV, Sullivan PM, Sondhi V, Umetani M, Geradts J, McDonnell DP (2013) 27-Hydroxycholesterol links hypercholesterolemia and breast cancer pathophysiology. Science 342:1094–1098

Ni YG, Di MS, Condra JH, Peterson LB, Wang W, Wang F, Pandit S, Hammond HA, Rosa R, Cummings RT, Wood DD, Liu X, Bottomley MJ, Shen X, Cubbon RM, Wang SP, Johns DG, Volpari C, Hamuro L, Chin J, Huang L, Zhao JZ, Vitelli S, Haytko P, Wisniewski D, Mitnaul LJ, Sparrow CP, Hubbard B, Carfi A, Sitlani A (2011) A PCSK9-binding antibody that structurally mimics the EGF(A) domain of LDL-receptor reduces LDL cholesterol in vivo. J Lipid Res 52:78–86

Nissen SE, Tuzcu EM, Brewer HB, Sipahi I, Nicholls SJ, Ganz P, Schoenhagen P, Waters DD, Pepine CJ, Crowe TD, Davidson MH, Deanfield JE, Wisniewski LM, Hanyok JJ, Kassalow LM (2006) Effect of ACAT inhibition on the progression of coronary atherosclerosis. N Engl J Med 354:1253–1263

Nohata N, Hanazawa T, Kikkawa N, Mutallip M, Fujimura L, Yoshino H, Kawakami K, Chiyomaru T, Enokida H, Nakagawa M, Okamoto Y, Seki N (2011) Caveolin-1 mediates tumor cell migration and invasion and its regulation by miR-133a in head and neck squamous cell carcinoma. Int J Oncol 38:209–217

Obad S, dos Santos CO, Petri A, Heidenblad M, Broom O, Ruse C, Fu C, Lindow M, Stenvang J, Straarup EM, Hansen HF, Koch T, Pappin D, Hannon GJ, Kauppinen S (2011) Silencing of microRNA families by seed-targeting tiny LNAs. Nat Genet 43:371–378

Oehlke J, Birth P, Klauschenz E, Wiesner B, Beyermann M, Oksche A, Bienert M (2002) Cellular uptake of antisense oligonucleotides after complexing or conjugation with cell-penetrating model peptides. Eur J Biochem 269:4025–4032

Onat A, Direskeneli H (2012) Excess cardiovascular risk in inflammatory rheumatic diseases: pathophysiology and targeted therapy. Curr Pharm Des 18:1465–1477

Orom UA, Nielsen FC, Lund AH (2008) MicroRNA-10a binds the 5′UTR of ribosomal protein mRNAs and enhances their translation. Mol Cell 30:460–471

Ouimet M, Franklin V, Mak E, Liao X, Tabas I, Marcel YL (2011) Autophagy regulates cholesterol efflux from macrophage foam cells via lysosomal acid lipase. Cell Metab 13:655–667

Palinski W, Miller E, Witztum JL (1995) Immunization of low density lipoprotein (LDL) receptor-deficient rabbits with homologous malondialdehyde-modified LDL reduces atherogenesis. Proc Natl Acad Sci USA 92:821–825

Pallan PS, Allerson CR, Berdeja A, Seth PP, Swayze EE, Prakash TP, Egli M (2012) Structure and nuclease resistance of 2′,4′-constrained 2′-O-methoxyethyl (cMOE) and 2′-O-ethyl (cEt) modified DNAs. Chem Commun (Camb) 48:8195–8197

Piao L, Zhang M, Datta J, Xie X, Su T, Li H, Teknos TN, Pan Q (2012) Lipid-based nanoparticle delivery of Pre-miR-107 inhibits the tumorigenicity of head and neck squamous cell carcinoma. Mol Ther 20:1261–1269

Poliseno L, Salmena L, Zhang J, Carver B, Haveman WJ, Pandolfi PP (2010) A coding-independent function of gene and pseudogene mRNAs regulates tumour biology. Nature 465:1033–1038

Popa C, van Tits LJ, Barrera P, Lemmers HL, van den Hoogen FH, van Riel PL, Radstake TR, Netea MG, Roest M, Stalenhoef AF (2009) Anti-inflammatory therapy with tumour necrosis factor alpha inhibitors improves high-density lipoprotein cholesterol antioxidative capacity in rheumatoid arthritis patients. Ann Rheum Dis 68:868–872

Popa CD, Arts E, Fransen J, van Riel PL (2012) Atherogenic index and high-density lipoprotein cholesterol as cardiovascular risk determinants in rheumatoid arthritis: the impact of therapy with biologicals. Mediat Inflamm 2012:785946

Prakash TP, Allerson CR, Dande P, Vickers TA, Sioufi N, Jarres R, Baker BF, Swayze EE, Griffey RH, Bhat B (2005) Positional effect of chemical modifications on short interference RNA activity in mammalian cells. J Med Chem 48:4247–4253

Qian P, Banerjee A, Wu ZS, Zhang X, Wang H, Pandey V, Zhang WJ, Lv XF, Tan S, Lobie PE, Zhu T (2012) Loss of SNAIL regulated miR-128-2 on chromosome 3p22.3 targets multiple stem cell factors to promote transformation of mammary epithelial cells. Cancer Res 72:6036–6050

Raal FJ, Santos RD, Blom DJ, Marais AD, Charng MJ, Cromwell WC, Lachmann RH, Gaudet D, Tan JL, Chasan-Taber S, Tribble DL, Flaim JD, Crooke ST (2010) Mipomersen, an apolipoprotein B synthesis inhibitor, for lowering of LDL cholesterol concentrations in patients with homozygous familial hypercholesterolaemia: a randomised, double-blind, placebo-controlled trial. Lancet 375:998–1006

Raal F, Scott R, Somaratne R, Bridges I, Li G, Wasserman SM, Stein EA (2012) Low-density lipoprotein cholesterol-lowering effects of AMG 145, a monoclonal antibody to proprotein convertase subtilisin/kexin type 9 serine protease in patients with heterozygous familial hypercholesterolemia: the Reduction of LDL-C with PCSK9 Inhibition in Heterozygous Familial Hypercholesterolemia Disorder (RUTHERFORD) randomized trial. Circulation 126:2408–2417

Ramirez CM, Davalos A, Goedeke L, Salerno AG, Warrier N, Cirera-Salinas D, Suarez Y, Fernandez-Hernando C (2011) MicroRNA-758 regulates cholesterol efflux through posttranscriptional repression of ATP-binding cassette transporter A1. Arterioscler Thromb Vasc Biol 31:2707–2714

Ramirez CM, Rotllan N, Vlassov AV, Davalos A, Li M, Goedeke L, Aranda JF, Cirera-Salinas D, Araldi E, Salerno A, Wanschel A, Zavadil J, Castrillo A, Kim J, Suarez Y, Fernandez-Hernando C (2013) Control of cholesterol metabolism and plasma high-density lipoprotein levels by microRNA-144. Circ Res 112:1592–1601

Raterman HG, Levels H, Voskuyl AE, Lems WF, Dijkmans BA, Nurmohamed MT (2013) HDL protein composition alters from proatherogenic into less atherogenic and proinflammatory in rheumatoid arthritis patients responding to rituximab. Ann Rheum Dis 72:560–565

Rayner KJ, Suarez Y, Davalos A, Parathath S, Fitzgerald ML, Tamehiro N, Fisher EA, Moore KJ, Fernandez-Hernando C (2010) MiR-33 contributes to the regulation of cholesterol homeostasis. Science 328:1570–1573

Rayner KJ, Esau CC, Hussain FN, McDaniel AL, Marshall SM, van Gils JM, Ray TD, Sheedy FJ, Goedeke L, Liu X, Khatsenko OG, Kaimal V, Lees CJ, Fernandez-Hernando C, Fisher EA, Temel RE, Moore KJ (2011a) Inhibition of miR-33a/b in non-human primates raises plasma HDL and lowers VLDL triglycerides. Nature 478:404–407

Rayner KJ, Sheedy FJ, Esau CC, Hussain FN, Temel RE, Parathath S, van Gils JM, Rayner AJ, Chang AN, Suarez Y, Fernandez-Hernando C, Fisher EA, Moore KJ (2011b) Antagonism of miR-33 in mice promotes reverse cholesterol transport and regression of atherosclerosis. J Clin Invest 121:2921–2931

Ridker PM, Thuren T, Zalewski A, Libby P (2011) Interleukin-1beta inhibition and the prevention of recurrent cardiovascular events: rationale and design of the canakinumab anti-inflammatory thrombosis outcomes study (CANTOS). Am Heart J 162:597–605

Ridker PM, Howard CP, Walter V, Everett B, Libby P, Hensen J, Thuren T (2012) Effects of interleukin-1beta inhibition with canakinumab on hemoglobin A1c, lipids, C-reactive protein, interleukin-6, and fibrinogen: a phase IIb randomized, placebo-controlled trial. Circulation 126:2739–2748

Roth EM, McKenney JM, Hanotin C, Asset G, Stein EA (2012) Atorvastatin with or without an antibody to PCSK9 in primary hypercholesterolemia. N Engl J Med 367:1891–1900

Rottiers V, Obad S, Petri A, McGarrah R, Lindholm MW, Black JC, Sinha S, Goody RJ, Lawrence MS, Delemos AS, Hansen HF, Whittaker S, Henry S, Brookes R, Najafi-Shoushtari SH, Chung RT, Whetstine JR, Gerszten RE, Kauppinen S, Naar AM (2013) Pharmacological inhibition of a MicroRNA family in nonhuman primates by a seed-targeting 8-Mer AntimiR. Sci Transl Med 5:212ra162

Ryan BM, Robles AI, Harris CC (2010) Genetic variation in microRNA networks: the implications for cancer research. Nat Rev Cancer 10:389–402

Salmena L, Poliseno L, Tay Y, Kats L, Pandolfi PP (2011) A ceRNA hypothesis: the Rosetta Stone of a hidden RNA language? Cell 146:353–358

Samson S, Mundkur L, Kakkar VV (2012) Immune response to lipoproteins in atherosclerosis. Cholesterol 2012:571846

Schaefer EJ (2013) Effects of cholesteryl ester transfer protein inhibitors on human lipoprotein metabolism: why have they failed in lowering coronary heart disease risk? Curr Opin Lipidol 24:259–264

Schnall-Levin M, Zhao Y, Perrimon N, Berger B (2010) Conserved microRNA targeting in Drosophila is as widespread in coding regions as in 3'UTRs. Proc Natl Acad Sci USA 107:15751–15756

Sehgal A, Vaishnaw A, Fitzgerald K (2013) Liver as a target for oligonucleotide therapeutics. J Hepatol 59:1354–1359

Selbach M, Schwanhausser B, Thierfelder N, Fang Z, Khanin R, Rajewsky N (2008) Widespread changes in protein synthesis induced by microRNAs. Nature 455:58–63

Seth PP, Vasquez G, Allerson CA, Berdeja A, Gaus H, Kinberger GA, Prakash TP, Migawa MT, Bhat B, Swayze EE (2010) Synthesis and biophysical evaluation of 2′,4′-constrained 2′O-methoxyethyl and 2′,4′-constrained 2′O-ethyl nucleic acid analogues. J Org Chem 75:1569–1581

Shaw PX, Horkko S, Chang MK, Curtiss LK, Palinski W, Silverman GJ, Witztum JL (2000) Natural antibodies with the T15 idiotype may act in atherosclerosis, apoptotic clearance, and protective immunity. J Clin Invest 105:1731–1740

Shi Q, Rafii S, Wu MH, Wijelath ES, Yu C, Ishida A, Fujita Y, Kothari S, Mohle R, Sauvage LR, Moore MA, Storb RF, Hammond WP (1998) Evidence for circulating bone marrow-derived endothelial cells. Blood 92:362–367

Singaraja RR, Sivapalaratnam S, Hovingh K, Dube MP, Castro-Perez J, Collins HL, Adelman SJ, Riwanto M, Manz J, Hubbard B, Tietjen I, Wong K, Mitnaul LJ, van Heek M, Lin L, Roddy TA, McEwen J, Dallinge-Thie G, van Vark-van der Zee L, Verwoert G, Winther M, van Duijn C, Hofman A, Trip MD, Marais AD, Asztalos B, Landmesser U, Sijbrands E, Kastelein JJ, Hayden MR (2013) The impact of partial and complete loss-of-function mutations in endothelial lipase on high-density lipoprotein levels and functionality in humans. Circ Cardiovasc Genet 6:54–62

Singh Y, Murat P, Defrancq E (2010) Recent developments in oligonucleotide conjugation. Chem Soc Rev 39:2054–2070

Small EM, Olson EN (2011) Pervasive roles of microRNAs in cardiovascular biology. Nature 469:336–342

Stein EA, Gipe D, Bergeron J, Gaudet D, Weiss R, Dufour R, Wu R, Pordy R (2012a) Effect of a monoclonal antibody to PCSK9, REGN727/SAR236553, to reduce low-density lipoprotein cholesterol in patients with heterozygous familial hypercholesterolaemia on stable statin dose with or without ezetimibe therapy: a phase 2 randomised controlled trial. Lancet 380:29–36

Stein EA, Mellis S, Yancopoulos GD, Stahl N, Logan D, Smith WB, Lisbon E, Gutierrez M, Webb C, Wu R, Du Y, Kranz T, Gasparino E, Swergold GD (2012b) Effect of a monoclonal antibody to PCSK9 on LDL cholesterol. N Engl J Med 366:1108–1118

Steinberg D, Witztum JL (2010) Oxidized low-density lipoprotein and atherosclerosis. Arterioscler Thromb Vasc Biol 30:2311–2316

Sugano M, Makino N (1996) Changes in plasma lipoprotein cholesterol levels by antisense oligodeoxynucleotides against cholesteryl ester transfer protein in cholesterol-fed rabbits. J Biol Chem 271:19080–19083

Sugano M, Makino N, Sawada S, Otsuka S, Watanabe M, Okamoto H, Kamada M, Mizushima A (1998) Effect of antisense oligonucleotides against cholesteryl ester transfer protein on the development of atherosclerosis in cholesterol-fed rabbits. J Biol Chem 273:5033–5036

Sullivan D, Olsson AG, Scott R, Kim JB, Xue A, Gebski V, Wasserman SM, Stein EA (2012) Effect of a monoclonal antibody to PCSK9 on low-density lipoprotein cholesterol levels in statin-intolerant patients: the GAUSS randomized trial. JAMA 308:2497–2506

Sun D, Zhang J, Xie J, Wei W, Chen M, Zhao X (2012) MiR-26 controls LXR-dependent cholesterol efflux by targeting ABCA1 and ARL7. FEBS Lett 586:1472–1479

Swayze EE, Siwkowski AM, Wancewicz EV, Migawa MT, Wyrzykiewicz TK, Hung G, Monia BP, Bennett CF (2007) Antisense oligonucleotides containing locked nucleic acid improve potency but cause significant hepatotoxicity in animals. Nucleic Acids Res 35:687–700

Tay Y, Zhang J, Thomson AM, Lim B, Rigoutsos I (2008) MicroRNAs to Nanog, Oct4 and Sox2 coding regions modulate embryonic stem cell differentiation. Nature 455:1124–1128

Thomas LJ, Hammond RA, Forsberg EM, Geoghegan-Barek KM, Karalius BH, Marsh HC Jr, Rittershaus CW (2009) Co-administration of a CpG adjuvant (VaxImmune, CPG 7909) with CETP vaccines increased immunogenicity in rabbits and mice. Hum Vaccin 5:79–84

Tissot AC, Spohn G, Jennings GT, Shamshiev A, Kurrer MO, Windak R, Meier M, Viesti M, Hersberger M, Kundig TM, Ricci R, Bachmann MF (2013) A VLP-based vaccine against interleukin-1alpha protects mice from atherosclerosis. Eur J Immunol 43:716–722

Tome-Carneiro J, Larrosa M, Yanez-Gascon MJ, Davalos A, Gil-Zamorano J, Gonzalvez M, Garcia-Almagro FJ, Ruiz Ros JA, Tomas-Barberan FA, Espin JC, Garcia-Conesa MT (2013) One-year supplementation with a grape extract containing resveratrol modulates inflammatory-related microRNAs and cytokines expression in peripheral blood mononuclear cells of type 2 diabetes and hypertensive patients with coronary artery disease. Pharmacol Res 72:69–82

Trajkovski M, Hausser J, Soutschek J, Bhat B, Akin A, Zavolan M, Heim MH, Stoffel M (2011) MicroRNAs 103 and 107 regulate insulin sensitivity. Nature 474:649–653

Trang P, Wiggins JF, Daige CL, Cho C, Omotola M, Brown D, Weidhaas JB, Bader AG, Slack FJ (2011) Systemic delivery of tumor suppressor microRNA mimics using a neutral lipid emulsion inhibits lung tumors in mice. Mol Ther 19:1116–1122

Truong TQ, Aubin D, Falstrault L, Brodeur MR, Brissette L (2010) SR-BI, CD36, and caveolin-1 contribute positively to cholesterol efflux in hepatic cells. Cell Biochem Funct 28:480–489

Tsukerman P, Stern-Ginossar N, Gur C, Glasner A, Nachmani D, Bauman Y, Yamin R, Vitenshtein A, Stanietsky N, Bar-Mag T, Lankry D, Mandelboim O (2012) MiR-10b downregulates the stress-induced cell surface molecule MICB, a critical ligand for cancer cell recognition by natural killer cells. Cancer Res 72:5463–5472

van Poelgeest EP, Swart RM, Betjes MG, Moerland M, Weening JJ, Tessier Y, Hodges MR, Levin AA, Burggraaf J (2013) Acute kidney injury during therapy with an antisense oligonucleotide directed against PCSK9. Am J Kidney Dis 62:796–800

van Puijvelde GH, Hauer AD, de Vos P, van den Heuvel R, van Herwijnen MJ, van der Zee R, van Eden W, van Berkel TJ, Kuiper J (2006) Induction of oral tolerance to oxidized low-density lipoprotein ameliorates atherosclerosis. Circulation 114:1968–1976

van Rooij E, Olson EN (2012) MicroRNA therapeutics for cardiovascular disease: opportunities and obstacles. Nat Rev Drug Discov 11:860–872

van Rooij E, Marshall WS, Olson EN (2008) Toward microRNA-based therapeutics for heart disease: the sense in antisense. Circ Res 103:919–928

van Rooij E, Purcell AL, Levin AA (2012) Developing microRNA therapeutics. Circ Res 110:496–507

van Wanrooij EJ, de Vos P, Bixel MG, Vestweber D, van Berkel TJ, Kuiper J (2008) Vaccination against CD99 inhibits atherogenesis in low-density lipoprotein receptor-deficient mice. Cardiovasc Res 78:590–596

Vasudevan S, Tong Y, Steitz JA (2007) Switching from repression to activation: microRNAs can up-regulate translation. Science 318:1931–1934

Veedu RN, Wengel J (2010) Locked nucleic acids: promising nucleic acid analogs for therapeutic applications. Chem Biodivers 7:536–542

Vester B, Wengel J (2004) LNA (locked nucleic acid): high-affinity targeting of complementary RNA and DNA. Biochemistry 43:13233–13241

Vickers KC, Rader DJ (2013) Nuclear receptors and microRNA-144 coordinately regulate cholesterol efflux. Circ Res 112:1529–1531

Vickers KC, Palmisano BT, Shoucri BM, Shamburek RD, Remaley AT (2011) MicroRNAs are transported in plasma and delivered to recipient cells by high-density lipoproteins. Nat Cell Biol 13:423–433

Visioli F, Davalos A (2011) Polyphenols and cardiovascular disease: a critical summary of the evidence. Mini Rev Med Chem 11:1186–1190

Visioli F, Giordano E, Nicod NM, Davalos A (2012) Molecular targets of omega 3 and conjugated linoleic fatty acids - "micromanaging" cellular response. Front Physiol 3:42

Voight BF, Peloso GM, Orho-Melander M, Frikke-Schmidt R, Barbalic M, Jensen MK, Hindy G, Holm H, Ding EL, Johnson T, Schunkert H, Samani NJ, Clarke R, Hopewell JC, Thompson JF, Li M, Thorleifsson G, Newton-Cheh C, Musunuru K, Pirruccello JP, Saleheen D, Chen L, Stewart A, Schillert A, Thorsteinsdottir U, Thorgeirsson G, Anand S, Engert JC, Morgan T, Spertus J, Stoll M, Berger K, Martinelli N, Girelli D, McKeown PP, Patterson CC, Epstein SE, Devaney J, Burnett MS, Mooser V, Ripatti S, Surakka I, Nieminen MS, Sinisalo J, Lokki ML,

Perola M, Havulinna A, de Faire U, Gigante B, Ingelsson E, Zeller T, Wild P, de Bakker PI, Klungel OH, Maitland-van der Zee AH, Peters BJ, de Boer A, Grobbee DE, Kamphuisen PW, Deneer VH, Elbers CC, Onland-Moret NC, Hofker MH, Wijmenga C, Verschuren WM, Boer JM, van der Schouw YT, Rasheed A, Frossard P, Demissie S, Willer C, Do R, Ordovas JM, Abecasis GR, Boehnke M, Mohlke KL, Daly MJ, Guiducci C, Burtt NP, Surti A, Gonzalez E, Purcell S, Gabriel S, Marrugat J, Peden J, Erdmann J, Diemert P, Willenborg C, Konig IR, Fischer M, Hengstenberg C, Ziegler A, Buysschaert I, Lambrechts D, Van de Werf F, Fox KA, El Mokhtari NE, Rubin D, Schrezenmeir J, Schreiber S, Schafer A, Danesh J, Blankenberg S, Roberts R, McPherson R, Watkins H, Hall AS, Overvad K, Rimm E, Boerwinkle E, Tybjaerg-Hansen A, Cupples LA, Reilly MP, Melander O, Mannucci PM, Ardissino D, Siscovick D, Elosua R, Stefansson K, O'Donnell CJ, Salomaa V, Rader DJ, Peltonen L, Schwartz SM, Altshuler D, Kathiresan S (2012) Plasma HDL cholesterol and risk of myocardial infarction: a mendelian randomisation study. Lancet 380:572–580

Wang D, Xia M, Yan X, Li D, Wang L, Xu Y, Jin T, Ling W (2012) Gut microbiota metabolism of anthocyanin promotes reverse cholesterol transport in mice via repressing miRNA-10b. Circ Res 111:967–981

Wang L, Jia XJ, Jiang HJ, Du Y, Yang F, Si SY, Hong B (2013) MicroRNAs 185, 96, and 223 repress selective high-density lipoprotein cholesterol uptake through posttranscriptional inhibition. Mol Cell Biol 33:1956–1964

Wiggins JF, Ruffino L, Kelnar K, Omotola M, Patrawala L, Brown D, Bader AG (2010) Development of a lung cancer therapeutic based on the tumor suppressor microRNA-34. Cancer Res 70:5923–5930

Wigren M, Kolbus D, Duner P, Ljungcrantz I, Soderberg I, Bjorkbacka H, Fredrikson GN, Nilsson J (2011) Evidence for a role of regulatory T cells in mediating the atheroprotective effect of apolipoprotein B peptide vaccine. J Intern Med 269:546–556

Xu Q, Dietrich H, Steiner HJ, Gown AM, Schoel B, Mikuz G, Kaufmann SH, Wick G (1992) Induction of arteriosclerosis in normocholesterolemic rabbits by immunization with heat shock protein 65. Arterioscler Thromb 12:789–799

Xu J, Wang Y, Tan X, Jing H (2012) MicroRNAs in autophagy and their emerging roles in crosstalk with apoptosis. Autophagy 8:873–882

Yasuda T, Ishida T, Rader DJ (2010) Update on the role of endothelial lipase in high-density lipoprotein metabolism, reverse cholesterol transport, and atherosclerosis. Circ J 74:2263–2270

Yoo BH, Bochkareva E, Bochkarev A, Mou TC, Gray DM (2004) 2′-O-methyl-modified phosphorothioate antisense oligonucleotides have reduced non-specific effects in vitro. Nucleic Acids Res 32:2008–2016

Yu RZ, Geary RS, Monteith DK, Matson J, Truong L, Fitchett J, Levin AA (2004) Tissue disposition of 2′-O-(2-methoxy) ethyl modified antisense oligonucleotides in monkeys. J Pharm Sci 93:48–59

Zamecnik PC, Stephenson ML (1978) Inhibition of Rous sarcoma virus replication and cell transformation by a specific oligodeoxynucleotide. Proc Natl Acad Sci USA 75:280–284

Zatsepin TS, Oretskaya TS (2004) Synthesis and applications of oligonucleotide-carbohydrate conjugates. Chem Biodivers 1:1401–1417

Zhang DW, Lagace TA, Garuti R, Zhao Z, McDonald M, Horton JD, Cohen JC, Hobbs HH (2007) Binding of proprotein convertase subtilisin/kexin type 9 to epidermal growth factor-like repeat A of low density lipoprotein receptor decreases receptor recycling and increases degradation. J Biol Chem 282:18602–18612

Zhang J, Kelley KL, Marshall SM, Davis MA, Wilson MD, Sawyer JK, Farese RV Jr, Brown JM, Rudel LL (2012a) Tissue-specific knockouts of ACAT2 reveal that intestinal depletion is sufficient to prevent diet-induced cholesterol accumulation in the liver and blood. J Lipid Res 53:1144–1152

Zhang J, Yu Y, Nakamura K, Koike T, Waqar AB, Zhang X, Liu E, Nishijima K, Kitajima S, Shiomi M, Qi Z, Yu J, Graham MJ, Crooke RM, Ishida T, Hirata K, Hurt-Camejo E, Chen YE,

Fan J (2012b) Endothelial lipase mediates HDL levels in normal and hyperlipidemic rabbits. J Atheroscler Thromb 19:213–226

Zhang L, McCabe T, Condra JH, Ni YG, Peterson LB, Wang W, Strack AM, Wang F, Pandit S, Hammond H, Wood D, Lewis D, Rosa R, Mendoza V, Cumiskey AM, Johns DG, Hansen BC, Shen X, Geoghagen N, Jensen K, Zhu L, Wietecha K, Wisniewski D, Huang L, Zhao JZ, Ernst R, Hampton R, Haytko P, Ansbro F, Chilewski S, Chin J, Mitnaul LJ, Pellacani A, Sparrow CP, An Z, Strohl W, Hubbard B, Plump AS, Blom D, Sitlani A (2012c) An anti-PCSK9 antibody reduces LDL-cholesterol on top of a statin and suppresses hepatocyte SREBP-regulated genes. Int J Biol Sci 8:310–327

Zhang Y, Xiong Q, Hu X, Sun Y, Tan X, Zhang H, Lu Y, Liu J (2012d) A novel atherogenic epitope from Mycobacterium tuberculosis heat shock protein 65 enhances atherosclerosis in rabbit and LDL receptor-deficient mice. Heart Vessel 27:411–418

Zhang LY, Ho-Fun Lee V, Wong AM, Kwong DL, Zhu YH, Dong SS, Kong KL, Chen J, Tsao SW, Guan XY, Fu L (2013) MicroRNA-144 promotes cell proliferation, migration and invasion in nasopharyngeal carcinoma through repression of PTEN. Carcinogenesis 34:454–463

Zhou X, Caligiuri G, Hamsten A, Lefvert AK, Hansson GK (2001) LDL immunization induces T-cell-dependent antibody formation and protection against atherosclerosis. Arterioscler Thromb Vasc Biol 21:108–114

Zincarelli C, Soltys S, Rengo G, Rabinowitz JE (2008) Analysis of AAV serotypes 1-9 mediated gene expression and tropism in mice after systemic injection. Mol Ther 16:1073–1080

Index

A

ABCG5. *See* ATP-binding cassette half-transporters G5 (ABCG5)
ABCG8. *See* ATP-binding cassette half-transporters G8 (ABCG8)
ABC transporters, 184–186, 238, 306, 320, 321, 598
Alcohol, 263–265, 579, 582–584
Animal models, v, 80, 132, 190–197, 199, 214, 215, 303, 381, 438, 461, 462, 499, 519, 533, 549, 624, 634, 638, 672
Antibodies, 5, 40, 59, 188, 214, 239, 347, 352, 378, 387, 467, 490, 494, 495, 497–499, 546, 649–675
Antisense, 135, 194, 196, 245, 649–675
Antiviral activity, 219, 498
Apolipoprotein(s) (Apo), 5, 58, 119, 184, 211, 237, 261, 303, 343, 408, 410, 413, 438–440, 458, 485, 517, 530, 572, 596, 621, 653
Apolipoprotein A-I (apoA-I)
  analogues, 641, 642
  infusion, 349, 634, 638–639
  mimetics, 197, 412, 493, 499, 631–643
  mutations, 23, 58, 61, 64–72, 74, 77, 84, 86, 189, 304, 634
Apolipoprotein A-IV (apoA-IV), 8, 10, 14, 17, 56, 57, 72–73, 87, 212, 213, 304, 313, 342, 343, 604
Apolipoprotein E (apoE), 8, 11, 17, 19, 56, 57, 60, 72–73, 79–81, 83–85, 120, 132–137, 186, 187, 190, 192, 195, 212, 219, 304, 307, 308, 310, 312–316, 318, 323, 342, 343, 349, 352, 353, 377, 384, 387, 415, 438, 439, 604, 624, 636, 637, 639, 642, 672–674
Apolipoprotein M (apoM), 8, 54, 136, 212, 243, 289, 374, 438, 466, 491, 535

Atherosclerosis, 20, 59, 120, 188, 209, 244, 259, 287, 301, 344, 376, 428, 457, 486, 512, 545, 574, 600, 624, 633, 654
Atherosclerotic lesion reduction, 634, 674
ATP-binding cassette half-transporters G5 (ABCG5), 122, 140–141, 149, 150, 157, 190–192, 388, 663, 664
ATP-binding cassette half-transporters G8 (ABCG8), 122, 140–141, 149, 157, 190–192, 388, 663, 664
ATP-binding cassette transporter A1 (ABCA1), 21, 56, 121, 184, 232–237, 287, 289, 306–309, 341, 372, 376, 410, 438, 461, 495, 514, 595, 599, 621, 624, 636, 662
ATP-binding cassette transporter G1 (ABCG1), 63, 130, 184, 185, 238, 289, 309–310, 341, 375, 410, 461, 598, 624, 636, 662
Autoimmune disease, 455–472

B

Bacteria, 14, 214, 216–219, 459, 464, 485, 488–492
Bacterial pathogen, 216–218
Bisphenol, 220

C

CAD. *See* Coronary artery disease (CAD)
Cancer, v, 118, 220, 221, 242, 247, 263, 295, 313, 344, 435, 553, 659, 664–666
CEC. *See* Cholesterol efflux capacity (CEC)
Cell cholesterol efflux, 198, 472, 600
CETP inhibitors, 8, 76, 82, 187, 194, 196, 287, 340, 372, 414, 426, 432, 619, 620, 626, 654, 670

CHD. *See* Coronary heart disease (CHD)
Cholesterol, v, 77, 187, 189–190, 197–199, 315, 342–343, 346, 439, 441, 466, 468–469, 574, 595, 598–600, 671
Cholesterol efflux, 20, 22, 58, 62, 63, 65, 68, 73, 76–80, 82–84, 89, 90, 128, 131, 132, 146–154, 181–199, 231, 233–236, 239–241, 270, 271, 292, 293, 307, 309, 312, 315, 318–320, 341, 342, 349, 376, 378, 385–388, 410, 441, 458, 460, 461, 466, 468, 469, 472, 495–497, 574–576, 580, 581, 583, 596, 598–600, 624, 625, 634, 636, 637, 661–666
Cholesterol efflux capacity (CEC), 77, 187, 189–190, 197–199, 315, 342–343, 346, 439, 441, 466, 468–469, 574, 595, 598–600, 671
Cholesteryl ester transfer protein (CETP), 8, 9, 75–76, 82–83, 139, 187, 264, 372, 409, 572, 619–621, 654–655, 670
Chronic kidney disease (CKD), 248, 347, 350, 355, 426–444, 494, 495, 601
Clinical phenotypes, 74
Composition, vi, 4, 6, 7, 19, 23, 28, 29, 35, 39, 40, 61, 75, 77, 81, 84, 90, 122, 186, 187, 189, 198, 217, 218, 221, 248, 262, 263, 266, 268–270, 272, 304, 316, 350, 353–355, 388, 426, 436, 438, 439, 443, 444, 460, 465–468, 470–472, 487, 489, 493, 498, 499, 511, 596, 626, 672
Coronary artery disease (CAD), 74, 78, 80, 82, 83, 138, 187, 198, 199, 244, 340, 342, 344, 345, 347, 350–355, 372, 373, 375, 385, 394, 407, 461, 496, 516, 535, 536, 553, 578, 603, 620, 625, 626, 672, 674
Coronary heart disease (CHD), 156, 248, 261–273, 287, 307, 371, 373, 393, 411, 510, 530, 571, 582, 583, 606, 619, 620, 622, 633, 635, 654, 669
Cubilin, 87

D
Diabetes, v, 20, 74, 80, 138, 193, 215, 216, 266, 268, 269, 271, 295, 296, 321, 340, 344, 345, 347, 349–351, 354, 355, 375, 391, 405–417, 427, 429–431, 433, 442, 462, 486, 517, 537–543, 548, 549, 553, 577, 578, 601, 607, 622, 623, 674
Diabetic ulcer, 548, 549
Drug development, 552
Dyslipidemia, 57, 70–72, 74, 155, 265–267, 271, 353, 408, 410, 426, 428–436, 495, 496, 597, 599, 605, 607, 623, 653–654, 666, 668, 669, 675

E
Ecto-F_1-ATPase, 86
Endothelial cells, 58, 76, 77, 83, 85–87, 154, 156, 196, 237, 238, 241, 242, 244–246, 248, 262, 264, 294, 314, 344–348, 351–354, 373–378, 380, 384–387, 389, 392, 412, 416, 467, 469, 492, 493, 511, 513–516, 518, 519, 534, 539–548, 552, 595, 600–603, 606, 674
Endothelial function, 198, 272, 346, 374–376, 443, 465, 516, 541, 542, 601–603, 606, 620
Endothelial lipase (EL), 75, 80–82, 88, 150, 190, 192, 246, 294, 372, 374, 386, 410, 487, 655, 671
Endothelium, 136, 193, 264, 340, 348, 373–377, 381, 465, 514, 516, 519, 542–546, 552, 601, 674
Exercise, 265, 347, 413, 443, 576–578

F
Fat, 90, 139, 263, 265, 273, 304, 305, 307, 311–314, 317, 318, 320, 322, 323, 377, 411–413, 538, 571, 573, 574, 637
Fibrates, v, 121, 122, 124, 138, 139, 141, 426, 431, 433, 594–597, 599, 602, 605–607, 622
Foam cells, 78, 80, 125, 128, 130, 153, 183–186, 188–190, 234, 235, 307, 308, 310, 314, 319, 320, 341, 343, 382, 383, 385, 386, 388, 599, 672

G
Gene, v, 12, 59, 113–157, 185, 216, 264, 287, 290, 303, 344, 372, 409, 460, 489, 512, 536, 574, 595, 619, 633, 651

H
HDL-C increase, 426, 471, 496, 577, 581
Heart failure, 271, 433, 443, 529, 530, 533, 536, 545, 550, 551, 553, 622, 669
Hepatic lipase (HL), 10, 75, 80–82, 88, 190, 192, 290, 293, 410, 411, 429, 577, 580, 597, 671
Hepatocyte nuclear factors (HNF), 120
Heterogeneity, 6, 16–19, 21, 36, 213–214, 267, 341, 355, 377, 438, 549, vi
High density lipoproteins (HDLs)
  antiatherogenic function improvement, 671
  biogenesis, 53–90, 121, 125, 150, 153, 157, 196, 306–309, 665
  catabolism, 53–90, 156, 350, 577, 596
  dysfunction, 442, 443, 467, 468, 656, 666

phenotypes, 56, 61, 62, 68, 69, 74, 84
remodeling, 53–90, 121, 122, 136, 190, 294, 441, 487
subclasses, 5–7, 88–90, 268, 443, 593–607, 626
therapy, 518, 530, 546–549
Hormone nuclear receptors, 118–120
Hyperalphalipoproteinemia, 82, 322
Hypertriglyceridemia, 10, 68–72, 74, 127, 140, 155, 311, 408–411, 429, 495, 496, 653, 656, v, vi
Hypoalphalipoproteinemia, 461, 490, 495, 496

I

Infections, 7, 72, 90, 137, 216–219, 309, 458, 464, 469, 485–499, 601, 661
Inflammation, 7, 14–17, 21–23, 39, 90, 119, 125, 130, 137, 189, 193, 218, 243, 263, 287, 292, 309, 314, 321, 349, 350, 376, 378, 381, 413, 439, 441, 455–472, 486, 488–491, 495, 496, 514, 530, 536, 537, 539–541, 546–548, 582, 595, 606, 620, 624, 635, 636, 662, 671, 672, 675, v
Innate immunity, 217–219, 245, 463, 485, 491, 492, 494, 499
Ischaemia/reperfusion injury, 393, 510, 513–516, 519, 530, 532–536
Ischaemic stroke, 510–512, 516, 518, 519

K

Knockout mice, 60, 155, 192–194, 216, 294, 304–321, 323, 375, 376, 379, 384, 387, 392, 413–415, 488, 493, 532, 534, 539, 635–637, 641, 642

L

Lecithin/cholesterol acyltransferase (LCAT), 8–10, 12, 17, 20, 23, 28, 31–34, 40, 56, 60–68, 71–75, 84, 88–90, 146, 157, 187, 190, 192, 198, 212–214, 231, 247, 248, 288, 289, 291, 304, 306, 311–313, 316, 319, 321, 322, 341, 342, 344, 345, 385, 386, 410, 436, 437, 458, 469, 471, 485, 487, 580, 583, 603, 604, 641
Lipidome, 23–28, 40, 214, 341, 355
Lipidomics, 23, 27, 28, 39, 272
Lipoprotein lipase (LPL), 8–10, 21, 70, 71, 157, 289, 290, 294, 296, 307, 311, 312, 317, 409, 429, 464, 465, 470, 486, 580, 583, 596, 621, 659
Lipoproteins, 4, 60, 118, 183, 209, 231, 232, 261, 265, 304, 342, 372, 379–388, 409, 425, 458, 485, 511, 530, 573, 596, 619, 655
Liver X receptor (LXR), 76, 119, 125–129, 131, 133, 135, 138–141, 147–149, 154, 157, 185, 186, 194, 196–197, 410, 599, 611, 619, 623–625, 663
Low-density lipoprotein cholesterol (LDL-C) reduction, 633, 667, 669
LPL. *See* Lipoprotein lipase (LPL)

M

Macrophages, 11, 57, 76, 125, 183, 215, 234, 271, 295, 305, 341, 379, 412, 439, 459, 485, 539, 547, 548, 580, 598, 599, 621, 633, 655, 662
Metabolic syndrome, 80, 138, 263, 340, 372, 405–417, 578, 599, 601, 603, 623
Metabonomics, 272
MicroRNAs (miRNAs), 90, 143–150, 185, 197, 247, 271, 378, 436, 649–675, vi
Modifications, 4, 19–23, 39, 64, 125, 186–188, 190, 199, 341, 342, 348, 350, 354, 355, 381–385, 426, 428, 429, 432, 436–444, 457, 459, 467–469, 472, 575, 581, 604, 620, 634, 651–653, 655, 660, 672

N

Nanoparticles, 219–220, 263, 643, 659
Niacin, v, 86, 197, 287, 340, 353, 372, 426, 429, 432, 433, 594–598, 600, 603, 606, 607

O

Obesity, 22, 138, 263, 265, 266, 268, 273, 340, 406, 407, 412–413, 415, 427, 428, 571, 573–575, 601, 605, 623
Oligonucleotides, 194, 196, 245, 649–675
Organophosphates (OP), 12, 220
Oxidation, 20, 78, 80, 187, 210–212, 214–216, 342–346, 348, 353, 354, 380, 382, 384, 385, 413, 439, 459, 467, 472, 514, 539, 575, 577, 580, 582, 604–606, 621, 672
Oxidative stress, 23, 26, 209–215, 351, 427, 440–442, 466, 495, 514, 518, 531, 536–540, 542, 575, 576, 605, 606

P

Paraoxonase, 9, 12, 90, 209, 212, 214–217, 260, 271, 343, 344, 348, 458, 467, 485, 514, 515, 539, 540, 575, 603, 671

Parasites, 217–219, 459, 485, 494, 495, 499
Peroxisome proliferator-activated receptors (PPAR), 126, 128, 138, 139, 157, 196, 197, 429, 432, 433, 619, 621–623
Phospholipid transfer protein (PLTP), 8, 9, 13, 14, 17, 40, 75, 76, 78–79, 90, 121, 122, 139–140, 190, 192, 214, 312–315, 321, 353, 583
Posttranscriptional, 85, 113–157, 185, 410, 414
Posttranslational, 4, 19–23, 39, 113–157, 185, 186, 233, 292, 341, 343, 436, 457, 459, 496
PPAR. *See* Peroxisome proliferator-activated receptors (PPAR)
Preβ- and α-HDL particles, 60, 62, 63, 67, 69, 71, 88–89, 136, 187, 188, 642
Protein stability, 156
Proteome, 7–23, 39, 90, 214, 341, 352, 353, 355, 436, 440, 488
Proteomics, 7–9, 16, 19, 22, 89, 90, 270–272, 293, 352–354, 440, 443, 515

R

Regulation, 7, 15, 16, 21, 78, 85, 117–157, 185, 191, 217, 236, 292, 313, 316, 346, 353, 377, 389, 411, 412, 415, 433, 458, 472, 517, 533, 538, 540, 651, 656, 657, 662–664
Reverse cholesterol transport (RCT), 77, 81, 82, 122, 147–150, 155–157, 183–199, 215, 217, 231, 261, 267, 268, 271, 303, 307, 308, 315, 316, 319, 341–343, 354, 385–388, 426, 438, 440, 442, 458, 495, 497, 511, 535, 575, 583, 621, 633, 636, 637, 641, 642, 661–665
RVX-208, 619, 625–626

S

Scavenger receptor class B type I (SR-BI), 63, 73, 75, 76, 81–86, 121, 122, 141–143, 146, 149–152, 155–156, 184–186, 188, 190, 192–197, 211, 218–220, 238–244, 248, 261, 313, 315–321, 341, 342, 346, 347, 374, 375, 377, 378, 386–388, 391, 392, 416, 438, 439, 467, 468, 489, 495, 497–499, 532, 533, 536, 543–548, 576, 595, 598–602, 665, 672
Signal transduction, 118, 229–249, 391, 460, 461, 533
Smoking, vi, 263–265, 287, 427, 577–581
Smooth muscle cells, 237, 243, 244, 246, 248, 314, 350, 377, 380, 390–391, 428, 541, 544, 665
Sphingosine-1-phosphate (S1P), 11, 24, 25, 27, 28, 136, 243–247, 269, 292, 347, 350, 352, 374–376, 378, 391, 393, 416, 441, 458, 460–463, 491, 493, 513, 532–533, 535–536, 601, 602
SR-BI. *See* Scavenger receptor class B type I (SR-BI)
Statins, v, 245, 262, 270, 303, 324, 340, 345, 353, 371, 372, 429–434, 461, 571, 594–595, 598–599, 601–604, 606, 607, 618–621, 624–626, 666, 667, 669
Steroidogenesis, 85, 149, 306, 317
Structure, vi, 3–41, 57, 71, 75, 77, 85, 89, 118, 119, 136, 186, 189, 220, 270, 272, 273, 296, 337–355, 389, 394, 425, 426, 458–460, 472, 533, 536, 545, 571, 578, 620, 634, 675
Systemic inflammation, 189, 217, 314, 455–472, 486, 489–491, 495, 635, 671

T

Thrombosis, 391, 392, 436, 546, 674
Tissue repair, 527–553
Transcriptional, 113–157, 185, 234, 343, 410, 489, 663, 664
Transcytosis, 58, 86–87, 193, 196, 238, 385–387, 518

U

Uremia, 436
Uremic toxin, 427, 428, 436, 439, 440, 442

V

Ventricular remodelling, 530, 550
Virus, 137, 219, 314, 485, 486, 492, 495–499

W

Weight, 241, 263, 266, 375, 412, 427, 573–576, 581, 623, 673
Wound healing, 530, 547–549, 552

MIX
Papier aus verantwortungsvollen Quellen
Paper from responsible sources
FSC® C105338

If you have any concerns about our products,
you can contact us on
ProductSafety@springernature.com

In case Publisher is established outside the EU,
the EU authorized representative is:
Springer Nature Customer Service Center GmbH
Europaplatz 3, 69115 Heidelberg, Germany

Printed by Libri Plureos GmbH
in Hamburg, Germany